HANDEL

HANDEL
A Celebration of his Life and Times
1685–1759

Edited by Jacob Simon

NATIONAL PORTRAIT GALLERY, LONDON

Published for the exhibition held from
8 November 1985 to 23 February 1986
at the National Portrait Gallery, London

Exhibition organiser   Jacob Simon
Exhibition designers   Brennan and Whalley

© National Portrait Gallery, 1985
Published by the National Portrait Gallery,
St Martin's Place, London WC2H 0HE

Handel: a celebration of his life and times
1685–1759.
1. Handel, George Frideric—Exhibitions
I. Simon, Jacob   II. National Portrait Gallery,
Great Britain
780'.92'4   ML410.H13

ISBN 0-904017-68-0
ISBN 0-904017-67-0 Pbk

Catalogue edited by Paula Iley
Designed by Trevor Vincent
Maps © Andras Bereznay
Printed in England by Jolly & Barber Ltd, Rugby, Warwickshire

FRONT COVER
Handel by Philip Mercier, c.1730 (no. 2, detail)
BACK COVER
Detail from frame of *Handel* by Thomas Hudson, 1756 (no. 12)

Contents

Foreword · 7

Handel: his Life and Work · 9
Donald Burrows

Handel: the Manuscripts and the Music · 18
Anthony Hicks

Handel and the Orchestra · 25
Jeremy Montagu

CATALOGUE · 29

Catalogue Notes and Abbreviations · 30

I
The Image of Handel · 33
Handel Described: a Selection of Accounts by his Contemporaries
Portraits of Handel

II
Handel and the Courts of Europe 1685–1712 · 49
Map: Europe after the Treaty of Utrecht (1713) showing the cities visited by Handel
Handel and his Family
Hamburg and the Opera
Handel and Italy
Handel and Hanover

III
London: the First Decade 1710–19 · 75
Map: Handel's London
Rinaldo: Handel's First Visit to London
Queen Anne's London
Handel and Lord Burlington, 'the Apollo of the Arts'
The Celebrated Water Musick
'Princ'ly Chandos': Handel at Cannons
The Orchestra at Cannons

IV
The Royal Academy of Music 1719–28 · 111
The Origins of the Academy
The Composers
Librettists and Designers
'The three best Voices in Europe'
The Operas
Competition and Rivalries

V
Italian Opera and English Oratorio 1729–41   139
The Court of George II
The Opera Revived
The Rise of Oratorio
Vauxhall Gardens

VI
From Messiah to the Fireworks Music 1741–9   173
Handel and his Friends
In Praise of Handel
The Instruments of Handel's Orchestra
Messiah: Dublin and London
War and Peace
The Royal Fireworks

VII
The Last Decade 1749–59   221
The Foundling Hospital
The Blind Composer
Death and Burial

VIII
The Commemoration of Handel   235
The Early Biographies
'No Vice but the Love of Music': Sir Samuel Hellier and the Sharp Family
The 1784 Handel Commemoration
The Apotheosis of Handel
Handel Festivals: the Nineteenth Century
Handel Editions
Handel in our Time

APPENDIX 1
Inventory of Musical Instruments at Cannons and in London,
belonging to the Duke of Chandos, 23 August 1720   283

APPENDIX 2
Handel's Naturalisation, 1727   284

APPENDIX 3
Handel's Library: the Evidence of Book Subscription Lists   286

APPENDIX 4
Handel's Collection of Paintings   289

Index   291

List of Lenders   296

Foreword

This celebration of the three hundredth anniversary of Handel's birth is another in the series of historical exhibitions organised by the National Portrait Gallery over the past twenty years, which has included tributes to Boswell, Pepys, Richard III and Thomas More. It is also one of many events mounted in Britain and abroad in the course of European Music Year, which marks the birth of three great composers, Handel, J.S. Bach and Domenico Scarlatti, in 1685. In the longer perspective, the Handel tercentenary is a further chapter in the long series of Handel commemorations which have continued ever since his death in 1759. In this degree of interest, Handel is different from Bach and Scarlatti, both of whom suffered periods of neglect before coming to public attention again.

The exhibition is an attempt to tell the story of Handel's life and of the creation of his music through the people and events of his time. Though the emphasis is on Handel's years in London where, over almost half a century, his path crossed that of many of the great figures of the period, his early life in Germany and Italy and his reputation and commemoration from his death to the present day are also treated.

The National Portrait Gallery has depended greatly, in putting the exhibition together, on the great generosity of lenders public and private in this country, Canada, Germany, Switzerland and the United States, to all of whom we are deeply obliged. Special acknowledgement should be made to the British Library and the Gerald Coke Handel Collection for their exceptionally generous response. Any exhibition organiser owes a great deal to help from many individuals. In particular I should like to thank David Alexander, Malcolm Baker, Charles Beare, Carol Blackett-Ord, Iain G. Brown, Elizabeth Einberg, David Freeman, Albert Gellman, Elizabeth Gibson, John Greenacombe, Philip Grimes, Ruth Hayden, Robin Herbert, Timothy Hudson, Douglas Johnson, John Kenworthy-Browne, Graf von Kielmansegg, Katharine Lochnan, John Lord, Nathalie McCance, Sir Robin Mackworth-Young, J.V.G. Mallet, Betty Matthews, Sir Oliver Millar, Oliver Neighbour, Mrs E. Nixon, J.W. Phillips, Judith Powling, Jane Roberts, Martin Roche, The Earl of Shaftesbury, Judith Sheppard, Barbara Small, Lindsay Stainton, Graham C.G. Thomas, Lord Christopher Thynne, Stephen Toms, Peter Ward-Jones, William Waterhouse, Timothy Wilson and Dr Percy Young. I add my thanks to the many others who have assisted or obliged me.

As an art historian rather than a musicologist I have relied heavily on advice from my fellow catalogue authors. In particular Anthony Hicks has taken a close interest in the exhibition from an early stage and has always been a fount of ideas and knowledge; he has often steered me from the pitfalls which await a novice in a new field. To him and to Donald Burrows and Jeremy Montagu I am most grateful. I also owe much to a small number of friends and colleagues who have been exceptionally kind: Gerald Coke, Oliver Davies, Terry Friedman and Ralph Hyde. At the National Portrait Gallery I thank all my colleagues, of whom I mention only those who have been most closely involved: Penny Dearsley, Paula Iley, Jean Liddiard and Jo Rickards. I am deeply grateful for their continuing support and understanding in helping me resolve the problems of a large exhibition. The Embassy of the Federal German Republic has kindly provided support for a programme of associated concerts and recitals; recorded music and stereo equipment have been supplied by Philips Electronics. The exhibition has been designed by Michael Whalley of Brennan and Whalley under a budget which was exceptionally restricting for a project on this scale. The catalogue has been produced by Trevor Vincent with great patience working to a very tight schedule. To all the above and to the many others who have helped, my sincere thanks.

Jacob Simon

2 (detail)

Handel: his Life and Work
Donald Burrows

Any biographer of Handel quickly discovers an imbalance in the surviving material about the composer. We can document Handel's public and professional life with fullness and precision: there are periods when we can even account for his movements day by day. Yet there is surprisingly little firm evidence about his private life and many aspects of his personality. That Handel had a strong personality is clear from the reactions of his contemporaries. In his youth he seems to have gained almost immediate favour with such diverse patrons as Roman cardinals, German princes and Electors, and Queen Anne of Great Britain. In his later years he emerges as rather excitable and impatient, yet generous and an entertaining companion. However, we know virtually nothing about his opinions on the political, religious or philosophical issues of his day. His letters, which are few in number in any case, are mostly businesslike, though a liveliness of style is evident enough. He was not, apparently, given to writing official memoranda, and we must conclude that a large part of the management of his career in the London theatres was conducted verbally. As to the identities and characters of his personal friends, the evidence is again surprisingly thin; furthermore, much modern speculation on his sexual proclivities has been based on a passage from Hawkins's *General History of the Science and Practice of Music* (written considerably after the composer's death), which is open to easy misinterpretation. The evidence of contemporary caricaturists must be considered, though with some caution. The various manifestations of the 'Charming Brute' (nos. 8–9) may have gained credibility from some streak of gluttony in Handel's personality, but the purpose behind the caricature has yet to be explained convincingly; it is likely that some wider social or political intention dictated the nature of the images. Similarly the literary caricatures of the heavily accented 'German' Handel need careful examination: some are fictional, others are obviously recorded in a deliberately overdrawn and ironic manner. In 1985 we can only try to imagine the sound of Handel's voice on the basis of his German upbringing, his known multi-lingual abilities and his fifty-year residence in England, for much of which he no doubt had to use Italian to converse with his singers and French in his dealings with the Hanoverian court in London. Possibly the most curious aspect of Handel's biography is that, while the composer has always been identified, quite correctly, in England with the House of Hanover, many of his strongest patrons and collaborators were, like Charles Jennens, librettist of *Saul* and *Messiah*, supporters of the Old Succession (without necessarily being Jacobites).

After three centuries we can perhaps accept the gaps, limitations and mysteries in the evidence. It was not so easy for the generations immediately following Handel's death to do so, for the man and his music quickly gained status as cultural objects. The Handelian image that developed in England during the later eighteenth century was, if not distorted, at least partial. 'The great and good Mr Handel' is a phrase penned by someone who had known the composer for the last two decades of his life, and we need not doubt that this description had had some basis in fact. But when later generations emphasised the grandeur of the 'oratorio' Handel, adding a moralistic garnish of idealised virtue, they left aside the more varied, humanistic Handel of the operatic repertory that had been the foundation-stone of his art. Perhaps this was inevitable as the memory of Handel's earlier operatic career receded. No doubt also, those who had known Mr Handel found themselves in later years with an attentive audience for stories about the composer. Thus myths and anecdotes proliferated, and bogus biography mingled with fact. The problem of separating fact from fiction is not confined to literary materials, for casual associative objects and bric-à-brac of all sorts soon acquired dubious pedigrees. By the nineteenth century the more credulous could have viewed scores that Handel had never played from, several harpsichords that Handel probably never touched, pens, watches and tuning forks that he probably never owned or used and, inevitably, some portraits of very dubious authenticity. Gaps in the genuine evidence left many a vacuum to be filled by the sentimental and, less commonly, the unscrupulous.

Distorted as the later Handelian image was in some respects, it was nevertheless founded on a correct instinct that Handel's music was outstanding and in some way special. With a longer historical perspective we can appreciate that in Handel's music, as in that of J.S. Bach, the techniques and forms that

Fig. 1 Handel's birthplace, Halle, in 1870
(photo: Händel-Haus, Halle)

had been developed during the seventeenth century are found in their most refined form. Paradoxically, it is in Handel's borrowings and adaptations from other composers that his technical genius as a composer is most clearly displayed. No historical barriers separate us from an appreciation of the emotional range of Handel's music, one we can share with his contemporaries and immediate successors: as Handel's first biographer John Mainwaring remarked in 1760, the composer was at his best when he had 'strong character to mark, or passion to express'. Curiously enough, the very qualities that put Handel in the forefront of dramatic composers also act as a barrier to an assessment of his own personality; from such a wide-ranging sympathy for different characters, moods and situations it would be difficult to draw narrow conclusions about Handel himself. In Handel's music the style is strong, the man recessive. By following through his life and career with the help of the associative objects presented here we can probably come as close to an awareness of Handel's personality as is possible.

The earliest documentary record of Handel's life is the entry in the baptism register of the Liebfrauenkirche, Halle. We know from the funeral sermon for Handel's mother, published in 1731 (no. 20), that Georg Friederich Handel was born on 23 February 1685, the day before his baptism. Presumably this was conducted at the Handel family's house in Halle (fig. 1); although the church was within easy walking distance, such an excursion would have been foolhardy with a newly born infant in February. The baptismal date is given in Old Style, according to the Julian calendar, which was also currently in use in England. (Handel, in common with all Britons, 'lost' 11 days in 1752 when the modern Gregorian calendar was adopted.) However, although the date is given in Old Style, the baptismal register records the year in New Style: in contemporary Britain the year would have been given as 1684, the new year 1685 commencing on 25 March. The date given on Handel's monument (see no. 17) is thus strictly consistent with British usage, although observers at the 1784 Commemoration may erroneously have believed that they were celebrating the centenary of Handel's birth.

By the revision of political boundaries in the Treaty of Westphalia, Halle had been transferred from Saxony to Prussia just before Handel's birth. Handel was therefore presumably born a Prussian citizen, and the frequent contemporary references to him as a Saxon may be taken as a geographical generalisation. His father, Georg Händel (see no. 19), nevertheless continued to work for a Saxon court, which had transferred from Halle to Weissenfels, and the encouragement of young Georg's musical talents is generally attributed to the intervention of his father's employer, Duke Johann Adolf of Saxony. In view of the relative seniority of Georg senior (Georg Friederich was the fruit of his second marriage), his death in Handel's twelfth year was, although untimely, probably taken as part of the natural course. No doubt Handel was following his father's wishes, revealed or implied, when he registered at Halle University in 1702 with a view to some safe professional career, probably in the law. In the same year, however, he also obtained an appointment as organist of the (Calvinist) Domkirche in Halle, and the next year we find him touring the organ-lofts of Hamburg and Lübeck in company with Johann Mattheson (see no. 27). The first major decision of his life came in 1703 when, at the age of 18, he left Halle and a stable provincial career in Middle Germany for a job in the Hamburg Opera House. His move may not have been unprepared: Keiser, the principal composer at the Opera House, had close professional and family connections with the Halle–Weissenfels area. There are good reasons for suspecting that each of Handel's subsequent changes of geographical centre was made on the basis of carefully prepared patronage and recommendation.

According to Mattheson, Handel entered the employ of the Hamburg Opera House as a back-desk violinist, but we soon find him in the director's seat at the harpsichord, and occupying this with a tenacity that led to a duel with Mattheson outside the theatre after a performance. Direction and composition went together at the Opera House, and *Almira*, Handel's first opera, was produced early in 1705. *Almira* is Handel's first complete surviving dateable composition, but his autographs of the opera and of keyboard music from the same period have not survived. Handel's earliest surviving musical autograph is unquestionably that of his F major setting of the Latin psalm *Laudate pueri*, but its date is uncertain. At the earliest, it might have been written at Halle or Hamburg in 1704; at the latest, it might date from Handel's first weeks in Italy in 1706.

Hamburg was, in a sense, a natural posting-station between Halle and Italy. Not only did the mercantile life of the city give Handel a broader international perspective and range of contacts, but the Opera House itself was an international institution. *Almira* was composed as a bilingual opera, with some arias in Italian and some in German. According to Mainwaring's biography, Handel was assured in Hamburg by Prince Ferdinand de' Medici that 'there was no country [other than Italy] in which a young proficient could spend his time to so much advantage; or in which every branch of his profession was cultivated with so much care'.

Although Mainwaring says that Handel resolved to go to Italy 'on his own bottom', it seems very likely that his contact with Ferdinand de' Medici gave him the security of an initial base in Florence. Many details of Handel's Italian itinerary are still obscure. The first certain evidence of his presence in Italy is his signature on a copy of some music by Agostino Steffani, accompanied by the endorsement 'Roma 1706', which may pinpoint it to the first months of 1707 according to our modern reckoning. From Rome in 1707 came two compositions that reveal a sudden and startling musical maturity: the Latin psalm *Dixit Dominus* (see no. 34), and the allegorical oratorio *Il Trionfo del Tempo e del Disinganno*. It was in Rome that Handel received encouragement from enlightened musical patronage – the Cardinals Pamphili, Ottoboni and Colonna (see nos. 30–1 and 33) and the Marquess Ruspoli. Pamphili was the instigator of *Il Trionfo del Tempo* and possibly its librettist, as he certainly was also the author of a cantata in praise of Handel, set to music, perhaps rather immodestly, by the composer. In Rome Handel met some of the world's leading instrumentalists, in particular the famous violinist-composer Arcangelo Corelli (see no. 32). For Colonna, Handel composed Latin church music for Carmelite services: for Ruspoli he composed a considerable number of Italian cantatas during 1707–8, some accompanied by *basso continuo* only, and some employing an instrumental ensemble. More spectacular was his oratorio *La Resurrezione*, produced on a lavish scale under Ruspoli's patronage on Easter Sunday, 1708 (see nos. 36–7).

Diverse and stimulating as Roman musical life was, it lacked a vital dimension for Handel because opera performance was under a papal ban. His first all-Italian opera, *Rodrigo*, was given in Florence during November 1707, presumably with Medici support, and the crowning success of his Italian visit was *Agrippina* (see no. 38) for Venice in 1709–10. At Naples he composed his extended cantata/serenata *Aci, Galatea e Polifemo* in 1708, apparently to adorn the entertainments of a wedding celebration.

Italy (and particularly Venice, the scene of Handel's latest Italian success) was a centre for diplomatic and social activity no less than for artistic experience. There Handel made contacts which advanced the next stages of his career with the opera composer-diplomat-Bishop Agostino Steffani, formerly Kapellmeister at Hanover, Johann Adolf von Kielmansegg, who held office at the Hanoverian court and was married to the half-sister of the Elector (see no. 63), and with Charles Montagu, Earl of Manchester (see no. 39). By early 1710 the 25-year-old Handel had evidently decided that his long-term future lay not in Italy but with a regular job at one of the German courts, and in June of that year he accepted the Kapellmeistership at Hanover. There are hints that he had considered the possibilities of employment at Berlin or Düsseldorf, where unquestionably the musical opportunities were more exciting, and it is quite possible that he did not regard the Hanoverian post as more than a short-term episode. Nevertheless, there was much to be said for Hanover. The Electoral family showed enthusiasm for Handel's harpsichord playing and for his compositions. Although the court opera was in abeyance for financial reasons, the great days of operatic activity under Steffani were still within memory and some informal hints may have been dropped that the opera would be re-started again when opportunity presented itself. In the meantime Handel's conditions of service as Kapellmeister seem to have been very liberal: apparently only light duties were expected of him at Hanover, and he was free to travel to other musical centres, provided due arrangements were made with the court authorities. Whether to fulfil some previous promise, or just out of curiosity as to the greenness of the grass elsewhere, Handel visited the court at Düsseldorf (see no. 43), and may have stopped briefly at other German courts as well. None of them, apparently, presented immediate allurement.

London offered greater prospects. By the time of Handel's first visit to London in the autumn of 1710, the infighting between the different interests that had characterised the introduction of Italian opera into the capital's theatrical life had given place to a more settled regime. In contrast to the situation at Hanover, the financial basis for an Italian opera company in London does not seem to have been affected critically by the demands of the War of Spanish Succession. At the Queen's Theatre, Haymarket, Handel found a regular opera company with a permanent orchestra, supported by an aristocratic audience sufficiently enthusiastic to ensure a measure of financial security. Italian opera was flourishing, yet still a fashionable novelty. In some ways the ambience, more stimulating than that at the German courts, must have reminded Handel of his days in Hamburg, although artistic provision in London was rather more lavish: Nicolini, the Haymarket's leading *castrato*, was as good an operatic performer as could be found anywhere in Italy. The artistic management of the opera company, although occasionally eccentric and over-enthusiastic, was intelligent and discriminating. The composer's need for an operatic outlet was matched exactly by the Haymarket opera's need for an outstanding composer. It is little wonder, then, that Handel's first London opera, *Rinaldo* (see nos. 45–7), was received with an enthusiasm indirectly reflected in the satirical comments of *The Spectator* (see p. 77). Other aspects of London's musical life were attractive also: Handel is said to have taken an early interest in the organ at St Paul's Cathedral (see no. 54) and, if not on his first visit to London, then soon after his return in 1712, he seems to have made contacts with the professional singers in the choirs of the Chapel Royal, Westminster Abbey and St Paul's Cathedral. By 1714 he was participating with the choir of St Paul's Cathedral in their musical evenings at the Queen's Arms tavern; possibly, also, he was a visitor to Thomas Britton's concerts (see no. 59). Between his visits to London, Handel mentioned in a letter that he hoped to receive some English words from the poet John Hughes (see no. 51) to set to music, in terms indicating that he had previously become acquainted with Hughes, and was making an effort to learn the English language.

Handel's presence in London was also to the advantage of his Electoral employer. Assuming no last-minute Jacobite intervention, the Elector of Hanover's eventual succession to the British throne was a certain and none-too-distant prospect, given the parlous state of Queen Anne's health by 1712. Anne supported this succession but would not allow any of the Hanoverian family to reside in her country, so the presence of a cultural representative in the form of the Hanoverian Kapellmeister was welcome to both sides. The Queen never attended the opera house herself, but Handel performed before her at St James's Palace with soloists from the opera house in 1711. Handel's recognised status at the British court a year later is manifest from his compositions, an ode for the Queen's birthday and church music for the thanksgiving service celebrating the Peace of Utrecht (see no. 55). The Utrecht *Te Deum* is Handel's first dated autograph setting of an English text. The protracted nature of the Peace negotiations meant that the Utrecht music was not performed until six months after composition; Handel's delay in returning to Hanover so that he could see this music to performance, coupled with Hanoverian official disapproval of the British unilateral diplomacy which the Peace represented, precipitated his dismissal from the Kapellmeistership. Queen Anne compensated the composer with an English pension at the end of 1713, and the clear underlying implication of the course of events during the year is that Handel had decided by spring 1713 that his future lay in London. As things turned out, the pattern that was already apparent by 1713 formed the basis of Handel's career for the rest of his life. This centred on activity as a composer and musical entrepreneur in the London theatres, supported by a wide circle of the nobility and gentry, and also by the court. Handel's royal pensions gave him a background of financial security and, of course, his association with the court led to the composition of music for state occasions.

Baron Kielmansegg, who had been the agent of the Hanoverian court's displeasure with Handel in 1713, is, with the violinist Geminiani, usually credited with effecting some reconciliation between Handel and the Elector when the latter became King George I of Great Britain in 1714. However, it is also possible that the Water Party on the Thames, which was the occasion for the *Water Music*, may have been a piece of stage management intended to demonstrate a reconciliation between the court of Hanover and the composer. In his role as King of Great Britain, George I had had no quarrel with Handel, and heard his music at the Chapel Royal on a couple of occasions during his first month in London.

As Kapellmeister at Hanover, Handel would no doubt have been entitled to accommodation in the court's property and meals at the appropriate table. A pension in London carried no such provision for the necessities of life. In his early years in London Handel apparently stayed for some time with a Mr Andrews of Barn-Elms (Barnes), and the dedication in the word-book for Handel's fourth London opera (*Amadigi*, 1715) strongly implies that he also stayed for a period with the Earl of Burlington: 'This Opera more immediately claims your Protection, as it was compos'd in Your own Family.'

Two years later the Haymarket opera house closed down – principally, it seems, because politics affecting the Royal Family had divided the patronage on which the opera depended. We know little about Handel's life as the opera project gradually ran out of steam: he may well have visited Germany for a while. What is certain, however, is that for some time after the closure of the opera house Handel enjoyed support and patronage from James Brydges, Earl of Carnarvon and later Duke of Chandos (see no. 68). During 1717–18 he composed *Acis and Galatea*, *Esther* and the Chandos Anthems for Brydges. The Earl was in the midst of creating a palace for himself at Cannons, Edgware; his domestic chapel was not completed until 1720 and the Chandos Anthems were first performed at the neighbouring parish church of St Lawrence (fig. 2), which Brydges had already rebuilt at his own expense and, it must be said, to his own gratification. Household accounts at Cannons were kept with care, and we know the room allocations of the singers and instrumentalists who served Brydges as domestic musicians. Handel does not appear among their number, perhaps because he was regarded as a guest rather than an employee. In spite of the copiousness of Brydges's domestic records many things remain uncertain; we do not even know whether *Esther* was first performed at Cannons or at Brydges's London house. The general impression of this period of Handel's life is, however, that it was a relaxing interlude from normal theatrical routine, which gave him the opportunity

Fig. 2 St Lawrence, Whitchurch, completed in 1716, where much of Handel's 'Chandos' music was first performed (photo: National Monuments Record)

to extend his contacts with English men of letters such as Gay, Pope, Hughes and Arbuthnot. During his time at Cannons, Handel seems also to have turned his attention to keyboard music. Apparently with publication in mind, he settled to writing down and tidying up the improvisations of his earlier years as well as to composing new pieces (see no. 74). The story of the 'Harmonious Blacksmith', duly commemorated in the churchyard at St Lawrence's, is one of the wilder fictions of Handelian biography, but the suite containing the famous movement was published just after Handel's Cannons period.

By 1719 the social and political divisions among London's potential operatic patrons were on the mend. The replacement of the rather hand-to-mouth management of the Haymarket opera of 1710–17 with the constitutional security of the Royal Academy of Music as proposed in 1719 (see nos. 84, 86) was one of the most significant developments for Handel's career: he quite rightly referred to the Academy proposal in a letter to his brother-in-law as 'affairs of the greatest moment, on which all my fortunes depend'. We tend to view the Academy with hindsight of the general wear and tear that ensued after a few years, when the first flush of enthusiasm had died away, and deduce from this that the enterprise was uncertain from the start. In fact, Handel could hardly have expected greater prospects for a secure operatic career than those provided by the Academy. Richard Steele ironically compared the Academy with a South Sea Company, but few of those involved can seriously have expected it to fulfil financial objectives; at best, it might not lose too much money for a couple of years. Finance aside, Handel's concerns, like those of the directors, were mainly practical. The important tasks were to gather patrons together, to set up proper theatre management, to collect singers from Europe, to establish a regular orchestra and to accumulate the necessary equipment in the way of costumes and stage sets. Handel was appointed 'Master of the Orchestra, with a salary'; this appointment no doubt carried the implication that, while the composition of operas in the early years was to be allocated variously to Handel, Bononcini, Ariosti, Porta and others, the day-to-day running of the opera seasons was in the hands of Handel in collaboration with manager-librettists such as Haym. Handel made the most of the opportunities and the relative luxury of the early years of the Academy. He had the best singers in Europe to work with, including Senesino and Francesca Cuzzoni (see nos. 93 and 95). Operas such as *Radamisto*, *Ottone* and *Giulio Cesare* are undoubtedly 'singers' operas', but only a mistaken and rather gullible belief in the propaganda against *opera seria* from a later generation would lead us to underestimate their dramatic worth. Working closely with his librettists, Handel produced coherent music-dramas, and his singers were not above contributing suggestions for ensuring the success of dramatic characterisation (see no. 92). If quality and coherence alone could have guaranteed the success of the Academy operas, then all might have been well. However, it was apparently considered necessary to maintain the patrons' interest with the constant novelty of new singers, in addition to the re-employment of proven favourite performers. The appearance of Faustina Bordoni (see no. 96) from 1726 as a rival prima donna to the already resident soprano Cuzzoni led fairly predictably to competition within the company of a particularly destructive kind, as well as needlessly adding to the budgetary problems. Rather than giving the Academy a new lease of life, these events finally exposed the inherent commercial vulnerability of the original arrangement. Handel and his collaborators were thereafter faced with the task of maintaining opera performances on a less exalted basis.

Following the death of King George I in 1727 Handel continued to receive the same support from his successor, encouraged by Queen Caroline and by Princess Anne, King George II's eldest daughter and, apparently, Handel's favourite pupil. The new reign began auspiciously for Handel as, probably at the King's own instigation, he contributed four magnificent anthems to the coronation service at Westminster Abbey (see no. 110). The service itself seems to have been rather disorganised, but Handel's music made its mark at the crowded Abbey (as it has done at every coronation since). Five years later Handel made a point of including a couple of the coronation anthems in his first theatre oratorio performance, and of noting in the advertisement that the music would be 'disposed after the Manner of the Coronation Service'. It is a fortunate coincidence that Handel had taken British nationality earlier in 1727 (see Appendix 2), because otherwise he might not have been eligible to contribute to the coronation music. In a sense his naturalisation was the final event in a series during the 1720s that reflected Handel's now settled life in London. From 1723 he rented a house in a new building development in Brook Street which was to remain his home for the rest of his life. The house was in the parish of St George's, Hanover Square, whose parish church was also newly constructed (see nos. 147–8). In 1723 Handel was given an appointment as Composer to the Chapel Royal, and from the next year we have our first evidence that he was also music master to the royal princesses. The pension granted by Queen Anne had been continued by her successors, so that overall Handel was receiving quite substantial personal patronage from the Royal Court by the mid-1720s.

The Royal Academy performed for one season into George II's reign, still with its two leading ladies and the famous *castrato* Senesino, but then collapsed, so there was no opera company for the 1728–9 season (the operatic seasons straddled the new year, with performances in winter and spring when the number of patrons 'in town' was at its greatest). However, early in 1729 Handel, in

HANDEL: HIS LIFE AND WORK

Fig. 3 Antonio Bernacchi (1685–1756), Handel's leading singer in his 1729/30 season; caricature from the studio of Marco Ricci (see no. 49) *Albert R. Gellman Collection, Courtesy of the Israel Museum, Jerusalem*

conjunction with his manager Heidegger, arranged to continue the operas for a further five years, and the composer toured the continent for new singers. He put together a good cast, including the *castrato* Bernacchi (fig. 3), the tenor Fabbri and the soprano Strada (see no. 115), and mounted a respectable season with two new operas, *Lotario* and *Partenope*. The operas of the next few years have been rather overshadowed in historical narrative by Handel's introduction of English works into his theatre season in 1732, but the years between 1730 and 1735 saw the production of some of Handel's most rounded operatic masterpieces – *Sosarme*, *Orlando*, *Arianna*, *Ariodante* and *Alcina*. To judge from the scores of the last three, Handel's principal *castrato* Carestini (see no. 116) was at least as fine a musician as the more famous Senesino. The inclusion of *Esther* and *Acis and Galatea* in Handel's theatre programme in 1732 was the culmination of a curious chain of events. For reasons still obscure, but which may well have been connected with internecine warfare among the cathedral choirs, Bernard Gates, the Master of the Children of the Chapel Royal, mounted a private performance of *Esther* on Handel's birthday at the Crown and Anchor tavern with members of his own choir (see no. 128). The performance was successfully repeated and it was apparently envisaged that the whole production might be transferred to the theatre, but the Dean forbade members of the Chapel from acting in costume in theatres 'even with the books in their hands'. Handel therefore mounted *Esther* with his own company, but adding some Chapel singers for the choruses and performing the work without actions (see no. 129). He adapted and added substantially to his 15-year-old Cannons oratorio, mixing Italian singers (and language) into the work; the same treatment was applied to his other dramatic Cannons work *Acis and Galatea*, thrown quickly into the programme as Handel discovered that a theatre across the road from his own in the Haymarket was mounting a rival production.

Handel's career as it developed in the 1730s was untidy and volatile. He added a visit to Oxford on to the end of his 1732/3 theatre season, but this was a diversion: his main energies were devoted to keeping some sort of permanent theatre season going in London. Once again Handel found himself in the midst of a competitive situation which proved self-destructive, as a rival opera company came into existence and divided the available patronage, which was already thinly spread. The Opera of the Nobility acquired the services of a greatly acclaimed *castrato*, Farinelli (see no. 120), and attracted soloists away from Handel's company. By the beginning of the 1734–5 season the Nobility company had forced Handel out of the premier opera theatre in the Haymarket; as his rivals moved into the Haymarket, Handel, at the end of his five-year agreement, made an arrangement with John Rich at the newer theatre in Covent Garden (see no. 131). Handel varied his programmes in an effort to attract and retain his audience, mounting Italian operas (incorporating, for a time, the additional spectacle provided by Madame Sallé's dancers), Italian serenatas, English and Italian oratorios, and English odes. He also introduced organ concertos into the intervals of his performances, an entirely original development (see no. 136). In about 1736 he seems to have come to an arrangement with John Walsh which put the publication of his music on a regular footing.

The late 1730s must have held little joy for Handel. 1737 was a particularly bad year, as his health broke

15

down in the last throes of the battle with the Nobility Opera; and hardly had Handel returned from convalescence than Queen Caroline, probably one of Handel's most faithful patrons through these difficult years, died. He managed to return to the scene of his former triumphs at the Haymarket for a couple of seasons, but by the autumn of 1739 he had to retrench again, and moved his activities to the more modest outpost of Lincoln's Inn Fields Theatre (see no. 140). Yet musical quality was not affected: *Saul*, arguably Handel's first masterpiece of English oratorio and his first collaboration with Charles Jennens as librettist, was produced in his last season at the Haymarket (see nos. 137–8); the masterly set of 12 *Concerti Grossi Opus 6* was composed between seasons in 1739; and *L'Allegro, il Penseroso ed il Moderato* received its first performance in the inauspicious surroundings at Lincoln's Inn Fields during the great freeze-up of 1740. Nor should one pass lightly over the operas from these years, such as *Serse* and *Imeneo*. Ironically, just at the time that Handel's mainstream career was in its greatest difficulty, Jonathan Tyers put up the monument to him in Vauxhall Gardens that seems to symbolise the beginning of a wider popular reputation (see no. 6).

By 1740 Handel, now in his mid-fifties, was committed by financial and artistic ties, as well as by routine, to a career in London. His visit to Dublin in 1741–2 must be seen against that background: it took him away from London at a period when his reputation and activities were growing stale with his regular audience, and yet it carried no implications of a complete break. Dublin was an elegant provincial centre (see p. 199), mirroring on a smaller scale the social and artistic life of London. In spite of Handel's hint that he was thinking of performing a second season in Dublin in 1742–3, it is clear that this trip was, like his previous visit to Oxford, an excursion. He never gave up his London house and he maintained contact with London affairs through correspondence. This is not to belittle the pleasure that Handel must have obtained from the success of his Dublin visit, which seems to have been altogether agreeable. *Messiah* may well have been the highlight of his Dublin programme, but it was produced at the end of a season that had in any case been successful and well received (see no. 186). He gave only one Italian performance in Dublin, of *Imeneo* (presumably without stage action), and concentrated on English odes and oratorios; after his return from Dublin he never again gave either a staged production, or a work in Italian, in London.

Messiah had been composed in London in 1741, before Handel set off for Dublin. *Samson*, also composed in 1741, was not included in the Dublin programme but was the major new work for Handel's all-English season at Covent Garden Theatre in 1743. A new opera management meanwhile mounted Italian operas at the Haymarket, but Handel had left the operatic world for good. Nevertheless, during 1743–5 Handel seems to have deliberately mixed oratorios with secular works such as *Hercules* and *Semele* when planning his programmes, and very soon terminological confusion arose because these secular works came to be described casually as 'oratorios'.

It is uncertain whether the Royal Family supported the oratorios as fervently as they had supported his operas, but Handel's continuing relationship with the court under George II is evident from his church music. During the 1730s Handel had supplied anthems for weddings of the royal children and for Queen Caroline's funeral. In the 1740s he provided the music for a service in the Chapel Royal following the King's return after the Victory of Dettingen (see no. 190), and again for the thanksgiving service for the Peace of Aix-la-Chapelle in 1749. The Fireworks Music was also composed for the Peace celebrations (see nos. 195–202). Handel's oratorio-type performances in the 1740s were affected by the course of national politics, which must have touched the audiences significantly. The '45 Jacobite rebellion so disturbed London's cultural life that Handel's programmes were dislocated completely. As London recovered, so the 'Victory Oratorios', *Judas Maccabaeus* and *Joshua*, restored Handel's career; 1749, the year of the Peace of Aix-la-Chapelle, saw the confident sunshine of *Susanna* and *Solomon*.

By 1750 Handel's English theatre performances, now regularly playing a Lenten season at Covent Garden, were an established institution. Handel's last oratorios, *Theodora* and *Jephtha*, concentrating on themes of endurance in the face of personal tragedy, seem to reflect the personal preoccupations of the aging composer, and blindness interrupted the composition of the second (see no. 206). In 1749–50 Handel had taken on expanding charitable commitments in support of the Foundling Hospital, and his annual *Messiah* performances in the Hospital's chapel from 1750 onwards were the foundation of the work's popularity in England. As Handel's blindness increased, the musical management of his performances effectively passed into the hands of the John Christopher Smiths, father and son. A policy of 'run and revive' kept his oratorios going successfully through the last decade of Handel's life, almost without any further intervention, and indeed Smith junior was able to continue the oratorios in the years after Handel's death with hardly any sense of significant interruption.

From the last two decades of Handel's life we can identify a few personal friends with certainty: among them were the Earl of Gainsborough, Bernard Granville, James Harris, and Anthony, 4th Earl of Shaftesbury (see nos. 157–8 and 160). To be a friend of Handel was also to be an enthusiast for his music: all four of these either promoted provincial performances or established private manuscript collections of his music for their own delectation. Handel's will (no. 212) names a complete range of the composer's acquaintances – relations with Halle connections, professional associates, personal servants and favoured friends. Perhaps the most attractive feature of Handel's personality is revealed in the will through his extensive charitable bequests. A paragraph from the final codicil to his will, added only a couple of days before his death, reveals the institutional image that Handel knew he was leaving behind: 'I hope to have the permission of the Dean and Chapter of Westminster to be buried in Westminster Abbey in a private manner at the discretion of my Executor, Mr. Amyand, and I desire that my Executor may have leave to erect a monument for me there.' In spite of the contingent language of this paragraph, it is difficult to believe that some informal arrangement had not already been made. The publication of Mainwaring's two hundred-page *Memoirs of the Late George Frederic Handel* a year later was a monument of a different kind, and no less revealing of Handel's status.

Few composers have enjoyed such an honoured position in their last years as Handel; in fact taken overall, the general pattern of Handel's life was fortunate, in spite of occasional storms and set-backs. As a young man he gained acceptance with kings and princes on the basis of his musical talents: in maturity he survived a career in the London theatres which was nearly twice as long as Bach's period of service at Leipzig, and he died in comfortable financial circumstances. One-sided as the enthusiasm for Handel was in Britain during the century and a half after his death, it nevertheless kept performances of his music within the common experience. Handel's music soon acquired 'classic' status and from 1776 formed a principal part of the programmes of the newly established 'Concerts of Ancient Music'. If the Handelian reputation was in danger of fading, the 1784 Commemoration put it firmly back into the centre of London's consciousness, supported by the personal enthusiasm of the King. Hardly less significant was the commencement three years later of Samuel Arnold's collected edition of Handel's works (no. 265). The mushrooming expansion of amateur choral singing in Britain during the mid-nineteenth century called upon Handel's music to provide the staple repertory for large singing festivals; the performances surrounding the 1859 Commemoration set the tone for the large-scale Handel performances at the Crystal Palace and at provincial festivals which continued into our present century.

Although not fêted in Germany on the same scale as in England, Handel was by no means forgotten there. Mattheson produced a German version of Mainwaring's biography of the composer, adding a few stories of his own. Handel's oratorios enjoyed occasional revivals in Germany, and Gottfried van Swieten's enthusiasm for 'Ancient Music' led Mozart to arrange *Messiah* and *Acis and Galatea* for performance in Vienna during the 1780s. Haydn carried the memory of the 1791 Handel festival at Westminster Abbey back with him to Vienna; Beethoven was pleased to receive printed editions of Handel's works; Mendelssohn performed Handel's works at Leipzig, along with his more famous revivals of Bach's music. In 1885, the bicentenary year, performances of Handel's music in Germany still outnumbered Bach performances. By 1885 Friedrich Chrysander's collected edition of Handel's works was nearing completion, with 83 volumes (out of an eventual total of 95) in print (see no. 270).

Chrysander's edition, although imperfect by the standards of modern scholarship, was a significant catalyst towards the wider appreciation of Handel's music that has taken place during the last half century. In 1819 Beethoven had encouraged his pupil and patron, the Archduke Rudolph, 'not to forget Handel's works, as they always offer the best nourishment for your ripe musical mind, and will at the same time lead to admiration for this great man'. That judgement, based on the knowledge of a relatively narrow repertory of Handel's music, has been confirmed and strengthened by wider experience. It is a testimonial well worth bearing in mind as Handel's life is retraced through the objects reproduced here.

Handel: the Manuscripts and the Music
Anthony Hicks

Handel's will, drawn up on 1 June 1750, included the following bequest: 'I give and bequeath to Mr Christopher Smith my large Harpsicord, my little House Organ, my Musick Books, and five hundred Pounds sterl:'[1] That modest description, 'Musick Books', referred to a collection of incalculable value: it contained the original composing scores (or 'autographs') of nearly all the works Handel had written since he left Germany around 1706, the working copies (or 'conducting scores') of most of the operas, oratorios and other large-scale works composed in London between 1711 and his death in 1759, and much else besides. It is our good fortune that almost all of the manuscripts in this collection survive today, albeit somewhat scattered. The bulk of the autographs, given by Smith's son to George III, now forms part of the Royal Music Library, donated to the British Library in 1957 by the present Queen, while a supplementary set, mainly comprising fragments and sketches, is to be found at the Fitzwilliam Museum, Cambridge.[2] The conducting scores are mostly at the Staats- und Universitäts Bibliothek, Hamburg,[3] having remained in Smith's family until the mid-nineteenth century,[4] though several isolated volumes survive in other collections. Here the autograph and the conducting score of *Jephtha* have been reunited (nos. 206–7).

Why are these manuscripts important, and how do they help performers who wish to play Handel's music in the closest possible accord with his own expectations and intentions? Clearly much of their value lies simply in their association with the composer himself. Even if perfectly accurate editions of a piece of music exist, there is always a special fascination about the composer's own original score. It reflects the moment when a musical idea is born, when the sounds heard in the composer's imagination are defined and become capable of being translated by others into actual sound. The very appearance of the page will have its own interest: the notes may run fluently without alteration, or additions and crossings-out may show that the final form of the music was achieved only with great labour.

More prosaically, the manuscripts were, and still are, working documents. In Handel's own time they were the sources from which were prepared the part-books used by the singers and instrumentalists for actual performances, and they remain today the main source – though by no means the sole one – for good modern editions of Handel's music. They have to be interpreted with care, however, and some knowledge of how Handel went about the composition of large-scale works such as operas and oratorios is necessary before the present contents of the manuscripts can be properly understood.

Handel's normal method of composing was to write a skeleton score of each act of an opera or oratorio, containing only the vocal lines of the arias and ensembles together with the instrumental bass and a few cues in the other instrumental parts. The words of the recitatives were written between blank staves but not set to music. In this way Handel could quickly establish the overall structure of each act, or even the whole work. He then went back over the score, writing the music for the recitatives and filling in the orchestration, and might indicate the completion of this process with the note 'ausgefüllet' ('filled out') and the date (fig. 4).[5] By this time

Fig. 4 Detail of final page of the score of *Arminio*; Handel writes (in Italian) that the opera was finished on 3 October 1736 and adds (in German) that the scoring was completed ('ausgefüllet') on the 14th British Library, RM 20.a.8., f. 80

preparation of the performing material would be well under way. Since the singers only required a vocal line and an instrumental bass from which to work, their parts in the arias could be copied from Handel's skeleton score even before the orchestration was complete; the recitatives and any revisions made by Handel during the filling-out process were entered later. As Handel completed the orchestration, each number was handed over to his copyists who transcribed it into the conducting score and part-books. By the time Handel entered the final date of completion at the end of the autograph, virtually the whole of the performing material was in existence, ready for full rehearsals.

This moment was not, however, the completion of the composition process, which is why Handel's dates in his autographs need to be treated with caution: they represent only particular moments in a continuing process of evolution. Once the singers had received their parts they might ask for changes, if they felt they were being denied the opportunity to give of their best. Anecdotes tell of Handel emphatically rejecting such demands, but these occasions were probably memorable only because they were exceptional.[6] Handel well understood that operas had to satisfy singers and audiences as well as himself, and there is evidence that he could be perfectly receptive to suggestions from a singer if they were soundly based and not merely the promptings of vanity. In 1723 the contralto Anastasia Robinson (see no. 92) had at least two arias in *Ottone* replaced because they did not suit her personality. A year later Handel made major revisions to *Tamerlano* (see no. 103) before the first performance at the instigation of the tenor Borosini, who had sung the role of Bajazet in an earlier setting of the same libretto by Gasparini. Borosini obviously wanted to ensure that the dramatic potential of the role should be fully developed in Handel's new setting – a desire the composer was pleased to satisfy. Further pre-performance revisions might be necessitated by the sudden unavailability of a singer or instrumentalist, or the discovery that a particular singer's vocal range or ability were not those expected. Nor did revision end with the first performance. Public reaction or further thoughts on the part of those involved might suggest further changes, and there might be special occasions – such as a benefit performance for a singer – when it was customary to add new arias.

More revisions would almost certainly be required if an opera or oratorio was revived. It was the custom for a group of singers to be engaged for a particular season (autumn to spring), and all major vocal works produced that season were written to fit that group. Some singers might be engaged for subsequent terms, but invariably others were replaced, and occasionally there might be an entirely new set of voices, as in 1733, with the defection of Handel's company to the Opera of the Nobility. Thus an opera conceived for one season would seldom fit the singers of another. At the very least a few transpositions, cuts and additions might be required for a revival, and sometimes a good deal more.

All these revisions, whether made before performance, during a run of performances or for revivals, were put into effect by changes in the autographs and conducting scores. Generally speaking, once the conducting score had been copied, it alone became the recipient of revisions of the simpler kind – cuts, changes to vocal lines, indications of transpositions and occasional re-ordering of numbers. When Handel had to write new music, however, he did so on fresh sheets of manuscript paper; a copy was then made and neatly inserted into the conducting score. The autograph of the additional music might be inserted into the original autograph – not necessarily in the place for which it had been composed – or it might remain separate among loose papers. As a consequence, the autograph scores often got into a highly disordered state, in which material rejected during the original process of composition might be found uncancelled alongside additions made many years later for a revival; the whole being further confused by later losses and muddles which occurred before the autographs were bound, in the late eighteenth century. To make matters still worse, material for one autograph was sometimes inadvertently bound up with another.[7] The conducting scores are, superficially, less confused, but they inevitably tend to show most prominently the last revisions made and it is often the first version of an opera or oratorio that is the most interesting and artistically coherent. In the case of oratorios the final revisions may have nothing to do with Handel, since the conducting scores continued to be used and altered after his death.

To some extent the layers of revisions in autographs and conducting scores can be separated out by detailed study. Here it is useful to have an understanding of the way the manuscripts were originally assembled. Paper was supplied in the form of gatherings of four folios, derived from folding and cutting a single large sheet. These would be ruled with inked staves by some kind of mechanical apparatus – the exact form is not known – and were then ready to receive music. Handel liked to use a fresh supply of paper for original compositions of complete works or numbers, although he might use up the odd scrap for shorter insertions. He frequently numbered the gatherings as he used them and wrote the music continuously, not necessarily taking a new sheet for the start of an aria. As a result, insertions made during any of the revision processes will tend to be obvious because they interrupt continuous sections of music, although to which process of revision they actually belong may not be immediately clear. More precise information can be obtained from a study of the water-marks and stave-rulings of the individual sheets. The gatherings of the original score will have a consistent pattern of water-marks, and will share this with the paper of other works written during the same period; the stave-rulings will also have special characteristics dictated by the particular setting of the instrument used to make them. These two kinds of evidence, taken together, will often give a date within a year or two, sufficient to determine the occasion for which the insertions were written.[8]

However, these techniques of analysis can only take us so far. If several alternative versions of a passage exist, all written on the same type of paper, the order in which they were written will not be obvious and it will be hard to tell which passage Handel actually used for his own performances. If sections of the manuscripts are simply missing, either because they have been accidentally lost or (as may be the case with the conducting scores) because they have been discarded during revision, then sources for the music must be sought elsewhere.

We might hope to find a third category of working documents, namely the instrumental and vocal part-books used by the musicians in Handel's performances, but unfortunately most of this material is lost. The part-books used in London were perhaps stored at one of the theatres used by Handel – the King's Theatre, Haymarket, or Covent Garden – and subsequently consumed in one of the fires that destroyed those buildings. A few fragments do survive, however, full of tantalising hints of the information and music that may have disappeared in the lost material. For example, the recently found part-books for Latin church music written for the Feast of Our Lady of Mount Carmel in Rome in July 1707 contain an antiphon setting not extant elsewhere, and have important corrections in Handel's own hand, while the continuo bass part (no. 135) of *Alexander's Feast* – the only major survivor of the material used in Handel's London performances – has several amendments reflecting Handel's revisions to the work, including the bass line of an otherwise lost solo, probably sung by Mrs Cibber (see no. 185) in Handel's Dublin performances of 1742.

Disappointment at the loss of this material is mitigated by the existence of many other sources. Broadly speaking these fall into three categories: word-books[9] (usually printed, though some important manuscript word-books exist for the oratorios); contemporary printed editions of the music; and 'secondary' musical manuscripts, so called because, though derived from Handel's own manuscripts, they were not used for performance and therefore may not reflect Handel's intentions in all respects.

Word-books are especially helpful in determining the form taken by a work on particular occasions. From the earliest days of opera and oratorio it had been customary to provide audiences with the full texts of what was being sung, a practice that faded away only in the late nineteenth century, when the custom of keeping the house lights on during performances declined in favour of the darkened auditorium. The word-book also served as a programme, usually providing a dedication, a cast-list and an 'argument' summarising the main elements of the story and indicating which features were historical and which invented. The books of Italian operas sung outside Italy (including Handel's operas for London) also provided complete translations into the native language, printed in parallel with the sung text. Word-books are usually dated, new editions normally being issued for revivals, so revisions in the text can usually be precisely associated with particular revivals. Fortunately, copies of the word-books for the first productions of all Handel's operas, oratorios and similar works survive, as do those for most of the revivals.

Sometimes other word-books will be useful. The librettos of Handel's operas are almost always based on earlier versions set by other composers, so it is often helpful to consult these sources to elucidate Handel's pre-performance revisions; they may also be useful in understanding parts of the story which have become obscured in Handel's version because of cuts in the recitative, introduced to make Italian opera more acceptable to English audiences. For the oratorios there is sometimes the additional bonus of a manuscript word-book provided for Handel by his

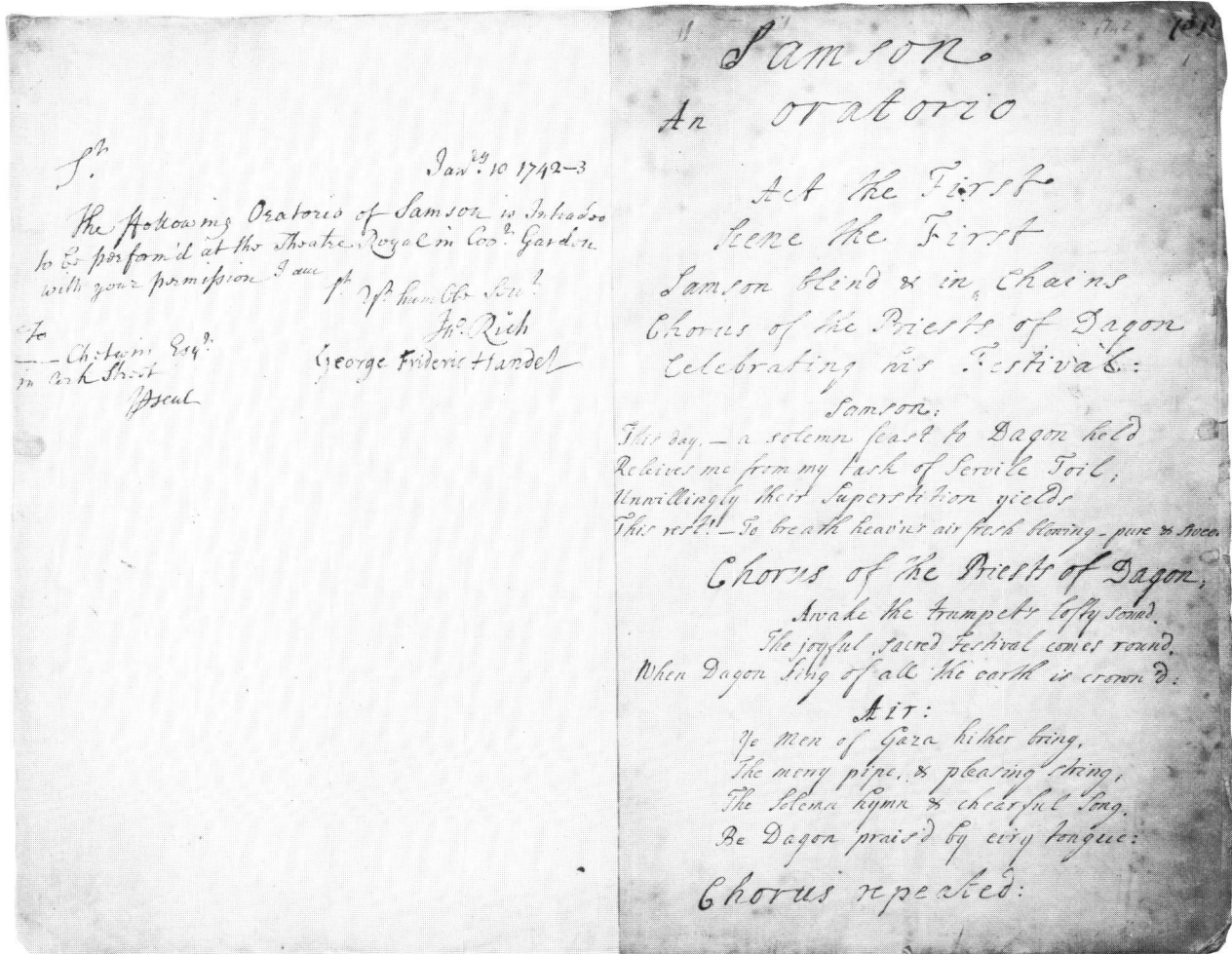

Fig. 5 The manuscript word-book for the oratorio, *Samson*, submitted to the inspector of stage-plays, William Chetwynd, MP, on 10 January 1743, showing the prefatory letter signed by Handel and John Rich
The Huntington Library, San Marino, California

librettist, which survives because it was later passed to the Inspector of Stage Plays for examination. In two cases – *Samson* (see fig. 5) and *Alexander Balus* – the manuscript is that actually used by Handel when composing; other manuscript word-books are fair copies, rather than original working documents, but may still contain amendments by Handel himself, reflecting any last-minute revisions made before performance.[10]

The contemporary printed editions of Handel's major works form a vast and highly varied collection, but the items of most interest for determining an accurate text of the music are the few full scores published in Handel's lifetime (*Alexander's Feast*, *Acis and Galatea* and some English church music), together with the much more common issues of selections from the operas and oratorios.[11] (Such publication only occurred in London; no music of Handel's was printed in connection with his operatic productions in Germany and Italy.) The selections invariably appeared under titles of the form 'Songs in the Opera [or Oratorio] of . . .', and contained most if not all the arias and solo vocal ensembles, as well as the overture and any major orchestral items played during the course of the work (see for instance nos. 47 and 189). Recitatives were excluded, however, and so were the choruses of the oratorios; these omissions did not seriously affect amateurs who wished to perform their favourite numbers in domestic circumstances, but they did hinder rival productions of complete works. Though prints are not reliable as primary sources for the music – inner parts and other orchestral details were often omitted – the fact that they were generally based on the conducting score as it existed during the first run of performances makes them useful for distinguishing between revisions in the manuscripts made before performance and those made for later revivals.

Our final sources of information are the vast range of secondary manuscripts, and particularly several great multi-volume collections of Handel's music (see nos. 137, 158 and 160). These are largely the product of commissions from the most faithful of Handel's wealthier supporters, who wished to have his music available to them in their libraries in a far more complete and accurate form than the contemporary prints could provide. One of the earliest collections, important for the preservation of several operas and other works in the versions of their first performances, is that formed by Elizabeth Legh of Adlington Hall, Cheshire, and maintained until her death in 1734. It seems to have been her intention that it 'be put in some library or publick Room at Cambridge there to be seen or copied',[12] but it eventually passed to Handel's friend Thomas Harris (see no. 160), and remains in the possession of the descendants of his brother James.[13] Another group of scores still preserved as a family heirloom is that formed by the 4th Earl of Shaftesbury, the full extent of which has only recently come to light;[14] it is unusual in its combination of printed scores and manuscripts, giving reasonably full coverage of Handel's output. Other collections have passed to public institutions. The origins of that known (after its nineteenth-century owner) as the Barrett-Lennard Collection, now at the Fitzwilliam Museum, Cambridge, are obscure, but it has been plausibly suggested that the collection is the one described by Charles Burney as formerly belonging to Handel's publisher John Walsh, which was otherwise thought to be lost.[15] The scores prepared for Bernard Granville (see no. 157), brother of Mary Delany, are now in the British Library.[16]

The greatest of all the collections was that formed by Handel's friend Charles Jennens (see no. 181), librettist of *Messiah* and other oratorios. On Jennens's death in 1773 it passed to his relative the Earl of Aylesford, and thus became known as the Aylesford Collection. After remaining at the family seat at Great Packington for many years it was sold at auction in 1918 and became dispersed; but the bulk of the Handel material is now happily preserved in the Royal Music Collection at the British Library and in the Newman Flower Collection at Manchester Central Library, and several other volumes can be traced elsewhere.[17] Quite apart from its comprehensiveness – it preserves much music, particularly instrumental, not otherwise extant – it is remarkable for containing sets of vocal and instrumental part-books. This is either because Jennens occasionally liked to arrange domestic performances of the music, or perhaps because he simply felt it grander to have 'score and parts' rather than the plain scores that seemed to satisfy everyone else. Though the part-books were not used in Handel's own performances and cannot therefore be assumed to be reliable in every detail, they preserve the general characteristics of the original performing material and shed light on aspects of contemporary performing practice. For example, where Handel has not clearly indicated the extent to which oboes and bassoons should double the violin and bass parts, the Aylesford parts will give more precise details. A point of particular interest occurs in the horn parts (fig. 6) of the *Occasional Oratorio*,[18] which double the trumpets in certain choruses though there is no indication in Handel's score that they should do so. Such a practice explains why horn players took part in the Foundling Hospital performances of *Messiah*, the score of which contains no horn parts; presumably the horns doubled the trumpets in the final choruses of Parts 2 and 3.

The annotations in Jennens's hand in the Aylesford scores and parts bear witness to his lively interest in the music – perhaps spilling over into arrogance at times, but never merely dilettante. There is evidence that Jennens closely supervised the work of the copyists working on his commissions. His copy of *Saul* (no. 137) – the work for which he first acted as

Fig. 6 *Occasional Oratorio*, part-book for the first horn, showing the doubling of the trumpets in the chorus, 'Let us break off'. From the Aylesford Collection *Henry Watson Music Library, Manchester, MS 130, Hd4, v.110*

Handel's librettist – was prepared at a very early stage in the composition, and takes account of annotations which Handel had allowed Jennens to make in the autograph score. (Later Jennens added further annotations to his copy.) As Handel revised the work substantially before it was performed, Jennens's score and part-books (the latter occasionally copied earlier than the score) provide valuable information about the history of the oratorio's composition.

This example shows that the secondary manuscripts may deviate substantially from the version of a work actually performed by Handel himself, and are generally more valuable for the individual items they contain than for their overall structure. Many copies of the operas were made in the 1740s or 1750s, long after Handel had ceased to perform them, and copyists of this period could be quite arbitrary in their choice of copy text (autograph, conducting score or an earlier secondary copy), or of the version to adopt if they were faced with alternative settings of an aria, or other ambiguities. The copyist of the score of *Scipione* in the Granville Collection,[19] for example, not only included items rejected by Handel before the first performance in 1726, but also adopted some of the additions made in 1730 when the opera was revised for a different set of singers; thus Scipio himself appears both as alto (as in 1726) and as tenor (as in 1730).

For secondary copies some independent evidence of the date of copying may be important in order to establish when a particular version of the score was current. Detailed studies of the handwriting of the various copyists associated with Handel have so far played a greater role in dating than examination of water-marks or stave-rulings, though the latter may be taken more into account as the data obtained from study of the primary manuscripts becomes available. For the period from about 1720 to the early 1760s the basic groundwork on copyists was laid down by Jens Peter Larsen,[20] who not only established the identification of the hands of Christopher Smith and his son, together with a number of anonymous copyists who worked with them, but also pointed out how the characteristics of particular hands (notably the elder Smith's) changed in the course of time and could themselves be used for dating purposes. Larsen's work has recently been refined and extended by Winton Dean and others, with the result that a wider circle of copyists has been revealed.[21] With particular regard to the early English period (1711 to the early 1720s), it is now clear that a viola player known only as D. Linike was the most important of Handel's copyists. Useful work has also been done on Handel's Italian copyists, particularly Antonio Giuseppe Angelini and others who worked for Handel when he was in Rome.[22]

This survey of musical sources, superficial though it is, serves to indicate why the task of establishing a reliable text for most of Handel's works is often complex. It is naïve to suppose that the best text can be obtained simply by consulting the composer's autograph score. In matters of detail the readings of an autograph may have been decisively repudiated by the composer's alterations in the working manuscripts used by the original players and singers; such changes may now be reflected only in secondary copies which in other respects are unreliable. In terms of overall structure the autographs are particularly unhelpful, since their present state may be confused and incomplete as a result of substantial revisions during and after composition, and losses and re-orderings caused by later careless handling. Indeed Handel's frequent revisions undermine the very concept of a single determinable text. In his time every opera or oratorio performance might be a unique event in terms of its content as well as of interpretation: at each performance under the composer an adjustment might be made to deal with some particular circumstance which had arisen. The work itself was an imagined abstraction, distilled out of the composer's constantly changing perception of what it ought to be and its actual realisation in performance, the latter compromised to a greater or lesser degree by ephemeral exigencies.

When a conscientious performer today asks the old question 'How can I realise the composer's intentions?', it is no answer to say 'Keep to the score', even if we understand that we are not thinking of the matters of interpretation – rhythm, tempo, embellishment and so on – which give life to performances of baroque music. What precisely is the score to which the performers should keep? Some of its features will be fixed beyond question; but for others it will be necessary to choose between the various versions which result from the composer's revisions. Here the performer will need to be guided on the one hand by the evidence of the sources, and on the other by the particular conditions of his own performance. Did Handel discard an aria merely because he had a singer incapable of doing it justice? If so, it may be right – that is, in accordance with what we perceive to be the composer's preferred intention – to restore it. Did he delete a section of aria or chorus because, in the light of several performances, he judged that the number was too long in its context? Then perhaps restoring the passage would be a retrograde step. If we believe that a new aria was added to an opera simply to pacify an insistent prima donna, we may feel justified in omitting it; but if we are convinced that it was inserted to improve the overall balance of an act, we should want to keep it. Investigation of the musical sources will not help us answer these questions: instead it will probably reveal previously unknown aspects of the history of a composition, prompting further questions and suggesting new choices of alternatives.

Such problems must be kept in perspective. Deciding upon a text is no more than the first step in preparing a musical performance; ultimately it will be the quality of the performance itself, not the particulars of the underlying text that will engage our emotions. But if performers have the sense that they are working from texts that have been thoughtfully and thoroughly prepared from a full consideration of all the sources, they will be more inclined to give of their best; and if the audience is conscious that the sounds they are hearing are in good agreement with those the composer imagined, they will listen the more acutely. Thus the labours of the musical scholar in locating, comparing and understanding the complexities hidden in Handel's musical manuscripts will – occasionally – have their reward.

NOTES

For abbreviations, see Catalogue Notes on page 30.

1. *Deutsch*, p. 691.
2. The autographs and other Handel manuscripts in the Royal Music Collection at the British Library are catalogued in *Squire*; those in the Fitzwilliam Museum (excluding additions this century) are catalogued by A.H. Mann in J.A. Fuller-Maitland and A.H. Mann, *Catalogue of the Music in the Fitzwilliam Museum, Cambridge*, 1893. A. Hyatt King's *Handel and his Autographs*, 1967, is a useful general guide. A new detailed catalogue of all Handel autographs is now in preparation (see note 8).
3. Hans Dieter Clausen's thesis *Händels Direktionspartituren ('Handexemplare')*, vol. VII of *Hamburger Beiträge zur Musikwissenschaft*, Hamburg, 1972, includes a catalogue of nearly all Handel's conducting scores, including those outside Hamburg (that of *Jephtha* is omitted, however, apparently by mistake).
4. It would appear that they were acquired as 'waste paper' by the Bristol book-dealer Thomas Kerslake shortly after the death of Sir Henry Rivers at Winchester in 1851. See H.J. Gauntlett's communication in *Notes and Queries*, 2nd series, no. 53, 3 January 1857, pp. 16–17, and Kerslake's reply (*ibid.*). A copy of the catalogue in which Kerslake offered the collection for £45 (*Books from the Libraries of the Rev. John Eagles ... and various others including Handel's Autograph* [sic] *Scores*) is in Bristol Public Library; the entry describing the scores is reprinted in *Clausen, op. cit.*, p. 17. The scores were bought by Handel's biographer, Victor Schoelcher. He offered them to the British Museum, but 'failing in this negotiation, he finally disposed of the collection to Dr Chrysander, for the public library in Hamburg'. (G.A. Macfarren in the preface to his edition of the vocal score of Handel's *Belshazzar*, Novello, Ewer and Co, n.d., p. iv).
5. Handel's datings are quoted by *Squire*.
6. Handel's threat to throw the soprano Cuzzoni out of the window for refusing to sing the aria 'Falsa immagine' in *Ottone* is related by *Mainwaring*, p. 110–11; his rage at Carestini's rejection of the aria 'Verdi prati' in *Alcina* is described in Burney, *Sketch*, p. [*24].
7. For example, f. 42 of RM 20.c.2 in the Royal Music Collection, the main autograph of *Riccardo Primo*, belongs to *Semele*; ff. 62–6 in RM 20.c.4, the main autograph of *Rodelinda*, are the autographs of additional arias for *Giulio Cesare*.
8. A catalogue of Handel's autographs, giving full details of stave-rulings and water-marks for every page, is currently being prepared by Martha Ronish and Donald Burrows.
9. 'Word-book' is here always used to indicate a physical document, printed or manuscript; 'libretto' indicates the verbal text of a musical work in an abstract sense.
10. The extant manuscript word-books of Handel's odes and oratorios are in the Larpent Collection at the Huntington Library and Art Gallery, San Marino, California, except for that of *Theodora*, which is at Manchester Central Library. The Larpent items are catalogued in D. MacMillan, *Catalogue of the Larpent Plays in the Huntington Library*, San Marino, 1939; that of *Susanna* should be entered under the date 1749, not 1759.
11. The early prints are thoroughly catalogued in *Smith*.
12. This instruction appears in Elizabeth Legh's draft will, preserved at Adlington (quoted in C. Legh, *The Organ at Adlington Hall*, n.d.). Her proved will has not been traced.
13. According to Malmesbury family tradition the manuscripts were returned to Handel on Elizabeth Legh's death, and were later given by him to Thomas Harris.
14. An inventory of the collection made in 1761 (copy in British Library, Add. MS 39864, ff. 98–102) agrees closely with the present contents of the collection at St Giles's House, Dorset. The present writer in collaboration with Winton Dean is currently preparing an account of the Shaftesbury collection for publication.
15. See Donald Burrows, 'Walsh's editions of Handel's Opera 1–5, *Music in Eighteenth Century England*, ed. C. Hogwood and R. Luckett, Cambridge, 1983, p. 95, n. 27.
16. Egerton MSS 2910–46. The Granville Collection was first offered for sale by Puttick and Simpson on 29 January 1858. The sale included the 'Father Smith' organ at Calwich Abbey, the country seat of Bernard Granville.
17. The Aylesford manuscripts added to the Royal Music Collection are included in Squire's catalogue. Those in the Flower Collection at Manchester are catalogued in A.D. Walker, *George Frideric Handel: The Newman Flower Collection*, Manchester, 1972; this however includes under the general heading 'The Aylesford Manuscripts' all the manuscript items formerly owned by Flower, several of which have no connection with the Aylesford Collection.
18. Manchester Central Library, MS 130, Hd 4 v. 110 and v. 111.
19. British Library, Egerton MS 2922.
20. J.P. Larsen, *Handel's Messiah: Origins, Composition, Sources*, 1957.
21. W. Dean, 'Handel's Early London Copyists', *Bach, Handel, Scarlatti: Tercentenary Essays*, Cambridge, 1985, pp. 75–97.
22. K. Watanabe, 'Die Kopisten der Handschriften von den Werken G.F. Händels in der Santini-Bibliothek, Münster', *Ongaku-gaku (Journal of the Japanese Musicological Society)*, XVI, 1970, pp. 225–62.

Handel and the Orchestra
Jeremy Montagu

'When I came hither first, I found, among the English, many good players, and no composers; but now, they are all composers, and no players': so Handel's *bon mot* to Charles Burney,[1] and yet the evidence that we have, from pay sheets and other lists of Handel's musicians, would seem to contradict him. In 1710, Uffenbach went to the opera at the Haymarket and wrote: 'The orchestra . . . are all foreigners, mostly German and then French . . .',[2] and the budget for the Royal Academy opera season of 1719/20 lists some twenty players with foreign names out of an orchestra of 33,[3] a proportion much the same as that at Cannons in the same years.[4] Quantz, in 1727, noted that the orchestra for Handel's *Admetus* at the Italian Opera consisted for the greater part of Germans, some Italians and a few Englishmen.[5] But by 1754, for the performance of *Messiah* at the Foundling Hospital, the majority of the players seem to have been English.[6]

The evidence, of course, is scanty. Such few lists of players that survive do so by chance, and much the same is true for those of the instruments which they played. Only two relevant and thorough inventories have come down to us with details of the instruments of an orchestra, and are of such importance that they are major features of this exhibition: Dr Pepusch's list of the Duke of Chandos's instruments at Cannons in 1720, and Samuel Hellier's catalogue of the instruments in his own possession (fig. 12, p. 282).

While Hellier shows a bias towards the Cremonese violin, Chandos is firmly in the English tradition of preference for the Austrian maker, Jacob Stainer, and both list most of their other string instruments from London makers. They share a preference for German horns, either London-made by a German immigrant, or imported from Germany, and while both include a trumpet by Harris of London (see no. 82), it would appear that the great English tradition of brass instrument making by such superb craftsmen as William Bull had either lapsed or was neglected. Certainly Harris, for example, was making horns in this period, for in 1717 he supplied a set to the Royal Buckhounds,[7] but this is where the lack of surviving instruments leaves us only with guesswork. It may have been that the English makers were still restricted to the type of horn suitable for the hunting field, and that only from Germany could one obtain the rather wider-bored and smoother-toned horns suitable for the orchestra. The Cannons Leichamschneider horns (see no. 81) have not survived, and whether they, like the Hellier Hofmasters, were provided with a set of crooks for different tonalities, we do not know.

The only woodwind instrument noted at Cannons was a German bassoon (it seems probable that the oboist, J.C. Kytch, was one of several players who used his own instrument), but Hellier went to the leading London makers for his woodwind, chiefly to Thomas Stanesby junior, a maker who was also associated with Handel and whose father was a member of the first generation of great English makers, in succession to foreign settlers such as Bressan. It was the Stanesbys and their contemporaries such as Bradbury who perfected the English oboe, bassoon, flute and recorder, making the flute in four joints instead of three, and introducing subtle changes in the bore profiles which affected the tone quality to the extent that the sound of the English woodwind is immediately recognisable as different from that of the contemporary French and German. Much of these changes took place during Handel's working life in London, and it is a continual frustration that we do not, and cannot, know what makes of instrument were used by the individual members of Handel's orchestra. Only occasionally, as with the Stanesby double bassoon (no. 173), is the evidence such that we can be reasonably certain that this instrument, and no other, was that which was used in Handel performances.

The main source of information that we have for the range of Handel's instruments is that of his scores. Occasionally he makes use of the bass viol, the lone survivor of that glory of English music, the chest of viols. The Cannons list is evidence that these were usually, at least, by the great English makers including Henry Jaye and Barak Norman (see nos. 78–9). The viol, while playing in much the same register as the violoncello, was an instrument with six strings as against the cello's four, strung at lower tension with a more intimate and yet a more biting tone quality, often more suitable for accompanying the voice than the cello. The viols, as a consort, had all-but died out in the latter part of the seventeenth century with the fashion for all things French brought back from his exile by Charles II, and certainly they would have been wholly unsuitable for the Handelian opera and oratorio orchestra. The bass, however, had lingered on, both as a solo instrument in the latter part of the seventeenth century and as one of the alternative bass voices for accompaniment.

The violins which King Charles had imported from France were, by Handel's time, more often modelled on those by Jacob Stainer of Absam near Innsbruck in Austria than on those of the Italian masters. The Stainer model, with its high arching of belly and back, had a sweeter, if less powerful, sound than those of Stradivari. Their popularity is reflected not only in the fact that the Duke of Chandos had four violins by Stainer at Cannons, but also that all the makers recorded by Hellier (apart from his Amati and his Stradivari) were London makers who worked on the Stainer model. It was only in the second half of the century, when orchestras grew larger in size, that the Italian violins really came into their own in Britain, for it was then that volume, rather than sweetness of tone came to be preferred. Today, with the increase in performance of eighteenth-century music on original instruments, or copies, we are again hearing the silvery sound of the high-arched violin, the sound that Handel would have heard in his own orchestra.

Now that we are becoming accustomed to the recreated baroque orchestra, it is probably unnecessary to say here that this is not the only change of instrumental sonority between Handel's period and the present day. Most of the basic changes took place in the latter part of the eighteenth century, the oboes becoming narrower in bore and more piercing in sound, the flutes becoming louder (and in the mid-nineteenth century louder still). Indeed the word 'flute' immediately after Handel's period changed its meaning: when Handel specified *flauto* he meant what we call the recorder, while what we now think of as the flute was always described as a 'traversa' by Handel, the 'transverse' or 'German' flute (see no. 178). It was the rise of expressiveness in music, the need to play more softly or loudly at will, which killed off the recorder and led to the popularity of the flute we know today. The bassoon also became louder, and the choke or constriction of the bore of the bell, so characteristic of the English bassoon of the first half of the eighteenth century, was opened out, with a resulting loss of tonal character. Curiously enough the horn became quieter, but this was because players found that they could obtain more notes in the middle register by placing and moving their hand in the bell of the instrument. More important to the music, this discovery meant that the horn's melodic range was an octave lower than in Handel's period. Thus a change can be traced: from a somewhat raucous tone, suitable for the hunting field, in the earlier years of Handel's career, to the broad, smooth quality of instruments by Hofmaster and others, typical of Handel's middle and later period, then to the velvety tone of the Mozart and Haydn period, which in turn deteriorated into the heavy sound we know in the modern orchestra.

The leading brass instrument of Handel's day was the natural trumpet, a magnificent instrument playing at the top-most part of its range. Handel wrote many superb solos for the trumpet, and there are few of his oratorios that lack a trumpet *obbligato* and one or more choruses accompanied by trumpets and drums. Few had the impact of the chorus in *Messiah* which caused a king to leap to his feet, but even today, with instruments which produce a pale shadow of the tone of the true natural trumpet, we stand for the Hallelujah Chorus in memory of that occasion. The trombones, those majestic instruments which came into the orchestra through the Church and, later, the opera, seem to have been rare in Handel's London. He seldom used them, and no contemporary instruments are known to survive in Britain. It is possible that a group of players visited London from abroad, probably from Germany, at the time that he was writing *Saul* and *Israel in Egypt* in 1738, and that the absence of players here on a continuing basis is why he used the instruments so seldom.

Handel's orchestration was always effective and normally economical, achieving the maximum impact and characterisation, and nowhere is this more apparent than in his use of brass instruments, especially the trumpets, and of the drums. In works such as *Messiah* they are reserved for the greatest climaxes, the timpani being heard only in the Hallelujah and in the two final choruses. Handel's drums were usually small, as can be seen from no. 223, but since they were played with wooden-headed sticks, their sound was clear and their rhythm precise. On special and rare occasions Handel would indent to the Royal Armouries for a larger pair of drums, known as double drums, for they were double the diameter of the normal size and an octave lower in pitch (see fig. 7).[8] Only one pair of such drums has survived, and can be seen in the Bate Collection at

HANDEL AND THE ORCHESTRA

Fig. 7 *Royal Artillery in the Low Countries* by David Morier; a detail showing the great kettledrums on their special carriage; Handel borrowed such drums from the Tower of London for use in his oratorio seasons *Reproduced by Gracious Permission of Her Majesty the Queen. On loan to the National Army Museum*

the Faculty of Music in Oxford.[9] Their sound is not only lower than that of the ordinary drums, but also much richer in tone.

Handel was himself famous as a player as well as a composer, and not only did he direct his performances from the harpsichord but he also frequently played an organ concerto between the parts of an opera or oratorio. He certainly owned both a large and a small organ, for in his will he left his small house organ to John Christopher Smith the elder, and in a codicil he bequeathed his large organ, which was then at the Theatre Royal in Covent Garden, to the manager John Rich. The latter was destroyed by fire in 1808, and whether the former has survived is unknown. Smith also received his 'large harpsichord', which must have been the one given by his son, John Christopher Smith junior, to George III in gratitude for receiving a pension. Although Hawkins's reference to Handel's owning a fine Ruckers harpsichord has always been associated with this instrument, recent research by Ann and Peter Mactaggart has brought new evidence to light.[10] A later royal inventory of 1825 mentions 'a harpsichord by Handell [*sic*] in a walnut tree case'. No maker of this name is known so it seems probable that this reference is to a harpsichord owned by Handel; if so, the type of case described can only apply to an English instrument, such as the Kew harpsichord by Shudi (see fig. 8); Smith's gift to George III may not, as has been thought in the past, apply to a Ruckers instrument.

Fig. 8 Double-manual harpsichord by Burkat Shudi, 1740. Made for Frederick Prince of Wales *Reproduced by Gracious Permission of Her Majesty the Queen. On display at Kew Palace*

27

Ruckers was the great name in harpsichord making in the sixteenth and seventeenth centuries, and it is noteworthy that there was still one of his instruments at Cannons in Handel's time, and another among the Hellier instruments. Also at Cannons was one by Tabel, a Fleming who had settled in London and who brought the Flemish tradition of the Ruckers family to England. It was Tabel, whose only known instrument is included here (no. 75), who established the later English school of harpsichord making, for he was the master of both Kirckman and Shudi, the two great makers active into the second half of the eighteenth century. All three were foreigners who settled in London, but in his turn Shudi was the master of John Broadwood, whose firm still exists today as makers of pianofortes. Another keyboard instrument at Cannons was a spinet by Hitchcock, who was the only English maker to adopt the German practice of making harpsichords with a rounded tail; thus it is possible that the design for a painted lid by Thornhill (no. 150), marked as for an instrument owned by Handel, was for a harpsichord of his make. Indeed it is possible, too, that the single-manual in the Mercier portrait (no. 2) is also a Hitchcock, for he was one of the few makers to inlay his 'black' notes with alternate strips of ivory and ebony, the so-called skunk-tail sharps.

It is difficult to characterise the English harpsichords in words. Although they were a continuation of the Flemish tradition, they were different in sound and effect, and different, too, from the German and French. The great French instruments, which are those most familiar in the concert hall today, have an incomparable richness, but the English instruments have a grandeur all of their own, especially in the bass, and a sharpness with the lute stop (aptly called the nazard, or nasal, in German) like no others. Each country had its own style and its own sound: Bach on the German instruments of Silbermann and Hass must have sounded as different from Couperin on the Blanchet and Taskin as both did from Handel on his Shudi, or at Cannons on the Tabel. Today few players can afford to own so many different harpsichords, and few concert platforms have the space to accommodate them, with the result that we hear all composers on the same instrument.

The same problem of space has meant that only the Tabel can be included here. The choice of instruments, which may be divided into three main groups, has depended very much on the chances of what has survived and is available. The earliest group is a selection based on the instruments listed at Cannons by Dr Pepusch (nos. 75–83). Then there are those catalogued by Samuel Hellier, together with the instruments of another enthusiast for Handel's music, Granville Sharp, who was portrayed by Zoffany with his family. These instruments (nos. 217–31) mainly belong to the end of Handel's life or the years after. An intermediate general section shows the type of woodwind and some of the stringed instruments used by Handel in his music (nos. 167–80). A few instruments, for example the carillon which Handel occasionally used, cannot be shown, since, like the trombone, none from his period has survived. Others, such as the large double drums, the triple harp and the theorbo, are excluded for reasons of space. One graphic description of Handel's orchestra, by Sir John Clerk of Penicuik, quoted at length on p. 145, describes a performance of *Orlando* in 1733, when the orchestra included '2 large bass violins each about 7 foot in length ... with strings not less than a $\frac{1}{4}$ of an inch diameter ...', but even if there were space, no double bass of the Handel period has survived in original condition. It has been attempted to show only those which have survived in original condition, or been restored to that condition. There is little point in showing a violin in the altered state in which they are normally used today, for such may be seen in any music shop or orchestra. Handel's music sounded very different from twentieth-century music simply because the instruments were different, and it is the instruments Handel could have known which are exhibited here.

NOTES

For abbreviations, see Catalogue Notes on page 30.

1. Burney, *Sketch*, p. 33 (footnote).
2. W.H. Quarnell and M. Mare (trans. and eds.), *London in 1710 from the Travels of Zacharias Conrad von Uffenbach*, 1934, pp. 17 f.
3. Judith Milhous and Robert D. Hume, 'New Light on Handel and The Royal Academy of Music in 1720', *Theatre Journal*, XXXV, 1983, pp. 149–67.
4. *Baker*, pp. 132–3.
5. *Deutsch*, p. 754.
6. *Deutsch*, pp. 750–1.
7. J.P. Hore, *The History of the Royal Buckhounds*, 1893, p. 289 (footnote).
8. H.G. Farmer, *Handel's Kettledrums and other Papers on Music*, 2nd edition, 1960, pp. 85–99.
9. Repr. in Jeremy Montagu, *The World of Baroque & Classical Musical Instruments*, 1979, p. 61, pl. 44.
10. Ann and Peter Mactaggart, 'A Royal Ruckers: Decorative and Documentary History', *The Organ Yearbook*, XIV, 1983, pp. 90–3.

CATALOGUE

Catalogue Notes and Abbreviations

Catalogue entries for Handel's autographs and other printed and manuscript scores, and word-books from Handel's lifetime, are by Anthony Hicks, who has also contributed to the introductory text on pp. 272–3. Catalogue entries for musical instruments are by Jeremy Montagu, together with the introductory texts on pp. 105, 191 and 239. Other entries and texts, together with the appendices, are by Jacob Simon.

TITLES
Titles for books and engravings, as given in inverted commas, are taken from their original form, though some are abbreviated. Works by Handel are otherwise titled in the form set out in the work-list by Anthony Hicks in Winton Dean, *The New Grove Handel*, 1982.

DIMENSIONS
Sizes are given in the order height, width, depth. For engravings, image size is given; for busts, height given includes the socle; for books, the closed size of the volume is given; and for stringed instruments, the width of the bouts is given in the order upper, middle, lower.

PROVENANCE
Provenance is not normally given for engravings and books.

LITERATURE AND EXHIBITIONS
References are selective and in particular, not all past exhibitions are given. Place of publication is only included where this is not London. *The Dictionary of National Biography* (1885–1901) and *The New Grove Dictionary of Music and Musicians* (1980) are not cited, although frequently consulted.

SQUARE BRACKETS
In some cases inscriptions on exhibits have been completed with information derived from a source other than the exhibit itself; this is indicated in square brackets, thus; 'S[enesin]o'. Similarly, details of works referred to under the heading 'Literature' are shown in square brackets when not actually given in the publication cited.

DATES
All dates for events in Britain are in Old Style until 14 September 1752, when Britain adopted the Gregorian calendar which had long been in force on much of the continent. Events abroad are dated in New Style (11 days ahead). Some dates on Handel's autograph scores and letter, and those of foreign treaties, are given in New Style followed by Old, thus: 21/10 August 1733 (this example is from a letter to Germany; see no. 23). The calendar year in Old Style began on 25 March, meaning that in the period from 1 January to 24 March some dates were written thus: 20 January 1746/7; this has usually been expressed in the modern form 20 January 1747.

ABBREVIATIONS
References to specific pages of manuscripts are given thus: f.16v (folio 16, verso), or f.16r (folio 16, recto). The most frequently used references are given in an abbreviated form, as shown below.

Baker
C.H. Collins Baker and Muriel I. Baker, *The Life and Circumstances of James Brydges First Duke of Chandos*, Oxford, 1949

Binyon
Laurence Binyon, *Catalogue of Drawings by British Artists . . . in the British Museum*, 4 vols., 1898–1907

Burney
Charles Burney, *An Account of the Musical Performances in Westminster-Abbey and the Pantheon, . . . 1784 in Commemoration of Handel*, 1785

Burney, Sketch
'Sketch of the Life of Handel' in *Burney* (see above)

Burrows
Donald Burrows, 'Handel and Hanover', *Bach, Handel, Scarlatti. Tercentenary Essays*, ed. Peter Williams, Cambridge, 1985

Coke
The Gerald Coke Handel Collection, catalogue of an exhibition at Jenkyn Place, Bentley, Hampshire, July 1985

Coxe
[William Coxe], *Anecdotes of George Frederick Handel and John Christopher Smith*, 1799

Croft-Murray
Edward Croft-Murray, *Decorative Painting in England 1537–1837*, 2 vols., 1962 and 1970

CS
John Chaloner Smith, *British Mezzotinto Portraits*, 4 vols., 1883

Dean
Winton Dean, *Handel's Dramatic Oratorios and Masques*, 1959

Deutsch
Otto Erich Deutsch, *Handel. A Documentary Biography*, 1955

Dorris
George Dorris, *Paulo Rolli and the Italian Circle in London*, The Hague, 1967

Esdaile
Katherine Esdaile, *The Life and Works of Louis François Roubiliac*, 1928

Gibson
Elizabeth Gibson, 'Owen Swiney and the Italian Opera in London', *Musical Times*, cxxv, 1984, pp. 82–6

Greenacombe
John Greenacombe, 'Handel's House: A History of No. 25 Brook Street, Mayfair', *London Topographical Record*, xxv, 1985, pp. 111–30

Hawkins
Sir John Hawkins, *A General History of the Science and Practice of Music*, first published in 4 vols., 1776; references given are to the reprint in 2 vols., 1875

Highfill
Philip H. Highfill, Jr, and others, *A Biographical Dictionary of Actors, Actresses, Musicians, Dancers, Managers and Other Stage Personnel in London, 1660–1800*, Carbondale and Edwardsville (USA), 10 vols. to date, 1973–

Hogwood
Christopher Hogwood, *Handel*, 1984

Ingamells and Raines
John Ingamells and Robert Raines, 'A Catalogue of the Paintings, Drawings and Etchings of Philip Mercier', *Walpole Society*, xlvi, 1978, pp. 1–70

Kerslake
John Kerslake, *Early Georgian Portraits*, National Portrait Gallery, 2 vols., 1977

Landon
H.C. Robbins Landon, *Handel and his World*, 1984

Mainwaring
[John Mainwaring], *Memoirs of the Life of the Late George Frederic Handel*, 1760

Mallet
J.V.G. Mallet, 'Some Portrait Medallions by Roubiliac', *Burlington Magazine*, civ, 1962, pp. 153–8

MI
Edward Hawkins, *Medallic Illustrations of the History of Great Britain and Ireland . . .*, 2 vols., 1885

Nicolson
Benedict Nicolson, *The Treasures of the Foundling Hospital*, Oxford, 1972

Niggl
Paul Niggl, *Musiker-Medaillen*, Hofheim am Taunus, 1965

Paulson
Ronald Paulson, *Hogarth: His Life, Art, and Times*, 2 vols., New Haven and London, 1971

Phillips
Hugh Phillips, *Mid-Georgian London*, 1964

Piper
David Piper, *Catalogue of Seventeenth Century Portraits in the National Portrait Gallery*, Cambridge, 1963

Poole
Mrs Reginald Lane Poole, *Catalogue of Portraits in the possession of The University Colleges, City, and County of Oxford*, vol. 1, Oxford, 1907

Rococo
Rococo. Art and Design in Hogarth's England, catalogue of an exhibition at the Victoria and Albert Museum, May–September, 1984

Rosenfeld and Croft-Murray
Sybil Rosenfeld and Edward Croft-Murray, 'A Checklist of Scene Painters working in Great Britain and Ireland in the 18th Century', *Theatre Notebook*, xix, xx, 1965

Satires
M.D. George and others, *Catalogue of Political and Personal Satires . . . in the British Museum*, 11 vols., 1857–1954

Schoelcher
Victor Schoelcher, *The Life of Handel*, 1857

Smith
William C. Smith, *Handel. A Descriptive Catalogue of the Early Editions*, 2nd ed., Oxford, 1970

Smith and Humphries
William C. Smith and Charles Humphries, *A Bibliography of the Musical Works Published by the Firm of John Walsh during the years 1721–1766*, 1968

Squire
William Barclay Squire, *Catalogue of the King's Music Library. Part I: The Handel Manuscripts*, 1927

Vertue
George Vertue, 'Vertue Notebooks', *Walpole Society*, 7 vols. (nos. xviii, xx, xxii, xxiv, xxvi, xxix, xxx), 1930–55

4

1 *The Image of Handel*

Images of Handel abound. He was portrayed more often than any other great composer of his century and perhaps of any century. This diversity stands in great contrast to the dearth of authentic portraits of Johann Sebastian Bach or Domenico Scarlatti, both born in the same year as Handel. The explanation is to be sought in their differing backgrounds. Unlike Bach or Scarlatti, Handel did not derive most of his patronage from a single source such as the Church or court. For much of his life he worked in a country in which his support was drawn from a relatively wide cross-section of society and where the habit of commissioning portraits was extremely widespread.

This is not the place to attempt a complete catalogue of Handel portraits, though almost all the major types are represented, including two images which are reproduced and discussed for the first time (nos. 3 and 16). There have been several recent attempts to list Handel's portraits (see p. 260) but the most useful and accurate survey is undoubtedly that by John Kerslake in his *Early Georgian Portraits* (1977). In this catalogue certain unifying themes are touched on: Handel's choice of artist, his interest in his portraits, and the incidence of ownership, although it must be said that the evidence is often too limited to draw firm conclusions.

Little is known of Handel's attitude to portraiture, either from his own pen or at the hands of his patrons or the artists commissioned. That Handel took an interest in paintings is apparent from the extensive and wide-ranging catalogue of his own collection (see Appendix 4). But for his portraits it seems that Handel preferred to look to foreign artists who, if not actually German by origin, almost all trained or worked in Germany at some stage: Balthasar Denner at Hamburg, Philip Mercier at Berlin and G.A. Wolfgang at Augsburg and Berlin (see nos. 1, 2 and 5). The print engraved by Jacobus Houbraken (no. 4) derives from a lost original done on the continent (though where is not known). Even the French-born sculptor, Louis François Roubiliac had probably worked at Dresden as a young man, coincidentally at the very time of Handel's long visit there in 1719. It is not until 1748 when he sat to Thomas Hudson (see no. 11) that Handel is known to have gone to an English artist.

Of Handel portraits, only the two grandest images – the Roubiliac statue for Vauxhall Gardens and the Hudson full-length (see nos. 6 and 12) – are known to be the result of commissions by friends. Three of the finest and most important were owned by Handel himself for many years: the oils by Denner and Mercier already mentioned and the Roubiliac marble bust of 1739 now in the Royal Collection. A fourth portrait, the earlier of the two paintings by Hudson (no. 11), also seems to have been commissioned by him. All four were given away to close personal friends: the Mercier to Thomas Harris in 1748, the Hudson to his family in Germany, perhaps on his last visit in 1750, and the Roubiliac and the Denner to J.C. Smith (the son it seems rather than the father), probably towards the end of Handel's life or at his death.

With the exception of the Vauxhall Gardens statue none of Handel's portraits was publicly exhibited during his lifetime. However one portrait in particular, the engraved head by Houbraken (no. 4), probably the most reticent in appearance, must have had his personal approval as an image by which he was willing to be known publicly. The print was a gift to subscribers to *Alexander's Feast*, a score which Handel himself published in 1738. At the opposite extreme is his understandably strong reaction to Goupy's caricature of him as a glutton (no. 8), which is said to have cost the artist a legacy.

After Handel's death the considerable market for his image was largely satisfied with casts of Roubiliac's busts and copies of Hudson's portraits, though the Handel Commemoration in 1784 began an expansion in Handel imagery which continued into the nineteenth century (see p. 260).

Handel Described: a Selection of Accounts by his Contemporaries

His Appearance

I went last Friday to 'Alexander's Feast'; but it was such a melancholy pleasure, as drew tears of sorrow to see the great though unhappy Handel, dejected, wan, and dark, sitting by, not playing on the harpsichord, and to think how his light had been spent by *being overplied in music's cause*.

Lady Shaftesbury in a letter to James Harris, 13 March 1753

He was in his person a large made and very portly man. His gait, which was ever sauntering, was rather ungraceful, as it had in it somewhat of that rocking motion, which distinguishes those whose legs are bowed. His features were finely marked, and the general cast of his countenance placid, bespeaking dignity attempered with benevolence.

Sir John Hawkins, *A General History of Music*, 1776

The figure of HANDEL was large, and he was somewhat corpulent, and unwieldy in his motions; but his countenance, which I remember as perfectly as that of any man I saw but yesterday, was full of fire and dignity; and such as impressed ideas of superiority and genius. . . .

HANDEL's general look was somewhat heavy and sour; but when he *did* smile, it was his fire and sun, bursting out of a black cloud. There was a sudden flash of intelligence, wit, and good humour, beaming in his countenance, which I hardly ever saw in any other. . . .

HANDEL wore an enormous white wig, and, when things went well at the Oratorio, it had a certain nod, or vibration, which manifested his pleasure and satisfaction. Without it, nice observers were certain that he was out of humour.

Charles Burney, 'Sketch of the Life of Handel', 1785

He was large in person, and his natural corpulency, which increased as he advanced in life, rendered his whole appearance of that bulky proportion, as to give rise to Quin's inelegant, but forcible expression; that his hands were feet, and his fingers toes. From a sedentary life, he had contracted a stiffness in his joints, which in addition to his great weight and weakness of body, rendered his gait awkward; still his countenance was open, manly, and animated; expressive of all that grandeur and benevolence, which were the prominent features of his character.

William Coxe, *Anecdotes of Handel*, 1799

His Manner and Speech

He had a way of speaking peculiar to himself, by which he made the gravest people laugh, without ever laughing himself.

Johann Mattheson, *George Friderich Handels Lebensbeschreibung*, 1761

He was impetuous, rough, and peremptory in his manners and conversation, but totally devoid of ill-nature or malevolence; indeed, there was an original humour and pleasantry in his most lively sallies of anger or impatience, which, with his broken English, were extremely risible. His natural propensity to wit and humour, and happy manner of relating common occurrences, in an uncommon way, enabled him to throw persons and things into very ridiculous attitudes. Had he been as great a master of the English language as Swift, his *bons mots* would have been as frequent, and somewhat of the same kind.

Charles Burney, 'Sketch of the Life of Handel', 1785

I had the pleasure of seeing and conversing with Mr Handel, who, with his other excellences, was possessed of a great stock of humour. No man ever told a story with more effect. But it was requisite for the hearer to have a competent knowledge of at least four languages – English, French, Italian, and German, for in his narrative he made use of them.

Dr W.C. Quinn in a letter to Charles Burney, 1788

In temper he was irascible, impatient of contradiction, but not vindictive; jealous of his musical pre-eminence, and tenacious in all points, which regarded his professional honour.

William Coxe, *Anecdotes of Handel*, 1799

14

Portraits of Handel

1 [*in colour between pp. 36–7*]

By Balthasar Denner (1685–1749)

Painted by the composer's exact contemporary and compatriot, Balthasar Denner, this is one of the earliest authentic portraits of Handel, if not the earliest. Like the Mercier (no. 2), another early image, it was apparently a gift from Handel to a friend, in this case to J.C. Smith the younger whose will, dated 13 May 1786, provides the first mention of the painting: 'I give to William Coxe Handel's pourtrait by Denner'.

Balthasar Denner was born in Hamburg but his career took him to various parts of Germany and to Denmark, Holland and England. He was in London from 1721 to 1725 and again from August 1726 to July 1728, and it must surely have been on one of these visits that Handel found time to sit to him, rather than on his hurried trip to the continent in 1729. As *Kerslake* has pointed out, a commission during Denner's last stay would seem reasonable on the grounds of costume (the looped ends of Handel's wig and the style of the steinkerk about his neck belong to the late 1720s).

Denner had painted the portraits of the family of Barthold Brockes, Handel's German patron, shortly before his first visit to England, but it is not possible to show that Handel knew of his work. He arrived in England bringing with him a highly finished portrait head of an old woman which was greatly admired for its minutely detailed finish, not least it seems by Handel himself, who acquired two similar portraits by him of an old man and an old woman (see Appendix 4). At the end of his second stay, Denner resolved to give up this very finished style and, according to George Vertue writing in 1728, 'to paint in a bolder & stronger manner & with more freedom'. If the Handel portrait is somewhat difficult to place, lying between his highly realistic style and the more painterly manner of his earlier work (such as the Brockes portraits), this is perhaps a reflection of Denner's changing taste. It may be compared to Denner's own self-portrait of 1730 in the Brunswick Museum.

An example of his freer late style is the portrait at Knole, signed and dated 1736, which has long been described as of Handel, perhaps since as early as 1778 when the Duke of Dorset is believed to have added it to his collection of musicians' portraits. However on grounds of likeness the identification seems improbable; in any case it does not seem possible for Denner and Handel to have met in 1736.

National Portrait Gallery, London
NPG 1976

Oil, 74.9 × 62.6 cms/$29\frac{1}{2}$ × $24\frac{5}{8}$ ins

Engraved by E. Harding for Coxe's *Anecdotes*, 1799

PROVENANCE
Not mentioned in Handel's will but said to have been given or bequeathed by him to J.C. Smith the younger (1712–95) according to Smith's step-son William Coxe (see *Anecdotes*, 1799); by then, under the terms of Smith's own will, drawn up in 1786, Coxe owned the portrait; it passed to his sister Martha (1749–*c*.1835), who had married, in 1768, Sir Peter Rivers; given by their daughter-in-law, Lady Rivers, to the Sacred Harmonic Society; sold by the 'late' Society at Christie's, 3 March 1883, lot 82, where purchased for Henry Littleton of Novello & Co; purchased from his son's executors by A.F. Hill, who gave it to the National Portrait Gallery, 1923.

VERSIONS
I) Anonymous (?later) oil, Civico Museo Bibliografico Musicale, Bologna, with the same head and shoulders but larger in format and showing Handel seated at a keyboard instrument; reproduced in *Connoisseur*, CCXV, February 1985, p. 94.
II) A lost oil, described only as Handel by Denner, about 71 × 58 cms, recorded in the Dutch Royal Collection from 1763 to 1798 (correspondence in NPG Archive).

LITERATURE
Vertue, III, p. 34; *Coxe* (advert.) and pp. 49, 55; *Schoelcher*, p. 356; *The Illustrated London News*, XXIV, 1859, pp. 577 (engraved), 580; W.C. Smith, 'More Handeliana', *Music and Letters*, XXXIV, 1953, p. 14; *Kerslake*, pp. 121–2.

Handel, miniature by G.A. Wolfgang the younger, c. 1737 (no. 5)

Handel by Balthasar Denner, c. 1726–8 (no. 1)

Handel by Thomas Hudson, 1748–9 (no. 11)

Handel by Thomas Hudson, 1756 (no. 12)

2 [detail in colour on front cover, in black and white on p. 8]

By Philip Mercier (1689–1760)

In this most informal of Handel's painted portraits, the greatest early image, Handel is seated at the harpsichord in the act of composing, quill pen in hand and an unfinished piece of music before him on the desk; the music is not identifiable.

Philip Mercier, born in Berlin of French Huguenot parents, had studied there and at Paris before settling in London by 1719. Until his move to York in 1739 he was one of the leading painters in London, introducing a style heavily influenced by the French rococo and by Watteau in particular. From 1729 to 1736 he was Principal Painter to Frederick Prince of Wales, and (as Handel did in music) gave lessons in painting to Frederick's sisters, the Princesses (see no. 112 for their group portrait by Mercier).

His portrait of Handel was dated stylistically to c.1730–5 in the 1969 Mercier exhibition and to c.1735–6 in the recent Royal Opera House exhibition, where its resemblance to his own self-portrait of 1735 is noted. This would make Handel fifty, which seems too late; as *Kerslake* points out, Handel looks younger here than in the majority of known portraits, and a date of about 1730 seems more likely. It is perhaps worth noting that Mercier was one of the subscribers towards the cost of building Covent Garden Theatre in 1733.

The portrait remained in Handel's possession until c.1748 when, according to an inscription on the back, it was given to his friend, Thomas Harris, a lawyer of Lincoln's Inn. He was called upon to witness Handel's will two years later, and was to witness all its subsequent codicils except the last one in which he himself was named as a beneficiary. Handel had known the Harris family for many years and the family correspondence shows a close concern for Handel's health and well-being as much as for his music. Thomas died childless and the picture passed to his brother, James (see no. 160), well known as a collector of Handel scores.

The Viscount FitzHarris

Oil, 127 × 101.6 cms/50 × 40 ins

Signed bottom centre *Phil. Mercier / pinxit An . . . .* (the date illegible)

PROVENANCE
Given by Handel to Thomas Harris, c.1748 (according to an inscription on the back of the re-lining canvas); by descent from Harris's nephew, the 1st Earl of Malmesbury, to the Viscount FitzHarris.

COPIES
I) By Miss Benson, c.1825, sold Christie's, 20 July 1872, lot 100, and again 18 January 1873, lot 75; subsequently acquired by the Händel-Haus, Halle, but destroyed 1943.
II) By Hans List, made since 1945 for the Händel-Haus from photographs.

LITERATURE
Kerslake, pp. 129–30, pls. 322–3; *Ingamells and Raines*, p. 24, no. 51.

EXHIBITED
National Portraits Exhibition, South Kensington Museum, 1867, no. 411; *Philip Mercier*, York City Art Gallery and Iveagh Bequest, Kenwood, 1969, no. 21; *Royal Opera House Retrospective 1732–1982*, Royal Academy, 1982–3, no. 188 (repr. in colour).

3 [illustrated on p. 34]

By an unknown engraver

This almost unknown portrait of Handel is an unfinished proof engraving, presumably made in England in view of the form of the early, probably contemporary, identification: 'Mr Handell' (the rest of the inscription, in French, was added at a much later date). The only other recorded example, also unfinished, is in the Bibliothèque Nationale, Paris (Schoelcher Collection). In the turn of the head and body and the style of the wig it is not unlike the oil portrait by Balthasar Denner (no. 1), though in reverse. It may date to about the same period or slightly later but cannot be identified with the print advertised on 2 March 1738, as the French inscription suggests, since this is now known to be that by Houbraken (no. 4). Who was responsible for this portrait is not known, but it does not appear to be in the style of Antoine Benoist (1721–70), mentioned in the inscription, who in any case did not come to London until the 1740s.

Royal College of Music

Engraving, unfinished, 30.2 × 21.9 cms/$11\frac{7}{8} \times 8\frac{5}{8}$ ins

Inscribed in an early eighteenth-century hand
Mr Handell, with a note in French added in the mid-nineteenth century
Ce portrait est entierem.^t [?] inconnu des marchands de gravures les plus experts de Londres et des conservateurs du cabinet des Estampes du British Museum. Mr Carpenter du British Museum trouve dans cette planche quelque ressemblance avec la manière de Benoist [?] graveur francais que vivait à Londres et qui produisit plusieurs ouvrages en 1747. Le portrait est probablement celui que Walsh annonce dans la London Gazette March 2^d 1738

LITERATURE
Julian Marshall, 'Handel', *A Dictionary of Music and Musicians*, ed. George Grove, I, 1879, p. 657.

4 [illustrated on p. 32]

By Jacobus Houbraken (1698–1780), the portrait after an unknown original, surround and scene underneath after Hubert François Gravelot (1699–1773)

Houbraken's engraving is the only portrait print known to have been published with Handel's approval. Though re-used in later publications, it was originally issued to subscribers to the score of his dramatic ode, *Alexander's Feast*, which was printed by John Walsh but published by Handel himself and available from his house in Brook Street. It was advertised in *The Country Journal* for 22 April 1738: 'This Day is publish'd (And are ready to be deliver'd to the Subscribers for Alexander's Feast) A Print of Mr. Handel. Engraved by the celebrated Mr Houbraken of Amsterdam. The Ornaments design'd by Mr Gravelot.' The print was not bound into the score but was delivered separately to the subscribers (who numbered 124 for 146 copies); it was first mentioned in an earlier advertisement, apparently contained in the *London Daily Post* for 15 June 1737, and then on 2 March 1738 was described as 'very near finish'd'.

Below the portrait is a scene from the opening of the first act of *Alexander's Feast*. The Greek Emperor Alexander is seated on the right with 'the lovely Thais by his side'; on the left is 'Timotheus plac'd on high', whose 'flying Fingers touch'd the Lyre', an image which subscribers to the score will readily have associated with Handel himself in view of the similarities to the almost exactly contemporary statue of Handel at Vauxhall Gardens.

Houbraken's print with its elaborate rococo surround and vignette beneath the portrait takes the same form as his engraved portraits for Birch's *Heads of the Illustrious Persons of Great Britain*, issued in parts from 1737. Handel is thus elevated to the ranks of the most distinguished company. The publication of the portrait in this particular format can be seen as an act of self-publicity, conceived at least as early as June 1737 at a desperate moment in his career. This was the time of his final struggle with the Opera of the Nobility, while he was struck down with illness.

Houbraken's portrait of Handel and a contemporary or slightly later French engraving by Schmidt (version II, listed below) both preserve the appearance of a lost original. While Sir John Hawkins, the only early commentator on Handel's likenesses, wrote of the Houbraken that 'the features are too prominent', he used the same image in the form of the Schmidt engraving as the source for the illustration in his *General History of Music* (1776), describing the portrait as having been painted abroad; it was one of the few pictures, he said, which were 'to any tolerable degree likenesses'. Hawkins's very different opinion of the two engraved versions of the same image may have been influenced by Houbraken's reversal of the original, his standard practice. This has the result of showing Handel's facial mole on his right cheek whereas Roubiliac shows it on his left (see no. 7).

Gerald Coke Handel Collection

Engraving, 35.6 × 22.1 cms/14 × 8¾ ins

Inscribed on the frame around the portrait
GEORGE FRIDERIC HANDEL, and at the bottom *J. Houbraken sculps. Amst.*

VERSIONS AND RELATED WORKS
I) An oil by Francis Kyte, 1742 (National Portrait Gallery), long considered to be the original until the Houbraken engraving was dated to 1738.
II) A related engraving with some differences by G.F. Schmidt at Paris, c.1738–44, perhaps published c.1741 with the French edition of Handel's concerto op. 4 no. 1 as suggested by William C. Smith (1953).
III) A pastel attributed to Susanna, Countess of Shaftesbury in the family collection.
IV) Other versions are listed by *Kerslake*, p. 123.

LITERATURE
Hawkins, II, p. 912; Keith Milnes, *Memoirs relating to the Portrait of Handel*, 1828 (where the Kyte oil, listed as I above, is proposed as the original); W.C. Smith, *Concerning Handel, his Life and Works*, 1948, pp. 111–44; W.C. Smith, 'More Handeliana', *Music and Letters*, XXXIV, 1953, pp. 23–4; *Smith*, p. 90, no. 1; *Kerslake*, pp. 122–3, pl. 33.

5 [in colour opposite p. 36]

By Georg Andreas Wolfgang the younger (1703–45)

This miniature appears to date to a visit to London in 1737 by the Augsburg-born painter and miniaturist, Georg Andreas Wolfgang. The evidence for this visit rests on two portraits which are closely related to the miniature: a drawing which, according to the signature, was made in London in 1737 and an inscribed engraving (see versions III and II below). Despite Handel's remarkably youthful appearance (he was 52 in 1737), a dating in the late 1730s, though queried by Smith, is not inconsistent with the style of the costume and wig and is confirmed by the presence of the score of *Alexander's Feast* (first performed in 1736 and revived in 1737) in the large painting attributed to Wolfgang in the Gerald Coke Handel Collection (see version I below).

Like the almost contemporary engraved portrait by Houbraken (no. 4), this miniature shows Handel with an organ in the background, an apt feature at any period of Handel's life given his prowess at the keyboard but particularly so in the late 1730s when his organ concertos had become a well-known feature of his oratorio performances. In 1738 his *Six Concertos Opus 4* were published, including two given when *Alexander's Feast* was first performed in 1736.

By Gracious Permission of Her Majesty The Queen

Miniature on ivory, 17.0 × 13.4 cms/6¾ × 5¼ ins

Signed bottom left *G A/W*

PROVENANCE
Perhaps acquired for the Royal Collection by the Prince Consort on 4 December 1860 (according to an annotated list of 1881 in the Royal Archives; information from Richard Walker).

VERSIONS
I) A very similar oil portrait, attributed to G.A. Wolfgang the younger, size 121.5 × 98 cms, showing a score marked *Alexanders Feast* in Handel's right hand, is in the Gerald Coke Handel Collection. This was in the K.E. Henrici sale, Berlin, 24 March 1928, lot 129, when it was described as from the collection of Dr Harry E. Smith of Streatham.
II) An engraving by J.G. Wolfgang (1662–1744) after his son, G.A. Wolfgang the younger, inscribed as painted in London and engraved in Berlin, is similar to the miniature but shows Handel at half-length against a plain background.
III) A drawing also by G.A. Wolfgang the younger, signed and dated *G.A. Wolffgang, Pinx. London 1737*, is in the Musikbibliothek Peters, Leipzig; this is closely related to the engraving.

LITERATURE
W.C. Smith, *Concerning Handel, his Life and Works*, 1948, p. 132; *Kerslake*, p. 130, pl. 325.

6 [*in colour between pp. 44–5*]

Model for the Vauxhall Gardens Statue

By Louis François Roubiliac (1702/5–62)

This model, one of Roubiliac's most satisfying terracottas, is for the life-size statue of Handel commissioned by Jonathan Tyers (see no. 142) for Vauxhall pleasure gardens, since 1965 in the Victoria and Albert Museum. The statue was first mentioned in the *London Daily Post* on 15 April 1738 and was unveiled at Vauxhall at the beginning of the summer season on 1 May. Its original setting in a niche surmounted by a group of 'Harmony and Genii', also by Roubiliac, is shown in an early engraving (no. 144).

A sculpture on this scale, Roubiliac's first major commission, must have taken a considerable time to complete, and was perhaps begun in the summer of 1737 before Handel's departure for the continent in September. His season at Covent Garden earlier in the year had ended with a performance of his setting of Dryden's ode, *Alexander's Feast*, on 25 June. It is this work, subtitled 'The Power of Music' that Roubiliac must have had in mind: he makes Handel the modern embodiment of the verses of the ode's opening scene, 'Timotheus plac'd on high . . . with flying Fingers touch'd the Lyre'. Furthermore *Alexander's Feast* is the only named score in the pile of volumes on which Handel leans.

Handel's contemporaries saw the statue as representing him in the character of Apollo or, more usually, that of Orpheus, as for example in these early verses:

Drawn by the Fame of these embower'd Retreats,
See Orpheus rising from th' Elysian Seats!
Lost to th' admiring World three thousand Years,
Beneath great Handel's Form he re-appears.

The sunburst on Handel's lyre in the full-scale marble would however favour an identification with Apollo. This is confirmed by the account of the circumstances in which Roubiliac, then an assistant of the sculptor Henry Cheere, received the commission, as recorded by J.T. Smith, the son of Roubiliac's own pupil Nathaniel Smith:

At the time Mr. Tyers had engaged in the Vauxhall-Garden speculation, he requested the advice of Mr. Cheere as to the best mode of decoration. 'I conclude you will have Music,' observed Cheere, 'therefore you cannot do better than to have a carving of an Apollo. What do you say to a figure of Handel?' 'Good,' replied Jonathan, 'but that will be too expensive, friend Cheere.' 'No,' answered the sculptor, 'I have an uncommonly clever fellow [this was Roubiliac] working for me now, . . . employ him, and he will produce you a fine statue.'

This terracotta is a work of great delicacy and is a masterpiece of the English rococo. In 1751 it was seen by George Vertue, who describes it very clearly as 'the Model in clay baked of Mr Handel done by Mr Roubilac – the same from which the Foxhall statue in Foxhall Gardens was done . . . this modell near 2 foot high is in posses of Mr Hudson painter'. Thomas Hudson painted two portraits of Handel (nos. 11 and 12); he was a friend of Roubiliac's and owned a number of his terracotta studies.

The Syndics of the Fitzwilliam Museum, Cambridge
M3–1922

Terracotta, 47 cms/18½ ins high

PROVENANCE
Thomas Hudson; his sale, Christie's, 26 February 1785, lot 37; purchased by Nathaniel Smith and sold by him to his master, Joseph Nollekens; Nollekens sale, 3 July 1823, lot 60 (as Carlini), purchased by 'Hamlet, the silversmith', presumably Thomas Hamlet; probably purchased for Northwick Park at Hamlet's sale held by Robins, 29 July 1833 (the only recorded copy of the sale catalogue derives from the Northwick Park Library); Captain E.G. Spencer-Churchill, Northwick Park, by whom given to the Fitzwilliam Museum, 1922.

LITERATURE
Vertue, III, pp. 84, 157; J.T. Smith, *Nollekens and his Times*, II, 1828, p. 93; *Esdaile*, pp. 41f, 172; Terence Hodgkinson, 'Handel at Vauxhall', *Victoria and Albert Museum Bulletin*, I, no. 4, 1965, pp. 1–14; *Kerslake*, p. 129.

EXHIBITED
Handel and the Fitzwilliam Museum, Fitzwilliam Museum, 1974, no. 2; *Rococo*, no. F9.

7 [*in colour opposite p. 44*]

By Louis François Roubiliac (1702/5–62)

The terracotta model for the marble bust of 1739 at Windsor. According to Coxe's *Anecdotes* the marble was owned by Handel who gave it to J.C. Smith; he in turn gave it, with Handel's music manuscripts, to King George III. The terracotta, one of Roubiliac's finest early works, is probably the 'Mr Handel' seen together with other busts by Vertue in Roubiliac's studio in 1741, and may have been the one in the sculptor's sale in 1762. It is closely related to the marble bust, compared with which it is marginally larger, and has the added naturalistic detail of the mole on Handel's left cheek; this feature is only found in one other portrait, the engraving by Houbraken (no. 4).

The Thomas Coram Foundation for Children

Terracotta, 72.4 cms/28½ ins high

PROVENANCE
According to Brownlow (1858) this terracotta belonged to Barrett, the proprietor of Vauxhall Gardens (perhaps George Rogers Barrett, d. 1818), and then to the bass singer James Bartleman (1769–1821), though it was not included in the Bartleman sale at White's, 20 February 1822; offered for sale, 1824, by H. Rodd of 9 Great Newport Street; subsequently acquired by Sir Frederick Pollock, 1st Bart (1783–1870) and given by him to the Foundling Hospital, 1844.

VERSIONS
I) An early plaster cast, formerly in the Gambier Parry collection, was acquired by the Huntington Art Gallery, San Marino, California in 1972.
II) Other plaster casts belong to the Thomas Coram Foundation and the Gerald Coke Handel Collection: Schoelcher (1857) wrote that 'a very good cast... has been taken, and copies may now be easily obtained'.
III) A bronze is in the Händel-Haus, Halle.
IV) Early saleroom references to busts of Handel by Roubiliac are difficult to relate to any particular bust and may just as well refer to the later terracotta (no. 16) as the 1739 type. In Roubiliac's own sale (1762) there were five plaster busts, a terracotta and two moulds. At the following auctions, Handel busts were offered for sale: John Blackwood, February 1778, lot 1326 ('Roubiliac, marble busto of Handel'); John Stanley, Christie's, 24 June 1786, lot 89 ('a remarkable fine bust, exquisitely modell'd, by Roubilliac'); anonymous (?John Belchier), Christie's, 29 March 1805 (a terracotta model for a bust of Handel); D. Lysons (*Environs of London*, II, 1795, p. 22) notes in the possession of John Bacon Esq (not the sculptor) 'the original cast of Roubiliac's bust of Handel'. Other Handel busts, probably but not necessarily based on Roubiliac, were advertised by F. Bull in 1758 (see no. 14), Charles Harris, *c.* 1780 (see no. 16) and by Domenico Brucciani & Company in the nineteenth century ('Handel, four sizes', 28, 22, 18 and 10 inches high; their catalogues are in the Victoria and Albert Museum Library).

LITERATURE
Schoelcher, p. 354; J. Brownlow, *History of the Foundling Hospital*, 1858, p. 137; *Esdaile*, pp. 51, 226; *Vertue*, III, p. 105; John Kerslake, 'Roubiliac's "Handel": a terracotta restored', *Burlington Magazine*, CVIII, 1966, p. 475, figs. 46–7; *Nicolson*, pp. 87–8, no. 102; *Kerslake*, pp. 127, 130, pls. 329–30.

8

'The Harmonious Boar'

By Joseph Goupy (*c.* 1680–*c.* 1770)

Joseph Goupy's notorious caricature of Handel's gross appetite is said to have ended their friendship and cost him a legacy in Handel's will. The drawing is a satire on Handel's gluttony (he is shown in the character of a fat pig or boar surrounded by food and drink), his vanity (the mirror), his fondness for loud orchestral effects (the drums, horn, trumpet and bassoon) and perhaps, as the anecdote repeated below suggests, his deceit (the owl on his head). Other features – the large white wig and the fat hands with tiny fingers (described by Burney as 'like those of a child') – serve to make the caricature more telling. As engraved by Goupy (see no. 9, version I, 'The Harmonious Boar'), the horse is turned into a braying donkey and the background arch is filled with smoking cannons – perhaps an allusion to the Fireworks Music of 1749 with its 101 cannons (see no. 201). The scroll trodden underfoot, 'Pension

Benefit Nobility Friendship', is less easy to interpret but may be a reference to patronage and friendship disdained.

Horace Walpole says that Goupy had quarrelled with Handel. A much later account given by Laetitia Hawkins, the daughter of the musical historian Sir John, in her *Anecdotes* of 1822 gives more details:

> At a time when Handel's circumstances were less prosperous than they had been, he invited Goupy to dine with him. The meal was plain and frugal, as he had warned his guest it must be; and for this Handel again apologised, adding that he would give him as hearty a welcome as when he could treat with claret and French dishes. Goupy returned a cordial reply; and they dined. Soon after dinner, Handel left the room, and his absence was so long, that Goupy at last, for want of other employ, strolled into the adjoining back-room, and walking up to the window, which looked diagonally on that of a small third room, he saw his host sitting at a table covered with such delicacies as he had lamented his inability to afford his friend. Goupy, to whom possibly such viands had little less relish than to his host, was so enraged, that he quitted the house abruptly, and published the engraving or etching ... in which Handel figures as a hog in the midst of dainties.

Laetitia Hawkins places this incident at the time of the first performance of Handel's oratorio *Deborah* in 1733, but a date in the later 1740s would seem much more likely, since Handel's friendship with Goupy (see no. 91) continued until at least 1743. The three engravings of this subject (see no. 9) all appear to date from the period 1749–54.

Henry Angelo in his *Reminiscences* reports that Goupy's pupil, Lady Burlington, was said to have had a hand in fabricating this satire. This may be an extension of the story associating her with Goupy's earlier caricature of Farinelli and Cuzzoni with Heidegger (see no. 121).

The Syndics of the Fitzwilliam Museum, Cambridge
No. 961

8

Gouache, black and coloured chalks on paper, laid on canvas, 39.4 × 32.4 cms/15½ × 12¾ ins

Inscribed on a scroll under Handel's foot
Pension Benefit Nobility Friendship and at bottom right *Pax*

PROVENANCE
Probably the picture in the artist's sale at Mr Hobbs, 23 March 1765, lot 39 (Goupy, 'An Hog playing on an Organ, with various Embellishments, all Emblematical'), where noted by Horace Walpole; Puttick and Simpson, 29 January 1858, lot 201, when it was acquired by Edward F. Rimbault; W.H. Cummings; given by six friends of the late R.A. Streatfield to the Fitzwilliam Museum, 1919.

LITERATURE
Laetitia-Matilda Hawkins, *Anecdotes, Biographical Sketches and Memoirs*, I, 1822, pp. 196f; Horace Walpole, *Anecdotes of Painting in England*, 1862, ed. J. Dallaway and R.N. Wornum, III, p. 752; E.F. Rimbault, 'Goupy's Caricature of Handel', *Notes and Queries*, 5th series, V, 1876, pp. 263–5; Henry Angelo, *The Reminiscences of Henry Angelo*, I, 1904, p. 316.

EXHIBITED
Handel and the Fitzwilliam, Fitzwilliam Museum, 1974, no. 3.

9

9
'The Charming Brute'

By an unknown engraver

Goupy's caricature of Handel was first engraved and published by Goupy himself, it seems, in an unsigned print entitled 'The true Representation and Character &c' which follows the original pastel (no. 8) fairly closely but in reverse. The accompanying verses describe Handel as a 'Harmonious Boar'. This engraving is one of two anonymous later derivatives, entitled 'The Charming Brute', where the emphasis both in the image and the accompanying verses ('His sole Devotion is to Eat') is placed more strongly on Handel's enormous appetite. A satirical poem, *The Scandalizade*, published in 1750, takes up the same theme of his great appetite.

Gerald Coke Handel Collection

Engraving, 30.2 × 23.5 cms/11⅞ × 9¼ ins

Inscribed below the image
THE *Charming* BRUTE− / *Pub according to Act of Parliam.!*
March 21ˢᵗ 1754,
and with eight lines of verse beginning
The Figure's odd; within the image on a scroll,
I AM MYSELF ALONE and with a bill of fare hanging from his pocket

VERSIONS
I) 'The true Representation and Character &c' or 'The Harmonious Boar', 38.4 × 30.5 cms, closely related to the Fitzwilliam Museum pastel (no. 8); *Satires*, no. 3272.
II) 'The Charming Brute', closely related to this engraving of 1754 but without the satyr holding up the mirror, and with *Messiah* engraved on the music at bottom right; it is not listed in *Satires*.

LITERATURE
Satires, no. 3273; Richard W. Wallace, 'Joseph Goupy's Satire of George Frideric Handel', *Apollo*, CXVII, 1983, pp. 104–5.

10

By an unknown artist

Once one of the best-known portraits of Handel and long described as by Thomas Hudson, this painting now appears to be the work of an altogether less familiar hand: perhaps, as tentatively suggested by *Kerslake*, that of Francis van der Myn (1719–83), one

10

of a family of Flemish artists who worked in England. It was formerly linked with Faber's mezzotint of Handel after Hudson, but this is now known to have been engraved from the Hamburg picture (no. 11).

The portrait came to the Royal Society of Musicians in 1851 after a dispute about the intentions of the first known owner, Redmond Simpson (see under 'Provenance'). The Society also owns an identical painting of the composer which has been in their possession since at least 1813. A close examination of the two side by side shows a greater liveliness in Simpson's portrait and suggests that the other may be a copy.

The Royal Society of Musicians of Great Britain

Oil, 77 × 64.3 cms/30¼ × 25¼ ins

PROVENANCE
The first known owner was Redmond Simpson (d. 1787), by whom given *c.* 1785 to the Concert of Ancient Music (together with portraits of Geminiani, Purcell and Corelli) and displayed in their then concert hall at the Tottenham Street Rooms (where listed in 1785 with the other three portraits); at the demise of the Ancient Concert, passed according to the conditions of Simpson's gift (but not without dispute with the Royal Academy of Music), to the Royal Society of Musicians, who received all four portraits in 1851.

VERSIONS
I) An almost identical portrait of the same size, also belonging to the Royal Society of Musicians, was in its possession by 3 April 1813, source unknown (it was then decided to clean and frame it to correspond with a portrait of George III).
II) A similar but smaller portrait, size 27.9 × 21.6 cms in the collection of the National Portrait Gallery (NPG 2151), comes from the W.H. Cummings Collection.

LITERATURE
Unpublished papers belonging to the Royal Society of Musicians (deed of gift, 25 April 1785, signed by Simpson and Sir Watkin Williams Wynn; letter, 17 October 1839, from W. Dance; Minute Book, entries for 3 April 1813, 1 September 1850, 6 July and 7 September 1851); Robert Elkin, *The Old Concert Rooms of London*, 1955, p. 89; Kerslake, pp. 125–6.

11 [*in colour between pp. 36–7*]

By Thomas Hudson (1701–79)

Thomas Hudson's strongly posed portrait of the composer in his maturity (he was 64 at the time of its completion) is a particularly revealing image. Set against a dark background, Handel's heavy face is framed in the bushy white wig for which he was known, and his corpulence is emphasised by the line of the coat. It is the portrait of a man of character and strong will, in which the whole pose is given force by the gesture of his right hand pressing down on his thigh (emphasised by the extended thumb) and his left holding up a score (perhaps a part of *Messiah*, which lies on the table).

Much speculation has centred on the dating of this portrait. In an effort to explain the existence of engravings dated 1748 (listed below), the signature on the portrait, which clearly registers as *T Hudson. Pinxit. 1749*, has been read as 1747. However the answer lies in changes Hudson made to the oil. The sequence of events appears to begin with the engraving of Hudson's apparently finished portrait in 1748. For some reason the artist then decided to make further changes: in particular he altered the embroidery on the composer's right-hand coat cuff and the decoration of the table leg to give us the portrait we have now. The traces of its original appearance can be made out in the paint, and are fully confirmed by the engravings, which show the coat cuff embroidered with a wide zig-zag band beneath a narrow straight one, and the table with a tightly coiled spiral scroll in the position of the present stylised rococo shell.

Thomas Hudson was apparently the only English-born artist to have the opportunity to paint not just one large portrait of Handel but two (see no. 12 for the later full-length). It seems likely that Handel will have gone to him not so much as the fashionable portrait painter of the moment (Hudson was at the height of his success in the late 1740s), but as an artist well known to him over many years from his portraits of his friends and associates: Senesino and Pepusch in the 1730s, Susannah Cibber in 1739, John Beard in about 1743 and Charles Jennens in 1747 (nos. 181, 185 and 187). Hudson had an interest in music, judging from the appearance of his name in subscribers' lists to new music publications (though only one of these, Boyce's *Twelve Sonatas for Two Violins* of 1747, specifies his address; so enabling one to be sure that it was the artist and not someone else of the same name who subscribed). Quite how well Hudson and Handel knew each other is difficult to say. They were both Governors of the Foundling Hospital, for which Hudson offered to paint the composer's portrait in 1750 (see p. 223). They shared an interest in collecting Old Master paintings, sometimes buying at the same auction sale. In addition, one of the treasures of Hudson's own collection was Roubiliac's terracotta model for the Handel statue at Vauxhall (no. 6).

Staats- und Universitätsbibliothek, Hamburg

Oil, 121.0 × 100.5 cms/47⅝ × 39½ ins

Signed to the left of the table leg in monogram *THudson. Pinxit./1749*. and inscribed on the paper on the table [ME]*SSIAH/AN/*[OR]*ATORIO*

ENGRAVED
I) In mezzotint by Andrew Miller as *Thos. Hudson pinxit London 1748* (CS 22a).
II) In mezzotint by John Faber (twice; it is assumed some accident may have happened to the first plate), firstly as *T. Hudson pinxt. I. Faber fecit. 1748.* and then again with the date as 1749 (CS 175 and 175a).
III) Most later engravings of the picture derive from the above.

PROVENANCE
Said to have been taken by Handel to his relatives in Halle on his last visit to Germany in 1750 and then (according to Förstermann, 1844) by descent to two grand-daughters of Handel's niece Johanna Friederika Flörke, Mademoiselle Collas du Bignon and Frau Professor Senff; sold for Professor Senff by Chrysander to a group of Hamburg art lovers with the idea that it should join the Handel manuscripts (see no. 129) in the Hamburg Library; recorded as presented to the Library by Friedrich Gültzow of Hamburg in 1883.

VERSIONS
I) The best known is the painting given by Samuel Howard to the University of Oxford in 1778 (*Poole*, I, no. 257) – a stiff copy, and certainly not by Thomas Hudson (*Landon*, p. 209, repr. in colour).
II) A drawing sold at Hôtel Drouot, Paris, 12 September 1920, lot 196, catalogued as dated 1 June 1743 but probably later in origin and perhaps from the mezzotint by Faber.

LITERATURE
K.E. Förstermann, *Georg Friedrich Haendel's Stammbaum*, Leipzig, 1844, p. 12; F. Chrysander, 'Händels Originalportrait von Hudson', *Allgemeine Musikalische Zeitung*, IV, 1869, pp. 126–7; *Kerslake*, p. 124, pl. 336.

EXHIBITED
Händel und Hamburg, Staats- und Universitätsbibliothek, Hamburg, May–June 1985, no. 66.

12 [*in colour opposite p. 37*]

By Thomas Hudson (1701–79)

This monumental full-length image is the only known painting of the composer in the blindness of his old age. Executed in 1756 for Charles Jennens, the librettist of *Messiah*, it is the work of Thomas Hudson, to whom Jennens had turned for his own portrait (no. 181) in 1747. Like Hudson's earlier portrait of Handel (no. 11), it includes the score of *Messiah*.

A preparatory sketch by Hudson, in the Royal Collection at Windsor, shows that in developing the composition the artist decided to give greater prominence to his subject by lowering the point of view and reducing the composition on both sides. The magnificent frame complete with musical trophies is in keeping with Jennens's reputation for grand effects; not for nothing was he nicknamed 'Solyman the magnificent'.

Although the portrait was kept in London at first, Jennens soon moved it to Gopsall, his palatial home in Leicestershire, and the portrait remained in his nephew's family until 1967 when it became the subject of the National Portrait Gallery's first national appeal. Hudson's earlier portrait of Handel, despite being sent to his family in Germany, was widely circulated through Faber's mezzotint, and remained the better known of the two images until recently. They are brought together here for the first time.

National Portrait Gallery, London
NPG 3970

Oil, 238.8 × 146.1 cms/94 × 57½ ins

Signed and dated bottom left *T. Hudson Pinx!/1756*, and inscribed *MESSIAH* on the open book

PROVENANCE
Charles Jennens (first recorded in his London home in 1761); bequeathed by him to his nephew Penn Assheton Curzon and thence by descent to the 6th Earl Howe;
his sale at Christie's, 7 July 1967, lot 104, purchased by Leggatt, from whom purchased by the National Portrait Gallery, 1968.

LITERATURE
R. and J. Dodsley, *London and its environs described*, V, 1761, p. 91; *Kerslake*, pp. 123–5.

EXHIBITED
Thomas Hudson 1701–1779, Iveagh Bequest, Kenwood, 1979, no. 56.

13

After Thomas Hudson (1701–79)

A head and shoulders version of Hudson's full-length of 1756 (no. 12), very similar in format to the oval portrait after Hudson in the Royal Collection and perhaps deriving from it. The portrait was given to Oxford University by the dramatist George Colman or his son.

Faculty of Music, Oxford University
(By Permission of the Heather Professor of Music)

Oil, 80 × 71.1 cms/31½ × 28 ins, oval

PROVENANCE
Given by George Colman to the Music School, University of Oxford, before 1796.

LITERATURE
Poole, I, no. 389.

Handel by L.F. Roubiliac, *c.* 1739 (no. 7)

Handel, model for the monument in Westminster Abbey, by L.F. Roubiliac, c. 1759 (no. 17)

Handel, model for the Vauxhall Gardens statue, by L.F. Roubiliac, *c.* 1737 (no. 6)

George I by J.M. Rysbrack (no. 41)

14 [*illustrated on p. 35*]

By an unknown sculptor

Formerly attributed to Roubiliac as a possible study for the Windsor marble or the Vauxhall statue (no. 6), this little bust was then optimistically equated with one of the terracottas in the sculptor's posthumous sale in 1762. But a comparison with the full-size terracotta model (no. 7) in the Thomas Coram Foundation shows that the present bust is the work of another, altogether less experienced, hand.

Although quite finely finished in its surface detailing, the bust lacks the sophistication of a life-sitting to a major sculptor. In view of its scale, it may perhaps have been intended as a prototype for one of the editions available in Handel's last years and after his death. It is however too small to be identified with the earliest of these, advertised by F. Bull (otherwise unknown) in an edition of thirty in April 1758: 'To the Lovers of Music . . . a Busto of Mr. Handel . . . which will make a rich and elegant Piece of Furniture . . . to be twenty three and a half high, and eighteen Inches broad.' Somewhat later, in a catalogue of about 1780, Charles Harris was offering three sizes (see under no. 16); this terracotta bust corresponds most closely to Harris's middle size but there is no evidence otherwise to substantiate a link with him.

There is a nineteenth-century cast of this bust by D. Brucciani & Co. in the Gerald Coke Handel Collection. A closely related bust of this kind is the almost identical small bronze acquired by the Fitzwilliam Museum in 1977. Until the discovery of an inscribed or documented example, it will be difficult to be more precise about the origins of this type.

National Portrait Gallery, London
NPG 878

Terracotta on a modern painted plaster socle,
43.8 cms/17¼ ins high

PROVENANCE
Said to have been acquired at either the sale of William Kitchener, an amateur musician (1775–1827), or that of the singer James Bartleman (1769–1821) by the conductor Sir George Smart (1776–1867; see no. 257); given by him to Richard Clark (1780–1856), from whom received *c.*1852 by W.F. Whitehall; given by him to the National Portrait Gallery, 1891.

LITERATURE
A Catalogue of the Statues, Bass Reliefs, Bustos, &c. of Charles Harris, Statuary, n.d., pp. 12, 14, 15 (unique copy in Victoria and Albert Museum Library); *Esdaile*, pp. 51–2, pl. 10a; *Kerslake*, pp. 126–7.

15

Lost Death-mask

Photograph of a cast of the mask taken by Louis François Roubiliac (1702/5–62)

The death-mask will have been taken as soon as possible after Handel's decease, which occurred at 8 am on Saturday 14 April 1759.

The earliest reference to the mask is in the weekly magazine, *The Mirror*, in 1834: 'Roubiliac after the death of Handel, produced a mould from his face, which he afterwards touched upon, by opening the eyes. From this mould a few impressions were taken . . .'. In an unpublished typescript of the Schoelcher Collection, now in the Bibliothèque Nationale in Paris, a plaster cast is described as having been made by J. Bishop & Son.

No example of the mask can now be traced. One cast was sold with the W.H. Cummings Collection at Sotheby's, 21 May 1917, lot 826, when it was bought by Maggs; two were listed in the *Musical Times* in August 1941; another was offered to the National Portrait Gallery by Thomas Armstrong in 1948 but, sadly, refused. A photograph of the Armstrong mask from the Gallery files is shown here.

The mask seems to have formed the basis for Roubiliac's monument in Westminster Abbey (see no. 17); in this the features are somewhat simplified and at the same time strengthened so that the memorial is effective seen from the transept below. Much closer to the mask is the recently identified posthumous bust of the composer from Grimsthorpe Castle (no. 16).

National Portrait Gallery, London

Photograph, 19.3 × 14.3 cms/7⅝ × 5⅝ ins

LITERATURE
The Mirror, XXIV, no. 873, 19 July 1834 (repr.); Schoelcher, pp. 355–6; Julian Marshall, 'Handel', *A Dictionary of Music and Musicians*, ed. George Grove, I, 1879, p. 656 (repr.); *Musical Times*, LXXXII, 1941, p. 307 (repr.).

16

Attributed to Louis François Roubiliac (1702/5–62)

In its mood of almost sad introspection, this bust of Handel is quite different from Roubiliac's earlier images (nos. 6 and 7). It is the only portrait other than the Westminster Abbey monument to show the composer bare-headed, with neither wig nor cap. From its close similarities to the monument, and more especially to Handel's death-mask (see no. 15), it would seem that this bust was made late in his life or, more probably, shortly after his death.

The bust has not previously been published and was unknown until its identification at Grimsthorpe Castle in Lincolnshire by both John Mallet and Malcolm Baker. No record has been found in the family papers of the Dukes of Ancaster as to how it came to Grimsthorpe.

In the absence of documentation, it is not possible to verify an attribution to Roubiliac beyond doubt. The face bears a striking similarity to the death-mask taken by Roubiliac while the form of the draperies (despite the somewhat patchy nature of the surface) may be compared with Roubiliac's contemporary bust of Joseph Wilton at the Royal Academy. The likelihood is that this bust of Handel was made in Roubiliac's studio while he was working on the monument. It should however be noted that the sculptor Charles Harris (d.1795?), who provided the monument to the 3rd and 4th Dukes of Ancaster (presumably shortly after the death of the latter in 1779), was advertising busts of Handel in his catalogue of c.1780, offering three different sizes ('large as life', 16 to 18 inches and 11 inches). These were probably derived from Roubiliac, though not specified as such, and were almost certainly in plaster, from their prices (2 guineas, 16/– and 7s. 6d). Other portrait busts

16 (before cleaning)

which Harris offered can be associated with Roubiliac. Perhaps he acquired this Handel terracotta from Roubiliac's studio sale with the intention of making casts and then supplied the original to Grimsthorpe.

The bust is also intriguing for the hair-line cracks which intersect Handel's face. In this less than fully presentable form it demonstrates the problems and techniques of the eighteenth-century sculptor in terracotta. Some sculptures, such as the earlier terracotta by Roubiliac (no. 7), have the appearance of being modelled directly in clay and then fired. This baking process was hazardous: at best it could lead to firing cracks and at worst the complete explosion of the work in the kiln, if the clay was not evenly packed. But it was standard practice to take a mould – probably, in the case of a terracotta, even before firing, as an insurance against disaster. The mould was like a skin, taken in as many separate pieces as the shape and projections of the original demanded for practical reasons. From this mould, once the pieces had been clamped together, a cast could be made by pressing clay or plaster into the shape: hence the term 'press-mould', generally applied to terracottas of this type. In the case of the bust, certain cracks in the surface such as that across the forehead seem to be firing marks, but perhaps following the joins in the press-mould, while others, such as faintly visible raised lines in the body of the sculpture may result from the process of taking a press-mould, during which clay would be squeezed upwards and out through the joins of the mould to form a faint ridge. Our knowledge of the technique of English eighteenth-century terracottas is developing as more examples are subject to close study, and it is still the subject of more than one point of view. The present analysis benefits from conversations with John Larsen, Malcolm Baker and Richard Cook.

A later marble of this type is in the Royal Collection at Kew Palace. It was at Buckingham House in 1817, where it was placed above the organ in the Queen's Japan Room, as shown in an engraving by Pyne. Another marble, recorded in the Alfred Morrison Collection last century as bare-headed, was perhaps of this type. The bust appears in a rather different guise in Sir William Beechey's full-length portrait, *Mrs Montague and her Sister Decorating the Bust of Handel*, shown at the Royal Academy in 1802 and sold recently at Sotheby's, New York, 17 June 1982, lot 80.

Grimsthorpe and Drummond Castle Trustees

Terracotta, 71 cms/28 ins high

PROVENANCE
Acquired by the Dukes of Ancaster at an unknown date.

17 [*in colour between pp. 44–5*]

Model for the Monument in Westminster Abbey

By Louis François Roubiliac (1702/5–62)

In his final illness, three days before his death on 14 April 1759, Handel signed a last codicil to his will saying that he hoped to be buried in Westminster Abbey and desiring that his executor, George Amyand, 'may have leave to erect a monument for me there and that any sum not Exceeding Six Hundred Pounds be expended for that purpose'. The Dean and Chapter of Westminster received a fee of 25 pounds in 1759 for a site for the monument in the south transept of the Abbey. No reference to Roubiliac occurs among Amyand's payments as Handel's executor.

The monument was 'opened', according to the *Gentleman's Magazine*, on 10 July 1762, six months after the sculptor's own death and more than three years after Handel's. It represents Handel in a moment of inspiration, with pen in hand, pointing to a figure above (described as David in the *Gentleman's Magazine*), and with part of *Messiah* before him, the immortal words 'I know that my Redeemer Liveth'. Just as Roubiliac's Vauxhall Gardens statue of Handel is his first major work, so the Westminster Abbey monument is his last.

Roubiliac took a death-mask of Handel (see no. 15) which, according to Schoelcher (1857), served as the basis for his likeness. This Hawkins described as 'the most perfect resemblance [in which] the true lineaments of his face are apparent'.

The Ashmolean model is on a particularly bold scale for a Roubiliac terracotta and is made up of two parts, with the architectural setting separate from the figure and table. It is close in detail to the monument as executed and must have served as the final model. A recently discovered terracotta, now in the Gerald Coke Handel Collection, is apparently an earlier sketch design showing the composer gazing to the right (repr. in colour, *Rococo*, pl. XXVI). Roubiliac's sale catalogue of 12 May 1762 does not list either of these terracottas but records a model (now lost) described as 'Mr Handell's monument' under 'Moulds in Plaister', lot 61 (second day); it includes four 'designs' as lot 70 which may also be for the monument.

A cast was made of the head of the monument in the last century, perhaps by Domenico Brucciani and Company. It should not be identified, as Esdaile proposed, with the designs listed in Roubiliac's sale catalogue. Examples are to be found in a number of major collections, including the Fitzwilliam Museum (which has two casts) and the British Library; the earliest appears to be that given to the Royal Society of Musicians in 1889.

The Visitors of the Ashmolean Museum, Oxford

Terracotta, 98.0 cms/38⅝ ins high

PROVENANCE
Given by James Wyatt to the University of Oxford, 1848.

LITERATURE
Gentleman's Magazine, XXXII, 1762, p. 340; *Schoelcher*, p. 355; *Esdaile*, p. 154; *Kerslake*, p. 131.

EXHIBITED
Rococo, no. S46 (repr.).

11 *Handel and the Courts of Europe* 1685–1712

The most extensive account of the early years of Handel's life and of his travels in Germany and Italy before his first visit to London at the age of 25 is given by John Mainwaring, whose biography of the composer was published in 1760, the year after his death (see no. 214). Towards the end of his account of Handel's time in Italy, Mainwaring tells us that the composer 'began to think of returning to his native country', adding, 'Not that he intended this to be the end of his travels: for his curiosity was not yet allay'd, nor likely to be so while there was any musical court which he had not seen'. Mainwaring devotes more than half his biography to these early years of Handel's on the continent, and the information and sentiments he provides must derive largely from the memories of Handel himself passed on by him to his amanuensis, John Christopher Smith the younger, under whose direction the biography was said to have been written.

Although other German musicians travelled widely, the range of experience Handel had gained by the time he was 25 must have been unrivalled. He had worked in four major German cities and four Italian, but most importantly he had had the opportunities to experiment and to realise a succession of highly ambitious compositions.

In Germany, then still a collection of independent states under the umbrella of the Holy Roman Empire, he worked at Halle and Hamburg before his Italian journey and at Hanover and Düsseldorf on his return. It was at Halle that he learnt the elements of composition and probably visited both Berlin and Leipzig where he will have had the opportunity to attend the opera. At the age of 18, in 1703, he moved to Hamburg where he had experience as an orchestral player before writing his first operas, *Almira* and *Nero*, for performance in 1705.

It was Italy which gave Handel the experience and confidence he needed as a composer. Between his arrival at about the age of 21 and his departure some three or four years later in 1710, he wrote his first large-scale religious works for performance in Rome, and two operas, *Rodrigo* for Florence and *Agrippina* for Venice, as well as many other cantatas and compositions. While he spent much of his time in Rome, he also worked in Naples and probably visited Siena.

When Handel returned to Germany in 1710 he spent some months at Hanover, where he was appointed Kapellmeister to the Elector, and a similar period at Düsseldorf. Both were courts where opera had thrived but was in abeyance due to the War of Spanish Succession (which Mainwaring described as the 'most prosperous war . . . which England had ever waged'). It was to London, the richest city in Europe but the last to succumb to the lures of Italian opera, that Handel went next.

LITERATURE
Mainwaring, pp. 69, 74.

Handel and his Family

Handel was born on 23 February 1685 at Halle, a town now in East Germany but then part of Brandenburg (the future Prussia). But Halle had belonged to Saxony until 1680, and old habits die hard: throughout his life Handel was known as a Saxon – 'viva il caro Sassone' was the cry at the Venice Opera in 1710; 'your petitioner was born at Hall in Saxony' is how Handel described himself for his naturalisation as a British citizen in 1727.

Handel was the son of a doctor, Georg Händel (see no. 19), by his second marriage, made when he was past sixty, to Dorothea Taust, the much younger daughter of the local pastor. For our knowledge of Handel's childhood, Mainwaring's biography is almost the only source, and presumably represents Handel's own memories as told by him to his amanuensis John Christopher Smith:

> From his very childhood Handel had discovered such a strong propensity to Music, that his father, who always intended him for the study of the Civil Law, had reason to be alarmed. Perceiving that this inclination still increased, he took every method to oppose it. He strictly forbad him to meddle with any musical instrument; nothing of that kind was suffered to remain in the house, now was he ever permitted to go to any other, where such kind of furniture was in use. All this caution and art, instead of restraining, did but augment his passion. He had found means to get a little clavichord privately convey'd to a room at the top of the house. To this room he constantly stole when the family was asleep.

Mainwaring continues with the story of how Handel, though still but a boy, attracted the attention of his father's employer, the Duke of Saxe-Weissenfels, by his playing on the organ. Through the Duke's persuasion, Handel was allowed to study music, and 'The first thing which his father did at his return to Hall, was to place him under one ZACKAW, who was organist to the cathedral church.' This was Friedrich Wilhelm Zachow, organist at the Liebfrauenkirche, who proved an effective teacher.

Mainwaring goes on to describe a visit by Handel to Berlin in 1698 and his meeting with the composers Giovanni Bononcini and Attilio Ariosti (see nos. 88–9); however perhaps Handel met them on separate occasions since Ariosti did not come to Berlin until 1702. This was the year Handel enrolled at Halle University and became organist of the Cathedral there. Handel was by now 17, and the following year left home for Hamburg (see no. 26). His visits to Halle in future were to be made as and when his travels allowed.

In 1708 Dr Michael Dietrich Michaelsen (see no. 21), a name often to feature in Handel's correspondence (see nos. 22–3), married Dorothea Sophia Händel, the composer's only surviving sister. Of their five children only one survived childhood: Johanna Friederika. It was fortunate for Handel that he was in Germany when she was born and was able to stand as a godparent at her baptism at Halle on 23 November 1711 – in the same church, the Liebfrauenkirche, where he himself had been baptised 26 years before. This favourite niece, who took her second name from his, was Handel's closest relative after his mother's death (see under no. 20) and was eventually to be made his residuary legatee.

18 [illustrated on p. 50]

Map and Views of Halle
By I. C. Homann
Published by Johann Baptista Homann (1664–1724)

This engraving offers a map of Halle and its suburbs and three views: in the centre the city of Halle as seen from the west; to left, the saltworks; and to right, the great lecture hall of the university. Under the Treaty of Westphalia, Halle passed from Saxony to Prussia in 1680.

Handel's parents were married in the church of St Bartholomew at Giebichenstein, marked on the map as a suburb of the city by the River Saale on the north side of Halle. The Liebfrauenkirche, where Handel was baptised and later learnt the organ under Friedrich Zachow, is the church with the twin towers, marked in the view as number 8 ('Die Markt, od Maria Kirch'). The cathedral or 'Dom', where Handel was appointed organist in 1702, is number 20 on the map. In the same year Handel enrolled at the newly founded University; the view of the lecture hall shows the installation ceremony for a new Pro-rector with the professors seated on the stage, the students in the body of the hall and musicians on the right in front of the windows.

The British Library Board
K. Top. 99.22

Engraving, 47 × 56.2 cms/18½ × 22⅛ ins

Inscribed DARSTELLUNG des GRUNDRISSES und PROSPECTES der Königl. Preussisch – Magdeburgischen und des Saal-Crayses Haupt Stadt HALLE | welcher daselbst auf Kosten und Verlag IOHANN BAPTISTAE HOMANNS . . . | von I.C. HOMANN . . . | zufinden in NÜRNBERG bei dem Authore.; the view of Halle titled *Prospect der Stadt HALLE wie solche von Abend anzusehen*; and the interior of the auditorium titled *Das grosse Auditorium | mit dem actu solemni der creation eines neuen Prorectoris academiae magnifici*.

19

Georg Händel (1622–97)

By J. Sandrart, probably Johann Jakob von Sandrart (1655–98), after Benjamin von Block (1631–90)

'The noble, honourable, greatly respected and renowned Herr Georg Hendel, duly appointed Valet to the Elector of Brandenburg' is how Handel's father is described in the register for his second marriage on 23 April 1683 to Dorothea Taust (see no. 20). From this union Handel was born nearly

19

two years later, and baptised as 'Georg Friederich' on 24 February 1685. Handel's father was a successful doctor working at the local court; since 1680 he had been 'barber-surgeon' (the two professions went together) to Johann Adolf I, Duke of Weissenfels, a short distance from Halle, and before that to Duke Augustus of Saxony. He was the son of Valentin Händel, a coppersmith, and had trained as a doctor at Hamburg and Lübeck before settling in Halle; his first wife, Anna, was the widow of the local surgeon Christoph Oettinger, and two of their sons followed in their father's profession.

The Trustees of the British Museum
B. b. 13–84

Engraving, 19.2 × 14.6 cms/7½ × 5¾ ins

Inscribed above the portrait *Georg Händel | Fürstl:Sächs:Magd: geheimbter Cammerdiner und Leib Chirurgus.*; below the portrait, four lines of verse signed *J.A.O.D.* and beginning *Artz-ney Kunst bringet gunst! die Wissenschaft besteht*; and signed at the bottom *B. Block pinxit. I. Sandrart sculpsit*

LITERATURE
Deutsch, p. 1.

20
Memorial Volume for Handel's Mother

Funeral sermon
Given by Johann Georg Francke
Printed by Johann Grunert, Halle, 1731

This traditional German memorial volume, a slim folio with black covers containing the sermon preached at the funeral of Handel's mother on 22 December 1730 together with other tributes, was printed in Halle at Handel's expense. On 10 August 1731 he writes to his brother-in-law, Michael Michaelsen (see no. 21), thanking him for sending copies of the volume and for having arranged the funeral in his absence.

In the sermon Handel is described as standing 'in especial grace, by reason of his exceptional knowledge of music, as Director of Music to the reigning Majesty in England and Elector of Hanover, George II'. His birth date is correctly stated, perhaps for the first time in print, as 23 February 1685. A suitably pious account of the life of Handel's mother is given. The daughter of Georg Taust, pastor of the parish of Giebichenstein near Halle, she was born in 1651 and christened Dorothea after her mother. In 1683 she was persuaded to marry the widower Georg Händel (see no. 19), 'Valet to H.R.H. August, Duke of Saxony', in a service 'which her father, to his greatest pleasure, could himself conduct'. In the seven years following her marriage to the 61-year-old Georg she had four children, of whom Handel was the second but the eldest to survive.

On his travels in Europe Handel seems to have visited his mother whenever he could, the last occasion being in June 1729 when he was seeking singers for the opera in London. In February 1731, two months after her death, Handel wrote to his brother-in-law: '. . . I note the care that you have taken in burying my Mother according to her last wishes. I cannot stop my tears here . . . her memory will always stay by me . . .'.

The British Library Board
C 70.i.11.(1)

Printed memorial volume, 28 pp., 36.5 × 23.2 cms/14⅜ × 9⅛ ins; bound with:
i) Funeral sermon for Handel's sister, Dorothea Sophia, died 11 August 1718, first wife of Dr Michael Michaelsen
ii) Funeral sermon for Christiane Sophia Dreysig, died 11 September 1725, second wife of Dr. Michaelsen
iii) Funeral oration for Maria Augusta Flörke, died 28 July 1730, unmarried sister-in-law of Johanna Friederika Flörcke
iv) Verses on the death of Johanna Friederika Flörcke, *née* Michaelsen, died 24 February 1771. The second child of Handel's sister, Dorothea Sophia, she was his niece and residuary legatee.

Titled *Die Wolthaten, welche Gott, durch einen seligen Todt, an seinen Gläubigen thut . . . Wurden, In dem, Der Weiland . . . Frau Dorotheen Taustin, Des Weiland . . . Herrn George Händels . . . Frau Witwe . . . Gehaltenen Leichen-Sermon Vorgestelt von Johann George Francken*

PROVENANCE
W.H. Cummings by 1881; his sale, Sotheby's, 21 May 1917, lot 823, purchased by the British Museum.

LITERATURE
W.H. Cummings, *Proceedings of the Royal Musical Association*, VII, 1881, p. 30; *Deutsch*, pp. 263–5 (with a translation of the sermon).

21

Michael Dietrich Michaelsen (1681–1748)

By Johann Christoph Sysang (1703–54)
after Christian Heinrich Sporleder (b. 1719)

Michaelsen married Handel's younger sister, Dorothea Sophia, in 1708; she gave birth to five children before her early death in 1718. The only one to survive childhood, Johanna Friederika, was described by Handel as his favourite niece; she received a ring from him for her wedding (see no. 23) and was subsequently made co-executor and chief beneficiary of his will.

Michaelsen was Doctor of Law and War Councillor to the King of Prussia at Halle. In correspondence Handel always addressed him as 'tres honoré frere', and relied on him to help his mother and to manage his family affairs in Halle during the long intervals between his occasional visits home.

This engraving was published in memory of Michaelsen after his death at Nuremberg in 1748 while a hostage for the Prussian side in the War of Austrian Succession.

Gerald Coke Handel Collection

Engraving, 30.8 × 19.4 cms/12¼ × 7⅝ ins

Inscribed *MICHAEL DIEDERICH MICHAELSEN, I.V.D. ET S.R.M. PORUSS. A CONSILIIS RERUM BELLICARUM. NATVS BREMAE DIE XXV. IVL. MDC. LXXXI. DENATVS DIE XX. IVL. MD.CCXLVIII*, and at the bottom *J.C.H. Sporleder pinxit Pict Acad. Frider. Hal./J.C. Syfang sc. Lips. Acad. Frider. Sculp*

EXHIBITED
Coke, p. 21, pl. 6d.

22

Handel sends Wedding Presents for his Niece and her Husband

Autograph letter from Handel in London
to his brother-in-law Michael Dietrich Michaelsen
in Halle, 28/17 August 1731

This letter in French from Handel to his brother-in-law, Michael Dietrich Michaelsen (see no. 21), was written in 1731 and not 1736 as previously published. It comes less than three weeks after Handel had written to Michaelsen thanking him for copies of the funeral oration given at his mother's death (see no. 20). He now writes on a happier note in response to news sent him of the forthcoming marriage of his favourite niece, Michaelsen's daughter, Johanna Friederika. She was later made an executor of his will. Handel is delighted that his niece is marrying a person 'd'un Caractere et d'un Merite si distingué', namely Dr Johann Ernst Flörke, Professor of Law at Halle University. The marriage took place on 6 December 1731, and the first of their eight children was born in September the following year.

Handel sends the groom 'une Montre d'Or de Decharmes avec une Chaine d'Or et deux Cachets un d'Amatiste et l'autre d'Onyx'. Neither this gold watch, probably the work of Simon Decharmes, a Huguenot watchmaker, nor the two seal rings referred to, are now known. To the bride, his niece, he gives a solitaire diamond ring (see no. 24).

Dr Bernhard Foss

Letter, single sheet, folded in half, 2 pp., 22.7 × 18.1 cms/8⅞ × 7⅛ ins page size

PROVENANCE
Michael Dietrich Michaelsen; by descent in the family of his youngest grand-daughter, Johanna Friederika Flörke (1747–1809), whose grand-daughter, Caroline Wilhelmine, married in 1828 Dr Heinrich Foss; thence by descent to Dr Bernhard Foss.

LITERATURE
Deutsch, pp. 413–15 (for a full transcript).

23
Handel's Family Affairs in Germany

Autograph letter from Handel in London to his brother-in-law Michael Dietrich Michaelsen in Halle, 21/10 August 1733

Handel seldom corresponded in German, as here, preferring to write in French even to friends in Germany. He thanks his brother-in-law for his news of relatives in Gotha and for the year's accounts for his late mother's house in Halle. This was now owned by Handel and seems to have been the one next door to his birthplace. Handel goes on to say that however much he wishes to pay a visit to his relatives, he finds the pressure of business in London quite overwhelming, a reference to his work for the forthcoming opera season during which a rival company, the Opera of the Nobility, was set up.

This is one of seven letters to survive from Handel's correspondence with Michaelsen between 1719 and 1736; it is the most extensive sequence known other than that with Charles Jennens (see nos. 186 and 191).

Dr Bernhard Foss

Letter, single sheet, folded in half, 3 pp., 22.6 × 18.5 cms/8⅞ × 7¼ ins page size

PROVENANCE
As no. 22.

LITERATURE
Deutsch, pp. 330–1 (for a full transcript).

24
Diamond Ring

By family tradition this piece was a gift from Handel to his niece Johanna Friederika Flörke (*née* Michaelsen), and perhaps that sent by him in August 1731 on the occasion of her wedding to Dr Johann Ernst Flörke. This ring however has one large and two smaller stones, which does not tally with Handel's description of his wedding gift in a letter (no. 22) as a diamond ring with a single stone of a little over 7½ grains in weight; he considered it 'of the first water and quite perfect'.

Dr Bernhard Foss

Gold set with one large and two small diamonds, 1.7 cms/⅝ ins diameter

PROVENANCE
Said to have been a gift from Handel to his niece, Johanna Friederika Flörke; by descent to Dr Bernhard Foss.

24

25
Chatelaine with Watch

The watch by John Ellicot (1706–72)

This watch, made by John Ellicot in London, can be dated to 1749 or 1750 from the number on the movement. It is attached to an elaborate chatelaine and, according to family tradition, was a gift from Handel to his family on his last visit to Germany in 1750. For other watches associated with Handel see nos. 22 and 151.

Dr Bernhard Foss

Gilt chatelaine with watch, overall length 18 cms/7⅛ ins; watch, 5 cms/2 ins diameter

Inscribed on the movement of the watch *Jn.º Ellicott*. LONDON *3119*, with the London hallmark for 1749

PROVENANCE
Said to have been a gift from Handel to his relatives in Germany; by descent in the family of his niece, Johanna Friederika Flörke, to Dr Bernhard Foss.

HANDEL

Hamburg and the Opera

'Next to the Opera of Berlin, that of Hamburgh was in the highest request. It was resolved to send [Handel] thither on his own bottom, and chiefly with a view to improvement.' So wrote Mainwaring many years later of Handel's move from Halle to Hamburg in 1703. A rather earlier account, by the friend of Handel's Hamburg days, Johann Mattheson (see no. 27), speaks of his coming there in the summer of 1703, 'rich in capability and goodwill. He made almost his first acquaintance with me, by means of which he had access to the organs and choirs of the town, to operas and concerts. . . . At first he played second violin in the Opera orchestra, and behaved as though he did not know how many beans made five, for he was inclined by nature to dry jokes.' Mattheson provides the chief source of information on these happy early years; he tells of their times together, adding revealingly, 'I know for certain, should he read this, he will laugh inwardly, for outwardly he laughs little.'

On 17 August 1703, not long after his arrival, Handel travelled with Mattheson to Lübeck, 'and in the coach we constructed many double fugues, *da mente, non da penna*'. There they both played the organ, competing for the position as successor to Dietrich Buxtehude. The job carried a sting in its tail: marriage to Buxtehude's daughter – 'Since, however, a marriage contract was proposed in connection with the affair, for which neither one of us showed the slightest desire, we left the place. . . .'

When in December the following year Mattheson's third opera was performed, he fought his famous duel with Handel over the right to harpsichordist's position in the final act (see under no. 27). Fortunately Handel had a lucky escape. On 8 January 1705 Handel's own first opera, *Almira*, a story of the Queen of Castile sung in a mixture of German and Italian (see fig. 9), was given with Mattheson perhaps in the leading tenor role of Fernando; it enjoyed a run of some twenty nights and was followed on 25 February by a second opera, *Nero*.

Though Handel must have known the opera at Leipzig where his friend Telemann lived from 1701 (see no. 28), and probably at Berlin, his participation at Hamburg, first as a humble orchestral player, then at the harpsichord and finally as composer, gave him first-hand experience of the theatre. The Hamburg Opera had been founded in 1678, and under its director Reinhard Keiser had a reputation as 'the most enterprising' of the German opera houses. Handel learnt much from Keiser and was willing to use his work as a source for his own compositions – as was noted by his contemporaries even within his own lifetime.

It was at Hamburg that Handel met the Prince of Tuscany, brother of the Grand Duke, who, according to Mainwaring, encouraged him to travel to Italy. This he did, probably travelling to Florence in 1706.

LITERATURE
Mainwaring, pp. 27–8, 39; *Deutsch*, pp. 502–3; *Hogwood*, pp. 26–8.

26

Map and View of Hamburg

Published by the heirs of Dr Homann

Handel came to Hamburg in 1703 at the age of 18. The ancient 'free and Hanseatic' city-state was a prosperous port on the river Elbe. It was not subject to a ruling prince as most other German cities but had its own elected town council. In this eighteenth-century engraving, the city is seen from the south side of the Elbe. In the centre is the Maria Magdalene church (marked as number 11 on the map) where Handel played the organ with the composer Johann Mattheson (see no. 27) shortly after his arrival in Hamburg. The Theater on the Gänsemarkt, home to Handel's first operas, *Almira* and *Nero*, in 1705, is not marked but lay just to the north of the city centre, not far from the site of the present opera house.

The British Library Board
Maps 30238 (3)

Engraving, coloured, 46.1 × 55.7 cms/$18\frac{1}{8}$ × $21\frac{7}{8}$ ins

Inscribed with the names of the buildings etc and at the top in German *Des Heil: Röms: Reichs freye Handels –u Hansee Stadt HAMBURG. in einem accuraten PLAN u: PROSP: entworfen u: edirt v: Homaenn: Erben. Cum Privil. S.C.M.*, and in French *PLAN de la Ville Imperiale d'HAMBURG. mis au jour par les Heritiers de feu Mons: le Doct. Homann.*

26

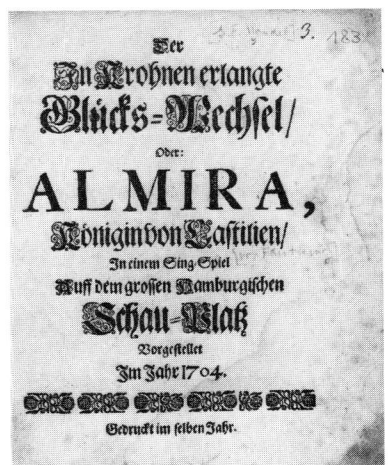

Fig. 9 *Almira*, Handel's first opera, performed at the Theater am Gänsemarkt, Hamburg on 8 January 1705; title-page from one of three editions of the word-book; dated 1704
Staats- und Universitäts Bibliothek, Hamburg

27

Johann Mattheson (1681–1764)

By Johann Jakob Haid (1704–c.1767)
after Johann Salomon Wahl (1689–1765)

One of Handel's earliest friends on coming to Hamburg in 1703 was Johann Mattheson, the Hamburg-born composer and in later life a prolific writer on music. His first opera, *Die Plejades*, was produced in 1699, and four years later he obtained for Handel a position in the opera orchestra as second violinist and subsequently harpsichordist. He later wrote a detailed account of their years together in Hamburg which was published in 1740. It included the story of his duel with Handel:

27

... when my third opera, *Cleopatra*, was performed and Händel sat at the harpsichord, a misunderstanding arose.... I, as composer, conducted, and at the same time personated Antonius, who, about half-an-hour before the end of the play commits suicide.... I had been accustomed, after this action, to go into the orchestra and accompanying the rest myself... this time, however, it was denied me. Instigated by several people, we thereupon, at the exit of the Opera House, on the public market place and with a multitude of onlookers, fought a duel, which might have passed off very unfortunately for both of us had God's guidance not graciously ordained that my blade, thrusting against the broad, metallic coat-button of my opponent, should be shattered.

However Mattheson goes on to say that they became better friends than ever before. In later years he was to be responsible for adapting Handel's *Radamisto* for the Hamburg stage in 1722.

Mattheson's many original writings on music, published between 1715 and 1740, make him a figure of exceptional importance in the study of the history and theory of late baroque music in Germany. It is not possible here to do more than pick out some of his many references to the music of Handel. In his *Critica Musica* (no. 104), in 1722, he comments on Handel's borrowing from his own work; in *Der vollkommene Capellmeister* (*The Perfect Kapellmeister*), 1739, perhaps his most important book on musical theory, Handel's playing on the harpsichord and organ is singled out for praise: 'One may say that Händel... is not easily surpassed by anyone in organ playing: unless it should be by Bach in Leipzig: therefore, these two... should stand in the forefront'; finally in his *Grundlage einer Ehren-Pforte* (*Basis for a Roll of Honour*) of 1740, he gives short biographies of 149 musicians, in many cases based on autobiographies received from the subjects (Telemann's is quoted below; see no. 28). But to Mattheson's displeasure Handel did not oblige him, so that his biography of Handel relies heavily on his personal experience.

This engraved portrait of Mattheson was made in or soon after 1744, following his appointment as Legation Councillor to the Duke of Holstein. Mattheson had gained extensive diplomatic experience from his former position as secretary to the English Resident in Hamburg; he knew English well, and in 1709 had taken an English wife, Catharina Jennings. Mattheson subsequently translated Mainwaring's biography of Handel (no. 214) into German as *Georg Friedrich Händels Lebensbeschreibung* (Hamburg, 1761).

Gerald Coke Handel Collection

Mezzotint engraving, 31.7 × 19.1 cms/12½ × 7½ ins

Inscribed *IOANNES MATTHESON* | *Celsitudinis Imperialis Magni Russiae Princip.* | *Supremi Holsatiae Ducis* | *Legationum Consiliarius.* | *cet.* | *nat. Hamburg d. 28. Sept. A. 1681.*, and at the bottom *Wahll pinxit.* | *V. Dec.* | *Ioh. Iac. Haid. sc. et exc. Aug. Vind.*

Included in Bruckers's *Pinacotheca Scriptorum nostra aetate litoris illustrium*, Augsburg, 1746

LITERATURE
Deutsch, pp. 135, 232, 393, 485, 501–3.

Georg Philipp Telemann (1681–1767)
By Georg Lichtensteger (1700–81)

Telemann was considered the leading composer of his day in Germany and was known for his prolific output of operas, oratorios, cantatas and instrumental works. He studied at Leipzig, becoming organist of the Neuekirche there in 1704 and writing several operas for the Leipzig Opera. After periods as Kapellmeister at Sorau, Eisenach and Frankfurt am Main, he moved to Hamburg in 1721 as music director of the five principal churches and Kantor of the Johanneum, positions he held until his death; he was also director of the Hamburg Opera from 1722–37.

Telemann and Handel were lifelong friends from the time of their first meeting as students. Like Handel he had been instructed to study law when his only interest was music. In his autobiography of 1739 he writes of meeting Handel in Halle on his way to Leipzig in 1701; once there he kept in close touch with so sympathetic a friend (the city was only twenty miles from Halle): 'in fashioning subjects of melody, Handel and I were constantly occupied, with frequent visits on both sides, and also with correspondence'.

Many of the operas Handel wrote for the Royal Academy of Music in the 1720s were adapted by Telemann for performance at the opera house at Hamburg, the same theatre which Handel had known as a young man writing his first opera. Here *Tamerlano*, *Ottone* and *Riccardo Primo* were all given between 1725 and 1729 under Telemann's direction, and over the period of his directorship 13 of Handel's London operas were staged. Handel for his part subscribed to Telemann's *Musique de Table*, published in Hamburg in 1733, and was not above borrowing from Telemann's work on occasion. Many years later, in December 1750, after Handel had arrived back in London from his last visit to the continent, he wrote to his old friend. He had received Telemann's 'most agreeable letter' while he was in The Hague and was touched by his kindness and the gift of Telemann's essay, *Das neue musikalische System*; in return he sent Telemann a crate of exotic plants to add to his collection, a gift which was repeated in 1754 after Handel learnt that reports of Telemann's death were false.

The Trustees of the British Museum
1944-10-14-419

28

Engraving, 23.4 × 15.5 cms/9¼ × 6⅛ ins

Inscribed GEORGIVS PHILIPPVS TELEMANN / *REIPVBLICAE HAMBVRGENSIS DIRECTOR* / *CHORI MUSICI* / *Natus Magdeburgi MDCLXXXI die 14. Martii*, and bottom left *G. Lichtensteger del. et sc.*

LITERATURE
Deutsch, pp. 340, 492, 696–7, 754–5; Willi Maertens, 'Händels Freundschaft mit Telemann', *Händel und Hamburg*, catalogue of an exhibition at Staats- und Universitätsbibliothek, Hamburg, 1985, pp. 109–21.

Handel and Italy

Handel probably reached Italy some time in 1706 and remained there until early in 1710. Until fairly recently his Italian years have been seen chiefly through the eyes of John Mainwaring (see no. 214). In his pages, the link between Germany and Italy is given as the Prince of Tuscany, Ferdinando de' Medici, who met Handel in Hamburg and recommended him to travel to Italy to further his musical knowledge, at the same time offering him his protection. There is no reason to doubt that Handel may have visited the Medici court on his first arrival in Italy, as Mainwaring suggests, but his successful Florentine opera *Rodrigo* (see no. 35) belongs a year later, to the autumn of 1707. Though much of the detail of Mainwaring's account can now be seen to be confused, his statement of Handel's main purpose is plausible enough: 'The nature of his design in travelling made it improper for him to stay long in any one place . . . it was his resolution to visit every part of Italy, which was any way famous for its musical performances.'

From early in 1707, and possibly before, Rome was the main centre of Handel's activity. Here he had three patrons mentioned by Mainwaring, the Cardinals Ottoboni, Pamphili and Colonna (see nos. 30, 31 and 33), and one who has only come to prominence in recent years with the discovery of his family archive: the Marchese Francesco Maria Ruspoli, who can now be seen as of great importance.

Little is known of Handel's music for Ottoboni, despite his renown as a patron. For Pamphili he wrote the oratorio *Il Trionfo del Tempo e del Disinganno* (later revised as *The Triumph of Time and Truth*) to a text by the Cardinal himself early in 1707 and certainly before mid-May. For Colonna he provided music for the feast day of Our Lady of Mount Carmel in July. But for the Marchese Ruspoli, created a Prince by the Pope in 1708, Handel acted as resident composer for much of his two years in Rome.

From May to October 1707 Handel stayed with Ruspoli, writing cantatas and sacred motets for performance by the soprano Margherita Durastanti (she was later to sing for Handel in London). He then travelled to Florence for the production of *Rodrigo* and from there a visit to Venice for the carnival of the winter of 1707/8 is possible; if so he is likely to have met the Earl of Manchester (see no. 39), the British ambassador, who is said to have invited him to London.

In 1708 Handel was again with Ruspoli, from late February until November, except for a visit to Naples in May, June and early July to write his cantata, *Aci, Galatea e Polifemo*, for performance at the wedding of the Duke of Alvito. For Ruspoli Handel's great oratorio *La Resurrezione* (see no. 36) occupied much of his time before Easter when it was first performed; on his return to Rome in July he provided a weekly cantata including the famous 'O come chiare e belle', with its topical references to Ruspoli himself, performed in September. (In the same month there are payments for ice for 'Monsù Endel', to help keep him cool.)

Handel's movements from December 1708 until November 1709 remain uncertain, though he may have composed music for a Good Friday performance in Siena in 1709 and seems to have spent some time in Florence. He was certainly in the city in early November, and travelled from there to Venice, his last stop in Italy, where in the new year his opera *Agrippina* (see no. 38), written to a libretto of Cardinal Vincenzo Grimani, was given a triumphant first performance; the audience, Mainwaring says, 'were thunderstruck with the grandeur and sublimity of his stile'.

LITERATURE
Mainwaring, pp. 40–1, 49, 51, 53–4, 62–3; Ursula Kirkendale, 'The Ruspoli Documents on Handel', *Journal of the American Musicological Society*, XX, 1967, pp. 222–73; Anthony Hicks, 'Handel's Early Musical Development', *Proceedings of the Royal Musical Association*, CIII, 1976–7, pp. 80–9.

29

29

The Piazza Navona, Rome

By Gomar Wouters (b. c.1649)

Rome in the early eighteenth century when Handel knew the city was still the goal of artists and noblemen on the Grand Tour; it was however a place of great contrasts, ranging from the mean dwellings of the mass of the populace and the decayed grandeur of the half-buried ruins of ancient Rome to the splendour of the baroque piazzas, palaces and churches. The city (and the surrounding Papal States) came under the personal rule of the Pope. Great power was held by certain rich families, among which the names of Aldobrandini, Borghese, Colonna and Pamphili were prominent. In Handel's time the Pope was Giovanni Francesco Albani, Clement XI, a great patron of the arts but one who disapproved of opera.

The Piazza Navona preserves the exact outline of the ancient Roman stadium, the Circus Agonalis. It owes its present form, as a showpiece of the Roman baroque, to the Pamphili Pope, Innocent X, who reigned from 1644–55. The spectacular fountain with the obelisk, the Fountain of the Four Rivers, is Bernini's masterpiece, unveiled in 1651. It faces the façade of the church of St Agnese in Agone, largely the work of Bernini's great rival, Borromini. To the left of the church lies the long façade of the Palazzo Pamphili built from 1644–50. Here in 1707 Handel's *Trionfo del Tempo e del Disinganno*, written for Cardinal Benedetto Pamphili (see no. 30), may have received its first performance in the Palace's magnificent gallery with its great ceiling painted in the florid high baroque style of Pietro da Cortona.

The Trustees of the British Museum
1981.u.3214

Engraving, 39.6 × 69.6 cms / 15⅝ × 27⅜ ins

Inscribed *PIAZZA NAVONA ANTICO CIRCO AGONALE DELL' IMPERATORE SEVERO ALESSANDRO EDIFICATA SV LE RVINE DI ESSO | ... | Data in Luce da Domenico de Rossi Erede di Gio. Giacomo de Rossi ... l'Anno 1693. il di 6. Maggio.*, with the names of the buildings and fountains; signed bottom right in the image *G. Wouters Caualier delin. et Sculp.*

30

Cardinal Benedetto Pamphili (1653–1730)

By Jacques Blondeau (1655–98)
after an unknown artist

Benedetto Pamphili (or Panfili), a relative of the Pamphili Pope, Innocent X, who reigned from 1644–55, was created a Cardinal at the age of 28 in 1681. He was an enthusiastic patron of music, at one time employing Corelli (see no. 32) as his chief musician. Cardinal Pamphili was perhaps Handel's first patron in Rome and provided the text for one of his earliest major Roman compositions, the oratorio *Il Trionfo del Tempo e del Disinganno*, probably performed during Lent 1707 and in any case written before 14 May when a copyist's bill was paid. (Fifty years later the music was adapted as *The Triumph of Time and Truth* to form Handel's last English oratorio.) *Il Trionfo* was presumably performed at the Pamphili Palace in the Piazza Navona (see no. 29). The substantial orchestra was led by Corelli.

Another work for Pamphili, written by summer 1708, was of a rather different sort: the cantata 'Hendel, non può mia Musa', written by Pamphili in praise of Handel himself. Many years later Handel described the Cardinal as 'an old Fool' for flattering him (this story is told by Charles Jennens; see no. 214).

The Trustees of the British Museum
1848–3–4–182

Engraving, 20 × 14.5 cms/7⅞ × 5¾ ins

Inscribed BENEDICTVS S.R.E. PRIMVS DIACONVS CARD. PAMPHILIVS ROMANVS CREATVS DIE. 1. SEPTEMBR. MDCLXXXI | Ia. Blondeau Sculp. | Io. Iacobi di Rubeis Formis Romae ad Templ: S.M. de Pace cum Priu. S. Pont.

31 [*in colour opposite p. 96*]

Cardinal Pietro Ottoboni (1667–1740)

By Francesco Trevisani (1656–1746)

Elevated to the purple at the age of 22 by his great-uncle, Pope Alexander VIII in 1689, Cardinal Pietro Ottoboni was the greatest patron of the arts in Rome at the turn of the century. He was vice-chancellor of the Catholic Church, and lived in great splendour in the Palazzo della Cancelleria. His huge household included the musician Arcangelo Corelli (see no. 32), who led his private orchestra, and the artist Francesco Trevisani, responsible for this portrait.

The weekly concerts held at the Cancelleria were well known and, according to Mainwaring, who describes Ottoboni as among Handel's 'greatest admirers', Handel 'was desired to furnish his quota' of pieces for performance – though unfortunately it is difficult now to identify his work for the Cardinal, since the Ottoboni archives for this period are missing. Mainwaring also tells of Handel's contest with Domenico Scarlatti, his exact contemporary and like him a great keyboard virtuoso:

> the Cardinal was resolved to bring him and Handel together for a trial of skill. The issue of the trial on the harpsichord has been differently reported. It has been said that some gave the preference to Scarlatti. However, when they came to the Organ there was not the least pretence for doubting to which of them it belonged. Scarlatti himself declared the superiority of his antagonist....

Trevisani's portrait of Ottoboni in red silk cardinal's dress and biretta was probably painted in the 1690s or early 1700s; it includes the Ottoboni double-headed eagle as decoration to the chair behind the

Cardinal. Ottoboni became the artist's main patron at the end of the seventeenth century and commissioned numerous pictures from him, including several portraits and a large set-piece of 'Cardinal Ottoboni's entrance into Rome' (reproduced in colour in *Landon*, p. 65).

The Bowes Museum, Barnard Castle, County Durham

Oil, 134.3 × 98.5 cms/52⅞ × 38¾ ins

Signed on the letter held by the sitter *All Emmo e Revmo Sig de / JL Sig. Card Otthoboni / re [?] / Vicecan / Per / francesco Treuisani*

PROVENANCE
Galliera Sampieri, Bologna, from whom purchased by John Bowes, 1874; part of his collection given to the Bowes Museum.

LITERATURE
Eric Young, *Catalogue of Spanish and Italian Paintings, The Bowes Museum*, Barnard Castle, 1970, pp. 126–7, pls. 67, 95.

32

Arcangelo Corelli (1653–1713)

By Hugh Howard (1675–1738)

Arcangelo Corelli was known throughout Europe as a composer of chamber music and as a virtuoso violinist; in England his concertos remained very popular throughout the eighteenth century. Corelli came to Rome at about the age of 22 and spent most of the rest of his life in the city. He was made director of music for Cardinal Pamphili (see no. 30) in 1687, and when Pamphili went to Bologna in 1690 he came under the protection of Cardinal Ottoboni (see no. 31), gaining his personal friendship.

During Handel's visit to Rome, Corelli performed in several of his works. The story is well known of how he complained of the difficulty he found in playing Handel's overtures. This seems to have been a matter of the style of performance rather than technique: Corelli's response to the much younger Handel, according to *Mainwaring*, the earliest source for this story, was to declare 'that he did not understand them; *i.e.* knew not how to execute them properly, and give them the strength and expression they required. When Handel appeared impatient, *Ma, caro Sassone* (said he) *questa Musica è nel stylo Francese, di ch'io non m'intendo* [But my dear Saxon, this music is in the French style, which I do not understand].' *Mainwaring* also tells that Handel adjusted the overture of *Il Trionfo del Tempo* (see under no. 30) to suit Corelli's style.

This small painting is one of a group of similar portraits of Corelli attributed to the Irish artist Hugh Howard, who studied in Rome under Carlo Maratta for a few years at the end of the seventeenth century. In 1700 he returned home to Ireland before settling in London. At least eight portraits of Corelli of this type, some rather larger, either in oil or chalk, have been listed by Michael Wynne, who discusses the strong tradition that Howard actually painted Corelli's portrait *ad vivum* as Sir John Hawkins claimed in his *General History of Music* (1776). However some of the more elaborate versions are of such poor quality as to be studio versions or later copies. They do not match the liveliness of this small oil on paper, which is undoubtedly the best of the surviving paintings (a fine version attributed to Carlo Maratta in the collection of Corelli's pupil, the Earl of Mount Edgcumbe, was destroyed in the 1939–45 War).

Faculty of Music, Oxford University (By Permission of the Heather Professor of Music)

Oil on paper mounted on board (?faced in canvas), 34.6 × 26.4 cms/13⅝ × 10⅜ ins

VERSIONS
See *Wynne*, below.

LITERATURE
Mainwaring, pp. 56–7; Michael Wynne, 'Hugh Howard: Irish Portrait Painter', *Apollo*, XC, 1969, pp. 314–17.

32

33

Cardinal Carlo Colonna 1665–1739

By Hieronymus Rossi after an unknown artist
Published by Domenico de Rossi

Carlo Colonna was created a Cardinal in May 1706. The following year he commissioned Handel to provide music for the celebration of the feast day of Our Lady of Mount Carmel, on 16 July 1707 in the church of Santa Maria di Monte Santo in the Piazza del Popolo. For this occasion Handel certainly wrote five pieces: two antiphons, *Haec est regina virginum* and *Te decus virgineum*, two psalm settings, *Laudate pueri* and *Nisi Dominus*, and a motet, *Saeviat tellus* (in praise of the Carmelites). That the Carmelite festival was regularly patronised by the Colonna family and was generally a great public occasion is clear from the record of the 1708 celebration made by the Roman diarist Francesco Valesio; unfortunately, he does not mention the celebrations of 1707.

The Trustees of the British Museum
B.b.9.294

Engraving, 20.3 × 15.3 cms/8 × 6 ins
Inscribed *CAROLVS S.R.E. DIACONVS CARDINALIS COLVMNA ROMANVS CREATVS DIE XVII MAIJ MDCCVI*, and at the bottom *Hieronymus Rossi Incidit | Dominicus de Rubeis formis Romae ad Templum S. Mariae de Pace cum priuil. Summi Pontifici*

33 (detail)

34

Dixit Dominus

Handel's autograph score

Handel's setting of Psalm 110, *Dixit Dominus Domine meo* (*The Lord said unto my Lord*), Psalm 109 in the Vulgate numbering, is the greatest of his choral works composed in Italy. It is the first of his compositions for which a dated autograph is extant (the only earlier autograph is that of the *Laudate pueri* in F major, which is undated). The date is unfortunately not precise: Handel was uncertain of the day of the month and, as shown on f. 82v, simply wrote '1707 li [blank] d'Aprile'. He seems also to have been confused over the year, first writing '1706' and then changing it to '1707'. This slip suggests he had been following the style of beginning the year on 25 March, and had completed the composition early in April just after the year had changed.

In the Roman Catholic liturgy *Dixit Dominus* is the first of the five psalms sung in the service of Vespers (at major feasts, when First Vespers is sung on the eve of the feast, and Second Vespers on the day itself, *Dixit Dominus* forms part of Second Vespers). The fact that the other psalms set by Handel are also proper to Vespers has led to the suggestion (first put forward by James S. Hall in 1959) that they were all sung in a single Vespers service for the feast of Our Lady of Mount Carmel on 16 July 1707; but while *Laudate pueri* and *Nisi Dominus* can be associated with this feast, there is no evidence that *Dixit Dominus*, written three months earlier, was used on this occasion. The circumstances of its composition and first performance thus remain obscure.

What is clear is that in *Dixit Dominus* Handel set out to show his Roman patrons his remarkable skill in writing large-scale choral works. The composition of cantatas for solo voice (his main creative activity in Italy) gave no opportunity to create the grand effects of counterpoint and block harmony which Handel loved but which could only be realised with a chorus and full orchestra. The expansiveness of *Dixit* (it plays for about 35 minutes) brings it into what seems to be a tradition of 'big' settings of the psalm – Vivaldi's are on the same scale – though the writing for the choral voices is unusually demanding, with many high B flats in the soprano lines and other difficulties.

Such boldness, typical of Handel's early Italian works, came to seem crude to the composer in his later years. In 1733 he re-used some of the music of the opening chorus of *Dixit* in his oratorio *Deborah*, making its harmonies more conventional and its counterpoint smoother-flowing.

The British Library Board
RM 20.f.1

Autograph score, 82 ff., 22.2 × 28.0 cms/8¾ × 11 ins; bound with the autograph of *Laudate pueri* in D major (1707)

Inscribed at the end of the score on f. 82v, *S.D.G. G.F. Hendel, 1707* [altered from *1706*] *li* [blank] *d'Aprile. Romae.*

PROVENANCE
Part of the collection of 'Musick Books' bequeathed by Handel to J.C. Smith the elder in 1759 and by him to his son, J.C. Smith the younger, in 1763; most of the autographs were given by him to King George III, *c.*1772; they were deposited by King George V with the Royal Music Library in the British Museum, 1911; given by Queen Elizabeth II, 1957.

LITERATURE
Squire, p. 61; J.S. Hall, 'The Problems of Handel's Latin Church Music', *Musical Times*, C, 1959, p. 197; J.S. Hall, 'Handel among the Carmelites', *Dublin Review*, no. 233, 1959, p. 121; Anthony Hicks, 'Handel's Early Musical Development', *Proceedings of the Royal Musical Association*, CIII, 1977, p. 80; *Landon*, pp. 41–2.

35
Rodrigo

Handel's autograph score

According to Mainwaring's *Memoirs*, Handel's first opera for Italy, *Rodrigo*, was produced at Florence with the support of the Grand Duke of Tuscany and his brother Prince Ferdinando de'Medici. Handel is said to have received '100 sequins and a service of plate' for it. That Handel did indeed write an opera on the subject of Rodrigo, last of the Visigothic Kings of Spain, was confirmed by the existence of a substantial portion of the autograph score, but it is only recently that evidence of the production has come to light. In 1974 Reinhard Strohm announced his discovery of copies of the word-book printed in Florence in 1707, previously unnoticed because the title was not *Rodrigo* but *Vincer se stesso è il maggior vittoria* (*Self-conquest is the Greatest Victory*). Later, Robert and Norma Weaver were able to show that the production was one of several promoted by a group of Florentine noblemen known as the Accademia degl'Infuocati, and probably took place in the theatre in the Via del Cocomero in November 1707. The libretto is based on Francesco Silvani's *Il Duello d'Amore e di Vendetta*, produced with M.A. Ziani's music in Venice in 1700; the highly melodramatic plot involves the attempt of Florinda, Rodrigo's former mistress, to avenge his refusal to make her his Queen, a fate averted by the loyalty and skill of his long-suffering wife, Esilena.

The autograph lacks the opening of Act 1 and the opening and closing sections of Act 3, but fortunately a complete version of Act 3 can be supplied from a copy made for the 4th Earl of Shaftesbury, recently found in the collection of the present Earl. A comparison with the 1707 word-book shows, however, that the autograph does not represent the form in which the opera was actually produced. Several changes, including the provision of seven new arias, were made before first performance, though there is no sign of them in the autograph; they were presumably made in a conducting score now lost. What seems to have happened is that Handel composed the opera in Rome in September or October 1707 and was uncertain of the abilities of the singers for whom he was writing. The changes were made in Florence when Handel finally got to know his cast. Evidence that the opera was written in Rome is provided by the aria 'Per dar pregio' (sung by Esilena at the end of Act 1), the opening of which at ff. 42v–43r is shown, with its introductory recitative, 'Cor mio non gir con fasto'. It is closely based on an aria in the cantata 'Il delirio amoroso', composed earlier in 1707, and therefore Handel was able to pass much of the writing to a copyist who can be identified as his Roman copyist Giuseppe Antonio Angelini. Handel himself wrote a new version of the opening bars, and Angelini then took over the task of copying the elaborate passages for solo violin.

The British Library Board
RM 20.c.5

Autograph score, i + 106ff., 22.2 × 27.9 cms/8¾ × 11 ins

PROVENANCE
As no. 34.

LITERATURE
Squire, p. 68; R. Strohm, 'Händel in Italia: nuovi contributi', *Rivista Italiana di Musicologia*, LX, 1974, p. 152; R.L. and N.W. Weaver, *A Chronology of Music in the Florentine Theatre 1590–1750*, Detroit, 1978, p. 210.

36

La Resurrezione

Handel's autograph score

After the production of *Rodrigo* at Florence in the autumn of 1707 Handel's movements are not clear, but by March 1708 he had returned to the Marchese Ruspoli at the Bonelli Palace, and was at work on his most important composition for Rome, the *Oratorio per la Resurrezione di Nostro Signor Gesù Cristo* (Oratorio for the Resurrection of Our Lord Jesus Christ). It was first performed on Easter Sunday (8 April), and was repeated the following day. Ruspoli spared no expense for the production. A large orchestra, led by Arcangelo Corelli (see no. 32), was engaged, and the performance was given in the Great Hall of the Palace in an elaborate setting, featuring backcloths depicting key moments in the Resurrection story. Carlo Sigismondo Capece's libretto operates on two levels. The theological points are brought out in a series of dramatic exchanges between Lucifer and an Angel, while events on earth are related in the conversations of Mary Magdalene, Mary Cleophas and St John the Evangelist. When the Angel appears in the empty tomb and gives the news of the resurrection to the women, the two planes of the drama are united.

The large array of instruments provided by Ruspoli enabled Handel to produce several novel effects of orchestral colour. There is some particularly prominent writing for the viola da gamba, not in common use in Italy. The part may have been intended for a visiting virtuoso – possibly (as Julie Anne Sadie has suggested) the German violist Ernst Christian Hesse, who had met Handel in Hamburg and is known to have spent most of 1708 in Italy. The autograph is open at ff. 19v–20r to show, after the end of an aria for Lucifer, the first recitative for Mary Magdalene ('Notte, notte funesta'), a melancholy reflection on the Crucifixion matched in mood by the attenuated accompaniment for recorders (*flauti*) and viola da gamba. The start of the aria 'Ferma l'ali', in which the same instruments appear, follows; the violins are muted (*sordi*) and instructed to play *appogiato*, ie the notes to be sustained or 'leaned upon'.

As in the case of *Rodrigo*, the autograph of *La Resurrezione* does not show the final form of the work defined by the printed word-book of 1708 (no. 37), and again the explanation is that changes were made in a conducting score before first performance.

Fortunately the conducting score of *La Resurrezione* is preserved, in the Santini Collection at Münster, and it contains the expected alterations, some in Handel's own hand. The most important differences from the autograph are the addition of overtures to both parts of the oratorio and the rearrangement of the first scene, so that the overture leads directly into the Angel's brilliant opening aria with trumpets—a stunning beginning to the work.

The British Library Board
RM 20.f.5

Autograph score, i + 79 ff., 21.6 × 27.9 cms/8½ × 11 ins

Inscribed at the end of the score on f. 79v, *Roma la Festa di Pasqua dal Marche Ruspoli li [blank] d'Aprile 1708*

PROVENANCE
As no. 34.

LITERATURE
Squire, pp. 62–3; R. Ewerhart, 'New Sources for Handel's *La Resurrezione*', *Music and Letters*, XLI, 1960, p. 127; R. Ewerhart, 'Die Händel–Handscriften der Santini-Bibliothek in Münster', *Händel-Jahrbuch*, 1960, p. 111; Ursula Kirkendale, 'The Ruspoli Documents on Handel', *Journal of the American Musicological Society*, XX, 1967, pp. 222–73, 518; Anthony Hicks, 'Handel's *La Resurrezione*', *Musical Times*, CX, 1969, p. 145.

37

'Oratorio per La Risurrettione di Nostro Signor Giesu Cristo ... L'Anno MCCVIII'

Word-book for *La Resurrezione*,
text by Carlo Sigismondo Capece
Printed for Antonio de' Rossi, Rome, 1708

The accounts for the production of *La Resurrezione* at the Bonelli Palace on 8 and 9 April 1708 include a bill from the printer Antonio de' Rossi for the printing of 1,500 word-books, one specially bound for presentation to the dedicatee of the oratorio, Cardinal Gualtierio. The number is remarkable but perhaps indicates that word-books were supplied to those attending the three rehearsals of the music which preceded the first performance.

In broad terms the text follows the final performing version of the oratorio as reflected in the conducting score now at Münster. The displayed page, as well as showing a decorative wood-cut of Christ rising from the tomb, indicates that the oratorio opened with the Angel's aria 'Disserratevi ò Porte d'Averno' ('Be unbarred, oh you gates of Avernus') and not the recitative for Lucifer found in Handel's

37

autograph (no. 36). In details, however, the text often differs slightly from that set by Handel, presumably as a result of last-minute revisions by Capece which Handel was unable or unwilling to incorporate into his score.

Manchester City Council, Cultural Services Department (Henry Watson Music Library)

Printed word-book, 22 pp., 24 × 17 cms/9½ × 7 ins

PROVENANCE
Sir Newman Flower; purchased with his collection for the Henry Watson Music Library, 1965.

LITERATURE
Anthony Hicks, 'Handel's *La Resurrezione*', *Musical Times*, CX, 1969, p. 145.

38

Agrippina

Handel's autograph score

Agrippina established Handel's reputation as a composer of Italian opera. Mainwaring relates that the production at Venice played for 'twenty-seven nights successively' and was received with 'acclamations of *viva il caro Sassone!* [long live the beloved Saxon!] and other expressions of approbation too extravagant to be mentioned'. The opera was performed at the San Giovanni Crisostomo theatre in the winter carnival of 1709–10, but probably did not open on the first night of the carnival (26 December 1709) as is sometimes stated. Copies of individual arias in the Coke and Schönborn-Wiesentheid Collections are headed 'S. Gio. Grisos.mo 2ª' (ie 'second [opera at] S[an] Gio[vanni] Grisostomo'). To some extent Handel ensured its success by drawing extensively upon earlier compositions which had already found favour in Rome, Florence and Naples.

The libretto, by Cardinal Vincenzo Grimani, Viceroy of Naples, is a highly satirical account of the efforts of the Empress Agrippina to ensure that her son Nero becomes the nominated successor of the Emperor Claudius. In Act 2 Claudius's successful conquest of Britain is celebrated in the chorus 'Di timpani e trombe al suono giulivo' ('To the joyful sound of drums and trumpets'), for which Handel has added the appropriate instruments to his score. The displayed pages (ff. 49v–50r), showing the entry of the voices, illustrate the usual way in which choruses were sung in operas of this period: not by a 'chorus' in the modern sense, but by an ensemble of all the available soloists. The names of the characters required to sing each line are carefully indicated by Handel. At this period he wrote his trumpet parts at actual pitch, but followed the convention of writing for the drums in C major as 'transposing' instruments. In his later scores the drums are notated at sounding pitch.

As in the cases of *Rodrigo* and *La Resurrezione*, the autograph of *Agrippina* does not reflect the performing versions of the opera. For that, revised versions of two of the arias and an additional aria (probably all sung in performance) were inserted into the score, and another additional aria (again used in performances, as the printed word-book confirms) appears in a separate volume of autograph fragments (British Library, RM 20.d.2, f.11). The extant manuscript copies of the opera, probably deriving from a lost conducting score, display several other revisions; one such copy, now also lost, was the source for

38

Samuel Arnold's edition of 1795, which is thus a more faithful reflection of Handel's performing version than the later Händel-Gesellschaft score, based mainly on the autograph.

The British Library Board
RM 20.a.3

Autograph score, i + 121 ff., 21.6 × 29.2 cms/8½ × 11½ ins

PROVENANCE
As no. 34.

LITERATURE
G.C. Bonlini, *Le Glorie della Poesia*, Venice, 1730, p. 158; *Mainwaring*, pp. 52–3; *Squire*, pp. 3, 93; F. Zoberly, 'Werke Händels in der Gräfl. von Schönbornschen Musikbibliothek', *Händel-Jahrbuch*, 1931, pp. 101–4; *Deutsch*, p. 27.

39
Charles Montagu,
1st Duke of Manchester (1662–1722)
By Sir Godfrey Kneller (1646–1723)

Kneller's Kit-cat Club portrait of the diplomat Charles Montagu, 4th Earl of Manchester (from 1719, 1st Duke) was probably painted about 1710, fairly soon after Manchester's return home from Venice where he had acted as British ambassador from June 1707 to October 1708. In Venice he is said to have met Handel and, according to *Mainwaring*, extended an invitation to him to visit London. If this is so it supports the theory that Handel visited Venice in the winter of 1707/8, travelling north from Florence after his opera *Rodrigo* had opened there in November 1707 and remaining in Venice for the winter carnival, until his return to Rome by the end of February 1708. Handel was certainly in Venice towards the end of 1709 preparing for the first performance of his celebrated opera *Agrippina* (see no. 38), given in January 1710, but this was some time after Lord Manchester had left.

Handel was not the only figure to receive Lord Manchester's encouragement while in Venice. At the request of Sir John Vanbrugh, still the manager of the Queen's Theatre, he arranged for the singer Nicolini (see no. 49) to come to London. He was also responsible for bringing over the Venetian painter Giovanni Antonio Pellegrini, who came with Marco Ricci in 1708 and undertook designs for the Italian opera as well as decorating Lord Manchester's town house with frescoes. Manchester was later deputy governor of the Royal Academy of Music (see no. 84).

39

National Portrait Gallery, London
NPG 3216

Oil, 90.8 × 70.5 cms/35¾ × 27¾ ins

Signed bottom left *GK*, with Manchester's coat of arms and earl's coronet top left

PROVENANCE
Jacob Tonson, secretary of the Kit-cat Club, and like other portraits in this set, by descent in the family until purchased by the National Art-Collections Fund and given by the Fund to the National Portrait Gallery, 1945.

LITERATURE
Mainwaring, p. 72; *Piper*, pp. 215, 398.

HANDEL

Handel and Hanover

Handel set off from Italy soon after the successful performance of his opera, *Agrippina*, in Venice in January 1710. Travelling by way of Innsbruck (which he had left by 9 March), he presumably visited his mother in Halle before going on to Hanover where he probably arrived in May. On 16 June he was formally appointed as Kapellmeister, so beginning his long connection with the Hanoverian Royal Family.

When Handel came to Hanover he was welcomed by Baron Kielmansegg (see no. 63) who had been Hanoverian ambassador in Venice and was later to commission the *Water Music* from Handel. According to Mainwaring's biography of the composer, it was Kielmansegg who introduced him at court and obtained his appointment. But Handel already had other plans; he told Kielmansegg that 'the favour intended him would hardly be consistent either with the promise he had actually made to visit the court of the Elector Palatine [see no. 43], or with the resolution he had long taken to pass over into England, for the sake of seeing that of London'. In fact Handel was granted permission to travel as he wished, and so 'on these easy conditions he thankfully accepted' his appointment.

Handel used Hanover as a secure base from which to explore the possibilities of finding a position which would suit his talents and his propensity towards opera and large-scale compositions. Within a few months he had travelled to Düsseldorf to visit the Elector Palatine and from there, by November 1710, to London. Though he returned to Hanover in June 1711 he remained for only a year and was encouraged to settle permanently in London by the success of his opera *Rinaldo*. In June 1713 the displeasure of the Elector was made clear by his dismissal from his position at Hanover – with the message however that 'he is by no means in disgrace . . . he will be quite all right when the elector comes [to London]'.

The importance of Handel's time in Hanover lies not so much in his music – the opera was in abeyance owing to the continental War and his identified works are few: the cantata 'Apollo e Dafne' and some chamber duets which *Mainwaring* describes as for the practice of the Electoral Princess – but rather for the relationships he struck up with his royal patrons-to-be in London: George I (then Elector; see no. 41) and George II and Queen Caroline (then Electoral Prince and Princess; see nos. 109 and 114). The Dowager Electress (the Elector's mother; see no. 42) reported of Caroline that she was 'entertained by the music of a Saxon who surpasses everyone who has ever been heard in harpsichord-playing and composition'. A musician of Handel's stature was an unusual pleasure for the court at Hanover, so it is no surprise to learn that the Electoral Prince and his wife were 'charmed with him'; they were both of Handel's own generation and had deep sympathies for music. Their appreciation was to be made immediately apparent years later, when Handel was the preferred choice over those who might normally have expected the commission for the anthems for George II's coronation in 1727 (see no. 110).

LITERATURE
Mainwaring, pp. 71–2; *Burrows*, pp. 35–59.

1. Catholische Kirch 4. Neustätter Kirch 7. Zum H.
2. Kleber Thor 5. Kalenbergische Thor 8. Da[f]s Ze
3. Reitstalle 6. Stein Thor 9. Da[f]s Le

40

40
View of Hanover from the South-west
By Friedrich Bernhard Werner (1690–1778)

Hanover, like many other continental cities in the eighteenth century, still retained its defensive walls. Though the main area of conflict in the War of Spanish Succession was on the Rhine and in the Low Countries, Hanover was still actively involved in the war against the French at the time Handel was appointed Kapellmeister in 1710. In this slightly later view, taken from the south-west of the city in 1729, the Royal Palace ('Fürst Schloss') is the low building (marked as number 12) in front of the market church (13). Herrenhausen, the summer residence of the Elector of Hanover, lay on the north-west side of the city, somewhat off to the left of this view.

The British Library Board
Maps 29895 (2)

Engraving, 29.7 × 99.0 cms/11¾ × 39 ins

Inscribed *HANNOVER | F.B. Werner Siles. Wratisl. delin. 1729. | Haered. Ier. Wolffij excud. Aug. V. | Cum. Pr. Sac. Caes. Maj.*, and with a key giving the names of the various numbered buildings in the view

41 [*in colour opposite p. 45*]
George I (1660–1727)
By John Michael Rysbrack (1694–1770)

Handel first met Georg Ludwig, Elector of Hanover, the future King George I, in 1710. (The various German Electors selected the Emperor who held sway over the union of German states which made up the Holy Roman Empire.) His appointment as Kapellmeister in June began his lifelong association with the House of Hanover. Despite Mainwaring's assertion that Handel's prolonged absences from Hanover meant that he 'did not dare to shew himself at court' when the Elector came to London as King in 1714, it is now apparent that Handel's dismissal from his post at Hanover in June the previous year was done in a manner designed to avoid such future difficulties: the Hanoverian Resident in London reported that he had 'let slip a few words to inform him [Handel] that, when some day His Highness comes here, he might re-enter his service'. That this was so is shown by George I's continued payment of Handel's pension from the British crown and even, in October 1715, a back payment of six months' arrears for his Hanoverian service in 1712. To the £200 paid by

Queen Anne, George I added a further £200 a year, and was of course to support the Royal Academy of Music with an annual payment of £1,000.

Handel's most famous composition for the King was the *Water Music* (see p. 95), commissioned through Baron Kielmansegg in 1717. In February 1723 he was appointed 'Composer of Musick for his Majesty's Chapel Royal', and on a number of occasions he wrote anthems to mark the King's return from visits to Hanover. In 1727, one of George I's last deeds before leaving for the continent where he died on 11 June was to give the Royal Assent, 'Soit fait comme il est desiré', to the Act for Handel's naturalisation (see Appendix 2). George I's son, grandson and great-grandson were all to follow him as enthusiastic patrons of the composer's music (see nos. 109, 112 and 234). This fine terracotta bust, by the Flemish-born sculptor John Michael Rysbrack, may date to the later part of George I's reign or is perhaps a posthumous work.

National Portrait Gallery, London
NPG 4156

Terracotta, 62.9 cms/24¾ ins high

PROVENANCE
Probably by descent in the family of the Earls Poulett at Hinton St George, Somerset; sold from there at auction by Wooley & Wallis, Salisbury, 1959, purchased by Roger Warner, from whom purchased by the National Portrait Gallery, 1960.

LITERATURE
Kerslake, pp. 86–7, pl. 235; *Burrows*, pp. 44–6.

42

Princess Sophia, Electress of Hanover (1630–1714)

By John Smith (c. 1652–1742) perhaps
after Friedrich Wilhelm Weidemann (1668–1750)

Under the terms of the Act of Settlement (1701), if Queen Anne died childless the British throne would pass to Princess Sophia, Dowager Electress of Hanover (she was grand-daughter to King James I) and her Protestant heirs. Though Queen Anne maintained a policy of not inviting the Hanoverian Royal Family to England, there was great public interest in the likely future holders of the British crown. This engraving of the Electress was probably published in 1706 and, as inscribed, was done from a 'painting lately brought from Hanover', perhaps the work of the Berlin court painter Friedrich Wilhelm Weidemann, although his original is not known. In the event the Electress died in June 1714, a few weeks before Queen Anne, so it was her son the Elector who became King as George I (see no. 41).

In three letters to her grand-daughter, Sophia Dorothea, Princess Royal of Prussia, written from Herrenhausen in June 1710, the Electress mentions Handel and his appointment as Kapellmeister: 'But there is not much to say from here except that the elector has taken on a Master of the Chapel named Handel, who plays marvellously on the harpsichord in which the electoral prince and princess take a great deal of pleasure' (the Electoral Prince was the future George II).

National Portrait Gallery, London

Mezzotint engraving, 29.9 × 25 cms/11¾ × 9¾ ins

Inscribed *The Most Illustrious Princess Sophia Electrice Dowager of Brunswick, | Successor to ye Crown of England &c. after her Majesty Queen Ann & her Royal issue. | Done by I. Smith from an Original painting lately brought from Hanover. & Sold by him at ye Lyon & Corwn* [sic] *in Russell Street Covent Garden*, with the scratch date, *1706*, bottom left

LITERATURE
CS, p. 1223, no. 237.

42

43

Johann Wilhelm, Elector Palatine (1658–1716)

By Elias Christoph Heiss (1660–1731)
after Jan Frans van Douven (1656–1727)

When Johann Wilhelm inherited the Palatinate in 1690, a scattered territory which lay on either side of the Rhine, it had been devastated by the occupying armies of Louis XIV. He moved his capital from Heidelberg to Düsseldorf and built up a reputation both as a military commander and an enlightened patron of the arts. This portrait dates to 1691, the year of his second marriage to Anna Maria Luisa, the last of the Medici, and was engraved from a work by Jan Frans van Douven, the Dutch court artist who painted a number of portraits for the Elector and who died at Düsseldorf in 1727.

Probably while still in Italy, Handel had agreed to visit Düsseldorf. He journeyed there from Hanover in the summer of 1710 and seems to have remained for most of August, September and October before departing for London. Mainwaring's biography of the composer describes how 'The Elector Palatine was much pleased with the punctual performance of his promise, but as much disappointed to find that he was engaged elsewhere.' The opera which the Dowager Electress of Hanover thought Handel was going to write at Düsseldorf did not materialise but Handel's excellence as a harpsichord player is reported by the Electress Palatine in letters to her brother. Handel re-visited Düsseldorf on his way back from London the following June.

The Trustees of the British Museum
B.b.13–222

Mezzotint engraving, 28.8 × 16.1 cms/11⅜ × 6⅜ ins

Inscribed beneath the portrait *SEREN.mo AC POTENTISSIMO / PRINCIPI / IOANNI WILHELMO / COMITI PALATINO / RHENI / ELECTORI ARCHIDAPIFERO / S.R.I. VICARIO.*, and at bottom right *E.C. Heiss Sc. Aug. Vind.*

VERSION
A larger engraved variant of this image gives the artist's name, J.F. van Douven, and the date 1691.

LITERATURE
Mainwaring, p. 74; H. Kühn-Steinhausen, 'Die Briefwechsel der Kurfürstin Anna Maria Luise von der Pfalz', *Düsseldorfer Jahrbuch*, XL, 1938, pp. 14–256.

43

III *London: the First Decade* 1710–19

Handel's first visit to London was quite short: from November 1710, when he arrived from Hanover via Düsseldorf, until perhaps May 1711, when he returned to Hanover the way he came. During this time he had written one opera, *Rinaldo*, and also some special music to mark the birthday of Queen Anne; both these compositions received their first performance in February. They form an indication of his future creative pattern: of work for the theatre interrupted by the occasional court commission.

When Handel returned to London and settled there, probably in October 1712, he is said to have stayed initially with a 'Mr Andrews, of Barn-Elms' who has not been identified. For much of the rest of the decade he seems to have come under the protection first of the young Earl of Burlington in London and then of the Earl of Carnarvon (later Duke of Chandos) at Cannons in Middlesex. According to the music historian Sir John Hawkins, Handel spent three years at Burlington House, perhaps from the winter of 1712/13 until 1716. It seems that there may have been an interval about which we know little, perhaps including a visit to Germany, before he moved to Cannons in the summer of 1717. By February 1719 Handel's thoughts had turned elsewhere, for he then wrote from London to his brother-in-law at Halle to say that he would soon be travelling to Germany – as he did later in the spring, in search of singers for the newly formed Royal Academy of Music.

Compared to the treadmill of the opera seasons of the Royal Academy of Music in the 1720s, or to Handel's own ventures in the years up to the mid-century, his early years in London seem not to have placed heavy demands on him musically. He provided three new operas for the Haymarket Theatre between 1712 and 1715; with his earlier *Rinaldo*, these were revived in succeeding seasons until 1717, after which the opera was in abeyance until 1720. For Queen Anne he provided his Utrecht *Te Deum* and a Birthday Ode, both in 1713; for George I his *Water Music* in 1717. Other than this music for the court, his chief compositions were for the future Duke of Chandos: the Chandos Anthems (1717–18) and two oratorios or masques to English words, *Esther* and *Acis and Galatea*. If there was a period of Handel's career in London without obvious financial problems, musical rivalries or personal ill-health, this was the time. Yet we know less about the detail of these early years than of any succeeding period of his life.

HANDEL'S LONDON Based on John Rocque's *Survey of London and the Country near Ten Miles Round*, 1746

1. Crown and Anchor tavern (*Esther* performed, 1732)
*2. John Walsh, father and son, music publishers
3. Whitehall Stairs (likely embarking point for Chelsea; *Water Music*, 1717)
4. Banqueting House Chapel (musical rehearsals for some state occasions)
5. Westminster Abbey (coronation, 1727; burial of Queen Caroline, 1737; Handel's burial, 1759 and monument, 1762)
6. Houses of Parliament (Handel's Act of Naturalisation, 1727)
7. Fireworks Machine (1749)
8. The Queen's Chapel, then known as the French Chapel (royal wedding, 1734)
9. Earl of Manchester, later Duke of Manchester (1710–13, 1715–22)
10. Duke of Chandos (1720–35)
11. Duke of Kent (to 1744)
12. Duke of Portland (1710–22); Frederick Prince of Wales (1743–51)
13. Earl of Egmont (from 1719)
14. Frederick Prince of Wales (from 1732)
15. The Queen's Theatre, later the King's Theatre, Haymarket (first used by Handel 1711)
16. Earl of Burlington, Burlington House (?Handel resident c.1715)
*17. Earl of Carnarvon, later Duke of Chandos (c.1710–20)
*18. Newburgh Hamilton, librettist
*19. William Bromfield, surgeon (1745–92)
20. St George's, Hanover Square (Handel's parish church)
21. Handel's house (1723–59)
22. Earl of Shaftesbury (c.1750)
23. Bernard Granville (from 1736)
24. Mrs Pendarves, later Mrs Delany (1733–41)
25. The Pantheon (1784 Handel Commemoration)
26. John Christopher Smith the elder (1723–50)
27. George Prince of Wales (1718–27); Frederick Prince of Wales (1743–51)
28. Louis François Roubiliac, sculptor (from 1741)
29. Theatre Royal, Covent Garden (first used by Handel 1734)
30. Thomas Hudson, painter (1746–64)
31. Duke of Newcastle
32. Theatre Royal, Lincoln's Inn Fields (used by Handel 1739–41)
*33. Charles Jennens, librettist (later in Great Ormond Street)
34. Foundling Hospital Chapel (supported by Handel from 1749)

* Precise location of house within street not established.

(From a reproduction by Harry Margary, Lympne Castle, Kent)

47

Rinaldo: Handel's First Visit to London

When the Elector of Hanover's newly appointed Kapellmeister arrived in London in November 1710, preceding his master by almost four years, he was coming to the largest city in Europe and probably already the wealthiest. A sign of London's expanding cultural ambitions was the newly built Queen's Theatre (the King's Theatre from 1714) in the Haymarket (see no. 48). For a time it was in direct competition with the Drury Lane Theatre, but it came to be used primarily for the performance of opera, and was to witness some of Handel's greatest successes.

Italian opera had been introduced in London as recently as 1705 with *Arsinoe, Queen of Cyprus* at Drury Lane: 'The design of this entertainment being to introduce the Italian manner of musick on the English stage', claimed Thomas Clayton, one of the three composers involved, who explained that 'The stile of this musick is to express the passions, which is the soul of musick.' Though actually sung in an English translation by an English cast, other operas soon followed into which Italian singers were introduced. When Bononcini's *Camilla*, one of the most successful operas of the century, was revived in 1709, the great *castrato* singer Nicolo Grimaldi, called Nicolini (see no. 49), joined the cast, leading the prompter at Drury Lane, William Chetwood, to describe the production as 'an odd Medley – Mrs *Tofts*, a mere *Englishwoman*, in the part of *Camilla*, courted by *Nicolini* in *Italian*, without understanding one single Syllable each other *said* or *sung*'. It was the arrival of a number of Italian singers, most especially Nicolini, that encouraged the change to opera sung in Italian by Italians, a process which was complete by the time of Handel's coming to London in 1710. The new fashion was attacked by Joseph Addison in *The Spectator* for 21 March 1711, soon after the successful staging of Handel's first opera, *Rinaldo*:

> At length the Audience grew tired of understanding Half the Opera, and therefore to ease themselves intirely of the Fatigue of Thinking, have so ordered it at present, that the whole opera is performed in an unknown Tongue.... I cannot forbear thinking how naturally an Historian who writes two or three hundred Years hence ... will make the following Reflection, *In the Beginning of the Eighteenth Century the* Italian *Tongue was so well understood in* England, *that Operas were acted on the publick stage in that Language*.

The taste for this new form of entertainment was in any case well established by the time of Handel's first visit to London.

Reports of Handel's 'uncommon abilities' had reached England before his arrival, according to his later biographer, John Mainwaring, who continues, 'Some persons here had seen him in Italy, and others during his residence at Hanover.' Mainwaring most obviously is speaking of the Earl of Manchester (see no. 39) who, when British ambassador in Venice, is reputed to have invited Handel to England. 'Many of the nobility were impatient for an Opera of his composing. To gratify this eagerness, *Rinaldo*, the first he made in England, was finished in a fortnight's time.' It was only by making extensive use of music from his earlier works that Handel was able to produce his autograph (no. 45) at such speed. For the first night, on 24 February 1711, the new theatre manager at the Haymarket, Aaron Hill (see no. 44), who had commissioned the opera from Handel produced a word-book in which he explains his approach to improving performances:

> The Deficiencies I found, or thought I found, in such ITALIAN OPERA'S as have hitherto been introduce'd among us, were, *First*, That they had been compos'd for Tastes and Voices, different from those who were to sing and hear them on the *English* Stage: And *Secondly*, That wanting the Machines and Decorations, which bestow so great a Beauty on their Appearance, they have been heard and seen to very considerable Disadvantage.

Rinaldo met with great success and ran for 15 performances. But ten days after the opera's opening, Joseph Addison found in it the ideal subject to ridicule in *The Spectator*, then but newly begun: 'An Opera may be allowed to be extravagantly lavish in its Decorations, as its only Design is to gratify the Senses, and keep up an indolent Attention in the Audience. Common Sense however requires, that there should be nothing in the Scenes and Machines which appear Childish and Absurd.' Addison takes to task the sight of 'Nicolini exposed to a Tempest in Robes of Ermin, and sailing in an open Boat upon a Sea of Paste-Board'; he also ridicules the 'Singing Birds' (actually sparrows) released towards the end of the first Act: 'there have been so many Flights of them let loose in this Opera, that it is feared the House will never get rid of them . . . besides the Inconveniences which the Heads of the Audience may sometimes suffer from them'. Interestingly it is the sets and stage effects, Hill's attempts at greater lavishness, which he attacks and not the singing and music.

Handel returned to Hanover before the season's last performance of *Rinaldo* in June. But his thoughts were still on England: the following month he wrote to a friend in London, asking if the poet John Hughes (see no. 51) could supply 'one of his charming poems in English', and saying that he had 'made some progress' in learning English.

LITERATURE
W.R. Chetwood, *A General History of the Stage*, 1749; Mainwaring, p. 78; Deutsch, pp. 32, 35, 44.

44

Aaron Hill (1685–1750)

By Hendrik Hulsbergh (d. 1729)
after an unknown original

Playwright, critic, poet and adventurer, Hill was one of the more colourful characters of his time, dabbling in many ventures. A schoolfellow of John Gay, he travelled to Constantinople at the age of 15, also visiting Greece, Egypt and Palestine. On his return he acted as secretary to Lord Peterborough for a number of years. In November 1709, at the age of 24 (when this portrait engraving was made), he became manager of the Drury Lane Theatre, despite his lack of any previous experience. After a stormy six months in which he had to tussle for control with his actors, on occasions coming to blows, the theatre was closed and he left Drury Lane. Despite his setbacks, the intervention of his patron, William Collier MP, enabled him to gain control of the operas at the Queen's Theatre in the Haymarket (see no. 48), so forcing out Owen McSwiny (see no. 56), who took over at Drury Lane.

One of Hill's first acts on taking over the Haymarket on 22 November 1710 must have been to commission Handel, then probably just arrived in London, to write a new opera, *Rinaldo* (see no. 45), which was performed with immense success on 24 February 1711. As at Drury Lane his management did not outlast his first season. Subsequently he wrote and produced various plays, including *Zara* in 1736, which was Mrs Cibber's first appearance on the stage, but he never regained control of a major theatre. He also indulged in schemes for extracting oil from beech nuts, for settling a colony in Georgia and for floating timber down Scottish rivers for the navy.

His very varied literary efforts were published in a collected posthumous edition (no. 155); this Handel subscribed to and it includes the text of Hill's letter of 1732 encouraging Handel to produce opera in English as well as his 'Ode on Occasion of Mr Handel's Great Te Deum' (1733).

National Portrait Gallery, London

Engraving, 26.5 × 17.1 cms/10½ × 6¾ ins

Inscribed around the portrait oval *Mr AARON HILL.* | *Ætatis Suae 24 Aº Domī 1709*, and lower right *H: Hulsbergh Sculp:*

45

Rinaldo

Handel's autograph score

Rinaldo was Handel's first opera for London, and opened at the Queen's Theatre in the Haymarket on 24 February 1711. The libretto, by Giacomo Rossi, followed a plan devised by the theatre manager Aaron Hill, whose declared purpose was 'to frame some Drama, that, by different Incidents and Passions, might afford the Musick Scope to vary and display its Excellence, and fill the Eye with more delightful Prospects, so at once to give two Senses equal Pleasure' (Preface to the 1711 word-book; see no. 46). Basically Hill was attempting to combine the stage spectacle of the older Purcellian semi-operas (he probably had the Dryden–Purcell *King Arthur* particularly in mind) with the new idioms of Italian music, for which London was then acquiring a taste. The result was a triumph for all concerned. *Rinaldo* was given 15 performances in 1711 and was frequently revived until 1717. In 1731 Handel gave it a further revival, though in a drastically amended version.

Rinaldo is the first of Handel's 'magic' operas, making much use of the scenic transformations which delighted eighteenth-century audiences (and can still delight twentieth-century ones, on the rare occasions when they are faithfully recreated). Hill's scenario was based on Tasso's great epic poem *La Gerusalemme Liberata* (*Jerusalem Delivered*), an elaborately mythologised version of the story of the first Crusade, when Godfrey of Bouillon freed Jerusalem from Saracen occupation. In the opera Godfrey is assisted by the young knight Rinaldo (who is betrothed to Godfrey's daughter Almirena), while Argantes, the Saracen King of Jerusalem, has an ally in the sorceress Armida. Her aim is to distract Rinaldo from the fight, first by kidnapping Almirena and then by attempting to seduce him. Needless to say, the Christian heroes finally triumph against all the odds.

What survives of the original autograph is divided between the British Library and the Fitzwilliam Museum, Cambridge (MS 254). The pages of the British Library fragment displayed (ff. 26v–27r) show an episode in Act I, in which Armida descends in a black cloud full of fire-breathing monsters and whisks Almirena away as Rinaldo looks on helplessly. After the stage directions comes the vigorous orchestral prelude which accompanies the action, and there follows the start of the famous aria 'Cara sposa', a powerful expression of Rinaldo's desperation and grief at the loss of his beloved. Throughout his life Handel made a habit of copying all stage directions into his autographs, even in the case of the dramatic oratorios, though these were never staged. The context of the music was always of vital importance to him.

The British Library Board
RM 20.c.3

Autograph score, ii + 57 ff., 23.5 × 29.2 cms/9¼ × 11½ ins; f. 54 contains sketches relating to the 1717 revival; ff. 55–6 contain an aria used in the 1731 revival, but actually written in 1730 as an addition to *Giulio Cesare*

PROVENANCE
As no. 34.

LITERATURE
Squire, p. 66; *Deutsch*, p. 32.

46

'Rinaldo, An Opera. As it is Perform'd At the Queens Theatre in London.'

Word-book, first edition, text by Giacomo Rossi, translated by Aaron Hill
Printed by Thomas Howlatt, 'Printer to the *House*', 1711

The word-book issued for the first performances of *Rinaldo* in February 1711 was in the bilingual form which was quickly established in the earliest days of Italian opera in London and continued well into the nineteenth century. Most of the book is occupied with Giacomo Rossi's Italian libretto, set out in parallel with an English verse translation by Aaron Hill, who had originally provided the scenario from which Rossi worked. The prefatory material is unusually extensive. Hill's dedication of the opera to Queen Anne is followed by his preface, explaining his aim of accommodating Italian opera to English taste. A further preface by Rossi ('Il Poeta al Lettore' – 'The Poet to the Reader') contains the following reference to Handel, which is not as flattering as it seems:

> ... il signor Hendel, Orfeo del nostro secolo, nel porla in musica, a pena mi diede tempo di scrivere, e viddi, con mio grande stupore, in due sole settimane armonizzata da quell' ingegno sublime, al maggior grado di perfezzione un opera intiera.

[Mr Hendel, the Orpheus of our century, when composing the music, scarcely gave me time to write, and I saw, to my great amazement, an entire opera set to music by that sublime genius, with the highest degree of perfection, in just two weeks.]

There was some private irony here, since Rossi must have been well aware that much of the music was taken from Handel's earlier works, so that it was indeed composed before Rossi had 'time to write'; in some cases the words used with the original music had to be incorporated into Rossi's text. After Rossi's preface comes the argument (a brief summary of the plot) and then the cast-list in both Italian and English; the music is credited to 'Signor Georgio Frederico Hendel, Maestro di Capella di S.A.E. d'Hanover'.

This edition of the word-book presumably continued in use for the revivals of *Rinaldo* in 1712, 1713 and 1714, despite the fact that several cast changes and alterations to the opera took place. New word-books were however issued for the substantially revised versions produced by Handel in 1717 and 1731; the latter was provided with an entirely new translation by Samuel Humphreys.

The British Library Board
1078.m.6.(8)

Printed word-book, x + 66 pp., 20.2 × 12.9 cms/8 × 5⅛ ins, bound with various other pamphlets

LITERATURE
Deutsch, pp. 32–4 (for a full transcript of the dedication and prefaces).

47 [*illustrated on p. 76*]

'Song's in the Opera of Rinaldo Compos'd by Mr Hendel'

Printed score, first edition
Printed for John Walsh senior and John Hare, 1711

John Walsh's score of the overture and arias in *Rinaldo* appeared in April 1711, two months after the first performance. It was the first music by Handel to be printed under his own name. (The music from *Rodrigo* used as act-tunes for a revival of Ben Jonson's *The Alchemist* in January 1710, and published by Walsh the following June, had been credited to 'an Italian Master'.)

As was often the case with the prints of Walsh and others only a few numbers are shown in their original scoring; often the accompaniments are reduced to treble and bass lines only, a considerable loss of effect in such pieces as the aria 'Or la tromba', originally accompanied by full orchestra with four trumpets. Far less usual is the transposition of the arias for alto voice to suit soprano or tenor range; presumably this was thought to be a useful convenience for amateur performers. Though the overture is included, the marches and the orchestral items in the opera are not.

The decorative title-page is an example of a passe-partout, printed from a plate whose centre can be replaced with different titles. This particular design had already appeared in Pepusch's *Six English Cantatas* in April 1710, and again in a score of the pastiche opera *Etearco*, before being used for *Rinaldo*. It was clearly engraved with the Pepusch cantatas in mind, as can be shown from the open book of music beneath the title. The right page can just be made out as the recitative 'As Beauty's Goddess from the Ocean Sprung', which comes at the beginning of the first of his cantatas, 'The Island of Beauty'. The quotation on the left page is from Horace (*Odes*, Book IV, no. 9, lines 3 and 4): 'Non Ante vulgatas per Artes / Verba loquor socianda Chordis' ('I utter words to be accompanied with strings, by skills not hitherto made known').

The British Library Board
RM 7.h.24

Printed score, 65 pp., 35.0 × 23.8 cms/13¾ × 9⅜ ins

LITERATURE
Smith, p. 56, no. 1.

48

The Queen's Theatre, later the King's Theatre, Haymarket

By William Capon (1765–1827)

The Queen's Theatre, from 1714 the King's Theatre and now known as Her Majesty's, is the oldest London theatre on its original site, though three times re-built. It was situated on the west side of the Haymarket, extending down as far as Pall Mall, and was the home of Italian opera in London for many years. The Theatre was first used by Handel in 1711 for his *Rinaldo*, and for all his subsequent operas until he was forced out by the Opera of the Nobility in 1734, whereupon he moved to Covent Garden (see no. 131). He returned to the Haymarket for the 1738 and 1739 seasons and again in 1744/5.

The Theatre was designed in 1704 by Sir John Vanbrugh, the soldier and playwright who was then just setting out on his new career as an architect. It opened in 1705 under the management of Vanbrugh and Congreve with an Italian opera but they both lost large sums of money on the venture and in 1708 disposed of the management to Owen McSwiny (see no. 56). Very few views of the Theatre are known; this watercolour of the exterior was made by the stage designer William Capon in 1783, six years before the Theatre's destruction by fire.

Inside, the auditorium was renowned for its magnificence. Colley Cibber, as an actor and manager, was highly critical of such display: 'Almost every proper Quality, and Convenience of a good Theatre had been sacrificed, or neglected, to shew the Spectator a vast, triumphal Piece of Architecture.... For what could their vast Columns, their gilded Cornices, their immoderate high Roofs avail when scarce one word in ten could be distinctly heard in it?' The Theatre was much more suitable for opera, as Cibber acknowledged: 'the Tone of a Trumpet, or the Swell of an Eunuch's holding note, 'tis true, might be sweeten'd by it'.

Guildhall Library

Watercolour, 18.2 × 21.0 cms/7⅛ × 8¼ ins

The Theatre façade bears the sign RIGAUT'S FENCING ACADEMY and four theatre posters, CORNWALL, LA REGINA di GOLCONDA. RAUZZINI, LA BUONA FIGLINOLA. RAUZZINI and SLINGSBY NIGHT; inscribed on the later card mount *The first Opera House in the Haymarket burnt down 1789. W. Capon.*

Engraved by Charles John Smith after the Theatre had been burnt in 1789

PROVENANCE
Given to the Guildhall Library in 1959.

VERSIONS
Several versions exist, including one in the British Museum dated 1783 (Crace Port. XI. 112. Sheet 51).

LITERATURE
Colley Cibber, *Apology for the life of C. Cibber, by himself*, 1740, pp. 183, 240–1.

48

49

Nicolo Grimaldi, called Nicolini (1673–1732)

From the album of Count Francesco Algarotti, probably made in the studio of Marco Ricci (1676–1730) or perhaps by Antonio Maria Zanetti (1679/80–1767)

Nicolini (on the right of this double sheet), a leading Neapolitan *castrato* singer and occasional librettist, came to England at the invitation of the Earl of Manchester in 1708, 'the first truely great singer who had ever sung in our theatre'. His earliest London appearance in an adaptation of Alessandro Scarlatti's *Pirro e Demetrio* (see under no. 50) excited Steele's favourable comment: Nicolini 'sets off the Character he bears in an Opera by his Action, as much as he does the Words of it by his Voice. Every Limb, and every Finger, contributes to the Part he acts . . .' (*Tatler*, 3 January 1709).

In February 1711, Nicolini created the title part in Handel's first London opera, *Rinaldo*, to great acclaim. He left London in the summer of 1712, but after seasons in Venice and Naples he returned to London in the spring of 1715, appearing in the revival of *Rinaldo* and taking the title role of Handel's latest opera, *Amadigi*, in May.

Like other drawings in the Algarotti album, this study will have been made in Venice where Nicolini frequently appeared after his final return to Italy in 1717; the original at Windsor has been attributed to Marco Ricci, who was also responsible for an earlier painted caricature which included Nicolini (no. 50).

The Algarotti album, or rather what is known of it, consists of 45 caricatures, mainly of opera singers who appeared in Venice during the first half of the eighteenth century. They are mounted in pairs on large sheets, size 37.4 × 52.8 cms, from a long-lost dismembered album, and are inscribed in the hand of Count Francesco Algarotti, the celebrated Venetian connoisseur and man of letters, who twice visited England, in 1736 and 1739.

Edward Croft-Murray has shown that they can be identified with a volume listed in the posthumous catalogue (1776) of the Algarotti Collection, with an attribution to Marco Ricci, as 'Caricatures de tous les Acteurs et Actrices qui de son [Ricci's] temps chantoient sur les Théatres d'Opéra de Venise; à la plume'. The caricatures are all of singers appearing in Venice at some time during the last 13 years of Marco Ricci's life, that is between his return from London in 1716 and his death early in 1730.

The Algarotti drawings give every appearance of being based on the caricatures in an album at Windsor Castle put together for Joseph Smith, British consul at Venice, and sold with his collection to George III in 1762. Most of the two hundred Windsor drawings are by Marco Ricci or his close friend the connoisseur Antonio Maria Zanetti. The Algarotti caricatures shown here (nos. 94, 115 and 117), according to Croft-Murray, are probably fair copies or 'worked-up tracings . . . executed either in Ricci's studio, or perhaps by Antonio Maria Zanetti'. A third such album, now in the Cini Foundation, Venice, belonged to Zanetti himself and is probably his own work. The presence of certain subjects in identical form in all three albums emphasises the difficulties of attribution for such caricatures.

Mounted with this is another drawing, no. 18, of an unidentified singer, inscribed *Benedettino*.

Albert R. Gellman Collection,
Courtesy of the Israel Museum, Jerusalem

Pen and brown ink over pencil traces,
24.8 × 17.7 cms/9¾ × 7 ins

Inscribed by Algarotti bottom centre *Nicolino*, and top left *19*

PROVENANCE
Count Francesco Algarotti (1712–64);
inherited by his brother, Bonomo (d. 1776), but then lost sight of; probably acquired on the continent before or after the 1914–18 War by W.S. Bayley, a Canadian collector; purchased from his widow, Mrs K. Bayley, by Albert R. Gellman, 1967.

LITERATURE
Contessa Maria Algarotti Corniani, *Catalogues des Tableaux, des Desseins et des Livres, – de la Galerie du feu Comte Algarotti*, 1776, no. 54 (as by Marco Ricci); Edward Croft-Murray, *An Album of Eighteenth Century Venetian Operatic Caricatures formerly in the Collection of Count Algarotti*, catalogue of exhibition at the Art Gallery of Ontario, Toronto, 1980 (as from the studio of Marco Ricci or by A.M. Zanetti).

50

50

The Rehearsal of an Opera

Attributed to Marco Ricci (1676–1730)

One of a group of caricature paintings of opera rehearsals, thought to be the work of Marco Ricci, the Venetian artist invited to London by the Earl of Manchester in 1708 (see no. 39). It has been suggested in a closely argued study by Eric White that the scene may represent the rehearsal of *Pyrrhus and Demetrius* as held in the practice room at the new Haymarket Theatre in 1709. White goes on to propose that Nicolini (Nicolo Grimaldi the *castrato*; see no. 49), in red, directing the orchestra, is singing a duet with Mrs Katherine Tofts (?1685–1756), who is dressed in white. The woman with the fan over her face may be 'the Baroness' (?Johanna Maria Lindelheim), and the singer with the muff, 'Margherita' (Francesca Margherita de l'Epine, *c.*1683–1746). If these identifications are to be trusted, and what evidence there is tends to support them, then the duet may be 'Caro(a), caro(a)' ('Charmer, charmer') – the only one sung by Nicolini and Mrs Tofts in *Pyrrhus and Demetrius*. This is one of the few operas for which Marco Ricci is known to have designed the scenery; his sets were for the revival staged on 2 April 1709. It had been adapted from an earlier opera by Nicola Haym, who may be the figure at the harpsichord. Among other less certain identifications are Pepusch (see no. 71), the future husband of Margherita, standing in a scarlet cloak on the right, and Heidegger (see no. 57), seated on the right with a music book.

This painting gives us a glimpse of the London opera world immediately before Handel's arrival in England. The use of both Italian and English singers in the same production was soon to disappear in favour of the Italians, who dominated the London opera stage for thirty years. Of the performers, Nicolini sang for Handel in *Rinaldo* (see no. 45) in 1711 and Margherita in various works in 1712 and 1713, including *Il Pastor Fido* and *Teseo*.

J. & A. Beare Ltd

Oil, 44.0 × 58.0 cms/$17\frac{1}{4} \times 22\frac{7}{8}$ ins

PROVENANCE
Anonymous sale, Sotheby's, 6 April 1977, lot 17, purchased by Colnaghi; acquired subsequently by J. & A. Beare Ltd.

VERSIONS
I) Sotheby's, 24 June 1970, lot 18 (from the collection of Christopher Turner), 41 × 54.5 cms (repr. *White*, pl. 2).
II) Yale Center for British Art, New Haven (from the collection of Sir Watkin Williams-Wynn, Bart), 48.5 × 58.0 cms (repr. *Hogwood*, pl. 5 in colour, and *White*, pl. 1).
III) Castle Howard, 46.2 × 54.3 cms; this version has an open window in place of the landscape on the wall (repr. *White*, pl. 3).

LITERATURE
E. W. White, 'The Rehearsal of an Opera', *Theatre Notebook*, XIV, 1960, pp. 79–91, pl. 2.

Queen Anne's London

By the time Handel returned to London in the autumn of 1712 on a further leave of absence from Hanover, the continental War was almost over after ten years of struggle. The memory of his *Rinaldo*, given in London 18 months earlier, and of the entertainment mounted by him for the Queen's birthday in February 1711 must have been fresh in the public mind when a few months later Handel received a remarkable commission from the Queen. It was to write a Te Deum to be sung in St Paul's Cathedral when peace was proclaimed. This was the 'Utrecht' *Te Deum* (see no. 55), which, though completed in January 1713, was not performed until July, two months after the official proclamation of peace.

It may seem strange that the commission went to a foreigner, as Mainwaring tried to explain: 'had our musicians been thought equal to the task, a foreigner would hardly have been applied to for the song of triumph and thanksgiving, which was now wanted'. William Croft, composer for the Chapel Royal, who had written an attractive Te Deum setting in 1709, would have been the obvious choice among native composers. It appears even stranger to have given the commission to Handel as a servant of the Hanoverian court since the Elector, as one of the Allies, was strongly opposed to England's unilateral decision to seek peace. Handel's part in the Utrecht celebrations, rather than his prolonged absence from Hanover, may well have been the real cause of his dismissal from the Elector's service in June, only a few weeks before the thanksgiving at St Paul's, as has been proposed in a recent study, 'Handel and Hanover', by Donald Burrows.

Another official commission came to Handel on the occasion of the Queen's birthday in February 1713. For this he prepared his Birthday Ode, 'Eternal source of light divine' (see no. 53), though it is not clear if Anne's state of health allowed it to be performed then. Later the same year, on 28 December, Handel's service was officially recognised by the award of a royal pension: 'unto Our Trusty & well-beloved George Frederick Hendel Esq . . . One Annuale Sum or Pencon of 200£'. On 1 August 1714 the Queen died.

If the court was an important source of recognition, the opera house absorbed much of Handel's energies. *Il Pastor Fido*, *Teseo* and *Silla* were all performed between November 1712 and June 1713. Of these *Teseo* was the most substantial piece, and one which met with both success and disaster: it had a long run of 13 performances but the manager, Owen McSwiny (see no. 56), ran off with the receipts for the first two nights. This opera saw the beginning of Handel's long co-operation with John James Heidegger (see no. 57), who now took on the management of the Haymarket Theatre where he was to remain for more than thirty years.

Handel's other occupation when he had no particular engagement was, according to Hawkins, to play the organ at St Paul's Cathedral (see no. 54): 'the tone of the instrument delighted Handel and a little intreaty was at any time sufficient to prevail on him to touch it . . . and he has been known, after evening service, to play to an audience as great as ever filled the choir'.

Early eighteenth-century London was a city in transformation. St Paul's was newly completed (though Thornhill's paintings in the dome were not begun until 1715); the opera house in the Haymarket had only opened in 1705; Burlington House where Handel was living for a period (see no. 61) was in the process of being transformed; and the area around Handel's future home in Brook Street had yet to be developed. No wonder Daniel Defoe wrote of 'New squares, and new streets rising up every day to such a prodigy of buildings, that nothing in the world does, or ever did equal it, except old Rome in Trajan's time' (*Tour through the Whole Island of Great Britain*, 1725).

LITERATURE
Mainwaring, p. 86; *Hawkins*, II, p. 589; *Burrows*, pp. 41–5.

51

John Hughes (1677–1720)

By Gerard van der Gucht (1697–1776)
after Sir Godfrey Kneller (1646–1723)

Hughes was an interesting but minor poet and historian ('what he wanted in genius he made up as a honest man'), who attracted Handel's attention on his first visit to London. Soon after the composer's return to Germany he wrote (in French) to a friend at London in July 1711: 'Please convey my best compliments to Mons. Hughes. I shall take the liberty of writing to him at the earliest opportunity. If, however, he would honour me with his commands, and if he would add one of his charming poems in English, he would give me great pleasure.' Hughes seems to have responded to this invitation with his cantata 'Venus and Adonis', the first English text set by Handel; he later provided part of the libretto for Handel's oratorio *Acis and Galatea* (no. 73).

Hughes was fond of the violin, which he sometimes played at the concerts given by Thomas Britton (see no. 59). He was a staunch supporter of the idea of opera in English, writing *Calypso and Telemachus*, which was set to music by Johann Ernst Galliard (see no. 153) and performed at the Haymarket Theatre in 1712 amidst opposition from the Italian singers; in the preface the case for opera to be 'perform'd in a Language understood by the Audience' is advanced – not for the last time, as a very similar polemic (see no. 155) addressed by Aaron Hill to Handel twenty years later shows.

This engraved portrait of Hughes formed the frontispiece to his *Poems on Several Occasions* (1735), a posthumously published collection of his work. Kneller's original oil of 1718 (which is reversed in the engraving) was given by Hughes to his patron the Lord Chancellor Cowper in January 1720, the month before the poet's early death, aged 43. (The painting appeared in the Cowper sale at Christie's, 16 October 1953, lot 62, bought in.)

National Portrait Gallery, London

Engraving, 13.7 × 8.1 cms/5$\frac{3}{8}$ × 3$\frac{1}{8}$ ins

Inscribed *J. HUGHES ESQ. | G. Kneller Eq: Bar: Pinx: 1718. | G. Vander Gucht Sculp.*

LITERATURE
Deutsch, pp. 44–5.

52

Queen Anne and the Knights of the Garter

By Peter Angellis (1685–1734)

This rare representation of the court of Queen Anne at the period Handel settled in England is thought to show the ceremony of the Chapter of the Order of the Garter held by Queen Anne at Kensington Palace on 4 August 1713, the year before her death. Though the six new knights then invested with the insignia of the Order cannot be individually identified in the picture, they were all of high rank. Three of them must have become known to Handel through his association with the Royal Academy of Music: the Duke of Kent, the 3rd Earl of Strafford (both were subscribers to the Academy in 1719; the Duke had been intended as governor) and the 3rd Earl of Peterborough, who married Handel's singer, Anastasia Robinson (see no. 92).

Queen Anne was Handel's first royal patron in England, awarding him a pension of £200 per annum on 28 December 1713. His music was first performed before the Queen at a splendid court occasion soon after his arrival in London:

> Tuesday, the 6th February [1711], being the Queen's Birth-day, the same was observed with great Solemnity: the Court was extream numerous and magnificent; the Officers of state, Foreign Ministers, Nobility, and Gentry, and particularly the Ladies, vying with each other, who should almost grace the Festival. Between one and Two in the Afternoon, was perform'd a fine Consort, being a Dialogue in Italian, in Her Majesty's Praise, set to excellent Musick by the famous Mr. Hendel, a Retainer to the Court of Hanover, in the Quality of Director of his Electoral Highness's Chapple, and sung by Cavaliero Nicolini Grimaldi, and the other Celebrated Voices of the Italian Opera: with which her Majesty was extreamly well pleas'd.

Handel's music for Queen Anne's birthday in 1711 is his first known performance in England, preceding the opera *Rinaldo* by a few weeks. The music is now lost but another composition, the well-known *Birthday Ode for Queen Anne*, survives, and was written for a similar occasion two years later (see no. 53).

National Portrait Gallery, London
NPG 624

Oil, 62.2 × 75 cms/24½ × 29½ ins

Signed bottom left *P. Angelis: F/17 . .*
(the last two digits of the date indistinct)

PROVENANCE
Perhaps Christie's sale, 27 May 1847, lot 296, purchased by Garcia; William Ruck of Cheltenham, from whom purchased by the National Portrait Gallery, 1881.

LITERATURE
Piper, p. 392, pl. 32a; *Burrows*, p. 41.

53

Birthday Ode for Queen Anne

Handel's autograph score

A tradition of celebrating New Year's Day and the birthday of the monarch with newly composed musical odes became established in the reign of James II (though its origins are much earlier) and continued until the death of George III in 1820. The words were generally the responsibility of the Poet Laureate, and the music was provided by the Master of the King's (or Queen's) Music. In the last years of Anne's reign, when the holders of these offices were Nahum Tate and John Eccles, the custom seems temporarily to have broken down. In 1711 the Queen's birthday, 6 February, was celebrated at St James's Palace by a 'dialogue in Italian in Her Majesty's praise, set to excellent Musick by the famous Mr Hendel', a work which seems to be lost. Handel's later English Birthday Ode ('Eternal source of light divine') is based on a text attributed by Charles Jennens to Ambrose Phillips, though it does not appear among Phillips's published work. No specific evidence of a performance of the Ode has been found; it is probable that it was composed for the 1713 celebration and revised for that of 1714, but the illnesses of the Queen may have prevented its performance on either occasion.

Each verse of the Ode is set for solo voice or voices and is concluded by a setting for full chorus of the refrain 'The day that gave great Anna birth / Who fix'd a lasting peace on Earth'. In the last verse Handel varies the scheme by following the alto solo 'United nations shall combine' with a further setting for soloists and chorus (the opening of which is displayed, ff. 34r–35v); the Ode is then concluded with a repetition of the first chorus. At the entry of the chorus in the last verse Handel marks the soprano and alto lines for *Mrs Robins*[on] and *Mr Eilfurt* (ie Anastasia Robinson and the counter tenor Richard Elford), and the chorus as *Co Eco* (ie 'Coro in Eco' or Echo Chorus).

The British Library Board
RM 20.g.2

Autograph score, i + 36 ff., 29.2 × 22.8 cms/11½ × 9 ins

PROVENANCE
As no. 34.

LITERATURE
Squire, p. 55; A. Boyer, *The Political State of Great Britain*, I, 1711, p. 227; Winton Dean, 'Charles Jennens's Marginalia to Mainwaring's Life of Handel', *Music and Letters*, LIII, 1972, p. 160; Donald Burrows, 'Handel and the English Chapel Royal', unpublished dissertation, Open University, 1981; *Burrows*, pp. 35–59 (especially p. 43).

54

'A Prospect of the Choir, of the Cathedral Church of St. Paul'

By Robert Trevitt (d. 1723)

This was the interior of St Paul's Cathedral as Handel first knew it, with the great organ by Father Smith straddling the choir. Wren's cathedral was still very new (the first service had been held in 1697) and was

53

LONDON: THE FIRST DECADE 1710–19

54

one of the wonders of London for visitors and Londoners alike.

The General Thanksgiving of 7 July 1713, at which Handel's Utrecht *Te Deum* (see no. 55) received its much delayed first public performance, is not known in any engraving. But it probably appeared much as in this view of an earlier thanksgiving, held on 31 December 1706 to mark the 'Great and Wonderful Successes' of that year.

Queen Anne and her husband are seated on the dais on the right (though the Prince was actually indisposed in 1706 and the Queen did not attend in 1713), with the bishops and clergy in the apse, the Lords on the benches in the aisles and the Commons occupying some of the choir stalls and the galleries above. Around the organ are ranged some of the choir men and musicians, mainly with stringed instruments, but two play oboes; the choir boys must have been placed further round the gallery.

Purcell's *Te Deum and Jubilate for St Cecilia's Day*, composed in 1694, had been the standard setting for thanksgivings at St Paul's until supplanted by Handel's in 1713. The General Thanksgiving of that year was the last of seven held in the Cathedral during Queen Anne's reign in celebration of Marlborough's military successes. It marked the rather dilatory conclusion of the War of Spanish Succession by the Peace of Utrecht on 31 March 1713; among the British gains were Gibraltar and Hudson's Bay. The Peace congress had however opened much earlier, so that Handel's *Te Deum* was already complete in January 1713.

During his early years in London, Handel made a habit of playing the organ in St Paul's after evensong, and is said to have attracted an audience as great as ever filled the choir.

Guildhall Library

Engraving on two sheets (coloured), 57.7 × 98.6 cms/ $22\frac{3}{4} \times 38\frac{3}{4}$ ins

Inscribed *A PROSPECT of the CHOIR, of the CATHEDRAL CHURCH of St PAUL, on the GENERAL THANKSGIVNG the 31.st of Decem.^r 1706. Her Majesty and both Houses of Parliament present*; a lengthy description follows, ending *The Choir and other Musicians filled the Organ Gallery and y^e Returns of the Galleries adjacent on either Side.*; and at bottom right *Robt Trevitt Fecit*

LITERATURE
Donald Burrows, 'Handel and the English Chapel Royal', unpublished dissertation, Open University, 1981.

55

Utrecht Te Deum and Jubilate

Handel's autograph score

The Te Deum in its English version ('We praise Thee, O God') is the main canticle of the Anglican morning service. Large-scale musical settings were traditional at times of celebration, and particularly as expressions of thanksgiving for victories. When Handel came to London the model setting was that of Henry Purcell (produced for the St Cecilia's Day celebrations of 1694), which William Croft's fine setting of 1709 did not entirely displace. It was probably with Purcell's design in mind that Handel wrote the first of his Te Deum settings (he eventually produced four others), completing it on 12 January 1713. That this was written anticipating the celebration of the end of the War of Spanish Succession, the details of which were hammered out in lengthy negotiations at Utrecht during 1712, is confirmed by a letter from the Hanoverian diplomat Thomas Grote to the Elector at Hanover; on 13 January he wrote that Queen Anne had commissioned a piece of music from Handel: 'This music is, as I understand, a Te Deum, which shall be sung in St Paul's Cathedral when peace is proclaimed.'

Handel was optimistic about the conclusion of the negotiations. The Treaty of Utrecht between England and France was not signed until 31 March and the Peace was not proclaimed officially until 5 May. Meanwhile, however, he managed to arrange 'rehearsals' (ie previews) of the *Te Deum* at the Banqueting House, Whitehall (then in use as a Royal Chapel) in the early part of March. The official thanksgiving service was further delayed by the poor health of Queen Anne, but eventually took place (in the Queen's absence) at St Paul's Cathedral on 7 July 1713, when Handel's *Te Deum* was joined by a complimentary setting of the Jubilate (Psalm 100: 'O be joyful in the Lord all ye lands'). The music swiftly rivalled Purcell's in popularity, and was no doubt the main reason for the Queen's grant of an annual pension of two hundred pounds to Handel the following December. His contribution to the celebrations may however have soured his relations with Hanover, since the Elector, the future George I of England, was opposed to the terms of the Treaty of Utrecht.

The Utrecht *Te Deum* is written on the grand scale, but Handel is always sensitive to the varying moods of the words and there are several passages of a quiet, expressive character to contrast with the brilliant writing for full chorus and orchestra (with trumpets) elsewhere. The verse 'We Believe that Thou shalt come to be our judge' is one such; the orchestral accompaniment (f. 28v) is reduced to solo flute (*Travers*[a]) and strings only (the organ is silenced by the instruction *senza Organ*[o]), and indications to use solo voices appear in the vocal lines, though Handel's exact intentions are not quite clear. '9' on the right-hand page indicates the start of the ninth 'gathering' of the score, or group of four sheets.

The British Library Board
RM 20.g.5

Autograph score, ii + 73 ff. (*Te Deum* ff. 1–42, *Jubilate* ff. 43–73), 28.6 × 22.8 cms/11¼ × 9 ins

Inscribed on f.42r (the end of the *Te Deum*) S.D.G. G.F.H. Londres ce 14 de Janv. v[ieux]. st[yle]. a 1712., and on f.72r (the end of the *Jubilate*) S.D.G. G.F. Hendel ... 1713. (the rest of the date cut away during binding)

PROVENANCE
As no. 34.

LITERATURE
Squire, p. 87; *Deutsch*, p. 61; M. Tilmouth, 'A Calendar of References to Music in Newspapers published in London and the Provinces (1660–1719)', *R.M.A. Research Chronicle*, 1, 1961, p. 84 (entry for 27 March); H. Becker, 'Die frühe Hamburgische Tagespresse, als musikgeschichtliche Quelle', *Beiträge zur Hamburgischen Musikgeschichte*, 1, 1956, pp. 22–45 (especially pp. 34–5); *Burrows*, pp. 35–59 (especially pp. 41–5).

56

56
Owen McSwiny (d. 1754)
By John Faber (d. 1756) after Jean Baptiste van Loo (1684–1745)

Owen McSwiny (or Swinny) was the rogue who ran off with the takings from Handel's third opera, *Teseo*, in January 1713 and absconded to Italy: 'after... Two Nights Mr Swiny Brakes & runs away & leaves ye Singers unpaid, ye Scenes & Habits also unpaid for' (Francis Colman's *Opera Register*). But McSwiny was a useful fellow, whom Handel forgave, and in the 1720s he is to be found scouring Italy for singers for the Royal Academy of Music. He returned to London in 1735, almost thirty years after his first arrival there from Ireland in 1706.

In his early years he had been a successful theatre manager both at Drury Lane and at the Queen's Theatre in the Haymarket. In May 1708 Sir John Vanbrugh gave up his interest in the Haymarket to McSwiny, writing to the Earl of Manchester (see no. 39): 'I have parted with my whole concern to Mr. Swiney, only reserving my rent: So that he is entire possessor of the Opera and most people think will manage it better than anybody.' McSwiny took on the *castrato* Nicolini (see no. 49), and produced a series of successful operas including *Pyrrhus and Demetrius* (see under no. 50) before running into trouble with Handel's *Teseo*.

In his self-imposed exile in Italy he devised a successful scheme, which was well under way by 1722, for a series of paintings by leading Italian artists; these are among the most intriguing images of their time, and depict allegorical tombs commemorating great men from England's recent history. His main patron in this venture was Charles, 2nd Duke of Richmond, who was in 1726 a deputy governor of the Royal Academy of Music. To him he also addressed his letters about opera productions in Italy in his capacity as a correspondent for the Academy, a position he had been offered in 1724. McSwiny sent libretti to London and also became involved with negotiations for the hire of singers including the celebrated soprano, Faustina Bordoni (see no. 96). After the dissolution of the Academy in 1728, Handel travelled to Italy in search of new singers, meeting McSwiny at Bologna in March 1729. McSwiny provided further help the following year when Handel was obliged to remain in London; he writes on 18 July 1730 from Bologna about the prospects of engaging Senesino or Carestini at 1,200 guineas – eventually Senesino was chosen but at a salary of 1,400 guineas.

In this engraving made after his return to London, McSwiny appears to be dressed in a theatrical role. According to George Vertue, McSwiny's was one of the first portraits painted by Jean Baptiste van Loo on his arrival in London from Paris in December 1737, though this engraving was not made until 1752. (The painting was sold at Christie's, 7 March 1927, lot 113.)

National Portrait Gallery, London

Mezzotint engraving, 30.7 × 25.4 cms/12⅛ × 10 ins

Inscribed *Owen M:Swiny Esq! | Vanloo Pinx! | J. Faber fecit 1752. | Price 2 Shillings Sold at the Golden Head near the Church Bloomsbury Square.*

LITERATURE
CS, p. 391, no. 231; *Deutsch*, pp. 53, 256–8; Francis Haskell, *Patrons and Painters*, 2nd edition, 1980, pp. 287–91; *Gibson*, pp. 82–6.

57
John James Heidegger (1666–1749)
By John Faber (d. 1756) after Jean Baptiste van Loo (1684–1745)

John James Heidegger was a central figure in the London theatre world for forty years, working closely with Handel on the production of opera at the King's Theatre in the Haymarket at one period or another. Born in Zurich, the son of a Professor of Theology, he left his wife and children as a result of an unfortunate love affair and travelled in Europe before settling permanently in London where he is first known in 1707. Sometimes known as 'Count' Heidegger, he was a man who attracted much public attention and his name crops up on numerous occasions. He was the 'Swiss Count whose Earldom lies in the land of the Moon', 'the most ugly man that was ever formed' (Mrs Delany), the 'strange bird from Switzerland' (Alexander Pope).

Heidegger was well established in the London opera world by the time Handel's *Rinaldo* was mounted at the Haymarket Theatre in 1711; he took over as manager there two years later when Owen McSwiny (see no. 56) ran off with the takings from *Teseo* in January 1713. The next new opera by Handel known to have been performed at the theatre was *Amadigi di Gaula* on 25 May 1715, with a libretto dedicated by Heidegger to Lord Burlington; this was revived in

57

1716 and again the following year.

When opera performances were resumed under the auspices of the Royal Academy of Music in 1720, Heidegger acted as manager while Handel held the position of Master of the Orchestra, an arrangement which lasted until the Academy's failure in 1728. Heidegger found a lucrative sideline to opera management in the immensely successful masquerades which he mounted at the King's Theatre, in the face of considerable opposition from the Bishop of London (see under nos. 105–6). After the demise of the Academy he entered into an agreement with Handel to produce opera at the Theatre; one of the former subscribers, Lord Percival, noted in his diary for 18 January 1729 the Academy's decision 'to permitt Hydeger and Hendle to carry on operas without disturbance for 5 years and to lend them for that time our scenes, machines, clothes, instruments, furniture, etc'. This arrangement ran its full term until 1734 when Heidegger must have felt he had little choice financially but to throw in his hand with Handel's rivals, the Opera of the Nobility. Handel subsequently returned to the Haymarket for some oratorio seasons but their close partnership was at an end.

Though Heidegger was now in his seventies, he continued his management of the Theatre. It was while the French artist, Jean Baptiste van Loo, was on his immensely successful visit to England that Heidegger sat to him for his portrait (which George Vertue thought was 'rather homelier than he was'). It was subsequently engraved in 1749 by Faber whose mezzotint, shown here, was probably intended as a memorial plate.

The Trustees of the British Museum
1902–10–11–1516

Mezzotint engraving, 31 × 25.4 cms/12¼ × 10 ins

Inscribed *John James Heidegger Esqr | Vanloo pinx^t! | J. Faber fecit 1749. | Price 1s. 6d. Sold by I Faber at ye Golden Head in Bloomsbury Square*

LITERATURE
CS, p. 367, no. 184; *Vertue*, III, p. 84; *Deutsch*, pp. 67, 234; *Highfill*, VII, p. 234.

58 [in colour opposite p. 232]

Richard Leveridge (1670/1–1758)

By Thomas Frye (c. 1710–62)

For more than half a century Richard Leveridge played a leading part in London's theatrical life, both as a composer of popular songs and as a fine bass singer (he holds a favourite song, 'Ghosts of every occupation' from *The Necromancer*). At his retirement in 1751 at the age of eighty a subscription was raised for 'honest Dick Leveridge, the Father of the English Stage'. Thomas Frye's sympathetic portrait, his most ambitious of the singer, belongs to this late period of Leveridge's life.

Leveridge had first come to prominence in Henry Purcell's theatre company in 1695, its last season. He took leading parts in many of the operas performed in London between 1705 and 1708, including *Arsinoe* and *Camilla*, but was overshadowed by the growing popularity of Italian singers. When Handel returned to London on his second visit, Leveridge was one of the leading singers employed for the winter season of 1712–13, appearing in both of Handel's new operas at the Queen's Theatre: *Il Pastor Fido* opening in November 1712 and the much more successful *Teseo*, first performed in January. Leveridge rarely sang for Handel in later seasons, turning instead to lighter musical work for the theatre and pantomime. He set English words to some of Handel's Italian songs and wrote many songs of his own, publishing them as single sheets and as collections.

Gerald Coke Handel Collection

Oil, 90.2 × 69.8 cms/35½ × 27½ ins

Signed at the bottom left of the music *Frye Pinx!*; inscribed on the music, near the top, *Ghosts of . . .* and *Sung by Mr Leveridge*

ENGRAVED
 I) In mezzotint by William Pether
    (*CS*, III, p. 985, no. 23).
 II) In line by J. Saunders for *The European Magazine*, 1793.

PROVENANCE
Probably the portrait painted for John Ellis (1698–1791), at whose sale bought for *The European Magazine*; Christie's sale, 17 March 1967, lot 31, purchased by Gooden & Fox for Gerald Coke.

LITERATURE
The European Magazine, XXIV, 1793, pp. 243, 364 (engraved, see above).

EXHIBITED
Coke, p. 29, pl. 4 (repr. in colour).

59

Thomas Britton (1644–1714)

By Thomas Johnson (fl. early 18th century) after John Woolaston (*c*. 1672–1749)

A most unusual participant in the cultural life of Queen Anne's London was the musical 'small coal-man' (or coal merchant) Thomas Britton, at whose concerts Handel is said to have played the harpsichord. His concert room was above his coal store; it was 'very long and narrow, and had a ceiling so low, that a tall man could but just stand upright'. His recitals were widely patronised: a 'Harmonious society of tickle-fiddle gentleman', according to his neighbour, Ned Ward, or in the more polite terms of the antiquarian Ralph Thoresby, in his diary on 5 June 1712, 'a noble concert of music, vocal and instrumental, the best in town, which for many years [Britton] has had weekly for his own entertainment, and of the gentry, &c., gratis, to which most foreigners of distinction, for the fancy of it, occasionally went'.

Britton was also an amateur chemist and an enthusiastic book collector. His wide-ranging taste in music may be judged from the sale catalogue of his library, which contained scores by many Italian and English composers, though not the work of Handel who had only been in London a few years at the time of Britton's death in 1714.

Britton's portrait was twice painted by his friend, the artist and amateur musician John Woolaston, who sometimes played the violin at his concerts. On one occasion, as Woolaston later told the historian Sir John Hawkins, on hearing the cry of 'small-coal', he invited Britton to come in and painted his portrait there and then wearing his blue linen frock with coal measure in hand; this is the painting in the National Portrait Gallery which appears to be dated 1703. The other portrait of Britton is known only through engravings such as this mezzotint. Both the image and Matthew Prior's verses point to the contrast between Britton's trade (his coal measure is behind him) and his devotion to music and books (he is tuning a harpsichord; a violin and bookshelves can be seen in the background).

National Portrait Gallery, London

Mezzotint engraving, 35.8 × 28.7 cms/14⅛ × 11¼ ins

Inscribed around the portrait THOMAS BRITTON SMALCOAL MAN, and beneath, *Thos. Iohnson fecit.*; also with eight lines of verse by Matthew Prior, beginning

Tho' doom'd to small-coal, yet, to Arts ally'd,
Rich without Wealth and Famous without Pride,
Musick's Best Patron, Judge of Books and Men,
Belov'd and Honour'd by Apollo's Train; . . .

LITERATURE
Hawkins, II, p. 790; *CS*, p. 736, no. 1; *Highfill*, II, p. 343.

Handel and Lord Burlington, 'the Apollo of the Arts'

During Handel's early years in London one of his most important patrons was the young and wealthy Earl of Burlington (see no. 60), who showed an early love of music and became a great enthusiast for all the arts and especially architecture. Little can be learnt of Handel's life at this time from contemporary documents but Sir John Hawkins in his *General History of Music*, written some sixty years later, describes Handel as staying at Burlington House for three years and attributes the operas *Il Pastor Fido*, *Teseo* and *Amadigi* (composed in the period 1712–15) to Handel's residence there.

Teseo was first performed at the Queen's Theatre on 10 January 1713 and was dedicated to Lord Burlington by Nicola Haym in his libretto. Another opera, *Silla*, was privately produced, receiving a single performance on 2 June 1713, either at the Queen's Theatre or possibly at Burlington House. On 17 May 1714 Lord Burlington left London to go to Italy on the Grand Tour and he did not return until 2 May 1715, almost a year later, bringing with him Filippo Amadei, cellist and composer, and the Castrucci brothers. Handel's next opera, *Amadigi di Gaula*, opened a few weeks later on 25 May; in the libretto Burlington is thanked for 'that Generous Concern Your Lordship has always shown for the promoting of Theatrical Musick, but this Opera more immediately claims Your Protection, as it is compos'd in Your own Family'.

It was at Burlington House, according to Hawkins, that Handel became acquainted with Alexander Pope, John Gay and Dr Arbuthnot (see no. 62). 'There Hendel strikes the Strings . . . There oft' I enter', wrote Gay in his *Trivia* (no. 152), first published early in 1716, adding, 'For Burlington's belov'd by ev'ry Muse.' By the summer of 1717 and probably earlier, Handel left Burlington House having, according to Hawkins, 'received a pressing invitation from the duke of Chandois to undertake the direction of the chapel at his superb mansion, Cannons'.

In 1719 Lord Burlington lent his name to the subscription list (no. 84) to the Royal Academy of Music, putting up £1,000 alongside the Duke of Chandos and the Duke of Newcastle. It was to Burlington that Handel wrote from Dresden reporting his progress in engaging Senesino and other singers to come to London (see no. 87). Burlington meanwhile set off to Italy on his second Grand Tour, and while there appears to have invited Bononcini (see no. 88) to come to London to compose for the Academy. His subsequent interest in music is less well documented but his wife, Dorothy, was well known for her enthusiasm for the theatre; in 1727 she took Faustina's side in the celebrated rivalry between the Academy's two sopranos (see no. 107) and was later associated with the artist Joseph Goupy in his caricatures of Cuzzoni and Farinelli (no. 121) and perhaps also of Handel (no. 8).

In 1733 Lord Burlington was among the subscribers to the rival Opera of the Nobility. His support for Handel continued however, and as late as 1750, three years before his death, there are entries in Burlington's accounts of payments 'for the use of Mr. Handel' and of subscription tickets for his oratorio seasons at Covent Garden.

LITERATURE
Hawkins, II, p. 859; *Deutsch*, pp. 67, 206, 255, 304; James Lees-Milne, *Earls of Creation*, 1962, p. 114; R.W. Williams, 'Lord Burlington and the "Gifts for 'Senr. Pope'": A False Dating', *Notes and Queries*, CCXVIII, 1973, pp. 7–8, 207.

60 [*in colour between pp. 96–7*]

Richard Boyle, 3rd Earl of Burlington (1694–1753)

Attributed to Jonathan Richardson (*c*.1665–1745)

The 'architect' Earl and early patron of Handel is shown here in his early twenties in the sort of informal undress of the period often associated with portraits of artists and writers. His architectural interests are indicated by the pair of dividers he holds and by the background view of the 'Casina' or 'Bagnio' at Chiswick which he himself designed in 1717 ('the first essay of his Lordship's happy invention'). The portrait was probably painted between then and his departure for Italy on his second Grand Tour in 1719.

Burlington was only ten years old when he succeeded his father as Earl and came into his vast estates in Ireland, Yorkshire and at Chiswick. He also owned a sizeable and profitable property in London, extending north from Burlington House in

61

Piccadilly towards the Cavendish estate. Burlington never took more than a passing interest in politics despite his position as a Privy Councillor and a member of the House of Lords. It was, in George Vertue's words, as 'the noble Maecenas of Arts' that he claimed attention. His enduring love of music is discussed above. In architecture, his patronage of Colen Campbell and William Kent, as well as his own designs, helped establish the Palladian style as the accepted manner in England during the mid-eighteenth century.

National Portrait Gallery, London
NPG 4818

Oil, 146.0 × 116.8 cms/57½ × 46 ins

Inscribed beneath the Burlington coat of arms *Earl of Burlington*

PROVENANCE
Sotheby's, 28 July 1954, lot 137, purchased by Colefax & Fowler; Mrs G.C. Lancaster, from whom purchased by the National Portrait Gallery, 1970

LITERATURE
Kerslake, pp. 29–30, pl. 88.

61

Burlington House

By Johannes Kip (1653–1722) after Leonard Knyff (1650–1722)

Burlington House in Piccadilly as Handel first knew it was a fairly plain Restoration house of brick with some stone ornamentation and a hipped roof. For much if not all of his stay it was in the hands of builders. The alterations which transformed it into a showpiece of the new Palladian style probably began with the interior in 1712 or early 1713 when Lord Burlington (see no. 60) was still only 18. Two Venetian artists in London, Giovanni Antonio Pellegrini and Sebastiano Ricci, provided wall paintings for the hall and staircase (perhaps the 'animated picture' of John Gay's poem; see no. 152). Structural work continued during the young Earl's first visit to Italy from May 1714 to May 1715 and was not complete until 1720; it included the re-facing of the south front by Colen

Campbell (incorporated, though with an additional storey, into the present façade of the Royal Academy of Arts at Burlington House).

Guildhall Library

Engraving, 32.0 × 46.9 cms/12⅝ × 18½ ins

From *Britannia Illustrata* (or the *Nouveau Theatre de la Grande Bretagne*) issued in various editions from 1707 onwards

Inscribed (on either side of the Burlington coat of arms) *Burlington House in Pickadilly Belonging to the R.! Honble Charles Boyle Baron Clifforde of Londesburgh, and Earle of Burlington . . .* [with details of his other titles and official positions] / *L. Knyff. Del. / I.Kip Sc.*

LITERATURE
Bernard Adams, *London Illustrated 1604–1851*, 1983, p. 40, no. 11; F.H.W. Sheppard (ed.), *St James, Westminster. Part II (North of Piccadilly)*, vol. XXXII of the *Survey of London*, 1963, pp. 395–402.

62

Alexander Pope (1688–1744)

By Jonathan Richardson (c. 1665–1745)

A late work by Jonathan Richardson, closely related to a drawing of 1737 at Windsor Castle showing Pope with a laurel wreath

62

It was perhaps at Burlington House that Pope first met Handel. Their paths then crossed at intervals during the thirty years before Pope's death in 1744, showing Pope as a supporter of music set to English words in the face of the fashion for Italian opera. The poet appears to have provided much of the text (in English) for Handel's *Esther* when first performed in 1718 (see no. 128); this was later adapted to become his first oratorio staged in public. In *The Dunciad* (1729), Pope enthusiastically supports Gay's *The Beggar's Opera*: it 'drove out of England the Italian Opera, which had carry'd all before it for ten years'. It was not until 1741 that Handel finally turned his back on producing opera in London but already in *The New Dunciad* (no. 164), published in March 1742, Pope places Handel as the leader of the rebellion against opera and refers to his being driven 'to th' Hibernian shore', a reference to his visit to Dublin in 1741–2.

There remains the question of how far Pope was able to appreciate music. As Brownell has made clear in a recent study, the long-standing tradition that Pope was no judge of music first appears in Mainwaring's biography of Handel (1760) with its story of the poet at the Earl of Burlington's: 'After [Handel] . . . had played some of the finest things he ever composed, Mr. Pope declared, that they gave him no sort of pleasure.' Later, Charles Burney in his *History of Music* (1776) described Pope as 'by nature wholly insensible to the charms of music'. However Brownell suggests that 'not a shred of evidence bears it out', and that more credence should be given to the earlier and more reliable report of an answer given by Pope to Joseph Spence in 1744; when asked, '"Did you ever learn anything of music?", Pope replied, "Never, but I had naturally a very good ear, and have often judged right of the best compositions in music by force of that"'.

National Portrait Gallery, London
NPG 1179

Oil, 61.3 × 45.8 cms/24⅛ × 18 ins

PROVENANCE
Given by Alfred A. de Pass to the National Portrait Gallery, 1898.

LITERATURE
W. K. Wimsatt, *The Portraits of Alexander Pope*, New Haven and London, 1965, p. 217, no. 54 (repr. in colour opposite p. 137); M. R. Brownell, 'Ears of an Untoward Make: Pope and Handel', *Musical Quarterly*, LXII, 1976, pp. 554–70; *Kerslake*, p. 214 (repr. in colour).

The Celebrated Water Musick

On a fine July evening in 1717 a splendid court entertainment took place on the Thames with music commissioned from Handel by Baron Kielmansegg (see no. 63). The occasion was described in the *Daily Courant* two days later:

> On Wednesday [the 17th] Evening, at about 8, the King took Water at Whitehall in an open Barge, wherein were also the Dutchess of Bolton, the Dutchess of Newcastle, the Countess of Godolphin, Madam Kilmanseck, and the Earl of Orkney. And went up the River towards Chelsea. Many other Barges with Persons of Quality attended, and so great a Number of Boats, that the whole River in a manner was cover'd; a City Company's Barge was employ'd for the Musick, wherein were 50 Instruments of all sorts, who play'd all the Way from Lambeth (while the Barges drove with the Tide without Rowing, as far as Chelsea) the finest Symphonies, compos'd express for this Occasion, by Mr Hendel; which his Majesty liked so well, that he caus'd it to be plaid over three times in going and returning.

Within Handel's own lifetime his *Water Music* had acquired a fame which it has kept ever since. The piece is also endowed with special status by Mainwaring and later biographers of Handel as the means by which he was restored to royal favour. But it is now known that although Handel had been dismissed in June 1713 from his position as Kapellmeister to the Elector of Hanover, it was done in terms to imply that he could re-enter the royal service when the Elector came to London as King, and indeed George I allowed the pension granted to Handel by Queen Anne to be continued.

Another contemporary account of the royal Water Party is given by Friedrich Bonet, the Prussian Resident in London, in a letter to his masters in Berlin. In this he attributes the organisation of the concert to Baron Kielmansegg, the King's close aide. He adds that 'the music had been composed specially by the famous Handel a native of Halle, and His Majesty's principal Court Composer'. The musicians 'played on all kinds of instruments, to wit trumpets, horns, hautboys [oboes], bassoons, German flutes, French flutes [recorders], violins and basses'. Bonet goes on to say that the 'concert cost Baron Kilmanseck £150 for the musicians alone'.

To gain an idea of the appearance of the occasion, with boats covering the whole river, one must turn to engravings of other events on the Thames (see no. 65), even though these do not include musicians. There are many reports of river parties accompanied by music during the eighteenth century, but the only view showing musicians with their instruments is the painting by Johann Zoffany of the Sharp family, now on loan to the National Portrait Gallery (no. 228).

Our knowledge of the *Water Music*, and indeed even of the original content and order of the movements, is hindered by the lack of either Handel's autograph or the conducting score. It seems possible that the work is an amalgam of pieces, only some of which were written specially for the summer evening in July 1717. The *Water Music* was not published until about 1733 and then in a partial and not wholly accurate form (see no. 64).

63 [*in colour opposite p. 97*]

Johann Adolf VI, Baron Kielmansegg (1668–1717)

By Michael Dahl (?1659–1743)

Baron Kielmansegg, the Hanoverian diplomat who accompanied George I to London in 1714, is well known for commissioning the *Water Music* from Handel in 1717. But in this portrait by Dahl, the Swedish painter working in London, it is his position as Master of the Horse to George I as Elector of Hanover which is emphasised; he is shown turning the pages of a folio illustrated with an engraving of a rider on a prancing horse. Kielmansegg's long service at Hanover had begun in 1692, his position at court being enhanced in 1701 by his marriage to Sophia Charlotte, the Elector's half-sister.

The Baron was an amateur musician with a deep sympathy for opera. He had studied the guitar in Paris while a young man and gave concerts with the best French singers at his villa near Hanover. He had known Handel for more than seven years at the time of his commission for the *Water Music*, having first met the young composer at Venice in 1709–10 while acting as Hanoverian ambassador to the Republic. It was Kielmansegg who, in Mainwaring's account, invited Handel to Hanover, where he took up the position of Kapellmeister in June 1710, and it was he who was directed to forward the news of Handel's dismissal from his post in 1713.

Kielmansegg's early death at the age of 49 in November 1717 came just over three years after he had settled in London, and within months of the *Water Music* commission.

Private Collection

Oil, 125.4 × 102.5 cms/49$\frac{3}{8}$ × 40$\frac{3}{8}$ ins

Signed *M Dahl pinx 1716* (the last two digits indistinct)

PROVENANCE
By descent in the Kielmansegg family.

VERSIONS
I) A very similar portrait, presumably also by Dahl, belongs to another descendant of the sitter; it has a column and curtain in the right background and further writing (14 lines instead of nine) on the left page of the folio on the table.
II) A copy of this by Countess G. St Genois-Stolberg belongs to a third branch of the family; Kielmansegg's portrait in armour by Kneller (*Familien Cronik*, repr. opposite p. 454), signed and dated 1714, is in the same collection.

LITERATURE
Mainwaring, p. 71; Erich Graf von Kielmansegg, *Familien Cronik der Herren Freiherren und Graffen von Kielmansegg*, 2nd edition, Vienna, 1910, opposite p. 454 (repr.); *Burrows*, pp. 35–59.

64

'The Celebrated Water Musick in Seven Parts'

Printed part-books, first edition
Printed for and sold by John Walsh, *c.* 1733

Though the *Water Music* must have circulated in manuscript quite soon after its first performance on the Thames on 17 July 1717 – a number of performances at theatres are advertised in the 1720s – it took some years to get into print; a complete version in full score did not appear before Arnold's edition of 1788. Walsh issued the overture in his third collection of *Overtures for Violins in all their Parts* around 1725, and some of the minuets appeared in a collection of

Cardinal Ottoboni by Francesco Trevisani (no. 31)

James Brydges, 1st Duke of Chandos by Herman van der Myn, c. 1721–5 (no. 68)

Richard Boyle, 3rd Earl of Burlington attributed to Jonathan Richardson, *c.* 1717–19 (no. 60)

Johann Adolf VI, Baron Kielmansegg by Michael Dahl, 1716 (no. 63)

1729. The first printing of a substantial part of the music was *The Celebrated Water Musick in Seven Parts*, issued by Walsh around 1733. As was then the custom for orchestral music, the publication was in the form of separate part-books, not in full score. The 'Seven Parts' comprised first and second horns, first and second violins with oboes, viola, cello and bassoon. Only about half the music was included. The fact that certain movements had already appeared elsewhere may explain some of the omissions, but it is not clear why Walsh omitted other items or made no attempt to expand the edition after 1743, when he published a virtually complete version of the music arranged for harpsichord.

Both the 1733 part-books and the 1743 harpsichord version reflect the now customary division of the *Water Music* into three separate suites, in F major, D major and G major and minor (though no contemporary source specifically divides the music in this way). However the earliest known sources for the *Water Music* as a whole are manuscript keyboard versions dating from the early 1720s (one in New York Public Library and one in the Earl of Malmesbury's collection), and these reflect a 'mixed' order in which the D major and G major/minor movements are mingled. This scheme is also followed in Arnold's score, and may represent the order in which the music was first performed. Sadly, the complete loss of the autographs of the music played in 1717 means that theories about the original form and order of the movements must remain speculative. Modern paper analyses of the autographs of two movements related to the *Water Music* – versions in F major of the first two movements of the D major Suite – have proved that they date from the period 1722–3, and are not survivors of an early version of the F major *Water Music* suite, as has been suggested.

Gerald Coke Handel Collection

Printed part-books, lacking two of the original seven, 33.0 × 24.0 cms/13 × 9½ ins

Bound with various Handel overtures including *First Treble Musick in the Play call'd the Alchemist by a* [sic] *Italian Master* (not in *Smith*)

LITERATURE
W.C. Smith, 'The Earliest Editions of the *Water Music*', *Concerning Handel*, 1948, pp. 269–87; *Smith*, p. 255, no. 1.

65

'The South East Prospect of Westminster Bridge'

By Benjamin Cole (b. 1723)
after Antonio Canaletto (1697–1768)

Something of the importance of the Thames as a thoroughfare at the time of the first performance of Handel's *Water Music* can be gained from this later view. In 1717 the King took a barge at Whitehall, just out of this view at the right; he was accompanied by musicians in a city barge, even perhaps one of those shown here. They went upstream past Westminster (on the right) and Lambeth (in the left distance) to Chelsea. At that time Westminster Bridge was no more than a future dream.

The occasion shown here is the annual procession of the Lord Mayor to be sworn into office at Westminster (still held each year though now no longer by water). The City barge, 'the finest in Europe' (it had been built in 1737), is seen broadside-on in the foreground; it is accompanied by the barges of nine of the city livery companies. Canaletto's view is of the 1746 procession which took place on 29 October: 'Last Wednesday Noon the Right Hon William Benn Esq. Lord Mayor of this City, went in great State from Guildhall to the Three Cranes, and from thence to Westminster by Water, where he took the usual Oaths of Qualification.' Canaletto's magnificent painting of the subject, now in the Mellon Collection, belongs to his first year in London. Engravings of it were being advertised in March 1747, with the title 'The South East Prospect of Westminster Bridge', but the view is actually from the north looking due south.

Guildhall Library

Engraving, coloured, 37.4 × 57.3 cms/14¾ × 22½ ins

Titled as above and inscribed with the names of various buildings and of the owners of the barges, among which, most prominently, is *G The City Barge, the finest in Europe* ... and also the barges of the Stationers (H), Goldsmiths (K), Skinners (L), Clothworkers (M), Vintners (N), Merchant Taylors (O), Mercers (P), Fishmongers (Q) and the Drapers (R); and at bottom centre *B. Cole sc*

LITERATURE
R.J.B. Walker, *Old Westminster Bridge*, 1979, pp. 167, 238, pl. 32 (the original oil).

66
Model for the Shipwrights' Company Barge
By an unknown modeller

Handel used a barge belonging to one of the great city livery companies, to carry the musicians for the *Water Music*. No barge of the period survives, but contemporary or slightly later prints, such as no. 65, confirm that this model gives a good idea of their design; the forward half was devoted to oarsmen and the richly decorated cabin for passengers. Accommodating Handel's orchestra – fifty instruments are mentioned in one newspaper report – cannot have been easy, especially as there is no evidence to suggest that the cabin roofs were designed to take passengers or musicians at this period (as later).

National Maritime Museum, London

Gilded, painted (red and blue) and varnished wood, 109 cms/43 ins long, scale about 1:17

Inscribed on the side with the Shipwrights' motto WITHIN THE ARK SAFE FOREVER; with the Shipwrights' coat of arms on the stern and animals from the Ark on the roof

PROVENANCE
Given by Sir James Caird to the National Maritime Museum, c. 1935.

LITERATURE
Peter Norton, *State Barges*, National Maritime Museum, 1972, p. 22 (repr.).

67
Model of a Royal Barge
By an unknown modeller

The Royal Barge which carried George I and his party of five members of the court from Whitehall to Chelsea while listening to the *Water Music* would have been rather larger than is suggested by this model, one of Queen Anne's state barges. This craft is closer to the sort of shallop used in coastal waters; the model can accommodate ten oarsmen wearing the uniform of the Royal Bargemen and a figure of the Queen in the stern.

The earliest Royal Barge to survive is that made for Frederick Prince of Wales in 1732, now in the National Maritime Museum. It is a particularly splendid craft, 67 feet in length, designed by William Kent, and was equipped, according to a contemporary bill, with seats for two French horn players. Any further musicians would presumably have been obliged to have their own craft, as at the first performance of the *Water Music*.

National Maritime Museum, London

Gilded and varnished wood, 18½ cms/47 ins long, scale about 1:24 (the lower planking is omitted in order to show the hull construction; the figure of Queen Anne is a modern replacement)

PROVENANCE
Part of the model collection of Charles Sergison (1654–1732), Clerk of the Acts to the Admiralty Board; kept at Cuckfield Park, Sussex, until purchased by Sir James Caird; given by him to the National Maritime Museum, c. 1935.

LITERATURE
Peter Norton, *State Barges*, National Maritime Museum, 1972, p. 14 (repr.).

66

'Princ'ly Chandos': Handel at Cannons

'No nobleman in England, and very few in Europe, lives in greater splendour, or maintains a grandeur and magnificence, equal to the Duke of Chandos', wrote Daniel Defoe in *A Tour through the Whole Island of Great Britain* in 1725. James Brydges, Paymaster-General of the forces abroad during the Marlborough wars, Earl of Carnarvon from 1714 and Duke of Chandos from 1719, made a fortune from his dealings as supplier to the army. He spent it on rebuilding and furnishing Cannons, his grand house near Edgware, long since demolished (and now probably best known from the adjoining Canons Park). Though he personally was the butt of poets ('Since all he got by fraud he lost by stocks' – Jonathan Swift), it was his house and the extravagant establishment he kept there which attracted most attention. Architecturally his greatest display was the remodelling of the two main fronts at Cannons (see no. 69) and the addition to the house of a lavish private chapel. Brydges also employed many of the leading decorative painters of the day: Sir James Thornhill and Antonio Bellucci in the house (see no. 70), and Louis Laguerre (fig. 2) at the nearby church of St Lawrence ('On painted Ceilings you devoutly stare, Where sprawl the Saints of Verrio or Laguerre' – Alexander Pope).

James Brydges's other great extravagance at Cannons was his private orchestra (see nos. 75–83): 'the Duke maintains there a full choir,' wrote Defoe of his chapel, 'and has the worship performed there with the best music, after the manner of the Chapel Royal, which is not done in any other noble man's chapel in Britain; no not the Prince of Wales's, though heir apparent to the crown.' The orchestra at Cannons numbered some two dozen musicians, many of whom doubled in some other role: the harpsichordist, George Monroe, was a waiting page and John Tetlow, violinist, was the Duke's valet.

Though J.C. Pepusch was Master of Music at Cannons (see no. 71), Handel was employed as composer-in-residence from about August 1717, probably spending much of his time there during the next 12 to 18 months. By 25 September 1717, when Brydges wrote to John Arbuthnot, Handel had written four of his Chandos Anthems (see no. 72), and two more were on the way: 'Mr Handle has made me two new Anthems very noble ones & most think they far exceed the two first. He is at work for 2 more & some Overtures to be plaied before the first lesson.' Two masques, set to English texts, seem to belong to the following year: *Acis and Galatea* to the words of John Gay and Alexander Pope (probably the 'little opera' described by a visitor on 27 May 1718; see no. 73), and *Esther*, perhaps by Pope (see no. 128). Handel probably left Cannons early in 1719, when plans were put in hand for the formation of the Royal Academy of Music. James Brydges, now Duke of Chandos, was one of only three subscribers to the Academy at £1,000, the highest rate (see no. 84).

Handel's music at Cannons is sometimes said to have been written for the magnificent chapel which James Gibbs added on to the house, but this was only dedicated in August 1720. Until then the nearby church of St Lawrence, Whitchurch, re-modelled for Brydges and in use by 1716, would have been used for worship. Music not only accompanied the chapel services at Cannons but as John Macky described (*Journey through England*, 1722) was performed at meal times: 'His musick also play when he is at table; he is served by gentlemen in the best order, and I must say that few German sovereign princes live with that magnificence, grandeur and good order . . . the dining room [is] very spacious . . . & at the end of it a room for his musick, which performs both vocal and instrumental, during the time he is at table.'

68 [*in colour between pp. 96–7*]

James Brydges, 1st Duke of Chandos (1674–1744)

By Herman van der Myn (1684–1741)

This is the left-hand portion of a once much larger double portrait of the Duke and Duchess of Chandos by the Dutch artist Herman van der Myn, who was working in England from about 1721. The picture was already cut down by the time of its gift to the British Museum in 1762 but had been seen intact at Cannons by Horace Walpole 18 years earlier: 'there is an admirable immense picture of her [the Duchess], drawing the Duke's portrait by one Vandernime

69

[sic]. He is in a Roman habit with buskins and cerulean stockings.' The Duchess's embroidered slipper and her easel can be made out on the right.

As a young man James Brydges played the flute, and at Oxford at the age of 16 or 17 he joined in the music meetings of the Mermaid Club which included Daniel Purcell, brother of the famous composer. Though there is some mention of music in the journal Brydges kept from 1697–1702 it is not until after his long period of office as Paymaster-General of the forces that his musical interests come to the fore with the foundation of the Cannons orchestra (see nos. 75–83).

National Portrait Gallery, London
NPG 530

Oil, 69¼ × 48½ cms/175.9 × 123.2 ins

PROVENANCE
Painted for the Duke; passed to James Farquharson (one of his secretaries and a beneficiary of his will); given by him to the British Museum, 1762; transferred to the National Portrait Gallery, 1879.

LITERATURE
Margaret Crum, 'An Oxford Music Club, 1690–1719', *The Bodleian Library Record*, IX, 1974, pp. 83–99; Kerslake, pp. 36–7, pl. 107; John Kerslake, 'The Duke of Chandos' Missing Duchess', *National Trust Studies 1981*, 1980, pp. 138–49 (repr. in colour).

69

'The South Front of Cannons in Middlesex'

By Hendrik Hulsbergh (d. 1729) after John Price (d. 1736)

James Brydges, the future Duke of Chandos (see no. 68), had purchased the Jacobean house of Cannons in 1713, and appointed William Talman as his first architect. Despite his reputation for extravagance, the Duke's indecisiveness and his reluctance to countenance certain of the more costly expenses in fitting out Cannons led to delays. It took two changes in architect and the appointment of James Gibbs in 1716 to decide on the design of the two main fronts, to the south and east. Gibbs was also responsible for the richly finished chapel which was built onto the house and in use by 1720.

The magnificent south front at Cannons, built in Portland stone and surmounted by statues by John van Nost, was designed by Gibbs but executed with some modifications by the builder John Price between 1720 and 1724. Cannons may not have been the largest of the grand new houses which were built on the fortunes of war, nor was it architecturally the finest, but in its richness and extravagance and its unification of the arts, it became a symbol of an age of ostentation. Much to the Duke's displeasure, Cannons was popularly identified with Pope's 'Timon's Villa' in his *Epistle to the Earl of Burlington*: 'At Timon's Villa let us pass a day. Where all cry out, "What sums are thrown away!"'

The British Library Board
K.Top.30.25a

Engraving, 37.8 × 63.3 cms/14⅞ × 24⅞ ins

Inscribed *The South Front of Cannons in Middlesex, the Seat of his Grace James Duke of Chandos, &c. To whom this Plate is most Humbly Inscrib'd by his Graces most Obed! Servant John Price. Architect. Built, Anno 1720/J. Price Delin:/H. Hulsbergh Sculp!*

LITERATURE
Terry Friedman, *James Gibbs*, 1984, pp. 51–3, 108–13, 315.

70

Aeneas before Dido in the Temple of Juno

By Sir James Thornhill (1675/6–1734)

Thornhill's fine sketch for the decoration of a hall or staircase depicts the Trojan hero, Aeneas (on the left) before Dido, Queen of Carthage (on the right with outstretched arms). It was associated with Cannons by Edward Croft-Murray on the basis both of its subject matter (there is a related drawing inscribed as for the Duke of Chandos at Cannons) and of the symbolism of the decoration beneath the main scene, which suggests a commission for a duke with military associations: the ducal coronet and the feigned relief of cupids forging arms are set between various military trophies.

Thornhill's association with Cannons probably began in January 1717, with his preparation of a design (not executed) for re-building the east front of the house. He received at least £300 for his work at Cannons which included elaborate decorative schemes for the staircase and the Saloon above, probably painted between 1719 and 1724. This sketch is an indication of the richness of his work and, in the absence of any surviving decoration from Cannons, of the sort of effect visitors must have received.

Private Collection

Oil, 58.4 × 41.8 cms/23 × 16½ ins
LITERATURE
Croft-Murray, 1, p. 266, pl. 134.

71

Dr Johann Christoph Pepusch (1667–1752)

By an unknown artist

Pepusch was born in Berlin but settled in London as a young man. As a composer and teacher he led a long and active career which has been rather overshadowed by the genius of Handel but, despite Charles Burney's later assertions, there is no contemporary evidence to suggest that there was any personal animosity between the two composers.

Pepusch was appointed 'Master of the Music' at Cannons (see no. 69) at a salary of £100 per annum in about 1716 and held the position until at least 1729. From what we know of his duties for the Duke of Chandos (see no. 68) they do not appear to have been onerous: responsibility for the Duke's small orchestra (see p. 105) and the composition of some new music. When Pepusch drew up an inventory of scores and musical instruments at Cannons in 1720 (see Appendix 1), his own cantatas and anthems were very much in evidence, including his *Six English Cantatas*, dedicated to Chandos, and two pieces written for the Duke's birthday.

71

Pepusch had been involved in the London theatre world from the time he arranged *Thomyris* for performance at Drury Lane in 1707 while he was the theatre's harpsichordist. He wrote several operatic masques for Drury Lane in 1715/16 including *Venus and Adonis*, and married one of the singers, Margherita de l'Epine, in about 1722 (see under no. 50). But his greatest success came with John Gay's *Beggar's Opera* (see no. 108) in 1728 for which he wrote the overture and arranged the music (which included more than one piece by Handel). In his later years he was the leading member of the Academy of Ancient Music and spent much time teaching and writing.

Faculty of Music, Oxford University
(By Permission of the Heather Professor of Music)

Oil, 75.6 × 62.8 cms/29¾ × 24¾ ins

PROVENANCE
Dr Philip Hayes, Professor of Music at Oxford; given by him to the Music School, Oxford, before 1795.

LITERATURE
Poole, I, p. 160, no. 388; Kerslake, p. 210, pl. 622.

72

Chandos Anthem: O Praise the Lord with one Consent

Handel's autograph score

The 11 anthems composed by Handel for James Brydges, Earl of Carnarvon and later Duke of Chandos, were probably all written between the summer of 1717 and the summer of 1718 and performed at the newly built church of St Lawrence, Whitchurch. Six of them are scored for a three-part chorus while the others require four- or even five-part choruses, and were almost certainly the last to be composed.

O Praise the Lord with one Consent is one of the most expansive anthems in this later group. The words are taken from the Psalms (no. 35, the whole of no. 117, and no. 148) but in this case entirely from the metrical versions of Nahum Tate and Nicholas Brady, rather than Handel's usual source, the Book of Common Prayer. Unusually, Handel does not begin the anthem with an overture but instead writes an extended orchestral introduction to the opening chorus. Both introduction and chorus are based on a theme resembling the opening phrase of the hymn-tune 'St Anne', probably written by William Croft and best known today as the tune of Isaac Watts's hymn 'O God, our help in ages past'; this resemblance may be coincidental, however, as the outline of the theme is suggested in the opening sonata of Handel's cantata 'Tu fedel? tu costante?', written in 1707 a year before the publication of the hymn-tune. On display is the first entry of the choral voices. The trebles sing the complete opening line, and the full chorus join in at 'with one consent' – a naïve but effective illustration of the words.

The Syndics of the Fitzwilliam Museum, Cambridge
MS 255

Autograph score, 36ff., 22.8 × 29.2 cms/9 × 11½ ins

Inscribed by Handel ψ [Psalm] *135.*, and by Lord Fitzwilliam *Fitzwilliam 1778*

PROVENANCE
This score cannot have been included among the autographs given by the younger Smith to King George III, but was presumably acquired by Richard, 7th Viscount Fitzwilliam in 1778, as indicated on the first page; subsequently part of his bequest founding the Fitzwilliam Museum, 1816. (See also no. 34.)

LITERATURE
A.H. Mann, 'Manuscripts and Sketches by Handel', *Catalogue of the Music in the Fitzwilliam Museum, Cambridge*, ed. J.A. Fuller-Maitland and A.H. Mann, 1893, p. 169.

73

Acis and Galatea

Handel's autograph score

On 27 May 1718 Sir David Dalrymple reported to the Earl of Loudoun some details of his recent visit to the Earl of Carnarvon at Cannons:

> He has a Choorus of his own, the Musick is made by himself and sung by his own servants, besides which there is a litle opera now a makeing for his diversion whereof the Musick will not be made publick. The words are to be furnished by Mrs Pope & Gay, the musick to be composed by Hendell.

There is no doubt that the 'little opera' was *Acis and Galatea*, a work quite different from Handel's earlier treatment of an Italian version of the same story composed for Naples ten years earlier. It is at once the most profound and most delightful of Handel's early English works. John Gay probably wrote the first version of the libretto (perhaps intended only for three singers), and Alexander Pope may have been responsible for its expansion into a substantial choral work. As in the case of the Italian operas, however, the choruses were almost certainly sung by an ensemble of solo voices: one of the early copies (in the collection of the Earl of Malmesbury) adds the names of the five characters to the choral lines, though this feature does not appear in the autograph. (The five characters are Acis, Galatea, Polyphemus, Damon and Coridon; the last, omitted from modern editions, sings the air 'Would you gain the tender creature', which Handel added after completing the rest of the score, using a text by John Hughes.)

The story is that related in Book XIII of Ovid's *Metamorphoses*. The sea-nymph Galatea and the shepherd Acis are happily in love, but the giant Polyphemus also falls in love with Galatea and becomes so jealous that he crushes Acis under a huge rock. Galatea cannot restore Acis to life but grants him immortality by changing him into a brook, in whose waters she can bathe for ever. Handel makes Polyphemus a slightly comic, not unsympathetic character. This element of caricature is most obvious in the famous air 'O ruddier than the cherry' (ff. 33v–34r), the giant's clumsy attempt to woo Galatea. In the accompaniment Handel includes a recorder (*Flauto*) as a jocular reference to the pipe made of 'a hundred reeds of decent growth' which Polyphemus says he is going to play. In many manuscript copies, including the Malmesbury score, this instrument is specified to be a high-pitched sopranino recorder (*Flauto piccolo ottavo*; see no. 176). Almost certainly this is Handel's own intensification of the joke; the alteration may have been made in a lost conducting score.

In 1732 Handel reacted to an unauthorised performance of *Acis and Galatea* by producing a new bilingual version at the King's Theatre, combining sections of the English and Italian versions with newly composed music. The mixture of languages, although artistically indefensible, suited the combination of English and Italian singers available to him at the time. This version was revived several times until 1736, but in 1739 Handel returned to the all-English version. He performed *Acis* for the last time at Dublin in 1742. With the publication of a complete full score of the work in 1743 – the only dramatic work of his to be published in its entirety in his lifetime – he could leave further performances to others.

The British Library Board
RM 20.a.2

Autograph score, 68 ff, 22.8 × 29.2 cms/9 × 11½ ins; ff. 66–68, containing the chorus 'Smiling Venus', are part of the 1732 additions to the work

PROVENANCE
As no. 34.

LITERATURE
Squire, p. 1; *Dean*, pp. 153–90;
P. Rogers, 'Dating "Acis and Galatea"',
Musical Times, CXLV, 1973, p. 792.

74

'Suites de Pieces pour le Clavecin Composées par G.F. Handel. Premier Volume.'

Printed score, first edition
Printed for Handel and sold by John Christopher Smith senior and Richard Meares, 1720

The first published volume of Handel's harpsichord suites, his *Suites de Pieces pour le Clavecin*, was 'printed for the Author' (ie on behalf of Handel himself) and issued jointly by John Christopher Smith and Richard Meares in November 1720. The engraving and printing were done at Cluer's Printing Office, Bow Church Yard. It was probably the only volume of Handel's music ever published under the direct supervision of the composer. Some copies show signs of proof amendments on the plates which go beyond mere corrections; these were almost certainly ordered by Handel himself.

In a prefatory note Handel explains the reason for issuing the volume:

> I have been obliged to publish Some of the following lessons because Surrepticious and incorrect copies of them have got abroad. I have added several new ones to make the Work more usefull which if it meets with a favourable reception: I will still proceed to publish more reckoning it my duty with my Small talent to Serve a Nation from which I have receiv'd so Generous a protection.

The reference to 'Surrepticious ... copies' has sometimes been taken to relate to a pirated publication by Jeanne Roger of Amsterdam entitled *Pieces à un & Deux Clavecins Composées par Mr. Hendel* (the plates for which were almost certainly prepared by John Walsh), but the plate number of Roger's edition (490) indicates a date of 1721. Handel thus seems to have been referring to the circulation of manuscript copies, but it is possible he had learnt of the preparation of the Roger edition and produced his own publication to forestall it.

Many of the individual movements that make up the eight suites in the volume were written much earlier than 1720; a few probably date from Handel's time in Hamburg (1704–6). The revisions made for the 1720 publication were wide-ranging: they include improvements of minor textual details, whole-scale re-composition of existing movements, the addition of entirely new movements and the compilation of a new suite (no. 7 in G minor) from a number of independently conceived pieces. Suite no. 5 in E major ends with the set of variations known as the 'Harmonious Blacksmith', but the story that purportedly gave rise to this nickname is highly suspect. The piece is in fact modelled closely on a *chaconne* in G major, ruling out the possibility that Handel was inspired to write it by the striking of an anvil sounding the note E.

The optimistic designation 'Premier Volume' was not followed up until about 1727, when Walsh prepared a 'Second Volume', almost certainly without Handel's authority; it consisted mostly of pieces written earlier than those in the 1720 volume. In 1733 Walsh re-issued the 'Second Volume' in an improved form, but several textual defects remaining in the music suggest that Handel himself can have taken little interest in the publication.

Gerald Coke Handel Collection

Printed score, 94pp., 24.0 × 30.5 cms/9½ × 12 ins
Titled as above and inscribed in pen inside the cover *NW*

LITERATURE
Smith, p. 248, no. 1; F. Lesure, *Bibliographie des éditions musicales publiées par Estienne Roger et Michel-Charles Le Cene (Amsterdam 1696–1743)*, Paris, 1969; Terence Best, 'Handel's Keyboard Music', *Musical Times*, CXII, 1971, p. 845.

74

The Orchestra at Cannons

However magnificent the Duke of Chandos's establishment, the available orchestra at Cannons was small, as might be expected in an English country house. We are fortunate that Johann Christoph Pepusch, who was Chandos's Master of the Music, has left us an inventory of the instruments (fig. 12, p. 282) which were under his care on 23 August 1720, a date near enough to Handel's residence at Cannons to give us a good idea of the instrumental resources. A transcript of the inventory is given in Appendix 1.

Only one of these instruments can be identified today as having then been at Cannons, and that one is of sufficient size to make it impossible to display here: the organ, now in Holy Trinity Church, Gosport. Pepusch's second item is most closely matched by a double-manual harpsichord with a virginals in the bentside made by Johannes Ruckers and dating from 1619, the only one of this sort to survive; it is now in the Instrument Museum of the Conservatoire Royal de Musique at Brussels (cat. no. 2935). Most of the other instruments listed by Pepusch are represented here by their closest surviving equivalents. The inventory does not however include quite the full range of instruments at Cannons, judging from Collins Baker's list of the Duke's household, which shows that there were players on the oboe, flute and, in 1733, the harp.

Cannons's sad fate of demolition was heralded by the sale held on the premises by Mr Cock the auctioneer on 1 June 1747 and the ten following days. Only the first four instruments in the 1720 inventory were included, and their descriptions provide some slight additional information. The dimensions of the 'large and fine organ' in the chapel are given as 18 ft 4 ins high, 9 ft 9 ins wide and 6 ft 6 ins deep. The Ruckers harpsichord, listed as 'with a spinet at one end', was then in His Grace's Study on the first floor, and the other harpsichord 'by Harmannus Tabel' was in His Grace's Dressing Room. The Hitchcock spinet 'in a walnut-tree case' was in the Little Saloon, also on the first floor.

LITERATURE
A Catalogue of the Genuine Household Furniture &c of His Grace James Duke of Chandos Deceased at his late Seat called Cannons, catalogue of sale held by Mr Cock, 1 June 1747 and ten following days, pp. 4, 14, 28, 41; *Baker*, pp. 131–40.

75 [*illustrated on p.* 106]

Double-manual Harpsichord
By Hermann Tabel (?1660s–1738)

Hermann Tabel came to London from the Low Countries *c.*1716 or before, having served his apprenticeship in Antwerp with the Couchets, who were successors to the Ruckers dynasty. He was the master of Jacob Kirckman, who came to London from Bischweiler, near Strasbourg, and of Burkat Shudi (see no. 162), who came from Glarus in Switzerland and was later to become Handel's friend. Thus Tabel formed the direct link between the Ruckers tradition and the eighteenth-century English harpsichord-making school. He is such a key figure in the history of English harpsichords that scholars have lamented that this instrument, of 1721, is his only harpsichord known to survive. The Cannons instrument is described as 'A Harpsichord with two Rows of Keys made at London by Hermanus Table' in the 1720 inventory.

Warwickshire Museum Collection

Inscribed on the nameboard HERMANUS TABEL; FECIT LONDINI 1721,
and signed on the uppermost key of the lower manual in Tabel's handwriting
N.° 43/Herm̄/Tabel/Fecit/Londi/–ni–/1721.;
the rose in the soundboard reproduces Tabel's seal

Oak case veneered with walnut (external) and maple (internal), pine soundboard and bottom boards;
5 octaves, FF–f³ without FF♯;
scale c² 34.3 cms/13½ ins for lower 8′; 15.1 cms/6 ins for 4′;
restored by John Broadwood & Sons, May, 1900, including new ebonies for the naturals;
length 247.0 cms/97¼ ins, width 94.3 cms/37⅛ ins, depth 28.6 cms/11¼ ins excluding the stand, which may be original

PROVENANCE
Helen, Countess of Radnor, by *c.*1900;
by descent to Mrs Pleydell-Bouverie; Captain Lane;
Messrs J.G. Morley, *c.*1936; Raymond Russell, 1945 or after; his sale, Sotheby's, 25 May 1951, lot 31;
Mrs Mary Edmonds from 1951;
her sale, Sotheby's, 2 July 1965, lot 151, purchased by the Warwickshire Museum.

LITERATURE
Donald Boalch, *Makers of the Harpsichord and Clavichord 1440–1840*, 2nd edition, Oxford, 1974, pp. 174–5, pls. 2, 3, 37;
Charles Mould, 'The Tabel Harpsichord',
Keyboard Instruments, Studies in Keyboard Organology 1500–1800, ed. Edwin Ripin, New York, 1977, pp. 59–65, pls. 46–52.

75

76

Violin

By John Barrett (fl. 1714–40)

The 'Mr Barrett' who made the Cannons double bass is almost certainly to be identified with John Barrett; he was the only maker of this name known to have been active in London at this period. No double bass by him is now known, and of his violins very few are in the original state, as this one is.

The instruments included here (nos. 77–9 and 169–70) are all in the state in which they would have been known to Handel, with short, straight necks, low bridges, short fingerboards and, internally, thin sound posts and small bass bars. Barrett's violins are modelled on those of Jacob Stainer (see no. 77) and seem to stand up well to the numerous alterations and modifications which were carried out shortly before or after 1800.

Sir Yehudi Menuhin

Labels, printed *John Barrett* and *Made by Iohn Barrett at y^e Harp | & Crown in Pickadilly London. 1722*

Overall length 59.3 cms/23$\frac{3}{8}$ ins, body length 35.5 cms/14 ins, width of bouts 16.8, 11.5 and 20.4 cms/6$\frac{5}{8}$, 4$\frac{1}{2}$ and 8 ins, string length 32.7 cms/12$\frac{7}{8}$ ins

77

Violin

By Jacob Stainer (*c.* 1617–83)

This violin, which was discovered very recently in Germany, is here shown for the first time. It is still in its original state.

There were four Jacob Stainer violins at Cannons in 1720, with labels dated 1660, 1665, 1676 and 1678. Stainer was born in Absam in the Tyrol. He seems to have spent most of his life there but, perhaps because much of his work was for the princely court at Innsbruck, his labels are marked *in Absom prope Oenipontum* ('at Absam near Innsbruck').

Stainer perfected the German model of the violin, which differed quite markedly in shape from the Italian model, and particularly from that established by Stradivarius. In his own day, and for more than a century afterwards, Stainer's instruments were pre-eminent and valued above all others for their full, silvery tone. They were highly popular in England, and many seventeenth- and eighteenth-century English makers, such as John Barrett (see no. 76), followed his model. Only towards the end of the eighteenth century, with the changing social conditions of orchestral performance, did the Cremonese model of violin, with its rounder tone and more powerful sound, become the more desirable. Today, with the rise of the baroque orchestra, the Stainer violin and those instruments which were influenced by it, are once more coming into their own.

Private Collection

Label, inscribed in pen and ink *Jacobus Stainer in Absom / propè oenipontum M Pia. 1679*
(*M Pia* = 'Manu Propria', or 'by my own hand');
with an early red wax seal on the back of the peg box impressed with a coat of arms of a rampant leopard (?) and a crest of another rampant animal, perhaps a deer or a horse

Overall length 57.9 cms/22¾ ins, body length 35.1 cms/13⅞ ins, width of bouts 16.3, 11.2 and 20.1 cms/6⅜, 4⅜ and 7⅞ ins, string length 32.4 cms/12¾ ins; the long neck is original; the fingerboard is perhaps late eighteenth century; one old, perhaps original, peg is in ebony tipped with ivory

77 (showing the inside)

78

Bass Viol

By Henry Jaye (fl. *c*.1610–67)

English viols were acknowledged to be the best in Europe in the late sixteenth and early seventeenth centuries. Following the sixteenth-century makers John Rose, father and son, Henry Jaye was the finest maker of the first half of the seventeenth century. His craftsmanship was superb and the tone of his instruments, many of which are still in use, was unrivalled in quality. His earliest known viol, once in Canon Galpin's collection, is dated 1611 and is now in the Gemeente Museum in The Hague. His last, dated 1667, is a tenor in the Victoria and Albert Museum. Both the instrument included here and the bass viol at Cannons (which was dated 1613) belong to Jaye's first decade.

Edward Withers Ltd

Label, inscribed in pen and ink *HENREE : IAY: IN: / SOVTHWARKE : / 1619.*,
and stamped on the bridge *D M KESSLER*,
and on the fingerboard *8000* (probably by Hills)

Overall length 133.4 cms/52½ ins, body length 72.3 cms/28½ ins, width of bouts 33.8, 24.2 and 40.1 cms/13¼, 9½ and 15¾ ins, string length 75.2 cms/29⅝ ins; the tailpiece and bridge are replacements

LITERATURE
F.W. Galpin, *Old English Instruments of Music*, 1910, pp. 89–90, pl. xvii (for the Hague viol);
Anthony Baines, *Catalogue of Musical Instruments: II – Non-Keyboard Instruments*, Victoria and Albert Museum, 1968, p. 3, pl. 1 (for the V & A tenor)

79

Bass Viol

By Barak Norman (*c*.1670–*c*.1740)

While Henry Jaye was recognised as one of the best English viol makers of the first half of the seventeenth century, Barak Norman reigned supreme at the end of that century and into the eighteenth; he was the last great English viol maker. This is an early instrument, with elaborately inlaid tailpiece and fingerboard in the style favoured by Norman. Another viol of the same year, 1692, is in the Royal College of Music.

Norman was also famous for his violins, and is said to have been one of the first English cello makers. Thus his work reflects the change of taste in English music which accompanied the Restoration of Charles II, when the viols of the old English style gave way to the violins of the French court. Only the bass viol survived with any frequency, and even that was heard only rarely.

Private Collection

Label, inscribed in pen and ink *Barak Norman / At the basse viol / in St Pauls Ally / London Fecit / 1692*; with two restorers' labels, *Restored by / G Saint George / May 1899 / New Neck fingerboard / No 24*, and *DIETRICH KESSLER / LONDON / RESTORED JAN 1953*; the bridge is stamped *D M KESSLER*

Overall length 115.6 cms/45½ ins, body length 65.5 cms/25¾ ins, width of bouts 28.9, 21.5 and 34.9 cms/11⅜, 8½ and 13¾ ins, string length 67.1 cms/26⅜ ins; the support of the inlay on the neck has been renewed

80
Bassoon
By Hermann Wietfelt (c.1663–c.1738)

Hermann Wietfelt was one of a dynasty of woodwind makers in Burgdorf, with a father and brother also making bassoons and oboes and a son and a grandson following him in the same trade. A number of instruments by members of the family survive, but this appears to be the only one by Hermann Wietfelt. Pastor Hart suggests that it may have been Hanoverian influence, perhaps even that of Handel, that led to the presence of this instrument, made in a Hanoverian town, at Cannons.

Boosey & Hawkes plc

Stamped on the butt .HARM./.WIETFELT., and on the wing and long joint .H./.WIETFELT.

Maple with brass mounts and four brass keys for B♭, D, F and A♭, with an added right-hand thumb lever and other alterations; the crook is probably not original;
overall length 123.2 cms/48½ ins,
tube length 172.3 cms/67¾ ins excluding crook

PROVENANCE
Probably acquired by David James Blaikley before 1930 for the Boosey and Hawkes Collection.

LITERATURE
Günter Hart, 'Die Holzblasinstrumentenmacher Wietfelt', *Das Musikinstrument*, XII, 1963, pp. 681–2;
Anthony Baines, *Antique Musical Instruments of Historical Interest of the Boosey & Hawkes Group*, 1972, p. 14, no. 250.

81
Pair of French Hunting Horns
By Johannes Leichamschneider (b. 1679)

The Leichamschneider brothers of Vienna, Michael and Johannes, were primarily responsible for creating the Viennese model of both hunting and orchestral horns in the early eighteenth century, producing instruments of different, and more 'modern', tone quality than the earlier English school of makers such as William Bull and Christian Bennett. There are still considerable gaps in our knowledge of instrument making in England at this period; it seems possible that there was a hiatus between the makers already referred to and those of the middle of the century such as Nicholas Winkings and John Christopher Hofmaster (see no. 220), thus compelling those who wanted the best instruments to look to Vienna or Nuremberg. It is certainly an indication in this direction that both the Duke of Chandos and Sir William Morgan owned horns by one of the leading Viennese makers. The Chandos pair is dated 1711 but these, belonging to 1725, are the last known instruments by Johannes Leichamschneider. His brother, however, continued to make horns and trumpets into the 1740s.

Private Collection

Engraved on the bells
IOHANNES | LEICHAM | SCHNEIDER | IN: WIENN 1725:
and ·I·/·L·/·IN·/WIENN

Silver, maximum width 56.0 cms/22 ins,
bell diameter 24.8 cms/9½ ins;
in F; triple-coiled; engraved on gilt panels on the bells with three scenes of stag, hare and bear hunts, probably by Michael Heinrich Renz, and also with the Morgan crest; mouthpieces missing

PROVENANCE
Sir William Morgan of Tredegar House, Newport;
by family descent until sold at Christie's, Geneva, 9 November 1976, lot 136, where purchased by the present owner.

LITERATURE
M.R. Apted, 'Social Conditions at Tredegar House, Newport in the 17th and 18th centuries', *The Monmouthshire Antiquary*, III, part 2, 1972–3, pp. 129–30; Horace Fitzpatrick, *The Horn and Horn-Playing*, 1970, pp. 42–3.

82
Pair of Trumpets
By John Harris (fl. 1716–21)

A John Harris, most probably the maker of this pair of trumpets, entered his mark at Goldsmiths Hall on 21 March 1716, less than four months before these trumpets, supplied by Seth Partridge to the Bristol City Council, were paid for on 7 July at a cost of £21. 17s. 6d. It seems likely that he was making instruments somewhat earlier because although both instruments are in the key of D, they are not an identical pair, suggesting that they were supplied from stock rather than made to order. They were purchased for the two troops of cavalry formed in 1715 by Bristol gentlemen in case of a Jacobite uprising. Harris also made horns, providing a pair for Tredegar in 1721 at a cost of £7. 1s. –, and ten for the Royal Buckhounds in 1717, again in brass at a cost of £26. 16s. –, none of which is known to have survived. His instruments are surprisingly rare, only three trumpets being known at present, these and one now in the Bate Collection, Oxford, which was converted to a slide trumpet by Köhler in the last century and is said to have been the model for all his slide trumpets. It then belonged to Thomas Harper, who also possessed another Harris trumpet, illustrated by Hipkins and Gibb, which has subsequently disappeared.

Bristol City Council

Inscribed on the bell JOHN HARRIS LONDINI FECIT, and engraved on the throat in 1800 *Chamber of Bristol*; not hallmarked

Silver, overall length 72 cms/28¾ ins,
tube length (exterior) 217 and 219 cms/85⅜ and 86¼ ins,
bell diameters 11.0 and 11.2 cms/about 4⅜ ins;
both instruments are in D

PROVENANCE
Supplied by Seth Partridge to the Bristol City Council in 1716.

LITERATURE
A.J. Hipkins & William Gibb, *Musical Instruments, Historic, Rare and Unique*, 1888, p. 85, pl. 36; J.P. Hore, *History of the Royal Buckhounds*, 1893, p. 289 (footnote); Joseph Wheeler, 'Further Notes on the Classic Trumpet', *Galpin Society Journal*, XVIII, 1965, pp. 15–22; Maurice Byrne, 'The Goldsmith-Trumpet-makers of the British Isles', *Galpin Society Journal*, XIX, 1966, pp. 80–1; M.E. Williams, *Civic Treasures of Bristol*, Bristol, 1984, p. 54 (repr.); unpublished papers for Tredegar House, National Library of Wales (Tredegar MSS 315–45 under 3 April 1721).

83
Oboe
By Joseph Bradbury (b. c.1670)

No oboes are listed by Dr Pepusch in the Cannons inventory, but Handel consistently wrote for them in his Chandos music; it is very likely that the oboe player employed at Cannons, Jean Christian Kytch, would have brought his own instrument with him. Only eight English oboes dating from the first quarter of the eighteenth century are known: two by Bradbury and six by the elder Stanesby, three of which are exhibited here (nos. 167–8 and 226). Of these, this instrument by Bradbury is one of the earliest and, because his instruments are so rare, one of the most important. Its tortoise-shell staining is characteristic of the late seventeenth- and early eighteenth-century English oboe.

All that is known of Bradbury is that he was born in about 1670, the son of a wood-turner, William Bradbury of Stapleford in Hertfordshire, and was apprenticed to Thomas Garrett in London on 5 November 1684. Because he did not become a freeman of the Turners' Company, it is presumed that he worked outside London. Nor is it known whether Garrett was an instrument maker or just a general turner, but it is significant that Bradbury's fellow apprentice, his senior in the craft by two years, was the elder Stanesby. Halfpenny has suggested that it is possible that Garrett was thus the founder of the native English oboe-making school, the successors of the generation of Bressan, who may have brought the instrument with him from France.

York Castle Museum
DA 592

Stamped on all three joints JO / BRADBURY / (crown)

Tortoise-shell stained boxwood, three brass keys for middle C and left- and right-hand E♭, ivory bell ring (probably a replacement); the E flat key was duplicated on the early oboe because players had not come to a decision on whether the left or right hand should be uppermost;
overall length 59.5 cms/23⅜ ins

PROVENANCE
De Rell (a music hall entertainer in Ireland); purchased from Mrs Dearden (whose stage name was De Rell) by Dr Kirk; given with his collection to the York Castle Museum, 1937.

LITERATURE
Eric Halfpenny, 'The English 2- and 3-Keyed Hautboy', *Galpin Society Journal*, II, 1949, pp. 10–26, pl. 2;
Eric Halfpenny, 'Biographical Notices of the Early English Woodwind-making School, c.1650–1750', *Galpin Society Journal*, XII, 1959, pp. 44–52; Mary Kirkpatrick, 'Another Oboe by Jonathan [sic] Bradbury: Restoration and Comparison', *Galpin Society Journal*, XXXIII, 1980, pp. 106–10, pl. 8c.

List of Subscribers to the Royal Academy of Musick.

Henry Duke of Kent. 200
Thomas Holles Duke of Newcastle. 1000
Charles Duke of Grafton. 200
Henry Duke of Portland. 600
Charles Duke of Manchester. 200
James Duke of Chandois. 1000
James Duke of Montrose. 200
Charles Earl of Sunderland. 200
Henry Earl of Rochester. 200
James Earl of Berkeley. 200
Richard Earl of Burlington. 1000
George Henry Earl of Litchfield. 200
Henry Earl of Lincoln. 200
Henry Visco.t Lonsdale. 200
Thomas Earl of Strafford. 200
William Earl Cadogan. 200
Talbot. Earl of Sussex. 200
Henry Earl of Thomond. 200
George Earl of Halifax. 200
David - Earl of Portmore. 200
Count Bothmer. 200
Allen Lord Bathurst, 200
Robert Lord Bingley. 200
George Lord Lansdowne. 200
John Lord Gower. 200

IV *The Royal Academy of Music* 1719–28

'I beg that you will not judge my eagerness to see you by the lateness of my departure', wrote Handel to his brother-in-law at Halle on 20 February 1719; 'it is greatly to my regret that I find myself kept here by affairs of the greatest moment, on which (I venture to say) all my fortunes depend; but they have continued much longer than I had anticipated.' These 'affairs' must have been the arrangements for the projected Royal Academy of Music in which Handel played the central part.

The King had shortly before received a petition from 'Severall Gentlemen who desire to be incorporated by Letters patents for Encouragement of Musick'. This project was described by Handel's biographer, John Mainwaring, some forty years later:

> During the last year of his residence at Cannons, a project was formed by the Nobility for erecting an academy at the Haymarket. The intention of this musical Society, was to secure to themselves a constant supply of Operas to be composed by Handel, and performed under his direction. For this end a subscription was set on foot: and as his late Majesty was pleased to let his name appear at the head of it, the Society was dignified with the title of the Royal Academy.

In fact Handel was not the only composer employed by the Academy but, from November 1719, he held the position of 'Master of the Orchestra'. This official appointment came some six months after he had acted on instructions (see no. 86) from the Academy's governor, the Duke of Newcastle, to travel abroad in search of singers. The project, though under the generous patronage of the King, seems to have been the initiative of a group of noblemen and gentlemen led, at least financially, by Handel's two greatest patrons, the Earl of Burlington and the Duke of Chandos. The subscribers to the Academy's shares, mostly of £200 but a few of £1,000, did not have to put their money down but agreed to pay it over in 5% 'calls' as and when the Academy gave due notice. A profit of 25% was optimistically envisaged but not guaranteed. Financially the Academy was organised as a joint stock company in a way which invited comparison with the South Sea Company: 'Yesterday South Sea was 174. Opera Company 83, and a half. No Transfer,' was the joke in *The Theatre*, a paper edited by Sir Richard Steele. This was taken up again a week later in the issue for 8 March 1720, introducing an imaginary Italian singer in the process: 'At the Rehearsal on Friday last, Signior Nihilini Beneditti rose half a Note above his Pitch formerly known. Opera Stock from 83 and a half, when he began; at 90 when he ended.' But the fun did not last; on 23 September Paolo Rolli, the Academy's Italian secretary, wrote to his friend Giuseppe Riva, 'My dear Riva, what ruination has the Southsea crash caused! The whole nobility is at its last gasp; only gloomy faces are to be seen,' and then again, on 18 October, 'They ought to gibbeted these South Sea Directors, who have ruined all my friends – and I very much fear that they will in consequence have ruined the Academy. God damn 'em. . . .'

The Academy survived, though it encountered difficulties of its own: arguments between Handel and his more temperamental singers such as Senesino and Cuzzoni (see nos. 93 and 95); the dismissal of Paolo Rolli, after his scheming led to a quarrel with the directors (see no. 90); disagreements among the directors themselves as time wore on, about the employment of expensive new singers such as Faustina (see no. 96); and squabbles on stage between the two rival prima donnas, Cuzzoni and Faustina (see under no. 107).

For all the Academy's problems its success was enormous. *Radamisto*, its first opera by Handel, given in April 1720, was extravagantly received and many others met with great success. The Academy established Italian opera on a firmer footing than it had previously enjoyed or was to experience for many years to come. It produced a stream of good, and at times great, operas, mostly by Handel or Bononcini (see no. 88). Its singers and its orchestra were highly regarded. Not least it could stand comparison with the Italian opera in other countries, as the comment of foreigners visiting London testify. During *Giulio Cesare*'s long run, one opera-goer, a Mr Lecoq, could write to Dresden on 31 March 1724: 'The passion for the opera here is getting beyond all belief. It is true that the music is beautiful and varied. There are three composers, including the famous Handel, each of whom writes two operas every winter. The orchestra, by and large, is of high quality and good care is taken to produce new voices at the theatre from time to time.' There was also a certain national pride in the Italian opera; it was in this light that the *Flying Post* for 4 February 1727 saw the Academy's decision to employ the three composers, Ariosti, Bononcini and Handel:

> Thus, as this Theatre can boast of the three best Voices in Europe, and the best Instruments; so the Town will have the Pleasure of having three different Stiles of composing. And, as Musick is part of Mathematicks, and was always ... esteemed a very rational and noble Entertainment; this Polite and Rich Nation will by Collecting what is perfect out of various Countries, become the Place where all Travellers will stay to be diverted and instructed in this Science, as well as in all others.

Why did the Academy collapse? In November 1727 Mrs Pendarves, one of Handel's most loyal supporters, wrote to her sister, 'I doubt operas will not survive longer than this winter, they are now at their last gasp; the subscription is expired and nobody will renew it. The directors are always squabbling, and they have so many divisions among themselves that I wonder they have not broke up before'. It would seem that after the disputes of the previous summer the audience had grown tired of opera and the subscribers were no longer willing to go on supporting the Academy financially. This view is confirmed by a much later account written by another of Handel's admirers, Lord Shaftesbury:

> These Animosities were very prejudicial to the Interest of the Academy and the Houses began to grow thinner upon it. The Beggar's Opera appearing soon after, gave such a Turn to the Town, the Opera's were generally neglected: ... the Fund for maintaining Opera's being exhausted they ceased of course – and the Singers left England. This is the real Cause of the discontinuance of Opera's at that time.

And so it was that on 4 January 1729, seven months after the last performance, a meeting was called 'in order to consider some Proposals that will then be offered for carrying on Operas; as also for disposing of the Effects belonging to the said Academy'. It was there decided to prosecute the subscribers who had not paid, and to let Heidegger and Handel take over the Academy's effects for a five-year term. 'It all past off in a great hurry, and there were not above 20 there,' noted Lord Percival, one of the few subscribers to attend.

LITERATURE
Mainwaring, pp. 96–7; *Deutsch,* pp. 85, 113, 115, 160, 218, 234, 844–5;
Judith Milhous and Robert D. Hume, 'New Light on Handel and The Royal Academy of Music in 1720', *Theatre Journal,* XXXV, 1983, pp. 149–67.

The Origins of the Academy

Within a few days of the King's ordering the incorporation of the Royal Academy of Music on 9 May 1719, Handel received his instructions from the Duke of Newcastle to travel abroad to seek singers for the Academy. This he did, spending much of the summer and autumn at Dresden where he hoped to engage the *castrato* Senesino, and other singers, to come to London.

84 [*illustrated on p. 110*]

'List of Subscribers to the Royal Academy of Musick'

This is the earliest list of subscribers to the Royal Academy of Music and may date to the second half of May 1719, according to *Deutsch*. It is headed by Henry, Duke of Kent, Lord Chamberlain until 1717, said to have been originally intended as governor of the Academy but in the event the Duke of Newcastle, his successor as Lord Chamberlain, was appointed. Each of the 63 subscribers guaranteed £200 of the joint stock of £10,000 except for five: Newcastle himself, the Duke of Chandos and the Earl of Burlington at £1,000 each (see nos. 85, 68 and 60), the Duke of Portland at £600 and Viscount Castlemaine at £400. Another notable Handel supporter was the Duke of Manchester (see no. 39), deputy governor of the Academy from November 1719.

Public Record Office, London
LC 7/3, no. 15, pp. 52–3

Single manuscript sheet folded in half, 4pp., 31.6 × 20.0 cms/12 × 7⅞ ins page size

LITERATURE
Deutsch, pp. 91–2 (for a full list of subscribers).

85

Thomas Pelham-Holles, 1st Duke of Newcastle (1693–1768)

By John Faber (d. 1756)
after Sir Godfrey Kneller (1646–1723)

The nominal governor of the Royal Academy of Music, but often absent from meetings, was the 26-

85

year-old Lord Chamberlain the Duke of Newcastle, statesman and future Prime Minister. He was responsible in January 1719 for seeking advice about the petition to found an opera company, and in May he signed the warrant (no. 86) authorising Handel to travel overseas in search of singers for the Academy. The subscribers' prospectus for the Academy laid down Newcastle's role: 'That the Lord Chamberlaine for the time being be declared perpetuall Governur of the said Company and he is to have a negative Voice in everything.' Newcastle resigned as Lord Chamberlain in 1724 and so ended his connection with the Academy. However he remained a supporter of Handel and his operas.

National Portrait Gallery, London

Mezzotint engraving, 30.3 × 25.1 cms/11⅞ × 9⅞ ins

Inscribed *The Most Noble* PRINCE THOMAS HOLLES, *Duke of* NEWCASTLE, *Marquis and Earl of* CLARE, *Viscount* HAUGHTON, *Baron* PELHAM *of* LAUGHTON *Lord Chamberlain of his Maj.*^ties *Household* . . . [with details of his other official positions] / *G. Kneller* S.R.I. et Mag. Brit. Baronet *pinx* / *I Faber junior fecit* / *sold by I Faber Jun.*^t *at the 2 Golden Balls near The Savoy in the Strand*

LITERATURE
CS, p. 400, no. 255.

86

'Warrant to Mr Hendel to procure Singers for the English Stage'

Copy warrant

On 14 May 1719, in his instructions to 'Mr Hendel Master of Musick', the Duke of Newcastle (see no. 85) as Lord Chamberlain and governor of the Royal Academy of Music directed Handel, 'by his Majestys Command ... forthwith to repair to Italy Germany or such other Place or Places as you shall think proper, there to make Contracts with such Singer or Singers as you shall judge fit to perform on the English Stage'. Handel's instructions are further detailed as to 'procure Voices to sing in the Opera', to issue contracts for periods of 'one Year and no more' and to 'engage Senezino as soon as possible ... for as many Years as may be'. Should he 'meet with an excellent Voice of the first rate', he was to inform the governor of the terms on which he or she might be engaged; he was also to send back copies of any agreements entered into.

Though Handel visited Italy in search of singers ten years later, on this occasion he chose to go to Dresden where, as he reported to Lord Burlington in July, he entered into negotiations for Senesino and other singers which proved largely successful (see no. 87).

Public Record Office, London
LC 5/157, pp. 233–5

Manuscript volume of copy warrants,
36.5 × 24.2 cms/14$\frac{3}{8}$ × 9$\frac{1}{2}$ ins

LITERATURE
Deutsch, pp. 89–90 (for a full transcript).

87

Handel's Visit to Dresden: the Hire of Singers for the Royal Academy of Music

Autograph letter from Handel in Dresden to the Earl of Burlington, 26/15 July 1719

Following his commission from the governors of the Royal Academy of Music to seek singers for their proposed opera company and in particular to engage Senesino (see no. 93), Handel set off to the continent. There he spent much of his time at Dresden, the home of a famous opera company and the court of Friedrich August I, Elector of Saxony (who was also August II, King of Poland).

Handel writes to Lord Burlington (see no. 60) on the prospect of a successful outcome to his long journey: 'Je suis icy à attendre que les engagements de Sinesino, Berselli, et Guizzardi, soyent finis, et que ces Messieurs – d'ailleurs bien disposés, s'engagent avec moy pour la Grande Bretagne.' Senesino came to London at the end of 1720 and Matteo Berselli a little before, but Guicciardi never arrived.

Everything will be decided in a few days' time, Handel writes, referring to Burlington as his benefactor and patron. It is apparent from this letter that Handel remained on good terms with his former protector, who was a very active director of the Royal Academy of Music. In fact while Handel was still abroad, Burlington set out for Italy, where he encouraged Bononcini (see no. 88), Handel's future rival, to come to London for the Academy. Handel remained at Dresden for much of the summer and autumn. In September he would have seen *Teofane*, the germ of his own future opera *Ottone*, performed in the splendour of the new opera house in the Zwinger Palace under the short-lived direction of Antonio Lotti.

This was the last production at Dresden for some time to come: Lotti returned to Venice, Handel left for London with many of the leading singers engaged to follow and the Dresden opera closed its doors.

The date on the letter, '26/15 de Juillet 1719', is written in both the New Style and the Old, a reminder that most of the continent used the Gregorian calendar which was eleven days ahead of England and was not adopted in Britain until 1752.

The Trustees of the Chatsworth Settlement

Letter, single sheet, folded in half,
3 pp., 23.5 × 18 cms/9$\frac{1}{4}$ × 7$\frac{1}{8}$ ins page size

Endorsed by the 6th Duke of Devonshire *Mr Handel to L^d Bn.*

PROVENANCE
By descent from the 3rd Earl of Burlington in the family of the Dukes of Devonshire; the Trustees of the Chatsworth Settlement.

LITERATURE
Deutsch, pp. 93–4 (for a full transcript and translation); Johannes Gress, 'Händel in Dresden (1719)', *Händel-Jahrbuch*, 1963, pp. 135–51.

The Composers

'Everybody is grown now as great a judge of music as they were in your time of poetry,' wrote Gay to Swift in 1723, 'and folks that could not distinguish one tune from another now daily dispute about the different Styles of Handel, Bononcini and Attilio.' This sense of competition received a certain degree of encouragement from the Academy's Directors. In *Muzio Scevola*, first given in 1721, each act was written by a different composer (see under no. 99), and this pattern was repeated in another form in 1727, when it was announced:

> The Directors of the Royal Academy of Musick have resolved, that after the Excellent Opera composed by Mr. Hendel, which is now performing: Signior Attilia shall compose one: And Signior Bononcini is to compose the next after that . . . the Town will have the Pleasure of having these three different Stiles of composing.

Unlike the rivalries between some of the Academy's singers, there is little to show that the employment of more than one composer was ever a matter of professional jealousy; it was rather a necessity, if the Academy was to mount as many as half a dozen new operas in a season. Of the 487 performances mounted by the Royal Academy of Music in the eight years between 1720 and 1728, the music of 245 was by Handel, 108 by Giovanni Bononcini, 55 by Attilio Ariosti and the remaining 79 by other composers or *pasticcii*, according to Chrysander's calculations.

Though Bononcini was a composer of much greater importance than Ariosti, they shared a somewhat similar background: both spent their early years in Bologna and subsequently gained considerable experience internationally; they also both came from an earlier generation than Handel, Bononcini being born in 1670 and Ariosti in 1666. Indeed Handel's first biographer, John Mainwaring, tells a story, impossible chronologically, of the 13-year-old Handel meeting both Bononcini and Ariosti in Berlin in 1698 (though such a meeting could have taken place a few years later). The peculiar merits of the three composers' music were characterised by the American author James Ralph in his pamphlet *The Touch-Stone* in 1728 (later re-issued as *The Taste of the Town*): Handel 'would furnish us with Airs expressive of the Rage of Tyrants, the Passion of Heroes and the Distresses of Lovers in the Heroick Stile'; Bononcini would 'sooth us with sighing Shepherds, bleating Flocks, chirping Birds, and purling Streams in the *Pastoral*'; while Ariosti would 'give us good Dungeon Scenes, Marches for a Battel, or Minuets for a Ball, in the *Miserere*'.

LITERATURE
Deutsch, pp. 149, 201, 224, 231–2, 237.

88

Giovanni Bononcini (1670–1747)

Attributed to Bartholomew Dandridge (1691–c.1755)

Bononcini was fifty when he came to London in 1720 to provide operas for the Royal Academy of Music. He was a composer of great experience and

88

was famous throughout Europe for his operas and cantatas; indeed his *Camilla* had achieved 64 performances in London between 1706 and 1709, and in 1707 plans were afoot to invite him to London but they came to nothing. From 1698 to 1712 he was employed at the Viennese court for a very high salary and from there he visited Berlin in 1702–3. This was presumably the occasion for his meeting, described by Mainwaring, with the youthful Handel.

Between 1714 and 1720 Bononcini was in Rome. He was encouraged to come to London by Lord Burlington, who had heard his work in Rome in 1715 on his first Grand Tour and who travelled to Italy again in August 1719. Burlington was present at the meeting of the directors of the Royal Academy on 30 November that year, when it was decided to write to Bononcini to settle terms for his coming to London 'for composing & performing in the Orchester'.

This portrait belongs to Bononcini's London years and therefore shows him in his fifties. He arrived in the autumn of 1720, and his *Astarto* opened the Academy's second season on 19 November with an immensely successful run of 24 performances; his operas provided the mainstay of the Academy's next two seasons. He also wrote a much-admired anthem for the funeral of the Duke of Marlborough in August 1722; but that autumn he was caught in a wave of anti-Catholic feeling in which his association with known Jacobites told for more than his music, and he seems not to have been formally re-engaged by the Academy for the 1722/3 season. However he was encouraged to remain in London by a most generous pension of £500 per annum, awarded him in May 1724 by the young Duchess of Marlborough, for whom he agreed to direct the two weekly concerts she gave at her London home; this was, according to Mrs Pendarves (later Mrs Delany; see no. 156), 'provided he will *not* compose any more for the ungrateful Academy, who do not deserve he should entertain them, since they don't know how to value his work as they ought'. Despite this Bononcini did provide a further opera for the Academy: *Astianatte* in May 1727, which was the occasion for the infamous battle between the supporters of the rival sopranos, Faustina and Cuzzoni (see under no. 107).

Though there is no evidence of personal animosity between Bononcini and Handel, the opera audience delighted in making comparisons between them. *Muzio Scevola* (see no. 99), given in April 1721, almost seems designed to provide just such an opportunity, for one Hanoverian in London, De Fabrice, reported a week after the first performance: 'Each act of this opera is by a different composer, – the first by a certain Pipo, the second by Bononcini and the third by Hendell, who easily triumphed over the others'. Later, on 15 January 1723, De Fabrice could write of the passions aroused by the opera: 'there exist two factions, the one supporting Hendell, the other Bononcini, the one for Cenesino [Senesino] and the other for Cossuna [Cuzzoni]. They are as much at loggerheads as the Whigs and Tories, and even on occasion sow dissension among the Directors.' The most famous embodiment of this notion of rivalry was John Byrom's 'Epigram on the Feuds between Handel and Bononcini' which began to circulate privately in May 1725:

> Some say, compar'd to Bononcini,
> That Mynheer Handel's but a Ninny;
> Others aver, that he to Handel
> Is scarcely fit to hold a Candle:
> Strange all this Difference should be
> 'Twixt Tweedle-dum and Tweedle-dee!

Bononcini's downfall came in 1732. He was accused of allowing a madrigal by Antonio Lotti 'In una siepe ombrosa,' to be presented as one of his own works at a performance of the Academy of Ancient Music; the ensuing controversy led to an exchange of letters with Lotti in Venice which appeared in print in 1732, despite Lotti's conciliatory words that the resentment raised over the madrigal was out of all proportion since 'Signor Buononcini can make others equal and much superior.' Bononcini also quarrelled with his patroness, the Duchess of Marlborough. He left London in October 1732 and eventually returned to Vienna where he died in 1747.

In comparing Bononcini's music with Handel's, Charles Burney's assessment is probably the most fair:

> The truth is, that Bononcini's peculiar merit in setting Italian words seems to have been out of reach of an English audience, and that Italians were alone competent to judge of it ... but in majesty, grandeur, force, fire, and invention, which are not local beauties, but striking and intelligible in all countries, Handel was infinitely his superior.

Royal College of Music

Oil, 127.0 × 101.5 cms/50 × 40 ins

Inscribed bottom right at a later date *Signor Bononcini*.

Engraved in mezzotint (head and shoulders) before 1736 by Joseph Sympson junior

PROVENANCE
Christie's, 1896 (as by Kneller);
given by William Barclay Squire to the Royal College of Music, 1898.

LITERATURE
Lady Llanover (ed.), *The Autobiography and Correspondence of Mary Granville, Mrs. Delany*, 1861, I, pp. 79–80; *Deutsch*, pp. 97, 126, 148; Lowell Lindgren, 'The Three Great Noises "Fatal to the interests of Bononcini"', *Music Quarterly*, LXI, 1975, pp. 560–83.

EXHIBITED
Apollo of the Arts: Lord Burlington and his Circle, Nottingham University Art Gallery, 1973, no. 61.

89

Attilio Ariosti (1666–1729?)

By John Simon (1675–1754)
after Enoch Seeman junior (*c*.1694–1745)

Attilio Ariosti spent the last 13 years of his life in London, where for a time he wrote operas for the Royal Academy of Music. Born in Bologna, Ariosti was known as a fine singer and instrumentalist as well as a composer. Often described as 'Padre Attilio' (as a young man he was invested as a Servite monk but soon left this order), he led a varied musical career which took him to Venice, and thence to Berlin from 1697 to 1703, where he was Kapellmeister to the Electress Sophia Charlotte. It was in Berlin in 1698 that he first met Handel, according to Mainwaring's rather confused account written many years after the event; the dubious story is told there of Ariosti's taking Handel, by then 13 years old, on his knee to play the harpsichord.

Before their paths crossed again, Ariosti moved to Vienna and then back to Italy; his position as a Catholic monk at the Protestant court of Berlin and his controversial behaviour in Vienna led, it is said, to his de-frocking. The course of events is not altogether clear but by July 1716 he was in London, where he was first heard playing a musical interlude between the acts of Handel's *Amadigi*. In 1719, the year of this portrait engraving, he was helping to negotiate with singers for the new Royal Academy of Music, but it was not until the 1722/3 season that he was engaged as a composer by the Academy. His first opera, *Coriolano*, opened in February 1723 with the newly

89

arrived soprano, Francesca Cuzzoni; it was described in the *British Journal* as 'said to exceed any thing of the kind ever seen upon the stage'. Though responsible for many fewer new productions than Handel or Bononcini, his work was attractive and generally met with public approval. He died in England two years after the performance of his last opera, *Teuzzone*, at the King's Theatre in 1727.

The Trustees of the British Museum
1874–8–8–1209

Mezzotint engraving, 31.8 × 25.3 cms/12½ × 10 ins

Inscribed *Attilius Ariosti Bononiensis* | *E. Seeman Jun: pinx.* | *J Simon fecit 1719*

LITERATURE
CS, p. 1067, no. 18; *Deutsch*, pp. 150, 215.

Librettists and Designers

When the Royal Academy of Music was formed, the writer and poet Paolo Rolli was made its Italian secretary and official librettist. He provided the texts for some of the Academy's earliest operas before a quarrel with the directors in 1722. He was replaced by the composer and librettist Nicola Haym, who had already provided Handel with texts for *Teseo* (1713) and *Radamisto* (1720). Over a period of six years he supplied seven or eight libretti for Handel including *Ottone*, *Giulio Cesare* and *Tamerlano* (see nos. 100–3) as well as others for Ariosti and Bononcini. The texts were nearly always adapted from those of existing Italian operas and to this end Owen McSwiny (see no. 56) was employed by the Academy to supply word-books from the operas he came across in Italy.

In 1726 Giuseppe Riva, a former member of the Italian circle in London, characterised the needs of the Academy at the Haymarket Theatre in response to an enquiry from the Italian historian Ludovico Antonio Muratori about obtaining a commission for a young friend: '... operas that come from Italy cannot be of service in this theatre. It is necessary to revise or rather to deform them in order to render them acceptable. Few verses of recitative and many arias is what they want over here'. The taste of the London audience for good songs and as little connecting recitative as possible may have made some of the plots difficult to follow, but it enabled Riva to lay down a formula for success:

> For this year and for the two following there must be two equal parts in the operas for Cuzzoni and Faustina; Senesino is the chief male character, and his part must be heroic; the other three male parts must proceed by degrees with three arias each, one in each Act. The duet should be at the end of the second Act, and between the two ladies. If the subject has in it three ladies, it can serve because there is a third singer here.

For all that is known of the librettists and their work, little can be said about the scene painters. The design of the sets for the opera does not appear to have received the same degree of attention as was the habit in Paris or Italy. Although on occasion the scenes were the subject of special comment (*Rinaldo* in 1711 is a notable example; see p. 77), more often than not they were ignored in contemporary accounts

90

of the opera, perhaps because the sets were often re-used from an earlier opera. Occasionally the libretto may mention the designer's name, as with *Admeto* and *Riccardo Primo* in 1727, when Joseph Goupy (see no. 91) is described as supplying the new scenes. In 1724 George Vertue refers to Goupy and another artist, Peter Tillemans, both foreigners long resident in London, as painting scenes 'which were *much* approved of'. However a French visitor in London in 1728, Pierre Jacques Fougeroux, perhaps more used to the Paris Opéra, was less complimentary.

As usual McSwiny had advice to give. In 1727 he recommended shipping costumes from Italy and suggested that the Valeriani brothers, Giuseppe and Domenico, two of the leading Italian scene painters, should be employed to come from Rome at a salary of £600. Nothing came of the idea but it would appear that on occasion their designs were adapted for the London stage by another artist living in London, John Devoto. When the Academy closed its doors it was resolved to lend to Handel and Heidegger (see no. 57) its 'scenes, machines, clothes, instruments, furniture, etc' for a term of five years.

LITERATURE
Deutsch, pp. 176, 185, 197, 234, 303; *Gibson*, pp. 84–5.

90
Paolo Rolli (1687–1767)
By Giovanni Battista Bellucci (fl. c. 1700–39)

Never a friend of Handel's but on occasion a useful collaborator, the Italian poet and librettist Paolo Rolli had come to London in 1715 at the invitation of the Earl of Pembroke (or, it is said, the Earl of Stair). He remained until 1744, his stay coinciding with the establishment and the subsequent decline of Italian opera in London. This portrait was painted by his fellow countryman Giovanni Battista Bellucci, a relation of the better known Antonio Bellucci who worked extensively for the Duke of Chandos.

Rolli was a prominent member of the Italian literary and cultural circle in London, and one whose letters are full of gossip and intrigue. Though hardly a poet of the first order (he does not compare, for instance, with his fellow student, Metastasio), he was an adept writer, whose libretti and translations helped enhance his reputation in England.

In 1719 he was made Italian secretary to the Royal Academy, and during its first three seasons he provided libretti for ten productions, including *Muzio Scevola* (see no. 99) and Handel's *Floridante*. Even before his dismissal in 1722 (he had become 'embroiled with the Directors') there was no love lost between him and Handel, as his extensive private correspondence makes only too clear. To Abbato Giuseppe Riva, the Modenese Representative in London, he wrote on 23 September 1720, shortly before the Academy's second season: 'Dear Riva, I submit myself to all shows of humility towards him within the limits of decorum, and we shall see whether that bristly nature of his will soften,' and a few weeks later on 18 October: 'But I am amused that he should suspect me and not trust my polite manners.' Nevertheless he provided libretti for three of Handel's later Academy operas (*Scipione*, *Alessandro* and *Riccardo Primo*) and, though he joined with the Opera of the Nobility from 1733–7, he prepared the text for Handel's last opera, *Deidamia*, in 1741 (see no. 127). His antagonism towards Handel remained; to Senesino on 4 February 1729 he wrote that he had 'always been and will always be most reserved towards him, nor did I wish him well on his departure', and to Riva on 12 June 1730 he took Handel's partnership with Heidegger to task, saying of 'their worthless operas' that 'in truth they succeed no better than they deserve'.

Private Collection

Oil, 90.8 × 71.5 cms/35¾ × 28⅛ ins

Inscribed (probably by Lady Grisel Baillie of Mellerstain, before 1731) *Sig.r Paolo Rolli / Belluci P.1723*

PROVENANCE
By descent at Mellerstain from George Baillie to the Earls of Haddington, and thence to the present owner.

LITERATURE
Deutsch, pp. 186, 237, 254–5; G.E. Dorris, *Paolo Rolli and the Italian Circle in London 1715–1744*, The Hague, 1967.

91
Joseph Goupy (c. 1680–c. 1770)
By Richard Bean (1792–1817)
perhaps after Michael Dahl (?1659–1743)

Joseph Goupy is one of those elusive figures: a close friend of Handel's for many years whose friendship is however revealed only by the occasional glimpse. He was a minor French artist who was in London by 1711 when he enlisted as a subscriber to Sir Godfrey Kneller's Academy. He specialised in pastel and gouache copies of Old Masters, so much so that by 1717, he was said to have been 'in such reputation that he could with the greatest ease get £600 a year and upwards'.

He was the nephew and pupil of Louis Goupy, who accompanied Lord Burlington to Italy. It may be through this connection that he found employment as a scene designer for the operas of the Royal Academy of Music in 1724, working with the painter Peter Tillemans according to George Vertue. The operas concerned may have been, as Deutsch suggests, Handel's *Tamerlano* or *Giulio Cesare*, both performed in 1724, or his *Rodelinda* (1725). In 1727 Goupy designed the sets for *Admeto* and *Riccardo Primo*. His etching, 'Mutius Scaevola burning his hand in the presence of Porsena', dated 1726, was inspired by the earlier Academy opera on that theme (see no. 99); his painting of 'Mucius Scaevola' was in Handel's sale in 1760. In 1730 Goupy was responsible for the publication of a caricature, 'Thou tunefull Scarecrow' (no. 121), featuring Farinelli, Cuzzoni and Heidegger. His portrait of Senesino was published in a mezzotint engraving by E. Kirkall (no. 93).

Goupy's close friendship with Handel probably came about through their work together for the Royal Academy of Music. It seems to have ended with Goupy's over-cruel satirical pastel of Handel (no. 8), probably done in the late 1740s, which is said to have cost him a legacy in Handel's will. A portrait

91

of Handel by Goupy was owned by Frederick Prince of Wales in 1742 (when a payment was made for its framing); the portrait is now lost but may have existed in more than one version, since the catalogue of a sale of pictures from Goupy's collection on 23 March 1765 included a 'Mr. Handel' as the first lot (this is not the pastel caricature which also featured in the sale).

Hawkins called Goupy an 'intimate' friend of Handel, one of the few who 'seemed to possess most of his confidence'; this would seem to be borne out by two widely spaced reports of Goupy's role as a mediator in delicate circumstances. In 1729, shortly before Handel's departure to look for new singers in Italy, Goupy was used to make an indirect approach to Handel's former collaborator Paolo Rolli (see no. 90), to discover his feelings about Handel's new proposals; he was given the task of explaining that 'Faustina had been the cause of the differences that had arisen' between Rolli and Handel (letter from Rolli to Senesino, 4 February 1729). And then in 1743, when John Christopher Smith the elder was asked by Lord Middlesex to persuade Handel to write two new operas for Middlesex's new venture (see p. 173), Smith chose Goupy as go-between: 'I wrote ... to Mr. Goupy with the desire to communicate it to Mr. Handel (for it seems he has taken an aversion to see me ...)' (letter from Smith to Lord Shaftesbury, 28 July 1743).

Handel owned various engravings and pictures by Goupy; the composer's posthumous sale in 1760 (see Appendix 4) featured Goupy's engravings after Salvator Rosa and Poussin, and three other works by him: a watercolour of Belisarius after Van Dyck (see fig. 14, p. 289) and two pictures, perhaps in gouache, of Hagar and Ishmael and 'Mucius Scaevola' (mentioned above).

Goupy's early patrons included Baron Kielmansegg (see no. 63) and the Duke of Chandos (see no. 68), both of whom paid high prices for his reduced-scale copies of the Raphael cartoons. He was appointed cabinet painter to Frederick Prince of Wales in 1736 and enjoyed a close association with the Prince's family, acting as drawing master to the Princess of Wales and her children. Prince Frederick's early death in 1751 appears to have been a great blow to Goupy, and he is said to have died in poverty.

This early nineteenth-century engraving, perhaps from the lost portrait by Michael Dahl (which is recorded as showing him holding a landscape), is the only known likeness of Goupy other than his representation in Gawen Hamilton's *A Conversation of Virtuosis* of 1735 (National Portrait Gallery).

National Portrait Gallery, London

Engraving, 18 × 17 cms/7⅛ × 6⅝ ins

Inscribed *JOSEPH GOUPY, Engraved by R. Bean, from the Original Picture. Pub. by C. Dyer, Compton Str! Soho.*

LITERATURE
A Catalogue of ... Pictures belonging to Joseph Goupy, ... sold by auction by Mr. Hobbs, 23 March 1765 (copy in British Library, C.131.h.4.(9)); C.R. Grundy, 'Documents Relating to an Action Brought against Joseph Goupy in 1738', *Walpole Society*, IX, 1921, pp. 77–88; *Vertue*, III, pp. 21, 31, and VI, p. 169; *Deutsch*, pp. 235–7; Betty Matthews, 'Unpublished Letters concerning Handel', *Music and Letters*, XL, 1959, pp. 261–8.

'The three best Voices in Europe'

Italian opera meant Italian singers. At the beginning of the Royal Academy of Music's second season in November 1720, Mrs Pendarves (later Mrs Delany) could write: 'The stage was never so well served as it is now, there is not one indifferent voice, they are all Italians.' Even the only English singer of any note to be engaged by the Academy, Mrs Anastasia Robinson (see no. 92), who appeared in its first five seasons, had been born in Italy.

When the *Flying Post* on 4 February 1727 claimed that the Academy could 'boast of the three best Voices in Europe' it was referring to the *alto castrato* Francesco Bernardi, called Senesino, and to the two sopranos Francesca Cuzzoni and Faustina Bordoni (see nos. 93–6). Others shared this opinion; the German composer Johann Quantz described them as 'all three virtuosos of the front rank'. Their fame has tended to overshadow some of the Academy's other fine singers such as Margherita Durastanti, the soprano who had first sung for Handel in Rome in 1707.

Italy was known for its thriving music and opera. It was also known for its *castrati* singers. The habit of castrating boys to retain their voices seems to have originated in the ancient interdiction on women's singing in church, and to have been perpetuated by the popularity of Italian opera. Not everyone liked the result. The Frenchman, De Brosses, was critical at first: 'As for the castrati, this sort of voice does not please me at all; except for one or two, all I have heard have seemed to me miserable. It is not worth while forfeiting one's effects for the right to chirp like that.' Later he was to write:

> One must be accustomed to the voices of the castrati in order to enjoy them. Their timbre is as clear and piercing as that of choirboys and much more powerful; they appear to sing an octave above the natural voice of women. Their voices have always something dry and harsh ... but they are brilliant, light, full of sparkle, very loud, and with a very wide range.

The fashion for *castrati* led to a prejudice against tenor and bass voices which Handel began to overcome only in his later operas. As Paolo Rolli (see no. 90) could write of Annibale Pio Fabri in 1729, then newly arrived in London: 'Fabri is a great success. Would you have believed that a tenor could have such a triumph here in England?'

Why were Italian singers willing to make the long journey north to England? London could of course boast a superb opera company, but the real attraction was money: the salaries paid were generally as high if not higher than anywhere else in Europe. A French visitor in 1728 thought the fees paid 'exorbitant', yet the means by which the Academy had been able to 'carry off the best that Italy possessed'. The foremost singers such as Faustina and Bordoni could command £2,000 a year plus the receipts of a benefit performance, which might bring as much as £1,800 on a single night. In the mid-1730s Farinelli is calculated to have been taking £5,000 in a year. These were huge sums. Senesino was said to have returned to Italy to build 'a fine house with an inscription over the door to let the world know 'twas the folly of the English had laid the foundation of it'. Even Margherita Durastanti's salary of a 'thousand pounds and more per year' was considerable, as Paolo Rolli indicated in August 1720: 'The half of this saved means a lot of money in Italy.' When Durastanti retired from the English stage in 1724 she sang a cantata in praise of the English nation at her benefit performance. As one foreigner in London observed with amazement, 'That one day brought her more than a thousand pounds sterling.... Have you ever heard ... of prodigality and favour to equal this towards a woman already old, whose voice is mediocre and worn out? That is what the English are like....'

LITERATURE
Deutsch, pp. 111, 118, 160, 242, 249, 754;
Angus Heriot, *The Castrati in Opera*, 1960, p. 14;
Winton Dean, 'A French Traveller's View of Handel's Operas', *Music and Letters*, LV, 1974, pp. 172–8.

92

92
Anastasia Robinson, Countess of Peterborough (c.1692–1755)
By John Vanderbank (1694–1739)

Anastasia Robinson was first given a role by Handel in his *Birthday Ode for Queen Anne* (no. 53), written for performance in February 1713. She soon started singing in his operas and was one of the few English singers to appear in the productions of the Royal Academy of Music. By then her voice had sunk from soprano to a fine contralto. For Handel she appeared in *Radamisto* (1720), *Muzio Scevola* (1721), *Floridante* (1721), *Ottone* (1723), *Flavio* (1723) and *Giulio Cesare* (1724) before her retirement from the stage in 1724.

She is seen here playing a double-manual harpsichord in a portrait by John Vanderbank painted in 1723. She was later described by Mrs Delany (see no. 156) as 'not handsome, but of a pleasing modest countenance, with large blue eyes; her deportment easy, unaffected, and graceful; her manner and address very engaging . . .'. In her own words, Mrs Robinson described herself in a letter to Giuseppe Riva about her part in *Ottone* (see no. 100): 'Nature design'd me as a peaceable Creature'.

The daughter of a portrait painter, Thomas Robinson, she was born in Italy where her father went to study soon after his marriage. Her talent for music was encouraged by him and on their return to England she took singing lessons from Dr Croft. She was obliged to take up music professionally with the advent of her father's blindness. In 1722 she entered into a secret marriage with Charles Mordaunt, the elderly Earl of Peterborough; this was said to be the first marriage of any stage personality to a peer. It was not for 13 years, however, that Lord Peterborough was willing publicly to acknowledge her as his wife, though in 1724 he went as far as to cane Senesino who had insulted her: the singer was obliged 'to confess upon his knees that Anastasia was a nonpareil of virtue and beauty'.

Private Collection

Oil, 127.0 × 101.9 cms/50 × 40⅛ ins

Engraved in mezzotint by John Faber, 1727, as after John Vanderbank, 1723

PROVENANCE
Perhaps by descent from the sitter's husband, the 3rd Earl of Peterborough, whose niece married Jonathan Shipley, tutor to Lord Peterborough's family and later Bishop of St Asaph; thence by descent through the Rowley-Conwy family to the present owner.

93
Francesco Bernardi, called Il Senesino (d. by 1759)
By Elisha Kirkall (1682–1742)
after Joseph Goupy (c.1680–c.1770)

The celebrated *alto castrato*, called Senesino from his birthplace, Siena, was the Royal Academy of Music's leading male singer. In the Academy's instructions to Handel in May 1719 (see no. 86), Senesino was the only artist singled out by name: he was to be engaged 'as soon as possible . . . and for as many years as possible'. It was for this reason that Handel travelled to Dresden, where Senesino had been performing since 1717 (his earliest stage appearance seems to have been in Venice ten years before). Negotiations dragged on, but Senesino eventually accepted the Academy's offer of three thousand guineas to come to London for its second season in September 1720.

On 23 September that year, the Academy's Italian secretary Paolo Rolli (see no. 90) wrote to his friend, Giuseppe Riva, of Senesino's arrival: 'I am delighted to find this famous artist a man well-mannered, well-read, extremely kind and endowed with the noblest sentiments.' However within a few weeks the seeds of personal animosity between Senesino and Handel were sown when Handel responded to an outburst from Senesino by calling him 'a damned fool'. Senesino already had a reputation for ill-mannered behaviour (at Dresden he had been dismissed for refusing to sing his part and tearing up that of another singer), and the imperious Handel would not brook opposition. Despite their personal disagreements an open break was avoided while Senesino sang for the Academy.

Senesino appeared in all the Academy's 32 operas in the eight years following his arrival, including 14 by Handel (listed on p. 127). His success was immense. After his first London appearance in Bononcini's *Astarto* in November 1720, Mrs Pendarves, later Delany, called him 'beyond Nicolini both in person and voice'; John Gay, writing to Dean Swift about the latest amusements of the town, described him as 'daily voted to be the greatest man that ever lived'; and the Hanoverian diplomat, Monsieur De Fabrice, thought him 'beyond all criticism' in *Giulio Cesare* in 1724.

In May 1726 public attention was diverted by the arrival of Faustina (see no. 96), who joined Cuzzoni (see no. 95) and Senesino as the Academy's star

93

August 1730 he was engaged at a salary of 1,400 guineas and welcomed back by the London audience in November. Before Senesino's final break with Handel, which came by June 1733, he performed in four new operas by the composer and in *Esther* (in English), and other oratorios. He then sang for the Opera of the Nobility until 1736, when he returned to Italy for good.

His departure was satirised in a songsheet, 'The Ladies Lamentation for ye Loss of Senesino' (no. 123), and he was also the subject of many caricatures by Marco Ricci and A.M. Zanetti (see no. 94). This portrait by Joseph Goupy, the Academy's scene designer in 1724 and 1727 (see no. 91), gives some hint of Senesino's majestic bearing and rather porcine features. In March 1740, when Horace Walpole met the great singer near Siena, he wrote, rather spitefully: 'We thought it a fat old woman; but it spoke in a shrill little pipe, and proved itself to be Senesini.'

National Portrait Gallery, London

Mezzotint engraving, 26.2 × 22.1 cms/$10\frac{1}{4}$ × $8\frac{3}{4}$ ins

Inscribed *Francesco Bernardi, detto il Senesino.*/ *J. Goupy pinx.*/ *E. Kirkall fe.*/, and with four lines of Italian verse beginning *Siena illustre gli diè Nascita e Nome*, with the English translation *Renown'd Siena gave him Birth and Name*

LITERATURE
CS, p. 788; *Deutsch*, pp. 112, 118, 149, 160; *Gibson*, p. 83.

attractions. It was probably no coincidence that, little over a month later, Senesino was reported as indisposed and returned to Italy for a stay which lasted six months – though when the Academy's correspondent, Owen McSwiny (see no. 56), met him in Bologna, he reported that he had never seen him looking in better health. McSwiny had received instructions to look at other singers including Carestini (see no. 116), and went so far as to recommend Minelli to take Senesino's place. However Senesino came back to London and sang in the Academy's last two seasons.

When Italian opera began again at the Haymarket Theatre in December 1729, it was for a five-year term under the direction of Handel and Heidegger. Senesino was not Handel's first choice for the new company and, as usual, had been made aware of this by Paolo Rolli; he gave Handel 'a cold reception' when the composer arrived in Venice in 1729, but in

94

Francesco Bernardi, called Il Senesino (d. by 1759)

From the album of Count Francesco Algarotti, probably made in the studio of Marco Ricci (1676–1730) or perhaps by Antonio Maria Zanetti (1679/80–1767)

Senesino (on the right of this double sheet) is recognisable by his snub nose (compare no. 93). He is shown out walking rather than dressed for the stage, but the 'majestic figure' he made as a singer comes across despite, or perhaps because of, the caricature. Like other drawings in the series (see no. 49) this study will have been made in Venice, where Senesino first sang in 1707–8. After his long stay in London, from 1720–8, he returned home, appearing at Venice again in 1729. Indeed Antonio Maria Zanetti, the Venetian caricaturist and connoisseur, mentions Senesino in a letter of 11 January 1729 (quoted by *Croft-Murray*); it is to Zanetti that the original of this drawing at Windsor has been attributed.

This is mounted with another drawing, no. 16, of Bartolomeo Bartolini (1685–after 1730), seen from behind. He sang in Venice from 1712–24.

Albert R. Gellman Collection,
Courtesy of the Israel Museum, Jerusalem

Pen and brown ink over pencil traces,
24.6 × 17.5 cms/9¾ × 6⅞ ins

Inscribed by Algarotti bottom centre *Senesino,* and top left *17*

PROVENANCE
As no. 49.

LITERATURE
As no. 49.

95

Francesca Cuzzoni (*c.* 1700–70)

By Philip Mercier (*c.* 1689–1760)

One of the most splendid sopranos of her century, Francesca Cuzzoni came to London in 1722 at the invitation of the Royal Academy of Music. This was the first of three visits: she later sang for the Opera of the Nobility from 1734–6, and in 1750–1 returned for a number of benefit performances, though sadly plagued by debt and in ill health.

Cuzzoni's engagement for the Academy's fourth season was heralded by an announcement in the *London Journal* in October 1722; according to the same paper on 22 December, shortly before her arrival, she had been paid an advance of £250 by Heidegger, the manager at the Haymarket (see no. 57). Her début was in *Ottone* on 12 January 1723, as was reported by the Hanoverian courtier, Monsieur de Fabrice, three days later:

> In the end the famous Cozzuna not merely arrived but even sang in a new opera by Hendell, called Othon ... with enormous success; the house was full to overflowing. Today is the second performance and there is such a run on it that tickets are already being sold at 2 and 3 guineas which are ordinarily half a guinea, so that it is like another Mississippi or South Sea Bubble.

It was during a difficult rehearsal for *Ottone,* when Cuzzoni refused to sing her aria 'Falsa imagine', that the famous incident occurred of Handel's seizing her by the waist and threatening to throw her out of the window (see under no. 100).

Rumours at the end of her second season in May 1724 that she was leaving London proved false; Cuzzoni continued to command attention as the Academy's leading soprano. In *Rodelinda* in February 1725, one of Handel's two new operas of the season, it was her costume which provoked most comment, as Horace Walpole noted: 'on her appearing in this opera, in a brown silk gown, trimmed with silver, with the vulgarity and indecorum of which all the old ladies were much scandalised, the young adopted it as a fashion, so universally, that it seemed a national uniform for youth and beauty.'

When Faustina Bordoni (see no. 96) arrived in 1726, Cuzzoni's undisputed reign was over; the new star, though not necessarily superior as a singer, was a far more attractive woman and had the advantage of novelty. The rivalry between the two great sopranos, encouraged by their supporters in the audience, led them to blows on stage in June 1727 (see under nos. 96 and 107), and Cuzzoni was discharged, though only temporarily, for she performed in the following season. With the closure of the Academy Cuzzoni and the other singers returned to Italy, and though Handel had thoughts of using her in his new company in 1729, she was by then engaged elsewhere.

Cuzzoni's success as a singer came despite her personal appearance, which was the subject of more than one caricature (see no. 121). Horace Walpole called her 'short and squat, with a doughy cross face'. Few portraits are known, and the identification of this drawing by Philip Mercier as Cuzzoni, though deriving from an inscription on the old mount (as recorded in 1876), is not altogether convincing. The drawing may be dated to 1723–5 and relates closely to a figure in a conversation piece, 'Lovers in a Park', belonging to the Duke of Northumberland (*Ingamells and Raines*, no. 218), and like Mercier's other early work demonstrates a profound knowledge of recent artistic developments in France. Mercier later painted Handel's portrait (no. 2), as well as that of Frederick Prince of Wales with his sisters (no. 112).

The Trustees of the British Museum
1895–9–15–953

Red, white and black chalks, 37.2 × 29.4 cms/14⅝ × 11⅝ ins

PROVENANCE
Sir J.C. Robinson; John Malcolm, from whom purchased by the British Museum, 1895.

LITERATURE
Mainwaring, p. 110; J.C. Robinson, *Descriptive Catalogue of Drawings ... of John Malcolm*, 1876, no. 500 (as by Pater); *Deutsch*, pp. 136, 138, 147, 207; *Ingamells and Raines*, p. 64, no. 278.

96 [in colour between pp. 204–5]

Faustina Bordoni (1700?–81)

By Bartolomeo Nazari (1699–1758)

Faustina Bordoni, generally known simply as Faustina, was one of the greatest of eighteenth-century sopranos. She had excited admiration by her performances in Italy from 1716 onwards, being dubbed 'La nuova sirena', and she also appeared abroad, first at Munich in 1723 and then at Vienna. Already in March 1723 the *London Journal* carried a rumour that she would be coming to London: 'we are well assured Faustina, the fine Songstress at Venice, is invited . . .'. Though not all the directors of the Royal Academy of Music were happy with the idea of bringing Faustina to London, the Academy's correspondent in Italy, Owen McSwiny (see no. 56), was instructed in April 1725 to continue his efforts to persuade her to come, and by 6 July he was able to report that she had contracted to do so. When in September the *London Journal* carried the news of her coming visit ('to rival Signiora Cuzzoni'), it gave her salary as £2,500, though in the event she was paid £2,000, the same as Cuzzoni (see no. 95).

With the arrival of Faustina the following spring the Royal Academy had, at vast cost, brought together three of the finest singers in Europe. In his new opera *Alessandro*, premiered at the Haymarket on 5 May 1726, Handel gave equal parts to Faustina and Cuzzoni: the same number of arias, a duet each with the hero, Senesino (see no. 93), and a duet sung together, with matching parts.

When the new season at the Haymarket began in January 1727, it was with an audience divided between the supporters of Faustina and those of Cuzzoni, led on the one side by Lady Burlington and Lady Cowper (who wrote of her favourite, 'she is the devil of a singer'), and on the other by Lady Pembroke and the Duke of Rutland. Handel's *Admeto*, opening on 31 January, met with enormous success ('every night fuller than ever was known at any Opera for so long together') but ended in April with the audience breaking up in disorder and the publication of recriminating pamphlets (see no. 107). The opera season was brought to a close with Bononcini's *Astianatte* on 6 June, as the *British Journal* reported: 'a great Disturbance happened at the Opera, occasioned by the partisans of the Two Celebrated Rival Ladies, Cuzzoni and Faustina. The Contention at first was carried on by Hissing on one Side, and Clapping on the other; but proceeded at length to Catcalls and other great Indecencies.'

This was too good an opportunity for the satirists to miss: *The Devil to Pay at St James's*, a pamphlet subtitled 'A Full and true account of a most horrible and bloody battle between Madame Faustina and Madame Cuzzoni', proclaimed that 'it is certainly an apparent Shame that two such well bred Ladies should call Bitch and Whore, should scold and fight like any Billingates'. Another such publication, *The Contre Temps; or, Rival Queans*, a skit set in the Temple of Discord, has Handel saying: 'I think 'tis best – to let 'em fight it out', with the directions: '*Faustina lays flat Cuzzoni's nose with a sceptre, Cuzzoni breaks her head with a gilt leather crown: Handel, desirous to see the end of the battle, animates them with a kettledrum.*'

The following season opened in October, soon after George II's coronation, and passed off rather more quietly; but a letter from Mrs Pendarves (later Mrs Delany) to her sister on 25 November 1727 correctly predicted that the operas were 'at their last gasp', adding that 'Senesino goes away next winter, and I believe Faustina, so you see harmony is almost out of fashion'. It was Faustina's illness in June 1728 which brought the Academy's final season to an early close. Though she later sang with great success in Italy and at Dresden, she never returned to London.

Faustina was one of the great singers of the age. Burney reported that 'she had the art of sustaining a note longer, in the opinion of the public, than any other singer, by taking her breath imperceptibly . . . her professional perfections were enhanced by a beautiful face, a symmetric figure, though of small stature, and a countenance and gesture on the stage which indicated an entire intelligence of the several parts she had to represent'. This portrait of her holding a lute was probably commissioned by Marshal Schulenburg, a successful military commander in the service of the Venetian republic. His fine collection of pictures included several works by the portrait painter Bartolomeo Nazari, who worked mainly at Venice (he also painted Farinelli; see no. 120).

Chaucer Fine Arts, London

Oil, 134 × 106 cms/52¾ × 41¾ ins

PROVENANCE
Field Marshal Count Johann Mathias von der Schulenburg; thence by descent; Sotheby's sale, 23 June 1982, lot 69; purchased by Chaucer Fine Arts.

LITERATURE
Deutsch, pp. 151, 207, 210, 212, 218; *Gibson*, p. 83.

The Operas

During most of the Royal Academy's nine annual seasons (which usually ran from October or November to May or June) Handel prepared one or two new operas for performance. Between 1720 and 1728 some 14 works were given for the first time and there were as many revivals, often involving substantial revisions. In most seasons he mounted three or four of his own operas. It is not possible to represent them all here, but the autographs or printed scores of five of the more famous are included and a complete table of those by Handel is given.

SEASON	OPERA (number of performances at first run)	DATE OF FIRST PERFORMANCE (revivals)
1	*Radamisto* (10)	27 April 1720 (1720, 1721, 1728)
2	*Muzio Scevola* (3) (only Act 3 by Handel)	15 April 1721 (1722)
3	*Floridante* (15)	9 December 1721 (1722, 1727, 1733)
4	*Ottone* (14)	12 January 1723 (1723, 1726, 1727, 1733)
	Flavio (8)	14 May 1723 (1732)
5	*Giulio Cesare* (13)	20 February 1724 (1725, 1730, 1732)
6	*Tamerlano* (12)	31 October 1724 (1731)
	Rodelinda (14)	13 February 1725 (1725, 1731)
7	*Scipione* (13)	12 March 1726 (1730)
	Alessandro (13)	5 May 1726 (1727, 1732)
8	*Admeto* (19)	31 January 1727 (1727, 1728, 1731)
9	*Riccardo Primo* (11)	11 November 1727
	Siroe (18)	17 February 1728
	Tolomeo (7)	30 April 1728 (1730, 1733)

97

'Radamistus. An Opera. As it is Perform'd At the King's Theatre in the Hay-Market, For the Royal Academy of Musick'

Word-book for *Radamisto*,
text probably by Nicola Haym
Printed for Thomas Wood, 1720

The first production of the newly formed Royal Academy of Music in 1720 was Giovanni Porta's *Numitore*, but that seems to have been something of a stop-gap. The main interest lay in the second opera of the short season, Handel's *Radamisto*, first performed on 27 April. Its production may have been delayed so that King George I and the Prince of Wales could appear together at the first night and thereby demonstrate their recent reconciliation. Handel himself dedicated the opera to the King, by no means a common gesture for a composer; his dedication is printed in this word-book.

According to Burney, the libretto was prepared by Nicola Haym (a plausible attribution, though curiously difficult to confirm from contemporary sources). It is adapted from Domenico Lalli's *L'amor tirannico*, first set by Gasparini in 1710, and is founded on an episode related by Tacitus (*Annals*, Book XII) concerning events in Asia Minor in AD 51. As usual, the full text of the opera is printed in the word-book with a parallel translation into English. For the first revival of the opera in December 1720, substantial revisions to both words and music were made to accommodate a very differently distributed cast (the alto Senesino took over the title role from the soprano Margherita Durastanti, while Durastanti sang the part of Zenobia, originally written for the contralto Anastasia Robinson).

The British Library Board
163.g.58

Printed word-book, viii + 71 pp.,
19.8 × 11.8 cms/$7\frac{3}{4} \times 4\frac{5}{8}$ ins

LITERATURE
Charles Burney, *A General History of Music*, Dover edition, New York, 1935, II, p. 701; *Deutsch*, p. 103.

98

'Il Radamisto Opera Rapresentata Nel Regio Teatro D'Hay Market Composta Dal Sig^re, Georgio Fredrico Handel.'

Printed score, first edition
Published by 'the Author' (ie Handel)
Printed by Richard Meares and
John Christopher Smith the elder, 1720

Richard Meares was originally an instrument maker who turned his attention to music publishing around 1713. He was later joined by his son, who bore the same name and who took over the business when the father died around 1722. The firm was associated with Handel in three major publications of 1720–1, all of which are included here (for the *Suites de Pieces*, see no. 74; for *Muzio Scevola*, see no. 99). The score of the overture, arias and part of the final *coro* of *Radamisto* – virtually the whole opera apart from recitatives – was published in mid-December 1720. The engraving of the copper plates was done for Meares by Thomas Cross, one of the best engravers of the time; he worked for several publishers as well as publishing some music on his own account.

As usual, the instrumentation of the arias is not always given in full, but the overture (displayed) is printed in its original scoring, and exemplifies the clarity and accuracy of the publication. The score also contains the text of the copyright privilege granted to Handel by the King on 14 June 1720. It would seem from the endorsement 'Sold by Jn? Walsh' on this copy that Walsh took over the younger Meares's stock after the latter ceased trading around 1726.

The British Library Board
RM 7.h.21

Printed score, 121 pp., 35.2×24.0 cms/$13\frac{7}{8} \times 9\frac{1}{2}$ ins

Titled as above and inscribed in manuscript on the title-page, presumably in Walsh's hand *Sold by Jn? Walsh*

LITERATURE
Smith, p. 53, no. 1; Charles Humphries and W.C. Smith, *Music Publishing in the British Isles*, 2nd edition, Oxford, 1970, illustration following p. 120, pp. 122, 230.

99

'The Most Favourite Songs in the Opera of Muzio Scaevola Compos'd by Three Famous Masters.'

Printed score, first edition
Printed for Richard Meares, 1722

Muzio Scevola, the sixth opera produced by the Royal Academy of Music, was first performed at the King's Theatre on 15 April 1721. It was a collaborative effort by three of the composers attached to the Academy – Filippo Amadei composed Act 1, Giovanni Bononcini Act 2 and Handel Act 3 – and as such, unique as far as London operatic productions were concerned, though similar collaborations were not unknown in Italy. Precisely how and why joint authorship was decided upon in this instance is not known: it may have been a deliberate attempt on the part of the Academy directors to stimulate musical rivalry, but expediency or a desire to experiment are equally plausible causes.

Paolo Rolli's libretto, derived from Livy's *History of Rome*, interweaves several fictional love tangles with the events surrounding Lars Porsenna's attempt to restore the Emperor Tarquinius to the Roman throne. Each act is spiced with a major dramatic incident. Act 1 ends spectacularly with Horatius Cocles's defence of the Sublician bridge and his leap into the Tiber (Macaulay's stirring account of the same incident – 'O Tiber, father Tiber, to whom the Romans pray' – is strikingly anticipated in the accompanied recitative 'O Padre Tevere'). At the heart of Act 2 is the scene in which Mutius Scaevola puts his hand into the altar flame to prove his Roman valour before Porsenna, and in Act 3 Mutius's warrior-maiden lover, Clelia, demonstrates her fidelity with another aquatic plunge.

Handel's music for Act 3 (never published in complete form) is one of his richest operatic scores of this period, suggesting that whether or not a sense of competition was intended, he was determined that any comparisons would be in his favour. Bononcini's act is by no means negligible, however; it includes a fine lament for Mutius and some excellent examples of his *affetuoso* style. Even the contribution of the lesser Amadei possesses a few expressive arias perfectly appropriate to their context.

Richard Meares issued the 'Most Favourite Songs' from *Muzio Scevola* in August, 1722. The selection was not generous: four arias by Bononcini and his overture to Act 2; three arias by Handel; and one by Amadei. (The overture was not new. It comes from Bononcini's serenata *Il fiore delle eroine*, produced in Vienna in 1704, and had already done duty as the overture to Peter Motteux's pasticcio opera *Thomyris, Queen of Scythia*, produced at Drury Lane in April 1707.) One of the Handel arias was 'A chi vive' (not included in the Händel-Gesellschaft edition). In his *Critica Musica* of July 1722, Johann Mattheson complained that Handel had borrowed its main theme from his opera *Porsenna*, produced in Hamburg in 1702.

The British Library Board
G 192(2)

Printed score, 22 pp., 33.5 × 21.8 cms/13¼ × 8⅝ ins; bound with songs from Bononcini's *Griselda* and *Crispus* and Handel's *Rodelinda* (*Smith*, p. 61, no. 4).

LITERATURE
Smith, p. 39, no. 1; *Deutsch*, pp. 125–6, 135; L.E. Lindgren, *A Bibliographical Scrutiny of Dramatic Works set by Giovanni and His Brother Antonio Maria Bononcini*, Harvard University dissertation, 1972, pp. 823–5, 986.

99

100

Ottone

Handel's autograph score

Ottone was first produced at the King's Theatre on 12 January 1723 and marked the début of the soprano Francesca Cuzzoni (see no. 95). Nicola Haym's libretto is adapted from that of Antonio Lotti's *Teofane*, an opera Handel had seen at Dresden in 1719; the original author was Stefano Pallavicino. The basis of the story is an historical event, the marriage of Otto II, King of Germany, to Princess Theophano, daughter of the Eastern Emperor, at Rome in 972. Handel subjected the opera to extensive revision during the course of composition and had some difficulties in meeting the requirements of his singers. Cuzzoni objected to her first aria, a short, quiet piece with simple *continuo* accompaniment; but Handel rightly judged it to be dramatically apt, and put his new prima donna firmly in her place:

> Having one day some words with CUZZONI on her refusing to sing *Falsa imagine* in Ottone; Oh! Madame, (said he) je sçais bien que Vous êtes une veritable Diablesse; mais je Vous ferai sçavoir, moi, que je suis Beelzebub le *Chéf* des Diables. With this he took her up by the waist, and, if she made any more words, swore that he would fling her out of the window.

Anastasia Robinson (see no. 92) was also unhappy with the arias in her role of Matilda, Otto's cousin, but was more subtle in her approach. Working through Giuseppe Riva, the Modenese ambassador in London and a good friend of Haym, she pointed out that she was unsuited to songs requiring 'fury and passion to express them':

> Nature design'd me a peaceable Creature [she wrote to Riva], and it is as true as strange, that I am a Woman and cannot scold; my request is, that if it be possible (as sure it is) the words of my second Song *Pensa spietata Madre* should be changed, and instead of reviling Gismonda with her cruelty, to keep on the thought of the Recitativo and perswade her to beg her Sons life of Ottone . . . a Short Melancholly Song would be proper. I have some diffcultys allso in the last I sing, but for fear that by asking too much I might be refus'd all, I dare not mention them. . . .

She got her way. Matilda's 'Pensa spietata Madre', the opening of which is displayed (f.53v), was replaced before first performance by 'Ah! tu non sei', conforming to her desire for a 'short melancholly song' (it follows on f.55, an insertion in the score). The unstated difficulties in her last song ('Nel suo sangue') were also noted, and Handel provided a new setting which, though fiery in style, as the drama required, was much easier to sing. Cuzzoni was later pacified by the addition of three new arias for her benefit performance in the opera on 26 March 1723.

This volume is one of five containing autograph material for *Ottone*, and consists mainly of the original draft; a missing section in Act 2 is in Fitzwilliam Museum MS 256, pp. 53–64. British Library RM 20.b.10 contains additional arias inserted before first performance, for Cuzzoni's benefit or for the 1726 revival; and odd leaves are found in Fitzwilliam MSS 262 (p. 17) and 263 (p. 67). For Handel's embellished versions of three arias, see no. 269.

The British Library Board
RM 20.b.9

Autograph score, 104 ff., 23.5 × 29.2 cms/9¼ × 11½ ins

Inscribed at the end of the score on f. 104 *Fine à Londres August, 21n/10v A.1722* (ie August 21 New ('nouveau') Style, 10 Old ('vieux') Style, Anno 1722)

PROVENANCE
As no. 34.

LITERATURE
Mainwaring, pp. 110–11; *Squire*, p. 56; G. Bertoni, 'Giuseppe Riva e l'Opera italiana a Londra', *Giornale storico della letteratura italiano*, LXXXIX, 1927, pp. 317–24; J.M. Knapp, 'The Autograph Manuscripts of Handel's *Ottone*', *Festskrift Jens Peter Larsen*, Copenhagen, 1972, p. 167.

101

Giulio Cesare in Egitto

Handel's autograph score

One of the greatest works of Handel's Royal Academy period, *Giulio Cesare in Egitto* opened at the King's Theatre on 20 February 1724. It was immediately successful and achieved 13 performances. Handel revived it in 1725, 1730 and 1732. The libretto is based on one by the Venetian poet G.F. Bussani, originally written for Antonio Sartorio in 1676. It tells of Julius Caesar's visit to Egypt in 47 BC, his flirtation with Cleopatra and his encounter with her treacherous brother Ptolemy: these roles were sung respectively by Senesino, Cuzzoni and Gaetano Berenstadt. The score is among Handel's richest, and includes a ravishing, exotic scene in which Cleopatra appears to Caesar in a representation of Mount Parnassus, surrounded by the Muses playing instruments (including harp, theorbo and viola da gamba) on stage.

The final scene is set in the port of Alexandria, where Caesar bids farewell to Cleopatra. It is introduced by a splendid Sinfonia demonstrating another of Handel's instrumental innovations: two pairs of horns crooked in different keys (D and G); the opening (ff. 146v–147r) is displayed.

In addition to the main autograph volume exhibited here, autograph material for *Giulio Cesare* is also found in British Library RM 20.e.4., ff. 62–6 (two arias composed for the revival of 1725), RM 20.c.3, ff. 55–7 (an aria added for Anna Strada's benefit performance of 21 March 1730, later used in *Rinaldo*, see

101

no. 45) and Fitzwilliam Museum MS 256, pp. 65–72 (additions of 1725 and 1730). The main autograph lacks the music for the recitatives; Handel must have written these in the first conducting score, which is now lost. It was replaced in 1725 by the present conducting score, now at Hamburg.

The British Library Board
RM 20.b.3

Autograph score, i + 154 ff., 23.5 × 29.2 cms/9¼ × 11½ ins
Inscribed in pencil at the end of the score on f. 154 *Anno 1723*

PROVENANCE
As no. 34.

LITERATURE
Squire, p. 34; J.M. Knapp, 'Handel's *Giulio Cesare in Egitto*', *Studies in Music History: Essays for Oliver Strunk*, Princeton, 1968, p. 389.

102

'Julius Caesar: An Opera. Compos'd by G. Frederick Handel, of London, Gent.'

Printed score, first edition
Published by John Cluer, 1724

Giulio Cesare was the first of Handel's operas to be published by the firm of John Cluer, marking the beginning of a collaboration which lasted from 1724 to 1730. The operas of 1720–3 had been issued by Richard Meares (see *Radamisto*, no. 98) and John Walsh (*Floridante*, *Ottone*, *Flavio*), nominally on Handel's own behalf; the formula 'published by the Author' appears on the title-pages. Cluer became the official publisher of nine of the ten subsequent operas (the exception was *Tolomeo* in 1728), though the irrepressible Walsh issued his own rival versions anonymously (they were simply described as 'Sold at the Musick Shops') and he continued to include Handel's music in the various serial publications which regularly appeared under his own imprint (*Overtures*, *Sonatas or Chamber Aires*, *Apollo's Feast*). It is possible that Handel quarrelled with Walsh around 1723, as has sometimes been suggested, but his change to Cluer may simply have been due to enthusiasm for Cluer's new methods of printing music from engraved copper plates rather than the stamped pewter plates used by Walsh. Handel's connection with Cluer's firm ended in 1730, the year in which Elizabeth Cluer (widow of John, who died in 1728) married her late husband's foreman Thomas Cobb, who then took over the business.

For *Julius Caesar: An Opera*, issued in July 1724, Cluer adopted the pocket-sized format he had just used for *A Pocket Companion for Gentlemen and Ladies*, the first volume of which contained several Handel arias. The beautiful engraving of the score gives it an attractive appearance, but the small size actually makes it rather inconvenient to play from; Cluer's subsequent scores (such as *Tamerlano*, no. 103) are in the usual quarto format. All the arias and ensembles of *Giulio Cesare* are included, together with the overture, but not the Act 3 Sinfonia (see no. 101) and other orchestral numbers.

The unusual design of the title-page is not entirely original: the central feature of a couple playing harpsichord and bass viol in a book-lined room is taken directly from an engraving ('Ici d'un concert tout divin . . .') by Bernard Picart (1673–1733) published in Paris in 1708. In common with Watteau and other French artists of the period Picart enjoyed depicting 'demi-caractères' in idealised, quasi-theatrical national costumes. The musical couple are 'Espagnolettes', dressed in a Spanish style particularly fashionable in the early part of the eighteenth century. Both in the original engraving and Cluer's version there is something odd about the treatment of the harpsichord keyboard, the black keys of which (corresponding to the white keys on a modern piano) seem far too long; possibly it is a clumsy attempt to represent a double manual. The grant of copyright for 14 years from 14 June 1720 reproduced opposite the title-page was made by James Craggs in his capacity as Secretary of State but seems to have done little to deter pirated editions.

The British Library Board
D310.b(1)

Printed score, 118 pp., 22.5 × 15.1 cms/8⅞ × 6 ins; bound with a flute reduction of the opera (*Smith*, p. 33, no. 13)

PROVENANCE
Thomas Greatorex (inscribed on the flyleaf with his initials, the name completed in another hand); Alfred Rosse (?) (inscribed on the title-page); acquired by the British Museum, 1857.

LITERATURE
Smith, p. 30, no. 1, and pp. xiv–xv; *Deutsch*, p. 162; Charles Humphries and W.C. Smith, *Music Publishing in the British Isles*, 2nd edition, Oxford, 1970, pp. 109–10; M.M. Grasselli and Pierre Rosenberg, *Watteau, 1684–1724*, catalogue of an exhibition at the National Gallery of Art, Washington, 1984, pp. 520–1.

103

'Tamerlane: an Opera, Compos'd by M.r Handel'

Printed score, first edition
Published by John Cluer, 1724

The opera *Tamerlano*, first produced at the King's Theatre on 31 October 1724, was the immediate successor to *Giulio Cesare*. It is no less masterly, but the contrast between the two works is striking. Though *Tamerlano* demands no exotic orchestral colours or scenic spectacle, it has a dramatic intensity unsurpassed by any comparable work of the period. It ends in tragic mood with the death of the captured Ottoman Emperor Bajazet, driven to suicide by the treatment he and his daughter Asteria receive at the hands of the Tartar conqueror Tamerlane (Marlowe's 'Tamburlaine').

The opera was drafted between 3 and 23 July 1724, but was substantially revised and re-written before performance. At first Handel worked from a libretto by Agostino Piovene, which Gasparini had set for Venice in 1710 (its source was the play *Tamerlan ou la Mort de Bajazet* by Jacques Pradon, produced in Paris in 1675). He then turned to a new version of the libretto produced (again with Gasparini's music) at Reggio Emilia in 1719 under the title *Il Bajazet*. The most striking feature of the new version was a revised final scene encompassing the death of Bajazet devised by the Modenese poet Ippolito Zanella and the tenor Francesco Borosini, who sang the role of Bajazet. Borosini also played Bajazet in Handel's version, and it was doubtless through him that Handel obtained the new libretto and was persuaded to revise his own opera on similar lines.

Cluer's score of the overture, arias and ensembles of *Tamerlano* was issued on 14 November 1724, in the middle of the opera's very successful opening run of 12 performances. He claimed that the music had been 'Correct'd and Figur'd' by Handel himself, and drew attention to a new feature: '... to render this Work more acceptable to Gentlemen & Ladies, every Song is *truly translated* into English Verse, & the Words Engrav'd to the Musick under the Italian, which was never done before in any OPERA. This score does indeed seem to have been unique in its provision of singable English translations of the arias (*pace Deutsch*, these were not the English versions given in the word-book, which do not fit the music); and it remained so. Cluer may well have found that the translations did little to increase sales, and indeed it is hard to imagine that they were widely used, since, though skilful and accurate, they make little sense outside the context of the opera.

The illustrated title-page was later used by Cluer for most of the Handel operas which appeared under his imprint. No immediate model is known, though the seated female figure playing a harp is reminiscent of one of Domenichino's sibyls or saints. She is perhaps intended to personify Music rather than to represent St Cecilia; one would have expected Cecilia to be playing the organ, but that instrument, though prominent, is placed in the background. The two wind instruments lying on top of a tambourine and a music book appear to be an oboe (in front) and an early clarinet. On a bench behind, there is a recorder and more music books, one inscribed with Senesino's name, while on the right, forming a somewhat awkward foot-rest for the harpist, is a tenor viol.

Gerald Coke Handel Collection

Printed score, 89 pp., 29.7 × 23.0 cms / 11¾ × 9 ins

LITERATURE
Smith, p. 72, no. 1; *Deutsch*, p. 175; J.M. Knapp, 'Handel's Tamerlano: the Creation of an Opera', *Musical Quarterly*, LVI, 1970, pp. 405–30; R. Strohm, 'Händel und seine Italienischen Operntexte', *Händel-Jahrbuch 1975/76*, 1977, pp. 118–20.

Competition and Rivalries

The opera world was a source of perpetual gossip and frequent intrigue, as is apparent from the newspaper columns and pamphlets of the period and the private correspondence of a number of the Italians in London, especially Paolo Rolli, the librettist (see no. 90). The singers were often the subject of the caricaturists' art both in Venice and England (see for instance nos. 94, 106 and 121); their squabbles were ridiculed by sixpenny pamphleteers (see no. 107); and their exorbitant rewards were attacked in Hogarth's prints (see nos. 106 and 122). The Academy also faced many rival attractions, of which Heidegger's masquerades at the opera house itself, and John Rich's production of *The Beggar's Opera* at Lincoln's Inn Fields (see nos. 105 and 108), were only two. Handel himself was criticised for borrowing one of his arias in *Muzio Scevola* from the music of a friend of his Hamburg days, Johann Mattheson (see no. 104).

104

'Critica Musica'

By Johann Mattheson (1681–1764)
Published by the author, Hamburg, 1722

The issue of Handel's extensive borrowings from other composers (as well as from his own work) has been discussed *ad nauseam* from his own day to this. Apparently the earliest public mention of the subject occurs in *Critica Musica*, the first music periodical, written and published by the friend of Handel's Hamburg days, Johann Mattheson (see no. 27). In the first volume (1722), Handel is chided by Mattheson for taking melodies from his work:

> In the opera *Porsenna*, of my composition, as it was performed here 20 years ago, and accompanied by Haendel under my Direction is found an Aria whose opening words run: *Diese Wangen will ich Küssen* ('These cheeks I would kiss'). It can well be that the melody may have seemed not unacceptable to Haendel: for he has not only in his *Agrippina*, which appeared in Italy, but also in another, new opera, recently performed in England and treating of *Mutio Scaevola* chosen just this same melody, almost note for note.... I will set out a fragment here; not indeed as if I were accusing the man of *plagii*: far be that from me....

The dividing line between borrowing from other works and plagiarism was something of a fine one but the issue was at times taken with great seriousness, as in the case of Bononcini, who in 1731 was accused of presenting a motet of Antonio Lotti's under his own name (see under no. 88). Handel's indebtedness to other contemporary composers such as Telemann (see no. 28), Reinhard Keiser and Gottlieb Muffat in Germany and Agostino Steffani, Domenico Scarlatti and Giovanni Porta among the Italians, has been the subject of many studies. In his own day the issue was discussed by Mattheson on more than one occasion and was also recognised in England. It is worth quoting from two of Charles Jennens's typically forthright letters written in 1743 about his collection of Italian music: 'Handel has borrow'd a dozen of the pieces & I dare say I shall catch him stealing from them; as I have formerly, both from Scarlatti & Vinci', and a month later, 'I am sorry I mention'd my Italian Musick to Handel, for I don't like to have him borrow from them who has so much a better a fund of his own.'

The changing attitudes to the subject have been nicely summarised in the preface to Christopher Hogwood's recent book on Handel:

> His technique of composing with borrowed material, ranging from small themes to complete movements, was noted during his lifetime (first by Mattheson in 1722) and has been a recurrent thorn to commentators ever since. To some nineteenth-century writers it was an issue of moral propriety which caused them a certain distress. Nowadays opinion is centred more on the musical value of such procedures, and the modern reader may more easily adopt the viewpoint of Handel's own century and perhaps accept the recommendation of John Potter, that a composer 'should examine every thing he can meet with and like the curious bee, suck sweets from every flower' (*Observations on the Present State of Music and Musicians*, 1762).

The British Library Board
556.b.6

Printed periodical, two volumes bound in one;
vol. 1 368 + xvii pp., vol. 2 380 + viii pp.,
21.0 × 18.0 cms/$8\frac{1}{4}$ × $7\frac{1}{8}$ ins

LITERATURE
Deutsch, p. 135 (for the translation of the passage quoted above);
Hogwood, pp. 7–8, 182–3.

105

Masquerade at the King's Theatre

Attributed to Giuseppe Grisoni (1699–1769)

This view has a double interest: it is one of the very few interiors of the King's Theatre and it shows one of Heidegger's notorious masquerades in progress. Soon after taking over the management of the Theatre in 1713 Heidegger decided on this novel and popular form of entertainment: 'For the news in London,' wrote Alexander Pope to Lady Mary Wortley Montagu in 1717, 'I'll sum it up in short; we have masquerades at the theatre in the Haymarket, of Mr. Heideker's institution.' These occasions were highly successful; they were also profitable, bringing Heidegger £300 or £400 an evening and, it was said, £5,000 a year.

The masquerade is seen here as from the front of the stalls or pit of the Theatre looking across the stage through an archway at the back to a large room beyond. Such an occasion was described by Montague Bacon in a letter to Lady Mary Wortley Montagu: 'There was a masquerade on Thursday last at the Haymarket Playhouse. By laying planks over the Pit, they made a continued floor as far as the Boxes, which were blocked up with pieces of fine painting, and two or three of the side Boxes left open for wine and other things.' He continues that tickets were sold by Heidegger for a guinea and a half, that is three times the cost of an opera seat. An earlier masquerade, probably held in the adjoining Long Room, was described in a press report on 15 February 1718:

> The Room is exceedingly large, beautifully adorned, and illuminated with five hundred Wax

105

Lights. . . . By the vast Variety of Dresses (many of them very rich) you would fancy it a Congress of principal Persons of all Nations in the World, as Turks, Italians, Indians, Polanders, Spaniards, Venetians, &c. There is an absolute Freedom of Speech, without the least Offence given thereby; while all appear better bred than to offer at any Thing profane, rude, or immodest, but Wit incessantly flashes about in Repartees, Honour, and good Humour, and all Kinds of Pleasantry.

When in 1724 the Bishop of London attacked the morality of such entertainments the only effect seems to have been to encourage interest: the King declared that 'whilst there were Masquerades, he wou'd go to them', Hogarth brought out his satirical print (no. 106), and Giuseppe Grisoni did a 'fine picture representing the Masquerade with the various habits', according to George Vertue – probably this painting or a version of it. However too little is known of Grisoni's work in England to make a firm attribution.

The Trustees of the Victoria and Albert Museum
P22-1948

Oil, 71.0 × 90.0 cms/28 × 35¾ ins

PROVENANCE
Acquired by the Victoria and Albert Museum, 1948.

VERSIONS
I) Sir Osbert Sitwell, Bart (repr. in *Country Life*, CVI, 1949, p.673.
II) Sotheby's, 10 July 1985, lot 69, size 101.5 × 109.0 cms.

LITERATURE
Vertue, III, p. 20; Edward Croft-Murray and Hugh Phillips, 'The Whole Humours of A Masquerade', *Country Life*, CVI, 1949, pp. 672–5, pl. 5; *Deutsch*, p. 80; *Highfill*, VII, p. 237.

106

The Bad Taste of the Town

By William Hogarth (1697–1764)

This is a caricature on the prevailing fashion for masquerades and operas at the expense of the legitimate theatre. It was Hogarth's first independent venture, being published in February 1724 as 'The Bad Taste of the Town' though it is often referred to as 'Masquerades and Operas'. On the left Heidegger (H . . .; see no. 57) looks out of a window of the King's Theatre in the Haymarket; he was the instigator of the highly lucrative fashion for masquerades, and watches a satyr and a fool, who is holding a purse marked *1000l*, lead a crowd of people in masquerade dress into his theatre. From the façade hangs a sign

106

(*Operas*) based on a recent caricature of the singers Cuzzoni, Berenstadt and Senesino (see nos. 95 and 93) but with one addition: the three kneeling noblemen on the right, one of whom, said to be the Earl of Peterborough, cries *Pray Accept 8000l.* as a heap of gold is poured out of a purse. On the other side a great crowd gathers outside a colonnade which probably represents the theatre at Lincoln's Inn Fields (see no. 140) where the pantomine *Dr Faustus* was first presented in 1715; over the arch Harlequin is pointing to a sign with a winged devil marked *Dr FAUSTUS is Here*. In front a woman crying *Waste paper for Shops* is pushing a wheelbarrow loaded up with the works of Congreve, Dryden, Otway and Shakespeare. Hogarth links this state of affairs with that arbiter of fashionable taste, the Earl of Burlington (see no. 60), by the inclusion of the gateway to Burlington House in the background. The two reclining figures of Raphael and Michelangelo look up at the statue of William Kent which surmounts the pediment, a barbed comment on Burlington's espousal of Kent as the universal artist of his age.

The Trustees of the British Museum
1857-5-9-13

Engraving, 12.7 × 17.2 cms/5 × 6¾ ins

Signed bottom right in the image *Wm Hogarth Inv: et Sculp!*; with various inscriptions including eight lines of verse beginning *Could new dumb Faustus, to reform the Age*, and at bottom centre *Price 1 Shilling. 1724*.

LITERATURE
Ronald Paulson, *Hogarth's Graphic Works*, 1965, I, no. 34.

107

'An Epistle from S[igno]r S[enesin]o to S[ignor]a F[austin]a'

By an anonymous pamphleteer
Published by James Roberts, 1727

This rare pamphlet is an example of the then prevailing fashion for satirical broadsides pretending to be the letters of one singer to another. It was published on 8 March 1727, at the height of the rivalries between the singers in Handel's *Admeto*, a new opera first performed at the King's Theatre on 31 January. The title-page includes a vignette of the last scene in the opera when the lovers, Admeto and Alceste, played by Senesino and Faustina Bordoni (see nos. 93 and 96), are united. Senesino is singing the famous aria, 'Si Caro, caro si' (the words are engraved beneath the scene); this is rendered in the original libretto as 'Yes my Life my loveliest Dear, / Thus I grasp and clasp thee here, / Close to my panting Heart.' The pamphlet is an obvious attack on Faustina; the sentiments attributed to Senesino in this spurious letter are anything but loving: '. . . at my Cost, you strive Applause to gain', he is made to accuse, and then, in direct reference to his final love scene, 'Tho' Head and Hands my Waste and Bosom Press; I curse the Little Song . . .'.

A response to this letter was sure. The London opera audience was split between the supporters of the two prima donnas, Faustina and Cuzzoni, the 'Rival Queens' of the pamphlet's opening line, who 'disturb the peaceful Stage'. More than one rejoinder was published; they were, the *British Journal* said on 25 March, either 'dull' or 'lewd'. Matters went from bad to worse. Easter Tuesday, 4 April, saw the sixteenth performance of a run in which 'the House [was] filled every night fuller than ever was known at any Opera for so long together'. With the young Princess Amelia present the audience's persistent hissing and shouting was no longer deemed acceptable. The letter of apology which Lady Pembroke, Cuzzoni's patron, felt obliged to send to court, gives some idea of the evening's events: 'the Cuzzoni had been publicly told . . . she was to be hissed off the stage'; she 'would have quitted the stage when they cat-called her to such a degree in one song that she was not heard one note'; afterwards the audience would not 'suffer the Faustina to speak'.

From then on, one party or the other was deliberately indisposed until the première of Bononcini's new opera *Astianatte* on 6 May; during its run a similar but worse disturbance occurred in Princess Caroline's presence, leading to fighting on stage and the abandonment of the performance. As another pamphlet, *The Devil to Pay at St James's*, put it soon after: 'it is not now (as formerly), i.e. are you High Church or Low, Whig or Tory; are you for Court or Country, King George, or the Pretender: but are you for *Faustina* or *Cuzzoni*, Handel or Bononcini. There's the Question.'

Private Collection

Printed pamphlet, 8 pp., 34.3 × 22.2 cms/13½ × 8¾ ins

Inscribed beneath the stage scene on the title-page, with the words of an aria from *Admeto*,   Si caro caro Si   Ti Stringo al fin cosi   Nel Seno amato

LITERATURE
Dorris, pp. 93–4; *Deutsch*, pp. 201, 205–11.

107

Covent Garden Piazza and Market by Joseph van Aken, c. 1726–30 (no. 132)

The Beggar's Opera by William Hogarth, 1729–31 (no. 108)

108 [*in colour opposite p. 137*]

The Beggar's Opera

By William Hogarth (1697–1764)

The Academy's last season was somewhat overshadowed by the unparalleled success of *The Beggar's Opera* which opened on 29 January 1728 at John Rich's theatre at Lincoln's Inn Fields and ran for 63 nights. John Gay's *Beggar's Opera* was something new: a ballad opera with a low-life setting (Newgate prison, Swift's suggestion) and a strongly satirical text, which attacked the affections of society and the political corruption of Sir Robert Walpole's government. Its novelty disconcerted the critics; as William Congreve declared, 'It will either take greatly or be damned confoundedly,' while Mrs Pendarves, Handel's constant supporter, thought 'the taste of the town is so depraved, that nothing will be approved of but the burlesque. The Beggar's Opera entirely triumphs over the Italian one.' Almost all the songs in *The Beggar's Opera* were taken from existing tunes (including those of Handel), and were adapted to their purpose by the composer J.C. Pepusch (he had been employed by the Duke of Chandos at Cannons at the same time as Handel; see no. 71). With the huge success of the venture – which made 'Gay rich and Rich gay' – John Rich was encouraged to build a new theatre at Covent Garden (see no. 131).

Hogarth's painting is set in Newgate prison; it shows a scene in Act 3 where Lucy Lockitt (left), the gaoler's daughter, and Polly Peachum (right), daughter of the devious lawyer, plead to their fathers for the life of the central figure, the manacled highwayman Macheath, who stands with his arms folded, at a loss to know which of his two wives to choose. The protagonists are watched by an aristocratic audience ranged in two stage boxes. On the extreme right the seated figure holding a word-book is the Duke of Bolton, who was so enamoured of Lavinia Fenton, the actress playing Polly, that he later married her. On the same side but further on the left, the two figures standing in conversation are said to be John Rich (on the left) and Mr Cock the auctioneer, and behind them John Gay.

Hogarth painted as many as six versions of this subject between 1728 and 1731; the first was bought by John Rich, who later commissioned a more elaborate and spacious working-up of the subject (now at the Yale Center for British Art), of which this painting is a version. It was made for the MP, Sir Archibald Grant, who also commissioned Hogarth's 'Committee of the House of Commons' but was declared bankrupt in 1732 before he could collect his paintings. The picture's frame, early if not original, is surmounted by a painted relief which includes a portrait of John Gay.

The Trustees of the Tate Gallery
N 02437

Oil, 57.1×75.9 cm/$22\frac{1}{2} \times 29\frac{7}{8}$ ins

Inscribed on the swags on either side of the coat of arms, VELUTI IN SPECULUM [as if in a mirror] / *UTILE DULCI* [instruction with delight]: from Horace's *Ars Poetica*, 1, 343

PROVENANCE
(based on information from Elizabeth Einberg): Commissioned by Sir Archibald Grant, MP, 2nd Bart, 1729 but unfinished in 1731 and not collected owing to Grant's bankruptcy in 1732; eventually acquired by William Huggins (1696–1761); by descent to his son-in-law, Sir Thomas Gatehouse; sold by him, *c.* 1776, to the Rev. Thomas Monkhouse (d. 1798), Fellow of Queen's College, Oxford; passed to his executor, the Hon. and Rev. Henry Jerome de Salis (d. 1810); by 1817 with Thomas Bowerbank of Lothbury; sold *c.* 1833–4 to Dr John Murray; by descent to his grandson, from whom purchased by the National Gallery, 1909.

VERSIONS
Reproduced and discussed in *Paulson* and elsewhere.

LITERATURE
Deutsch, p. 220; *Paulson*, 1, pp. 183–91.

Georgio II.do Mag: Brit: Franc: & Hiber: Regi

Nuptias Ceremoniales inter Annam Mag: Brit: Principissam Regalem et Gulielmum Principem Arausionensem habitas in Capella Regia S.ti Jacobi apud Londinum Martis 14.o An: 1733.
Devotissimus et obligatissimus servus Gul: Kent.
Humillime offert, dicat, dedicatque

V *Italian Opera and English Oratorio* 1729–41

In 1728, when the Royal Academy of Music's performances came to a close, Handel was 43 years old, a mature and highly experienced composer, dedicated to the art of opera and, since the previous year, a naturalised British subject. Despite the problems which the Academy had met, its eight seasons of new operas and full houses must have seemed blissfully successful years in retrospect. For during the mid-1730s Handel faced a financially ruinous four-year battle with a rival company, the Opera of the Nobility, and, though he continued with opera for a time, in the end he gave it up for good with the production of *Deidamia* in 1741. The decline of Italian opera coincided with a willingness on Handel's part to experiment with oratorio in English, a form which in due course led to *Messiah* and several other great successes.

In a number of other ways it was not a quiet period. It saw the death of Handel's 79-year-old mother in 1730 (see no. 20 for her memorial volume) and, also at Halle, the marriage of the composer's niece Johanna Friederika the following year (and not in 1736 as is usually said; see no. 22 for Handel's letter sending a ring as a present). An occasion of a rather different kind was the offer of a doctorate from Oxford University in the summer of 1733, but although Handel travelled to Oxford and saw his music performed there he did not proceed to take his degree, for reasons which are not entirely clear. The period also witnessed Handel's first serious illness, in May 1737: 'a Paraletick Disorder, he having at present no Use of his Right Hand'; from this complaint he is said to have recovered by taking the vapour baths at Aix-la-Chapelle. It was at this time that Jonathan Tyers, proprietor at Vauxhall Gardens, chose to have the young sculptor Louis François Roubiliac carve a life-size statue of Handel (see under no. 6): a commission without precedent, and a sign of the esteem in which Handel was widely held.

Musically Handel's output was very varied. There was of course his work for the court, often written at short notice in response to events such as the coronation in 1727 or the funeral of Queen Caroline in 1737 (see p. 140). Though opera remained Handel's favoured form throughout the period (see p. 145), he felt obliged to answer the challenge of his rivals by putting on performances of oratorios from 1732 onwards, at first relying on earlier works, composed for the Duke of Chandos, but in the course of the decade writing eight further oratorios or dramatic odes in English. By the later 1730s these had found an audience which was often enthusiastic. Another feature of these entertainments, intended to fill out his evenings and to add to their attraction in the face of competition elsewhere, was the introduction of organ concertos played by Handel himself as an interval feature or a prelude to an act. He felt encouraged to publish two of his infrequent instrumental collections, *Six Concertos for the Harpsichord or Organ* (1738) and *Twelve Grand Concertos* (1740). In the season of 1740/1, his last before taking *Messiah* to Dublin, Handel produced two new operas, revived two of his oratorios and repeated two of his dramatic odes in his attempt to satisfy his audience.

LITERATURE

Deutsch, p. 434.

HANDEL

The Court of George II

Though there was no love lost between George II and his son, Frederick Prince of Wales, they shared an enthusiasm for the opera. Both father and son subscribed large sums to Handel's operas, and later to his oratorios, though Frederick temporarily deserted the composer when the Opera of the Nobility was formed in 1733 (see p. 145).

It was to Handel that George II turned for music for state occasions: the coronation in 1727, the weddings of his eldest daughter in 1734 and his son in 1736, the funeral of his wife Queen Caroline in 1737, the celebration of his own safe return after victory at Dettingen in 1743 and the celebration of the Peace of Aix-la-Chapelle in 1749. From the King and his family Handel received pensions in one form or another of at least £600 a year: £200 as a continuation of the royal pension first granted by Queen Anne in 1713, £200 for composing for the Chapel Royal and £200 for teaching the royal princesses. This was quite separate from any sums Handel received for support of his operas or oratorios, and formed a security for Handel in times of financial hardship when his public ventures were losing money.

109

George II (1683–1760)

By George Vertue (1684–1756) after
Sir Godfrey Kneller (1646–1723)

George II first met Handel in Hanover in 1710 when he was 27 and the composer 25. He was to become Handel's lifelong supporter and patron. As Prince of Wales he was a frequent visitor to the opera (Handel's *Floridante* was dedicated to him in 1721). As King, he seems personally to have made the decision that Handel should compose the music for his coronation in 1727, as was announced in the newspapers on 14 September: 'the famous Mr Handel is appointed to compose the Antheim, which is to be sung at that Grand Ceremony'. Handel's four anthems included *Zadok the Priest* (see no. 110), performed at every coronation since; it was described by George II's grandson, George III, as 'probably the most perfect if possible of all his superb compositions'.

Throughout Handel's years of struggle George II went out of his way to lend his support. In October 1734, when the Opera of the Nobility was threatening to destroy the viability of Handel's operas, a note appears in the Treasury minute books that 'the King intends that the 1,000*l* for the undertakers of the Opera shall be paid to Mr Hendell and not to the Academy of Music, as the last 1,000*l* was'. Burney describes George II as 'a steady patron of Handel' during the difficulties of his seasons of the mid-1740s; the King 'constantly attended his Oratorios, when they were abandoned by the rest of his court'. For his part, Handel continued to provide music for state occasions such as the service held in November 1743 to celebrate the King's safe return after the Battle of Dettingen (see no. 190).

National Portrait Gallery, London

Engraving, 45.3 × 34.3 cms/17⅞ × 13½ ins

Inscribed *His most Excellent Majesty GEORGE II, KING OF GREAT BRITAIN, FRANCE and IRELAND, &c,&c,&c.* / *G. Kneller Eq : Baronet pinx: 1717/Geo Vertue Londini Sculp: 1727.*, and on a ribbon at the top *DIEU ET MON DROIT*

LITERATURE
Burney, *Sketch*, p. 29; *Deutsch*, pp. 370–1; Donald Burrows, 'Handel and the 1727 Coronation', *Musical Times*, CXVIII, 1977, pp. 469–73.

Coronation Anthems

Handel's autograph score

On 25 February 1723 Handel was appointed 'Composer of Musick for His Majesty's Chappel Royal'. The appointment seems to have been made to allow him to compose occasional anthems for royal services, particularly those connected with the return of King George I from his visits to Hanover. It carried no direct remuneration and indeed could not, as aliens were forbidden to hold offices of profit under the crown. (This ban continued even after Handel's naturalisation; see Appendix 2.) Handel thus maintained a link with the Chapel Royal and English church music.

It happened that after the death of George I on 11 June 1727 and the succession of his son as George II, a change in the Chapel Royal establishment also took place. William Croft, organist and composer to the Chapel Royal and Master of the Children, died at Bath on 14 August, and on 4 September Maurice Greene became the new organist and composer, while Handel's old friend Bernard Gates took over as Master of the Children. Greene might reasonably have expected to compose the music for the coronation of George II and Queen Caroline, but – presumably through the personal intervention of the new King – the task fell to Handel. He duly produced a splendid set of four grand ceremonial anthems (*Let thy Hand be Strengthened*, *Zadok the Priest*, *The King shall Rejoice* and *My Heart is Inditing*), one of which – *Zadok the Priest* – has been sung at every subsequent coronation.

The actual performance of the anthems at Westminster Abbey during the ceremony on 11 October 1727 seems not to have gone smoothly (see under no. 111), and it was probably only when Handel incorporated them in his oratorios *Esther* (1732) and *Deborah* (1733) that he secured a satisfactory hearing for the music.

On the displayed opening, ff. 19r–20v, of the autograph of *The King shall Rejoice*, the indications for voices show the numbers of singers that Handel had in mind: C[anto].*12*; H[ughes]. *et 6*; Fr[eeman]. *et 6*; Church *et 6*; W[heely] *et 6*; Gates *et 6* (the names are spelt out more fully on an earlier page) – ie 12 trebles, 7 first altos, 7 second altos, 7 tenors, 7 first basses and 7 second basses, representing 'approximately what might have been expected from a full combined muster of the Chapel Royal and Westminster Abbey

110

choirs, with appropriate deputies' (Burrows). The unusual size of the paper, with its 18 staves, suggests the expansive nature of the music. At the bottom of the page Handel has written a new set of words ('The great King of Kings . . .') and indicated appropriate adjustments to the vocal lines. This addition was made in 1733, when Handel adapted the chorus for use in *Deborah*.

The British Library Board
RM 20.h.5

Autograph score, 50 ff., 38.1 × 24.2 cms/15 × 9½ ins

Inscribed on f. 1r, *Anthem. 1 Kings 1.48* [*Zadok the Priest*], on f. 18r, ψ [Psalm] *21.v.1. et v.3* [*The King shall Rejoice*], and on f. 36r, ψ [Psalm] *45.v.1, 10, 12* [*My heart is Inditing*], and *Is.*[Isaiah] *49.v.23*

PROVENANCE
As no. 34.

LITERATURE
Squire, p. 12; Donald Burrows, 'Handel and the 1727 Coronation', *Musical Times*, CXVIII, 1977, pp. 469–73.

111

'The Form and Order of the Service... to be Observed in the Coronation of Their Majesties'

Order of Service
Printed by John Baskett, 'Printer to the King's most Excellent Majesty', 1727

The Order of Service for the coronation of King George II and Queen Caroline, held at Westminster Abbey on 11 October 1727, was drawn up by the Archbishop of Canterbury, William Wake, from the work of his predecessors in office. The form devised by Archbishop Thomas Tenison for the crowning of George I in 1714 was the model for the new King's coronation, while the form for the Queen's coronation was taken from the office drawn up in 1685 by Archbishop William Sancroft for the coronation of James II and Queen Mary. Wake's final version was approved by the Privy Council on 20 September, and an order was made for printing one hundred copies, a surprisingly small number.

Wake's own copy, shown here, bears annotations in his hand recording, as a note on the title-page indicates, 'W[ha]t was done or omitted at that Coronation'. The notes amount to a catalogue of disasters. The first anthem, Purcell's unaccompanied setting of *I was Glad*, 'was omitted... by the Negligence of the Choir of Westm[inste]r'. At the Recognition Handel's *The King shall Rejoice* was 'in confusion: All irregular in the Musick' – perhaps because the Chapel Royal Order of Service had *Let thy Hand be Strengthened* at this point, and some musicians may have begun that instead. At the Anointing the hymn 'Come Holy Ghost our Souls Inspire', intended to be chanted, was 'by the mistake of the Music not sung', and the next anthem, Handel's *Zadok the Priest*, was sung in its place. As Winton Dean has observed, 'it was evidently a typical Georgian occasion'.

Lambeth Palace Library
Cod. Misc. 1079B

Printed order of service with manuscript notes in Archbishop Wake's hand, 89 pp., 24.4 × 19.5 cms/$9\frac{5}{8}$ × $7\frac{3}{4}$ ins

Titled in full *The Form and Order of the Service that is to be Performed, and of the Ceremonies that are to be Observed, in the Coronation of Their Majesties, King George II. and Queen Caroline, in the Abby Church of S. Peter, Westminster, on Wednesday the 11th of October, 1727.*, and inscribed in manuscript on the title-page *With notes of w^t was done or omitted at that Coronation*

PROVENANCE
William Wake, Archbishop of Canterbury; Dr Head, Archdeacon of Canterbury, by whom presented 1766, to Thomas Secker, Archbishop of Canterbury; given by him to Lambeth Palace Library, 1767.

LITERATURE
Winton Dean, notes to recording of Coronation Anthems on Decca Argo ZRG 5369 (1963); Donald Burrows, 'Handel and the 1727 Coronation', *Musical Times*, CXVIII, 1977, pp. 469–73.

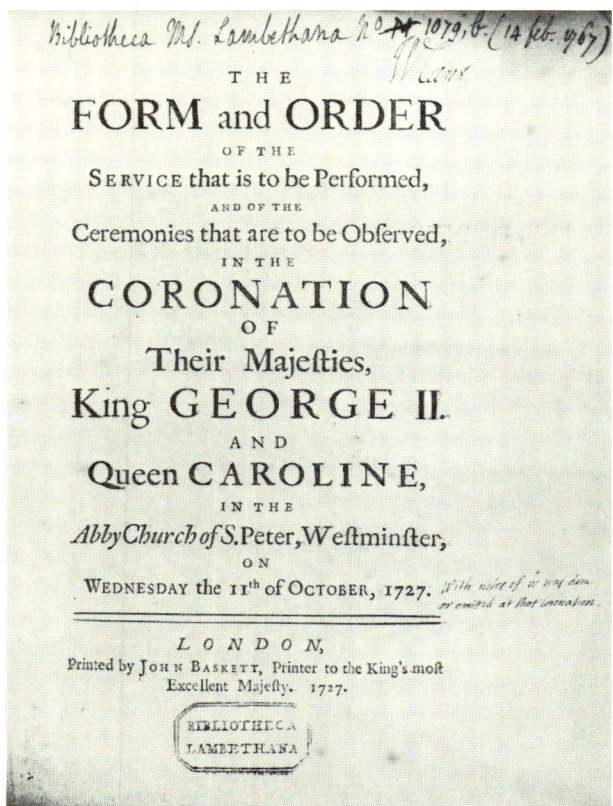

111

112 [in colour opposite p. 152]

Frederick Prince of Wales and his Sisters

By Philip Mercier (1689–1760)

'The Music Party', to use this picture's modern title, is the best of the three versions of the subject, though smallest in format. It shows Frederick Prince of Wales (1707–51) playing the cello, with three of the royal Princesses, his three eldest sisters. In the background is the Dutch House at Kew where Anne, the Princess Royal, was living before her marriage to Prince William of Orange in 1734 (see no. 113). The painting is dated 1733 when Anne would have been 24; she is usually identified with the figure seated at the harpsichord with, behind her, her younger sister, Princess Caroline, aged 20, playing the mandora and on the right, Princess Amelia aged 22 reading from Milton.

The artist, Philip Mercier (who also painted Handel's portrait; see no. 2), had been appointed principal painter to the Prince of Wales in 1729, shortly after the Prince's arrival in London from Hanover; he was also employed to teach the Princesses to paint and draw. Though this conversation piece is the most successful of Mercier's royal portraits, the presence of the Prince in the company of his sisters presents some difficulties, since by 1733 the Prince is said by Lord Hervey to have been on poor terms with his sisters, especially Anne. He had not spoken to her 'this twelve-month', according to Lord Egmont writing in December 1733, who also describes the Prince's enthusiasm for his bass viol and Anne's playing on the harpsichord.

From as early as 1724 Handel had acted as music master to the royal Princesses; in that year Princess Anne and Princess Caroline, then aged 15 and 11 respectively, made a visit to St Paul's Cathedral where they 'heard the famous Mr. Hendel, (their Musick Master) perform upon the Organ'. The Princesses, most especially the eldest, Anne the Princess Royal, remained lifelong supporters of Handel, attending his operas and subscribing to his music. For Anne's marriage in 1734 to the Prince of Orange Handel wrote a special anthem and a serenade, *Parnasso in Festa*.

In 1736, to mark the occasion of the Prince of Wales's wedding, Handel produced his opera *Atalanta*. Three years previously Frederick had set himself up as patron of Handel's rivals, the Opera of the Nobility, though his action seems to have been as much a demonstration of his antipathy to his father and his sister Anne, as a move against Handel. Lord Hervey in his memoirs bears this out: 'the Prince, in the beginning of his enmity to his sister, set himself at the head of the other opera to irritate her'. He goes on that the King and Queen 'were both Handelists, and sat freezing constantly at his empty Haymarket Opera, whilst the Prince with all the chief of the nobility went as constantly to that of Lincoln's Inn Fields'.

However this may be, recent research by Carole Taylor in the Prince of Wales's household accounts shows that Frederick supported Handel in 8 of the 11 seasons he mounted in London between 1733/4 and 1744/5. Of the £1120.10s.- received by Handel from the Prince, a total of £750 was subscribed for opera in the three seasons 1733/4, 1736/7 and 1737/8, while smaller sums were paid for attendance at his oratorios starting in 1738/9. The Prince provided £1,000 to the Opera of the Nobility over the four seasons from 1733/4, and a similar sum to Lord Middlesex's opera company but spread over just two seasons, 1742/3 and 1743/4. It would seem that the Prince unsuccessfully tried to persuade Handel to co-operate with Lord Middlesex; a letter of John Christopher Smith the elder written in July 1743 refers to several approaches from the Prince of Wales desiring him to compose for the opera directors. The absence of any payment from the Prince to Handel for the 1743/4 season is perhaps an indication of some displeasure on the Prince's part. It may be to this that Mrs Delany referred when on 11 February she wrote to her sister that 'Mr. Handel and the Prince had quarrelled, which I am sorry for. Handel says that the *Prince* is quite out of *his* good graces!.'

National Portrait Gallery, London
NPG 1556

Oil, 45.6 × 57.8 cms/17¾ × 22¾ ins

Signed and dated bottom right *Ph. Mercier fecit / 1733*

PROVENANCE
Bought from Frank T. Sabin Ltd by the National Portrait Gallery, 1909.

VERSIONS
i) National Trust (Cliveden House), size 78.0 × 56.5 cms; set in the grounds of Kew with the Dutch House behind.
ii) Royal Collection (Windsor Castle), size 79.0 × 58.0 cms; set in the Banqueting House at Hampton Court.

LITERATURE
Ralph Edwards, 'Mercier's Music Party', *Burlington Magazine*, XC, 1948, pp. 308–12; *Deutsch*, pp. 173, 380, 582; Betty Matthews, 'Unpublished Letters concerning Handel', *Music and Letters*, XL, 1959, p. 263; *Kerslake*, pp. 338–40, pl. 950; Ingamells and Raines, p. 22, no. 39; Carole Taylor, 'Handel and Frederick, Prince of Wales', *Musical Times*, CXXV, 1984, pp. 89–92.

EXHIBITED
Philip Mercier, York City Art Gallery and Iveagh Bequest Kenwood, 1969, no. 24.

113 [*illustrated on p. 138*]

The Marriage of the Princess Royal to the Prince of Orange

By Jacques Rigaud (*c.* 1681–1754?)
after William Kent (1685?–1748)

For the marriage of Princess Anne, Handel's pupil and patron, to Prince William of Orange on 14 March

1734, Handel wrote his Wedding Anthem, *This is the Day* (mostly arranged from *Athalia*, first performed in Oxford the previous year). His serenata for the occasion, *Parnasso in Festa*, was given at the Haymarket Theatre in the presence of the Royal Family on the previous evening.

The marriage ceremony took place at the 'French Chapel Royal', adjacent to St James's Palace, known to us now as the Queen's Chapel, built by Inigo Jones from 1623–5. Kent's engraving of the ceremony is of especial interest for showing an occasion on which a work by Handel was performed; no other authentic view is known from the composer's lifetime. Handel's Anthem was sung soon after the joining of hands by the Bishop of London, shown here. The gallery above the altar, built for the occasion, is crowded with musicians and singers; if Handel is to be identified, it is with the keyboard player near the left end of the second row. Some of the choir boys can be made out ranged along the front, and on the right two violinists and a flautist are just to be distinguished from the many other figures.

George II spared no expense for the marriage of his eldest daughter; the fitting up of the Chapel by William Kent showed 'an extreme good taste', according to the master of ceremonies Lord Hervey; it had 'as much finery as velvets, gold and silver tissue, galloons, fringes, tassels, gilt lustres, and sconces could give'. Kent made a pen drawing of the Royal couple at the altar with their attendants which has all the appearance of a quick sketch made on the spot during the ceremony; it is now in the British Museum. The publication of Rigaud's engraving was announced in the newspapers on 24 October, more than six months after the wedding.

National Portrait Gallery, London

Engraving, 34.3 × 25.6 cms/13½ × 10 ins

Inscribed *Georgio II^{do} Mag: Brit: Franc: & Hiber.Regi | Nuptias Ceremoniales inter Annam Mag: Brit: Principissam Regalem et Gulielmum Principem Aransionensem habitas in Capella Regia S^{ti} Jacobi apud Londinum Martis 14°: An^{o} 1733 | Devotissimus et obligatissimus servus Gul: Kent. Humillime offert, dicat, dedicatque | Guliemus Kent Decoravit et Delineavit | J. Rigaud Sculp.*

LITERATURE
Lord John Hervey, *Memoirs of the Reign of George II*, ed. J.W. Croker, I, 1884, p. 316;
Vertue, III, p. 68; Deutsch, pp. 360–1;
Paulson, I, p. 318; H.M. Colvin (ed.), *The History of the The King's Works, 1660–1782*, V, 1976, p. 243.

114

Queen Caroline (1683–1737)

By P. van Gunst
after Sir Godfrey Kneller (1646–1723)

When Handel was appointed as Kapellmeister to the Elector of Hanover in June 1710 it was the Electoral Princess, Caroline, and her husband, the future George II, who were reported to be particularly charmed with the composer and entertained by his music; for her, Handel wrote a number of duets while he was at Hanover. In London Caroline frequently attended his operas and appointed him as music master to her daughters. At her death on 20 November 1737 the King commissioned a funeral anthem which Handel was obliged to compose in great haste; his tribute to a faithful patron, *The Ways of Zion do Mourn*, was ready in time for rehearsal on 14 December. After its first performance the Duke of Chandos described the composition as 'exceeding fine, and adapted very properly to the melancholly occasion of it'.

National Portrait Gallery, London

Engraving, 35.4 × 27 cms/14 × 10⅝ ins

Inscribed *CAROLINE QUEEN, OF GREAT BRITAIN FRANCE, AND IRELAND &c | Phil: a Gunst Sculpsit. | M. Marrebeek excudit*

LITERATURE
Deutsch, p. 443; Burrows, p. 39.

114 (detail)

The Opera Revived

Soon after the collapse of the Royal Academy of Music, moves were made to continue the production of opera in London. This new project was headed by Lord Bingley, a former deputy governor of the Academy; Handel was to have £1,000 'for the composition, whether it will be by himself or by whomsoever else he may choose', according to Paolo Rolli (who was writing to Senesino), while the manager of the King's Theatre, J.J. Heidegger, was to receive £2,200 for the theatre, scenery and costumes. On 18 January 1729 this arrangement was blessed by the Academy's subscribers, as Lord Percival recorded in his diary: 'I . . . went to a meeting of ye members of ye Royal Accademy of Musick; where we agreed . . . to permitt Hydeger and Hendle to carry on operas without disturbance for 5 years and to lend them for that time our scenes, machines, clothes, instruments, furniture, etc.'

Handel's later operas may conveniently be divided into two groups: seven composed during his five-year term with Heidegger at the King's Theatre in the Haymarket from 1729–34; and a later group, produced at his own risk, at first at Covent Garden (some six new operas between January 1735 and May 1737), then at the Haymarket again for two seasons, and finally at Lincoln's Inn Fields in 1740/1. For four seasons, from 1733–7, he was in competition with a rival company, the Opera of the Nobility, the origins of which are discussed below.

Twice Handel had to re-build his company of singers almost from scratch: in 1729, and again in 1733, when the Opera of the Nobility stole almost all his performers. Each time Handel looked to Italy for replacements. He considered Farinelli (see no. 120) in 1729, but instead Senesino came back again in 1730 as his leading *castrato*. In 1733 he turned to Giovanni Carestini for two seasons, and subsequently to Gioacchino Conti in 1736/7 (see nos. 116–17 and 124). Handel's prima donna from 1729–37 was the soprano Anna Maria Strada del Pò (see no. 115); she was the only one of his singers not to go over to the Nobility Opera in 1733. On her return to Italy she was succeeded by another fine soprano, La Francesina (see no. 125), French by birth but Italian by training, who appeared from 1737–41 and 1744–6.

The last opera staged by Handel to star Senesino was *Orlando*, first performed on 27 January 1733; a Scottish visitor, Sir John Clerk of Penicuik, gives a graphic account of the tenth and last night on Saturday 5 May:

> I never in all my life heard a better piece of musick nor better perform'd – the famous Castrato, Senesino made the principal Actor the rest were all Italians who sung with very good grace and action, however, the Audience was very thin so that I believe they get not enough to pay the Instruments in the orchestra. I was surprised to see the number of Instrumental Masters for there were 2 Harpsichords, 2 large basse violins each about 7 foot in length at least with strings proportionable that cou'd not be less than a $\frac{1}{4}$ of an inch diameter, 4 violoncellos, 4 bassoons, 2 Hautbois, 1 Theorbo lute & above 24 violins. These made a terrible noise & often drown'd the voices. One Signior Montagnania sung the bass with a voice like a Canon. I never remember to have heard any thing like him. amongst the violins were 2 Brothers of the name Castrucci who play'd with great dexterity.

The bass Antonio Montagnana sang for Handel from 1731–3 and in 1737/8; Pietro Castrucci was leader of the orchestra and had originally been brought to England with his brother Prospero by Lord Burlington in 1715. The description of the orchestra is not dissimilar to one of the few other accounts, that given by a French visitor, Pierre Fougeroux, in 1728.

It was in the summer of 1733 that the general sense of dissatisfaction with Handel's management of the opera, due perhaps to his attitude to his singers but more especially to his attempt to double prices earlier in the year, showed itself in a plan to set up the rival company which came to be called the Opera of the Nobility. The whole scheme was explained in a plea for support sent by Lord Delawarr to the Duke of Richmond:

> There is a spirit got up against the Dominion of Mr Handel, a Subscription carry'd on, and Directors chosen, who have contracted with Senisino, and have sent for Cuzzoni, and Farinelli, it is hoped he will come as soon as the Carneval of Venice is over, if not sooner. The

General Court gave power to contract with any Singer Except Strada, so that it is Thought Handel must fling up, which the Poor Count [Heidegger] will not be sorry for, There being no one but what declares as much for him, as against the Other, so that we have a Chance of seeing Operas once more on a good foot. Porpora is also sent for. We doubt not but we shall have your Graces Name in our Subscription list.

This letter, it should be noted, does not date to January 1733 as published but to 16 June, the day following a meeting of subscribers to the new venture and a week after the end of Handel's opera season.

It is not quite clear where the initiative for the new scheme arose, but it had the support of Frederick Prince of Wales (see no. 112) in the form of a £250 subscription. An important part was played by Senesino, who had apparently been dismissed by Handel at the end of May; later in the year the news appeared in another form in one of the Paris journals, *Le Pour et Contre*: 'there was an irreconcilable rupture between Senesino and Handel, and . . . the former produced a schism in the company and hired a separate theatre for himself and his partisans'. Elsewhere it was said that Handel's rivals had resolved never to compromise with Handel: 'Point d'accommodement à jamais avec le Sr Handel'. Though the Opera of the Nobility collapsed in 1737, the absurdity of the competition between Handel and his rivals had already been expressed by Mrs Pendarves, Handel's great supporter, in a letter of 16 May 1735 to Dean Swift: 'Our Operas have given much cause of dissension; men and women have been deeply engaged; and no debate in the House of Commons has been urged with more warmth; the dispute of the merits of the composers and singers is carried to so great a height, that it is much feared, by all true lovers of music, that operas will be quite overturned. I own I think we make a very silly figure about it.'

LITERATURE
Unpublished diary of Sir John Clerk of Penicuik, Scottish Record Office (GD 18/2110/1, pp. 16–17), quoted with the permission of Sir John Clerk of Penicuik, Bart; *Deutsch*, pp. 234 (Lord Percival's diary is in the British Library, Add. MS 17720), pp. 235, 303–4 (Lord Delawarr's letter to the Duke of Richmond is Goodwood MS 103, f. 173 in the Goodwood Archives, at the West Sussex Record Office), and pp. 315–16, 335, 341, 390.

115

115

Anna Maria Strada del Pò (fl. 1720–40)

From the album of Count Francesco Algarotti, probably made in the studio of Marco Ricci (1676–1730) or perhaps by Antonio Maria Zanetti (1679/80–1767)

Anna Maria Strada del Pò (seen on the right) was Handel's leading soprano in all his operas and oratorios from 1729–37. She seems to have begun her career in Venice in 1720. She sang in Naples from 1724–6 and while there married the theatre manager Aurelio del Pò, supposedly because he could find no other way of satisfying his debts to her.

Handel presumably heard her sing on his visit to Italy in the spring of 1729 to look for singers, and had by May engaged her for London. Her first stage appearance was as Adelaide in *Lotario* on 2 December 1729. A few days earlier we find Mrs Pendarves writing to her sister about Handel's new singers: 'La Strada is the first woman; her voice is without exception fine, her manner perfection, but her person is *very bad*, and she makes *frightful mouths*.' She was nicknamed 'the Pig', according to Burney, who attributed her success to Handel: 'she . . . sung herself into favour, particularly with the friends of Handel, who used to say that by the care he took in composing for her, and his instructions, from a coarse singer with a fine voice, he rendered her equal at least to the finest performer in London'.

Unlike the other members of Handel's company she alone did not go over to the rival Opera of the Nobility in the summer of 1733 but continued to perform for Handel until 1737 when, according to Burney, her husband sued the composer for the arrears in her salary. She left England in June 1738, going to Breda at the invitation of Princess Anne, Handel's former pupil, and from there to Italy. She is not known to have sung after 1740 and appears to have retired to her native Bergamo.

Like other drawings in this series (see no. 49) this study will have been made in Venice, probably at the beginning of her career; the original sketch at Windsor has been attributed to Marco Ricci. It is mounted with another drawing, inscribed *L'Ambreville*, either of Anna d'Ambreville-Peroni who sang in Italy, Vienna and Prague from 1711–28, or of her younger sister, Rosa D'Ambreville-Borosini, who sang in Venice, Turin and Vienna from 1715–25.

Albert R. Gellman Collection,
Courtesy of the Israel Museum, Jerusalem

Pen and brown ink over pencil traces,
24.7 × 17.5 cms/9¾ × 6⅞ ins

Inscribed by Algarotti bottom centre *La Stradina*, and top left *39*

PROVENANCE
As no. 49.

LITERATURE
As no. 49; *Deutsch*, p. 247.

116

Giovanni Carestini (*c.* 1705–60)

By John Faber (d. 1756) after George Knapton (1698–1778)

The *castrato*, Giovanni Carestini was brought to London by Handel for the 1733/4 and 1734/5 seasons to replace Senesino, who had led the breakaway to the rival Opera of the Nobility. Handel had heard Carestini in Italy on his visit in 1729 and tried to engage him then. The following year he was willing to offer the singer a fee of 1,200 guineas – the same as for Senesino – but it was not until October 1733 that he was able to entice him to London. Once there Carestini was well received, though somewhat overshadowed by Senesino and then Farinelli who sang for Handel's rivals. In two seasons he created parts in several Handel operas including *Ariodante* and *Alcina* (the occasion for his rejection of the aria 'Verdi prati'; see no. 119) and also sang in some of Handel's oratorios. Carestini left England in July 1735 but returned in 1739 to sing for Lord Middlesex's opera company. In subsequent seasons he sang in Italy, Dresden, Berlin and, as late as 1754, St Petersburg.

This mezzotint was published towards the end of Carestini's second season in London. Burney described him as 'tall, beautiful and majestic . . . a very animated and intelligent actor', and characterised his voice as 'a powerful and clear soprano, which afterwards changed into the fullest, finest and deepest countertenor that has perhaps ever been heard'.

The Trustees of the British Museum
1902-10-11-1266

Mezzotint engraving, 29.0 × 25.4 cms/11⅜ × 10 ins

Inscribed *Joannes Carestini | George Knapton Pinx! | J. Faber Fecit 1735 | Sold by J. Faber at the Golden Head in Bloomsbury Square.*

LITERATURE
CS, p. 321, no. 55; *Deutsch*, p. 258.

116

117

Giovanni Carestini (c. 1705–60)

Two drawings from the album of Count Francesco Algarotti, probably made in the studio of
Marco Ricci (1676–1730)
or perhaps by Antonio Maria Zanetti (1679/80–1767)

Both of these drawings show Carestini; a comparison of the full-face study with Knapton's portrait (no. 116) shows how extreme is the caricature, with only the fullness of face and the heavy, set mouth readily recognisable. Like other drawings in this series (see no. 49), these studies will have been made in Venice where Carestini sang on various occasions between 1724 and 1749, that is both before and after his London visit of 1733–5. Similar drawings at Windsor are attributed to Marco Ricci.

Albert R. Gellman Collection, Courtesy of the Israel Museum, Jerusalem

Pen and brown ink over pencil traces,
24.9 × 17.6 cms/9¾ × 7 ins

Inscribed by Algarotti bottom centre *Carestini*, and top left *14* and *15*

PROVENANCE
As no. 49.

LITERATURE
As no. 49.

118

Marie Sallé (1707–56)

By Gilles Petit (1694–1760) after Paul Fenouil

Marie Sallé was one of the leading French dancers of the early eighteenth century. Her uninhibited performances at Covent Garden in her ballet *Pigmalion*, given from January to March 1734, excited much comment: 'She ventured to appear without skirt, without a dress, in her natural hair, and with no ornament on her head. She wore nothing in addition to her bodice and under petticoat but a simple robe of muslin arranged in drapery after the model of a Greek statue.'

It was to her, the 'divine Sallé', that Handel turned to find a counter-attraction to Farinelli (see no. 120) who opened the season for his rivals, the Opera of the Nobility, at the end of October 1734. This was just a few days before Handel revived *Il Pastor Fido* as the first opera of Sallé's season with him; he added a prologue specifically for her: 'a new Dramatic Entertainment (in Musick) call'd, *Terpsichore*'. Later in the season she appeared in *Ariodante* and *Alcina* (see no. 119). This last was particularly successful, receiving 18 performances, but unfortunately towards the end of its run she found herself so hissed at by an organised claque that she left London never to return:

> Mademoiselle Sallé, who had at first been as favourably received by the English as Farinelli (however, in due proportion to her talents), found herself afterwards bitterly attacked both in verse and prose. . . . The opera *Alcine* was given, the story of which is taken from *Ariosto*. Mlle Sallé had composed a ballet, in which she cast herself for the role of Cupid and took upon herself to dance in male attire. This, it is said, suits her very ill and was apparently the cause of her disgrace.

Such was the sad end to her career on the London stage, which had begun almost twenty years before, when at the age of nine she had appeared with her brother in the 1716/17 season at Lincoln's Inn Fields; they had also provided 'entertainments of dancing' at a revival of Handel's *Rinaldo* on 5 June 1717 at the Haymarket Theatre.

The Trustees of the British Museum
1873-7-12-274

118

Engraving, 25 × 20.2 cms/9⅞ × 8 ins

Inscribed *L APRÈS-DINÉ*/*La Dame à la Promenade*/Melle *Sallé peint par Fenouil*/*Gravé par Petit*

LITERATURE
Deutsch, pp. 75, 387, 390; *Royal Opera House Retrospective 1732–1982*, catalogue of an exhibition at the Royal Academy, 1982, pp. 123, 126.

119

Alcina

Handel's autograph score

In 1734 Handel relinquished his old base at the King's Theatre to his rivals, the Opera of the Nobility. Fortunately he was able to contract with John Rich to use the latter's newly built theatre at Covent Garden for two nights each week, and, stimulated by the competition, he produced one of his most sumptuous seasons of opera and oratorio. The operas all featured ballet sequences specially written for the French dancer Marie Sallé and her company, whom Rich – by coincidence or design – had also engaged for the same season.

Alcina, the final opera of the season, opened on 16 April 1735 and achieved a remarkable run of 18 performances. It is the last and perhaps the greatest of Handel's 'magic' operas. The libretto, loosely derived from episodes in Ariosto's *Orlando Furioso*, is based on that of Riccardo Broschi's *L'isola d'Alcina*, produced in Rome in 1728 (the author is not known). It tells how the young knight Ruggiero, seduced from his true heroic calling by the sorceress Alcina, is rescued by his fiancée Bradamante. In the process Alcina and her magic island are destroyed; she falls truly in love with Ruggiero and becomes unable to use her evil powers against him.

'Verdi prati' (see ff. 57v–58r), one of the most famous numbers in the opera, is sung by Ruggiero in Act 2 as he bids farewell to the 'verdant meadows' and the beauties of Alcina's island; he has learnt that they are mere illusions concealing a barren desert, but the hauntingly nostalgic melody of the aria tells us that he will never forget them. Burney states that the original singer, the *castrato* Giovanni Carestini (see no. 116), at first refused to sing the aria – he may have initially been unaware of its dramatic context – but was persuaded after the composer unleashed the full force of his wrath. The autograph shows Handel's careful marking *pianiss*[imo] for the orchestra at the entry of the voice. *Ritornello ut supra*, at the end of the page, is an instruction to repeat the opening orchestral introduction at the close of the aria.

The only date in the autograph is that of completion, 8 April 1735, which, being only eight days before the first performance, must represent the point at which Handel completed his final revision of the score. The clefs and words of the recitatives were prepared for Handel by his copyist John Christopher Smith the elder – an unusual procedure suggesting haste in the composition, though there is no hint of perfunctoriness in the music.

The British Library Board
RM 20.a.4

119

Autograph score, 90 ff., 22.8 × 29.2 cms/9 × 11½ ins

Inscribed at the end of the score on f. 84 *Fine dell' Opera G F Handel April 8 1735*

PROVENANCE
As no. 34.

LITERATURE
Burney, *Sketch*, p. [*24]; *Squire*, p. 4;
R. Strohm, 'Händel und seine italienischen Opern', *Händel-Jahrbuch 1975/76*, 1977, pp. 101–59.

120

Carlo Broschi, called Farinelli (1705–82)

By Bartolomeo Nazari (1699–1758)

'One God, one Farinelli' was the cry when the *soprano castrato* Carlo Broschi, called Farinelli, first appeared on the London stage in 1734. Perhaps the greatest eighteenth-century singer and certainly the most widely known, Farinelli never sang for Handel, though some years before he came to London for the Opera of the Nobility, he had been considered by Handel. Paolo Rolli wrote to Senesino in February 1729 that Handel was 'for sure, going directly to Venice for Farinello' and later reported that he 'returned from his travels very full of Farinello and extremely loud in his praises'. When Farinelli finally came to London in October 1734, it was to sing for Handel's rivals. Rolli then reported to Giuseppe Riva that 'Farinello was a revelation to me, for I realise that till then I had heard only a small part of what human song can achieve, whereas I now conceive I have heard all there is to hear.'

This portrait by Bartolomeo Nazari (who also painted Faustina; see no. 96) was painted in Italy shortly before Farinelli's departure for London. It was almost certainly commissioned by the 3rd Earl of Essex, the British ambassador in Venice who helped negotiate terms for the singer's visit.

Farinelli's reception in London was ecstatic; at first rehearsal the band is said to have been unable to follow him, 'gaping with wonder, as if thunderstruck'; at his first public performance, the opera *Artaserse* by Hasse given on 29 October 1734, Senesino was so taken with his singing that, 'forgetting his stage character, [he] ran to Farinelli and embraced him'. His benefit performance in March 1735 brought him a huge sum, reported to be £2,000. The gifts that were showered on him were the subject of Hogarth's caricature in one of his prints of *A Rake's Progress* (see no. 122).

Farinelli sang in London for three seasons. Though his voice was without compare, his acting was thought to be awkward. One later account, Pickering's *Reflections upon Theatrical Expressions in Tragedy* (1755), describes his set poses, one of which is closely echoed in this portrait: 'his left hand settled upon his hip, in a beautiful Bend like that of the Handle of an old fashion'd Caudle Cup'.

Royal College of Music

Oil, 143.2 × 119.4 cms/56⅜ × 47 ins

Signed and dated on the harpsichord *Bartholomeus Nazarij Berg:/Pinxit. Venetus An. 1734*

PROVENANCE
Probably the portrait commissioned by William Cappel, 3rd Earl of Essex, in 1734; Arthur F. Hill, by whom given to the Royal College of Music, 1933.

LITERATURE
F.J.B. Watson, 'A Portrait of Farinelli by Bartolommeo Nazari', *Burlington Magazine*, XCII, 1950, pp. 266–9; *Deutsch*, pp. 229, 238, 374; Highfill, V, p. 149; Fernando Noris, 'Bartolomeo Nazari', *I Pittori Bergamaschi Dal XII al XIX secolo: Il Settecento*, Bergamo, 1982, pp. 222 (repr. in colour) and 232, no. 26.

120

ITALIAN OPERA AND ENGLISH ORATORIO 1729–41

121

121

'Thou tunefull Scarecrow'

By Joseph Goupy (*c.*1680–*c.*1770)
and Dorothy, Countess of Burlington (1699–1758)
after Marco Ricci (1676–1730)

This caricature shows Francesca Cuzzoni and Farinelli (see nos. 95 and 120) singing a duet with the opera impresario Heidegger (see no. 57) seated in the background. In June 1730 Paolo Rolli wrote to his friend Giuseppe Riva that Marco Ricci 'a few days before his death sent to this Goupy a caricature of Cuzzona and Farinello singing a duet: Goupy added the figure of Eidegar seated with his face turned up, and it was printed in honour and glory of that great band of rogues, the singers'. The caricature is often found with Lady Burlington's name added and, though probably etched by Goupy, the grouping of the figures is hers according to Henry Angelo's much later account. That of Heidegger is taken from a drawing by Goupy of 'Heidegger in a rage' (in the British Museum). The figure of Farinelli corresponds closely to a drawing by Marco Ricci at Windsor.

The Trustees of the British Museum
1868–8–8–3572

Etching, 23.0 × 20.2 cms/9 × 8 ins

Inscribed with eight lines of verse beginning *Thou tunefull Scarecrow, & thou warbling Bird*, in pen *css Burlington des / Goupy sc*, and with the names of the singers, *Cuzzoni / Farinelli or Senesino / Heidegger*

LITERATURE
Henry Angelo, *Reminiscences*, I, 1828, pp. 406–7; *Satires*, no. 2022; Anthony Blunt and Edward Croft-Murray, *Venetian Drawings . . . in the Royal Collection*, 1957, pp. 163–6, fig. 27; Deutsch, p. 255.

122

A Rake's Progress

By Louis Gérard Scotin (1690–after 1755)
after William Hogarth (1697–1764)

This second scene from Hogarth's *Rake's Progress*, the 'Morning Levée' (the original oil of 1733–4 is in the Soane Museum), is a satire on the many ways for a rich young man to squander his money. One of the most expensive of such indulgences was the Italian opera, hence the long scroll shown hanging over the back of the chair, 'A List of the rich Presents Signor Farinelli the Italian Singer Condescended to Accept of yᵉ English Nobility & Gentry'. At the end of the scroll an engraving includes Farinelli himself on a pedestal, as the subject of worship, being offered flaming hearts by adoring women. The scene is inscribed with the cry *One God, one Farinelli*, which is usually attributed to Lady Bingley's rapture of delight at first hearing the celebrated *castrato* sing.

The musician seated at the harpsichord was presumably intended to evoke memories of Handel, who wore just such a long wig (compare his portrait by Denner, no. 1). The music however is from an imaginary opera, *The Rape of the Sabines*, the cast of which is given: Senesino (see no. 93) and Carestini (see no. 116) are ironically (for *castrati*) described as 'Ravishers', and among the other names are those of Cuzzoni (see no. 95) and Strada.

The Trustees of the British Museum
S 2–37

HANDEL

Engraving, 31.5 × 38.7 cms/12⅜ × 15¼ ins

Inscribed *Invented &c by Wm Hogarth & Publish'd According to Act of Parliament June ye 25. 1735. | Plate. 2^d*, and with sixteen lines of verse beginning *Prosperity, (with Horlot's smiles, Most pleasing, when she most beguiles.)*

LITERATURE
Ronald Paulson, *Hogarth's Graphic Works*, 1965, I, no. 133, pl. 245.

123
'The Ladies Lamentation for ye Loss of Senesino'

Songsheet, headpiece engraved by George Bickham junior (1706?–71)

The headpiece to this songsheet from Bickham's celebrated *Musical Entertainer* is a parody on the extraordinary emotions aroused by the great Italian singers; here Senesino (see no. 93) is about to embark for Italy but is begged to stay by weeping ladies and bowing gentlemen; two servants carry a hand-barrow laden with 'Ready Money', his vast fees for singing.

The verses of this popular song seek the cause of the ladies' unhappiness: 'I gently requested the Cause of her moan, / She told me her sweet Senisino was flown. . . .'/'Why who is this Mortal so Cruel said I' . . . 'Tis neither for Man, nor for Woman said she, / That thus in Lamenting I water the lee, / My Warbler

123

Frederick Prince of Wales and his Sisters by Philip Mercier, 1733 (no. 112)

Burkat Shudi and his Family by C.M. Tuscher, c. 1742 (no. 162)

The Sharp Family by Johann Zoffany, 1779 (no. 228)

Celestial sweet Darling of fame, / Is a Shadow of something, a Sex without Name' – an allusion to the fact that Senesino was a *castrato*.

The engraved headpiece is said to be a copy, reversed, from an earlier caricature of the arrival of Senesino in 1720, but it is likely that this dates not to his first visit but to his second in 1730, since the women to either side are identified as 'The Rival Queens, Signora Faustina and Signora Cuzzoni' (see under no. 107).

The Trustees of the British Museum
1868–8–8–3526

Etching and engraving, 32.6 × 20.0 cms/12$\frac{7}{8}$ × 7$\frac{7}{8}$ ins plate size
From George Bickham's *The Musical Entertainer*, I, c.1737, no. 38
Titled as above and inscribed with music, verses and, beneath the headpiece, *G. Bickham jun sc*.

LITERATURE
Satires, no. 2147.

124

Gioacchino Conti, called Gizziello (1714–61)

By Alexander van Aken (c.1699–1749) after Charles Lucy (1692–c.1736)

This is one of three portraits of *castrato* singers engraved by Alexander van Aken (or Haecken) in 1735–6, the others being of Senesino and Farinelli. Brought over by Handel in April 1736 as an attraction to counteract Farinelli's success with the rival Opera of the Nobility, Conti (or Gizziello as he was known, from the name of his master, Domenico Gizzi) had sung in Rome, Naples and Venice before coming to London. During his short stay in 1736 and 1737 he appeared in leading parts in several new operas including *Ariodante*, *Atalanta* and *Berenice*. His singing was greatly liked by the poet Thomas Gray – except for the shape of his mouth which, 'when open, made an exact square'. He was considered 'near equal in merit to the late Carestini' by another contemporary observer, Benjamin Victor, while many years later Burney described his style as 'remarkable for pathos, delicacy, and refinement', and also commented on how he was 'so modest and diffident, that when he first heard Farinelli, at a private rehearsal, he burst into tears, and fainted away with despondency'.

National Portrait Gallery, London

Mezzotint engraving, 35.6 × 25.5 cms/14 × 10$\frac{1}{8}$ ins
Inscribed around the feigned oval *GIOACCHINO CONTI GIZZIELLO*, and below *CH: LUCY PINXIT ALEX. VANHAECKEN FE.1736*

LITERATURE
Deutsch, pp. 409–10.

125

Elizabetta Du Parc, called La Francesina (d. 1778?)

By John Faber (d. 1756) after George Knapton (1698–1778)

The French-born but Italian-trained soprano, usually known as Francesina, came to London in the autumn of 1736 for the Opera of the Nobility but in the following season appeared in Handel's *Faramondo* and *Serse*, so beginning a long connection which lasted until 1746 but for a break centred on the period when Handel was in Dublin. Though she sang in his last two operas, *Imeneo* and *Deidamia* (see no. 127), given in the 1740/1 season, it is as an oratorio singer, usually in English, that she became well known, appearing in almost all Handel's works of this period. In February 1744 she took the title role in *Semele* following an absence of more than two years from the London stage. After the first night Mrs Delany wrote to her sister: 'Francesina is extremely improved, her notes are more distinct, and there is something in her running-divisions that is quite surprising. She was much applauded.'

Francesina's first arrival in London in 1736 was soon followed by the publication of this mezzotint, her only known portrait. She holds a score open at Cleopatra's aria, 'Tu la mia stella sei amabile speranza', from Act 1 of Handel's *Giulio Cesare* (see no. 101). Perhaps she sang this aria at a benefit or concert performance, since it was not an opera she ever performed. The aria is one which would have been well suited to her style: 'that natural warble and agility of voice, which Handel . . . seems to have had great pleasure in displaying', in the words of Burney.

National Portrait Gallery, London

Mezzotint engraving, 28.9 × 22.2 cms/11$\frac{3}{8}$ × 8$\frac{3}{4}$ ins
Inscribed *Sign.^{r} Lisabetta Du Parc detta La Francesina. / George Knapton Pinx^t/ J. Faber fecit 1737. / Sold by J. Faber at the Golden Head in Bloomsbury Square*.

LITERATURE
Deutsch, p. 582.

125

126

Serse

Handel's autograph score

Serse (*Xerxes*) was completed on 14 February 1738 and produced at the King's Theatre on 15 April. The libretto is derived from that provided by Niccolo' Minato for Cavalli in 1655 via a revised version made by Silvio Stampiglia for Giovanni Bononcini in 1694. (Handel knew Bononcini's score and to some extent modelled his own setting upon it, as well as using several of its themes in other works.) Even as revised, the Venetian origins of the libretto are apparent in the broadly satirical treatment of the leading character and in the appearance of the comic servant Elviro. These features can however be found in other operas (*Almira*, *Agrippina*, *Flavio*, *Deidamia*), and so the not uncommon description of *Serse* as Handel's only comic opera is highly misleading. The story centres on the love of King Xerxes of Persia for the Princess Romilda, who is herself in love with the King's brother Arsamene. Eventually the King's intentions are foiled by the misunderstanding of Romilda's father Ariodate, and he returns to his former fiancée Amastre while Romilda and Arsamene are happily betrothed. The mischievous behaviour of Romilda's sister Atalanta comes to naught.

The opening scene of the opera contains one of Handel's best-known melodies. After a short recitative accompanied by the orchestra, Xerxes sings the aria 'Ombra mai fu', a loving tribute to the shade of a plane tree. In countless instrumental and orchestral arrangements this aria has become known as 'Handel's Largo', though as can be seen from the displayed opening (f. 5r) Handel's actual marking is the faster tempo of *Larghetto*. The composer's meticulous stage directions are again in evidence: the scene shows 'a summer house at the side of a most beautiful garden, at the centre of which is a plane tree; Xerxes under the plane tree.' Dynamic contrasts in the aria (*Pian*[o], *Fort*[e]) are also clearly marked.

The British Library Board
RM 20.c.7

Autograph score, 91ff., 24.2 × 27.9 cms/9½ × 11 ins

Inscribed on f.1 *angefangen den 26 December 1737 od Montag, den 2 X tag*; on f.34r *Fine dell' Atto primo Jan. 9 1738*; on f.59v *Fine dell' Atto 2do Jan. 25. 1738*; and at the end of the score, on f.91r, *Fine dell' Opera G.F. Handel. London Februar. 6. 1738. geendiget auszufüllen den 14 dieses Febr 1738*.

PROVENANCE
As no. 34.

LITERATURE
Squire, p. 79; H.S. Powers, 'Il Serse trasformato', *Musical Quarterly*, XLVII, 1961, p. 481 and XLVIII, 1962, p. 73.

127

'Deidamia an Opera Compos'd by Mr. Handel.'

Printed score, first edition
Printed for and sold by John Walsh, 1741

Deidamia, Handel's last opera, has a libretto by Paolo Rolli which, as far as is known, is an original work rather than an adaptation. It is based on an early episode in the mythical life of Achilles. The boy has been sent to the island of Scyros to avoid a prediction that he will be killed if he should ever take up arms against Troy. Disguised as a girl, he has been brought up with the daughters of King Lycomedes, one of whom, Deidamia, discovers his secret and becomes his lover. The Greeks, however, have been told that they cannot defeat Troy without Achilles, and the wily Odysseus eventually tricks the young hero into revealing his true nature by offering him as a present the choice of a sword and armour or a chest of fine silk dresses; Achilles eagerly opts for the former.

Handel wrote the music between 29 October and 20 November 1740 and directed the first performance at Lincoln's Inn Fields on 10 January 1741. Like *Imeneo*, Handel's other operatic production of this season, *Deidamia* failed miserably, achieving only two further performances. Apart from the revival of *Imeneo* as a 'serenata' in Dublin in 1742, Handel never returned to Italian opera again.

Despite the failure of the work on the stage, Walsh's score of *Deidamia* included all the arias and ensembles of the opera as well as the overture and the march which follows it. The frame of the engraved title-page, surmounted by an eagle and featuring three winged cupids with conventional representations of a music book, a cello, a violin and a flute, was derived from a plate first used for Pietro Castrucci's *Sonate . . . Opera Seconda* (1734) and modified for Weideman's *XII Sonatas or Solos for a German Flute* (1737).

Gerald Coke Handel Collection

Printed score, 22 pp., 32.8 × 24.6 cms/12⅞ × 9¾ ins

Titled as above and inscribed on the label opposite the title-page *Arie in Deidamia & Olympia. Pr: 12 Shill: / R. Howman 1740*; another label bears a coat of arms and the motto, *Labile quod opportunum*, of the family of Howman of Bexwell, Norfolk

PROVENANCE
R. Howman, 1740.

LITERATURE
Smith, p. 22, no. 1; *Deutsch*, pp. 509–11.

EXHIBITED
Coke, p. 25, pl. 7b.

127

The Rise of Oratorio

Handel's commitment over many years had been to opera. It was only the performance of his Cannons oratorio, *Esther*, at the Crown and Anchor tavern in the Strand (see nos. 128–30) which put in hand the events which led to the rise in popularity of oratorio over opera. This private occasion was organised by Bernard Gates, Master of the Children of the Chapel Royal, in February 1732, but was followed on 20 April by an unauthorised public performance: 'Esther an Oratorio or, Sacred Drama . . . the Musick by Mr. Handel'. Handel felt obliged to respond by putting on his own revised version of the oratorio, which had been written some 14 years before for the Duke of Chandos; this he announced for performance on 2 May at the King's Theatre. Though performed at the opera house, the presentation was kept deliberately simple to suit a work on a religious theme, apparently at the insistence of the Bishop of London: 'N.B. There will be no Action on the Stage, but the House will be fitted up in a decent Manner, for the Audience.' Francis Colman described the oratorio as having 'all ye Opera Singers in a sort Gallery, no acting'.

Such was the success of *Esther* that within a few weeks Handel faced a pirated version of his other Cannons composition, the masque *Acis and Galatea*; again he responded by revising the work extensively, adding music from a work of his Italian period, *Aci, Galatea e Polifemo*, and announced the first performance for 10 June. He soon proceeded to write two new oratorios, *Deborah* and *Athalia*, both of which were first performed the following year, and now began to include oratorios towards the end of his opera seasons, usually in the spring or during Lent. Even his rivals in the Opera of the Nobility began to copy him, as Mrs Pendarves (later Mrs Delany), a committed Handelian, reported on 28 March 1734: 'I went to the oratorio at Lincoln's Inn, composed by Porpora . . . some of the choruses and recitatives are extremely fine and touching, but they say it is not equal to Mr. Handel's oratorio of Esther or Deborah.'

Handel had written two oratorios many years before, *La Resurrezione* and *Il Trionfo del Tempo e del Disinganno*, both performed in Rome (the word 'oratorio' derives from the performances given by the Roman society, the Congregazione dell'Oratorio, an order of priests founded by St Philip Neri in the sixteenth century). His English oratorios were however perceived as something quite novel, and in the course of the 1730s he developed the oratorio as his own individual contribution to dramatic art in London. His subjects were taken from the Old Testament such as *Deborah*, *Saul* (see no. 137) and *Israel in Egypt*, or, like *Athalia*, treated a Jewish theme. Handel also experimented with a number of other works performed in English, including two of Dryden's St Cecilia's Day odes – *Alexander's Feast* (see nos. 133–5) and the *Ode for St Cecilia's Day* (1739) – and a work inspired by Milton, *L'Allegro, il Penseroso ed il Moderato*, in 1740. Despite their success, Handel's future lay almost exclusively with oratorio, as his composing activity in the ten years succeeding *Messiah* was to show.

In his oratorios Handel's personal role was much

128

more conspicuous than with opera; in 1735 Mrs Pendarves described his concertos on the organ in *Esther* as 'the finest things I ever heard in my life', and later a French visitor, the poet Anne-Marie Fiquet du Bocage, remarked on a visit to London in 1750:

> The Oratorio, or pious concert, pleases us highly. *English* words are sung by *Italian* performers, and accompanied by a variety of instruments. Handel is the soul of it: when he makes his appearance, two wax lights are carried before him, which are laid upon his organ. Amidst a loud clapping of hands he seats himself, and the whole band of music strikes up exactly at the same moment. At the interludes he plays concertos of his own composition, either alone or accompanied by the orchestra. These are equally admirable for the harmony and the execution.

Why did oratorio succeed where opera failed? Of his early biographers, Mainwaring described the oratorio as 'more suited to the native gravity and solidity of the English', while Burney stated that, 'his Oratorios . . . being in our own language, have chiefly endeared him to the nation'. Hawkins saw Handel's knowledge of the Scriptures as being such that 'the sublime sentiments with which they abound would give opportunities for displaying his greatest talents'; he also pointed out how an oratorio had the practical advantage of being less expensive to mount than opera, having no costly scenery or costumes. Whatever the case, the heyday of Italian opera had passed.

LITERATURE
Mainwaring, p.126; Burney, *Sketch*, p. 21; *Hawkins*, II, p. 889; *Deutsch*, pp. 288–9, 292–3, 361, 384, 686.

128

Esther

Manuscript score

Esther, the first English oratorio, was the most substantial work composed by Handel for the Earl of Carnarvon, later Duke of Chandos (see no. 68). Unfortunately the circumstances of its composition and first performance are obscure. The first and last pages of the autograph are missing, so any dates which Handel may have noted are lost. When the work was performed in London in 1732 by the choir of the Chapel Royal, the word-book printed for the performance stated that it had been 'composed for the Most Noble James Duke of Chandos . . . In the Year 1720', but the earliest extant copy (see no. 161) has the title 'The Oratorium composed by George Frederick Handel Esquire in London 1718'. The earlier date seems the more likely, since it brings the work into the period when Handel is definitely known to have been working for the Earl of Carnarvon; there is no evidence that this connection continued beyond 1718.

The libretto of *Esther* is derived from Jean Racine's play of 1689, via an English version by Thomas Brereton published in 1715; the story ultimately comes from the Biblical book. It is not certain who prepared the text actually set by Handel; both Alexander Pope and his friend John Arbuthnot are mentioned as authors in contemporary documents, and it may be that both were involved. Little attempt was made to give dramatic coherence to the libretto, but Handel had enough scope to paint a convincing portrait of the villainous Haman, instigator of a Jewish persecution prevented by the brave intervention of Esther. Much of the music is powerful and original, though a certain amount of it is taken from Handel's setting of B.H. Brockes's Passion Oratorio, completed a couple of years earlier. Horns are used for the first time in an English choral work.

Little attention seems to have been paid to *Esther* at the time of its first performance, but in 1732 Bernard Gates arranged a series of three performances by the boys and men of the Chapel Royal choir, the first of which was held at the Crown and Anchor tavern in the Strand (see no. 130) on 23 February 1732, Handel's forty-seventh birthday. The exhibited score is clearly connected with these performances. The title-page is copied from Gates's word-book, and details of the performances are recorded inside. Except for the addition of the instrumental introduction of *Zadok the Priest* at the start of the final chorus, the version of the score is that of 1718. (See also nos. 129 and 161.)

Gerald Coke Handel Collection

Manuscript score, 197pp., 23.5 × 30.0 cms/$9\frac{1}{4}$ × $11\frac{7}{8}$ ins

Titled in full ESTHER. an ORATORIO; or Sacred Drama. the MUSICK as it was Composed for the MOST NOBLE JAMES DUKE OF CHANDOS. by George Frederick Handel, in the year 1720. And Perform'd by the CHILDREN OF His MAJESTY's Chapel, on Wednesday 23: February 1731. LONDON 1732

LITERATURE
W.C. Smith, 'Handeliana', *Music and Letters*, XXXI, 1950, pp.126–7; *Dean*, pp. 203–4.

129
Esther

Conducting score with autograph additions by Handel

Bernard Gates's performances of the original 1718 version of *Esther* in 1732 (see no. 128), though ostensibly private, were undoubtedly attended by many influential musical connoisseurs and were much enjoyed (Viscount Percival noted in his diary that 'the company were highly pleased'). It may seem odd that Handel himself was not immediately encouraged to present a performance of his own, but he was fully engaged in his usual season of Italian opera, and English oratorio was not only untried as a means of public entertainment but also poorly suited to a company largely of Italian singers. Others, seeking profit as well as pleasure, were not prepared to see *Esther* lie dormant. In April 1732 an anonymous group of musicians announced a performance in the Great Room of York Buildings, Villiers Street, emphasising that the work was 'Never Perform'd in Publick before'. This act of piracy stimulated Handel into action. His response was to produce a substantially revised version of the score to suit his company. It was announced in these terms:

> At the King's Theatre in the Hay-Market, on Tuesday the 2nd Day of May, will be performed, *The Sacred Story* of ESTHER: an *Oratorio* in *English*. Formerly composed by Mr. Handel, and now revised by him, with several Additions, and to be performed by a great Number of the best Voices and Instruments.
> NB There will be no Action on the Stage, but the House will be fitted up in a decent Manner, for the Audience. The Musick to be disposed after the Manner of the Coronation Service.

The 'best Voices' included the Chapel Royal singers for the choruses, and the 'Additions' included two of the Coronation Anthems (*Zadok the Priest* and *My Heart is Inditing*), as well as many new pieces. Dramatically the work was considerably improved by the provision of a new opening scene; Samuel Humphreys supplied the words for this and the other newly composed items. Six performances were given, launching the great series of English oratorios which were to displace opera as Handel's prime means of musical expression. Handel revived the new version several times up to 1757, often with further revisions.

To create the conducting score of this new version of *Esther*, Handel had John Christopher Smith interpolate the revisions into an earlier copy of the original version; several amendments and additions were also made in his own hand. As the displayed left-hand page shows, Smith began the final chorus ('The Lord our enemy has slain') by copying the original opening but leaving blank staves for second and third trumpets and timpani; Handel filled in these extra parts himself. The continuation, with elaborate 'Allelujah' solos for Senesino, was copied from an autograph now partly surviving in the Fitzwilliam Museum (MS 251, pp. 37–8). Later (probably in 1740) Handel made major changes to the form of the chorus which he had to continue by writing an entirely new autograph; the start of this is the right-hand page displayed. These further amendments to the conducting score necessitated the removal of some of the pages added in 1732; six of these survive in the Fitzwilliam Museum (MS 251, pp. 31–6) together with fragments of additions made for later revivals (MS 251, pp. 39–42).

Staats- und Universitätsbibliothek, Hamburg
M. C/261

Manuscript score, 126 ff., 44.4 × 28.9 cms/17½ × 11⅜ ins

PROVENANCE
Bequeathed by Handel to J.C. Smith the elder, 1759, and by him to his son J.C. Smith the younger, 1763; by him in turn to his step-daughter, Lady Rivers; perhaps sold locally at Winchester by the Rivers family; purchased by Thomas Kerslake by 1857, from who purchased by Victor Schoelcher; sold by him to Friedrich Chrysander for the Hamburg Library.

LITERATURE
Deutsch, pp. 288–9; Werner Rackwitz and Helmut Steffens, *George Frideric Handel: A Biography in Pictures*, pls. 116a, b; Dean, pp. 222–4; H.D. Clausen, *Händels Direktionspartituren ('Handexemplare')*, Hamburg, 1972, pp. 139–43.

130
St Clement Danes with the Crown and Anchor Tavern

By John Maurer (fl. 1713–61)

In this view east along the Strand, the Crown and Anchor tavern is the four-storey building on the right at the corner with Arundel Street (marked by its crown and anchor sign). Here Handel's *Esther* (see no. 128) was performed in February 1732; here too the Academy of Ancient Music met, from its foundation in 1726 (when it was known as the Academy of Vocal Music) until 1784, in what Hawkins

ITALIAN OPERA AND ENGLISH ORATORIO 1729-41

130

described as 'a spacious room, in every respect proper for musical performances'. The Academy performed music by the 'old masters of every country' as well as oratorios by Handel' This engraving of 1753 appears to be the only view of the tavern before it was pulled down and re-built in 1790.

The Church of St Clement Danes (built by Wren, 1680-2; the tower by Gibbs, 1719-20; bombed 1941 and re-built) is all that now survives of this view. The east end of the Strand with its many shops and inns was a prosperous residential district, but one which quickly became less salubrious as one neared Temple Bar, which is just to be seen in the far distance here.

Guildhall Library

Engraving, 16.1 × 26.6 cms/6⅜ × 10½ ins

Inscribed *J. Maurer del. et sculp. 1753. According to Act of Parliament* / *A Perspective View of S.t Clements-Church in the Strand* [repeated in French] / *London Printed for Rob.t Sayer at the Golden Buck facing Fetter Lane End. Fleet Street*

LITERATURE
Hawkins, II, p. 805; Hugh Phillips, *Mid-Georgian London*, 1964, p.181 (repr., but misidentified); Bryant Lillywhite, *London Signs*, 1972, p.142, no. 5581 (Lillywhite's much more detailed typescript is in the Guildhall Library).

131

'Rich's Glory or his Triumphant Entry into Covent-Garden'

By an unknown engraver
perhaps after William Hogarth (1697-1764)

In this satirical view of the proceedings at the opening of John Rich's new theatre on 7 December 1732, the procession is seen crossing Covent Garden with St Paul's church in the background. The print is a reflection of the success of John Gay's *Beggar's Opera* (see no. 108), which was said to have made Rich gay and Gay rich. Near the front is 'Johnny Gay on Porters Shoulders' just entering the colonnade leading to the theatre. Next 'we see gay Rich in gilded Chariot ride'. Behind in the crowd 'The players follow... Drest in each Character for which they're fam'd', including James Quin and Lacy Ryan. Looking on in disgust (in the right corner marked P) is Alexander Pope, 'a bright man of Taste', with a torn-up copy of *The Beggar's Opera*.

John Rich had managed the small playhouse in Lincoln's Inn Fields (see no. 140) since 1714. His new theatre was erected in 1731-2 to the design of

131

Edward Shepherd. Rich advertised for subscribers to take up fifty shares, each of £300, all of which were sold; they gave the shareholder the right of free admission as well as a two-shilling rent for each night the theatre was used, and they proved highly popular, often changing hands at a premium. Among the subscribers were the Dukes of Chandos and Richmond but many were of modest standing, including Philip Mercier the artist, who painted Handel's portrait at around this period (no. 2).

Covent Garden was not used exclusively for opera until the nineteenth century, though Handel's *Alcina* (1735) and *Berenice* (1737) among others had their first performance here. It is however for its association with his oratorios, and other works in English, including *Alexander's Feast* (1736), *Messiah* (its first London performance was in 1743), *Judas Maccabaeus* (1747) and *Jephtha* (1751), that the theatre is particularly memorable.

Guildhall Library

Engraving, 16.9 × 30.9 cms/6⅝ × 12⅛ ins

Titled at the top as above; and inscribed in the image bottom left *WHIE SCULP* (ie W. Hogarth inv. et sculp.), and below, thirty lines of verse beginning *Not with more glory through the Streets of Rome*; at bottom left *WHIE SULP*, and bottom centre *London Printed 7. Oct. 1811 by Robert Wilkinson, N°. 58 Cornhill* (this print is a re-issue from the original plate)

LITERATURE
Satires, no. 1899; Ronald Paulson, *Hogarth's Graphic Works*, 1965, 1, no. 278.

132 [*in colour opposite p. 136*]

Covent Garden Piazza and Market

By Joseph van Aken (*c.* 1699–1749)

A view of Covent Garden looking east from beside the church of St Paul's. Inigo Jones's Covent Garden piazza, laid out in the 1630s, was used as a market from the mid-seventeenth century. This painting is dated by John Hayes to 1726–30, shortly before the new Theatre Royal was built behind the houses of the piazza on part of the site of the present opera house. Its main entrance was approached along the north-east colonnade of the piazza (on the far left as seen from the church).

No view of either the exterior or the interior of the theatre – the one as modest as the other is grand – is known from the period of Handel's occupation. After he was forced out of the King's Theatre in the Haymarket by his latest rivals, the Opera of the Nobility, in 1734, he was offered house-room at the new Theatre at Covent Garden which had been open less than two years. With the exception of a few seasons, Handel used the Theatre for the last 25 years of his life, at first for his operas but then for the performance of oratorio. His spring oratorio seasons, ending shortly before Easter with a performance of *Messiah*, became an accepted feature of London life.

The Museum of London

Oil, 96 × 100.6 cms/38 × 40$\frac{1}{8}$ ins

PROVENANCE
B.T. Batsford, sold at Christie's, 20 December 1911, lot 153, purchased by Knoedler; purchased by the London Museum, 1912.

VERSION
A very similar view (mentioned by *Hayes*), size 62.2 × 75.0 cms, signed in monogram by Joseph van Aken, was in 1978 with the Leger Galleries.

LITERATURE
John Hayes, *Catalogue of Oil Paintings in the London Museum*, 1970, pp. 21–2, no. 1, pl. 1.

133

Alexander's Feast

Handel's autograph score

For the winter of 1735/6 Handel left operatic ventures to his rivals at the King's Theatre, though he gave a short opera season at Covent Garden in the spring of 1736 to celebrate the marriage of the Prince of Wales. He turned his thoughts towards an English choral work, but a little different from the form of the English oratorio which he had just brought into being with *Esther*, *Deborah* and *Athalia*. Instead he looked back to the old-established form of the ode. He had already produced a fine example of the English court ode in the *Birthday Ode for Queen Anne* (see no. 53), and he now looked to the somewhat more distinguished tradition of the odes produced between 1683 and the early 1700s to celebrate

St Cecilia's Day, 22 November. (The entirely fictitious Cecilia had been established in medieval times as the patron saint of music and inventor of the organ.) Musically the greatest Cecilian odes were those composed by Purcell, especially his setting of the 1692 ode by Nicholas Brady, *Hail! Bright Cecilia*. The best poetic texts were however those by John Dryden, *A Song for St Cecilia's Day* (set by G.B. Draghi in 1687), and *Alexander's Feast, or the Power of Music* (set in 1697 by Jeremiah Clarke, whose music is unfortunately lost).

It seems that Handel's friend, Newburgh Hamilton (later librettist of *Samson* and the *Occasional Oratorio*) drew the composer's attention to *Alexander's Feast* (see no. 134). It is easy to see why the poem appealed to Handel. Dryden gives a vigorous description of the feast held by Alexander the Great to celebrate the conquest of Persia, at which the musician Timotheus shows the power of his art by changing Alexander's mood with his playing and singing. Eventually he inspires the monarch to burn Persepolis to avenge the Greeks slain by the Persians. In the final stanza of the poem Dryden introduces Cecilia for the first time. She is credited with enlarging music's 'former narrow bounds . . . with Nature's mother-wit and arts unknown before'. But Dryden, having exalted Timotheus, cannot easily abandon him, and so it is suggested he should 'divide the crown' with Cecilia. Handel sets the final lines of the ode as a complex fugue on four subjects, illustrating the new dimensions of music which Cecilia may be supposed to have inspired.

The opening of the final stanza – 'Thus long ago, e'er heaving bellows learnt to blow' – is set as a tenor recitative eked out with short interpolations from recorders and violas; the veiled orchestral sound suggests the vision of Alexander and Timotheus fading into the mists of antiquity. Handel originally thought to use a harp, and (as the displayed opening, ff. 66v–67r, shows) prepared staves for it, but he never filled them in. The left-hand page shows the end of the chorus 'Thais led the way', with the cue *concerto per l'organo* cancelled in pencil. It seems to have been Handel's first intention to play the organ concerto op. 4 no. 1 at this point, but he finally decided to place it after the chorus 'Let old Timotheus yield the prize', concluding the ode and leading to an additional final chorus with words by Newburgh Hamilton.

British Library
RM 20. d. 4

Autograph score, 88 ff., 22.8 × 29.2 cms/9 × 11½ ins
Inscribed on f. 47v, *Fine della parte prima January ye 5. 1736*, on f. 80r, *12 Jan. 1736*, and at the end of the score on f. 88r, *Fine. 17 January 1736*.

PROVENANCE
As no. 34.

LITERATURE
Squire, p. 7.

134

'Alexander's Feast; or, The Power of Musick. An Ode Wrote in Honour of St. Cecilia, By Mr. Dryden.'

Word-book, first edition, text adapted by Newburgh Hamilton from John Dryden
Printed for J. and R. Tonson, 1736

As well as providing the full text of Dryden's Ode, the word-book issued for the first performances of *Alexander's Feast* in February 1736 carries a preface by Newburgh Hamilton, who brought the Ode to Handel's attention and advised on how it might be set to music. Describing the poem as one 'which no Man can add to, or abridge, in any thing material, without injuring it', Hamilton declared that he had confined himself 'to a plain Division of it into *Airs*, *Recitative*, or *Chorus's*; looking upon the Words in general so sacred, as scarcely to violate one in the Order of its first Place'. He also commended Handel's music in a poem of his own ('To Mr Handel: On his setting to Musick Mr. *Dryden's* "Feast of Alexander"'), and this was not the only verse of his to appear in the word-book. Given his view on additions to Dryden, it is surprising to find the final stanza of the Ode followed by an 'Additional Chorus' taken from Hamilton's own Cecilian ode of 1720, *The Power of Musick*.

It is likely however that it was Handel rather than Hamilton who demanded the addition. The setting of the Ode itself turned out to be too short to provide the entertainment of about three hours of music which Handel's audiences expected, and the added chorus was just one of the 'extras' by which Handel extended the music to an appropriate length. The other additions are indicated in the word-book: a concerto 'for the Harp, Lute, Lyricord, and other Instruments' (a version of the Harp Concerto, op. 4, no. 6) in Part 1; a concerto 'for two Violins, Violoncello, &c.' (the so-called '*Alexander's Feast* Concerto') and an Italian cantata ('Cecilia, volgi un sguardo') before Part 2; and an organ concerto (op. 4, no. 1) before the 'Additional Chorus'. (The 'Lyricord', a sort of harpsichord with bowed strings, probably never materialised; contemporary manuscripts of the music indicate that just harp and lute were the solo instruments in the first concerto.)

The exhibited copy of the word-book contains pencilled annotations by a member of the audience at one of the first performances, which provide interesting information about the performers. Entries variously given as *argono* and *argony* prove that the Italian lutenist, singer and composer Carlo Arrigoni was a prominent soloist, taking the tenor part in the Italian cantata and playing the lute in the first concerto; he was active in London from about 1732 to 1736. A more heart-felt note appears on page 17 (displayed), in connection with the well-known bass aria 'Revenge, Revenge, Timotheus cries'. This was sung by a Mr Erard, about whom nothing is known; he was not, one suspects, a singer of the first rank. The annotator, presumably ignorant of his name, writes *German* at the start of the aria, and against the lugubrious middle section adds *O for Seneseno*: obviously not all of Handel's followers were happy with the move away from Italian opera. *B* against the accompanied recitatives 'Now strike the Golden Lyre again' and 'Give the Vengeance due' indicates the tenor John Beard (see no. 187).

Winton Dean Esq.

Printed word-book, 20 pp., 23 × 18.5 cms/9 × 7¼ ins

LITERATURE
R.M. Myers, *Handel, Dryden and Milton*, 1956, pp. 31–4, 95–104; Winton Dean, 'An unrecognised Handel singer: Carlo Arrigoni', *Musical Times*, CXVIII, 1977, pp. 556–8.

135

Alexander's Feast

Manuscript part-book for violoncello

This violoncello part for *Alexander's Feast* is the only part-book actually used in Handel's London performances which survives in complete form. It was originally copied by the elder Smith from the earliest state of the autograph (no. 133), and was substantially amended both before the first performance in February 1736 and subsequently for the many revivals of the work. Cuts are indicated

135

by pencilled cancellations or, more commonly, by pasted-in sheets of blank paper, and the new versions of passages altered are also pasted in. (Experienced modern orchestral players will note that methods have changed little over 250 years.) At least one of the pasted-over passages is unique to this manuscript – the bass line of an otherwise lost solo added for Mrs Cibber in the Dublin performances of 1742.

The displayed pages (ff. 11v–12r) show the end of the recitative 'He chose a mournful muse' and the start of the soprano solo 'He sung Darius great and good', which was considerably shortened by Handel at an early stage, probably before first performance. The cancelled passage is carefully covered by pasted-in sheets. It will be noted that the singer's words and vocal line are shown as well as the instrumental bass, so that the player can follow the singer even if some liberty is taken with the rhythm.

Royal College of Music
MS 900

Manuscript part-book, iii + 32 + ii ff., 30 × 23 cms/11¾ × 9 ins

Inscribed on the cover *Alexander's Feast. N.º 1 Violoncello* and on f. 1 *Harpsichord. Sig.ʳ Pasqualini.*

LITERATURE
Donald Burrows, 'Handel and *Alexander's Feast*', *Musical Times*, CXXII, 1982, pp. 252–5.

136

'Six Concertos For the Harpsicord or Organ Compos'd by Mr: Handel'

Printed keyboard score, first edition
Printed for and sold by John Walsh, 1738

The six concertos first published by Walsh in 1738 and designated as Handel's Opus 4 were all written in 1735 or 1736 for performance with oratorios and odes. Nos. 2, 3, 4 and 5 were composed for Handel's oratorios at Covent Garden in March and April 1735: no. 2 in B flat and no. 3 in G minor were probably first performed with *Esther* on 5 March, no. 5 in F with *Deborah* on 26 March and no. 4 in F (completed on 25 March) with *Athalia* on 1 April. Mrs Delany found the concertos in *Esther* 'the finest things I ever heard in my life', an opinion that was widely shared. English oratorio, itself an innovation, thus became coupled with a new instrumental form designed to display another facet of Handel's genius – his skill as a keyboard player. The other concertos of the op. 4 set, no. 1 in G minor and no. 6 in B flat, were composed the following year and first performed with *Alexander's Feast* on 19 February 1736; no. 6 was originally written for harp and was apparently performed with an additional lute part (now lost) played by Carlo Arrigoni.

Walsh published a solo keyboard arrangement of no. 2 in *The Ladies Entertainment 5th Book* at the beginning of September 1738. On 25 September an advertisement in the *London Daily Post* declared that six concertos were being prepared for publication, 'some of which have been already printed by Mr. Walsh and the others done without the knowledge or Consent of Mr. Handel.' Two days later Walsh announced:

> Whereas there is a spurious and incorrect Edition of six Concertos of Mr. Handel's for the Harpsichord and Organ, publish'd without the Knowledge, or Consent of the Author. This is to give Notice, (That the Publick may not be imposed upon by that mangled Edition) That these are now printed from Mr. Handel's original Manuscripts, and corrected by himself, the same Six Concertos, the Copy of which I have purchased from Mr. Handel, and shall be publish'd in a few days.

Whether the 'mangled Edition' ever existed in fact (no copy has been traced) or whether it was a fantasy invented by Walsh to force Handel into publication is not clear, but Handel's own note on the title-page of Walsh's edition (reproduced under the details of inscription below) is undoubtedly genuine, and confirms that he had supplied an authoritative score to Walsh. (Handel did not supervise the preparation of the print, however; it has several major errors.) The edition was issued in the form of a keyboard score – including a reduction of the orchestral passages as well as the solo organ part – which could be played on its own or in conjunction with the separately issued orchestral parts.

Gerald Coke Handel Collection

Printed keyboard score, 48 pp., 30.1 × 22.5 cms/$11\frac{7}{8} \times 8\frac{7}{8}$ ins

Inscribed on the title-page *These Six Concertos were Publish'd by Mr. Walsh from my own Copy Corrected by my Self, and to Him only I have given my Right therein. George Frideric Handel.*

LITERATURE
Smith, p. 224, no. 1; Deutsch, pp. 466–9.

137
Saul
Manuscript score of Act 1

Charles Jennens (see no. 181) had begun collecting Handel's music by 1725, when he was one of the subscribers to Cluer's score of *Rodelinda* (the first Handel score to carry a subscription list). He soon formed a plan of collecting all the composer's music in manuscript full scores and part-books, and he seems to have ordered Handel's copyists to copy every note of the music, including items removed from major works before performance and uncompleted fragments. Jennens's collection eventually became the largest and most comprehensive of all those assembled in Handel's lifetime, and preserved many pieces which might subsequently have become lost. It also included music by other composers, particularly scores and other manuscripts sent from Italy by Jennens's friend Edward Holdsworth. The collection was inherited by the 3rd Earl of Aylesford (Jennens's great-nephew), and hence became known as the 'Aylesford Collection'.

Many of the scores copied for Jennens carry annotations in his hand, bearing witness to his lively interest in the music. He figured the bass lines of many pieces, which suggests that he played them through on the harpsichord; he sometimes entered the dates of composition found in Handel's autographs (which the copyists invariably ignored); and, more controversially, he occasionally 'improved' Handel's word-setting.

Jennens's interest in the preparation of the score and parts of *Saul*, the first work on which he collaborated with Handel, was particularly intense. The oratorio was drafted between 23 July and 15 August 1738, but was subjected to substantial revision between 18 September (when Jennens visited Handel and went over the score with him) and 27 September; further revisions were probably made before the first performance at the King's Theatre on 16 January 1739. At the time of his visit to Handel, Jennens was allowed to annotate the autograph of *Saul* with directives for his copyist, all of which were faithfully followed in Jennens's score, though Handel himself subsequently cancelled the annotations and ignored them. Volumes 1 and 2 of the score (covering Acts 1 and 2 of the work) were prepared before Handel's major revisions, and thus preserve a version of the oratorio never performed by Handel. The corresponding vocal parts seem to have been prepared at an even earlier stage, since they include bars which Handel cancelled before they had been orchestrated. As a substantial portion of the autograph of *Saul* is lost, the 'Aylesford' score and parts are of vital importance in determining the composition history of the oratorio.

The displayed page of volume 1 of the score shows part of the original setting of the opening chorus, called 'An Epinicion, or Song of Triumph' by Jennens. Handel set verses 2 and 4 of the five-verse text to the same music, and it happened that the words of verse 4 ('The youth inspir'd by thee, O Lord') fitted the music more aptly than did those of verse 2 ('An infant rais'd by thy command'). Jennens tried to deal with the problem by providing a new text for verse 2 ('When thou to quell the rebel host'), which, as can be seen, he entered in his own hand in his score. He also supplied the same text to the printer of the word-book issued for the first performance. Meanwhile, however, Handel had provided his own solution to the problem by re-setting verse 2 to entirely different music which was almost certainly used at the first and subsequent performances; to that extent, therefore, the word-book does not reflect the version of first performance as is usually the case. The figures on the bass line of the displayed page are also Jennens's.

Manchester City Council, Cultural Services Department (Henry Watson Music Library)
MS 130 Hd. 4 v. 269

Manuscript score, 328 ff., 29 × 24 cms/11½ × 9½ ins

Inscribed by the copyist *Saul an Oratorio. Compos'd by G.F. Handel Esqr.*, and at end of volume, in Jennens's hand, *Aug 1. 1738*

PROVENANCE
Charles Jennens (1700–73); bequeathed to his great-nephew the 3rd Earl of Aylesford (son of Jennens's niece Mary Fisher who married the 2nd Earl of Aylesford in 1712); by descent to the 8th Earl; sold Sotheby's, 13 May 1918, as part of lot 265; subsequently acquired by Sir Newman Flower (1879–1964), whose collection was purchased for the Henry Watson Music Library in 1965.

LITERATURE
[N. Flower], *Catalogue of a Handel Collection Formed by Newman Flower*, Sevenoaks [*c*.1920], p. 10; L. Duck, 'The Aylesford Handel Manuscripts', *The Manchester Review*, x, 1965, pp. 228–32; A.D. Walker, *George Frideric Handel: The Newman Flower Collection in the Henry Watson Music Library*, Manchester, 1972, p. 45.

138

Saul: Receipt for Kettledrums from the Tower of London

Signed by Handel

In *Saul*, Handel experimented with a wide range of orchestral effects, which required among other instruments three trombones, then still a novelty in the orchestra, a carillon and the great kettledrums from the Tower (see fig. 7, p. 27).

Although this indenture or receipt for borrowing the drums is dated 18 January 1738/9 (ie 1739 in the New Style), it would appear from a letter of 13 January from Lord Wentworth to his father the Earl of Strafford that Handel had in fact received the drums some days earlier, in time for the first performance of *Saul* on the 16th: 'I hear Mr. Handell has borrow'd of the Duke of Argylle a pair of the largest kettledrums in the Tower, so to be sure it will be most excessive noisy with a bad set off singers. . . .' These splendid instruments were said to have been taken by the Duke of Marlborough at the Battle of Malplaquet in 1709. They sounded an octave lower than the standard kettledrums and were used by Handel in the Hallelujah Chorus in Act 1 of *Saul* and in the Dead March in Act 3, and subsequently in other oratorios.

The text of the receipt is previously unpublished and is set out in full on p. 166.

Gerald Coke Handel Collection

138

Single sheet, 19.5 × 28.1 cms/7¾ × 11⅛ ins, now mounted in a volume of its own

LITERATURE
Deutsch, p. 472; H.G. Farmer, *Handel's Kettledrums and Other Papers on Military Music*, 2nd edition, 1960, pp. 85–96.

EXHIBITED
Coke, p. 32, pl. 126.

This Indenture made the *Eighteenth* day of *January 173⁸/₉* in the *Twelfth* Year of the Reign of our Sovereign Lord GEORGE 2ᵈ by the Grace of God, KING of *Great Britain*, *France* and *Ireland*, Defender of the Faith, *&c.* Between *His Grace John Duke of Argyll & Greenwich* Master General of His MAJESTY's Ordnance, and the Principal Officers of the same, on the Behalf of the King's most Excellent MAJESTY on the one Part; And *George Frederick Hendal Esq[r].* on the other Part; Witnesseth, That the said *George Frederick Hendel* hath received out of His MAJESTY's Stores, within the Office of the Ordnance, *at the Tower the Kettle Drums undermentioned the same being lent to him for the use of the Oratorios at the Kings Theatre in the HayMarkett, By Order of the Board dated the 16 Instant which he's to return in as good Order as when receiv'd.*

   Large Kettle Drums to beat before a Train } One Pair
   of Artillery with Heads Compleat

                     G.F. Handel

[on the back in pen:]
Large Kettle Drums lent to M[r]. Handel for the use use [sic] *of the Oratorioes* [sic] *at the Kings Theatre in the Hay Markett himself for Ditto 18 Janry 1738/9*

139

'Twelve Grand Concertos in Seven Parts for Four Violins, a Tenor Violin, A Violoncello with a Through Bass for the Harpsichord.'

Printed part-books, first edition
Published by 'the Author' (ie Handel)
Printed for John Walsh, 1740

The 'Twelve Grand Concertos' were composed in a single burst of creative energy in the autumn of 1739. Handel completed the first concerto, in G major, on 29 September and immediately went on to compose the others, the last (in A major, eventually published as no. 11) being finished on 30 October. Walsh published the complete set by subscription in February 1740 as the composer's Opus 6. The composition of orchestral or instrumental music with immediate publication in mind was by no means usual for Handel (all his previous 'Opera' had been compilations of past work), and there can be little doubt that in this instance he was deliberately emulating an earlier 'Opus 6' – the twelve *concerti grossi* of Arcangelo Corelli, published in 1714 and long regarded as exemplifying the elegance and fiery invention of the best music in the Italian style. Handel's concertos are generally scored, like Corelli's, for an orchestra of strings and *continuo* featuring a 'concertino' or solo group of two violins and a cello. The possibility that Handel was consciously creating a new model of excellence in orchestral writing for a new generation cannot be dismissed, but the composition of the concertos was also a highly practical exercise, since it provided a stock of music which Handel could use in addition to his organ concertos during his oratorio concerts. In the winter season of 1739/40 most of the Opus 6 concertos were given their first performances in that context.

As was common practice in the publication of orchestral music, Walsh issued the 'Grand Concertos' in separate part-books, not in full score. (Handel added oboe parts to four of the concertos, but these were not published, and probably had not been written when Walsh received his copy of the music.) The *violino primo concertino* part carries the new copyright privilege Handel had taken out in October 1739 and an impressive list of subscribers' names, headed by six members of the Royal Family. Several provincial musical societies also subscribed, confirming the widespread enthusiasm for Handel's music throughout the British Isles.

The British Library Board
g.274.a

Printed part-books, 31.5 × 23.7 cms/12⅜ × 9¼ ins

PROVENANCE
Rev. Robert Ashe (with his bookplate).

LITERATURE
Smith, p. 222, no. 1; *Deutsch*, pp. 487–9, 498–9.

140

Lincoln's Inn Fields Theatre

By George Shepherd (fl. 1800–30)

The Theatre Royal at Lincoln's Inn Fields stood in Portugal Street, on the south side of the square. The Theatre, converted from a tennis court in the Restoration period, was re-fitted by John Rich in 1716 and used by him until he moved to Covent Garden after the success of *The Beggar's Opera* (see nos. 108 and 131). It was here that Handel was obliged to come in the bitter-cold winter of 1739/40, when his *Ode for St Cecilia's Day* and *L'Allegro, il Penseroso ed il Moderato* were first performed, and in 1740/1 for

Imeneo and *Deidamia*. The Theatre later became Spode's china warehouse and was demolished in 1848.

Guildhall Library

Watercolour, 21.5 × 29.6 cms/8½ × 11⅝ ins

Signed bottom right *Shepherd. 1811.*

PROVENANCE
Bequeathed by Viscount Wakefield to the Guildhall Library, 1938.

141

A Concert at Montagu House

By Marcellus Laroon (1679–1772)

In the 1730s, after his retirement from military service, Marcellus Laroon produced a number of large finished drawings of musical parties; this example is described by Horace Walpole (writing at least thirty years after the event) as representing John, 2nd Duke of Montagu (1690–1749) and his daughter Mary, Countess of Cardigan, later Duchess of Montagu (1711–75), who are the standing figures on the extreme left in the doorway. The harpsichordist is usually identified with Heidegger (see no. 57), manager of the opera with Handel; however he is not known as a harpsichord player. Montagu was a director of the Royal Academy of Music in 1719, and was later responsible for the Royal Fireworks in 1749 (see p. 212) in his capacity as Master General of the Ordnance.

The Trustees of the British Museum
1848–7–8–207

Pen and brown ink with grey wash over pencil, 45.4 × 33.3 cms/17⅞ × 13⅛ ins

Signed bottom right *Marcellus. Laroon Fecit. 1736*, and inscribed by Horace Walpole on the mount *A Concert, by Captain Laroon, The gentleman on the left under the door is John 2nd Duke of Montagu, the lady standing by him is his 2nd daughter Mary Countess of Cardigan afterwards Duchess of Montagu. H.W.*

PROVENANCE
Horace Walpole; Strawberry Hill sale, 17 May 1842, lot 121, purchased by J.W.B. Tiffin, from whom purchased by the British Museum, 1848.

LITERATURE
Horace Walpole, *A Description of Strawberry Hill*, 1784, p. 91; *Binyon*, III, p. 36, no. 4; Robert Raines, *Marcellus Laroon*, 1967, p. 131, no. 55, pl. 24.

140

Vauxhall Gardens

The great Handel statue at Vauxhall (see no. 6), commissioned by Jonathan Tyers from the rococo sculptor Louis François Roubiliac, was an attraction and an advertisement for the Gardens. For Handel it must have seemed a magnificent gesture of recognition, coming quite unexpectedly but at a critical moment in his fortunes. Unveiled in May 1738 but perhaps begun the summer before, it idealises – one might say idolises – the composer as the god Apollo playing on his lyre, or as Orpheus performing his lays; both allusions were commonly made at the period (see nos. 143 and 146).

The Gardens at Vauxhall were not new: they had begun as the New Spring Gardens in 1661 and were conveniently reached by river. However, it was not until the grounds were reopened in 1732 by the young entrepreneur, Jonathan Tyers (see no. 142), that they acquired the celebrity which they held in London life in the eighteenth and early nineteenth centuries. Over nearly thirty years, Tyers added a series of elaborate walks, arches, fountains, temples and pavilions in the Chinese, Moorish, Gothic and Classical tastes. The whole ensemble was brilliantly lit at night – a further innovation.

Tyers made Vauxhall into a place of popular resort by providing an elaborate and changing spectacle, yet at the same time maintaining an air of quality which attracted audiences from many walks of life. The Gardens were lent distinction by the remarkable range of rococo sculptures and paintings which he commissioned, including Roubiliac's statue of Handel, the earliest item of any consequence. It was followed in the early 1740s by the paintings which decorated the supper boxes; these were largely the work of Francis Hayman, who had already painted a delightful conversation piece (no. 142) of Tyers and his family.

Tyers is not known to have had any special interest in Handel's music before his commission to Roubiliac, which seems to have come as the inspired action of a man with an eye for the coming fashion, perhaps prompted by a suggestion made by Roubiliac's then employer, Henry Cheere. As if the statue was not enough, Tyers made the magnificent gesture of subscribing for fifty tickets at Handel's benefit night at the King's Theatre on 28 March 1738.

Music was of course one of the delights of Vauxhall. The orchestra building (see no. 144) was among the first erected by Tyers; a newspaper report in May 1736 speaks of 'a Band consisting of above thirty of the ablest Performers', and in 1737 an organ was installed. Concerts seem at first to have been chiefly instrumental. It took some persuasion on Tyers's part to overcome the reluctance of the finest singers to perform in the open air, but by 1745 both John Beard (see no. 187) and Thomas Lowe were appearing at Vauxhall; in 1746 Lowe sang Handel's 'Song on the Victory obtained over the Rebels', which was published as a patriotic songsheet.

Early accounts of Vauxhall suggest that Handel's music was quite frequently performed there; the *Daily Post* (18 April 1738) speaks of 'his Harmony [which] has so often charm'd even the greatest Crouds into the profoundest Calm'. Unfortunately, few reports specify the pieces played, though it is known that Handel's setting of 'God Save the King' from the Coronation Anthem, *Zadok the Priest*, closed the season in 1738 and 1745. Handel did not normally conduct or write music for Vauxhall or any of the London pleasure gardens, but a 'Hornpipe compos'd for the Concert at Vauxhall, 1740' is known from manuscript scores in the Gerald Coke Handel Collection and the Royal Music Library. Handel's season ticket to the Gardens (see under no. 146) will have given him the opportunity to attend when he pleased.

His greatest success at Vauxhall – and perhaps Vauxhall's greatest attendance ever – was the rehearsal of the *Music for the Royal Fireworks* (see p. 212) on 21 April 1749, which he was reluctantly persuaded to give at Vauxhall. It brought twelve thousand spectators at 2s. 6d. each, clogging London Bridge for three hours at a time – the only way of reaching Vauxhall other than by river.

LITERATURE
Deutsch, pp. 465, 622–3; David Coke, *The Muse's Bower. Vauxhall Gardens 1728–1786*, catalogue of an exhibition held at Gainsborough's House, Sudbury, 1978.

142

142

Jonathan Tyers (1702–67) and his Family

By Francis Hayman (1708–76)

Two years after the unveiling of Roubiliac's statue of Handel (see no. 6) at Vauxhall Gardens, Jonathan Tyers the proprietor went to Francis Hayman, artist of the supper box paintings at Vauxhall, for this conversation piece, the first of several he commissioned from the artist. Tyers is seen taking tea with his family; the panelled room setting, almost certainly Hayman's creation, includes, significantly, a profile medallion portrait of Frederick Prince of Wales, Tyers's most prestigious patron at Vauxhall and also his ground landlord. From the left the other figures are his 14-year-old eldest son, Thomas (in gown and mortar-board), who later took over the management of the Gardens; pouring tea, his younger daughter, Elizabeth junior; his wife Mrs Elizabeth Tyers, cup in hand; and on the right Margaret, the eldest child.

This is a picture of success: of the 38-year-old Tyers seated confidently, surrounded by his family, with the portrait of his patron in the background.

National Portrait Gallery, London
NPG 5588

Oil, 77.8 × 106.2 cms/$30\frac{5}{8} \times 41\frac{3}{4}$ ins

Signed lower left in monogram *FHayman 1740 P*

PROVENANCE
Jonathan Tyers; probably by descent through his younger daughter, Elizabeth, who married John Wood *c.* 1750; by descent to Mrs Skelton; Mrs Dorothy Scobell Wood; Morton Lee (dealer), from whom purchased by Robert Tritton, 1949; Christie's, 15 July 1983, lot 65 (withdrawn); purchased from the estate of Mrs Robert Tritton by the National Portrait Gallery, 1983.

LITERATURE
John Ruch, 'A Hayman Portrait of Jonathan Tyers's Family', *Burlington Magazine*, CXII, 1970, pp. 493–7, fig. 8.

EXHIBITED
Francis Hayman, Iveagh Bequest, Kenwood, 1960, no. 5; *Rococo*, no. F2 (repr.).

143

'The Invitation to Mira Requesting Her Company to Vauxhall Garden'

Songsheet, headpiece by George Bickham junior (1706?–71) after Hubert François Gravelot (1699–1773)

This songsheet, from George Bickham's *Musical Entertainer*, was published in June 1738, the month after the unveiling of Roubiliac's statue of Handel at Vauxhall (see no. 6). The headpiece, with Handel reigning supreme in the form of Apollo over an assembly of musicians, dancers and putti, was designed by Hubert Gravelot, who was also responsible for the decorative surround of the Houbraken engraving of the composer (no. 4) published earlier the same year.

The songsheet is the earliest engraved view of the statue. Written for the flute by Thomas Gladwin, the verses by John Lockman are an invitation to Vauxhall: 'Come, Mira, Idol of ye Swains ... To Bowers where heav'n-born Flora reigns, & Handel warbles Airs divine.'

London Borough of Lambeth Archives Department

Etching and engraving, 32.8 × 19.9 cms/12⅞ × 7⅞ ins plate size
From George Bickham's *The Musical Entertainer*, II, 1738, no. 2, pl. 5
Titled as above and inscribed with music, verses and *Gravelot inv. G. Bickham jun. sc. The Words by M. Lockman. The Musick by M. Gladwin. According to ye Late Act, 6 June, 1738*

LITERATURE
Deutsch, p. 463.

EXHIBITED
Rococo, no. F11 (repr.).

144

'A Perspective View of Vaux Hall Garden'

By John Maurer (fl. 1713–61)

This shows the original setting of Roubiliac's statue, in an arched niche in the Grand South Walk, with the figures of 'Harmony and Genie' also by Roubiliac, surmounting the arch. In an anonymous 'Letter from a foreigner to his friend at Paris' of 1 July 1742, published in the *Gentleman's Magazine*, the scene at Vauxhall Gardens is set:

> Right before me extended a long and regular Vista; on my Right Hand, I stepp'd into a delightful Grove ... Farther on, still on my right, thro' a noble Triumphal Arch, with a grand Curtain, still in the picturesque Stile, artificially thrown over it, an excellent Statue of *Handel* appears, in the Action of playing upon the Lyre, which is finely set off by various Greens, which form, in Miniature, a Sort of woody Theatre.

The two-storey pavilion on the left of this view housed the orchestra and the organ, installed in 1737.

London Borough of Lambeth Archives Department

Etching, coloured, 23.5 × 40.7 cms/9¼ × 16 ins

Titled as above and inscribed *J. Maurer delin et sculp.*; the remainder of the inscription, part of which is mounted on the back of the print, reads *Printed for John Bowles, at the Black Horse in Cornhill./ London Publish'd according to Act of Parliament 1744*

LITERATURE
Gentleman's Magazine, XII, August 1742, pp. 419–20 (as from *Champion*, no. 424, 5 August 1742).

EXHIBITED
Rococo, no. F13 (repr.).

144

145

'The Triumphal Arches, Mr. Handels Statue &c in the South Walk of Vauxhall Gardens'

By Johann Sebastian Müller (c.1715–c.1789) after Samuel Wale (1721–86)

This shows the new setting of the Handel statue. At some time between the publication of the earlier view in 1744 (no. 144) and this one in 1751, the statue was placed in the open, at the focal point of a semicircle of supper boxes, with Roubiliac's figure of Harmony surmounting a large niche some way behind the statue.

Though seen from almost the same point as the earlier view, the series of triumphal arches across the South Walk have transformed the appearance of the vista. They were among the many improvements made at Vauxhall towards the end of the 1740s.

Dr Terry Friedman

Engraving and etching, 25.8 × 40.2 cms/10⅛ × 15⅞ ins

Titled in English and French as above and inscribed *S. Wale delin!/ Publish'd according to Act of Parliament 1751/ I.S. Muller sculp./ Printed for Tho Bowles in St Pauls Church Yard & John Bowles in Cornhill.*

LITERATURE
Rococo, no. F18 (repr.).

146

Season Ticket for Vauxhall Gardens

By an unknown silversmith

The season tickets for Vauxhall Gardens are among the most delightful mid-eighteenth-century tokens. They usually incorporate a classical figure and a motto within a rococo surround: here Handel, in the form of his statue at Vauxhall Gardens (see no. 6), is likened to 'caressing Orpheus'. Though the design of some of these tickets has sometimes been attributed to Hogarth there is no reason to see his hand at work here.

Handel's own silver ticket to the Gardens, last known in the collection of George J. Powell in the late nineteenth century, is unlocated; it is decorated with a figure of Arion holding a lyre astride a dolphin and bears the engraved name *Geo: Fr: Handell, Esq* on the reverse.

The Trustees of the British Museum

Silver, 4.4 cms/1¾ ins high

Inscribed BLANDIUS ORPHEO (Horace, *Odes*, Book 1, no. 24, line 13), and engraved on the reverse S^r: Joⁿ: English Barn^t/ 592

PROVENANCE
Sir John English, Bart (not recorded in *The Complete Baronetage*); Smith; Montague Guest, by whom given to the British Museum, 1907.

LITERATURE
Warwick Wroth, 'Tickets of Vauxhall Gardens', *Numismatic Chronicle*, 3rd series, XVIII, 1898, pp. 80, 83; *Catalogue of the Montague Guest Collection of Badges, Tokens and Passes*, 1930, p. 77, no. 680, pl. III.

EXHIBITED
Rococo, no. F24b (repr.).

Songs in
MESSIAH
an
ORATORIO
Set to Musick by
MR HANDEL

London. Printed for I. Walsh, in Catharine Street, in the Strand. of whom may be had. The Works of Mr Handel, Geminiani, Corelli, and all the most Eminent Authors of Musick.

VI *From Messiah to the Fireworks Music* 1741–9

The 1740s saw two of Handel's best-known works, *Messiah* and the *Music for the Royal Fireworks*. It was also the decade when he abandoned opera and was rumoured to be returning home – in fact he went to Dublin for a long season. Two years after his return to London, when the subscriptions to his oratorio season failed, he was obliged to return subscribers' monies; afterwards he never relied on subscriptions again. Despite these trials and his increasing concern with his health, he had, by the end of the 1740s, evolved a successful pattern of oratorio seasons each spring at Covent Garden Theatre and regular charity concerts in support of the Foundling Hospital.

The season of 1740/1 was not an easy one for Handel. It saw the last of his Italian operas, *Deidamia* (see no. 127), which received only three performances and seems to have persuaded him finally to relinquish opera in favour of oratorio. He also appears to have been so disenchanted with his audience that he was tempted to leave the country in 1741. An anonymous letter in the *London Daily Post*, published on 8 April, a few days before a performance of *L'Allegro ed il Penseroso*, urged support for Handel: '. . . this Oratorio of Wednesday next is his last for this Season, and if Report be true, probably his last for ever in this Country, let them . . . fill this his last House, and shew him on his Departure, that London, the greatest and richest city in the World, is great and rich in Virtue, as well as in Money . . .'.

Though Handel had abandoned opera, a group of young noblemen were determined to try their hand where he had failed. Their leader was Charles Sackville, Earl of Middlesex, later Duke of Dorset, whose taste for opera had been encouraged by the years he had spent in Italy on the Grand Tour. Handel's reaction to the scheme was reported by Thomas Dampier to his friends in Geneva on 30 July 1741: 'He laughs very much at the opera which is preparing for next winter. He has refused to have anything to do in the matter. There are eight subscribers, each one 1,000 *l*. I can remember the names of some of them: Lord Middlesex, Lord Brooke, Lord Conway, Lord Holderness, Mr Conway, Mr. Frederick &c. Lord Middlesex it seems is the chief manager in the affair. . . .' Later in the year, immediately before departing for Ireland, Handel attended the opening night of Lord Middlesex's opera season: a pasticcio, *Alessandro in Persia*, which, he wrote to Charles Jennens after his arrival in Dublin, 'made me very merry all along my journey'. The following year, on his return to London in September, Handel wrote again to Jennens denying that he was directing the opera next winter, adding that 'The gentlemen who have undertaken to middle with Harmony can not agree, and are quite in a confusion' (see no. 186). However Handel may have wavered in his opposition, as emerges in a letter from J.C. Smith the elder, written in July 1743: 'It seems that Mr. Handel promised my Lord Middlesex that if he would give him for two new operas 1000 guineas. . . . He would compose for him next Season, after which he declined his promise and said that He could – or would do nothing . . .'. In fact Handel had decided to continue with his oratorios: he mounted *Samson* in February 1743 with a mainly English cast including John Beard, Mrs Cibber and Mrs Clive (see nos. 185 and 187–8). A few days after the opening, Horace Walpole, a subscriber to Middlesex's operas, wrote to Horace Mann: 'Handel has set up an Oratorio against the Operas, and succeeds. He has hired all the goddesses from farces, and the singers of *Roast Beef* from between the acts at both theatres, with a man with one note in his voice, and a girl without ever a one. . . .' On 2 March the Royal Family attended, drawing a further comment from Walpole: 'The Oratorios thrive abundantly – for my part, they give me an idea of heaven, where everybody is to sing whether they have voices or not.'

Handel's later oratorios established his reputation in this field. But it was not an easy process; the introduction of *Messiah* to Handel's London audience on 23 March 1743 was not without controversy, demonstrating the doubts some people had as to the propriety of performing works on a sacred theme in the theatre (see p. 199). Quite a different approach was taken by another observer, Miss Catherine Talbot, after hearing a performance of *Samson*: 'I really cannot help thinking this kind of entertainment must necessarily have some effect in correcting or moderating at least the levity of the age.'

LITERAURE
Deutsch, pp. 518, 520, 531, 554, 560–1, 577;
Betty Matthews, 'Unpublished Letters concerning Handel', *Music and Letters*, XL, 1959, p. 263.

147

Handel and his Friends

By the 1740s the pattern of Handel's life, his friendships and his habits were largely set, though he was still to branch out in some new directions, for example with his visit to Dublin in 1741–2 and his support for the Foundling Hospital in the 1750s (see p. 223).

Handel had been living in Brook Street at what is now No. 25, just on the Grosvenor Square side of Bond Street, since 1723. Though the neighbourhood was a good one, his house was a modest three storeys, as is shown in a recently discovered drawing by John Buckler (no. 147) which, though dating to the 1830s, is the earliest known. Many of Handel's friends lived nearby: in or around Grosvenor Square, in the Bond Street area or in Soho (see map, p. 74).

Handel's friends belonged to several groups. There were his close supporters from the nobility and gentry, many of long standing. Mrs Delany (no. 156) had first met Handel when she was a girl, and her brother, Bernard Granville, was another close friend. The correspondence of the Earl of Shaftesbury and his cousins the Harrises (see nos. 158 and 160) provides a sympathetic view of the problems Handel faced as a composer. These families also formed collections of Handel scores (see nos. 159 and 161); not only the printed editions but also unpublished manuscript music. A second group were his professional acquaintances; these included his librettists Thomas Morell, Newburgh Hamilton and Charles Jennens (see nos. 181 and 194), all of whom were remembered in his will, and his scene designer Joseph Goupy, with whom he seems to have quarrelled after a long friendship (see no. 8). But he had other companions, drawn from quite different walks of life, about whose friendship little is known: the surgeon John Belchier whom Handel remembered in his will and who helped Burney with stories of the composer; the 'scarlet-dyer at Old Ford, near Bow', one 'Hunter', whom Hawkins describes as having copies made of all Handel's music (where is his collection now?); and the perfumer of Bond Street, James Smyth, who received £500 in Handel's will, more than any other friend excepting J.C. Smith the elder.

Concerning Handel's own life, we know more of the externals: how much his pew cost to rent at his parish church, St George's Hanover Square (see no. 148), or the way he voted in the 1749 Westminster Byelection (Whig of course – the Tories were too closely associated with the Stuart succession for Handel; see no. 149). Of his personal life – his habits and possessions – we know little. Such relics as 'Handel's watch' (no. 151) or 'Handel's bookcase', though traditionally associated with his name (indeed the watch is so inscribed) cannot with any real certainty be shown to have been his. We are on better ground with his portraits – indeed those by Denner, Mercier and Roubiliac (nos. 1 and 2; see also no. 7) were all at one time in his possession – or with his own collection of pictures (see Appendix 4): his version of Goupy's reduction of Van Dyck's *Belisarius* must have looked very much like that in the Royal Collection (fig. 14, p. 289).

In a similar way, though no printed book or score which belonged to Handel is known, we can glimpse something of his library through contemporary book subscription lists (see Appendix 3) and find other copies of these same books. Among them are works by contemporary literary figures such as John Gay and Aaron Hill (nos. 152 and 155), and finely printed volumes such as Pine's edition of Horace, or Galliard's score of his *Hymn of Adam and Eve* (nos. 153–4).

Perhaps naturally the associational items which have been subject to most discussion, from his century to this day, are the harpsichords linked with his name. Two harpsichords by the Ruckers family – at Fenton House on loan from the Royal Collection and in the Victoria and Albert Museum – have been associated with him, but as Russell has shown in his detailed appendix to his catalogue of the Victoria and Albert Museum's keyboard instruments, neither can be traced back before the second half of the nineteenth century. Indeed this is the problem with so many Handel relics: to bridge the gap in documentation of almost two generations which separates Handel's own day from the last century. There are three harpsichord makers who have been associated with Handel. To summarise the evidence, both Mercier's portrait of Handel (no. 2) showing a single-manual 'skunk tail' harpsichord and Thornhill's drawing of a harpsichord lid (no. 150) with its double curve to the case are consistent with an instrument by a member of the Hitchcock family, the leading English maker of the early eighteenth century. The Swiss immigré, Burkat Shudi (see no. 162 for the portrait of him tuning a harpsichord), received Handel's patronage according to his grandson; Shudi produced fine walnut case instruments such as that owned by Frederick Prince of Wales (fig. 8, p. 27), and it is not impossible as Jeremy Montagu discusses (p. 27) that an instrument linked with Handel's name which was in the Royal Collection in 1825 may have been his. Handel's Ruckers is first mentioned by Hawkins ('He had a favourite Ruckers harpsichord, the keys whereof, by incessant practice, were hollowed like the bowl of a spoon'). But it is not possible to prove that this is the 'large Harpsichord' which Handel bequeathed to J.C. Smith the elder. No doubt Handel used more than one harpsichord and over the space of almost five decades in London he is quite likely to have changed his instruments, as the inscription on the Thornhill drawing implies: 'Mr. Handel has or had the Harpsichord this design was made for'.

Fig. 10 Plans of Handel's House, No. 25 Brook Street, as built 1721–2; the room names are from the inventory made after Handel's death in 1759
(Drawn by John Sambrook, Survey of London)

147 [*illustrated on p. 174*]

Handel's House, No. 25 Brook Street

By John Buckler (1770–1851)

This is the earliest view of Handel's house in Brook Street, and gives a very good idea of its original appearance. Here he lived from 1723 until his death in 1759. The drawing is from a volume of London buildings, one of a large series by John Buckler, the prolific architectural and topographical draughtsman.

Brook Street was one of the many new developments of the 1720s. It was described in 1736 as 'For the most part nobly built and inhabited by People of Quality'. Handel's was one of four adjoining houses (the others being Nos. 23, 27/29 and 31) built in 1721–2 by George Baines, a slater, who had taken a building lease of the site from Huntley Bigg, a scrivener. Recent research by John Greenacombe has produced extensive new information on the history of the house. Handel must have moved in between 3 July and 8 August 1723, as two ratebooks show; he held a short-term lease (its length is not known) at a rent of £60 per annum.

The house was on three floors with a basement and attic. It was three windows wide and two rooms deep. The original floor plan has been reconstructed by Greenacombe (fig. 10), the names being taken from the inventory made some months after Handel's death. On the Brook Street side, as seen in Buckler's drawing, the attics were probably originally servants' rooms and the second floor a good-sized bedroom for Handel. The first-floor front room, the Dining Room, with the Handel-like figure peering out of the window, was the largest in the house, extending the full width of the façade, and so the most likely to have been used for private rehearsals of his operas and oratorios; a few privileged friends would sometimes attend, as Mrs Pendarves (later Mrs Delany, see no. 156) did for *Alcina* in 1735: 'Yesterday morning my sister [Ann Granville] and I went with Mrs. Donellan to Mr. Handel's house to hear the first rehearsal of the new opera Alcina. . . . Whilst Mr. Handel was playing his part, I could not help thinking him a necromancer in the midst of his own enchantments.' At the front on the ground floor the Fore Parlour, as the most readily accessible room from the street, would probably have been used for selling subscriptions to concerts and to his scores. The Back Parlour on the first floor, with its single south-facing window, is the one described by Cum-

148

mings in 1893 as traditionally Handel's 'composition room', where he would have written his music.

In Handel's day the houses were not numbered; subsequently his became No. 57 and from 1867, No. 25. Over the years it has been radically altered, so much so that externally only the second floor is original; the top floor probably replaced the garrets shown in the drawing as early as the 1830s; the ground floor was the subject of a very crude alteration in 1905–6 and the first is a restoration of 1953–4. Internally the only substantial feature to remain from Handel's day is the wooden staircase. A blue plaque on the façade commemorates Handel's residence; it was put up in 1952 and replaces an earlier tablet of 1870. At the time of writing the house is under threat of re-development and its final fate is still uncertain.

The British Library Board
Add. MS 36370 f. 37

Pencil, 26.8 × 18.5 cms/10½ × 7¼ ins

Inscribed at the bottom (over traces of an earlier inscription) *The house in Lower Brook Street at present, 1839, numbered 57, in which Handel lived and died; – as it appeared before the front of the Attic Story was raised* / JB [in monogram] *Aug! 5th 1839*.

PROVENANCE
Bequeathed by the artist's grandson, Charles Alban Buckler, to the British Museum in 1905.

LITERATURE
Deutsch, pp. 385–6; *Greenacombe*, p. 114, fig. 1.

148

'A View of St. Georges Church Hanover Square'

By an unknown engraver
First published by T. Jefferys, *c.*1751

Erected under the Act for Fifty New Churches of 1711, St George's Hanover Square was built by John James to a budget of £10,000 and completed in March 1725, not long after Handel had moved into his new house in nearby Brook Street (see no. 147).

For most of his life Handel does not appear to have played much part in the affairs of the church but, according to Hawkins writing in 1776, 'For the last two or three years of his life he was used to attend divine service in his own parish church of St George, Hanover Square . . . expressing by his looks and gesticulations the utmost fervour of devotion'. This is confirmed by the recently published record which shows that he rented a place in one of the pews in the 'Middle Isle South' from 1755.

Guildhall Library

Engraving, 24.6 × 40.2 cms/9¾ × 15⅞ ins

Inscribed *A View of S!. Georges Church Hanover Square from Conduit Street London* [repeated in French]/ *London Printed for Jn? Boydell Engraver Cheapside*

LITERATURE
Hawkins, II, p. 910; *Greenacombe*, p. 118, n. 18.

149

Poll-book for the Westminster By-election of 1749

The 1749 Westminster By-election, one of the most bitterly fought of the century, was a contest between Granville Leveson Gower, Viscount Trentham, a Whig, who was obliged to seek re-election on his appointment as a Lord of the Admiralty, and Sir George Vandeput, Bart, a wealthy Huguenot, who had been persuaded to stand against him for the Tories.

Handel's naturalisation in 1727 meant that as a male householder and rate-payer, he was entitled to vote; this he did on 1 December (the poll remained open from 24 November to 8 December), being entered as 'George Frederick Handell Brook Street Esqr'. He cast his vote for the sitting Whig, Lord Trentham, who won by a narrow margin, receiving 4,811 votes to Vandeput's 4,654. A scrutiny of the poll was demanded by the Tories but the result, declared on 15 May 1750, was to give Lord Trentham a slightly increased majority of 170.

In those days the ballot was neither secret nor universal; the poll was printed as a 260-page book, *A Copy of the Poll for a Citizen for the City and Liberty of Westminster*, giving the names of voters and their votes. Threats of eviction had been made by the Duke of Bedford against those of his tenants who did not vote for Lord Trentham, son of his brother-in-law, the Earl Gower. Handel would not have been subject to such pressures but he would in any case have felt a natural sympathy for the Whigs – as the party of government and as the supporters both of the Hanoverians and the Protestant succession.

Greater London Council, Greater London Record Office

Card-covered manuscript volume, 38.3 × 25.3 cms/15$\frac{1}{8}$ × 10 ins, 95 pp. (the last 48 blank)

Inscribed on the cover *(D) 5 St. George near Hanover-Square Pro*[duce]*d by Mr Corbett 22d April*

LITERATURE
Romney Sedgwick, *The History of Parliament. The House of Commons 1715–1754*, I, 1970, pp. 286–7; *Greenacombe*, p. 118.

149 (detail)

150

150

Design for the Decoration of Handel's Harpsichord

By Sir James Thornhill (1675/6–1734)

'Mr Handel has (or had) the Harpsichord this design was made for': these words are the earliest mention of any harpsichord owned by Handel.

Thornhill's design for the decoration of the lid of a harpsichord, long since lost, presents some fascinating problems. The shape of the lid's curve led Russell to propose as possible maker either one of the Hamburg families (Hass or Zell), or the Hitchcocks of London. Russell makes the further suggestion that the single-manual harpsichord to be seen in Philip Mercier's approximately contemporary portrait of Handel (no. 2) may be the instrument in question, in view of its elaborate use of ivory and ebony, a feature used by the Hitchcocks for their keyboards. However Thornhill's drawing is for an instrument of about 7 feet 8 inches as measured by the scale beneath the design, and is most likely a double manual, too large for the Mercier.

Thornhill has chosen as his main subject the Arts and Sciences sacrificing to Apollo; some of the subsidiary themes are taken from Ovid, his favourite source, and are explained in his typically copious annotations. No harpsichord with decoration by Thornhill is known, but such an instrument was advertised for sale in 1754 in the *Public Advertiser* (2 March):

> To be sold at Della Corbett's at the Naked Boy in Tavistock Street, Covent Garden. A very fine ton'd Double row'd Original Harpsichord of Hans Rhekers, beautifully japann'd and the Lid painted in Figures by Sir James Thornhill, which was late in the Possession of Mr Robinson, Organist.

'Mr Robinson' is presumably John Robinson, organist of Westminster Abbey. Both his wife and daughter sang for Handel in his oratorios. Despite this link, Robinson's instrument, as a Hans Ruckers, is unlikely to be one and the same as Handel's, since Ruckers is not known to have made harpsichords with a curved tail.

The Syndics of the Fitzwilliam Museum, Cambridge
PD 7-1954

Red chalk, the outline of the lid in pen and brown ink over traces of pencil and incised lines, 21.0 × 44.1 cms/8¼ × 17⅜ ins

Inscribed in pen and brown ink by Thornhill *Mr Handel has or had the Harpsichord this design was made for* (the words *or had* inserted); with long explanations of the subject matter on both sides of the drawing, starting on the recto *Peace introduces Arts & Sciences, who sacrifice to Apollo &c.*

PROVENANCE
H.S. Reitlinger; sold Sotheby's, 26 May 1954, lot 591, to the Fitzwilliam Museum.

LITERATURE
Raymond Russell, *The Harpsichord and Clavichord*, 2nd edition, 1973, p. 168, pl. 62.

151

'Handel's Watch'

By Golling of Augsburg, perhaps Andreas Golling (fl. 1712–46)

Handel's supposed personal possessions are difficult to authenticate. This watch is engraved with his name: perhaps at a later date, since inscriptions of ownership are not normally found on watches of this period. Its claim to have belonged to Handel would be strengthened if its early history could be verified; it is said to have passed with the duplicate copy of Handel's will from the descendants of Handel's executor, Sir George Amyand, to William Snoxell (d. 1879). From there its provenance is clear enough; though separated from the will, it can be traced through different ownerships to its gift to the British Museum in 1935.

In style the silver case belongs to the years c. 1730–50. A receipt for repairs, dated 1745, apparently with the watch in 1879, probably referred to the movement which, like most watches at the period, was made separately. Although the movement was ostensibly produced at Augsburg, the arcaded minutering round the edge of the dial is possibly an indication of Dutch manufacture, perhaps of about 1720.

Two other watches, by Decharmes and Ellicot, have been associated with Handel (see nos. 22 and 25).

The Trustees of the British Museum
1935, 10–24.1

Silver (the case, with a *repoussé* design of musicians playing instruments, houses the movement which is separate), 4.7 cms/1⅞ ins case diameter

Engraved on the case beneath the musicians, *G.F. Handel*; marked on the inside *MV* (beneath a coronet?); engraved on the face of the watch *Golling/Augspur*, and on the backplate of the movement *Golling Augspurg*

PROVENANCE
Said to have belonged to George Amyand, Handel's executor, and to have passed from his heirs to William Snoxell; his sale, Puttick and Simpson, 11 June 1879, lot 486 (offered for sale with a receipt for repairs dated 1745) but withdrawn and entered again, 21 July 1879, lot 82 (without the receipt), purchased by a Mr Adam; Miss Henrietta M. Mackenzie; given by F.E.A. Mackenzie to the British Museum, 1935.

LITERATURE
Musical Times, XXXIV, 14 December 1893, pp. 8 (repr.), 28; unpublished letter from A.H. King, 29 March 1951, quoted in British Museum Departmental Register; *Deutsch*, pp. 414–15; Cecil Clutton and George Daniels, *Watches*, 1965, pp. 132–3, figs. 249, 260 (for watches with similar faces).

152

'Poems on Several Occasions'

By John Gay (1685–1732)
Printed for Jacob Tonson and Bernard Lintot, 1720

John Gay's *Poems on Several Occasions*, a finely printed collected edition, is the first book to feature Handel's name among the subscribers (see Appendix 3). It includes Gay's long and affectionate poem about London, originally published in 1716, *Trivia or, the Art of Walking the Streets of London*, with its well-known verses on Burlington House and Handel:

> Yet *Burlington's fair palace still remains;*
> *Beauty within, without proportion reigns,*
> *Beneath his eye declining art revives,*
> *The wall with animated picture lives;*
> *There* Hendel *strikes the strings, the melting strain*
> *Transports the soul, and thrills through ev'ry vein;*

The Earl of Burlington (see no. 60), a patron of both Gay and Handel, led the subscribers' list with fifty copies to his name.

The British Library Board
83.k.21

Printed book, volume 1 (of two), viii + 268 pp., 30.2 × 24.5 cms/11⅞ × 9⅝ ins

153

153

'The Hymn of Adam and Eve, Out of the Fifth Book of Milton's Paradise-Lost'

By Johann Ernst Galliard (1687?–1747)
Published 1728

This setting of Milton's 'Morning Hymn' from *Paradise Lost* (1667), with its handsome engraved title-page by John Pine, is the work of Johann Ernst Galliard. A German composer and oboist, he settled in London in about 1706, producing music for the stage and in particular for John Rich's very successful pantomimes at Lincoln's Inn and then Covent Garden.

Galliard's *Hymn of Adam and Eve*, a setting for two voices and strings, met with considerable success, despite its rather old-fashioned music. Among the subscribers were 'Mr. Handel 4 books', Bononcini, Colley Cibber and Maurice Greene, each taking two books, Dr Pepusch taking twelve books, and Mrs Anastasia Robinson.

The great vogue for Milton in Handel's day did not pass him by: both *L'Allegro, il Penseroso, ed il Moderato* (1740) and *Samson* (1743) derive from Milton's work and were admired for their union of the talents of two great men.

The British Library Board
B.440.(1)

Volume of printed music, 30 pp., 17.4 × 25.5 cms/$6\frac{7}{8}$ × 10 ins; bound with John Valentine's *Thirty Psalm Tunes* and other late eighteenth-century music

Inscribed at the bottom of the engraved title-page
J. Pine inv: & Sculp:, and on the last page of the music *Engraven by Tho: Atkins*.

LITERATURE
R.M. Myers, *Handel, Dryden & Milton*, 1946, pp. 45–80;
Deutsch, p. 230.

154

The Work of Horace

Published by John Pine, 1733 (volume 2, 1737)

One of the finest illustrated editions of the classical poets, this volume of Horace's works was published by John Pine in 1733. He used a separate engraved plate for each page, so allowing the Latin text and the finely detailed illustrations to be integrated to best effect. The long list of eminent subscribers is led by Frederick Prince of Wales; Handel's name follows that of Dr Edmund Halley. Unusually for subscription lists of this period there are three pages of subscribers from abroad: Vienna, Paris, Berlin, Amsterdam, Geneva and Leipzig.

This first volume contains the Odes and Epodes of Horace, written between about 40 and 23 BC. Ode 24 of Book I, addressed to his friend Virgil, is the source of the inscription *Blandius Orpheo* beneath the image of Handel on some Vauxhall Gardens

154

season tickets (see no. 146). Handel's duet 'Beato in ver chi può', written in 1742, is an Italian version of Horace's 'Beatus ille', Epode 2.

Dr Terry Friedman

Printed book, volume 1 (of two), xxviii + 265 pp., 23.3 × 15.6 cms/9¼ × 6⅛ ins
Titled in full *Quinti Horatii Flacci Opera (The Work of Quintus Horatius Flaccus)*
LITERATURE
Deutsch, p. 555.

155

'The Works of the late Aaron Hill, Esq'

Printed for the benefit of his family, 1753

When Aaron Hill (see no. 44) mounted Handel's first London opera, *Rinaldo*, in 1711, he used his dedication of the libretto to put the case for opera in English. The theme is taken up again over twenty years later in his letter to Handel of 5 December 1732, the text of which we owe to this posthumously published edition of his work (to which Handel was a subscriber). His meaning, he explains to Handel, is that 'you would be resolute enough to deliver us from our *Italian bondage*; and demonstrate, that *English* is soft enough for Opera, when compos'd by poets who know how to distinguish the *sweetness* of our tongue, from the *strength* of it. . . .' Handel refused to succumb to opera in English; but the performance of *Esther* (see no. 128) given earlier in 1732 was a foretaste of the future, which lay increasingly in English oratorio at the expense of Italian opera: this change was complete by the end of the decade.

The third volume of this collected edition of Hill's work includes his poem, *An Ode on Occasion of Mr. Handel's great Te Deum at the Feast of the Sons of the Clergy*, which refers to the performance at St Paul's Cathedral on 1 February 1733.

Gerald Coke Handel Collection

Printed book, volume 1 (of four), xxxii + 351 pp., 20.7 × 13.3 cms/8⅛ × 5¼ ins
LITERATURE
Deutsch, pp. 299, 306–7, 745.

156

Mary Granville, Mrs Pendarves, later Mrs Delany (1700–88)

By an unknown artist

No other friend of Handel could claim to have met him on his first visit to London in 1710–11, as Mrs Delany did, and to have attended the Handel Commemoration concerts in Westminster Abbey in 1784. Her frequent letters to her sister from 1720 onwards afford many glimpses of the London opera world and of her long friendship with Handel. She was the wife firstly of Alexander Pendarves, from 1718 until his death in 1724, and then from 1743 of Dr Patrick Delany, Dean of Down in Ireland, who died in 1768. This little-known painting, though an anonymous work of perhaps the 1720s, is the only known oil portrait from her early years.

As Mrs Delany later told, she first met Handel when he played at her uncle's house on his first visit to London. She was only ten years old at the time: 'My uncle archly asked me if I thought I should ever play as well as Mr. Handel. "If I did not think I should", cried I, "I would burn my instrument!"'

Her close friendship with Handel dates to the 1730s and 1740s. In a delightful letter of 12 April 1734, written from her house in Brook Street a few hundred yards from Handel's, she provides a rare glimpse of a social occasion at which Handel was present:

I must tell you of a little entertainment of music I had last week.... I had Lady Rich and her daughter, Lady Cath. Hanmer and her husband, Mr. and Mrs. Percival, Sir John Stanley and my brother, Mrs Donellan, Strada and Mr. Coot. Lord Shaftesbury begged of Mr. Percival to bring him, and being a *profess'd friend* of Mr. Handel (who was here also) *was admitted;* I never was *so well* entertained at *an opera!* Mr. Handel was in the best humour in the world, and played lessons and accompanied Strada and all the ladies that sung from seven o' the clock till eleven. I gave them tea and coffee, and about half an hour after nine had a salver brought in of chocolate, mulled white wine and biscuits. Everybody was easy and seemed pleased....

Exactly a year later, on 12 April 1735, she writes to her mother about an occasion when she went to Handel's:

Yesterday morning my sister and I went with Mrs. Donellan to Mr. Handel's house to hear the first rehearsal of the new opera Alcina. I think it is the best he has ever made, but I have thought so of so many, that I will not say positively 'tis the finest, but 'tis so fine I have not words to describe it.... Whilst Mr. Handel was playing his part, I could not help thinking him a necromancer in the midst of his own enchantments.

Following her marriage to Dr Delany she was for much of the year in Ireland, but pursued her old enthusiasms whenever she came to London, attending rehearsals at Handel's house or giving small parties of her own, as on 3 April 1744: 'Handel, my brother and Donnellan *dine here*, and we are to be entertained with Handel's playing over Joseph to us....' Dr Delany was not always willing to accompany her to the theatre: 'Semele is charming.... But it being a profane story D.D. does not think it proper for him to go....' Mrs Delany even ventured to put herself forward as a librettist for Handel, as she tells her sister on 10 March 1744: 'and how do you think I have lately been employed? Why, I have made a drama for an oratorio, out of Milton's Paradise Lost, to give Mr. Handel to compose to; it has cost me a great deal of thought and contrivance....' But nothing came of this.

Her later letters from Ireland show a touching concern for Handel's blindness but she was now

156

more often than not at a distance. After her second husband's death in 1768 she returned to London, taking up a position at court. She attended the 1784 Westminster Abbey Handel Commemoration, reporting with that enthusiasm which had lasted a lifetime: 'the effect was wonderful, and I had the courage ... of hearing it *four* times'.

Mrs M. McNair-Wilson

Oil, 62.7 × 53.0 cms/24⅝ × 20⅞ ins

PROVENANCE
By descent in the family of Mrs Delany's sister, Anne D'Ewes, to Mrs M. McNair-Wilson.

LITERATURE
Deutsch, pp. 31, 363, 385, 587, 589; Ruth Hayden, *Mrs Delany, her life and her flowers*, 1980, p. 37 (detail repr.).

157

157

Bernard Granville (1698/9–1775)

Attributed to Sir Godfrey Kneller (1646–1723)

Bernard Granville's friendship with Handel is more elusive than that of his younger sister, Mrs Delany (see no. 156), though he appears to have been no less an enthusiast for his music. His more settled circumstances (he inherited more than one fortune and acquired an estate at Calwich Abbey in Staffordshire) allowed him to build up a substantial collection of manuscript scores of Handel's music, now in the British Library.

Both Granville's uncles had a pronounced taste for music. It was Sir John Stanley, Bart, a commissioner of customs, for whom Handel played in front of the family in the winter of 1710/11, according to Mrs Delany's recollections many years later (see under no. 156), while George Granville, playwright, poet and politician, created Lord Lansdowne in 1711, was among the 63 original subscribers to the Royal Academy of Music in 1719 (see no. 84).

Bernard Granville lacked his sister's even temperament of mind: after his death she wrote of his 'unbridled will' and his 'unhappy temper, meeting in early youth with many disappointments . . . having had too little contradiction and government in childhood'. Perhaps his proud aloofness is the dominant emotion of this his only known portrait, which may be dated to about 1720. Bernard Granville was however often in his sister's company and is mentioned time and again in her letters (usually as 'Bunny') with obvious affection. He was at her party with Handel in 1734: 'Bunny staid with me after the company was gone, eat a cold chick with me, and we chatted till one o' the clock.' On 27 November 1736, 'Bunny came from the Haymarket Opera . . . [he] has made me a present of Handel's Book of Lessons.'

Granville installed an organ in his music room in 1756, traditionally on Handel's advice, though the story that Handel played this organ and was often his guest seems unlikely. In the third codicil to his will, signed on 4 August 1757, Handel left Granville two Rembrandt landscapes, one of which had been a present from Granville to Handel. It was to Granville that James Smyth wrote on 17 April 1759: 'According to your request to me when you left London that I would let you know when our good friend departed this life, on Saturday last . . . died the great and good Mr. Handel . . . he has left to you the two pictures you formerly gave him'.

Mrs M. McNair-Wilson

Oil, 75.5 × 62.9 cms/29¾ × 24¾ ins

PROVENANCE
As no. 156.

LITERATURE
Roger Granville, *The History of the Granville Family*, Exeter, 1895, pp. 442–8; Mary Teresa Fortescue, *The History of Calwich Abbey*, 1914, pp. 110–37 (repr.).

158

Anthony Ashley Cooper, 4th Earl of Shaftesbury (1711–71)

Attributed to Peter Scheemakers (1691–1781) or his son Thomas (1740–1808)

Lord Shaftesbury came into his title before he was two years old, on the death of his father, the celebrated 'philosopher' Earl. At the age of 14 he married Susanna Noel, daughter of the 3rd Earl of

Gainsborough. As 4th Earl of Shaftesbury he took his seat in the House of Lords in 1732 on his coming of age, but rarely took an active part in politics. 'His delights', wrote Edwin Hodder, biographer of his grandson the philanthropist 7th Earl, 'were in literature and art, and society'.

He was one of the youngest members of that small circle of Handel's patrons and supporters who were also personal friends. During Handel's last 25 years Shaftesbury and his wife seem to have taken the greatest personal interest in both Handel's music and his well-being. Mrs Pendarves allows us the earliest glimpse of that friendship when she wrote to her sister on 12 April 1734 about a 'little entertainment of music' she had held the previous week: 'Lord Shaftesbury begged of Mr. Percival to bring him, and being a *profess'd friend* of Mr. Handel (who was here also) was admitted'. From 1737 Shaftesbury's name features among the subscribers to Handel's published scores and he seems to have begun to build up his sizeable collection of manuscript scores at this period (see no. 159). Handel singles out the Shaftesburys for mention in a letter of 29 December 1741 to Charles Jennens: 'You know how much Their kind Protection is precious to me.'

Lord Shaftesbury's correspondence, often with his cousin, James Harris, also a close friend of Handel's (see no. 160), provides us with a measure of his concern for Handel's health and well-being: 'Poor Handel looks something better. I hope he will entirely recover in due time, though he has been a good deal disordered in his head,' he wrote to Harris from London on 24 October 1745. Again, some years later: 'I have seen Handel several times since I came hither and think I never saw him so cool and well. He is quite easy in his behaivour, and has been pleasing himself in the purchase of several fine pictures. . . . We have scarce talked at all about musical subjects, though enough to find his performances will go off incomparably' (13 February 1750). Such references occur in almost all Shaftesbury's letters.

After Handel's death, Lord Shaftesbury prepared a lengthy private note on Handel's life and affairs; this was in response to the extracts from Mainwaring's biography, which were published by the *London Chronicle* in June 1760.

This bust, a pair to the posthumous marble of Shaftesbury's father, the 3rd Earl, also in the Dorset County Museum, was attributed by Gunnis to Peter Scheemakers. It is very similar to the one on the monument to the 4th Earl by Scheemakers's son Thomas, erected in the parish church at Wimbourne St Giles following his death in 1771.

Dorset Natural History and Archaeological Society

Marble, 79 cms/31⅛ ins high

Inscribed on the socle *A.A.C. / 4th Earl of Shaftesbury*

PROVENANCE
By descent to the 10th Earl of Shaftesbury; his sale Christie's, 26 June 1980, lot 12, purchased by the Dorset Natural History and Archaeological Society for the Dorset County Museum.

LITERATURE
Rupert Gunnis, *Dictionary of British Sculptors 1660–1851*, 1953, pp. 343–4; *Deutsch*, pp. 624, 680, 844–8.

158

159

Italian Cantatas

Manuscript score

The 4th Earl of Shaftesbury (see no. 158) has secured a claim to fame as a loyal patron of Handel; by 1734 he was, according to Mrs Delany, 'a profess'd friend of Mr. Handel' and his name appears on the list of subscribers to printed scores from 1737 onwards. He also commissioned manuscript scores of Handel's works not available in reasonably complete printed editions and built up a fine collection which is still at St Giles's House in Dorset, the Shaftesbury family seat. The full extent of the collection became known only in 1983, though its contents could be inferred from an inventory made in 1761, a copy of which is in the British Library (Add. MS 39564, ff. 98–102).

Handel himself is known to have visited St Giles's House occasionally, and his personal interest in the 4th Earl's collection is attested by the displayed page of a volume of 49 Italian cantatas (strictly speaking, there are 48 cantatas in the volume, since no. 39 is merely a partial copy of 'Ne tuoi lumi', which also appears complete as no. 7). In many such anthologies, the cantatas 'Venne voglia ad amore' and 'Vedendo amor' appear together as if they were parts of a single cantata, and they so appear in the Shaftesbury volume, though the copyist has added *Fine* at the end of 'Venne voglia'. That 'Vedendo amor' is indeed a continuation of 'Venne voglia' is confirmed by Handel's own pencilled note at the start of 'Vedendo amor': *continuatione della cantata precedente*.

The Earl of Shaftesbury

Manuscript score, 422 pp., 23 × 30.5 cms/9 × 12 ins
Inscribed *Handell's Cantatas*

PROVENANCE
By descent from the original owner, the 4th Earl of Shaftesbury.

160

James Harris (1709–80)

By Charles Bestland (1763–c.1837)
after Joseph Highmore (1692–1780)

James Harris was a close friend and patron of Handel and a passionate musical enthusiast; he was also a man of more intellectual authority than any of the composer's other admirers. His works on theoretical criticism and grammar, of which *Hermes* (1751) is the chief, were preceded by his *Three Treatises* published in 1744; in the second of these, devoted to 'Music, Painting and Poetry', he describes Handel's genius as 'being it self far the sublimest and most universal now known'.

Harris was a lifelong enthusiast for Handel's work; he was already buying his music at the age of twenty while still a law student (a copy of Handel's *Suites de Pieces pour le Clavecin* dated 1729 is in the Gerald Coke Handel Collection). On 5 May 1737, in a letter to his cousin Anthony, 4th Earl of Shaftesbury (see no. 158), he wrote that his consolation, if Handel gave up opera, would be 'to look over his Scores, and recollect past Events. . . . This was an Excell[t] Chorus and that a Charming peice [sic] of Recitative.' His brother, Thomas Harris, was the recipient of Handel's portrait by Mercier (no. 2) in 1748. After Handel's death he is said to have contributed the list of the composer's works to Mainwaring's biography of 1760 (no. 214).

Highmore's portrait of James Harris, a sensitive image still in the Malmesbury family collection, was probably painted soon after the publication of the *Three Treatises* (perhaps indicated by Harris's gesture to the volumes on the table). It is unlikely to be as early as 1740, the date inscribed on this engraving from the posthumous edition of Harris's works published by his son, Lord Malmesbury, in 1801. This is confirmed by an old inscription on the oil original which bears the date 1745, the year of Harris's marriage to Elizabeth Clarke. Her portrait, also by Highmore, is in the Malmesbury Collection, leading one to suppose that the two images were painted as a marriage pair in 1745. If so Harris would have been doing no more than following the lead of his friend and cousin, Lord Shaftesbury, who had the previous year commissioned a pair of grand full-lengths of himself and his wife from Highmore on the occasion of his own marriage.

National Portrait Gallery, London

Stipple engraving, 17.3 × 13.7 cms/6¾ × 5⅜ ins
From *The Works of James Harris Esq*, 1801 (frontispiece)

Inscribed on either side of the Harris coat of arms
JAMES HARRIS. MDCCXL AET. SV.XXXI/ Jos. Highmore Pinx!/ C. Bestland sculp!

LITERATURE
Deutsch, pp. 432–3, 599; A.S. Lewis, *Joseph Highmore: 1692–1780*, Harvard University PhD, 1975, published by University Microfilms International, Ann Arbor, Michigan, and London, 1979, II, p. 426, no. 79.

160

161

Esther

Manuscript score

Elizabeth Legh of Adlington Hall, Cheshire seems to have been the first of Handel's English supporters to attempt to assemble a reasonably complete collection of his works. As well as subscribing to the printed scores of the operas she commissioned manuscript copies of all Handel's major works as soon as they were composed. By 1734, when she died, she had collected at least 38 manuscript volumes, including composite volumes of shorter instrumental and keyboard pieces in addition to operas and choral works. Two volumes of cantatas are now in the Bodleian Library (Mus.d. 61, 62); the rest of the scores, a uniformly bound set of 36, were apparently returned to Handel after Elizabeth Legh's death and were later presented by him to Thomas Harris, brother of James (see no. 160), and recipient of the Mercier portrait (no. 2). Together with the portrait the scores passed to Thomas's nephew the 1st Earl of Malmesbury.

The fact that Elizabeth Legh's copies were made at an early stage in the history of the works concerned gives them special value in determining the versions of first performance, and some volumes contain music not extant elsewhere. Another important feature is the dating of many of the manuscripts, both by the copyists and Elizabeth Legh herself, which often helps establish dates of composition which otherwise could only be guessed. In this respect this copy of *Esther* is of particular significance since the date of 1718, clearly recorded on the title-page (endorsed by Elizabeth Legh's less clear inscription), is not found elsewhere. The surprising statement that the work was composed in London gives rise to speculation that the first performance might also have taken place there, rather than at Cannons. (See also no. 128.)

The Earl of Malmesbury

Manuscript score, ii + 197 + ii ff., 29.7 × 23.5 cms/11¾ × 9¼ ins

Inscribed on the spine *Oratorium*, on the fly-leaf *Eliz. Legh 1718* (the last digit obscured by a blot), and on the title-page *The Oratorium. Composed by George Frederick Handel. Esquire, in London 1718* (Elizabeth Legh's index to the volume is written on the penultimate fly-leaf)

PROVENANCE
Part of a 36-volume set prepared for Elizabeth Legh of Adlington Hall, Cheshire; ?returned to Handel after her death in 1734; given to Thomas Harris; by descent from his nephew the 1st Earl of Malmesbury.

LITERATURE
Deutsch, p. 200; H.E. Smither, *A History of the Oratorio*, Chapel Hill, 1977, II, p. 189; Winton Dean, 'Handel's early London Copyists', *Bach, Handel, Scarlatti: Tercentenary Essays*, Cambridge, 1985, pp. 75–97.

162 [*in colour opposite p. 152*]

Burkat Shudi (1702–73) and his Family

By Carl Marcus Tuscher (1705–51)

Burkat Shudi, the Swiss harpsichord maker and friend of Handel, moved to new premises in Great Pulteney Street, Soho, in 1742. This delightful conversation piece was probably painted then or very soon after, and is traditionally said to depict the parlour of the new premises. Burkat's eldest surviving son, Joshua (1736–54), stands by his father as he tunes a splendid harpsichord. His wife Catherine (1707?–58), a protective arm around the younger boy Burkat (1737–92?), is reading the newspaper, perhaps even the issue of the *Daily Advertiser* for 5 October 1742 which announced that Shudi had 'removed from Meard's Street in Dean Street, Soho to Great Pulteney Street, Golden Square'. The walls of the room are hung with a landscape painting, traditionally of Shudi's birthplace, and portraits of Shudi's great patron, Frederick Prince of Wales, and his wife Augusta Princess of Wales; though only partially shown they can be identified with a pair of mezzotint engravings by John Faber after portraits by Bartholomew Dandridge and Charles Philips. The truncated appearance of the prints and painting has led to the suggestion that the painting has been cut down, but an examination of the top paint edge does not support this theory.

Burkat Shudi, or Tschudi as he was born, came to London from Glarus in Switzerland at the age of 16 as a journeyman joiner. He worked for the Flemish harpsichord maker Herman Tabel (see no. 75) for a time, but by 1728 he had set up his own business, which for some years was conducted from premises in Meards Street, Soho (a few doors from John Christopher Smith the elder). Shudi became one of the most important harpsichord makers working in London in the mid-century. By the time of the marriage of his daughter, Barbara, to John Broadwood, his apprentice and successor who gave his name to the famous firm of piano makers, English instruments were well known throughout Europe for their outstanding qualities.

Like so many of Handel's friendships, that with Shudi is difficult to document. His grandson, James Shudi Broadwood, writing in 1838, went so far as to attribute the very success of his business to 'the powerful patronage of his friend Handel' who is described as 'a constant guest at his table, ever well covered with German dishes and German wines'. Only some ten of the 380 instruments made by Shudi in Handel's lifetime survive. The earliest is of 1729 with the penned inscription, *Questo cimbalo è de l^a Sig^{ra} Anna Strada 1731, London*; it belonged to Anna Maria Strada del Pò (see no. 115) and is sometimes said, without any real evidence, to have been a gift from Handel to his singer; another, of 1740, was the property of Mrs Delany (see no. 156); while a very fine harpsichord of the same year at Kew (fig. 8, p. 27) was apparently originally owned by Frederick Prince of Wales. A double manual was sold at auction at Oxford on 26 June 1788, optimistically described as 'made for the immortal Handel and used by him till his death'. The instrument shown in this painting was allegedly presented to Frederick the Great of Prussia to mark the capture of Prague in 1744.

Until very recently the artist of this important conversation piece had defied identification, with names such as Zoffany, Mercier and Du Pan being tentatively proposed. However it has now been established, through the keen eye of Gavin Graham and the research of Brian Allan, that the picture is the work of the German painter and architect Carl Marcus Tuscher, who was born at Nuremberg and spent many years in Italy before coming to London in 1741 for a two-year stay. The varied poses and the distinctive, rather flat faces, the bright colours and the frieze-like composition can be closely paralleled in signed or documented works by Tuscher. Furthermore the date of about 1742 already proposed for this family group (when the boys would have been five and six; they were eighteen months apart) fits perfectly with Tuscher's short stay in London.

National Portrait Gallery, London
NPG 5776

Oil, 83.4×141.5 cms / $32\frac{3}{4} \times 55\frac{3}{4}$ ins

Inscribed beneath the engraved portraits shown on the wall
His Royal Highness Frederick Prince of Wales &c and
Her Royal Highness Augusta Princess of Wales &c

PROVENANCE
Painted for Burkat Shudi; bequeathed by him to his daughter Barbara (1749–76), who married John Broadwood in 1769; thence by descent to Captain Evelyn Broadwood (1889–1975); the Broadwood Trust, by which sold Sotheby's, 12 December 1984, lot 32; purchased by the National Portrait Gallery.

LITERATURE
H.F. Broadwood, *Some Notes made by J.S. Broadwood, 1838, with Observations & Elucidations by H.F. Broadwood, 1862*, 1862, pp. 6–7; William Dale, *Tschudi, The Harpsichord Maker*, 1913, pp. 33–4, 42–5 and 65 (repr. as frontispiece); David Wainwright, *Broadwood By Appointment: A History*, 1982, pp. 27–8, 81; Brian Allan, 'Carl Marcus Tuscher. A German Artist in London 1741–1743', *Apollo*, CXXII, July 1985, pp. 32–5, fig. 281.

EXHIBITED
Rococo, no. B14 (repr. in colour on cover).

In Praise of Handel

Of the many contemporary poems to mention Handel perhaps the finest and one of the very earliest is John Gay's *Trivia* of 1716 (no. 152). Handel is the central subject of poems by Daniel Prat (1722) and Aaron Hill (1733) (see nos. 163 and 155), and features in the work of Alexander Pope (1742), Tobias Smollett (1746) and Henry Fielding (1751). He also inspired outpourings by numerous other minor and anonymous authors, of which perhaps the most famous is John Byrom's 'Epigram on the Feuds between Handel and Bononcini' (see no. 88) of 1725.

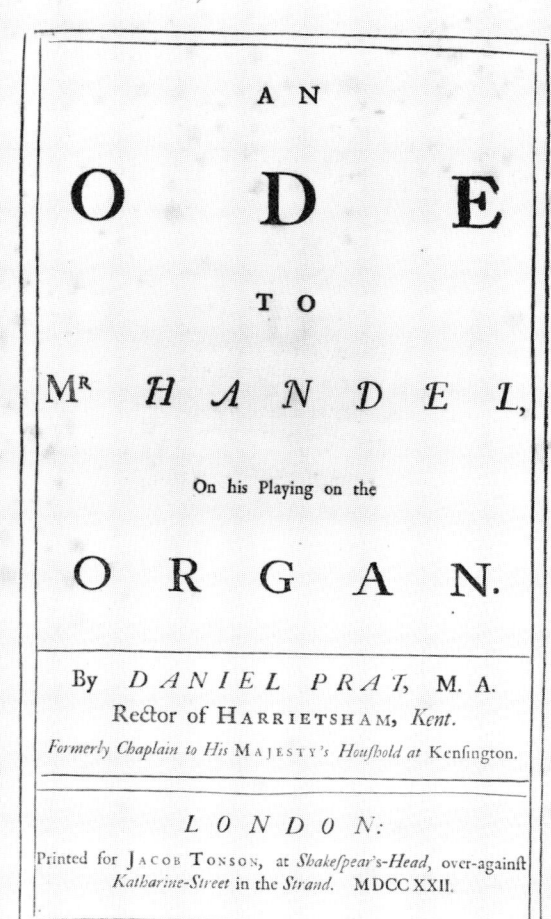

163

163

'An Ode to Mr Handel
On his Playing on the Organ'

By Daniel Prat
Printed for Jacob Tonson, 1722

This poem is the earliest to have been dedicated to Handel. It takes as its theme his celebrated performance upon the organ:

> *We hear, inspiring sacred Dread,*
> *The deep majestic Organ blow,*
> *Symbol of Sounds that rowse the dead!*

Daniel Prat proclaims himself on the title-page as Rector of Harrietsham, Kent, and formerly Chaplain to His Majesty's Household at Kensington.

Gerald Coke Handel Collection

Printed pamphlet, 9 pp., 31.9 × 20.9 cms/12½ × 8¼ ins

Titled as above and inscribed in pen at the top of the title-page *David Polhill Esq.*

LITERATURE
Deutsch, pp. 139–44.

EXHIBITED
Coke, p. 17.

164

'The New Dunciad: As it was found
In the Year MDCCXLI.'

By Alexander Pope (1688–1744)
Printed for Thomas Cooper, 1742

Alexander Pope had known Handel for the best part of thirty years when his famous description of the composer was published in Book IV of *The Dunciad* in March 1742 (Books I to III had been published in 1728):

> *Strong in new Arms, lo Giant Handel stands,*
> *Like bold Briareus, with a hundred hands;*
> *To stir, to rowze, to shake the soul he comes,*
> *And Jove's own Thunders follow Mars's Drums.*
> *Arrest him, Empress; or you sleep no more –*
> *She heard, and drove him to th' Hibernian shore.*

Pope's earlier association with Handel is discussed above (see under no. 62); here, the allusions in *The Dunciad* are to contemporary events. The 'hundred hands' of the monster Briareus which 'shake the soul' are explained in Pope's own footnote: 'Mr Handel had introduced a greater number of Hands and more variety of instruments into the Orchestra, and employed even Drums and Cannon to make a fuller Chorus.' Pope uses 'Cannon', as others were to do, to refer to kettledrums, which were used in *Saul* (see no. 138) and other oratorios. This richer sound, Pope tells us, 'proved so much too manly for the fine gentlemen of his age, that he was obliged to remove his Musick into Ireland'. Handel had been in Dublin for four months at the time of the publication of this poem, and a few weeks later gave *Messiah* its first performance.

The British Library Board
11630.f.17

Printed book, Book VI of *The Dunciad*, iv + 44 pp., 26.9×21.0 cms/$10\frac{5}{8} \times 8\frac{1}{4}$ ins

LITERATURE
Deutsch, p. 543.

165

'Advice: A Satire'

By Tobias Smollett (1721–71)
Printed for Mary Cooper, 1746

Published anonymously, this satire on London life by the novelist Tobias Smollett takes the form of a series of interchanges between the Poet and his Friend; here the Poet speaks:

> ... *Again shall* Handel *raise his laurel'd brow,*
> *Again shall harmony with rapture glow!*
> *The spells disolve, the combination breaks,*
> *And rival Punch no more in terror squeaks.*
> *Lo, R—ss—l falls a sacrifice to whim,*
> *And starts amaz'd in* Newgate *from his dream:*

Smollett's own footnote explains the rivalry between Handel's oratorios and the puppet shows of the unfortunate Russel, possibly the tenor who sang in *Saul* on 16 January 1739.

In 1749 Handel provided the incidental music for Smollett's tragedy, *Alceste*, but this never reached production. Smollett later described Handel (in the *Continuation of the Complete History of England*, 1762) as 'the most celebrated master in music which this age produced'.

Gerald Coke Handel Collection

Printed pamphlet, 16 pp., 33.7×22.0 cms/$13\frac{1}{4} \times 8\frac{5}{8}$ ins

LITERATURE
Deutsch, pp. 634–5, 657, 679, 850.

166

'Amelia'

By Henry Fielding (1704–54), first edition
Printed for Andrew Millar, 1751

In this story of the virtuous Amelia's adventures, Fielding includes a visit to the oratorio with her landlady, Mrs Ellison:

> Upon the Evening ... the two Ladies went to the Oratorio, and were there time enough to get a first Row in the Galery ... Amelia ... being a great Lover of Music, and particularly of Mr. Handel's Compositions. Mrs. Ellison was, I suppose, a great Lover likewise of Music. ...
>
> Tho' our Ladies arrived full two Hours before they saw the Back of Mr. Handel; yet this Time of Expectation did not hang extremely heavy on their Hands. ...

By the time Henry Fielding's novel of *Amelia* first appeared in 1751, an evening at a Handel oratorio had become an accepted part of London's cultural life. That the two ladies 'saw the Back' of Mr Handel was entirely characteristic of his performance at the organ when he faced away from the audience.

The British Library Board
12614.b.25

Printed book, volume 2 (of four), viii + 262 pp., 17.0×10.4 cms/$6\frac{3}{4} \times 4\frac{1}{8}$ ins

LITERATURE
Deutsch, p. 714.

The Instruments of Handel's Orchestra

This selection of instruments is designed to show as far as possible, in the space available, the instruments which Handel employed in his orchestra throughout his life in England.

The treble lines of Handel's orchestra were normally played by violins in two, and often three, lines, plus at least two oboes and quite frequently four. The oboes played along with both the violin parts as a general rule, certainly whenever the violin lines are marked *tutti*, falling silent whenever the violins are marked *solo*, and of course frequently have their own parts. Two violins are shown among the Cannons instruments (nos. 76–7).

The middle line was covered by violas, and sometimes by the third violins. Violas were occasionally divided, since the older concept of alto and tenor parts was still to some extent alive. A tenor viola is shown here, for this size is seldom seen today. Players find it a physical strain to support so large an instrument and to cover the necessary stretch, even though some are aware that the tone quality of so large but shallow-bodied an instrument has a sonority that cannot be approached by the more popular, smaller but deeper, alto viola.

The bass parts were taken by cello and double bass accompanied by bassoons, again normally two and occasionally four, playing with them just as the oboes played with the violins, and again often having their own independent lines. A cello of very slightly later date is shown here; regrettably it has not been possible to find a contemporary double bass in original condition which is not in active use in one of the ensembles now recreating the sounds of the eighteenth century. We are particularly fortunate in being able to display the only two known surviving Stanesby bassoons and the unique Stanesby junior double bassoon which was directly associated with Handel's music. The bass line was supported by the continuo instruments, those capable of chordal accompaniment such as the organ, harpsichord (see no. 75), bass lute or theorbo and harp.

In many of his works, Handel wrote parts for the flute (by which he and all his contemporaries meant the recorder), normally the treble but occasionally smaller and larger sizes, and also for the *traversa*, or transverse flute, the instrument which we call the flute today. Parts for the horns (see nos. 220 and 230) are frequent, as are those for trumpets (see nos. 82 and 221–2) and drums (see no. 224). Finally, one must remember that orchestras need a pitch standard for tuning, and thus the pitch pipe was an important adjunct to music-making.

167

Oboe

By Thomas Stanesby senior (*c*.1668–1734)

Only six of the elder Stanesby's oboes are now known, three of which are included here (see nos. 168 and 226). He served his apprenticeship with Joseph Bradbury (see no. 83), and became the leading maker of recorders and oboes in the London of his day. His and Bradbury's are the earliest English oboes with makers' names to have come down to us. The other leading woodwind maker, Bressan, who came to London from France at the end of the seventeenth century, is known to have made oboes, but none survives.

Edinburgh University Collection of Historic Musical Instruments (By Permission of the Curator)

Stamped on each joint T/STANESBY/(eight-pointed star)

Stained boxwood, with three silver keys for middle C and left- and right-hand E♭;
overall length 59.0 cms/23¼ ins

PROVENANCE
F.G. Rendall (1890–1953); lent by his widow, 1967, and eventually given to Edinburgh University, 1980.

LITERATURE
Eric Halfpenny, 'The English 2- and 3-Keyed Hautboy', *Galpin Society Journal*, II, 1949, p. 10, no. 5;
Philip Bate, *The Oboe*, 1956, pl. II;
Lyndesay Langwill, *A Check-List of the Double-Reeded Instruments in the Edinburgh University Collection of Historic Musical Instruments*, Edinburgh, 1981, p. 2, no. 62.

EXHIBITED
British Musical Instruments, Galpin Society, London, 1951, no. 66;
European Musical Instruments, Galpin Society, Edinburgh, 1968, no. 82.

HANDEL

167   168

168

Oboe

By Thomas Stanesby senior (c. 1668–1734)

This is a more elaborate example than no. 167, with ivory mounts.

Horniman Museum
(Carse Collection) 14.5.1947–277

Stamped on each joint T / STANESBY / (eight-pointed star)

Stained boxwood, with ivory mounts at the top of the bell and lower body joints, and three silver keys for middle C and left- and right-hand E♭;
overall length 59.6 cms/23½ ins

PROVENANCE
Nettlefold Collection; sold by Puttick & Simpson, 25 April 1946, purchased by Adam Carse; given by him to the Horniman Museum, 1947.

LITERATURE
Eric Halfpenny, 'The English 2- and 3-Keyed Hautboy', *Galpin Society Journal*, II, 1949, p. 10, no. 6.

EXHIBITED
Eighteenth Century Musical Instruments: France and Britain, Victoria and Albert Museum and elsewhere, 1973, no. 96.

169

Tenor Viola

By Antonio Ungarini (1696–1771)

This instrument is typical of the large, tenor-size viola found all over Europe in the seventeenth century. It was made in 1741 by Antonio Ungarini of Fabriano in Italy. Though the small viola was the more commonly found size in England in the eighteenth century, Handel's frequent two-part writing for the viola strongly suggests he expected to find at least some instruments of the tenor size in his orchestra. Both kinds were, of course, tuned to the same pitch, but the larger body of the tenor adds greater resonance to the lower notes.

The majority of string instruments made before the end of the eighteenth century have been modernised (see under no. 76), with consequent changes to the tone quality. This example was neglected for two centuries, being used as a nest for mice (hence the enlargement of the bass sound hole), but as a result it retains its original neck and fingerboard and has not been reduced in size.

J. & A. Beare Ltd

Label, inscribed in pen and ink
FABRIANO 1741 / ANTONIO VNGARINI

Overall length 70.5 cms/27¾ ins, body length 45.1 cms/17¾ ins, width of bouts 21.8, 15.0 and 26.2 cms/8⅝, 5⅞ and 10¼ ins, string length 38.0 cms/15 ins; one-piece back;
the left sound hole has been damaged;
the pegs, bridge, tail-piece and tail button are missing

170

Violoncello

By Thomas Smith (?fl. 1740–90)

Thomas Smith was a pupil and the successor of Peter Wamsley, working from the same address of the Harp and Hautboy in Piccadilly. Like Wamsley he worked mainly on the Stainer model (see no. 77), but whereas his master made examples of all bowed string instruments, Smith appears to have specialised in violoncellos, becoming the leading maker of his day. This example, though not labelled, probably dates from the latter part of his working life, perhaps from the 1770s.

The history of cello-making in eighteenth-century England goes back to Barak Norman (see no. 79),

FROM MESSIAH TO THE FIREWORKS MUSIC 1741–9

who is generally thought to have been the first maker; one dated 1718 is recorded. Wamsley, his slightly later contemporary, was making cellos from at least 1727 to 1754, while Smith spans the next generation. Two group portraits included here show eighteenth-century cellos: *Frederick Prince of Wales with his Sisters* (no. 112) and *The Sharp Family* (no. 228).

Private Collection

Overall length 120.6 cms/47½ ins, body length 73.7 cms/29 ins, width of bouts 34.5, 23.1 and 42.5 cms/13⅝, 9⅛ and 16¾ ins, string length 66.7 cms/26¼ ins;
the neck, fingerboard, bridge and end piece belong to the early nineteenth century, but are set up in a manner similar to that of Handel's time.

171

Bassoon

By Thomas Stanesby senior (*c.* 1668–1734); the wing joint by his son, Thomas Stanesby junior (1692–1754)

This is both the only known bassoon by the elder Stanesby and the oldest surviving English bassoon. The wing joint by the younger Stanesby is presumably a replacement, perhaps for a damaged original or possibly to bring the pitch up to that common in the second quarter of the century: English pitch at the beginning of the century was around A = 409, and by the 1730s it had risen to A = 415.

The bassoon was known in England from the 1680s onwards, but so far no reference has been found for any English makers earlier than the maker of the present instrument. However, its general appearance and the heavy choking of the bore of the bell joint both suggest that the English school of bassoon-making was already well established.

William Waterhouse Esq

Stamped on the butt, long joint and bell T.STANESBY/(star), and on the wing joint (star)/STANESBY/IUNIOR

Stained maple with brass mounts and four brass keys on saddles for B♭, D, F and A♭, the F key with a forked touch; overall length 126.3 cms/49¾ ins, tube length 219.0 cms/86¼ ins without crook

PROVENANCE
Not known before 1979, when sold at Teynham Auction Rooms, Sittingbourne, Kent, to Nigel Peverett; offered by him at Sotheby's, 8 September 1979, lot 123; bought in and sold to William Waterhouse.

EXHIBITED
The Proud Bassoon, Edinburgh University Collection of Historic Musical Instruments, 1983, no. 5.

HANDEL

171

172   173

172

Bassoon

By Thomas Stanesby junior (1692–1754)

Stanesby junior was apprenticed to his father in 1706 and was particularly noted for his flutes, probably only producing recorders, oboes and bassoons after his father's death or retirement (he may also perhaps have produced clarinets; only a barrel joint survives bearing his name). His pupil and successor was Caleb Gedney, many of whose instruments were in Sir Samuel Hellier's orchestra (see p. 239).

This is the second-oldest surviving English bassoon and is unusual in that it bears the name of its purchaser (monograms are quite common); unfortunately there is no record of to whom it was given. Ralph Jenison (1696–1758) was a Member of Parliament twice (1724–41 and 1749–58) and also twice Master of the Royal Buckhounds (1737–44 and 1746–57). He was a frequent subscriber to music, especially that of Handel, his name appearing in subscription lists as early as 1727 with the publication of *Admeto*. Muraeus was a minor instrument maker or dealer in London; his stamp on three of the bassoon's joints may indicate a repair or a later sale, though it is noteworthy that the instrument has had very little use.

William Waterhouse Esq

Stamped on each joint *STANESBY / IUNIOR*, on the butt *LONDON / 1747*, and on three joints *MURAEUS /* (rampant lion); engraved on the bell mount, *Ex Dono, R:* [coat of arms] *Jenison, Arm: 1747*, with the Jenison arms repeated on the brass band at the top of the butt

Maple with brass mounts and four brass keys for B^b, D, F and A^b;
the touch of the F key may be a replacement;
the key head is modern;
the wing joint may have been shortened; pitch: A = 415;
overall length 122.5 cms/48¼ ins,
tube length 213.3 cms/84 ins without crook

PROVENANCE
Given by Ralph Jenison to an unspecified recipient, 1747;
Canon Francis W. Galpin by the mid-1880s;
purchased from him by Rev. C.J. Sharp, 1946;
Sotheby's, 18 April 1968, lot 3M,
purchased by William Waterhouse.

LITERATURE
F.W. Galpin, *Old English Instruments of Music*, 1910, pl. 34;
Eric Halfpenny, 'The Evolution of the Bassoon in England, 1750–1800', *Galpin Society Journal*, X, 1957, p. 30;
Eric Halfpenny, 'Further Light on the Stanesby Family', *Galpin Society Journal*, XIII, 1960, pp. 67–8;
Brian Galpin, 'Canon Galpin's Check Lists', *Galpin Society Journal*, XXV, 1972, p. 4.

EXHIBITED
Royal Military Exhibition, London, 1890, no. 151;
British Musical Instruments, Galpin Society, London, 1951, no. 92;
European Musical Instruments, Galpin Society, Edinburgh, 1968, no. 240;
The Proud Bassoon, Edinburgh University Collection of Historic Musical Instruments, 1983, no. 6.

173

Double Bassoon

By Thomas Stanesby junior (1692–1754)

The only surviving English double or contrabassoon from the first half of the eighteenth century (dated 1739), this is exceptional in that it descends a full octave below the normal bassoon, unlike almost all other eighteenth- and early nineteenth-century double bassoons. Bach, for example, wrote for the smaller quart- or quintfagott, descending only a fourth or a fifth below the ordinary bassoon.

This instrument is unusual in that it is built exactly as a normal bassoon but almost double the size, with a long joint, or bass joint, some four feet long. Double bassoons are more often constructed with shorter joints, often with extra upward and downward turns, so that the total height is little more than that of the ordinary bassoon and the player less inconvenienced in handling the instrument. The size of this example is such that the player may have had to perform standing rather than sitting, especially if the original crook, probably about two feet long, followed the normal shape of a bassoon crook.

Double bassoons of any shape or size are rarely found in the music of Handel's period, which is doubtless why the advent of this instrument was considered newsworthy. The *London Daily Post and General Advertiser* for Monday 6 August 1739, and on subsequent days in the same month, carried an advertisement which must surely have referred to this very instrument, made in that same year:

> Marybone-Gardens, This day, and to be continued, The Usual Concert. To which will be added, (for the better Entertainment of Gentlemen and Ladies, the Undertaker being desirous to make the Concert the most complete,) Two Grand or Double Bassoons, made by Mr. Stanesby, jun. the Greatness of whose Sound surpasses that of any other Base Instrument whatsoever. Never perform'd before.

It must be stressed that this advertisement refers specifically to Stanesby junior. Canon Galpin misquoted it as Stanesby senior, and all subsequent authors, with the sole exception of Langwill, have followed him, causing much fruitless speculation in the attempt to reconcile a 1739 advertisement with the elder Stanesby's death in 1734 when this surviving instrument, dated 1739, bears the younger Stanesby's stamp. Burney had probably started the confusion when he wrote of the double bassoon used in the 1784 Handel Commemoration, 'It was made with the approbation of Mr. HANDEL, by Stainsby, the Flute-maker, for the coronation of his late majesty, George the Second.', that is, in 1727. It is probable that Burney, who was writing in 1784, simply got his facts wrong. He was also wrong when he went on 'nor . . . was it ever introduced into any band in England, till now . . .', for what is certain is that Handel, willing as ever to experiment with novel effects in his orchestration, scored for it in *L'Allegro* in January 1740, marking *bassons e basson grosso* against the air 'Let the pealing organ', and employing it again in 1749 in the *Music for the Royal Fireworks*. The performances in 1739 and 1740 aroused sufficient interest to be lampooned in a concert advertisement in the *General Advertiser* for 20 October 1740, announcing 'twenty-four bassoons accompanied on the violoncello, intermixed with duets by four double bassoons accompanied by a German flute': surely a spoof rather than an eighteenth-century Hoffnung concert.

Whether two such instruments as this really existed is not known. Handel appears only to have written for one, and Burney and the later references cited in detail by Halfpenny only mention a single instrument. It is most probable, therefore, that it was this identical instrument 'which was so conspicuous in the Orchestra and powerful in its effect' at the 1784 Commemoration, to quote Burney again. It apparently towered over the singers in the very front of the orchestra, immediately behind the conductor who was seated at the keyboard of the organ, according to the 'Plan of the Orchestra' which Burney prints (curiously, it is not visible in his picture of the orchestra).

Halfpenny traces the further history of the instrument used in the 1784 concerts (ie presumably the history of this instrument) down to 1803, though with a gap between 1788 and that date. Thereafter it vanished. An instrument which may have been this one, was sold at Sotheby's on 8 April 1878, lot 380, and purchased by Fenton (information from Graham Wells of Messrs Sotheby). Thereafter the history is sufficiently clear.

National Museum of Ireland, Dublin

Stamped on all joints *STANESBY / IUNIOR* and on the butt joint also *LONDON / 1739*

Cherrywood with brass mounts and crook, with four brass keys for B♭, D, F and A♭;
overall length 252.4 cms/99 3/8 ins,
tube length 439.2 cms/173 ins excluding the crook, which is not original; the pitch is an octave below the normal bassoon

PROVENANCE
Seen by F.W. Galpin at Fenton's shop in New Oxford Street in 1887 and subsequently purchased by W. Ringrose Atkins (according to a letter of Galpin's belonging to William Waterhouse), and certainly in Atkins's possession at the time of the Royal Military Exhibition in 1890; bequeathed by him to the National Museum of Ireland, 1907.

LITERATURE
Burney, p. 7; F.W. Galpin, *Old English Instruments of Music*, 1910, p. 169; Eric Halfpenny, 'Stanesby, Major and Minor', *Music & Letters*, xxxiv, 1953, pp. 41–7; Lyndesay G. Langwill, *The Bassoon and Contrabassoon*, 1965, pp. 114–17.

EXHIBITED
Royal Military Exhibition, London, 1890, no. 169;
British Musical Instruments, Galpin Society, London, 1951, no. 104;
European Musical Instruments, Galpin Society, Edinburgh, 1968, no. 259.

174

Treble Recorder

By Thomas Stanesby senior (*c.* 1668–1734)

The recorder was the common flute of the early eighteenth century and appears frequently in Handel's scores – normally, as here, in pairs.

Horniman Museum
(*Carse Collection*) 14.5.47–319

Stamped on each of three joints *T/STANESBY/* (eight-pointed star)

Ebony with ivory head and mounts and foot ring;
overall length 51.1 cms/20 1/8 ins,
sounding length 44.8 cms/17 5/8 ins

PROVENANCE
Given by Adam Carse to the Horniman Museum, 1947.

LITERATURE
Eric Halfpenny, 'The English Baroque Treble Recorder', *Galpin Society Journal*, ix, 1956, pp. 82–90, no. 4.

EXHIBITED
Eighteenth Century Musical Instruments: France and Britain, Victoria and Albert Museum and elsewhere, 1973, no. 71.

175

Treble Recorder

By Thomas Stanesby junior (1692–1754)

It is generally presumed that the younger Stanesby did not make recorders until after his father's death in 1734 (see under no. 172).

Horniman Museum
(*Carse Collection*) 14.5.47–275

Stamped *STANESBY / IUNIOR*

Stained boxwood without mounts;
overall length 50.0 cms/19 3/4 ins,
sounding length 43.2 cms/17 ins

PROVENANCE
Given by Adam Carse to the Horniman Museum, 1947.

LITERATURE
Eric Halfpenny, 'The English Baroque Treble Recorder', *Galpin Society Journal*, ix, 1956, pp. 82–90, no. 8.

EXHIBITED
Eighteenth Century Musical Instruments: France and Britain, Victoria and Albert Museum and elsewhere, 1973, no. 72.

176

Descant Recorder

By Richard Haka (1645/6–1709)

This is the size of instrument that Handel indicated as *Flauto Piccolo*.

Richard Haka said, on his marriage, that he was born in London, but he spent all his working life in Amsterdam. As yet, the precise date of his birth and the details of his early life have not been traced, and it is not known whether Haka was originally his name (if so, perhaps of Bohemian origin) or whether he was a 'Harker' who changed the spelling of his name to maintain the correct sound in Dutch. He was particularly noted for his recorders, oboes and *deutsche Schalmeien* (a form of shawm). This is the only example of his work in Britain, and was made in about 1700, towards the end of his working life.

Edinburgh University Collection of Historic Musical Instruments (By Permission of the Curator)

Stamped *HAKA*

Ivory, made in one piece and therefore with the lowest finger holes duplicated to be accessible to the little finger of either hand; the unwanted hole was stopped with wax;
overall length 33.6 cms/13 1/4 ins, sounding length 29.5 cms/11 5/8 ins

175

176

177

PROVENANCE
Probably acquired by H.I. Brackenbury (1869–1938); part of the C.H. Brackenbury Memorial Collection, lent by Mark Brackenbury to the Edinburgh University Collection of Historic Musical Instruments.

LITERATURE
John Dick and Arnold Myers, *A Checklist of the Flutes and Whistles in the Edinburgh University Collection of Historic Musical Instruments*, Edinburgh, 1982, p. 6, no. 1037.

EXHIBITED
European Musical Instruments, Galpin Society, Edinburgh, 1968, no. 23.

177

Bass Recorder

Perhaps by a member of the Stanesby family

Handel wrote for *basso di flauti* in his opera *Giustino*, performed at Covent Garden in 1737. This instrument was attributed by Halfpenny to Thomas Stanesby senior on stylistic grounds, with the suggestion that the stamped initials, *M.S.*, might stand for Mary Stanesby, his wife.

Warwickshire Museum Collection

Stamped *M.S.* in a scroll

Pearwood (according to Halfpenny) or maple (according to the Warwickshire Museum), brass key, crook and head-cap ring; pitch: F at $A = 409$ (which confirms a date contemporary with Stanesby senior);
overall length 107.4 cms/$42\frac{1}{4}$ ins including crook, body length 100.6 cms/$39\frac{5}{8}$ ins, sounding length 89.7 cms/$35\frac{3}{8}$ ins; very heavy brass crook, 1.32 cms/$\frac{1}{2}$ in outside diameter

PROVENANCE
Canon Francis W. Galpin, after 1895; Dr Rosamond Harding, 1946; Eric Halfpenny, 1951–62; Warwickshire Museum, 1973.

LITERATURE
Eric Halfpenny, 'Further Light on the Stanesby Family', *Galpin Society Journal*, XIII, 1960, p. 66;
Eric Halfpenny, 'Technology of a Bass Recorder', *Galpin Society Journal*, XV, 1962, pp. 49–54, pl. 6.

EXHIBITED
British Musical Instruments, Galpin Society, London, 1951, no. 10.

178

Transverse Flute

By Thomas Stanesby junior (1692–1754)

Handel frequently wrote for *traversa*, as he normally referred to this instrument – usually, as with the recorders, in pairs. This instrument is very similar in the size of its ivory mounts to the flûte d'amour (no. 179), suggesting an early date, very possibly before the death of the elder Stanesby.

Horniman Museum
(*Carse Collection*) 14.5.47–241

Stamped *STANESBY / IUNIOR*

Dark brown wood with heavy ivory head-cap and mounts and silver key for D♯;
overall length 61.6 cms/$24\frac{1}{4}$ ins, sounding length 55.1 cms/$21\frac{3}{4}$ ins

PROVENANCE
Given by Adam Carse to the Horniman Museum, 1947.

EXHIBITED
Eighteenth Century Musical Instruments: France and Britain, Victoria and Albert Museum and elsewhere, 1973, no. 89.

179
Flûte d'Amour
By Thomas Stanesby junior (1692–1754)

When Handel wrote for *traversa bassa* in *Riccardo Primo* for the Royal Academy in 1727, he scored it as a transposing instrument in B flat, thus indicating that he had in mind this size of instrument, commonly called the flûte d'amour, with B flat as the lowest note. Nowadays, the description 'bass flute' implies a larger instrument, either in G or an octave below the normal flute.

This instrument by Stanesby junior may have been made early in the eighteenth century while his father was still alive. Before its embouchure (or mouthhole) was enlarged, thus raising the pitch to B♭ at A = 415, it must have been pitched at the early eighteenth-century level of around A = 409. The only other Stanesby junior flute bearing the same mark as this instrument is in the Museo Civico at Modena; it retains the small embouchure and is pitched well below A = 415. This evidence that the younger Stanesby made flutes at the pitch used in his father's lifetime strongly suggests that there may have been a division of labour: that the father made recorders, oboes and bassoons (no flute by him has ever been found) and the son specialised in flutes, perhaps only making the other instruments after his father died or retired.

Bate Collection of Historical Instruments, Faculty of Music, University of Oxford

Stamped STANESBY / IUNIOR / ♭B

Stained boxwood, with heavy ivory head-cap and mounts and a silver key for D♯;
overall length 75.4 cms/29¾ ins,
sounding length 69.6 cms/27⅜ ins

PROVENANCE
Purchased for the Bate Collection, 1984.

LITERATURE
A detailed drawing of the instrument, with all measurements, has been published by the Bate Collection.

180
'Handel's Pitch Pipe'
By an unknown maker

A pitch pipe resembles an organ pipe but is designed normally to be mouth-blown. A long movable plunger acts as a stopper and is marked with a series of note names; when the plunger is drawn out so that a certain note name is level with the end of the pipe, the pipe will produce that note. Before the invention of the tuning fork in 1711 by Handel's trumpet player, John Shore, the pitch pipe was used as a tuning standard from which to tune other instruments. The fork was much more accurate, and also more portable, but the pipe continued in use as a convenient device for giving a note to a choir when there was no other instrument present; it has the advantage over a fork that it can produce any note within its range, usually one octave, whereas a fork can only produce a single note.

This pitch pipe is not documented before 1842 when it was given to the Sacred Harmonic Society: though of the type used in the eighteenth century, its condition and the relatively high level of its pitch mean that it must date to after Handel's lifetime.

The Royal Society of Musicians of Great Britain

Mahogany, square section, in a later leather-covered case in a mahogany box with five silver ownership plaques; scale marked from D to D; pitch: an octave above A = 434; overall length 21 cms/8¾ ins, section 2 × 2 cms/¾ × ¾ ins

PROVENANCE
According to the silver plaques attached to the box of the pitch pipe, given by Handel to Dr Burney, whose son gave it to his cousin, Mr Sansom; given by him to Mr Harker, 1842, who the same year gave it to the Sacred Harmonic Society, Exeter Hall; sold by the Society to Messrs Edward & Sons for George Mence Smith, 1882; at his death, purchased by C.T. Johnson, 1896; at his death, purchased by William Bradford of Leominster, 1924; purchased from him by Margaret Godfrey (Mrs J.H. Bennett) and given to the Royal Society of Musicians, 1929.

LITERATURE
A.E. Ellis, 'On the History of Musical Pitch', *Journal of the Society of Arts*, 1880.

178

179

Messiah: Dublin and London

Messiah has gained such a special reputation among Handel's works that it is not easy for us now to understand why it took so long for the oratorio to find public favour. The idea for *Messiah* belongs to Charles Jennens (see no. 181), with whom Handel had already worked on *Saul* (see no. 137) and *L'Allegro*, first performed in 1740. Jennens's exalted plans are explained in a revealing letter (no. 182) to his friend Edward Holdsworth on 10 July 1741; he has produced a 'Scripture Collection' and he hopes Handel 'will lay out his whole Genius & Skill upon it, that the Composition may excell all his former Compositions, as the Subject excells every other Subject'.

Handel's score (no. 183) was completed in the space of only 24 days: starting work on 22 August, he finished Part I on 28 August, Part II on 6 September and Part III on 12 September; the final orchestration took the next two days. It appears that the work was written with Dublin in mind, where Handel had been invited to give a season by the Lord-Lieutenant, William Cavendish, 3rd Duke of Devonshire. Handel left London early in November and travelled to Dublin via Chester and Holyhead, arriving on 18 November. Dublin was then the largest provincial city in the British Isles, and counted a lively society of its own with the Lord-Lieutenant, Governor of Ireland, as the apex to the social pyramid. A special service with a *Te Deum* and *Jubilate* by Handel, was held on 10 December at St Andrew's church and then on the 12th an announcement appeared in *Faulkner's Dublin Journal*: 'On Monday next . . . the 14th of December (and every Day following) Attendance will be given at Mr Handel's House in Abby-street, near Lyffey-street . . . to receive the Subscription Money for his Six Musical Entertainments in the New Musick-Hall in Fishamble street.'

The first concert was held on 19 December and the immediate success of the Dublin season was reported by Handel to Charles Jennens on 29 December:

> The Nobility did me the Honour to make amongst themselves a Subscription for 6 Nights, which did fill a Room of 600 Persons, so that I needed not sell one single Ticket at the Door, and without Vanity the Performance was received with a general Approbation. . . . I opened with the Allegro, Penseroso, & Moderato and I assure you that the Words of the Moderato are vastly admired.

Jennens was of course the author of *Il Moderato*.

Handel's attempts to augment his chorus with members of the choir of St Patrick's Cathedral met with a temporary rebuff at the hands of Jonathan Swift, Dean of St Patrick's Cathedral, who was already beginning to show signs of insanity:

> And whereas it have been reported, that I gave a licence to certain vicars to assist at a club of fiddlers in Fishamble Street, I do hearby declare that I remember no such licence to have been ever signed or sealed by me; . . . intreating my said Sub-Dean and Chapter to punish such vicars as shall ever appear there as songsters, fiddlers, pipers, trumpeters, drummers, drum-majors, or in any sonal quality, according to the flagitious aggravations of their respective disobedience, rebellion, perfidy, and ingratitude.

It was only at the end of March, after a successful season of oratorios including *Alexander's Feast*, that *Messiah* was announced in an advertisement in *Faulkner's Dublin Journal*: 'For the Relief of Prisoners in the several Gaols, and for the Support of Mercer's Hospital . . . and of the Charitable Infirmary . . . on Monday the 12th April, will be performed at the Musick Hall in Fishamble Street, Mr. *Handel's new Grand Oratorio, called the* MESSIAH, in which the Gentlemen of the Choirs of both Cathedrals will assist. . . .' After a rehearsal on 9 April followed by a first performance on 13 April, *Faulkner's Dublin Journal* declared of *Messiah* that 'Words are wanting to express the exquisite Delight it afforded to the admiring crowded Audience. The Sublime, the Grand, and the Tender, adapted to the most elevated, majestick and moving Words, conspired to transport and charm the ravished Heart and Ear.' Handel gave *Messiah* again on 3 June, his last performance before leaving Dublin on 13 August, well satisfied with the success of his long stay.

In contrast to the welcome given to *Messiah* in Dublin, it did not reach the London stage without a certain amount of controversy as to whether a work on such a theme should be given in a theatre. Charles Jennens wrote on 21 February 1743, a month before the performance, 'there is a clamour about Town, said to arise from the B[rs] [Brothers, ie the Methodists] against performing it'. As Jennens makes clear,

Handel felt obliged to avoid the title *Messiah*, instead advertising the work on 19 March as 'A New Sacred Oratorio'. On the same day a letter in the *Universal Spectator* asked the public 'to consider . . . the Impropriety of Oratorios, as they are now performed. . . . An Oratorio either is an Act of Religion, or it is not; if it is, I ask if the Playhouse is a fit Temple to perform it in, or a Company of Players fit Ministers of God's Word. . . .' Despite this controversy, public reaction to *Messiah* was remarkably muted, and there were no reports in the newspapers when it was given the first of three performances at Covent Garden Theatre on 23 March, with a cast which now included Kitty Clive and John Beard (see nos. 187 and 188).

Handel gave two performances of *Messiah* in April 1745 but it was not until 1750, with the first of the annual performances at Foundling Hospital Chapel, that the work began to find favour (see p. 223). By the time of Handel's death nine years later, *Messiah* had achieved its position of pre-eminence among Handel's works, and was given each spring at both the Foundling Hospital and at Covent Garden Theatre. Although the very first performance of *Messiah* in Dublin was given in aid of three charitable causes, it was through the Foundling Hospital performances that the work became associated with the idea of raising funds for charity, a pattern which continued into the nineteenth century (see nos. 255–6).

LITERATURE
Deutsch, pp. 527, 530, 536–7, 542, 546, 563; *Hogwood*, p. 183.

181

181

Charles Jennens (1700–73)

By Thomas Hudson (1701–79)

It was Charles Jennens who provided the text of *Messiah* for Handel. A rich man, he came from a very different walk of life to Handel's other librettists and, though a friend of the composer's, he always allowed his superiority to affect their dealings. This portrait was painted by Thomas Hudson in 1747, the year Jennens inherited his father's estate at Gopsall in Leicestershire. Jennens had a reputation for great wealth and extravagance: he was nicknamed 'Solyman the Magnificent' and is said to have used a coach and four, complete with four footmen to go but a few minutes' walk. Though his literary talent was modest – his edition of Shakespeare's tragedies (1770–4) was much criticised – the very real merit of the texts he prepared for Handel has sometimes been overlooked.

Jennens's personal collaboration with Handel extended to four works: *Saul*, based on the Old Testament books of Samuel, which appears to have been received by Handel as early as 1735 but not set to music until September 1738 (see no. 137); *L'Allegro, il Penseroso ed il Moderato*, an expanded version of Milton's poems which received its first performance in 1740; *Messiah*, written in 1741 and performed in 1742 (see Handel's and Jennens's letters, nos. 182 and 186); and *Belshazzar*, the subject of several letters from Handel to Jennens in the summer of 1744 (see no. 191), in which he eagerly presses his librettist for the text of the new oratorio.

In his correspondence with Jennens, Handel is often at pains to assure him of his esteem and of the wider regard of the public. For his part Jennens's unbearable arrogance is abundantly clear from his private correspondence with his friend Edward Holdsworth. Early in 1743 he writes: 'His Messiah has disappointed me. . . . I shall put no more Sacred Words into his hands, to be thus abus'd,' though he relents a little after hearing the work for the first time: ''Tis after all, in the main, a fine Composition, not withstanding some weak parts, which he was too idle & too obstinate to retouch. . . .' Nevertheless Jennens continued his attacks on Handel, so much so that Holdsworth felt obliged to restrain him; he wrote from Florence on 28 October that year, 'You have contributed, by yr. own confession, to give poor Handel a fever, and now He is pretty well recover'd, you seem resolv'd to attack him again; for you say you have not yet done with him. This is really ungenerous, & not like Mr. Jennens. Pray be merciful; and dont you turn Samson, & use him like a Philistine. . . .'

Jennens was a subscriber to all Handel's printed scores and also owned a substantial collection of manuscript scores of Handel's music, the largest of its kind; though this is now widely dispersed there are substantial holdings in the British Library and Manchester Public Library (see no. 137). His extensively annotated copy of Mainwaring's *Life of Handel* (no. 214) demonstrates his wide-ranging knowledge of Handel's music. Though the composer chose to work with other librettists in his last years, he seems to have remained in touch with Jennens, supplying him with a specification for a new organ for Gopsall in September 1749 and sitting to Thomas Hudson for the full-length portrait commissioned by Jennens in 1756 (no. 12). Handel remembered Jennens in his will in 1757, not by leaving him money as he did his less wealthy friends but by the bequest of two pictures, 'the Old Man's head and the Old Woman's head done by Denner'.

Private Collection

Oil, 125.7 × 99.0 cms/49½ × 39 ins

Signed and dated on the right *Thos. Hudson pinxit / 1747*

PROVENANCE
Presumably by descent from the sitter through his great-nephew Heneage Finch, 3rd Earl of Aylesford, and thence by way of his second son Charles Finch to the present owner.

LITERATURE
John Steegman, *A Survey of Portraits in Welsh Houses*, I, Cardiff, 1957, p. 129, no. 35, pl. 19c; *Deutsch*, p. 394; *Hogwood*, p. 184.

EXHIBITED
Thomas Hudson 1701–1779, Iveagh Bequest, Kenwood, 1979, no. 33.

182

'The Subject is Messiah'

Autograph letter from Charles Jennens at Gopsall to Edward Holdsworth (who was travelling on the continent), care of the banker Monsieur Waters at Paris, 10 July 1741

Charles Jennens's extensive correspondence with his close friend the classical scholar Edward Holdsworth (1684–1746) covers many topics of classical, literary and musical interest. In this letter, Jennens does not turn to music until the third page, requesting

HANDEL

Holdsworth to purchase various compositions by D'Astorga, Latilla and Jommelli. He then goes on to tell of his plans to encourage Handel to write a new oratorio, *Messiah*:

> Handel says he will do nothing next Winter, but I hope I shall persuade him to set another Scripture Collection I have made for him, & perform it for his own Benefit in Passion week. I hope he will lay out his whole Genius & Skill upon it, that the Composition may excell all his former Compositions, as the Subject excells every other Subject. The Subject is Messiah.

Handel began work on his autograph (no. 183) the next month and rapidly completed it. But as Jennens's subsequent correspondence with Holdsworth makes clear, he was far from pleased with the result; see under no. 181.

Gerald Coke Handel Collection

Letter, single sheet, folded in half, 4 pp., 22.8 × 18.2 cms/9 × 7$\frac{1}{8}$ ins page size; bound into the second of two volumes of unpublished correspondence between Jennens and Holdsworth, 1729–46

PROVENANCE
By descent to Earl Howe as no. 12; his sale at Christie's, 4 July 1973, lot 3, purchased for Gerald Coke.

LITERATURE
Coke, p. 31.

183
Messiah
Handel's autograph score

Handel began what was to become his most famous work on 22 August 1741 and completed the score on 14 September. It was first performed in a slightly revised form at Neal's Music Hall in Dublin on 13 April 1742, the proceeds of the concert being divided between three local charities. One press report (*Faulkner's Dublin Journal*, 10 April 1742) claimed that the oratorio had been composed specifically for the Dublin charities. There is some plausibility in this statement. Uniquely among Handel's oratorios, the orchestra of *Messiah* as originally composed contains no oboes or bassoons, and there is no indication of singers' names against the vocal solos. The implication is that Handel was uncertain about what orchestral forces and what singers would be available to him, and this suggests he was writing for Dublin rather than London. (*Samson*, completed in draft only six weeks after Handel had finished *Messiah*, and first performed in London in 1743, is scored in Handel's usual manner). For his later London revivals, Handel added oboes, bassoons and horns (probably doubling the trumpets) to his score.

The displayed pages (ff. 99v–100r) show the end

of the tenor solo 'Thou shalt break them' and the beginning of the chorus 'Hallelujah! for the Lord God omnipotent reigneth!' – universally known as the Hallelujah Chorus, though there are many other choral settings of 'Allelujah' or 'Hallelujah' in Handel's works. It will be seen that though Handel has prepared staves for the trumpets (*T1 et 2*) and timpani (*Tymp.*) he uses only the string orchestra at first, allowing the trumpets and drums to make an exciting entry a little later on. To save himself the work of copying out all the verbal repetitions Handel writes the words in the treble and bass staves only, and uses symbols to indicate repeats.

This volume contains all the autograph material that survives from Handel's original score (the end of the overture and the start of the recitative 'Comfort ye' are lost), but other autograph material exists elsewhere. The Fitzwilliam Museum has a page of sketches (MS 263, p. 58), and the conducting score formerly at St Michael's College, Tenbury contains the autographs of Handel's revised settings of 'Rejoice greatly' (probably written for Francesina in 1745), 'But who may abide' (written for the *alto castrato* Gaetano Guadagni in 1750) and 'Thou art gone up' (also for Guadagni).

The British Library Board
RM 20.f.2

Autograph score, i + 140 ff., 23.5 × 29.8 cms/9¼ × 11¾ ins; f. 21 is a nineteenth-century insertion containing a re-copied version of f. 22r (disfigured by a blot); ff. 133–138r were added for the Dublin performance, 1742; f. 138v contains a copy in the elder Smith's hand of a solo added for the London performance of 1743; ff. 139–40 were added for the London revival of 1745

Inscribed on f. 1r ♄ *angefangen den 22. August 1741*, on f. 52v (the end of Part I) *August 28* ♀ *1741*, on f. 106r (the end of Part II) *An* ☉ *Septembr. 6. 1741.*, and on f. 132v *Fine dell'Oratorio. G.F. Handel.* ♄ *Septembr. 12. 1741. ausgefüllet den 14. dieses.*

PROVENANCE
As no. 34.

LITERATURE
Squire, p. 52; *Deutsch*, pp. 544–6; Watkins Shaw, *A Textual and Historical Companion to Handel's 'Messiah'*, 1965, pp. 23–30.

184

184

'Messiah. An Oratorio Compos'd by Mr. Handel.'

Word-book, first edition, text by Charles Jennens (compiled from the Bible and Prayer Book Psalter) Printed by George Faulkner, Dublin, 1742 ('Price a British Six-pence')

In a letter of 29 December 1741 Handel thanked Jennens for 'the Lines You was pleased to send me, in Order to be prefix'd to Your Oratorio Messiah, which I set to Musick before I left England'. The lines in question appeared on the title-page of the word-book issued for the first performance of *Messiah* at Neal's Music Hall in Dublin on 13 April 1742:

MAJORA CANAMUS.
And without Controversy, great is the Mystery of Godliness: God was manifested in the Flesh, justified by the Spirit, seen of Angels, preached among the Gentiles, believed on in the World, received up in Glory.
In whom are hid all the Treasures of Wisdom and Knowledge.

The Latin phrase ('Let us sing of greater things') is aptly taken from Virgil's Fourth Eclogue, in which a prediction of a return to the Golden Age ushered in by 'a new offspring sent down from high heaven' suggests a pagan perception of the coming of Christ. The Biblical quotations (I Timothy iii, 16 and Colossians ii, 3) summarise the themes of the oratorio.

As is usually the case, the word-book indicates the nature of the musical setting of each of the numbers ('Recitative', 'Recitative accompany'd', 'Song', 'Chorus', etc). It thus provides important evidence concerning the form in which the oratorio was sung at the first performance. It shows, for example, that 'But who may abide' was sung as a recitative (though in Handel's autograph it is set as a bass aria or 'Song'), that 'How beautiful are the feet' was sung in the duet-and-chorus version, and that 'Their sound is gone out' was omitted. The British Library copy has singers' names entered by someone who attended one of the early performances, giving additional useful information; it appears, for example, that Mrs Cibber sang the alto version of the recitative 'Then shall the eyes of the blind be opened' and the whole of the aria 'He shall feed his flock' in an alto version not represented in modern editions.

Gerald Coke Handel Collection

Printed word-book, original blue paper covers, 16 pp., 24.0 × 19.5 cms/9½ × 7⅝ ins

Titled as above and inscribed in pen on the title-page in two different hands *Henry Paget / Moses Brown's b'ot at Vendue M^r M . . . ts*(?)

LITERATURE
J.C. Culwick, *Handel's Messiah: Discovery of the Original Word-Book used at The First Performance in Dublin, April 13, 1742*, Dublin, 1891; Watkins Shaw, *A Textual and Historical Companion to Handel's 'Messiah'*, 1965, pp. 109–13.

185 [*in colour opposite p. 204*]

Susannah Cibber (1714–66)

By Thomas Hudson (1701–79)

Handel, according to Burney, 'was very fond of Mrs. Cibber, whose voice and manners . . . softened his severity for her want of musical knowledge'. Her success in Handel's oratorios was by all accounts due to the intelligence and feeling she brought to her roles rather than the quality of her singing voice, which was not in itself remarkable. She later became a great tragic actress, appearing in partnership with David Garrick. This portrait, a soft and delicate image which is arguably one of Thomas Hudson's finest works, was not engraved until 1749 but may be rather earlier, so similar in appearance is it to her other portrait by Hudson, the Garrick Club picture of 1739.

Susannah Maria Cibber was the sister of the composer Thomas Arne in whose pirated performance of Handel's *Acis and Galatea* she appeared in May 1732 at the age of 18. She married Theophilus Cibber, son of the actor and dramatist Colley Cibber, in 1734 but separated in 1738 in particularly public and bitter circumstances; her husband sued her lover, John Sloper, for £5,000 damages but was awarded only £10 on grounds of his collusion in the matter. Her next public appearance was in Dublin during the winter of 1741/2 where she acted in James Quin's theatre season and sang for Handel in many of his concerts. On 13 April 1742, after a period of illness, she appeared in the first performance of *Messiah*. Her singing of 'He was despised' deeply moved the audience, creating such an effect that the Rev. Dr Delany (later to marry Handel's close friend, Mrs Pendarves; see no. 156) rose to his feet exclaiming, 'Woman, for this, be all thy sins forgiven!'. According to Thomas Sheridan she joined force of expression with the charms of music: 'No person of sensibility, who has had the good fortune to hear Mrs Cibber sing in the oratorio of the Messiah, will find it very difficult to give credit to accounts of the most wonderful effects produced from so powerful an union.'

In London she sang again in *Messiah* and took the roles of Micah in *Samson* in 1743 and Lichas in *Hercules* (1745); she was cast for the part of Daniel in *Belshazzar* in 1744 (see no. 191), but does not seem to have appeared. From then on she entered into the partnership with David Garrick on the stage at Drury Lane which continued almost unbroken until her death.

National Portrait Gallery, London
NPG 4526

Oil, 76.2 × 62.9 cms/30 × 24¾ ins

Engraved in reverse by J. Marchand, 1749

PROVENANCE
By descent to T.G.B. Forster of Holt, Wiltshire; sold Sotheby's, 24 November 1965, lot 90 and again, 11 January 1967, lot 140, where purchased by Leggatt's for the National Portrait Gallery.

LITERATURE
Burney, *Sketch*, p. 35; *Kerslake*, pp. 55–6 (repr. in colour).

Susannah Cibber by Thomas Hudson, before 1749 (no. 185)

Kitty Clive by Jeremiah Davison, *c.* 1735 (no. 188)

Faustina Bordoni by Bartolomeo Nazari (no. 96)

John Beard by Thomas Hudson, c. 1743 (no. 187)

186

Messiah: 'how zealous they are in Ireland for Oratorio's'

Autograph letter from Handel in London to Charles Jennens at Gopsall, 9 September 1742

Handel left Dublin on 13 August 1742, having been away from London for more than nine months. In this letter written soon after his return home he compliments Jennens on 'how well Your Messiah was received' in Ireland and encloses a very favourable report from the Bishop of Elphin, Dr Edmund Synge, which begins prophetically: 'As Mr Handel in his oratorio's greatly excells all other Composers I am acquainted with, so in the famous one, called The Messiah he seems to have excelled himself. . . .' To this Handel adds his own note, presumably addressed to Jennens: 'I send you this S[r] only to show how zealous they are in Ireland for Oratorio's. I could send you a number of Instances more from others in Print and in writing.'

Handel also promises to send Jennens 'the printed Book of the Messiah' (see no. 184) and describes his plans for the coming season: 'The report that the Direction of the Opera next winter is comitted to my

Care, is groundless. The gentlemen who have undertaken to middle with Harmony cannot agree, and are quite in a Confusion. Whether I shall do something in the Oratorio way (as several of my friends desire) I cannot determine as yet.' In the event Handel did mount a season of oratorios at Covent Garden during February and March 1743, including the first London performance of *Messiah* and a new work, *Samson*.

Despite his confident assertion, towards the end of his letter, that 'Certain it is that this time 12 month I shall continue my Oratorio's in Ireland,' Handel never returned there.

Gerald Coke Handel Collection

Letter, single sheet, folded in half, 3 pp., 22.7 × 19.0 cms/9 × 7½ ins page size

PROVENANCE
By descent to Earl Howe as no. 12; his sale at Christie's, 4 July 1973, lot 1 (with eight other letters); subsequently purchased by Gerald Coke.

LITERATURE
Deutsch, pp. 554–5 (for a full transcript).

EXHIBITED
Coke, p. 31, pl. 11a.

187 [*in colour opposite p. 205*]

John Beard (*c*. 1717–91)

By Thomas Hudson (1701–79)

Hudson's striking portrait of the singer and actor John Beard in his red coat trimmed in gold has been dated stylistically by Ellen Miles to about 1743; this was the year when Beard appeared in the first London performance of *Messiah*.

Beard was Handel's greatest English singer, and his talents were developed and encouraged by the composer. He first performed an important part in a work by Handel while still a choir boy of the Chapel Royal, when in 1732 he took the treble part of the Israelite priest in the revival of *Esther* (see no. 128). Beard left the Chapel Royal in October 1734 and immediately joined Handel's opera company. He was noticed by a visitor to *Il Pastor Fido* in November: 'Beard... shines in the Opera of Covent Garden and Mr Hendell is so full of his Praises that he says he will surprise the Town with his performances before the Winter is over.' By then Beard's voice had deepened to the strong tenor which made him the leading English singer of his generation. In the course of the next 25 years, though not exclusively a Handel singer, he performed for him on numerous occasions, appearing in some ten operas and almost all his oratorios, creating the title roles in *Samson* (1743), *Belshazzar* (1745), *Judas Maccabaeus* (1746) and *Jephtha* (1752).

In 1739 Beard married the recently widowed Lady Henrietta Herbert, daughter of the Earl of Waldegrave, in the face of opposition from her family and much gossip ('there is no prudence below the girdle', wrote Lord Egmont of her five days after the wedding). Beard's marriage, as the first of a male player of any degree to a peer's daughter, was the subject of exceptional comment, just as Anastasia Robinson's to Lord Peterborough (see no. 92) had been considered remarkable a few years before as the first of a singer or actress to a peer. Lady Henrietta died in 1753, and six years later Beard married Charlotte, the daughter of John Rich, the manager of Covent Garden Theatre, taking over the management of the theatre at Rich's death in 1761. Beard retired from the stage in 1767.

Gerald Coke Handel Collection

Oil, 74.2 × 62.2 cms/29¼ × 24½ ins

Engraved by James McArdell, *c*. 1764 (*CS*, p. 841, no. 14).

PROVENANCE
Sotheby's, 15 July 1959, lot 138, purchased by Gooden & Fox for Gerald Coke.

LITERATURE
Ellen Miles, *Thomas Hudson (1701–1779): Portraitist to the British Establishment*, Yale University PhD, 1976, published by University Microfilms International, Ann Arbor, Michigan and London, 1978, I, p. 31, no. 12.

188 [*in colour between pp. 204–5*]

Kitty Clive (1711–85)

By Jeremiah Davison (*c*. 1695–1745)

Like Mrs Cibber, another Handel singer of the 1740s, Kitty Clive was better known in the theatre than the opera house. She usually appeared at Drury Lane, at first under Colley Cibber and later with David Garrick, and it is as one of the great comic actresses of the middle of the eighteenth century that she is perhaps best known. Though Handel had written songs for her in the later 1730s it was only in the short season of 1743 that she sang in his oratorios, creating the role of Dalila in *Samson* and singing in the first London performance of *Messiah* and in *L'Allegro*. After the first performance of *Samson* in February 1743, Horace Walpole (later to own this painting) wrote derisively of Handel's hiring 'all the goddesses from farces'; he probably had Mrs Clive particularly in mind, for he had noticed her the previous May in one of her famous imitations of Italian opera singers: 'There is a little simple farce at Drury Lane . . . in which Mrs Clive mimics the Muscovita admirably, and Beard, Amorevoli tolerably.' Indeed Burney picked on this lighter side of her talent to praise: 'her singing, which was intolerable when she meant it to be fine, in ballad farces and songs of humour was, like her comic acting, every thing it should be'.

This portrait of the mid-1730s was commissioned by Kitty Clive's brother from Jeremiah Davison, a successful portrait painter of Scottish origins.

Sixth Marquess of Bath

Oil, 125.8 × 100.2 cms/49½ × 39½ ins

Engraved by Alexander van Aken, 1735, as after his brother, Joseph van Aken, who may have been responsible for the draperies in the portrait

PROVENANCE
Given to Kitty Clive by her brother, James Raftor; Horace Walpole; Strawberry Hill sale, 18 May 1842, lot 13, purchased by P. Botfield; acquired by the Marquess of Bath.

LITERATURE
Horace Walpole, *A Description of Strawberry Hill*, 1784, p.92; *Deutsch*, pp. 549, 560.

189 [*illustrated on p.172*]

'Songs in Messiah an Oratorio Set to Musick by Mr. Handel.'

Printed score, first edition
Printed for John Walsh by 1763

Songs in Messiah, the earliest published collection of music from the oratorio, was not issued immediately following the first performance, as was usually the case; instead publication was delayed until after the composer's death. This untypical cautiousness on the part of Walsh may have been the result of the hostility with which *Messiah* was received by certain religious factions when it was first performed in London in 1743. Some of the music did however emerge in Walsh's various serial publications. The overture was published in 1743, first in the instrumental part-books of *Six Overtures for Violins, etc . . . Eighth Collection* and then in a keyboard arrangement in *Six Overtures fitted to the Harpsichord or Spinnet . . . Eighth Collection*. In April 1744 an instrumental version of 'He was despised' (described as 'Air . . . in the Sacred Oratorio') was included in Part 3 of Volume V of *Sonatas or Chamber Aires*, a long-running series of arrangements of opera and oratorio arias. In another series, *Handel's Songs selected from his Oratorios*, the arias of *Messiah* gradually appeared over the period 1749–59.

Songs in Messiah, containing the overture, Pastoral Symphony and all the arias of the oratorio, seems not to have been issued until 1763; no advertisement for such a publication can be traced before that date. There is evidence, however, that Walsh had prepared the edition in Handel's lifetime. Though the earliest state of the print (of which this exhibit is an exemplar) has added figurings on the bass line in a style not used by Walsh before 1760, the main engraving style is consistent with an earlier date. The passe-partout title-page was first used for *Judas Maccabaeus* in 1747, and William C. Smith concluded from an examination of other uses of the plate that the *Songs in Messiah* version belonged to about 1749. The music plates may originally have been engraved even earlier, and not from an authoritative source, since in some cases the items reflect the London performances of 1743 and sometimes go back to versions discarded by Handel before the first performance in Dublin in 1742. The music text is often corrupt. A curious feature also suggesting an early date is the printing of the alto voice parts of 'O though that tellest' and 'He was despised' an octave higher than sounding pitch, as if they were high tenor parts; 'He was despised' is marked *To be sung an Octave lower*. Despite its importance as the first substantial printing of a very famous work, *Songs in Messiah* is of very little textual value; modern editions are based almost entirely on the more reliable manuscript sources.

Gerald Coke Handel Collection

Printed score, 70 pp., 33 × 24.5 cms/13 × 9⅝ ins; bound with *The Triumph of Time and Truth*, 1757 (*Smith*, p.146, no.1)

Titled as above and inscribed inside the front cover *Richard Warner 1765*, and with the bookplates of Richard Warner of Woodford Row, Essex and of W.C. Smith

PROVENANCE
Richard Warner, 1765; W.C. Smith, from whom acquired by Gerald Coke.

LITERATURE
W.C. Smith, 'The Earliest Edition of "Messiah"' *Concerning Handel*, 1948, pp. 67–110, pl. 4; *Smith*, p.116, no.1; Watkins Shaw, *A Textual and Historical Companion to Handel's 'Messiah'*, 1965, pp. 127–31.

HANDEL

War and Peace

The war with Spain which broke out in October 1739 gave way to a wider conflict during the 1740s, the War of Austrian Succession, in which Great Britain, Holland and Austria were opposed to France and Prussia. The conflict dragged on until near the end of the decade, with British defeats outweighing successes, on land at least.

In London the winter season of 1739/40 opened with the first performance of Handel's *Ode for St Cecilia's Day*. But the extremely cold weather – the Thames froze over – and the excitement of war kept the audience away: 'Plays we have none, or damned ones', complained Richard West to Horace Walpole in a letter on 13 December, 'Handel had had a concerto this winter. No opera, no nothing. All for war and Admiral Haddock.'

The War did not stop Handel visiting the continent in the summer of 1740, perhaps to seek further singers for the opera. In fact the major conflict did not begin until 1741 and it was only in 1742 that Britain became substantially involved; the following June George II led his troops to an indecisive victory at Dettingen (see no. 190), subsequently marked by Handel's new *Te Deum* setting and an anthem, first performed in November. Handel's winter seasons at Covent Garden or, occasionally, the Haymarket, continued as usual but were now, since his return from Dublin, exclusively given over to oratorio (see no. 191 for *Belshazzar*).

The second conflict of the 1740s, the '45 Jacobite rebellion, had a much more critical effect on London life. The Young Pretender's rapid progress from his landing in the Hebrides on 21 July 1745, to Derby, the furthest point in his march south, on 4 December, was a threat alarmingly closer to home than a continental war. Handel turned his talents to such overtly patriotic works as his *Song for the Gentlemen Volunteers of the City of London*, first performed on 14 November, and his *Occasional Oratorio*, hurriedly produced in the winter months to mark the retreat of the rebels and given its first performance at Covent Garden on 14 February 1746.

His next three major works, *Judas Maccabaeus*, *Alexander Balus* and *Joshua*, treat military themes from the Old Testament and Apocrypha and can be seen in the context of contemporary events: *Judas Maccabaeus* is dedicated to the Duke of Cumberland (see no. 192), and the other two works, first performed in 1748, were intended to capitalise on the success of this oratorio. The continental War had by this time reached an impasse; preliminary articles of peace were signed in June 1748 and the final treaty, the Peace of Aix-la-Chapelle, on 7 October that year. It was to mark the official proclamation of peace the following spring that Handel composed his Peace Anthem and his *Music for the Royal Fireworks* (no. 195), his last works written for such national celebrations.

190

George II at the Battle of Dettingen

By John Wootton (c.1682–1764)

John Wootton's grandiose scene of the victorious George II at Dettingen in June 1743 shows the King (the last British sovereign to lead his troops into battle) with, it seems, his 22-year-old son the Duke of Cumberland (see no. 192) and, in profile on the right, the youthful 4th Earl of Holderness, for whom the picture may have been painted. In the background the British infantry can be seen firing on a French cavalry charge. The battle took place at Dettingen, south-east of Frankfurt, close to the River Main; it saw the triumph of combined British, Hanoverian and Austrian forces over a superior French army.

To mark the King's safe return from battle, one of the few British military victories of the War of Austrian Succession, Handel composed his Dettingen *Te Deum*, first performed at the Chapel Royal at St James's Palace in November 1743.

Since George II's coronation in 1727 (see no. 109), Handel had produced music for many state occasions, including the marriage of the King's eldest daughter in 1734 (see no. 113) and the funeral of his wife, Queen Caroline, in 1737. Handel died one year before George II, having known him for almost fifty years, first as Electoral Prince at Hanover, then as Prince of Wales and finally as King. Throughout this time Handel found him 'a steady patron', to use Burney's words, and the King would attend his oratorios 'when they were abandoned by the rest of his court'.

National Army Museum, London

Oil, 141.5 × 158.7 cms/55¾ × 62½ ins

Signed bottom right *J. Wootton/Fecit*; inscribed bottom left *THE BATTLE OF DETENGEN/ Gained by the Allies of the Queen of/ Hungary under the Command of his Majesty/ the King of Great Brittain over the French/ Army (Allies to the Emperor) commanded/ by the Mareshall Duc de Noailles./ June the 16/27 1743*

PROVENANCE
Perhaps painted for Robert, 4th Earl of Holderness; presumably by descent through his only surviving child, Amelia, wife of the 5th Duke of Leeds, and thence to the 11th Duke; his sale at Sotheby's, 14 June 1961, lot 17, purchased by the National Army Museum.

VERSION
At Hopetoun House, dated 1753, slightly smaller with different accompanying figures.

LITERATURE
Hornby Castle, Catalogue of Pictures, 1898, p. 18, no. 196; *Kerslake*, p. 95, pl. 251.

191

Belshazzar: Handel's Plans for the 1744/5 Season

Autograph letter from Handel in London to Charles Jennens 'in the Country', perhaps at Gopsall, 9 June 1744

After a successful oratorio season at Covent Garden in the spring of 1744, Handel's mind turned to the following winter; on 9 June, shortly before leaving London, he is to be found writing to his librettist Charles Jennens (see no. 181) about his plans for the new season: 'I have taken the Opera House in the Haymarketh. engaged, as Singers, Sig.ra Francesina, Miss Robinson, Beard, Reinhold, Mr Gates with his Boyes's and several of the best Chorus Singers from the Choirs, and I have some hopes that Mrs Cibber will sing for me.' He goes on to say that Mrs Cibber would need the permission of John Rich, the manager at Covent Garden, but in the event she did not appear. Handel then goes on to the point of his letter, his next collaboration with Jennens, the oratorio *Belshazzar*, which was not performed until 27 March 1745: 'Now should I be extreamly glad to receive the first Act, or what is ready, of the new Oratorio with which you intend to favour me, that I might employ all my attention and time, in order to answer in some measure the great Obligation I lay under.'

This is the first of five letters from Handel to Jennens in a summer of feverish activity in which he wrote *Belshazzar* and another oratorio, *Hercules* (to a text by Thomas Broughton). These months saw 'the peak of Handel's creative life', according to Winton Dean in his monumental survey of Handel's oratorios: 'The creative stamina of an artist who could compose a work of the physical and spiritual dimensions of *Belshazzar* immediately after completing *Hercules* is little short of miraculous, especially as he seems to have had no time to ponder either in advance.'

On 19 July, on Handel's return to London, he wrote to Jennens that the first act of *Belshazzar* 'gives me great Pleasure' and requested the remaining acts; by 21 August he had received the second act and began work on the music while he waited 'impatiently' for the third act; on 13 September Handel again begged Jennens 'with anxiety' to complete the oratorio, expressing his 'great Delight in setting it to Musick'; finally on 2 October he had received the third act and within three weeks had completed the composition. It is sad then to remark that when *Belshazzar* was performed in March it received only three performances.

Gerald Coke Handel Collection

Letter, single sheet folded in half, 3 pp., 22.7 × 18.7 cms/9 × 7⅜ ins page size

PROVENANCE
As no. 186.

LITERATURE
Deutsch, pp. 590–1 (for a full transcript); *Dean*, pp. 434–5.

EXHIBITED
Coke, p. 32, pl. 12a.

192

William, Duke of Cumberland (1721–65)

By Bernard Baron (1696–1762)
after John Wootton (c. 1682–1764)

When Bonnie Prince Charlie landed in the Hebrides in July 1745, the 24-year-old Duke of Cumberland, who had served at Dettingen with his father, George II, was chosen to lead the hastily assembled English army north to meet the threat. His single-mindedness in victory in 1746 brought him the name 'Butcher Cumberland' among his foes but a hero's welcome in London, a city which had come close to panic the previous year with the news that Prince Charles had reached Derby on his march south.

It is as the proud victor of Culloden that Wootton has portrayed Cumberland, just as he did his father at Dettingen (no. 190). 'See the conquering hero comes' are the words which spring to mind – from Handel's Victory Oratorio, *Joshua*.

Handel's *Occasional Oratorio*, quickly put together and first performed in February 1746 as Cumberland pressed on north, was his own contribution to the tide of patriotic feeling in London. But victory had not yet been won: the words were 'expressive of the rebels' flight and our pursuit of them', wrote the Rev. William Harris to his sister-in-law on 8 February. His second piece, *Judas Maccabaeus*, comes after Cumberland's success at the Battle of Culloden in April where he viciously slaughtered the remnants of Bonnie Prince Charlie's army and put an end to the Jacobite rising. According to Handel's librettist, Thomas Morell, writing many years later, its plan was 'designed as a compliment to the Duke of Cumberland', and indeed the word-book (no. 193) is so dedicated. The first performance in April 1747 preceded the publication of this engraving by a matter of months.

National Portrait Gallery, London

Engraving, 50.8 × 39.8 cms/20 × 15⅝ ins

Inscribed *His Royal Highness William Duke of Cumberland. &c. &c. &c. With a View of the routed Rebel Army near Culloden | J. Wootton Pinxit | Publish'd According to Act of Parliament August the 10.th 1747. | B. Baron Sculp | Printed for and Sold by Bern.d Baron in Panton Square Pickadilly. Tho.s Bowles in S.t Pauls Church Yard, & John Bowles at y.e Black Horse in Cornhill.*

LITERATURE
Deutsch, p. 629; Dean, p. 461.

193

'Judas Macchabaeus. A Sacred Drama. As it is Perform'd at the Theatre-Royal in Covent Garden. The Musick by Mr. Handel.'

Word-book, probably for the first performance, text by Thomas Morell
Printed for John Watts, 1747

Judas Maccabaeus was the first collaboration between Handel and the Rev. Thomas Morell. It was first produced at Covent Garden on 1 April 1747, but had been composed eight months earlier, in July and August 1746. Morell's libretto seems to have been conceived even sooner, before the end of 1745. In a letter written around 1770 Morell stated that he had written it at Kew in 1746, at Handel's request, and that it 'was designed as a compliment to the Duke of Cumberland, upon his returning victorious from

193

Scotland'. However, the first edition of the word-book contains the air 'O liberty' in a footnote, with Morell's explanation that the air 'was design'd and wrote for this Place but it got, I know not how, into the *Occasional Oratorio* . . .'. Since 'O liberty' was certainly included in the *Occasional Oratorio* in February 1746 it seems that Handel must have had at least part of the libretto of *Judas Maccabaeus* before the end of 1745. Morell and Handel may well have wished to have a 'victory' oratorio ready at the earliest opportunity, but could only complete their plans when the Jacobite forces had been decisively defeated.

Morell's text wisely avoids specific topical references, but in dedicating the work to the Duke of Cumberland he invites his readers and listeners to equate the Duke with Judas Maccabaeus, the leader of the Jewish resistance to Syrian domination around 160 BC. Judas is shown rallying the Israelites 'in defence of your nation, religion and laws', a sentiment that appealed to English Protestants resisting the claims of a Catholic pretender to the British throne. The incidents of the story are taken from the First Book of Maccabees, in the Apocrypha, though Morell also added a few details from the Jewish historian Josephus. The easy tunefulness of much of the music and the reference to the temporarily popular Duke caught the mood of the moment in 1747, and enabled Handel to put behind him the losses of his disastrous season of 1744/5. The popularity of *Judas Maccabaeus* helped to establish Handel's Lenten oratorio seasons at Covent Garden as a permanent feature of London life. Unlike several other oratorios, *Judas Maccabaeus*, with *Messiah*, *Israel in Egypt* and *Samson*, sustained its popularity throughout the nineteenth century, and is still one of the more frequently performed.

Gerald Coke Handel Collection

Printed word-book, iv + 16 pp., 24.3 × 19.5 cms/9$\frac{1}{2}$ × 7$\frac{5}{8}$ ins

LITERATURE
Dean, pp. 461–76.

194

Thomas Morell (1703–84)

By James Basire (1730–1802)
after William Hogarth (1697–1764)

The classical scholar, the Reverend Thomas Morell, was the favoured librettist of Handel's old age, preparing the texts for his oratorios between 1746 and 1757, including *Judas Maccabaeus* (1747), *Theodora* (1750), *Jephtha* (1752) and *The Triumph of Time and Truth* (1757). In a letter written some years after Handel's death, Morell explains the origins of their collaboration: '. . . Mr Handell applied to me, when at Kew, in 1746, and added to his request the honour of a recommendation from Prince Frederic. Upon this I thought I could do as well as some that had gone before me, and within 2 or 3 days carried him the first Act of *Judas Maccabaeus* which he approved of.'

Morell was 'a warm friend and a cheerful companion, who loved a jest, told a good story and sang a good song' (*Dictionary of National Biography*). Unlike the aloof Jennens, Morell seems to have developed close links with Handel. He accompanied the composer on his visit to Tunbridge Wells in August 1758 when in search of a cure for his penultimate illness, and was remembered in Handel's will, in a codicil dated 6 August 1756: 'I give to Doctor Morell of Turnham Green Two Hundred Pounds.'

In this, his only known portrait, Morell is depicted 'in the character of a cynic philosopher' by his friend William Hogarth whom he had helped with the drafting of his *Analysis of Beauty*. Hogarth's original drawing (in the Horne Collection, Florence) is followed closely but for the introduction of an organ, Morell's favourite instrument. The engraving was published in February 1762 as the frontispiece to Morell's two-volume Greek thesaurus. A related oil, based on the print, is in the Gerald Coke Handel Collection.

National Portrait Gallery, London

Engraving, 19.7 × 14.9 cms/7$\frac{3}{4}$ × 5$\frac{7}{8}$ ins

Inscribed at the bottom *T. MORELL, S.T.P–S.S.A.* [ie Professor of Sacred Theology – Secretary of the Society of Antiquaries]/ *W^m. Hogarth del.* / *James Basire sculp.*; on one of the books *Æ t. 60.*; and on the paper on which he writes *THESARUS.*; and above the image at top right *to face Morelli Dedicatio*

LITERATURE
Deutsch, pp. 737, 851; Dean, pp. 462–3; Paulson, I, p. 344, pl. 292 (the original drawing).

194

The Royal Fireworks

The signing of the Treaty of Aix-la-Chapelle on 7 October 1748 brought the long years of the War of Austrian Succession to an inconclusive close, with little advantage for Britain.

Plans to mark the Peace with a grand display of fireworks were put in hand on 18 June 1748, the day after the signing of the Treaty preliminaries. The Duke of Montagu, a supporter of Handel from thirty years before (see no. 141), wrote in his capacity as Master General of the Ordnance to the Duke of Newcastle, saying that the officers in charge of the fireworks 'tell me they will take a considerable time to prepare'. Charles Frederick (see no. 196), an official in the Ordnance and Comptroller of the Woolwich Laboratory, was given responsibility, and already by 5 August was being paid quite substantial sums on account of the intended occasion. The site for the display was marked out in Green Park on 4 November.

If the proposed celebrations were an obvious object of fun for Horace Walpole's sharp pen, his is the first mention of music for the occasion:

> Charles Frederick has turned all his virtu into fireworks, and ... has prepared such a spectacle for the proclamation of the peace as is to surpass all its predecessors of bouncing memory. It is to open with a concert of fifteen hundred hands, and conclude with so many hundred thousand crackers all set to music, that all the men killed in the war are to be wakened with the crash, as if it was the day of judgment ... I wish you could see him making squibs of his papillotes, and bronzed over with a patina of gunpowder, and talking himself still hoarser on the superiority that his firework will have over the Roman naumachia.

Despite Walpole's mocking tone in this letter of 6 October, there seems no reason for his even mentioning music unless the idea was under discussion. Perhaps Handel was already involved, but his name does not occur until January.

To act as architect and designer of the fireworks, the well-known scene designer, the Chevalier Servandoni (see no. 197), was chosen, at the wish of the Duke of Montagu. He used four Italians as assistants: two artists, Andrea Casali (see no. 196) and Andrea Soldi, who produced the paintings and sculpture which decorated the 'Fireworks Machine', as it became known; and two specialists in fireworks, Gaetano Ruggieri and Giuseppe Sarti of Bologna.

Frederick took everything very seriously by all accounts; he was 'in the Green Park from 8 in the morning till 4 in the afternoon, has an office built there for him. The rest of the day he gives audiences and worries his spirits and his person till 'tis reduced to a shadow' (according to Mrs Boscawen in a letter to her husband, Admiral Boscawen, who was Frederick's brother-in-law). Frederick had much to contend with in the early months of 1749. There was the excitable Servandoni to keep in check who, already on 5 January, was advertising a book of the project 'without consulting the Duke of Montagu, the Controller Mr Frederick or anybody'; in April the Duke was complaining of 'being plagued almost every day' by Servandoni, and on the night of the fireworks there was the well-known incident of his arrest for drawing his sword on Frederick. Another minor nuisance was Andrea Casali's unofficial scheme to produce his own engravings, which led to the Board's chosen engraver, George Vertue (see no. 199), complaining of the Italians, perhaps with some justice, that he was 'a tool to their designs'.

This is not to forget Handel, who is described in the *London Magazine* for December 1748 as composing music for the occasion. However, he does not really enter the story until the month before the fireworks. 'Now Hendel proposes to lessen the number of trumpets, &c and to have violeens', writes the Duke of Montagu to Frederick on 28 March, 'I dont at all doubt but when the King hears it he will be very much displeased.' And on 9 April, 'Mr Hendel has hetherto refused to let it be at Foxhall [a rehearsal was proposed at Vauxhall], which his Majesty seems to think he was in the wrong of.'

The rehearsal of the music finally took place on Friday 21 April, having been postponed from Monday 17th to Monday 24th and then brought forward again at the Duke of Cumberland's wish ('Munday is a drawing-room day and therefore, may be, would not be agreeable to the Duke'). From the 36 lbs of powder delivered to Frederick at Vauxhall the day before the rehearsal, it would seem that more than the music was being rehearsed. Despite the changes in date the rehearsal was very well attended: 'Yesterday there was the brightest and most numerous

Assembly ever known at Spring Gardens, Vauxhall', according to the *General Advertiser*; 'an audience of above 12,000 persons (tickets 2s. 6d.). So great a resort occasioned such a stoppage on London Bridge [still the only bridge at that time], that no carriage could pass for 3 hours'.

On the night, Thursday 27 April, it was not the music but the fireworks, perhaps inevitably, which attracted attention – and then not always in favourable terms. For Horace Walpole, as usual a lively witness of events, the fireworks

> by no means answered the expense, the length of preparation, and the expectation that had been raised:. . . . The rockets and whatever was thrown up into the air succeeded mighty well, but the wheels and all that was to compose the principal part, were pitiful and ill-conducted, with no changes of coloured fires and shapes: the illumination was mean, and lighted so slowly that scarce anybody had patience to wait the finishing; and then what contributed to the awkwardness of the whole, was the right pavilion catching fire, and being burned down in the middle of the show.

Fireworks had formed a part of previous peace celebrations, but on a less ambitious scale. After Utrecht (see no. 55) a 'Royal Fire-work' was held on the Thames at Whitehall in 1713, but the focus of public celebration appears to have been the General Thanksgiving in St Paul's; then and after Dettingen in 1743, a special setting of the Te Deum was commissioned from Handel. In 1749 it was the *Music for the Royal Fireworks* which attracted most attention, and not Handel's new Peace Anthem given at the Chapel Royal in St James's Palace two days earlier.

LITERATURE
Unpublished manuscripts in the Public Record Office (Board of Ordnance, WO 47/34 and 48/89) and the British Library (notebooks by George Vertue, Add. MS 23090); *London Magazine*, XVII, November 1748, p. 522; *Deutsch*, pp. 661–9; W.S. Lewis (ed.), *The Yale Edition of Horace Walpole's Correspondence*, XIX, 1955, p. 151, n. 9 (quoting British Library, Add. MS 32715, f. 210) and XX, 1960, pp. 47–8.

195
Music for the Royal Fireworks
Handel's autograph score

The music Handel provided for the Royal Fireworks of 27 April 1749 consists of a grand overture and five shorter movements: a *bourrée*, a siciliana called 'La Paix' ('The Peace'), a fast movement called 'La Réjouissance' ('The Rejoicing') and two minuets which may have been played in alternation. The titles suggest that Handel may originally have been working to a specific programme and expected the music to accompany certain set-pieces in the display. A cancelled and largely illegible note near the start of the allegro of the overture containing the word *guerre* is a hint that this piece, dominated by fanfares exchanged between trumpets and horns, may have been intended as a 'battle' piece. In the event the music was played before the fireworks began, while it was still light; there were obvious practical objections to performing in the dark with fireworks exploding all around.

A better-documented change of plan concerned the instrumental forces. As can be seen from the opening page of the autograph (f. 16r), Handel originally conceived the music for a massive wind ensemble: 3 first trumpets, 3 second trumpets, 3 third trumpets ('Principale' was a term used for low-register trumpet playing) and 3 pairs of kettledrums; 3 first horns, 3 second horns, 3 third horns; 12 first oboes, 8 second oboes, 4 third oboes; 8 first bassoons, 4 second bassoons and a double bassoon (the cancelled words after 'Contra Bassone' are 'et Serpente'; the fairly soft-toned serpent could have added little to the effect). But by the time Handel came to compose 'La Réjouissance' he had decided to include strings in his score, and he went back over the other movements adding instructions for their usage. The additions *e Violino 1, e Violino 2, e Viola* and *e li Violoncelli e Contra Bassi* can be seen here. The Duke of Montagu reported Handel's second thoughts to Charles Frederick on 28 March 1749:

> I think Hendel now proposes to have but 12 trumpets and 12 French horns; at first there was to have been sixteen of each, and I remember I told the King so, who, at that time, objected to their being any musick; but when I told him the quantity and number of martial musick there was to be, he was better satisfied, and said he hoped there would be no fidles. Now Hendel proposes

to lessen the number of trumpets, &c. and to have violeens. I dont at all doubt but when the King hears it he will be very much displeased.

Handel's worry was almost certainly the difficulty of obtaining enough competent wind players to do justice to music which is often technically demanding: strings would help cover slips and lapses of intonation. The number of instruments reported as having taken part ('a band of 100 musicians') suggests that Handel got his way.

The British Library Board
RM 20.g.7

Autograph score, 29 ff., 29.8 × 24.2 cms/$11\frac{3}{4}$ × $9\frac{1}{2}$ ins consisting of Concerto in D major, HWV 335a (ff. 1–10); Concerto in F major, HWV 335b (ff. 11–15); and the Fireworks Music (ff. 16–29); the concertos (c.1746) contain earlier versions of the Fireworks overture

PROVENANCE
As no. 34.

LITERATURE
Squire, p. 44; *Deutsch*, pp. 661–9.

196 [*in colour opposite p. 216*]

Sir Charles Frederick (1709–85)

By Andrea Casali (c.1700–84)

Charles Frederick (from 1761 Sir Charles), 'Comptroller of his Majesty's Fireworks as well as for War as for Triumph', was the guiding hand in the preparations for the Royal Fireworks.

A man of diverse interests – antiquary, scholar, architect and Member of Parliament – Frederick emerges as one of those energetic eighteenth-century amateurs of the arts and sciences. As a young man, on the Grand Tour from 1737–9, Frederick met the Italian history and portrait painter Andrea Casali in Rome, where he sat to him in 1738 for the portrait now at the Ashmolean Museum, Oxford. Soon afterwards, in about 1741, Casali 'proposed to Come over to England to try his fortune', according to George Vertue, 'having had some encouragement by Mr. Frederick & his Friends at Rome'. He was therefore well known to Frederick before being commissioned to provide several of the decorations for the Royal Fireworks in 1748–9.

Casali's second portrait of Frederick is unusually austere for a work of this period: it shows him in almost monkish garb, poring over a large folio from which he has been momentarily interrupted; at hand is a mortar-bomb and a sketch of a mortar in action.

The portrait may date to the period of his appointment to the Ordnance in 1746 as Comptroller of His Majesty's Laboratory at Woolwich. It was in this capacity that he set about organising the Royal Fireworks, with an abundance of enthusiasm which rose above the problems of the occasion.

Perhaps as a tribute to his success, Frederick was made Surveyor General of the Ordnance in 1750, a post he held until 1782. As a Member of Parliament for more than forty years (from 1741 to 1785), he was noted more for his willingness to vote for the government than for any contribution to debate. He was made a Knight of the Bath in 1761.

Somerville & Simpson Limited

Oil, 127.0 × 101.6 cms/50 × 40 ins

Inscribed on the mortar in the drawing on the table *G. R* (below a crown) and on the relief on the pedestal, . . . *UCTIS X/ET/X.X.*

PROVENANCE
Christie's, 23 December 1954, lot 272; offered by J. Singer to the National Portrait Gallery, 1955, but refused; Somerville & Simpson Limited by 1980.

LITERATURE
Edward H. Fellowes, *The Family of Frederick*, Windsor, 1932, pp. 60–4; *National Art-Collections Fund Annual Report*, 1980, under no. 2871, repr. p. xlvii (advert.); Stella Rudolph, *La Pittura del '700 a Roma*, Milan, 1983, pl. 133.

197

Giovanni Niccolò Servandoni (1695–1766)

By Simon Charles Miger (1736–1820) after Jean-François Colson (1733–1803) (or his father Jean-Baptiste Colson, 1686–1762)

The Chevalier Servandoni, the designer of the Fireworks, had an international reputation for his stage designs and firework set-pieces. Florentine by birth but French by adoption, he was a pupil of the painter G. P. Panini at Rome, and later of the architect G.I. Rossi. His first visit to London lasted from 1721 to 1723, during which time he was employed as a scene painter for the Italian Opera at the King's Theatre in the Haymarket. On his return to England in 1747 he produced sets for a number of John Rich's Shakespeare productions at Covent Garden Theatre and also for Smollett's ill-fated masque *Alceste*, which was to have been accompanied by music, written by Handel, in the winter of 1749/50.

Servandoni's earlier forays into the rather specialised field of firework design had included spectacular set-pieces marking the birth of the Dauphin in 1729 and the Paris Peace celebrations of 1739. For the Royal Fireworks in London in 1749 he adopted a compromise between a triumphal arch and a temple as centre-piece, which was linked by colonnades to pavilions on each side. One of these caught fire during the performance; the enraged Servandoni, known for his quick temper, drew his sword 'affronting Charles Frederick . . . he was disarmed and taken into custody, but discharg'd the next day on asking pardon before the D[uke] of Cumberland'.

This French engraving, apparently the only known portrait of Servandoni, was not published until some years after his death in Paris in 1766.

National Portrait Gallery, London

Engraving, 23.8 × 17.1 cms/9 5/16 × 6 11/16 ins

From Gautier Dagoty, *Galerie Françoise, ou Des Hommes Et Des Femmes Célèbres*, Paris, 1771, no. VI

Inscribed *SERVANDONI/Colson pinx./Miger sculp.*

LITERATURE
Ambroise Firmin-Didot, *Les Graveurs de Portraits en France*, Paris, 1877, vol. 2, p. 211, no. 1562; *Croft-Murray*, II, pp. 274–6; *Rosenfeld and Croft-Murray*, pp. 36–7.

198
'An Horizontal View of the Public Fireworks'

By an unknown engraver

The structure for the fireworks, shown here before the arcades and side pavilions had received their decoration, was a great attraction during the months preceding the display on 27 April 1749, and was the subject of a number of popular engravings.

This view, taken from the Library at St James's Palace, from which the King viewed the proceedings, is unusual in marking the 'grand Area or Music Gallery' (4) where the orchestra was to be placed. It is also noteworthy for showing the setting particularly well. On the left is Buckingham House (14), past the wall of which runs Constitution Hill (6). This was then a track, which led behind the fireworks up to Hyde Park Corner (7) and Chesterfield House in Mayfair (8). Above the firework structure are the Pothouse (10), with smoke rising from the kiln of the brickworks, and Chelsea College (11), apparently used for the newly founded Lock Hospital for venereal diseases (see no. 209). In the distance is Chelsea Church (12) and the Surrey Hills (13), their prominence emphasised by the intervening countryside, still very rural at this time.

National Portrait Gallery, London

Engraving, 15.1 × 25.7 cms/6 × 10 1/8 ins

In *The Universal Magazine*, IV, March 1749, opposite p. 138, 'Of Gun-powder and Fire-works'

Inscribed at the top *An Horizontal View of the* PUBLIC FIREWORKS *ordered to be exhibited on occasion of the* GENERAL PEACE *concluded at Aix le Chapelle on November, 7th 1748.*, and underneath, THE EXPLANATION, a key to the numbered features on view

LITERATURE
Phillips, p. 44 (reproducing a similar engraving).

199
'A Plan and Elevation of the Royal Fire-Works'

By George Vertue (1684–1756) based on designs by Giovanni Niccolò Servandoni (1695–1766)

The official view of the fireworks, an architectural elevation and plan, engraved by George Vertue for the Board of Ordnance. The structure was built of

Sir Charles Frederick by Andrea Casali, late 1740s (no. 196)

John Christopher Smith the Younger by Johann Zoffany, *c.* 1763 (no. 210)

wood and canvas, the allegorical statues in plaster. It extended 410 feet across Green Park, running at right angles to The Mall. The central temple was 144 feet in length and 114 feet high (to the top of the large sun, 176 feet).

This engraving was the only part published of a much more grandiose scheme for a volume of twenty plates of the Firework Machine and its decorations: 'all on the largest & most sumptuous paper finely printed'. It was to be 'equal to such works done beyond [the] seas especially the noble illuminations & fireworks at Paris on the marriage of the Infanta of Spain 1738'. In an unpublished notebook George Vertue records the sad details of the frustration of a noble enterprise. The project was first discussed at meetings on 10 and 14 November 1748, when Vertue records that 'Mr Frederick proposes and is willing I only should be concerned to my benefit and advantage'. Nearly five months later he was to write of this single 'fine plate' that it was 'to be published at the expence of the fireworks Italians & to their profitt'.

Sadly he adds, 'But me – nothing – only a tool to their designs' (though he does acknowledge that he was recompensed by Charles Frederick). In the interim he had had to battle with the Italians: first with the Chevalier Servandoni, who published advertisements for the proposed book without permission and was 'underhand' enough to supply a rival publisher with a drawing; and then with Andrea Casali, who 'proposed to engrave the four great designs at his own expense', which, as Vertue comments, 'will interfere with the design of a book'. On the publication of an unauthorised engraving in the *London Magazine* for December 1748, the Ordnance Board decided to produce a single plate – the engraving exhibited here – 'as expeditiously as possible'. But even in June the idea of a book, now of 25 plates, lingered on; it would depend on 250 subscriptions being received, but it is clear that the moment to engage public support for the project had passed.

Guildhall Library

Engraving, 38.5 × 54.0 cms/15⅛ × 21¼ ins

Inscribed at the top *A PLAN and ELEVATION of the ROYAL FIRE-WORKS to be performed in S.t JAMES's PARK, April the 27.th 1749, on Account of the GENERAL PEACE, signed at Aix la Chapelle, Oct.r 7. MDCCXLVIII.* and with the dimensions of the machine and numbers of fireworks; beneath the plan *The Machine invented and design'd by the Chev.r Servandoni.| The Fireworks principally perform'd by the Sign.ti Gaetano, Ruggieri, and Giuseppe Sarti of Bologna.| G. Vertue Sculp. LONDINI.| Drawn, Engrav'd and Publish'd 5.th of April 1749, according to Act of Parliament,* and at the bottom *By Order of his Grace the DUKE of MONTAGU Master General, and the Right Hon.ble the L.t General and the rest of the principal OFFICERS of his Majesty's Ordnance; and performed by the Direction of Charles Frederick Esq.r Comptroller of his Majesty's Laboratory at Woolwich.| price 1 shilling*; in the sun over the central arch *VIVAT REX*

LITERATURE
Notebook by George Vertue, British Library, Add. MS 23090, ff. 97–104.

200

'A Description of the Machine for the Fireworks ... and a Detail of the Manner in which they are to be exhibited in St James's Park, Thursday, April 27, 1749'

Published by Gaetano Ruggieri and Giuseppe Sarti by order of the Board of Ordnance
Printed by William Bowyer, 1749

The official programme published for the Board of Ordnance and sold for sixpence. The 'Machine for the Fireworks' is described in great detail; its centre-piece, a 'magnificent Doric Temple', lay some 500 feet from the Royal Library adjoining St James's Palace. It was from here that the King was to watch on the night.

There was considerable public interest in the nature of the proposed display in the months beforehand, and the papers and magazines published whatever descriptions or engraved views they could come by. It was for this reason that the publication of George Vertue's official engraving (no. 199) was brought forward. This programme seems to have met similar problems. A reference in Vertue's notebook, for 18 April, speaks of '35 guineas demanded & paid for the written coppy of the Fireworks to be printed', seeming to suggest some sort of clandestine agreement. Whatever the case, it may have been in response to unofficial competition that the Board's programme, published three days later, contained a notice threatening to prosecute anyone else who printed or abridged the description.

Most of the programme is taken up with the fireworks, but the music is mentioned in passing: 'After a Grand Overture of Warlike Instruments composed by Mr. Handel, a Signal is given for the Commencement of the Firework, which opens by a Royal Salute of 101 Brass Ordnance ... The Salute ended, the Firework begins to display itself....' The climax of the fireworks came with the eleventh of the 13 displays: 'The Grand fixed Sun on the top of the Machine composed of 3 Circles of Rays whose diameter of fire is 70 feet, in it's Center appears VIVAT REX in bright fire'; this was followed by 576 rockets in 12 flights. By eleven o'clock the grandest display of fireworks to be seen in eighteenth-century London was all but over.

A manuscript presentation copy of the programme is in the Gerald Coke Handel Collection. Illustrated with Vertue's engraving, it may well have been made for the King's own use; it bears the bookplate of George III. Another small pamphlet was published by the Ordnance after the event; much of it is a straightforward reprint of the official programme but it adds that during Handel's overture 'eighteen small Cannon were fired' and mentions the 'Band of Musick' performing during the King's inspection of the Machine some time before the fireworks began.

Gerald Coke Handel Collection

201

Bound pamphlet, 16 pp., 23.0 × 17.6 cms/9 × 6⅞ ins

Titled as above and inscribed in pen on the title-page *J. Lee. Dec! 1831 found at Totteridge Park*

LITERATURE
Deutsch, p. 667.

EXHIBITED
Coke, p. 29.

201

'A View of the Public Fire Works'

By an unknown engraver
Published by Henry Overton (fl. 1701–51)

Of the few views of the fireworks actually in progress this is the most impressive, and includes details not found elsewhere, such as the soldiers who were employed to control the crowds, the 'Scaffolding for the Quality' on the left, with tickets at 10s. 6d. for the best seats, and, in the background, the '101 Brass Ordnance' which fired the opening Royal Salute.

The Trustees of the British Museum
1877–6–9–2060

Engraving, 45.4 × 116.2 cms/17⅞ × 45¾ ins

Inscribed *A View of the Public* FIRE WORKS *Exhibited on Occasion of the* GENERAL PEACE *Concluded at Aix La Chapelle, October 7. 1748. Shewing the various Statues, Trophies and Paintings with which that Curious Piece of Architecture is Ornamented, also the 3 Fire Suns; &c &c &c./ Printed for and Sold by H. Overton at the White Horse without Newgate London. According to Act of Parliament*, with a lengthy description of the fireworks

202

'A Perspective View of the Building for the Fireworks in the Green Park, taken from the Reservoir'

By P. Augier after P. Brookes

The use of back lighting, cut-outs and coloured materials has here transformed a daytime scene of strolling couples into a dramatic re-creation of the effect of fireworks going off. The spectacle is as seen from the reservoir on the north side of Green Park, close to Piccadilly, with Buckingham House on the right.

Gerald Coke Handel Collection

Engraving, coloured, 23.0 × 43.1 cms/9 × 17 ins, the outlines of the structure pin-pricked and the main fireworks cut away and backed with patches of various coloured fabrics; the rockets in the sky also added

Titled as above and inscribed *P. Brookes. del./ P. Augier sculp./ Printed for R. Sayer opposite Fetter Lane Fleet Street/41*

In the Name of God Amen.

I George Frideric Handel considering the Uncertainty of human Life doe make this my Will in manner following. viz⸺

I give and bequeath unto my Servant Peter le Blond, my Cloathes and Linnen, and three hundred Pounds Sterl: and to my other Servants a Year Wages.

I give and bequeath to Mr Christopher Smith my large Harpsicord, my little House Organ, my Musick Books, and five hundred Pounds Sterl:

Item I give and bequeath to Mr James Hunter ~~⸺⸺⸺⸺⸺⸺⸺⸺~~ five hundred Pounds Sterl:

I give and bequeath to my Cousin Christian Gottlieb Handel of Coppenhagen one hundred Pounds Sterl:

212 Handel's will (first page)

VII *The Last Decade* 1749–59

In April 1750, *Messiah* was performed at the Foundling Hospital for the first time. Handel was by now 65. In the previous ten years his stature had grown in the eyes of his contemporaries to the point at which he was an accepted national figure. 'The Great and Good Mr Handel' is how one friend, James Smyth, was to describe him immediately after his death.

Before going abroad in the summer of 1750 he made his will (no. 212); all we know of his movements on this last journey to the continent is that his coach met with an accident in Holland on his way to visit family and friends in Germany; apparently he took with him his recently completed portrait by Thomas Hudson (no. 11). Earlier in the year it is reported that he had 'been pleasing himself in the purchase of several fine pictures'. What evidence there is suggests that the collecting of Old Master paintings was an activity Handel came to late in life (see Appendix 4) but which, sadly, was cut short by his blindness, already apparent by February 1751 when composing *Jephtha* (no. 206).

This oratorio, completed in August 1751 and performed at Covent Garden in February the following year, was his last substantial original work. From now on, his theatre seasons each Lent were taken up by revivals of favourite oratorios, though sometimes with new arias added or other amendments. His repertory had two mainstays: *Judas Maccabaeus*, which was performed every year during the 1750s, and *Messiah*, which was always the concluding performance of the oratorio season, sometimes being given as late as Easter Thursday. A further, charity performance then followed some weeks after Easter at the Foundling Hospital Chapel (see no. 203).

With the onset of blindness Handel had to seek assistance. He turned to his former pupil, John Christopher Smith the younger, ignoring a suggestion that the blind John Stanley could help (see nos. 210 and 208). Mainwaring, writing in 1760, says that Handel desired that Smith 'would play for him and assist him in conducting the Oratorios'. The performance of *Messiah* at the Foundling Hospital on 15 May 1754 was probably the last given under Handel's direct supervision; on 25 June his approval of J.C. Smith's appointment as organist of the Chapel is reported in the Hospital minutes with a note that 'on Account of his Health [Handel] excused himself from giving any further instructions relating to the Performers'.

Even after Handel's eyesight failed, 'he continued to play concertos and voluntaries between the parts of his Oratorios to the last, with the same vigour of thought and touch, for which he was ever so justly renowned', according to Burney's *Sketch of the Life of Handel*; 'at last, however, he rather chose to trust to his inventive powers, than those of reminiscence: for, giving the band only the skeleton, or ritornels of each movement, he played all the solo parts *extempore*, while the other instruments left him, *ad libitum*; waiting for the signal of a shake, before they played such fragments of symphony as they found in their books'. Other contemporary reports give a mixed view: on 9 March 1753 he seemed to Lady Shaftesbury to be 'dejected, wan and dark, sitting by, not playing on the harpsichord', while less than a month later, on 3 April, her husband was writing that 'Handel's playing is beyond what even *He* ever did.'

Lord Shaftesbury was also a witness to Handel's late work as a composer; in 1757, on 8 February he reported that 'Mr Handel is better than he has been for some years and finds he can compose Chorus's as well as other music to his own (and consequently to the hearers) satisfaction.' The chorus referred to is probably 'Sion now her head shall raise', added to *Esther* when it was performed on 25 February that year and described by Thomas Morell (see no. 194) as 'his last chorus'. Burney refers to the piece as having been dictated to J.C. Smith, but it seems that most other late additions to Handel's oratorios, if not actually copied from earlier material, may be in good part Smith's own work, acting only under the general supervision of Handel.

The late portrait of Handel by Thomas Hudson (no. 12), a full-length painted for his librettist, Charles Jennens, does not suggest a man in decline so much as in the fullness of old age. In these last years, with more time on his hands, Handel's attendance at his parish church, St George's, Hanover Square (see no. 148), seems to have become a more regular feature of his life. In the four codicils to his will written during his final two and a half years, his thoughts evidently turned to the future and the past, as he remembered more than thirty friends in England and relatives abroad with bequests, and asked to be buried in Westminster Abbey. Both his funeral, attended by some three thousand people, and his monument (see nos. 213 and 17), bear witness to the high regard in which 'the Great and Good Mr Handel' was held.

203

The Foundling Hospital

Eighteenth-century London was a cruel city, in which illegitimate children were frequently abandoned to their fate, often to die. It was as a result of the energies of the former merchant sailor, Captain Thomas Coram, that in 1739 the King granted a royal charter 'for establishing an Hospital for exposed children'. The Foundling Hospital opened in temporary accommodation in 1741 and was met with a flood of applications from destitute or deserted mothers. Having been accepted, children were sent to foster-parents in the country until the age of four or five when they were brought back to the Hospital to be educated.

The Hospital found a permanent site in Lamb's Conduit Fields, then on the very edge of London near what is now Brunswick Square (see map, p. 74). The need to raise funds for the project, one of the foremost philanthropic enterprises in mid-eighteenth-century London, led to the appointment of a large roll of governors. Under the encouragement of William Hogarth, who had painted a splendid portrait of Coram in 1740, several leading painters and sculptors donated their work to the Hospital and became governors. Thomas Hudson gave two full-length portraits, including that of the Hospital's architect Theodore Jacobsen, and, in April 1750, offered to paint Handel's portrait as well. Although it was noted that 'Mr Handel had consented to sit for it', the project came to nothing, perhaps as a result of Handel's absence on the continent that summer and the onset of blindness the next winter. Hudson had already produced one portrait of the composer in 1748–9 and Handel was to sit for another in 1756 (nos. 11–12).

Handel's first association with the Hospital came in May 1749 with his offer of a recital of vocal and instrumental music in aid of the Chapel; in May the following year his gift of an organ was marked by a special performance of *Messiah*. The performances of Handel's work, and in particular *Messiah*, helped to attract the attention of society to the Hospital; 'vast numbers of the nobility and gentry were drawn to the hospital [and] . . . were afterwards warmly engaged in promoting it', according to Handel's biographer, John Mainwaring. In the course of ten years Handel brought in some £7,000 for the Hospital, while the highest figure raised in a single year was £969. These were substantial sums, so it was quite natural for the governors to wish to ensure the composer's goodwill and continued support. As usual, however, Handel did not like to be taken for granted. When he was first offered a position as a governor in 1749 he refused, saying that 'he should Serve the Charity with more Pleasure in his Way, than being a Member of the Corporation'. When later the Hospital misunderstood Handel's intentions and proposed to petition for exclusive performing rights to *Messiah* Handel burst into a rage, exclaiming that *his* music would not go to Parliament. Nevertheless Handel was well disposed and in the third codicil to his will, in 1757, he bequeathed a copy of the score of *Messiah* and performing parts (no. 205) to the Hospital.

LITERATURE
R.H. Nichols and F.A. Wray, *The History of the Foundling Hospital*, 1935, pp. 20, 202–6; *Deutsch*, pp. 669–70; *Nicolson*, p. 29, n. 25; Donald Burrows, 'Handel and the Foundling Hospital', *Music and Letters*, LVIII, 1977, pp. 269–84.

203
'A Perspective View of the Foundling Hospital, with Emblematic Figures.'

By Charles Grignion (1721–1810) and Pierre-Charles Canot (1710–77) after Samuel Wale (1721–86)

This unusual view of the Foundling Hospital is one of a pair of engravings published in April 1749, the month before Handel's first charity concert in the Hospital Chapel. It is to the design of the painter and prolific book-illustrator Samuel Wale, one of the first artists to be appointed as a governor of the Hospital in 1746. He gave three small painted views of London hospitals to decorate the Foundling's court room. The meaning of the 'Emblematic Figures' of the title is not entirely clear but the verses beneath the view refer both to the foreground statue of Venus and the three figures of sneering Frenchmen on the right.

The Hospital was built to the design of Theodore Jacobsen, a merchant and amateur architect who gave his services free. The boys' wing, on the left, had been completed in 1745; the Chapel with its large arched windows, the centre-piece of the design, was substantially complete by Spring 1749 but at the time of this engraving awaited fitting-out – it was without windows and had to be 'sash'd and made commodious' before being used for the first time at Handel's concert on 27 May. As well as his recently completed music for the Royal Fireworks (see no. 195), a new anthem was given, *Blessed are They that Considereth the Poor*, composed specially for the occasion.

The immense success of the annual performances of *Messiah* at the Foundling Hospital in the 1750s helped establish the pre-eminence which this work has enjoyed ever since. By the same token these occasions greatly benefited a still very young charitable institution – so much so that Handel has been called 'the greatest benefactor of the Chapel' (*Nichols and Wray*). In his later years Handel left the direction of the Foundling Hospital performances to John Christopher Smith (see no. 210), appointed organist of the Chapel in June 1754 with Handel's approval.

Guildhall Library

Engraving, 29.2 × 42.9 cms/11½ × 16⅞ ins

Inscribed at the top *A PERSPECTIVE VIEW OF THE FOUNDLING HOSPITAL, WITH EMBLEMATIC FIGURES.*, and at the bottom *S. Vale delin./ Publish'd 14.th of April, 1749 according to Act of Parliament/ Grignion & Canot Sculp*; also inscribed with verses beginning *Myriads of Proselites sustain thy Cause, . . .* , and dedicated *To his Grace John Duke of Bedford, This Plate is humbly Inscribd/ . . . by his Graces most Dutiful and Obedient Servant Margrett Granville.*

LITERATURE
R.H. Nichols and F.A. Wray, *The History of the Foundling Hospital*, 1935, p. 202; *Deutsch*, pp. 670–1.

204

Foundling Hospital Anthem

Conducting score with autograph additions by Handel

On 7 May 1749 Handel attended a meeting of the General Committee of the Foundling Hospital and, as the minute book records, 'offered a performance of vocal and instrumental musick to be held at the Hospital', the proceeds of which 'should be applied to the finishing the chapel of the Hospital'. The concert took place on 27 May and, according to the word-book (a copy of which is in the Rowe Library, King's College, Cambridge), was divided into three parts. The first contained the two pieces Handel had contributed to the celebrations of the Peace of Aix-la-Chapelle the previous month, the Music for the Royal Fireworks and the Anthem on the Peace. Part II consisted of excerpts from *Solomon* 'relating to the *Dedication of the Temple*'. Part III opened with an unidentified concerto and concluded with 'The Anthem Composed on this Occasion'.

The new anthem was *Blessed are They that Considereth the Poor*, now generally known as the Foundling Hospital Anthem. No complete autograph of the work was ever written, because Handel took much of the music from other works (the Funeral Anthem, *Susanna* and *Messiah*) and simply gave John Christopher Smith the elder instructions to copy the relevant sections into the conducting score, which is thus the earliest source for the complete Anthem. For a repeat performance Handel expanded the work with three new sections for solo voices and adapted sections of one of the former Funeral Anthem choruses in order to make use of the soloists. On display is a section of the chorus 'They deliver the Poor that cry', showing Handel's alterations. Pencil marks indicate deletion of the original vocal bass line and the oboe parts, while in ink Handel has written a new part for the second sopranos (*C*[anto].*2 soli*) and added *pian* ('softly') to the bass and violin lines. Further pencil notes pointing out Handel's handwriting (eg 'The Second Voice & *all* the pencil marks are in *Handel* [sic] *writing*') were written in the 1850s by Rophino Lacy, who assisted Victor Schoelcher in the writing of the latter's *Life of Handel*.

The Thomas Coram Foundation for Children

Manuscript score, i + 51 ff., 23 × 28.5 cms/9 × 11¼ ins

Inscribed on f. 1r *Anthem for the Foundling Hospital*, and on f. 5r *Part 3d* (Chorus 'Blessed are they', originally the start of Part 3 of the concert of 27 May 1749)

PROVENANCE
Presumably deposited at the Foundling Hospital by the composer and retained after his death.

LITERATURE
Donald Burrows, 'Handel and the Foundling Hospital', *Music and Letters*, LVIII, 1977, pp. 269–84;
G.F. Handel, *Foundling Hospital Anthem*, ed. Donald Burrows, Edition Peters, London, Frankfurt and New York, 1983.

205

Messiah

Manuscript part-book for second soprano

In the third codicil to his will, dated 4 August 1757 (see no. 212), Handel bequeathed to the Foundling Hospital 'a fair copy of the Score and all Parts of my Oratorio called The Messiah'. This bequest was duly fulfilled after Handel's death, as is acknowledged in the Hospital minutes for 13 June 1759; a manuscript full score and a set of instrumental and vocal parts was delivered to the charity by J.C. Smith the elder. Apart from the loss of one or two part-books, this material still survives in the archives of the Hospital, now known as the Thomas Coram Foundation. Although it appears that the score and parts were never actually used, they were certainly prepared under the supervision of the elder Smith as genuine performing material would have been, and thus provide important evidence of certain contemporary practices. Since *Messiah* is for the most part scored only for strings (see under no. 183), the way in which oboes and bassoons were added to the score for Handel's London performances cannot be deduced from the original score alone; the Foundling Hospital part-books, however, provide appropriate parts for these instruments throughout the oratorio.

A significant feature of the vocal part-books intended for the soloists is the inclusion of the music for the choruses. The idea of the 'chorus' being an ensemble of the principal singers was carried over from opera into oratorio; the chorus in the modern sense (ie the choir of the Chapel Royal in Handel's performances) was a reinforcement of the principal singers, not a separate body.

For reasons which are not clear, the material copied for the Foundling Hospital in 1759 reflected the version of *Messiah* prepared in 1754 (possibly a new set of performing material for the work had been made in that year, and the copyists of 1759 simply reproduced it, ignoring all later revisions). Handel's company in 1754 included the soprano Christina Passerini, whom Telemann had recommended to him, and the second soprano part-book of the Foundling Hospital material reproduces the music she sang in *Messiah* at Covent Garden and at the Foundling Hospital Chapel that year. To create a part for her in *Messiah* without diminishing the role of his leading soprano Giulia Frasi, Handel transposed the two arias originally added to the oratorio in 1750 for the *alto castrato* Guadagni. The opening of the first, 'But who may abide', is displayed; it is transposed to the high key of A minor (a version presently unpublished, though it may be heard on the Decca L'Oiseau-Lyre recording). On the left-hand page is the conclusion of the chorus 'And the Glory of the Lord'.

The Thomas Coram Foundation for Children

Manuscript part-book, 24 ff., 23.8 × 30.3 cms/9$\frac{3}{8}$ × 12 ins

Inscribed on the cover *Messiah an Oratorio Second Soprano N° 28*, and on f. 1 *Messiah an Oratorio Sig^{ra} Passerini*

PROVENANCE
Bequeathed by Handel to the Foundling Hospital, 1759.

LITERATURE
Watkins Shaw, *A Textual Companion to Handel's 'Messiah'*, 1965, pp. 89–108; *Nicolson*, p. 49.

The Blind Composer

When Handel had to break off from composing *Jephtha* on 13 February 1751 with his deeply touching note in the score about his failing sight (see no. 206), the prospect of blindness faced him. He was almost 66 and, despite a history of occasional chronic illness probably caused by overwork, he had another eight years to live. Although his sight recovered sufficiently for him to complete all but Act 3 of *Jephtha* by 27 February he decided to seek advice from Samuel Sharp of Guy's Hospital, the first of three specialists he consulted about his blindness. 'Noble Handel hath lost an eye', wrote Sir Edward Turner on 14 March. The following year he went to Dr William Bromfield (see no. 209), an eminent surgeon at St George's Hospital whose operation produced a short-lived improvement in his sight. Then, in the year before his death, Handel was treated by the rather less reputable, though highly reputed 'opthalmiater', the so-called Chevalier John Taylor (see no. 211), again without success.

In the early years of his blindness he decided to take on J.C. Smith the younger (see no. 210) as his assistant and to recommend him as organist for the Foundling Hospital (see p. 223). With his help the oratorio seasons at Covent Garden continued without interruption; and yet the audience's reaction could not be the same, as both Coxe and Burney relate. William Coxe, Smith's step-son, tells how

> When Smith played the organ at the Theatre, during the first year of Handel's blindness, Samson was performed, and Beard sung, with great feeling,
>
> > 'Total eclipse – no sun, no moon,
> > 'All dark amid the blaze of noon.' –
>
> The recollection that Handel had set this air to music, with the view of the blind composer then sitting by the organ, affected the audience so forcibly, that many persons present were moved to tears.

'Poor Handel! how feelingly must he recollect the *"total eclipse"*', wrote his friend of many years, Mrs Delany, in November 1752. And Handel's own performance was found to be just as moving. Burney recalled that 'To see him, however, led to the organ, after this calamity, at upwards of seventy years of age, and then conducted towards the audience to make his accustomed obeisance, was a sight so truly afflicting and deplorable to persons of sensibility, as greatly diminished their pleasure, in hearing him perform.'

LITERATURE
Burney, *Sketch*, pp. 29–30; *Coxe*, p. 45; *Deutsch*, p. 703.

206

Jephtha

Handel's autograph score

Handel's last oratorio, *Jephtha*, has a libretto by Thomas Morell (see no. 194) based on the incident related in the Biblical Book of Judges. The Israelite chieftain Jephtha vows that if he is victorious over the Ammonites he will sacrifice the first living thing to greet him on his return. It is his daughter who greets him, his only child, and in the Bible it would appear that the chastened Jephtha grimly carries out his vow. The relevant text has been subject to other interpretations, however, since human sacrifice is contrary to Jewish law, and Morell took advantage of these to introduce an Angel in the final scene who commutes the fate of the daughter to live 'in pure and virgin state for ever'. In addition to his Biblical source Morell took hints from the Latin play *Jephthes sive Votum (Jephtha, or the Vow)* by the Scottish humanist George Buchanan (1506–82), including the names of Jephtha's wife and daughter (Storgè and Iphis) and a scene in which Storgè is prey to dire forebodings.

The broad theme of the oratorio – the destruction of human happiness by divine ordinance – had a tragic resonance with the composer's own situation when he began composition in January 1751: he was going blind. The weakening of his sight, the physical labour of writing the score and the psychological implications of the story brought Handel to a mental crisis while setting to music the final chorus of Act 2; this is sung just after the famous recitative 'Deeper and deeper still', in which Jephtha breaks down as, compelled by his vow, he resolves to carry out the terrible sacrifice. The opening words of the chorus ('How dark O Lord are thy decrees, all hid from mortal sight') proved too much for Handel. Scarcely able to fill in the orchestration (in the upper string parts only the note heads are written, the rhythm of the bass line being presumed) he managed to complete the section and noted at the end of the page: 'biss hierher komen den 13 Febr. ☿ 1751 verhindert worden wegen ['relaxation' cancelled] so relaxt des gesichts meines linken auges' ('got as far as this on Wednesday 13 February 1751, unable to continue because the sight of my left eye is so weakened'). Ten days later, on his sixty-sixth birthday, he was fortunately able to record: 'den 23 ♄ dieses etwas besser worden wird angegangen' ('on Saturday the 23rd of this month a little better, resumed work'). Nevertheless composition remained a struggle, and the work was not completed until 30 August. The first performance took place at Covent Garden on 1 March 1752. For the revival the following year Handel added a quintet and an air with music taken from *Agrippina*; the younger Smith may have assisted with the adaptations.

The displayed pages of the autograph (ff. 91v–92r), showing the notes quoted above, are one of music's most moving documents.

The British Library Board
RM 20.e.9

Autograph score, i + 134 ff., 23.5 × 29.8 cms/9¼ × 11¾ ins

Inscribed on f. 5r; *Oratorio Jephtha angefangen den 21 Janr 1751* ☽; on f. 49r (the end of Act 1) *geendiget den 2 Febr ♄ 1751. völlig ♂ agost 13.1751*: *Fine della Parte Seconda. geendiget den 27 dieses Febr* ☿ *1751*; f. 97v: *Act 3. Jephtha ♂ Juin 18.*; on f. 122v (the end of the chorus 'Theme sublime') *July 15 or 17 1751* ☽ or ☿; and on f. 134v, *G.F. Handel aetatis 66. Finis.* ♀ *Agost 30. 1751.*

PROVENANCE
As no. 34.

LITERATURE
Squire, p. 49; *Dean*, pp. 588–624.

208

207

Jephtha

Conducting score with autograph additions by Handel

The conducting score of *Jephtha*, in accordance with Handel's normal practice, is basically a fair copy by the elder Smith made directly from the autograph; and, as was also customary, Handel avoided writing out for a second time any music which he had decided to take from an earlier work. There is no sign in the main autograph (no. 206) of the sinfonia that heralds the appearance of the Angel in Act 3: instead Handel chose to use the finale of his violin sonata in D major (a late work, written only a year or two before *Jephtha*). He made some pencilled amendments on the autograph of the sonata (British Library RM 20.g.13, ff. 5–8), cutting or simplifying passages unsuitable for orchestral playing. Smith then copied the revised movement into the conducting score, leaving blank staves in which (as shown) Handel himself filled in a viola part to complete the movement.

Staats- und Universitätsbibliothek, Hamburg
MA/1024.3

Volume 3 (of three) of manuscript score, 64 ff., 23.8 × 30.0 cms/9⅜ × 12 ins

PROVENANCE
As no. 129.

208

John Stanley (1712–86)

By James McArdell (1728–65) perhaps after John Michael Williams (fl. 1743–60)

When Handel consulted Samuel Sharp at Guy's Hospital about his failing eyesight in 1751, the surgeon recommended the blind Stanley to play the organ at his oratorios, as a person whose memory never failed. This brought a swift retort from Handel, who burst out laughing, 'Mr Sharp, have you never read the Scriptures? Do you not remember, if the blind lead the blind, they fall into the ditch?' Instead Handel turned to John Christopher Smith (see no. 210). Nevertheless Stanley was to conduct *Alexander's Feast* in March 1753, and a number of other oratorios within Handel's own lifetime. After Handel's death he entered into an agreement with Smith to carry on the Handel oratorio seasons held at Covent Garden each Lent.

Unlike Handel, Stanley was blind from childhood as a result of an accident. He was well known both as a virtuoso organist, whose playing is said to have attracted Handel, and as a composer of cantatas and organ music. He relied on his phenomenal memory and acutely developed hearing to lead as full a life as possible; he was heard to remark that he would not receive his sight, if it was in his power: he felt perfectly happy under his present circumstances.

Stanley owned two portraits of Handel: a Roubiliac bust ('A remarkable fine bust, exquisitely modell'd', lot 89 in his posthumous sale, Christie's, 24 June 1786) and an oil painting, derived from Hudson's famous image of the blind Handel, which was engraved as in his collection.

The Trustees of the British Museum
1887–4–6–81

Mezzotint engraving, 31.0 × 24.8 cms/12¼ × 9¾ ins

Inscribed *Mr Stanley, Organist. | Ja^s Mc Ardell fecit. | Sold by E. Fisher, Engraver, at the Golden head in Leicester Square, & by Ryland & Bryer at the King's Arms in Cornhill, London.*

LITERATURE
Coxe, p. 44; *CS*, p. 897, no. 170, state III.

209

Dr William Bromfield (1713–92)

By Francis Cotes (1726–70)

Dr William Bromfield, founder of the Lock Hospital for venereal diseases and surgeon at St George's Hospital, was one of three specialists who treated Handel's failing eyesight. On 4 November 1752 the *General Advertiser* reported that 'Yesterday George-Frederick Handel, Esq: was couch'd [a cataract operation] by William Bromfield, Esq; Surgeon to her Royal Highness the Princess of Wales, when it was thought there was all imaginable Hope of Success by the Operation, which must give the greatest Pleasure to all Lovers of Musick.' But any improvement in Handel's eyesight was short-lived, for on 27 January 1753 it was reported that 'Mr. Handel has at length, unhappily, quite lost his sight. Upon his being couch'd sometime since, he saw so well, that his friends flattered themselves his sight was restored for a continuance; but a few days have entirely put an end to their hope.'

Later the same year, on 7 May, a performance of Handel's *Judas Maccabaeus* was staged at the King's Theatre for the benefit of the Lock Hospital, probably at Bromfield's instigation. That Handel was well disposed to the Hospital is confirmed by his acceptance of the appointment as a Governor of the Hospital.

Francis Cotes's pastel was probably drawn a few years later, in about 1757, at the period when the artist was producing his most strikingly realistic images. It was bequeathed by Bromfield to Sir James Wright, Bart (d. 1803), formerly the British Resident at Venice. By coincidence he had lived in Handel's house (see no. 147) in Brook Street from 1772–82, subsequently sub-letting the property.

The Royal Academy of Arts

Pastel, 69.2 × 54.3 cms/27¼ × 21⅜ ins

PROVENANCE
Bequeathed by the sitter to Sir James Wright, Bart, 1792; given by him to the Royal Academy of Arts, 1796.

LITERATURE
Royal Academy Council Minute Book, entry for 17 December 1796 (unpublished); George C. Peachey, 'William Bromfield 1713–1792', *Proceedings of the Royal Society of Medicine*, VIII, 1915, pp. 103–26; Carl Winter, 'Francis Cotes, R.A', *Connoisseur*, LXXXVIII, 1931, p. 174 (repr.); *Deutsch*, pp. 727, 731, 743; Edward Mead Johnson, *Francis Cotes*, Oxford, 1976, p. 62, no. 81.

210 [*in colour opposite p. 217*]

John Christopher Smith (1712–95)

By Johann Zoffany (1733–1810)

The only known portrait of the organist and composer John Christopher Smith the younger. In his childhood he was a pupil of Handel's; in his maturity he acted as Handel's amanuensis during the composer's blindness. He came to London as a boy of eight in 1720 to join his father, Johann Christoph Schmidt (1683–1763) of Ansbach, who had left his native city a year or two previously to act as Handel's principal copyist and treasurer.

209

Smith's education was unusual, as his early biographer, his step-son William Coxe (see no. 216), makes clear. At the age of 13 he was 'placed under the tuition of Handel', turning later to Dr Pepusch (no. 71), and more especially to the organist Thomas Roseingrave, for his instruction when Handel showed no inclination to take his pupil beyond the rudiments of music. Coxe reports that he also had the 'good fortune to become acquainted with Dr Arbuthnot' at whose house he frequently met Swift, Pope, Gay and Congreve, 'a society highly improving to a young man'.

Despite Smith's exceptional musical background, it was his 'peculiar turn of disposition, not to live much with the professors of music . . . and the friendships he formed in life were with men of a different profession'. In any case Smith's occasional early works for the stage, such as his English opera *Ulysses* (1733), met with only limited success. In 1746–8, on a long visit abroad as tutor to an invalid youth, he spent some time in Geneva, where he fell in with the literary circle of such like-minded Englishmen abroad as Robert Price of Foxley, an amateur musician and composer, and Benjamin Stillingfleet. They served as an introduction to David Garrick, who in 1755 commissioned from Smith a full-length Shakespearean opera, *The Fairies*, the first of three works for the stage which he wrote at this period.

With the onset of blindness Handel called on Smith to act as his assistant. He took over some of the organisation for the oratorio seasons held at Covent Garden each spring. He also assumed responsibility for the annual performance of *Messiah* at the Foundling Hospital and in 1754 was appointed as their organist with the approval of Handel (see p. 223). It would seem that many of the late additions to Handel's oratorios, as Anthony Hicks has shown, owe more to Smith than has previously been supposed.

For his portrait, painted within a few years of Handel's death, Smith turned to Johann Zoffany, a German artist who had settled in London in 1760. He may have been encouraged to do so by the example of his friend Richard Neville Neville, whose own portrait by Zoffany (at Audley End), painted in 1761, is one of the artist's earliest English commissions. But Neville was only the first of Smith's friends to go to Zoffany; David Garrick followed in 1762 and also his two literary friends, Robert Price and Benjamin Stillingfleet, both of whom had recently prepared oratorio texts for him. Smith's portrait, perhaps painted at the time of his marriage to Martha Coxe in 1763, shows him in his early fifties; Smith was 'naturally grave', his step-son wrote, with 'hands delicately formed, blue and impressive eyes, and his whole countenance remarkably open'.

After Handel's death Smith carried on the Lenten oratorio seasons at Covent Garden working with the blind organist, John Stanley (see no. 208). It appears that he gave lessons on the harpsichord to Augusta, the dowager Princess of Wales (he was well known as a teacher); and it is possible that the music shown in his portrait, marked *varia* . . . , may have been intended for such a lesson, if not the bass part to one of his works. In gratitude to George III, who awarded him a pension after Augusta's death, Smith gave the King his collection of Handel's autographs which had come to him via his father from the composer himself; it seems that he also gave the marble bust of Handel by Roubiliac now at Windsor (see no. 7 for the terracotta model). In 1784, now living in Bath, he declined an invitation from the King to attend the Westminster Abbey Handel Commemoration, pleading old age as his excuse. Smith still retained the oil portrait of Handel by Denner (no. 1). His name will for ever be linked with Handel's by his step-son's short biography, published four years after his death in 1795: *Anecdotes of George Frederick Handel and John Christopher Smith.*

Gerald Coke Handel Collection

Oil, 88.9 × 67.9 cms/35 × 26¾ ins

Inscribed on the music *varia* . . .

Engraved by E. Harding for William Coxe's *Anecdotes*, 1799 (see no. 216); the engraving was made from a watercolour reduction of the portrait by W.N. Gardiner (contained in an extra-illustrated copy of the *Anecdotes* in the Gerald Coke Handel Collection)

PROVENANCE
Probably part of John Christopher Smith's bequest to his step-son, Peter Coxe; Sotheby's sale, 30 November 1960, lot 131, purchased by Gooden and Fox for Gerald Coke.

LITERATURE
Coxe (advert.), repr. p. 36; Anthony Hicks, 'The late additions to Handel's oratorios and the role of the younger Smith', *Music in Eighteenth-Century England*, ed. C. Hogwood and R. Luckett, Cambridge, 1982, pp. 147–69.

211

John Taylor (1703–72)
By Richard Cooper (d. 1764)
after William De Nune (fl. 1729–50)

The quack oculist and self-styled 'Chevalier', John Taylor, enjoyed European-wide success through his powers of self-publicity. This is the most elaborate of several engraved portraits which he saw fit to encourage and is full of references to his wide-ranging exploits. The format adopted should be compared to an earlier engraving of Handel (no. 4); it is closely based on Houbraken's *Heads of the Illustrious Persons of Great Britain* (that of George Buchanan), and was presumably intended to excite comparison.

If Taylor travelled extensively it was because the beneficial effects of his operations were usually only temporary; by the time this was apparent he had moved on. It appears that Taylor met Handel in Tunbridge Wells in August 1758; his operation on the aging composer was lauded in an anonymous poem published as a puff in the *London Chronicle* on 24 August, 'On the Recovery of the Sight of the Celebrated Mr. Handel, by the Chevalier Taylor'. However Handel's sight was *not* restored, as Taylor himself later acknowledged in his *History of the Travels and Adventures of the Chevalier John Taylor, Ophthalmiater* (1761); in the same mention he describes his treatment in 1750 of J.S. Bach, but does not reveal that Bach's initial recovery of sight was followed by blindness and death within a few months.

The Trustees of the British Museum
1858-4-17-269

Engraving, 35.2 × 22.2 cms/13⅞ × 8¾ ins

Inscribed on the tablet in English, Latin, Italian, French and Spanish *John Taylor* MD *Oculist to the King of Great Britain Fellow of Several Colleges of Physicians. &c*, followed by eight lines in Latin, commencing *Effigiem Taylor*; on the paper at bottom right LECTURES *On the Means of Preserving Sight and on the Nature & Method of Removing the Several defects of it call'd Weakness of Sight: In a Stile design'd to be understood by the Ladies, for whom they were chiefly intended*, with various volumes of Taylor's international lectures, opinions and books bearing dates from 1727 to 1744; and at the bottom of the plate, *W. De Nune Pinxit. R. Cooper Sculpsit*.

LITERATURE
R.R. James, *Studies in the History of Ophthalmology in England prior to the year 1800*, Cambridge, 1933; *Deutsch*, pp. 804–6, 849; David M. Jackson, 'Bach, Handel and the Chevalier Taylor', *Medical History*, XII, 1968, pp. 385–93.

211

HANDEL

Death and Burial

On Easter Saturday 14 April 1759, at the age of 74, Handel died at his house in Brook Street. He had lived in London for nearly fifty years, 36 in this one house. It had been his hope to die, like his Saviour, on Good Friday. Three days after his death James Smyth, a friend who lived close by in Bond Street, wrote to Bernard Granville (see no. 157), Handel's long-standing acquaintance:

> According to your request to me when you left London, that I would let you know when our good friend departed this life, *on Saturday last at 8 o'clock in the morn died the great and good Mr. Handel.* He was sensible to the last moment; made a codicil to his will on Tuesday, ordered to be buried privately in Westminster Abbey, and a monument not to exceed £600 for him. . . . He died as he lived – a good *Christian*, with a true sense of his duty to God and man, and in perfect charity with all the world.

His funeral in Westminster Abbey on Friday 20 April was a solemn occasion, with the choirs of the Chapel Royal, St Paul's Cathedral and Westminster Abbey who sang William Croft's *Burial Service*. His monument by Roubiliac was completed in 1762 (see no. 17). In his will (no. 212) he made numerous personal and charitable bequests from his large fortune. Quoting again from Smyth's letter: 'I had the pleasure to reconcile him to his old friends; he saw them and forgave them, and let all their legacies stand! . . . He has died worth £20,000, and left legacies with his charities to nearly £6,000.'

Handel was the subject of numerous obituaries in the press and of a volume of memoirs (no. 214) by John Mainwaring, Fellow of St John's College, Cambridge, who produced in 1760 what was the first book-length biography of any musician. But this was only the beginning of the commemoration of Handel.

LITERATURE
Deutsch, pp. 818–19.

Fig. 11 Sir George Amyand, Bart, MP (1720–66), executor of Handel's will, by Antoine Pesne, 1745
Christie's, South Kensington, 8 January 1985

212   [*illustrated on p. 220*]

Handel's Will and Four Codicils

The will in Handel's own hand, the codicils (except the third) probably written out by the lawyer, John Hetherington, and signed by Handel

'I George Frideric Handel considering the Uncertainty of human life doe make this my will in manner following.' So wrote Handel on 1 June 1750, shortly before his last visit to Germany. There are four codicils to his will, the first made on 6 August 1756 and the last on 11 April 1759, three days before his death. Handel made more than thirty specific bequests to his friends, relatives and servants.

Of particular interest is the mention of certain of Handel's possessions: 'I give and bequeath to Mr Christopher Smith my large Harpsicord, my little House Organ, my Musick Books, and five hundred Pounds sterl:' (another £1,500 was added in the first codicil in 1756). This bequest was to Smith senior, though his son (see no. 210) inherited in due course. The large harpsichord was given by Smith junior to George III (see no. 234) together with Handel's autograph scores; the little house organ is not known.

Richard Leveridge by Thomas Frye (no. 58)

Friedrich Chrysander by Leopold Kalckreuth, 1901 (no. 269)

'I give to John Rich Esq.' my Great Organ that stands at the Theatre Royal in Covent Garden' (third codicil, 4 August 1757): this organ was destroyed with the theatre in the fire of 1808.

'I give a fair copy of the Score and all the parts of my Oratorio called The Messiah to the Foundling Hospital': the second soprano part is shown here (no. 205).

Handel's bequests of pictures to Charles Jennens and Bernard Granville are discussed in Appendix 4.

Handel appointed two executors to his will: his niece and residuary legatee, Johanna Friederika Flörke, and his friend George Amyand, MP (see fig. 11, p. 232). It was Amyand who proved the will on 26 April and put its provisions into effect, not without protests from some of Handel's German relatives, as documents in the Royal College of Music show. Amyand was a wealthy merchant in the German trade, a banker, and from 1754 to his early death in 1766 a Member of Parliament. An almost identical duplicate copy of the will and its codicils in the Gerald Coke Handel Collection, evidently written out at the same time as the original, appears to have come to Amyand in his capacity as executor.

Public Record Office, London

The will is written on both sides of a single sheet, 24×19 cms/$9\frac{1}{2} \times 7\frac{1}{2}$ ins; the four codicils are bound into a volume, 35×21.5 cms/$13\frac{3}{4} \times 8\frac{1}{2}$ ins

LITERATURE
Deutsch, pp. 691–2, 776, 784, 788–9, 814–15; W.C. Smith, 'More Handeliana', *Music and Letters*, XXXIV, 1953, pp. 15–18.

213

'Fees for the Funeral of George Frederick Handel Esq.'

Funeral fee book for Westminster Abbey

Handel's funeral expenses, here published for the first time, came to the considerable total of £45. 5s. 6d. and were paid by his executor, George Amyand (see fig. 11, p. 232).

Despite the wish expressed in his will to be buried 'in a private manner', Handel was given a particularly grand funeral:

> On Friday Night [20 April 1759] the Remains of the late Mr. Handel were deposited at the Foot of the Duke of Argyle's Monument in Westminster-Abbey; the Bishop, Prebendaries, and the whole Choir attended, to pay the last Honours due to his Memory; and it is computed there were not fewer than 3000 Persons present on this Occasion [*London Evening Post*, 24 April 1759].

The service was taken by the Dean, Dr Zachary Pearce, Bishop of Rochester, who is recorded as receiving £2 in his capacity as Dean and another £1 for officiating. The £10 paid 'To the Fabrick' and the last item of £6 for a 'Leaden Coffin' were standard charges, while the £17. 17s. 2d. paid by John Merest, Receiver General at the Abbey, to Anselm Bayly, the Precentor, would have represented the fees of several participants in the funeral.

Further charges of 6 guineas for a gravestone and £25 for the site of the monument are recorded in the Abbey Treasurer's account book for 1759 (Westminster Abbey Muniment 33790).

The Dean and Chapter of Westminster

Manuscript fee book, ii + 134 + iii pp., 19.8×15.6 cms/$7\frac{3}{4} \times 6\frac{1}{8}$ ins

[Transcript of the fee account, on p. 125 verso:]

Fees for the Funeral of George Frederick Handel Esq.' in the South Cross of Westminster Abby.

		£.	s.	d
	To the Fabrick	10.	0.	0
	To the Dean	2.	0.	0
	To the Subdean	0.	13.	4
	To the 12 Prebendaries	6.	0.	0
The Dean.	To the Minister officiating	1.	0.	0
	To the Chantor & Choir	8.	3.	4
	To the Organ Blower	0.	5.	0
	To the Receiver			
	To the Register	0.	10.	0
	To the Clerk of the Works	0.	13.	4
	To the Gravemaker	0.	10.	0
	To the Mason	0.	5.	0
	To the two Sacrists	0.	13.	4
	To the two Virgers	0.	13.	4
	To the eight Bearers	1.	0.	0
	To the 4 Bellringers	0.	13.	4
	To the 12 Almsmen	1.	10.	0
	To the 2 Porters	0.	12.	6
	To the Beadle	0.	5.	0
	For 9.^{lb} of white Wax Candles & 3 doz.ⁿ of Torches	2.	8.	0
	For the Church Pall	0.	10.	0
	Leaden Coffin	6.	0.	0

April 14.th 1759 Rec^{ed} of George Amyand Esq.' 45. 5. 6

April 21.st 1759 Rec^{ed} then of John Merest Esq.' the Sum of— Seventeen Pounds Seventeen Shillings & two pence, for the Funeral of George Frederick Handel Esq.', for my self & others therein concerned—by me

£17. 17. 2   *Anselm Bayly*

[on the previous page, before the main fee account:]

May 1.st 1759 Rec^{ed} then of Mr.' Merest the Sum of three Pounds for my Fees for the Funeral of George Frederick Handel Esq.', by me

3..0..0   *Z^y Roch.' Dean*

VIII *The Commemoration of Handel*

Within his own lifetime Handel had achieved classic status; this he has never lost – unlike most of his contemporaries, who at one time or another have been out of fashion. The commemoration of the composer from his own time to this day is a subject worthy of an exhibition in itself. The work of the image-makers (see nos. 246–54), the biographers (nos. 214–16) and the editors of his music (nos. 264–70) is gathered here with souvenirs of some of the great Handel Festivals and performances from his own century to the present (see nos. 232–44, 255–63 and 271–7).

240

The Early Biographies

To Handel falls the honour of the first biography of a musician to be published as a book (no. 214). Written by the Rev. John Mainwaring, who had apparently never met the composer, it appeared in 1760 and was followed in 1761 by a German edition, the work of Handel's former friend, Johann Mattheson, who embellished his translation with stories deriving from his own experience. Sir John Hawkins's *General History of Music* (1776) and Charles Burney's work of the same title (1776–89) provide biographical information on Handel but of greater importance are Burney's 'Sketch of the Life of Handel' of 1785 (no. 242) and the Rev. William Coxe's *Anecdotes of George Frederick Handel* of 1799 (no. 216). These volumes are all based on personal knowledge of Handel or the reports of those who knew him; they are anecdotal in approach. The first critical biographies in the modern sense follow after a gap of more than fifty years: Victor Schoelcher's *Life of Handel* (1857) and Friedrich Chrysander's three-volume *G.F. Händel* (1858–67).

214

'Memoirs of the Life of the Late George Frederic Handel. To which is added, A Catalogue of his Works and Observations upon them.'

By John Mainwaring (1724–1807)
Printed for Robert and James Dodsley, 1760

John Mainwaring, Handel's first biographer, was a Fellow of St John's College Cambridge, rector of Church Stretton, Shropshire, and, later, Professor of Divinity at Cambridge. His biography is in three parts, the 'Memoirs of the Life', which relies heavily on biographical information from John Christopher Smith the younger; a 'Catalogue' of Handel's works (pp. 145–55) contributed by James Harris (see no. 160); and the 'Observations' on these works (pp. 159–208) provided by another close friend of Smith, Robert Price of Foxley. Mainwaring's authorship of the volume, though sometimes doubted, is confirmed by two letters from him to Dodsley, claiming two thirds of the profit arising from his work (these letters, dated 13 June 1776 and 2 September 1787, are in the Gerald Coke Handel Collection).

As a volume, Mainwaring's *Memoirs* is the first extended biography of a European composer in any language. As a source for Handel's life it provides essential information, even though much of it was compiled many years after the events of which it tells; more than half of the biographical section is concerned with Handel's German and Italian years and the information here must come, if indirectly, from Handel himself. As a critical survey of Handel's work the 'Observations' of the final section of the book, however derivative in approach, have been seen as noteworthy as a pioneering attempt to provide a coherent critical point of view in discussing the music.

This copy of the *Memoirs* belonged to Charles Jennens (see no. 181), and his annotations, on 22 of the 208 pages, provide otherwise unknown information on the chronology of two of Handel's works, *Saul* and *Belshazzar*, for which he himself provided the libretti. His corrections to Mainwaring are terse and often quite pointed: *Semele* he describes as 'No Oratorio, but a baudy Opera'; when Mainwaring accuses Handel of 'consulting his interest at the expence of his art', Jennens demands: 'Explain your self, if you can!' His longest annotation (on p. 63) relates to Mainwaring's account of Cardinal Pamphili, who in about 1707 wrote *Il Trionfo del Tempo* for Handel to set and also the verses of a cantata in honour of Handel himself: 'Handel told me that the words of Il Trionfo &c. were written by Cardinal Pamphilii, & added "an old Fool!" I asked "why Fool? because he wrote an Oratorio? perhaps you will call *me* fool for the same reason!" He answered "So I would, if you flatter'd me, as He did."'

Mainwaring's biography was translated into German with critical additions by Johann Mattheson (see no. 27) and published at Hamburg in 1761. An abridged French translation was published at Paris in 1768.

Robin Golding Esq

Printed book, 208 pp., 20.4 × 13.6 cms/8 × 5⅜ ins

LITERATURE
Deutsch, pp. 842–3; Peter Kivy, 'Mainwaring's *Handel*: its Relation to English Aesthetics', *Journal of the American Musicological Society*, XVII, 1964, pp. 170–8;
Winton Dean, 'Charles Jennens's Marginalia to Mainwaring's Life of Handel', *Music and Letters*, LIII, 1972, pp. 160–4;
William Weber, 'Intellectual Bases of the Handelian Tradition 1759–1800', *Proceedings of the Royal Musical Association*, CVIII, 1981, pp. 100–14; Burrows, pp. 36–8.

214

[Facsimile of page 63 with handwritten marginal annotations:]

and others extempore. † One of these was in honour of HANDEL himself. He was compared to ORPHEUS, and exalted above the rank of mortals. Whether his Eminence chose this subject as most likely to inspire him with fine conceptions, or with a view to discover how far so great an Artist was proof against the assaults of vanity, it is not material to determine. HANDEL's modesty was not however so excessive, as to hinder him from com-

† The ABBE DU BOS, speaking of that general turn for Music for which the Italians from the highest to the lowest have ever been remarkable, continues thus,— Ils sçavent encore chanter leurs amours dans des vers qu'ils composent sur le champ, & qu'ils accompagnent du son de leurs instruments. Ils les touchent, si non avec délicatesse, du moins avec assez de justesse : c'est ce qui s'apelle *improviser*.

Manuscript annotations: Cantata the 19th in my collection, which contains only 51 in all. Handel told me that the words of Il Trionfo &c. were written by Cardinal Pamphili, and as an old fool!! I ask'd 'Why fool?' 'Because he wrote an Oratorio'. 'Perhaps you will call me fool for the same reason.' 'He answer'd if I would, if you flatter'd me as He did.'

215

Sir John Hawkins (1719–89)

By James Roberts (1753–c.1809)

Sir John Hawkins published his celebrated five-volume magnum opus *A General History of the Science and Practice of Music* in 1776, shortly after the first volume of Charles Burney's work on the same subject had appeared. Hawkins's book is a compendium of information, paying special attention to the music of the sixteenth, seventeenth and early eighteenth centuries up to the death of Handel, at the expense of the composers of Hawkins's own day. This bias towards 'ancient music', as it was called, and the book's unanalytical approach and digressive style attracted criticism from the partisans of Burney's better-organised and more wide-ranging survey; but for the biographer of Handel the book provides extensive information on his life not available elsewhere, drawing on Hawkins's personal recollections of the composer.

Hawkins had studied as an attorney, later acting as chairman of the Middlesex Quarter Sessions, but he devoted much time and energy to his literary and antiquarian interests. He was a close friend of the blind organist John Stanley (see no. 208) and also knew Handel, Boyce and Pepusch. Through his contributions to the *Gentleman's Magazine* he became well acquainted with Samuel Johnson, and acted as one of his literary executors; he retained Johnson's friendship despite Johnson's gibe, 'a most *unclubable* man' – a result of his rudeness at meetings of Johnson's Club of friends.

In this the only portrait of Hawkins, he is shown with his *History of Music* lined up volume by volume in the background. Hawkins was reluctant to sit for his portrait, considering the wish for this form of immortality to be 'foppery', but according to his daughter, Laetitia Matilda Hawkins, he was persuaded to sit by Dr Hayes, Professor of Music at Oxford, so that his portrait might be placed in the

215

music school there. The result, though described by the artist as a 'capital likeness', was disliked by his daughter: it showed him, she claimed, 'as he never looked, dressed as he never dressed, and employed as never was employed; for he is represented smirking in a velvet coat, with a volume in his hand which ... must be one of the last new novels then printed'.

Faculty of Music, Oxford University
(By Permission of the Heather Professor of Music)

Oil, 76.5 × 63.8 cms/30$\frac{1}{8}$ × 25$\frac{1}{8}$ ins

Inscribed below the books *Sr J. Hawkins. | 1786* (the last digit indistinct); the books are labelled HISTORY OF MUSIC VOL. I etc

Engraved by R. Clamp, 1794 for E. and S. Harding's *Biographical Mirrour*, I, 1796, p. 158

PROVENANCE
Given by the artist to the Music School at Oxford, probably in 1788 when living in Oxford.

LITERATURE
Poole, I, no. 396; Percy A. Scholes, *The Life and Activities of Sir John Hawkins*, 1953, pp. 183–4.

EXHIBITED
Samuel Johnson 1709–84, Arts Council, London, 1984, no. 16.

216

'Anecdotes of George Frederick Handel, and John Christopher Smith'

By William Coxe (1747–1828)
Printed by W. Bulmer & Co, 1799

Like Mainwaring's earlier biography, the authorship of the 'Anecdotes' has been doubted, but it is certainly the work of the Rev. William Coxe, Archdeacon of Wiltshire and step-son of one of the two subjects of the book, John Christopher Smith the younger (see no. 210). Coxe was a Fellow of King's College Cambridge and author of many volumes including *Memoirs of Sir Robert Walpole* (1798) and *Memoirs of John, Duke of Marlborough* (1818–19).

Coxe provides short lives of both Handel (in 34 pages) and Smith (in 25 pages), to which he adds a list of Smith's compositions and 34 pages of music taken from three of his works. His life of Handel, as he himself states in the 'Advertisement' which precedes the text, draws on Mainwaring's *Memoirs* (he does not give him as author but refers to it as being 'written under the inspection of Mr. Smith'), Hawkins's *History of Music*, and the works of Burney, to which, Coxe claims, were added some anecdotes of 'unquestionable authority'. It is however in his life of Smith, his step-father, that Coxe gathers together much biographical information that is otherwise unpublished.

The list of subscribers which precedes the volume includes John Mainwaring, Handel's biographer of forty years before, Lord Malmesbury, son of Handel's friend James Harris, and Lord Fitzwilliam, whose Handel collections formed part of the founding bequest of the Fitzwilliam Museum. Many of Smith's family and friends subscribed, including Lord Braybrooke, Bishop Thomas Dampier and Uvedale Price, all sons of Smith's Geneva friends.

The British Library Board
G 2182

Printed book, iv + 64 pp. + 34 pp. of music, 28.0 × 21.8 cms/11 × 8$\frac{5}{8}$ ins

PROVENANCE
Rt Hon Thomas Grenville (1755–1840), by whom bequeathed to the British Museum, 1840.

LITERATURE
Percy Young, introduction to the reprint of this volume by Da Capo Press, New York, 1979.

'No Vice but the Love of Music':
Sir Samuel Hellier and the Sharp Family

Two men of the latter part of the eighteenth century not only were enthusiasts for the music of Handel but also have left us solid evidence of the means by which they indulged their enthusiasm. One was Sir Samuel Hellier (1737–84) of the Wodehouse, Wombourne, Staffordshire, and the other was Granville Sharp (1735–1813), one of the leaders in the fight for the abolition of slavery. Both left records of their practical musicianship, Hellier a list in his own hand of the instruments which he possessed for his private band, and Sharp a family portrait by Zoffany (no. 228), commissioned by his brother, showing a number of his instruments. By great good fortune, some small part of both these collections of instruments has survived in the hands of the two families.

Hellier's instruments seem normally to have been played at the Wodehouse, where a music room was built for that purpose, whereas Sharp's were often heard, to the great delight of the Royal Family and many others, from the family barge, *The Apollo*, as may be seen in the painting. There are numerous references to the family water parties in the diaries of Elizabeth Prowse, one of Granville Sharp's sisters and the lady seated at the keyboard of the piano holding a sheet of music.

The Sharp instruments have passed down to us through the Lloyd-Baker family, the descendants of the baby seated on her mother's lap, and are now on loan to the Bate Collection of Historical Instruments in the Faculty of Music of the University of Oxford. The Hellier orchestra passed first to Thomas Shaw, a friend of Hellier's and his heir, and then to Shaw's great-grandson, Thomas Shaw-Hellier, later Colonel Commandant of the Royal Military School of Music (1883–93), who greatly amplified the collection and was partly responsible for three major exhibitions in which a number of the instruments appeared; subsequently they passed to a cousin of his great-niece, Mr J.W. Phillips, and most are now on loan to the Warwickshire Museum.

Of the instruments in the Zoffany painting which are no longer in the family's possession, the keyboard is one of the earliest depictions of a fortepiano in England, as pointed out by Laurence Libin, Curator of Musical Instruments at the Metropolitan Museum of Art, who suggested also that the large lute played by Judith Sharp was an angelica, a form of theorbo in which the strings are tuned diatonically rather than in wider intervals, and well suited to the skilled amateur. The one instrument which might eventually be identified is the serpent, for the decorative medallion on the bell, from which a chain is suspended, leading presumably to another medallion on the other side, is an unusual feature.

Of the Hellier instruments no longer owned by the family, the later travels of one are well known: one of the pair of violins 'Foreign' heading the list is the famous Hellier Stradivarius, one of the most beautiful the Italian master ever made. The others cannot be identified, though one might compare the 'Mettle Trumpet wth Silver Ornam^{ts} by Harris' with the one on loan to the Bate Collection (see under no. 82), and the Stanesby Junior bassoon with that in the Waterhouse Collection (no. 172).

The Hellier instruments are listed here in the order given by Sir Samuel quoting his own descriptions of them; the Sharp instruments are as they appear from left to right in the painting.

LITERATURE
The diary of Elizabeth Prowse, manuscripts in the Gloucestershire County Record Office (D 3549, Box 73).

217

Sir Samuel Hellier (1737–84)

Attributed to James Shaw (fl. 1769–84)

Samuel Hellier had a passionate fondness for music, and for the work of Handel in particular: he had 'no vice ... but the love of music', to use his own words. His portrait shows him in Van Dyck costume holding a scroll of music, and is attributed to a local artist, James Shaw, a pupil of Edward Penny and a student at the Royal Academy in 1769.

Hellier went up to Exeter College, Oxford, at the age of 16 in 1753; he is noted as studying the violin in 1754 and was buying Handel's music in 1756 (see no. 219). He took his degree as MA in 1758 and became a barrister of the Inner Temple. In 1762 he was knighted on his appointment as High Sheriff of Worcestershire.

Our knowledge of Hellier's later musical activities is derived mainly from his letters written from London to his bailiff, John Rogers, at the Wodehouse. There are many references in the late 1760s to his purchases of musical instruments: an organ for the parish church which was fitted by Mr Adcock in 1767; various wind instruments of which he writes, 'I hope to send you Horns & Hautboys but Bassoon's are so much call'd for I can get none just yet' (15 January 1767); and a double bass in May 1768. He even suggests a performance of *Messiah* to inaugurate the new organ.

Hellier's father, also Samuel, had died in 1751 when his son was 14. His interest in music is less well documented than that of his son, but his name is found in some music subscription lists and it is possible that he acquired certain of the Hellier instruments, such as the serpent (no. 227).

J.W. Phillips Esq

Oil, 76.2 × 63.5 cms/30 × 25 ins

PROVENANCE
Sir Samuel Hellier (1737–84); bequeathed to his friend the Rev. Thomas Shaw who changed his name to Shaw-Hellier; by descent at the Wodehouse to his great-grandson, Col Thomas Shaw-Hellier, commandant of the Royal Military School of Music; by descent to J.W. Phillips.

LITERATURE
Hellier papers, Staffordshire Record Office, temporary deposit, Tp.1226/Add. K/1, 1322/2.

218

217

218

Handel's Temple at the Wodehouse, Wombourne, Staffordshire

By J. Hughes (fl. 1773)

Among Sir Samuel Hellier's garden buildings at the Wodehouse, Wombourne (to the south of Wolverhampton), was a temple to Handel, with a figure of the composer, based on the Westminster Abbey monument, as its centre-piece. It was built to the design of the young James Gandon, his 'Elevation of a mausoleum to the memory of Handel' being exhibited at the Society of Arts in 1768. Sir Samuel wrote to his bailiff on 20 April that year: 'As Handel's Temple is an entire new design and as Mr Gandon has drawn it and exhibited it to public view I am very anxious to have it completed out of hand; for which reason pray forward it with great expedition.'

The drawing was made in 1773 by the little-known artist, J. Hughes, as one of a series of small views of the garden at the Wodehouse, including a music room pavilion and a hermitage. These buildings, like Handel's Temple, are all long since destroyed.

J.W. Phillips Esq

Pen and black ink and grey wash, 11.5 × 15.6 cms/4½ × 6⅛ ins

Signed bottom left *J: Hughes delin*.

PROVENANCE
As no. 217.

LITERATURE
Hellier papers, Staffordshire Record Office, temporary deposit, Tp.1226/Add. K/1; M.W. Greenslade (ed.), vol. xx of *A History of the County of Stafford*, in *Victoria History of the Counties of England*, 1984, pp.193 (repr.), 206; Edward McParland, *James Gandon, Vitruvius Hibernicus*, 1985, pp.15, 207, pl.13.

219

'L'Allegro, Il Penseroso, ed Il Moderato.
The Words taken from Milton.
Set to Musick by Mr Handel.'

Printed score
Printed for John Walsh, 1747

This edition of Handel's *L'Allegro, il Penseroso ed il Moderato*, his setting of Milton's poems as expanded by Charles Jennens, was acquired by Samuel Hellier in 1756 while still at Oxford. To the printed songs, the other sections have been added in manuscript (in more than one hand), a prerequisite for giving a full performance. Manuscript additions of this sort are found in a number of other volumes in Hellier's substantial collection of Handel printed editions.

J.W. Phillips Esq

Printed score with manuscript additions, 221 pp., 29.8 × 24.7 cms/11¾ × 9¾ ins
Titled as above and inscribed in pen on the title-page *S. Hellier. Exon, Coll, Oxon. 1756.*

PROVENANCE
As no. 217.

LITERATURE
Smith, p. 95, no. 9.

220

Pair of French Horns
By John Christopher Hofmaster (fl. 1751–d. 1764)
'One Pair of Concert French Horns wth Crooks &c Complete in painted Boxes by Hofmaster London'

Hofmaster, presumably originally Johann Christoph Hofmeister, settled in London some time before the mid-century and introduced the new orchestral horns, built with detachable crooks. Their great advantage was that players could use the same instrument in different keys, merely changing the couplers, whereas with the older models such as those by Leichamschneider (no. 81), a different horn would have to be used for each key. Hofmaster's origin is as yet unknown, but it is clear that he introduced the wide-bore Eastern European type of horn, which remained the norm in this country until superseded in the first third of the nineteenth century by the French model.

Both these horns are right-handed, unlike the Sharp pair (see no. 230); this follows the more normal eighteenth-century practice of having one coiled right-handed and the other left-handed, which looks more impressive in the orchestra. One has only a master crook and the smallest coupler: the other has the master crook, which accepts the mouthpiece, and four couplers which fit between the master crook and the horn and produce the harmonic series of

220

different keys. The couplers can be used individually and in combination.

A mute has survived, the earliest horn mute to have done so in this country. When held in position, for it does not fit snugly into the bell like a modern mute, the tone is modified as with mutes today, and the pitch remains unaffected, unlike the early trumpet mutes which raise the pitch by a semitone. Another rare survival is the pair of red painted wooden cases for the horns, probably the originals.

These horns are probably those referred to in a letter from London sent by Sir Samuel Hellier to his bailiff John Rogers on 18 January 1767. He writes, 'I have now sent you a French Horn in the key of D. & mouthpiece to it. . . . When you see how you are I will send down an other Horn but till you are a little Settled one will do.'

J.W. Phillips Esq

Inscribed on the bells
IOHN | CHRISTOPHER . HOFMASTER | IN . PICCADILLY | LONDON | 17

Brass, the inside of the bells painted in ox-blood,
maximum width 51 cms/20 ins,
tube length 235 cms/92½ ins without crook or couplers,
master crook length 51 cms/20 ins,
bell diameter 24 cms/9⅜ ins,
mouthpiece socket diameter 0.8 cms/⅓ in;
pitches (at A = 430 hz):
with master crook alone 484 hz (B),
with smallest coupler 437 hz,
with next coupler 429 hz (A),
with next coupler 409 hz (G♯),
with longest coupler 343 hz (F);
the tuning bits and the other crooks listed in the Royal Military Exhibition catalogue have all disappeared;
mute 13 × 8.3 × 7.4 cms/5⅛ × 3¼ × 2⅞ ins

PROVENANCE
As no. 217; now on loan to the Warwickshire Museum.

LITERATURE
See no. 217.

EXHIBITED
Royal Military Exhibition, London, 1890, no. 308.

221

Trumpet

By John Christopher Hofmaster (fl. 1751–d. 1764)
'One Silver Trumpet by Hofmaster London in a Red Leather Case'

This handsome instrument with its finely embossed bell is still in its original state. The mouthyard passes through a groove in the ball and is lashed with cord to the bellyard. The front bow is tied to the bell with thin cord which passes through a small hole in the rim of the bell. This construction is typically English: German trumpets did not have a groove in the ball, which was anyway smaller, and the mouthyard was held away from the bellyard by a wooden spacing block.

A trumpet banner embroidered with the royal coat of arms as used from 1714–1801, known from old photographs but long thought to be lost, has recently been rediscovered and is now reunited with the trumpet. The red leather case listed in the Hellier inventory is not now known.

J.W. Phillips Esq

Inscribed on the bell *HOFMASTER LONDON*, and engraved with the Hellier coat of arms and motto
PRO RE PUBLICA SEMPER

Silver, overall length 69.2 cms/27¼ ins,
tube length 212.8 cms/83¾ ins,
bell diameter 11.1 cms/4⅜ ins;
catalogued at the Royal Military Exhibition as in D, but the present pitch is over a tone above that note,
E at A = 435

PROVENANCE
As no. 220.

LITERATURE
Eric Halfpenny, 'Notes on Two Later British Trumpets', *Galpin Society Journal*, XXIV, 1971, pp. 79–83, pls. 9–12, 14a.

EXHIBITED
Royal Military Exhibition, London, 1890, no. 294.

221

222

222

Trumpet

By an unknown maker, early eighteenth century
'One Brass Trumpet with Crooks Mouthpieces &c'

No mouthpiece to this trumpet now survives, but there is a crook which might be original. The bell is unusually conical, perhaps indicating an early eighteenth-century date, and has a wide garland. The trumpet is in original state, with the mouthyard passing through a groove in the ball and lashed with silver wire to the bellyard. The front bow is tied to the bell with wire. The metal of the bell is badly corroded through age under the garland.

J.W. Phillips Esq

Brass, overall length 67.5 cms/26½ ins without crook,
71.4 cms/28 ins with crook,
tube length 207.6 cms/81¾ ins,
crook tube length 41.5 cms/16⅜ ins,
bell diameter 10.7 cms/4¼ ins

PROVENANCE
As no. 220.

EXHIBITED
International Loan Exhibition, Crystal Palace, 1900, no. 273.

223

Pair of Kettledrums

By an unknown maker, mid- or early eighteenth century
'One Pair of Kettle Drums Complete in Red Leather Cases'

Timpani of this period are sufficiently rare in England to make it very difficult to judge their date or origin. Certainly these are of eighteenth-century date, probably no later than the mid-century and quite possibly earlier. They are probably of English manufacture in that they do not appear to be German and are unlikely to be French. The small difference of diameter between the two drums is typical of timpani of the first half of the century; drums of this size, but with a larger difference between them, can still be seen today in use by the Household Cavalry, dating from the latter part of the last century. Since there is no trace of any fittings for horse harness, it is clear that these timpani were made for orchestral use. The red leather cases listed in the inventory are not known.

J.W. Phillips Esq

Brass shells, iron feet, goat-skin heads;
overall diameters 60.2 cms/23¾ ins and 58.5 cms/23 ins,
head diameters 52.5 cms/20⅝ ins and 50.5 cms/19⅞ ins,
total height 38.2 cms/15 ins and 36.7 cms/14½ ins,
shell depth 33.6 cms/13¼ ins and 31.5 cms/12⅜ ins;
the shells are each hammered from a single sheet of brass; scribed lines on the under side of the shell 120° apart mark the positions of the three small, red-painted feet, which were necessary for storing the drums (placing the shells directly on the ground was liable to dent them); in performance they would be supported on stands or trestles; the heads are reverse-lapped on deep, narrow hoops (25 cms high, 3 cms thick) and tensioned by eight screws with small square heads, tuned with a key similar in size to a clock key, which pass through lugs fixed to the hoops and piercing the heads in the normal way for the period.

PROVENANCE
As no. 220.

EXHIBITED
International Loan Exhibition, Crystal Palace, 1900, no. 5.

223

224

225

Tambourine and Case

By an unknown maker, mid-eighteenth century
'One Tambourine Drum in a Red Leather Case'

Few tambourines of this period are known. It is perhaps associated with the side drum (no. 224) and, like it, dating to about 1750, as suggested by Colonel Shaw-Hellier. What appears to be its original case survives with the tambourine.

J.W. Phillips Esq

Red-painted wood shell,
diameter 51 cms/20 ins, depth 14.3 cms/5⅝ ins;
calf skin stitched to an iron wire hoop which is secured by nine tension hooks which pierce the skin so as to bear on the hoop, and pass through L-shaped metal hasps screwed on to the shell, beyond which they are tensioned with wing nuts;
the shell is made of a single piece of red-painted wood formed into a hoop with a scarf joint which overlaps by 23 cms and is nailed at random (there is no pattern of nails such as was common with side drums); there is a thumb-hole in the joint and there are eight pairs of high-domed brass jingles let into the side of the shell; original case, with the top hinged on a chord halfway between the diameter and the edge, and a folding handle on the lid, diameter 57.4 cms/22½ ins, depth 20 cms/7⅞ ins

PROVENANCE
As no. 220.

224

Side Drum

By an unknown maker, perhaps by Nicholson
'One Common Drum Mahog.ʸ by Nicholson, London in a Red Leather Box'

A mahogany side drum of mid-eighteenth-century date. The attribution in Hellier's list to Nicholson, an unrecorded maker, has not been verified. There are a pair of eighteenth-century side drum sticks in the same collection. As is not unusual for this period, they differ slightly in size; the shorter has been broken and repaired. The wood appears to be beefwood and there are seven incised lines at the butt end of both. The red leather box listed in the inventory is not known.

J.W. Phillips Esq

Mahogany side drum with red-painted counter hoops,
diameter 44.5 cms/17½ ins, depth 47 cms/18½ ins

PROVENANCE
As no. 220.

225

226

226
Oboe

By Thomas Stanesby senior (*c.*1668–1734)

Both the Gedney and the Stanesby junior oboes listed by Sir Samuel Hellier have disappeared, and this instrument, though from the previous generation, was presumably acquired later perhaps by Colonel Shaw-Hellier, who was actively collecting towards the end of the nineteenth century. Stanesby junior served his apprenticeship with his father, and Caleb Gedney with Stanesby junior; thus all three represent a direct chain of tradition.

J.W. Phillips Esq

Stamped T / STANESBY / (8-pointed star)
Tortoise-shell stained boxwood, with two silver keys for C and E♭; overall length 59.1 cms/23¼ ins
PROVENANCE
Perhaps acquired by Colonel Shaw-Hellier in the late nineteenth century; by descent to J.W. Phillips; now on loan to the Warwickshire Museum.
EXHIBITED
International Loan Exhibition, Crystal Palace, 1900, no. 166.

227
Serpent

French, by an unknown maker, eighteenth century
'One Serpent silver ornaments made at Paris in a Red Leather Case'

This type of key-less serpent is usually referred to as a *serpent d'église*; by the time that the instrument was adopted into the English military band it had usually acquired three keys. Presumably Sir Samuel Hellier used it to bulk out the bass parts of his orchestra, though with three cellos, a double bass and two bassoons this would hardly seem necessary. The red leather case listed in the inventory is not known.

J.W. Phillips Esq

Wood, leather stained black, with silver-plated brass mounts and crook, and ivory mouthpiece; the bell mount is engraved with the Hellier crest; there are six finger holes, no keys; overall length 97 cms/38¼ ins, tube length about 233 cms/91¾ ins
PROVENANCE
As no. 220.
EXHIBITED
Royal Military Exhibition, London, 1890, no. 329.

227

228 [*in colour opposite p.153*]

The Sharp Family
By Johann Zoffany (1733–1810)

Music-making on the Thames was a feature of London life throughout the eighteenth century, from the time of the performance of Handel's *Water Music* for George I in 1717 (see p. 95) and before. This noble tradition was continued by the remarkably talented sons and daughters of Thomas Sharp (1693–1758), Archdeacon of Northumberland; here they are gathered for one of their celebrated water parties which were held on the Thames each summer from as early as 1755. This very diverse family, united by its love of music, is commemorated in this masterpiece by Johann Zoffany, which was painted soon after his return to London from Italy in 1779. It shows the Sharps on their sailing barge *The Apollo*, at Fulham, with their instruments, some of which still survive, namely the French horns and the clarinet (lying on the piano) and the flageolets (held up above the serpent) (nos. 229–31).

Zoffany takes the spectator on board *The Apollo* so that one is looking along the length of the vessel from its right side to the stern, where the four Sharp brothers and three sisters are gathered with their families. The painting was commissioned by William Sharp (1729–1810), who is seen standing at the tiller with hat raised; he was surgeon to George III and wears Windsor uniform with its distinctive red collar. His wife and baby daughter, Mary (with the kitten), are in front of him; his are the French horns on the piano. Of his three brothers, each distinguished in his own way, Dr John Sharp (d. 1792), Archdeacon of Northumberland like his father, is on the right and has laid aside his cello for the moment; his wife is behind him and their daughter, Jemima, wearing a miniature round her neck, is at top right. On the left is James Sharp, an engineer, who played the serpent; his daughter, Catherine, is beside him, and her mother, in the black shawl, looks down on them. Granville Sharp (1735–1813), the famous philanthropist and slavery abolitionist, holds his favoured flageolets in one hand, his clarinet being near to reach on the piano. Granville's three sisters are to his left; Mrs Elizabeth Prowse is at the piano (her diaries are quoted below), Judith Sharp with music in hand is to her left, and above and behind, Frances Sharp, the youngest sister, plays the theorbo or perhaps the angelica. The remaining figures, on the extreme left, are the cabin boy and boatmaster. In the background, William Sharp's yacht, *The Union*, is moored in front of Fulham church, next to which is his balconied cottage; on the extreme right is old Fulham Bridge.

The progress of the painting is recorded in Mrs Prowse's diary: sittings for the 'Family Picture' were taking place in November 1779; Zoffany travelled to Northamptonshire in December 1780 to complete her portrait in time for the Royal Academy exhibition in the spring. Several accounts of the Sharp family water parties survive in her diaries and in the memoirs of her brother, Granville Sharp. Mrs Prowse gives a particularly detailed account of a 'water scheem', as she calls it, on 5 August 1770; this was she thinks the first occasion when they were noticed by the Royal Family. They took boat at Kew Bridge with 'a good band on board' and as they passed the Prince of Wales's House, 'all the 4 Princes ... came down to the water side & stud for near half an hour ... & sent to request 2 or 3 different songs'. A few days later, starting above London Bridge, they went upstream on the tide to Kew where the King and Queen admired their performance; the first piece, Mrs Prowse writes, was 'God Save the King', then 'My 2 Sis[rs] sang let Ceser & Uranio live, Bro:[r] G[ranville] a concerto on the 2 flutes [the flageolets], & accompanied by Bro:[r] James on the Surpent'; the performance was interrupted by a thunderstorm but a special awning 'was drawn over the performers, & Instruments in a moment'.

Among the instruments which Granville lists for the barge are: 'Violins, Horns, Kettledrums, and sticks, Boat whistle, Serpents, Harpsichord, and Stand, Flute box, Violincello, Clarinets, Hautboys'. In 1777 he received the King on board to the tune of 'See the conquering hero comes'; the following year at the request of the Prime Minister, Lord North, they entertained 'six Ambassadors and Foreign Ministers' as well as North himself and his family.

Granville Sharp is best known for his unfailing efforts in the fight for the abolition of slavery. But his musical abilities were remarkable: he was particularly interested in the theoretical side, publishing an introduction to vocal music, and enjoyed singing and playing at sight and had a good bass voice; he played the flute, clarinet, oboe, flageolet and kettledrums. He liked to sign himself by the musical notation G♯, ie G. Sharp. The Sharp brothers, Granville, William and James, also gave fortnightly concerts of sacred music on Sunday evenings, augmented by professional performers. Although few details of their programmes survive, their evenings were said to end with the Hallelujah Chorus, and their strong interest in Handel's music may be judged from their substantial collections of his scores, both printed and manuscript.

The Trustees of the Lloyd-Baker Settled Estates

Oil, 115.5 × 125.7 cms/45½ × 49½ ins

PROVENANCE
Painted for William Sharp; by descent in the family of his only daughter, Mary (1778–1812), who married in 1800 Thomas Lloyd-Baker (1774–1841) of Hardwicke Court, Gloucestershire; thence to his great-great-grand-daughter, the late Miss Olive Lloyd-Baker.

LITERATURE
Prince Hoare, *Memoirs of Granville Sharp, Esq.*, 2nd edition, 1828, I, pp. 207–19, and II, pp. xviii–xxiv; E.C.P. Lascelles, *Granville Sharp and the Freedom of Slaves in England*, 1928, pp. 119–26; James Lees-Milne, 'Stouts Hill, Uley and Hardwicke Court, Gloucestershire', *Country Life*, CLIV, 1973, pp. 17–18, fig. 9 (showing the painting and instruments *in situ* at Hardwicke Court); John Kerslake, 'A Note on Zoffany's "Sharp Family"', *Burlington Magazine*, CXX, 1978, pp. 753–4, figs. 32–4.

EXHIBITED
Royal Academy, 1781, no. 85;
Johann Zoffany, National Portrait Gallery, 1976–7, no. 87; now on loan to the National Portrait Gallery.

229

229
Pair of One-hand Flageolets
By John Mason (fl. *c.* 1755)

While the French flageolet was a professional musician's instrument, with two thumb and four finger holes, and was used from at least the seventeenth century into the twentieth, the English flageolet, originally with six finger holes and no thumb hole, was an amateur's simple substitute for the recorder. In the early nineteenth century it was considerably elaborated, with a double, even occasionally a triple, tube; it nevertheless remained an amateur's instrument and was eventually replaced by the tin or penny whistle. The instruments displayed here are an early and unique attempt at elaboration. Each flageolet has a thumb and four finger holes and three keys, leaving it difficult to conceive how they were played with one in each hand. There are traces of cork blocks in three places, suggesting that they may at one time have been strutted together; both caps are shaved down on their sides to make them fit more closely. They are held by Granville Sharp in the Zoffany painting and it appears that performance on these flageolets was his speciality.

John Mason is a shadowy figure, few of whose instruments have survived. The only contemporary references to him and his work are those published by *Byrne*, recording his residence in King Street, Westminster from 1754–6.

C.G.M. Lloyd-Baker Esq

Stamped with (rampant lion) / J. MASON. LONDON / (two rampant lions); *A* is scratched on both instruments; R is scratched on one, presumably for the right hand, and // is scratched on the other

Boxwood with ivory mounts and mouthpieces and three brass keys;
overall length 55.0 cms/21⅝ ins,
sounding length 31.3 cms/12⅜ ins

PROVENANCE
Perhaps given by Granville Sharp, who played these instruments, to his brother William to go with the Zoffany painting; then as no. 228; now on loan to the Bate Collection of Historical Instruments, Faculty of Music, University of Oxford.

LITERATURE
Maurice Byrne, 'Schuchart and the Extended Foot-Joint', *Galpin Society Journal*, XVIII, 1965, pp. 7–13.

230

Pair of French Horns

By John Christopher Hofmaster (fl. 1751–d. 1764)

Save that this is a true pair, one left-handed and the other right-handed, there is no practical difference between these instruments, played by William Sharp, and those of Sir Samuel Hellier (no. 220). Both orchestras were using the best horns available in their day. The master crooks and four couplers are reproductions, made by Keith Burdett, Sheffield, in 1980, and based on those in the Hellier Collection.

C.G.M. Lloyd-Baker Esq

Inscribed on the bells *JOHN | CHRISTOPHER . HOFMASTER | IN . PICCADILLY | LONDON | 17*

Brass, the insides of the bells painted in ox-blood, maximum width 51.0 cms/20 ins, tube length 235.0 cms/92½ ins without crook or coupler, bell diameter 24.0 cms/9⅜ ins

PROVENANCE
As no. 228; now on loan to the Bate Collection of Historical Instruments, Faculty of Music, University of Oxford.

EXHIBITED
British Musical Instruments, Galpin Society, London, 1951, no. 210.

231

Pair of Clarinets

By George Miller (fl. before 1777–1814)

While it would appear that the clarinet was little known in Britain before the middle of the eighteenth century, we have the evidence of Handel's *Tamerlano* (1724) and *Riccardo Primo* (1727) that at least two players possessed such instruments. In addition there exists, in the Dayton Miller Collection in the Library of Congress, a clarinet barrel by Stanesby junior, which must be earlier than 1754, and the Hellier list includes a pair of clarinets by Stanesby's apprentice and successor, Caleb Gedney.

However, the first English maker whose instruments have survived is George Miller, and Halfpenny has suggested that this pair of clarinets are not only his finest work but also his earliest, dating from perhaps 1760, earlier than Miller's first known appearance in the Westminster rate books in 1777. The bells are the early English conical pattern, instead of the later flared style, and the bell and lowest joint are made in one piece, both features characteristic of Miller's earlier instruments. They must in any case be earlier than 1779, the year in which Zoffany painted the Sharp family picture, in which one may be seen lying on the pianoforte.

C.G.M. Lloyd-Baker Esq

Stamped with (sunburst) | *G.MILLER* | *LONDON* | (five-petalled flower)

Tortoise-shell stained boxwood, with ivory mounts and brass keys; originally made with the five keys usual at this period (E, F♯, A♭, A and the speaker); later a sixth key was added for shakes and these may have been the first English clarinets to adopt this key which is on the left side of the instrument instead of the usual right side; at a subsequent date the hole beneath the key on one of the instruments has been plugged; on the same instrument the original one-piece construction of barrel and mouthpiece has been replaced by shortening the barrel by 2 mm and providing a new separate mouthpiece;
overall length 67.5 and 67.3 cms/26⅝ and 26½ ins

PROVENANCE
As no. 229.

LITERATURE
R.B. Chatwin, 'Handel and the Clarinet', *Galpin Society Journal*, III, 1950, pp. 3–8; Eric Halfpenny, 'Early English Clarinets', *Galpin Society Journal*, XVIII, 1965, pp. 42–56, pls. 5a and b; Maurice Byrne, 'George Miller and John Cramer', *Galpin Society Journal*, XIX, 1966, pp. 134–5.

EXHIBITED
British Musical Instruments, Galpin Society, London, 1951, no. 114.

The 1784 Handel Commemoration

The great Handel Commemoration of 1784, with five large-scale performances in Westminster Abbey and the Pantheon, was an event without precedent in musical history and helped establish the fashion for such festivals. According to the introduction of Charles Burney's record of the celebrations, his book, *An Account of the Musical Performances . . . in Commemoration of Handel*, the 'great idea' for this occasion began in a conversation between Viscount Fitzwilliam, Sir Watkin Williams Wynn and Joah Bates in 1783. They lamented that there was no occasion to bring together London's exceptional wealth of musical performers for 'a performance . . . on so grand and magnificent a scale as no other part of the world could equal'. Burney continues: 'The birth and death of HANDEL naturally occurred to three such enthusiastic admirers of that great master, and it was immediately recollected, that the next (now the present) year, would be a proper time for the introduction of such a custom: as it formed a *complete century* since his birth, and an exact *quarter of a century* since his decease'. In calling 1784 the centenary of Handel's birth they were following the inscription on Handel's Westminster Abbey monument which gives his year of birth as 1684, a confusion perhaps arising from the use of the Old Style calendar in England until 1752.

If there was a precedent for the Handel Commemoration, it was the Shakespeare Jubilee held by David Garrick at Stratford upon Avon in 1769 – an earlier celebration, amounting almost to an act of worship, of the memory of a great national figure. Indeed it was the solitary eminence of Shakespeare that was often invoked in writing of Handel in the generation after his death; an example from the pen of William Coxe occurs in his *Anecdotes of Handel*: 'The distinction due to Shakespeare in energy of poetry, to Michael Angelo in sculpture and painting, Handel may justly claim in the sister art; to him belongs the Majesty of Music.'

The Handel Commemoration was originally intended to have begun on 20 April 1784, the anniversary of his funeral 25 years before (hence the sarcophagus design of the ticket for the first day's performance; see no. 236); but the sudden dissolution of Parliament led to its postponement by five weeks. All the performances were held in Westminster Abbey except for the second, which included arias from Handel's Italian operas and was deemed more suitable for the Pantheon in Oxford Street with its large assembly room (see no. 233). Both it and the Abbey were fitted out for the purpose under the direction of James Wyatt, the architect of the Pantheon and the Abbey's surveyor. His plans for the nave of the Abbey were, according to Burney, intended 'to produce the effect of a royal musical chapel, with the orchestra terminating one end, and the accommodations for the Royal Family, the other'. The King lent his enthusiastic patronage to the festival, attending all five performances. Such was the popularity of the occasion that two extra concerts were arranged for Thursday 3 and Saturday 5 June in addition to the original three held on Wednesday 26, Thursday 27 and Saturday 29 May. Each day crowds gathered at the Abbey at nine o'clock in the morning, though the performances did not start until noon. In all, some 4,500 people could squeeze into the nave and the temporary galleries in the aisles to listen to an orchestra and choir numbering more than five hundred. On the first day, when the Earl of Sandwich, one of the commemoration directors, gave the conductor Joah Bates the signal to begin, the orchestra led off with the Coronation Anthem *Zadok the Priest*. Many descriptions of the occasion survive; one recently published account, that of the young composer John Marsh, gives an idea of the effect of the occasion:

> The first movement of this or [*sic*] introductory symphony I had always before thought for want of a sufficient body of harmony, to support the arpeggioes of the fiddles, to be rather tedious, and after the first ten or twelve bars uninteresting and monotonous. But with the band now engaged in it, I thought nothing could exceed the grandeur of the effect. This symphony with its beautiful modulation having continued some time, at length the whole force of the orchestra with all the voices, full organ, trumpets, trombones, double drums, etc. burst upon us all at once in the words – 'Zadok, the priest' – the force and effect of which almost took me off my legs, and caused the blood to forsake my cheeks. As to the

succeeding movement – 'And [all] the people rejoiced' – with the following chorus – 'God save the king!' – it was all ecstasy.

The five concerts raised a total of £12,736. After the payment of expenses, £6,000 went to the Society of Musicians' Fund for decayed Musicians and £1,000 to Westminster Hospital. Such was the success of the Commemoration that it was repeated every year until 1791 (except when George III was ill in 1788 and 1789).

LITERATURE
Burney, pp. 3–4; H.D. Johnstone, 'A Ringside Seat at the Handel Commemoration', *Musical Times*, CXXV, 1984, pp. 632–5.

232 [*illustrated on p. 234*]

Westminster Abbey Handel Commemoration

By Edward Edwards (1738–1806)

This is very much the view the King had of the Handel Commemoration in Westminster Abbey, looking along the nave to the west end from a special structure sited in front of the choir screen. It shows the temporary balconies lining the aisles, the audience seated on benches in the body of the nave, and at the far end the steeply tiered orchestra and choir rising high up to either side of the organ (itself installed specially for the occasion and surmounted by a portrait of Handel).

When exhibited in 1793 the painting was described as '... the commemoration of Handel, taken from the manager's box', with the added information that many of the figures were portraits (though they are not now recognisable). The artist, Edward Edwards, teacher of perspective at the Royal Academy, described the painting as having 'occupied much of his time', and it seems possible that his work may be based on more than one of the Handel Commemorations, of which that in 1791 was the last.

Yale Center for British Art (Paul Mellon Collection)

Oil, 128.0 × 101.5 cms/50$\frac{3}{8}$ × 40 ins

Signed on the balcony at right *E. Edwards/93*(?)

PROVENANCE
Apparently by descent from J. Smith, a violoncellist at the 1784 Commemoration, to F.Y. Smith, 1904; Arthur F. Hill; John Baskett, 1967; Paul Mellon, by whom given to the Yale Center, 1981.

LITERATURE
Edward Edwards, *Anecdotes of Painters*, 1808, p. vi.

EXHIBITED
Royal Academy, 1793, no. 198; *Music Loan Exhibition*, 1904 (illustrated catalogue, 1909, p. 229).

233

'An Inside View of the Pantheon'

By William Angus (1752–1821) after John Dixon (fl. 1772–1813)

The Pantheon in Oxford Street ('the most beautiful edifice in England', according to Horace Walpole) was built as a grand assembly room, for masquerades and entertainments, by James Wyatt. Opened in 1772, it was fitted out by Wyatt with a special royal box for the 1784 Handel Commemoration. The Pantheon was destroyed by fire in 1792 (it stood on the site of what is now the 'Pantheon' branch of Marks & Spencer; see map, p. 74).

The Pantheon was chosen for the second performance of the 1784 Handel Commemoration in preference to the Abbey since the programme included various secular works, and in particular airs from several of Handel's operas. The concert took place on 27 May and in Burney's opinion, expectations seemed generally disappointed: 'The extreme heat of the weather, augmented by the animal heat of more than Sixteen hundred people, closely wedged together, must have considerably diminished the delight which the lovers of Music expected to receive....' The royal box, seen here, was 'lined with crimson sattin, and ornamented with looking glass.... The box was covered with a dome, in which were placed the royal supporters in gold.' The organ was in the gallery, 'decorated with a transparent portrait of Handel, from an original painting presented to the Concert of Ancient Music by Mr. Redmond Simpson' (this was no. 10; the transparency taken from it was by Robert Smirke).

This view of the interior was engraved from a drawing by Wyatt's assistant, John Dixon, for the proprietors of *The European Magazine* at a cost of 13 guineas. Four thousand copies were printed, the magazine being sold at the Abbey and the Pantheon by special arrangement.

National Portrait Gallery, London

An Inside View of the PANTHEON exhibiting their MAJESTIES Box &c as fitted up under the direction of MR. JAMES WYATT, for the Commemoration of HANDEL.
Published as the Act directs by J. Sewell, May 25 1784.

233

Engraving, 17.5 × 25.3 cms/6⅞ × 10 ins
In *The European Magazine*, v, May 1784, opposite p. 325

Inscribed *An Inside View of the PANTHEON exhibiting their MAJESTIES Box &c. as fitted up under the direction of MR. JAMES WYATT, for the Commemoration of HANDEL.* / *Published as the Act directs by J. Sewell, May 25 1784.* / J. Dixon delᵗ / Angus sculp / EUROPEAN MAGAZINE

LITERATURE
Editorial minutes of *The European Magazine* for 1784 (unpublished manuscript in British Library, Add. MS 38278, ff. 79, 85); *Burney*, pp. 45–7.

234

George III (1738–1820)

By Joseph Nollekens (1737–1823)

'While that boy lives, my music will never want a protector,' said Handel of the future King, George III. This story was told by George III himself in conversation with Charles Burney's daughter in 1789. His enthusiasm for Handel's music was deeply felt and almost obsessive. He was the fortunate owner from about 1772 of the vast bulk of the composer's autographs (a gift from J.C. Smith the younger) and he provided his support for the idea of a complete edition of Handel's work, subscribing to 25 copies of Samuel Arnold's scores when publication began in 1787.

George III was also a frequent concert-goer, lending his patronage to the Handel Commemoration in Westminster Abbey in 1784 and subsequent years and subscribing to the Concert of Ancient Music from 1785. This concert society, dedicated to such 'old masters' as Handel and Purcell, came to be known as the King's Concert once George III began to be actively involved; he helped select programmes (mainly works by Handel, judging from the lists he made, some of which survive in the Royal Library and elsewhere); and more significantly he enlisted the help of the choirs of the Chapel Royal.

George III's personal copy of Mainwaring's biography of Handel (in the British Library) is extensively annotated by the King in English and German. He considered Handel's Coronation Anthems 'the most perfect if possible of all His superb Compositions', and in this context he takes Maurice Greene to task as 'that wretched little crooked ill natured insignificant writer player and musician'. Alexander Pope is called 'a little rimer'. Handel was superior to 'flimsy Bononcini'. George III's opinions on the subject were very strongly held, as Charles Burney found to his exasperation when writing his account of the 1784 Commemoration (see no. 243). The King was the fourth generation of the Hanoverian Royal Family to support Handel's music, and his conviction did much to keep alive his memory.

The Royal Society of London

Marble, 85 cms/33½ ins high including socle and square plinth

Signed on the right side of plinth
Nollekens F!/A. MDCCLXXIII and inscribed on the front
GEORGIVS III PATRONVS MUNIFICVS

PROVENANCE
Commissioned by the Royal Society, 1773.

LITERATURE
W.C. Smith, 'George III, Handel and Mainwaring', *Musical Times*, LXV, 1924, pp. 789–95; Norman H. Robinson, *The Royal Society. Catalogue of Portraits*, 1980, p.132.

235

Joah Bates (1740–99)

By William Daniell (1769–1837)
after George Dance (1741–1825)

Joah Bates was famed as the conductor of the great Westminster Abbey Handel Commemorations, the first of which was held in 1784. He had early shown his enthusiasm for Handel's music when he organised two performances of *Messiah* in 1766 to inaugurate the new organ in the parish church of his native Halifax; this is said to have been the first occasion on which the oratorio was given in the north of England. Bates was a fine organist and, rather in the manner of Handel, he played a concerto on the new organ between acts each day. At this time he was just 26 years old, an Eton and Cambridge scholar, whose position as tutor to the son of the Earl of Sandwich led to his appointment as the Earl's private secretary at the Admiralty. Bates and Sandwich's shared musical enthusiasms led to the establishment of the Concert of Ancient Music in 1776, which Bates conducted; it was dedicated to the 'productions of venerable old masters, particularly those of Purcell and Handel'.

In 1783, in company with Viscount Fitzwilliam and Sir Watkin Williams Wynn, Bates planned the Handel Commemoration which was held in Westminster Abbey the following year. His performance on this occasion was conducted from the harpsichord, which had been specially linked, at a distance of 19 feet, to the organ, so that he could see the whole

234

235

orchestra. The *Gentleman's Magazine* for May 1784 thought that at the first performance he

> appeared to be so agitated and inflamed by the subject during performance – his mind was so involved, and his powers so roused, that his instrument, though immense in its tones, could hardly give utterance to his sentiments. Driven along by the torrent so powerfully, he was at times too rapid in the movement, but his judgement quickly corrected his feelings; and a band more easily directed, more distinct in its impressions, or more perfect in its harmony, we never saw.

Others were more free with their praise: the *Daily Advertiser* on 27 May called him 'the most accomplished Dilettante in Europe', while his sister-in-law wrote after the close of the Commemoration that 'Joah is at present the wonder of every living creature.'

National Portrait Gallery, London

Soft-ground etching, 20.2 × 15.3 cms/8 × 6 ins

Inscribed *Joah Bates/Geo. Dance del Decr 20.1794/Published by Willm. Daniell N°. 9 Cleveland Street, Fitzroy Square, London July 1. 1809.*

LITERATURE
'Batesiana', *Musical Times*, XCVI, 1905, pp. 99–100; unpublished letter from Mrs Harry Bates to her sister-in-law, 7 June 1734 (communicated by Betty Matthews).

236

Admission Ticket to the First Performance, Westminster Abbey, 26 May 1784

By John Keyse Sherwin (1751–90) after Biagio Rebecca (1735–1808)

The illustrated admission tickets to the Handel performances at Westminster Abbey in 1784, rare though they are in their original form (as no. 238), are well known from Burney's Account of the commemoration (no. 242) where they are all reprinted. This ticket formed plate III to his volume where the design, 'a Sarcophagus, with a Medallion of the great Musician over it', is explained by the original intention to hold the performance in April to coincide with Handel's funeral. The engraving served as the source for various later profile portraits of the composer, including the early nineteenth-century relief on the organ at Belton House, Lincolnshire.

National Portrait Gallery, London

Stipple engraving, printed in brown, 16.4 × 11.1 cms/6½ × 4⅜ ins

Inscribed on the sarcophagus *WESTMINSTER ABBEY COMMEMORATION OF HANDEL* and below, *MAY XXVI MDCCLXXXIV / B. Rebecca del! / I.K. Sherwin, Engraver to his Majesty & his Royal Highness the Prince of Wales, Sculpsit.*

237

Admission Ticket to the Second Performance, the Pantheon, 27 May 1784

By Francesco Bartolozzi (1727–1815) after Giovanni Battista Cipriani (1727–85)

This ticket is as reprinted by Burney as plate IV to his volume (no. 242), where the design is described as 'Handel composing sacred Music; the Genius of Harmony crowning him, and a Seraph wafting his Name to Heaven'.

237

National Portrait Gallery, London

Engraving, 19.8 × 16.5 cms/8⅞ × 6½ ins

Inscribed on the ribbon at the top with the opening lines of Dryden's *Song for St Cecilia's Day* (1687; set by Handel in 1739), FROM HARMONY, FROM HEAVENLY HARMONY THIS UNIVERSAL FRAME BEGAN; on the scroll held by the 'Seraph' HANDEL; and below the image *I.B. Cipriani R.A. Inveni!* | *F. Bartolozzi RA. Engraver to his Majesty Sculp!* | *Pub.d as the Act directs April 1.st 1784.*

238

Admission Ticket to the Third Performance, Westminster Abbey, 29 May 1784

By Francis Haward (1759–97)
after Robert Smirke (1753–1845)

A rare example of the original *Messiah* ticket which is better known in its reprinted form as plate v in Burney's book (no. 242). The design is there described as 'BRITANNIA pointing to a Pyramid, on which the Name of HANDEL is engraved; a Genius offering the First-fruits of a Sacrifice to his Memory; and on the Back-ground, a perspective View of Westminster-Abbey'.

A preparatory design is in the Victoria and Albert Museum (E 1228.1948). It is horizontal in format and, among other differences, shows a bust of Handel before the pyramid. The presence of the Pantheon (on the left) in place of Westminster Abbey suggests that the design was originally intended for the second performance in the Handel Commemoration, the only one to be held at the Pantheon.

Gerald Coke Handel Collection

Engraving, 18.5 × 14.5 cms/7¼ × 5¾ ins

Inscribed on the pyramid HANDEL; around the top THE DEAD SHALL LIVE. THE LIVING DIE (from the last chorus in Dryden's *Song for St Cecilia's Day*, 1687; set by Handel in 1739) and at the bottom *MAY 29 MESSIAH, R: SMIRK, pinx!, F. HAWARD sculp! RA.*; below the plate-mark *SUBSCRIBERS TICKET*, *West-Door*; and in red within a laurel wreath WESTMINSTER ABBEY

EXHIBITED
Coke, p. 45, pl. 15b.

238 (detail)

239

Two Medals for the 1784 Handel Festival

By an unknown medallist

This design was struck in both gold and silver. It is reproduced as the frontispiece (no. 242) of Burney's *Account of the Musical Performances in Westminster Abbey* (1785), and is there described as 'struck on occasion of the Commemoration of HANDEL, and worn by their Majesties and the Directors, on the Days of Performance'.

The British Museum medal is probably that worn by King George III himself at the performances of the 1784 Festival, since it came to the museum with the royal medal collection in 1825. No provenance is recorded for the example from the Fitzwilliam Museum, which might suggest an early acquisition date; it would be nice to think that it was part of the Founder's Bequest and therefore the property of Richard, Viscount Fitzwilliam, one of the directors of the 1784 Festival and a great Handel enthusiast.

According to Dr Edmund Ayrton, one of the sub-directors at the commemoration, the medal was first struck with the initial G omitted before Handel's name. Of the 17 specimens made before the die was broken (only one of which has been traced, that shown here from the British Museum), eight in gold went to the King and Queen, the five directors and the conductor, Joah Bates (see no. 235), while the remainder in silver were given to the eight sub-directors (which included Ayrton himself) and the architect, James Wyatt. A further five hundred medals from a new die were given to the principal performers. Ayrton owned two commemoration medals, but only a correctly struck specimen of slightly later date has been traced (no. 240).

a) *The Trustees of the British Museum*
b) *The Syndics of the Fitzwilliam Museum, Cambridge*

Gold, each 3.2 cms/1¼ ins diameter

Inscribed around a profile draped bust of Handel, left, COMM. G.F. HANDEL MDCCLXXXIV. (but the British Museum medal is without the letter 'G'), and on the reverse within an oak wreath SUB./AUSP./G.III. ('under the auspices of George III')

PROVENANCE
a) George III?; given by George IV to the British Museum in part settlement of his debts, 1825.
b) Not known.

LITERATURE
Unpublished statement by Dr Edmund Ayrton, 26 July 1808 (Gerald Coke Handel Collection); *MI*, II, no. 422; *Niggl*, no. 813; Laurence Brown, *A Catalogue of British Historical Medals 1760–1960*, I, 1980, p. 62, no. 259.

239a   239b

240 [*illustrated on p. 235*]

Commemoration Medal and Ribbon, 1791

By an unknown medallist

This medal was one of two which belonged to Dr Edmund Ayrton, the organist and composer, who was prominent in the London musical world in the late eighteenth century (for details of the other medal see under no. 239). Such medals were awarded to all performers at the 1784 Handel Commemoration but, from the inscription on this example, the circulation was more restricted at the 1791 festival, this medal being given 'by permission of His Majesty' on 1 June, when the last of the four festival performances was held.

Ayrton records that the King wished that medal holders might wear their medals 'on all occasions, & when ever they pleased, on the left Breast' (letter of 1808, Gerald Coke Handel Collection). A drawing of the 1790 Commemoration (no. 244) shows another of the assistant directors, John Jones, wearing a medal in this fashion.

Gerald Coke Handel Collection

Silver, glazed and set in a gilt mount, overall diameter 3.2 cms/1¼ ins, joined to a faded off-white silk ribbon

Inscribed on the edge of the mount BY PERMISSION OF HIS MAJESTY 1ST OF JUNE 1791 TO EDM^D AYRTON MUS.DOC. SUB DIRECTOR (the medal is inscribed as no. 240)

PROVENANCE
Presumably one of the 'two Commemoration Medals' bequeathed by Dr Edmund Ayrton (1734–1808) to his eldest son, Edward Edmund Ayrton; Gerald Coke.

LITERATURE
Unpublished statement by Dr Edmund Ayrton, 26 July 1808 (Gerald Coke Handel Collection); *Highfill*, I, pp. 185–7.

EXHIBITED
Coke, p. 45, pl. 15a.

241

Handel Commemoration Ring, 1784

By an unknown silversmith

Eight or ten of these rings were said to have been made for the directors of the 1784 Handel Commemoration. In 1905 one was in the possession of Dr J. Curling Bates, descendant of Joah Bates (see no. 235), conductor at the Commemoration. The portrait of Handel, apparently printed on satin, is a miniature version of that by Hudson, 1748 (no. 11).

Gerald Coke Handel Collection

Gold, inset with a portrait of Handel printed in black on white satin, 3.4×2.0 cms/$1\frac{5}{16} \times \frac{3}{4}$ ins, oval, nearly pointed at each end

Inscribed beneath the portrait
G.F. HANDEL/Ob! April 14/1759./Æt 75

LITERATURE
'The Handel Commemoration Rings', *The Athenaeum*, no. 2279, 1 July 1871, p. 16; 'Handel Commemoration (1784) Rings', *Notes and Queries*, 4th series, VIII, 1871, p. 39, and 5th series, VI, 1875, p. 315; 'Batesiana', *Musical Times*, XLVI, 1905, p. 99 (reproduces the Bates ring).

EXHIBITED
Coke, p. 43.

241

242

'An Account of the Musical Performances in Westminster-Abbey, and the Pantheon ... in Commemoration of Handel.'

By Charles Burney (1726–1814)
Published by T. Payne & Son and
G. Robinson, 1785
('for the Benefit of the Musical Fund')

Charles Burney's 125-page account of the five concerts of the 1784 Handel Commemoration was illustrated with eight plates (see nos. 236–8) and prefaced by his 'Sketch of the Life of Handel'. It was published early in 1785 after many difficulties. As the King remarked when Burney presented him with his dedication copy, 'You have made, Dr. Burney, a much more considerable book of this Commemoration than I had expected: or, perhaps, than you had expected yourself.'

Burney's first setback came when he was obliged to allow the proceeds from his book to be used for the benefit of the Fund for Decayed Musicians. His disappointment is apparent from his description of a meeting with the Directors of the Commemoration: 'you intend giving the profits of this acc^t to the Fund. Dr B. – don't you?' enquired Lord Sandwich; 'No, my Ld – I had no such intension – It will be a drop in the Ocean, after the thousands it has gained by the Commemoration – & yet though little for them to receive, it will be a great deal for me to give.' But Burney had no choice: 'His Majesty expects it,' was Sandwich's response.

His next problem was over the approach to be adopted in the text. The King read the book 'in MS, sheet by sheet, in fact as it was written'. Both the King and the conductor Joah Bates had their views, which Burney, with his more critical mind, strongly disliked: '*one key* of Panegyric is all they want. – fine! – very fine! – charming! exquisite! grand! Sublime!!! These are all the notes (a Hexachord) I must use.'

Once the text was ready Burney was faced with delays in publishing the volume over the 'business of the plates'. One of the directors, Sir Watkin Williams Wynn, wished Francesco Bartolozzi to engrave the frontispiece using the Commemoration medal (no. 239) as his subject. But as Burney wrote to Lord Sandwich on 27 November 1784: 'Bartolozzi, whose hands are always full, had not made a beginning 10 Days ago ... he very justly, I think, objected that the Medal only with a naked head of Handel, neither very remarkable for likeness nor workmanship ... would cut but a small, & no very interesting, figure, on a whole 4.to page. He therefore directed my nephew [E.F. Burney] to draw a design for ornamenting the plate & in a very short time he carried 3 for Bartolozzi's choice.' The design which was chosen shows the obverse and reverse of the medal in laurel wreaths held up on a ribbon by an angel.

The Society of Musicians paid Burney's expenses for the volume and thanked him for 'the very Elegant History of Handel &c'. Burney was to be given thirty books 'for his Friends' and five hundred were to be delivered to the performers. This copy is

THE COMMEMORATION OF HANDEL

242

243
Charles Burney (1726–1814)
By Joseph Nollekens (1737–1823)

By the time Charles Burney, the music historian, published his *Account of the Musical Performances in Westminster-Abbey . . . in Commemoration of Handel* (no. 222) in 1785, he was already well known in England and abroad for three earlier publications: *The Present State of Music in France and Italy* (1771), *The Present State of Music in Germany, the Netherlands and the United Provinces* (1773) and *A General History of Music* (1776; completed in five volumes in 1789).

Burney was apprenticed to Thomas Arne in 1744, and came to know Handel in the late 1740s through attending oratorio rehearsals at the composer's house in Brook Street. As he records in his 'Sketch of the Life of Handel' (see no. 242), the experience of

inscribed with the name of Mr Rawson (of Nottingham), one of the 84 bass singers in the choir at the Commemoration.

British Library
Hirsch I.644a

Printed book, xvi + 70 + 139 pp.,
28.0 × 22.5 cms / 11 × 8⅞ ins

Inscribed in pen on the title-page *Presented by the Directors to Mr Rawson*

LITERATURE
Unpublished draft of a letter from Charles Burney to Lord Sandwich, 27 November 1784, in the Gerald Coke Handel Collection; Percy A. Scholes, *The Great Dr. Burney*, II, 1948, pp. 72–3; Roger Lonsdale, *Dr. Charles Burney: A Literary Biography*, Oxford, 1965, pp. 292–314; Pippa Drummond, 'The Royal Society of Musicians in the Eighteenth Century', *Music and Letters*, LIX, 1978, p. 280; Betty Matthews, 'Handel and the Royal Society of Musicians', *Musical Times*, CXXV, 1984, pp. 79–80; *Hogwood*, p. 240.

243

rehearsing *Judas Maccabaeus* in 1748 was a memorable one: 'unfortunately, something went wrong, and Handel, with his normal impetuosity, grew violent: a circumstance very terrific to a young musician. At length, however, recovering from my fright, I ventured to say that I fancied there was a mistake in the writing; which, upon examining, Handel discovered to be the case. . . .'

Burney's attitude to Handel and his music was ambivalent. He agreed that 'Handel was superior in the strength and boldness of his style', and in public he was willing to concur with the general uncritical praise of Handel's work. But in private, as his comments at the time of the 1784 Handel Commemoration make clear (see no. 242), he held very different opinions. It was only in 1811 that he tentatively advanced his real view in the Rees *Cyclopedia*: 'We will not assert that his vocal melodies were more polished and graceful than those of his countryman and contemporary Hasse; or his recitatives, or musical declaration, superior to that of his rivals, Bononcini and Porpora.'

In this bust Burney is shown by Nollekens in his Oxford robes as a Doctor of Music. It was exhibited at the Royal Academy in 1802 and was probably begun the previous year, a part payment of forty guineas having been made on 25 September 1801 ('Received of Doc^r Charles Burney the sum of forty guineas on account for a Bust of Doc^r Burney'). Several casts were taken, including one which was in the Bodleian Library by 1806.

The Trustees of the British Museum
1944, 7–4, 1

Marble, 69.0 cms/27¼ ins high

Inscribed *Nollekens F^t 1802*

PROVENANCE
Given by Edward Burney to the British Museum, 1944.

LITERATURE
Burney, *Sketch*, p. 35; unpublished receipt signed by Nollekens and dated 25 September 1801 in the Pierpont Morgan Library, New York (information from John Kenworthy-Browne); John Kenworthy-Browne, 'Taken from the Marble: Portraits in Plaster, by Nollekens', *Country Life*, CLVII, 1975, pp. 1509–10, fig. 6.

EXHIBITED
Royal Academy, 1802, no. 1059; *Eighteenth Century Portrait Busts*, Iveagh Bequest, Kenwood, 1959, no. 20.

244

244

Spectators at the 1790 Commemoration of Handel

By John Nixon (fl. 1781–1818)

The Commemoration of Handel took place in Westminster Abbey on five occasions after 1784: from 1785–7 and again in 1790 and 1791. In this view of the 1790 Commemoration, the Earl of Sandwich, one of the five directors, is flanked by the formidable-looking Lady M. Barker, otherwise not known, and John Jones, one of the assistant directors and organist of St Paul's Cathedral. He is wearing a Commemoration medal in the manner laid down by the King (see under no. 240).

John Nixon made many sketches of London's musical and theatrical life in the 1780s and 1790s. A watercolour of spectators at the 1787 Handel Commemoration is also in the Gerald Coke Handel Collection (reproduced in *Coke*, pl. VIII, in colour).

Gerald Coke Handel Collection

Pen and grey and brown washes,
15.5 × 19.9 cms/6⅛ × 8 ins

Initialled bottom right *JN:1790*, and inscribed in pen along the balustrade *Lady M Barker   Lord Sandwich   Jones (Organist at St Paul's)*
Engraved in 1790 by Nixon as 'N . . . Gallery Westminster Abbey Com. of HANDEL'

PROVENANCE
The French Hospital of La Providence; sold at Christie's, *c*.1977.

EXHIBITED
Coke, p. 11, pl. VIII (in colour).

245
Joseph Haydn (1732–1809)
By Thomas Hardy (1757–c.1805)

'Hardy, Otto, Guttenbrun, Hoppner, Dassie. The first four gentlemen painted my portrait, Dassie in wax'. Of the four portraits Haydn noted as having been painted of him in London during his 18-month stay in 1791/2, Thomas Hardy's is perhaps the finest. It was almost certainly commissioned by the music publisher John Bland, who had visited Vienna in 1787 to invite Haydn to England and with whom the composer stayed for a short time on his arrival in London on 1 January 1791. Hardy was an artist much patronised by Bland, for whom he made an engraved version of this portrait which was published on 13 February 1792.

'He is the master of us all', Haydn is said to have cried on hearing the Hallelujah Chorus; he was present at the large-scale performance of *Messiah* given on 1 June 1791 during the last of the great Handel Commemorations in Westminster Abbey. To his biographer Giuseppe Carpani he admitted 'that when he heard the music of Hendl in London, he was struck as if he had been put back to the beginning of his studies and known nothing up to that moment. He meditated on every note and drew from those most learned scores the essence of true musical grandeur.'

Haydn's own great oratorio, *The Creation*, is based on a libretto which he took with him from London at the end of his second visit in 1795; this was presented to him 'to compose in the spirit and manner of Handel's'. When it was first performed in London at Covent Garden in March 1800, the *Morning Herald* was slightly grudging in its praise: 'it is certainly a fine composition, in every respect worthy of its great author... and although not equal in grandeur to the divine compositions of the immortal Handel, is, nevertheless, on the whole, a very charming production'.

The Royal College of Music

Oil, 76.2 × 63.5 cms/30 × 25 ins

Engraved by the artist, 1792 (as in the possession of J. Bland)

PROVENANCE
John Bland; Arthur F. Hill by 1904; given by him to the Royal College of Music, 1933.

LITERATURE
Manfred Huss, *Joseph Haydn*, Eisenstadt, 1984, p.174 (repr. in colour); *Hogwood*, pp. 243, 247.

EXHIBITED
Royal Academy, 1792, no. 368; *Music Loan Exhibition*, 1904 (illustrated catalogue, 1909, p. 230, repr.); *Joseph Haydn in Seiner Zeit*, Eisenstadt, 1982, no. 556 (repr. in colour).

245

The Apotheosis of Handel

'Apotheosis of Handel' was the title chosen by Dr Samuel Arnold for a fanciful engraving by Biagio Rebecca (no. 249), the first to be issued with his edition of Handel's scores in 1787. Another volume from the same series included a rather solid depiction of the Vauxhall Gardens Handel statue as its frontispiece (no. 265). These engravings exemplify the two approaches to the production of portraits of Handel in the succeeding century: the imaginative and the derivative.

Within Handel's own lifetime and in the years immediately after his death, the demand for his image, if not satisfied by engravings such as those by Jacobus Houbraken or John Faber, was met with painted copies of one or other of the two portraits by Thomas Hudson (that given to Oxford University by George Colman, no. 13, is an example), or else with casts of Roubiliac's busts (several are listed under no. 7). However the 1784 Handel Commemoration seems to have opened up a new genre, the fanciful portrait or imaginative reconstruction, often retaining only a hint of Handel's appearance: the profile portrait on the ticket for the first performance in Westminster Abbey in 1784 (no. 236) seems quite recognisable if compared with Cipriani's romantic vision (no. 237).

The early nineteenth century saw the supply of 'Handel portraits' grow to meet the demand, with his name being attached to portraits other than his own. Though long since discredited these die hard, as illustrations to recent Handel biographies show; they include the signed and dated Denner of 1736 probably acquired for Knole by the Duke of Dorset as early as 1776 (see under no. 1), the so-called 'Hogarth' which was engraved by Charles Turner in 1821 (the original is now in Leipzig), the 'Thornhill' in the Fitzwilliam Museum which was first written up by Richard Clark in 1830, and the 'Kneller' given to the Foundling Hospital in 1843. Many of these make their first appearance in the early or mid-nineteenth century, when such legends as the 'Harmonious Blacksmith' (also the work of Richard Clark) began to gain force and other relics emerged – 'The Handel Bookcase' in the Fitzwilliam Museum is first documented in 1832, 'Handel's pitch pipe' (no. 180) in 1842 and 'Handel's watch' (no. 151) not until 1879. More public forms of remembrance were the grand monuments which began to appear in several countries: the bust by Rudolf Schadow was included in Ludwig I's Walhalla of German heroes built overlooking the Danube near Regensburg in 1842, a full-size statue was unveiled in the market-place at Halle in 1859 and Handel took his place at last in the Paris Opéra in the guise of a seated statue by Jules Salmson. He also appears on the Albert Memorial in London.

The passing of the generation who knew Handel's likeness from personal experience also saw the beginnings of a concern to document and catalogue the authentic portraits. Keith Milnes's *Memoir relating to the Portrait of Handel by Francis Kyte* (1829) is an early example of a serious study of Handel's iconography while in Austria the publication of Aloys Fuchs's listing of 32 Handel portraits in 1839 seems to have been the first attempt at a comprehensive catalogue. Later listings include that by Vogel (1896) in German and those of Marshall (1879), Coopersmith (1932), Smith (c.1977), Kerslake (1977) and Highfill (1982) in English. With the exception of *Kerslake* all the recent surveys suffer from the same fault of allowing the apparent number of images to multiply by failing to connect different references to the same portrait and by accepting optimistic old identifications at face value.

LITERATURE
Keith Milnes, *Memoir relating to the Portrait of Handel by Francis Kyte and containing Proofs of its Authenticity*, 1829; Aloys Fuchs, 'Verzeichniss der sämmtlichen Porträts des ... Georg Friedrich Händl', *Allgemeiner Musikalischer Anzeiger*, Vienna, XI, 1839, pp. 227–30; Julian Marshall, 'Handel', *A Dictionary of Music and Musicians*, ed. George Grove, I, 1879, pp. 656–7; Emil Vogel, 'Händelportraits', *Jahrbuch der Musikbibliothek Peters*, III, 1896, pp. 22–32; J.C. Coopersmith, 'A List of Portraits, Sculptures, etc of G.F. Händel', *Music and Letters*, XIII, 1932, pp. 156–67; W.C. Smith, *Handel, Iconography*, microfilm of typescript in Gerald Coke Handel Collection, c.1977; *Kerslake*, pp. 120–32; *Highfill*, VII, pp. 89–95.

246

246

Handel Clock-case

Made by Bow China Works, 1759

Several Bow pieces commemorating Handel, including three other clock-cases, were made soon after his death in April 1759. This example, with the opening to display a watch near the top, is surmounted by a figure of Father Time. The body of the case is decorated with sheets of music bearing Handel's name and the date, 5 November 1759. That these Bow pieces were intended as a tribute to the composer is demonstrated by another clock-case (see *Tait*, fig. 1), which is inscribed on the back: 'To great handle the god of music'. Somewhat strangely, where the music decorating these clock-cases can be identified it has been found not to be by Handel but to have been chosen purely for effect. The minuet on the front of this case is a correct representation of the score of the 'Brede Minuet' from an anonymous collection, *Apollo's Banquet*, 1754 (*Smith and Humphries*, no. 40).

The Geoffrey Freeman Collection of Bow Porcelain, on loan to the Pallant House Gallery, Chichester

Soft-paste bone-ash porcelain, enamelled in colours, 31.2 cms/12¼ ins high

Impressed mark T^o; decorated with sheet music etc inscribed *Minuetto* (front of case), *A Cantata, Handle by Lorv*(?) and *Haymakers* (left side), and *A Song, Sonata IV* and *Nov 5 1759* (right side)

PROVENANCE
Mrs Donald J. Morrison of Nova Scotia;
Sotheby's sale, 3 April 1973, lot 194, purchased by Geoffrey Freeman.

VERSIONS
The three other clock-cases of this type illustrated by Tait are:
I) at the Cecil Higgins Museum, Bedford;
II) formerly with Messrs D.M. & P. Mannheim;
III) formerly with the Antique Porcelain Company.

LITERATURE
Hugh Tait, 'Handel and Bow', *Apollo*, LXXVI, 1962, pp. 384–90, fig. 5; Anton Gabszewicz and Geoffrey Freeman, *Bow Porcelain, the Collection formed by Geoffrey Freeman*, 1982, no. 139

EXHIBITED
English Pottery and Porcelain 1300–1850, Detroit Institute of Arts, 1954, no. 205; *Rococo*, no. O20.

247

George Frideric Handel

After Louis François Roubiliac (1702/5–62)

This bronze bust is closely related to a number of oval bronze reliefs of Handel in later life, attributed to Roubiliac. A good example of this type belonging to Cheng Huan was shown in the Rococo exhibition. So little is known about Roubiliac's work in bronze that it is difficult to be precise about authorship or dating. It seems that towards the end of his career he made reliefs of this kind, of Handel and a number of other contemporary and historical figures including Alexander Pope, David Garrick, Inigo Jones, Isaac Newton and Oliver Cromwell. Of these, Handel featured among the *basso relievos* included in Roubiliac's posthumous sale in 1762, and in another early Christie's sale in 1766 or '67. This bronze bust, though not of a quality to have been made by Roubiliac, comes from the same mould as the reliefs, differing only in its surface ornamentation of scroll work on the cloak and the embroidery to the border of the coat. A very similar bust in bronze is in the collection of Sir Brinsley Ford; however, the type is much more commonly found in ceramic – two examples are shown here, no. 248a and b – which would suggest that the mould circulated quite widely after Roubiliac's death.

Gerald Coke Handel Collection

Bronze, 23.6 cms/9¼ ins high

LITERATURE
J.V.G. Mallet, 'Some Portrait Medallions by Roubiliac', *Burlington Magazine*, CIV, 1962, pp.153–8; *Rococo*, no. S45.

248

George Frideric Handel

By Ralph Wood the younger (1748–95)

These two ceramic busts are closely related to the bronze based on Roubiliac (no. 247), and probably derive from the same mould. They are not uncommon (there are seven in the Coke Collection), but are seldom so finely modelled as the plain bust shown here, which is fully marked by Ralph Wood of Burslem, Staffordshire.

While these figures remained in production at the Wood factory over a period of years, it is suggested by Irving Smith that the type was first produced to mark the Handel Commemoration of 1784, using a model by Enoch Wood (1759–1849). An earlier date should however be considered in view of the origins of this type in the work of Roubiliac. Given their illustrious prototype, it is not surprising that these Handel busts were described by Frank Falkner as among Wood's finest, both in modelling and in portraiture.

Gerald Coke Handel Collection

Ceramic, a) white (blanc de chine), b) coloured (primrose cloak, spotted and lined in purple over a mauve coat), each 23.4 cms/9¼ ins high

Impressed inside socle, on
a) with *80/Ra. Wood/Burslem* and on b) with *80*; inscribed on the front of the socle of b, HANDEL

LITERATURE
Frank Falkner, 'The George Stoner Collection of Figures and Groups by the Ralph Woods of Staffordshire', *Connoisseur*, XXVII, 1910, pp. 23–9; J.I. Smith, 'Mimicry in Staffordshire figures', *Connoisseur*, CLXV, 1967, pp.52–5.

EXHIBITED
Coke, p.11, pl. 5d.

248

249

'Apotheosis of Handel'

By James Heath (1757–1834)
after Biagio Rebecca (1735–1808)

Delivered to the subscribers to Dr Arnold's Handel edition (no. 265) with the twelfth number, *Messiah*, although loose prints were also being advertised on 4 June 1787 as available from Arnold. The print was described as having been taken from a picture by Hudson belonging to Arnold (perhaps the one in his sale, 27 May 1803, lot 17); but the portrait, if one of the usual Hudson types, can hardly be the source of inspiration, which lies closer to Roubiliac's last bust of Handel (no. 16). Rebecca's original pen sketch (in the collection of D.L.T. Oppé) shows an additional putto holding Handel aloft, and the two angels in slightly differing positions. For an earlier but hardly more faithful image of Handel by this artist, see no. 236.

National Portrait Gallery, London

Engraving, 32.0 × 22.2 cms / 12$\frac{3}{8}$ × 8$\frac{3}{4}$ ins

Inscribed *APOTHEOSIS of HANDEL / The Portrait from an original Picture of Hudson's in the possession of D^r. ARNOLD. / Publish'd the 26th of May 1787, being the Anniversary of the Commemoration of HANDEL / Designed by Rebecca / Engraved by Heath / Printed* [By Cox]

LITERATURE
Smith, p.1, no.1; Kerslake, p.124.

250

George Frideric Handel

By J. Ridgway Bates & Co from a model by
William Theed the younger (1804–91)
after Louis François Roubiliac (1702/5–62)

One of three sizes modelled by Theed from the Roubiliac bust at Windsor (see no. 7), and produced by the Staffordshire firm of J. Ridgway Bates, who traded under this name from 1855 to 1858. *The Art Journal* in 1859 records their successors in the business, Bates, Brown-Westhead, Moore & Co, producing this type of bust for 'Mr Hawkins of the China Court, Crystal Palace'. The bust would 'become a cherished possession in homes where music is a source of happiness'. Its publication shortly before the Handel Festival in 1859 was 'auspicious – in a few weeks tens of thousands of people will assemble at the Crystal Palace to commemorate the centenary'.

249

Royal College of Music

Parian ware, 36.9 cms / 14$\frac{1}{2}$ ins high

Impressed *J. RIDGWAY . BATES . & Co.*

PROVENANCE
Purchased from Richard Dennis by the Royal College of Music, 1984.

LITERATURE
The Art Journal, v, 1859, p.195; Charles and Dorrie Shinn, *The Illustrated Guide to Victorian Parian China*, 1971, p.62.

EXHIBITED
The Parian Phenomenon, held by Richard Dennis at Chelsea Town Hall, 1984, no.403.

251

'G.F. Haendel.'

Continental, made by an unknown factory

A continental biscuit-ware figure of Handel, perhaps produced in Germany in the second half of the nineteenth century. But for the inscription and the token presence of sheets of music, including the bound score of *Samson*, the figure would not be immediately recognisable; it does not obviously derive from an earlier portrait of the composer. *Samson* was one of the first of the oratorios to be published in Chrysander's Handel edition for the German Handel Society in 1861. The growing interest in Handel's music in Germany in the late nineteenth century is also indicated by the unveiling of Hermann R. Heidel's life-size statue of Handel in the market-place at Halle in 1859, and the purchase of Handel's conducting scores for the Hamburg Library.

Anthony Gasson Esq

Biscuit-ware figure, 41 cms/16⅛ ins high

Impressed on the front of the base *G.F. HÆNDEL.*;
on the volume at Handel's feet *SAMSON*;
and on the underside of the base *1213* and *13*

252

George Frideric Handel

By Johann Bankel (1837–1906)
after Carl Jäger (1833–87)

Jäger's romanticised portrait of Handel as the noble composer was the most reproduced image of Handel in late nineteenth-century Germany. It is based on Hudson's portrait of 1748 (no. 11), acquired by the Hamburg Library in 1883 but already well known through engravings such as that by Faber. Carl Jäger was a Nuremberg artist who specialised in illustrative historical paintings; this portrait originated in a project of the publisher, Frederick Bruckmann, for a series of twelve portraits of German composers, his *Galerie deutscher Tondichter* (1870–1). They were subsequently engraved for E.F. Rimbault's edition, *Gallery of German Composers* (1873).

Gerald Coke Handel Collection

Steel engraving, 18.2 × 12.6 cms/7⅛ × 5 ins
Inscribed *C. Jäger pinx.*/*J. Bankel sculps.*

253

'The Child Haendel'

By the Berlin Photographic Company
after Margaret Isabel Dicksee (1858–1903)

When Margaret Dicksee's painting was shown at the Royal Academy in 1893, it was thus explained in the catalogue:

> Handel's father, objecting to his son's absorbing devotion to music, forbade his following his bent, and banished all musical instruments to an attic, where, however, the little musician discovered them, and under cover of night resumed his beloved pursuit. The sounds this produced, and the flitting of the little white-clad figure, started the notion that the house was haunted, until the truth was revealed.

This is a variation on the story first retailed in Mainwaring's memoirs of Handel (1760); there, instruments were not allowed in the house at all, but the child Handel contrived to have a clavichord smuggled into the attic.

Photogravure prints on India paper were sold by the Berlin Photographic Company at £1. 1s. – each, and a limited edition of two hundred 'Artists Proofs' at £3. 3s. –.

This kind of historical scene depicting men of genius in childhood or at the moment of artistic creation was a speciality of Margaret Dicksee, and indeed of the extensive Dicksee family of artists; her brother, Sir Frank, became President of the Royal Academy.

The Trustees of the British Museum
1908-4-14-138

Photogravure print, 39 × 52.4 cms/15⅜ × 20⅝ ins

Titled as above and inscribed within the image, bottom right *Margaret Isabel Dicksee/1893*;
below the image *Margaret J. Dicksee:/The Child Haendel/ No 2314./Société Photographique 10 Rue Vivienne Paris./ Photographische Gesellschaft Berlin/Berlin Photographic Company 133 New Bond Street London. 14 East 23rd Street, New York.*; and above the image *Copyright, 1893, by Photographische Gesellschaft.*

LITERATURE
Musical Times, XXXIV, 14 December 1893, p. 47 (advert.).

254

254

The Waiting Handel

By Batt, pseudonym for
Oswald Charles Barrett (1892–1945)

First published in the issue of the *Radio Times* for 21 February 1935, and subsequently as 'The Waiting Handel' in Percy Scholes's *Oxford Companion to Music* (1938), to which 'Batt' contributed many portraits (these were described as being intended to 'penetrate to the mind of the character represented and express his personality').

'The Waiting Handel' is explained by the artist in the following terms:

> The blind old man is seen smoking as he awaits the arrival of his amanuensis. Behind him hangs the portrait of himself, painted for him in earlier days by Balthasar Denner. It now hangs in the National Portrait Gallery.
>
> His harpsichord (to be seen in the South Kensington Museum) stands open near at hand. His wig and stock, symbols of a world in which he is no longer much interested, are thrown off for his greater ease. His food is left untouched, for his immense appetite of former days is now gone.
>
> Thus Handel must have waited many times for John Christopher Smith to come and commit his last music to paper.

Gerald Coke Handel Collection

Wood engraving, 15.3 × 20.4 cms/6 × 8 ins
Inscribed below and also in the image *BATT*

Handel Festivals: the Nineteenth Century

The nineteenth-century obsession with large-scale Handel performances had its origins in the 1784 Handel Commemoration in Westminster Abbey (see p. 249). Another carry-over from the eighteenth century was the preference for Handel's English oratorios at the expense of most of the rest of his music; ultimately this reflects Handel's own concentration late in life on an annual season of oratorios.

The expansion in the movement for public choral performances, marked by the growing number and scale of provincial festivals (see nos. 255–6), reached a climax in the great Handel Festivals given at the Crystal Palace in 1857 and 1859 and then triennially until 1926 (see nos. 259–63). The number of performers grew from the 500 brought together for the 1784 Festival, to the 700 often employed by the Sacred Harmonic Society for concerts during their later seasons (see no. 258), and then at the Crystal Palace to 2,500 in 1857, more than 3,000 in 1859 and 4,500 in 1883.

To contemporary observers (led by Queen Victoria; see under no. 259), it was the power of so many voices in chorus which was the most striking effect. To the more percipient it was the emotion behind such events which was the abiding memory: *Messiah* in Westminster Abbey in 1784 was described as 'a great act of national assent to the fundamental truths of religion' (Rev. Robert Hall) and, in the mid-nineteenth century, it was asserted that 'The Handel Commemorations have become identified with national feelings' (Dean Ramsay, 1862). As Christopher Hogwood has recently written, 'The massiveness of such forces, quite disproportionate to the scale of Handel's music, was justified in the opinion of the audiences by its effects. It was not to render Handel's score that 4,500 gathered to perform *Messiah*, but rather to participate in a ceremony whose basis was an assertion of power.'

LITERATURE
Hogwood, p. 256.

255 Three Tickets for *Messiah*

a) Birmingham Musical Festival, 1811
  By William Radclyffe (1780–1855) after
  a member of the Barber family of Birmingham
b) Chapel Royal, Whitehall, 5 July 1814
  By John Samuel Agar (fl. 1796–1835)
  after John Thurston (1774–1822)
c) Place and date not stated
  By J. Bottomley after
  William Marshall Craig (*c*. 1765–*c*. 1828)

These elaborate illustrated tickets for Handel's *Messiah* continue the tradition begun at the 1784 Handel Commemoration but show a romantic vision of events which stands in contrast to the gentler classical approach of the earlier tickets (nos. 236–8).

The triennial Birmingham Musical Festival was held at the church of St Philip's for the benefit of the Birmingham General Hospital. Radclyffe and Barber produced tickets for a number of the festivals; in the ticket for the 1811 *Messiah*, perhaps their earliest such partnership, a figure holding a lyre and wreath is seated on a cloud which hides the sun; beams of sunlight shine down on a distant view of Birmingham with St Philip's on the right and the Hospital at left. For a later festival at Birmingham see under no. 256.

The performances of *Messiah* in 1814 organised by the 'Committee for the relief of the Suffering Germans' were held at the Banqueting House, then used as a Chapel Royal. The splendid ticket for the occasion depicts an allegorical scene with a helmeted soldier rescuing Liberty (the female figure holding up the cap of liberty) from the shackles of Tyranny, with his flaming torch.

The third ticket (the occasion is not indicated) is the most dramatic. An angel is blowing the Last Trumpet on the Day of Judgement against a background of lightning and pyramids; a skull and crossbones can be seen at her feet and she supports a large tablet with the words, 'The Trumpet shall sound and the dead shall be raised', from the aria near the end of *Messiah*.

Gerald Coke Handel Collection

Engravings,
a) 13.9 × 19.8 cms/5½ × 7¾ ins,
b) 20.4 × 25.0 cms/8 × 9⅞ ins,
c) 20.7 × 15.2 cms/8⅛ × 6 ins

255c

Inscribed on a), BIRMINGHAM MUSICAL FESTIVAL MESSIAH 1811 | Barber Delt | W. Radclyffe Sc^t
with a blindstamp of the General Hospital bottom left;
b), *Chapel Royal Whitehall* | *Second Day, Tuesday, July 5th 1814.* | THE MESSIAH. | *Doors open at Eleven—begins at One* | *For the Benefit of the SUFFERERS by the War in Germany* | *J. Thurston del!* | *R. Ackermann Excudit* | *J. Agar Sculp!*;
c), MESSIAH | W.M. Craig del.. | J Bottomley Sculp! | THE TRUMPET SHALL SOUND AND THE DEAD SHALL BE RAISED

LITERATURE
For b) see *The Favour of Your Company*, Museum of London, 1980, no. 32.

256

Three Programmes for Messiah at Provincial Musical Festivals

a) Leicester Musical Festival, 1827
   (St Margaret's Church, 6 September)
b) Norfolk and Norwich Musical Festival, 1830
   (St Andrew's Hall (?), 23 September)
c) Birmingham Musical Festival, 1834
   (New Town Hall, 9 October)

The centre-piece of most provincial musical festivals in the early nineteenth century was a performance of *Messiah*, usually held in aid of the local hospital or a similar institution. The version of *Messiah* with Mozart's accompaniments, which became popular at this period, was used by both Thomas Greatorex at Leicester in 1827 and Sir George Smart at Norwich in 1830. The festival at Birmingham in 1834, previously held in the church (see under no. 255), was the first to be given in the still incomplete new town hall. Two singers, Deborah Knyvett and the celebrated John Braham, appeared at all three venues (the Norwich programme is Braham's personal copy).

Percy M. Young Esq

Programmes,
a) 20 pp., 21.5 × 13.5 cms/$8\frac{1}{2} \times 5\frac{1}{4}$ ins,
b) 19 pp., 23.3 × 19.0 cms/$9\frac{1}{8} \times 7\frac{1}{2}$ ins,
c) 12 pp., 22.8 × 18.2 cms/$9 \times 7\frac{1}{8}$ ins

Inscribed in pen on the title-pages,
a) *no 5 / I.H.*;
b) *Mr. Braham*;
c) *J. Hedges*

257

Sir George Smart (1776–1867)

By William Bradley (1801–57)

Sir George Smart conducted at the 1834 Handel Commemoration in Westminster Abbey, held to mark the supposed 150th anniversary of the composer's birth. He continued the traditions of the earlier Handel Festivals held at the Abbey from 1784 to 1791, the first of which he attended as a chorister at the Chapel Royal and turned the pages of the score for the conductor Joah Bates.

Smart was known as a teacher of the traditional manner of singing Handel's songs; he was also the

257

conductor of many provincial festivals in which Handel's music featured prominently, and he edited the Dettingen *Te Deum* for the Handel Society in 1846–7. As a composer he was known for his vocal music (he holds the music of one of his glees in this portrait). As a conductor his range extended to much contemporary music, including in 1826 the first English performance of Beethoven's Ninth Symphony.

National Portrait Gallery, London
NPG 1326

Oil, 75.8 × 63.7 cms/30 × 25 ins

Signed and dated bottom left *WB/1829*, and inscribed on the music *Glee 4 Voices/G T Smart*

PROVENANCE
By descent to the sitter's daughter, Miss Ann Caroline Smart; purchased from her executors by the National Portrait Gallery, 1902.

LITERATURE
'Mr and Mrs Joah Bates', *Musical Times*, XLVI, 1905, p. 19;
Julian Herbage, *Messiah*, 1948, pp. 58–62;
Richard Walker, *Regency Portraits*, National Portrait Gallery, 1985, p. 456.

EXHIBITED
National Portraits, South Kensington Museum, 1868, no. 105.

258

Three Handbills for Handel Performances by the Sacred Harmonic Society

a) The Coronation Anthems and other works by Mendelssohn and Haydn, 23 February 1844
b) *Solomon*, 25 February 1859
c) *Deborah*, 10 December 1869

The Sacred Harmonic Society was a leading amateur choral society for much of Queen Victoria's reign. It was founded in 1832 with Joseph Surman as conductor and was dedicated to large-scale religious choral works. In 1836 it moved to Exeter Hall in the Strand, where the following year it gave the first London performance of Mendelssohn's *St Paul*. But while such composers as Haydn, Mozart and Beethoven featured in the Society's programmes, it was the music of Handel which was given pride of place.

The fashion for large-scale performances is evident from the Society's handbills: there were five hundred performers under Surman in 1844 and for *Solomon* in 1859 nearly seven hundred, 'the largest in Exeter Hall'. When Wagner attended the Society's performance of *Messiah* in 1855 he was struck by the ethos behind such occasions:

> It was here that I came to understand the true spirit of English musical culture, which is bound up with the spirit of English Protestantism. This accounts for the fact that an oratorio attracts the public far more than an opera. A further advantage is secured by the feeling among the audience that an evening spent in listening to an oratorio may be regarded as a sort of service, and is almost as good as going to church.

Under Sir Michael Costa, who had been appointed conductor in 1848, the Society organised the great Handel Festivals at the Crystal Palace in 1857 and 1859 (see nos. 259 and 261). The Society was disbanded in 1882 and its library was acquired by the Royal College of Music. Among its other treasures were several portraits of Handel, including the Vauxhall Gardens statue (see under no. 6) which stood in the entrance to Exeter Hall.

Royal College of Music

Printed handbills,
a) 23×19 cms/$9 \times 7\frac{1}{2}$ ins,
b) 21.5×19 cms/$8\frac{1}{2} \times 7\frac{1}{2}$ ins,
c) 25.5×20.3 cms/10×8 ins

LITERATURE
Hogwood, p. 251.

259

'Great Handel Festival, at the Crystal Palace, June 15th, 17th and 19th, 1857'

Programme book
Published by J. Alfred Novello, 1857

After the close of the Great Exhibition in 1851, Joseph Paxton's great glass conservatory, the 'Crystal Palace', was moved to Sydenham where it reopened in an enlarged form in June 1854. Three years later the first of the great Handel Festivals was held there as a preliminary to the 1859 centenary; it was organised by the Sacred Harmonic Society under Sir Michael Costa as conductor and drew on performers from all over the country for the choir of two thousand and the orchestra of five hundred. Three oratorios were given: *Messiah*, *Judas Maccabaeus* and *Israel in Egypt*. The Queen, the Prince Consort and their four eldest children attended *Judas Maccabaeus* on 17 June; in her journal she shows the not unusual preference for the great choruses rather than the solos:

> An immense number of people there, & an Orchestra & Chorus, of 2,000, conducted by Costa. The heat was awful & such a glare, it was like being in a hothouse. The Choruses magnificent, but none equal to the 'Conquering Hero', which begins softly, then gradually increases in loudness with the drumers, ending with the whole strength of the Chorus & Orchestra, the effect of which was quite overpowering & moving, bringing back such recollections of last year, & the return of our Heroes, – also of the dear Old Duke of Wellington. The songs are tedious, & I do not care for them.

Royal College of Music

Programme book with three programmes;
Judas Maccabaeus, 44pp., 24×19.7 cms/$9\frac{1}{2} \times 7\frac{3}{4}$ ins;
44pp., 24×19.7 cms/$9\frac{1}{2} \times 7\frac{3}{4}$ ins;
bound into the first volume (1857–9) of a four-volume set of programmes for the Handel Festivals at the Crystal Palace from 1857–80

260

Medal for the 1857 Crystal Palace Handel Festival

By John Pinches (d. 1905)

Struck for the Sacred Harmonic Society on the occasion of the Handel Festival at the Crystal Palace in 1857 and awarded to each of the two thousand five hundred participants. The bust of Handel is taken from the Vauxhall Gardens statue (see under no. 6) by Roubiliac, which the Society had purchased three years previously. The medal is attributed by L. Forrer to John Pinches, a pupil of W.J. Taylor, who produced the 1859 Festival medal (no. 262).

Private Collection

Bronze, 4.2 cms/1⅝ ins diameter

Inscribed on the obverse behind a profile bust of Handel
HANDEL; below, *PINCHES. LONDON.*;
on the reverse around a lyre,
CRYSTAL.PALACE. HANDEL. FESTIVAL. JUNE 1857; and
on the edge of the medal; *W. GLYDON. PERFORMER*

LITERATURE
MI, p. 696, no. 425; L. Forrer, *Biographical Dictionary of Medallists*, IV, 1909, p. 548.

261   [*illustrated on p. 266*]

'The Great Handel Festival at the Crystal Palace'

By an unknown engraver
From *The Illustrated London News*, XXXIV, 2 July 1859, pp. 20–1.

The 1859 Handel Festival at the Crystal Palace was a grander affair than any that had gone before. This view from *The Illustrated London News* is of *Israel in Egypt*, given on 24 June under the baton of Sir Michael Costa. At this concluding concert, 'the multitude congregated . . . was unprecedented on any previous occasion amounting to the enormous number of *twenty-six thousand eight hundred and twenty-six persons*'. By comparison the 1784 celebration 'was in truth a puny affair . . . at none of the performances did the number of the audience exceed three thousand, a number not equal to that of the performers assembled last week in the orchestra'. The attendance figures came to a grand total of 81,319 and with takings of some £33,000. From then on, Handel Festivals were held triennially at the Crystal Palace.

Gerald Coke Handel Collection

Wood engraving, 34.5 × 51.7 cms/13½ × 20⅜ ins

Titled as above and inscribed with the date; the names of Handel's oratorios are visible around the gallery to either side of the organ (on which Handel's portrait hangs)

262

Medal for the 1859 Crystal Palace Centenary Commemoration

By William Joseph Taylor (1802–85)

Struck on the occasion of the Centenary Handel Festival at the Crystal Palace in 1859 and awarded to each of the three thousand participants. Like the 1857 medal (no. 260), the bust of Handel is based on the statue from Vauxhall Gardens.

Private Collection

Bronze, 5.1 cms/2 ins diameter

Inscribed on the obverse around a profile bust of Handel
GEORGE FREDERIC HANDEL;
below, W.J. TAYLOR. LONDON.;
on the reverse within a wreath of palm and olive,
CENTENARY COMMEMORATION – CRYSTAL PALACE MDCCCLIX;
and on the edge of the medal, MR T.W. CONSTANTINE. STEWARD.

LITERATURE
MI, p. 697, no. 426.

263

'The Great Triennial Handel Festival, at the Crystal Palace in 1862'

Programme
Published by Robert R. Burt, 1862

The fold-out at the back of the 1862 Handel Festival programme exemplifies once more the Victorians' delight in detailing the grand scale of their Handel performances. Here on a single sheet the roof sections are given of the large interiors used in the great musical festivals: halls are shown at Birmingham, Bradford, Leeds and Liverpool, the cathedrals at Worcester and Gloucester, the Minster at York and, in London, Exeter Hall, Westminster Hall, St Paul's Cathedral and Westminster Abbey. Largest of all are the 1862 International Exhibition in Hyde Park and the Crystal Palace, which before re-building had been used to house the 1851 Great Exhibition.

The frontispiece of the programme shows the audience and musicians in the vast space of the Crystal Palace while the title-page reproduces a ring based on a print of Hudson's first portrait of Handel (no. 11).

Royal College of Music

Programme, 28 pp. + fold-out,
21.4 × 14.0 cms/8 3/8 × 5 1/2 ins

263

Handel Editions

Probably more of Handel's music was published in his own lifetime than for any other great composer of the period. However these early editions, produced in London by John Walsh and others (see for example nos. 64 and 102), suffer from being incomplete and sometimes inaccurate. The operas and oratorios, for instance, were usually printed as collections of songs omitting the recitatives and choruses; only *Alexander's Feast* and *Acis and Galatea* were published complete in Handel's day. Walsh issued a complete *Samson* in 1763, and complete scores of most of the other oratorios were issued by his successors between 1767 and 1787.

Since then there have been four attempts at the production of a collected edition of Handel's entire output. The earlier two, Samuel Arnold's (1787–97) and the London Handel Society's (1843–58), were published in London and were never completed. The German Handel Society (Händel-Gesellschaft) edition (1858–1902) was the first to cover all of Handel's major works; the Halle Handel Edition (Hallische Händel-Ausgabe or 'HHA'), begun in 1955, is still in progress.

Arnold's edition, the first attempt at a collected edition of any composer's works, gave reasonably thorough coverage of the oratorios and instrumental music but included only five operas. It was issued to subscribers in 180 numbers or fascicles, which were usually bound into about fifty volumes. (It is discussed below; see no. 265). In 1843 the Handel Society of London was formed, with the express intention of producing 'a superior and standard edition of the works of Handel'. Like Arnold, the Society failed to attract enough subscriptions, so that its publishers, Cramer, Addison and Beale, took over the edition in 1848 and continued issuing volumes for a further ten years. In all 16 were produced under such distinguished editors as Mendelssohn, Moscheles and Sir George Smart.

The Deutsche Händel-Gesellschaft (German Handel Society) was founded in 1856 to produce 'a critical and uniform edition of the whole of Handel's works in full score, with pianoforte arrangement and German translation of the text'. The first volume, *Susanna*, appeared in 1858. By 1860, however, the Society was no more than a name, and the production of the edition over the next forty years became the sole responsibility of the editor, Friedrich Chrysander (see no. 269). At the time of his death in 1901 only two of the projected 96 volumes (including two facsimile editions) remained unpublished: volume 45 (*Messiah*) was seen through the press by Max Seiffert, but volume 49 (miscellaneous vocal music) was never issued. Six supplementary volumes, containing music by other composers from which Handel borrowed, were issued between 1888 and 1902. This edition is still the standard printed source for most of Handel's operas and several oratorios, and has been reprinted twice (Gregg Press, 1965–8; E. Kalmus, c.1980, without the supplements).

The new Halle Handel Edition, prepared under the auspices of the George-Friedrich-Händel-Gesellschaft at Halle, began in 1955 as a practical edition based on Chrysander, but this policy met with considerable criticism and in 1958 the edition was converted into a full critical edition. As Anthony Hicks indicates in his essay (pp. 18–24), the problems of editing Handel's major works are complex both in principle and practice, and in some of the earlier Halle volumes the challenge was not adequately met; new editorial arrangements (involving an editorial board of scholars from both German states, from Britain and the USA) have recently been instituted to ensure that future volumes are prepared to the highest scholarly standards. To date 31 volumes have been issued.

The old collected editions in full score were intended for scholarly reference and professional usage. Other forms of publication met the needs of amateurs. In Handel's own time Walsh and others issued several kinds of instrumental arrangements of the opera and oratorio songs, in addition to the printed scores. Later, while Arnold's edition was appearing, the firm of Harrison & Co. produced vocal scores of the songs in the English choral works, though the choruses themselves were only occasionally included.

A decisive change of emphasis came in the nineteenth century as a result of the establishment and growth of amateur choral societies, especially in the great industrial centres. Vincent Novello (1781–1861) and his son Alfred (1810–96), founders of the firm of Novello & Co., were among the first to recognise the commercial potential of publishing vocal scores of choruses and complete choral works in cheap and convenient octavo format, clearly laid out and using treble clefs instead of the old C clefs for the alto and tenor voices. In the 1840s, the Handel oratorios most favoured by choral societies – *Messiah*, *Israel in Egypt*, *Samson* and *Judas Maccabaeus* – were issued in company with the oratorios of Haydn and Mendelssohn and became best sellers.

Vincent Novello's original editions were replaced in due course by others, not always with advantage (some presented cut versions, others incorporated errors derived from Chrysander's editions). The most famous of all was Ebenezer Prout's edition of *Messiah* (1902), which, though now disregarded, was thoughtfully prepared and most enlightened for its time. Novello & Co. eventually published all of Handel's English oratorios except *Joseph* and the *Occasional Oratorio*, together with other choral works, but the dramatic oratorios on lesser-known subjects, often issued with bowdlerised words, never became popular. The vocal scores were generally made available in three bindings: paper covers, board covers and red boards with gilt lettering (see nos. 267a, b, and d). Certain scores were allowed their own special binding (such as John West's edition of *Judas Maccabaeus*; see no. 267e) but the standard cream-and-brown Novello design, familiar to many generations of choral singers until it was superseded in the late 1960s, was used for most of the paper-covered and plain-board scores (see no. 267c).

Some of the older Novello scores are still in print (notably Vincent Novello's edition of *Jephtha*, 1851); others have now been replaced by the newly edited scores of the Novello Handel Edition which, under the general editorship of Watkins Shaw, maintains the firm's traditional association with Handel. Another important series of modern practical Handel editions, covering orchestral and chamber music as well as choral works, is that published by Bärenreiter in conjunction with the Halle Handel Edition.

264

Dr Samuel Arnold (1740–1802)

By William Daniell (1769–1837)
after George Dance (1741–1825)

Samuel Arnold, the editor of the first collected edition of Handel's work (see no. 265), was one of the best-known musicians of his time. He learnt music at the Chapel Royal under Bernard Gates and James Nares and succeeded Nares as its organist and composer in 1783. But his most lasting success was as a composer of opera and other theatrical music, at first at Covent Garden Theatre, where he was engaged by John Beard in 1764, and then from 1777 at the Little Theatre in the Haymarket.

Arnold was one of the eight assistant directors at the Handel Commemoration concerts in 1784; he also conducted performances of *Messiah* at the Foundling Hospital and Drury Lane, and at the death of John Stanley took over the oratorio seasons at Drury Lane. A prolific composer over many years and an active theatre manager, he is now best remembered for his Handel edition.

National Portrait Gallery, London

Soft-ground etching, 26.8 × 20.0 cms/$10\frac{1}{2}$ × $7\frac{7}{8}$ ins plate size

Inscribed *Samuel Arnold Mus. D. | Geo. Dance del! Jan.ʸ 25. 1795. | Published by Wilᵐ. Daniell Nº 9 Cleveland Street Fitzroy Square London July 1, 1812. | Wᵐ. Daniell Fecit.*

264

265

'Israel in Egypt. A Sacred Oratorio in Score Composed in the Year 1738 By G.F. Handel.'

Printed score, the Samuel Arnold edition

Samuel Arnold's is the first collected edition of Handel's music and indeed of any composer's work. The idea for such an edition seems to have originated with a proposal by the music publisher Robert Birchall, dated 22 June 1783. This was inserted by Burney in his Account of the 1784 Handel Commemoration with a note that the King and many other admirers of Handel's music wished that 'a uniform and complete edition of all his various works . . . might be engraved, in score'. But Birchall's edition did not materialise.

Samuel Arnold (see no. 264) was 46 when he published his own prospectus in 1786, and although he did not complete the series (claiming as an excuse that subscriptions had fallen off), he produced 180 parts, which made up about 50 volumes, at close to the advertised rate of one part a fortnight, from May 1787 until September 1797. The King led the first published list of 380 subscribers in 1788 with 25 copies to his name. Given the then prevailing fashion for Handel's oratorios, Arnold chose to concentrate on these works and on Handel's instrumental music at the expense of his Italian operas, of which he only managed to publish five. In the year of his death in 1802 he wrote how 'the progress of the work was impeded by frequent interruption . . . which required the strongest exertions, perseverance and resolution to overcome'. Despite the edition's inaccuracies, which Arnold himself acknowledged, it long remained the sole printed source for many of Handel's compositions and was not altogether superseded until the completion of Chrysander's edition more than one hundred years later.

This score, *Israel in Egypt*, was published as parts 92 to 98 in about 1791 and includes the second of the illustrated frontispieces which appeared from time

265

to time; it was advertised in Arnold's prospectus as 'a Capital Engraving by Bartolozzi, from the Celebrated Statue of Handel in Vauxhall Gardens.' (The first frontispiece to appear was Rebecca's 'Apotheosis of Handel'; see no. 249.)

Gerald Coke Handel Collection

Printed score, 282 pp., 40.2 × 30 cms/15¾ × 11¾ ins

Titled as above, the frontispiece engraving of Roubiliac's statue in Vauxhall Gardens inscribed *For D^R. ARNOLD'S Edition of HANDEL'S Works | From the Statue in Vauxhall Gardens. | Pub^d. Jany. 1. 1789 by S Arnold. | B. Rebecca del. | Printed by C.W. Richards | F. Bartolozzi sculp.*

LITERATURE
Paul Hirsch, 'Dr Arnold's Handel Edition (1787–1797)', *Music Review*, VIII, 1947, pp. 106–16; *Smith*, p. 2, no. 2 and p. 109, no. 4.

266

Vincent Novello (1781–1861)

By Edward Novello (1813–36)

Vincent Novello edited the first cheap edition of *Messiah* published by his son, Alfred, in 1846, as well as various other Handel oratorios (see no. 267). Though he had long been an editor and publisher in his own right, he had begun his career at the age of 17 as organist to the Portuguese Embassy Chapel where his playing of Haydn's and Mozart's masses, then new to London, soon made him well known. In this portrait by his son Edward he holds a Mozart Requiem Mass; the painting was probably done in the early 1830s at the time he was collecting material for his proposed biography of Mozart and while his edition of Mozart masses was still in course of publication. In 1834 he was organist at the Handel Festival in Westminster Abbey.

The Novello edition of *Messiah* was announced in highly enthusiastic terms in July 1846: 'The cheapest Musical Publication ever offered to the public in respect both to quality and quantity!!! Handel's Sacred Oratorio "THE MESSIAH" in Vocal Score. With separate accompaniment for the Organ or Pianoforte. Arranged by VINCENT NOVELLO.' It was published in twelve monthly parts at 6d each and at this price met with enormous success; in the *Morning Herald* less than three years later it was claimed that 'some 20,000 copies of "The Messiah" alone have already been sold'. As a result a series of cheap editions of the work of Handel and other composers were brought out by Novello's in a variety of inexpensive editions.

266

In the 1860s the basic octavo edition of *Messiah* was selling an average of 6,500 copies a year, almost twice as many as the second favourite, Haydn's *Creation*.

National Portrait Gallery, London
NPG 5864

Oil, 76.5 × 64.1 cms/30⅛ × 25¼ ins

Inscribed on the fore-edge of the volume held by Novello, MOZ^T REQUIEM

PROVENANCE
By descent to the sitter's great-grand-daughter, Contessa Bona Gigliucci, by whom bequeathed to the National Portrait Gallery, 1983

LITERATURE
Michael Hurd, *Vincent Novello – and Company*, 1981, pp. 1–11 (repr.)

HANDEL

267

Five Scores of Handel Oratorios published by Novello's

a) *Israel in Egypt*, c.1857–9
   Novello's Centenary Edition (ocatavo), price 4/–
   Scarlet cloth binding embossed in gold
b) *Alexander's Feast*, c.1876–98
   Folio edition, price 10/–
   Scarlet cloth binding embossed in gold
c) *Dettingen Te Deum*, c.1884
   Novello's Original Octavo Edition, price 1s. 6d.
   Paper boards, printed in brown on pink
d) *Solomon*, c.1893
   Novello's Octavo Edition, price 4/–
   Scarlet cloth binding embossed in gold
e) *Judas Maccabaeus*, 1906 (or after)
   Published by Novello & Co Ltd
   Blue cloth boards

The cheap editions of Handel's work published by Joseph Alfred Novello, starting with the Messiah in 1846, were an immediate success. In the foreword of the earliest volume included here, *Israel in Egypt*, Novello advanced two reasons other than price for the success of his venture: 'an extended use for these books was invented ... viz., the use of such books by the Audience, instead of being confined, as previously to the performers'. Secondly, they were also 'the first type works which had the *words* (or text) of a proportionate size to *the music*, so that both could be conveniently read at the same distance from the eye'.

Novello's method of printing, stereotyping, allowed him to duplicate his printing plates relatively easily; it combined clarity with economy and enabled large print runs to be achieved. With the growth in his sales Novello was able to reduce his prices in 1849 ('Good printed copies can thus be obtained ... for less than the blank [music] paper'), and again for the most popular works in 1854. In March 1857 he advertised his Centenary Edition of 'Handel's Oratorios and Cantatas' at 1s. 6d. or 2/– for a vocal score; this was intended, he claimed, 'to facilitate the universal celebration of the Centenary commemoration of the great Composer's death (in 1759), which appears likely to absorb the chief share of attention in musical society between the present period and the year 1859'.

a) *Royal College of Music*
b) *Private Collection*
c) to e) *Anthony Hicks Esq*

a) iii + 172 pp., 26 × 17.8 cms/$10\frac{1}{4}$ × 7 ins,
b) 95 pp., 36 × 26.2 cms/$14\frac{1}{8}$ × $10\frac{1}{4}$ ins,
c) 70 pp.,
d) vii + 218 pp.,
e) vii + 202 pp.,
size of c–e as a

PROVENANCE
a) Joseph James Curling (his bookplate, dated 1859).

LITERATURE
Miriam Miller, 'The early Novello octavo editions', *Music and Bibliography. Essays in honour of Alec Hyatt King*, 1980, pp.160–9.

267b

268

'Bagster's Gem Edition of Handel's Messiah'

Printed score, miniature edition
Published by S. Bagster & Sons Ltd, 1899

This little volume with its green paper covers is the ultimate pocket edition. The minute print is only just large enough to read; it was presumably intended for spectators rather than performers and is small enough to take anywhere.

Royal College of Music

Printed score, 181pp., 13.3 × 10 cms/5¼ × 4 ins

269 [*in colour opposite p. 233*]

Friedrich Chrysander (1826–1901)

By Leopold Kalckreuth, Graf von Kalckreuth (1855–1928)

Chrysander's three-volume Handel biography, and more especially his great Handel edition for the Händel-Gesellschaft (no. 270), took the greater part of his life's effort from 1856 until his death in 1901. After the loss of his annual grant from the Hanoverian crown in 1866, he transferred the production of the edition to a printing shop set up for the purpose in his garden at Bergedorf near Hamburg. To obtain funding he took on the editorship of the *Allgemeine musikalische Zeitung* and sought other editorial work. On his frequent visits to London to examine Handel's autographs, then still kept at Buckingham Palace, he would stay with his daughter and her husband, Charles Volkert, the music publisher.

   This portrait was painted in his printing workshop during the last months of his life. Chrysander took a strong interest in its progress, expressing a dislike of the emphasis given to the engraver and the printer at work in the background. Chrysander is holding a volume now in the Bodleian Library (MS Don.c.69) which contains copies by the elder Smith of five soprano arias from *Ottone* (1722); it is open at the aria sung by Teofane, 'Benchè mi sia crudele'. While this appears on the two sides of a single leaf (folio 9) in the score itself, the artist has felt obliged to depict the aria as facing pages in the portrait. Of some interest is the transposition of the aria from A minor to D minor and the indication of the vocal ornamentation of the part; this is one of the few instances where such ornamentation of an existing text can be linked with Handel himself.

Hamburger Kunsthalle

Oil, 102.0 × 119.5 cms/40⅛ × 47 ins
Signed and dated top right *Kalckreuth d.j./H.1901*.

PROVENANCE
Purchased from a legacy of John R. Warburg, 1901, by the Hamburger Kunsthalle

LITERATURE
Eva Maria Kraft and Carl-Wolfgang Schümann, *Katalog der Meister des 19. Jahrhunderts in der Hamburger Kunsthalle*, Hamburg, 1969, p.139, no.1661; Winton Dean (editor), *G.F. Handel. Three Ornamented Arias*, 1976, preface.

EXHIBITED
Händel und Hamburg, Staats- und Universitätsbibliothek, Hamburg, 1985, no.58.

270

'Sieg der Zeit und Wahrheit. Oratorium von Georg Friedrich Händel.'

Printed score of *The Triumph of Time and Truth*, volume 20 of the German Handel Society edition
Published by the Deutsche Händel-Gesellschaft, Leipzig, c.1865

Friedrich Chrysander, the leading Handel scholar of his generation (see no. 269), was among the 35 signatories to the original prospectus of the German Handel Society issued in 1856. He himself took over the production of the edition when Breitkopf, the original publishers, withdrew in 1864 after the publication of only 18 of the 96 volumes.

   Like Arnold's edition the early volumes were devoted to the oratorios and instrumental and church music at the expense of operas in Italian. This volume, *The Triumph of Time and Truth*, the revison made in 1757 of an oratorio written in Rome fifty years earlier, was one of the first published under Chrysander's auspices; it has a preface by him dated 8 July 1865 discussing Thomas Morell's alterations to the text in 1757, and includes a libretto in English and German. The scores appear to have been bound up when needed; the exhibited copy, though first published almost thirty years previously, has an advertisement on the inside cover bearing the date 1894.

Private Collection

Printed score, paper covers, viii + 181pp.,
30.6 × 28.0 cms/14 × 11 ins

Handel in our Time

'If I were a member of the House of Commons, I would propose a law making it a capital offence to perform an oratorio by Handel with more than eighty performers in the chorus and orchestra, allowing forty-eight singers and thirty-two instrumentalists.' George Bernard Shaw's strongly held feelings on the performing of Handel's oratorios is an early sign of the reaction against the 'monster performances' so prevalent in the last century and into our own.

The two facets of modern interest in Handel's music, a willingness to recreate the style of performance originally intended, and a wish to extend the current repertory to include all his works, mean that Handel's music is more widely available than ever before. Recordings allow the modern listener the opportunity to hear much of Handel's music on request. An important prerequisite to a better knowledge of Handel's work, especially of his operas, was Chrysander's Handel edition, completed in 1902 (see no. 270). This meant that scores of almost all Handel's work, including his operas, were readily available, but it was not until 1920 at Göttingen that the movement to bring Handel's operas back to public attention was initiated after 180 years of neglect. At Göttingen (see no. 271) and then at Handel's birthplace, Halle, and other German cities, some 19 of the 39 operas were staged before the outbreak of war in 1939. In England, the Cambridge University Musical Society led the way, starting with *Semele* in 1925, but it was only with the formation of the Handel Opera Society in 1955 that a more consistent attempt was made at the renaissance of Handel's Italian operas in the country of their first performance (see no. 274).

But though much of the operatic repertory has now been revived, the interpretation of Handel's operas, and their adaptation to the modern stage, remain a subject of debate. The growing understanding of the conventions of baroque opera means that adapting Handel's operas to make them acceptable to modern audiences is no longer necessary.

271

Two Programmes for the Göttingen Handel Festivals

Published by W.H. Lange, Göttingen, 1921 and 1922

The productions mounted at Göttingen in the 1920s under Oskar Hagen were the first complete performances of Handel's operas anywhere since his lifetime. These two festival programmes with essays by Oskar Hagen and Hermann Albert were published to mark the 1921 Handel Festival which featured *Rodelinda* and the 1922 Festival during which *Ottone* was given (as *Otto und Theophano*) with Thura Hagen-Leisner as Teofane (her striking costume is illustrated in the programme). These performances were heavily adapted with the *castrato* parts transposed down an octave and the plots rewritten to suit an audience unfamiliar with the conventions of Handel's opera. Subsequently productions were given at festivals all over Germany, with great success: Hagen's arrangement of *Giulio Cesare* which led the 1922 Festival had 222 performances in 34 different cities, while *Rodelinda* was given 136 times over a seven-year period.

271b

Gerald Coke Handel Collection

Two souvenir programmes,
a) 28 pp., 22.2 × 15.2 cms/8¾ × 6 ins,
b) 28 pp., 24.1 × 17.0 cms/9½ × 6¾ ins

LITERATURE
W.C. Smith, 'Handel Festivals and Operas in Germany from 1920–62', *A Handelian's Notebook*, 1965, pp. 134–5.

272

'G.F. Handel and Halle. In Memory of the 250th Anniversary of the Composer's Birth'

Edited by Dr Lore Liebenam
Published by the Deutsch-Englischer Kulturaustausch, Halle, 1935

The frontispiece of this pamphlet, redolent of its period, shows the midnight celebration at the Handel monument in Halle on 23 February 1935, with 36 uniformed youths holding flaming torches aloft against a backcloth of the floodlit Liebfrauenkirche (it was here that Handel had learnt the organ). Halle was one of the main centres of the Handel revival in Germany between the Wars, and has held an annual Handel Festival since 1952.

National Portrait Gallery, London.

Printed pamphlet, 20 pp., 21 × 14.8 cms/8¼ × 5¾ ins

PROVENANCE
Given by the editor, Dr Lore Liebenam, to the National Portrait Gallery, 1935.

LITERATURE
Hogwood, pp. 271–2.

273

Costume Designs for a Projected Co-operative Society Production of Belshazzar, 1937

By David Bomberg (1890–1957)

A dramatic, boldly coloured design for the costumes of the chorus of Babylonian Jews in *Belshazzar*, made for the London Co-operative Society in autumn 1937 but never carried out. In December Bomberg wrote to a friend that he had got as far as a large-scale model, the planning of the seven scenes and seven designs for the costumes of the chorus, before giving up the project.

Belshazzar was mounted by the Workers' Musical Association at the Scala Theatre in May the following year, the first modern production in London, but Bomberg's designs were not used.

The Trustees of the Victoria and Albert Museum
E2975–1962

Charcoal, Indian ink and watercolour, 50.2 × 61.0 cms/19¾ × 24 ins sheet size

Signed and dated *Bomberg 37* and inscribed *Opera – Costumes for Handel's 'Belshazzar' Chorus men & women together 90 voices, Babylonian Jew with shawl or cloak Jewess Babylonian Jew.*

PROVENANCE
David Bomberg; purchased by the Victoria and Albert Museum, 1962.

LITERATURE
Dean, pp. 456, 663; William Lipke, *David Bomberg*, 1967, p. 82 (pl. 28 reproduces two related designs).

EXHIBITED
David Bomberg, Herbert Art Gallery, Coventry, 1960, no. 135.

273

HANDEL

274

Five Productions of the Handel Opera Society

a) *Deidamia*, St Pancras Town Hall, 3 June 1955
   Poster and programme for the Society's first performance
b) *Semele*, Sadler's Wells Theatre, 23 June 1959
   Programme for the Society's first performance at Sadler's Wells
c) *Rodelinda*, Sadler's Wells Theatre, 24 June 1959
   Photograph by Houston Rogers showing Janet Baker as Eduige with Alfred Hallett and Raimund Herincx
d) *Saul*, Sadler's Wells Theatre, 23 June 1965
   Photograph by Donald Sotheran showing Geraint Evans as Saul and Marjorie Biggar as David
e) *Esther*, Sadler's Wells Theatre, 5 November 1980
   Stage design by Peter Rice

The Handel Opera Society was organised by Charles Farncombe with the intention of bringing Handel's operas to the stage in England. His aim, as expressed in the Society's published history, was 'to revive an authentic score as Handel himself had written it and at the same time to adapt the eighteenth-century style to twentieth-century convention in such a way as to satisfy an audience largely untutored in that style'.

Over the past thirty years the Society has staged more than twenty different operas and a few oratorios, in some cases bringing to the theatre works not seen since the eighteenth century. Such an opera was *Deidamia*, sung in English for the Society's début performance in June 1955 at St Pancras Town Hall under the baton of Charles Farncombe. Two years later Joan Sutherland, then at the beginning of her career, took the lead part in *Alcina*, to great acclaim; she appeared again in *Rodelinda* in 1959, as part of the Handel bicentenary celebrations. By then the Society had progressed to Sadler's Wells Theatre, where *Semele* was the first work to be staged, with Heather Harper singing the title role.

During the Handel tercentenary the Society has staged *Rodrigo* at Sadler's Wells but has also had to face the news of the withdrawal of its Arts Council grant. The continuation of the Handel Opera Society beyond its thirtieth birthday is thus very doubtful.

a), b) *Royal College of Music*
c), d) *Handel Opera Society*
   e) *Tom Hawkes Esq*

a) Poster, 50.8 × 38 cms/20 × 15 ins and programme, 4 pp., 21.0 × 13.3 cms/8¼ × 5¼ ins;
b) programme, 16 pp., 18.5 × 12.7 cms/7¼ × 5 ins;
c) photograph, 20.3 × 25.4 cms/8 × 10 ins;
d) photograph, 19.0 × 25.4 cms/7½ × 10 ins;
e) design in pen and brown ink over pencil, squared, 36.8 × 55.4 cms/14½ × 21¾ ins

LITERATURE
The story of the Handel Opera Society [1975], and supplement for 1975–80.

275

The Bicentenary Celebrations, 1959

a) 'Purcell–Handel Festival London June 1959'
   The Festival souvenir programme book with essays on 'Handel's Dramatic Works' by Winton Dean and 'Handel in May Fair' by Gwen Ferguson
b) 'Henry Purcell. George Frideric Handel. Catalogue of a Commemorative Exhibition', British Museum, May–August 1959
c) Sir Arthur Bliss laying a wreath at Handel's tomb in Westminster Abbey on 14 April 1959, the day of the bicentenary of Handel's death
   Photograph taken by *The Times*

The celebrations in 1959 to mark the bicentenary of Handel's death coincided with the tercentenary of Purcell's birth. The occasion saw the first modern Handel performance at Covent Garden with the staging of *Samson*, a substantial historical exhibition at the British Museum and a 'Purcell–Handel Festival' in June which included performances by the Handel Opera Society of *Semele* and *Rodelinda* (see no. 274).

a) *Royal College of Music*
b) *National Portrait Gallery, London*
c) *Handel Opera Society*

a) Programme, 44 pp., 28.5 × 23 cms/11¼ × 9 ins;
b) Catalogue, 47 pp., 24.8 × 18.5 cms/9¾ × 7¼ ins;
c) Photograph, 15.2 × 15.2 cms/6 × 6 ins

276
The National Portrait Gallery's Handel Appeal, 1968
Handbill and programme

Following the sale by the Earl Howe in July 1967 of the great full-length portrait of Handel by Thomas Hudson (no.12), which had been in the family since commissioned, the National Portrait Gallery launched its first national appeal to help raise the £9,450 required to purchase the portrait for the nation. The appeal committee, with Lord Harewood as chairman, included Benjamin Britten, Yehudi Menuhin, Malcolm Sargent, Joan Sutherland and Michael Tippett. This handbill was for the special fund-raising concert, appropriately of *Messiah* (the score appears in the painting), given at the Albert Hall on 10 July 1968 with Joan Sutherland and the London Philharmonic Orchestra under Richard Bonynge.

National Portrait Gallery, London

Printed handbill, 29.0 × 22.6 cms/11$\frac{3}{8}$ × 8$\frac{7}{8}$ ins; programme, 16pp., 25.3 × 20.4 cms/10 × 8 ins

277
Four Recent Handel Festival Programmes

a) Halle Handel Festival, 1971
   Programme for *Ariodante*, Landes Theater, Halle, 30 May 1971
b) Göttingen Handel Festival, 23 to 28 June 1973
   Programme book, open at a portrait of William C. Smith (1881–1972), the eminent Handel scholar particularly noted for his definitive work *Handel. A Descriptive Catalogue of the Early Editions* (1960, 2nd edition 1970), with an appreciation of his life by Percy Young
c) Handel in London, 1984
   Programme book of the festival organised by the Royal Society of Musicians
d) London Handel Festival, 1985
   Programme book for the eighth London Handel Festival at St George's, Hanover Square, 20–7 April 1985, with the London Handel Orchestra, formed by Denys Darlow in 1978

a), b) *Percy M. Young Esq*
c), d) *Private Collection*

Four programmes,
a) 16pp. + 4pp. insert, 18 × 10 cms/11 × 4 ins,
b) 67pp., 21.0 × 15.0 cms/8$\frac{1}{4}$ × 5$\frac{7}{8}$ ins,
c) 52pp., 29.5 × 21.0 cms/11$\frac{5}{8}$ × 8$\frac{1}{4}$ ins,
d) 40pp., 29.8 × 21.0 cms/11$\frac{3}{4}$ × 8$\frac{1}{4}$ ins

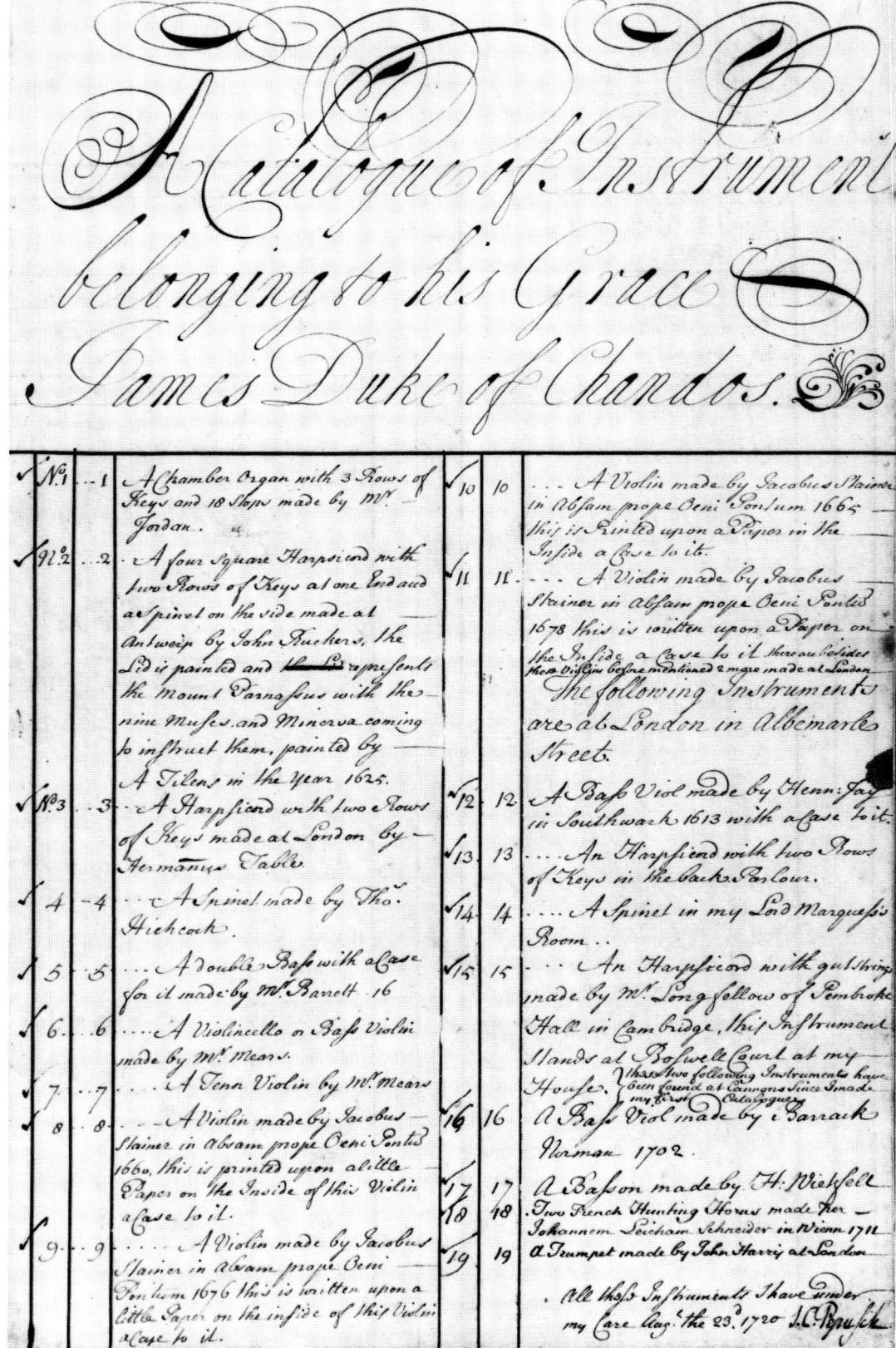

Fig. 12 The 'Catalogue of Instruments', belonging to the Duke of Chandos, 23 August 1720
Stowe Papers, The Huntington Library, San Marino, California

Appendix 1

Inventory of Musical Instruments at Cannons and in London, belonging to the Duke of Chandos, 23 August 1720

The inventory records most of the instruments at the Duke of Chandos's seat, Cannons near Edgware, in 1720 shortly after Handel left the Duke's employ. It is discussed above on p. 105 in more detail. First published by Collins Baker in 1949 in slightly abbreviated form and repeated by Deutsch, the text is given here in full.

The inventory is on the first leaf of a twenty-page folio manuscript, bound in marbled paper covers, size 31.8 × 20.0 cms, in the Stowe Papers (ST66) at the Huntington Library, San Marino, California. Most of the remaining pages list the Duke's manuscript music books including several works by Handel. The inventory is in the hand of a Mr Noland and is countersigned at the end by Dr Pepusch, with later annotations (on page 9) made in 1729 by the chaplain, Dr G. Baxter. It is in two columns, with the numbering on the left, from 1 to 19, given twice.

[p. 1]

A Catalogue of Instrument belonging to his Grace James Duke of Chandos.

No. 1 *A Chamber Organ with 3 Rows of Keys and 18 Stops made by Mr. Jordan*

No. 2 *A four square Harpsichord with two Rows of Keys at one End and a Spinet on the side made at Antwerp by John Ruckers, the Lid is painted and the Lid* [struck through] *represents the Mount Parnassus with the nine Muses, and Minerva coming to instruct them, painted by A Tilens in the year 1625.*

No. 3 *A Harpsichord with two Rows of keys made at London by Hermanus Table*

4 *A Spinet made by Tho: Hichcock.*

5 *A double Bass with a Case for it made by M:r Barrett – 16*

6 *A Violincello or Bass Violin/made by M:r Mears*

7 *A Tenor Violin by M:r Mears*

8 *A Violin made by Jacobus Stainer in Absam prope Oeni Pontus 1660, this is printed upon a little Paper on the Inside of this Violin a Case to it.*

9 *A Violin made by Jacobus Stainer in Absam prope Oeni Pontum 1676 this is written upon a little Paper on the inside of this Violin a Case to it.*

10 *A Violin made by Jacobus Stainer in Absam prope Oeni Pontum 1665 this is Printed upon a Paper in the Inside a Case to it.*

12 *A Violin made by Jacobus Stainer in Absam prope Oeni Pontus 1678 this is written upon a Paper on the inside a case to it.* [Added later:] *there are besides these violins before mentioned 2 more made at London.*

The following Instruments are at London in Albemarle Street

12 *A Bass Viol made by Henn: Jay in Southwark 1613 with a case to it.*

13 *An Harpsicord with two Rows of keys in the back Parlour.*

14 *A Spinet in my Lord Marquess's Room*

15 *An Harpsicord with gutstrings made by M:r Longfellow of Pembroke Hall in Cambridge, this Instrument stands at Boswell Court at my House.* [Added later:] *these two following Instruments have been found at Cannons since I made my first Catalogue.*

16 *A Bass Viol made by Barrack Norman 1702*

17 *A Bassoon made by H. Wietfelt*

18 *Two French Hunting Horns made per Johannem Leicham Schneider in Wienn 1711*

19 *A Trumpet made by John Harris at London*

All these Instruments I have under my Care Aug:t the 23:d 1720
J.C. Pepusch

[p. 9]

Observations upon comparing two Catalogues of his Grace's Musical Instruments One wrote by Mr Noland and subscrib'd by Dr Pepusch in 1720. Aug 23.

The other wrote by the D:r himself and deliver'd to his Grace Octob: 23:1721.

From N:o 1 to N:o 11 inclusive there is an exact agreem't. But at the end of N:o 11 in Nolands Catalogue there are these words added.

There are besides the violins before mention'd 2 more made at London.

The D:rs Catalogue owns [?] one of these made at London by John Brown 1709 a Quere what is become of the other?

Appendix 2

Handel's Naturalisation, 1727

The Act for Handel's naturalisation is in the House of Lords Record Office and has never been published. It is written in a legal hand in brown ink on two parchments, both 30 cms wide, and 49 cms and 46 cms long respectively, joined following line 44 to form a single scroll, with a third, 68 cms long, acting as a wrapper. Quite often, rather than commence one's own bill, the method of naturalisation at the period, a petitioner might seek to have his name added to an existing bill. In this case Handel and two others have joined their names to Louis Sechehaye's of Metz.

The Act takes the standard form, allowing all property and legal rights but excluding the petitioners from holding civil or military office and from being Members of Parliament or of the Privy Council. Handel's personal petition, and the committee proceedings of the two Houses of Parliament, are set out in *Deutsch*, pp. 202–5.

Fig. 13 The Act for Handel's naturalisation, 1727 *House of Lords Record Office*

The scroll is inscribed on the back:

> *An Act for Naturalizing*
> *Louis Sechehaye George Frideric Handel and others*
> *A°, 13°, Georgij Regis*

At the head of the Act in another more rounded hand is the usual statement of royal assent:

> *Soit Baillié aux Seignieurs.*
> *Soit fait comme il est Desiré. Aceste Bille (avec les*
> *Amendements les Seignieures sont assentis*

In the Act itself there are various amendments; these are indicated here by underlining, except for deletions of the indefinite article which are shown by an asterisk. The marginal numbers are only given in the Act for those lines including amendments.

1. Humbly Beseech
   Your most Excellent Majesty the Lords Spiritual
   and Temporal and Commons in this present
   Parliament assembled Louis Sechehaye Son
   of Peter Sechehaye by Elizabeth his wife born
6. at Metz in the Bishoprick of Metz and Verdun

[on a separate parchment, to follow line 6:]

> press 1, lin: 6.
> after (Verdun) insert –
> George Frideric Handel Son of George Handel by Dorothy his Wife
> born at Hall in Saxony Anthony Fursteneau Son of Gerhard
>   Anthony
> Fursteneau by Clarata Beaute his Wife born at Osnabruck in
>   Westphalia
> and Michael Schlegel Son of Michael Schlegel by Anne his Wife
>   born at
> Gorlitz in Saxony

[end of separate parchment]

> out of Your Majesties Allegiance having constantly
> professed the Protestant Religion and give
9. Testimony thereof and of their Loyalty and
   Fidelity to Your Majesty and the Good of the
   Kingdom of Great Britain That it may be Enacted
   And be it Enacted by the Kings most Excellent
   Majesty by and with the Advice and Consent of the
   Lords Spiritual and Temporal and Commons
   in this present parliament assembled and by
   the Authority of the same That the said Louis

[line inserted:]

17. Sechehaye George Frideric Handel Anthony Fursteneau
    and Michael Schlegel shall be and they are hereby from hence=
    =forth Naturalized and shall be adjudged and
    taken to be to all intents and purposes naturaliz=
20. =ed and as * freeborn Subjects of this Kingdom
21. of Great Britain and they are and shall be from
    henceforth adjudged reputed and taken to be in
    every condition Respect and degree free to all

24. Intents purposes and Constructions as if they had
25. been born * National Subjects within the kingdom
    of Great Britain And be it further Enacted
    declared and Ordained by the Authority aforesaid
28. That the said Louis Sechehaye George Frideric Handel
    Anthony Fursteneau and Michael Schlegel shall be and they
29. are hereby Enabled and adjudged able to all intents
    purposes and Constructions whatsoever to inherit
    and be inheritable and inherited and to demand
    Challenge Ask Take retain have and enjoy all or
    any Manors Lands Tenements hereditaments
    Goods Chattels Debts Estates Trusts and all other
    privileges and Immunities Benefit and Ad=
    =vantage in Law or Equity belonging to the Leige
    people and Natural born Subjects of this Kingdom
38. and to make their Resort or Pedigree as heirs to their
    Ancestors Lineal or Collaterall by reason of any
    Dissent Remainder or Reverter Right or Title or
    Conveyance Legacy or Bequest whatsoever which
    hath may or shall from henceforth discend
43. remain revert Accrue or grow due to them as also
    from henceforth to take have retain keep and

[here the two parchments are joined]

> Enjoy all Manors Lands Tenements and here=
2. =ditaments which they have may or shall have by
   way of purchase or Gift of any person or persons
   whatsoever as also to prosecute pursue maintain
   avow justifye and defend all and all manner of
   Actions Suites and Causes and all other things to
7. do as lawfully liberally freely and surely as if they
   had been born of British parents within this
   kingdom of Great Britain and as any other
   person or persons born or derived from British
   parents within this kingdom may lawfully in
   any wise do and in all things and to all intents
   and purposes shall be taken to be and shall be
14. * Naturall Leige Subjects of this kingdom of
    Great Britain any Law Statute provision Custom
    Ordinance or other thing whatsoever had made
    done promulged proclaimed or provided to the
    contrary thereof in any wise notwithstanding
    Provided always And it is declared and
    Enacted by the Authority aforesaid that the
21. said Louis Sechehaye George Frideric Handel
    Anthony Fursteneau and Michael Schlegel
    or either of them shall not hereby be Enabled
22. to be of the Privy Council or a Member or Members of either
    House of Parliament or to take any Office or
    Place of Trust either Civil or Military or to have
    any Grant of Lands Tenements or hereditaments
26. from the Crown to themselves or any other person
27. in trust for them or either of them anything therein contained
    to the contrary thereof in anywise not with=
    =standing.

Appendix 3

Handel's Library: the Evidence of Book Subscription Lists

Handel's own autographs and conducting scores are well known (see p. 18), and inevitably overshadow the complete loss of his collection of printed books. No catalogue or inventory of his library survives or is known to have been made; however, by using contemporary book subscription lists (they are normally found bound at the front of a volume), it is possible to draw up a short catalogue of 24 books and scores to which Handel subscribed between 1720 and 1758, the year preceding his death.[1]

Nevertheless there are inherent faults in this method of approach: at best it can provide only a very partial view of a library; at worst it may be misleading if a well-known name has been listed as a subscriber without permission in order to give a work an extra cachet of approval. On some occasions a name may have been received too late to include on the list; the absence then of Handel's name from a subscription may not necessarily indicate that he did not own the work. Many other books were of course published without resorting to the device of seeking subscriptions.

In the catalogue given below all the items except Telemann's *Musique de Table* were published in England. Several of the earlier volumes are the work of Handel's friends or associates, in particular John Gay (1720), Henry Carey (1729), J.C. Smith the younger (1732) and James Miller (1741). The later music books are mostly by young or obscure composers whom Handel may have felt some duty to encourage; even in his final years he seems to have been sufficiently interested to subscribe to volumes of this kind by Elizabeth Turner (1756) and John Bennett (1758).

How was Handel's support solicited? Not perhaps by the advertisements for subscribers which regularly appeared in the press, but by a more direct approach, if the example of William Felton is typical. The story is given by Burney:[2] for his second organ concertos, Felton 'was extremely ambitious of the honour of [Handel's] name and acceptance of a book, merely to grace his list, without involving him in any kind of expense'. But he met with no success, Handel bursting into one of his famous rages, even though he had subscribed to Felton's first volume of concertos in 1744. Of the other volumes listed here, two are known to have been advertised in advance in the press, Gay's in 1720 and Carey's in 1726, three years before publication.

In view of his extensive borrowings from composers such as Keiser, Steffani, Muffat and Telemann, Handel's library must have been rich in earlier printed and manuscript music, some of which he probably brought with him to England with his own early autographs. Only the occasional reference is known to such volumes, such as the manuscript copy of Steffani's *Duets*, which Handel acquired soon after his arrival in Italy in 1706, or Krieger's *Keyboard Exercises* which he had while still a student.[3] One would also expect that in London Handel accumulated a library of the English poets, Milton, Dryden and Congreve, whose works were adapted for his oratorios and operas.

It is likely that the listing below can be expanded; its publication at this stage is intended to encourage a more detailed examination of the use which Handel may have made of these volumes. We know that he borrowed heavily from Telemann; it is of interest that the subscription to Miller immediately precedes his employment as a librettist; but the presence of Handel's name among the subscribers to Rolt or Rowell remains to be explained. However, in the absence of a contemporary catalogue, we are unlikely ever to discover as much of his library as of those of his contemporaries and successors – including in this country Thomas Britton, J.C. Pepusch and William Boyce, and on the continent J.S. Bach and Joseph Haydn.[4]

What happened to Handel's library at his death? His 'musick books' were bequeathed to John Christopher Smith the elder; he in turn left to his son, the younger Smith (see no. 210), 'All my Musick Books and Pieces of Musick whether Manuscript or otherwise ... left to me ... by ... my Friend George Frederick Handel', together with his own music books.[5] Smith the younger died in 1795, having given the Handel autographs to

George III (see no. 234). He bequeathed his library to his step-children: his English and theological books to Peter Coxe and George Coxe respectively, his French books to Emilia Coxe, while his music books went to his other step-daughter, Lady Rivers.[6] Here the trail becomes less clear. Lady Rivers died in 1835, stipulating in her will that except for specific bequests her possessions should be auctioned. Her portrait of Handel by Denner (no. 1) was given by her daughter-in-law, Lady Rivers, to the Sacred Harmonic Society in 1857, but it would seem that Handel's conducting scores were sold locally in Winchester; they turn up again in the catalogue of a Bristol bookseller, Thomas Kerslake, and were the subject of a short correspondence in *Notes and Queries*.[7] No reference is to be found to printed music in his catalogue.

NOTES

1. Some of these volumes have been noted by *Deutsch* (pp. 121, 230, 299, 340, 421, 577, 599, 643, 745 and 781); others are recorded in *Book Subscription Lists. A Revised Guide* by F. J. G. Robinson and P. J. Wallis, Newcastle upon Tyne, 1975 (with four supplements), and have been traced with the generous help of Mr Peter Wallis.
2. Burney, *Sketch*, p. 32.
3. Agostino Steffani's *Duets* are in the British Library (Add. MS 37,779). Handel's copy of Johann Krieger, *Anmuthige Clavier Ubung*, Nuremberg, 1699, was subsequently owned by Bernard Granville (see no. 157), and was among the Granville manuscripts offered at auction by Puttick and Simpson, 29 January 1858, lot 185. Krieger was organist at Halle and later at Weissenfels during Handel's youth.
4. See A. Hyatt King, *Some British Collectors of Music c. 1600–1960*, Cambridge, 1963, pp. 11–12, 15–16 and 20; Robin A. Leaver, *Bach's Theological Library*, 1983; Maria Hörwarthner, 'Joseph Haydns Bibliothek', *Joseph Haydn in Seiner Zeit*, exhibition catalogue, Eisenstadt, 1982, pp. 172–4.
5. James S. Hall, 'John Christopher Smith. Handel's Friend and Secretary', *Musical Times*, XXX, 1955, pp. 132–4.
6. William C. Smith, 'More Handeliana', *Music and Letters*, XXXIV, 1953, pp. 11–24.
7. H. J. Gauntlett, 'Did Handel possess a Musical Library?', *Notes and Queries*, 2nd series, no. 53, 3 January 1857, pp. 16–17 (with a response from Thomas Kerslake); see also Hans Dieter Clausen, *Händels Direktionspartituren ('Handexemplare')*, vol. VII of *Hamburger Beiträge zur Musikwissenschaft*, Hamburg, 1972, pp. 16–20.

Books to which Handel Subscribed

HENRY CAREY
Poems on Several Occasions
London, E. Say, 1729 (3rd edition)
Dedicated to the Countess of Burlington
Subscribers: 'Mr George Frederick Handel'
Note: Among the poems are tributes to Handel (p. 108), Pepusch, Geminiani, Galliard, Roseingrave, Lampe, Stanley and Dubourg

JOHN GAY
Poems on Several Occasions
London, Jacob Tonson and Bernard Lintot, 1720 (2 vols.)
Subscribers: 'Mr Handel', also the Earl of Burlington and the Duke of Chandos (fifty copies each)
See no. 152

AARON HILL
The Works of the late Aaron Hill, Esq
London, for the benefit of Hill's family, 1753 (4 vols.)
Dedicated to HRH The Princess Dowager of Wales
Subscribers: 'George Frederick Handel, Esq'
See no. 155

JAMES MILLER
Miscellaneous Works in Verse and Prose
London, J. Watts, 1741
Note: Not examined; listed by *Robinson and Wallis*, p. 23

JOSEPH MITCHELL
Poems on Several Occasions
London, the author, 1729 (2 vols.)
Dedicated to John, Earl of Stair
Subscribers: 'Mr Handell', also the Earl of Burlington, the Duke of Chandos and Aaron Hill
Note: The second volume includes two poems on Francesca Cuzzoni: 'Of Seigniora Cuzzoni's Voice and Face' (p. 222), and 'To Seigniora Cuzzoni' (p. 225)

JOHN PINE (publisher)
Quinti Horatii Flacci Opera
London, John Pine, 1733 (2 vols.)
Dedicated to Frederick Prince of Wales
Subscribers: 'Geo: Frederick Handel Esq:'
See no. 154

RICHARD ROLT
Memoirs of the Life of the late Right Honourable John Lindesay, Earl of Craufurd and Lindesay
London, Henry Kopp, 1753
Dedicated to James, Duke of Hamilton and Brandon
Subscribers: 'George Frederick Handel, Esq'

J. ROWELL (translator)
The Marchioness de Lambert's Letters to her Son and Daughter, on True Education, &c.
London, M. Cooper, 1749
Dedicated to the Rt Hon Lady Betty Cecil
Subscribers: 'George Frederick Handel, Esq'

JOSEPH STANGLINI
Nouvelle Methode pour apprendre La Langue Italienne
London, J. Read, 1724
Dedicated to Sir Richard Grosvenor, Bart, MP
Subscribers: 'Monsieur Handel', also Bononcini, Senesino and Heidegger

Scores to which Handel Subscribed

JOHN BENNETT
Ten Voluntaries for the Organ or Harpsichord
London, the author, 1758
Subscribers: 'George Frederic Handel, Esq.'

WILLIAM BOYCE
Solomon. A Serenata. In Score, Taken from the Canticles
London, John Walsh for the author, 1743
Subscribers: 'George Frederick Handel, Esq.'

WILLIAM BOYCE
Twelve Sonatas for Two Violins with a Bass for the Violoncello or Harpsichord
London, the author (sold by John Walsh), 1747
Subscribers: 'George-Frederick Handel, Esq;'

THOMAS CHILCOT
Six Suites of Lessons for the Harpsichord or Spinet
London, William Smith, 1734
Dedicated to Viscountess Bateman
Subscribers: 'Mr George Frederick Handel'

THOMAS CHILCOT
Twelve English Songs With their Symphonies. The Words by Shakespeare, and other Celebrated Poets
London, John Johnson, 1744
Dedicated to Samuel Strode Esq
Subscribers: 'George Frederick Handel, Esq.'

WILLIAM FELTON
Six Concerto's for the Organ or Harpsichord with Instrumental Parts
London, John Johnson, 1744
Subscribers: 'George Frederick Handell, Esq'
Note: See above for Felton's approach to Handel concerning his second organ concertos

JOHANN ERNST GALLIARD
The Hymn of Adam and Eve. Out of the Fifth Book of Milton's Paradise Lost
London, 1728
Subscribers: 'Mr Handel 4 books'
See no. 153

ELISABETTA DE GAMBARINI
Six Sets of Lessons for the Harpsichord
London, the author, 1748
Dedicated to Viscountess Howe
Subscribers: 'Sig. G.F. Handel'
Note: Handel's name is absent from the list of subscribers to her second set of lessons for the harpsichord issued shortly afterwards

BARNABUS GUNN
Two Cantata's and Six Songs
Gloucester, Robert Raikes, 1736
Subscribers: 'Mr. Handel'

MUSGRAVE HEIGHINGTON
Six Select Odes of Anacreon, in Greek and Six of Horace in Latin
London, John Simpson for the author, 1744
Dedicated to Robert Walpole, Earl of Orford
Subscribers: 'George Frederick Handel, Esq'

JOHN CHRISTIAN MANTEL
Six Setts of Lessons For the Harpsichord or Organ ... Opera Prima
London, William Smith, 1752
Subscribers: 'Mr Handel'

JAMES NARES
Eight Setts of Lessons for the Harpsichord
London, John Johnson for the author, 1747
Dedicated to the Earl of Abingdon
Subscribers: 'George Frederick Handel, Esq'

JOHN CHRISTOPHER SMITH THE YOUNGER
Suites de Pièces pour le Clavecin
London, the author, 1732
Dedicated to the Rt Hon the Lady Walpole
Subscribers: 'Geo: Frederick Handel Esq:'
Note: Vol. 2 was published by John Walsh, 1737

GEORG PHILIPP TELEMANN
Musique de Table
Hamburg, 1733
Subscribers: 'Mr Hendel, Docteur en Musique, London'
(Handel was the only subscriber in Britain)
Note: Not examined; listed by *Deutsch*, p. 340

ELIZABETH TURNER
A Collection of Songs with Symphonies and a Thorough Bass, With Six Lessons for the Harpsichord
London, the author, 1756
Subscribers: 'Geor. Fred. Handel Esq'
Note: With a vignette of Apollo listening to a woman at the harpsichord; her music volumes on the floor are labelled Corelli, Purcell, Handel and Boyce

Appendix 4

Handel's Collection of Paintings

Brought together here for the first time are various references to Handel as a collector. Some of these are unpublished, in particular the record of his saleroom purchases and the catalogue of his own posthumous sale in 1760. It has not proved possible to trace any of his paintings although equivalent works or other versions can be found (see fig. 14). Some of the engravings listed in the sale catalogue as lots 1 to 9, for example those by Joseph Goupy, can be identified.

The Making of a Collection

Handel appears to have acquired paintings from three sources:

His own portraits,
commissioned or given to him, discussed in more detail on p. 33. At one time or another he owned portraits by Denner, Mercier, Roubiliac and Hudson (nos. 1, 2, 7 and 11).

Paintings by his contemporaries in England,
perhaps acquired directly from the artists. These include works by or after Peter Angellis (lot 20 in the sale catalogue below), Joseph Goupy (lots 6, 8, 35, 43 and 58), Giovanni Antonio Pellegrini (lots 15 and 26), Marco Ricci (lots 51–3), Sebastiano Ricci (lot 65), G.N. Servandoni (lots 12, 25 and 67), Peter Tillemans (lot 56) and John Wootton (lot 39). Several of these artists designed sets for the opera in London: Pellegrini and Marco Ricci in 1709–10, Goupy and Tillemans in 1724, and Servandoni in 1721–3 and again in 1749. Another contemporary artist represented was Balthasar Denner whose pair of heads of an old man and woman are discussed below.

Old Master paintings,
most of which will have been bought at auction. However one of his two paintings attributed to Rembrandt was a gift from Bernard Granville (see no. 157) and is mentioned in Handel's will (no. 212). Some saleroom purchases are contained in a manuscript volume, *Sale Catalogues of the Principal Collections 1711–59*, in the Victoria and Albert Museum Library (86.00.18–19); these are set out below. In each case only the year, 1749/50, is specified and not the precise date of the sale.

Mr Edwin's Sale 1749/50
  Lot 2   'A small history'   Carracci   Bt. Handel £2.3s.–

Major & Gouijn's Sale 1749/50   2nd Day
Lot 37   'A Landskip with Moses in the Bullrushes'
                    Parocelle Sen[r]   Bt. Handel £12.12s.–
Lot 45   'A Landschape'   Sal. Rysdael   Bt. Handel £6.16s.6d.

Lady Sunderland's Sale 1749/50
Lot 19   'The Sacking of Troy'   P. da Cortona
                    Bt. Handel £7.0s.–
Lot 27   'A Landskip with a Bacchanalian'   Prelemburch
                    Bt. Handel £10.10s.–

Mr Cosein's Sale 1749/50   1st Day
Lot 39   'A large landscape & Figs'   Rembrandt
                    Bt. Handel £39.18s.–

These purchases, in the order given above, can probably be identified with lots 66, 36, 24, 55(?) and 64 in Handel's sale. The last item, the *Large Landscape with Figures* by Rembrandt, was his most costly acquisition: it was auctioned at Anthony Cousein's sale on 8 February 1750 for 38 guineas (hardly expensive compared to the price of half a guinea for the best opera seats at that period).

Fig. 14 Belisarius, a copy by Joseph Goupy of a picture attributed to Van Dyck in the collection of the 3rd Earl of Burlington. Handel owned another copy of this subject by Goupy
By Gracious Permission of Her Majesty the Queen

Lord Shaftesbury noted this purchase in a letter of 13 February (*Deutsch*, p. 680): Handel 'has been pleasing himself in the purchase of several fine pictures, particularly a large Rembrandt, which is indeed excellent'. In his will, a 'Landskip, a view of the Rhine' was one of two Rembrandts bequeathed by Handel to Bernard Granville. These remained at the Granville house, Calwich Abbey, until 1844 when they were sold to the Reverend Walter Bromley-Davenport of Capesthorne.

The Dispersal of Handel's Collection

Handel made certain gifts and bequests: his own portraits by Denner and Roubiliac, which he gave to J.C. Smith the younger, according to Coxe's *Anecdotes* (1799); his 'Old Man's head and the Old Woman's head done by Denner', which he bequeathed to Charles Jennens (still at Gopsall in 1857: see *Schoelcher*, p. 342, n. 2); and the two Rembrandts, discussed above, which went to Granville.

His sale catalogue is known through a unique copy in the Frick Art Library, New York, by whose kind permission it is here transcribed as it stands. The first day's sale was devoted to dispersing another collection, that of John Ellys (c. 1701–57), intriguingly described as portrait painter and 'keeper of the Lions in the Tower'.

A CATALOGUE Of the GENUINE and ENTIRE COLLECTIONS Of *Italian, Flemish* and *Dutch* PICTURES, of JOHN ELLYS, Esq; AND GEORGE FREDERICK HANDEL, Esq; Both lately Deceased; Which, by Order of their respective Executors, Will be sold by AUCTION, By Mr. LANGFORD, At his House in the *Great Piazza, Covent Garden*, On *Wednesday* and *Thursday*, the 27th and 28th of this Instant *February* 1760.

The same may be viewed on *Monday* the 25th, and every Day after till the Time of Sale, which will begin each Day *punctually* at *Twelve o'Clock*

CATALOGUE of which may be had *gratis* at Mr. LANGFORD's in the *Great Piazza* aforesaid.

CONDITIONS of SALE as usual.

Second Day's SALE, *Thursday, February* 28, 1760.
Mr. *HANDEL*'s COLLECTION.

LOT PRINTS, &c.
1. EIGHT of the Planets, by *N. Dorigny*
2. The Flight into *Egypt*, by *Solimene*, and 8 others
3. Seven Prints of the Angles and Crucifixion, by *N. Dorigny*
4. Eight Landscapes, by *N. Poussin*
5. Six Sea Pieces, by Mr. *Scott* and Mr. *Lambert*
6. Four Landscapes, by *Goupy*, after *Poussin*, &c.
7. Four Views of *Malta*, and 3 others by *Dorigny*
8. Eight Landscapes, by *Goupy*, after *Sal. Rosa*
9. *Rubens* Stag hunting, by *ditto*, and 5 others
10. Eight Maps on Rollers
11. Two burnished Gold Frames

PICTURES.
12. Two Door Pieces, after *Servandoni*, and 2 others
13. A Landscape with Buildings, by *Vandiest*, a small Landscape, by *Mompert*, and a Sea Piece — PERCELLES
14. A *Dutch* Conversation, the Style of *Ostade*; a Landscape, the Stile of *Housman*, and 2 Portraits
15. Two History Pieces — PELLIGRINI
16. Two Flower Pieces, *Ardine*, and 1 by *Montingo*
17. A fresh Gale, *Percelles*, a Calm, and 2 others
18. A Town on Fire, by *Ab. Van Bassan*, 2 Heads in one Picture, and a Landscape — VANDIEST
19. A Conversation of Boors — D. TENIERS
20. Two Monkeys in Friars Habits, by *Angelles*, and a small Landscape — VANGOEN
21. *Narcissus* in a Landscape, *Fr. Mola*, and a Man playing on a German Flute, Stile of — TITIAN
22. A Flower Piece, *Ardine*, and a Landscape
23. Two [?] Landscapes of [torn] — [Torn]
24. A Landscape and Cattle, after *Berghem*, by *Old Griffier*, and a ditto by *Sol. Ruysdael*
25. A Piece of Ruins — SERVANDONI
26. *Sampson* and *Dalilah* — PELLIGRINI
27. A Head, the Stile of — REMBRANDT
28. A Conversation — WATTEAU
29. Its Companion — DITTO
30. *Jupiter* and *Leda*, in Water Colours
31. *Jupiter* and *Danaë*, its Companion
32. A Landscape — MOMPERT
33. A Lamp-light — S. ANDREA
34. The Rape of Proserpine
35. *Hagar* and *Ishmael* — GOUPY
36. A Landscape with a Bridge, and the finding of *Moses* — PAROCELLE
37. Two Angels Heads in an oval, and its Companion — CANTARINI
38. A Landscape — SWANEVELT
39. A Sun-set — WOOTON
40. A Landscape — N. POUSSIN
41. An upright ditto — HORIZONTI
42. Its Companion
43. *Belisarius*, after *Vandyck*, in Water Colours — GOUPY
44. An upright Landscape — MICHAU
45. Ditto with Vulcan's Forge — HONDIUS
46. A Landscape — ROWLAND SAVORY
47. *Jupiter* and *Iö*, in Water Colours
48. *Jupiter* and *Ixion*, its Companion
49. A Hunting Piece — HONDIUS
50. A Landscape — HORIZONTI
51. Ditto and Buildings, in Water Colours — M. RICCI
52. Two ditto — DITTO
53. Two ditto, small — DITTO
54. A large Piece of Architecture and Ruins, after — PANINI
55. The Siege of *Troy* — N. POUSSIN
56. Two Battle Pieces — TILLEMANS
57. A small Landscape — V. BRUEGHELL
58. *Mucius Scaevola* — GOUPY
59. *Venus* with her Attendants — ANDREA DEL SARTO
60. A Landscape and Figures, on Copper — FERG
61. Its Companion — DITTO
62. A Landscape — LUCATELLI
63. The Doge's Palace — CANNALETTI
64. *Bacchus* and *Ariadne* — POELENBURGH
65. *Venus* and *Adonis* — SEB. RICCI
66. *Venus* and *Cupid*, in an oval — CARRACCI
67. A large Piece of Ruins and Figures — SERVANDONI

Index of Exhibit Titles

HWV numbers refer to the Handel-Werke Verzeichnis (ie catalogue of Handel's works) in B. Baselt, *Thematisch-systematisches Verzeichnis, Händel-Handbuch*, vols. I to III Leipzig, 1978, 1984, vol. III in preparation.

'Account of the Musical Performances in Westminster-Abbey, and the Pantheon . . . in Commemoration of Handel.' 242
Admission Tickets: to the First Performance, Westminster Abbey, 26 May 1784 236; to the Second Performance, the Pantheon, 27 May 1784 237; to the Third Performance, Westminster Abbey, 29 May 1784 238; *see also* Season Ticket for Vauxhall Gardens 146; Three Tickets for Messiah 255
'Advice: A Satire' 165
Aeneas before Dido in the Temple of Juno 70
'Amelia' 166
'Anecdotes of George Frederick Handel, and John Christopher Smith.' 216
Anne, Queen, and the Knights of the Garter 52
'Apotheosis of Handel' 249
Ariosti, Attilio 89
Arnold, Dr Samuel 264

Bad Taste of the Town, The 106
'Bagster's Gem Edition of Handel's Messiah' 268
Baker, Photograph of Janet, as Eduige in Rodelinda, Handel Opera Society, 24 June 1959 274c
Barges: Model for the Shipwrights' Company 65; Model of a Royal 66
Bass Recorder 177
Bassoons: Double 173; by Stanesby junior 172; by Stanesby senior and junior 171; by Wietfelt 80
Bass Viols: by Jaye 78; by Norman 79
Bates, Joah 235
Battle of Dettingen, George II at the 190
Beard, John 187
Beggar's Opera, The 108
Bernardi, Francesco, called Il Senesino: from album of Count Algarotti 94; by Kirkall 93; *see also* pamphlet, 'An Epistle from S[igno]r S[enesin]o to S[ignor]a F[austin]a' 107
Bicentenary Celebrations, 1959, The 275
Biggar, Photograph of Marjorie, as David, Handel Opera Society, 23 June 1965 274d
Bliss, Photograph of Sir Arthur, laying a wreath, The Bicentenary Celebrations, 1959 275c
Bononcini, Giovanni 88
Bordoni, Faustina: by Nazari 96; *see also* pamphlet, 'An Epistle from S[igno]r S[enesin]o to S[ignor]a F[austin]a' 107
Boyle, Richard, 3rd Earl of Burlington 60
Britton, Thomas 59
Bromfield, Dr William 209
Brook Street, Handel's House, No. 25 147
Brydges, James, 1st Duke of Chandos 68
Burlington House 61
Burlington, Richard Boyle, 3rd Earl of 60
Burney, Charles 243

'Cannons in Middlesex, The South Front of' 69
Cantatas, Italian 159
Carestini, Giovanni: from album of Count Algarotti 117; by Faber 116
Caroline, Queen 114
'Celebrated Water Musick in Seven Parts, The' (HWV 348–50) 64

Chandos, James Brydges, 1st Duke of 68
'Charming Brute, The' 9
'Child Haendel, The' 253
Chrysander, Friedrich 269
Cibber, Susannah 185
Clarinets, Pair of 231
'Clavecin. Suites de Pieces pour le, Composées par G.F. Handel. Premier Volume.' (HWV 426–33) 74
Clive, Kitty 188
Clock-case, Handel 246
Codicils, Handel's Will and Four 212
Colonna, Cardinal Carlo 33
Commemoration Medal and Ribbon, 1791 240
Commemoration Ring, 1784 241
Commemorations: of Handel, Spectators at the 1790 244; Westminster Abbey Handel 232
Concert at Montagu House, A 141
Concertos: 'Six C. For the Harpsicord or Organ Compos'd by Mr: Handel' [Opus 4] (HWV 289–94) 136; 'Twelve Grand C. in Seven Parts for Four Violins, A Tenor Violin, A Violoncello with a Through Bass for the Harpsichord' [Opus 6] (HWV 319–30) 139
Conti, Gioacchino, called Gizziello 124
Cooper, Anthony Ashley, 4th Earl of Shaftesbury 158
Co-operative Society Production of Belshazzar, 1937, Costume Designs for a Projected 273
Corelli, Arcangelo 32
'Coronation of their Majesties, The Form and Order of the Service . . . to be Observed in the' 111
Costume Designs for a Projected Co-operative Society Production of Belshazzar, 1937 273
Covent Garden: Piazza and Market 132; 'Rich's Glory or his Triumphant Entry into Covent-Garden' 131
'Critica Musica' 104
Crown and Anchor Tavern, St Clement Danes with the 130
Crystal Palace: engraving, 'The Great Handel Festival at the C.P.' 261; Medal for the 1859 Centenary Commemoration 262; programme, 'Great Triennial Handel Festival, C.P., 1862' 263; programme book, 'Great Handel Festival, at the C.P., June 15th, 17th and 19th, 1857' 259
Cumberland, William, Duke of 192
Cuzzoni, Francesca 95

Death-mask, Lost 15
Delany, Mrs (previously Mary Granville, Mrs Pendarves) 156
Descant Recorder 176
'A Description of the Machine for the Fireworks . . . and a Detail of the Manner in which they are to be exhibited in St James's Park, Thursday, April 27, 1749' 200
Design for the Decoration of Handel's Harpsichord 150
Dettingen, George II at the Battle of 190
Diamond Ring 24
Double Bassoon 173
Double-manual Harpsichord 75
Drums: Pair of Kettle- 223; Saul: Receipt for Kettledrums from the Tower of London 138; Side 224
'Dunciad, The New: As it was found In the Year MDCCXLI.' 164

'Epistle from S[igno]r S[enesin]o to S[ignor]a F[austin]a, An' 107

Evans, Photograph of Geraint, as Saul, Handel Opera Society, 23 June 1965 274d

Farinelli (Carlo Broschi) 120
Faustina Bordoni: by Nazari 96; *see also* pamphlet, 'An Epistle from S[igno]r S[enesin]o to S[ignor]a F[austin]a' 107
'Fees for the Funeral of George Frederick Handel Esq.' 213
Festivals: Medals for the 1784 Handel, Two 239; Medal for the 1857 Crystal Palace Handel 260; Programmes for Messiah at Provincial Musical, Three 256; programme book, 'Great Handel F., at the Crystal Palace, June 15th, 17th and 19th, 1857' 259; programme, 'Great Triennial Handel F., Crystal Palace, 1862' 263; Programmes for the Göttingen Handel, Two 271; Programmes, Four Recent Handel 277; wood engraving, 'The Great Handel F. at the Crystal Palace' 261
Fireworks: 'A Description of the Machine for the F. . . . and a Detail of the Manner in which they are to be exhibited in St James's Park, Thursday, April 27, 1749' 200; 'An Horizontal View of the Public F.' 198; Music for the Royal (HWV 351) 195; 'A Perspective View of the Building for the F., in the Green Park, taken from the Reservoir' 202; 'A Plan and Elevation of the Royal F.' 199; 'A View of the Public F.' 201
Flageolets, Pair of One-hand 229
Flutes: d'Amour 179; Transverse 178
'Form and Order of the Service . . . to be Observed in the Coronation of their Majesties' 111
Foundling Hospital: Anthem (HWV 268) 204; 'A Perspective View of the F.H. with Emblematic Figures' 203
La Francesina (Elizabetta Du Parc) 125
Frederick Prince of Wales and his Sisters 112
Frederick, Sir Charles 196
French Horns, Pairs of: right-handed (Hellier) 220; true pair (Sharp) 230
French Hunting Horns, Pair of 81
Funerals: 'F. of George Frederick Handel Esq., Fees for the' 213; *see also* Memorial Volume for Handel's Mother 20

George I 41
George II: 109; at the Battle of Dettingen 190
George III 234
Gizziello (Gioacchino Conti) 124
Göttingen Handel Festivals: 1921 and 1922, Two Programmes for the 271; 1973, programme book 277b
Goupy, Joseph 91
Granville, Bernard 157
Granville, Mary, Mrs Pendarves, later Mrs Delany 156
Great Handel Festival: programme book, 'G.H.F. at the Crystal Palace, June 15th, 17th and 19th, 1857' 259; engraving, 'The G.H.F. at the Crystal Palace' 261
'Great Triennial Handel Festival, Crystal Palace, 1862' 263
Grimaldi, Nicolo, called Nicolini 49

Hagen-Leisner, Photograph of Thura as Teofane 271a
Halle: Map and Views of 18; pamphlet, 'G.F. Handel and H. In Memory of the 250th Anniversary of the Composer's Birth' 272; programme, Handel Festival, 1971 277a
Hamburg, Map and Views of 26
Händel, Georg 19
Handel, George Frideric:
  Correspondence from: letter to the Earl of Burlington, Handel's Visit to Dresden: the Hire of Singers for the Royal Academy of Music 87; letter to Jennens, 1742, Messiah: 'how zealous they are in Ireland for Oratorio's' 186; letter to Jennens, 1744, Belshazzar: Handel's Plans for the 1744/5 Season 191; letter to Michaelsen, 1731, Handel Sends Wedding Presents to his Niece and her Husband 22; letter to Michaelsen, 1733, Handel's Family Affairs in Germany 23
  Correspondence to: from the Duke of Newcastle, 'Warrant to Mr Hendel to procure Singers for the English Stage' 86
  Gifts from: Chatelaine with Watch 25; Diamond Ring 24; Portraits: by Denner 1, by Mercier 2, by Hudson 11
  House, Handel's, No. 25 Brook Street 147
  Mother, Memorial Volume for Handel's 20
  Personal effects: Harpsichord, Design for the Decoration of Handel's 150; 'Pitch Pipe, Handel's' 180; 'Watch, Handel's' 151
  Portraits of: by Bankel (steel engraving) 252; by Batt (wood engraving), The Waiting Handel 254; by Berlin Photographic Company (photogravure print), 'The Child Haendel' 253; by Denner 1; Continental, unknown factory (biscuit-ware figure), 'G.F. Haendel' 251; by Goupy (gouache), 'The Harmonious Boar' 8; by Heath (engraving), 'Apotheosis of Handel' 249; by Houbraken (engraving) 4; by Hudson, 1748–9 11; by Hudson, 1756 12; by Mercier 2; by Ridgway Bates & Co. (Parian ware bust) 250; by Roubiliac, Lost Death-mask, Photograph of 15; by Roubiliac (terracotta), Model for the Monument in Westminster Abbey 17; by Roubiliac (terracotta), Model for the Vauxhall Gardens Statue 6; by Roubiliac (terracotta model for marble bust, Windsor) 7; after Roubiliac (bronze bust) 247; attributed to Roubiliac (terracotta bust) 16; by an unknown artist 10; by an unknown engraver 3; by an unknown engraver, 'The Charming Brute' 9; by an unknown sculptor (terracotta bust) 14; by Wolfgang (miniature) 5; by Wood (two ceramic busts) 248
  Publications concerning: by Burney, 'An Account of the Musical Performances in Westminster-Abbey, and the Pantheon . . . in Commemoration of Handel' 242; by Coxe, 'Anecdotes of George Frederick Handel, and John Christopher Smith' 216; by Fielding, 'Amelia' 166; by Mainwaring, 'Memoirs of the Life of the Late George Frederic Handel. To which is added, A Catalogue of his Works and Observations upon them.' 214; by Pope, 'The New Dunciad: As it was found In the Year MDCCXLI.' 164; by Pratt, 'An Ode to Mr Handel On his Playing on the Organ' 163; by Smollett, 'Advice: A Satire' 165
  Receipt, Saul: Receipt for Kettledrums from the Tower of London 138
  Temple at the Wodehouse, Wombourne, Staffordshire, Handel's 218
  Will and Four Codicils, Handel's 212
  Works by:
    Acis and Galatea (HWV 49a) 73
    Agrippina (HWV 6) 38
    Alcina (HWV 34) 119
    Alexander's Feast (HWV 75): autograph score 133; manuscript part-book for violoncello 135; published by Novello's 267b; word-book, 'A.F. or, The Power of Musick. An Ode Wrote in Honour of St. Cecilia, By Mr. Dryden.' 134
    'L'Allegro, Il Penseroso, ed Il Moderato. The Words taken from Milton. Set to Musick by Mr. Handel.' (HWV 55) 219

Belshazzar (HWV 61): Costume Designs for a Projected Co-operative Society Production of, 1937 273; letter, Handel's Plans for the 1744/5 Season 191
Birthday Ode for Queen Anne (HWV 74) 53
Chandos Anthem: O Praise the Lord with one Consent (HWV 254) 72
Coronation Anthems (HWV 258–61): autograph score 110; and other works by Mendelssohn and Haydn, Handbill for Handel Performances by the Sacred Harmonic Society 258a
Deborah (HWV 51), Handbill for Handel Performance by the Sacred Harmonic Society 258c
Deidamia (HWV 42): Poster and programme for the Handel Opera Society, 3 June 1955 274a; printed score, 'D. an Opera compos'd by M.r Handel.' 127
Dettingen Te Deum (HWV 283) 267
Dixit Dominus (HWV 232) 34
Esther (HWV 50a & b): conducting score with autograph additions 129; manuscript score 128; manuscript score belonging to James Harris 161; stage design by Peter Rice, Handel Opera Society, 5 November 1980 274e
Foundling Hospital Anthem (HWV 268) 204
Giulio Cesare in Egitto (HWV 17): autograph score 101; printed score, 'Julius Caesar: An Opera. Compos'd by G. Frederick Handel, of London, Gent.' 102
Israel in Egypt (HWV 54): printed score, 'A Sacred Oratorio in Score Composed in the Year 1738 By G.F. Handel.' 265; published by Novello's 267a
Italian Cantatas 159
Jephtha (HWV 70): autograph score 206; conducting score 207
Judas Maccabaeus (HWV 63): published by Novello's 267e; word-book, 'Judas Macchabaeus. A Sacred Drama. As it is Perform'd at the Theatre-Royal in Covent Garden. The Musick by Mr. Handel.' 193
'Julius Caesar. An Opera. Compos'd by G. Frederick Handel, of London, Gent.' (HWV 17) 102; see also Giulio Cesare in Egitto
Messiah (HWV 56): autograph score 183; 'Bagster's Gem Edition of Handel's M.' 268; letter from Handel to Jennens, 'how zealous they are in Ireland for Oratorio's' 186; letter from Jennens to Holdsworth, 'The Subject is M.' 182; manuscript part-book for second soprano 205; printed score, 'Songs in M. an Oratorio Set to Musick by M.r Handel' 189; Three Programmes for, at Provincial Musical Festivals 256; Three Tickets for 255
Music for the Royal Fireworks (HWV 351) 195
'Muzio Scaevola (HWV 13), The Most Favourite Songs in the Opera of, Compos'd by Three Famous Masters.' 99
Ottone (HWV 15): 100; photograph of Thura Hagen-Leisner as Teofane 271a
Radamisto (HWV 12a): printed score, 'Il Radamisto Opera Rapresentata Nel Regio Teatro D'Hay Market Composta Dal Sig.re, Georgio Fredrico Handel.' 98; word-book, 'Radamisto. An Opera. As it is Perform'd At the King's Theatre in the Hay-Market, For the Royal Academy of Musick.' 97
Resurrezione, La (HWV 47): autograph score 36; word-book, 'Oratorio per la Risurrettione di Nostro Signor 'Giesu Cristo ... L'Anno MCCVIII' 37
Rinaldo (HWV 7a): autograph score 45; printed score, 'Song's in the Opera of Rinaldo Compos'd by Mr Hendel' 47; word-book, 'Rinaldo, An Opera. As it is Perform'd At the Queens Theatre in London.' 46

Rodelinda (HWV 19), Photograph of Janet Baker as Eduige, Handel Opera Society, 24 June 1959 274c
Rodrigo (HWV 5) 35
Saul (HWV 53): manuscript score of Act 1 137; Photograph of Geraint Evans as Saul and Marjorie Biggar as David, Handel Opera Society, 23 June 1965 274d; Receipt for Kettledrums from the Tower of London 138
Semele (HWV 58), Programme for the Handel Opera Society, 23 June 1959 274b
Serse (HWV 40) 126
'Sieg der Zeit und Wahrheit. Oratorium von Georg Friedrich Händel.' ['The Triumph of Time and Truth'] (HWV 71) 270
'Six Concertos For the Harpsichord or Organ compos'd by Mr. Handel' [Opus 4] (HWV 289–94) 136
Solomon (HWV 67): Handbill for Handel Performance by the Sacred Harmonic Society 258b; published by Novello's 267d
'Suites de Pieces pour le Clavecin Composées par G.F. Handel. Premier Volume.' (HWV 426–33) 74
'Tamerlane: an Opera, compos'd by M.r Handel' (HWV 18) 103
Triumph of Time and Truth, The (HWV 71) 270
'Twelve Grand Concertos in Seven Parts for Four Violins, a Tenor Violin, A Violoncello with a Through Bass for the Harpsichord. [Opus 6] (HWV 319–30) 139
Utrecht Te Deum and Jubilate (HWV 278, 279) 55
'Water Musick, The Celebrated, in Seven Parts' (HWV 348–50) 64
Handbills: for Handel Performances by the Sacred Harmonic Society, Three 258; see also The National Portrait Gallery's Handel Appeal, 1968 276
Handel Opera Society, Five Productions of the 274
Hanover from the South-west, View of 40
'Harmonious Boar, The' 8
Harpsichords: Design for the Decoration of Handel's 150; Double-manual 75; 'Six Concertos For the Harpsicord or Organ Compos'd by Mr: Handel' [Opus 4] (HWV 289–94) 136; 'Twelve Grand Concertos in Seven Parts for Four Violins, A Tenor Violin, A Violoncello with a Through Bass for the H.' [Opus 6] (HWV 319–30) 139
Harris, James 160
Hawkins, Sir John 215
Haydn, Joseph 245
Heidegger, John James 57
Hellier, Sir Samuel 217
Hill, Aaron: 44; 'The Works of the late Aaron Hill, Esq' 155
Horace, The Work of 154
'Horizontal View of the Public Fireworks, An' 198
Horns: Pair of French Hunting 81; Pair of French (right-handed, Hellier) 220; Pair of French (true pair, Sharp) 230
House, Handel's, No. 25 Brook Street 147
Hughes, John 51
Hunting Horns, Pair of French 81

'Inside View of the Pantheon, An' 233
'Invitation to Mira Requesting Her Company to Vauxhall Garden, The' 143

Jennens, Charles 181
Johann Adolf VI, Baron Kielmansegg 63
Johann Wilhelm, Elector Palatine 43

Kettledrums: Pair of 223; Saul: Receipt for, from the Tower of London 138
Kielmansegg, Baron, Johann Adolf VI 63
King's Theatre: Masquerade at the 105; previously the Queen's Theatre 48

'Ladies Lamentation for ye Loss of Senesino, The' 123
Leveridge, Richard 58
Lincoln's Inn Fields Theatre 140
London Handel Orchestra, Programme book for the London Handel Festival, 1985 277d

'Machine for the Fireworks, A Description of ... and a Detail of the Manner in which they are to be exhibited in St James's Park, Thursday, April 27, 1749' 200
McSwiny, Owen 56
Manchester, Charles Montagu, 1st Duke of 39
Maps and Views: of Halle 18; of Hamburg 26
Marriage of the Princess Royal to the Prince of Orange, The 113
Masquerade at the King's Theatre 105
Mattheson, Johann 27
Medals: for the 1784 Handel Festival, Two 239; and Ribbon, Commemoration, 1791 240; for the 1857 Crystal Palace Handel Festival 260; for the 1859 Crystal Palace Centenary Commemoration 262
'Memoirs of the Life of the Late George Frederic Handel. To which is added, A Catalogue of his Works and Observations upon them.' 214
Memorial Volume for Handel's Mother 20
Michaelsen, Michael Dietrich 21
Models: for the Monument in Westminster Abbey 17; of a Royal Barge 66; for the Shipwrights' Company Barge 65; for the Vauxhall Gardens Statue 6
Montagu, Charles, 1st Duke of Manchester 39
Montagu House, A Concert at 141
Morell, Thomas 194

National Portrait Gallery's Handel Appeal, 1968, The 276
Newcastle, Thomas Pelham-Holles, 1st Duke of 85
'New Dunciad: As it was found In the Year MDCCXLI., The' 164
Nicolini (Nicolo Grimaldi) 49
Novello, Vincent 266
Novello's, Five Scores of Handel Oratorios published by 267

Oboes: by Bradbury 83; by Stanesby senior (without ivory mounts) 167; by Stanesby senior (with ivory mounts) 168; by Stanesby senior (tortoise-shell stained boxwood) 226
'Ode to Mr Handel On his Playing on the Organ, An' 163
One-hand Flageolets, Pair of 229
O Praise the Lord with one Consent (Chandos Anthem) (HWV 254) 72
'Oratorio per La Risurettione di Nostro Signor Giesu Cristo ... L'Anno MCCVIII' (HWV 47) 37
'Order of the Service, The Form and, ... to be Observed in the Coronation of their Majesties' 111
'Organ, Six Concertos For the Harpsicord or, Compos'd by Mr: Handel' [Opus 4] (HWV 289–94) 136
Ottoboni, Cardinal Pietro 31

Pamphili, Cardinal Benedetto 30
Pantheon: Admission Ticket to the Second Performance, 27 May 1784 237; 'An Inside View of the P.' 233
Du Parc, Elizabetta, called La Francesina 125

Pelham-Holles, Thomas, 1st Duke of Newcastle 85
Pendarves, Mrs (Mary Granville), later Mrs Delany 156
Pepusch, Dr Johann Christoph 71
Perspective Views: 'A P.V. of the Building for the Fireworks in the Green Park, taken from the Reservoir' 202; 'A P.V. of the Foundling Hospital with Emblematic Figures' 203; 'A P.V. of Vaux Hall Garden' 144
Peterborough, Anastasia Robinson, Countess of 92
Piazza Navona, Rome, The 29
'Pitch Pipe, Handel's' 180
'Plan and Elevation of the Royal Fire-Works, A' 199
del Pò, Anna Maria Strada 115
'Poems on Several Occasions' 152
Poll-book for the Westminster By-election of 1749 149
Pope, Alexander 62
Productions of the Handel Opera Society, Five 274
Programmes: for the Göttingen Handel Festivals, Two 271; for Messiah at Provincial Musical Festivals, Three 256; Recent Handel Festival, Four 277; see also 'Great Triennial Handel Festival, Crystal Palace, 1862' 263; Five Productions of the Handel Opera Society 274a, b; The Bicentenary Celebrations, 1959 275a
'Prospect of the Choir, of the Cathedral Church of St. Paul, A' 54
Public Fireworks: 'An Horizontal View of the P.F.' 198; 'A View of the P.F.' 201; see also Fireworks
Purcell-Handel: 'Catalogue of a Commemorative Exhibition', The Bicentenary Celebrations, 1959 275b; Festival programme book, The Bicentenary Celebrations, 1959 275a

Queen's Theatre, later the King's Theatre, The 48

Rake's Progress, A 122
Recorders: Bass 177; Descant 176; Treble by Stanesby junior 175; Treble by Stanesby senior 174
Rehearsal of an Opera, The 50
'Rich's Glory or his Triumphant Entry into Covent-Garden' 131
Rings: Diamond 24; Handel Commemoration, 1784 241
Robinson, Anastasia, Countess of Peterborough 92
Rolli, Paolo 90
Royal Academy of Music: Handel's Visit to Dresden: the Hire of Singers for the 87; 'List of Subscribers to the Royal Academy of Musick' 84
Royal Barge, Model of a 66
Royal Fireworks: 'A Plan and Elevation of the R.F.' 199; see also Fireworks
Royal Society of Musicians, 'Handel in London' festival programme book, 1984 277c

Sacred Harmonic Society, Three Handbills for Handel Performances by the 258
St Clement Danes with the Crown and Anchor Tavern 130
'St. Georges Church Hanover Square, A View of' 148
'St. Paul, A Prospect of the Choir, of the Cathedral Church of' 54
Sallé, Marie 118
Season Ticket for Vauxhall Gardens 146
Il Senesino (Francesco Bernardi): from album of Count Algarotti 94; by Kirkall 93; see also pamphlet, 'An Epistle from S[igno]r S[enesin]o to S[ignor]a F[austin]a' 107
Serpent 227
Servandoni, Giovanni Niccolò 197

Shaftesbury, Anthony Ashley Cooper, 4th Earl of 158
Sharp Family, The 228
Shipwrights' Company Barge, Model for the 65
Shudi, Burkat, and his Family 162
Side Drum 224
Smart, Sir George 257
Smith, John Christopher the Younger: 210; 'Anecdotes of George Frederick Handel, and J.C.S.' 216
Smith, William C., Portrait in Göttingen Handel Festival programme book, 1973 277b
'Songs in Messiah an Oratorio Set to Musick by Mr. Handel.' (HWV 56) 189
'Song's in the Opera of Rinaldo Compos'd by Mr Hendel' (HWV 7a) 47
Sophia, Princess, Electress of Hanover 142
'South East Prospect of Westminster Bridge, The' 67
'South Front of Cannons in Middlesex, The' 69
Spectators at the 1790 Commemoration of Handel 244
Stanley, John 208
Strada del Pò, Anna Maria 115
'Subject is Messiah, The' 182

Tambourine and Case 225
Taylor, John 211
Te Deum, Dettingen (HWV 283) 267
Telemann, Georg Philipp 28
Temple at the Wodehouse, Wombourne, Staffordshire, Handel's 218
Tenor Viola 169
Tenor Violin: 'Twelve Grand Concertos in Seven Parts for four Violins, a T.V., a Violoncello with a Through Bass for the Harpsichord.' [Opus 6] (HWV 319–30) 139
Tickets: Admission, to the First Performance, Westminster Abbey, 26 May 1784 236; Admission, to the Second Performance, the Pantheon, 27 May 1784 237; Admission, to the Third Performance, Westminster Abbey, 29 May 1784 238; for Messiah, Three 255; Season, for Vauxhall Gardens 146
Transverse Flute 178
Treble Recorders: by Stanesby junior 175; by Stanesby senior 174

'Triumphal Arches, Mr. Handels Statue &c in the South Walk of Vauxhall Gardens, The' 145
Trumpets: Pair of, by Harris 82; by Hofmaster 221; by an unknown maker 222
'tunefull Scarecrow, Thou' 121
Tyers, Jonathan, and his Family 142

Vauxhall Gardens: 'A Perspective View of Vaux Hall Garden' 144; Model for the Statue 6; Season Ticket for 146; 'The Invitation to Mira Requesting Her Company to Vauxhall Garden' 143; 'The Triumphal Arches, Mr. Handels Statue &c in the South Walk of V.G.' 145
Views: of Hanover from the South-west, 40; 'V. of St. Georges Church Hanover Square, A' 148; 'V. of the Public Fireworks, A' 201
Viola, Tenor 169
Violins: by Barrett 76; by Stainer 77; 'Twelve Grand Concertos in Seven Parts for Four V., a Tenor V., a Violoncello with a Through Bass for the Harpsichord.' [Opus 6] (HWV 319–30) 139
Violoncello: 170; 'Twelve Grand Concertos in Seven Parts for Four Violins, A Tenor Violin, A.V. with a Through Bass for the Harpsichord.' [Opus 6] (HWV 319–30) 139
Viols, Bass: by Jaye 78; by Norman 79

Waiting Handel, The 254
'Warrant to Mr Hendel to procure Singers for the English Stage' 86
Watches: Chatelaine with 25; 'Handel's W.' 151
Westminster Abbey: Admission Ticket to the First Performance, 26 May 1784 236; Admission Ticket to the Third Performance, 29 May 1784 238; Handel Commemoration 232; Model for the Monument in 17; see also 'Fees for the Funeral . . .' 213
'Westminster Bridge, The South East Prospect of' 67
Westminster By-election of 1749, Poll-book for the 149
Will and Four Codicils, Handel's 212
Wodehouse, Wombourne, Staffordshire, Handel's Temple at the 218
Works of Horace, The 154
'Works of the late Aaron Hill, Esq, The' 155

List of Lenders

Her Majesty The Queen 5

The Visitors of the Ashmolean Museum, Oxford 17

Bate Collection of Historical Instruments, University of Oxford 179
Sixth Marquess of Bath 188
J. & A. Beare Ltd 50, 169
Boosey & Hawkes plc 80
The Bowes Museum, Barnard Castle, County Durham 31
Bristol City Council 82
The British Library Board 18, 20, 26, 34, 35, 36, 38, 40, 45, 46, 47, 53, 55, 69, 73, 97, 98, 99, 100, 101, 102, 104, 110, 119, 126, 133, 139, 147, 152, 153, 164, 166, 183, 195, 206, 216, 242
The Trustees of the British Museum 19, 28, 29, 30, 33, 43, 57, 89, 95, 106, 116, 118, 121, 122, 123, 141, 146, 151, 201, 208, 211, 239a, 243, 253

The Trustees of the Chatsworth Settlement 87
Chaucer Fine Arts, London 96
Gerald Coke Handel Collection 4, 9, 21, 27, 58, 64, 74, 103, 127, 128, 136, 138, 155, 163, 165, 182, 184, 186, 187, 189, 191, 193, 200, 202, 210, 238, 240, 241, 244, 247, 248, 252, 254, 255, 261, 265, 271
The Thomas Coram Foundation for Children 7, 204, 205

Winton Dean Esq 134
Dorset Natural History and Archaeological Society 158

Edinburgh University Collection of Historic Musical Instruments 167, 176

The Viscount FitzHarris 2
The Syndics of the Fitzwilliam Museum, Cambridge 6, 8, 72, 150, 239b
Dr Bernhard Foss 22, 23, 24, 25
The Geoffrey Freeman Collection of Bow Porcelain (on loan to the Pallant House Gallery, Chichester) 246
Dr Terry Friedman 145, 154

Anthony Gasson Esq 251
Albert R. Gellman Collection, Courtesy of the Israel Museum, Jerusalem 49, 94, 115, 117
Robin Golding Esq 214
Greater London Council, Greater London Record Office 149
Grimsthorpe and Drummond Castle Trustees 16
Guildhall Library 48, 54, 61, 67, 130, 131, 140, 148, 199, 203

Hamburger Kunsthalle 269
Staats- und Universitätsbibliothek, Hamburg 11, 129, 207
Handel Opera Society 274c, d, 275c
Tom Hawkes Esq 274e
Anthony Hicks Esq 276c–e
Horniman Museum (Carse Collection) 168, 174, 175, 178

Lambeth Palace Library 111
London Borough of Lambeth Archives Department 143, 144
C.G.M. Lloyd-Baker Esq 228, 229, 230, 231

Mrs M. McNair-Wilson 156, 157
The Earl of Malmesbury 161
Manchester City Council, Cultural Services Department (Henry Watson Music Library) 37, 137
Sir Yehudi Menuhin 76
The Museum of London 132

National Army Museum, London 190
National Maritime Museum, London 65, 66
National Museum of Ireland, Dublin 173
National Portrait Gallery, London 1, 12, 14, 15, 39, 41, 42, 44, 51, 52, 56, 59, 60, 62, 68, 85, 91, 93, 109, 112, 113, 114, 124, 125, 142, 160, 162, 185, 192, 194, 197, 198, 233, 235, 236, 237, 249, 257, 264, 266, 272, 275b, 276

University of Oxford (Bate Collection of Historical Instruments) 179
University of Oxford (Faculty of Music) 13, 32, 71, 215

Pallant House Gallery, Chichester (on loan from the Geoffrey Freeman Collection of Bow Porcelain) 246
J.W. Phillips Esq 217, 218, 219, 220, 221, 222, 223, 224, 225, 226, 227
Private Collections 63, 70, 77, 79, 81, 90, 92, 107, 170, 181, 260, 262, 267b, 270, 277c, d
Public Record Office, London 84, 86, 212

Royal Academy of Arts 209
Royal College of Music 3, 88, 120, 135, 245, 250, 258, 259, 263, 267a, 268, 274a, b, 275a
The Royal Society of London 234
The Royal Society of Musicians of Great Britain 10, 180

The Earl of Shaftesbury 159
Somerville & Simpson Limited 196

The Trustees of the Tate Gallery 108

The Trustees of the Victoria and Albert Museum 105, 273

Warwickshire Museum Collection 75, 177
William Waterhouse Esq 171, 172
Henry Watson Music Library, Manchester City Council, Cultural Services Department 37, 137
The Dean and Chapter of Westminster 213
Edward Withers Ltd 78

Yale Center for British Art (Paul Mellon Collection) 232
York Castle Museum 83
Percy M. Young Esq 256, 277a, b

Raranimus 229
Repenomamus 229
Reticulitermes 409n197
Revueltosaurus 198
Rhinaderma 220
Rhipsalis 395n131
Rhizanthes 164
Rhizobiales 405n88
Rhizonema 153
Rhynchohyalus 360n157
Roridula 351n83
Rousettus 137

Sahelanthropus 71
Saimiri 445n77
Saltoblattella 345n132
Sapajus 455n164
Sapria 396n158
Sarracenia 86
Sauripterus 224–25
Saxidomus 36
Sceloporus 346n156
Scoliodon 219
Scolodecus 89
Scytonema 153
Sharovipteryx 191
Shuvosaurus 198
Siamosorex 62
Silene 210; 423n27
Silesaurus 416n196
Similicaudipteryx 415n144
Simoselaps 64
Simosuchus 59; 338n127, 128
Sinornis 420n106
Sinornithosaurus 62, 195; 414n135
Sinosauropteryx 414n135
Sivapithecus 344n78
Smilodon 54, 55; 335ns57, 66; 336n79
Solanum 403n5
Solenodon 62
Speophyes 116
Sphaerolobus 41; 306n1
Sphaerotermes 186, 187
Sphagesaurus 59
Sphagnum 328n167
Sphinx 201
Spizaetus 422n137
Stegastes 177
Stegosaurus 197
Stenopterygius 78
Stopharia 385n4
Stylites 157
Styloichthys 222
Sulcia 173
Suminia 67; 333n122
Symplocarpus 163

Talpanas 130
Taricha 36, 37

Tarsipes 434n154
Temnothorax 408n162
Teratornis 208
Teredo 188
Termitomyces 187
Tetramorium 273
Thalassocnus 434n160
Thamnophis 37
Thyroptera 83
Tianyulong 414n137
Tikisuchus 198
Tiktaalik 222, 225
Titanichthys 50
Toxoplasma 395n144
Toxotes 271
Trachylepis 217
Trachymyrmex 184; 406n101; 408n184
Tremblaya 174
Trichosaurus 342n34
Trochita 28
Troodon 199
Typhlonectes 220
Tyrannosaurus 193, 198;
    338n128

Upucerthia 323n13
Uromastyx 6n115
Utricularia 85

Varanops 232
Varanus 232n84
Velociraptor 194
Ventostega 222
Verrucaria 152
Volaticotherium 230
Volvox 305–6n1; 371n69
Vorona 194

Watasenia 364n1

Xenocranium 231
Xenops 32
Xenopus 325n58
Xenosmilus 56
Xianglong 191
Xiaotingia 414n131
Xylocopa 113
Xylophaga 188

Yacarerani 60; 436n204
Yalkparidon 32
Yulong 415n146

Zanabazar 199
Zanclorhynchus 72
Zinderia 174
Zoophagus 150
Zygaena 326n113

Monodelphis 334n18
Mononykus 52, 196
Monopterus 430n9
Monorhaphis 96
Monosiga 315n11
Monotropsis 161
Morchella 177
Mosacaulis 390n10
Mycena 159
Mycetarotes 182
Mycocepurus 182
Myotis 324n38
Myotragus 253–54; 432n81
Myrmecophaga 51
Myzopoda 83

Nandinia 336n82
Nautilus 320n94
Necrolestes 232
Nehvizdyella 390n11
Neofelis 56
Neomys 62
Nepenthes 351n99; 352n101
Nereis 66; 341n258
Neuquenornis 433n117
Nipponaphis 172
Nomamyrmex 185
Nostoc 152, 311n8; 313n10
Notaden 352n131
Notiomystis 424n40
Notoplana 248
Notoryctes 232
Numbigilga 435n171

Ocepechelon 46
Octopus 19; 320ns103, 111
Ocymyrmex 368n149
Ocythoe 12
Odobenocetops 346n162
Oedipina 76
Ohmdenia 50
Onychoselache 73
Ophiocordyceps 183
Ophiocoma 103
Opthalmosaurus 77
Oreomystis 418n25
Oreopithecus 70–71, 281
Ormia 135
Orrorin 71
Oryctodromeus 415n160
Osmia 250
Oxyrrhis 357n48

Pachyrhizodus 348n229
Pakasuchus 59, 233
Pandanus 275, 276, 277
Panderichthys 222, 224
Panderodus 339n172
Pantodon 105; 410n9

Pantophthalmus 169
Paralouatta 281
Paramachairodus 335n74
Paramecium 306n1; 314n10; 443n22
Parasphaeria 186
Parastizopus 397n12
Paratrechalea 89
Paulinella 313n10
Peneteius 59
Pereskia 156
Petaurus 68
Petroderma 152
Philcoxia 85
Philodendron 163
Photuris 109
Phyllomedusa 70
Physarum 237
Phytophora 151
Picoides 323n1
Pinna 355n229
Pitohui 35; 325n75
Plagiosuchus 330n32
Plasmodium 395n144
Platecarpus 78
Plectropomus 289
Plioplatecarpus 429n173
Plotosaurus 78
Poiana 336n82
Polyboroides 422n137
Polymita 425n74
Polysiphonia 177
Poposaurus 198
Portiera 401n148
Postosuchus 416n189
Potos 336n82
Prasinohaema 350n58
Prionodon 336n82
Prognathodon 347n186
Prophaethon 260
Prosthecobacter 312n9
Protopithecus 281; 455n152
Psettodes 348n219
Pseudemoia 217
Pseudocarcharias 443n16
Pseudoeurycea 76
Pseudopanax 158
Pseudothelphusa 166
Psittacosaurus 196
Pterodaustro 51
Pteronura 268
Ptomacanthus 223, 225
Ptychozoon 192
Puccinia 153, 154

Qianosuchusmixtus 416n193

Rafflesia 162, 164
Rahonavis 194, 196; 413n113
Rana 423n22

INDEX OF GENERA   491

Glycera 66
Glyphorynchus 32
Gnathonemus 147
Goganasus 222
Gooloogongia 224
Guiyu 223, 225
Gunnera 311n8

Haldanodon 230
Halitherium 178
Halobates 165
Haplozoon 310n3
Harmothoe 247
Hatena 314n10
Hegetotherium 32
Heliamphora 352n101
Heliconius 442n167
Heliophobius 337n95
Heliotropium 392n48
Hemilepistus 166
Hemiscyllium 72
Heteroteuthis 110
Hirtella 355n205
Histioteuthus 362n223
Hodgkinia 173
Hogna 353n153
Holaspis 192
Homo 166, 213, 282, 284; 445n109
Homotherium 55; 335ns63, 69; 336n79
Homunculus 258
Hospitalitermes 86; 351n86
Hydnora 164
Hylecoetus 180
Hymenochirus 45
Hymenomonas 315n11
Hyopsodus 377n158
Hypsomocona 90

Ibla 321n28; 355n237
Icarosaurus 412n64
Ichthyostega 222
Idiurus 190
Iharkutosuchus 60
Ifrita 35
Incisivosaurus 195
Indicator 289
Indohyus 380n234
Ishikawaella 401n151
Isoetes 157
Istiodactylus 422n135

Janjucetus 337n114
Jinfengopteryx 415n142

Kahaono 91
Kallikaike 258
Keichousaurus 219
Kentrophoros 403n2
Klebsiella 181; 410n219

Kraterokheirodon 433n126
Kryptobaatar 218
Kuehneosaurus 412n64
Kuwajimalla 337n119

Lactarius 159
Lagothrix 114
Lasius 405n85; 409n194
Leonardoxa 186
Leptolyngbya 95
Lialis 64
Limusaurus 197
Linckia 361n200
Lineatriton 76; 411n63
Lithops 387n37
Litoria 324n44; 325n58
Littoraria 179
Livoniana 222
Longirostravis 374n31
Longisquama 414n137
Lottia 441n114
Lucihormetica 365n31
Lymnaea 457n24
Lystrosaurus 226

Mabuya 216, 217
Macaranga 86
Macrauchenia 335n66
Mandageria 224
Marchantia 210
Mariliasaurus 60
Masiakasaurus 413n110
Mastophora 89
Mecistotrachelos 191
Megalancosaurus 411n51
Megalania 232
Megalomyrmex 185
Megalopta 250
Megaphragma 443n22
Megarhyssa 341n271
Megaxantho 63
Mei 415n156
Meiacanthus 339n184
Melanophila 117-18
Merimna 119
Mesodinium 314n10
Mesosaurus 215
Metarhizium 183; 315n11; 400n88
Methylcoccus 312n9
Metopaulias 167
Microbotyrum 387n57
Microraptor 194, 195, 196
Microspathodon 177
Miracinonyx 57
Miragaia 197
Mitrastema 162
Mnierpes 105
Mola 109
Monilina 154

Caudipteryx 195
Cayaponia 417n20
Cebus 257, 280–82; 342n20; 445n77; 455n164
Cephalorhynchus 142
Cephalotes 192
Chaetopterus 365n51
Chalcides 217
Changyuraptor 413n123
Channa 430n9
Channallabes 430n5
Chelodina 46
Cheiromeles 73
Chilecebus 258
Chimaerasuchus 59
Chironectes 131
Chironex 97
Chiropotes 445n77
Chlamydomonas 95
Choloepus 231
Chromidina 315
Chrysopelea 192
Clarias 221
Clymenella 404n41
Coelurosauravus 412n64
Coenobita 249
Colisa 272
Conchoraptor 198
Confuciusornis 414n130
Coprinus 385, n4
Cornus 41
Costus 417n20
Cosymbotus 412n75
Cretaceogekko 83
Crocodylus 416n193
Cryptocercus 186; 409n200
Cryptococcus 210
Cryptoprocta 336n82
Cuscuta 162
Cyclodomorphus 63
Cynomorium 162
Cyphomyrmex 182; 406n117

Dactylopsila 32, 69
Dakosaurus 338n128
Dalechampia 28
Danionella 57
Dasypeltis 65
Daubentonia 32
Dermophis 77
Dialommus 105
Diatryma 334n35
Dictyostelium 177; 306n1
Dilong 195
Diploptera 175
Dolichopteryx 102
Doubledaya 405n74
Dracaena 47
Draco 192
Drosophila 9, 124, 127, 136, 246; 457n23

Dryopithecus 71
Dyakia 365n56

Echinops 39
Ecteninion 229
Effigia 197, 198
Elachistodon 340n237
Elaeocarpus 393n65
Elanus 205
Eleutherodactylus (*E. coqui*) 215
Ellobius 210
Emblemasoma 135
Emydocephalus 64; 340n232
Enchytraeus 247
Enhydrina 340, n225
Epichloë 154
Epifagus 161; 395n142; 396n149
Equatorius 71
Ernanodon 434n158
Eryx 216
Escovopsis 183, 184, 185
Etmopterus 110
Euchambersia 60
Eudibamus 344n89
Eudorina 313n9
Euhadra 63, 213
Eukrohnia 102
Eulamprus 217
Eulemur 115
Eumecia 217
Euops 180
Eurhinosaurus 77; 346n166
Eurotamandua 434n158
Eusthenopteron 224
Everettia 213
Excalibosaurus 346n166

Falcarius 415n142
Falculea 31
Fecenia 88
Fossa 336n82
Frankia 311n8
Fruitafossor 229, 231
Fulgora 82

Galerita 326n118
Galleria 125
Gasteropelecus 410n9
Gastrotheca 220; 430n222
Gemmata 312, n9
Geosaurus 338n128
Geranospiza 422n137
Gerrothorax 330n32
Gigantocypris 102
Gilliesia 393n89
Gingko 390n11
Glanosuchus 226
Globidens 78
Glomus 387n31

Index of Genera

With ns, pagination refers to start of n, although in some cases word is on a following page.

Abarenicola 404n40
Acanthocnemus 119
Acanthostega 222
Acionyx 57
Acokanthera 39
Acrecebus 455n151
Acrobates 190
Acromyrmex 181–85; 406n121; 408n159
Actinocoryne 236
Adamantinasuchus 60
Adenia 155–56
Aegyptopithecus 257
Aipysurus 340n226, 232
Albinykus 415n157
Alloposus 11
Alouatta 114; 445n77
Alticus 430n13
Amphioctopus 20
Anableps 105, 219
Aneura 395n142
Anolis 147
Anthracomedusa 287
Antillothrix 455n142
Aotus 115
Aplysia 15
Apterostigma 182, 183, 84
Arachnodes 399n79
Aradus 119
Archaeopteryx 193–94, 195, 198, 206, 225; 414n130; 414n131
Arctictis 336n82
Argentavis 422n138
Armadillosuchus 59
Armillaria 365n56
Aspergillus 183
Astyanax 128
Ateles 114
Atretochoana 45
Atta 181, 183–84; 408n184
Attercopus 87
Aurelia 359n122
Australodelphis 346n162

Azendohsaurus 416n166
Azolla 152; 311n8; 313n10

Babiana 202
Barbaturex 232
Barbourofelis 55
Bathylychnops 105
Baumannia 173
Beipiaosaurus 415n140
Benthalbella 110
Birgus 249
Bisonalveus 339n173
Bitis 340n230
Blarina 61
Blanzyopteris 159
Botanophila 154
Brachaluteres 325n81
Brachiosaurus 198
Bradypus 231
Branisella 454n139
Brenthia 90
Brycon 272
Buchnera 402n162
Buckholderia 183

Cacajao 445n77
Cactospiza 275
Caecilata 45
Caenorhabditis (C. elegans) 126
Caipora 281
Calappa 63
Calliarthron 388n7
Caluromys 69
Camponotus 250
Cancer 242
Carsonella 402n162
Cartelles 455n151
Casarea 64
Castorocauda 230
Cataglyphis 250; 368n149
Catasetum 328n169
Cataulacus 193

(n) Vision (*continued*)
schizochroal 103-104
slit eyes 99
spherical aberration 12, 98
stemmata 104; 363, n246
stereoscopic vision 105
strabismus 105
superpositional compound eye 113
tapetum lucidum 107, 110, 114
telephoto 106
tetrachromatic 114
transparency 106-107, 108-109
trichromacy 111, 114-115, 296
ultra-violet 111-112, 113, 115, 117, 296;
   353, n160
ultra-violet vision 111

Sex
aphrodisiac 175
asexuality 27, 182
ejaculation 212; 393, n89; 424, n46
love darts 213
love tokens 213
monogamy 166
orgasms 212
penis 212
penile stylet 425, n73
phalloid organ 212
pseudocopulation 85; 393, n89
safe sex 211
scrotum 206
seducin 175
sex chromosomes 209-210
sex pheromones 89
sexual attraction 451, n17
sexual selection 114, 212, 264, 266
sexually transmitted disease 154
sperm 124-125, 168, 209, 212, 213; 424, ns 45,
   46; 425, ns 52, 81; 426, n91
testes 209, 258; 356, n22
XX-YY sex chromosomes 209-210
ZZ-ZW sex determination 209

Sociality 29
fission-fusion system 142
matriarchs 268, 293
social carnivores 290
social play 19, 283-285
social structure 276-277
soldiers 171, 172; 409, n197
(a) Eusociality 29, 168, 170-171, 179-180
eusocial mammal 175
synalpheid eusociality 174

Sponges 96-97, 174, 238-239, 279
glass sponges 96, 238

Symbiosis
bryosymbionts 395, n136
primary endosymbiosis 313, n10
symbiotic algae 96, 101
symbiotic bacteria 173
tertiary endosymbiosis 357, n35

Tools 19, 274-283
anvils 276, 279, 282; 451, n8; 456,
   ns 184, 188
flinging objects 279
glove 279
hammering 280, 282
hooked tools 275
lance 275
lithic tools 280
manipulative ability 70
metatools 277-278
probes 274, 275-277; 452, n36; 454, n131;
   456, n191
rake 452, n37
sticks (and twigs) 274, 281
stone tools 281-282, 290
tents 87
throwing 273, 274, 282; 452 n34;
   455, n174
tool use 19-20, 268, 270, 273-283
weaving 90, 91

Vocalization (and language) 265-268
antiphony 264
babbling stage 265, 266
birdsong 264-268
dialects 265, 266
drumming 129, 275; 452, n37
duets 264
formants 265, 267
language 263-268
lingual articulation 265
music 294-295
phonic lips 141
purring 267
rumbles 131
sign language 265
signature whistles 263
sonations 264
songs 264, 294
synthetic songs 265
techno music 264
universal music 294-295
vowels 267

electrogeneration 145
electroreceptors 144-149
  Schnauzenorgan 147
  tuberous receptors 146
(e) Gustation 126-127; 372, n85
  umami 127
(f) Hearing (and acoustics)
  acoustic fovea 135, 138
  acoustic lens 141
  acoustic mirror 139
  auditory fovea 134
  cochlea 134, 135, 136, 137, 143
  deaf children 265
  ears 56, 118, 129, 133-137, 141, 142, 144,
   228
  impedance-matching hearing 133
  Johnston's organ 135-136; 377, n151
  Lombard effect 140; 378, n180; 380, ns
   227, 228
  pinnae 137
  semicircular canals 71, 133, 141
  stereo 123
  stridulation 264
  tympanic ears 125, 133, 135, 136; 376, ns
   100, 127; 398, n33
  tympanal muscles 448, n55
  ultrasound 136, 137; 377, ns 153, 159; 378,
   n171; 447, n23
(g) Infra-red (and thermal)
  bolometer 116
  infra-red organs 118
  thermosensation 94
  warmth 116
(h) Mechanosensory (mechanoreceptors) 118,
  131-132, 135-136, 149; 368, n166; 374, ns
  33, 36; 381, n250
  chordotonal organ 135
  Meissner corpuscles 131
  Pacinian corpuscles 131; 374, ns 48, 53;
   381, n250
  scolopedial cells 135, 136
(i) Olfaction 101, 166
  anosmia 122
  esthetascs 248
  olfactory organs 249
(j) Pressure
  hydrodynamic image 128
  lateral-line organ system 75, 128
(k) Seismic
  seismic communication 32, 131; 374, n53
(l) Tactile 130-133
  Eimer's organs 132
  Herbst corpuscles 130
  tactile fovea 131, 132
  tactile sensitivity 242
(m) Vibration 173, 175
  vibrational communication 129-130; 409,
   n197

(n) Vision
  adaptive optics 99
  appositional compound eye 113
  bifoveal vision 147
  binocular vision (steropsis) 105; 368, n163
  blue eyes 115
  blue vision 113
  camera eyes 12-13, 93, 98, 100, 106-107,
   288
  cerebral eyes 100
  chromatic aberration 364, n1
  chromatic adaptation 96
  ciliary photoreceptors 100
  ciliary retina 438, n3
  circular polarized light 112
  colour-blind 297
  colour vision 95-96, 98, 111; 358, n60; 360,
   n164
  compound eyes 101-104, 112, 200
  cornea 12, 102, 105
  cyclopean eye 103
  de-focussing 358, n95
  dichromatic 113-114
  dim light 112
  dinoflagellate eyes 95
  extra-ocular muscles 12
  extrinsic eye muscles 145
  eye 128, 132, 133, 136, 140, 237; 373, n2;
   380, n224; 384, ns 363, 378
  fibre optics 96-97, 118
  five-site rule 112
  four-eyed fish 219
  fovea 106, 131, 139, 147; 361, n169
  lens 12, 95, 97, 98, 99, 100, 101, 104, 106,
   110, 111, 113, 129; 357, n49; 359, ns 101,
   115; 362, ns 201, 210
  mirror optics 101-102
  monochromatic 101-102, 113, 115; 366,
   n69
  Müller cells 97
  ocellus 95; 359, n116
  ocular feathers 107
  optical fibres 96-97
  photoreceptors 359, n112
  phototransduction cascade 100
  pin-hole (camera) eye 101-102, 117; 362,
   n210
  pit eyes 99; 359, n115
  polarized light 108, 112; 357, n49; 360,
   n164; 361, n197
  pupils 12, 98, 101, 105, 113, 288; 335, n59
  red/green colour blindness 114-115
  retina 13, 93, 97, 98, 102, 106, 110, 111,
   241, 272
  retinitis pigmentosa 94
  reverse eye 96, 110
  rhabdomeric photoreceptors 100; 438, n3
  saccades 106, 132, 147

(b) Dinosaurs (*continued*)
stegosaurs 197
therizinosaurians 196; 414, n135; 415, ns 140, 142
theropods 52, 62, 72, 194, 260; 333, n122
troodontids 194, 195, 199; 414, n135; 415, ns 142, 146, 156
tyrannosaurids 195; 347, n186
(c) Ichthyosaurs 46, 47, 77-78, 142, 210, 219
(d) Mammal-like
anomodonts 59, 67, 226; 333, n122
cynodonts 229
dicynodonts 333, n122
galesaurids 433, n138
varanopid 196
sphenacodontid 229
synapsid 60, 67, 228-229; 333, n122
therapsids 226, 229
theriodonts 149
therocephalian 226
Tritylodontidae 59
(e) Pterosaurs 51, 193; 374, n31; 421, n135
azhdarchids 412, n94
ctenochasmatids 412, n9
(f) Squamates 62, 75, 216
(i) Lizards 64, 75, 133, 175, 192, 216-217, 275, 290
acrodontans 191, 232
amphisbaenians (worm-lizards) 75
anguid lizards 75
anolid lizards 33, 68, 84, 147, 270
Australian dragon-lizard 423, n14
basilisk lizard 344, n89
borioteiids 59
Burton's legless lizard 64, 75
butterfly lizard 412, n65
carphodactyline geckos 84
chameleons 45, 68, 105
flying dragon 192
geckos 68, 83-84, 175, 192; 366, n79; 367, n111
horned lizards 52, 84
iguanians 59; 337, n119; 427, n125
Komodo Dragon 62, 232, 284
lacertids 192
mastigure 290
Mexican beaded lizard 61
monitor lizards (varanids) 219, 232, 284; 336, n84
phyllodactylids 350, n47
scleroglossans 337, n119
shingleback 350, n58
skinks 84, 216-217; 342, n12
teiids 47
thorny devil lizard 52, 84
(ii) Mosasaurs 47, 78, 210, 219; 346, n172; 347, ns 183, 186, 188, 191
aigialosaur 429, n173

globidentids 47
(iii) Snakes 50, 64-65, 75-76, 116, 119, 210, 216
atractaspidids 62
black mamba 64
boids 116-117, 119, 216; 368, n158
bolyerines 64
colubrids 65, 216
coral snake 64
corn snake 426, n103
crotalids (rattlesnakes, etc) 116-117, 119; 448, ns 34, 55
dipsadines 63
elapids 62, 64
fer-de-lance 282
garter snake 37
kukri snake 340, n233
natricines 65
paradise tree snake 192
pythons 116
scolecophidians 65
sea-snakes 15, 64, 122
taipan 64
turtle-headed sea-snake 64
vipers 62, 216
(g) Turtles 26, 46, 78, 130, 212, 284; 333, n122; 370, n27; 371, n63
amphichelydians 26
cryptodire 26
emydid 284
leatherback turtles 109; 346, n172; 433, n121
matamata 46
Nile soft shelled turtles 284
pleurodires 26, 46
sea turtles 370, n27
tortoises 274

Sensory systems
jamming avoidance response (JAR) 138, 148
sensilla 116, 118, 119, 126
(a) Anemometer 67
(b) Balance
statocyst 133
statoliths 133
(c) Echolocation 111, 137-143, 267
buzz 139, 142; 382, n284; 448, n55
constant frequency (CF) 138
Doppler shift compensation 139
Frequency Modulation (FM) 138
phonic lips 141
ultrasound 377, n153; 447, n23
(d) Electroreception (and electrogeneration)
ampullae of Lorenzini 144
ampullary electroreceptors 146, 149; 385, n412
electric fields 144
electric fovea 147

aphrodisiacs 175
areolae 217
biparental care 49, 166
blastocyst 217
brood care 179, 186
bursa copulatrix 213
chorioallantois 217
chorion 214, 217
cooperative breeders 268
courtship 12, 25, 211, 213, 214, 232, 264, 274
dwarf males 12
eggs 196, 208, 214-215, 274
endometrium 220
endotheliochorial placenta 218
epitheliochorial placenta 217, 218
extra-embryonic membranes 214
extra-pair copulation 424, n40
foam nests 39-40
gonads 209
hemochorial placenta 218
hemipenes 425, n60
hermaphrodites 213; 425, n73
heterogametic sex 209
lekking 15, 169
love darts 213
mammary glands 175
maternal care 167
maternal-fetal conflict 428, n165
mating plug 426, n91
matrotrophy 215, 220
milk 124, 175, 219; 403, n185
miscarriages 167
monogamy 166
mouthbrooding 49
nests 87, 90, 196, 226
nuptial dance 15
nuptial gift 89, 90, 213
omphaloplacenta 217
ovaries 209
oviduct 215
oviparity 216-220
ovipositors 66, 180, 181
ovules 163
parental care 228, 232, 276
parental imprinting 211
penis 14, 16, 212
phalloid organ 212
placenta 216-220
placentotrophy 216
pollen tube 154, 163
precocial breeding 260
pseudocopulation 85; 393, n89
pubic nipples 33
safe sex 211
scrotum 206
seducin 175
self-fertilization 28
serosa 426, n93

sex chromosomes 209-210
sex determination 209
sex linkage 114
sexual dimorphism 354, n181
sexual mimicry 15, 109; 393, n89
sexual selection 114, 212, 264, 266
sperm 124-125, 168, 212, 213
testes 124, 209, 258; 356, n22
trophotaeniae 220
umbilical cord (appendiculae) 219
unisexuality 322, n48
viviparity 175, 216-220
XX-YY sex chromosomes 209-210
yolk 215
ZZ-ZW sex determination 209

Reptiles 191, 215, 284
archosaurians 62, 112, 197; 338, n128
choristoderans 219
diapsids 191; 411, n51
kuehneosaurids 191
nothosaurids 219
pachypleurosaurids 219
parareptiles 133
pareiasaurs 59
placodonts 47, 178
plesiosaurs 78, 219
pliosaurid 337, n114
protorosaurs 46
pseudosuchians 197
reptilomorphs 149
silesaurids 416, n196
sphenodont 62
(a) Crocodiles 47, 132, 196, 210, 212
    caiman 82, 284
    Chinese alligator 47
    notosuchians 59, 232
(b) Dinosaurs 62, 173, 196-199; 399, n73; 403, n185
    allosauroids 196
    alvarezsaurids 196
    ankylosaurs 59
    ceratopsids 197; 333, n122
    coelurosauravians 191
    deinonychosaurians 195
    dromaeosaurids 62, 194, 195
    hadrosaurs 57-58
    heterodontosaurid 414, n137
    maniraptorian 196
    megalosauroid 414, n137
    megaraptorans 196
    ornithischians 196
    ornithomimosaurs 197
    oviraptorids 195, 198
    oviraptorosaurians 195, 196; 414, n135
    psittacosaurid 196; 414, n137
    saurischians 197
    sauropods 196

(d) Flowers
   anthers 41; 387, n57; 389, n10
   carrion flowers 162-163
   chamber blossoms 389, n10
   floral gigantism 162
   nectar 127, 200-203; 417, ns 6, 20; 418, ns
      31, 34, 65
   nectaries 86
(e) Gymnosperms 157, 161
   conifers 119
   cycads 163; 310, n8
   pine cones 88
   pines (pinaceans) 369, n182
(f) Leaves 155, 158, 161; 389 n9; 390, n13;
      393, ns 65, 79; 394, n96
   megaphylls 389, n9
   microphylls 389, n9
   stomata 157
   venation 390, n13
(g) Monocotyledons (monocots) 161, 275;
      389, n10; 390, n13; 392, n48; 396, n168
   alismatids 389, n10
   alliacean 393, n89
   arums 163-164
   bamboo 235
   bromeliaceans 157, 167
   dead-horse arum 162, 163, 164
   grasses 154; 389, n10; 390, n13; 392, n48;
      393, n68
   marsh grass 179
   oats 307, n2
   philodendron 163, 164
   sea-grasses 178; 389, n10
   skunk cabbage 163
   Solomon's lily 396, n170
   titan arum 163
   voodoo lily 164
   wild oat 160
(h) Parasites 160-162
   beechdrop 161
   cytinaceans 162
   dodder 162
   haustoria 161, 162
   Maltese mushroom 162
   mistletoe 161
   Orobanchaceae 389, n140
   Rafflesiales 162
(i) Pollination 28
   anemophily (wind pollination) 389, n10
   buzz-pollination (vibratory sonication)
      322, n60; 389, n10
   cantharophily 163; 389, n10
   chiropterophilous syndrome 379, n206
   hummingbird pollination 417, n20
   hydrophily (water-pollination) 389, n10
   insect pollination 111
   mammal pollination 396, n176
   ornithophily 417, n20

pollination syndromes 139, 162, 169; 393,
   n89
pseudo-pollination 154
(j) Seeds 41, 161; 395, ns 123, 131
   dust seeds 395, n123
   elaiosome 390, n12
   endoderm (plants) 42
   endosperm 423, n34
   maple seed 419, n55
   myrmecochory 155
   pappus-like parachute 389, n10
   seed dispersal 74
(k) Succulents
   agaves 156; 379, n206
   cacti 156-157; 391, n33
   euphorbiacean 86, 156, 162
   portulacineans 156
   succulent tissue 156-157
(l) Vascular system
   xylem 155, 173; 389, n8

Platyhelminthes (Flatworms) 27, 36, 101, 234,
   248; 357, n49 (flatworms) and 248; 307, n2;
   310, n3 (platyhelminthes)
   acoels 248; 439 n59; 441, n119
   monogenean platyhelminthes 27
   polyclads 248
   tapeworms 216; 307, n2; 310, n3; 426, n111
   trematodes 400, n111; 401, n128; 407, n134
   triclads 248

Protistans (see also Algae) 4, 6, 7; 305, n1
   amoeba 305, n1; 307, n3; 313, n10
   apicomplexans 161
   cercozoans 313, n10
   choanoflagellates 6, 235, 238; 305, n1; 315,
      n11; 371, n80; 437, ns 53, 54
   coccolithophorids 315, n11
   diatoms 236; 313, n10
   heliozoans 437, n31
   slime moulds 237-238; 305, n1
   social amoeba 177
   trypanosome 117
   (a) Ciliates 7; 313, n10; 315, n13; 327, n149;
      403, n2
      chromidinids 315, n13
   (b) Dinoflagellates 4, 36, 37, 95, 237; 310, n3;
      313, n10; 327, n149
      polykrikids 357, n46
      warnowiids 95, 237

Reproduction and care (see also Embryology
   and Sex)
   afterbirth 217
   allantois 214
   altricial breeding 260
   amnion 214-215
   amniotic egg 214-215

gigantothermy 433, n121
heterothermic 226
metabolic budget 241
non-shivering thermogenesis 227
photosynthesis 155
poikilothermy 375, n85
reverse temperature effect 80
salt secretion 390, n11
thermogenesis 163-164, 168-169; 389, n10;
    396, n177
thermoregulation 163, 193, 226
vernalization 391, n15
warm bloodedness 79-80, 169

Plants 4, 202, 206, 210, 234, 235; 305, n1; 307,
    n2; 310, ns 3, 8; 417, n20; 418, n31; 420,
    n106; 421, n111; 423, n34; 436, ns 2, 6
autumnal colours 155; 391, n15
bipolar growth 389, n9
colour space 391, n14
cones 163
euphyllotes 389, n9
extrafloral nectaries 388, n58
ferns 388, n58
fruits 155; 390 n13; 391, n14
heterospory 391, n16
lactifers 159
leguminous nodules 310, n8
lianas 155, 159, 190; 336, n82
lignin 388, n7; 389, n8
lycophytes 389, n9
lycopsids 157
mangroves 98, 155, 271, 281; 390, n11; 398,
    n49
megaspores 391, n16
microspores 391, n16
mixotrophy 160
mycoheterotrophy 161
mycophycobioses 152
nitrogen-fixation 4, 181, 187; 307, n2; 310, n8;
    313, n10
nitrogen-fixing clade 310, n8
periderm 157
phytoliths 58
plant galls 172
pollen 41, 154, 163; 387, n57; 389, n10; 391,
    n16; 393, n89; 417, n20
pollen tube 154, 163
quillworts 157
resins 159
roots 156, 157, 161; 389 n9; 390, n11
spores 41
storage roots 156
Sudden Oak Death 151
thermogenic plants 163-164
trichomes 86, 90, 158-159; 354, n200; 355,
    n205; 394, n96; 397, n187
water-ferns 310, n8; 313, n10

wind dispersal 389, n10
wire plants 158
wood 155; 388, n7
xerophytes 156
(a) Angiosperms (flowering plants; see also
    flowers and monocots) 24, 155-164,
    210; 310, n8; 389, n10; 390, ns 11, 13; 397,
    n187
    amaranthaceans 392, n48
    Amazon waterlily 163
    aspen 158
    aviceniaceans 390, n11
    balsams 201
    bertram palm 203
    blueberry 154
    bunchberry dogwood 41
    campanulids 161, 201
    candlenuts 275-276
    carrots 156
    caryophyllids 161; 387, n57
    coconuts 20
    cucurbitacean 417, n20
    daisy 393, n89
    dicots 389, n10
    Ericales (ericaceans) 160, 161, 162
    gum trees 180
    iris 202; 393, n89
    lamiids 161
    magnolias 389, n10
    malpighiaceans (Malpighiales)162
    Malvales 162
    milk-weeds 394, n96
    oil-seed rape 162
    orchids 157, 160; 328, n169; 379, n206;
        393, n89
    palm fruits 281
    piassava 281
    privet 395, n118
    pyroloids 160
    penstemons 417, n20
    rat's tail 202
    rhizophoraceans 390, n11
    sacred lotus 163; 351, n92; 397, n206
    white mulberry 41
(b) Bryophytes 155, 161, 210; 310, n8; 328,
    n167; 388, n5; 389, n8; 395, n136
    ghostwort 161
    liverworts 161, 210
    metzgerialean liverworts 389, n8
    polytrichaceous mosses 389, n8
    splachnacean mosses 162
(c) Carnivorous 85-86, 151
    bladderwort 85
    epiascidiation 85
    pitcher-plants 85-86
    sarraceniaceans 85
    sundews 85
    Venus fly-trap 85

Organic compounds (*continued*)
dopamine (DOPA) 317, n21; 355, n233
avenacin 307, n2
benzoquinones 38; 327, ns 119, 121, 148
bilirubin 350, n58
biliverdin 350, n58
candicidine 407, n144
capsaicin 368, ns 148, 158
cardenolides 394, n96
carotenoid 421, n122
DDT 160
dieldrin 160
dimethyl sulphides 121
floral volatiles 379, n206
formic acid 326, n118
guanine 37, 110
hydrazine (N_2H_4) 311, n9
hydrocarbons 29, 174
hydrogen peroxide 15, 38; 307, n2
ink 15, 110
ladderanes 311, n9
latex 159; 394, ns 96, 100, 101
lipids 141, 164; 390, n12
malic acid 157
methyl (CH_3) molecule (methylation) 211
monosodium glutamate 127
mucopolysaccharides 91
mucus 209, 213, 214
opium 159
ouabain 39
phenols 118
phosphoglycerate (BPG) 227
phospholipids 236
phosphorylated inositols (IPs) 227
pyrazines 121
pyrethroids 160
quinine 127
ranaspumins 40
resins 159; 394, n101
rubber 159
scytonemin 364, n7
serotonin (5-hydroxytryptamine) 121;
    315, n13
steroids 159
sterols 179, 181; 311, n9
sugars 127, 200; 371, n76; 372, n114
vitamin E 327, n148
volatiles 162-163; 393, n89; 396, ns 170, 176

Parasites (see also parasitic plants, parasitic
    wasps) 7, 28, 95, 104, 135; 307, n2; 310, n3
brood-parasitism 28
hemiparasites 161
holoparasitic 161, 164

People
Aeschylus 274
Boran people 289

Jerome "Dizzy" Dean 89
Mortimer 9-10, 120, 127, 301-303
Polynesians 262
sea-gypsies (Moken people) 113
Shivaji 349, n38

Physics
angular acceleration 133
atmospheric CO_2 156, 158; 392, n48
ballistics 45, 48
ballistospores 41
ballistospory 386, ns 27, 28
birefringence 104
boundary layer 128
carbon dioxide 127, 166, 170
cavitation 44, 81, 143-144
Coanda effect 38
coherent light scattering 206
colours 295-297
constant frequency (CF)   138
Doppler shift compensation 139
elastic spring 144
enantiomers 122
frequency modulation (FM) 138
gravity 133
Helmholtz resonator 382, n302
hygoscopic 88
hypodermic syringe 62, 213
infra-red 116-118
infra-red thermography 163
magnetic 125
microwaves 116
Milky Way 169
osmosis 88
photon 125
Rayleigh scattering 206
refractive index 12, 108-109, 271
starlight 113
suction 44-47
turgor pressure 40, 41, 151
ultra-violet radiation 111-112
ultra-violet screening 15, 112
van der Waals forces 83, 86, 88
vibratory sonication 389, n10
vortex 203; 418, n20; 419, n55

Physiology 194
artificial sweeteners 127
blood 212
blood-clotting 92
blushing 207
circadian rhythms 97; 356, n22; 457, n7
cranial endothermy 80
C_4 photosynthesis 156, 157; 392, n48
Crassulacean Acid Metabolism (CAM)
    photosynthesis 157-158
discontinuous gas exchange (DGC) 442, n169
endothermy 163-164, 200, 226-227

GENERAL INDEX   481

(e) Toxins and venom 60-62, 66
    alkaloids 34-35, 86; 388, n70
    batrachotoxin 35
    caerulein 325, n58
    pumiliotoxins 327, n148
    saxitoxin 36
    tetrodotoxin 36

Molluscs 178
  chitons 65, 104; 317, n29
  scaphopods 316, n9
  (a) Bivalves
    ark-shells 101
    bivalve mollusks 91; 426, n91
    corbiculids 426, n111
    fan mussel 91
    giant clams 101
    oysters 280, 281
    Pectinidae 319, n77
    scallops 101; 319, n77
    shipworm (bivalve mollusk) 188
  (b) Cephalopods 7, 11, 93, 97, 105, 108, 109,
    112, 129, 133, 244, 245, 247; 315, n13
    cuttlefish 11, 14, 15, 18, 100, 108
    Pearly nautilus 101
    (i) Octopus 11, 72, 83, 101, 288, 294; 425,
      n59
      blue-ringed octopus 36
    (ii) Squid 11, 13, 101, 109-110, 292; 358,
      n95; 362, n223
      bobtail squid 110
      dumpling squid 18
      firefly squid 97; 364, n1
      oval squid 19
      vampire squid 110
  (c) Gastropods (snails) 15, 28, 106 (gastro-
    pods) and 63, 178-179 (snails)
    calyptraeids 28
    cerithioids 426, n111
    heteropods 106
    land snails 213
    limpets 65, 178
    littorinids 106
    pond snails 288
    predatory snails 26
    prosobranchs 106
    pulmonate snails 25, 106; 329, n15; 365,
      n56; 398, n41
    Samuri snail 213
    sea-hares 38
    slugs 213
    strombids 106
    viviparids 363, n236

Named animals
  Alex 292
  Betty 275, 278
  Calimero 19

Celeste 19
Clever Hans 269, 290, 292
Emily Dickinson 19
Fiona 269
Ginger 269
Griffin 292
Guillem 270
Lucretia McEvil 19
Mike 293
Pigface 284
Shelob 89

Nematodes 121, 150, 288; 407, n134; 457, n25
  steinernematids 382, n316

Nervous systems (see also Brains) 96, 98, 103,
  234-240
  acetylcholine 235
  action potentials 242
  axons 242
  bipolar neurons 123
  calcium action potentials 436, n23
  cortical neurons 242
  dorsal nodes 243
  fenestrae 243
  GABA (γ-amino butyric acid) 235
  ganglia 246
  giant axons 17, 242
  glial cells 242, 245
  globuli cells 248
  glomeruli 124
  glutamate reception 239
  intrinsic neurons 246
  ionotropic glutamate receptors 371, n70
  Kenyon cells 246
  myelin 17, 242-243
  nerve net 244
  neuropil 250
  neurons 98
  neurotransmission 234-235
  nodes of Ranvier 243
  oculomotor reflex 133
  pace-maker 99
  pain sensors 131
  pyramidal neurons 261
  synapses 235, 239, 242, 250
  synaptic scaffold proteins 239
  ventral nerve cord 246
  Von Economo neurons (VENs) 259, 294

Oomycetes (and water molds) 151; 307, n2

Organic compounds
  3-hydroxykynurenine 112
  acetic acid 326, n118
  aldehyde waxes 86
  ammonia 11, 16; 311, n9; 315, n13
  anthocyanins 120

Molecular biology (*continued*)
(d) Proteins
abductin 42
actins 311, n9
adenine nucleotide translocator (ANT)
227; 432, n104
arrestin 327, n149; 377, n144
calcium channel 235
collagen 206, 212; 315, ns 11, 13; 421,
n122; 425, n59
crystallin 12, 93, 94, 100, 110; 356, ns 8, 10,
13; 357, n 49; 359, n115; 366, n79
cytokines 218
decarboxylases 436, n6
DM domain factors 423, n37
Down Syndrome (DS) cell adhesion molecule *dscam* 307, n2
elastic proteins 16, 42
elastin 16, 42, 91, 131, 192
flagellin 307, n2
gelsolin 356, n8
gonadotrophins 428, n164
growth factor receptor 16
heat-sensitive ion channel protein (TRPA1)
116
immunoglobulins 307, n2
insulin-growth factor II 428, n170
interleukins 218
intermediate filament 311, n9
ion channels 235-236, 242
lipocalin fold 369, n174
lipocalins 61, 175
luciferin 109
minicollagen 327, n149
nitrophorins 369, n174
non-selective cation channel proteins 125
PAQR proteins 125
polycomb group 423, n34
potassium channel 235
prolactin 403, n185
proton channels 437, n25
proton pump 94
receptor kinases 305, n1
reflectins 110
resilin 42
seven-helix transmembrane proteins
(7-TM) 94, 124, 125, 127
SNARES (N-ethylmalemide-sensitive factor
attachment protein receptors) 305, n1
sodium channels 35, 37-38, 160
synaptic scaffold proteins 239
syncytins 218
toll-like receptors (TLRs) 307, n2
TRP (transient receptor potential) ion
channels 368, n148
tubulin 311, n9
ubiquitin 426, n90
uncoupling protein (ANT) 227

uncoupling protein (UCP1) 226-227
uncoupling proteins (UCPs) 164, 226
voltage-gated sodium channels 235-236,
243
(i) Enzymes
acetylcholinesterase 160, 235
AOX (alternative oxidase) 164
β-fructofuranosidase 394, n100
carbonic anhydrase 341, n258; 372,
n114; 398, n29
catalytic (serine) triads 151; 386, n17
cellulases 187
chitinases 351, n80
cutinases 151
esterase (acetylcholinesterase) 160
glucanases 188
kinases 305, n1; 307, n2; 311, n9; 315,
n11
luciferase 109
lysozyme 206
nitrogenases 310, n8
PAMs (peptidylglycine α-amidating
monooxygenases) 437, n53
PEPC (phosphoenolpyruvate
carboxylase) 392, n48
peroxidases 307, n2
phosphodiesterase (PDE) 360, n136
RuBisCO (ribulose-1, 5-bisphosphate
carboxylase/oxygenase) 162
serine proteases 61, 92, 151
synthase 327, n149
tyrosine kinases 315, n11
U-box ubiquitin ligases 438, n67
xylanases 187
(ii) Opsins 124; 327, n149
bacteriorhodopsin 94, 95, 126
c-opsins 100
encephalopsins 356, n22; 360, n134
halorhodopsin 94
melanopsins 94; 356, n22; 360,
n134
neuropsins 100, n22
panopsin 100; 356, n22
pteropsins 100; 356, n22
r-opsins 100
rhodopsin 94, 112
(iii) Respiratory proteins
hemerythrin 397, n204
hemocyanin 13, 166; 317, n29
hemoglobin 80
myoglobins 380, n229
(iv) Silk 40, 87-92; 328, n184; 426, n91;
439, n38;
cribellar silk 88
dragline silk 88, 129
marine silk 91
orb webs 88
sea silk 91

naked mole rats 134, 175
porcupines 158, 188, 231
rat 235, 270
ricefield rat 39
Siberian chipmunk 39
squirrels 39, 117, 256

Mathematics 292, 301-302; 460, n3; 461, n8
algebra 292
algorithms 148, 238
numbers 291
numerical distance 291
numerical magnitude 291
numerosity 290-292
zero 292

Medicine (see also Physiology)
antibiotic resistance 183-184
antibiotics 181, 183; 307, n2
Chagas disease 117
chlamydids 311, n9
cyanosis 317, n2
dental plaque 402, n153
fever 169
gangrene 158; 388, n70
herbal medicine 162
Irukandji Syndrome 97
kidney stones 157
narcolepsy 288
pathogens 183
pathogenic bacteria 121
pathogenic fungus 183
pregnancy clays 187
rabies 117
retinitis pigmentosa 94
sexually transmitted disease 154
suicide 294
summit disease 407, n134

Metazoans (also Arthropods, Chordates, Cnidarians, Echinoderms, etc)
Articulata 247
brachiopods 100
bryozoans 426, n111
chaetognaths (arrow worms) 36, 66, 102, 129, 244; 365, n51
ctenophores 239; 356, n2
deuterostomes 103; 310, n8; 359, n104
dicyemids (mesozoans) 7; 315, n13; 360, n141
diploblasts 239
ecdysozoans 126, 247
hemichordates 439, n59
Lophotrochozoa 247
phoronids 242
placozoans 239
rotifers 27, 150

Mimicry and crypsis , etc 29-30, 82, 89, 158,
162-163, 174, 265, 266; 402, n173; 407, n134;
409, ns, 197, 199; 422, n137
acoustic crypsis 142
acoustic mimicry 137
aposematism 34, 35; 346, n160; 388, n70; 393, n71
Batesian mimicry 325, n75; 377, n153; 409, n199; 422, n137
camouflage 14-15, 30, 102-103, 105, 107, 158
chemical mimicry 154, 174; 354, ns 175, 181
countershading 110
egg mimicry 409, n197
eye-spots 14, 95; 363, n254
floral mimicry 153-154
fruit mimic 407, n134
fungal mimics 153-154
molecular mimic 162
Müllerian mimicry 377, n153
olfactory mimicry 86
phagomimicry 15
sexual mimicry 15, 109; 393, n89

Minerals
aragonite 104
atacamite 341, n260
calcite 103-104; 362, n201; 382, n302
calcium oxalate 157
goethite 65
iron oxides 273
magnetite 65; 311, n9
optic axis 103-104

Molecular biology
draculin 73
endocrine 427, n138
endorphin 235
ergosterol 179
fermentation 206, 231
GABA (γ-amino butyric acid)160, 235
(a) Amino acids
alanine 87, 160
glycine 87
histidines 91, 317
isovaleric acid 141
leucine 236; 307, n2
mycosporine-like amino acids 15, 112; 364, n7
serine 160, 236
tryptophan 121
(b) Hormones 160, 235
allohormones 213-214
ecdysteroids 159
gonadotrophins 428, n164
lactogens 428, n164
seducin 175
(c) Pheromones 115, 122, 125, 170, 175; 393, n89
alarm pheromones 172

(d) Elephants 115, 131, 202, 258-259, 270, 279-280, 293-294; 336, n82
deinotheriids 454, n124
woolly mammoth 279
(e) Felids (cats) 57, 126, 141, 256, 267, 288; 444, n28; 449, ns 94, 95
African cheetah 57
American cheetah 57
barbourofelids 55
clouded leopards 56
Diard's clouded leopard 57
lynx 28
machairodonts 55
nimravids 55
sabretoothed cats 54-57, 230
(f) Marsupials 54, 64, 114, 131, 190, 204, 207, 218, 230, 231-232, 255-256, 257; 417, n212
antechinus 432, n97
bandicoot 428, n158
?bat 231
borhyaenids 54, 231
brown antechinus 417, n212
dasyuroids 432, n97
diprontodonts 343, n41
fat-tailed dunnart 432, n97
feathertail gliders 83, 190
kangaroos 231, 283
koala 267
marsupial "lion" 54
marsupial mole 232
marsupial "wolf" 54
marsupium 231
phalangeroids 69
pygmy rock wallaby 337, n95
Tasmanian devil 54
Tasmanian wombat 257
thylacines 54, 231
thylacosmilids 54, 56
(i) Opossums 49, 54, 69
didelphids 54, 69, 131
greater glider 190
honey possum 69, 114, 203; 419, n65; 434, n154
mahogany glider 190
mouse opossum 206-207
petaurids 190
possum 256; 432, n97
striped possum 257
sugar gliders 68, 69, 190
woolly opossum 69
yellow bellied glider 190; 323, n20
(g) Monotremes 60, 114, 132, 256
duck-billed platypus 60-61, 114, 132, 148-149, 283; 423, n11
echidna (spiny anteater) 51, 132, 256; 338, n152
long-beaked echidna 149

short-beaked echidna 149
(h) Primates 114, 190, 207, 213, 255, 257, 260, 267, 291
aye-aye 115
black lemur 115
Coquerel's sifaka 367, n118
coucang (slow loris) 338, n149
lemurs 32, 68, 114
red ruffed lemur 367, n118
sifakas 344, n88
slender loris 68
tarsier 205
(i) Apes and hominids 292 (apes) and 259 (hominids)
anthropoids 258
australopithecines 267
chimpanzees 277, 281, 293
great apes 127, 259
Neanderthals 293
orangutans 72, 269
(ii) Monkeys 127, 256, 260
atelines 280
bonnet macaque 280
callitrichids 260
capuchin monkeys 32, 39, 70, 257, 281-283; 342, n20
catarrhines 257
colobine monkeys 206
howler monkeys 114-115, 182, 281, 295
lion-tailed macaques 454, n131
long-tailed macaques 280
mandrill 206, 207
marmosets 266
New World monkeys (platyrrhines) 114, 127, 213, 257, 264, 267, 281
Old World monkeys 281
owl monkey 115; 327, n122
rhesus monkeys 291
spider monkeys 114, 131, 281
squirrel monkey 450, n183
vervet monkey 206
woolly monkeys 114
xenotrichinids 281
(i) Squirrels and other rodents 39, 115, 134, 190, 213, 230, 256; 428, n164
African crested rat 39; 417, n212
anomalurids 190
beavers 188, 230
dormouse 190
eomyids 190
flying squirrels 190, 191
ground squirrels 268
hamster 175
mice 342, n39
microtine voles 175
mole rats 56, 134; 337, n95
mole vole 210

microsyopids 257
moles 231
Mongolian gazelles 449, n97
multituberculates 58, 218
musk shrews 417, n212
nesophontids 62
notoungulates 32, 58; 434, n146
otter shrew 231
palaeanodontans 231
pangolins 51, 52, 131, 231
pantolestid 339, n173
paromomyids 190
pigmy hippo 445, n102
pigs 226-227, 231
placentals 54, 204, 230, 255
plesiadapiforms 68, 257
saiga 454, n119
short-tailed shrew 61; 377, n158
shrews 61, 86, 137, 231; 375, n89; 377, n158
sloths 206, 231, 281
star-nosed mole 131-132; 370, n32
tapirs 231, 279
tenrecs 39, 137, 138, 226, 231; 445, n92
tree shrews 85, 203, 256
triconodonts 229
ungulates 57
water shrew 370, n32
xenarthrans 51, 58, 230, 231, 283
zebra 445, n102
(a) Bats 33, 73, 83, 85, 111, 137-139, 141, 191,
    193, 203, 213, 254, 255, 258, 266, 267; 417,
    ns 20, 212; 448, n55
    fruit bats (flying foxes) 137-138
    glossophagine nectar (or flower) bats 333,
       n131; 366, n69
    greater mustached bat 138-139
    horseshoe bats (rhinolophids) 137,
       138-139
    megachiropterans 203
    molossians 73
    murinines 255
    myotisids 33, 73, 138
    mystacinids 33, 73-74
    naked bulldog bat 73
    nectar-feeding (nectarivorous) bats 52, 139
    noctule bats 138
    phyllostomids 33, 203
    pipistrellids 138
    short-tailed bat 74
    sucker-footed bats 83
    tomb bat 378, n175
    vampire bats 73, 117
    vespertilionids 83
(b) Carnivores (except felids) 29, 62, 212, 232,
    256
    African linsangs 336, n82
    Asian linsangs 336, n82
    badgers 268

bearded seals 46
binturong (bearcats) 336, n82
bears 56; 449, n95
brown bear 454, n110
canids 54-55, 258
dogs 293
euplerids 336, n82
Giant otters 268
giant panda 449, n95
harbor seals 122, 266
honey badger (ratel) 289
hyenas 54, 213
kinkajou 115; 336, n82
lion 267, 279
mustelids (weasels) 435, n168
otter 231
palm civet 336, n82
pandas 55
racoon 270
sea lions 370, n27
sea otters 122, 279
seals 113, 129, 230, 266; 380, n225
social carnivores 290
viverrids 336, n82
walrus 346, n162; 445, n102
(c) Cetaceans 254, 259
    aetiocetids 50
    archaeocetes 259
    balaenopterans 50, 51
    beaked whales 142; 346, n162
    belugas 293
    Blainville's beaked whales 381, n281
    blue whale 141
    bottlenose dolphins 259, 293; 458, n47
    bowhead whales 266
    cetartiodactyls 140
    delphinids 142
    dolphins 81, 141, 142, 149, 259-260, 263,
       266, 279, 290, 292; 433, n121; 459, n114
    dwarf sperm whale 142; 318, n56
    false killer whales 458, n47
    finless porpoise 458, n47
    Guiana dolphin 149
    hourglass dolphins 381, n280
    humpback dolphins 458, n47
    humpback whales 50, 266, 294
    killer whales 142, 259
    mysticetes 50, 122
    pakicetids 141
    porpoises 142
    Risso's dolphins 293
    river dolphins 142
    roqual whale 50
    rough-toothed dolphin 293
    sperm whales 141, 142
    squalodonts 370, n31
    toothed whales (odontocetes) 122, 140,
       259; 370, n31

Geological intervals (*continued*)
    Paleocene 208
    Permian 67, 133, 149, 196, 215, 228;
      390, n11
    Permo-Triassic 60; 333, n122; 392, n64
    Pleistocene 56, 71, 232; 421, n135
    Pliocene 54, 156; 346, n162
    Silurian 223, 225
    Triassic 46, 47, 62, 178, 189, 191, 197, 198,
      219, 220, 229; 389, n10

Immune systems (and other defences) 169,
   220, 235
  adaptive immune system 307, n2
  antifungal compounds 179
  antigens 307, n2
  antimicrobial defences 34-35, 180, 184, 187,
    206; 307, n2
  B-cells 307, n2
  bacteriocins 307, n2
  bacteriocytes 188
  defensinlike peptides 61
  dentigerumycin 407, n144
  fungicides 160
  innate immune system 151; 307, n2
  insecticide 160
  leucine-rich repeats (LRR) 307, n2
  lymphocytes 307, n2
  organochlorine cyclodiene insecticides 160
  organophosphates (OPs) 160; 395, n114
  T-cells 307, n2
  thorns 158
  toll-like receptors (TLRs) 307, n2
  trichomes 86, 90, 158-159; 354, n200; 355,
    n205; 394, n96; 397, n187

Lichens 152-153, 155, 235; 310, n8; 313, n10
  ascolocular lichens 387, n50
  peltigeralean lichens 153
  pyrenocarpous lichens 387, n50

Locomotion
  anguilliforms 332, n96
  aquaplaning 86
  ballooning 87
  bipedality 68-69, 71-72; 343 ns 60, 72; 344, ns
    88, 89
  C-type rapid start 271-272
  deep-diving 380, n229
  foot-propelled divers 204
  glider (gliding)   189-193, 194-195, 230; 319,
    n77; 336, n82
  jet-propulsion 97, 189
  knuckle-walking 72
  leapfrogging 285; 457, n226
  parasagittal gait 229
  punting 73
  rolling toads 399, n72

S-type rapid start 271
tandem-running 272
thunniform locomotion 77
(a) Flying
  aerofoil 191, 193
  delta-wing 191
  flightlessness 284; 345, n125
  flying fish (exocoetids) 105, 189
  hovering flight 201, 203
  soaring 422, ns 136, 138, 139
  stalling 191
  winglet 191

Mammals 3, 5, 67, 68-72, 74, 80, 122, 125, 126,
  127, 129, 130, 133, 134, 135, 137, 140, 148,
  158, 202, 203, 206, 208, 210, 211, 212, 215,
  217, 218, 219, 220, 221-233, 255-260, 279-
  283, 287; 307, n2; 343, n41; 345, n119; 347,
  n202; 372, n85; 375, ns 84, 89; 376, ns111,
  118; 377, n151; 380, ns 235, 249; 381, n268;
  383, n360; 396, n176; 419, n65; 420, n86; 423,
  ns 11, 34; 427, ns 136, 138; 428, n156; 432,
  n105; 433, n122; 434, ns 154, 157; 443, ns 7,
  11, 13; 445, n92; 446, ns 117, 151; 454, n119;
  457, ns 7, 15, 23; 458, n40
  aardvarks 230
  Afrotheria 230, 231; 445, n92
  ant-eaters (see also myrmecophagy) 51-52, 72,
    196, 230-231; 333, n122
  antelope 454, n119
  apatemyids 32
  armadillos 230, 231
  artiodactyls 256
  borophagine 54
  bovids 280; 432, n81; 456, n202
  camel 342, n1
  chalicothere 435, n170
  cow 206
  deer 267; 417, n212
  desmans 129, 230
  docodonts 230
  dugongs 48, 177-178
  elephant shrews 396, n176
  euarchontoglires 324, n21
  flying lemurs (dermopterans) 190
  gazelle 267
  goitred gazelles 449, n97
  golden moles (chrysochlorids) 231
  gondwanatherians 58
  Haemothermia 227
  hedgehogs 39, 60, 158, 231; 375, n77
  hippopotamus (hippo) 230
  horses 218, 231, 254, 269; 428, n164
  hyraxes 231
  Laurasiathera 230
  litoptern 335, n66
  manatees 122; 445, n55; 445, ns 92, 102;
    446, n118

parental imprinting 211
plastome 161
Wnt/β-catenin pathway 199

Geography
  Argentina 51, 232
  Australia 32, 35, 46, 52, 54, 111, 180, 225, 232, 279; 324, n44
  Berlin 194
  Borneo 45, 85, 86, 205
  Brazil 45, 59, 139, 200, 281; 421, n135
  Burma 113
  Caatinga 282
  Caribbean 281
  Cebu 166
  Cerrado 282
  Chad 187
  China 46, 59, 191, 194, 195, 222, 223
  Christmas Island 166
  Clipperton Island 166
  Colorado 230
  Costa Rican 272
  Cuba 281
  Cyprus 400, n93
  Dorking 427, n112
  Drem airfield 302
  England 207
  Florida 56, 61
  France 159
  Galápagos 105; 353, n153
  Germany 194
  Greece 274
  Haiti 62
  Hawaii 31, 88, 90, 130, 166, 201
  Hungary 60
  India 113, 202
  Inner Mongolia 195
  Jamaica 167
  Kenya 289
  Kerguelen 444, n28
  Kyrgystan 191
  Lake Malawi 49, 148
  Lake Tanganyika 49; 426, n111
  Liaoning 194
  Liberia 167 7
  Madagascar (Malagasy) 28, 31, 32, 34, 59, 83, 158, 175, 194, 211, 231; 323, ns3, 4; 336, n82; 422, n137
  Majorca 253
  Malaysia 83, 202
  Mauritius 84
  Mexico 49
  Minorca 390, n11
  Mongolia 196, 208
  Myanmar 232
  Namibia 397, n12; 399, n72
  New Caledonia 84, 161, 275-278
  New Guinea 35, 149; 324, n44

  New Zealand 33, 74, 84, 130, 158, 208, 262, 264, 283; 393, ns 65, 68
  Nicaragua 265
  North Sea 335, n69
  Oxford 275
  Papua New Guinea 84
  Patagonia 186
  Pennsylvania 224
  Peru 66, 268
  Phillipines 166
  Poland 191, 225
  Puerto Rica 215
  Roma Aquincum (Budapest) 92
  Round Island 64
  San Francesco del Deserto 302
  Sargasso Sea 189
  Shark Bay 279
  Sicily 274
  Solomon Islands 84
  South Africa 202
  South America 32
  Sri Lanka 265
  St. Paul's Rocks 207
  Sunda Island 57
  Tanzania 167, 187, 233
  Thailand 113
  Tobago 166
  Trinidad 166
  United States 62
  Venice 120, 302
  Virginia 191

Geological intervals
  Jehol Group 194
  Maevarano Formation 194
  Mazon Creek 287
  Morrison Formation 229
  Posidonia Shale 219
  Rhynie Chert 153
  Solnhofen Limestone 193
  (a) Periods
    Cambrian 112
    Carboniferous 73, 159, 260, 287
    Cretaceous 46, 47, 51, 58, 63, 83, 88, 145, 150, 171,186, 191, 194, 196, 197, 216, 218, 219, 232; 333, n122; 389, n10; 390, n11; 395, n128; 420, n106; 421, n135; 422, n146
    Devonian 50, 87, 153, 219, 222, 224
    Eocene 32, 58, 137, 140, 207-208, 232, 260; 349, n35; 417, n19
    Jurassic 90, 197, 219, 229; 447, n23
    Miocene 32, 64, 70, 71, 74, 156, 182, 190, 201, 208, 232, 258, 281; 345, n119; 370, n31; 455, n151
    Oligocene 62, 140, 178, 190, 201, 207, 231, 280; 337, n114
    Ordovician 152

Fish (*continued*)
  (e) Sharks (and relatives) 47, 110, 144-145,
      219, 223, 253; 426, n98; 428, n167
      alopiids 253
      bamboo shark 44
      basking shark 50
      bonnet head 47
      cat sharks 44
      chimaeroids 385, ns 412, 422
      common thresher shark 79
      crocodile shark 443, n16
      dalatiids 365, n44
      epaulette shark 72-73
      etmopterids 365, n44
      great white sharks 285
      hammerhead shark (sphyrinids) 47, 253
      horn shark 44, 47
      lamnids 79
      megamouth shark 50
      nurse shark 44, 47
      orectolobiformids 44
      salmon shark 79
      scyliorhinids 44
      spadenose shark 219
      swell shark 325, n81
      whale shark 50; 443, n16

Fires 117-118
  serotiny 369, n182
  wild-fires 369, n182

Food
  bird's nest soup 139
  bread 43
  gluten 43
  Moa Twizzlers 262
  Roquefort cheese 116
  sausages 29, 122
  umami 127

Fossils
  amber 83, 84, 88, 150, 159, 171; 323,
      n12; 349, n35; 414, n135
  graphoglyptid 179
  La Brea tar-pits 208
  Mazon Creek biota 287
  trace fossils 179, 186, 225
  trackways 225

Fungi 40-41, 85, 94, 95, 109, 111, 150-154, 159,
    179, 180, 181-182; 355, n205; 365, n56; 369,
    n200; 385, n4; 386, ns 13, 21, 24, 27; 387,
    n36; 388, ns 7, 70; 395, n136
  arbuscular mycorrhizae 152; 310, n8
  asci 41
  ascomycetes 41, 150, 151, 152, 153, 154,
      179, 180, 181; 386, ns 13, 24; 387, n30;
      395, n136

basidiomycetes 40, 150, 151, 152, 153,
    181, 182, 210; 305, n1; 365, n56;
    385, n4; 386, n 27
black yeasts 184
coral-mushroom (pterulaceans) 182
cyphelloids 386, n28
ectomycorrhizal associations 152
entomopathogenic fungi 183
fruiting body 150, 183, 187; 386, ns 27, 28
fungal spores 40-41
gasteromycetes 41; 386, ns 27, 28
glomeromycotans 152; 310, n8
hyphomycetes 150
lepiotaceans 182
milk-caps 159
morels 177
mushrooms 150, 186, 187
mycorrhizal fungi 160; 386, n24
nematophagous fungi 150
ophiostomatoids 179
pathogenic fungus 183-184
psathyrellids 386, n28
resupinates 386, n27
rusts 153-154
secotioid syndrome 386, n28
stinkhorn 162-163
truffles 387, n30
yeasts 162, 203, 234; 305, n1; 396, n177
zygomycetes 41

Genes
  amnesiac 9
  ASPM (abnormal spindle-like microcephaly)
      445, n108
  atonal 136
  barfly 9
  cheapdate 9
  hangover 316, n3
  lethargus 457, n25
  mt *coxI* gene 395, n134
  Notch 440, n67
  Pax 97
  Pax6 95, 99-100; 360, n141
  PaxA 99-100
  PaxB 99-100
  PaxC 99
  Pjvk 382, n295
  prestin 137, 143
  promotor region 356, n13
  pseudogenes 113, 115
  reelin 261
  Six 1/2 97
  Tmc1 382, n295
  TRPV1 117

Genome
  genome size 173-174, 193
  haplodiploidy 179

scavenging 208
suction feeding 224
tongue-bite apparatus 48
xylophagy 187-188
(a) Predation
  ambush predators 53, 55, 97, 145,
    223-224; 335, n59; 384, n378
  co-operative hunting 289-290; 450,
    n120
  group hunting 148
  hypercarnivores 55, 258; 338, n128

Fish 5, 33, 104-105, 108, 109, 110, 111,
  122, 177, 206, 215, 252, 271-272, 273,
  285, 290; 311, n9; 368, n154
acanthodians 223
actinopterygians 252
amphibious fish 105
anabantids 222, 272
angler fish 110
antennariids 44, 72
archerfish 271-272
arthrodires 219
billfish 80
blenny 271; 339, n184
blue-striped fangblenny 318, n50
butterfly fish 221; 410, n9
callorhinchids 385, n412
carangid 375, n99
carp 355, n234
catfish 45, 146; 332, n96
cave fish 128, 271
centrarchids 49
clariids 221
cleaner-fish 318, n50
comb grouper 49
coral trout 289
cyprinids 57
cyprinodontiforms 219
damselfish 177
diphycercal tails 224
flatfish 15, 80; 423, n14
four-eyed fish 219
fourhorn poacher 344, n91
fugu fish 36, 37
galaxids 332, n96
goby 36, 271; 325, n81
goldfish 252
gouramis 222, 272
gymnotiforms 38, 145-146
hagfish 307, n2
ichthyodectiforms 81
knife-fish (xenomystids) 146
labrids 48; 443, n16
labrisomids 105, 221
lamprey 307, n2; 447, n153
lampridiform 80
lepidotrichia 225

marlins 80
mormyriforms (elephant-nose fish) 38,
  145-146, 221, 254, 285; 438, n1
mosquito-fish 290
ocean sunfish 109
opah 80
osteoglossomorphs 48
pachycormids (pachycormiforms) 50, 81
pachyrhizodontids 81
paddlefish 50, 145, 147, 148
pigfish 72
pipefish 362, n235
placoderms 50, 219
poeciliids 219
pomacentrids 177
porcupine-fish 158
puffer-fish 35-36, 37; 325, ns 78, 81, 84
Queen trigger-fish 325, n78
ratfish 47
salmon 48
sandlance 105-106
sawfishes 383, n332
scombroids 80
scopelarchids 110
sea robins 72
snail fishes 444, n27
sparids 47
stargazers 145
sturgeon 145
sunfish 35
swim-bladder 11, 221
swordfish 80
teleosts 44, 110
tetraodontiforms 36
thoracopterid 189
toad-fish 448, n55
tuna 78-80
tusk fish 451, n8
wrasse 451, n8
yellowhead jawfish 320, n115
zenarchopterid half-beaks 429, n204
(a) Cichlids 48-49, 293
  heroines 47
(b) Eels
  electric eel 144, 221
  moray eels 49, 289
  synbranchids 222
(c) Rays 47, 48, 144, 253; 317, n28
  cownose rays 47-48
  electric rays 48
  manta rays 48
  mobulid rays 80
  skates 73, 145
(d) Sarcopterygians 73, 220, 222
  coelacanths 191, 220, 222, 225
  lungfish 144, 221, 222, 225; 344, n103
  rhizodontids 224
  tristichopterids 224

Disease
  anthrax 315, n11
  avascular necrosis 78
  bends 380, n229
  Caisson disease 78

Dung, etc 19, 41, 169; 399, n73 (dung) and
  30, 85, 90, 162, 169, 179, 181, 187; 351, n83
  (feces)
  bird-dropping 354, n181
  coprolites 58
  latrines 85

Echinoderms 103-104; 426, n111; 439, n59
  brittle-stars (ophuiroids)   103
  echinoids 103; 307, n2; 349, n30; 382, n302
  holothurians 426, n111
  sand-dollars 331, n67
  starfish 361, n200
  tube feet 103

Ecology
  arboreality 131, 169
  caves 139
  chemotaxis 120, 124-125, 236
  deserts 116, 166
  zooplankton 147

Elements
  calcium 103, 157, 167, 227
  copper 66, 91, 206
  iron 65
  manganese 341, ns 267, 272
  nitrogen 85; 310, n8
  oxygen 155
  silica 96
  zinc 66, 182

Embryology
  direct development 25, 215
  embryos 214-220; 305, n1
  gastrulation 305, n1
  larvae 111, 123, 179, 180, 185
  meroblastic 215

Evolution
  anachronistic adaptations 158
  Aquatic ape hypothesis 71
  attractors 24, 33
  Cambrian explosion 307, n2
  cladistics 157
  clones 172
  concerted convergence 390, n13; 394, n96
  co-operation 289-290
  correlated progression 228-229
  Darwinian trench 191
  deep homology 95, 97, 99-100, 123, 218, 244,
    261, 288; 368, n148; 392, n48; 450, n102

Dollo's "law" 28, 216; 345, n139
ecomorphs 33, 49, 68, 76, 84, 88, 230
evolutionary inherency 156
evolutionary irreversibility 25
Fortean Bladders 45
gigantism 416, n194; 422, n146; 435, n188
inherency 288
innovation 276
law of homologous series 23
logic 126, 239; 307, n2
mammalness 208, 232
miniaturization 195, 242
molecular inherency 235-236
mosaic evolution 223
multilevel selection theory (MLS) 305, n1
Non-Darwinian Mechanisms 112
optimal solutions (optimizations) 41, 46, 57,
  63, 82, 83, 88, 124, 125, 132, 140, 201, 235,
  241; 310, n8; 335, n62; 339, n188; 375, n89;
  437, ns 25, 47
ornithization 195
search engine 295
self-organization 93
sensitivities 125
sexual selection 114, 212, 264, 266, 267
terrestrialization 25, 46
universal scaling laws 271

Feeding
  asymmetric feeding 63
  bait-fishing 274, 281
  bat nectarivory 379, n206
  breakfast 229
  carnivory 26, 150
  carrion 275
  coprophagy 435, n175
  durophagy (molluscivory) 35-36, 47, 49
  fishing 274
  flowerpiercers 418, n31
  folivorous 206, 231, 281
  geophagy 187
  grazing 45
  hematophagy 61, 73
  herbivory 29, 177-178, 197
  honeydew 174
  insectivores 255
  matrotrophy 215, 220
  mixotrophy 160
  mycoheterotrophy 161
  myrmecophagy 51-52, 65
  nectarivory (nectar feeding) 200-202, 255
  oophagy 65
  phytophagous 250
  phytophagous insects 28, 29
  planktivory 48, 50
  ruminant 206
  saliva 87, 173; 401, n139; 407, n137
  saprotrophy 150

GENERAL INDEX 471

Cnidarians 40, 93, 239; 327, n149
  Bigmessidae 25
  corals 25, 33
  hydromedusans 109
  hydrozoans 99
  nematocysts 40
  siphonophores 110; 438, n77
  (a) Cubozoans 93, 97-99, 287
    carybdeids 97
    rhopalia 98-99; 369, n205

Cognition
  abstractions 263-264, 290, 295-298
  analogical reasoning 277
  associative learning 270
  autism 294
  cognitive map 32
  cognitive capacities 243
  cognitive landscapes 268
  consciousness 241, 261, 271, 299-300
  death 293
  dreams 287, 288
  duffers 269
  emotional intelligence 252
  extra-terrestrial intelligence 22
  grieving 459, n109
  idiot-savant syndrome 294
  intelligence 17, 238, 286
  learning 18, 121, 272, 276
  memory 18, 237-238, 250; 459, n114
  mental image 277
  mind 12, 20, 278-279, 286, 294-300
  mirror self-recognition 19, 292-293
  myrmecoid pedagogy 272
  plant neurobiology 286
  Platonic reality 295
  psychophysics 290-292, 296-297
  qualia 94, 115, 120, 121, 122, 126, 127,
    292, 295; 368, n148
  rapid eye movement (REM) 287
  recognize faces 289
  REM sleep 287-288
  saliency 272
  sleep 287-289
  sleep deprivation 288
  slow wave sleep 287
  string-pulling task 269, 277, 278
  suicide 294
  synesthia 126
  transfer experiments 278
  trap-tube test 270
  unihemispheric sleep 259
  Weber-Fechner law 291

Colours
  red 86, 114
  red/green colour blindness 115

Culture, etc
  Aesop's fable 278
  Assumption of the Virgin 120
  bamboo traps 166
  bassoon 214
  Battle of Britain 271
  Beagle 36, 207; 388, n7
  bicycle 245
  blowpipes 35
  Boran people 289
  cravats 91
  cultural transmission 279
  dancing 381, n269
  domestication 254
  Eiffel Tower 125, 135
  Frari 120
  Gilbert and George 121
  Golden Fleece 92
  Golden House (Domus Aurea) 95
  Haber process 310, n8
  Hamlet 286
  hats 83, 91
  H.M.S. *Astræa* 302
  inebractometer 9
  International Festival of Worm
    Charming 129
  Magister monoplanes 302
  Mr Weston's Good Wine 299
  Monet 269
  Namagiri 302
  NASA 191
  ochre 114, 273-274
  Origin 93
  paint and paintings 273-274, 294
  Paleolithic art 56
  Parsifal 115
  Picasso 269
  Pied Piper of Hamelin 270
  Pillenwerfer 143
  Platonic reality 295
  San Francesco del Deserto 302
  Spitfire 272
  symbolism 273-274
  technology 276
  The hitchhiker's guide to the galaxy 10
  tiepins 213
  Tiepolo 274
  Titian 120
  trebuchet 41
  U-boats 143
  worm grunting 129

Developmental biology (Evo-Devo)
  embryonic hourglass 305, n1
  Fork-head box gene (*FoxP2*) 267
  GEM 305, n1
  geminin 305, n1

Brains (*continued*)
    auditory cortex 139
    barrel field 256
    blood-brain barrier 17, 245
    brain size 32, 69, 253
    cerebellar foliation 260; 443, n16
    cerebellum 205, 247, 252, 253, 255, 258,
        285; 452, n40
    cerebellum-like structures 443, n11
    cerebral cortex 247
    cerebrotypes 253, 254-255, 257, 276
    cochlear nucleus 134
    corpus callosum 255, 259
    cortex 246, 255, 256
    deuterocerebrum 248, 249
    dorsal ventricular ridge 261
    electrosensory lateral-line lobe 146
    encephalization 252-262
    encephalization quotient (EQ) 199; 445,
        n109
    endocasts 198, 257, 260; 334, n37;
        335, n63
    forebrains (telencephalon) 198, 252,
        253, 261
    gyrification 256, 258
    gyrus 251, 259
    hemiellipsoidal bodies 248-249
    hindbrain 148
    hippocampus 247, 250, 252
    hyperpallium 261
    intraparietal sulcus 291
    King Cortex 246, 283
    lateral pallium 252
    lissencephaly 445, n92
    median pallium 252
    mesopallium 276
    midbrain 146, 148
    mushroom bodies 246-251
    neocortex 259, 261
    neuropil 250
    nidopallium 261
    olfactory lobe 248
    optic lobe 13
    parasol cells 248
    principal sensory trigeminal nucleus
        (PrV) 130
    protocerebrum 246, 248
    somatosensory cortex 133
    subventricular zone (SVZ) 255, 261-262
    sulcus 251
    torus circularis 146
    tripartite brain 244
    tritocerebrum 244
    ventricular zone 261
    vertical lobes 18, 247
    visual cortex 149, 256, 272
    vocal nuclei 266
    Wulst 261

Cells
    acidocalcisomes 311, n9
    adipocytes 227
    anammoxosomes 311, n9
    apicoplast 395, n144
    axopodia 437, n31
    bioluminescence 96, 97, 109-110
    carboxysomes 311, n9
    cell adhesion 305, n1
    cell-cell communication 305, n1
    chloroplasts 313, n10
    choanocytes 437, n61
    chromatophores 14; 318, n40
    cilia 123, 128
    cryptochromes 97
    division of labour 276-277
    endocytosis 311, n9
    endothelial cells 16, 245
    epithelial cells 239
    flask cells 239
    futile cycling 80, 164, 227
    gas vesicles 311, n9
    Herbst corpuscles 130
    heterocysts 313, n10
    holoblastic cleavage 215
    magnetosomes 311, n9
    Mauthner cells 272
    Meissner corpuscles 131
    melanophores 356, n22
    melanosomes 346, n172
    melatonin 420, n91
    mitochondria 80, 95, 99, 116, 119, 164;
        311, n9; 357, n49; 395, n134
    Müller cells 97
    multicellularity 305, n1; 311, n9
    myeloid bodies 360, n143
    nematocysts 327, n149
    nuclear membrane 311, n9
    organelle 402, n165
    Pacinian corpuscles 131
    parasol cells 248
    photophores 110
    phytoliths 58
    plastids 161
    programmed cell death (PCD) 305, n1
    tight junctions 245
    volutin granules 311, n9

Chordates (also Birds, Mammals,
        Reptiles, etc)
    conodonts 336, n89; 339, n172
    doliolids 109
    salps 426, n111
    sea-squirts 426, n111
    tetrapods 222-226
    tunicate 39, 109
    urochordates 359, n104
    vertebrates 93, 133, 136

oilbirds 113, 207–8
orioles 323n1; 418n23
ornithurine birds 414n137
palaeognathans 130, 208, 262
passerines 204, 264
pelican 339n188
penguins 122, 204
pigeons 147, 269; 447n12
plain-tailed wren 264
plotopterids 204
rails 208
ratites 208, 210
regulids 378n186
rhea 208
roseate spoonbill 204
sandgrouse 205
sandpipers 130
scolopacids 130
sea horses 220; 342n12
seedsnipe 205
shorebirds 130
siskins 269
songbirds (oscines) 126, 265–66
speckled mousebird 421n111
spoonbills 204
Sri Lankan drongo 265
stinking pheasant 206
stitchbird 424ns40,46
streamertail 264
strong-billed woodcreeper 327n122
sunbirds 202
sunbittern 205
swiftlets 138, 139
swifts 204, 211
thrashers 420n96
tinamou 210
turaco (or touraco) 206
vangids 31
water-fowl 130
waxwings 201
woodcreepers 31, 32
woodpeckers 22, 31, 255; 447n174
yellow-bellied sapsucker 32; 324n20
yellow-breasted chat 424n40
zebra finches 265
(a) Bowerbirds 28, 211; 403n5
    painting 274, 294
    satin bowerbird 274
(b) Corvids (crows) 255, 262, 264, 270, 275–
    78, 290, 292
    brown-necked ravens 290
    fan-tailed crow 452n29
    magpie 292
    New Caledonian crows 262, 275–78;
      323n2
    northern blue jays 452n42
    ravens 284
    rooks 270, 278
    Steller's jay 275

(c) Ducks
    Argentine lake duck 212
    Muscovy duck 212
(d) Eagles and other raptors 147
    accipitrids 422n137
    bat hawk 420n90
    buzzards 452n29
    condor 208; 421n135
    crane hawk 422n137
    Golden eagles 274
    gymnogene 422n137
    Harris's hawk 290; 450n120
    hawk eagles 422n137
    kites 205; 422n137
    sea eagle 422n137
    serpent eagle 422n137
(e) Sea-birds 323n1
    gannets 205
    giant auk 262
    gulls 452ns28, 29
    herring gulls 130
    skuas 290
    tropicbirds 205
    whiskered auklet 130
(f) Hummingbirds 53, 80, 200–2, 204, 207,
    227, 264, 266; 377n159; 417n20; 419n65;
    449n86
    Anna's hummingbird 264
(g) Owls 105, 134, 205, 254, 262; 448n34
    barn owl 148
    burrowing owls 19
(h) Parrots 130, 255, 262, 264, 265, 270, 283;
    421n111
    African grey 265, 268, 292
    budgie 427n112
    Goffin's cockatoo 452n37
    green-rumped parrotlet 421n111
    hyacinth macaws 450n115
    kakapo 262, 284; 421n111
    kea 264, 283–84; 453n95
    lorikeets 418n23
    macaws 447n174; 450n120
    palm cockatoo 450n115; 452n37
    parakeets 265
    Senegal parrots 374n36
    Vasa parrots 211
(i) Vultures
    bearded vultures 273
    Egyptian vulture 208, 274, 290
    New World vultures 208
    Old World vultures 208
    palm-nut vulture 421n135; 422n136

Brains (see also Nervous Systems) 98, 130, 194,
    198, 228, 241–51, 252–62, 265–66, 276
accessory lobe 248
anterior commissures 255
amygdala 252
arcopallium 261

Bacteria (*continued*)
  verrucomicrobia 311n9
  vibrionaceans 109
  (a) Cyanobacteria 36, 95; 310n8; 311n9;
      313n10; 366n76
      prochlorophytes 313n10
  (b) Proteobacteria
      α-proteobacteria 173
      β-proteobacteria 174
      γ-proteobacteria 109, 173

Behaviour
  alarm calls 101, 129, 131, 265
  anointing 39, 117
  bucket-brigades 182
  burying the dead 279; 407n137
  cosmetics 273, 282
  courtship 145, 211, 213, 232, 264, 274
  defense syndrome 394n96
  eavesdropping 142, 143
  escape manoeuvre 26
  ethotypes 88
  facial flushing 207
  grave-yards 183, 293
  gregarious 393n71
  grooming 184
  juvenile delinquency 283–84
  knotting 50
  lekking 15, 169
  lethal assaults 281
  maze 237–38
  migratory syndrome 204
  navigation 250
  personality 18–19
  Pharaoh approach 170
  phototaxis 95
  play (playfulness) 283–85
  quorum sensing 289; 311n9
  schooling 253
  skiing 284
  sneezing 238
  tail-waving 117
  thanatosis (drops dead) 49, 89
  torpor 200, 226, 287; 419n76
  waggle-dances 457n19
  waste disposal 172, 184–85
  weeding 177, 181
  wool gathering 274
  worker policing 171
  zombies 183

Birds 125, 133, 134, 139, 193, 209,
    211, 227–28, 254–55, 260–62, 264–66,
    273, 283–84, 287; 403n185
  'akiapola'au 32
  antarctic prion 122
  arabian babblers 284
  argus pheasant 107
  bar-tailed godwit 204

bittern 205
black wheatear 320n115
blue-winged mountain-tanager 107
boobies 205
broad-billed prions (whale birds) 51
brown pelican 50–51
buffalo weaver bird 212
carib grackle 450n125
Coronaves 204
cowbirds 28, 265
cuckoos 28; 422n137
curlews 452n29
currawongs 290
Darwin's finches 275
doves 205
ducks 28, 130, 212, 254
egret 452n28
emu 194
enantiornithines 323n12; 374n31;
    413n121; 433n117
fairy wrens 425
feathers 194, 199, 205, 264; 323n1
flamingos 51, 204
flightless birds 262
foot-propelled divers 204
frogmouths 105, 205, 254; 421n133
furnariids 32
goatsuckers (caprimulgiforms) 205
goldfinches 269
gray flycatchers 452n36
great grey shrike 424n40
greater honeyguide 289
grebes 204
green-backed heron 274
hammerkop 420n96
herons 274; 378n186
hihi 424n40
hoatzin 205–6; 421n133
honeycreepers 201
hornbills 262, 283
house sparrows 269
ibises 130
jacana 421n119
Java sparrows 447n12
kagu 420n96
kiwi 130, 208, 262
loggerhead shrikes 290
loons 204
malachite sunbirds 202
malaysian peacock 107
manakins 264
meliphagids 201
Metaves 204
moa 158, 208, 262
monias 420n96
mousebird 421n111
neoavians 204
night-jars 205
ocular feathers 107

fruit-flies 9, 94, 123, 126, 127, 295;
   376n128; 396n170
fungus gnats 90, 111; 393n89
gall midges 181
hilarinids 90
lasiopterines 181
maggots 42
midges (chironomids) 90, 119
mosquitoes 29, 39, 127, 135, 160;
   307n2; 325n54
mycetophilids 90
parasitoid flies 28
sarcophagids 135
syrphids (flower flies) 169; 389n10
tachinids 135
tsetse fly 175
worm-lion 21
(viii) Hemipterans (including aphids)
   82, 91, 160, 171–73 (aphids) and 53,
   119, 172, 204; 307n2; 351n83; 369n174
   (hemipterans)
   assassin bugs 30
   bed bugs (reduviids) 117; 350n64;
      401n140
   cerataphidinids 171
   cicadas 169, 172, 173, 175; 361n176;
      448n55
   froghoppers (spittlebug) 40, 42, 174
   fulgoroids 30, 82
   gerromorphs 165
   leafhoppers 42, 91, 173
   margarodid aphids 363n246
   mealybugs 174–75
   phymatine bugs 53
   pseudococcid 174
   pyrophilus flat-bug 119
   sharpshooters (cicadellinids) 173
   stinkbugs 401n151
   treehoppers 173
   western conifer seed bug 119
   whitefly 160; 401n148
(ix) Moths (lepidopterans, in part) 89,
   102–3, 119; 360n168; 379n206
   castanet moth 448n25
   choreutids 90
   greater wax moth 125
   hawkmoths (or sphinx-moths) 103,
      113, 127, 136–37, 169, 200–201;
      347n202; 377n153; 379n206; 417n20
   hooktip moths 12, 12
   silkworm 91
   tiger-moth 138
(x) Odonata (dragon-flies) 42; 359n116
(xi) Orthopterans 53, 83
   cricket 135
   katydids 136; 447n23
   locusts 400n88; 419n70; 442n163
   weta 74

(xii) Termites (and cockroaches) 39, 82,
   171, 175, 186, 246; 399n73; 365ns31, 56;
   457n23(cockroaches) and 66, 85–86,
   186–87, 281; 409n200; 415n160; 452n36
   (termites)
   madagascan hissing cockroach 175
(xiii) Wasps 113, 171, 28
   beewolves 407n154
   ichneumon wasps 341n271;
      393n89
   paper wasp 250
   parasitic wasps 66, 117, 129;
      355n218
   vespids 171
   woodwasps 181
(d) Spiders 29, 66, 86–89, 129, 166, 245;
   317n29; 399n72
   aphantochilids 323n94
   bolas spiders 89
   cribellum 88
   cyrtarachnids 89
   lycopsids 353n153
   nursery web spiders 89
   ogre-faced spider 104
   orb webs 88–89
   pisaurids 89
   salticids 29, 86–87, 106, 129; 358n95;
      458n64
   scytodids 87
   spinnerets 87
   stegodyphids 322n80
   tarantulas 38
   tetragnathids 88
   theridiids 29; 353n150
   thermophilic ants 116
   wolf spiders 129
   zodariids 323n94
(e) Trilobites 104
   phacopids 103

Bacteria (also Disease and Medicine) 120, 177,
   178, 236; 307n2
actinomycete 179, 184; 325n85
archael bacteria 94
eubacteria 315n11
extremophiles 236
flagellar motor 307n2
gram-negative bacteria 188
gram-negative rhizobia 310n8
gram-positive actinorhizal 310n8
magnetotactic bacteria 311n9
myxobacteria 311n9
nitrogenous bacterium 187
pathogenic bacteria 121
planctomycetes 311n9
plasmids 311n9
streptococcal bacteria 315n11
symbiotic bacteria 173

Arthropods, (continued)
    Yeti crab 404n25
    (ii) Shrimps 96, 243
      alpheids 143
      euphausiids 360n168
      mantis shrimps 42, 112; 362n235
      opossum shrimp 439n42
      pandarids 83
      penaeids 129
      snapping shrimps 143
      synalpheids 176
(c) Insects 122, 124, 135, 168, 204, 245–47,
  250–51; 426n93
  ant lion 21
  archaeognathans 248
  argids 399n61
  bristle-tails 193
  caddisflies 90
  embiopterans 90
  ephemopterans 329n14
  gall thrips 171
  heteropterans 53, 82, 119
  homopterans 354n201
  hymenopterans 82, 90, 91, 111, 168
  lacewings 402n173
  lepidopterans 90, 204
  mantispids 53; 384n378
  neuropterans 21
  phasmatoids 345n125
  plectopterans 329n14
  praying mantis 53
  ranatrid water-stick insects 53
  rhachiberothids 53
  springtails 248; 385n4
  strepsipterans 104
  thrips 172
  titanopterans 53
  tree lobsters 345n125
  trichopterans 90
  water-beetles 63
  water stick insects 384n378
  webspinners 90
    (i) Ants 90, 174–75, 181–86, 192–93, 250,
      272, 273; 326n118; 401n144; 458n64
      ant rain 412n88
      army ants 168, 185; 351n86
      cartons 409n194
      crazy ant 166
      fire-ants 70; 332n92
      formicids 273
      leaf-cutter ants (attines) 27, 66, 121,
        181–86, 202; 368n148
      weaver ants 29, 82
    (ii) Bees and hornets 168, 169–70, 202,
      289; 311n9; 364n2; 389n10; 393n89;
      417n20; 418n37; 450n127 (bees) and
      170; 439n38 (hornets)
      bumble-bees 169; 442n157

      carpenter bee 113
      halictids (sweat bees) 171, 250, 273
      honeybees 169–70, 250, 287–88, 290;
        400n121
      stingless bees 170
  (iii) Beetles (coleopterans) 53, 87, 117,
    119, 169, 171, 179–80, 275; 389n10;
    423n12
    acanthocnemids 119
    ambrosia beetles 179–80
    bark beetles 29
    bombardier beetle 15, 38
    buprestids 117, 118
    carabids 409n199
    cave beetles 116
    cerambycid 66
    chrysomelids 30; 350n64
    dermestids 164
    diving beetle (dytiscids) 363n246
    dung beetles 169; 396n168
    dynastine scarabs 169
    fire-flies 109, 111
    flea beetles (alticinids) 42
    flour beetle 160
    glowworms 109
    ladybirds 173
    leather-winged sailor (lymexylids)
      180
    little ash beetle 119
    melyrids 35
    platypodids 179
    rhinoceros beetles 169
    scarab beetles 250
    scolytids 179
    ship-worm (beetle) 180
    soldier beetle 327n148
    stag-beetles 180, 188
    staphylinids 53
    weevils 30, 180
  (iv) Butterflies (lepidopterans, in part)
    102; 363n254; 379n206; 393n71
    caterpillars 90, 129, 168, 399n72
    ithomiids 90
    nymphalids 28; 442n167
  (v) Earwigs (Dermapterans) 345n122;
    348n4
  (vi) Fleas 42; 424n50
  (vii) Flies (Dipterans) 21, 53, 90, 111,
    180, 245; 354n201; 423n12 (dipterans)
    and 135, 162, 168; 288 (flies)
    asphondylines 181
    batfly 345n122
    blow-fly 241
    brachycerid flies 53
    carrion flies 163
    dance flies 89, 90
    empidids 53, 90

syrinx 139, 265–66
tracheal systems 166, 168, 180; 433n121
(g) Skeleton 293
  exoskeleton 144
  hydrostatic skeleton 77
  intramandibular joint 340n220
  lumbrosacral 375n91
  maxilloturbinals 226
  metatarsal bones 196
  nacre 321n28
  osteoderms 338n127
  pachyostosis 403n21
  pharyngeal jaws 49–50
  pygostyle 415n140
  sclerotic ring 346n167
  secondary palate 229
  skeleton space 24
  skulls 203, 208, 228
  subarctometatarsus 196
  turbinals 433n121
(h) Teeth (Dentition) 57–59, 190, 198 (dentition) and 197; 382n302 (teeth)
  canines 53–57, 62
  cellular cementum 346n176
  edentulous 46, 52; 340n237
  fangs 57, 62, 66, 224; 339n184
  heterodonty 59; 413n110; 415n142
  hinged teeth 64
  homodont dentition 142
  hypsodonty 57–58
  incisors 54, 58
  ivory 293
  labral tooth 26
  mandibles 382n309
  molars 58–59
  plagiaulacoid dentition 58
  plicidentine 336n84
  pre-maxillary teeth 426n91
  propaliny 333n122
  rhamphothecae 414n137
  supernumerary teeth 28
  trabecular bone 48
  tribosphenic 59
  zalambdodonty 435n184
  ziphodonty 338n128

Annelids 242, 244, 247–48
  alciopids 106–7
  earthworms 129–30, 214, 242
  echiuroids 440n67
  fan worm 17
  fireworm 97
  flabelligeridids 364n9
  leech 73
  lugworms 130; 404n40
  maldanids 310n3
  nereids 178, 244
  poeobids 109

polychaetes 66, 178; 404n41
sabellariids 355n233
sabellids 16, 101
serpulids 101
sipunculans (peanut worms) 440n67
syllids 97

Arthropods 245–51; 317n29; 319n69
  centipedes 399n57
  millipedes 39; 448n24
  onychophorans 220, 244, 245, 247; 352n131; 427n136
  tardigrades 150
  velvet worms 317n29; 352n131
(a) Chelicerates (except spiders) 16, 66
  harvestmen 166
  limulids 245
  mites 34
  opilionid harvestmen 373n3
  oribatid mites 27
  pseudoscorpions 87
  scorpions 66, 245, 288
  spider-mites 87; 401n148
  teneriffiid mites 448n55
  uropygids 326n118
  vinegaroon 326n118
(b) Crustaceans 27, 102, 105, 110, 120, 165, 242–43, 245, 248–49; 317n29
  amphipods 91
  barnacles 92; 424n50
  cephalocarids 248–49
  cirripedes 310n3
  copepods 98, 242–43; 373n13; 410n8
  crayfish 17, 144, 288
  cycloids 168
  darwinulid ostracods 27
  decapods 129; 339n204
  Eumalacostraca 249
  isopods 165–66, 176, 249
  krill 51
  lobster 448n55
  myodocopid ostracods 102
  ostracods 27, 28
  remipedes 249
  water-flea 372n85
  (i) Crabs 30, 63, 166, 167–68, 242; 344n99; 375n91
    anomurans 249
    blue crab 331n67
    brachyurans 166
    coconut (robber) crabs 122, 166, 249
    coenobitids 166
    grapsids 167
    hermit crabs 166, 249
    land-crabs 166–67
    ocypodids 398n33
    potamoid brachyurans 167

Anatomy (*continued*)
gelatinous composition 109
gills 25, 166, 188, 220
glossal tube 379n206
hairy pads 82
hammer 129, 144, 280
hands 131, 279
hearts 16
hyobranchial 45–46
hyolingual complex 46
hyomandibular complex 65; 431n59
infrabuccal pocket 184
insulation 169–70
integument 13
iridophores 108; 392n40
kidneys 16; 315n13; 454n121
Kranz anatomy 392n48
laterality 123, 277
malpighian tubules 40; 318n69
manipulative ability 70
metapleural glands 183–84, 185
mycangia (mycetangia) 180
nanostructure 206
neck 224
nipple arrays 102–3
noses (nostrils) 123, 131–32, 226, 279
opaline gland 15
patagium 190, 191–92
phonic lips 141
pinnae 137
plastron 397n2
precision grip 69–70
prehensile feet 68
prehensile tails 68, 131, 190, 280; 336n82;
    342n12
proboscis 279; 379n206
projectile tongue 45, 105
protofeathers 195
quill knobs 194
radula 65–66, 179; 329n15
reticulum 421n112
salt glands 340n222
scopula 86-87
segmentation 440n67
self-cleaning 102–3; 351n92
semicircular canals 71, 133, 141; 375n91
sensilla 116
shock-absorbers 205
skin 116
smooth pads 82
snorkel 279
spicules 96
spines 158
spinnerets 87
spiracles 40, 166
stylets 171, 172–73
suckers 14
swim-bladder 11–12, 221

tagmosis 27
tail 68
thorns 158
tongue 52, 137–38, 203, 219; 341n243;
    379n206
transparency 106–7, 108–9
trunk 131, 279
uropygidial gland 273
vomeronasal organ 115
whiskers 33, 55, 130, 132
(a) Fingers (and claws) 70, 83, 88, 143, 171,
    194, 206, 242; 335n62; 404n25 (claws) and
    14, 224-225 (fingers)
    hyperphalangy 350n47
    opposable thumb 68, 73
    pharyngeal formula 229
    thumb 225
(b) Gut 356n7
    oesophagus 109
(c) Hair
    fur 199
    keratin 199
    osmetrichia 417n212
(d) Muscle
    flight muscle 227
    oxidative muscle 436n199
    red aerobic muscle 13, 79
    sonic musculature 448n55
    sternoglossus 52
    superfast muscles 448n55
    tendons 48, 79
    tympanal muscles 448n55
    white anaerobic muscle 13, 79
(e) Reproductive
    baculum 212–13
    baubellum 212
    gonads 209
    mammary glands 175
    ovaries 209
    oviduct 215
    ovipositors 66, 181
    penis 16, 212
    phalloid organ 212
    placenta 216–20
    placentotrophy 216
    pubic nipples 33
    scrotum 206
    testes 124, 209, 258; 356n22
    trophotaeniae 220
(f) Respiratory
    air-breathing organs (ABO) 221
    air-sacs 205, 227, 267; 380n249
    book-lungs 398n18
    descended larynx 267
    larynx 137, 265–66
    lung 166, 221, 232
    lunglessness 45
    pulmonary system 193

GENERAL INDEX

Because of the range of topics and subjects they have been assigned to a series of more inclusive categories.

Agriculture 177–87
 agro-predators 185
 algal farms 177–79
 bacterial farms 404n25
 cattle-husbandry 174
 domatia 186
 fungal farming 179–87
 gardening 178
 monoculture 177

Alcohol 9, 150, 154; 371n70; 396n177
 alcoholism 9–10
 beer 203
 English pub 157
 gin and tonic 10, 189
 [Chateau] Petrus 115
 tequila 379n206
 Vouvray 123

Algae (see also protistans) 94, 231; 305n1; 371n80
 brown algae 152; 305n1; 389n8
 green algae 95, 152, 153, 235; 305n1
 microalgae 178
 prymnesiophytes 313n10
 red algae 15, 177, 178; 305n1; 388n7
 seaweeds 389n8
 symbiotic algae 96, 101
 volvocaleans 305n1

Amphibians 25, 34, 45–46, 76, 144, 215, 220, 290; 335n62
 tadpoles 25, 45
 temnospondyls 224
 (a) Caecilians 45, 77, 220
  typhlonectids 45, 220
 (b) Frogs 4, 25, 34, 45, 57, 70, 83, 117, 158, 192; 335n62; 377n159
  arboreal (tree) frogs 34, 70, 83
  arthroleptids 335n62
  burrowing frogs 34

 cane toads 35
 dendrobatid 34, 35, 36
 eleutherodactylids 34, 215
 hemiphractids 321n34
 hylids 192, 220
 leptodactylids 39
 mantellids 34
 microhylids 324n44
 myobatrachids 39
 pipids 220; 329n18
 reed frog 426n99
 rhacophorids 39, 192
 toad 35, 39; 325n63; 333n122; 339n72
 tree frog 83, 215
 (c) Salamanders 25, 45, 77, 83, 290; 339n72
  bolitoglossines 76
  desmognathids 329n22
  hemidactyllines 329n28
  mountain dusky salamander 426n91
  newt 36; 335n62
  plethodontids 25, 45, 76
  tiger salamander 36

Anatomy
 acoustic fats 141–42
 adhesion 30, 85, 91, 98, 143, 154, 271
 aorta 16
 arms 13–14
 baleen 50–51
 beaks 197, 208, 255; 414n137
 biominerals 321n28
 blood-air barrier 398n26
 blood vessels 79, 207, 215, 220
 blubber 433n121
 brown adipose tissue (BAT) 226, 227
 byssus 91, 92; 426n91
 calotte 315n13
 cartilage 16, 48
 chaetae 214
 clubs 213
 countercurrent system 79, 226; 433n121

NOTES 461

129. Shepard (1992), p. 512; note 128.

130. Shepard (1992), pp. 517–518; note 128.

131. See their paper in *Psychological Science* 3, 97–104 (1992).

132. Shepard's analysis of how we interpret the three-dimensional euclidean world is both analogous and equally compelling (Shepard [2001]; note 128). Here, too, the perception of objects and their movement reflect general laws of psychology that are embedded in a Newtonian world but as crucially interpret it. If, for example, an observer is presented with two equivalent two-dimensional images, but only one of which is illuminated at any one time, the individual will insist that they move from one position to another even though in reality they are being illuminated on an alternative basis. Motion is evidently perceived because in the Newtonian world, the same object is far more likely to move from one position to another rather than dematerializing and promptly popping back into existence. Corresponding experiments involving similar two-dimensional and then three-dimensional images again illustrate the necessary assumptions of surviving in a Newtonian world. Take the two images again, but now place them at 90˚ to each other. Illuminated in alternate fashion the observer will insist they rotate. Present these images again, but now in mirror form, and the observer perceives both rotation and the image moving out of the plane; otherwise, how could it reverse? Reality finally collapses, however, when the image is seen as three-dimensional and in mirror form: the invitation to enter a four-dimensional space is, for the brain, impossible.

133. Shepard (1987), p. 262; note 126.

134. See R.N. Shepard in *The science of the mind: 2001 and beyond* (R.L. Solso and D.W. Massaro, eds.), pp. 50–62 (Oxford; 1995).

135. Shepard (1995), p. 59; note 134.

136. Shepard (1995), p. 51; note 134.

137. Shepard (1987), p. 267; note 126.

138. From his *The story and the fable: An autobiography* (Harrap; 1940), pp. 57–58.

139. From his *Skin for skin* (Bradley Head; 1948), pp. 85–86.

140. See *Science* 160, 1308–1312 (1968).

141. The title of Polanyi (1968); note 140.

142. Polanyi (1968), p. 1310; note 140.

143. See *Journal of Applied Physiology* 104, 1844–1846 (2008); also the generally supportive, even enthusiastic, correspondence on pp. 1847–1850 and Macklem's generous reply (p. 1851).

144. Macklem (2008), p. 1846; note 143.

145. Macklem (2008), p. 1844; note 143.

146. Polanyi (1968), p. 1310; note 140.

147. Polanyi (1968), p. 1311; note 140.

CHAPTER 27

1. See J.J. Kripal *Authors of the impossible: The paranormal and the sacred* (Chicago; 2011).

2. E. Wigner *Communications in Pure and Applied Mathematics* 13, 1–14 (1960).

3. Complex numbers are described as $a + bi$ where a and b "exist" but i is "imaginary," so that $i^2 = \sqrt{-1}$. To the man in the street, complex numbers are fictional, but they have immense utility in mathematics.

4. Wigner (1960), p. 7; note 2.

5. See R. Kanigel *The man who knew infinity: A life of the genius Ramanujan* (Scribners; 1991).

6. Kanigel (1991; note 5), see pp. 281–282.

7. Kanigel (1991; note 5), p. 281.

8. Wigner (1960) makes several references to" miracles" in his article. For example, "The miracle of the appropriateness of the language of mathematics for the formulation of the laws of physics is a wonderful gift which we neither understand nor deserve" (p. 14).

9. Here quoting the Polish mathematician Mark Kac (p. 281), but Kanigel (1991; note 5) makes it abundantly clear that there is something very mysterious about Ramanujan's gifts.

10. See his autobiography *Jolly green giant: The autobiography of David J Bellamy OBE, Hon FLS, an Englishman* (Century; 2002), p. 88.

11. H. Owen *Journey from obscurity: Wilfred Owen: Memoirs of the Owen family 1893–1918* (Oxford; 1965); see vol. 3, *War*, pp. 198–201.

12. V. Goddard *Flight towards reality* (Turnstone; 1975).

13. J. O'Connor *Father Brown on Chesterton* (Frederick Muller; 1937).

93. See D.M. Palacios and D. Day *Marine Mammal Science* 11, 593–594 (1995).

94. See F. Ritter *Marine Mammal Science* 23, 429–433 (2007).

95. See L. Lodi *Marine Mammal Science* 8, 284–287 (1992).

96. See K.M. Dudzinski et al. *Aquatic Mammals* 29, 108–116 (2003).

97. See T.J. Smith and G.A. Sleno *Canadian Journal of Zoology* 64, 1581–1582 (1986); in this report one beluga was carrying a neonate, but it was not possible to observe whether it had actually died.

98. See, for example, K. McComb et al. *Science* 292, 491–494 (2001); also commentary on pp. 417–418 by E. Pennisi.

99. See, for example, K. Payne in *Animal social complexity: Intelligence, culture, and individualized societies* (F.B.M. de Waal and P.L. Tyack, eds.), pp. 57–85 (Harvard; 2003).

100. See, for example, C.A. Spinage *Elephants* (T & AD Poyer; 1994), pp. 108–109.

101. See I. Douglas-Hamilton et al. *Applied Animal and Behaviour Science* 100, 87–102 (2006).

102. Douglas-Hamilton et al. (2006), p. 87 [Abstract]; note 101.

103. Spinage (1994), p. 109; note 100.

104. See K. McComb et al. *Biology Letters* 2, 26–28 (2006).

105. See his *Africa's elephant: A biography* (Hodder & Stoughton; 2001). His chap. 22 on death details many other aspects of the elephants' reactions to death, including close examination of skeletal remains, attempts at resuscitation, and what for all the world looks like grief.

106. Meredith (2001), p. 186; note 105.

107. Spinage (1994), p. 109; note 100.

108. Spinage (1994), p. 110; note 100.

109. For an overview, see L.A. Bates et al. *Current Biology* 18, R544–R546 (2008). See also C. Braden's article "Not so dumbo" (*BBC Wildlife* 21[8], 32–38; 2003), which draws attention to Joyce Poole's observations of elephants grieving, as well as possible evidence for mirror self-recognition among captive (in a Las Vegas casino!) Asian elephants.

110. See N.G. Pillai *Journal of the Bombay Natural History Society* 42, 927–928 (1942).

111. See B. Gilbert *Smithsonian* 21, 40–51 (1990); an interesting piece that touches on aspects of elephant intelligence. See also M. Tennesen *Wildlife Conservation* 101(5), 66–67 (1998), on the artistic accomplishments of Mamie.

112. See J.H. Williams *Elephant Bill* (Rupert Hart-Davis; 1954), p. 86.

113. See U. Toke Gale *Burmese timber elephant* (Trade Corporation [9]; n.d. [c. 1974]); see p. 109 for a description of self-strangulation achieved by placing a foot on the trunk. This remarkable book provides a wonderful glimpse into the life of these elephants and like Williams's deserved classic (note 112) reports an analogous case of muffling a bell to avoid detection, but this time by coiling the trunk around the offending object (p. 126).

114. See, for example, J.M. Plotnik et al. (*Proceedings of the National Academy of Sciences, USA* 108, 5116–5121; 2011), who describe experiments that assess the cognitive capacity of Asian elephants in the context of cooperative pulling (specifically two individuals each hauling a rope to drag a table with food), and remark, "Through convergent evolution, elephants may have reached a cooperative skill level on a par with that of chimpanzees" (p. 5116 [Abstract]). This result is all the more interesting because while Plotnik et al. provide good evidence that the two elephants are aware of each other's needs (to the extent that in one case a female elephant simply stood on the rope while her partner did all the work), in the cases of rooks (otherwise celebrated for their cognitive capacities) it is not clear that they really understand what their partner is doing (see T. Bugnyar *Current Biology* 18, R530–R532 [2008]). Proverbial also are the powers of elephant memory, and this finds an important parallel in dolphins; see J.N. Bruck *Proceedings of the Royal Society of London, B* 280, 20131726 (2013).

115. See their paper in *Neuroscience & Biobehavioral Reviews* 32, 86–98 (2008).

116. Hart et al. (2008), p. 96; note 115. Others are less convinced; see G. Vallortigara et al.'s (*PLoS Biology* 6, e42; 2008) critique of T. Grandin and C. Johnson's *Animals in translation: Using the mysteries of autism to decode animal behavior* (Bloomsbury; 2005).

117. See, for example, D.A. Treffert's *Extraordinary people* (Black Swan; 1990).

118. See *Science* 291, 52–54 (2001).

119. See, for example, E. Mercado et al. *Animal Cognition* 8, 93–102 (2005).

120. Gray et al. (2001), p. 52; note 118.

121. Gray et al. (2001), p. 54; note 118.

122. See, for example, G. Roth and U. Dicke *Trends in Cognitive Sciences* 9, 250–257 (2005).

123. Roth and Dickie (2005), p. 256; note 122.

124. The title of John Horgan's influential book; in full *The end of science: Facing the limits of knowledge in the twilight of the scientific age* (Helix; 1996). For a more recent contribution along similar lines, see Russell Stannard's *The end of discovery: are we approaching the boundaries of the knowable?* (Oxford; 2010).

125. For one critique of Horgan's thesis (note 124), see D. Hoffman *Notices of the AMS* 45, 260–266 (1998).

126. See R.N. Shepard in *The latest on the best: Essays on evolution and optimality* (J. Dupré, ed.), pp. 251–275 (MIT Press; 1987).

127. Shepard (1987), p. 251; see note 126.

128. See R.N. Shepard in *The adapted mind: Evolutionary psychology and the generation of culture* (J.H. Barlow et al., eds.), pp. 495–532 (Oxford; 1992), and *Behavioral and Brain Sciences* 24, 581–601 (2001).

phins by J.R. Zaeschmar et al. *Marine Mammal Science* 29, 556–562 (2013), and assistance of a young finless porpoise calf by a pod of humpback dolphins by X.-Y. Wang et al. *Journal of Mammology* 94, 1123–1130 (2013).

48. See H. Friedman in *Symbiosis*, vol. 2, *Associations of invertebrates, birds, ruminants and other biota* (S.M. Henry, ed.), pp. 291–316 (Academic; 1967).

49. See L.L. Short and J.F.M. Horne in *Handbook of the birds of the world*, vol. 7, *Jacamars to woodpeckers* (J. del Hoyo et al., eds.), pp. 274–295 (Lynx; 2002).

50. Friedman (1967), p. 294; note 48.

51. See H.A. Isack and H.U. Reyer *Science* 243, 1343–1346 (1989).

52. See, for example, C.M. Drea and A.N. Carter *Animal Behaviour* 78, 967–977 (2009).

53. See, for example, S.K. Gazda et al. *Proceedings of the Royal Society of London, B* 272, 135–140 (2005).

54. See G.G. Frye and R.P. Gerhardt *Wilson Bulletin* 113, 462–464 (2001).

55. See M. Schulz *Notornis* 51, 167 (2004).

56. See M.G. O'Neil and R.J. Taylor *Corella* 8, 95–96 (1984).

57. See J.C. Bednarz *Science* 239, 1525–1527 (1988).

58. See R. Bowman *Wilson Bulletin* 115, 197–199 (2003).

59. See R. Yosef et al. *Naturwissenschaften* 98, 443–446 (2011).

60. See R. Yosef and N. Yosef *Journal of Ethology* 28, 385–388 (2010).

61. Yosef and Yosef (2010), p. 387; note 60.

62. See C. Uller et al. *Animal Cognition* 6, 105–112 (2003); also P. Krusche et al. *Journal of Experimental Biology* 213, 1822–1828 (2010).

63. See, for example, C. Agrillo et al. *Animal Cognition* 11, 495–503 (2008).

64. See M. Dacke and M.V. Srinivasan *Animal Cognition* 11, 683–689 (2008). For a possible case in the beetle *Tenebrio*, see P. Carazo et al. *Animal Cognition* 12, 463–470 (2009). Z. Reznikova and B. Ryabko (*Behaviour* 148, 405–434; 2011) infer an arithmetical ability in ants, while beyond the insects a numeric capacity appears to exist in a salticid spider (X.J. Nelson and R.R. Jackson *Animal Cognition* 15, 699–710; 2012).

65. See, for example, A. Nieder's commentary in *Neuron* 44, 407–409 (2004).

66. See A. Nieder and E.K. Miller *Proceedings of the National Academy of Sciences, USA* 101, 7457–7462 (2004).

67. See their paper in *Neuron* 37, 149–157 (2003), and commentary by S. Dehaene *Trends in Cognitive Sciences* 7, 145–147 (2003).

68. See, for example, the case of the bee odometer, K. Cheng et al. *Animal Cognition* 2, 11–16 (1999).

69. See their paper in *Neuron* 37, 4–6 (2003).

70. Pessoa and Desimone (2003), p. 5; note 69.

71. See, for example, S.T. Boysen and G.G. Berntson *Journal of Comparative Psychology* 103, 23–31 (1989).

72. For an overview, see I.M. Pepperberg *The Alex studies: Cognitive and communicative abilities of grey parrots* (Harvard; 1999).

73. See papers by I.M. Pepperberg in *Animal Cognition* (vol. 9, pp. 377–391 [2006], and vol. 15, pp. 711–717; 2012).

74. See, for example, M.J. Beran *PLoS Biology* 6, e19 (2008).

75. See J.M. Plotnick et al. *Proceedings of the National Academy of Sciences, USA* 103, 17053–17057 (2006), and *Zoo Biology* 29, 179–191 (2010).

76. See D. Reiss and L. Marino *Proceedings of the National Academy of Sciences, USA* 98, 5937–5942 (2001).

77. See F. Delfour and K. Marten *Behavioural Processes* 53, 181–190 (2001).

78. Reiss and Marino (2001), p. 5942; note 76.

79. See H. Prior et al. *PLoS Biology* 6, e202 (2008); see also commentary F.B.M. de Waal in e201. Curiously in the case of the New Caledonian crow, while they can employ mirrors to find concealed food, their capacity to self-recognition was open to question; see F.S. Medina et al. *Animal Behaviour* 82, 981–993 (2011).

80. Prior et al. (2008), p. 1648; note 79.

81. See A. Heschl and J. Burkart *Primates* 47, 187–198 (2006); also T. Suddendorf and E. Collier-Baker *Proceedings of the Royal Society of London, B* 276, 1671–1677 (2009).

82. See I.M. Pepperberg et al. *Journal of Comparative Psychology* 109, 182–195 (1995).

83. Plotnick et al. (2010), p. 184; note 75; see also F.B.M. de Waal *Proceedings of the National Academy of Sciences, USA* 102, 11140–11147 (2005).

84. See J.K. Desjardins and R.D. Fernald *Biology Letters* 6, 744–747 (2010).

85. Desjardins and Fernald (2010), p. 744 [Abstract]; note 84.

86. Desjardins and Fernald (2010), p. 745; note 84.

87. See P. Rochat and D. Zahavi *Consciousness and Cognition* 20, 204–213 (2011); as well as drawing attention to the varying cultural reactions to mirrors they also address important questions as to the ability to determine mirror self-recognition.

88. Titled *Blue-remembered hills: A recollection* (Bodley Head; 1983).

89. Sutcliff (1983), p. 83; note 88.

90. Although quite widely documented (see, for example, A. Leroi-Gourhan *Science* 190, 562–565 [1975; with commentary on pp. 880–881 by R.S. Solecki], and T. Akazawa et al. *Nature* 377, 585–586 [1995]), more skeptical views are offered, for example, by R.H. Gargett *Journal of Human Evolution* 37, 27–90 (1999), and D.M. Sandgathe et al. *Journal of Human Evolution* 61, 243–253 (2011).

91. See K.A. Cronin et al. *American Journal of Primatology* 73, 415–421 (2011).

92. See S. Harzen and M.E. dos Santos *Aquatic Mammals* 18, 49–55 (1992), and J. Mann and H. Barnett *Marine Mammal Science* 15, 568–575 (1999).

13. Rattenborg et al. (2009), p. 253 [Abstract]; note 12.

14. See, for example, M. Giurfa et al. *Nature* 410, 930–933 (2001), and papers in *Journal of Experimental Biology* by A.G. Dyer et al. (vol. 208, pp. 4709–4714; 2005) and S.-W. Zhang et al. (vol. 209, pp. 4420–4428; 2006). So, too, with novelty seeking, where Z.-Z.S. Liang et al. (*Science* 335, 1225–1228; 2012) demonstrate "intriguing parallels between honey bees and humans" (p. 1227).

15. Interestingly, the stages of sleep of the young bees who stay in the hive are much the same as the foragers (there is also evidence that when foraging is not possible, the workers will take opportunistic daytime naps; see B.A. Klein and T.D. Seeley *Animal Behaviour* 82, 77–83; 2011) out in the sunshine, but the architectural patterns do differ. Such a plasticity of behavior again echoes mammalian sleep (see A.D. Eban-Rothschild and G. Bloch *Journal of Experimental Biology* 211, 2408–2416 [2008]; also B.A. Klein et al. on pp. 3028–3040).

16. See his paper in *Journal of Comparative Physiology* 163, 565–584 (1988).

17. See their paper in *Journal of Comparative Physiology, A* 189, 599–607 (2003).

18. Sauer et al. (2003), pp. 604–605; note 17.

19. This includes their waggle dances, which become less precise in terms of directional information (see B.A. Klein et al. *Proceedings of the National Academy of Sciences, USA* 107, 22705–22709 [2010]) and consolidation of navigational memory (see L. Beyaert et al. *Journal of Experimental Biology* 215, 3981–3988 [2012]).

20. Sauer et al. *Journal of Sleep Research* 13, 145–152 (2004).

21. See W. Kaiser and J. Steiner-Kaiser *Nature* 301, 707–709 (1983).

22. See I. Tobler and J. Stalder *Journal of Comparative Physiology, A* 163, 227–235 (1988).

23. In the latter case they evidently take a little time to drop off (see K. Mendoza-Angeles et al. *Journal of Neuroscience Methods* 162, 264–271 [2007]), and they also display a form of sleep with slow electrical waves reminiscent of that found in mammals (see F. Ramón et al. *Proceedings of the National Academy of Sciences, USA* 101, 11857–11861 [2004], and K. Mendoza-Angeles et al. *Journal of Experimental Biology* 213, 2154–2164 [2010]). Apart from the bee, sleep has been identified in the fruit fly *Drosophila* (see D.A. Nitz et al. *Current Biology* 12, 1934–1940 [2002], and W.J. Joiner et al. *Nature* 441, 757–760; 2006), as well as the cockroach (see R. Stephenson et al. *Journal of Experimental Biology* 210, 2540–2547 [2007]).

24. For sleeplike states in the snail *Lymnaea*, see R. Stephenson and V. Lewis *Journal of Experimental Biology* 214, 747–756 (2011).

25. And indeed the humble nematode with its spells of lethargus (see D.M. Raizen et al. *Nature* 451, 569–572 [2008]).

26. See, for example, P.J. Shaw et al. *Science* 287, 1834–1837 (2000), and P.J. Shaw and P. Franken *Journal of Neurobiology* 54, 179–202 (2003).

27. See C. Cirelli et al. *Nature* 434, 1087–1092 (2005).

28. See their paper in *Current Biology* 18, R204–R206 (2008), which is commentary on Raizen et al. (2008; note 25) and related papers.

29. Olofsson and Bono (2008), p. R204 [Abstract]; note 28.

30. See J.M. Siegel *Trends in Neurosciences* 31, 208–213 (2008). Not only is the capacity for sleep largely lost in cave fish, but this has arisen convergently and may employ different genetic mechanisms; see E.R. Duboué et al. *Current Biology* 21, 671–676 (2011).

31. See their paper in *Current Biology* 18, R670–R679 (2008).

32. Allada and Siegel (2008), p. R676; note 31.

33. Allada and Siegel (2008), p. R675; note 31.

34. See E.R. Brown *Behavioural Brain Research* 172, 355–359 (2006).

35. See her paper in *Consciousness and Cognition* 17, 37–48 (2008).

36. Mather (2008), p. 39; note 35.

37. See A. Avarguès-Weber et al. *Journal of Experimental Biology* 213, 593–601 (2010); also Dyer et al. (2005); note 14.

38. See W. Gronenberg et al. *Brain, Behavior and Evolution* 71, 1–14 (2008).

39. See E.A. Tibbetts *Proceedings of the Royal Society of London, B* 271, 1955–1960 (2004).

40. See M.J. Sheehan and E.A. Tibbetts *Science* 334, 1272–1275 (2011), who not only document the ability of closely related wasp species to engage in facial recognition, but remark, "specialized face learning provides a remarkable example of convergent evolution between wasps and mammals. Although mammals and wasps have dramatically different eyes and neural structures . . . specializations for recognizing conspecific faces arisen independently in both groups" (p. 1274); see also commentary by A. Avarguès-Weber *Current Biology* 22, R91–R93 (2012). Even within the mammals, facial recognition may have arisen independently. As D.A. Leopold and G. Rhodes (*Journal of Comparative Psychology* 124, 233–251; 2010) remark, while a more ancient origin cannot be dismissed, when one considers the capacities in diurnal and social animals like sheep and macaques, "their face-selective cortical machinery might have evolved independently, as parallel adaptations of a general visual processing system, driven by similar sources of natural and sexual selection" (p. 241).

41. Dyer et al. (2005), p. 4713; note 14.

42. Zhang et al. (2006), p. 4428; note 14.

43. See their paper in *Current Biology* 18, R851–R852 (2008).

44. Sheehan and Tibbetts (2008), p. R852; note 43.

45. See, for example, T.D. Seeley and P.K. Visscher *Behavioral Ecology and Sociobiology* 56, 594–601 (2004).

46. See R. Bshary et al. *PLoS Biology* 4, e431 (2006).

47. See also the documentation of cooperative feeding involving false killer whales and bottlenose dol-

NOTES 457

192. See Q. Liu et al. *Animal Behaviour* 81, 297–305 (2011).

193. See D. Fragaszy et al. *Animal Behaviour* 79, 321–332 (2010).

194. Fragaszy et al. (2010), p. 331; note 193.

195. See E.B. Ottoni et al. *Animal Cognition* 8, 215–219 (2005); also Visalberghi et al. (2009); note 191.

196. For a magisterial and equally enjoyable overview, see G.M. Burghardt *The genesis of animal play: Testing the limits* (MIT Press; 2005).

197. Burghardt (2005), p. 103; note 196.

198. Burghardt (2005), p. 103; note 196.

199. Burghardt (2005), p. 194; note 196.

200. See, for example, D.M. Watson in *Animal play: Evolutionary, comparative, and ecological perspectives* (M. Bekoff and J.A. Byers, eds.), pp. 61–95 (Cambridge; 1998); for an overview, see chap. 9 of Burghardt (2005); note 196.

201. For an overview, see chap. 10 of Burghardt (2005); note 196.

202. For example, writing of "Kangaroos at play" (Watson [1998], part of the title; note 200), Duncan Watson reminds us, "The closest placental equivalent to macropodoids are the Artiodactyla, particularly the Cervidae and Bovidae. Despite some 130 million years of phyletic separation, macropodoids and artiodactyls display striking behavioural and ecological convergence" (p. 61). Nor is this surprising, because as Watson also remarks, these two groups "share very similar evolutionary histories . . . [and] the broad sweep of change remains remarkably convergent" (p. 82).

203. See their paper in *Behaviour* 140, 1091–1115 (2003).

204. Diamond and Bond (2003), p. 1093; note 203.

205. For an excellent and absorbing introduction, see J. Diamond and A.B. Bond *Kea, bird of paradox: The evolution and behavior of a New Zealand parrot* (University of California; 1999).

206. Diamond and Bond (1999), p. 78; note 205.

207. Diamond and Bond (1999), pp. 80–81; note 205.

208. Diamond and Bond (1999), p. 100; note 205.

209. Diamond and Bond (1999), p. 24; note 205.

210. See R.G. Powlesland et al. *Notornis* 53, 3–26 (2006).

211. See T.F. Wright et al. *Molecular Biology and Evolution* 25, 2141–2156 (2008).

212. See J. Diamond et al. *Behaviour* 143, 1397–1423 (2006).

213. Diamond and Bond (2003), p. 1098; note 203.

214. See B. Heinrich and R. Smolker in *Animal play: Evolutionary, comparative and ecological perspectives* (M. Bekoff and J.A. Byers, eds.), pp. 27–44 (Cambridge; 1998).

215. Heinrich and Smolker (1998), p. 36; note 214.

216. See O. Pozis-Francois et al. *Behaviour* 141, 425–450 (2004).

217. Pozis-Francois et al. (2004), p. 436; note 216.

218. Burghardt (2005), p. 286; note 196.

219. See B. Heinbuch and T. Wiegmann *Crocodile Specialist Group Newsletter* 19, 14–15 (2000).

220. See M. Kramer and G.M. Burghardt *Ethology* 104, 38–56 (1998).

221. See G.M. Burghardt et al. *Zoo Biology* 15, 223–238 (1996).

222. Burghardt et al. (1996), p. 235; note 221.

223. Kramer and Burghardt (1998), p. 53; note 220.

224. Burghardt (2005), p. 154; note 196.

225. Burghardt (2005), part of title of chap. 13; note 196.

226. Burghardt (2005), p. 318; note 196; see his fig. 13.1 with a needlefish leapfrogging a turtle.

227. Burghardt (2005), p. 340; note 196.

228. Burghardt (2005), p. 330; note 196.

229. Burghardt (2005), p. 351; note 196.

230. See my *Life's solution*, pp. 194–196 and fig. 7.11.

231. Burghardt (2005), p. 316; note 196.

CHAPTER 26

1. W. Shakespeare *Hamlet, Prince of Denmark*, act 3, scene 1.

2. See, for example, W.O. Stevens *The mystery of dreams* (George Allen & Unwin; 1950).

3. See, A. Trewavas *Plant behaviour and intelligence* (Oxford; 2014).

4. See their paper in *Medical Journal of Australia* 181, 707 (2004). The claim by L-A. Gershwin and P. Dawes (*Biological Bulletin* 215, 57–62; 2008) that the evidence for sleep "was rapidly shown to be erroneous" (p. 60) by B.J. Currie and S.P. Jacups (*Medical Journal of Australia* 183, 631–636; 2005) is without foundation; these latter authors did not dispute the existence of "rest" but merely pointed out that it did not occur at the time noted by Seymour et al.

5. Seymour et al. (2004), p. 707; note 4.

6. See S.D. Sroka in *Richardson's guide to the fossil fauna of Mazon Creek* (C.W. Shabica and A.A. Hay, eds.), pp. 57–63 (Northeastern Illinois University Press; 1997), and M.W. Foster in *Mazon Creek fossils* (M.H. Nitecki, ed.), pp. 191–267 (Academic; 1979).

7. As have the circadian clocks of mammals and insects; see papers by C. Helfrich-Förster in *Journal of Comparative Physiology, A* 190, 601–613 (2004), and *Biochemical Society Transactions* 33, 957–961 (2005).

8. See N.C. Rattenborg *Brain Research Bulletin* 69, 20–29 (2006).

9. See J.L. Kavanau *Neuroscience and Biobehavioral Reviews* 26, 889–906 (2002).

10. See P.S. Low et al. *Proceedings of the National Academy of Sciences, USA* 105, 9081–9086 (2008).

11. Low et al. (2008), p. 9085; note 10.

12. See N.C. Rattenborg et al. *Neuroscience and Biobehavioral Reviews* 33, 253–270 (2009); also J.A. Lesku et al. *Proceedings of the Royal Society of London, B* 278, 2419–2428 (2011).

155. See E.W. Heymann *Journal of Human Evolution* 34, 99–101 (1998); see also consecutive papers (pp. 2024–2047 and pp. 2048–2063) by I.B. Halenar in vol. 294 *Anatomical Record* (2011), who is more cautious regarding their terrestriality.

156. See, for example, A.L. Rosenberger et al. *Proceedings of the Royal Society of London, B* 278, 67–74 (2011), and S.B. Cooke et al. *Proceedings of the National Academy of Sciences, USA* 108, 2699–2704 (2010).

157. See their paper *American Museum Novitates* 3516, 1–65 (2006).

158. MacPhee and Meldrum (2006), p. 3; note 157.

159. See, for example, D.P. Watts et al. *American Journal of Primatology* 68, 161–180 (2006).

160. See F. Aureli et al. *American Journal of Physical Anthropology* 131, 486–497 (2006).

161. See S.M. Lindshield and M.A. Rodrigues *Primates* 50, 269–272 (2009).

162. See their paper in *Primates* 39, 545–548 (1998).

163. Richard-Hansen et al. (1998), p. 545 [Abstract]; note 162.

164. See, for example, W.C. McGrew and L.F. Marchant *International Journal of Primatology* 18, 787–810 (1997). Principal focus has been on *Cebus libidinosus*, but this and related species are sometimes put in the genus *Sapajus*; see, for example, J.W. Lynch Alfaro et al. *Journal of Biogeography* 39, 272–288 (2012).

165. See their paper in *International Journal of Primatology* 18, 787–810 (1997).

166. McGrew and Marchant (1997), p. 805; note 165.

167. See M.E.B. Fernandes *Primates* 32, 529–531 (1991); unlike their macaque cousins, however, these monkeys do not use stone (or snail) hammers, but rather other pieces of oyster.

168. See F.D.C. Mendes *Folia Primatologica* 71, 350–352 (2000).

169. See A. Souto et al. *Biology Letters* 7, 532–535 (2011).

170. Not to mention some memorable observations of a young male not only employing half an orange as a water cup but even using a stick as a cane while gazing at some nearby birds (see B. Urbani *Folia Primatologica* 70, 172–174 [1999]).

171. See E. Visalberghi et al. *American Journal of Primatology* 70, 884–891 (2008).

172. See G.C. Westergaard and S.J. Suomi *Journal of Human Evolution* 27, 399–404 (1994).

173. See G.C. Westergaard and S.J. Suomi *Current Anthropology* 35, 75–77 (1994).

174. See G.C. Westergaard et al. *International Journal of Primatology* 19, 123–131 (1998); these authors documented chucking food, while A. Cleveland et al. *American Journal of Primatology* 61, 159–172 (2003), describe captive capuchins trained to throw stones.

175. See A.C. de A. Moura and P.C. Lee *Science* 306, 1909 (2004).

176. See I.C. Waga et al. *Folia Primatologica* 77, 337–344 (2006).

177. See also E.B. Ottoni and P. Izar *Evolutionary Anthropology* 17, 171–178 (2008), who provide a very useful overview of capuchin tool use. N. Spagnoletti et al. (*Animal Behaviour* 83, 1285–1294; 2012), however, see tool use as more opportunistic.

178. See G.C. Westergaard *Journal of Material Culture* 3, 5–19 (1998).

179. Westergaard (1998), p. 13; note 178; for details of this "artistry," see G.C. Westergaard and S.J. Suomi *International Journal of Primatology* 18, 455–467 (1997).

180. See their paper in *American Journal of Primatology* 71, 242–251 (2009).

181. Mannu and Ottoni (2009), p. 246; note 180.

182. Mannu and Ottoni (2009), p. 248; note 180.

183. See S. Boinski *American Journal of Primatology* 14, 177–179 (1988); this author also reports various objects being employed by capuchins against a variety of animals, including not only sticks and fruit but on one occasion a squirrel monkey hurled at the author.

184. They are also capable of bipedally carrying a load to the anvil; see M. Duarte et al. *Journal of Human Evolution* 63, 851–858 (2012).

185. See, for example, D. Fragaszy et al. *American Journal of Primatology* 64, 359–366 (2004). Not only that, but a propitious positioning of the nut ensures a stable orientation; see D.M. Fragaszy et al. *PLoS ONE* 8, e56182 (2013).

186. See their paper in *American Journal of Physical Anthropology* 132, 426–444 (2007).

187. Visalberghi et al. (2007), p. 436; note 186.

188. See also E. Visalberghi et al. *Primates* 50, 95–104 (2009), who not only provide additional evidence for transport but also other features, including the capuchins waiting their turn to use an anvil and the likelihood of the young learning by observation. In addition Spagnoletti et al. (*Journal of Human Evolution* 61, 97–107; 2011) document sexual dimorphic differences in tool use, with the males generally using more tools and cracking more resilient nuts.

189. Even so, on occasion capuchins will bang stones together in what seems to be an aggressive display to deter predators; see A.C. de A. Moura *Folia Primatologica* 78, 36–45 (2007).

190. See G.R. Canale et al. *American Journal of Primatology* 71, 366–372 (2009), who note hammer stones in excess of 3 kg. Their report is important because it documents tool use in a different species (yellow-breasted capuchins) and notes again the connection between tool use and dry rain forests.

191. See E. Visalberghi et al. *Current Biology* 19, 213–217 (2009), and D.M. Fragaszy et al. *Animal Behaviour* 80, 205–214 (2010); also R.G. Ferreira et al. *American Journal of Primatology* 72, 270–275 (2010). Similar powers of discrimination are evident when captive capuchins are given probes of varying rigidity; see H.M. Manrique et al. *Animal Cognition* 14, 775–786 (2011).

(see A.B. Clifford and L.M. Witmer *Journal of Zoology, London* 264, 217–230 [2004]). Its flexible structure evidently evolved as a dust filter and has been co-opted by the males for some impressive roaring (see R. Frey et al. *Journal of Anatomy* 211, 717–736 [2007]).

120. See J.B. West *Respiration Physiology* 126, 1–8 (2001); also J.B. West et al. *Respiratory Physiology & Neurobiology* 138, 325–333 (2003).

121. See papers in *Proceedings of the National Academy of Sciences, USA* by A.P. Gaeth et al. (vol. 96, pp. 5555–5558; 1999) and A.G.S.C. Liu et al. (vol. 105, pp. 5786–5791 [2008]) assessing, respectively, fetal anatomy (especially kidney structure) and stable isotope values consistent with an aquatic mode of life early in the history of elephants.

122. See J. Shoshani *Trends in Ecology & Evolution* 13, 480–487 (1998).

123. See their paper in *Compte Rendus Palevol.* 8, 281–294 (2009; in French, with English summary).

124. Gheerbrant and Tassy (2009), pp. 290–291; note 123 (my translation). Most striking in this regard are the extinct deinotheriids, otherwise famous for their shovel-like lower tusks; see J.M. Harris *Zoological Journal of the Linnean Society* 56, 331–362 (1975).

125. See W. Wickler and U. Seibt *Ethology* 103, 365–368 (1997); also J.H. Williams *Elephant Bill* (Rupert Hart-Davis; 1954), pp. 151–152, and J. Poole *Elephants* (Colin Baxter Photography; 1997). See also the account by J.A. Gordon (*African Wildlife* 20, 75–79; 1966) of elephants constructing plugs for waterholes they have excavated.

126. For an overview, see S. Chevalier-Skolnikoff and J. Liska *Animal Behaviour* 46, 209–219 (1993).

127. There we find a positive correlation between the size of the species and the length of the tail (see M.S. Mooring et al. *Biological Journal of the Linnean Society* 91, 383–392 [2007]).

128. See B.L. Hart and L.A. Hart *Animal Behaviour* 48, 35–45 (1994); also U. Toke Gale *Burmese timber elephant* (Trade Corporation [9]; n.d. [c. 1974]), p. 130.

129. See B.L. Hart et al. *Animal Behaviour* 62, 839–847 (2001).

130. Hart et al. (2001), p. 847; note 129.

131. These include examples in captivity, such as lion-tailed macaques spontaneously manufacturing probes to extract syrup; see G.C. Westergaard *Journal of Comparative Psychology* 102, 152–159 (1988).

132. See A. Sinha *Folia Primatologica* 68, 23–25 (1997).

133. The title of the paper by A. Carpenter *Nature* 36, 53 (1887).

134. See M.D. Gumert et al. *American Journal of Primatology* 71, 594–608 (2009).

135. Gumert et al. (2009), p. 606; note 134.

136. See J. Padberg et al. *Journal of Neuroscience* 27, 10106–10115 (2007).

137. Padberg et al. (2007), p. 10108; note 136.

138. See, for example, C. Poux et al. *Systematic Biol-*ogy 55, 228–244 (2006), and C.G. Schrago *American Journal of Physical Anthropology* 132, 344–354 (2007).

139. M. Takai et al. *American Journal of Physical Anthropology* 111, 263–281 (2000), documents the earliest-known representative (*Branisella*). This form may have been terrestrial, and included grasses in its diet. Takai et al. also speculate that the initial radiation of the platyrrhines may have begun in Africa.

140. Takai et al. (2000), p. 278; note 139.

141. See J.A. Hodgson et al. *Proceedings of the National Academy of Sciences, USA* 106, 5534–5539 (2009).

142. See, for example, discussions revolving around *Antillothrix* by R.F. Kay et al. (*Journal of Human Evolution* 60, 124–128; 2011) and A.L. Rosenberger et al. (*Proceedings of the Royal Society of London, B* 278, 67–74; 2011).

143. See C.A. Lockwood *American Journal of Physical Anthropology* 108, 459–482 (1999).

144. See J.M. Organ *Anatomical Record* 293, 730–745 (2010); also A.L. Rosenberger *American Journal of Physical Anthropology* 60, 103–107 (1983).

145. See also J.G. Robinson and C.H. Janson in *Primate societies* (B.B. Smuts et al., eds.), pp. 69–82 (Chicago; 1987), who draw attention to four sets of socio-ecological convergence between various platyrrhines and Old World equivalents, as well as with the chimpanzees, but note that the arrangements are by no means identical.

146. See P.M. Kappeler and E.W. Heymann *Biological Journal of the Linnean Society* 59, 297–326 (1996).

147. See also J.U. Ganzhorn et al. *PLoS ONE* 4, e8253 (2009).

148. Kappeler and Heymann (1996), p. 310; note 146.

149. See M.D. Rose *Folia Primatologica* 66, 7–14 (1996).

150. See C. Cartelle and W.-C. Hartwig *Proceedings of the National Academy of Sciences, USA* 93, 6405–6409 (1996).

151. See W.C. Hartwig *Journal of Human Evolution* 28, 189–195 (1995); also L.B. Halenar and A.L. Rosenberger *Journal of Human Evolution* 65, 374–390 (2013), who assign some material to *Cartelles*. Although known only from a tooth, the Miocene cebid *Acrecebus* appears also to have been very large; see R.F. Kay and M.A. Cozzuol *Journal of Human Evolution* 50, 673–686 (2006).

152. These giant monkeys may have lived as recently as ten thousand years ago, and Indian legends hint at more recent survival. *Protopithecus* is interesting because it combines cranial features of howler monkeys and postcranial aspects of spider monkeys, but its convergence is probably with the howlers.

153. See W.C. Hartwig and C. Cartelle *Nature* 381, 307–311 (1996).

154. See this suggestion in E.W. Heymann *American Journal of Physical Anthropology* 55, 171–175 (2001), in his review for the rarity of platyrrhine folivores.

79. See G.R. Hunt et al. *Proceedings of the Royal Society of London, B* 273, 1127–1133 (2006).

80. See R. Rutledge and G.R. Hunt *Animal Behaviour* 67, 327–332 (2004).

81. See A.A.S. Weir et al. *Proceedings of the Royal Society of London, B* (Supplement) 271, S344–S346 (2004).

82. See A.H. Taylor et al. *PLoS ONE* 5, e9345 (2010); the authors explain the important distinction between experienced and naïve crows, and by and large the latter cohort were unimpressive and learned by trial and error.

83. Taylor (2010), p. 6; note 82.

84. See A.H. Taylor et al. *Proceedings of the Royal Society of London, B* 276, 247–254 (2009).

85. Taylor et al. (2009), p. 253; note 84.

86. Taylor et al. (2009), p. 252; note 84.

87. See, for example, the ingenious experiments reported by A.M.P. Bayern et al. *Current Biology* 19, 1965–1968; 2009 (with commentary on pp. R1039–R1040 by S.J. Shettleworth). These involved dropping stones to release food and are consistent with considerable cognitive capacities, but may fall short of true comprehension.

88. See their paper in *Comparative Cognition & Behavior Reviews* 2, 1–25 (2007).

89. Bluff et al. (2007), p. 6; note 88.

90. Bluff et al. (2007), p. 19; note 88.

91. See A.A.S. Weir and A. Kacelnik *Animal Cognition* 9, 317–334 (2006).

92. Weir and Kacelnik (2006), p. 331; note 91.

93. See J. Chappell and A. Kacelnik *Animal Cognition* 5, 71–78 (2002); also their subsequent paper (vol. 7, pp. 121–127; 2004) assessing their capacity to select tools of varying diameters.

94. Such metatool use can be arrived at spontaneously; see A.H. Taylor et al. *Current Biology* 17, 1504–1507 (2007); also their follow-up paper in *Communicative and Integrative Biology* 2, 311–312 (2009).

95. See also comparable experiments with a parrot, a kea, reported by A.M.I. Auersperg et al. *Animal Behaviour* 80, 783–789 (2010).

96. See J.H. Wimpenny et al. *PLoS ONE* 4, e6471 (2009); also A.H. Taylor et al. *Proceedings of the Royal Society of London, B* 279, 4977–4981 (2012), and commentary by A.M. Seed and N.J. Boogert *Current Biology* 23, R67–R69 (2013).

97. See papers in *Proceedings of the Royal Society of London, B* by A.H. Taylor et al. (vol. 277, pp. 2637–2643; 2010) and T. Bugnyar (vol. 278, pp. 634–640; 2011).

98. See Bluff et al. (2010), p. 1384; note 66.

99. See their paper in *Ethology* 111, 962–976 (2005).

100. Heinrich and Bugnyar (2005), p. 973; note 99.

101. See S. Tebbich et al. *Animal Cognition* 10, 225–231 (2007).

102. See C.D. Bird and N.J. Emery *Current Biology* 19, 1410–1414 (2009).

103. See C.D. Bird and N.J. Emery *Proceedings of the National Academy of Sciences, USA* 106, 10370–10375 (2009).

104. See his commentary in *Proceedings of the National Academy of Sciences, USA* 106, 10071–10072 (2009).

105. Kacelnik (2009), p. 10072; note 104.

106. See their commentary in *Current Biology* 19, R731–R732 (2009).

107. Taylor and Gray (2009), p. R731; note 106.

108. For an overview, see S.J. Shettleworth *Trends in Cognitive Sciences* 14, 477–481 (2010).

109. See D.M. Bauer *African Journal of Ecology* 39, 317 (2001).

110. See, for example, K.R.L. Hall and G.B. Schaller *Journal of Mammalogy* 45, 287–298 (1964); also V.B. Deecke (*Animal Cognition* 15, 725–730; 2012), who reports a brown bear in Alaska manipulating a rock, evidently for cleaning or grooming purposes.

111. See R. Smolker et al. *Ethology* 103, 454–464 (1997).

112. E.M. Patterson and J. Mann (*PLoS ONE* 6, e22243; 2011) point out that at first sight it is curious that the dolphins do not employ their echolocation to find the benthic prey as they can do elsewhere, and present evidence that the sponging is useful for finding fish without swim bladders that, along with the rough seafloor, would make echolocation more problematic.

113. See G.J. Parra *Mammalia* 71, 147–149 (2007).

114. See papers by M. Krützen et al. *Proceedings of the National Academy of Sciences, USA* 102, 8939–8943 (2005) and *Proceedings of the Royal Society of London, B* 281, 20140374 (2014); see also J.A. Tyne et al. *Marine Ecology Progress Series* 444, 143–153 (2012), who give this sponge carrying a more ecological slant. Links to genetic structure are detailed by A.M. Kopps et al. *Proceedings of the Royal Society of London, B* 281, 20133245 (2014).

115. See J. Mann et al. *PLoS ONE* 3, e3868 (2008); also A.M. Kopps et al. *Marine Mammal Science* 30, 847–863 (2014).

116. Shark Bay dolphins have also been observed lifting large conch shells to the surface, evidently because they contain a fish; see S.J. Allen et al. *Marine Mammal Science* 27, 449–454 (2011).

117. See, for example, L. Marino *Brain, Behavior and Evolution* 59, 21–32 (2002).

118. See S. Onodera and T.P. Hicks *Neuroscientist* 4, 217–226 (1999); also the popular article by J.(H.) Shoshani in *Natural History* 106(10), 36–45 (1997), titled "It's a nose! It's a hand! It's an elephant's trunk!"

119. See A.V. Milewski and E.S. Dierenfeld *Integrative Zoology* 8, 84–94 (2013). These authors provide a useful overview of the various proboscislike structures found among the mammals. What they define as a "prorhiscis" characterizes the proboscislike structure in that rather extraordinary antelope, the saiga

30. See, for example, papers in vol. 10 (1995) of *Behavioral Ecology* by P.V. Switzer and D.A. Cristol (pp. 213–219) and D.A. Cristol and P.V. Switzer (pp. 220–226).

31. Boswell (1983), p. 170; note 29.

32. See his *The book of lost books: An incomplete history of all the great books you'll never read* (Polygon; 2010).

33. Kelly (2010), p. 36; note 32.

34. See J. van Lawick-Goodall and H. van Lawick-Goodall *Nature* 212, 1468–1469 (1966); also J. van Lawick-Goodall *Advances in the Study of Behavior* 3, 195–249 (1970). The origins of this stone throwing have been discussed, and while the Egyptian vulture has a "particular propensity" to chuck things around, there is no evidence that this behavior is culturally transmitted; see C.R. Thouless et al. *Ibis* 131, 9–15 (1989).

35. See Y. Stoyanova et al. *Journal of Raptor Research* 44, 154–156 (2010).

36. In addition to the classic examples discussed in the text, other instances include gray flycatchers that employ the probe to feed on termites (see Boswell 1983; note 28).

37. See G.A. Wood *Corella* 8, 94–95 (1984); the drummer was a palm cockatoo. So, too, a captive Goffin's cockatoo employed a stick as a rake; see A.M.I. Auersperg et al. *Current Biology* 22, R903–R904 (2012).

38. See S. Tebbich et al. *Ecology Letters* 5, 656–664 (2002).

39. See I. Teschke et al. *Animal Behaviour* 82, 945–956 (2011), and I. Teschke and S. Tebbich *Animal Cognition* 14, 555–563 (2011); also S. Tebbich and R. Bshary *Animal Behaviour* 67, 689–697 (2004).

40. For an overview of the connection between brain size and tool use in birds, see L. Lefebvre et al. *Behaviour* 139, 939–973 (2002). A.N. Iwaniuk et al. (*Canadian Journal of Experimental Psychology* 63, 150–159; 2009) point out that such a correlation is not found when it comes to the relative size of the cerebellum, but the relationship does hold in terms of the degree of foliation, at least in terms of tool use as against employment of proto-tools.

41. Lefebvre et al. (2002), p. 939 [Abstract]; note 40.

42. For an early report see T.B. Jones and A.C. Kamil *Science* 180, 1076–1078 (1973), where captive and hungry northern blue jays shredded newspaper and used it to scoop food from outside their cage.

43. See R.P. Balda *Wilson Journal of Ornithology* 119, 100–102 (2007); his intriguing account includes the narrative of the Steller's jay first breaking off the stick and employing it as a lance before dropping it, and the crow then taking the initiative.

44. See C. Caffrey *Wilson Bulletin* 112, 283–284 (2000); also J. McCormick *Notornis* 54, 116–117 (2007).

45. See G.R. Hunt *Emu* 100, 109–114 (2000).

46. See their paper in *Emu* 102, 349–353 (2002).

47. Hunt and Gray (2002), p. 352; note 46.

48. See T. Troscianko et al. *Ardea* 96, 283–285 (2008); also C. Rutz et al. *Science* 318, 765 (2007).

49. Hunt and Gray (2002), p. 352; note 46.

50. See G.R. Hunt and R.D. Gray *Proceedings of the Royal Society of London, B* 271, S88–S90 (2004).

51. See A.A.C. Weir et al. *Science* 297, 981 (2002).

52. See the overview by C. Rutz and J.J.H. St Clair *Behavioural Processes* 89, 153–165 (2012).

53. See B. Kenward et al. *Ibis* 146, 652–660 (2004).

54. See C. Rutz et al. *Science* 329, 1523–1526 (2010).

55. Rutz et al. (2010), p. 1525; note 54.

56. Such a capacity is also evident from their willingness to use tools taken from a recently introduced climbing vine; see G.R. Hunt *New Zealand Journal of Zoology* 35, 115–118 (2008).

57. See G.R. Hunt et al. *Emu* 102, 283–290 (2002).

58. See J. Mehlhorn *Brain, Behavior and Evolution* 75, 63–70 (2010).

59. This extends to the crows using tools to assess unfamiliar objects (such as a metal frog or flashing bicycle light) that the experimenter has placed in the cage (see J.H. Wimpenny et al. *Animal Cognition* 14, 489–464 [2011]) or to prefer to use a tool when a model snake is too close for comfort (see A.H. Taylor et al. *Biology Letters* 8, 205–207 [2012]).

60. See B. Kenward et al. *Nature* 433, 121 (2005).

61. See B. Kenward et al. *Animal Behaviour* 72, 1329–1343 (2006); also G.R. Hunt and R.D. Gray *Behavioral and Brain Sciences* 30, 412–413 (2007), and G.R. Hunt et al. *New Zealand Journal of Zoology* 34, 1–7 (2007).

62. For an overview of the evolution of social structure and especially cooperative breeding, see J. Ekman and P.G.P. Ericson *Proceedings of the Royal Society of London, B* 273, 1117–1125 (2006).

63. See J.C. Holzaider et al. *Animal Behaviour* 81, 83–92 (2011).

64. See G.R. Hunt et al. *Ethology* 118, 423–430 (2012).

65. See J.C. Holzaider et al. *Behaviour* 147, 553–586 (2010).

66. See L.A. Bluff et al. *Proceedings of the Royal Society of London, B* 277, 1377–1385 (2010).

67. See G.R. Hunt and R.D. Gray *Animal Cognition* 7, 114–120 (2004).

68. See G.R. Hunt and R.D. Gray *Proceedings of the Royal Society of London, B* 270, 867–874 (2003).

69. See G.R. Hunt et al. *Australian Journal of Zoology* 55, 291–298 (2007).

70. Hunt and Gray (2003), pp. 873–874; note 68.

71. Hunt and Gray (2003), p. 873; note 68.

72. Hunt and Gray (2003), p. 873; note 68.

73. See G.R. Hunt and R.D. Gray *Biology Letters* 3, 173–175 (2007).

74. Hunt and Gray (2007), p. 174; note 73.

75. See, for example, T. Humle and T. Matsuzawa *American Journal of Primatology* 71, 40–48 (2009).

76. See L.F. Marchant and W.C. McGrew *Annals of the New York Academy of Sciences* 1288, 1–8 (2013).

77. See A. Bisazza et al. *Neuroscience and Biobehavioral Reviews* 22, 411–426 (1998).

78. See G.R. Hunt *Proceedings of the Royal Society of London, B* 267, 403–413 (2000).

171. See P. Krupczynski and S. Schuster *Current Biology* 18, 1961–1965 (2010).

172. See N.R. Franks and T. Richardson *Nature* 439, 153 (2006).

173. Tom Richardson and colleagues (see their paper in *Current Biology* 17, 1520–1526 [2007]) go on to remark, "By focusing on the underlying similarities among different taxa that achieve functionally similar outcomes, we would gain a deeper understanding both of the minimal criteria and the effects of the presence of more complex augmenting features" (p. 1524).

CHAPTER 25

1. For a valuable overview, see V.K. Bentley-Condit and E.O. Smith *Behaviour* 47, 185–221 and A1–A32 (2010).

2. See W.L. Merrill *Florida Entomologist* 55, 59–60 (1972), and J.T. Barber et al. *Animal Behaviour* 38, 550–552 (1989).

3. See V.S. Banschbach *Insectes Sociaux* 53, 463–471 (2006).

4. See G.W. Schultz *Journal of the Kansas Entomological Society* 55, 277–282 (1982). A somewhat similar case is described by M.H.J. Möglich and G.D. Alpert (*Behavioral Ecology and Sociobiology* 6, 105–113; 1979), involving ants chucking rocks into the entrance of nests of other ant species to discourage them from foraging. This, however, seems to be more a method of mechanical signaling rather than brute deterrence.

5. Schultz (1982), p. 278; note 4.

6. See my *Life's solution*, pp. 263–264.

7. These and other examples (include those of ant tool use) are reviewed by J.D. Pierce *Florida Entomologist* 69, 95–104 (1986).

8. See reports of food being broken against anvils by wrasse (Ł. Paśko *Zoo Biology* 29, 767–773; 2010) and tusk fish (A.M. Jones et al. *Coral Reefs* 30, 865; 2011).

9. See K. Delhey et al. *American Naturalist* 169 (Supplement), S145–S158 (2007).

10. Delhey et al. (2007), p. S154; note 9.

11. See J.J. Negro et al. *Animal Behaviour* 58, F14–F17 (1999); also commentary in vol. 64 (2002) by R. Arlettaz et al. (pp. F1–F3) and reply by Negro et al. (pp. F5–F7).

12. Negro et al. (1999), part of title; note 11.

13. Negro et al. (1999), p. F15; note 11.

14. See, for example, H.G. Vevers *Symposia of the Institute of Biology* 12, 133–139 (1964).

15. See L.S. Barham *Current Anthropology* 43, 181–190 (2002); also W. Roebroeks et al. *Proceedings of the National Academy of Sciences, USA* 109, 1889–1894 (2012).

16. See C.W. Curtis et al. *Nature* 449, 905–908 (2007); also commentary on pp. 793–794 by S. McBrearty and C. Stringer.

17. The color red not only has a powerful effect on human activities, but has widespread counterparts across the animal kingdom; see A.J. Elliot et al. *Journal of Experimental Psychology: General* 139, 399–417 (2010), who review this area in the context of sexual attraction. Also P. Prokop and M. Hromada *Ethology* 119, 605–613 (2013).

18. See C.S. Henshilwood et al. *Journal of Human Evolution* 57, 27–47 (2009).

19. See D.E. Bar-Yosef et al. *Journal of Human Evolution* 56, 307–314 (2009), and C.S. Henshilwood et al. *Science* 334, 219–222 (2011).

20. See, for example, papers in *Emu* by R.A. Gannan (vol. 30, pp. 39–41; 1930) and N. Chaffer (vol. 30, pp. 277–285; 1931).

21. See B.D. Bravery *Journal of Avian Biology* 37, 77–83 (2006); also R.E. Hicks et al. *Animal Behaviour* 85, 1209–1215 (2013).

22. See J.A. Endler et al. *Current Biology* 20, 1679–1684 (2010); also L.A. Kelley and J.A. Endler *Proceedings of the National Academy of Sciences, USA* 109, 20980–20985 (2012).

23. Endler et al. (2010), p. 1679 [Abstract]; note 22.

24. This is where crocodiles evidently use twigs to lure unsuspecting water-birds; see V. Dinets et al. *Ethology, Ecology & Evolution* 25, 174–184 (2013).

25. See S.K. Robinson *Wilson Bulletin* 106, 567–569 (1994).

26. See his paper in *Ibis* 128, 285–290 (1986).

27. The occurrences of such bait and lure fishing are reviewed by G.D. Ruxton and M.H. Hansell (*Ethology* 117, 1–19; 2011). While acknowledging that not only does this qualify as true tool use, evidently combining some degree of decision making and flexibility of behavior, they remind us that the cognitive demands of this process are yet to be established.

28. For two cases reported in *Waterbirds*, and specifically a gull in Paris and an egret in Kenya, see, respectively, the reports by P.-Y. Henry and J.-C. Aznar (vol. 29, pp. 233–234; 2006) and R.J. Post et al. (vol. 32, pp. 450–452; 2009), while the latter also report on a heron employed in passive bait fishing.

29. These include crows (reported by J. Boswell *Avicultural Magazine* 89, 170–181; 1983), buzzards (see A.H. Chisholm *Ibis* 96, 380–383 [1954]; also T. Aumann *Emu* 90, 141–144; 1990, although this report involves a captive bird), and curlews (see J.S. Marks and E.S. Hall *Condor* 94, 1032–1034 [1992]). See also S. Andersson *Condor* 91, 999 [1989]), who reports a fan-tailed crow mistaking a Ping-Pong ball for an egg, attempted to break it using several techniques, and ending up by holding a rock in its beak as a hammer. According to one report this activity extends to interfering humans; see S.W. Jones *Condor* 78, 409 (1976), who reports ravens employing rocks to defend their nest, while elsewhere these birds (as well as crows) have been observed on bombing missions (but using vegetation) to displace sitting gulls and evidently obtain any eggs (see W.A. Montevecchi *Condor* 80, 349 [1978]).

122. Cartmill (1990), p. 182; note 121.

123. Cartmill (1990), p. 183; note 121.

124. See R.E. Brockie and B. O'Brien *Notornis* 51, 52–53 (2004).

125. See S.J. Webster and L. Lefebvre *Animal Behaviour* 62, 23–32 (2001), who in their study of Barbados birds single out the carib grackle for its innovativeness.

126. See S. Watanabe et al. *Journal of the Experimental Analysis of Behavior* 63, 165–174 (1995).

127. For a series of cautionary remarks on Watanabe et al. (1995; note 126) that have a wider bearing on how far human perception can be extrapolated, see J. Monen et al. *Journal of the Experimental Analysis of Behavior* 69, 223–226 (1998). On the other hand, using the same set of paintings by Monet and Picasso, bees also show discrimination; see W. Wu et al. *Journal of Comparative Physiology, A* 199, 45–55 (2013).

128. Here the basis of the experiment involves a tube containing a piece of food that from one end can be recovered by using a stick (allowing a subset of experiments revolving around stick length, etc.), but if pushed from the other end results in the peanut (or whatever other morsel) falling into a trap.

129. See N. Mendes et al. *Biology Letters* 3, 453–455 (2007).

130. Mendes et al. (2007), p. 454; note 129.

131. Emery and Clayton (2009), p. 28; note 114.

132. See A.M. Seed et al. *Current Biology* 16, 697–701 (2006).

133. See her overview in *Current Biology* 16, R244–R245 (2006).

134. Chappell (2006), p. R244; note 133.

135. See J. Liedtke et al. *Animal Cognition* 14, 143–149 (2011). While failing the standard trap-tube puzzle, once a slot was made, the success rate was considerably better. The authors also point out aspects of parrot behavior that might militate against this specific cognitive challenge.

136. Liedtke et al. (2011), p. 147; note 135.

137. See, for example, C.D. Bird and N.J. Emery *Proceedings of the Royal Society of London, B* 277, 147–151 (2009).

138. See M. Nissani *Journal of Experimental Psychology: Animal Behavior Processes* 32, 91–96 (2006).

139. Nissani (2006), p. 91; note 138.

140. Seed et al. (2006), p. 700; note 132.

141. See fig. 3 in F.A. Beach's classic analysis of equating animal behavior with the rat in *American Psychologist* 5, 115–124 (1950).

142. See M. Pettit *British Journal for the History of Science* 43, 391–421 (2010).

143. See L.W. Cole *Journal of Animal Behavior* 2, 299–309 (1912); also L.F. Whitney *Journal of Mammalogy* 14, 108–114 (1933).

144. See M. Leal and B.J. Powell *Biology Letters* 8, 28–30 (2012).

145. Leal and Powell (2012), p. 28 [Abstract]; note 144.

146. See M. Vasconcelos et al. *Biology Letters* 8, 42–43 (2012), and on pp. 44–45 the reply by Leal and Powell.

147. See R. Bshary et al. *Animal Cognition* 5, 1–13 (2002); see also the interesting commentaries by J. Call (pp. 15–16) and A. Miklósi (pp. 17–18).

148. See K.P. Chandroo et al. *Fish and Fisheries* 5, 281–295 (2004).

149. See J.D. Rose *Diseases of Aquatic Organisms* 75, 139–154 (2007).

150. Rose (2007), p. 151; note 149.

151. See W. Wickler *Zeitschrift für Tierpsychologie* 14, 393–428 (1957; in German with English summary).

152. Wickler (1957), p. 427; note 151.

153. See papers by L.R. Aronson in *American Museum Novitates* 1486, 1–22 (1951), and *Annals of the New York Academy of Sciences* 188, 378–392 (1971).

154. Such jumping seems to be unrelated to a tidal pool becoming inhospitable by virtue of salinity or temperature, but may be a result of intraspecific competition.

155. See the helpful overview by S. Schuster *Current Biology* 17, R494–R495 (2007).

156. See, for example, S. Schuster et al. *Current Biology*, 16, 1565–1568 (2004).

157. Schuster (2007), p. R494; note 155.

158. See T. Schlegel et al. *Current Biology* 16, R836–R837 (2006).

159. See E. Arzt et al. *Proceedings of the National Academy of Sciences, USA* 100, 10603–10606 (2003).

160. See S. Schuster et al. *Current Biology* 16, 378–383 (2006).

161. See S. Wöhl and S. Schuster *Journal of Experimental Biology* 210, 311–324 (2007).

162. See Wöhl and Schuster (2007); note 161. They point out that this capacity for acceleration is all the more remarkable because, necessarily living close to the air-water interface, the amount of drag is greatly increased.

163. See S. Temple et al. *Proceedings of the Royal Society of London, B* 277, 2607–2615 (2010).

164. See A. Mokeichev et al. *Proceedings of the National Academy of Sciences, USA* 107, 16726–16731 (2010); also I. Rischawy and S. Schuster *Journal of Experimental Biology* 216, 3096–3103 (2013).

165. Mokeichev et al. (2010), p. 16728; note 164.

166. Mokeichev et al. (2010), p. 16729; note 164.

167. Wöhl and Schuster (2007), p. 323; note 161.

168. See T. Schlegel and S. Schuster *Science* 319, 104–106 (2008).

169. See J. Vrerke *Aquarien Magazin* 5, 66–69 (1971); also K.H. Lüling *Bonner Zoologische Beiträge* 20, 416–422 (1969; in German).

170. For the phylogenetic position of *Colisa* within the anabantoids, see L. Rüber et al. *Systematic Biology* 55, 374–397 (2006).

83. See K.M. Bohn et al. *PLoS ONE* 4, e6746 (2009).

84. See K.M. Stafford et al. *Journal of the Acoustical Society of America* 124, 3315–3323 (2008).

85. See R.S. Payne and S. McVay *Science* 173, 585–597 (1971); also, for example, N. Eriksen et al. *Behaviour* 142, 305–328 (2005).

86. While crows and parrots tend to take center stage with respect to cognitive capacity, the recognition of an episodic-like memory in hummingbirds is in line with this connection; see P.L. González-Gómez et al. *Animal Behaviour* 81, 1257–1262 (2011).

87. Kuhl (2003), p. 9646; note 45.

88. See, for example, R. Frey et al. *Journal of Anatomy* 210, 131–159 (2007).

89. See G. Hewitt et al. *Folia Primatologica* 73, 70–94 (2002).

90. See, for example, B. de Boer *Journal of the Acoustical Society of America* 126, 3329–3343 (2009).

91. See Z. Alemseged et al. *Nature* 443, 296–301 (2006); also commentary on pp. 278–281 by B. Wood.

92. See B. de Boer *Journal of Human Evolution* 62, 1–6 (2012).

93. See G.E. Weissengruber et al. *Journal of Anatomy* 201, 195–209 (2002).

94. See G. Peters *Mammal Review* 32, 245–271 (2002). Peters also draws attention to apparently contradictory examples of purring in cats, as when in labor or severely injured, and speculates this is a sort of autocommunication, a sort of equivalent of humming to oneself.

95. Apart from purring cats and monkeys (and at this juncture let us mention those humming bears; see G. Peters et al. *Acta Theriologica* 52, 379–389; 2007), oddly the giant panda is the exception, again indicative of "comfort or contentment" (p. 385); this sound also emanates from the civets (or viverrids).

96. See W.T. Fitch and D. Reby *Proceedings of the Royal Society of London, B* 268, 1669–1675 (2001).

97. See papers by R. Frey et al. in *Journal of Morphology* 269, 1223–1237 (2008), and *Journal of Anatomy* 218, 566–585 (2011), addressing, respectively, the Mongolian and goitred gazelles; also K.O. Efremova et al. *Naturwissenschaften* 98, 919–931 (2011).

98. See papers by B.D. Charlton et al. in *Journal of Experimental Biology* (vol. 214, pp. 3414–3422; 2011) and *Animal Behaviour* (vol. 84, pp. 1565–1571; 2012). The low notes, however, depend specifically on particular vocal folds; see B.D. Charlton et al. *Current Biology* 23, R1035–R1036 (2013).

99. See, for example, D. Reby et al. *Proceedings of the Royal Society of London, B* 272, 941–947 (2005); and A.G. McElligott et al. *Journal of Zoology London* 270, 340–345 (2006).

100. See A. MacLarnon and G. Hewitt *Evolutionary Anthropology* 13, 181–197 (2004).

101. See, for example, C.S.L. Lai et al. *Nature* 413, 519–523 (2001), with commentary on pp. 465–466 by S. Pinker and F. Liégeois et al. *Nature Neuroscience* 6, 1230–1237 (2003).

102. For an overview, see C. Scharff and J. Petri *Philosophical Transactions of the Royal Society of London, B* 366, 2124–2140 (2011), albeit from the stance of deep homology (and perhaps a dash of special pleading?).

103. See G. Li et al. *PLoS ONE* 2, e.900 (2007).

104. See S.E. Fisher and C. Scharff *Trends in Genetics* 25, 166–177 (2009).

105. See S. Haesler et al. *PLoS Biology* 5, e321 (2007); also C. Scharff and S. Haesler *Current Opinion in Neurobiology* 15, 694–703 (2005).

106. Haesler (2007), p. 2893; note 105.

107. See D.M. Webb and J. Zhang *Journal of Heredity* 96, 212–216 (2005).

108. See S.M.H. Gobes and J.J. Bolhuis *Current Biology* 17, 789–793 (2007).

109. Gobes and Bolhuis (2007), p. 791; note 108.

110. Brainard and Doupe (2000), p. 32 (box 1); note 40.

111. See L.C. Davenport *PLoS ONE* 5, e11385 (2010).

112. See G.R. Michener *Journal of Mammalogy* 85, 1019–1027 (2004).

113. Michener (2004), p. 1026; note 112.

114. See N.J. Emery and N.S. Clayton *Current Opinion in Neurobiology* 19, 27–33 (2009).

115. See J. Boswell *Avicultural Magazine* 89, 170–181 (1983); this author also documents a series of other tool uses by birds (p. 452; notes 23–36). In general, and despite their obvious cognitive capacities, the parrots are not known for their tool use. A. Borsari and E.B. Ottoni (*Animal Cognition* 8, 48–52; 2005), however, note the employment of wood to wedge nuts during feeding by captive hyacinth macaws from Brazil and also draw attention to its convergent appearance in Polynesian palm cockatoos.

116. See L.A. Bates and R.W. Byrne *Methods* 42, 12–21 (2007).

117. See their paper in *Ethology* 112, 493–502 (2006).

118. Seibt and Wickler (2006), p. 496; note 117.

119. See A. Krasheninnikova and R. Wanker *Behaviour* 147, 725–739 (2010).

120. See also the investigation by C. Shuck-Paim et al. (*Animal Cognition* 12, 287–301; 2009) of string pulling in three species of macaw. Once again, at least one parrot in each species succeeded in the string-pulling task, but a more challenging arrangement (involving crossed strings) defeated all of them; while their cognitive capacities were self-evident, it was not possible to demonstrate that these macaws grasped the causal connections. On the other hand a Harris's hawk showed a degree of competence that rivals corvids and parrots, although perhaps significantly this raptor is a social species and practices cooperative hunting (see E.N. Colbert-White et al. *Ibis* 155, 611–615 [2013]).

121. See his paper in *International Journal of Primatology* 11, 173–191 (1990).

(2003); also D. Lipkind et al. *Nature* 498, 104–108 (2013).

45. See her commentary in *Proceedings of the National Academy of Sciences, USA* 100, 9645–9646 (2003).

46. Kuhl (2003), p. 9645; note 45.

47. See A.P. King et al. *Ethology* 111, 101–117 (2005).

48. King et al. (2005), part of title; note 47.

49. See, for example, T.F. Wright and G.S. Wilkinson *Proceedings of the Royal Society of London, B* 268, 609–616 (2001).

50. See their paper in *Nature* 459, 564–568 (2009); also commentary on pp. 519–520 by W.T. Fitch.

51. Fehér et al. (2009), p. 567; note 50.

52. See, for example, M.S. Fee et al. *Nature* 395, 67–71 (1998); also C.P.H. Elemans et al. *Journal of Experimental Biology* 212, 1212–1224 (2009).

53. See T. Riede and F. Goller *Brain and Language* 115, 69–80 (2010); also D.N. Düring et al. *BMC Biology*, 11, e1 (2013).

54. Riede and Goller (2010), p. 77; note 53.

55. This includes employment of so-called superfast muscles that are essential for the rapid and high-frequency modulations (see C.P.H. Elemans et al. *PLoS ONE* 3, e2581 [2008]). Nor are they alone, because this remarkable physiological system (see L.C. Rome *Annual Review of Physiology* 68, 193–221 [2006] has also evolved in the bats, where it is essential for the terminal echolocatory buzz (see C.P.H. Elemans et al. *Science* 333, 1885–1888 [2011]). Further afield it has evolved independently in the toad-fish (such sonic musculature is convergent: see, for example, K.S. Boyle et al. *Journal of Morphology* 274, 377–394 [2013], and S. Millot and E. Parmentier *BMC Evolutionary Biology* 14, e24 [2014], while H.-K. Mok et al. (*Frontiers in Zoology* 8, e31; 2012) document a transitional case and rattlesnake to provide, respectively, impressive mating sounds and the canonical rattle (see L.C. Rome et al. *Proceedings of the National Academy of Sciences, USA* 93, 8095–8100 [1996]). Nor is this capacity restricted to the vertebrates; superfast sonic muscles have evolved independently in the lobster (where the extremely rapid contraction of an antennal muscle produces a characteristic buzzing sound; see papers in vol. 42 of *Journal of Cell Biology* [1969] by J. Rosenbluth [pp. 534–547] and M. Mendelson [pp. 548–563], as well as a letter from the latter in *Science* 334, 1202; 2011) and perhaps less surprisingly in the tympanal muscles of the very noisy cicada (see P.C. Nahirney et al. *FASEB Journal* 20, 2017–2026 [2006]; also R.K. Josephson and D. Young *Journal of Experimental Biology* 118, 185–208 [1985]). Superfast muscles are most likely employed in the very-fast-moving teneriffiid mites, at least to judge from their kinematics (see G.C. Wu et al. *Journal of Experimental Biology* 213, 2551–2556 [2010]).

56. See G.J.L. Beckers et al. *Current Biology* 14, 1592–1597 (2004); see also V.R. Ohms et al. (*Journal of Experimental Biology* 215, 85–92; 2012), who extend this work and draw attention to this articulation being

controlled also by the gape of the beak and length of the trachea in addition to lingual positioning.

57. Beckers et al. (2004), p. 1595; note 56.

58. For an overview, see A.J. Doupe et al. *Trends in Neurosciences* 28, 353–363 (2005).

59. See D. Aronov et al. *Science* 320, 630–634 (2008).

60. See, for example, the overview by R. Mooney *Learning & Memory* 16, 655–669 (2009).

61. See E.D. Jarvis and C.V. Mello *Journal of Comparative Neurology* 419, 1–31 (2000). A. Suh et al. (*Nature Communications* 2, e443; 2011) present evidence that parrots are the closest relatives of the passerines, with the implication that their capacity for vocalization is ancestral rather than convergent. Much depends on the paleoneurology of their common ancestors.

62. See M. Gahr *Journal of Comparative Neurology* 426, 182–196 (2000).

63. See E.D. Jarvis et al. *Nature* 406, 628–632 (2000).

64. See G. Feenders *PLoS ONE* 3, e1768 (2008).

65. See E.D. Jarvis *Journal of Ornithology* 148 (Supplement, 1), S35–S44 (2007).

66. See his paper in *Annals of the New York Academy of Sciences* 1016, 61–76 (2004).

67. Farries (2004), p. 70; note 66.

68. A tension that could throw light on the origins of human language and unrestricted by ethical constraints (such as decapitation) that apply to humans but not—for better or worse—to zebra finches.

69. See, for example, J.H. Goldberg and M.S. Fee *Journal of Neurophysiology* 103, 2002–2014 (2010), and J.H. Goldberg et al. *Journal of Neuroscience* 30, 7088–7098 (2010).

70. Jarvis (2007), p. S35 [Abstract]; note 65.

71. Jarvis (2007), p. S36; note 65.

72. See E.D. Jarvis *Annals of the New York Academy of Sciences* 1016, 749–777 (2004); this article also provides an excellent overview of the evolution of birdsong.

73. Jarvis (2004), p. 749 [Abstract]; note 72.

74. See, for example, papers in *Behaviour* by A.M. Elowson et al. (vol. 135, pp. 643–664; 1998) and C.T. Snowdon and A.M. Elowson (vol. 138, pp. 1235–1248; 2001).

75. See M. Knörnschild et al. *Naturwissenschaften* 93, 451–454 (2006).

76. See B. McCowan and D. Reiss *Journal of Comparative Psychology* 109, 242–260 (1995).

77. See V.B. Deecke et al. *Animal Behaviour* 60, 629–638 (2000); also L. Rendell and H. Whitehead *Animal Behaviour* 70, 191–198 (2005).

78. See B.J. Le Boeuf and L.F. Petrinovich *Animal Behaviour* 22, 656–663 (1974).

79. See D. Reiss and B. McCowan *Journal of Comparative Psychology* 107, 301–312 (1993).

80. See A.D. Foote et al. *Biology Letters* 2, 509–512 (2006); also R.L. Eaton *Carnivore* 2, 22–23 (1979).

81. See K. Ralls et al. *Canadian Journal of Zoology* 63, 1050–1056 (1985).

82. See M. Knörnschild et al. *Biology Letters* 6, 156–159 (2010).

448   NOTES

CHAPTER 24

1. See, for example, W.T. Fitch *The evolution of language* (Cambridge; 2010).
2. See his paper in *Neuron* 65, 795–814 (2010).
3. Fitch (2010), p. 796; note 2.
4. See, for example, papers by L.M. Herman et al. *Cognition* 16, 129–219 (1984), and *Journal of Experimental Psychology: General* 122, 184–194 (1993).
5. See papers in *Proceedings of the National Academy of Sciences, USA* by V.M. Janik et al. (vol. 103, pp. 8293–8297; 2006) and S.L. King and V.M. Janik (vol. 110, pp. 13216–13221; 2013).
6. Janik et al. (2006), p. 8295; note 5.
7. See his paper in *Current Biology* 16, R598–R599 (2006).
8. Barton (2006), p. R599; note 7.
9. See R. Ferrer-i-Cancho and B. McCowan *Entropy* 11, 688–701 (2009).
10. See E. Mercado et al. *Behavioural Processes* 50, 79–94 (2000).
11. See J. McDermott and M.D. Hauser *Cognition* 104, 654–668 (2007).
12. See, for example, D. Porter and A. Neuringer *Journal of Experimental Psychology: Animal Behavior Processes* 10, 138–148 (1984), and S. Watanabe and K. Sato *Behavioural Processes* 47, 53–57 (1999), who address discriminative capacities of pigeons and Java sparrows, respectively.
13. See, for example, C.K. Catchpole and P.J.B. Slater *Bird song: biological themes and variations* (Cambridge; 1995).
14. See C.J. Clark and T.J. Feo *Proceedings of the Royal Society of London, B* 275, 955–962 (2008).
15. See C.J. Clark *Biology Letters* 4, 341–344 (2008).
16. See K.S. Bostwick and R.O. Prum *Journal of Experimental Biology* 206, 3693–3706 (2003).
17. See K.S. Bostwick and R.O. Prum *Science* 309, 736 (2005).
18. See K.S. Bostwick et al. *Proceedings of the Royal Society of London, B* 277, 835–841 (2010); nor is it accidental that in contrast to typical avian bones, the ulna is now massively constructed; see K.S. Bostwick et al. *Biology Letters* 8, 760–763 (2012).
19. Bostwick and Prum (2003), p. 3703; note 16.
20. See also C.J. Clark et al. *Science* 333, 1430–1433 (2011), who in describing sonation in hummingbirds remark, "we hypothesize that acoustic communication signals produced by aeroelastic flutter have evolved many times in birds" (p. 1433).
21. See R.O. Prum *Animal Behaviour* 55, 977–994 (1998).
22. Bostwick and Prum (2005), p. 736; note 17.
23. In the insects, such an ability has evolved numerous times, and in the katydids includes the production of ultrasound (see F. Montealegre-Z et al. *Journal of Experimental Biology* 209, 4923–4937 [2006]) and dates back to at least the Jurassic (see J.-J. Gu et al. *Proceedings of the National Academy of Sciences, USA* 109,

3868–3873 [2012]; also commentary on pp. 3606–3607 by J. Rust).
24. Familiar as are insect stridulations—think of the cricket and cicada—among the arthropods it has evolved independently in giant millipedes; see T. Wesener et al. *Naturwissenschaften* 98, 967–975 (2011).
25. Bostwick and colleagues (2010; note 18) also remark, "Particularly intriguing is the convergence of this mechanism of sound production to the Castanet moth" (p. 841).
26. See N.I. Mann et al. *Biology Letters* 2, 1–4 (2006).
27. Mann et al. (2006), p. 1 [Abstract]; note 26.
28. See N.I. Mann et al. *Behaviour* 146, 1–43 (2009).
29. See, for example, E.A. Brenowitz *Journal of Neurobiology* 33, 517–531 (1997).
30. See, for example, P. Laiolo and A. Rolando *Evolutionary Ecology* 17, 111–123 (2003); also P. Enggist-Dueblin and U. Pfister *Animal Behaviour* 64, 831–841 (2002), who document cultural transmission of their vocalizations.
31. See, for example, N.E. Collias and E.C. Collias *Behaviour* 141, 1151–1171 (2004), who document convergence of calls in African finches, specifically with corvids. For other examples of acoustic convergence between birds, see, for example, the case of a penguin and shearwater documented by P. Jouventin and T. Aubin *Ibis* 142, 645–656 (2006).
32. See, for example, A.R.J. Ferreira *Auk* 123, 1129–1148 (2006); also M. Araya-Salas and T. Wright *Biology Letters* 9, 20130625 (2013).
33. See A.B. Bond and J. Diamond *Behaviour* 142, 1–20 (2005).
34. Extending to the hissing of owls, a sound that evidently serves to imitate rattlesnakes and so deter squirrels from moving into their burrows, thus qualifying as an example of "Acoustic Batesian Mimicry"; see M.P. Rowe et al. *Ethology* 72, 53–71 (1986; part of the title).
35. See E. Goodale and S.W. Kotagama *Proceedings of the Royal Society of London, B* 273, 875–880 (2006); also E. Goodale et al. *Ethology* 120, 266–274 (2014).
36. Goodale and Kotagama (2006), p. 878; note 35.
37. For reports of corresponding mimicry in the wild, see A.J. Cruickshank et al. *Ibis* 135, 293–299 (1993).
38. See L. Bottoni et al. *Ethology, Ecology and Evolution* 15, 133–141 (2003).
39. See, for example, A.J. Doupe and P.K. Kuhl *Annual Review of Neuroscience* 22, 567–631 (1999), and J.J. Bolhuis *Nature Reviews Neuroscience* 11, 747–759 (2010).
40. See, for example, M.S. Brainard and A.-J. Doupe *Nature Reviews Neuroscience* 1, 31–40 (2000).
41. See T.J. Gardner et al. *Science* 308, 1046–1049 (2005).
42. See W.-C. Liu et al. *Proceedings of the National Academy of Sciences, USA* 101, 18177–18182 (2004).
43. Liu et al. (2004), p. 18182; note 42.
44. See M.H. Goldstein et al. *Proceedings of the National Academy of Sciences, USA* 100, 8030–8035

of the neoavians, precocial development has emerged a number of times and from which altricial breeding has then reevolved, notably in some seabirds. See R.E. Ricklefs and J.M. Starck in *Avian growth and development: Evolution within the altricial-precocial spectrum* (J.M. Starck and R.E. Ricklefs, eds.), pp. 366–380 (Oxford; 1998).

135. Ricklefs and Starck (1998), p. 379; note 134.

136. See A.N. Iwaniuk et al. *Brain, Behavior and Evolution* 68, 45–62 (2006); the list of avian groups with foliated cerebelli shows that there is a phylogenetic signal, as well as body size, but cognitive capacity is probably also a factor.

137. See J.M. Burkart et al. *Evolutionary Anthropology* 18, 175–186 (2009).

138. Brain endocasts of some Cretaceous birds also suggest they were crepuscular hunters; see E.N. Kurochkin et al. *Biology Letters* 3, 309–313 (2007).

139. See A.C. Milner and S.A. Walsh *Zoological Journal of the Linnean Society* 155, 198–219 (2009), and S. Walsh and A. Milner *Journal of Systematic Palaeontology* 9, 173–181 (2011).

140. See E.D. Jarvis et al. [and The Avian Brain Nomenclature Consortium] *Nature Reviews Neuroscience* 6, 151–159 (2005).

141. See their paper in *Anatomical Record, A* 287, 1080–1102 (2005).

142. Reiner et al. (2005), p. 1088; note 141.

143. For a concise overview, see A. Reiner *Biology Letters* 5, 122–124 (2009); also Reiner et al. (2005); note 141.

144. See N.J. Emery and N.S. Clayton *Science* 306, 1903–1907 (2004); also A. Seed et al. *Ethology* 115, 401–420 (2009).

145. See A.B. Butler and R.M.S. Cotterill *Biological Bulletin* 211, 106–127 (2006), and A.B. Butler *Brain Research Bulletin* 75, 442–449 (2008).

146. See, for example, L. Puelles et al. *Journal of Comparative Neurology* 424, 409–438 (2000).

147. See L. Medina and A. Reiner *Trends in Neuroscience* 23, 1–12 (2000), who are supportive of the homology arguments.

148. See, for example, A. Abellán et al. *Journal of Comparative Neurology* 518, 3512–3528 (2010a), addressing the expression of LIM-homeodomain genes *Lhx1* and *Lhx5*; also Abellán et al. *Cerebral Cortex* 20, 1788–1798 (2010b).

149. See T. Nomura et al. *PLoS ONE* 3, e1454 (2008).

150. Nomura et al. (2008), p. 3; note 149.

151. Nomura et al. (2008), p. 8; note 149; as the authors acknowledge this is a near-quote from the illuminating review of cortical development in mammals by P. Rakic (*Brain Research Reviews* 55, 204–219; 2007, see p. 214).

152. See F. Tissis et al. *Brazilian Journal of Medical and Biological Research* 35, 1473–1484 (2002); also A.M. Goffinet et al. *Journal of Comparative Neurology* 414, 533–550 (1999).

153. See E. Pérez-Costas et al. *Journal of Chemical*

Neuroanatomy 27, 7–21 (2004), for its occurrence in lamprey.

154. The case for the SVZ in the monotremes is not clear; note 51.

155. See C.J. Charvet et al. *Brain, Behavior and Evolution* 73, 285–294 (2009).

156. See G.F. Striedter and C.J. Charvet *Biology Letters* 5, 134–137 (2009).

157. See Y. Wang et al. *Proceedings of the National Academy of Sciences, USA* 107, 12676–12681 (2010).

158. Wang et al. (2010), p. 12679; note 157.

159. See his paper in *Current Opinion in Neurobiology* 15, 686–693 (2005a).

160. Güntürkün (2005a), p. 691; note 159.

161. See also O. Güntürkün *Brain Research Bulletin* 66, 311–316 (2005b), where the same thesis is set out, and with emphasis on dopaminergic functions.

162. Güntürkün (2005a), p. 686 [Abstract]; note 159.

163. Güntürkün (2005a), p. 691; note 159.

164. See A.N. Iwaniuk et al. *Journal of Zoology, London* 263, 317–327 (2004).

165. Iwaniuk et al. (2004); note 164, suggest that in either case the demands of folivory (p. 206) and a long intestine for microbial fermentation in the kakapo, and thermoregulatory requirements (or a heavier skeleton) in the Great auk led to the familiar energetic trade-off between the brain and other needs.

166. See K.W.S. Ashwell and R.P. Scofield *Brain, Behavior and Evolution* 71, 151–166 (2008).

167. See J.R. Corfield et al. *Brain, Behavior and Evolution* 71, 87–99 (2008); also Ashwell and Scofield (2008); note 166.

168. For an overview, see L. Lefebvre and D. Sol *Brain, Behavior and Evolution* 72, 135–144 (2008); their survey extends beyond birds to the convergent similarities with primates, but also emphasizes the complexity of this area.

169. See their paper in *Brain, Behavior and Evolution* 65, 40–59 (2004).

170. Iwaniuk et al. (2004), p. 53; note 169.

171. C. Schuck-Paim et al. (*Brain, Behavior and Evolution* 71, 200–215; 2008), for example, find a correlation between brain size and environmental variability in Neotropical parrots; interestingly, "larger-brained species are more tolerant to climatic variability" (p. 210).

172. See their paper in *Journal of Comparative Neurology* 507, 1663–1675 (2008).

173. Striedter and Charvet (2008), p. 1664; note 172.

174. See J. Cnotka et al. *Neuroscience Letters* 433, 241–245 (2008); these authors place the New Caledonian crow as fourth in the pecking order (so to speak), exceeded by two macaws and a woodpecker, although in the latter case it has a large cerebellum; see also K.A. Jønsson et al. *BMC Evolutionary Biology* 12, e72 (2012), who argue that corvids as a whole are encephalized.

93. Shoshani et al. (2006), p. 154; note 92.

94. See M. Goodman et al. *Proceedings of the National Academy of Sciences, USA* 106, 20824–20829 (2009); also vol. 107, p. 8498 (2010).

95. See, for example, A.Y. Hakeem et al. *Anatomical Record, A* 287, 1117–1127 (2005); also P.R. Manger et al. *Neuroscience* 167, 815–824 (2010).

96. See B.L. Hart and L.A. Hart in *Evolution of the nervous system: A comprehensive reference*, vol. 3, *Mammals* (J.H. Kaas and L.A. Krubitzer, eds.), pp. 491–497. (Academic; 2007).

97. See B. Jacobs et al. *Brain, Structure & Evolution* 215, 273–298 (2011).

98. Jacobs et al. (2011), p. 294; note 97.

99. See A.Y. Hakeem et al. *Anatomical Record* 292, 242–248 (2009).

100. Hakeem et al. (2009), p. 242; note 99.

101. See J.M. Allman et al. *Brain Structure & Evolution* 214, 495–517 (2010).

102. See C. Butti and P.R. Hof *Brain Structure & Evolution* 214, 477–493 (2010); also C. Butti et al. *Cortex* 49, 312–326 (2013); significantly, occurrences of VENs include the pigmy hippo (which is aquatic and closely related to the cetaceans), as well as the manatee (quite close to the elephants), walrus, and perhaps more surprisingly the zebra.

103. See C. Butti et al. *Journal of Comparative Neurology* 515, 243–259 (2009); also P.R. Hof and E. van der Gucht *Anatomical Record* 290, 1–31 (2007), and Butti and Hof (2010); note 102.

104. For an overview, see H.H.A. Oelschläger and J.S. Oelschläger in *Encyclopedia of marine mammals* (W.F. Perrin et al., eds.), pp. 133–158 (Academic; 2002).

105. See his paper in *Brain Research Bulletin* 75, 450–459 (2008).

106. Oelschläger (2008), part of title; note 105.

107. Oelschläger (2008), p. 455; note 105.

108. S.-X. Xu et al. (*Proceedings of the Royal Society of London, B* 279, 4433–4440; 2012) documented evidence for convergent evolution with respect to positive selection of the ASPM gene (abnormal spindlelike microcephaly) that is already implicated with encephalization in primates. A number of amino-acid sites also show evidence for positive selection, but unlike *prestin* (p. 137), here there is no evidence for specific molecular convergence. See, however, a critique by S.H. Montgomery et al. in a subsequent issue (vol. 281, 20131743; 2014).

109. See S.H. Montgomery et al. (*Evolution* 67, 3339–3353; 2013), who remind us that although cetaceans had the highest encephalization quotients (EQ) on the planet until the arrival of *Homo*, the stories of cetacean and primate encephalization show important differences, not least relative decreases of EQ among the mysticetes.

110. See L. Marino *Brain, Behavior and Evolution* 56, 204–211 (2000).

111. See also L. Marino et al. *Brain, Behavior and Evolution* 51, 230–238 (1998).

112. See H.H.A. Oelschläger et al. *Brain, Behavior and Evolution* 71, 68–86 (2008). The extent of gyrification in cetaceans such as the dolphins and pilot whales substantially exceeds that of humans, although this may be in part because of correlations between volume of brain and surface area, and between surface area and length of exposed gyri; see H. Elias and D. Schwartz *Science* 166, 111–113 (1969). Although the emphasis is here on the odontocetes, the mysticete brain is also highly gyrified; see N. Eriksen and B. Pakkenberg *Anatomical Record* 290, 83–95 (2007).

113. Oelschläger et al. (2008), p. 77; note 112.

114. See L. Marino et al. *Anatomical Record, A* 281, 1256–1263 (2004).

115. See R.J. Tarpley and S.H. Ridgway *Brain, Behavior and Evolution* 44, 156–165 (1994).

116. For further discussion, see my *Life's solution*, p. 254, and note 144 on p. 422; also S. Ridgway et al. *Journal of Experimental Biology* 209, 2902–2910 (2006).

117. See O.L. Lyamin et al. *Neuroscience & Biobehavioral Reviews* 32, 1451–1484 (2008), who remind us that "this unusual form of mammalian sleep" (part of the title) applies to all cetaceans, and while most likely of ancient origin may have evolved more than once.

118. Notably in the eared seal and manatees; see N.C. Rattenborg et al. *Neuroscience & Biobehavioral Reviews* 24, 817–842 (2000).

119. See, for example, N.C. Rattenborg et al. *Nature* 397, 397–398 (1999); also Rattenborg et al. (2000); note 118, and T. Fuchs et al. *Biology Letters* 5, 77–80 (2009).

120. See L. Marino *Anatomical Record, A* 290, 694–700 (2007).

121. See L. Marino et al. *Journal of Mammalian Evolution* 7, 81–94 (2000).

122. See L. Marino et al. *Anatomical Record, A* 281, 1247–1256 (2004).

123. See, for example, L. Marino *Evolutionary Anthropology* 5, 81–85 (1996).

124. See, for example, I.I. Glezer and P.J. Morgane *Brain Research Bulletin* 24, 401–427 (1990).

125. See, for example, P.R. Hof et al. *Anatomical Record, A* 287, 1142–1152 (2005).

126. Hof et al. (2005) p. 1151; note 125.

127. See J. Huggenberger *Journal of the Marine Biological Association, UK* 88, 1103–1108 (2008); also corrigendum in vol. 90, p. 213 (2010).

128. Huggenberger (2008), p. 1103; note 127.

129. Huggenberger (2008), p. 1105; note 127.

130. See, for example, N.J. Emery and N.S. Clayton *Science* 306, 1903–1907 (2004).

131. See their paper in *Biology Letters* 5, 125–129 (2009).

132. Isler and van Schaik (2009), title of their paper; note 131.

133. See also S. Schultz and R.I.M. Dunbar *Biological Journal of the Linnean Society* 100, 111–123 (2010).

134. This conclusion is based on the observation that while altricial breeding occurs in even basal members

41. See A.N. Iwaniuk and P.L. Hurd *Brain, Behavior and Evolution* 65, 215–230 (2005).

42. Clark et al. (2001), p. 191; note 38; these authors point out there are "distinct changes in the distribution of brain volume among other regions" (p. 191).

43. Iwaniuk and Hurd (2005), p. 228; note 41.

44. Iwaniuk and Hurd (2005), p. 228; note 41.

45. See his paper in *Current Biology* 15, R649–R650 (2005).

46. Sultan (2005), p. R650; note 45.

47. Sultan (2005), p. R649; note 45.

48. Sultan (2005), p. R649; note 45.

49. See W. de Winter and C. Oxnard *Nature* 409, 710–714 (2001); also commentary by W.M. Brown in *Trends in Ecology & Evolution* 16, 471–473 (2001).

50. de Winter and Oxnard (2001), p. 712; note 49.

51. See A.F.P. Cheung et al. *Cerebral Cortex* 20, 1071–1081 (2010).

52. Cheung et al. (2010), p. 1079; note 51.

53. See, for example, M.L. Reynolds *Anatomy and Embryology* 173, 81–94 (1985).

54. See their paper in *Progress in Neurobiology* 82, 122–141 (2007).

55. Karlen and Krubitzer (2007), p. 122 [Abstract]; note 54.

56. Karlen and Krubitzer (2007; note 54) point out that in this case of a somatosensory convergence, the packing arrangement of central and peripheral cells is reversed in the placentals.

57. See J.H. Kaas *Brain, Behavior and Evolution* 59, 262–272 (2002).

58. See J.E. Janečka et al. *Science* 318, 792–794 (2007).

59. Kaas (2002), p. 270; note 57.

60. See J.H. Kaas *Brain and Mind* 1, 7–23 (2000).

61. See K.W.S. Ashwell and C.D. Hardman *Brain, Behavior and Evolution* 79, 57–72 (2012).

62. See M. Hassiotis et al. *Comparative Biochemistry and Physiology, A* 136, 827–850 (2003).

63. Hassiotis et al. (2003), p. 844; note 62.

64. Hassiotis et al. (2003), p. 849; note 62.

65. Hassiotis et al. (2003), p. 844; note 62.

66. See his chapter in *The biology of marsupials* (D. Hunsaker, ed.), pp. 157–278 (Academic; 1977); also Karlen and Krubitzer (2007); note 54.

67. Johnson (1977), p. 262; note 66.

68. For an overview, see J.E. Nelson and H. Stephan in *Carnivorous marsupials* (M. Archer, ed.), vol. 2, pp. 699–706 (Royal Zoological Society of New South Wales; 1982).

69. Karlen and Krubitzer (2007), p. 123; note 54.

70. See A.N. Iwaniuk et al. (2001); they quote unpublished data (table 3) giving this animal an EQ of 320.

71. Karlen and Krubitzer (2007), p. 123; note 54. In addition K.W.S. Ashwell *Brain, Behavior and Evolution* 71, 181–191 (2008), notes that the EQ of the honey possum and sugar glider are comparable to prosimians and may reflect the fact that each are arboreal.

72. See M.T. Silcox et al. *Journal of Human Evolution* 58, 505–521 (2010).

73. Silcox et al. (2010), p. 518; note 72.

74. Silcox et al. (2010), p. 519; note 72.

75. Silcox et al. (2010), p. 519; note 72.

76. Silcox et al. (2010), p. 519; note 72.

77. See W. Hartwig et al. *Anatomical Record* 294, 2207–2221 (2011), who document larger brain sizes in *Cebus* (EQ of 3.08 in *C. apella*) and *Saimiri*, as well as relatively large brains in *Cacajao* and *Chiropotes*. In contrast the howler (*Alouatta*) has a conspicuously smaller brain, which the authors suggest might reflect a shift in the face to make room for an expanded throat.

78. See E.L. Simons et al. *Proceedings of the National Academy of Sciences, USA* 104, 8731–8736 (2007).

79. Simons et al. (2007), p. 8735; note 78.

80. See K.E. Sears et al. *American Museum Novitates* 3617, 1–29 (2008).

81. See R. Kay et al. *Journal of Vertebrate Paleontology* 26 (Supplement, 3), 83A–84A (2007).

82. See M.F. Tejedor et al. *Proceedings of the National Academy of Sciences, USA* 103, 5437–5441 (2006).

83. See their paper in *Journal of Vertebrate Paleontology* 28 (Supplement, 3), 99A (2008).

84. Kay et al. (2008), p. 99A; note 83. The four instances they list are the stem-group platyrrhines, cebids, atelines, and pithecids, and they also note that brain reduction has occurred at least once in the cebids.

85. Kay et al. (2008), p. 99A; note 83.

86. See L. Krubitzer *Annals of the New York Academy of Sciences* 1156, 44–67 (2009).

87. See his paper in *Scripta Geologica* 139, 1–93 (2009).

88. Lyras (2009), p. 74; note 87.

89. Lyras (2009), p. 73; note 87.

90. See S. Pitnick et al. *Proceedings of the Royal Society of London, B* 273, 719–724 (2006).

91. Yet in the case of the hypercarnivorous canids, they are evidently more vulnerable to extinction (see B. van Valkenburgh et al. *Science* 306, 101–104 [2004]), while in the bats in general the more demanding their ecology, such as hunting in more complex environments (see J.M. Ratcliffe et al. *Brain, Behavior and Evolution* 67, 165–176 [2006]) or enhancing their spatial memory (see J.M. Ratcliffe *Behavioural Processes* 80, 247–251 [2009]), then the larger the brain.

92. See their paper in *Brain Research Bulletin* 70, 124–157 (2006); also J. Benoit et al. (*Brain, Behavior and Evolution* 81, 154–169; 2013), who point out some aspects (e.g., gyrification) are primitive among the elephant's relatives in the Afrotheria, with the corollary that the "simple" brain in tenrecs may be secondary. More generally, I. Kelava et al. (*Frontiers in Neuroanatomy* 7, e16; 2013) suggest that mammals initially showed gyrification, and a reversion to so-called lissencephaly (and so a smoother brain) took place several times independently, including in manatees.

cerebellum-like structures have evidently evolved independently, including the medial octavolateral nucleus associated with the lateral-line system (p. 146), the electrosensory lobe of weakly electric fish (p. 134), and the dorsal cochlear nucleus (pp. 128–129) of mammals. In describing these and other examples C.C. Bell (*Brain, Behavior and Evolution* 59, 312–326 [2002]; also N.B. Sawtell and C.C. Bell *Journal of Physiology: Paris* 102, 223–232; 2008) emphasizes that disentangling independent evolution and homology is not straightforward, but suggests that these structures will emerge when the appropriate developmental machinery for the somatosensory region is duly activated.

12. For an overview, see K.E. Yopak *Journal of Fish Biology* 80, 1968–2033 (2012).

13. See K.E. Yopak et al. *Proceedings of the National Academy of Sciences, USA* 107, 12946–12951 (2010), who noted that the allometry of the shark brains is strikingly similar to that of mammals; in addition there are interesting parallels in gyrification.

14. See K.E. Yopak et al. *Brain, Behavior and Evolution* 69, 280–300 (2007).

15. See T.J. Lisney et al. *Brain, Behavior and Evolution* 72, 262–282 (2009).

16. This might entail either the topological complexities of coral reefs or the challenges of dimensionality in open water. So, too, there are some convergences between the pelagic sharks and equivalent teleosts in the form of tuna. Nevertheless, there is a different emphasis on sensory modalities, and although the tuna has a large brain, it is outstripped by the sharks, with the palm going to the pelagic crocodile shark (*Pseudocarcharias*); see T.J. Lisney and S.P. Collin *Journal of Fish Biology* 68, 532–554 (2006). Thus, the highly foliated cerebellum of the whale shark has evidently evolved independently of open-water lamnids; see K.E. Yopak and L.R. Frank *Brain, Behavior and Evolution* 74, 121–142 (2009). With respect to coral-reef habitats, the largest brains in the teleost fish are found in such groups as the labrids, although here also the demands to remember the local geography are not necessarily easy to disentangle from social factors such as schooling and ranking; see R. Bauchot et al. *Copeia* 1977, 42–46 (1977).

17. In the case of the sharks G.E. Nilsson et al. (*Proceedings of the Royal Society of London, B,* 267, 1335–1339; 2000) raise the fascinating possibility that by employing a more efficient biochemistry (linked to the otherwise energetically demanding pumping of sodium and potassium across the ion channels of nerve cells), these fish have stumbled on a more effective neural architecture.

18. See T.J. Lisney and S.P. Collin, *Journal of Fish Biology* 68, 532–554 (2006).

19. See his chapter (pp. 117–193) in *Sensory biology of sharks, skates and rays* (E.S. Hodgson and R.F. Mathewson, eds.; Office of Naval Research; 1978).

20. Northcutt (1978), p. 118; note 19.

21. See, for example, A.P. Klimley *Zeitschrift für Tierpsychologie* 70, 297–319 (1985).

22. In this section I only address vertebrates, but the prize for brain reduction surely goes to a parasitic wasp (*Megaphragma*), which is much the same size as a *Paramecium* (c. 200 μm) and has a correspondingly minute brain composed of fewer than four hundred neurons, not surprisingly lacking nuclei. As A.A. Polilov (*Arthropod Structure & Development* 41, 29–34; 2012) points out, despite this extraordinary miniaturization, the wasp still can fly and find its hosts. For an overview of the implications for miniaturized nervous systems, see J.E. Niven and S.M. Farris *Current Biology* 22, R323–R329 (2012).

23. See their paper in *Biology Letters* 1, 283–286 (2005).

24. Safi et al. (2005), title of their paper; note 23.

25. Safi et al. (2005), p. 285; note 23.

26. Safi et al. (2005), p. 285; note 23.

27. For another interesting example, but this time in deep cold-water snail fishes, see M.J. Lannoo et al. (*Copeia* 2009, 732–739; 2009), where diminution in some sensory modalities (but not olfaction) has evidently led to significant brain loss.

28. See M. Köhler and S. Moyà-Solà *Brain, Behavior and Evolution* 63, 125–140 (2004); also M.R. Palombo et al. *Quaternary International* 182, 160–183 (2008). See also Ph. Derenne *Mammalia* 36, 459–481; 1972 (in French with English summary) for a similar example among feral cats in Kerguelen.

29. These have a number of other interesting consequences, not least changes in dentition, smaller eyes set in orbits that now allowed binocular vision, and loss of the shock absorbers in the legs; see Köhler and Moyà-Solà (2004) and Palombo et al. (2008); note 28.

30. Köhler and Moyà-Solà (2004), p. 136; note 28.

31. See, for example, M. Röhrs and P. Ebinger *Zeitschrift für zoologische Systematik und Evolutionsforschung* 31, 233–239 (1993; in German, with English summary).

32. See, for example, P. Ebinger *Brain, Behavior and Evolution* 45, 286–300 (1995).

33. Ebinger (1995), p. 298; note 32.

34. See, for example, M.W. Bruford et al. *Nature Reviews Genetics* 4, 900–910 (2003).

35. Ebinger (1995), p. 298; note 32.

36. See R.A. Barton and P.H. Harvey *Nature* 405, 1055–1058 (2000).

37. See A.N. Iwaniuk et al. *Proceedings of the Royal Society of London B (Supplement)* 271, S148–S151 (2004).

38. See their paper in *Nature* 411, 189–193 (2001); also commentary on pp. 141–142 by J.H. Kaas and C.C. Collins.

39. See also L. Marino et al. *Brain, Behavior and Evolution* 56, 204–211 (2000).

40. Clark et al. (2001), p. 190; note 38.

156. See S.E. Fahrbach et al. *Journal of Neurobiology* 57, 141–151 (2003).

157. In the bumblebee, and probably because of its rather different social structure, there is again a correlation with brain size and foraging, but this time with respect to the medial calyx of the mushroom body; see A.J. Riveros and W. Gronenberg *Brain, Behavior and Evolution* 75, 138–148; 2010.

158. See S. O'Donnell et al. *Neuroscience Letters* 356, 159–162 (2004).

159. See T.A. Jones et al. *Neurobiology of Learning and Memory* 92, 485–495 (2009).

160. See W. Gronenberg et al. *Journal of Experimental Biology* 199, 2011–2019 (1996).

161. See S. Kühn-Bühlmann and R. Wehner *Journal of Neurobiology* 66, 511–521 (2006).

162. See A.R. Adam et al. *Proceedings of the Royal Society of London, B* 277, 2157–2163 (2010).

163. In a comparable way the brains (and notably the mushroom body) of gregarious locusts are larger than their solitary counterparts; see S.R. Ott and S.M. Rodgers *Proceedings of the Royal Society of London, B* 277, 3087–3096 (2010).

164. Somewhat curiously, in a study of the acceleration of genes linked to bee eusociality, although there are unsurprising correlations with such areas as gland development and carbohydrate metabolism in the case of those associated with the brain, the acceleration is much more pronounced in bees with a primitive eusociality; see S.H. Woodard et al. *Proceedings of the National Academy of Sciences, USA* 108, 7472–7477 (2011). They suggest that the more fluid social environment in comparison with the advanced eusociality of the honeybee might be the driver.

165. See G.S. Withers et al. *Developmental Neurobiology* 68, 73–82 (2008).

166. See S.M. Farris and N.S. Roberts *Proceedings of the National Academy of Sciences, USA* 102, 17394–17399 (2005).

167. Although seemingly opposite, the fact that among a suite of nymphalid butterflies the size of the mushroom body varies by an order of magnitude, with the largest being found in a species of *Heliconius* that specializes on passionflowers, could indicate particular cognitive demands; see J. Sivinski *Journal of Insect Behavior* 2, 277–283 (1989).

168. Farris and Roberts (2005), p. 17398; note 166.

169. See S.M. Farris *Brain, Behavior and Evolution* 72, 1–15 (2008*b*); also Farris (2008*a*; note 85). Note also an interesting correlation between the ability of insects (and other arthropods) to engage in discontinuous gas exchange (DGC) of the trachea and the possession of an enlarged and convoluted mushroom body. DGC has evolved multiple times, but as P.G.D. Matthews and C.R. White (*American Naturalist* 177, 130–134; 2011) point out, it is in just such groups where respiration need not be under central control that decapitation does not prevent ventilation.

170. See S.M. Farris and N.J. Strausfeld *Journal of Comparative Neurology* 456, 305–320 (2003).

171. Farris and Strausfeld (2003), p. 319; note 170.

172. See her paper in *Arthropod Structure & Development* 34, 211–234 (2005); also Farris (2008*a*); note 85.

173. Farris (2005), p. 212; note 172.

174. For example, Farris (2005; note 172) remarks, "The Hymenoptera thus provide a robust example of two motifs that have been repeatedly demonstrated in the evolution of the vertebrate cortex: the elaboration of specific sensory inputs and a concomitant expansion of their target regions in the higher processing centers" (p. 224).

175. Continuing study of the annelidan equivalent of the mushroom bodies will be instructive. R. Tomer et al. *Cell* 142, 800–809 (2010; also commentary on pp. 679–681 by L.B. Sweeney and L.-Q. Luo) argue, however, for a "deep homology" of annelidan mushroom body and vertebrate pallium, even though they concede that "it is unlikely that a mushroom body-like shape was already in place in the protostome-deuterostome ancestor . . . and that higher degrees of histological (and thus, functional) complexity were acquired independently in divergent evolutionary lineages" (p. 807). Much hinges, of course, on what one means by "higher degrees."

176. See their paper in *Journal of Neuroscience* 21, 6395–6404 (2001).

177. Farris et al. (2001), p. 6404; note 176.

CHAPTER 23

1. See the work on the lungfish telencephalon by A. González and R.G. Northcutt *Brain, Behavior and Evolution* 74, 43–55 (2009).

2. See, for example, H. Ito and N. Yamamoto *Biology Letters* 5, 117–121 (2009).

3. See I. Rodríguez-Molder *Brain, Behavior and Evolution* 74, 20–29 (2009), for a similar argument applying to the chondrichthyeans.

4. See C. Broglio et al. *Brain Research Bulletin* 66, 277–281 (2005); also T. Mueller and M.F. Wullimann *Brain, Behavior and Evolution* 74, 30–42 (2009).

5. Broglio et al. (2005), p. 280; note 4.

6. See, for example, M. Portavella et al. *Journal of Neuroscience* 24, 2335–2342 (2004).

7. As Broglio et al. (2005; note 4) note, such work supports the notion "that the [median pallium] of goldfish, like the pallial amygdala of mammals, is essential for emotional learning and memory" (p. 280).

8. See their paper in *Brain Research Bulletin* 66, 365–370 (2005).

9. Rodríguez et al. (2005), p. 365 [Abstract]; note 8.

10. Broglio et al. (2005), pp. 277–278; note 4.

11. While the cerebellum is evidently a fundamental feature of the vertebrate brain, a number of

116. See S. Watanabe et al. *Learning & Memory* 15, 633–642 (2008).

117. See, for example, L.F. Sempere et al. *Evolution & Development* 9, 409–415 (2007).

118. See E. Ruiz-Trillo et al. *Molecular Phylogenetics and Evolution* 33, 321–332 (2004).

119. See, for example, papers by M. Reuter et al. in *Tissue and Cell* (vol. 30, pp. 57–63; 1998, and vol. 33, pp. 119–128; 2001) and A. Bery et al. *Development, Genes and Evolution* 220, 61–76 (2010). O.I. Raikova et al. (*Zoologica Scripta* 33, 71–88; 2004) draw attention to various parallelisms in the evolution of acoel brains, noting that although this group is no longer allied to the other flatworms, "they seem to share the main trends in nervous system evolution," notably a "concentration of the nerve net fibres into several longitudinal and transversal nerves and the sinking of the nervous system into the parenchyma" (p. 86).

120. See, for example, M. Nakazawa et al. *Molecular Biology and Evolution* 20, 784–791 (2003).

121. See, for example, H. Koopowitz and L. Keenan *Trends in Neurosciences* 5, 77–79 (1982).

122. See his paper in *Hydrobiologia* 321, 79–87 (1986).

123. Koopowitz (1986), p. 80; note 122.

124. See D. Hadenfeldt *Zeitschrift für wissenschaftliche Zoologie* 133, 586–638 (1929).

125. See T.H. Bullock and G.A. Horridge in *Structure and function in the nervous systems of invertebrates*, vol. 1, pp. 535–577 (Freeman; 1965).

126. See their paper in *Journal of Comparative Neurology* 195, 697–716 (1981).

127. Keenan et al. (1981), p. 699 (emphasis in original); note 126.

128. See J. Baguñà and M. Riutort *Canadian Journal of Zoology* 82, 168–193 (2004); also J. Paps *Molecular Biology and Evolution* 26, 2397–2406 (2009).

129. See, for example, C. Mißbach et al. *Arthropod Structure & Development* 40, 317–333 (2011).

130. See N.J. Strausfeld et al. *Journal of Comparative Neurology* 513, 265–291 (2009).

131. See M. Kollmann et al. *Arthropod Structure & Development* 40, 304–316 (2011).

132. See N.J. Strausfeld *Proceedings of the Royal Society of London, B* 276, 1929–1937 (2009); and N.J. Strausfeld and D.R. Andrew in *Arthropod Structure & Development* 40, 276–288 (2011).

133. See, for example, D.F. Mellon et al. *Journal of Comparative Neurology* 321, 93–111 (1992).

134. See, for example, M.E. McKinzie et al. *Journal of Comparative Neurology* 462, 168–179 (2003). The comparison between the parasol cells and possible equivalents in the mushroom body is, however, queried by G. Wolff et al. (*Journal of Comparative Neurology* 520, 2824–2846; 2012), who suggest that the equivalent lies with the basiophilic cell bodies rather than the Kenyon cells.

135. See S. Koenemann et al. *Arthropod Structure & Development* 39, 88–110 (2010).

136. See M.E.J. Stegner and S. Richter *Arthropod Structure & Development* 40, 221–243; 2011.

137. See his paper in *Arthropod Structure & Development* 39, 143–153 (2010).

138. Jenner (2010), p. 145; note 137.

139. See also J. Kubrakiewicz et al. *Zoology* 115, 261–269 (2012), but see T. Stemme et al. (*BMC Evolutionary Biology* 12, e168; 2012), and M.E.J. Stegner et al. (*Journal of Morphology* 275, 269–294; 2014), who are less convinced about a remipede-cephalocarid connection.

140. Again long thought to be primitive, possession of accessory lobes with olfactory neuropil and hemiellipsoidal bodies forming part of an "unexpectedly" complex brain supports a malacostracan-remipede connection; see M. Fanenbruck and S. Harzsch *Arthropod Structure & Development* 34, 343–378 (2005); on the other hand B. von Reumont et al. (*Molecular Biology and Evolution* 29, 1031–1045; 2012) suggest the remipedes are sister group to the insects.

141. See their paper in *Journal of Comparative Neurology* 470, 25–38 (2004).

142. Sullivan and Beltz (2004), p. 32; note 141.

143. See his paper in *Brain, Behavior and Evolution* 52, 186–206 (1998); see also Strausfeld (2009); note 130.

144. Strausfeld (1998), p. 202; note 143.

145. See, for example, J.M. Sullivan and B.S. Beltz *Journal of Comparative Neurology* 481, 118–126 (2005); also Stemme et al. (2012; note 139).

146. See S. Brown and G. Wolff et al. *Journal of Comparative Neurology* 520, 2847–2863 (2012).

147. This has been compensated for by the development of other sensory modalities; see S. Harzsch et al. *Arthropod Structure & Development* 40, 244–257 (2011).

148. See S. Harzsch and B.S. Hansson *BMC Neuroscience* 9, e58 (2008). They note in terms of arrangement in the olfactory lobe, "The tall, narrow, and elongate shape of the glomeruli . . . [of *Coenobita*] seem to represent one extreme end of [the crustacean] variability" (p. 27), but more intriguingly in both this hermit crab and the robber crab, the "hemiellipsoid body also displays a lamellar organization" (p. 29).

149. See G. Scholtz and S. Richter *Zoological Journal of the Linnean Society* 113, 289–328 (1995).

150. See M.C. Stensmyr et al. *Current Biology* 15, 116–121 (2005).

151. Loesel and Heuer (2010), p. 32 (emphasis added); note 104.

152. Strausfeld et al. (1998), p. 28; note 102.

153. See, for example, papers in *Journal of Comparative Neurology* by N.J. Strausfeld et al. (vol. 424, pp. 179–195; 2000) and W. Gronenberg (vol. 435, 474–489; 2001).

154. See their paper in *Journal of Neuroscience* 30, 6461–6465 (2010).

155. Hourcade et al. (2010), p. 6461 [Abstract]; note 154.

Zrzavy et al. *BMC Evolutionary Biology* 9, e189 [2009]; also T.H. Struck vol. 7, e57; 2007), but they exhibit a sort of agriculture (p. 178) and behaviors that have been compared with the octopus (see S.M. Evans *Nature* 211, 945–948 [1966]; also his papers in *Animal Behaviour* [vol. 14, pp. 102–106, and pp. 107–119; 1966]).

69. See A.S. Denes et al. *Cell* 129, 277–288 (2007); also D. Arendt et al. *Philosophical Transactions of the Royal Society of London, B* 363, 1523–1528 (2008). See also K. Tessmar-Raible et al. *Cell* 129, 1389–1400 (2007).

70. Nishikawa (2002), p. 247; note 7.

71. See their paper in *Philosophical Transactions of the Royal Society of London, B* 337, 261–269 (1992); also M. Mizunami et al. *Zoological Science* 21, 1141–1151 (2004).

72. Wolf et al. (1992), title of the paper; note 71.

73. Wolf et al. (1992), p. 262; note 71.

74. See, for example, N.J. Abbott *Cellular and Molecular Neurobiology* 25, 5–23 (2005).

75. See M. Bundgaard and N.J. Abbott *Glia* 56, 699–708 (2008).

76. See S. Banerjee and M.A. Bhat *Annual Review of Neuroscience* 30, 235–258 (2007).

77. Bundgaard and Abbott (2008), p. 707; note 75.

78. Banerjee and Bhat (2007), p. 242; note 76.

79. See N.J. Lane and S.S. Campiglia *Journal of Neurocytology* 16, 93–104 (1987).

80. See N.J. Lane and H.J. Chandler *Journal of Cell Biology* 86, 765–774 (1980).

81. See N.J. Lane et al. *Tissue and Cell* 13, 557–576 (1981).

82. See J.B. Harrison and N.J. Lane *Journal of Neurocytology* 10, 233–250 (1981).

83. See N.J. Lane and N.J. Abbott *Cell and Tissue Research* 156, 173–187 (1975), and N.J. Abbott et al. *Experimental Eye Research* (Supplement, 1) 25, 259–271 (1977).

84. Thus F. Mayer et al. (*Journal of Neuroscience* 29, 3538–3550; 2009) identify a homology with transporter molecules in the blood-brain barrier of insects and mammals. But to refer to this as "Evolutionary conservation" (part of the title) is really to miss the point if we accept that the two barriers have arisen independently. What matters is how, why, and when these proteins were co-opted.

85. See S.M. Farris *Brain, Behavior and Evolution* 72, 106–122 (2008a).

86. Farris (2008a), p. 116; note 85.

87. See, for example, G. Laurent in *23 problems in systems neuroscience* (J.L. van Hemmen and T.J. Sejnowski, eds.), pp. 3–21 (Oxford; 2006).

88. Laurent (2006), title of his paper; note 87.

89. Laurent (2006), p. 4; note 87.

90. Laurent (2006), p. 3; note 87.

91. See G. Mayer and P.M. Whitington *Developmental Biology* 335, 263–275 (2009).

92. See G. Mayer and S. Harzsch *Journal of Comparative Neurology* 507, 1196–1208 (2008).

93. See J.E. Niven et al. *Annual Review of Entomology* 53, 253–271 (2008).

94. For an overview, see S.E. Fahrbach *Annual Review of Entomology* 51, 209–232 (2006).

95. See, for example, S.M. Farris and N.J. Strausfeld *Journal of Comparative Neurology* 439, 331–351 (2001).

96. See M. Mizunami et al. *Neuroscience Letters* 229, 153–156 (1997).

97. For a detailed series of comparisons between the vertebrate cerebellum-like structures (p. 443; note 11) and the insect mushroom body, see S. Farris *Arthropod Structure & Development* 40, 368–379 (2011).

98. See N.J. Strausfeld and Y.-S. Li *Journal of Comparative Neurology* 409, 626–646 (1999).

99. Farris and Strausfeld (2001), pp. 349–350; note 95.

100 Farris and Strausfeld (2001), p. 350; note 95.

101. See N.J. Strausfeld et al. *Arthropod Structure & Development* 35, 169–196 (2006b); also N.J. Strausfeld et al. (2006a); note 62.

102. See N.J. Strausfeld et al. *Learning & Memory* 5, 11–37 (1998).

103. See, for example, L. Orrhage and M.C.M. Müller *Hydrobiologia* 535, 79–111 (2005), and M.C.M. Müller *Integrative and Comparative Biology* 46, 125–133 (2006).

104. See, for example, R. Loesel and C.M. Heuer *Acta Zoologica* 91, 29–34 (2010).

105. See papers by C.M. Heuer and R. Loesel in *Cell and Tissue Research* 331, 713–724 (2008), and *Zoomorphology* 128, 219–226 (2009); also Strausfeld (1998); note 102.

106. See, for example, M. Myohara et al. *Developmental Dynamics* 235, 2051–2070 (2006).

107. See C. Yoshida-Noro et al. *Development, Genes and Evolution* 210, 311–319 (2000).

108. Yoshida-Noro et al. (2000), p. 314 (emphasis added); note 107.

109. See C.M. Heuer et al. *Frontiers in Zoology* 7, e13 (2010).

110. See C.J. Winchell et al. *Development, Genes and Evolution* 220, 275–295 (2010).

111. See his overview in *Current Biology* 18, R897–R898 (2008).

112. Hochner (2008), p. R897; note 111.

113. Hochner (2008), p. R898; note 111.

114. See, for example, R. Chase *Microscopy Research and Technique* 49, 511–520 (2000); also S. Ratté and R. Chase *Journal of Comparative Neurology* 417, 366–384 (2000). N. Higuchi et al. (*BMC Biology* 7, e21; 2009) argue that the occurrence of a specific gene (PsGEF) in the snail *Lottia* (a limpet), which plays an important role in the mushroom body of the fly, thereby points to a deep ancestry for the mushroom body (or equivalent). Yet as noted in the text, no mollusk appears to possess any sort of mushroom body, and as Higuchi et al. also note, in the fly PsGEF is employed elsewhere in the brain.

115. See S. Ratté and R. Chase *Journal of Comparative Neurology* 384, 359–372 (1997).

and C.F. Doggenweiler *Journal of Cell Biology* 30, 381–403 [1966]).

37. See K. Xu and J. Terakawa *Journal of Experimental Biology* 202, 1979–1989 (1999).

38. The similarities do not stop here, because some silk, such as that which enwraps the pupa of the hornet, also has electrical properties and has a number of intriguing similarities to myelinated nerves (see J.S. Ishay and S. Kirschboim *Journal of Optoelectronics and Advanced Materials* 2, 281–291 [2000]). These similarities, of course, are not exact, and the silk does not transmit nervous impulses; rather the electrical current appears to serve to release heat and so assist in the thermoregulation of the larva.

39. See, for example, M.R. Kaplan et al. *Neuron* 30, 105–119 (2001).

40. See J. Günther *Journal of Comparative Neurology* 168, 505–531 (1976).

41. See McAlear et al. (1958); note 25, and Heuser and Doggenweiler (1966); note 36.

42. Although as Roots (2008; note 21) notes, in the case of the opossum shrimp, these openings "bear a striking resemblance to those of vertebrates" (p. 105).

43. See their paper in *Current Biology* 19, R995–R1008 (2009).

44. Chittka and Niven (2009), p. R1006; note 43.

45. See his paper in *International Journal of Developmental Biology* 47, 555–562 (2003).

46. Ghysen (2003), p. 556; note 45.

47. See their paper in *Development, Growth & Differentiation* 51, 167–183 (2009).

48. Watanabe et al. (2009), p. 173; note 47.

49. See also B. Galliot et al. *Developmental Biology* 332, 2–24 (2009).

50. Chittka and Niven (2009), title of paper; note 43.

51. Chittka and Niven (2009), p. R1004; note 43.

52. Chittka and Niven (2009), p. R1004; note 43.

53. See his paper in *Brain, Behavior and Research* 74, 177–190 (2009).

54. Moroz (2009), p. 177; note 53.

55. Moroz (2009), p. 186; note 53.

56. See their paper in *Brain, Behavior and Evolution* 68, 191–195 (2006).

57. Jaaro and Fainzilber (2006), p. 194; note 56.

58. See, for example, H. Reichert *Biology Letters* 5, 112–116 (2009).

59. This has played a major role in discussion of the central nervous system (CNS) of the deuterostomes; see, for example, N.D. Holland *Nature Reviews Neuroscience* 4, 617–627 (2003). In the case of the hemichordates, however, it now seems likely that they have effectively a CNS; see M. Nomaksteinsky et al. *Current Biology* 19, 1264–1269 (2009); also commentary on pp. R640–R642 by E. Benito-Gutiérrez and D. Arendt. This suggests that the nerve net of echinoderms (the sister group of hemichordates) is the result of "simplification," but more importantly the most primitive extant bilaterians in the form of the acoels have a nerve net.

60. See, for example, R. Lichtnecker and H. Reichert *Heredity* 94, 465–477 (2005).

61. So in vertebrates the tripartite brain is formed from anterior to posterior of telencephalon, mesencephalon, and rhombencephalon, while the supposed equivalents in the insects comprise the protocerebrum, deuterocerebrum, and tritocerebrum.

62. See N.J. Strausfeld et al. *Proceedings of the Royal Society of London, B* 273, 1857–1866 (2006a).

63. See G. Mayer et al. *BMC Evolutionary Biology* 10, e255 (2010); also P.M. Whitington and G. Mayer *Arthropod Structure & Development* 40, 193–209 (2011).

64. See V. Rieger et al. *Invertebrate Biology* 129, 77–104 (2010).

65. See their paper in *Acta Zoologica* 91, 35–43 (2010).

66. Harsch and Wanninger (2010), p. 35 [Abstract]; note 65.

67. Exactly the same arguments apply to metazoan segmentation. Here, too, the commonality of developmental mechanisms at first sight would suggest a single origin, most obviously in the phyla annelids, arthropods, and chordates. Ariel Chipman (*BioEssays* 32, 60–70; 2010) explores the details of embryology, the most likely phylogenies, and the fossil record (see also J.P. Couso *International Journal of Developmental Biology* 53, 1305–1316 [2009]), with the apparent conundrum of evidence pointing against the common ancestor being segmented as against the obvious developmental similarities (see also E.C. Seaver *International Journal of Developmental Biology* 47, 583–595 [2003]). The problem is resolved if, as Chipman remarks, one assumes "parallel recruitment of pre-existing gene regulatory networks in the three phyla" (p. 65). As importantly at least part of this regulatory network, notably the gene *Notch*, was most likely co-opted from its role in determining the body axis and "could have been recruited several times throughout evolution because of *inherent characteristics* of the network that make them pre-adapted for a role in segment generation" (p. 66; emphasis added). Just the same principal of co-option seems to apply to the genes that now specify the segmentation of arthropod limbs, despite the fact that all of them are expressed in the more primitive and undifferentiated onychophoran leg (see R. Janssen et al. *Evolution & Development* 12, 363–372 [2010]). It is also evident that segmentation has been lost on multiple occasions, including among the annelids independently in the sipunculans and echiuroids; see J. Dordel et al. *Journal of Zoological Systematics and Evolutionary Research* 48, 197–207 (2010). Looking further afield, a similar story evidently applies to the *Hox* genes, which despite their canonical role in axial pattern in bilaterians were evidently co-opted from the cnidarians, where they were employed in a diversity of roles; see R. Chiori et al. *PLoS ONE* 4, e4231 (2009).

68. Not only are they highly derived annelids (see J.

Schierwater et al. *PLoS Biology* 7, e1000020 (2009*a*) with commentary by N.W. Blackstone in e100007.

81. To be sure, the former group has a bona fide nervous system. Irrespective of whether the placozoans once had a nervous system but lost it, or alternatively are like the sponges with "a nervous system in waiting," the genomic architecture of the placozoans has what is effectively a neural component with many of the proteins that are key to the operation of a nervous system (see M. Srivastava et al. *Nature* 454, 955–960 [2008]; also commentary by D.J. Miller and E.E. Ball in *Current Biology* 18, R1003–R1005; 2008; also C.L. Smith et al. *Current Biology* 24, 1565–1572 (2014), with commentary on pp. R655–R658 by E.M. Jorgensen).

82. See also papers by H.-J. Osigus et al. in *Molecular Phylogenetics and Evolution* vol. 66, pp. 551–557, and vol. 69, pp. 339–351 (both 2013).

83. Schierwater et al. (2009), p. 0041; note 80.

84. See B. Schierwater et al. *Communicative and Integrative Biology* 2, 403–405 (2009*b*).

85. Schierwater et al. (2009), part of title; note 84.

86. Schierwater et al. (2009), p. 404; note 84.

87. Schierwater et al. (2009), p. 404; note 84.

CHAPTER 22

1. Oxygen consumption in a mormyriform fish with an enormous brain tips in at about 60 percent; see G.E. Nilsson *Journal of Experimental Biology* 199, 603–607 (1996).

2. See J.W. Mink et al. *American Journal of Physiology: Regulatory, Integrative and Comparative Physiology* 241, R203–R212 (1981).

3. See S.B. Laughlin et al. *Nature Neuroscience* 1, 36–41 (1998). However, because of important differences between the rhabdomeric retina of insects and ciliary retina of vertebrates (see G.L. Fain et al. *Current Biology* 20, R114–R124 [2010]), the energetics are not always directly comparable. In particular, in daylight mammalian rods consume substantially less energy, although cones remain very expensive. This appears to explain why rods normally predominate except in a fovea; see H. Okawa et al. *Current Biology* 18, 1917–1921 (2008), and commentary on pp. R69–R71 of vol. 19 (2009) by E.J. Warrant.

4. Laughlin et al. (1998), p. 36 [Abstract]; note 3.

5. See their paper in *Journal of Experimental Biology* 211, 1792–1804 (2008).

6. Niven and Laughlin (2008), p. 1793; note 5.

7. See, for example, K.C. Nishikawa *Brain, Behavior and Evolution* 59, 240–249 (2002).

8. See, for example, S. Itzovitz et al. *Proceedings of the National Academy of Sciences, USA* 105, 9278–9283 (2008).

9. See, for example, D. Attwell and S.B. Laughlin *Journal of Cerebral Blood Flow and Metabolism* 21, 1133–1145 (2001).

10. See their paper in *Science* 301, 1870–1874 (2003).

11. Laughlin and Sejnowski (2003), p. 1872; note 10.

12. Attwell and Laughlin (2001), p. 1143; note 9.

13. See P. Lennie *Current Biology* 13, 493–497 (2003).

14. See D. Attwell and A. Gibb *Nature Reviews Neuroscience* 6, 841–849 (2005).

15. For an excellent overview, see J.E. Niven and S.M. Farris *Current Biology* 22, R323–R329 (2012).

16. See A.A. Faisal *Current Biology* 15, 1143–1149 (2005).

17. Attwell and Gibb (2005), p. 841; note 14.

18. Glial cells are often referred to as a sort of cellular "glue" in the brain, but as D.K. Hartline (*Glia* 59, 1215–1236; 2011) stresses, this largely misses the point in what is an exceedingly complex system. As importantly he reviews the evidence that glia are convergent.

19. For example, see P.M. Pereyra and B.I. Roots *Neurochemical Research* 13, 893–901 (1988).

20. See I. Fernández et al. *Journal of Morphology* 230, 265–281 (1996).

21. See her paper in *Neuron Glia Biology* 4, 101–109 (2008).

22. Roots (2008), p. 101 [Abstract], pp. 106 and 107; note 21.

23. Roots (2008), p. 106; note 21.

24. See C.K. Govind and J. Pearce *Journal of Comparative Neurology* 268, 121–130 (1988).

25. See J.H. McAlear et al. *Journal of Ultrastructural Research* 2, 171–176 (1958).

26. See A.D. Davis et al. *Nature* 398, 571 (1999); also Wilson and Hartline (2011); note 33.

27. See P.H. Lenz et al. *Journal of Comparative Physiology, A* 186, 337–345 (2000).

28. See T.M. Weatherby and P.H. Lenz *Arthropod Structure & Development* 29, 275–288 (2000).

29. See, for example, E.J. Buskey et al. *Adaptive Behavior* 20, 57–66 (2012).

30. See R.J. Waggett and E.J. Buskey *Journal of Experimental Marine Biology and Ecology* 361, 111–118 (2008).

31. See T.M. Weatherby et al. *Journal of Comparative Physiology* 186, 347–357 (2000).

32. See B.I. Roots and N.J. Lane *Tissue and Cell* 15, 695–709 (1983).

33. See C.H. Wilson and D.K. Hartline *Journal of Comparative Neurology* 519, 3259–3280 (2011).

34. See C.H. Wilson and D.K. Hartline *Journal of Comparative Neurology* 519, 3281–3305 (2011).

35. McAlear (1958), p. 176; note 25.

36. This point was reinforced by Roots (2008; note 21), who emphasized how the arrangement of the myelin "is exceptional in several respects and bears a striking resemblance to that of vertebrates" (p. 105). The exact degree of similarity is yet to be determined, not least whether the myelin sheath has a concentric or spiral arrangement. In any event such tight arrangements are not universal, and in other groups the packing is generally more open, with the layers of myelin separated by cytoplasm (see, for example, J.E. Heuser when foraging different nutrient patches, the slime mold does so to optimize its intake. Thus this "extraordinary organism can make complex nutritional decisions despite lacking a coordination center and comprising only a single vast multinucleate cell" (p. 4607 [Abstract]). As these workers also note, their "result has strong parallels with observations in other distributed systems, such as bacterial and fungal colonies, in which the overall pattern of growth is influenced by varying the concentration of nutrient" (p. 4608). See also C.R. Reid et al. *Proceedings of the National Academy of Sciences, USA* 109, 17490–17494 (2012).

48. See T. Nakagaki et al. *Physical Review Letters*, e068104 (2007).

49. See T. Saigusa et al. *Physical Review Letters*, e018101 (2008); also P. Ball *Nature* 451, 385 (2008).

50. Nakagaki (2001); p. 770; note 46.

51. Nakagaki et al. (2000), p. 470; note 44.

52. Nakagaki et al. (2004a), p. 2308; note 47.

53. But inherency versus convergences? For example, enzymes involved with the operation of neurotransmitters (PAMs = peptidylglycine α-amidating monooxygenases) are also found in algae but not in organisms (choanoflagellates, fungi) more closely related to animals. R.M.F. Attenborough et al. (*Molecular Biology and Evolution* 29, 3095–3109; 2012) suggest a secondary loss, but they do not exclude convergence.

54. See P. Burkhardt et al. (*Proceedings of the National Academy of Sciences, USA* 108, 15264–15269 [2011]), who report that an essential part of neural secretions that involve SNARE proteins (p. 306, note 1) had evolved in the choanoflagellates.

55. See A. Alié and M. Manuel *BMC Evolutionary Biology* 10, e34 (2010).

56. Alié and Manuel (2010), p. 1 [Abstract]; note 55.

57. See their paper in *Nature Reviews Neuroscience* 10, 701–712 (2009).

58. Ryan and Grant (2009), p. 701 [Abstract]; note 57.

59. See, for example, S.P. Leys and R.W. Meech *Canadian Journal of Zoology* 84, 288–306 (2006), and C. Bond *Invertebrate Biology* 132, 283–290 (2013).

60. See G.R.D. Elliot and S.P. Leys *Journal of Experimental Biology* 210, 3736–3748 (2007), and commentary by R.W. Meech *Current Biology* 18, R70–R72 (2008).

61. Attention has also been drawn to the diagnostic choanocytes. As Dave Jacobs and colleagues (*Integrative and Comparative Biology* 47, 712–723; 2007) note, "In adult sponges [they] also bear some similarity to eumetazoan sensory structures" (p. 720). While this comparison may be superficial, not only do these workers draw attention to a raft of molecules that subsequently find employment in the nervous system, but the subsequent recognition of one class (antennapedia homeobox genes) in the sponges is noteworthy. The specific genes (NK6 and NK7) play a central role in neurogenesis but in one particular sponge (a homoscleromorph) are only expressed in the choanocyte

(see E. Gazave et al. *Development, Genes and Evolution* 218, 479–489 [2008]).

62. See G.J. Tompkins-MacDonald and S.P. Leys *Marine Biology* 154, 973–984 (2008).

63. See S.P. Leys et al. *Journal of Experimental Biology* 202, 1139–1150 (1999); also M. Nickel *Invertebrate Biology* 129, 1–16 (2010).

64. Leys et al. (1999), p. 1148; note 63.

65. See O. Sakarya et al. *PLoS ONE* 2(6), e506 (2007).

66. See G.S. Richards et al. *Current Biology* 18, 1156–1161 (2008); also G.R.D. Elliott *Journal of Experimental Biology* 213, 2310–2321 (2010).

67. See also I. Marín (*BMC Evolutionary Biology* 10, e331; 2010), who documents the phylogeny of U-box ubiquitin ligases and notes that their occurrence in sponges and placozoans "suggests that many conditions required for the generation of a complex nervous system were already present in the last common ancestor of all animals" (p. 12).

68. See K.S. Kosik *Biology Letters* 5, 108–111 (2009); also R.D. Emes et al. *Nature Neuroscience* 11, 799–806 (2008).

69. Kosik (2009), p. 109; note 68.

70. Crucial though the consequences were for the emergence of a functioning synapse, Kosik (2009; note 68) also notes how the proposed "domain expansion through shuffling, duplication, and changes in protein expression level [that] are critical drivers in the evolution of cellular machines" (p. 108 [Abstract]) can hardly be regarded as exceptional evolutionary mechanisms.

71. See S. Shaham *Nature Reviews Neuroscience* 11, 212–217 (2010).

72. Shaham (2010), p. 212; note 71.

73. Shaham (2010), p. 215; note 71.

74. See V. Croset et al. *PLoS Genetics* 6, e1001064 (2011).

75. See A.F. Silbering and R. Benton *EMBO Reports* 11, 173–179 (2010).

76. See M. Srivastava et al. *Nature* 466, 720–726 (2010), and commentary on p. 673 by A. Mann.

77. Some hydrozoan areas of the epithelium lack nerves yet show highly effective conduction of impulses (see G.O. Mackie and L.M. Passano *Journal of General Physiology* 52, 600–621 [1968]), which, as George Mackie notes in the siphonophore epithelia (see G.O. Mackie *American Zoologist* 5, 439–453 [1965]), "conducts at a velocity and with a refractory period comparable to unmyelinated nerve fibers in vertebrates" (p. 450). This worker also notes that in one area of epithelium it acts like "a single giant axon spread out over the whole exumbrellar surface" (p. 450).

78. See M. Nickel *Invertebrate Biology* 129, 1–16 (2010).

79. For an overview, see M.J. Telford *Current Biology* 19, R339–R341 (2009).

80. See L.L. Moroz et al. *Nature* 510, 109–114 (2014), with commentary on pp. 38–39 by A. Hejnol. Also

57) note that "over 21% of the MASC [membrane-associated guanylate kinase signaling complex] genes and 25% of the PSD [post-synaptic density] genes were found to have direct orthologues in protosynaptic organisms" (p. 701 [Abstract]), such as yeast.

5. See A.O. Noda et al. *Gene* 365, 130–136 (2006).

6. In another case also we see the employment in very different ways of the same enzymes (known as decarboxylases) in animals (with various roles including neurotransmission) and plants (see L.E. Sáenz-de-Miera and F.J. Ayala *Journal of Evolutionary Biology* 17, 55–66 [2004]), where unsurprisingly they are specific to very different substrates.

7. See P. Thagard *Philosophy of Science* 69, 429–446 (2002).

8. Mineta et al. (2003); note 3.

9. See, for example, I. Wessler et al. *Japanese Journal of Pharmacology* 85, 2–10 (2001); also I. Wessler et al. *Clinical and Experimental Pharmacology and Physiology* 26, 198–205 (1999).

10. See G. Schiechl et al. *Plant Science* 175, 262–266 (2008).

11. See K. Kawashima et al. *Life Sciences* 80, 2206–2209 (2007).

12. See M. Raineri and P. Modenesi *Histochemical Journal* 18, 647–657 (1986).

13. See, for example, K. Kawashima and T. Fujii *Life Sciences* 74, 675–696 (2003).

14. It draws upon the familiar enzymatic theme of a rampantly convergent catalytic triad; see K. Sangane et al. *Plant Physiology* 138, 1359–1371 (2005).

15. These in turn belong to a much larger class of G protein–coupled receptors, which not only have deep phylogenetic origins but could be turned to neurotransmission with only minor reconfigurations of the binding pocket; see D.-H. Kuang et al. *Proceedings of the National Academy of Sciences, USA* 103, 14050–14055 (2006).

16. See P. Ramoino et al. *Journal of Experimental Biology* 213, 1251–1258 (2010).

17. Ramoino et al. (2010), p. 1257; note 16.

18. See his paper in *American Zoologist* 30, 907–920 (1990); also W.E.S. Carr et al. *Advances in Comparative and Environmental Physiology* 5, 25–52 (1989).

19. Mackie (1990), p. 909 [subtitle of section]; note 18.

20. See his paper in *L'Anneé Biologique* 58, 89–98 (1954).

21. Haldane (1954), p. 91 (my translation); note 20.

22. See, for example, G. Csaba et al. *Cell Biology International* 31, 924–928 (2007).

23. While propagation speeds associated with sodium channels are very variable and with differences equivalent to three orders of magnitude, the equivalent speeds associated with calcium action potentials show a remarkably restricted range, be it in cnidarians or heart muscle. L.F. Jaffe (*Biology of the Cell* 95, 343–355; 2003) reviews this area and suggests that we see a universal constraint reflecting the need to avoid the cell being poisoned by the calcium.

24. See, for example, B. Roux et al. *Biochemistry* 39, 13295–13306 (2000).

25. And not only their various ion channels, but even proton channels. Not surprisingly these play a key role in pH regulation (see T.E. DeCoursey *Cellular and Molecular Life Sciences* 65, 2554–2573 [2008]). Thomas DeCoursey remarks on how these channels both exhibit "perfect proton selectivity" (p. 2557) and also "their exquisite sensitivity" (p. 2559) to the pH on either side of the cell membrane.

26. See also W. Zhou et al. *Neuron* 42, 101–112 (2004), for evidence of a particular Na$^+$ channel that might have originated from the common Na$^+$/Ca$^+$ channel.

27. See P.A.V. Anderson and B.M. Greenberg *Comparative Biochemistry and Physiology, B* 129, 17–28 (2001).

28. See B.J. Liebeskind et al. *Proceedings of the National Academy of Sciences, USA* 108, 9154–9159 (2011); for aspects of the still-deeper history of these channels see B.J. Liebeskind et al. *Molecular Biology and Evolution* 29, 3613–3616 (2012).

29. See Liebeskind et al. (2011), p. 9155; note 28.

30. See A.L. Goldin *Journal of Experimental Biology* 205, 575–584 (2002).

31. This is a member of the heliozoans, which in fact are polyphyletic with convergent evolution of the diagnostic feeding structures known as axopodia (see S.I. Nikolaev et al. *Proceedings of the National Academy of Sciences, USA* 101, 8066–8071 [2004]).

32. See C. Febvre-Chevalier et al. *Journal of Experimental Biology* 122, 177–192 (1986).

33. See A.R. Taylor *PLoS ONE* 4(3), e4966 (2009).

34. Taylor (2009), p. 3; note 33.

35. See W.A. Catterall *Science* 294, 2306–2308 (2001), and R. Koishi et al. *Journal of Biological Chemistry* 279, 9532–9538 (2004). See also J. Payandeh et al. *Nature* 475, 353–358 (2011), and commentary on pp. 305–306 by R. Horn. Also B.J. Liebeskind et al. *Current Biology* 23, R948–R949 (2013).

36. See D.-J. Ren et al. *Science* 294, 2372–2375 (2001).

37. See M. Ito et al. *Proceedings of the National Academy of Sciences, USA* 101, 10566–10571 (2004).

38. See J.D. Lear et al. *Science* 240, 1177–1181 (1988).

39. See their paper in *Astrobiology* 5, 1–17 (2005).

40. Pohorille et al. (2005), p. 11; note 39.

41. See their paper in *Bollettino di Zoologia* 42, 57–79 (1975).

42. Piccinni and Omodeo (1975), p. 72; note 41.

43. See I. Walker *Acta Amazonica* 8, 423–438 (1978).

44. See their paper in *Nature* 407, 470 (2000).

45. Nakagaki et al. (2000), p. 470; note 44.

46. See T. Nakagaki *Research in Microbiology* 152, 767–770 (2001); also T. Nakagaki et al. (2000; note 44) and *Biophysical Chemistry* 92, 47–52 (2001).

47. See T. Nakagaki et al. *Proceedings of the Royal Society of London, B* 271, 2305–2310 (2004*a*), and *Biophysical Chemistry* 1107, 1–5 (2004*b*). See A. Dussutour et al. (*Proceedings of the National Academy of Sciences, USA* 107, 4607–4611; 2010), who demonstrate how

is to help the young in the delicate task of coprophagy, a habit that allows the acquisition of the gut flora necessary to help digest the gum leaves (p. 147).

176. See also B. Figueridio and C.M. Janis (*Biology Letters* 7, 937–940; 2011), who while agreeing on the striking convergences on wolves/dogs, draw attention to evidence for thylacine being an ambush predator and in this respect more similar to, say, a tiger.

177. See M.S. Springer et al. in *Molecular evolution and adaptive radiation* (T.J. Givnish and K.J. Sytsma, eds.), pp. 129–161 (Cambridge; 1997).

178. See his chapter (pp. 1–46) in *Kangaroos, wallabies and rat-kangaroos* (G. Grigg et al., eds.; Surrey Beatty & Sons; 1989).

179. Flannery (1989), p. 43; note 178.

180. Natalie Warburton (2006; note 181) draws attention as to how in the marsupial mole, the "use of the tail as a fifth limb during locomotion is analogous to the pentapedal rôle that the tail of a kangaroo . . . plays when moving forwards on all fours or fives. It is remarkable that two marsupials of such different histories and lifestyles have independently co-opted the tail for an active rôle in locomotion. It would be interesting to know whether there are any similarities in the tail musculature and vertebral joints, given this apparently convergent evolution of function between the marsupial moles and kangaroos" (p. 140).

181. See N.M. Warburton *Verhandlungen des Naturwissenschaftlichen Vereins in Hamburg* 42, 39–149 (2006). They also have similar diets; see C.R. Pavey et al. *Journal of Zoology, London* 287, 115–123 (2012).

182. See J.P. Gasc et al. *Journal of Zoology, London* 208, 9–35 (1986); Warburton (2006; note 181) discusses the possible degrees of convergence (p. 142).

183. See M. Archer et al. *Proceedings of the Royal Society, B* 278, 1498–1506 (2011).

184. The dentition (known as zalambdodonty and representing a form of molar that has evolved multiple times [see R.J. Asher and M.R. Villagra *Journal of Mammalian Evolution* 12, 265–282; (2005)]) of this Miocene animal may be distinctive, but as Mike Archer and colleagues (2011; note 183) point out, such "highly specialized morphological patterns . . . can be achieved in very different ways in different mammalian clades and can result from convergent rather than parallel evolution" (p. 1503).

185. See S. Ladevèze et al. *Journal of Anatomy* 213, 686–697 (2008).

186. See G.W. Rougier et al. *Proceedings of the National Academy of Sciences, USA* 109, 20053–20058 (2012).

187. See S.P. Modesto et al. *Naturwissenschaften* 98, 1027–1034 (2011).

188. See M.K. Hecht *Proceedings of the Royal Society of Victoria* 87, 239–250 (1975), and J.J. Head et al. *Zoological Journal of the Linnean Society* 155, 445–457 (2009); the latter authors point out that such gigantism has evolved at least four times among the varanids (see also J. Conrad et al. *PLoS ONE* 7, e41767 [2012]).

189. See G.M. Erickson et al. *Journal of Vertebrate Paleontology* 23, 966–970 (2003); for a more skeptical view, see S. Wroe *Australian Journal of Zoology* 50, 1–24 (2002).

190. See J.J. Head et al. *Proceedings of the Royal Society of London, B* 280, 20130665 (2013).

191. See their paper in *Biological Journal of the Linnean Society* 35, 379–407 (1988).

192. Losos and Greene (1988), p. 398; note 191.

193. See R.S. Seymour et al. *Proceedings of the Royal Society of London, B* 279, 451–456 (2012).

194. See, for example, F.H. Pough *Ecology* 54, 836–844 (1973), and C.J. Clemente et al. *Biological Journal of the Linnean Society* 97, 664–676 (2009).

195. Even so, if monitors are given a very good lunch they do not show any evidence of burning off the excess food via elevated body temperature, but merely increase their metabolic rate (see A.F. Bennett *Evolution* 54, 1768–1773 [2000]).

196. See A.F. Bennett *Journal of Comparative Physiology* 79, 259–280 (1972); also P.B. Frappell et al. *Comparative Biochemistry and Physiology, A* 133, 239–258 (2002).

197. Yet monitors have not taken the road to endothermy, and as importantly, unlike many lizards, neither have they to viviparity.

198. See H.-G. Horn *Mertensiella* 11, 167–180 (1999).

199. These lizards are also capable of a variety of bipedal postures (see G.W. Schuett et al. *Biological Journal of the Linnean Society* 97, 652–663 [2009]). Also S. Moritz and N. Schilling (*Journal of Morphology* 274, 294–306; 2013), who draw attention to the large volumes of oxidative muscle fibers in the trunk.

200. See J.D. Manrod et al. *Animal Cognition* 11, 267–273 (2008).

201. See P.M. O'Connor et al. *Nature* 466, 748–751 (2010).

202. O'Connor et al. (2010), p. 749; note 201.

203. O'Connor et al. (2010), p. 751; note 201.

204. So far as we know, however, no notosuchian, including *Pakasuchus*, became viviparous. Neither are there data that point to possible endothermy in notosuchians, and the evidence suggests that at least *Yacarerani* lay eggs; see F.E. Novas et al. *Journal of Vertebrate Paleontology* 29, 1316–1320 (2009).

CHAPTER 21

1. Ostensibly this is a relatively primitive animal, but in this specific case the type in question (a planarian) belongs to the lophotrochozoans (i.e., annelids, mollusks, etc.) and is anatomically simplified. This, however, has little bearing on the argument given here.

2. Some 37 percent of them in plants, and 30 percent in yeast.

3. See K. Mineta et al. *Proceedings of the National Academy of Sciences, USA* 100, 7666–7671 (2003).

4. In a similar vein Ryan and Grant (2009; note

certainly easy to point to differences and absences. For instance, the honey possum (*Tarsipes*) has no obvious counterpart among the placentals (Springer et al. [1997]; note 177), yet it does not escape the net of convergence, because not only is it nectar feeding like the hummingbirds (p. 201), but it also has independently evolved trichromatic color vision (p. 114).

155. See O. Madsen et al. *Nature* 409, 610–614 (2001).

156. See, for example, W. J. Murphy et al. *Nature* 409, 614–618 (2001); for a paleontological perspective, see R. Tabuce et al. *Proceedings of the Royal Society of London, B* 274, 1159–1166 (2007).

157. Just as with the afrotherian-laurasiatherian parallels (p. 231), there are analogues among the Mesozoic faunas (see J.F. Bonaparte *Actas IV Congreso Argentino Paleontologia y Biostratigraphia*, vol. 4, pp. 63–95 [1986]), notably with the mammals, where there is again a north-south distinction between the denizens of Gondwana and Laurasia (for an overview, see G.P. Wilson *Journal of Vertebrate Paleontology* 27, 521–531 [2007]).

158. See also pp. 353–354 in my *Life's solution*, where I discuss this convergence as well as the strikingly similar extinct taxa *Ernanodon* and *Eurotamandua*, both presumed anteaters and unlikely to be closely related to any xenarthran.

159. These may be closer to the afrotherians than the laurasiatherians (see B.M. Hallström et al. *Molecular Biology and Evolution* 24, 2059–2068 [2007]), and D.E. Wildman *Proceedings of the National Academy of Sciences, USA* 104, 14395–14400 [2007]).

160. Ponder indeed, because when we consider their fossil record, among the various surprises are the aquatic sloths feeding on seaweed (see C. de Muizon and H.G. McDonald *Nature* 375, 224–227 [1995], and C. de Muizon et al. *Journal of Vertebrate Paleontology* 24, 398–410 [2004]). *Thalassocnus* evidently was not an agile swimmer, but probably had powerful lips like those of the manatees to tear at seaweed and seagrass. Evidently, to judge by tooth abrasions, these sloths also ingested a good deal of sand. Other excursions in sloth form include the better-known giant sloths, and although the giant sloths are typically depicted as shambling vegetarians, R.A. Farina and R.E. Blanco (in *Proceedings of the Royal Society of London B*, 263, 1725–1729 [1996]) have argued for them being stabbing carnivores (although limited isotopic data [N and C] reaffirm a herbivorous lifestyle; see C.A.M. France et al. *Palaeogeography, Palaeoclimatology, Palaeoecology* 249, 271–282 [2007]).

161. See W. J. Foley et al. *Journal of Zoology, London* 236, 681–696 (1995).

162. See his paper in *Zoological Journal of the Linnean Society* 140, 255–305 (2004).

163. Gaudin (2004), p. 283; note 162.

164. See C.R. Dickman in *The encyclopedia of mammals* (D.W. Macdonald, ed.), pp. 120–123 (Oxford; 2009).

165. See K.D. Rose and R.J. Emry *Journal of Morphology* 175, 33–56 (1983).

166. See O.F. Chernova and R.S. Hoffmann *Zoologicheskii Zhurnal* 83, 159–165 (2004; in Russian, with English abstract).

167. See P. Vogel *Revue d'Écologie* (*Terre et Vie*) 38, 37–49 (1983; in French, with English summary).

168. This in turn finds another convergence with a Miocene mustelid (see C. Grohé et al. *Naturwissenschaften* 97, 1003–1015 [2010]).

169. See D.T. Rasmussen in *The evolution of perissodactyls* (D.R. Prothero and R.M. Schoch, eds.), pp. 57–78 (Oxford; 1989).

170. For example, the cosmic paleontologist Teilhard de Chardin described from northern China what he believed to be a chalicothere (see his paper with E. Licent in *Bulletin of the Geological Society of China* 15, 421–427 [1936]). This is a major group of extinct mammals belonging the perissodactyls, thus related to such groups as the horses and tapirs. They were obviously very puzzled with their discovery of this "aberrant type of chalicotherid" and even remarked on it "strangely converging to the Notungulata of South America" (p. 426). By 1939, and reporting in the same journal (vol. 19, pp. 257–267), their doubt was evidently growing as we read "this strange Chalicotherid(?)" (p. 257). Subsequently this fossil transpired to be yet another hyracoid.

171. Consider, for example, the extraordinary fossil *Numbigilga* from the Pliocene of Australia (see R.M.D. Beck et al. *Journal of Paleontology* 82, 749–762 [2008]). So odd is this marsupial that it conceivably represents a new order, or alternatively may be a relic of a much more primitive group (specifically the polydolopimorphs), but with the available material (mostly teeth) it is clear that any phylogenetic analysis is torpedoed by convergence of dental "features [that] are most likely homoplastic within marsupials, reflecting convergent adaptations to a frugivorous-omnivorous diet" (p. 749 [Abstract]).

172. In this context, consider highly enigmatic but fragmentary material from the Eocene of Egypt; although most likely marsupial, these fossils also have a number of batlike features (see M.R. Sánchez-Villagra et al. *Paläontologische Zeitschrift* 81, 406–415 [2007]).

173. Thus, the Australasian marsupials show convergences in the anatomy of the hand (see V. Weisbecker and M. Archer *Palaeontology* 51, 321–338 [2008]), and also have learned to glide at least three times (p. 190).

174. Springer et al. (1997), see p. 147; note 177.

175. A deep-seated pocket makes sense for marsupials either bounding across the outback or climbing trees, while a marsupium pointing backward is particularly associated with burrowing or, in one case, an aquatic mode of life. So why this configuration in the folivorous koala "bear" (actually a wombat)? Mark Springer and colleagues (1997; note 177) suggest that it

[2008]). Such dives rival those of various mammals and perhaps extinct reptiles, but in this case the most likely explanation is that the turtle is undertaking a reconnaissance for its gelatinous prey, which at night migrates upward to more accessible depths. Intriguingly, the trachea has a vascular lining that not only has a countercurrent system but it is evidently functionally equivalent to the turbinals of birds and mammals (see papers by J. Davenport et al. in *Journal of Experimental Biology* [vol. 212, pp. 3440–3447 (2009)] and *Journal of Experimental Marine Biology and Ecology* 450, 40–46 [2014]). These authors also point out that, unusually for mammals, dolphins lack turbinals, but so too the trachea has a vascular lining. Evidence also exists for thermoregulation in the soft-shelled turtle. Most probably this involves control of the blood supply (see E.N. Smith et al. *Physiological Zoology* 54, 74–80; 1981), but body temperature is maintained and regulated in a rather distinctive fashion (see B.P. Wallace and T.T. Jones *Journal of Experimental Marine Biology and Ecology* 356, 8–24 [2008]; also D.N. Penick et al. *Comparative Biochemistry and Physiology, A* 120–399–403 [1998]).

122. Rob Asher reminds me that one curious difference is that despite many birds entering torpor, a seasonal hibernation analogous to some mammals is almost unknown.

123. See C.G. Farmer *American Naturalist* 155, 326–334 (2000); also subsequent discussion in vol. 162 (2003) by M.J. Angilletta and M.W. Sears (pp. 821–825) and reply by Farmer (pp. 826–840).

124. Farmer (2000), p. 330; note 123.

125. For an overview, see B.S. Rubidge and C.A. Sidor *Annual Review of Ecology and Systematics* 32, 449–480 (2001).

126. Not to forget that not only does the fossil record continue to yield surprises, but in the immediate context of Triassic faunas the bizarre teeth that R.B. Irmis and W.G. Parker describe (*Canadian Journal of Earth Sciences* 42, 1339–1345; 2005) as *Kraterokheirodon* that *might* be some sort of synapsid is a pertinent reminder that we best keep our minds open.

127. See, for example, his paper in *Acta Zoologica* 88, 3–22 (2007a).

128. See his paper in *Journal of Evolutionary Biology* 19, 1231–1247 (2006).

129. Kemp (2006), p. 1232; note 128.

130. See C.A. Sidor *Evolution* 55, 1419–1442 (2001).

131. Kemp (2007a), p. 3 [Abstract]; note 127.

132. This concept of correlated progression can be applied more widely, as in the case of the origin of the tetrapods from the sarcopterygians (see T.S. Kemp *Proceedings of the Royal Society of London, B* 274, 1667–1673 [2007b]).

133. Kemp (2006), p. 1232; note 128.

134. See J. Liu et al. *Acta Palaeontologica Polonica* 54, 393–400 (2009).

135. Liu et al. (2009), p. 393 [Abstract]; note 134.

136. See his paper in *Journal of Vertebrate Paleontology* 16, 271–284 (1996).

137. Martinez (1996), p. 281; note 136.

138. The galesaurids (see F. Abdala *South African Journal of Science* 99, 95–96 [2003]) also show "an unusual combination of characters" (p. 95 [Abstract]), leading Fernando Abdala to conclude that his study "represent[s] an increment in the already high level of homoplasy acknowledged in cynodont phylogeny" (p. 96).

139. See, for example, H.-D. Sues *Zoological Journal of the Linnean Society* 85, 205–217 (1985).

140. See, for example, J.A. Hopson *Journal of Vertebrate Paleontology* 11 (Supplement, 3), 36A (1991); also A. Huttenlocker *Zoological Journal of the Linnean Society* 157, 865–891 (2009).

141. See J.A. Hopson *Journal of Vertebrate Paleontology* 15, 615–639 (1995).

142. Hopson (1995), p. 636; note 141.

143. See C.A. Sidor *Journal of Paleontology* 77, 977–984 (2003).

144. See Y.-M. Hu et al. *Nature* 433, 149–152 (2005).

145. See Z.-X. Luo *Nature* 450, 1011–1019 (2007).

146. See Z.-X. Luo and J. R. Wible *Science* 308, 103–107 (2005). In a somewhat similar vein B.J. Shockey et al. *Paleobiology* 33, 227–247 (2007), argue that the extinct South American mesotheriid notoungulates show fossorial adaptations convergent with the xenarthran anteaters (and aardvarks).

147. See T. Martin *Zoological Journal of the Linnean Society* 145, 219–248 (2005).

148. See Q. Ji et al. *Science* 311, 1123–1127 (2006); also commentary on pp. 1109–1110 by T. Martin.

149. Luo (2007), p. 1012; note 145.

150. See, for example, M.T. Clementz et al. *Palaios* 23, 574–585 (2008).

151. See L.D. Martin and T.J. Meehan *Naturwissenschaften* 92, 1–19 (2005); also T.J. Meehan and L.D. Martin *Naturwissenschaften* 90, 131–135 (2003), and L.D. Martin *Institute for Tertiary-Quaternary Studies—TER-QUA Symposium Series* 1, 33–40 (1985).

152. The title of the paper by Martin and Meehan (2005), note 151.

153. There is also evidence that toward the end of one of these community cycles, body size of such animals as the saber-tooths can rapidly increase (see L.D. Martin *Special Publications of the Carnegie Museum of Natural History* 8, 526–538 [1984]), esp. fig. 8). Could this have a parallel in the remarkable increase in the brain size of hominids (Martin and Meehan [2005]; note 151)?

154. In the world of convergence it would be very strange if the similarities are precise (although they can be pretty exact), nor should we expect a one-to-one correspondence. Thus, among the placental afrotherians, flying mammals have not evolved; neither have they in the marsupials (but see note 172). Nor have the latter group ever invaded the oceans. It is linked to a diagnostic polyunsaturation (i.e., more C=C bonds) in the lipids (see P.L. Else et al. *Physiological and Biochemical Zoology* 77, 950–958 [2004]).

92. See, for example, F. Bouillaud et al. *Biochimica et Biophysica Acta: Bioenergetics* 1504, 107–119 (2001).

93. See F. Berg et al. *PLoS Genetics* 6, e129 (2006).

94. See, for example, J. Borecký et al. *Bioscience Reports* 21, 201–212 (2001).

95. See, for example, S. Saito et al. *Gene* 408, 37–44 (2008).

96. See M. Jastroch et al. *Physiological Genomics* 22, 150–156 (2005).

97. Even in the case of the marsupials, which like their eutherian cousins are endothermic and adept thermoregulators, whether this capacity stems from a common ancestor is less clear. UCP1 is certainly present and involved in adipose tissue, yet in some marsupials (possum, antechinus) its expression is transient (see M. Jastroch *Physiological Genomics* 32, 161–169 [2008], while in antechinus there is no obvious connection between proton leakage and thermogenesis [see M. Jastroch et al. *Physiological and Biochemical Zoology* 82, 447–454; 2009]). To be sure, the fat-tailed dunnart does show a eutherian-style UCP1 activity, but this group (the dasyuroids) occupies quite a distal position in marsupial phylogeny (see R.M.D. Beck *Journal of Mammalogy* 89, 175–189 [2008]), suggesting that the capacity to employ UCP1 for thermogenesis has arisen in parallel (also D.A. Hughes et al. *BMC Evolutionary Biology* 9, e4; 2009).

98. See B.A. Block and J.R. Finnerty *Environmental Biology of Fishes* 40, 283–302 (1994).

99. See B.A. Block *News in Physiological Sciences* 2, 208–213 (1987); also Jastroch et al. (2005); note 96.

100. See, for example, J.E.P.W. Bicudo et al. *Bioscience Reports* 21, 181–188 (2001).

101. See Y. Emre et al. *Journal of Molecular Evolution* 65, 392–402 (2007).

102. See J. Mozo et al. *Bioscience Reports* 25, 227–249 (2005).

103. These also converge in their degree of polyunsaturation (see A.J. Hulbert et al. *Journal of Experimental Biology* 205, 3561–3569 [2002]).

104. See I. Walter and F. Seebacher *Journal of Experimental Biology* 212, 2328–2336 (2009); ANT is short for adenine nucleotide translocator and, like UCP1, serves as mitochondrial carrier (see E. Rial and R. Zardoya *Journal of Biology* 8, e58 [2009]).

105. See N.V. Mezentseva et al. *BMC Biology* 6, e17 (2008). These authors suggest that the adipocytes stem from a common ancestor with the mammals, and also point out that the pathway of differentiation is very similar. Common ancestry, or once again convergence?

106. See their paper in *Journal of Experimental Zoology, B* 291, 317–338 (2001).

107. Like mammals many birds exhibit torpor, notably the hummingbirds.

108. Schweitzer and Marshall (2001), p. 331; note 106.

109. See G.R. Bartlett *American Zoologist* 20, 103–114 (1980).

110. See R.E. Isaacks and D.R. Harkness *American Zoologist* 20, 115–129 (1980).

111. See J.R. Villar et al. *Comparative Biochemistry and Physiology, B* 135, 169–175 (2003).

112. See R.H. Michell *Nature Reviews Molecular Cell Biology* 9, 151–161 (2008).

113. See their overview in *Science* 206, 649–653 (1979).

114. Bennett and Ruben (1979), p. 649; note 113.

115. See their paper in *Physiological and Biochemical Zoology* 77, 1019–1042 (2004).

116. Hillenius and Ruben (2004), part of the title; note 115.

117. Hillenius and Ruben (2004), p. 1038; note 115, suggest, for example, that the Cretaceous enantiornithine bird *Neuquenornis* (see L.M. Chiappe and J.O. Calvo *Journal of Vertebrate Paleontology* 14, 230–246 [1994]) might "represent an independent development of aerobic flight musculature."

118. See B.G. Gardiner *Zoological Journal of the Linnean Society* 74, 207–232 (1982).

119. See his paper in *Zoological Journal of the Linnean Society* 92, 67–104 (1988).

120. Kemp (1988), p. 69; note 119.

121. To this roster can be added the scrolled turbinals within the nasal passage, albeit formed of cartilage rather than bone (see N.R. Geist *Physiological and Biochemical Zoology* 73, 581–589 [2000]). These avian turbinals operate in very much the same way as their mammalian equivalents. It is also worth drawing attention to another striking example of endothermic convergence, but this time in the remarkable jellyfish-chomping leatherback turtles (see W. Frair et al. *Science* 177, 791–793 [1972]). Not only do they have an elevated body temperature (along with some pythons; see P. Harlow and G. Grigg *Copeia* 1984, 959–965 [1984]), but these are one of the very few reptilian endotherms. The large size of these turtles has been used to suggest they might provide a useful analogy to the possible "gigantothermy" of the dinosaurs (see F.V. Paladino et al. *Nature* 344, 858–860 [1990]). Leatherback turtles can thrive in near-freezing waters (see M.C. James et al. *Journal of Experimental Marine Biology and Ecology* 335, 221–226 [2006]), and have been doing so since at least the Eocene (see L.B. Albright et al. *Journal of Vertebrate Paleontology* 23, 945–949 [2003]). These turtles also possess a direct analogy to the blubber of marine mammals (see papers by J. Davenport et al. in *Journal of the Marine Biological Association, UK* 70, 33–41 [1990], and *Journal of Experimental Biology* 212, 2753–2759 [2009]; also G.P. Goff and G.B. Stenson *Copeia* 1988, 1071–1075 [1988]). On occasion these turtles dive to depths in excess of a kilometer (see J.D.R. Houghton et al. *Journal of Experimental Biology* 211, 2566–2575

Z. Johanson and P.E. Ahlberg *Transactions of the Royal Society of Edinburgh: Earth Sciences* 92, 43–74 (2001).

50. See D. Snitting *Acta Zoologica* 90 (Supplement, 1), 273–284 (2009).

51. See his paper in *Palaeontology* 45, 577–593 (2002).

52. Clement (2002), pp. 589–590; note 51.

53. See their paper in *Journal of Vertebrate Paleontology* 17, 653–673 (1997).

54. Ahlberg and Johanson (1997; note 53) add, "Interestingly, the morphology and dentition of the most derived tristichopterids approaches that of rhizodonts. . . . Our phylogenetic analysis indicates that these similarities are homoplasies, presumably resulting from a similar mode of life" (p. 672).

55. See Z. Johanson et al. *Palaeontology* 46, 271–293 (2003).

56. Johanson et al. (2003), p. 289; note 55.

57. See, for example, T. Holland et al. *Journal of Vertebrate Paleontology* 27, 295–315 (2007).

58. See Z. Johanson and P.E. Ahlberg *Nature* 394, 569–573 (1998); also Johanson and Ahlberg (2001); note 49.

59. As reflected for example in the jaws; see J.E. Jefferey *Transactions of the Royal Society of Edinburgh: Earth Sciences* 93, 255–276 (2003), and M.D. Brazeau *Canadian Journal of Earth Sciences* 42, 1481–1499 (2005); see also M.D. Brazeau and J.E. Jeffery *Journal of Morphology* 269, 654–665 (2008), for remarks abut the hyomandibular and convergences.

60. See, for example, J.M. Garvey et al. *Journal of Vertebrate Paleontology* 25, 8–18 (2005).

61. See J.E. Jeffery *Biological Journal of the Linnean Society* 74, 217–236 (2001).

62. See E.B. Daeschler and N. Shubin *Nature* 391, 133 (1998); also M.C. Davis et al. *Bulletin of the Museum of Comparative Zoology at Harvard College* 156(1), 171–187 (2001).

63. See M.C. Davis et al. *Journal of Vertebrate Paleontology* 24, 26–40 (2004).

64. Davis et al. (2004), p. 39; note 63.

65. Title of paper by Daeschler and Shubin (1998); note 62.

66. See their paper in *Modularity in development and evolution* (G. Schlosser and G.P. Wagner, eds.), pp. 429–440 (Chicago; 2004).

67. Shubin and Davis (2004), p. 435; note 66.

68. See N.H. Shubin et al. *Nature* 440, 764–771 (2006).

69. Shubin et al. (2006), p. 764; note 68.

70. Shubin et al. (2006), p. 768; note 68.

71. See Z. Johanson et al. *Journal of Experimental Zoology B: Molecular and Developmental Evolution* 308, 757–768 (2007).

72. See M. Friedman et al. *Evolution & Development* 9, 329–337 (2007).

73. See G.P. Wagner and A.O. Vargas *Genome Biology* 9(3), e213 (2008).

74. See J.W. Warren and N.A. Wakefield *Nature* 238, 469–470 (1972); also G.C. Young *Australian Journal of Earth Sciences* 54, 991–1008 (2007).

75. See G. Niedźwiedzki et al. *Nature* 463, 43–48 (2010); also commentary on pp. 40–41 by P. Janvier and G. Clément.

76. See A. Warren et al. *Alcheringa* 10, 183–186 (1986).

77. See J.A. Clack *Palaeogeography, Palaeoclimatology, Palaeoecology* 130, 227–250 (1997); also G. Gouramanis *Australian Journal of Earth Sciences* 50, 811–825 (2003).

78. See G.C. Young *Alcheringa, Special Issue* 1, 409–428 (2006).

79. See his paper in *Zoological Journal of the Linnean Society* 147, 473–488 (2006).

80. See B.G. Lovegrove and F. Génin *Journal of Comparative Physiology, B* 178, 691–698 (2008).

81. Fascinatingly on the basis of bone microstructure, M. Köhler and S. Moyà-Solà (*Proceedings of the National Academy of Sciences, USA* 106, 20354–20358; 2009) argue that an insular bovid (*Myotragus*, pp. 253–254) lived in such a resource-poor environment that it effectively reverted to the physiology of an ectothermic reptile, growing more like a crocodile than a self-respecting cow (see, respectively, critique by S. Meiri and P. Raia and reply in vol. 107 [2010], pp. E27 and E28).

82. See, for example, J.A. Ruben et al. *Physiological and Biochemical Zoology* 76, 141–164 (2003); also J.F. Gillooly et al. *PLoS Biology* 4, e248 (2006), and for a crisp overview, A. Louchard and L. Viriot *Trends in Ecology & Evolution* 26, 663–673 (2011); see their box 1.

83. Other turbinate bones are coated with olfactory tissue. See B. Van Valkenburgh et al. (*Journal of Anatomy* 218, 298–310; 2011), who document how in the mammalian carnivores in terrestrial taxa, the relative proportions of olfactory to respiratory surfaces (70:30) are almost exactly opposite to that of the aquatic pinnipeds and sea otters. Not only has this evolved independently (and is consistent with the correlations of aquatic modes of life and diminished olfactory capacity), but suggests that this trade-off is inevitable.

84. See, for example, K. Schmidt-Nielsen et al. *Respiration Physiology* 9, 263–276 (1970).

85. See M. Laaß et al. *Acta Zoologica* 92, 363–371 (2011).

86. See W.J. Hillenius *Paleobiology* 18, 17–29 (1992).

87. See J. Ruben *Annual Review of Physiology* 57, 69–95 (1995).

88. Ruben (1995), p. 85; note 87.

89. See W.J. Hillenius *Evolution* 48, 207–229 (1994).

90. As Ruben (1995; note 87) also reminds us, this emergence "may have initially involved more substantive quantitative, rather than qualitative, changes in many aspects of cellular physiology" (p. 85).

91. Notably the central role of an elevated enzymatic activity in the context of mitochondrial membranes

Naturelle 1, 271–272 (1895); also S. Van Wassenberg et al. (*Nature* 440, 881; 2006), who report on the related *Channallabes* preying on insects.

6. See T. Moritz and K.E. Linsenmair *Journal of Fish Biology* 71, 279–283 (2007).

7. See K. Johansen et al. *Zeitschrift für Vergleichende Physiologie* 61, 137–163 (1968).

8. Graham (1997), p. 130; note 4.

9. To give just one example Graham (1997; note 4) draws our attention to how "The vascular epithelium of the air sacs and the buccopharynx of *Monopterus* [a swamp-eel] is highly convergent with that in *Channa*" [a snakehead fish] (p. 97).

10. Graham (1997), p. 253; note 4.

11. See K.F. Liem *Fieldiana: Zoology* (*New Series*) 37, 1–29 (1987).

12. Liem (1987), p. 24; note 11.

13. See, for example, the description by T.J. Ord and S.T. Hsieh (*Ethology* 117, 918–927; 2011) of the remarkable leaping blenny (*Alticus*), which as long as it keeps itself moist (it has no lung) is effectively entirely terrestrial.

14. See B.K. Das *Philosophical Transactions of the Royal Society of London, B* 216, 183–219 (1927).

15. The ability to gulp air evolved independently in the lungfish, a capacity that originated in a marine setting; see A.M. Clement and J.A. Long *Biology Letters* 6, 509–512 (2010).

16. See his paper in *Transactions of the Kansas Academy of Science* 63, 115–120 (1960); also T.H. Eaton *American Midland Naturalist* 46, 245–251 (1951).

17. Eaton (1960), p. 119; note 16.

18. See J.A. Clack *Annual Review of Earth and Planetary Sciences* 37, 163–179 (2009); also M.I. Coates *Annual Review of Ecology, Evolution, and Systematics* 39, 571–592 (2008).

19. See E.B. Daeschler et al. *Nature* 440, 757–763 (2006), with commentary on pp. 747–749 by P.E. Ahlberg and J.A. Clack; also J.D. Downs et al. *Nature* 455, 925–929 (2008).

20. See P.E. Ahlberg et al. *Nature* 453, 1199–1204 (2008).

21. See M.D. Brazeau and P.E. Ahlberg *Nature* 439, 318–321 (2006).

22. See M. Zhu and P.E. Ahlberg *Nature* 432, 94–96 (2004).

23. See, for example, E.B. Daeschler et al. *Acta Zoologica* 90 (Supplement, 1), 306–317 (2009).

24. One such example is a humerus from the Upper Devonian of Pennsylvania, which is assigned to tetrapods (see N.H. Shubin *Science* 304, 90–93 [2004]), but displays particular types of muscle insertion. These, in the words of Neil Shubin and colleagues, point to "a diversity of limb design in the earliest tetrapods" (p. 92). So, too, with a description of a new tetrapod from the Devonian of Greenland, which entertainly almost entirely demolishes some cherished cladograms, J.A. Clack et al. (*Palaeontology* 55, 73–86 [2012])

note how the emerging fossil evidence indicates "that by the Fammenian, diversity among tetrapods was already substantial" (p. 84).

25. See also V. Callier et al. (*Science* 324, 364–367 [2009]) who in their documentation of the ontogeny of the forelimbs of *Acanthostega* and *Ichthyostega* identify what they euphemistically refer to as "morphological experimentation" (p. 366), although one gets the impression that to the cladistically minded, this is profoundly irritating rather than deeply informative.

26. See C.A. Boisvert et al. *Nature* 456, 636–638 (2008); also C.A. Boisvert *Acta Zoologica* (Supplement, 1), 297–305 (2009).

27. See P. Ahlberg et al. *Palaeontology* 43, 533–548 (2000).

28. See T. Holland and J.A. Long *Acta Zoologica* (Supplement, 1) 285–296 (2009).

29. See their paper in *Nature* 444, 199–202 (2006).

30. Long et al. (2006), p. 199 [Abstract]; note 29.

31. See M. Zhu and X.-B. Yu *Nature* 418, 767–770 (2002).

32. See M. Zhu et al. *Nature* 458, 469–474 (2009); also M. Zhu et al. *Nature* 441, 77–80 (2006).

33. Zhu et al. (2009), p. 469 [Abstract]; note 32.

34. Zhu et al. (2009), p. 473; note 32.

35. See M.D. Brazeau *Nature* 457, 305–308 (2009).

36. See, for example, M. Friedman *Journal of Systematic Palaeontology* 5, 289–343 (2007).

37. See, for example, the example from the lungfish given by T. Qiao and M. Zhu *Acta Zoologica* 90 (Supplement, 1), 236–252 (2009).

38. See, for example, D. Snitting *Journal of Vertebrate Paleontology* 28, 637–655 (2008).

39. See A. Blieck et al. in *Devonian events and correlations* (R.T. Becker and W.T. Kirchgasser, eds.), pp. 219–235 (Geological Society of London, Special Publications 278; 2007).

40. See, for example, M.J. Markey and C.R. Marshall *Proceedings of the National Academy of Sciences, USA* 104, 7134–7138 (2007).

41. See her paper in *Paleontological Journal* 37, 449–460 (2003; a translation from *Paleontologicheskii Zhurnal* 2003[5], 3–14 [2003]).

42. Vorobyeva (2003), p. 455; note 41.

43. Vorobyeva (2003), p. 457; note 41 (both quotations).

44. See their paper in *Nature* 395, 792–794 (1998); with commentary on pp. 748–749 by P. Janvier.

45. Ahlberg and Johanson (1998), p. 794; note 44.

46. Ahlberg and Johanson (1998), p. 792 [Abstract]; note 44.

47. See G.C. Young *Alcheringa* 32, 321–336 (2008).

48. See F.J. Meunier and M. Laurin *Acta Zoologica* 93, 88–97 (2011), for evidence that the bones of the aquatic animal were already very tetrapodlike and mechanically preadapted for terrestrial life.

49. See, for example, J.E. Jeffery *Biological Journal of the Linnean Society* 74, 217–236 (2001), and

192. See, for example, papers in *Journal of Morphology* by J.I. Castro and J.R. Wourms (vol. 218, pp. 257–280; 1993) and L. Fishelson and A. Baranes (vol. 236, pp. 151–165; 1998).

193. See J.P. Wourms *Environmental Biology of Fishes* 38, 269–294 (1993).

194. Wourms (1993), p. 269 [Abstract]; note 193.

195. These, as Wourms (1993; note 193) also notes, are "virtually identical with that observed in trophonemata, villous extensions of the uterine wall in stingrays . . . [and so represent an] extraordinary convergence" (p. 278).

196. Others, however, are not convinced, and argue that to describe this type of placenta as epitheliochorial is an overinterpretation (see W.C. Hamlett et al. in *Reproductive biology and phylogeny of Chondrichthyes: Sharks, batoids and chimaeras* [W.C. Hamlett and B.G.M. Jamieson, eds.]), pp. 463–502 [Science Publishers; 2005]). Even so, William Hamlett and colleagues note that in a few "species the egg envelope degenerates and is not a constituent of the [uteroplacental] interface" (p. 484).

197. This interface also employs interleukins (p. 218), and significantly glycans mediate the attachment of the placenta (see C.J.P. Jones and W.C. Hamlett *Placenta* 25, 820–828 [2004]; also Hamlett et al. [2005]; note 196).

198. See A.N. Haines et al. *Placenta* 27, 1114–1123 (2006).

199. See J.E. Mank et al. *Evolution* 59, 1570–1578 (2005).

200. Typically this is by hypertrophy of the epithelium associated with the ovarian follicle (see the classic paper by C.L. Turner *Journal of Morphology* 67, 59–87 [1940]; also, for example, B.D. Grove and J.P. Wourms *Journal of Morphology* 220, 167–184; 1994).

201. See F.M. Knight et al. *Journal of Morphology* 185, 131–142 (1985).

202. See D.N. Reznick et al. *Science* 298, 1018–1020 (2002); also, for example, papers in *Biological Journal of the Linnean Society* by D. Reznick et al. (vol. 92, pp. 77–85; 2007), M.N. Pires et al. (vol. 99, pp. 784–796; 2010), and Wourms (1994; note 187).

203. Reznick et al. (2002), p. 1018; note 202.

204. These include the zenarchopterid half-beaks (beloniforms) (see D. Reznick et al. *Evolution* 61, 2570–2583 [2007]; also A.D. Meisner and J.R. Burns *Journal of Morphology* 234, 295–317; 1997), which are related to the flying fish.

205. See, for example, J.P. Wourms and D.M. Cohen *Journal of Morphology* 147, 385–402 (1975).

206. Wourms and Cohen (1975), p. 385 [Abstract]; note 205.

207. See note 136.

208. Wourms and Cohen (1975), p. 394 [emphasis in original]; note 205.

209. See K.N. Stölting and A.B. Wilson *BioEssays* 29, 884–896 (2007).

210. Stölting and Wilson (2007), p. 889; note 209. So, too, for example, lectins are employed, possibly with an antimicrobial function (see P. Melamed et al. *FEBS Journal* 272, 1221–1235 [2005]).

211. Stölting and Wilson (2007), p. 893; note 209.

212. See J.P. Wourms et al. *Environmental Biology of Fishes* 32, 225–248 (1991); also C.L. Smith et al. *Science* 190, 1105–1106 (1975).

213. See W. Wen et al. *Acta Palaeontologica Polonica* 58, 175–193 (2013).

214. See M.H. Wake *Journal of Experimental Zoology* 266, 394–413 (1993); also M.H. Wake and R. Dickie in vol. 282, pp. 477–506 (1998).

215. See, for example, D.J. Gower et al. *Journal of Evolutionary Biology* 21, 1220–1226 (2008).

216. See M. Delsol et al. *Compte Rendus des Séances de l'Academie des Sciences, série III*, 293, 281–285 (1981).

217. See M. Delsol et al. *Mémoires de la Société Zoologique de France* 413, 39–54 (1984).

218. See J. Hanken in *The origin and evolution of larval forms* (B.K. Hall and M.H. Wake, eds.), pp. 61–108 (Academic; 1999).

219. See D. Toews and D. Macintyre *Nature* 266, 464–465 (1977).

220. See, for example, M. Wilkinson et al. *Biology Letters* 4, 358–361 (2008); also Wake and Dickie (1998); note 214.

221. See O. Goicoechea et al. *Journal of Herpetology* 20, 168–178; also K. Busse *Copeia* 1970, 395 (1970).

222. There is some evidence in *Gastrotheca* suggesting that its tadpole stage reevolved from frogs with a direct development (see J.J. Wiens et al. *Evolution* 61, 1886–1899 [2007]), so echoing a comparable example in the salamanders (p. 25).

223. See R.E. Jones et al. *Journal of Experimental Zoology* 184, 177–184 (1973).

224. See H. Greven and S. Richter *Journal of Morphology* 270, 1311–1319 (2009).

225. Greven and Richter (2009), p. 1311; note 224.

CHAPTER 20

1. This, of course, echoes what D'Arcy Wentworth Thompson set out in his magisterial *On growth and form* (Cambridge; 1942), and has provided in equal parts inspiration and frustration to subsequent generations.

2. The title of his book (Penguin; 2009).

3. Not that all fish have this structure. "Twenty ways to lose your bladder" is the main title of the paper by A.R. McCune and R.L. Carlson (*Evolution & Development* 6, 246–259; 2004), but they suggest that the actual total would well be above fifty and for the most part is linked to either a benthic or deep-sea habit.

4. See J.B. Graham *Air-breathing fishes: Evolution, diversity, and adaptation* (Academic; 1997).

5. See L. Vaillant *Bulletin du Muséum d'Histoire*

157. See Z. Kielan-Jaworowska *Nature* 277, 402–403 (1979).

158. See U. Zeller *Zoologischer Anzeiger* 238, 117–130 (1999). In any event what is clear that in the bandicoot the placenta is strikingly convergent with that of the placentals; see H.A. Padykala and J.M. Taylor *Anatomical Record* 186, 357–386 (1976).

159. See A.M. Carter *Placenta* 29, 930–931 (2008); also his overview in *Physiological Reviews* 92, 1543–1576 (2012).

160. See C.C. O'Harra *Science* 71, 341–342 (1930).

161. See J.L. Franzen *Palaeontographica, Abteilungen. A* 278, 27–35 (2006).

162. See M.G. Elliot and B.J. Crespi *Placenta* 30, 949–967 (2009), and G. Cornelis et al. *Proceedings of the National Academy of Sciences, USA* 110, E828–E837 (2013); also A.C. Enders and A.M. Carter *Placenta* 33, 319–326 (2012).

163. See, for example, O. Heidmann et al. *Retrovirology* 6, e107 (2009), and G. Cornelis et al. *Proceedings of the National Academy of Sciences, USA* 109, E432–E441 (and pp. 2206–2207; 2012), with commentary on pp. 2184–2185 by H.S. Malik.

164. Nor are these the only convergences at a molecular level, because the placental lactogens (see I.A. Forsyth *Experimental and Clinical Endocrinology* 102, 244–251 [1994]) and chorionic gonadotrophins (see F. Stewart and W.R. Allen *Reproduction in Domestic Animals* 30, 231–239 [1995]) of primates, respectively, have evolved independently in such groups as the rodents and horses.

165. See M. Garratt et al. *Proceedings of the National Academy of Sciences, USA* 110, 7760–7765 (2013); these authors also discuss links to number of offspring and resolution of potential maternal-fetal conflict. They also caution, however, that conceivably placental type drove life history rather than vice versa.

166. See B. Crespi and C. Semeniuk *American Naturalist* 163, 635–653 (2004).

167. This is even more true where the litter is the product of multiple fathers, as in some viviparous sharks (see, for example, K.A. Feldheim et al. *Copeia* 2001, 781–786 [2001]).

168. See L. Paulesu et al. *Placenta* 16, 193–205 (1995).

169. See, for example, C. Cateni et al. *Reproductive Biology and Endocrinology* 1, e25 (2003).

170. We find other molecules, such as insulin growth factor II, being employed in placental teleosts (see M.J. O'Neill et al. *Proceedings of the National Academy of Sciences, USA* 104, 12404–12409 [2007]) and a key gene (Hβ58) for placental development in the lizard *Chalcides* (see L. Paulesu et al. *Placenta* 22, 735–741 [2001]).

171. Intriguingly they have a connection to the Toll receptors of the innate immune system (see L. Paulesu et al. *Evolution & Development* 10, 778–788; 2008).

172. For similar observations with respect to HoxA10-like proteins in lizards, see M. Thomson et al. *Comparative Biochemistry and Physiology, B* 142, 123–127 (2005).

173. Specifically the examples include an aigialosaur (see M.W. Caldwell and M.S.Y. Lee *Proceedings of the Royal Society of London, B* 268, 2397–2401 [2001]), as well as the more advanced (see, for example, M.W. Caldwell and A. Palci *Journal of Vertebrate Paleontology* 27, 863–880 [2007]) *Plioplatecarpus* (see G.L. Bell et al. *Journal of Vertebrate Paleontology* 16 [Supplement, 3], 21A–22A [1996]). The agialosaurs were relatively primitive, and giving birth on land cannot be ruled out, but that was surely not an option for the more advanced taxa that were evidently wholly aquatic.

174. See M. DeBraga and R.L. Carroll *Evolutionary Biology* 27, 245–322 (1993), and O. Rieppel and H. Zaher *Zoological Journal of the Linnean Society* 129, 489–514 (2000).

175. See A.S. Schulp et al. *Netherlands Journal of Geology* 84, 359–371 (2005).

176. See D.C. Deeming et al. *Modern Geology* 18, 423–442 (1993) and E.E. Maxwell and M.W. Caldwell *Proceedings of the Royal Society B (Supplement)* 270, S104–S107 (2003).

177. See R. Böttcher *Stuttgarter Beiträge zur Naturkunde, Serie B (Geologie und Paläontologie)* 164, 1–51 (1990) (in German, with English abstract).

178. See R. Motani et al. *PLoS ONE* 9, e88640 (2014).

179. See F.R. O'Keefe and L.M. Chiappe *Science* 333, 870–873 (2011); these authors also suggest a K-selected life history and possible parallels to the cetaceans.

180. See Y.-n. Cheng et al. *Nature* 432, 383–386 (2004).

181. See S. Renesto et al. *Journal of Vertebrate Paleontology* 23, 957–960 (2003); for a more cautious view, see P.M. Sander *Science* 239, 780–783 (1988).

182. See Q. Ji et al. *Naturwissenschaften* 97, 423–428 (2010).

183. See note 106.

184. The enclosed embryos had been spotted by earlier investigators, but they had been misidentified—in the case of the arthrodire, as its dinner.

185. See J.A. Long et al. *Nature* 453, 650–652 (2008), and a short item on p. 575 by C. Dennis.

186. See J.A. Long et al. *Nature* 457, 1124–1127, with commentary on pp. 1094–1095 by P.E. Ahlberg; also P. Ahlberg *Nature* 460, 888–889 (2009).

187. See J.P. Wourms *Israel Journal of Zoology* 40, 551–568 (1994).

188. For a record from the Carboniferous, see E.D. Grogan and R. Lund *Zoological Journal of the Linnean Society* 161, 587–594 (2011).

189. See N.K. Dulvy and J.D. Reynolds *Proceedings of the Royal Society of London, B* 264, 1309–1315 (1997); also, for example, J.A. López et al. *Molecular Phylogenetics and Evolution* 40, 50–60 (2006).

190. Dulvy and Reynolds (1997), p. 1314; note 189.

191. See, for example, W.C. Hamlett and M.K. Hysell *Journal of Experimental Zoology* 282, 438–459 (1998).

131. See, for example, J.R. Stewart and M.B. Thompson *Journal of Experimental Zoology, B* 312, 590–602 (2009).

132. Blackburn (2006), p. 138; note 113.

133. See C.R. Murphy et al. *Comparative Biochemistry and Physiology, A* 127, 433–439 (2000).

134. Published in *Journal of Morphology* 258, 346–357 (2003); also S.M. Adams et al. in vol. 268, pp. 385–400 (2007).

135. Hosie et al. (2003), p. 355; note 134.

136. Among the roster of invertebrate viviparity, particular attention should be paid to the arthropods, including scorpions (see O.F. Francke *Revue Arachnologique* 4, 27–37 [1982]) and also the onychophorans (see S.S. Campiglia and M.H. Walker *Journal of Morphology* 224, 179–198 [1995]). In the latter, the embryos are not only housed in sacs filled with fluid but they are attached to a syncitial placenta by a stalk. As the authors note, this arrangement is "analogous to the mammalian non-invasive epitheliochorial placenta" (p. 179 [Abstract]).

137. Hosie and her colleagues (2003; note 134) conclude, "Indeed, this is precisely the point made by the concept of "plasma membrane transformation",i.e.,despite the disparate modes of placentation . . . the behavior of the plasma membrane . . . and the initial stages of contact between maternal and fetal tissues [show] remarkable commonalities across species" (p. 355).

138. This includes an endocrine function reminiscent of mammals; see Guarino et al. *General and Comparative Endocrinology* 111, 261–270 (1998), while in the related *C. ocellatus*, gene regulation across a wide series of functions associated with pregnancy (e.g., tissue remodeling, blood supply, immune system, transport) show "clear parallels" with mammals; see M.C. Brandley et al. *Genome Biology and Evolution* 4, 394–411 (2012); also commentary on pp. 372–373 by D. Venton.

139. See their paper in *Placenta* 24, 489–500 (2003).

140. Jones et al. (2003), p. 489 [Abstract]; note 139.

141. Blackburn (1998), p. 589; note 107.

142. See A.F. Flemming and W.R. Branch *Journal of Morphology* 247, 264–287 (2001).

143. See D.G. Blackburn and A.F. Flemming *Journal of Experimental Zoology, B* 312, 579–589 (2009).

144. Blackburn and Flemming (2009), p. 586; note 143.

145. For further documentation of this remarkable structure, see D.G. Blackburn and A.F. Flemming *Journal of Morphology* 273, 137–159 (2012), the intimacy of which with the maternal tissue raises questions about immune response and hormonal control.

146. Blackburn and Flemming (2009), p. 587; note 143.

147. See S.M. Adams et al. *Journal of Morphology* 264, 264–276 (2005); also J.R. Stewart and M.B. Thompson *Journal of Experimental Zoology, A* 299, 13–32 (2003),

and O.W. Griffith et al. *Journal of Experimental Zoology, B* 320, 465–470 (2013).

148. Adams et al. (2005), p. 275; note 147.

149. Here too mammalianlike biochemistries connected to the permeability of membranes (involving the protein occludin [see J.M. Biazik et al. *Journal of Comparative Physiology, B* 177, 935–943 [2007]) and a lysosomal system (see J.M. Biazik *Journal of Experimental Zoology, B* 312, 817–826 [2009]) underline the complexity of this type of reptilian placentation (see B.F. Murphy and M.B. Thompson *Journal of Comparative Physiology, B* 181, 575–594 [2011], for an overview of the molecular dimension of squamate placentation).

150. See, for example, A. Herrel and M.P. Ramírez-Pinilla *Journal of Morphology* 249, 132–146 (2001), and F.B. Wooding et al. *Placenta* 31, 675–685 (2010).

151. See D.G. Blackburn et al. *Proceedings of the National Academy of Sciences, USA* 81, 4860–4863 (1984), and A.F. Flemming and D.G. Blackburn *Journal of Experimental Zoology, A* 299, 33–47 (2003); also Blackburn (1998); note 107, and Blackburn and Flemming (2009); note 143.

152. See S. Vieira et al. *Journal of Morphology* 271, 738–749 (2010). As these authors note, although in many ways of a squamate type, the ovaries of *Mabuya* show one intriguing difference inasmuch as, instead of having yolk platelets, "the oocyte's membrane forms numerous pits associated with coated vesicles allowing lipoprotein endocytosis. . . . [These] seem to be similar to those described for marsupials . . . [and] suggest a very interesting convergence between these two clades" (p. 748), and reflecting an advanced level of matrotrophy.

153. See her paper in *Herpetological Monographs* 20, 194–204 (2006).

154. Ramírez-Pinilla (2006), p. 194; note 153.

155. See D.G. Blackburn and L.J. Vitt *Journal of Morphology* 254, 121–131 (2002), and M.P. Ramírez-Pinilla et al. *Journal of Morphology* 267, 1227–1247 (2006); also Blackburn and Flemming (2009); note 143.

156. As Daniel Blackburn and Laurie Vitt note, these "chorionic areolae . . . [also] occur in several mammalian groups, including carnivores and primates" (p. 128; note 155). So, too, the epithelium of the chorion in *Mabuya* includes specialized cells with two very large nuclei, a feature again found in mammals. In terms of interpenetration, quite how close the placenta of *Mabuya* is to the equivalents of various mammals, notably the ungulates, is debatable. It is evident that by the neurula stage no remnant of the eggshell occurs (see A. Jerez and M.P. Ramírez-Pinilla *Journal of Morphology* 258, 158–178 [2003]), but it is now clear that, like *Trachylepis*, the placenta is close to a true endotheliochorial arrangement, with the chorion cells interpenetrating with the syncitial tissue of the uterine wall and its capillaries (see S. Vieira et al. *Anatomical Record* 290, 1508–1518 [2007]).

pendently—not only in the amniotes but in such groups as the sharks (see A. Collazo et al. *American Naturalist* 144, 133–152 [1994]).

99. For example, in an African reed frog the eggs are yolk-rich and show a sort of meroblasty (see A.D. Chipman et al. *Evolution & Development* 1, 49–61 [1999]). This frog, however, not only has a small egg, but a biphasic life cycle—that is, with an aquatic tadpole.

100. See E.M. Callery et al. *BioEssays* 23, 233–241 (2001), and for direct development see, for example, R.P. Elinson *Genesis* 29, 91–95 (2001).

101. See R.P. Elinson and Y. Beckham *Zoology* 105, 105–117 (2002).

102. See D.R. Buchholz et al. *Developmental Dynamics* 236, 1259–1272 (2007).

103. But in the corn snake the yolk mass is not only cellularized but is associated with a blood vessel; see R.P. Elinson and J.R. Stewart *Biology Letters* 10, 20130870 (2014).

104. Buchholz et al. (2007), p. 1269; note 103.

105. See S. Renesto and R. Stockar *Swiss Journal of Geosciences* 102, 323–330 (2009).

106. See G. Piñeiro et al. *Historical Biology* 24, 620–630 (2012).

107. See, for example, D.G. Blackburn *Journal of Experimental Zoology* 282, 560–617 (1998).

108. See their paper in *Philosophical Transactions of the Royal Society of London, B* 357, 269–281 (2002).

109. Reynolds et al. (2002), p. 280; note 108.

110. Reynolds et al. (2002), p. 279; note 108.

111. See Zh.V. Korneva *Journal of Evolutionary Biochemistry and Physiology* 41, 552–560 (2005), who documents a placentalike interaction between some of the eggs in the uterus of the tapeworm, and also notes that "the appearance of placental type interactions occurred repeatedly in phylogenetically remote groups of cestodes" (p. 559). Nor are these the only examples of viviparity among the invertebrates (and I exclude the many examples of brooding). Apart from the bryozoans (see A.N. Ostrovsky *Evolution* 67, 1368–1382 [2013], sea squirts [in the salps; see J.E.A. Godeaux in *Reproductive biology of invertebrates*, vol. 4, part B (eds. K.G. Adiyodi and R.G. Adiyodi), pp. 453–469 (Wiley Interscience; 1990)], and arthropods [see note 136], matrotrophy has been identified in various echinoderms, including some holothurians [see M.A. Sewell et al. *Invertebrate Reproduction and Development* 49, 225–236; 2006]). Viviparity in freshwater mollusks has evolved multiple times, notably in the cerithioid snails (including those of Southeast Asia [see F. Köhler et al. *Evolution* 58, 2215–2226; 2004], and also Lake Tanganyika (see papers by E.E. Strong and M. Glaubrecht in *Zoologica Scripta* 31, 167–184 [2002], and *Organisms, Diversity & Evolution* 7, 81–105; 2007), while among the bivalves matrotrophic examples include the corbiculids (see A.V. Korniushin and M. Glaubrecht *Acta Zoologica*

84, 293–315 [2003], and M. Glaubrecht et al. *Zoologica Scripta* 35, 641–654; 2006), sphaeriids (see C. Meier-Brook *Archiv für Hydrobiologie, Supplement-Band* 38, 73–150 [1970; in German, with English summary], and W.H. Heard *Malacologia* 16, 421–455 [1997]), and the unionids (see M.L. Schwartz and R.V. Dimock *Invertebrate Biology* 120, 227–236 [2001]). The snails show a variety of brooding locations, including uterine, while the bivalves employ the gills. Viviparity evidently confers other benefits, because a number of species are highly invasive.

112. But not oddly the birds that with one possible exception (this was a budgie living in Dorking: see my *Life's solution* (2003), p. 219 and note 137 [p. 410]) eschew viviparity.

113. See D.G. Blackburn *Herpetological Monographs* 20, 131–146 (2006).

114. See D.G. Blackburn *Amphibia-Reptilia* 3, 185–205 (1982).

115. Blackburn (1982), p. 189; note 114.

116. Blackburn *Amphibia-Reptilia* 6, 259–241 (1985).

117. See Y. Wang and S.E. Evans *Naturwissenschaften* 98, 739–743 (2011).

118. See M.S.Y. Lee and R. Shine *Evolution* 52, 1441–1450 (1998).

119. See V.J. Lynch and G.P. Wagner *Evolution* 64, 207–216 (2010); see also Y. Surget-Groba et al. *Biological Journal of the Linnean Society* 87, 1–11 (2006). For possible examples in the vipers, see A.M. Fenwick et al. *Journal of Zoological Systematics and Evolutionary Research* 50, 59–66 (2012).

120. See his paper in *American Zoologist* 32, 313–321 (1992).

121. Blackburn (1992), p. 316; note 120.

122. Blackburn (2006), p. 137; note 113.

123. Blackburn (2006), p. 137; note 113.

124. Blackburn (2006), p. 131 [Abstract]; note 113.

125. Although the skinks have been a particular focus of attention, as might be expected from its rampant convergence, various other lizard groups show placentation, such as the iguanians (see, for example, D.G. Blackburn et al. *Journal of Morphology* 271, 1153–1175 [2010]).

126. See J.R. Stewart and M.B. Thompson *Comparative Biochemistry and Physiology, A* 127, 411–431 (2000); see also pioneering earlier work such as by H.C. Weekes *Proceedings of the Zoological Society of London* 1935, 625–645 (1935).

127. See, for example, M.B. Thompson and B.K. Speake *Journal of Comparative Physiology, B* 176, 179–189 (2006).

128. Other varieties are also known (see, for example, J.R. Stewart and K.R. Brasch *Journal of Morphology* 255, 177–201 [2003]).

129. See, for example, M. Villagrán et al. *Journal of Morphology* 264, 286–297 (2005).

130. See D.G. Blackburn *Journal of Experimental Zoology* 266, 414–430 (1993).

the Royal Society of London, B 273, 1471–1475 (2006). A hermaphrodite sea slug (Siphopteron) goes one better by plunging its penile stylet into the forehead of the partner and then transfers prostrate fluids in a series of pulses; see R. Lange et al. Proceedings of the Royal Society of London, B 281, 20132424 (2013).

74. See J.M. Koene and S. Chiba American Naturalist 168, 553–555 (2006); also B. Reyes-Tur and J.M. Koene (in Animal Biology 57, 261–266; 2007) for a similar report in a Cuban snail, Polymita muscarum. In both cases the multiple habit means the dart is reused, whereas in many other snails it is lost and has to be resecreted. The dart of the Cuban snail is remarkably stilettolike.

75. See, for example, E.A. Fedoseeva Ruthenica 4, 103–110 (1994; in Russian, with English abstract), and J.M. Koene and H. Schulenburg BMC Evolutionary Biology 5, e25 (2005).

76. See A. Davison et al. Journal of Zoology, London 267, 329–338 (2005); also Koene and Schulenburg (2005), note 75.

77. The competing hypotheses are reviewed widely; see, for example, Koene (2006), note 71.

78. See H. Lind Journal of Zoology, London 169, 39–64 (1973).

79. See J.M. Koene and A. ter Maat Journal of Comparative Physiology, A 187, 323–326 (2001).

80. See J.M. Koene and R. Chase Journal of Experimental Biology 201, 2313–2319 (1998); also D.W. Rogers and R. Chase Behavioral Ecology and Sociobiology 50, 122–127 (2001).

81. Much has also been made of elements of sexual competition and an "arms race" between allohormones reconfiguring the female reproductive tract and attempted counteraction. Yet, while exchange of sperm may be simultaneous or sequential, both partners potentially stand to gain, and each snail can be very much treated as an independent actor (see R. Chase and K. Vaga Behavioral Ecology and Sociobiology 59, 732–739 [2006]).

82. See J.M. Koene et al. PLoS ONE 8, e69968 (2013).

83. See, for example, N.K. Michiels and L.J. Newman Nature 391, 647 (1998).

84. See The formation of vegetable mould through the action of worms, with observations on their habits (Murray; 1883).

85. See V. Nuutinen and K.R. Butt Journal of Zoology, London 242, 783–798 (1997).

86. Of the two populations studied (see V. Nuutinen and K.R. Butt Soil Biology and Biochemistry 29, 307–308 [1997]), one was Finnish and the other English. Oddly, the latter, "when undisturbed, had a much shorter and less complicated pre-mating phase" (p. 308). The authors are careful to note that several explanations may be possible, including laboratory versus field conditions or environmental factors. Always good to know that in the decorum of an English garden, order extends underground.

87. See, for example, J. Feldkamp Zoologische Jahrbücher: Abteilung für Anatomie und Ontogenie der Tiere 46, 609–632 (1924), pl. 25, fig. 12.

88. See A.J. Grove Quarterly Journal of Microscopical Science 69, 245–290 (1925).

89. See J.M. Koene et al. Invertebrate Reproduction and Development 41, 35–40 (2002).

90. See J.M. Koene et al. Behavioral Ecology and Sociobiology 59, 243–249 (2005); rather curiously the mucus contains remarkably high concentrations of a small protein, ubiquitin, which as its name suggests is very widespread. See S. König et al. Invertebrate Reproduction and Development 49, 103–111 (2006), but the exact role of the ubiquitin is not clear.

91. Darts and sharp chaetae are not the only way to inject an allohormone, and among the salamanders some species (e.g., mountain dusky salamander) engage in a courtship (see L.D. Houck and N.L. Reagan Animal Behaviour 39, 729–734 [1990]) that entails the male scraping the back of his partner with the so-called premaxillary teeth and then stroking it with the chin, which contains allohormone-secreting glands. In a different way, secretions of the accessory reproductive glands in the insects mingle with the sperm and can hijack the recipient female by, for example, inducing a disinclination to mate again or rendering her decidedly less attractive (see E.M. Adams and M.F. Wolfner Journal of Insect Physiology 53, 319–331 [2007]). These secretions also play an antimicrobial role, helping to protect the reproductive tract against infection (see C. Gillott Annual Review of Entomology 48, 163–184 [2003]). Another trick is to insert a mating plug to block anybody else's sperm. Interestingly it has a molecular structure that is convergent on both silk and the attachment threads (byssus) of some bivalve molluscs (p. 91), and with good reason, given that the plug must rapidly polymerize to be effective (see O. Lung and M.F. Wolfner Insect Biochemistry and Molecular Biology 31, 543–551 [2001]).

92. Darwin (1883), p. 26; note 84.

93. See J. Skulan Zoological Journal of the Linnean Society 130, 235–261 (2000). In this context, however, it is worth noting that the serosa of the insect egg provides an extraembryonic membrane that not only is analogous to the amnion but has a demonstrable role in preventing desiccation; see C.G.C. Jacobs et al. Proceedings of the Royal Society of London, B 280, 20131082 (2013).

94. Skulan (2000), p. 235 [Abstract] and elsewhere; note 93.

95. Skulan (2000), p. 249; note 93.

96. See M. Laurin Journal of Natural History 39, 3151–3161 (2005).

97. See M.J. Packard and R.S. Seymour in Amniote origins: Completing the transition to land (S.J. Sumida and K.L.M. Martin, eds.), pp. 265–290 (Academic; 1997).

98. Meroblastic cleavage has evolved five times inde-

behave surreptitiously but copulate in seclusion, in the bushes.

41. In terms of developmental explanations, see A.M. Herrera et al. *Current Biology* 23, 1065–1074 (2013), and commentary on pp. R523–R525 by P.L.R. Brennan.

42. See their paper in *Journal of Avian Biology* 39, 487–492 (2008).

43. Brennan et al. (2008), p. 491; note 42.

44. See D.L. Lack *Swifts in a tower* (Methuen; 1956), pp. 43–47.

45. Nor are these the only suggestions, and explanations revolving around sexual selection in such forms as sperm competition or female choice have merit. See the six hypotheses discussed by J.V. Briskie and R. Montgomerie *Journal of Avian Biology* 28, 73–86 (1997), and ensuing debate with T. Wesolowski in the same journal (vol. 30, pp. 483–485 [1999] and vol. 32, p. 188 [2001] and response [vol. 32, pp. 184–187; 2001]).

46. In the stitchbird, not only does the male engage in aggressive copulation, but in the breeding season the cloacal protuberance grows to an immense size, evidently to store sperm that enables either multiple or large ejaculations (see M. Low et al. *Behavioral Ecology and Sociobiology* 58, 247–255 [2005]). Nor are the birds alone, because in a frog (*Ascaphus*) the cloaca of the male is engorged by blood during copulation, which can continue for days; see D.E. Metter *Copeia* 1964, 710–711 (1964), and B. Stephenson and P. Verrell *Journal of Zoology, London* 259, 15–22 (2003).

47. See R. Wilkinson and T.R. Birkhead *Ibis* 137, 117–119 (1995).

48. See their paper in *Proceedings of the Royal Society of London, B* 277, 1309–1314 (2010).

49. Brennan et al. (2010), p. 1309 [Abstract]; note 48.

50. See K.G. McCracken *Auk* 117, 820–825 (2000). Let us also doff our hats to the barnacles. For their size they possess a penis the length of which is truly colossal, but remains a vital prerequisite given the need to explore as wide an area as possible for potential mates (see W. Klepal *Oceanography and Marine Biology: An Annual Review* 28, 353–379 [1990]). Living in calm waters as against pounding surf presents its own risks to such exploration, and accordingly the barnacle can shrink or grow the penis in response to the prevailing hydrodynamic conditions (see C.J. Neufeld and A.R. Palmer *Philosophical Transactions of the Royal Society of London, B* 275, 1081–1087 [2008]). The penis of the barnacles has other peculiarities, and indeed many such structures represent the "genitalic extravagance" (p. 12) of W.G. Eberhard's *Sexual selection and animal genitalia* (Harvard; 1985). See, for example, his fig. 1.9 displaying the male genitalia of the chicken flea, which, as Eberhard notes, is "One of the marvels of organic engineering."

51. See M. Winterbottom and T.R. Birkhead *Ostrich* 74, 237–240 (2003).

52. A somewhat similar structure, the clocal tip, characterizes some fairy wrens and may assist with sperm transfer and stimulating the female (see M. Rowe et al. *Journal of Avian Biology* 39, 348–354 [2008]).

53. See T.R. Birkhead et al. *Ibis* 135, 326–331 (1993); also W. Hoesch *Journal für Ornithologie* 93, 362–363 (1952).

54. See M. Winterbottom et al. *Nature* 399, 28 (1999); also *Behavioral Ecology and Sociobiology* 50, 474–482 (2001).

55. Winterbottom et al. (1999), p. 28; note 54.

56. See P.L.R. Brennan and R.O. Prum *Journal of Zoology* 286, 140–144 (2012) who demonstrate that this must be primitive to the birds; interestingly, however, the cloacal protrusions of the Vasa parrots evidently also employ blood (note 47).

57. See her paper in *Integrative and Comparative Biology* 42, 216–221 (2002); also *Proceedings of the Royal Society of London, B* 271, S293–S295 (2004).

58. Kelly (2002), p. 219; note 57.

59. In the males of some species of octopus there is a structure that is interestingly convergent on the mammalian penis (see M.J. Thompson & J.R. Voigt *Journal of Zoology, London* 261, 101–108 [2003]). Convergence is not exact inasmuch as although there is a striking anatomical congruence, complete with hydrostatic inflation via blood vessels and collagen fibrils to resist bending, the sperm (packaged in spermatophores) is transported along an external groove, and, as far as can be assessed, does not penetrate the female octopus.

60. See A.S. King in *Form and function in birds*, vol. 2 (A.S. King and J. McLelland, eds.), pp. 107–147 (Academic; 1981); see p. 112. Kelly (2004; note 57) also draws attention to an inflatable penis evolving independently in the crocodiles, as well as the snakes and lizards (squamates), although in the latter the so-called hemipenes typically erect by a sort of eversion (see, for example, H.G. Dowling and J.M. Savage *Zoologica* 45, 17–27 [1960]).

61. See G.F. Baryshnikov et al. *Journal of Mammalogy* 84, 673–690 (2003).

62. See, for example, C.E. Friley *Journal of Mammalogy* 30, 102–110 (1949).

63. See D.A. Kelly *Journal of Morphology* 244, 69–77 (2000).

64. See their paper in *Mammal Review* 32, 283–294 (2002).

65. Lariviere and Ferguson (2002), p. 288; note 64.

66. See S.A. Ramm *American Naturalist* 169, 360–369 (2007).

67. See A.J. Hobday *Mankind Quarterly* 41, 43–58 (2000).

68. See J. Lloyd and J. Mitchinson *The book of general ignorance* (Faber; 2006), p. 74.

69. See M. A. Landolfa *Veliger* 45, 231–249 (2002).

70. Landolfa (2002), p. 232; note 69.

71. See his paper in *Integrative and Comparative Biology* 46, 419–429 (2006).

72. Koene (2006), p. 420; note 71.

73. See R. Chase and K.C. Blanchard *Proceedings of*

primitive fish and amphibians not only have both XY and ZW systems evolved multiple times but either method can also evolve into the other (see T. Ezaz et al. *Current Biology* 16, R736–R743 [2006]). Most striking is a Japanese frog (*Rana rugosa*), where not only are both types found but the ZZ-ZW system has evolved twice (see M. Ogata et al. *Heredity* 100, 92–99 [2008]).

23. See D.W. Bellott et al. *Nature* 466, 612–616 (2010).

24. As Daniel Bellott (2010; note 23) and colleagues observe, "This convergent specialization for Z and X chromosomes evolved with female heterogamety and the X chromosome evolved in the opposite direction, with male heterogamety," but they go on to observe, "in amniotes, selective pressures to preserve or enhance male reproductive functions have trumped the differences between ZW and XY systems" (p. 615).

25. See J.A. Fraser and J. Heitman *Current Opinion in Genetics & Development* 15, 645–651 (2005).

26. See M. Nicolas et al. *PLoS Biology* 3 e4 (2004).

27. Given, however, that the karyotype of chromosome number in all species of *Silene* is the same, one might expect this system to only have evolved once in this plant. No, the data suggest that it happened at least twice, given that the sex chromosomes in one species (*S. colpophylla*) evidently derived from different autosomes (see M. Mrackova et al. *Genetics* 179, 1129–1133 [2008]).

28. See S. Okada et al. *Proceedings of the National Academy of Sciences, USA* 98, 9454–9459 (2001); also K.T. Yamato et al. in vol. 104, pp. 6472–6477 (2007).

29. Okada et al. (2001), p. 9459; note 28.

30. See J.A. Foster et al. *PLoS Biology* 2, e384 (2004).

31. Foster et al. (2004), p. 2253; note 30.

32. See R. Feil and F. Berger *Trends in Genetics* 23, 192–199 (2007); also R.J. Scott and M. Spielman *Cytogenetic and Genome Research* 113, 53–67 (2006), as well as T.M. Vu et al. *Development* 140, 2953–2960 (2013).

33. Feil and Berger (2007), p. 192; note 32.

34. Specifically this involves in plants the endosperm and the placenta of mammals, and here a complex of proteins known as the Polycomb group evidently play an important role. See, in particular, J.F. Gutierrez-Marcos et al. *Placenta* 33, e3–e10 (2012).

35. See C. Köhler et al. *Nature: Genetics* 37, 28–30 (2005); also L. Hennig and M. Derkacheva *Trends in Genetics* 25, 414–423 (2009).

36. Köhler et al. (2005), p. 28 [Abstract]; note 35.

37. Consider, for example, the so-called DM domain factors, involved with DNA binding and equipped with a characteristic cysteine-rich motif. These factors play an important role in sex determination, especially with regards to the male. Is this an ancient function (see, for example, its identification in a lophotrochozoan by A. Naimi et al. *Comparative Biochemistry and Physiology, A* 152, 189–196 [2009]), so reflecting an expected conservation, or could it be convergence? "ancestry or aptitude?" as Jonathan Hodgkin asks (part of the title in *Genes & Development* 16, 2322–2326

[2002], which provides a commentary on the paper by R. Lints and S.W. Emmons on pp. 2390–2402). Given its ubiquity of function, the simplest assumption is that it is very ancient and evolution employs what has been always available. A closer inspection, however, reveals difficulties with this view because the proteins it interacts with are not homologous. As Hodgkin (2002) writes, "An alternative hypothesis is that the recruitment of DM-domain factors to male-specific functions in development has occurred multiple times independently, as a result of convergent evolution, and that there is no link to a primordial sexual differentiation mechanism" (p. 2325)—a timely reminder that the same in biology need not be automatically equated with homology. Hodgkin continues by addressing a more profound issue: "Why then should this have happened so often—why should members of this family turn up so frequently as necessary for male development, when there are dozens of other kinds of transcription factors that might seem equally qualified? Is there even something special about male development that *attracts* this particular protein family?" (p. 2325; emphasis added). He reminds us also that there are interesting parallels to other binding proteins with the requisite properties, not least *Pax6*. As importantly, just because this DM domain factor appears to be a "molecule of choice," this does not exclude its involvement in other important developmental roles (see C.-S. Hong et al. *Developmental Biology* 310, 1–9 [2007]). This versatility is not only underappreciated but begs the question as to whether it also is just accidental. I doubt it.

38. See his paper in *Ibis* 96, 380–383 (1954).

39. Chisholm (1954), p. 380; note 38.

40. Evolution always allows exceptions. In the New Zealand stitchbird (or hihi, *Notiomystis cincta*), copulation is often face-to-face with the female thrown on her back (see S. Anderson *Notornis* 40, 14; 1993), and often in the context of male force majeur (see M. Low *Journal of Avian Biology* 36, 436–448 [2005]). Nor does the avian sexual landscape lack other—to some, depressing—similarities to the human condition. A number of bird species are monogamous, but in many cases the sexes also engage in "extra-pair copulation." Despite the best efforts of the "official" male (in the case of a Neotropical passerine, the yellow-breasted chat), not only do the males show the usual mate guarding, but there is evidence they will herd the errant female back to home territory: see H.L. Mays and G. Ritchison *Naturwissenschaften* 91, 195–198; 2004). Not only may the intruder male move "quietly and stealthily . . . using vegetation as a cover" (see P. Tryjanowski et al. *Behaviour* 144, 23–31 [2007]), but in the case of the great grey shrike (P. Tryjanowski et al. [2007], p. 24) once contact is made, the precopulatory gifts of food are more lavish (see P. Tryjanowski and M. Hromada *Animal Behaviour* 69, 529–533 [2005]), and not only do both sexes then

influencing the behavior of the potential host bird may not be easy to distinguish. There is no doubt the mimicry works because tits, which are not parasitized by cuckoos, respond as if they have seen a sparrow hawk (see N.B. Davies and J.A. Welbergen *Proceedings of the Royal Society of London, B* 275, 1817–1822 [2008]).

138. For the aerodynamics of the related *Argentavis*, the size of the Cessna light aircraft, see S. Chatterjee et al. *Proceedings of the National Academy of Sciences, USA* 104, 12398–12403 (2007).

139. The soaring abilities of not only the vultures but other birds of prey was surely one inspiration for humans learning how to glide. Given the many obvious differences between our reliance on carefully crafted sailplanes as against the millions of years of accumulative biological design when it comes to comparing soaring strategies, then as Zsuzsa Ákos and colleagues (see their paper in *Proceedings of the National Academy of Sciences, USA* 105, 4139–4143 [2008]) ask, "Are the common tricks the same or are there alternative successful tactics?" (p. 4139). In the context of the all-important process of exploiting thermals, they conclude, "All of the parameters we determined were nearly the same for both humans and birds. . . . evolving flight strategies of bird and human calculations lead to virtually the same outcome" (p. 4142).

140. See C. Stock *Rancho La Brea: A record of Pleistocene life in California* (Los Angeles County Museum of Natural History, Science Series, 20; 1956).

141. See Darwin (1890), see pp. 60–61, 115–120; note 2.

142. See, for example, D.A. Roff *Evolutionary Ecology* 8, 639–657 (1994).

143. See D.W. Steadman *Extinction & biogeography of tropical Pacific birds* (University of Chicago Press; 2006).

144. See J.J. Kirchman *Biological Journal of the Linnean Society* 96, 601–616 (2009).

145. Steadman (2006), p. 316; note 143.

146. Gigantism has evolved a number of times in the birds, and perhaps surprisingly even independently in the Cretaceous; see D. Naish et al. *Biology Letters* 8, 97–100 (2012).

147. See pp. 218–220 in my *Life's solution*.

148. See note 82.

149. See J. Harshman et al. *Proceedings of the National Academy of Sciences, USA* 105, 13462–13467 (2008); also papers in *Systematic Biology* by M.J. Phillips et al. (vol. 59, pp. 90–107; 2010), and J.V. Smith et al. (vol. 62, pp. 35–49; 2013).

150. Harshman et al. (2008), p. 13462 [Abstract]; note 149.

CHAPTER 19

1. See his paper in *Symposia of the Society for Experimental Biology* 34, 359–376 (1980).

2. Pain (1980), p. 359; note 1.

3. See T. DeFalco and B. Capel *Annual Review of Cell and Developmental Biology* 25, 457–482 (2009).

4. DeFalco and Capel (2009), p. 475; note 3.

5. DeFalco and Capel (2009), p. 457; note 3.

6. J.A.M. Graves in *Annual Review of Genetics* 42, 565–586 (2008).

7. Graves (2008), p. 566; note 6.

8. See C.L. Organ et al. *Nature* 461, 389–392 (2009).

9. See their paper in *Annual Review of Genomics and Human Genetics* 10, 333–354 (2009).

10 Wilson and Makova (2009), p. 334; note 9.

11. When the platypus genome was sequenced (giving, of course, other insights into convergence such as toxins, p. 61), not only did it reveal a remarkably complex arrangement with ten sex chromosomes, but despite being a paternal XY system it is evidently derived from a birdlike model, further reinforcing the evidence that the mammalian sex chromosomes only arose in the Cretaceous (see F. Veyrunes et al. *Genome Research* 18, R557–R559 [2008]; also L. Potrzebowski et al. *PLoS Biology* 6, e80; 2008). In addition the arrangements in the monotremes and other mammals (therians) arose effectively independently; see D. Cortez et al. *Nature* 508, 488–493 (2014), with commentary on pp. 463–465 by A.G. Clark.

12. As they have in the insects, where the XY system has evolved independently in at least the dipterans and coleopterans; see J.B. Pease and M.W. Hahn *Molecular Biology and Evolution* 29, 1645–1653 (2012).

13. See K. Matsubara et al. *Proceedings of the National Academy of Sciences, USA* 103, 18190–18195 (2006); also commentary on pp. 18031–18032 by E.J. Vallender and B.T. Lahn.

14. More surprising is evidence that the same arrangement in a lizard (Australian dragon lizard) is also convergent (see T. Ezaz et al. *Chromosome Research* 17, 965–973 [2009]), even though as squamates the lizards and snakes are close relatives. Or is it surprising, given that this convergence extends to a flatfish. See S.-L. Chen et al. *Nature Genetics* 46, 253–260 (2014).

15. See their paper in *BioEssays* 26, 159–169 (2004).

16. Vallender and Lahn (2004), p. 160; note 15.

17. See, for example, W. Just et al. *Sexual Development* 1, 211–221 (2007).

18. See Y. Tsuda et al. *Chromosoma* 116, 159–173 (2007).

19. See their paper in *Trends in Ecology & Evolution* 22, 389–391 (2007).

20. Mank and Ellegren (2007), p. 390; note 19.

21. Mank and Ellegren (2007), pp. 390–391; note 19.

22. It will be gratifying to request swabs from our extraterrestrial equivalents to confirm that the sex of their children is determined by either an XX-XY or ZZ-ZW system, with a runt of a Y or W chromosome. No choice, I am afraid. But such stability emerged from a much more labile system, and among the more

119. This has evolved independently in a wading bird, the jacana (see J. Dyck *Auk* 109, 293–301 [1992]).

120. See, for example, R.O. Prum and R.H. Torres *Integrative and Comparative Biology* 43, 591–602 (2003).

121. For an overview of blue coloration in animals, see K.D.L. Umbers *Journal of Zoology* 289, 229–242 (2013); see also S. Vignolini et al. (*Proceedings of the National Academy of Sciences, USA* 109, 15712–15715; 2012) for a convergent example from plants.

122. See R.O. Prum and R. Torres *Journal of Experimental Biology* 206, 2409–2429 (2003). They document how the collagen arrays are typically underlain by a layer rich in melanin granules, but this is not to confer color but serves to absorb any light that might penetrate this far into the skin and so produce incoherent scattering. In addition, in at least some cases a yellow coloration—in birds often due to carotenoid pigments—is again a structural rather than pigmentary color.

123. See R.O. Prum and R.H. Torres *Journal of Experimental Biology* 207, 2157–2172 (2004).

124. Prum and Torres (2004), p. 2167 (my italics); note 123.

125. So, too, Prum and Torres (2004; note 123) suggest that it is unlikely to be coincidental that this skin coloration is associated with groups that have evolved trichromatic vision, itself convergent (pp. 113–115).

126. See his paper in *Comparative Biochemistry and Physiology: A Molecular and Integrative Physiology* 143, 78–84 (2006).

127. Negro (2006), p. 143 [Abstract]; note 126.

128. See, for example, A.J. Bamford et al. *Ethology* 116, 1163–1170 (2010), who also discuss the likely role of this flushing in interactions.

129. See W.C.O. Hill *Zeitschrift für Morphologie und Anthropologie* 61, 18–32 (1969); also his paper in *Säugetierkunde Mitteilungen* 3, 145–151 (1955).

130. See chap. 13 in *The expression of the emotions in man and animals* (Murray; 1872).

131. Darwin (1872), p. 139; note 130.

132. Reviewed by C. Mourer-Chauviré *Bulletin de la Société Géologique de France* 170, 85–90 (1999).

133. A similar story applies to the hoatzin; see G. Mayr and V.L. De Pietri *Naturwissenschaften* 101, 143–148 (2014). The related frogmouths, now restricted to Australia, also had a much wider distribution in the Eocene; see S.J. Nesbitt et al. *PLoS ONE* 6, e26350 (2011).

134. See F. Hertel *Ecology* 75, 1074–1084 (1994), and *Journal of Theoretical Biology* 190, 51–61 (1998).

135. The fossil record can only give snapshots, to be sure sometimes of crystal clarity but as often blurred and in black and white. When, therefore, we draw up the roster of convergences, the fossil record, despite its limitations, can contribute. Consider the tiny condor from the Pleistocene of Brazil (see H.M.F. Alverenga and S.L. Olson *Proceedings of the Biological Society of Washington* 117, 1–9 [2004]). These authors speculate that this diminutive condor may have been convergent on the Old World palm-nut vulture. Looking further afield and deeper in time, the Cretaceous pterosaur *Istiodactylus* is interpreted as a scavenger with various convergences with its avian counterparts; see M.P. Witton *PLoS ONE* 7, e33170 (2012).

136. Good arguments exist to suggest that as obligate scavengers, such soaring fliers will outcompete any rivals (see G.D. Ruxton and D.C. Houston *Journal of Theoretical Biology* 228, 431–436 [2004]) that choose to remain earthbound. As Graeme Ruxton and David Houston note of the latter theoretical possibility, in fact, "Such a beast is unlikely to have evolved" (p. 431 [Abstract]). In other words, vultures will be your answer.

137. Nor do the convergences end here. Even within the Old World accipitrids (see H.R.L. Lerner and D.P. Mindell *Molecular Phylogenetics and Evolution* 37, 327–346 [2005]), the group that includes eagles and kites, the vulture morph has evolved twice. Among the accipitrids the story of convergences continues, such as between the palm-nut vulture (which, as its name suggests, is unusual because it eats nuts, as well as a variety of other foods) and the sea eagle (also documented by Lerner and Mindell [2005]). These authors also draw attention to other convergences within the accipiterids, notably the gymnogene (*Polyboroides*) and crane hawk (*Geranospiza*), as well as among the hawk eagles. This entails a previously unrecognized convergence in the supposed genus *Spizaetus* of Asia and South America that very much echoes the Old World / New World dichotomy of the vultures (see E. Haring et al. *Journal of Zoological Systematics and Evolutionary Research* 45, 353–365 [2007]). Two species of serpent eagle, one from West Africa and the other from Madagascar, also show striking similarities to hawks, possibly to reduce mobbing by other birds (see J.J. Negro *Ibis* 150, 307–314 [2008]). Nor does the roster end there, because since classical times the similarity between accipitrid hawks and cuckoos has excited attention. Aristotle (see Aristotle's *Historia animalium*, trans. D'Arcy Wentworth Thompson [vol. 4 of *The works of Aristotle*: J.A. Smith and W.D. Ross, eds.; Oxford; 1910]) was perhaps a more astute observer than the villager who had to be dissuaded from fetching his gun by Edgar Chance who had been engaged in a meticulous study of the local cuckoos (see his *The truth about the cuckoo* [Country Life; 1940, p. 88]). The yeomen Wilkes fell into the same type of trap as Henry Bates's blasting hawkmoths to bits. But it was an understandable mistake because this mimicry extends not only to the plumage, but size, shape and soaring manner of flight. Mimicry can evolve for several reasons, and in this case competing hypotheses such as the cuckoo resembling a dangerous predator (classic Batesian mimicry; see T.-L. Gluckman and N. I. Mundy *Animal Behaviour* 86, 1165–1181; 2013) or eyes and nocturnal vision drive changes in the skull architecture that serve to isolate the eyes from other movements in the skull (see R.A. Menegaz and E.C. Kirk *Journal of Human Evolution* 57, 672–687 [2009]).

94. See his paper in *Fortschritte der Zoologie* 30, 667–670 (1985); the quotation is the first part of the title.

95. Niemitz (1985), p. 670; note 94.

96. Fain and Houde (2004; note 84) also draw attention to the plumage and foraging behavior in ground litter of the metavian monias and coronavian thrashers, along with the respective dancing displays of the kagu and hammerkop.

97. As the title of M.D. Sorenson et al.'s paper (in *Molecular Biology and Evolution* 20, 1484–1498; 2003) says, "The hoatzin problem is still unresolved."

98. See J.M. Hughes and A.J. Baker *Molecular Biology and Evolution* 16, 1300–1307 (1999).

99. For example, W.P. Pycraft in *Ibis*, 1 (7th ser.), 345–373 (1895), drew attention to the striking similarity in the arrangement of the feathers, while R. Verheyen *Bulletin, Institut royal des Sciences naturelles de Belgique* 32(32), 1–8 (1956), emphasized the osteological similarities, albeit noting the likelihood of convergence as well.

100. See B.T. Thomas in *Handbook of the birds of the world*, vol. 3, *Hoatzin to auk* (J. del Hoyo et al.), pp. 23–32 (Lynx; 1996).

101. See D.A. Turner in *Handbook of the birds of the world*, vol. 4, *Sandgrouse to cuckoos* (J. del Hoyo et al., eds.), pp. 480–506 (Lynx; 1997).

102. The fossil record suggests that this has been for a very long time, and—given the hoatzins appear to have an African origin—suggests their arrival in South America was by rafting rather than flight and that their folivory might, so to speak, have saved their bacon during the westward passage across the Atlantic; see G. Mayr et al. *Naturwissenschaften* 98, 961–966 (2011).

103. See his paper in *Ibis*, 2 (6th ser.), 327–335 (1890).

104. Quelch (1890), p. 329; note 103.

105. See B.C. Stegmann *Publications of the Nuttall Ornithological Club* 17(6), 1–119 (1978). Stegmann's description of the hoatzins and turacos, and his emphasis on their similarities, is very much driven by his assumption of a close relationship rather than convergence. It is also overshadowed by assumptions of supposed primitiveness and a reluctance to believe that particular complex character states can reevolve (see, in particular, his comment on p. 57).

106. The primitive Cretaceous bird *Sinornis* was not only an adept arborealist, but Alan Feduccia has suggested that its skeletal configuration could be consistent with a large crop and so a hoatzinlike folivory (see p. 150 in his *The origin and evolution of birds* [Yale; 1996]). He notes, however, that the available plants would be "far less appealing," and while not dismissing this intriguing suggestion, L.M. Chiappe and C.A. Walker (in *Mesozoic birds: Above the heads of*

the dinosaurs [L.M. Chiappe and L.M. Wittmer, eds.], pp. 240–267 [University of California Press; 2002]) suggest insectivory is at least as likely.

107. See L.P. Korzun et al. *Ostrich* 74, 48–57 (2003).

108. See A. Grajal *Auk* 112, 20–28 (1998).

109. See P.J. Weldon and J.H. Rappole *Journal of Chemical Ecology* 23, 2609–2633 (1997).

110. Quelch (1890); note 103.

111. These appear to include the Neotropical green-rumped parrotlet, which feeds almost exclusively on seeds (see M.A. Pacheco et al. *Condor* 106, 139–143 [2004]), the speckled mousebird from sub-Saharan Africa (see C.T. Downs et al. *Auk* 117, 791–794 [2000]), and possibly the New Zealand kakapo (for this suggestion see p. 142 of Pacheco et al. [2004]). In at least the hoatzin and green-rumped parrotlet the very low oxygen levels in the plant-stuffed crop facilitate the activity of nitrogen-reducing bacteria and thereby, as in a very different way do the leguminous plants (pp. 310–311, note 8), obtain a valuable nutrient (see P. Gueneau et al. *Journal of Ornithology* 147 [Supplement 1], 176 [abstract]). In addition G. Mayr (*Ibis* 155, 384–396 [2013]) draws attention to the convergence between an Oligocene mousebird and parrots in terms of the skull.

112. Note that the portion of the forestomach in ruminants that is referred to as the reticulum not only has a honeycomblike mucosal lining, which is evidently important for sorting particles, but the type of diet also imposes "a strong convergent evolution of the internal muscosa [*sic*] of the reticulum"; see M. Clauss et al. *Journal of Zoology, London* 281, 26–38 (2010), p. 31.

113. See, respectively, W. Messier and C.-B. Stewart *Nature* 385, 151–154 (1997), and M.A. Pacheco et al. *Comparative Biochemistry and Physiology, A* 147, 808–819 (2007).

114. See K.E. Sullam et al. *Molecular Ecology* 21, 3363–3378 (2012).

115. See, for example, F. Godoy-Vitorino et al. *Applied and Environmental Microbiology* 74, 5905–5912 (2008), and M.G. Dominguez-Bello et al. *Physiological Zoology* 66, 374–383 (1993).

116. See A.-D.G. Wright et al. *ISME Journal* 3, 1120–1126 (2009). Employment of micro-organisms to assist in gut fermentation almost certainly evolved in the Paleozoic and in association with the multiple evolution of herbivory; see H.-D. Sues and R.R. Reisz *Trends in Ecology & Evolution* 13, 141–145 (1998).

117. In the case of fermentative digestion by the sloths, M.A. Pacheco et al. (2007; note 113) have suggested that the primary function of the lysozyme is not in a digestive capacity but to provide protection against pathogenic bacteria that would otherwise compete with the "friendly" microbial symbionts upon which this lethargic animal relies.

118. This topic is discussed in my *Life's solution* (2003), pp. 298 and 440–441 (notes 84–91).

[1984]) and some at least evidently function as scent bearers, or osmetrichia (p. 417, note 212).

67. See K.C. Welch et al. *Journal of Experimental Biology* 211, 310–316 (2008).

68. See papers by R.K. Suarez et al. in *Comparative Biochemistry and Physiology, A* 153, 136–140 (2009), and *Journal of Experimental Biology* 214, 172–178 (2011).

69. Such identifications presuppose a stable phylogeny for the birds, which unfortunately has remained somewhat elusive. Nevertheless, despite competing proposals (and genuine advances; see, for example, S.J. Hackett et al. *Science* 320, 1763–1768; 2008), the convergences here remain robust.

70. See J.W. Chapman et al. *Science* 327, 682–685 (2010). In addition to the famous long-distance migrations by some moths and butterflies, locusts can also engage in trans-Atlantic crossings; see M.W. Lorenz *Quaternary International* 196, 4–12 (2009).

71. Chapman et al. (2010), p. 682 [Abstract]; note 70.

72. Chapman et al. (2010), p. 684; note 70.

73. See also T. Alerstam et al. *Proceedings of the Royal Society of London, B* 278, 3074–3080 (2011).

74. See his paper in *Journal of Ornithology* 147, 212–220 (2006); also B.J. Frost and H. Mouritsen *Current Opinion in Neurobiology* 16, 481–488 (2006).

75. See Dingle (2006), part of title; note 74.

76. See also C.G. Guglielmo (*Integrative and Comparative Biology* 50, 336–345; 2010), who, in drawing attention to the specific metabolic and physiological demands of burning fat during the migration of birds, reminds us of the likelihood of convergences with migratory bats. Much remains to be learned, but in addition to convergences there are also differences, such as the role of torpor in the bats.

77. See the papers by R.E. Gill et al. *Condor* 107, 1–20 (2005), and *Proceedings of the Royal Society of London, B* 276, 447–457 (2009).

78. See R.K. Suarez and C.L. Gass *Comparative Biochemistry and Physiology, A* 133, 335–343 (2002).

79. See S.L. Olson and Y. Hasegawa *Science* 206, 688–689 (1979), and N.D. Smith *PLoS ONE* 5, e13354 (2010). G.J. Dyke et al. (*PLoS ONE* 6, e25672; 2011) prefer to draw the convergent comparison to the auks, while S. Kawabe et al. (*Zoological Journal of the Linnean Society* 170, 467–493; 2013) argue against convergence on the basis of brain structure.

80. See M. van Tuinen et al. *Proceedings of the Royal Society of London, B* 268, 1345–1350 (2001); also G. Mayr *Zoological Journal of the Linnean Society* 140, 157–169 (2004).

81. For the spoonbills, see E. Matheu and J. del Hoyo in *Handbook of the birds of the world*, vol. 1, *Ostrich to ducks* (J. del Hoyo, ed.), pp. 472–506 (Lynx; 1992), and for flamingos J. del Hoyo in the same volume (pp. 508–526). Although they both behave in much the same way in feeding—and catch much the same food—the flamingo not only differs in the method of feeding but in doing so shows a remarkable convergence with the baleen whales (p. 51).

82. By way of background, all living birds (the Neornithes) are divided into the palaeognathans (these include the large flightless birds) and the neognathans. In turn the latter comprise a basal group (including, for example, the ducks and pheasants) and a larger assemblage of metavians and coronavians that together form the Neoaves.

83. See S.J. Hackett et al. *Science* 320, 1763–1768 (2008); also M. Morgan-Richards et al. *BMC Evolutionary Biology* 8, e20 (2008); G. Mayr *Journal of Zoological Systematics and Evolutionary Research* 49, 58–76 (2011); and T. Pásko et al. *Molecular Phylogenetics and Evolution* 61, 760–771 (2011).

84. See their paper in *Evolution* 58, 2558–2573 (2004); so swifts and grebes are metavians, and their respective counterparts, the swallows and loons, are coronavians. See also P.G.P. Ericson (*Journal of Biogeography* 39, 813–824; 2012) for a somewhat similar analysis, but this time revolving around "biogeography and parallel radiations" [part of title].

85. On the sunbittern, so called because of its spectacular display, see B.T. Thomas in *Handbook of the birds of the world*, vol. 3. *Hoatzin to auk* (J. del Hoyo et al., eds.), pp. 226–233 (Lynx; 1996). Not only is its mode of stalking remarkably similar to the bittern, but it too can be heard across wide distances.

86. Despite the many evolutionary paths to diving in birds and mammals (see, for example, I.L. Boyd *Trends in Ecology & Evolution* 12, 213–217 [1997]), striking generalities emerge in terms of their physiology and behavior (see L.G. Halsey et al. *Functional Ecology* 20, 889–899 [2006]).

87. For the boobies, see C. Carboneras (pp. 312–325) and for the tropicbirds J. Orta (pp. 280–289) in *Handbook of the birds of the world*, vol. 1, *Ostrich to ducks* (J. del Hoyo et al., eds.; Lynx; 1992).

88. See D.T. Holyoak in *Handbook of the birds of the world*, vol. 5, *Barn-owls to hummingbirds* (J. del Hoyo et al., eds.), pp. 266–287 (Lynx; 1999).

89. See A.N. Iwaniuk et al. *Brain, Behavior and Evolution* 67, 53–68 (2006).

90. Another convergence is the wide gape that finds parallels with swift and swallows, and even more remarkably the bat hawk that, as its name suggests, pursues bats and swiftlets; see L.R. Jones et al. *Biotropica* 44, 386–393 (2012).

91. This crepuscular life is possibly linked to control of melatonin activity, and in both the caprimulgiforms and owls, the same genetic mechanism may be employed (see A.E. Fidler et al. *Molecular Phylogenetics and Evolution* 33, 908–921 [2004]).

92. See J.J. Negro et al. *Journal of Raptor Research* 40, 222–225 (2006).

93. The convergences extend further because of parallel developments in the owls and haplorhine primates (tarsiers and anthropoids), where large

flower and rob the plant of its nectar. In doing so, of course, they frustrate the plant's strategy of the visiting bird being covered in pollen that it is then meant to transport to an adjacent flower awaiting fertilization. An interesting convergence arises, however, because as with the hawkmoths the flowerpiercers have arrived at closely similar digestive strategies to the hummingbirds in order to deal with the high carbohydrate diet (see J.E. Schondube and C. Martinez del Rio *Journal of Comparative Physiology, B* 174, 263–273 [2004]).

32. See B. Anderson et al. *Nature* 435, 41–42 (2007).

33. See S.D. Johnson and S.W. Nicolson *Biology Letters* 4, 49–52 (2008); also M. Brown et al. *Journal of Avian Biology* 39, 479–483 (2008).

34. Johnson and Nicolson (2008), p. 50; note 33. The main exception to this convergence is the sucrose activity in hummingbirds, which unsurprisingly is much higher. The similarities in digestive strategy, e.g., dealing with low nitrogen levels, and high water content (hence dilute) are also emphasized by Nicolson and Flemming (2003; note 23), although they stress the contrast between the sucrose-rich nectars obtained by hummingbirds and hexose-rich nectars for sunbirds.

35. For one such exception, see A. Ortega-Olivencia et al. *Oikos* 110, 578–590 (2005).

36. As he points out in *Oikos* 44, 127–131 (1985), this hardly explains why nectarivorous birds and bees coexist in the Americas, and an alternative explanation is climatic, notably glaciations.

37. Interestingly (along with some honey wasps) the bees also show a striking convergence for specialized pollen harvesting (see A. Müller *Biological Journal of the Linnean Society* 57, 235–252 [1996]).

38. See T.H. Fleming and N. Muchhala *Journal of Biogeography* 35, 764–780 (2008).

39. Fleming and Muchhala (2008), pp. 772–773; note 38.

40. Fleming and Muchhala (2008), p. 776; note 38.

41. See their paper in *Journal of Biogeography* 30, 1357–1380 (2003). See also R.T. Corlett and R.B. Primack *Trends in Ecology & Evolution* 21, 104–110 (2006), who reach similar conclusions.

42. See, for example, B.J. MacFadden, *Trends in Ecology & Evolution* 12, 182–187 (1997).

43. See M. de Vivo and A.P. Carmignotto *Journal of Biogeography* 31, 943–957 (2004).

44. See Nicolson and Fleming (2003), note 23. A similar point was made by G.H. Pyke *Australian Journal of Ecology* 5, 343–369 (1980), who, while agreeing that hummingbirds and honeyeaters have similarities, they are "quite dissimilar" with the latter—for example, eating fruit and showing flocking behavior.

45. See F. Wiens et al. *Proceedings of the National Academy of Sciences, USA* 105, 10426–10431 (2008).

46. See H.M. Vose *Journal of Mammalogy* 54, 245–247 (1973).

47. See H.I. Rosenberg and K.C. Richardson *Journal of Morphology* 223, 303–323 (1995).

48. See K.C. Richardson et al. *Journal of Zoology, London* 208, 285–297 (1986); also note 65.

49. See J.A.W. Kirsch and F.-J. Lapointe in *Molecular evolution and adaptive radiation* (T.J. Givnish and K.J. Systma, eds.), pp. 313–330 (Cambridge; 1997).

50. See F.T. Muijres et al. *Science* 319, 1250–1253 (2008).

51. See R.J. Bomphrey et al. *Journal of Experimental Biology* 208, 1079–1094 (2005).

52. See his article in *Current Biology* 18, R468–R470 (2008).

53. Dickinson (2008), p. R470; note 52.

54. See D.R. Warrick et al. *Nature* 435, 1094–1097 (2005).

55. Finding these similarities among hovering animals is not so surprising, but the same employment of a leading-edge vortex in the descending maple seed (see D. Lentink et al. *Science* 324, 1438–1440 [2009]) indicates "a convergent aerodynamic solution in the evolution of high-performance flight of both animal wings and plant seeds" (p. 1440). A universal principle of physics explains how wing seeds move from plummet mode to extraordinarily effective dispersal.

56. Dickinson (2008), p. R470; note 52.

57. See C.C. Voigt and Y. Winter *Journal of Comparative Physiology, B* 169, 38–48 (1999).

58. See, for example, J.E. Hill and J.P. Smith *Bats: A natural history* (British Museum [Natural History]; 1984), pp. 66–69.

59. See their paper in *Annals of Botany* 104, 1017–1043 (2009).

60. Fleming et al. (2009), p. 1018; note 59.

61. See Y.J. Alvarez et al. *Molecular Biology and Evolution* 16, 1061–1067 (1999); also L.J. Hollar and M.S. Springer *Proceedings of the National Academy of Sciences, USA* 94, 5716–5721 (1997).

62. See her chapter (pp. 398–429) in *Bat ecology* (T.H. Kunz and M.B. Fenton, eds.; University of Chicago Press; 2003).

63. Dumont (2003), p. 399; note 62.

64. See P.W. Freeman *Biological Journal of the Linnean Society* 56, 439–463 (1995).

65. The papillose surfaces of the tongue tips in a glossophagine bat also serve as a nectar mop and have important parallels not only to the hummingbirds and honey possum but also nectar-feeding bees (see C.J. Harper et al. *Proceedings of the National Academy of Sciences, USA* 110, 8852–8857 [2013]). Patricia Freeman (1995; note 64) also notes, "There are many convergent features between nectarivorous bats and several mammalian myrmecophages" (cf. anteaters, pp. 51–52) (p. 458).

66. At least among the bats they obtain vital proteins from licking their fur of the pollen grains. It has been suggested (see D.J. Howell and N. Hodgkin *Journal of Morphology* 148, 329–336 [1976]) that the nectarivorous bats have hair specialized to trap this pollen, but this idea does not stand up to scrutiny (see D.W. Thomas et al. *Journal of Mammalogy* 65, 481–484

10. Bartholomew (1981), p. 47; note 9.

11. Bartholomew (1981), p. 56; note 9.

12. See N.S. Church *Journal of Experimental Biology* 37, 186–212 (1960); also G.A. Bartholomew and R.J. Epting *Journal of Experimental Biology* 63, 603–613 (1975).

13. Church (1960), p. 203; note 12.

14. Church (1960), p. 204; note 12.

15. Welch et al. (2006); note 7.

16. Also addressed in *Life's solution*, note 171 (p. 368), drawing on the work of G.A. Bartholomew.

17. See, for example, N. Sapir and R. Dudley *Journal of Experimental Biology* 215, 3603–3611 (2012).

18. See A.J. Ross and E.A. Jarzembowski in *The fossil record 2* (M.J. Benton, ed.), pp. 363–426 (Chapman & Hall; 1993), see p. 414, fig. 21.19. See also J.C. Regier et al. *Molecular Phylogenetics and Evolution* 20, 311–316 (2001), who discuss their molecular phylogeny.

19. See G. Mayr *Science* 304, 861–864 (2004), and *Journal of Ornithology* 148, 105–111 (2006). Bird nectarivory, however, had appeared by the Eocene; see G. Mayr and V. Wilde *Biology Letters* 10, 20140223 (2014).

20. See A. Louchart et al. *Naturwissenschaften* 95, 171–175 (2008). In the Americas hummingbirds are familiar visitors to flowers, seeking the nectar, of course, but unwittingly serving to ferry the pollen from gaping stamen to expectant stigma. And as ever there is convergence. In the case of those hummingbirds active in temperate regions, the flower type in six separate families has repeatedly converged toward a tubular shape (typically arranged to allow only the bill and tongue actual access), bright red colors, and nectar rewards (see J.K. Brown and A. Kodrick-Brown *Ecology* 60, 1022–1035 [1979]). Some plants can also get on the act because they mimic the hummingbird-frequented flowers, enjoying the benefits of pollination but craftily providing no nectar. In some groups, such as the Neotropical *Costus*, a shift to pollination by hummingbirds (from the bees) has occurred at least seven times (see K.M. Kay et al. *American Journal of Botany* 92, 1899–1910 [2005]; for similar examples in legumes see A. Bruneau in vol. 84, pp. 54–71 [1997]), while in the penstemons it might be as many as twenty-one times (see P. Wilson et al. *New Phytologist* 176, 883–890 [2007]). In the latter case Paul Wilson and colleagues intriguingly employ the term "attractor," and they also remark, "the number of origins of ornithophily [by hummingbirds] is astonishing, and indicates to us that the hummingbird pollination niche is *just waiting to be claimed* by plants with flowers like penstemons in a way that other pollination niches are not" (p. 889; emphasis added). An alternative metaphor to the term "attractor" is that of the vortex, where the organism is almost helplessly swept into configuration from which escape is very difficult. This is used to good effect by James Thomson and Paul Wilson in their overview of hummingbird pollination (see their paper in *International Journal of Plant Sciences*, 169, 23–38 [2008]). Here they nicely capture the sense of a

sort of invisible morphogenetic field when they write of how the "phenotypes tend to be drawn toward conformity with particular pollination syndromes by a coherent set of changes in a predictable set of characters" (p. 26). The associations between not only hummingbirds but bats and hawkmoths have evolved multiple times, but in the case of the latter two groups they seem to navigate repeatedly to dead ends. P. Duchen and S.S. Renner (*American Journal of Botany* 97, 1129–1141; 2010) argue that as far as the cucurbitacean *Cayaponia* is concerned there is good evidence of reversals from bat pollination back to bee pollination. In the case of the hummingbirds it remains an open-ended association so that bees and other insects may take up the baton of probing for the nectar (see E.A. Tripp and P.S. Manos *Evolution* 62, 1712–1737]2008]).

21. Mayr (2004); note 19.

22. This point draws on C. Westerkamp *Botanica Acta* 103, 366–371 (1990).

23. Many other birds are opportunistic nectar feeders, including the lorikeets and orioles (see S.W. Nicolson and P.A. Fleming *Plant Systematics & Evolution* 238, 139–153 [2003]).

24. See K.J. Burns et al. *Journal of Avian Biology* 34, 360–370 (2003); also commentary on pp. 321–323 by J.V. Remsen.

25. Among the spectacular radiation of the indigenous honeycreepers are two examples of a "nuthatch" type, the Hawaii and Kauai creepers. They are very similar, in terms of a very distinctive type of tongue (which, in contrast to the tubular tongue of the typical nectar feeders, is employed to catch insects; see H.D. Pratt *Condor* 94, 836–846 [1992]), as well as plumage and behavior. Unsurprisingly they were placed in the same genus (*Oreomystis*), but it now transpires this is a remarkable example of convergence (see D.M. Reding et al. *Biology Letters* 5, 221–224 [2009]). Dawn Reding and colleagues call it "extreme" (p. 221), a reasonable conclusion because even the "juvenile begging calls are nearly identical" (p. 221), and they conclude by noting that their findings "call into question the view that evolution is inherently contingent and unpredictable" (p. 223).

26. See R.C. Fleischer et al. *Current Biology* 18, 1927–1931 (2008); also commentary on pp. R1132–R1134 by I.J. Lovette.

27. Fleischer et al. (2008), p. 1928; note 26.

28. Fleischer et al. (2008), p. 1927; note 26.

29. Fleischer et al. (2008), p. 1929; note 26.

30. Westerkamp (1990; note 22) emphasizes that this division, although "dogma," is less rigid than sometimes supposed, with hummingbirds occasionally perching and other birds seen to hover. The hummingbirds, however, are masters at hovering, and on account of limb proportions and wing shape can maintain a constant position.

31. While various birds either hover or perch as they seek the sugar-rich nectars, the flowerpiercers take the short cut and, as their name suggests, slice open the

(2012); also J.A. Gauthier et al. *Bulletin of the Peabody Museum of Natural History* 52, 107–126 (2011).

187. See S. Chatterjee *National Geographic Research & Exploration* 9(3), 274–285 (1993). In this paper *Shuvosaurus* is interpreted as a theropod rather than a rauisuchid; see Nesbitt (2007), note 180.

188. See S. Chatterjee and P.K. Majumdar *Journal of Paleontology* 61, 787–793 (1987).

189. So too the tyrannosaur-like appearance of *Postosuchus* (see S. Chatterjee *Philosophical Transactions of the Royal Society of London, B* 309, 395–460 [1985]) is again a convergence (see, for example, S.L. Brusatte et al. *Acta Palaeontologica Polonica* 54, 221–230 [2009]).

190. Chatterjee (1993), p. 285; note 187.

191. See W.G. Parker et al. *Proceedings of the Royal Society of London, B* 272, 963–969 (2005); also A.P. Hunt et al. *New Mexico Museum of Natural History and Science, Bulletin* 29, 67–76 (2005).

192. Nesbitt (2011), p. 24; note 184.

193. The similarities are not restricted to the dinosaurs, because a marine archosaurian from the Triassic of China (*Qianosuchusmixtus*) is convergent on some of the Jurassic crocodyliforms. As Chun Li and colleagues (see their paper in *Naturwissenschaften* 93, 200–206 [2006]) note, this animal "explored and adapted to a marine environment in the Triassic, which was independently exploited by Jurassic marine crocodyliforms more than 40 million years later and the saltwater species of extant *Crocodylus*" (p. 205).

194. See S. Lautenschlager and J.B. Desojo *Paläontologisches Zeitschrift* 85, 357–381 (2011); also R. Redelstorff et al. (*Acta Palaeontologica Polonica* 59, 607–615; 2014) for a possible convergence in terms of gigantism.

. **195** See his article (a review of my *Life's solution*) in *Journal of Vertebrate Paleontology* 28, 586–587 (2008).

196. Molnar (2008), p. 586; note 195. Even amongst the silesaurids, the sister group of the dinosaurs, *Silesaurus* converges with the ornithischians (with respect to teeth) and theropods (in terms of ankle structure); see S.J. Nesbitt *Nature* 464, 95–98 (2010).

197. See papers by S.L. Brusatte et al. in *Science* 321, 1485–1488 (2008a), and *Biology Letters* 4, 733–736 (2008b).

198. Brusatte et al. (2008a), p. 1486; note 197.

199. See F. Knoll and D. Schwarz-Wings *Annales de Paleontologie* 95, 165–175 (2009).

200. See H. Osmolska *Acta Palaeontologica Polonica* 49, 321–324 (2004).

201. See, for example, H.-D. Sues *Zoological Journal of the Linnean Society* 62, 381–400 (1978).

202. See H.C.E. Larsson et al. *Journal of Vertebrate Paleontology* 20, 615–618 (2000).

203. See M. Kundrát *Naturwissenschaften* 94, 499–504 (2007); also A.M. Balanoff et al. (*Nature* 50, 93–96; 2013), who in their overview of preavian encephalization point out that the nonavian theropods are effectively "flight-ready."

204. See M.A. Norell *American Museum Novitates* 3654, 1–63 (2009).

205. See P.J. Currie and X.-J. Zhao *Canadian Journal of Earth Sciences* 30, 2231–2247 (1993); also D.A. Russell's paper in vol. 6, pp. 595–612; 1969.

206. See their paper in *Syllogeus* 37, 1–43 (1982).

207. See K.A. Stevens *Journal of Vertebrate Paleontology* 26, 321–330 (2006).

208. Russell and Séguin (1982), p. 35; note 206.

209. Russell and Séguin (1982), p. 35; note 206.

210. See D. Dhouailly et al. *Journal of Anatomy* 214, 587–606 (2009).

211. Dhouailly et al. (2009), p. 602; note 210.

212. Among the most interesting of convergences involving mammalian hair are the so-called osmetrichia that serve to store or disperse scents of varying degree of pungency. Identified in various deer (see D. Müller-Schwarze et al. *Journal of Ultrastructural Research* 59, 223–230 [1977], and M.T. Ajmat et al. *Biocell* 23, 171–176; 1999), appropriately musk shrews (see M. Balakrishnan *Journal of Zoology, London* 213, 213–220; 1987), the crested rat (see D.M. Stoddart *Journal of Zoology, London* 189, 551–553 [1979]), and bats (see M.B.C. Hickey and M.B. Fenton *Journal of Mammalogy* 68, 381–384 [1987]), while an example in the marsupials is the Brown antechinus (C.L. Toftegaard and A.J. Bradley *Journal of Zoology, London* 248, 27–30 [1999]). While a recurrent motif is scales or similar extensions, convergence is self-evident because of specific adaptations to store the scents that include medullary cells in the musk shrew, and perhaps the most strikingly in the crested rat, where the osmetrichia are highly vacuolated, and as Michael Stoddard (1979) remarks, "[resemble] a sponge" (p. 553).

CHAPTER 18

1. See his *The naturalist on the River Amazons* (Murray; 1892).

2. See his *Journal of researches into the natural history and geology . . . H.M.S. Beagle, etc.* (Nelson; 1890).

3. Darwin (1890), p. 49; note 2.

4. Bates (1892), p. 94; note 1.

5. See E. Warrant et al. *Journal of Experimental Biology* 202, 497–511 (1999).

6. There is also an interesting set of connections between optimal sugar content of a nectar versus its viscosity according to whether the fluid is obtained by capillary suction (e.g., hummingbirds), active suction (e.g., moths) or viscous dipping (e.g., bats); see W.-J. Kim et al. *Proceedings of the National Academy of Sciences, USA* 108, 16618–16621 (2011); also W.-J. Kim et al. *Proceedings of the Royal Society of London, B* 279, 4990–4996 (2012).

7. See K.C. Welch et al. *Physiological and Biochemical Zoology* 79, 1082–1087 (2006).

8. See, for example, J.E.P.W. Bicudo et al. *Journal of Experimental Biology* 205, 2267–2273 (2002).

9. See his article in *Insect thermoregulation* (B. Heinrich, ed.), pp. 46–78 (Wiley; 1981).

148. See P.M. O'Connor and L.P.A.M. Claessens *Nature* 436, 253–256 (2005).

149. See J.M. Clark et al. *American Museum Novitates* 3265, 1–36 (1999).

150. Clark et al. (1999), p. 16; note 149.

151. See, for example, A.L. Vergne et al. *Naturwissenschaften* 94, 49–54 (2007).

152. See Q.-J. Meng et al. *Nature* 431, 145–146 (2004).

153. Meng et al. (2004), p. 145; note 152.

154. See J. Botha-Brink and S.P. Modesto *Proceedings of the Royal Society of London, B* 274, 2829–2834 (2007).

155. Botha-Brink and Modesto (2007), p. 2833 (both quotations); note 154.

156. In this context at least one troodontid (*Mei*) adopted a sleeping posture like that of a bird (see S. Xu and M.A. Norell *Nature* 431, 838–841 [2004]), while her eggs must have been brooded (see D.J. Varriccho et al. *Journal of Vertebrate Paleontology* 22, 564–576 [2002]; also D.K. Zelenitsky and F. Therrien for a possible dromaeosaurid example [*Palaeontology* 51, 1253–1259; 2008]). The eggshells are well enough preserved to reveal their microstructure, but whether their avianlike structure derived from a common ancestor or is convergent remains uncertain. What is striking, however, are tiny theropod eggs from the Upper Cretaceous of Thailand, much the same size as those of a goldfinch, that have a structure that reveals once again an evolutionary mosaicism by combining a shell reminiscent of both birds and saurischian dinosaurs (see E. Buffetaut et al. *Naturwissenschaften* 92, 477–482 [2005]).

157. This is particularly striking in *Albinykus*, which not only is remarkably small, but shows three other trends coincident with the avialians including bone vascularization; see S.J. Nesbitt et al. *Journal of Vertebrate Paleontology* 31, 144–153 (2011). Among the interesting points they make is why this set of convergences should be found here, given the alvarezsaurids in question were ground-based insectivores.

158. See P.M. Barrett *Palaeontology* 52, 681–688 (2009).

159. See P. Senter *Paleobiology* 31, 373–381 (2005).

160. Ripping open wood in pursuit of termites. See N.R. Longrich and P.J. Currie *Cretaceous Research* 30, 239–252 (2009); nor were digging activities restricted to the theropods, because a hypsilodont-like ornithiscian (*Oryctodromeus*) excavated burrows, evidently for denning; see D.J. Varricchio et al. *Proceedings of the Royal Society of London, B* 274, 1361–1368 (2007), and commentary on pp. 1359–1360 by K. Padian; also A.J. Martin *Cretaceous Research* 30, 1223–1237 (2009).

161. Reviewed in N.R. Longrich et al. *Palaeontology* 53, 945–960 (2010). See also X. Xu et al. *Proceedings of the National Academy of Sciences, USA* 108, 2338–2342 (2011), for a striking case of forelimb reduction in the alvaresaurids, again most likely connected to digging.

162. Until recently what had been interpreted as trackways of theropods from the late Triassic or early Jurassic of Argentina (see R.N. Melchor et al. *Nature* 417, 936–938 [2002]; also S. De Valais and R.N. Melchor *Journal of Vertebrate Paleontology* 28, 145–159; 2008) were an intriguing hint that this group had gone aloft even earlier, but it now transpires that the date was erroneous and the true age is late Eocene (see R.N. Melchor et al. *Nature* 495, E1–E2 [2013]).

163. See R.B.J. Benson et al. *Naturwissenschaften* 97, 71–78 (2010).

164. Benson et al. (2010), p. 71 [Abstract]; note 163.

165. Benson et al. (2010), p. 76; note 163.

166. J.J. Flynn et al. *Palaeontology* 53, 669–688 (2010), remind us, in their description of a basal archosauromorph (*Azendohsaurus*), that in this assemblage of archosaurians, herbivory may have evolved at least six times with concomitant convergences in the dentition.

167. See P.C. Sereno et al. *Proceedings of the Royal Society of London, B* 277, 199–209 (2010).

168. See O. Mateus et al. *Proceedings of the Royal Society of London, B* 276, 1815–1821 (2009).

169. Mateus et al. (2009), p. 1815; note 168.

170. See his paper in *Paleobiology* 26, 489–512 (2000).

171. Carrano (2000), p. 506; note 170.

172. Carrano (2000), p. 489 [Abstract]; note 170.

173. See S.J. Nesbitt et al. *Journal of Systematic Palaeontology* 5, 209–243 (2007). Yes, there are theropods (see S.J. Nesbitt et al. *Science* 326, 1530–1533 [2009]), but no convincing remains of either sauropods or ornithischians are known (see also R.J. Butler et al. *Neues Jahrbuch für Geologie und Paläontologie, Abhandlungen* 249, 143–156 [2008]).

174. As well as related forms such as dinosauromorphs (see, for example, R.B. Irmis et al. *Science* 317, 358–361 [2007]) and dinosauriforms (see S.J. Nesbitt and S. Chatterjee *Neues Jahrbuch für Geologie und Paläontologie, Abhandlungen* 249, 143–156 [2008]).

175. For a brief overview, see, for example, S.J. Nesbitt et al. *Journal of Vertebrate Paleontology* 29, 498–516 (2009).

176. See D.M. Henderson and D.B. Weishampel *Senckenbergiana lethaea* 82, 77–91 (2002).

177. Henderson and Weishampel (2002), p. 89; note 176.

178. See their paper in *Nature* 459, 940–944 (2009).

179. Xu et al. (2009), p. 941; note 178.

180. See his paper in *Bulletin of the American Museum of Natural History* 302, 1–84 (2007).

181. Nesbitt (2007), p. 71; note 180.

182. See S.J. Nesbitt and M.A. Norell *Proceedings of the Royal Society of London, B* 273, 1045–1048 (2006).

183. Nesbitt and Norell (2006), part of title; note 182, and Nesbitt (2007), p. 1 [Abstract]; note 180.

184. See his paper in *Bulletin of the American Museum of Natural History* 352, 1–292 (2011).

185. Nesbitt (2011), p. 31; note 184.

186. See E.R. Schacher et al. *Journal of Morphology* 272, 1464–1491 (2011), and K.T. Bates and E.R. Schuchner *Journal of the Royal Society Interface* 9, 1334–1353

NOTES 415

137. Jocular professors (and yes, they do exist) like to draw the attention of students to a bird fooling around outside the office and remind them that dinosaurs never died out, they just turned into birds. From this perspective it is difficult to think of the evolution of theropods other than how supposedly scales (p. 199) turned into feathers and teethed jaws into horny beaks. The horny beaks have the technical name of rhamphothecae. T.L. Hieronymus and L.M. Witmer (*Auk* 127, 590–604; 2010) point out that among the coelurosaurian dinosaurs, "The evolution of rhamphothecae in ornithurine birds represent only one of at least seven independent occurrences" (p. 602). This integumentary story, however, is not unique to this group of dinosaurs. This is because representatives of the other major group of dinosaurs, the ornithischians, have representatives in the form of a heterodontosaurid (*Tianyulong*) (see X.-T. Zheng et al. *Nature* 458, 333–336 [2009]; also commentary on pp. 293–295 by L.M. Witmer), a horned psittacosaurid (see G. Mayr et al. *Naturwissenschaften* 89, 361–365 [2002]); also details of the skin reported by T. Lingham-Soliar (*Proceedings of the Royal Society of London, B* 275, 775–780 [2008]) and a megalosauroid (O.W.M. Rauhut et al. *Proceedings of the National Academy of Sciences, USA* 109, 11746–11751 [2012]). Overall, this suggests that some sort of integumentary structure is ancestral to both ornthischians and saurischians (including the theropods), but it is too early to discount independent acquisition that display filamentous arrays. Still more tantalizing, but correspondingly more controversial, is the strange reptile from the Triassic known as *Longisquama*. It is a striking beast, with a series of elongate integumentary plates. An earlier proposal that these were a type of feather (see T.A. Jones et al. *Science* 288, 2202–2205 [2000]) received a dusty reception (see R.O. Prum et al. *Science* 291, 1899–1902 [2001]; also R.R. Reisz and H.-D. Sues *Nature* 408, 428; 2000), but evidence that these extensions of the skin have developmental similarities to avian feathers combined with the phylogenetic distance between *Longisquama* and any dinosaur makes this a strong candidate for convergence (see S. Voigt et al. *Naturwissenschaften* 96, 81–86 [2009]; also M. Buchwitz and S. Voigt *Paläontologische Zeitschrift* 86, 313–331; 2012).

138. See, for example, P.J. Currie *Journal of Vertebrate Paleontology* 7, 72–81 (1987).

139. See, for example, T. Maryańska et al. *Acta Palaeontologica Polonica* 47, 97–116 (2002).

140. See A.M. Balanoff and M.A. Norell *Bulletin of the American Museum of Natural History* 372, 1–77 (2012). Convergences include loss of teeth, pneumatization of the bones, and brooding of eggs, but perhaps the most striking example is the evolution of a series of fused vertebra at the end of the tail, the pygostyle (see papers by R. Barsbold et al. in *Nature* [vol. 403, pp. 155–156 (2000a), and *Acta Palaeontologica Polonica* (vol. 45, pp. 97–106; 2000b)]); also W.S. Persons et

al. *Acta Palaeontological Polonica* 59, 553–567 (2014). A similar arrangement has evolved yet again in the more primitive therizinosaurian *Beipiaosaurus* (see X. Xu et al. *Acta Geologica Sinica* [English version] 77, 294–298 [2003]).

141. See X. Xu et al. *Nature* 419, 291–293 (2002); also A.M. Balanoff et al. *American Museum Novitates* 3651, 1–36 (2009).

142. As its name suggests, it has distinctive teeth, making this not only a remarkable case of dinosaurian heterodonty (and possibly convergent with such therizinosaurians as *Falcarius* [see L.F. Zanno *Zoological Journal of the Linnean Society* 158, 196–230; 2010]), but as Xing Xu and colleagues (2002; note 141) note, the "paired first premaxillary teeth are very similar to the incisors found in a few specialized mammalian lineages" (p. 293), most obviously the rodents and so suggesting a herbivorous diet. Other analyses (see L.E. Zanno and P.J. Makovicky *Proceedings of the National Academy of Sciences, USA* 108, 232–237 [2011]) indicate that herbivory also evolved independently in five other theropod lineages, including the ornithomimosaurs, therizinosaurs, and even a troodontid (*Jinfengopteryx*). Their analysis reveals both convergences and specific differences, and they make the important point that the endpoint of an edentulous beak may well have allowed the occupation of new niches and related shifts in diets.

143. See papers in *Vertebrata PalAsiatica* by Z.-H. Zhou and X.-L. Wang (vol. 38, pp. 111–127; 2000) and Z.-H. Zhou et al. (vol. 38, pp. 241–254; 2000).

144. See, for example, the striking feathers in *Similicaudipteryx*, although there is no indication these were employed for flight; see X. Xu et al. *Nature* 464, 1338–1341 (2010). See also subsequent critique by R.O. Prum in vol. 468, p. E1 (2010) and reply by X. Xu et al. (p. E2).

145. See X. Xu et al. *Nature* 447, 844–847 (2007).

146. J.-C. Lü et al. (*Naturwissenschaften* 100, 165–175; 2013) document juveniles of *Yulong* that were more or less chicken-sized. In addition, many of the more advanced theropods were primitively small (a necessary prerequisite for any aerial adventures), and increases in body size occurred not only in the oviraptorids (and also the more primitive ornithomimosaurs; see P.J. Makovicky et al. *Proceedings of the Royal Society of London, B* 277, 191–198; 2010; evidence also exists that ornithomimosaurs had feathers [see D.K. Zelenitsky et al. *Science* 338, 510–514; 2012]) but independently three times in the dromaeosaurids and at least once in the troodontids (see A.H. Turner et al. *Science* 317, 1378–1381 [2007]; also X. Xu et al. *Nature* 415, 780–784; 2002), and subsequently several times in the birds (see, for example, B.C. Livezey in *On Evolutionary change and heterochrony* (K.J. McNamara, ed.), pp. 169–193 (Wiley; 1995).

147. See E. Snively et al. *Zoological Journal of the Linnean Society* 142, 525–553 (2004), and M.A. White *Alcheringa* 33, 1–21 (2009).

(see P.L. Manning et al. *Biology Letters* 2, 110–112 [2006]). As Phillip Manning and colleagues note, not only were the wounds inflicted by the teeth, but the "fatal embrace is analogous to the hunting technique used by many species of big cat" (p. 112).

120. See X. Xu et al. *Nature* 421, 335–340 (2003); also commentary on pp. 323–324 by R.O. Prum. Turner et al. (2012; note 114) synonymize the two species of *Microraptor* (*M. gui* and *M. zhaoianus*, the latter being senior).

121. See J.-M. O'Connor et al. *Proceedings of the National Academy of Sciences, USA* 108, 19662–19665 (2011), who document the gut contents in the form of an enantiornithine bird which on other evidence is interpreted as also arboreal. Even so this animal had an eclectic diet, as is evident from gut contents in the form of fish bones; see L. Xing et al. *Evolution* 67, 2441–2445 (2013).

122. See K. Padian and K.P. Dial *Nature* 438, E3–E4 (2005). E-P. Gong et al. (*Palaeoworld* 21, 81–91; 2012) describe another species and also reconstruct it as a glider.

123. See papers by S. Chatterjee and R.J. Templin and D.E. Alexander et al. in *Proceedings of the National Academy of Sciences, USA*, respectively, in vol. 104, pp. 1576–1580 (2007), and vol. 107, pp. 2972–2976 (2010), with commentary on pp. 2733–2734 by J. Ruben; see also subsequent discussion and reply in vol. 107 (2010) by J. Brougham and S.L. Brusatte (p. E155) and D.E. Alexander et al. (p. E156). See also M.A.R. Koehl et al. *Integrative and Comparative Biology* 51, 1002–1018 (2012), and D. Evangelista et al. *PLoS ONE* 9, e85203 (2014). The closely related *Changyuraptor* has even longer feathers and was evidently aerodynamic; see G. Han et al. *Nature Communications* 5, e4382 (2014)

124. See X. Xu et al. *Nature* 410, 200–204 (2001).

125. See S. Chatterjee and R.J. Templin in *Feathered dragons: Studies on the transition from dinosaurs to birds* (P.J. Currie et al., eds.), pp. 251–281 (Indiana University Press; 2004).

126. See R.L. Nudds and G.J. Dyke *Science* 328, 887–889 (2010).

127. See, for example, papers by X. Wang et al. in *Journal of Evolutionary Biology* (vol. 24, pp. 1226–1231 [2011], and vol. 25, pp. 547–555 [2012], also correction on p. 2376) and discussion in *Science* (vol. 320; 2010) arising from Nudds and Dyke (2010; note 126) by G.S. Paul (p. 320-b) and X.-T. Zheng et al. (p. 320-c), plus reply (p. 320-d).

128. See J.R. Hutchinson and V. Allen *Naturwissenschaften* 96, 423–448 (2009).

129. Hutchinson and Allen (2009), p. 437; note 128.

130. Note the flying abilities of *Archaeopteryx* and *Confuciusornis* are questioned by R.L. Nudds and G.J. Dyke *Science* 328, 887–889 (2010); see also X. Wang et al. *Journal of Evolutionary Biology* 25, 547–555 (2012), and N.R. Longrich et al. *Current Biology* 22, 2262–2267 (2013), with commentary on pp. R992–R994 by X. Xu.

131. And maybe four times if new discoveries—in the form of yet another theropod (*Xiaotingia*) from the Upper Jurassic of China—are taken into account. This work (see X. Xu et al. *Nature* 475, 465–470 [2011], with commentary on pp. 601–602 by L.M. Witmer) has been taken to suggest that *Archaeopteryx* is part of a distinct lineage, closer to the other dromaeosaurids but crucially no longer on the branch leading to the birds themselves. Or at least that is the suggestion, because as Xing Xu and colleagues note, "our phylogenetic hypothesis is only weakly supported by the available data. . . . This low support is partly caused by various homoplasies, many of which are functionally significant" (p. 467). Their find echoes some earlier thoughts in the same direction (see, for example, D.-G. Hu et al. *Nature* 461, 640–643 [2009]), but the matter remains open; see M.S.Y. Lee and T.H. Worthy *Biology Letters* 8, 299–303; 2012 (also Turner et al. 2012; note 114; and P. Godefroit et al. *Nature* 498, 359–362 [2013], with commentary [doi: 10.1038/nature.2013.13088] by C. Woolston). See also C. Foth et al. *Nature* 511, 79–82 (2014).

132. Kurochkin (2006), p. S49; note 106.

133. See X. Xu et al. *Nature* 431, 680–684 (2004).

134. Xu et al. (2004), mentioned in title and text; note 133.

135. A few skeptical voices have been raised, arguing that in at least two cases that neatly span the theropod tree, from the primitive compsognathan *Sinosauropteryx* (see P.-J. Chen et al. *Nature* 391, 147–152 [1998], also commentary on pp. 119–120 by D.M. Unwin) to a dromaeosaurid in the form of *Sinornithosaurus*, the supposed feathers are no more than rotted fibers of collagen (see T. Lingham-Soliar et al. *Proceedings of the Royal Society of London, B* 274, 1823–1829 [2007], and T. Lingham-Soliar *Naturwissenschaften* 90, 428–432 [2003]; also T. Lingham-Soliar *Journal of Ornithology* 151, 193–200; 2010). Such taphonomic pitfalls need to be taken seriously, but there is little to support this view. Primitive feathers, possibly from a troodontid, are preserved in amber; see V. Perrichot et al. *Proceedings of the Royal Society of London, B* 275, 1197–1202; 2008). This arrangement around the body (see, for example, P.J. Currie and P.-J. Chen *Canadian Journal of Earth Sciences* 38, 1705–1727 [2001]), details of structure (see, for example, Q. Ji et al. *Nature* 410, 1084–1088 [2001]), and color (see F.-C. Zhang et al. *Nature* 463, 1075–1078 [2010], and Q.-G. Li *Science* 327, 1369–1372 [2010]). Occurrences in not only primitive theropods but also the therizinosaurs (see X. Xu et al. *Proceedings of the National Academy of Sciences, USA* 106, 832–834 [2009]) and oviraptorosaurians (see Q. Ji et al. *Nature* 393, 753–761 [1998]; also commentary on pp. 729–730 by K. Padian) indicates that at least the smaller theropods had a dense covering of integumentary extensions.

136. See, for example, X. Xu and F.-C. Zhang *Naturwissenschaften* 92, 173–177 (2005), and F.-C. Zhang et al. *Nature* 455, 1105–1108 (2009).

86. See S.P. Yanoviak and R. Dudley *Journal of Experimental Biology* 209, 1777–1783 (2006).

87. See S.P. Yanoviak et al. *Journal of Insect Behavior* 21, 164–171 (2008).

88. How many other species contribute to the "ant rain" (see, for example, P.D. Haemig *Animal Behaviour* 54, 89–97 [1997]), which applies as much to dislodged ants that have no gliding capacity but will land safely owing to their negligible weight, remains to be established.

89. See S.P. Yanoviak et al. *Biology Letters* 5, 510–512 (2009).

90. See E. Caviedes-Vidal et al. *Proceedings of the National Academy of Sciences, USA* 104, 19132–19137 (2007).

91. See A.L. Hughes and M.K. Hughes *Nature* 377, 391 (1995); also C.L. Organ et al. *Nature* 446, 180–184 (2007).

92. See, for example, J.D.L. Smith and T.R. Gregory *Biology Letters* 5, 347–351 (2009).

93. See C.L. Organ and A.M. Shedlock *Biology Letters* 5, 47–50 (2008).

94. The pterosaurs show their own convergences, as among the giant azhdarchids and ctenochasmatids (see B. Andres and Q. Ji *Palaeontology* 51, 453–469 [2008]).

95. See J.-C. Lü et al. *Acta Geologica Sinica* 79, 766–769 (2005).

96. See L.P.A.M. Claessens et al. *PLoS ONE* 4, e4497 (2009).

97. See P. Chambers *Bones of contention: The Archaeopteryx scandals* (John Murray; 2002).

98. See K.W. Barthel et al. *Solnhofen: A study in Mesozoic palaeontology* (Cambridge; 1990), and K.A. Frickhinger *The fossils of Solnhofen: Documenting the animals and plants known from Plattenkalks* (Goldschneck-Verlag; 1994; parallel German and English texts).

99. See G. Mayr et al. *Zoological Journal of the Linnean Society* 149, 97–116 (2007).

100. See, for example, U. Bergmann et al. *Proceedings of the National Academy of Sciences, USA* 107, 9060–9065 (2010).

101. See P. Christiansen and N. Bonde *Comptes Rendus Paleovol* 3, 99–118 (2004), and N. Longrich *Paleobiology* 32, 417–431 (2006).

102. See, for example, Z.-H. Zhou et al. *Journal of Anatomy* 212, 565–577 (2008); E.M. Morschhauser et al. *Journal of Vertebrate Paleontology* 29, 545–554 (2009); and Z.-H. Zhou et al. *Proceedings of the Royal Society of London, B* 277, 219–227 (2010).

103. See G.M. Erickson et al. *PLoS ONE* 4, e7390 (2009); also I.M. Chiappe et al. *Biology Letters* 4, 719–723 (2008), for dinosaurian-like life history in even the first birds.

104. See E.N. Kurochkin et al. *Biology Letters* 3, 309–313 (2007).

105. See P. Senter *Journal of Systematic Palaeontology* 5, 429–463 (2007); also, for example, A.H. Turner et al. *American Museum Novitates* 3557, 1–27 (2007).

106. See E.N. Kurochkin *Entomological Review* 86 (Supplement, 1), S45–S58 (2006; originally published in Russian in *Zoologischeskii Zhurnal* 85, 283–297 [2006]).

107. See X. Xu and M.A. Norell *Geological Journal* 41, 419–438 (2006); also M.-M. Chang et al. (eds.) *The Jehol Biota: The emergence of feathered dinosaurs, beaked birds and flowering plants* (Shanghai Scientific & Technical Publishers; 2003).

108. See, for example, F.E. Novas et al. *Proceedings of the Royal Society of London, B* 276, 1101–1107 (2009); in this case it is a bizarre dromaeosaurid with much shorter arms. For a comparable case from Liaoning see X.-T. Zheng et al. *Proceedings of the Royal Society of London, B* 277, 211–217 (2010).

109. See, for example, hints for dromaeosaurids from Madagascar (F. Fanti and F. Therrien *Acta Palaeontologica Polonica* 52, 155–166; 2007) and Chinese trackways that suggest a considerable diversity of body sizes (R.-H. Li et al. *Naturwissenschaften* 95, 185–191; 2008).

110. These include abelisaurid theropods (see S.D. Sampson and D.W. Krause [eds.] *Journal of Vertebrate Palaeontology* 27 [Supplement, 2; Memoir 8], 1–184; 2007), one of which (*Masiakasaurus*) displays a remarkable heterodont dentition (see S.D. Sampson et al. *Nature* 409, 504–506 [2001]). Here we see a procumbent set of teeth at the anterior, and as Matthew Carrano and colleagues (see their paper in *Journal of Vertebrate Paleontology* 22, 510–534 [2002]) note, "somewhat analogous anterior tooth morphologies can be observed in certain mammals" (p. 528).

111. See C.A. Forster and P.M. O'Connor *Journal of Vertebrate Paleontology* 20 (Supplement, 3), 41A–42A [Abstract] (2000).

112. See C.A. Forster et al. *Nature* 382, 532–534 (1996).

113. See C.A. Forster et al. *Science* 279, 1915–1919 (1998*a*); in this paper the fossil was named *Rahona*, but this was corrected to *Rahonavis* (vol. 280, p. 179; 1998*b*).

114. See P.J. Makovicky et al. *Nature* 437, 1007–1011 (2005); also A.H. Turner et al. *Bulletin of the American Museum of Natural History* 371, 1–206 (2012), who support its assignment to the dromaeosaurids as against avialians.

115. See also P.M. O'Connor and C.A. Forster *Journal of Paleontology* 30, 1178–1201 (2010).

116. Forster et al. (1998*a*), p. 1916; note 113.

117. See A.H. Turner et al. *Science* 317, 1721 (2007).

118. See M.H. Schweitzer et al. *Journal of Vertebrate Paleontology* 19, 712–722 (1999).

119. While these claws have been seen in the light of the dromaeosaurids merrily disemboweling their prey, in fact a slashing capacity is unlikely and their principal role appears to have been to grip their prey

47. See his paper in *Malayan Nature Journal* 13, 31 (1958).

48. Adams (1958), p. 31; note 47.

49. See papers in the *Zoological Journal of the Linnean Society* by S.E. Evans (vol. 76, pp. 97–123; 1982) and S.E. Evans and H. Haubold (vol. 90, pp. 275–303; 1987); also contributions by V.V. Bulanov and A.G. Sennikov *Paleontological Journal* 40 (Supplement, 5), S567–S570 (2006), and *Paleontogicheskii Zhurnal* 2010 (6), 81–93 (2010; in Russian, with English abstract).

50. See E. Frey et al. *Science* 275, 1450–1452 (1997), with commentary on p. 1419 by B. Wuethrich; also G. Schaumberg et al. *Paläontologische Zeitschrift* 81, 160–173 (2007).

51. What may be a third instance involves the remarkable diapsid *Megalancosaurus*, that was evidently an adept arboreal climber and could possibly glide; see S. Renesto *Rivista Italiana di Paleontologia e Stratigrafia* 106, 157–180 (2000).

52. See, for example, P.L. Robinson *Proceedings of the Geological Society of London* 1601, 137–146 (1962).

53. See S.E. Evans *Palaeontologia Polonica* 65, 145–178 (2009).

54. One possibility is that while the males glided, the females parachuted; see K. Stein et al. *Palaeontology* 51, 967–981 (2008).

55. See N.C. Fraser et al. *Journal of Vertebrate Paleontology* 27, 261–265 (2007).

56. See P.-P. Li et al. *Proceedings of the National Academy of Sciences, USA* 104, 5507–5509 (2007).

57. See G.J. Dyke et al. *Journal of Evolutionary Biology* 19, 1040–1043 (2006).

58. See D. Peters *Rivista Italiana di Paleontologia e Stratigrafia* 106, 293–336 (2000).

59. See, for example, D.W.E. Hone and M.J. Benton *Journal of Systematic Palaeontology* 5, 465–469 (2007).

60. See, for example, A.W. Heere *Copeia* 1958, 338–339 (1958).

61. See A.P. Russell and L.D. Dijkstra *Journal of Zoology, London* 253, 457–471 (2001).

62. See J.A. McGuire *American Naturalist* 161, 337–349 (2003).

63. This process is far from random, because a large body size (see J.A. McGuire and R. Dudley *American Naturalist* 166, 93–106 [2005]) has evolved at least four times. Moreover, in a way that finds a fascinating parallel to the "*Lineatriton* phenomenon" (p. 76), so in a study of the phylogenetics of *Draco* it transpires that one "species" (specifically *D. spilopterus*) has evolved three times (see J.A. McGuire and B.H. Kiew *Biological Journal of the Linnean Society* 72, 203–229 [2001]).

64. See J.A. McGuire and R. Dudley *Integrative and Comparative Biology* 51, 983–990 (2012); they also draw comparisons with the Permo-Triassic gliders and conclude that although *Coelurosauravus* and *Kuehneosaurus* were less competent, *Icarosaurus* "was

the best nonflapping . . . terrestrial vertebrate glider yet discovered" (p. 989).

65. If dropped, the butterfly lizard will parachute, yet in life the animals are entirely terrestrial (see J.B. Losos et al. *Journal of Zoology, London* 217, 559–568 [1989]).

66. See A. Schiøtz and H. Volsøe *Copeia* 1959, 259–260 (1959).

67. See B. Vanhooydonck et al. *Journal of Experimental Biology* 212, 2475–2482 (2009).

68. See his paper in *Bulletin of the Natural History Museum of London (Zoology)* 68, 155–163 (2002).

69. Arnold (2002), part of the title; note 68.

70. Arnold (2002), p. 155 [Abstract]; note 68.

71. Arnold (2002), part of the title; note 68.

72. McGuire and Dudley (2012; note 64) review the literature on a number of other gliding lizards; see p. 984.

73. Arnold (2002), p. 160; note 68.

74. See A.P. Russell et al. *Journal of Morphology* 247, 252–263 (2001).

75. The evolutionary transformations that led to the patagium remain of great interest. It is likely that initially the body folds evolved to store fat (and so remain helpful as shock absorbers on landing). Subsequently the extensions of the body made the lizard more cryptic, and only at this stage did a controlled descent become viable (see A.P. Russell *Zoological Journal of the Linnean Society* 65, 233–249 [1979]). This descent displays remarkable maneuverability, employing both the tail (as in the gecko *Cosymbotus*; see A. Jusufi et al. *Proceedings of the National Academy of Sciences, USA* 105, 4215–4219 [2008]) and the webbed feet (see B.A. Young et al. *Journal of Herpetology* 36, 412–418 [2002]).

76. See J.J. Socha *Journal of Experimental Biology* 209, 3358–3369 (2006); also J.J. Socha and M. LaBarbera in vol. 208, pp. 1835–1847 (2005).

77. See J.J. Socha et al. *Journal of Experimental Biology* 208, 1817–1833 (2005), and J.J. Socha *Integrative and Comparative Biology* 51, 969–982 (2012).

78. See J.J. Socha *Nature* 418, 603–604 (2002).

79. See S.B. Emerson and M.A.R. Koehl *Evolution* 44, 1931–1946 (1990).

80. See M.G. McCay *Journal of Experimental Biology* 204, 2817–2826 (2001).

81. See M.G. McCay *Biotropica* 35, 94–102 (2003).

82. See W.E. Roberts *Journal of Herpetology* 28, 193–199 (1994). She graphically describes this as "a dense writhing mass of frogs" (p. 194), a frenetic scene where not all the details were visible, but there was a good deal of kicking.

83. See S.P. Yanoviak et al. *Proceedings of the Royal Society of London, B* 277, 2199–2204 (2010).

84. For an overview, see S.P. Yanoviak et al. *Integrative and Comparative Biology* 51, 944–956 (2012).

85. See S.P. Yanoviak et al. *Nature* 433, 624–626 (2005).

5. See J. Davenport *Reviews in Fish Biology and Fisheries* 4, 184–214 (1994).

6. Their aerodynamic capacities are addressed by H. Park and H. Choi *Journal of Experimental Biology* 213, 3269–3279 (2010).

7. See E.R. Baylor *Nature* 214, 307–309 (1967).

8. See J. Davenport *Journal of Fish Biology* 62, 455–463 (2003). This point has also occurred to the copepods, which can engage in ballistic leaps; see B.J. Gemmell et al. *Proceedings of the Royal Society of London, B* 279, 2786–2792 (2012).

9. The African butterfly fish (*Pantodon*), with its powerful pectoral musculature, has been thought of as a flying fish (see P.H. Greenwood and K.S. Thomson *Proceedings of the Zoological Society of London* 135, 283–301 [1960]), in fact it is a highly competent ballistic leaper (see W.M. Saidel et al. *Environmental Biology of Fishes* 71, 63–72 [2004]). On the other hand the Brazilian fish *Gasteropelecus* can travel a considerable distance and is powered by quite enormous pectoral muscles (see W.G. Ridewood *Annals of the Magazine of Natural History* 12 [8th ser.], 544–548 [1913]). Although this fish is popular in the aquarium trade it appears to have received much less scientific attention, so it is particularly interesting that Walter Ridewood remarked that "there is no doubt that the pectoral fins are flapped vigorously during the passage of the fish through the air" (p. 545).

10. See M. Silviá et al. *Journal of Molluscan Studies* 70, 297–299 (2004).

11. Silviá et al. (2004), p. 297; note 10; also K.S. Cole and D.L. Gilbert *Biological Bulletin* 138, 245–246 (1970), where they note a "continuing ejection of water . . . provides a striking illustration of amphibious jet propulsion" (p. 246).

12. Silviá et al. (2004), p. 297; note 10.

13. See A.R. Ennos *Journal of Zoology, London* 219, 61–69 (1989).

14. Not surprisingly the savings in such energy expenditure find commonalities among at least mammalian gliders in different parts of the world, and interestingly these also extend to the types of diet (see R. Dial *Evolutionary Ecology Research* 5, 1151–1162 [2003]).

15. Dial (2003), p. 1159; note 14.

16. See L.H. Emmons and A.H. Gentry *American Naturalist* 121, 513–524 (1983).

17. See R. Dial et al. *Forest Science* 50, 312–325 (2004).

18. See R. Dudley and P. De Vries *Biotropica* 22, 432–434 (1990); also M.P. Heinicke et al. *Biology Letters* 8, 994–997 (2012).

19. For an overview, see G. Byrnes and A.J. Spence *Integrative and Comparative Biology* 51, 991–1001 (2012).

20. See, for example, R.W. Thorington et al. *Journal of Mammalian Evolution* 9, 99–135 (2002), and R.W. Thorington and E.M. Santana *Journal of Mammal-*ogy 88, 882–896 (2007); also B.S. Arbogast *Journal of Mammalogy* 88, 840–849 (2007).

21. See J.D. Kingdon in *East African mammals: An atlas of evolution in Africa*, vol. 2B, pp. 445–464 (Academic; 1974).

22. Kingdon (1974), p. 460; note 21.

23. See, for example, R.M. Nowak in *Walker's mammals of the world*, 6th ed., vol. 1, pp. 250–252 (Johns Hopkins; 1999).

24. See V. Fahlbusch in *Evolutionary relationships among rodents: A multidisciplinary analysis* (P.W. Luckett and J.-L. Hartenberger, eds.), pp. 617–629 (Plenum; 1985).

25. See G. Storch et al. *Nature* 379, 439–441 (1996), and B. Engesser and G. Storch *Ecologae Geologicae Helvetiae* 92, 483–493 (1999; in German, with English abstract).

26. See J.L. Johnson-Murray *Australian Journal of Zoology* 35, 101–113 (1987).

27. See R.W. Thorington *Science* 225, 1048–1050 (1984).

28. See P. Mein and J.-P. Romaggi *GeoBios Mémoire Spéciale* 13, 45–50 (1991; in French, with English abstract).

29. See K.C. Beard *Nature* 345, 340–341 (1990); also accompanying paper by R.F. Kay et al., pp. 342–344.

30. See J.A. Bunestad and C.B. Ruff *American Journal of Physical Anthropology* 98, 101–119 (1995).

31. See M.W. Hamrick et al. *American Journal of Physical Anthropology* 109, 397–413 (1999).

32. See J.I. Block et al. *Proceedings of the National Academy of Sciences, USA* 104, 1159–1164 (2007).

33. See R.F. Kay and M. Cartmill *Journal of Human Evolution* 6, 19–53 (1977).

34. See J. Meng et al. *Nature* 444, 889–893 (2006); also vol. 102, p. 446 (2007).

35. See S.M. Jackson *Mammal Review* 30, 9–30 (2000).

36. See, for example, R.W. Thorington and L.R. Heaney *Journal of Mammalogy* 62, 101–114 (1981).

37. Jackson (2000), p. 16; note 35.

38. See K.L. Bishop *Journal of Experimental Biology* 210, 2593–2606 (2007).

39. Bishop (2007), p. 2604; note 38.

40. Bishop (2007), p. 2594; note 38.

41. Bishop (2007), p. 2605; note 38.

42. See R.L. Essner *Journal of Experimental Biology* 205, 2469–2477 (2002).

43. See K.L. Bishop and W. Brim-Deforest *Journal of Experimental Zoology: Ecological Genetics and Physiology, A* 309, 225–242 (2008).

44. See K.L. Bishop *Journal of Experimental Biology* 209, 689–701 (2006).

45. See K.E. Paskins et al. *Journal of Experimental Biology* 210, 1413–1423 (2007).

46. See R.W. Thorington et al. *Journal of Mammalogy* 79, 245–250 (1998).

203. See, for example, J. Korb *Naturwissenschaften* 90, 212–219 (2003).

204. See J.P.E.C. Darlington *Journal of Zoology, London* 198, 237–247 (1982).

205. See A.J. Mills et al. *Journal of Zoology, London* 278, 24–35 (2009).

206. See J.M. Hunter *Journal of Cultural Geography* 14, 69–92 (1993).

207. See R.G. Ruggiero and J.M. Fay *African Journal of Ecology* 32, 222–232 (1994).

208. See papers by P. Duringer et al. in *Naturwissenschaften* 93, 610–615 (2006), and *Palaeogeography, Palaeoclimatology, Palaeoecology* 251, 323–353 (2007).

209. Molecular data, however, suggest a substantially earlier origination, at c. 30 Ma; see T. Nobre et al. *Molecular Ecology* 20, 2619–2627 (2011).

210. See D.K. Aanen and P. Eggleton *Current Biology* 15, 851–855 (2005); also D.K. Aanen et al. *Proceedings of the National Academy of Sciences, USA* 99, 14887–14892 (2002).

211. See C. Rouland-Lefevre et al. *Molecular Phylogenetics and Evolution* 22, 423–429 (2002).

212. See G.D. Piearce *Mycologist* 1, 111–116 (1987).

213. See T.G. Frøslev et al. *Mycological Research* 107, 1277–1286 (2003).

214. There is some evidence that the production of the mushrooms is synchronized with the activity of the termites (including the reproductive alates; see R.A. Johnson et al. *Journal of Natural History* 15, 751–756; 1981), and this underscores one important contrast with the attine ants. As Duur Aanen and colleagues (2002; note 210) note, although vertical transmission of a fungal clone has "two independent origins" (p. 14890), the usual (and original) method entails a horizontal transmission whereby the fungal spores are carried into the nest (see J. Korb and D.K. Aanen *Behavioral Ecology and Sociobiology* 53, 65–71 [2003]; also H.H. De Fine Licht et al. *Mycological Research* 109, 314–318 [2005], and *Molecular Ecology* 15, 3131–3138; 2006).

215. See D.K. Aanen et al. *BMC Evolutionary Biology* 7, e115 (2007).

216. See D.K. Aanen et al. *Science* 326, 1103–1106 (2009).

217. Aanen et al. (2009), p. 1105; note 216.

218. See E. Garnier-Sillam et al. *Insectes Sociaux* 36, 293–312 (1989).

219. Some wood-eating termites include the bacterium *Klebsiella* (see M. Doolittle et al. *Bioresource Technology* 99, 3297–3300 [2008]) as part of their complex gut flora, and like the attine ants this microbe helps to fix nitrogen (see J.A. Breznak et al. *Nature* 244, 577–580 [1973]).

220. See, for example, the overviews by D.K. Aanen and J.J. Boomsma in *Insect-fungal associations: ecology and evolution* (F. Vega and M. Blackwell, eds.), pp. 191–211 (Blackwell; 2005), and T.G. Wood and R.J. Thomas in *Insect-fungus interactions* (N. Wilding et al., eds.), pp. 69–92 (Academic; 1989).

221. See D.K. Aanen *Biology Letters* 2, 209–212 (2006).

222. See J.P.E.C. Darlington in *Nourishment and evolution in insect societies* (J.H. Hunt and C.A. Nalepa, eds.), pp. 105–130 (Westview; 1994).

223. See H.H. De Fine Licht et al. *Ecological Entomology* 32, 76–81 (2007).

224. See F. Hyodo et al. *Functional Ecology* 17, 186–193 (2003).

225. See P. Mora and C. Lattaud *Insect Science and its Application* 19, 51–55 (1999).

226. See M. Matoub and C. Rouland *Comparative Biochemistry and Physiology, B* 112, 629–635 (1995).

227. See C. Rouland et al. *Comparative Biochemistry and Physiology, B* 91, 459–465 (1988).

228. See M. Lamberty et al. *Journal of Biological Chemistry* 276, 4085–4092 (2001). Essential as this capacity is, the association of most termites with wood means that in the group as a whole, microbial infestation is a recurrent threat. See also C. Hamilton et al. (*Journal of Insect Physiology* 57, 1259–1266; 2011), who, although documenting secretion of antifungal defenses in the more primitive termites, also remind us of the likely analogues between the salivary glands of termites and the metapleural glands of the ants.

229. See T. Kuranouchi et al. *Journal of Applied Entomology* 130, 471–472 (2006).

230. See C. Vispo and I.D. Hume *Canadian Journal of Zoology* 73, 967–974 (1995).

231. See R.C. Dean *Biological Bulletin* 155, 297–316 (1978), and S.M. Gallager et al. *Journal of Experimental Marine Biology and Ecology* 52, 63–77 (1981).

232. See P.-N. Xu and D.L. Distel *Marine Biology* 144, 947–953 (2004).

233. See D.L. Distel et al. *Applied and Environmental Microbiology* 57, 2376–2382 (1991); see also Xu and Distel (2004); note 232.

234. See D.L. Distel and S.T. Roberts *Biological Bulletin* 192, 253–261 (1997).

CHAPTER 17

1. See R. Dudley et al. *Annual Review of Ecology, Evolution, and Systematics* 38, 179–201 (2007), and R. Dudley and S.P. Yanoviak *Integrative and Comparative Biology* 51, 926–936 (2012).

2. See G.-H. Xu et al. *Proceedings of the Royal Society of London, B* 280, 20122261 (2013); also A. Tintori and D. Sassi *Journal of Vertebrate Paleontology* 12, 265–283 (1992).

3. See J. Davenport *Journal of the Marine Biological Association, UK* 72, 25–39 (1992).

4. See E.A. Lewallen et al. *Biological Journal of the Linnean Society* 102, 161–174 (2011).

conflict or one where mutual sabotage turned out to be counterproductive? Although declaring themselves to be surprised, these workers concluded, "There is no evidence of conflict—in fact the evidence points towards complete cooperation between ants and fungi" (p. 995).

185. See S. Powell and E. Clark *Insectes Sociaux* 51, 342–351 (2004).

186. See M.B. Dijkstra and J.J. Boomsma *Naturwissenschaften* 90, 568–571 (2003), unless, that is, they run into "mercenaries" in the form of *Megalomyrmex* whose "guest" status is employed to deal with invaders by administering toxic alkaloids and thereby saving the fungus farms of the host *Sericomyrmex* (see R.M.M. Adams et al. *Proceedings of the National Academy of Sciences, USA* 110, 15752–15757 [2013]).

187. See R.M.M. Adams et al. *Naturwissenschaften* 87, 549–554 (2000).

188. Adams et al. (2000), p. 551; note 187.

189. As can the attine ants, although as B.E. Lechner and R. Josens (*Insectes Sociaux* 59, 285–288; 2012) point out, this example poses a number of interesting questions.

190. See V. Witte and U. Maschwitz *Naturwissenschaften* 95, 1049–1054 (2008).

191. In the case of leaf domatia it is evident they have evolved independently several times; see C. Leroy et al. *American Journal of Botany* 97, 557–565 (2010). Also R. Blatrix et al. *PLoS ONE* 8, e68101 (2013).

192. See E. Defossez et al. *New Phytologist* 182, 942–949 (2009); also overview on pp. 785–788 by M. Poulsen and C.R. Currie.

193. See E. Defossez et al. *Proceedings of the Royal Society of London, B* 278, 1419–1426 (2011).

194. As far as interactions between ants and fungi are concerned, mention should be made of the so-called cartons (see V.E. Mayer and H. Voglmayr *Proceedings of the Royal Society of London, B* 276, 3265–3273 [2009]). These typically are nests of earth and/or vegetation that are specifically reinforced with fungi. Not only do the ants (members of the *Lasius* group) culture the fungi with honeydew (see U. Maschwitz and B. Hölldobler *Zeitschrift für vergleichende Physiologie* 66, 176–189; 1970), but there is a specificity of association (see B.C. Schlick-Steiner et al. *Proceedings of the National Academy of Sciences, USA* 105, 940–943 [2008]). Most likely this employment of fungi arose twice (see M. Maruyama et al. *BMC Evolutionary Biology* 8, e237 [2008]), and as Munetoshi Maruyama and colleagues note, "the *Lasius* situation might thus indicate a stronger predisposition to evolve fungiculture for ants generally" (p. 8). To be sure the care and maintenance of the fungi seems very largely to revolve around their structural role in the cartons, but some workers have suggested that the fungus conceivably supplements the ants' diet (see, for example, J.S.B. Elliott *Transactions of the British Mycological Society* 5, 138–142 [1915]).

195. See J.F. Genise et al. *Palaeogeography, Palaeoclimatology, Palaeoecology* 287, 128–142 (2010).

196. Genise et al. (2010), p. 140; note 195.

197. This is most obvious in the convergent acquisition of eusociality and a caste system, complex vibrational communication, and a soldier caste (see B.L. Thorne et al. *Proceedings of the National Academy of Sciences, USA* 100, 12808–12813 [2003]; also E.A. Roux and J. Korb *Journal of Evolutionary Biology* 17, 869–875; 2004). Termites also reveal specific convergences, such as battering horn–like projections and snapping mandibles (see O.I. Scholtz et al. *Zoological Journal of the Linnean Society* 153, 631–650 [2008]). It is seldom a sensible idea to tackle an occupied nest, but its vigorous defense by the soldier caste is only one aspect of a complex social order. Another area includes care of the brood, but here the termites can be fooled by a remarkable mimicry. Termite eggs are small, round, and very smooth, and so is a fungus that forms brown balls which are industriously collected by the unwitting termites (see K. Matsuura et al. *Ecological Research* 15, 405–414 [2000]). When this mimicry was first discovered, a symbiotic association seemed possible, with perhaps the fungi helping to protect the eggs, but it now seems likely that the benefit is to the fungus (see K. Matsuura *Proceedings of the Royal Society of London, B* 273, 1203–1209 [2006]). While this association is best known among species of *Reticulitermes* (see T. Yashiro and K. Matsuura *Annals of the Entomological Society of America* 100, 532–538 [2007]), there is evidence that this mimicry not only entails a chemical identity (involving a cellulose; see K. Matsuura et al. *Current Biology* 19, 30–36; 2009), but as Kenji Matsuura and colleagues note, this "egg mimicry has evolved at least twice independently" (p. 34).

198. See D. Inward et al. *Biology Letters* 3, 331–335 (2007).

199. Inward et al. (2007), p. 333; note 198. Cockroaches also provide their own insights into convergence such as milk, as well as a remarkable case of Batesian mimicry with a noxious carabid beetle; see H. Schmied et al. *Entomological Science* 16, 119–121 (2013).

200. See N. Lo et al. *Current Biology* 10, 801–804 (2000); also D. Grimaldi and M.S. Engel *Evolution of the insects* (Cambridge; 2005), pp. 235–238. See also S.M. Farris and N.J. Strausfeld *Journal of Comparative Neurology* 456, 305–320 (2003), for the significance of mushroom bodies (pp. 246–251) in *Cryptocercus* and the termites, not least in germs of the evolution of sociality.

201. See R. Pellens et al. *Molecular Phylogenetics and Evolution* 43, 616–626 (2007); also K. Maekawa et al. *Insectes Sociaux* 55, 107–114 (2008), who emphasize the likely importance of oviparity.

202. See, for example, J.E. Francis and B.M. Harland *Cretaceous Research* 27, 773–777 (2006); also Grimaldi and Engel (2005); note 200.

153. Mueller et al. (2008), p. 2908; note 152.

154. We should note that employing antibiotic actinomycetes has evolved independently in other fungus-growing ants (*Allomerus*; see R.F. Seipke et al. *Antonie van Leeuwenhoek* 101, 443–447; 2012) and also in the beewolves (see W. Goettler et al. *Arthropod Structure & Development* 36, 1–9 [2007]; also M. Kaltenpoth and E. Strohm *Comparative Biochemistry and Physiology A* 146, S66–S67 [2007], and M. Kaltenpoth et al. *Physiological Entomology* 35, 196–200; 2010). In the beewolves, the female houses the bacteria in an antennal gland and evidently employs the secretions to inhibit the growth of microbes in the brood chamber, where paralyzed honeybees await their grisly fate as the wasp larvae mature. Wolfgang Goettler and colleagues not only draw attention to the way in which "fungus-growing ants and beewolves use analogous alliances with bacteria to combat the fungal menace" (p. 8), but in the spirit of this book also look to the future when, they write, "we predict that other Hymenoptera might have evolved comparable symbioses" (p. 8).

155. See A.E.F. Little and C.R. Currie *Ecology* 89, 1216–1222 (2008).

156. See H. Fernández-Marin et al. *Proceedings of the Royal Society of London, B* 276, 2263–2269 (2009).

157. See M. Poulsen et al. in *Behavioral Ecology and Sociobiology* 52, 151–157 (2002).

158. See H. Fernández-Marin et al. *Proceedings of the Royal Society of London, B* 273, 1689–1695 (2006).

159. Further evidence for the intricacy of the system is evident from the observation that in *Acromyrmex* the smaller workers not only have disproportionately large metapleural glands (see M. Poulsen et al. *Insectes Sociaux* 53, 349–355 [2006]), but evidently rely less on the antibiotics provided by the bacteria (see M. Poulsen et al. *Naturwissenschaften* 90, 406–409 [2003]).

160. See T.N. Walker and W.O.H. Hughes *Biology Letters* 5, 446–448 (2009).

161. Walker and Hughes (2009), p. 448; note 160.

162. Significantly, among ants that live in relatively simple societies, ants dying of fungal infection wander off to meet their end in isolation; see J. Heinze and B. Walter *Current Biology* 20, 249–252 (2010). Nor may this be restricted to the species in question (*Temnothorax unifasciatus*), because as these authors note, "scattered reports about dying bees, wasps and ants leaving their nests . . . indicate that social withdrawal is not restricted to *Temnothorax* but is a more widespread phenomenon" (p. 251). In his commentary (pp. R104–R105) Michel Chapuisat makes the interesting point that while such self-exclusion makes sense in small colonies, in cases like the leaf-cutter ants, not only do we see much more sophisticated behaviors revolving around hygiene and pathogen control, but, as he remarks, "Over evolutionary time, diseases might therefore become a factor contribut-

ing to the maintenance of advanced social behaviour" (p. R105).

163. See A.E.F. Little et al. *Naturwissenschaften* 90, 558–562 (2003).

164. Little et al. (2003), p. 558 [Abstract]; note 163.

165. See A.E.F. Little et al. *Biology Letters* 2, 12–16 (2006).

166. See A.G. Hart and F.L.W. Ratnieks *Behavioral Ecology* 13, 224–231 (2002).

167. See J.C.M. Jonkman *Zeitschrift für Angewandte Entomologie* 89, 217–246 (1980); see Fig. 20C.

168. See A.N.M. Bot et al. *Ethology, Ecology & Evolution* 13, 225–237 (2001).

169. See J.A. Zeh et al. *Biotropica* 31, 368–371 (1999).

170. See A.G. Hart et al. *Naturwissenschaften* 89, 275–277 (2002).

171. See P. Villesen et al. *Proceedings of the Royal Society of London, B* 269, 1541–1548 (2002).

172. As Palle Villesen and colleagues (2002; note 171) note, this has "interesting parallels with phylogenetic patterns of multiple mating found in bees and wasps" (p. 1545), specifically honeybees and vespines.

173. See W.O.H. Hughes and J.J. Boomsma *Evolution* 58, 1251–1260 (2004).

174. See W.O.H. Hughes et al. *Evolution* 62, 1252–1257 (2008).

175. See P. D'Ettorre et al. *Journal of Comparative Physiology, B* 172, 169–176 (2002).

176. See F.-J. Richard et al. *Journal of Comparative Physiology, B* 175, 297–303 (2005); also H.H. De Fine Licht et al. *Proceedings of the National Academy of Sciences, USA* 110, 583–587 (2013).

177. See S. Rønhede et al. *Mycological Research* 108, 101–106 (2004); also M. Schiøtt et al. *BMC Biology* 8, e156 (2010), who not only document the importance of these fungally derived enzymes (pectinases) but draw attention to a likely convergence with those of phytopathogenic microbes.

178. See F.-J. Richard *Behavioral Ecology and Sociobiology* 61, 1637–1649 (2007).

179. See J.F.S. Lopes et al. *Insectes Sociaux* 52, 333–338 (2005).

180. Lopes et al. (2005), p. 333 [Abstract]; note 179.

181. See W.O.H. Hughes and J.J. Boomsma *Proceedings of the Royal Society of London, B* 274, 1625–1630 (2007); also W.O.H. Hughes et al. *Proceedings of the National Academy of Sciences, USA* 100, 9394–9397 (2003).

182. Hughes and Boomsma (2007), p. 1629; note 181.

183. See H. Herz et al. *Behavioral Ecology* 19, 575–582 (2008).

184. Consider remarks by Jon Seal and Walter Tschinkel (*Functional Ecology* 21, 988–997; 2007). They were interested to see what the reaction might be if the cultivars were swapped, specifically from *Atta* to *Trachymyrmex*. Given that the genetics and physiology of the introduced strain were different, the question is whether such a study would reveal a world of accept novel strains, suggesting that when the nest is first established a replacement strategy potentially exists in case something goes wrong with the imported strain (see M. Poulsen et al. *Evolution* 63, 2235–2247 [2009]). Evidence for some form of sexual reproduction among the fungi again suggests that the crops are far from being monotonous clones (see A.S. Mikheyev et al. *Proceedings of the National Academy of Sciences, USA* 103, 10702–10706 [2006]; also K.R. Doherty et al. *Mycologia* 95, 19–23 [2003]).

122. See H. Fernández-Marin et al. *Insectes Sociaux* 54, 64–69 (2007); also vol. 50, pp. 304–308 (2003).

123. See, for example, H. Fernández-Marin et al. *Biological Journal of the Linnean Society* 81, 39–48 (2004).

124. See, for example, C.R. Currie et al. *Science* 299, 386–388 (2003).

125. See H.T. Reynolds and C.R. Currie *Mycologia* 96, 955–959 (2004).

126. See N.M. Gerado et al. *Proceedings of the Royal Society of London, B* 271, 1791–1798 (2004).

127. See S.J. Taerum et al. *Proceedings of the Royal Society of London, B* 274, 1971–1978 (2007).

128. See N.M. Gerardo et al. *BMC Evolutionary Biology* 6, e88 (2006).

129. See A. Rodrigues et al. *Microbial Ecology* 56, 604–614 (2008).

130. See A.P.P. de Andrade et al. *Sociobiology* 40, 293–306 (2002).

131. Especially by the tiny workers known as minims (see E.K.M. Vieira-Neto et al. *Insectes Sociaux* 53, 326–332 [2006], and H.M. Griffiths and W.O. Hughes *Ecological Entomology* 35, 529–537; 2010). In the case of those fungi living within the plant tissue (as so-called endophytes) then they pose a special risk. As Sunshine Van Bael and colleagues (*Proceedings of the Royal Society of London, B* 276, 2419–2426; 2009) remark, these "endophytic fungi are not welcome" (part of the subtitle). Not only the behavior of the ant but its fungal cultivar serve to exclude these unwelcome visitors.

132. See D.P. Hughes et al. *Ecological Entomology* 34, 214–220 (2009).

133. See their paper in *BMC Ecology* 11, e13 (2011).

134. Nor is this the only example of behavior control of an insect by the infective fungus. B.A. Roy et al. *Annual Review of Entomology* 51, 331–357 (2006), review many other examples of what they aptly call summit disease whereby the doomed insect dies with a view. In the case of ants alone, a mandibular grip has evolved independently in temperate ants infected not only by fungi (see P.I. Marikovsky *Insectes Sociaux* 9, 173–179 [1962]) but also trematodes (see M.Y. Manga-González et al. *Parasitology* 123, S91–S114 [2001]). Apart from fungal infestation, another striking example is found in cases where a nematode-infected ant (in this case a Neotropical species that has also learned to glide, pp. 192–193) has the terminal part of its abdomen (the gaster) transformed into a red, berrylike object that evidently fools birds into eating this fruit

mimic so the nematode can complete its life cycle; see S.P. Yanoviak et al. *American Naturalist* 171, 536–544 (2008); also commentary by D.P. Hughes et al. *Current Biology* 18, R294–R295 (2008). See also G.D. Ruxton and H.M. Schaefer (*Journal of Ecology* 99, 899–904; 2011), who query this case of visual mimicry.

135. See D.P. Hughes et al. *Biology Letters* 7, 67–70 (2011).

136. See, for example, papers by H.C. Evans and R.A. Samson in *Transactions of the British Mycological Society* vol. 79, pp. 431–453 (1982), and vol. 82, pp. 127–150 (1984).

137. It is worth mentioning those ants that bury their dead; see M. Renucci et al. *Insectes Sociaux* 58, 9–16 (2011). Such a propensity has also evolved independently in a fungal-farming termite, albeit with the additional feature of covering the corpse with saliva; see T. Chouvenc et al. *Actes des Colloques Insectes Sociaux* 16, 16–20 (2004). For a wider view of corpse management among the insects, see Q. Sun and X.-G. Zhou *International Journal of Biological Sciences* 9, 313–321 (2013).

138. See M.B. Pontoppidan et al. *PLoS ONE* 4, e4835 (2009). In the cases of the fungally infected ants from western Siberia, however, Marikovsky (1962; note 134) notes that the healthy ants go to considerable lengths to dismember the infected corpses.

139. See W.O.H. Hughes and J.J. Boomsma *Proceedings of the Royal Society of London, B* 271, S104–S106 (2004).

140. Hughes and Boomsma (2004), p. S106; note 139.

141. See A.V. Santos *FEMS Microbiology Letters* 239, 319–323 (2004).

142. See, for example, J. Barke et al. *BMC Biology* 8, e109 (2010).

143. See C.R. Currie et al. (1999; note 85).

144. These include dentigerumycin (see D.-C. Oh *Nature. Chemical Biology* 5, 391–393 [2009]) and candicidine (see S. Haeder et al. *Proceedings of the National Academy of Sciences, USA* 106, 4742–4746 [2009]).

145. Including *Metarhizium*; see T.C. Mattoso et al. *Biology Letters* 8, 461–464 (2012).

146. See C.R. Currie et al. *Science* 311, 81–83 (2006).

147. See their paper in *Proceedings of the National Academy of Sciences, USA* 106, 17805–17810 (2009).

148. Perhaps the antibiotics are only specifically applied to infected parts of the farm, or as likely they may serve as a general prophylactic.

149. See M. Poulsen et al. *Molecular Ecology* 14, 3597–3604 (2005).

150. See M.M. Zhang et al. *ISME Journal* 1, 313–320 (2007).

151. See M.J. Cafaro and C.R. Currie *Canadian Journal of Microbiology* 51, 441–446 (2005), and C. Kost et al. *Naturwissenschaften* 94, 821–828 (2007).

152. See U.G. Mueller et al. *Evolution* 62, 2894–2912 (2008).

and commentary on pp. 916–917 by J.H. Hunt; also S. Eilmas and M. Heil *Applied and Environmental Microbiology* 75, 4324–4332; 2009). This is certainly the case with the turtle ants, and here a symbiotic association with the bacterium *Rhizobiales* has evolved at least five times independently (see J.A. Russell et al. *Proceedings of the National Academy of Sciences, USA* 106, 21236–21241 [2009]).

89. See their paper in *Proceedings of the National Academy of Sciences, USA* 105, 5435–5440 (2008).

90. See also S.T. Meyer et al. *Ecological Entomology* 36, 14–24 (2011), who describe their role in "ecosystem engineering," with major effects on the understory and microclimate.

91. Schultz and Brady (2008), p. 5435; note 89.

92. See L.L. Rockwood and K.E. Glander *Biotropica* 11, 1–10 (1979); also A. Estrada and R. Coates-Estrada *American Journal of Primatology* 10, 51–66 (1986).

93. Rockwood and Glander (1979), p. 1; note 92.

94. See S.P. Hubbell *Biotropica* 12, 210–213 (1980).

95. See A.G. Hart and F.L.W. Ratnieks *Animal Behaviour* 59, 587–591 (2000); also their subsequent paper in vol. 62, pp. 227–234 (2001).

96. See A. Dussutour et al. *Animal Cognition* 12, 21–30 (2009).

97. See J.J. Howard *Behavioral Ecology and Sociobiology* 49, 348–356 (2001).

98. When the mandibles become blunted the ants are more likely to be found carrying the leaves rather than cutting them; see R.M.S. Schofield *Behavioral Ecology and Sociobiology* 65, 969–982 (2011).

99. See F. Roces and J.R.B. Lighton *Nature* 373, 392–393 (1995).

100. See U.G. Mueller et al. *Proceedings of the National Academy of Sciences, USA* 108, 4053–4056 (2011).

101. See F. Roces *Biological Bulletin* 202, 306–313 (2002). In the less advanced *Trachymyrmex*, decision making seems to be more at the colony level; see J.N. Seal and W.R. Tschinkel *Behavioral Ecology and Sociobiology* 61, 1151–1160 (2007).

102. See I.H. Chapela et al. *Science* 266, 1691–1694 (1994). Jordal and Cognato (2012; note 47) point out that this is much the same time the ambrosia beetles started to cultivate their fungi, although in contrast to the attine ants they repeatedly reevolved this agriculture.

103. See S.E. Solomon et al. *Insectes Sociaux* 51, 333–338 (2004).

104. See H. Fernández-Marin et al. *Journal of Natural History* 39, 1735–1743 (2005).

105. Mueller (2002), p. S67; note 80.

106. Schultz and Brady (2008); note 89.

107. Schultz and Brady (2008), p. 5438; note 89.

108. See J.F. Genise et al. *Palaeogeography, Palaeoclimatology, Palaeoecology* 386, 349–363 (2013).

109. See R.R. Snelling and J.T. Longino in *Insects of Panama and Mesoamerica: Selected studies* (D. Quintero and A. Aiello, eds.), pp. 479–494 (Oxford; 1992).

110. See A.B. Munkacsi et al. *Proceedings of the Royal Society of London, B* 271, 1777–1782 (2004).

111. See P. Villesen et al. *Evolution* 58, 2252–2265 (2004).

112. See B.T.M. Dentiger et al. *Evolution* 63, 2172–2178 (2009).

113. See A.G. Himler et al. *Proceedings of the Royal Society of London B* 276, 2611–2616 (2009), and C. Rabeling et al. *PLoS ONE* 4, e6781 (2009); also commentary by D. Fournier and S. Aron *Current Biology* 19, R738–R740 (2009).

114. See, for example, D. Fournier et al. *Nature* 435, 1230–1234 (2005) (and commentary on pp. 1167–1168 by D. Queller), and K. Ohkawara *Biology Letters* 2, 359–363 (2006).

115. Himler et al. (2009), p. 2615; note 113; also K. Kellner et al. *Journal of Evolutionary Biology* 26, 1353–1362 (2013).

116. Himler et al. (2009), p. 2615; note 113.

117. While perhaps unique in its capacity to switch crops, closely related species of *Cyphomyrmex* cultivate distantly related fungi (see T.R. Schultz et al. *Insectes Sociaux* 49, 331–343 [2002]). If, however, they are presented with a novel cultivar, the colony will reject it in favor of the tried and tested (see U.G. Mueller et al. *Behavioral Ecology* 15, 357–364 [2004]), employing what Ulrich Mueller and colleagues refer to as an "acute ability" (p. 362). So, too, in this ant, if cultivars are deliberately swapped the effects are not necessarily beneficial (see N.J. Mehdiabadi et al. *Behavioral Ecology* 17, 291–296 [2006]). Evidence that cultivation is not a locked-in process also comes from other lines of evidence. Thus, in two species, again of *Cyphomyrmex*, exchange of cultivars in either direction can occur (see A.M. Green et al. *Molecular Ecology* 11, 191–195 [2002]). Abigail Green and colleagues make an important general point when they observe that their documentation of this "ant-cultivar association may not be fundamentally different from other specialized mutualisms with connections to free-living populations" (p. 194), of which the lichens (pp. 152–153) are one obvious example.

118. See M. Poulsen and J.J. Boomsma *Science* 307, 741–744 (2005).

119. See A.B.F. Ivens et al. *Behavioral Ecology* 20, 378–384 (2009).

120. Ivens et al. (2009), p. 383; note 119.

121. Ivens et al. (2009; note 119) stress that although having quite separate origins "there may be remarkable parallels between symbiont policing and hygienic reactions against pathogens" (p. 383). Despite, however, these levels of scrutiny, other evidence suggests that distinct strains (if not species) can coexist (see A.B. Abril and E.H. Bucher *Microbial Ecology* 54, 417–423 [2007]) and exchange of cultivars is possible (see A.S. Mikheyev et al. *Molecular Ecology* 16, 209–216 [2007]). These observations are complemented by experiments whereby queens of *Acromyrmex* will view by D.K. Aanen et al. *Trends in Microbiology* 17, 179–182 (2009).

53. See L.R. Kirkendall et al. in *The evolution of social behaviour in insects and arachnids* (J.C. Choe and B.J. Crespi, eds.), pp. 181–215 (Cambridge; 1997); also Haanstad and Norris (1985); note 49.

54. See P.H.W. Biedermann and M. Taborsky *Proceedings of the National Academy of Sciences, USA* 108, 17064–17069 (2011).

55. See B.K. Mable and S.P. Otto *BioEssays* 20, 453–463 (1998).

56. See B.H. Jordal et al. *Biological Journal of the Linnean Society* 71, 483–489 (2000); also K. Peer and M. Taborsky *Behavioral Ecology and Sociobiology* 61, 729–739 (2007).

57. See D.S. Kent and J.A. Simpson *Naturwissenschaften* 79, 86–87 (1992).

58. See S.M. Smith et al. *Insectes Sociaux* 56, 285–288 (2009).

59. William Foster (personal communication) reminds me that the role of haplodiploidy versus diploidy in the evolution of eusociality may be overplayed inasmuch as this syndrome will only emerge if the genetic benefits that accrue are greater than the costs. Such thinking typically revolves around inclusive fitness and kin selection, which as a general explanation has been criticized by M.A. Nowak et al. *Nature* 460, 1057–1062 (2010), provoking a general uproar (see critiques in vol. 471 E1–E9 and reply p. E9–E10; also A.F.G. Bourke *Proceedings of the Royal Society of London, B* 278, 3313–3320 [2011]).

60. See L.R. Kirkendall *Annals of the Entomological Society of America* 99, 211–217 (2006); see also vol. 99(3), p. iii.

61. Farrell et al. (2001), p. 2022; note 46.

62. See W.D. Stone et al. *Canadian Journal of Zoology* 85, 232–238 (2007); see also C. Yuceer et al. *Acta Zoologica* 92, 216–224 (2011).

63. Berryman (1989), p. 149; note 45.

64. See J. Hulcr and A.I. Cognato *Evolution* 64, 3205–3212 (2010).

65. Hulcr and Cognato (2010), p. 3206; note 64.

66. See K. Sakurai *Journal of Ethology* 3, 151–156 (1985).

67. See X.-Q. Li et al. *Journal of Insect Physiology* 58, 867–873 (2012).

68. See C. Kobayashi et al. *Evolutionary Ecology* 22, 711–722 (2008).

69. See Q.D. Wheeler *Bulletin of the American Museum of Natural History* 183, 113–120 (1986).

70. See H. Francke-Grosmann in *Symbiosis*, vol. 2 (S.M. Henry, ed.), pp. 141–205 (Academic; 1967), and A. Egger *Anzeiger für Schädlingskunde Pflanzen- und Umweltschutz* 47, 7–11 (1974).

71. See L.R. Batra and H. Francke-Grosmann *American Journal of Botany* 48, 453–456 (1961).

72. See M. Tanahashi et al. *Naturwissenschaften* 97, 311–317 (2010).

73. The distribution of exoskeletal pockets that might serve as mycangia in the beetles is reviewed by V.V. Grebennikov and R.A.B. Leschen *Entomological Science* 13, 81–98 (2010). In addition to the curculionoids (weevils), they identify four families in the sphindids as strong contenders, suggesting multiple origins of coleopteran mycangia. They caution, however, that many other proposed examples are at best inconclusive.

74. See J. Bissett and A. Borkent in *Coevolution of fungi with plants and animals* (K.A. Pirozynski and D.L. Hawkesworth, eds.), pp. 203–225 (Academic; 1988). For an example in a non-social beetle (*Doubledaya*), see papers in *PLoS ONE* by W. Toki et al. (7, e41893 [2012] and 8, e79515 [2013])

. 75 See also J.J. Heath and J.O. Stireman *Entomologia Experimentalis et Applicata* 137, 36–49 (2010), and J.B. Joy *Proceedings of the Royal Society of London, B* 280, 20122820 (2013).

76. See O. Rohfritsch *Entomologia Experimentalis et Applicata* 128, 208–216 (2008); also R.J. Adair et al. *Fungal Ecology* 2, 121–134 (2009).

77. See H. Kajimura *Annals of the Entomological Society of America* 93, 312–317 (2000).

78. See E.M. Janson et al. *Journal of Chemical Ecology* 35, 1309–1319 (2009).

79. See J.J. Kukor and M.M. Martin *Science* 220, 1161–1163 (1983).

80. For excellent overviews, see, for example, C.R. Currie *Annual Reviews in Microbiology* 55, 357–380 (2001), and U.G. Mueller *American Naturalist* 160 (Supplement), S67–S98 (2002); also Mueller et al. (2005); note 44.

81. Although adults can supplement their diet with sap (see M. Littledyke and J.M. Cherrett *Bulletin of Entomological Research* 66, 205–217 [1976], and R.J. Quinlan and J.M. Cherrett *Ecological Entomology* 4, 151–160; 1979).

82. See C.R. Currie and A.E. Stuart *Proceedings of the Royal Society of London, B* 268, 1033–1039 (2001).

83. See M. Bass and J.M. Cherrett *Functional Ecology* 10, 55–61 (1996).

84. See N.D. Boyd and M.M. Martin *Journal of Insect Physiology* 21, 1815–1820 (1975).

85. C.R. Currie et al. *Nature* 398, 701–704 (1999); also vol. 423, p. 461 (2003). The ant *Lasius* also disinfects its fungally infected larvae using an acid/hydrocarbon mix; see S. Tragust et al. *Current Biology* 23, 76–82 (2013).

86. See D.M. Mangone and C.R. Currie *Canadian Entomologist* 139, 841–849 (2007).

87. See A.A. Pinto-Tomás et al. *Science* 326, 1120–1123 (2009).

88. Given the nitrogen-poor nature of either plant or hemipteran exudates (pp. 174–175), the vast number of ants in tropical rainforests look to a variety of nitrogenous sources including bacterial symbionts (see D.W. Davidson et al. *Science* 300, 969–972 [2003], also J.W. Cooper and M.W. Westneat *BMC Evolutionary Biology* 9, e24 (2009).

11 See H. Hata and M. Kato *Marine Ecology Progress Series* 237, 227–231 (2002).

12 See, for example, D.R. Lassuy *Bulletin of Marine Science* 30, 304–312 (1980).

13 See H. Hata and M. Kato *Marine Ecology Progress Series* 263, 159–167 (2003).

14. See H. Hata and M. Kato *Journal of Experimental Marine Biology and Ecology* 313, 285–296 (2004).

15. See, respectively, H. Hata et al. *Journal of Experimental Marine Biology and Ecology* 280, 95–116 (2002), and W.L. Montgomery *Bulletin of Marine Science* 30, 290–303 (1980).

16. See H. Hata and M. Kato *Biology Letters* 2, 593–596 (2006); also H. Halà et al. *BMC Evolutionary Biology* 10, e185 (2010), who document the geographically widespread nature of this cultivation, albeit with some differences with respect to the cultivars of *Polysiphonia*.

17. See his paper in *Marine Ecology Progress Series* 124, 201–213 (1995); also M.T. Clementz *Journal of Vertebrate Paleontology* 26, 355–370 (2006).

18. Preen (1995), part of the title; note 17.

19. See C.G. Diedrich *Palaeogeography, Palaeoclimatology, Palaeoecology* 285, 287–306 (2010).

20. Diedrich (2010), p. 293; note 19.

21. Thickening of bones in aquatic vertebrates as ballast, known as pachyostosis, has evolved multiple times; see A. Houssaye *Integrative Zoology* 4, 325–340 (2009).

22. A useful overview is provided by R.N. Hughes and C.J. Gliddon *Philosophical Transactions of the Royal Society of London, B* 333, 231–239 (1991), who also touch on some of the earlier work on fish agriculture.

23. See their chapter in *Plant-animal interactions in the marine benthos* (D.M. John et al., eds.), pp. 405–423 (Clarendon; 1992).

24. Branch et al. (1992), p. 407; note 23.

25. Consider also the bacterial farms that the Yeti crab, a hydrothermal vent-dweller, maintains on its claws and appears to nourish by waving them in zones of seepage; see A.R. Thurber et al. *PLoS ONE* 6, e26243 (2011). A number of other hydrothermal animals likely maintain similar associations that can be classified as a sort of agriculture, including a limpet; see S.K. Goffredi *Environmental Microbiology Reports* 2, 479–488 (2010).

26. See S.A. Woodin *Marine Biology* 44, 39–42 (1977); also M.C. Gambi et al. *Ophelia* 53, 189–202 (2000).

27. See P. Calow *Oecologia* 16, 149–161 (1974), and L.H. Kofoed *Journal of Experimental Marine Biology and Ecology* 19, 233–241 (1975).

28. See V.M. Connor and J.F. Quinn *Science* 225, 843–844 (1984).

29. See M.R. Langer and C.A. Gehring *Journal of Foraminiferal Research* 23, 40–46 (1993).

30. Connor and Quinn (1984), p. 844; note 28.

31. See V.M. Connor *Biological Bulletin* 171, 548–564 (1986).

32. See G.M. Branch *Animal Behaviour* 27, 408–410 (1979).

33. Branch (1979), p. 408; note 32.

34. See G.M. Branch *Oceanography and Marine Biology: An Annual Review* 19, 235–380 (1981).

35. See D.R. Lindberg *Bulletin of Marine Science* 81, 219–234 (2007); also C.D. McQuail and P.W. Froneman *Oecologia* 96, 128–133 (1993).

36. See J. Stimson *Ecology* 51, 113–118 (1970).

37. See O.E.E. Plaganyi and G.M. Branch *Marine Ecology Progress Series* 194, 113–122 (2000).

38. Branch (1981), p. 367; note 34.

39. See B.R. Silliman and S.Y. Newell *Proceedings of the National Academy of Sciences, USA* 100, 15643–15648 (2003); also R.D. Sieg et al. *Journal of Experimental Marine Biology and Ecology* 446, 122–130 (2013).

40. T. Hylleberg (*Ophelia* 14, 113–137; 1975) suggests that the lugworm (*Abarenicola*) can engage in a variety of feeding modes, including gardening, whereby it alters sediment properties to encourage microbial growth that then provides a food source.

41. The same may apply to an infaunal polychaete (*Clymenella*), which appears to transport bacteria from the surface to its feeding area, possibly to serve as a garden; see F.C. Dobbs and R.D. Whitlach *Ophelia* 21, 159–166 (1982).

42. See chapters by A. Seilacher in *Trace Fossils 2* (T.P. Crimes and J.C. Harper, eds.), pp. 289–334 (Geological Journal Special Issue 9; 1977), and *Evolution of animal behaviour: Paleontological and field approaches* (M.H. Nitecki and J.A. Kitchell, eds.), pp. 62–87 (Oxford; 1986).

43. See C. Guillard and D. Olivero *Palaios* 24, 257–270 (2009).

44. See R.A. Beaver in *Insect-fungus interactions* (N. Wilding et al., eds.), pp. 121–143 (Academic; 1989); also U.G. Mueller et al. *Annual Review of Ecology, Evolution, and Systematics* 36, 563–595 (2005).

45. See A.A. Berryman in *Insect-fungus interactions* (N. Wilding et al., eds.), pp. 145–159 (Academic; 1989).

46. See B.D. Farrell et al. *Evolution* 55, 2011–2027 (2001).

47. See B.H. Jordal and A.I. Cognato *BMC Evolutionary Biology* 12, e133 (2012).

48. See, for example, H. Gebhardt et al. *Mycological Research* 109, 687–696 (2005), and J. Hulcr et al. *Symbiosis* 43, 151–159 (2007); also Farrell et al. (2001); note 46.

49. See, for example, J.O. Haanstad and D.M. Norris *Microbial Ecology* 11, 267–276 (1985).

50. See L.T. Kok et al. *Nature* 225, 661–662 (1970).

51. See Y.J. Cardoza et al. *Ecological Entomology* 31, 6636–6645 (2006).

52. See J.J. Scott et al. *Science* 322, 63 (2008); also D.-C. Oh *Organic Letters* 11, 633–636 (2009), and over-

170. See J.D. Styrsky and M.D. Eubanks *Proceedings of the Royal Society of London, B* 274, 151–164 (2006).
171. C. Nielsen et al. *Biology Letters* 6, 205–208 (2010), document ants quickly removing fungally infected cadavers. They suggest that this activity may serve to reinforce the aphid-ant mutualism in a comparable way to the attine ant-fungus symbiosis (pp. 180–185).
172. See K. Matsuura and T. Yashiro *Naturwissenschaften* 93, 506–510 (2006).
173. See also T. Eisner et al. *Science* 199, 790–794 (1978), who document a remarkable mimicry whereby the larvae of a lacewing appropriate wax scales from the aphids upon which they feed, and thereby escape the attention of the otherwise vigilant attendant ants.
174. See D.J. Lohman et al. *Ecological Entomology* 31, 41–51 (2006).
175. Lohman et al. (2006), p. 44; note 174.
176. Lohman et al. (2006), p. 47; note 174.
177. See their paper in *BMC Evolutionary Biology* 12, e106 (2012); also A.J. Pontin *Ecological Entomology* 3, 203–207 (1978).
178. Ivens et al. (2012), p. 8; note 177.
179. See U. Maschwitz and U. Hänel *Behavioral Ecology and Sociobiology* 17, 171–184 (1985).
180. See M. Fölling et al. *Journal of Natural History* 35, 279–284 (2001).
181. See J. Perry and C.A. Nalepa *Insectes Sociaux* 50, 245–247 (2003).
182. See B. Stay and A.C. Coop *Tissue and Cell* 6, 669–693 (1974).
183. See also J.B. Benoit et al. *Journal of Insect Physiology* 57, 1553–1556 (2011), who document the role of lipophorins and comment on the similarities and differences with mammalian lactation.
184. See G.M. Attardo et al. *Journal of Insect Physiology* 52, 1128–1136 (2006).
185. Stay and Coop (1974); note 182. So, too, consider the various sort of nutritive milk produced by some birds, not least with the convergent involvement of prolactin (see, for example, N.D. Horseman and J.D. Buntin *Annual Review of Nutrition* 15, 213–238 [1995]), together with the intriguing possibility of avian lactation evolving in their predecessors, the dinosaurs (see P.L. Else *Journal of Experimental Biology* 216, 347–351 [2013]).
186. See A. Williford et al. *Evolution & Development* 6, 67–77 (2004).
187. Attardo et al. (2006); note 184.
188. Attardo et al. (2006), p. 1135; note 184.
189. See, for example, L. Sreng *Journal of Chemical Ecology* 16, 2899–2912 (1990).
190. See A. Korchi et al. *FEBS Letters* 449, 125–128 (1999).
191. See L. Sreng *Journal of Morphology* 182, 279–294 (1984).
192. See, for example, L. Briand et al. *Peptides* 25, 1545–1552 (2004).
193. See my *Life's solution*, p. 142.

194. See J.U.M. Jarvis *Science* 212, 571–573 (1981).
195. See J.U.M. Jarvis and N.C. Bennett *Behavioral Ecology and Sociobiology* 33, 253–260 (1993); also H. Burda et al. *Behavioral Ecology and Sociobiology* 47, 293–303 (2000).
196. See N.G. Solomon *Trends in Ecology & Evolution* 9, 264 (1994).
197. See their paper in *Oikos* 67, 573–576 (1993).
198. Spanier et al. (1993), title of paper; note 197.
199. See, for example, J.E. Duffy *Nature* 381, 512–514 (1996), and K.S. Macdonald et al. *Diversity and Distributions* 12, 165–178 (2006).
200. See J.E. Duffy et al. *Evolution* 54, 503–516 (2000); also C.L. Morrison et al. *Molecular Phylogenetics and Evolution* 30, 563–581 (2004), and errata, vol. 31, pp. 810–813 (2004).
201. See J.E. Duffy and K.S. Macdonald *Proceedings of the Royal Society of London, B* 277, 575–584 (2010).
202. Duffy and Macdonald (2010), p. 582; note 201.
203. See J.E. Duffy et al. *Behavioral Ecology and Sociobiology* 51, 488–495 (2002).
204. Duffy et al. (2002), p. 493; note 203.
205. See E. Toth and J.E. Duffy *Biology Letters* 1, 49–52 (2005).
206. See A. Anker et al. *Evolution* 60, 2507–2528 (2006).
207. Spanier et al. (1993); note 197.
208. See R. Pellens et al. *Molecular Phylogenetics and Evolution* 43, 616–626 (2007).

CHAPTER 16

1. See D.A. Brock et al. *Nature* 469, 393–396 (2011); also commentary on pp. 308–309 by J.J. Boomsma.
2. One should also draw attention to the sand-dwelling mouthless ciliate *Kentrophoros*, which is extremely flattened and on its "ventral" side maintains a garden of chemolithotrophic bacteria that it phagocytoses; see T. Fenchel and B.J. Bland *Ophelia* 30, 75–93 (1989).
3. Brock et al. (2011), p. 394; note 1.
4. Brock et al. (2011), p. 393 [Abstract]; note 1.
5. Agriculture usually implies food production, but humans also engage in "useless" cultivation, notably flowers. They are not alone because bowerbirds cultivate a *Solanum*. The plant benefits by the relative fertility of the soil and open aspect, while the fruit produced are greener and serve as decoration; see J.R. Madden et al. *Current Biology* 22, R264–R265 (2012).
6. See M. Pion et al. *Proceedings of the Royal Society of London, B* 280, 20132242 (2013).
7. See B. Frédérich et al. *Journal of Morphology* 269, 175–188 (2008).
8. See B. Frédérich et al. *American Naturalist* 181, 94–113 (2013).
9. Frédérich et al. (2013), p. 105; note 8.
10. See C.E.L. Ferreira et al. *Journal of Experimental Marine Biology and Ecology* 229, 241–264 (1998); see

for example, have an endosymbiotic association, but this time in association with gut (and ovary) tissue (see C. Ratzka et al. *Journal of Insect Physiology* 59, 611–623 [2013]).

145. S. Shigenobu and D.L. Stern *Proceedings of the Royal Society of London, B* 280, 20121952 (2013).

146. Shigenobu and Stern (2013), p. 5; note 145.

147. See L.A. Weinert et al. *Biology Letters* 3, 678–681 (2007).

148. This has evolved at least twice in the hemipterans, and in one case involved horizontal gene transfer from a bacterium to the endosymbiotic *Portiera* of whitefly (see D.B. Sloan and N.A. Moran *Biology Letters* 8, 986–989 [2012]), and the other acquisition by the aphid genome of a fungal gene (see E. Nováková and N.A. Moran *Molecular Biology and Evolution* 29, 313–323 [2012]), but this in turn finds a further convergence with a spider-mite (see B. Altincicek et al. *Biology Letters* 8, 253–257 [2012]).

149. These xylophagous insects are of more than arcane interest, because they are very serious agricultural pests acting as vectors for bacterial pathogens (see R.A. Redak et al. *Annual Review of Entomology* 49, 243–270 [2004]).

150. See J.A. Raven *Advances in Ecological Research* 13, 135–234 (1983).

151. For the most part the context of these symbioses is in the form of intimate endocellular associations located in the so-called bacteriome. Rather astonishingly, given the obvious differences of environment, the γ-proteobacteria *Ishikawaella* that inhabits the midgut of stinkbugs not only is an essential symbiont but shows strikingly convergent features, including reduction in genome size, with its endocellular counterparts; see N. Nikoh et al. *Genome Biology and Evolution* 3, 702–714 (2011).

152. See F. Husník et al. *BMC Biology* 9, e87 (2011), also commentary (e91) by H. Philippe and B. Roure. Mutualisms involving various proteobacteria extend far beyond these insects, and such associations have evolved more than thirty times with an apparent preference to derive from a parasitic ancestor; see J.L. Sachs et al. *Proceedings of the Royal Society of London, B* 281, 20132146 (2013).

153. *Sulcia* is a member of the so-called Bacteroides, which are probably closer than you realize should you have any dental plaque (see, for example, S.S. Socransky et al. *Journal of Clinical Periodontology* 25, 134–144 [1998]).

154. This is a very ancient association (see D.M. Takiya et al. *Molecular Ecology* 15, 4175–4191 [2006]).

155. See J.P. McCutcheon and N.A. Moran *Proceedings of the National Academy of Sciences, USA* 104, 19392–19397 (2007).

156. See J.P. McCutcheon et al. *Proceedings of the National Academy of Sciences, USA* 106, 15394–15399 (2009*a*).

157. See J.P. McCutcheon et al. *PLoS Genetics* 5, e1000565 (2009*b*).

158. The convergences are not precise—for example, with respect to vitamin biosynthesis. However, as John McCutcheon and colleagues (2009*a*; note 156) conclude, "a comparison of [*Hodgkinia*'s] amino acid biosynthetic capabilities with *Baumannia*'s reveals a remarkable case of convergent evolution, especially considering the vast differences between the two genomic architectures" (p. 15398).

159. See J.P. McCutcheon and N.A. Moran *Genome Biology and Evolution* 2, 708–718 (2010).

160. McCutcheon and Moran (2010), p. 716; note 159.

161. See J.P. McCutcheon and C.D. von Dohlen *Current Biology* 21, 1366–1372 (2011).

162. Mealybugs are sap-feeders, as are the aphids and psyllids. To survive on an almost pure sugar diet lacking many key compounds demands much the same solution. Best known is the employment of bacteria such as *Buchnera* (see E.C.H.J. van Ham et al., *Proceedings of the National Academy of Sciences, USA* 100, 581–586 [2003]) and *Carsonella* (which like *Baumannia* belong to the γ-proteobacteria), and here endosymbiotic associations have evolved independently many times (see J.T. Herbeck et al. *Molecular Biology and Evolution* 22, 520–532 [2005]; also P.H. Degnan et al. *Genome Research* 15, 1023–1033; 2005). So, too, these bacteria show striking and parallel reductions in genome size (see, for example, A. Nakabachi et al. *Science* 314, 267 [2006], and papers in *Molecular Biology and Evolution* by D.B. Sloan and N.A. Moran (vol. 29, pp. 3781–3792; (2012) and D.B. Sloan et al. (vol. 31, pp. 857–871; 2014), yielding a streamlined genetic apparatus that has thrown to the winds a whole suite of genes that in any other context would be regarded as essential. A further twist is found in a psyllid that employs *Carsonella* and another endosymbiont (again with a greatly reduced genome), but the latter also produces polyketide toxins that may serve in defense; see A. Nakabachi et al. *Current Biology* 23, 1478–1484 (2013); also commentary on pp. R657–R658 by J.P. McCutcheon.

163. See C.D. von Dohlen et al. *Nature* 412, 433–436 (2001); also F. Husnik et al. *Cell* 153, 1567–1578 (2013).

164. McCutcheon and Dohlen (2011), p. 1366 [Abstract]; note 161.

165. Not only is this degree of intricacy remarkable, but as Patrick Keeling (see his commentary in *Current Biology* 21, R623–R624 [2011]) points out, the minute genome size of *Tremblaya* begs the question of is it "an endosymbiont or an organelle?" (p. R624).

166. See P.J. Gullan and M. Kosztarab *Annual Review of Entomology* 42, 23–50 (1997).

167. See B. Stadler and A.F.G. Dixon *Annual Review of Ecology, Evolution, and Systematics* 36, 345–372 (2005).

168. See, for example, A.W. Shingleton and D.L. Stern *Molecular Phylogenetics and Evolution* 26, 26–35 (2003).

169. See A.W. Shingleton et al. *Evolution* 59, 921–926 (2005).

176), the remarkable case of the subterranean mole rats (see my *Life's solution*, pp. 141–143), and perhaps most extraordinarily the trematode parasites; see R.F. Hechinger et al. *Proceedings of the Royal Society of London, B* 278, 656–665 (2011), with commentary in *Current Biology* (vol. 20, pp. R985–R987; 2010) by P. Newey and L. Keller; also O. Miura *Marine Ecology Progress Series* 465, 119–127 (2012).

112. See their paper in *Biology Letters* 3, 331–335 (2007).

113. See, for example, B.J. Crespi *Nature* 359, 724–726 (1992), and L.A. Mound *Annual Review of Entomology* 50, 247–269 (2005).

114. See D.S. Kent and J.A. Simpson *Naturwissenschaften* 79, 86–87 (1992).

115. See his article in *Trends in Ecology & Evolution* 17, 199–200 (2002).

116. Foster (2002), p. 199; note 115.

117. See, for example, papers by D.L. Stern and W.A. Foster in *Biological Reviews* 71, 27–79 (1996), and in *The evolution of social behavior in insects and arachnids* (J.C. Choe and B.J. Crespi, eds.), pp. 150–165 (Cambridge; 1997).

118. See D.L. Stern *Evolution* 52, 155–165 (1998).

119. See D.L. Stern *Proceedings of the Royal Society of London, B,* 256, 203–209 (1994); also P.K. Rhoden and W.A. Foster *Insectes Sociaux* 49, 257–263 (2002).

120. As, of course, they have in other eusocial groups such as termites (pp. 185–177) and perhaps more surprisingly in the stingless bees; see C. Grüter et al. *Proceedings of the National Academy of Sciences, USA* 109, 1182–1186 (2012).

121. Not that all eusocial animals are reclusive; an Asian honeybee, for example, builds its comb out in the open. As M.S. Sarma et al. (in *Naturwissenschaften* 87, 241–243; 2000) explain, the worker bees can arrange themselves in a tilelike fashion to keep off the rain (as indeed do honeybee swarms in inclement weather; see S.M. Cully and T.D. Seeley *Insectes Sociaux* 51, 317–324; 2004). They also have a novel head-pushing method of levering debris such as leaves and sticks off the comb; some workers are apparently "particularly eager and persistent head-pushers" (p. 242) with certain individuals verging on the heroic.

122. See D.L. Stern and W.A. Foster (1997; note 117).

123. Indeed this asexuality has evolved multiple times (see F. Delmotte et al. *Proceedings of the Royal Society of London, B* 268, 2291–2299 [2001]).

124. See, for example, T.W. Chapman et al. *Behavioral Ecology* 13, 519–525 (2002), and M.J. McLeish and T.W. Chapman *Australian Journal of Entomology* 46, 300–304 (2007).

125. See, for example, L.G. Cook and P.J. Gullan *Biological Journal of the Linnean Society* 83, 441–452 (2004).

126. See P. Abbot et al. *Proceedings of the National Academy of Sciences, USA* 98, 12068–12071 (2001).

127. The title of the paper; see note 115.

128. Not only have soldiers evolved multiple times

among the eusocial insects, but more remarkably so, too, in clonal colonies of trematode parasites; see R.F. Hechinger et al. (note 111). These workers remark, "The division of labour recognized and described here for trematodes has strong parallels to other social systems with a soldier caste," notably with gall thrips, termites, and snapping shrimps, so that "All of these systems . . . have traits that fit the 'fortress-defence' model of sociality" (p. 663; both quotations).

129. See N. Pike and W. Foster *Animal Behaviour* 67, 909–914 (2004).

130. See M. Kutsukake et al. *Proceedings of the Royal Society of London, B* 276, 1555–1563 (2009).

131. See U. Kurosu et al. *Proceedings of the Royal Society of London, B* 270 (Supplement), S12–S14 (2003).

132. See T.G. Benton and W.A. Foster *Proceedings of the Royal Society of London, B* 247, 199–202 (1992).

133. See N. Pike et al. *Proceedings of the Royal Society of London, B* 269, 1211–1215 (2002).

134. See J.A. Pickett and D.C. Griffiths *Journal of Chemical Ecology* 6, 349–360 (1980); also A.F.G. Dixon *Aphid ecology: An optimization approach*, 2nd ed. (Chapman & Hall; 1998).

135. Kutsukake et al. (2009), p. 1562; note 130.

136. See A.F. Sanborn et al. *Biological Journal of the Linnean Society* 83, 281–288 (2004).

137. Dolling, W.R. *The Hemiptera* (Oxford; 1991).

138. See, for example, M. Uzest *Arthropod Structure & Development* 39, 221–229 (2010).

139. These include not only a series of convoluted pathways as the stylet seeks out the sap (see W.F. Tjallingii and T.H. Hogen Esch *Physiological Entomology* 18, 317–328 [1993]), but the production of two types of saliva, one of which is watery and helps to draw in the food while the other quickly hardens and serves to form an effective sheath, presumably to prevent leakage (see P.W. Miles *Nature* 183, 756 [1959]; also overview by D.G. Pollard *Bulletin of Entomological Research* 62, 631–714 [1973]).

140. This comparison is not entirely fanciful. While the great majority of hemipterans are sap or xylem suckers, in the bedbugs (which have independently evolved infrared detection, p. 117) the stylets seek for blood, and the intolerable itch of the weary traveler is a result of the reaction against the anticoagulants.

141. See T. Will et al. *Proceedings of the National Academy of Sciences, USA* 104, 10536–10541 (2007).

142. This is apparently by scavenging calcium that would trigger the response that would otherwise quickly shut down the punctured area and so prevent further leakage of the sap.

143. See R.B. Cocroft *Proceedings of the Royal Society of London, B* 272, 1023–1029 (2005).

144. See A.E. Douglas *Journal of Experimental Botany* 57, 747–754 (2006). Other bacterial symbionts can confer other advantages to the aphid, such as protection against parasitic wasps (see K.M. Oliver et al. *Proceedings of the National Academy of Sciences, USA* 100, 1803–1807 [2003]). Nor are they alone; ants,

73. Paradoxically not in the time of the dinosaurs, when in places the landscape would have been a scatological nightmare. A. Arillo and V.M. Ortuño (*Journal of Natural History* 42, 1405–1408 [2008]) explain how the cloacal mixing of feces and urea in these reptiles would make the highly nitrogenous dung much less attractive to the dung-beetles, a point supported by their lack of interest in the heaps of guano, including that produced by the cave-dwelling oilbirds (p. 139–140). On the other hand, these do receive the attention of cockroaches (see W.J. Bell et al. *Cockroaches: Ecology, behavior and natural history* (Johns Hopkins [2007]), and this chimes with some evidence that their Mesozoic counterparts tackled dinosaur dung (see P. Vršanský et al. *PLoS ONE* 8, e80560 [2013]).

74. See M. Dacke et al. *Current Biology* 23, 298–300 (2013); also commentary on pp. R149–R150 by J.L. Gould.

75. Bartholomew and Heinrich (1978), p. 81; note 71.

76. For an overview, see C. Pérez-Íñigo *Revista de Entomologos Ibéricos: Graellsia* 27, 161–176 (1971).

77. See P. Holter *European Journal of Entomology* 101, 365–372 (2004).

78. See, for example, T.H. Larsen et al. (*The Coleopterists' Bulletin* 60, 315–324; 2006), who document a series of striking specializations among Peruvian dung beetles, including those that inhabit attine nests (p. 182) and even prey on millipedes.

79. See A.J. Davis et al. in *Canopy arthropods* (N.E. Stork et al., eds.), pp. 417–432 (Chapman & Hall; 1997); also H. Wirta et al. *Molecular Phylogenetics and Evolution* 57, 710–727 (2010), who document "the *Arachnodes* clade . . . in which species have shifted (multiple times) to arboreal foraging" (p. 718).

80. See A.F. Sanborn *Denisia* 4, 455–470 (2002).

81. See A.F. Sanborn *Naturwissenschaften* 90, 305–308 (2003).

82. Sanborn (2003), part of the title; note 81.

83. With respect to the muscles connected to sound production in the insects, Barbara Block (see her paper in *News in the Physiological Sciences* 2, 208–213 [1987]) notes how these "tymbal muscles have a remarkable similarity to the heater cells [of the billfish, p. 80]" (p. 212) when it comes to such factors as concentration of mitochondria and loss of myofibrils.

84. This capacity even extends to ectotherms, because in the caterpillar of the gypsy moth, which engages in thermoregulation, its covering of setae assists with insulation; see T.M. Casey and J.R. Hegel et al. *Science* 214, 1131–1133 (1981).

85. See N.S. Church *Journal of Experimental Biology* 37, 186–212 (1960).

86. See T. Seeley and B. Heinrich in *Insect thermoregulation* (B. Heinrich, ed.), pp. 159–234 (Wiley; 1981).

87. See M.J. Kluger *Fever: Its biology, evolution, and function* (Princeton; 1979).

88. See P.T. Starks et al. *Naturwissenschaften* 87, 229–231 (2000); also V.L. Hunt et al. (*Journal of Thermal Biology* 36, 443–451; 2011), who describe an environmentally induced fever in locusts against fungal (*Metarhizium*) infection.

89. Starks et al. (2000), p. 229; note 88.

90. See A. Stabentheiner et al. *Ethology* 113, 995–1006 (2007).

91. See M. Ono et al. *Nature* 377, 334–336 (1995).

92. See M. Sugahara and F. Sakamoto *Naturwissenschaften* 96, 1133–1136 (2009).

93. In what is presumably an independent example from Cyprus, the bees mob the hornet, again engulfing it in a ball that asphyxiates the unwelcome guest (see A. Papachristoforou et al. *Current Biology* 18, R795–R796 [2007]).

94. See their paper in *Naturwissenschaften* 97, 319–323 (2010).

95. Greco et al. (2010), part of the title; note 94.

96. Greco et al. (2010), p. 322; note 94.

97. See S.H. Woodard et al. *Proceedings of the National Academy of Sciences, USA* 108, 7472–7477 (2011).

98. Woodard et al. (2011), p. 7475; note 97.

99. Woodard et al. (2011), p. 7476; note 97.

100. See their paper in *Naturwissenschaften* 95, 85–90 (2008).

101. Perrichot et al. (2008), p. 89 (emphasis added); note 100.

102. See, for example, B.N. Danforth *Proceedings of the National Academy of Sciences, USA* 99, 286–290 (2002); B.N. Danforth et al. *Systematic Biology* 52, 23–36 (2003); and M. P. Schwarz et al. *Annual Review of Entomology* 52, 127–150 (2007).

103. See S.G. Brady et al. *Proceedings of the Royal Society of London, B*, 273, 1643–1649 (2006). J. Gibbs et al. (*Molecular Phylogenetics and Evolution* 65, 926–939; 2012), however, propose an earlier origination time and also fewer acquisitions of eusociality.

104. See H.M. Hines et al. *Proceedings of the National Academy of Sciences, USA* 104, 3295–3299 (2007).

105. See, for example, S. Dreir et al. *Biology Letters* 3, 459–462 (2007).

106. See T. Wenseleers and F.L.W. Ratnieks *Nature* 444, 50 (2006).

107. See, for example, J.S. van Zweden et al. *Proceedings of the Royal Society of London, B* 274, 1421–1428 (2007); also commentary by A.F.G. Bourke in *Current Biology* 17, R519–R520 (2007).

108. But oddly often a failure to prefer one's nearest and dearest, odd because a key element in the discussion of the origins of eusociality has revolved around genetic relatedness.

109. See J.C. Jones and B.P. Oldroyd *Advances in Insect Physiology* 33, 153–191 (2007).

110. See, for example, instances in bees (A. Stow et al. *Biology Letters* 3, 422–424; 2007) and ants (M. Chapuisat et al. *Proceedings of the Royal Society of London, B* 274, 2013–2017 [2007], and G. Castella et al. *Animal Behaviour* 75, 1591–1596; [2008]).

111. And not only among the insects, because in the arthropods there are the synalpheid shrimps (pp. 175–

pp. 1–146 (CABI Publishing; 2001); also W.E. Holznagel et al. *Molecular Phylogenetics and Evolution* 57, 1017–1025 (2010).

42. See R. Diesel *Animal Behaviour* 38, 561–575 (1989).

43. See R. Diesel and M. Schuh *Behavioral Ecology and Sociobiology* 32, 11–15 (1993).

44. See R. Diesel *Proceedings of the Royal Society of London, B* 264, 1403–1406 (1997).

45. Diesel (1989), p. 573; note 42.

46. This is with abortive eggs (see P. Wegoldt *Behavioral Ecology and Sociobiology* 7, 329–332 [1980]), a strategy that has evolved in many other groups of animals (see J.C. Perry and B.D. Roitberg *Oikos* 112, 706–714 [2006]).

47. See J. Bayliss *African Journal of Ecology* 40, 26–34 (2002).

48. See also N. Cumberlidge et al. *Journal of Crustacean Biology* 25, 302–308 (2005).

49. Particularly striking are the inhabitants of mangroves (p. 390, note 11), some of which spend much of their time in the canopy, largely subsisting on fresh leaves (see, for example, A.A. Erickson et al. *Journal of Experimental Marine Biology and Ecology* 289, 123–138 [2003]; also S.M. Linton and P. Greenaway *Journal of Comparative Physiology, B* 177, 269–286; 2007). In coasts around the tropics similar crabs pursue this existence. Although it had been assumed they were related, Sara Fratini and colleagues (see their paper in *Evolutionary Ecology Research* 7, 219–233 [2005]) note that this "very specialized arboreal lifestyle evolved several times . . . providing another striking example of the likelihood of convergence in evolutionary biology and the degree of phenetic and ecological potential to be found among marine organisms" (p. 219).

50. See G. Scholtz *Contributions to Zoology* 83, 87–105 (2014); also C.L. Morrison et al. *Proceedings of the Royal Society of London, B* 269, 345–350 (2002) and L.M. Tsang et al. *Systematic Biology* 60, 616–629 (2011).

51. Morrison et al. (2002), p. 345; note 50.

52. J. Dzik (*Journal of Morphology* 269, 1501–1519; 2008) suggests a relationship to the branchiurans.

53. See F.R. Schram et al. *Journal of Paleontology* 71, 261–284 (1997).

54. Schram et al. (1997), p. 279; note 53.

55. The degree of geological overlap of cycloids and crabs is, however, more considerable than once realized (it now transpires that cycloids survived into the Cretaceous), thus making arguments for preemption less compelling; see R.H.B. Fraaije et al. *Journal of Paleontology* 77, 386–388 (2003). However, as we saw with the turtles (pp. 26–27) less effective forms can survive for very long periods, and even if the emergence of a crab-form was an inevitability (as their convergence strongly suggests), this still needs time.

56. For an assessment of this popular hypothesis, see R. Okajima *Lethaia* 41, 423–430 (2008).

57. See E. Marais et al. *Journal of Experimental Biology* 208, 4495–4507 (2005); in addition, discontinuous

gas exchange has evolved independently in several other groups of arthropod, including the centipedes (see, for example, C.J. Klok et al. *Journal of Experimental Biology* 205, 1019–1029 [2002]).

58. See their paper in *Science* 299, 558–560 (2003).

59. Westneat et al. (2003), p. 558; note 58.

60. See M. Locke *Journal of Insect Physiology* 44, 1–20 (1998); also commentary in *Nature* (vol. 391, pp. 129–130; 1998) by P.J. Mill.

61. Such lunglike systems are not only widespread in the lepidopterans, but appear to have evolved independently in the argid hymenopterans (see Y.-P. Li et al. *Acta Entomologica Sinica* 53, 110–117 [2010]; in Chinese, with English abstract).

62. See S. Powell and N.R. Franks *Animal Behaviour* 73, 1067–1076 (2007).

63. Incidentally in the army ants (see D.J.C. Kronauer et al. *Evolution* 61, 413–422 [2007]) this has evolved convergently "under strikingly similar selective pressures" (p. 413).

64. See, for example, J.C. Jones et al. *Science* 305, 402–404 (2004), who point out the more genetically diverse colonies are more adept at thermoregulation of the nest on account of their capacity to handle a spectrum of environmental shifts.

65. See, for example, G.A. Bartholomew in *Insect thermoregulation* (B. Heinrich, ed.), pp. 46–78 (Wiley; 1981); also B. Heinrich *Science* 185, 747–756 (1974).

66. See G.A. Bartholomew and J.R.B. Lighton *Journal of Comparative Physiology, B* 156, 461–467 (1986).

67. See B. Heinrich and C. Pantle *Journal of Experimental Biology* 62, 599–610 (1975).

68. See K.R. Morgan and B. Heinrich *Journal of Experimental Biology* 133, 59–71 (1987).

69. See G.A. Bartholomew and T.M. Casey *Science* 195, 882–883 (1977).

70. See R.S. Seymour et al. *Journal of Experimental Biology* 212, 2960–2968 (2009).

71. See G.A. Bartholomew and B. Heinrich *Journal of Experimental Biology* 73, 65–83 (1978).

72. Gerhard Scholtz (*Contributions to Zoology* 77, 139–148; 2008) wonders if the inspiration for the wheel came from the ancients observing these industrious beetles. Of course, the reason that nature never invented the wheel, other than in the convergent evolution of the bacterial flagellum is an old canard that I review in my *Life's solution* (p. 112). It is only appropriate to draw attention to wheeling spiders in Namibia (see J.R. Henschel *South African Journal of Science* 86, 151–152 [1990], caterpillars freewheeling in reverse (see J. Brackenbury *Journal of Insect Physiology* 48, 525–533 [1999]), and even among the amphibians in the form of rolling toads (see R.W. McDiarmid and S. Gorzula *Copeia* 1989, 445–451 [1989]) and salamanders (see M. García-París and S.M. Deban *Journal of Herpetology* 29, 149–151 [1995]). In all these cases this tactic is not an expression of joie de vivre, but a disinclination to be turned into lunch.

Colmer (*Journal of Experimental Biology* 215, 705–709; 2012) note, "Two kingdoms, one solution" (p. 707; subtitle), because in addition to insects and spiders, independently some aquatic plants have arrived at the same solution.

3. See, for example, W.A. Foster *Proceedings of the Third European Congress of Entomology* (H.H.W. Velthuis, ed.), pp. 209–216 (1986). He points out that in terms of intertidal existence, insects show an impressive range of adaptations, but when it comes to a wholly marine existence, insects would have to throw away nearly all that allowed their triumphant excursion in the opposite direction.

4. See G.D. Ruxton and S. Humphries *Marine Ecology* 29, 72–75 (2008).

5. See N.M. Andersen and L. Cheng *Oceanography and Marine Biology: An Annual Review* 42, 119–179 (2005).

6. See N.M. Anderson *Ecography* 22, 98–111 (1999).

7. See R.M. May *Philosophical Transactions of the Royal Society of London, B* 343, 105–111 (1994).

8. See G.J. Vermeij and R.K. Grosberg *Integrative and Comparative Biology* 50, 675–682 (2010).

9. See M.R. Warburg *Advances in Ecological Research* 17, 187–242 (1987).

10. See I. Tabacaru and D.L. Danielopol *Vie et Milieu* 46, 171–181 (1996; in French, with English abstract).

11. See S. Harzsch et al. *Arthropod Structure & Development* 40, 244–257 (2011).

12. See K.E. Lisenmair in *Evolutionary ecology of social and sexual systems: Crustaceans as model organisms* (J.E. Duffy and M. Thiel, eds.), pp. 339–364 (Oxford; 2007). This author also points out that this surprising degree of social complexity finds a close parallel in an insect, the Namibian desert beetle (*Parastizopus*).

13. See C. Schmidt and J.W. Wägele *Acta Zoologica* 82, 315–330 (2001).

14. See B. Hoese *Zoologische Jahrbücher Abteilung für Anatomie und Ontogenie der Tiere* 107, 396–422 (1982).

15. See P. Paoli et al. *Journal of Morphology* 253, 272–289 (2002).

16. See, for example, C. Bromhall *Tissue and Cell* 19, 793–807 (1987).

17. See A.M. Höfer et al. *Arthropod Structure & Development* 39, 13–21 (2000); they point out that in the case of harvestmen the similarities are significantly greater with the trachea of insects than with the spiders.

18. See, for example, J.A. Dunlop and M. Webster *Journal of Arachnology* 27, 86–93 (1999). Their suggestion, however, that the diagnostic book-lungs of scorpions and spiders are convergent is not, however, supported by more recent work; see, for example, C. Kámenez et al. *Biology Letters* 4, 212–215 (2008).

19. See K. Anger *Journal of Experimental Marine Biology and Ecology* 193, 119–145 (1995); also R. Diesel *Journal of Zoology* 250, 141–160 (2000). F. Giomi et al. (*Proceedings of the Royal Society of London, B* 281,

20132927; 2014) reiterate the degree of convergence and suggest the driving force toward air breathing is linked to temperature.

20. See D.P. Maitland *Journal of Crustacean Biology* 22, 497–501 (2002).

21. See G. Paulay and J. Starmer *PLoS ONE* 6, e19916 (2011).

22. See their paper in *Ecology Letters* 6, 812–817 (2003).

23. O'Dowd et al. (2003), part of title; note 22.

24. Paulay and Starmer (2011), p. 15; note 21.

25. Paulay and Starmer (2011), p. 16; note 21.

26. See V. Storch and U. Welsch *Zoologischer Anzeiger* 212, 73–84 (1984), who draw attention to similarities with vertebrate lungs in terms of the thinness of the blood-air barrier, and also secretory glands.

27. See C.A. Farrelly and P. Greenaway *Arthropod Structure & Development* 34, 63–87 (2005).

28. See P. Greenaway et al. *Journal of Experimental Biology* 140, 493–509 (1988).

29. Importantly the latter process involves carbonic anhydrase (see R.P. Henry *Journal of Experimental Zoology* 259, 294–303 [1991]; also S. Morris and P. Greenaway *Journal of Comparative Physiology B*, 160, 217–221; 1990).

30. See P. Greenaway *Memoirs of Museum Victoria* 60, 13–26 (2003).

31. See C.A. Farrelly and P. Greenaway *Zoomorphology* 112, 39–49 (1992).

32. See C.A. Farrelly and P. Greenaway *Journal of Experimental Biology* 187, 113–130 (1994).

33. Here we see two basic arrangements, one of which involves a complex and very effective portal system (see P. Greenaway and C.A. Farrelly *Physiological Zoology* 63, 117–139 [1990]). As is the way of the world, although many terrestrial crabs use gills and lungs, others have adopted novel solutions. Consider some shore-dwelling ocypodids that employ thin, windowlike membranes located on their legs (see D.P. Maitland *Nature* 319, 493–495 [1986]). Although these look strikingly similar to the tympanic membranes employed by many insects for hearing (p. 126), the associated vascularization points to oxygen exchange.

34. See E.W. Taylor and A.J. Innes *Biological Journal of the Linnean Society* 34, 229–247 (1988).

35. Taylor and Innes (1988), p. 242; note 34.

36. Taylor and Innes (1988), p. 243; note 34.

37. Taylor and Innes (1988), p. 243; note 34.

38. See A.J. Innes et al. *Comparative Biochemistry and Physiology A* 87, 1–8 (1987).

39. Innes et al. (1987), p. 5; note 38.

40. Innes et al. (1987), part of the subtitle; note 38.

41. While the focus is largely on arthropods and vertebrates, let us not forget the mollusks and notably the pulmonate snails. These actually occupy a wide variety of habitats, but terrestriality has probably occurred a number of times. For an overview, see G.M. Barker in *The biology of terrestrial molluscs* (G.M. Barker, ed.),

NOTES 397

et al. *Current Biology* 20, 1846–1852 (2010), and commentary on pp. R891–R893 by R. Benton.

171. See M.C. Stensmyr et al. *Nature* 420, 625–626 (2002).

172. See P. Marino et al. *Symbiosis* 47, 61–76 (2009).

173. See S.D. Johnson and A. Jürgens *South African Journal of Botany* 76, 796–807 (2010).

174. Davis et al. (2008), p. 54; note 165.

175. See S. Lev-Yadun et al. *BioEssays* 31, 84–88 (2009).

176. On the other hand, in the case of a cytinid parasitic plant, the volatiles it produces evidently serve to attract elephant shrews; see S.D. Johnson et al. *Proceedings of the Royal Society of London, B* 278, 2303–2310 (2011). As these workers note, one of the ketones in question "is also known from some bat-pollinated species, suggesting independent evolution of plant signals in derived, highly specialized mammal-pollination systems" (p. 2303 [Abstract]).

177. See, for example, R.S. Seymour and P. Schultze-Motel *Endeavour* 21, 125–129 (1997). Although the account given in this chapter explores the innate thermogenesis of plants that employ particular molecular mechanisms, differential heating can arise by solar absorption (see, for example, Y. Sapir et al. *Oecologia* 147, 53–59 [2006]). More remarkable is heat generated by yeasts growing on the nectar sugars (see C.M. Herrera and M.I. Pozo *Proceedings of the Royal Society of London, B* 277, 1827–1834 [2010]), which also provide the alcohol for tree shrews.

178. See his paper in *Scientific American* 276 (3, March), 104–109 (1997).

179. Seymour (1997), p. 104; note 178.

180. See their paper in *Thermochimica Acta* 391, 107–118 (2002).

181. Lamprecht et al. (2002), part of the title; note 180.

182. See R.S. Seymour *Plant, Cell and Environment* 27, 1014–1022 (2004).

183. See W. Barthlott et al. *Plant Biology* 11, 499–505 (2009).

184. See their paper in *Proceedings of the Royal Society of London, B* 271, S13–S15 (2004). With respect to the dragon lily, see R.S. Seymour and P. Schultze-Motel vol. 266, pp. 1975–1983 (1999).

185. Angioy et al. (2004), p. S13; note 184.

186. See R.S. Seymour et al. *Nature* 426, 243–244 (2003).

187. Some early-flowering plants have convergently developed a very dense covering of trichomes to confer a downy insulation to the buds; see H. Tsukaya and T. Tsuge *Plant Biology* 3, 536–543 (2001).

188. See R.S. Seymour et al. *Biology Letters* 5, 568–570 (2009).

189. See also I. Silberbauer-Gottsberger et al. *Dissertationes Botanicae* 346, 165–183 (2001).

190. See R.S. Seymour et al. *Functional Ecology* 18, 925–930 (2004), and J. Terry et al. *Plant Systematics and Evolution* 243, 233–247 (2004).

191. See L.B. Thien et al. *American Journal of Botany* 96, 166–182 (2009).

192. See R.S. Seymour and P.G.D. Matthews *Annals of Botany* 98, 1129–1135 (2006).

193. See R.S. Seymour and P. Schultze-Motel *Nature* 383, 305 (1996).

194. Seymour and Schultze-Motel (1996), p. 305; note 193.

195. See, for example, L.A. Núñez *Plant Systematics and Evolution* 254, 149–171 (2005).

196. See S. Patiño et al. *New Phytologist* 154, 429–437 (2002).

197. See S. Patiño et al. *Oecologia* 124, 149–155 (2000).

198. See R.S. Seymour et al. *Annals of Botany* 104, 823–832 (2009).

199. See R.S. Seymour *Bioscience Reports* 21, 223–236 (2001).

200. See K. Ito et al. *Journal of Experimental Botany* 54, 1113–1114 (2003).

201. See A.E. Vercesi et al. *Annual Review of Plant Biology* 57, 383–404 (2006).

202. See L.J. Sweetlove et al. *Proceedings of the National Academy of Sciences, USA* 103, 19587–19592 (2006).

203. See K. Ito and R.S. Seymour *Biology Letters* 1, 427–430 (2005).

204. This enzyme may be related to haemerythrin; see A.S. Albury et al. *Physiologia Planatarum* 137, 316–327; 2009.

205. See J.R. Watling et al. *Plant Signalling & Behavior* 3, 595–597 (2008); also A.M. Wagner et al. *Biochimica et Biophysica Acta: Bioenergetics* 1777, 993–1000 (2008), and R.E. Miller et al. *New Phytologist* 189, 1013–1026 (2011).

206. See, for example, N.M. Grant et al. *Journal of Experimental Botany* 59, 705–714 (2008), who document the role of AOX in the sacred lotus, but also provide evidence for likely posttranslational activity.

207. See Y. Onda et al. *Plant Physiology* 146, 636–645 (2008); also Y. Ito-Inaba et al. *Journal of Experimental Botany* 60, 3909–3922 (2009).

208. See Y. Ito-Inaba *Planta* 231, 121–130 (2009), and Y. Ito-Inaba et al. *Plant, Cell and Environment* 35, 544–566 (2012).

CHAPTER 15

1. See G.J. Vermeij and R. Dudley *Biological Journal of the Linnean Society* 70, 541–554 (2000).

2. In this context consider those organisms that have successfully developed a plastronlike structure where a superhydrophobic surface allows a thin film of air to stick to the body and permit gas diffusion from surrounding water. As O. Pedersen and T.D.

138. See H.E. Neuhaus and M.J. Emes *Annual Review of Plant Physiology and Plant Molecular Biology* 51, 111–140 (2000).

139. For an overview, see K. Krause in *Organelle genetics* (C.E. Bullerwell, ed.), pp. 79–103 (Springer; 2012).

140. In other parasitic groups, such as members of the Orobanchaceae, these pathways are retained; see N.J. Wickett et al. *Current Biology* 21, 2098–2104 (2012). In this group, parasitism per se has a single origin, but holoparasitism has evolved three times; see J.R. McNeal et al. *American Journal of Botany* 100, 971–983 (2013).

141. See K.H. Wolfe et al. *Proceedings of the National Academy of Sciences, USA* 89, 10648–10652 (1992).

142. The plastome of the ghostwort (see N.J. Wickett et al. *Molecular Biology and Evolution* 25, 393–401 [2008]) also shows gene loss. In this case the reduction in genome size is less dramatic than some parasitic angiosperms, but as Norman Wickett et al. note, not only is there "a common directionality of these losses," but as importantly in this liverwort (*Aneura*) the losses "are concentrated in genes and functional groups that are completely lost in [*Epifagus*]" (p. 397; both quotations).

143. See E. Delannoy et al. *Molecular Biology and Evolution* 28, 2077–2086 (2011).

144. These are an entirely parasitic group (including the malarial *Plasmodium* and other vicious parasites, such as the odious *Toxoplasma*) that derive from algae and still possess an echo of their life in the sunshine with a remnant plastid referred to as the apicoplast; see X.-M. Cai et al. *Gene* 321, 39–46 (2003); also M. Matsuzaki et al. *Molecular Biology and Evolution* 25, 1167–1179 (2008).

145. Wolfe et al. (1992), p. 10648 [Abstract]; note 141.

146. See K. Krause *Current Genetics* 54, 111–121 (2008).

147. See J. Schwender *Nature* 432, 779–782 (2004); also commentary on p. 684 by C. Surridge.

148. It provides its own examples of convergence (see D.L. Nickrent et al. *BMC Evolutionary Biology* 5, e38 [2005]).

149. Once again it is a member of the lamiids, and as with other parasitic plants the dodder plastid (see J.R. McNeal et al. *BMC Biology* 5, e55 [2007]) exhibits gene loss (with interesting parallels to *Epifagus*, notably with respect to the loss of introns in a maturase [*matk*]; see J.R. McNeal et al. *PLoS ONE* 4, e5982; 2009).

150. See J.H. Dawson et al. *Reviews of Weed Science* 6, 265–317 (1994).

151. See I. Dörr in *Sieve elements: Comparative structure, induction and development* (H.-D. Behnke and R.D. Sjölund, eds.), pp. 239–256 (Springer; 1990).

152. See K.C. Vaughn *Protoplasma* 219, 227–237 (2002).

153. In the words of Inge Dörr (1990; note 151) they are among "the most peculiar and highly differentiated cell types within the plant kingdom" (p. 249).

154. It is worth mentioning that despite reference to "haustoria" and "hyphae," when it comes to comparisons with the parasitic plants, for the most part the manner in which pathogenic fungi invade their hosts is marked more by differences than similarities (see A.M. Mayer *Phytoparasitica* 34, 3–16 [2006]).

155. See A.M. Koch et al. *New Phytologist* 162, 147–155 (2004).

156. See J.B. Runyon et al. *Science* 313, 1964–1967 (2006); also commentary on p. 1867 by E. Pennisi.

157. Runyon et al. (2006), p. 1967; note 156.

158. See J. Nais *Rafflesia of the world* (Natural History Publications Kota, Kinabalu; 2001). The extent of its parasitism is revealed by the fact that all trace of the chloroplast genome appears to have been lost; see J. Molina et al. *Molecular Biology and Evolution* 31, 793–803 (2014). L.A. Nikolov et al. (*Proceedings of the National Academy of Sciences, USA* 110, 18578–18583; 2013) demonstrate that even in the closely related *Rafflesia* and *Sapria* the floral chambers are not constructed in the same way and thus cannot be homologous.

159. See T.J. Barkman et al. *Proceedings of the National Academy of Sciences, USA* 101, 787–792 (2004), and D.L. Nickrent et al. *BMC Evolutionary Biology* 4, e40 (2004).

160. See C.C. Davis et al. *Science* 315, 1812 (2007).

161. See, for example, C. De Vega et al. *Annals of Botany* 100, 1209–1217 (2007).

162. See W. Meijer and J.F. Velkamp *Blumea: Journal of Plant Taxonomy and Plant Geography* 38, 221–229 (1993).

163. See T.J. Barkman et al. *Current Biology* 18, 1508–1513 (2008); also commentary on pp. R1102–R1104 by C.C. Davis.

164. Barkman et al. (2008), pp. 1508–1509; note 163; also M. Bendiksby et al. *Molecular Phylogenetics and Evolution* 57, 620–633 (2010), for further remarks on morphological convergences in the rafflesiaceans.

165. See C.C. Davis et al. *Current Opinion in Plant Biology* 11, 49–57 (2008).

166. See R.S. Beamen et al. *American Journal of Botany* 75, 1148–1162 (1988).

167. See A. Jürgens et al. *Ecology Letters* 16, 1157–1167 (2013); also for a helpful overview of "carrion and dung mimicry in angiosperms" (p. 730), see Vereecken and McNeil (2010); note 89.

168. So a tropical monocot attracts dung beetles; see S. Sakai and T. Inoue *American Journal of Botany* 86, 56–61 (1999).

169. See, for example, A. Jürgens et al. *New Phytologist* 172, 452–468 (2006).

170. In this case the volatiles are produced by Solomon's lily and serve to attract fruit flies; see J. Stökl is otherwise unknown in animals and was presumably imported by lateral gene transfer from a bacterial source (see T. Daimon et al. *Journal of Biological Chemistry* 283, 15271–15279 [2008]).

101. This type of sabotage has evidently evolved independently (see D.E. Dussourd and T. Eisner *Science* 237, 898–901 [1987]; also S.G. Compton *Ecological Entomology* 12, 115–118 [1987]), notably in some beetles and lepidopteran caterpillars. Nor are such tactics restricted to bypassing the latex, but can also be employed in diverting dangerous resins and chemicals.

102. See their paper in *Genes & Development* 16, 3113–3129 (2002).

103. Thummel and Chory (2002), p. 3124; note 102.

104. See, for example, L. Dinan *Russian Journal of Plant Physiology* 45, 296–305 (1998).

105. For an overview, see, for example, R.H. ffrench-Constant et al. *Philosophical Transactions of the Royal Society of London, B* 353, 1685–1693 (1998).

106. See R.H. ffrench-Constant et al. *Annual Review of Entomology* 48, 449–466 (2000).

107. See M. Thompson et al. *Insect Molecular Biology* 2, 149–154 (1993).

108. See D. Andreev et al. *Journal of Molecular Evolution* 48, 615–624 (1999).

109. See W. Du et al. *Insect Molecular Biology* 14, 179–183 (2005).

110. See J.A. Anstead et al. *Insect Biochemistry and Molecular Biology* 35, 249–256 (2005).

111. See M. Alon et al. *Insect Biochemistry and Molecular Biology* 36, 71–79 (2006).

112. See R.D. Newcomb et al. *Proceedings of the National Academy of Sciences, USA* 94, 7464–7468 (1997).

113. See, for example, C. Claudianos et al. *Insect Biochemistry and Molecular Biology* 29, 675–686 (1999).

114. Their paper is in *Proceedings of the National Academy of Sciences, USA* 103, 8757–8762 (2006); a key finding is that in one case of resistance (to malathion) the mutation was "ready and waiting," predating the use of the organophosphate (see also R.H. ffrench-Constant *Trends in Genetics* 23, 1–4 [2007]).

115. Hartley et al. (2006), p. 8761; note 114.

116. See J.W. Chen et al. *Applied and Environmental Microbiology* 73, 5162–5172 (2007).

117. See W. Liu et al. *Proceedings of the National Academy of Sciences, USA* 104, 3627–3632 (2007).

118. It is, however, by no means a one-way battle, and the countermeasures adopted by insects again involve convergences. Consider, for example, the repeated evolution of molecular strategies to allow the digestive juices of various dipterans and lepidopterans to tackle privet (see K. Konno et al. *Journal of Chemical Ecology* 36, 983–991 [2010]).

119. See, for example, N.A. Hynson and T.D. Bruns *Proceedings of the Royal Society of London, B* 276, 4053–4059 (2009); also L. Tedersoo et al. *Oecologia* 151, 206–217 (2007).

120. See M.A. Selosse and M. Roy *Trends in Plant Science* 14, 64–70 (2009).

121. See J.R. Leake *New Phytologist* 127, 171–216 (1994); also V. Merckx and J.V. Freudenstein in vol. 185, pp. 605–609 (2010).

122. Leake (1994), p. 186; note 121.

123. These so-called dust seeds have evolved at least fifteen times, and are especially characteristic of parasitic plants; see O. Eriksson and K. Kainulainen *Perspectives in Plant Ecology, Evolution and Systematics* 13, 73–87 (2011).

124. Leake (1994), p. 186; note 121.

125. Leake (1994), p. 172 [Abstract]; note 121.

126. See M.R. Klooster et al. *American Journal of Botany* 96, 2197–2205 (2009).

127. Leake (1994), p. 171 [Abstract]; note 121.

128. For an overview, see J. Kuijt *The biology of parasitic flowering plants* (University of California Press; 1969). At least some groups date back to the Cretaceous, and interestingly these older lineages tend to be the more specialized; see J. Naumann et al. *PLoS ONE* 8, e79204 (2013).

129. See J. Kuijt *Annual Review of Phytopathology* 15, 91–118 (1977).

130. See T.J. Barkman et al. *BMC Evolutionary Biology* 7, e248 (2007).

131. The sticky and viscid seeds of mistletoe are familiar, and not surprisingly dispersal by birds seems to have evolved multiple times (see C. Restrepo et al. in *Seed dispersal and frugivory: Ecology, evolution and conservation* [D.J. Levey et al., eds.], pp. 83–98 [CABI [2002], and even by marsupials (G. Amico and M.A. Aizen *Nature* 408, 929–930 [2000]). More strikingly, however, is the convergence between the mistletoe seeds and those of the Neotropical epiphytic cactus *Rhipsalis* (see A. de C. Guaraldo et al. *Biotropica* 45, 465–473 [2013]).

132. Barkman et al. (2007), p. 10; note 130.

133. Barkman et al. (2007), p. 10; note 130.

134. A clue might come from the mitochondrial genome and in particular the observation "that the mt *coxI* gene in nearly every parasitic lineage has been invaded by a group I intron" (Barkman et al. [2007], p. 7; note 130). At first sight this might seem rather trivial and as an explanation it is still tentative, but it may not be coincidental that this intron may have a fungal origin. If so, this suggests that mycoheterotrophy is the most likely route to parasitism.

135. See T.S. Field and T.J. Brodribb *Plant, Cell and Environment* 28, 1316–1325 (2005).

136. It is worth noting that symbiotic associations between fungi, especially ascomycetes, and various bryophytes are blatantly polyphyletic (see S. Stenroos et al. *Cladistics* 26, 281–300 [2010]). As Soili Stenroos et al. (2010) note, in these so-called bryosymbionts, "even highly specialized life strategies can be adopted multiple times during evolution" (p. 281 [Abstract]).

137. See M.I. Bidartondo et al. *Proceedings of the Royal Society of London, B* 270, 835–842 (2003).

89. In this context one should also draw attention to those orchids with flowers that provide an irresistible sexual lure (for a useful overview of this topic, see N.J. Vereecken and J.N. McNeil [*Canadian Journal of Zoology* 88, 725–752; 2010] aptly titled, "Cheaters and liars: chemical mimicry at its finest") to the male insect (often bees), typically employing pheromones (see F.D. Schiestl et al. in *Nature* 399, 421–422 [1999] and *Science* 302, 437–438 [2003] [or other volatiles; see J. Brodmann et al. *Current Biology* 18, 740–744; 2008] and a most encouraging (hairy) appearance. This engenders manifest sexual excitement in the insect, culminating in pseudocopulation. This pollination syndrome has evolved independently several times among the orchids (see, for example, R.B. Singer et al. *Annals of Botany* 93, 755–762 [2004], and L. Ciotek et al. *Flora* 201, 365–369; 2006) and beyond the orchids, notably in an iris (see N.J. Vereecken et al. [*Proceedings of the Royal Society of London, B* 279, 4786–4794; 2012] and a South African daisy [see A.G. Ellis and S.D. Johnson *American Naturalist* 176, E143–E151 (2010); also commentary on pp. R1020–R1022 by F.P. Schiestl in *Current Biology* (20; 2010)]). Nor is this the only example, because there is also a good case for this pollination syndrome in an alliacean (*Gilliesia*) (see P.J. Rudall et al. *American Journal of Botany* 89, 1867–1883 [2002]), the flowers of which have in the words of Paula Rudall and colleagues a "remarkable insect-like appearance" (p. 1878). Hugging orchids and receiving a good dose of pollen is one thing, but the further reaches of pseudocopulation enter some pretty strange territory. Here the insect engages in specific mating maneuvers with the flower. These are integral to the successful transfer of pollen but may also lead to ejaculation by the visitor (see A.C. Gaskett et al. *American Naturalist* 171[6], E206–E212 [2008]). This climax of pseudocopulation is also convergent because it has evolved in association with an ichneumonid wasp and a fungus gnat (see M.A. Blanco and G. Barboza *Annals of Botany* 95, 763–772 [2005]). In the former insect, having been drawn in by seductive pheromones (see F.P. Schiestl et al. *Botanical Journal of the Linnean Society* 144, 199–205;2[004]), the wasp then displays a tremendous eagerness (see E. Coleman *Transactions of the Entomological Society of London* 76, 533–539 [1928]; also *Proceedings of the Royal Entomological Society of London* 13, 82–83; 1938).

90. See their article in *Annual Review of Plant Biology* 62, 549–566 (2011).

91. Pichersky and Lewinsohn (2011), p. 549 [Abstract]; note 90.

92. See W.F. Pickard *New Phytologist* 177, 877–888 (2008).

93. See their paper in *Trends in Plant Science* 13, 631–639 (2008).

94. Hagel et al. (2008), p. 631 [Abstract]; note 93.

95. See T.M. Lewinsohn *Chemoecology* 2, 64–68 (1991).

96. Coevolutionary arms races between herbivorous insects and plants (and evidently equivalent fungivores) are not only intrinsically fascinating because of the way conflicting adaptations slug it out, but intuitively one might envisage an almost indefinite escalation. Consider the appropriately named milkweeds. As their name suggests, these plants are copious latex producers; although this feature is surprisingly labile in terms of evolution (see A.A. Agrawal et al. *Entomologia Experimentalis et Applicata* 128, 126–138 [2008]), milkweed also employ other lines of defense, notably trichomes and highly toxic chemicals such as cardenolides. For a striking example of molecular convergence that confers resistance to this compound and has arisen at least four times among insects, see S. Dobler et al. *Proceedings of the National Academy of Sciences, USA* 109, 13040–13045; [2012]). Intriguingly, despite this battery, there is a trend away from these methods of defense toward a "simpler" expedient of enhancing growth (see A.A. Agrawal and M. Fishbein *Proceedings of the National Academy of Sciences, USA* 105, 10057–10060 [2008]). An important driving force may be the specialization by the chomping insects, but this case is an important reminder that ecological interactions can be complex. At first sight this combinatorial richness whereby various strategies can be adopted might appear to speak against the likelihood of recurrent solutions evolving. That this need not be the case is, however, evident from an analysis as to how plants in the tropics manage to defend their young leaves, given that they are particularly attractive to herbivores on account of their tenderness and higher protein contents. It turns out that certain combinations of defense that in principle might be ideal are actually excluded because they make conflicting physiological demands (see T.A. Kursar and P.D. Coley *Biochemical Systematics and Ecology* 31, 929–949 [2003]). For example, fast growth of the leaf is incompatible with either low levels of protein or incorporation of defensive chemicals. The net result is that while a wide variety of options for defense in principle can be envisaged, only a few particular strategies are workable. These combinations evolve repeatedly, and as Thomas Kursar and Phyllis Coley (2003) note, "These tradeoffs have apparently led to the independent evolution of similar defense syndromes in many lineages" (p. 942). Their conclusions are an important reminder of concerted convergence. This is where entire adaptive systems must evolve in tandem, further restricting the evolutionary options.

97. See, for example, K. De Gussum et al. *Phytochemistry* 67, 2580–2589 (2006).

98. See, for example, S. Peters et al. *European Journal of Organic Chemistry* 2008, 1187–1194 (2008).

99. See P. Spiteller *Chemistry—A European Journal* 14, 9100–9110 (2008).

100. In some cases, such as the silkworm, the alkaloids of the mulberry latex are rendered harmless by a specific enzyme (a β-fructofuranosidase) that *and evolution.* Ecological Studies 114 (Springer; 1996); also Silvera et al. (2010); note 47.

51. See E.A. Nelson et al. *Functional Plant Biology* 32, 409–419 (2005).

52. See S.L. Martin et al. *International Journal of Plant Sciences* 166, 623–630 (2005).

53. See A.P. Vovides *Botanical Journal of the Linnean Society* 138, 155–162 (2002); see also D.J. von Willert et al. *Functional Plant Biology* 32, 389–395 (2005), for the interesting case of the gnetalean *Welwitschia*.

54. See D.M. Crayn et al. *Proceedings of the National Academy of Sciences, USA* 101, 3703–3708 (2004); also I.M. Quezada and E. Gianoli *Biological Journal of the Linnean Society* 104, 480–486 (2011), who argue that acquisition of CAM physiology ranks as a key adaptation in terms of capacity to diversify.

55. See, for example, H. Motomura et al. *Journal of Plant Research* 121, 163–177 (2008), and K. Silvera et al. *Plant Physiology* 149, 1838–1847 (2009).

56. Although the seedlings start with the C_3 method; see A. Altesor et al. *Acta Oecologica* 13, 777–785; 1992.

57. See U. Lüttge *Trees: Structure and Function* 22, 139–148 (2008).

58. See U. Lüttge *Annals of Botany* 93, 629–652 (2004).

59. See J.E. Keeley et al. *Nature* 310, 694–695 (1984).

60. See J.E. Keeley *Botanical Review* 64, 121–175 (1998).

61. See, for example, O. Pedersen et al. *New Phytologist* 190, 332–339 (2011).

62. See W.A. Green *Proceedings of the Royal Society of London, B* 277, 2257–2267 (2010).

63. Green (2010), p. 2263; note 62.

64. Given the polyphyly of CAM photosynthesis, a crisis in CO_2 levels need not be the only explanation, and others find a possible correlation to the massive desertification that occurred across the Permo-Triassic boundary (see J.E. Decker and M.J. de Wit *Terra Nova* 18, 9–17 [2006]).

65. A similar argument is given for another New Zealand plant (*Elaeocarpus*) where the color of leaves within browsing reach of the now-vanished moa are cryptic; see N. Fadzly and K.C. Barns *International Journal of Plant Sciences* 171, 828–833 (2010).

66. See N. Fadzly et al. *New Phytologist* 184, 495–501 (2009); also commentary by D. Lee and K. Gould on pp. 282–284.

67. See P.J. Grubb *Perspectives in Plant Ecology, Evolution and Systematics* 6, 125–146 (2003); W.J. Bond et al. *Oikos* 104, 500–508 (2004); and W.J. Bond and J.A. Silander *Proceedings of the Royal Society of London, B* 274, 1985–1992 (2007); also commentary by H. Dempewolf and L.H. Riesenberg *Current Biology* 17, R773–R774 (2007).

68. Correspondingly their absence appears to have had a significant impact on the evolution of grasses in New Zealand; see A. Antonelli et al. *Proceedings of the Royal Society of London, B* 278, 695–701 (2011).

69. See P.J. Grubb's assessment (*Journal of Ecology* 80,

585–610; 1992) with the important reminder of "A positive distrust of simplicity" [part of title].

70. See N. Halpern et al. *Environmental Microbiology* 9, 584–592 (2007).

71. In the larvae of butterflies alone aposematism has emerged perhaps twenty-three times (see B. Sillén-Tullberg *Evolution* 42, 293–305 [1988]). In many lepidopterans the larvae are also highly gregarious. It is clear that not only has this habit evolved many times (see B. Sillén-Tullberg *Evolution* 42, 293–305 [1988]), but it too has an important link to aposematism (see B.S. Tullberg and A.F. Hunter *Biological Journal of the Linnean Society* 57, 253–276 (1996), and A. Gagliardo and T. Guilford *Proceedings of the Royal Society of London, B* 251, 69–74 [1993]).

72. See J.F.V. Vincent and P. Owers *Journal of Zoology, London* 210, 55–75 (1986).

73. See M. Inbar and S. Lev-Yadun *Naturwissenschaften* 92, 170–172 (2005); also M.P. Speed and G.D. Ruxton *Evolution* 59, 2499–2508 (2005).

74. See S. Lev-Yadun *Journal of Theoretical Biology* 210, 385–388 (2001).

75. See S. Lev-Yadun *Journal of Theoretical Biology* 244, 183–188 (2003).

76. See S. Lev-Yadun *Biological Journal of the Linnean Society* 81, 413–416 (2004).

77. See S. Lev-Yadun *Journal of Theoretical Biology* 224, 483–489 (2003).

78. See S. Lev-Yadun and M. Inbar *Biological Journal of the Linnean Society* 77, 393–398 (2002).

79. See K. Yamazaki *Biological Journal of the Linnean Society* 104, 738–747 (2011); this worker suggests that the rustling of the leaves might also interfere with the vibrational communication used by many insects.

80. See M.E. Hanley et al. *Perspectives in Plant Ecology, Evolution and Systematics* 8, 157–178 (2007).

81. See L. Serna and C. Martin *Trends in Plant Science* 11, 274–280 (2006).

82. See C.E. Jeffree in *Insects and the plant surface* (B.E. Juniper and T.R.E. Southwood, eds.), pp. 23–64 (Edward Arnold; 1981); M. Krings et al. *Botanical Journal of the Linnean Society* 141, 133–149 (2003); and M.E. Hanley et al. *Perspectives in Plant Ecology, Evolution and Systematics* 8, 157–178 (2007).

83. See M. Krings et al. *Botanical Review* 69, 204–224 (2003).

84. See D.H. Gentry in *The biology of vines* (F.E. Putz and H.A. Mooney, eds.), pp. 3–49 (Cambridge; 1991); also E. Gianoli *Proceedings of the Royal Society of London, B* 271, 2011–2015 (2004), and R.J. Burham *Revista Brasileira de Paleontologia* 12, 149–160 (2009).

85. See M. Krings et al. *Evolutionary Ecology and Research* 4, 779–786, and *Botanical Journal of the Linnean Society* 141, 133–149 (2003).

86. See P.S. Bray and K.B. Anderson *Science* 326, 132–134 (2009).

87. See D. Grimaldi *Science* 326, 51–52 (2009).

88. Grimaldi (2009), p. 52; note 87.

out that an exception of sorts is found in Australia, where a quasi-cactus/euphorbacean has not evolved.

27. See M. Arakaki et al. *Proceedings of the National Academy of Sciences, USA* 108, 8379–8384 (2011).

28. See R. Nyffeler et al. *Haseltonia* 14, 26–36 (2008).

29. See J.D. Mauseth *International Journal of Plant Sciences* 165, 1–9 (2004); also R.M. Ogburn and E.J. Edwards *Current Biology* 23, 722–726 (2013).

30. This in turn allowed for a remarkable diversity of forms to evolve, with frequent convergence; see T. Hernández-Hernández et al. *American Journal of Botany* 98, 44–61 (2011).

31. See R. Nyffeler *American Journal of Botany* 94, 89–101 (2007).

32. See E.J. Edwards et al. *American Journal of Botany* 92, 1177–1188 (2005).

33. Peter Grubb (personal communication) draws my attention to the fact that although many xerophytes are spiny in the same manner as nearly all cacti, this is by no means a general rule, and many succulents, such as the so-called bottle trees, are not so equipped.

34. See E.J. Edwards and M.J. Donoghue *American Naturalist* 167, 777–793; also vol. 168, p. 132 (2006).

35. Edwards and Donoghue (2006), p. 780; note 34.

36. Edwards and Donoghue (2006), p. 789; note 34.

37. Edwards and Donoghue (2006), p. 789; note 34.

38. See R.M. Ogburn and E.J. Edwards *American Journal of Botany* 96, 391–408 (2009).

39. Ogburn and Edwards (2009), p. 402; note 38.

40. Ogburn and Edwards (2009), p. 406; note 38.

41. See W.P. Hartl et al. *Canadian Journal of Botany* 85, 501–517 (2007).

42. For plants see V.R. Franceschi and P.A. Nakata *Annual Review of Plant Biology* 56, 41–71 (2005).

43. See L.P. Ruiz and T.A. Mansfield *New Phytologist* 127, 473–481 (1994).

44. See, for example, S.-C. Han et al. *Nature* 425, 196–200 (2003); also commentary by C. Brownlee (*Current Biology* 13, R923–R942; 2003).

45. See J.E. Keeley and P.W. Rundel *International Journal of Plant Sciences* 164 (3, Supplement), S55–S77 (2003).

46. Not only is aluminium resistance in plants convergent (and a vital necessity in many acidic soils; see R.R. Ryan and E. Delhaize *Functional Plant Biology* 37, 275–284; 2010), but the two gene families responsible act by releasing either malate or citrate. So why, ask Peter Ryan and Emmanuel Delhaize, "are malate and citrate the common currencies of Al^{3+} resistance?" They suggest, "The answer may lie in the economy of these small organic compounds . . . [that] are ubiquitous in living cells and metabolically cheap to synthesise" (p. 277).

47. See K. Silvera et al. *Functional Plant Biology* 37, 995–1010 (2010).

48. In land plants the evolutionary impetus for C_4 photosynthesis, so called because the first product in the process to making sugars has four carbon atoms, was evidently declining levels of atmospheric carbon dioxide during geological time (see, for example, P.-A. Christin et al. *Current Biology* 18, 37–43 [2008]; also commentary on pp. R167–R168 by E.H. Roalson). This strategy is rampantly convergent (see, for example, R.F. Sage et al. *Journal of Experimental Botany* 62, 3155–3169 [2011]). As Rowan Sage and colleagues remark, "With 62 or more distinct origins, the C_4 pathway must be considered one of the most convergent of complex evolutionary phenomena in the living world" (p. 3165). They remind us that although such a frequency might indicate the relative ease of this transition (for a particularly compelling example in *Heliotropium* of transitional forms that include species with a proto-Kranz anatomy, see R. Muhaidat et al. [*Plant, Cell & Environment* 34, 1723–1736; 2011]; also D. Heckmann et al. [*Cell* 153, 1579–1588 (2013), with commentary on pp. 1427–1429 by K.D. Beer et al.]), the fact that different evolutionary routes are employed suggests that it is "genuinely convergent, in the sense that it has independently emerged from deep within many of the major angiosperm clades" (p. 3165). While C_4 photosynthesis is classically associated with a distinctive cellular configuration (Kranz anatomy), not only have alternative arrangements evolved but they are just as effective (see J.L. King et al. *Plant, Cell and Environment* 35, 513–523 [2012]). Where common ground does exist is a key ingredient in the C_4 process, a carboxylase enzyme known as PEPC (in full phospho*enol*pyruvate carboxylase). Significantly C_4 photosynthesis in the eudicots, notably among the amaranthaceans (see, for example, G. Kadereit et al. *Proceedings of the Royal Society of London, B* 279, 3304–3311 [2012]), seems to have first emerged at much the same time as it did in the monocots (see P.-A. Christin et al. *Journal of Experimental Botany* 62, 3171–3181 [2011]). In the latter case the best-known examples occur among the grasses. Even in this group the C_4 photosynthetic mechanism has evolved more than twenty times (see Grass Phylogeny Working Group II *New Phytologist* 193, 304–312 [2012]). Perhaps not so surprisingly, in at least some of the cases the same sites within the PEPC enzyme are subject to very similar changes and very often involve substitution with the same amino acids (see P.-A. Christin et al. *Current Biology* 17, 1241–1247 [2007]), but the change is adaptive, and given that at one amino acid site the invariable replacement (of serine by alanine) is evidently associated with catalytic properties of PEPC, so the convergence makes good sense. Not only that, but when the comparisons are extended to other C_4 plants, a striking analogy to the "five-site rule" of vision (p. 112) is found (see G. Besnard et al. *Molecular Biology and Evolution* 26, 1909–1919 [2009]).

49. See, for example, U. Lüttge *New Phytologist* 171, 7–25 (2006).

50. See K. Winter and J.A.C. Smith (eds.) *Crassulacean acid metabolism: Biochemistry, ecophysiology* *ular Phylogenetics and Evolution* 34, 159–166 [2005]), who also comment on likely convergence in seed viviparity), while in the Cretaceous world a relative of *Gingko* provided a Mesozoic counterpart. This is the gymnospermous ginkgoalean *Nehvizdyella* (see J. Kvaček et al. *American Journal of Botany* 92, 1958–1969 [2005]) that H.J. Falcon-Lang et al. (*Geoscientist* 16 [4, April], 4–6; 2006) place in a specifically mangrove setting. Still earlier, from the Permian of Minorca, are fossil soils with mangrovelike roots, albeit of an unknown plant; see A. Bercovici et al. *Review of Palaeobotany and Palynology* 158, 14–28 (2009).

12. Key to the success of myrmecochory is the evolution in the seed of a reward in the form of an appendage rich in lipids (the elaiosome). These have evolved multiple times (see, for example, J.L. Bronstein et al. *New Phytologist* 172, 412–428 [2006]), but even more impressive is the recurrent evolution of myrmecochory. As Szaboles Lengyel and colleagues (*Perspectives in Plant Ecology, Evolution and Systematics* 12, 43–55 [2010]) note, this syndrome "provides one of the best examples to date for convergent evolution in general and for the repeated evolution of plant-animal mutualisms more specifically" (p. 51). Careful to note that the roster of examples is incomplete, they "identified at least 101, but possibly up to a 147, independent origins of myrmecochory" (p. 43 [Abstract]).

13. A particularly striking example involves the division of flowering plants known as the monocotyledons (or monocots). Typically this group, of which the grasses are probably the most familiar, have narrow leaves with parallel venation and they rely on wind dispersal. Moving into shady habitats, such as a forest, presents problems because the light is dimmer and wind currents are muted. The evolutionary solution is twofold: move toward wider leaves (reinforced by a netlike venation) and develop fleshy fruits that will attract animals for dispersal. The main fruit types have also evolved multiple times, in part due to the selective pressures for dispersal (see C.M. Lorts et al. *Journal of Systematics and Evolution* 46, 396–404 [2008]). Even for enthusiasts of convergence these results are striking because the linkage of leaf type and fruit has evolved independently (see T.J. Givnish et al. *Proceedings of the Royal Society of London, B* 272, 1481–1490 [2005]) some twenty times. Givnish et al. (2005) also note that this concerted convergence works in the opposite direction if the monocot subsequently comes to occupy a more open habitat.

14. Curiously the "color space" of fruit is not only far more restricted than that of flowers, but with few exceptions the former almost entirely lies within the color space of flowers (see K.E. Stournaras et al. *New Phytologist* 198, 617–629 [2013]). Fruit color is rampantly convergent, and its failure to occupy more than 80 percent of the available color space points to strong constraints. One fact may well be that fruit needs to attract dispersers, while flowers are magnets for pollinators.

15. One must be flint-hearted not to be moved by the vivid reds and yellow that herald the winter days, yet curiously a wealth of adaptive explanations have failed to find a consensus (see M. Archetti et al. *Trends in Ecology & Evolution* 24, 166–173 [2009]). Only a minority of plants exhibit autumnal colors, but those that do are rampantly polyphyletic (see M. Archetti *Annals of Botany* 103, 703–713 [2009]) and, as Marco Archetti remarks, have "evolved independently at least 25 times" (p. 707). If seasonal cooling leads to autumnal colors, so too the process of vernalization (where a cold interval is necessary for spring flowering) is also convergent; see T.S. Ream et al. *Cold Spring Harbor Symposia on Quantitative Biology* 77, 105–115 (2012).

16. Bill Chaloner and John Pettitt (*Bulletin de la Société Botanique de France, Actualités Botaniques* 134[2], 39–49; 1987) spell this out when they write how they "see this evolutionary progression as a continuous process, followed independently and inevitably by several lines of vascular plant evolution" (p. 41). Confidence that this view is correct rests on several lines of evidence. One crucial precondition is the phenomenon known as heterospory. As the name suggests, this entails the evolution of microspores and megaspores: ultimately these will form, respectively, pollen and ovules. Yet heterospory is rampantly convergent, having evolved at least eleven times (see R.M. Bateman and W.A. DiMichele *Biological Reviews* 69, 345–417 [1994]). This is certainly not to imply that the transition was either simple or involved identical evolutionary trajectories, but Richard Bateman and William DiMichele note that the sheer frequency of this convergence in a number of distinct genomic and ecological contexts provides an "unparalleled opportunity to infer mechanisms that drove . . . [this] major evolutionary innovation" (p. 386).

17. See L.J. Rothschild *Philosophical Transactions of the Royal Society of London, B* 363, 2787–2801 (2008).

18. See R. Milo *Photosynthesis Research* 101, 59–67 (2009); also my *Life's solution*, pp. 109–111.

19. See his paper in *Systematic Botany* 31, 805–821 (2006).

20. See D.J. Hearn *American Journal of Botany* 96, 1941–1956 (2009).

21. Hearn (2009), p. 1950; note 20.

22. Hearn (2009), p. 1954; note 20.

23. Nor are they quite alone, because among the extinct cheirolepidiaceans, a group of Mesozoic conifers, one species (*Frenelopsis ramosissima*) has evidence for succulent shoots; see B.J. Axsmith and B.F. Jacobs *International Journal of Plant Sciences* 166, 327–337 (2005).

24. See, for example, R. Felger and J. Henrickson *Haseltonia* (5), 77–85 (1997).

25. For convergences within *Euphorbia*, see J.W. Horn et al. (*Molecular Phylogenetics and Evolution* 63, 305–326; 2012), who emphasize both the remarkable range of growth habits and rampant xeromorphy.

26. Peter Grubb (personal communication) points

types labile but they have evolved independently numerous times (see A.M.F. Tomescu *Trends in Plant Science* 14, 5–12 [2009]). So, too, it appears that the key developmental genes show a combinatorial complexity that allows "many independent origins of determinacy of leaf precursor structures" (p. 10). Alexandrou Tomescu further notes the implication that even within the euphyllotes there may have been "as many as nine independent origins of leaves" (p. 11). It remains significant that in terms of the occupation of morphospace, although by no means all plants evolved in the same way, the fact remains that during the Palaeozoic the boundaries of possible forms were not only reached relatively quickly but as significantly they could not be transgressed (see C.K. Boyce and A.H. Knoll *Paleobiology* 28, 70–100 [2002]). As Kevin Boyce and Andy Knoll note, there does appear to be a "limited number of ways that plants can form laminate photosynthetic surfaces" (p. 78). To be sure these authors dwell on the constraints imposed by preexisting ancestry, notably the nature of the proliferative zone or meristem, but here too it seems that prior geometries of cell proliferation in this region were in turn constrained by the simple exigencies of being any sort of land plant at all.

10. Many flowering plants employ wind pollination (anemophily). This has evolved more than sixty times and importantly is recurrently associated with features such as flower type and absence of nectar (see J. Friedman and S.C.H. Barrett *International Journal of Plant Sciences* 169, 49–58 [2008]). Wind dispersal per se had, of course, emerged much earlier and winged propagules have evolved multiple times (see D. Dilcher et al. *Review of Palaeobotany and Palynology* 98, 247–256 [1997]). Angiosperms are past masters of this technique, yet even so the specialized pappuslike parachute mechanism had evolved independently in the late Triassic (see B.J. Axsmith et al. *Palaeontology* 56, 1173–1177 [2013]). Anemophily may also be the usual route to what is arguably the strangest of pollination strategies, where water is the medium of dispersal, the aptly named hydrophily (see, for example, P.A. Cox *Annual Review of Ecology and Systematics* 19, 261–280 [1988]). Paul Alan Cox (*Scientific American* 269 [4; October], 50–56 [1993]) rightly insists that these "water-based pollination systems cannot be considered mere quirks of natural history as previously thought. Rather water-pollinated plants should now be viewed as compelling cases of convergent evolution" (p. 50). Most remarkable are those flowering plants that are permanently submerged and may be most familiar in the form of the sea grasses. Although a monocotyledon, they are not true grasses, but the group to which they belong (the alismatids) is not only widespread in freshwater (see D.H. Les et al. *Systematic Botany* 22, 443–463 [1997]) but has invaded the marine realm three times—and possibly more. Consider the peculiar *Mosacaulis* from the late Cretaceous, which, as R.W.I.M. van der

Ham et al. (*Review of Palaeobotany and Palynology* 168, 51–67 [2011]) demonstrate, was certainly aquatic and sea grass–like, but might be either a lycopsid or angiosperm. In angiosperm animal-mediated or biotic pollination, many intricate systems are known, but among the most remarkable is the so-called buzz pollination. This arrangement epitomizes the integration of insect and flower because in this system the pollen can only exit via pores in the anthers while their release is effected by a vibratory sonication generated by the "shivering" of the visiting bee (see, for example, M.J. King and S.L. Buchmann *Functional Ecology* 10, 449–456 [1998]; also P.A. De Luca and M. Vallejo-Marín [*Current Opinion in Plant Biology* 16, 429–435; 2013], who emphasize its convergence, especially in the solenoid flower-type). As Stephen Buchmann (see his chapter [pp. 73–113] in *Handbook of experimental pollination biology* [C.E. Jones and R.J. Little, eds.; Van Nostrand Reinhold; 1983]) remarks, "One of the most striking features of the "syndrome" of vibratile pollination is the independent evolution in floral morphology . . . and the resultant buzz pollination in so many phylogenetically unrelated and diverse families" (p. 85). Given the manner in which the pollen is released, industrious probing for nectar while receiving a full dusting is hardly an option, so instead of the customary nectar the reward comes in the form of energy-rich pollen grains. Buchmann (1983) also draws attention to a remarkable mimic in the form of a syrphid fly that not only looks like a bee but also releases the pollen by the same vibratory action. Syrphids or flower flies are especially important, but the more general importance of flies in pollination (see, for example, A. Ssymank et al. *Biodiversity* 9, 86–89 [2008]) is often overlooked. Even less familiar is the role of beetles. Yet beetle-mediated pollination (cantharophily) has evolved independently at least fourteen times in the dicots and six times among the monocots, with the so-called chamber blossoms (see P. Bernhardt *Plant Systematics and Evolution* 222, 293–320 [2000]) (as exemplified in the magnolias) being particularly characteristic. We should note the association between the flowers' thermogenesis (itself convergent, pp. 163–164 and the benefit incurred by the beetles (see, for example, R.S. Seymour et al. *Nature* 426, 243–244 [2003], and R.S. Seymour and P.G.D. Matthews *Annals of Botany* 98, 1129–1135 [2006]).

11. Among the flowering plants Robert Ricklefs and Roger Latham (see their chapter [pp. 215–229] in *Species diversity in ecological communities: historical and geographical perspectives* [R.E. Ricklefs and D. Schluter, eds.; Chicago; 1993]) note, "mangrove attributes appear to have evolved independently at least 15 times in 9 orders and 15 families" (p. 217). As these authors point out, the greatest diversity of mangroves resides in the rhizophoraceans (rosids) and aviceniaceans (asterids). Other convergences also emerge in features such as salt secretion (see S.H. Shi et al. *Molec-*

and Clint Chapple (2010) when they remark how the coincidence of complex biosynthetic pathways make it "of great importance to interrogate the genome of [*Calliarthron*] to determine definitively whether it encodes the enzymes necessary for monolignol synthesis" (p. 278). Again we encounter a profound point of tension in evolution: Is the system derived from a remote common ancestor, or is it inherent in the available molecular machinery? The evidence, however, continues to point to convergence. D.-M. Guo et al. (*Journal of Molecular Evolution* 71, 202-218; 2010) remind us, "it is very common that a specific character appeared much later than the origin of the gene families responsible for it" (p. 214). As Patrick Martone and colleagues note, because this "synthesis is exceptionally complex . . . it seems unlikely that *Calliarthron* and terrestrial plants evolved monolignol biosynthesis and polymerization completely independently" (p. 170). It would not be at all surprising if some components of the lignin pathway turn out to be very ancient. Evolution can only work on what is available, but it is crucial to determine not only the "original" function (so far as that is ever possible) but reasons for subsequent versatility. In the land plants the first step in the biosynthetic pathway of lignin may have revolved around the lateral gene transfer of an enzyme (a lysase) (see G. Emiliani et al. *Biology Direct* 4, e7 [2009]). If correct, the donor was most likely from among the fungi, and the transfer might have been mediated via the mycorrhizal association (p. 152).

8. Like lignin, xylem is convergent. Nor should this come as a surprise. Even among the giant seaweeds the capacity for long-distance transport, albeit for solutes rather than water, has evolved multiple times (see J.A. Raven *Plant, Cell and Environment* 26, 73-85 [2003]). In the case of the laminarid brown algae these convergences become more compelling (see K. Schmitz *Marine Biology* 78, 209-214 [1984]; also his chapter [pp. 1-18] in *Sieve elements: Comparative structure, induction, and development* [H.D. Behnke and R.D. Sjolund, eds.; Springer; 1990]). Klaus Schmitz notes how these algae "have evolved a very efficient long-distance transport system comparable to the sieve-element system of vascular plants" (p. 15). In the context of land plants, these convergences find a particular relevance because although the equivalent water-conducting cells (WCCs) in the bryophytes are biochemically complex, they have evolved independently (see R. Ligrone et al. *New Phytologist* 156, 491-508 [2002]) of the other vascular plants. The diversity of WCCs among the bryophytes indicates that even within this group they have multiple origins (see R. Ligrone et al. *Philosophical Transactions of the Royal Society of London, B* 355, 795-813 [2000]). At one level this is to be expected, given that any WCC has to involve a tubular construction and the organized removal of the cytoplasmic contents. Even so, the convergences are interesting. For example, Roberto Ligrone and colleagues remark how the "general appearance of WCCs in metzgerialean liverworts . . . is strongly reminiscent of tracheids" (p. 809), while the "striking similarity of leptoids [another type of WCC] in polytrichaceous mosses to sieve cells is probably an instance of homoplasy" (p. 811). Although bryophytes can synthesize lignin, it does not appear to be employed in its xylemlike tissue, but related chemicals in the form of polyphenols evidently provide the necessary sealant; see D.C. Sheirer *Bulletin of the Torrey Botanical Club* 107, 298-307 [1980]).

9. While there is necessarily an underlying set of developmental mechanisms in leaves connected to cell proliferation and vascularization, these mechanisms have been recruited independently (see C.J. Harrison et al. *Nature* 434, 509-514 [2005]), as exemplified in the group known as the lycophytes (see S.K. Floyd and J. L. Bowman *Current Biology* 16, 1911-1917 [2006]). In what William Friedman (see his paper in *Current Biology* 21, R554-R556 [2011]) aptly calls "Homoplasy Heaven" (part of title), he reminds us of not only this convergence between the lycophytes and the so-called euphyllotes (a concept embracing the ferns and seed-plants) but a raft of other homoplasies (such as megaspory (note 16), roots (see J.A. Raven and D. Edwards [*Journal of Experimental Botany* 52, 381-401; 2001], who suggest true roots [with a diagnostic root cap] have evolved at least twice), and bipolar growth (see, for example, H. Sanders et al. *Plant Systematics and Evolution* 291, 221-225 [2011]). Friedman writes, "Perhaps even more remarkably, these homoplasious developmental and structural innovations occurred in both lineages . . . within the evolutionary instant of a mere twenty or so million years. In a world of potentially infinite morphospace, it is striking that the two major lineages of vascular plants have independently arrived at fundamentally identical bauplans" (p. R555). While it is conjectural what, if anything, spurred the appearance of land plants, in the context of the evolution of leaves an important constraining factor may have been the greenhouse world in which these early plants were evolving. This is because of the heat stress any large photosynthetic surface, such as a leaf, would have experienced (see C.P. Osborne et al. *Proceedings of the National Academy of Sciences, USA* 101, 10360-10362 [2004]). It is striking that not only were the first plants leafless, but so it seems even the earliest trees (see C.P. Osborne et al. *Proceedings of the National Academy of Sciences, USA* 101, 10360-10362 [2004]). As global temperatures declined, leaves promptly emerged. This evidence chimes with the classic distinction between the so-called microphylls and megaphylls (with specific reference to convergence of megaphyllous leaves, see J. Galtier *International Journal of Plant Sciences* 171, 641-661 [2010]), except that it is now clear that not only are these leaf evolution of particular growth forms (see M. Grube and D.L. Hawksworth *Mycological Research* 111, 1116–1132 [2007]; also A. Tehler and M. Irestedt *Cladistics* 23, 432–454; 2007) and emergence of complex morphological characters, as in the parmelioid lichens; see O. Blanco et al. *Molecular Phylogenetics and Evolution* 39, 52–69 (2006).

51. See J. Miadlikowska and F. Lutzoni *American Journal of Botany* 91, 449–464 (2004).

52. See, for example, M.P. Nelsen et al. *American Journal of Botany* 94, 1289–1296 (2007).

53. See J.D. Lawrey et al. *Molecular Phylogenetics and Evolution* 44, 778–789 (2007), and D. Ertz et al. *American Journal of Botany* 95, 1548–1556 (2008).

54. Ertz et al. (2008), p. 1553; note 53.

55. For an overview, see H.K. Ngugi and H. Scherm *FEMS Microbiology Letters* 257, 171–176 (2006).

56. See B.A. Roy *Nature* 362, 56–58 (1993).

57. So, too, are more trained eyes—in yet another example of such mimicry, in this case where a smut (*Microbotyrum*) resembles the pollen-laden anthers of flowers (see, for example, P.H. Thrall et al. *Journal of Ecology* 81, 489–498 [1993], and J.A. Shykoff and E. Bucheli *Journal of Ecology* 83, 189–198 [1995]; also S.M. Altizer et al. *American Midland Naturalist* 139, 147–163; 1998). Closer examination of a number of supposed species of pinks (caryophyllaceans) revealed taxonomic misidentifications based on supposedly distinct anthers (see M.E. Hood and J. Antonovics *Proceedings of the Royal Society of London, B* 270, S156–S158 [2003]).

58. Nectaries are, of course, otherwise associated with flowers. The story, however, does not stop there, because so-called extrafloral nectaries (which appear to serve as a sort of defense by deflecting the attention of herbivores) are rampantly convergent and have evolved at least 457 times, including in the ferns (see M.G. Weber and K.H. Keeler *Annals of Botany* 111, 1251–1261 [2013]).

59. See B.A. Roy and R.A. Raguso *Oecologia* 109, 414–426 (1997).

60. See R.A. Raguso and B.A. Roy *Molecular Ecology* 7, 1127–1136 (1998).

61. See A. Naef et al. *New Phytologist* 154, 717–730 (2002).

62. Note also that the reverse situation can occur—that is, when a plant appears to be stricken by a fungal infection, thereby luring pollinating insects; see Z.-X. Ren et al. *Proceedings of the National Academy of Sciences, USA* 108, 7478–7480 (2011); also correction (vol. 109, p. 20774; 2012).

63. See L.R. Batra and S.W.T. Batra *Science* 228, 1011–1013 (1985).

64. See H.K. Ngugi and H. Scherm *Physiological and Molecular Plant Pathology* 64, 113–123 (2004).

65. See note 57.

66. Shykoff and Bucheli (1995), part of the title; note 57.

67. See F.P. Schiestl et al. *Biology Letters* 2, 401–404 (2006).

68. Schiestl et al. (2006), p. 401 [Abstract]; note 67.

69. See R. Kaiser *Science* 311, 806–807 (2006).

70. Despite the specificity of the association involving *Botanophila* there is no evidence for a tight coevolutionary lock (see A. Leuchtmann *Entomologia Experimentalis Applicata* 123, 13–23 [2007]); rather it is dynamic and has probably evolved more than once. And there may be other examples of fungi helping grasses to ward off attack. Thus the ergot fungus, whose unwitting consumption of its alkaloids when mixed with cereals such as rye can lead to disastrous psychotic effects (not to mention gangrene; see, for example, P.W.J. van Dongen and A.N.J.A. de Groot *European Journal of Obstetrics & Gynaecology and Reproductive Biology* 60, 109–116; 1995), may provide aposematic colors that in a manner directly parallel to the warning splodges on plants serve to warn off potential herbivores (see S. Lev-Yadun and M. Halpern *Symbiosis* 43, 105–108 [2007]).

CHAPTER 14

1. See P. Knauth and M.J. Kennedy *Nature* 460, 728–732 (2009); also commentary on pp. 698–699 by M.A. Arthur.

2. See, for example, M. Kennedy et al. *Science* 311, 1446–1449 (2006).

3. See P. Ward et al. *Proceedings of the National Academy of Sciences, USA* 103, 16818–16822 (2006).

4. See D.S. Heckman et al. *Science* 293, 1129–1133 (2001).

5. One curiosity is that in contrast to the higher plants where the obvious vegetative mass is the sporophyte (composed of diploid tissue) and the haploid gametophyte tissue is confined to reproductive structures, in the bryophytes it is the other way round.

6. Not least in terms of their tropical ecomorphs; see H. Kürschner *Nova Hedwigia* 69, 73–99 (1999).

7. "Touch wood," as we apotropaically murmur. Lignin (for an overview, see J.-K. Weng and C. Chapple *New Phytologist* 187, 273–285; 2010) is a remarkable material, both in terms of its biomechanical capacities (transformed by human ingenuity into ships like the *Beagle* and astounding feats of architectural construction) and its complex biosynthesis (see W. Boerjan et al. *Annual Review of Plant Biology* 54, 519–546 [2003]). Yet it has evolved convergently, in a red algae known as *Calliarthron* (see P.T. Martone et al. *Current Biology* 19, 169–175 [2009]). These plants, of course, are marine, and here the lignin provides a key structural role—not of course to resist the tyranny of gravity, but to strengthen the joints (or genicula) that allow the calcified thallus to bend in the water. But is it a genuine convergence? With respect to this convergence eyebrows are raised by Jing-Ke Weng multiple times (within particular groups there is a recurrence of form, as in the psathyrellids [see M. Padamsee et al. *Molecular Phylogenetics and Evolution* 46, 415–429; 2008]), and rampant convergence also occurs in what has been termed the secotioid syndrome (see H.D. Thiers *Mycologia* 76, 1–8 [1984]). Here, instead of the classic gills, the lamellae become contorted and exhibit a deliquescence that precludes ballistospory and so resembles the situation found in the gasteromycetes. Not only has the secotioid arrangement evolved several times (see J.S. Hopple and P. Vilgalys *Mycologia* 86, 96–107 [1994]), but despite a radical rearrangement of anatomy it may depend on a single mutation (see D.S. Hibbett et al. *American Journal of Botany* 81, 466–478 [1994]). Its recurrence may be unsurprising, but it reveals some general evolution principles. Most obvious is that what appears to be a significant evolutionary innovation may arise with minimal genetic change. Second, as Gerry Baura and colleagues (see their paper in *Mycologia* 84, 592–597 [1992]) remark, "secotioid forms are easy to produce but are often strongly selected against" (p. 595). Not only does this suggest that these sorts of things will, like the poor, always be with us, but a gasteromycete morphology (in which the secotioid syndrome may be a necessary intermediary step) not only evolved numerous times from mushroom ancestors (see, for example, U. Peintner et al. *American Journal of Botany* 88, 2168–2179 [2001]; also J.-M. Moncalvo et al. *Molecular Phylogenetics and Evolution* 23, 357–400; 2002), but the very instability of the linking forms may help to "precipitate" the more stable evolutionary solution (see T.D. Bruns et al. *Nature* 339, 140–142 [1989]; also S.P. Albee-Scott *Mycological Research* 111, 1030–1039; 2007). The evolutionary adventures by no means end here. One striking example is the so-called cyphelloids, which involve major simplification associated with tiny fruiting bodies. As Philomena Bodensteiner and colleagues (*Molecular Phylogenetics and Evolution* 33, 501–515; 2004) note, "the multiple independent origins of cyphelloid forms represent striking cases of parallel evolutionary reduction of complex fungal morphology" (p. 501).

29. See D.S. Hibbert et al. *Nature* 407, 506–508 (2000).

30. Not least with respect to the highly prized truffles (an ascomycete), yet a genomic investigation of this symbiosis (see F. Martin et al. *Nature* 464, 1033–1038 [2010]) reveals this ectomycorrhizal "symbiosis appears as an ancient innovation that developed several times during the course of Mycota evolution using different 'molecular toolkits'" (p. 1037).

31. They are evidently related to the other zygomycotans (Hibbert et al. *Mycological Research* 111, 509–547; 2007) and appear to be monophyletic (although the type form *Glomus* is evidently polyphyletic [see D. Schwarzolt et al. *Molecular Phylogenetics and Evolution* 21, 190–197; 2001]).

32. See, for example, V. Gianinazzi-Pearson *Plant Cell* 8, 1871–1883 (1996).

33. See B. Wang and Y.-L. Qiu *Mycorrhiza* 16, 299–363 (2005).

34. See T.N. Taylor et al. *Mycologia* 87, 560–573 (1995); also papers in *Mycological Progress* by N. Dotzler et al. vol. 5, pp. 178–184 (2006), and vol. 8, pp. 9–18 (2009).

35. See D. Redecker et al. *Science* 289, 1920–1921 (2000); also on pp. 1884–1885 commentary by M. Blackwell. See also D. Redecker et al. *Mycotaxon* 84, 33–37 (2002).

36. See B. Wang et al. *New Phytologist* 186, 514–525 (2010); also on pp. 267–270 commentary by P. Bonfante and M.-A. Selosse. See also M.I. Bidartondo et al. (*Biology Letters* 7, 574–577; 2011), who suggest that the earliest associations with land plants involves the mucoromycotan fungi rather than the more familiar glomeromycotans.

37. See C.L. Schoch et al. *Systematic Biology* 58, 224–239 (2009), see p. 236; also G.K. Mugambi and S.M. Huhndorf *Systematics and Biodiversity* 7, 453–464 (2009). Not only that, but while the standard arrangement of a fungal cortex enveloping an algal layer is the norm, an arrangement whereby the two components form vertical stacks has evolved multiple times. J. Vondrák and J. Kubásek (*Lichenologist* 45, 115–124; 2013) link this to high levels of solar irradiance, but also draw attention to similarities with the window-leaved plants such as the well-known living stone (*Lithops*).

38. See papers by W.B. Sanders et al. in *American Journal of Botany* 91, 511–522 (2004), and *European Journal of Phycology* 40, 353–361 (2005).

39. See, for example, H.-X. Xu et al. *Botany (Canadian Journal of Botany)* 86, 185–193 (2008).

40. Sanders et al. (2004), p. 516; note 38.

41. See J. Rikkinen et al. *Science* 297, 357 (2002).

42. See D. Papaefthimiou et al. *International Journal of Systematic and Evolutionary Microbiology* 58, 553–564 (2008).

43. See R. Lücking et al. *American Journal of Botany* 96, 1409–1418 (2009).

44. Lücking et al. (2009), p. 1416; note 43.

45. Lücking et al. (2009), p. 1416; note 43.

46. See T.N. Taylor et al. *American Journal of Botany* 84, 992–1004 (1997); see also R. Honegger et al. *New Phytologist* 197, 264–275 (2013).

47. See X.-I. Yuan et al. *Science* 308, 1017–1020 (2005).

48. See A. Gargas et al. *Science* 268, 1492–1495 (1995).

49. See, for example, M. Kluge et al. *Planta* 185, 311–315 (1991).

50. For example in the ascolocular lichens; see Y.-J. Liu and B.D. Hall (*Proceedings of the National Academy of Sciences, USA* 101, 4507–4512; 2004), who also address other convergences including those of yeast forms and types of asci. In pyrenocarpous lichens (see also L. Schmitt et al. *Mycologia* 97, 362–374 [2005]; also H.T. Lumbsch and S.M. Huhndorf *Mycological Research* 111, 1064–1074 [2007]), we see both recurrent

Environmental Microbiology 73, 3916–3923 [2007]). Nor is this a unique solution, because in another fungus (*Stopharia*) extraordinary stellate cells act like tiny caltrops that impale the nematode (see H. Luo et al. *Applied and Environmental Microbiology* 72, 2982–2987 [2006]). In both cases immobilization inexorably lends to digestion.

5. See, for example, G.L. Barron *Canadian Journal of Botany* 47, 1899–1902 (1969), and G.Y. Liou and S.S. Tzean *Mycologia* 89, 876–884 (1997).

6. See G.L. Barron and Y. Dierkes *Canadian Journal of Botany* 55, 3054–3062 (1977); also R.G. Thorn et al. *Mycologia* 92, 241–252 (2000), and A.T.E. Koziak et al. *Canadian Journal of Botany* 85, 762–773 (2007).

7. See C. Morikawa et al. *Mycological Research* 97, 421–428 (1993), and Y. Tanabe et al. *Mycologia* 91, 830–835 (1999).

8. Y. Li et al. (*Mycologia* 97, 1034–1046; 2005) suggest that these are the most primitive type of trapping device.

9. See, for example, D. Ahren and A. Tunlid *Journal of Nematology* 35, 194–197 (2003).

10. See M.L. Higgins and D. Pramer *Science* 155, 345–346 (1967).

11. See A.R. Schmidt et al. *Science* 318, 1743 (2007). In addition, E. Yang et al. (*Proceedings of the National Academy of Sciences, USA* 109, 10960–10965; 2012) suggest that active trapping of prey emerged close to the Permian-Triassic boundary.

12. See A.R. Schmidt et al. *American Journal of Botany* 95, 1328–1334 (2008).

13. No such association is known in living carnivorous fungi, and the convergent evolution of the lasso is also consistent with this sophisticated arrangement in living ascomycetes apparently being relatively derived (see Y. Yang et al. *Proceedings of the National Academy of Sciences, USA* 104, 8379–8384 [2007]).

14. See R.G. Thorn and G.L. Barron *Science* 224, 76–78 (1984).

15. Yang et al. (2007), p. 8383; note 13.

16. See X.-W. Huang et al. *Research in Microbiology* 155, 811–816 (2004).

17. Sure enough they possess one of the hallmarks of enzymatic convergence in the form of the classic catalytic triad of aspartic acid–histidine–serine (see, for example, M. Wang et al. *Canadian Journal of Microbiology* 52, 130–139 [2006]).

18. See M. Latijnhouwers et al. *Trends in Microbiology* 11, 462–469 (2003).

19. See P.N. Dodds et al. *New Phytologist* 183, 993–1000 (2009).

20. See N.P. Money et al. *Fungal Genetics and Biology* 41, 872–876 (2004).

21. See T.A. Richards et al. *Current Biology* 16, 1857–1864 (2006); also T.A. Richards et al. (*Proceedings of the National Academy of Sciences, USA* 108, 15258–15263 [2011]) for additional examples of lateral gene transfer from the fungi.

22. See C. Martinez et al. *Biochemistry* 33, 83–89 (1994).

23. See L. Belbahri et al. *Gene* 408, 1–8 (2008). A similar story is evident in nematode parasites on plants where various enzymes involved in the destruction of the cell wall were transferred from bacteria; see E.G.J. Danchin et al. *Proceedings of the National Academy of Sciences, USA* 107, 17651–17656 (2010).

24. In this case a possible source are bacteria (and in the case of some of the osmotrophic enzymes identified by Richards et al. [2006; note 21] they were earlier transferred to the ascomycetes from bacteria), but because these cutinases also occur in the fungi the precise directions of respective interchange of the genes remain uncertain. A further twist in the tale is the usurpation by the oomycetes of a process employed for colonization by the mycorrhizal fungi; see E. Wang et al. *Current Biology* 22, 2242–2246 (2013), with commentary on pp. R997–R999 by R. Geurts and V.G.A.A. Vleeshouwers.

25. See, for example, J.O. Andersson *Cellular and Molecular Life Sciences* 62, 1182–1197 (2005).

26. See R.R. Lew et al. *Fungal Genetics and Biology* 41, 1007–1015 (2004).

27. The fruiting bodies of basidiomycetes show a remarkable diversity. They include the classic mushroom cap bearing gills and supported on a stalk (and evidently with an aerodynamic shape to assist spore dispersal [see R. Deering et al. *Mycologia* 93, 732–736; 2001] after their ballistic blastoff, pp. 40–42), puffballs and other so-called gasteromycetes such as the earth stars and encrusting (or resupinate) varieties (see D.S. Hibbert *Systematic Biology* 53, 889–903 [2004]). All forms are convergent, although interestingly some trends are more prevalent than others, as for example the repeated evolution (at least twelve times) of the resupinates (see D.S. Hibbett and M. Binder *Proceedings of the Royal Society of London, B* 269, 1963–1969 [2002], and M. Binder et al. *Systematics and Biodiversity* 3, 113–157; 2005). So, too, mushrooms with gills have evolved at least six times and puffballs some four times (see J.M. Moncalvo et al. *Molecular Phylogenetics and Evolution* 23, 357–400 [2002]; also D.S. Hibbett et al. *Proceedings of the National Academy of Sciences, USA* 94, 12002–12006; 1997). In these and other gasteromycetes, some of which show an extraordinary degree of histological specialization into distinct tissues (see C.T. Ingold *Transactions of the British Mycological Society* 58, 179–185 [1972]), the capacity to explosively discharge the spores (ballistospory) has been lost. This may be an irreversible state, but it hasn't prevented repeated experimentation in this direction. As John Webster and Roland Weber (see their *Introduction to fungi* [Cambridge; 2007]), remark, "it still comes as a shock to most mycologists to realize just how strongly convergent the evolution of these [gasteromycetes] has been" (p. 579).

28. What of the mushrooms? This design has evolved

387. See R. Pusch et al. *Journal of Experimental Biology* 211, 921–934 (2008), and J. Engelmann et al. *Frontiers in Zoology* 6, e21 (2009).

388. Pusch et al. (2008), p. 933; note 387.

389. See his paper in *Journal of Comparative Physiology, A* 192, 601–612 (2006).

390. von der Emde (2006), p. 606; note 389.

391. See her paper in *Journal of Comparative Neurology* 472, 131–133 (2004).

392. Carr (2004), part of title; note 391.

393. Konishi (1993), p. 41; note 105.

394. See, for example, M. Kawasaki *Biological Bulletin* 191, 103–108 (1996).

395. See R.L. Green and G.J. Rose *Brain, Behavior and Evolution* 64, 85–103 (2004).

396. See, for example, M. Kawasaki and Y.-X. Guo *Journal of Neuroscience* 16, 380–391 (1996).

397. Kawasaki (1996), p. 107; note 394.

398. See M. Kawasaki *Journal of Comparative Physiology, A* 173, 9–22 (1993).

399. See, for example, M. Kawasaki *Current Opinion in Neurobiology* 7, 473–479 (1997).

400. See, for example, P. Cain and S. Malwal *Journal of Experimental Biology* 205, 3915–3923 (2002).

401. See, for example, M. Zhou and G.T. Smith *Journal of Experimental Biology* 209, 4809–4818 (2006).

402. See S.K. Tallarovic and H.H. Zakon *Animal Behaviour* 70, 1355–1365 (2005).

403. See M.E. Arnegard and B.A. Carlson *Proceedings of the Royal Society of London, B* 272, 1305–1314 (2005).

404. Arnegard and Carlson (2005), p. 1313; note 403.

405. See B.A. Carlson *Science* 332, 583–586 (2010).

406. See J.S. Albert et al. *Proceedings of the 5ᵗʰ Indo-Pacific Fish Conference, Nouméa, 1997*, pp. 647–656 (Société de française d'ichtyologie; 1999).

407. See chapters (pp. 107–134 and 135–169) by R. Nieuwenhuys and C. Nicholson in *Neurobiology of cerebellar evolution and development* (R. Llinos, ed.; Chicago; 1969). See also my *Life's solution*, pp. 194–195.

408. See, for example, U. Proske et al. *Philosophical Transactions of the Royal Society of London, B* 353, 1187–1198 (1998), and J.D. Pettigrew *Journal of Experimental Biology* 202, 1447–1454 (1999).

409. See J.D. Pettigrew and L. Wilkens in *Sensory processing in aquatic environments* (S.P. Collin et al.), pp. 420–433 (Springer; 2003).

410. Pettigrew and Wilkens (2003), p. 427; note 409.

411. See N.U. Czech-Damal et al. *Proceedings of the Royal Society of London, B* 279, 663–668 (2012).

412. See N.U. Czech-Damal et al. *Journal of Comparative Physiology, A* 199, 555–563 (2013). This analogy may also extend to some remarkable chimaeroids. Not surprisingly as chondrichthyeans they possess dense arrays of electroreceptors (see K.H. Andres and M. von Düring *Progress in Brain Research* 74, 113–131 [1988]), but much less is known of their functional capacities. Nevertheless, callorhinchids possess prominent snouts that have a high density of ampullary

receptors (for an overview, see T.J. Lisney *Reviews in Fish Biology and Fisheries* 20, 571–590; 2010), which may reflect "some kind of electrosensory specialization, perhaps relating to feeding" (p. 583).

413. Pettigrew (1999), p. 1452; note 408.

414. Pettigrew and Wilkens (2003), p. 420 [Abstract]; note 409.

415. See P.R. Manger et al. *Proceedings of the Royal Society of London, B* 263, 611–617 (1996).

416. See U. Proske and E. Gregory *Comparative Biochemistry and Physiology: Molecular & Integrative Physiology, A* 136, 821–825 (2003).

417. See their paper in *Philosophical Transactions of the Royal Society of London, B* 353, 1199–1210 (1998).

418. Pettigrew et al. (1998), p. 1199 [Abstract]; note 417.

419. See P.R. Manger et al. *Proceedings of the Royal Society of London, B* 264, 165–172 (1997).

420. See J.E. Gregory et al. *Journal of Physiology* 414, 521–538 (1989).

421. Proske et al. (1998), p. 1195; note 408.

422. Andres and von Düring (1988; note 412) speculate epidermal glands associated with electroreceptors in chimaeroids might be "involved in a kind of electrolyte secretion" (p. 122).

423. Proske et al. (1998), p. 1195; note 408.

424. See M.J. Phillips et al. *Proceedings of the National Academy of Sciences, USA* 106, 17089–17094 (2009).

425. See commentary on Phillips et al. (2009; note 424) in vol. 107 (2010) by A.B. Camens (p. E12) and reply by the former (p. E13).

426. See J. Klembara *Palaeontology* 37, 609–626 (1994).

427. See L.P. Tatarinov *Paleontological Journal* 31, 655–661 (1997).

CHAPTER 13

1. Such toxins have been repeatedly gained and lost; see T.N. Sherratt et al. *American Naturalist* 166, 767–775 (2005).

2. The nematodes are not inevitably doomed, and a touch response can allow escape; see S.M. Maguire et al. *Current Biology* 21, 1326–1330 (2011).

3. See, for example, G.L. Barron et al. *Canadian Journal of Botany* 68, 685–690 (1990), and S.J. McInnes *Polar Biology* 26, 79–82 (2003).

4. Larger victims are also found, including soil-dwelling crustaceans (see G.L. Barron *Canadian Journal of Botany* 68, 691–696 [1990]) and even insects in the form of springtails (see C. Drechsler *Mycologia* 36, 382–399 [1944]). Nor is the only way nematodes at least can meet a grisly end in the "hands" of fungi. In a basidiomycete (*Coprinus*), the hyphae bear a profusion of spiny balls (see H. Luo et al. *Mycologia* 96, 1218–1225 [2004]) that act like miniature burrs, lacerating the body of the nematode and allowing access of toxins (see H. Luo et al. *Applied and* principally the bulldog. The solution evidently lies in the fact that discharges emitted by the sexually mature male bulldogs have a much longer duration.

352. See T.H. Bullock and R.G. Northcutt *Journal of Comparative Physiology, A* 148, 345–352 (1982); see also M.R. Bradford *in* Bullock and Heiligenberg (1986), pp. 453–464; note 315.

353. See J.M. Jørgensen and T.H. Bullock *Journal of Neurocytology* 16, 311–315 (1987).

354. See M.R. Bradford et al. *Neuroscience Letters* 32, 35–39 (1982).

355. See S. Lavoué and J.P. Sullivan *Molecular Phylogenetics and Evolution* 33, 171–185 (2004).

356. Bullock et al. (1983), p. 41; note 335.

357. See, for example, papers in *Environmental Biology of Fishes* addressing convergences in reproductive patterns (F. Kirschbaum vol. 10, pp. 3–14; 1984) and ecological niches (including feeding) in floodplain settings (K.O. Winemiller and A. Adite vol. 49, pp. 175–186; 1997).

358. See, for example, T.H. Bullock *in* Bullock and Heiligenberg (1986), pp. 651–674; note 314; also J.R. Gallant et al. (*Journal of Experimental Biology* 215, 2479–2494; 2012) with respect to gene expression patterns in skeletal muscle and the derived electric organ.

359. Finger et al. (1986), p. 472; note 346.

360. The contrast of laminated versus nuclear regions of the torus circularis in these electric fish is echoed by a comparable distinction in the brains of mammals and birds; see pp. 260–262.

361. See their Table 1 (pp. 470–471); note 346.

362. Finger et al. (1986), p. 466; note 346.

363. In general, although equipped with eyes, the consensus has been that they serve little purpose in many weakly electric fish. This may not be correct because M. Landsberger et al. (*Journal of Physiologie, Paris* vol. 102, pp. 291–303; 2008) document what appears to be an exceptionally sensitive system in a mormyriform. In addition, J.A. Stevens et al. (*Brain, Behavior and Evolution* 82, 185–198; 2013) identify a mormyriform clade of petrocephalines where the sensory dedication to vision is conspicuously greater than to electroreception.

364. See W.G.R. Crampton *Anais da Academia Brasileira de Ciencias* 70, 805–847 (1998).

365. See C. Marrero and K.O. Winemiller *Environmental Biology of Fishes* 38, 299–309 (1993).

366. See, for example, A.A. Shirgaonka et al. *Journal of Experimental Biology* 211, 3490–3503 (2008).

367. See M.J. Lannoo and S.J. Lannoo *Environmental Biology of Fishes* 36, 157–169 (1993).

368. See his article in *Scientific American* 208 (3, March), 50–59 (1963).

369. See L.A. Wilkens et al. *Journal of Physiology, Paris* 96, 363–377 (2002); also L.A. Wilkens and M.H. Hofmann *Bioscience* 57, 399–407 (2007).

370. See, for example, A.A. Caputi and R. Budelli *Journal of Comparative Physiology, A* 192, 587–600 (2006).

371. See, for example, G. von der Emde and S. Fetz *Journal of Experimental Biology* 210, 3082–3095 (2007).

372. von der Emde and Fetz (2007), p. 3090; note 371.

373. von der Emde and Fetz (2007), p. 3093; note 371.

374. See their paper in *Current Biology* 14, 818–823 (2004).

375. Graff et al. (2004), p. 832; note 374.

376. See M.E. Castelló et al. *Journal of Experimental Biology* 203, 3279–3287 (2000).

377. See, for example, M. Hollmann et al. *Journal of Zoology* 276, 149–158 (2008).

378. Although most familiar in the vertebrates, especially the birds, they can help to confer an acuity in insects such as the mantids (p. 53). At first sight this might seem surprising given that the eyes of the mantids are, of course, compound. But as ever, similar challenges are met in similar ways, and it transpires that mantids possess a fovea (see J.C. Barros-Pita and H. Maldonado *Zeitschrift für vergleichende Physiologie* 67, 79–92 [1970], and G.A. Horridge and P. Duelli *Journal of Experimental Biology* 80, 165–190; 1979), which is a smaller area within the eye that provides especially acute perception. The convergence between the praying mantids and mantispids is all the more striking because although both are insects with compound eyes, the detailed arrangements are different (respectively simple apposition versus superposition), but mantispids lack a fovea (see K. Kral et al. *Journal of Experimental Biology* 203, 2117–2123 [2000]). Despite this, not only do both groups hunt and kill in much the same way, but the visual behavior of fixing the prey and use of triangulation is very much the same. The water stick insects do not have a visual acuity to match the mantids but are adept ambushers and quite capable of catching two prey simultaneously, one in each foreleg (see A. Cloarec *Journal of Experimental Biology* 120, 59–77 [1986]).

379. See G. von der Emde et al. *Journal of Physiology, Paris* 102, 279–290 (2008).

380. See Y. Galifret *Zeitschrift für Zellforschung und mikroskopische Anatomie* 86, 535–545 (1968); also A. Querubin et al. *Journal of Comparative Neurology* 517, 711–722 (2009).

381. See K.V. Fite and S. Rosenfield-Wessels *Brain, Behavior and Evolution* 12, 97–115 (1975).

382. See K.V. Fite and B.L. Lister *Brain, Behavior and Evolution* 19, 144–154 (1981).

383. Fite and Lister (1981), p. 150; note 382.

384. See P.M. Blough in *Avian visual cognition* (R.G. Cook, ed.; Cyberbook [see www.pigeon.psy.tufts.edu/avc/]; 2001).

385. In the Schnauzenorgan the mormyromasts occur at a very high density, and although those associated with the nasal region are broadly similar to their equivalents elsewhere, their arrangement is consistent with a foveal function; see M. Amey-Özel et al. *Journal of Morphology* 273, 629–638 (2012).

386. See, for example, J. Bacelo et al. *Journal of Comparative Neurology* 511, 342–359 (2008).

309. When it comes to extremely rapid release of an appendage, the mantis shrimps are by no means alone. Consider the mandibles of termite soldiers. These can not only inflict a lethal blow, but employ a mechanism that is convergent with the mandible strike of some ants; see M.A. Seid et al. *Current Biology* 18, R1049–R1050 (2008).

310. Exactly how this operates is still not quite clear. One component is the finger pointed at the part of the claw that forms a saddle-back, an arrangement also found in engineering and architectural solutions, but more recently other components of this superbly designed structure are believed to be central to the springlike action (see papers in *Journal of Experimental Biology* by S.N. Patek et al. (vol. 210, pp. 3677–3688 [2007]) and T.I. Zack et al. (vol. 212, pp. 4002–4009; 2009); also T. Claverie et al. *Evolution* 65, 443–461 (2011), who explore the modular construction of this remarkable appendage).

311. See, for example, M.V.L. Bennett in *Fish physiology*, vol. 5 (W.S. Hoar and D.J. Randall, eds.), pp. 347–491 (Academic; 1971), where the electric eel is discussed along with various other electric organs of fish.

312. See my *Life's solution*, pp. 182–184.

313. See his *Electroreception and communication in fishes*, forming vol. 42 (1996) of *Progress in Zoology*.

314. Kramer (1996), p. 17; note 313.

315. For an excellent overview, see T.H. Bullock and W. Heiligenberg *Electroreception* (Wiley; 1986).

316. These include a group of parasitic nematodes, the steinernematids; see D.I. Shapiro-Ilan et al. *Journal of Invertebrate Pathology* 100, 134–137 (2009).

317. See B.W. Patullo and D.L. Macmillan *Current Biology* 17, R83–R84 (2007).

318. See B.W. Patullo and D.L. Macmillan *Journal of Experimental Biology* 213, 651–657 (2010).

319. See P. Steullet et al. *Biological Bulletin* 213, 16–20 (2007).

320. See, for example, C.V.H. Baker et al. *Journal of Experimental Biology* 216, 2515–2522 (2013); also B. Fritzsch *Journal of Comparative Physiology, A* 173, 710–712 (1993).

321. See M. Watt et al. *Animal Behaviour* 58, 1039–1045 (1999).

322. See, for example, B. Fritzsch and U. Wahnschafte *Cell and Tissue Research* 229, 483–503 (1983). For an overview, see B. Fritzsch and H. Münz *in* T.H. Bullock and W. Heiligenberg (1986), pp. 483–496 (chap. 16); note 315.

323. See, for example, R.W. Murray *Journal of Experimental Biology* 39, 119–128 (1962).

324. See J.A. Sisneros and T.C. Tricas *Journal of Physiology, Paris* 96, 379–389 (2002).

325. See their paper in *Journal of Fish Biology* 80, 2055–2088 (2012).

326. Kempster et al. (2012), p. 2084; note 325.

327. See, for example, papers in *Journal of Experimental Biology* by S.M. Kajiura and K.N. Holland (vol. 205; pp. 3609–3621; 2002), and L.K. Jordan et al. (vol.

212, pp. 3044–3050; 2009); also C.N. Bedore and S.M. Kajiura *Physiological and Biochemical Zoology* 86, 298–311 (2013).

328. See his paper in *Interdisciplinary Science Reviews* 13, 27–39 (1988).

329. See Autrum (1988), pp. 34–35; note 328.

330. See, for example, L.J. Rosenberger *Journal of Experimental Biology* 204, 379–394 (2001).

331. See M. Camperi et al. *PLoS Computational Biology* 3(6), e113 (2007).

332. But convergent, inasmuch as molecular phylogeny of the batoids indicates independent evolution of the skate and sting-ray morphs, as well as the toothed rostrum of the sawfishes and sawsharks; see N.C. Aschliman et al. *Molecular Phylogenetics and Evolution* 63, 28–42 (2012).

333. See L.A. Wilkens et al. *Journal of Experimental Biology* 204, 1381–1389 (2001).

334. See J.H. Teeter et al. *Journal of Comparative Physiology, A* 138, 213–223 (1980).

335. See T.H. Bullock et al. *Brain Research Reviews* 6, 25–46 (1983).

336. See, for example, R.B. Leonard and W.R. Willis *Journal of Comparative Neurology* 183, 397–413 (1979); also U. Dahlgren and C.F. Silvester *Anatomischer Anzeiger* 29, 387–403 (1906).

337. Some reports speak of prey being paralyzed; see P. Moller *Electric fishes: History and behavior* (Chapman & Hall; 1995).

338. See V.D. Baron *Journal of Ichthyology* 49, 1065–1072 (2009).

339. See S. Lavoué et al. *PLoS ONE* 7, e36278 (2012); these authors also point out that the time taken to move from a general electroreception to the fully fledged electric organs is also much the same in either group.

340. See my *Life's solution*, pp. 184–188.

341. See his paper in *Current Opinion in Neurobiology* 5, 769–777 (1995).

342. Hopkins (1995), p. 775; note 341.

343. See J.G. New *Brain, Behavior and Evolution* 50, 244–252 (1997).

344. See his paper in *Journal of Fish Biology* 58, 1489–1511 (2001).

345. Alves-Gomes (2001), p. 1495; note 344.

346. See, for example, T.E. Finger et al. *in* Bullock and Heiligenberg (1986), pp. 465–481; note 315.

347. Alves-Gomes (2001), p. 1498; note 344.

348. See, for example, R.G. Northcutt et al. *Journal of Comparative Neurology* 421, 570–592 (2000).

349. See D.L. Whitehead *Journal of Morphology* 255, 253–260 (2003).

350. See K.H. Andres et al. *Anatomy and Embryology* 177, 523–535 (1988).

351. See S. Hanika and B. Kramer *Behavioral Ecology and Sociobiology* 48, 218–228 (2000); interestingly in this case the clarid in question has not evolved tuberous electroreceptors. This raises the question of how this catfish manages to detect the mormyrids,

Amundin (pp. 27–35), and W.M. Masters and H.E. Harley (pp. 249–259).

272. See V. Teloni et al. *Journal of Experimental Marine Biology and Ecology* 354, 119–131 (2008).

273. See P.W. Moore et al. *Journal of the Acoustical Society of America* 124, 3324–3332 (2008).

274. See M. Linneschmidt et al. *Proceedings of the Royal Society of London, B* 279, 2237–2245 (2012).

275. See L. Jakobsen and A. Surlykke *Proceedings of the National Academy of Sciences, USA* 107, 13930–13935 (2010).

276. See, for example, D.K.N. Dechmann et al. *Proceedings of the Royal Society of London, B* 276, 2721–2728 (2009).

277. See P.T. Madsen et al. *Bioacoustics* 15, 195–206 (2005).

278. See also S.M. Van Parijs et al. *Journal of the Acoustical Society of America* 108, 1938–1945 (2000), who document vocal similarities with the delphinids.

279. See T. Morisaka and R.C. Connor *Journal of Evolutionary Biology* 20, 1439–1458 (2007).

280. See J. Tougaard and L.A. Kyhn (*Marine Mammal Science* 26, 239–245; 2010), who document the likely convergence of echolocation sounds between *Cephalorhynchus* and the hourglass dolphins.

281. Blainville's beaked whales adopt a somewhat different strategy: keeping quiet in shallow water and only echolocating at depth; see N. Aguilar de Soto et al. *Marine Mammal Science* 28, E75–E92 (2012).

282. As Peter Madsen and colleagues (2005; note 277) stress, the example of the dwarf sperm whale is particularly interesting. Not only is it an inhabitant of deep waters, but with only a single pair of phonic lips its "sound production apparatus is very different from those of [dolphins], which in turn supports the notion of functional convergence from two very different morphological starting points" (p. 202).

283. See P.E. Nachtigall and A.Y. Supin *Journal of Experimental Biology* 211, 1714–1718 (2008), and W.W.L. Au and K.J. Benoit-Bird *Nature* 423, 861–863 (2003).

284. See S.L. DeRuiter et al. *Journal of Experimental Biology* 212, 3100–3107 (2009); in this case the authors note that unlike the bats the terminal buzzing continues after the prey is caught, but they remind us that the sound production system in cetaceans is entirely detached from the mouth.

285. See P.J.O. Miller et al. *Proceedings of the Royal Society of London, B* 271, 2239–2247 (2004).

286. See M. Johnson et al. *Journal of Experimental Biology* 209, 5038–5050 (2006).

287. See M. Johnson et al. *Proceedings of the Royal Society of London, B* 275, 133–139 (2008).

288. See their paper in *Journal of Experimental Biology* 208, 181–194 (2005).

289. Madsen et al. (2005), p. 190; note 288.

290. See L.M. Herman et al. *Journal of Comparative Psychology* 112, 292–305 (1998); also H.E. Harley et al. *Nature* 424, 667–669 (2003).

291. See M.J. Xitco and H.L. Roitblat *Animal Learning & Behavior* 24, 355–365 (1996).

292. See A.A. Pack and L.M. Herman *Journal of the Acoustical Society of America* 98, 722–733 (1995), and H.E. Harley et al. *Journal of Experimental Psychology. Animal Behavior Processes* 22, 164–174 (1996).

293. See Y. Liu et al. *Current Biology* 20, R53–R54 (2010); also S.J. Rossiter et al. *Communicative & Integrative Biology* 4, 236–239 (2011).

294. This entails a remarkable concentration near the carboxyl terminus, which appears to be especially sensitive (see Y. Li et al. *Current Biology* 20, R55–R56 [2010]) to high-frequency sound. The degree of molecular convergence between bats and whales is developed considerably further by J. Parker et al. (*Nature* 502, 228–231; 2013) and Z. Liu et al. (*Molecular Biology and Evolution* 31, 2415–2424 (2014).

295. See K.T.J. Davies et al. *Heredity* 108, 480–489 (2012); these convergences which involve the *Tmc1* and *Pjvk* genes are even more striking within the bats themselves, notably in two species that have independently evolved CF echolocation.

296. See G. Jones *Current Biology* 20, R62–R64 (2010).

297. Jones (2010), p. R62; note 296.

298. See M. Vater and M. Kössl in *Echolocation in bats and dolphins* (J.A. Thomas et al., eds.), pp. 89–99 (Chicago; 2004).

299. See E.G. Wever et al. *Proceedings of the National Academy of Sciences, USA* 68, 2381–2385 (1971a).

300. See E.G. Wever et al. *Proceedings of the National Academy of Sciences, USA* 68, 2908–2912 (1971b).

301. See M. Braun (1994); note 121.

302. Who would have thought this includes the sea urchins? However, their gnawing of the substrate, employing teeth that despite being made of calcite show exquisite bioengineering (see Y.-R. Ma et al. *Proceedings of the National Academy of Sciences, USA* 106, 6048–6053 [2009]), induces the approximately spherical test to behave as a so-called Helmholtz resonator (familiar from blowing across the neck of an empty bottle). So it is that millions of grinding teeth contribute to a dawn and evening chorus (see C. Radford et al. *Marine Ecology Progress Series* 362, 37–43 [2008]) around the coasts of New Zealand.

303. See M. Versluis et al. *Science* 289, 2114–2117 (2000).

304. See T. Devereux *Messenger gods of battle. Radio, radar, sonar: The story of electronics in war* (Brasseys; 1991).

305. See D. Lohse et al. *Nature* 413, 477–478 (2001).

306. See S.N. Patek et al. *Nature* 428, 819–820 (2004).

307. See S.N. Patek and R.L. Caldwell *Journal of Experimental Biology* 208, 3655–3664 (2005).

308. In ritualized fighting the potentially lethal blows are directed at the opponent's telson, the structure of which evidently is designed to absorb punishing but nonlethal treatment; see J.R.A. Taylor and S.N. Patek *Journal of Experimental Biology* 213, 3496–3504 (2010).

236. See, for example, M.D. Uhen *Annual Review of Earth and Planetary Sciences* 38, 189–219 (2010).

237. See F. Spoor et al. *Nature* 417, 163–166 (2002).

238. See her chapter in *Hearing by whales and dolphins* (W.W.L. Au et al., eds.), pp. 43–108 (Springer; 2000); but see B.M. Kandel and T.E. Hullar *Journal of Experimental Biology* 213, 1175–1181 (2010).

239. Ketten (2000), p. 76; note 238.

240. Spoor et al. (2002), p. 165; note 237.

241. See, for example, S. Nummela et al. *Anatomical Record, A* 290, 716–733 (2007) and J.H. Geisler et al. *Nature* 508, 383–386 (2014).

242. See, for example, W.S. Martin et al. *Journal of the Acoustical Society of America* 117, 2301–2307 (2005).

243. See J. Podos et al. *Ethology* 108, 601–612 (2002).

244. See, for example, M.H. Rasmussen and L.A. Miller in *Echolocation in bats and dolphins* (J.A. Thomas et al., eds.), pp. 50–53 (Chicago; 2004).

245. See, for example, K. Nakamura and T. Akamatsu in *Echolocation in bats and dolphins* (J.A. Thomas et al., eds.), pp. 36–40 (Chicago; 2004).

246. See B. Møhl et al. *Journal of the Acoustical Society of America* 114, 1143–1154 (2003).

247. For an overview, see T.W. Cranford et al. *Journal of Morphology* 228, 223–285 (1996).

248. See, for example, S. Huggenberger et al. *Anatomical Record* 292, 902–920 (2009), and P.T. Madsen et al. *Journal of Experimental Biology* 213, 3105–3110 (2010); also comment (pp. 1403–1404) and reply (pp. 1404–1405) by T.W. Cranford et al. and Madsen et al., respectively; also Cranford et al. (1996); note 247. See also P.T. Madsen et al. (*Journal of Experimental Biology* 216, 4091–4102; 2012), who describe how the whistling and clicking are produced on either side.

249. These air-sacs are central to sound production and of course need special adaptations to function at great depths, yet they are unlikely to be homologous with the air-sacs (p. 267) of other mammals; see J.S. Reidenberg and J.T. Laitman *Anatomical Record* 291, 1389–1396 (2008).

250. See S. Prahl et al. *Journal of Morphology* 270, 1320–1337 (2009), who draw attention to Pacinian-like mechanoreceptors (p. 131) that probably play an important role in detecting the vibrations.

251. This, however, is evidently a complex process and even allows these lips to serve as two sonar sources (see T.W. Cranford et al. *Journal of Experimental Marine Biology and Ecology* 407, 81–96 [2011]).

252. See, for example, papers by T.W. Cranford et al. in *Anatomical Record* (291, 353–378; 2008a) and *Bioinspiration & Biomimetics* (3, e016001; 2008b).

253. See S. Hemilä et al. *Journal of Comparative Physiology, A* 196, 165–179 (2010).

254. Not to mention the nose-leaf; see R. Kuc *Journal of the Acoustical Society of America* 128, 3190–3199 (2010).

255. See D.P. Phillips et al. *Hearing Research* 8, 13–28 (1982).

256. See U. Varanasi et al. *Nature* 255, 340–343 (1975).

257. M.F. McKenna et al. (*Marine Mammal Science* 28, 690–713; 2012) document melon anatomy in a variety of odontocetes, suggesting links between shape and type of sound production, focusing, and perhaps filtering of lower frequencies.

258. See papers by H.N. Koopman et al. *Journal of Comparative Physiology, B* 173, 247–261 (2003), and *Proceedings of the Royal Society of London, B* 275, 2327–2334 (2008).

259. See C. Litchfield et al. *Marine Biology* 52, 285–290 (1979); also Z.P.Z. Duggan et al. *Journal of Comparative Physiology, B* 179, 783–798 (2009).

260. See also C. Litchfield et al. *Marine Biology* 23, 165–169 (1973).

261. See C.J. Harper et al. *Journal of Morphology* 269, 820–839 (2008).

262. See T.W. Cranford et al. *PLoS ONE* 5, e11927 (2010); also Cranford (2008a, b); note 252.

263. See C.D. MacLeod et al. *Marine Ecology Progress Series* 326, 295–307 (2006).

264. See S.A. Dible et al. *Bioinspiration & Biomimetics* 4, e015005 (2009).

265. See, for example, L.A. Dankiewicz et al. *Journal of the Acoustical Society of America* 112, 1702–1708 (2002), who investigates the ability of dolphins to discriminate echo trains that are amplitude modulated.

266. This interesting story is also a rather melancholy one because the riverine dolphins are highly endangered. They appear to have invaded freshwater habits a number of times independently (see I. Cassens et al. *Proceedings of the National Academy of Sciences, USA* 97, 11343–11347 [2000]), but such events may have been favored by marine transgressions (see H. Hamilton et al. *Proceedings of the Royal Society of London, B* 268, 549–556 [2001]). The dolphins show a number of striking morphological similarities, notably a narrow rostrum. While some of these are probably convergent, phylogenetic analysis paints a more complicated picture, and not only may some features be more or less ancestral, but the extent to which life in freshwater has provided major adaptive pressures is less obvious (see J.H. Geisler et al. *BMC Evolutionary Biology* 11, e112 [2011]).

267. See T. Akamatsu et al. *Proceedings of the Royal Society of London, B* 272, 797–801 (2005).

268. For an overview of delphinid social systems see S. Gowans et al. *Advances in Marine Biology* 53, 195–294 (2007). They mention, of course, the classic dolphin-chimp comparison, but throw the net considerably wider to encompass various other mammals.

269. See, for example, R.C. Connor et al. *Animal Behaviour* 72, 1371–1378 (2006), who argue that this synchrony reflects male alliances. They point out that it is strikingly convergent to human alliances, such as coordinated dancing.

270. See T. Götz et al. *Biology Letters* 2, 5–7 (2006).

271. See chapters in *Echolocation in bats and dolphins* (J.A. Thomas et al., eds. [Chicago; 2004]) by W.W.L. Au (pp. xiii–xxvii), T.W. Cranford and M.

to manoeuvre through a dense tangle of leaves . . . and to adjust their fast approach flight exactly to within a few millimetres of the opening of a flower" (p. 1033).

208. See D. von Helversen and O. von Helversen *Nature* 398, 759–760 (1999).

209. See R. Simon et al. *Science* 333, 631–633 (2011).

210. See D. von Helversen *Journal of Comparative Physiology, A* 190, 515–521 (2004).

211. For an outstanding overview, see D.R. Griffin *Listening in the dark: The acoustic orientation of bats and men* (Yale; 1958); see also my *Life's solution*, pp. 181–182.

212. See, for example, M. Konishi and E.I. Knudsen *Science* 204, 425–427 (1979).

213. See, for example, J.J. Price et al. *Journal of Avian Biology* 35, 135–143 (2004).

214. See J.J. Price et al. *Ibis* 147, 790–796 (2005), and H.A. Thomassen et al. *Molecular Phylogenetics and Evolution* 37, 264–277 (2005).

215. See P. Gnaspini and E. Trajano in *Ecosystems of the world 30: Subterranean ecosystems* (H. Wilkens et al., eds.), pp. 251–268 (Elsevier; 2000).

216. See, for example, H.A. Thomassen et al. *Ibis* 146, 173–174 (2004).

217. See R.A. Suthers and D.H. Hector *Journal of Comparative Physiology, A* 156, 243–266 (1985).

218. See R.A. Suthers and D.H. Hector *Journal of Comparative Physiology, A* 148, 457–470 (1982).

219. See H.A. Thomassen et al. *Hearing Research* 225, 25–37 (2007).

220. See A.N. Iwaniuk et al. *Behavioural Brain Research* 167, 305–317 (2006).

221. See H.A. Thomassen and G.D.E. Povel *Biological Journal of the Linnean Society* 88, 631–643 (2006).

222. See G. Martin et al. *Naturwissenschaften* 91, 26–29 (2004).

223. The subtitle of the paper by Martin et al. (2004); note 222.

224. This ability recalls those spiders with eyes like head-lamps (p. 104), but the overengineering is assisted by a banking of the photoreceptors, up to three deep, in an arrangement that finds a parallel in some deep-sea fish (see L.M. Rojas *Brain, Behavior and Evolution* 64, 19–33 [2004]).

225. While other groups, notably the seals (pinnipeds) actively vocalize and have acute hearing (not to mention enhanced tactility, p. 129), they do not appear to employ echolocation; see R.J. Schusterman et al. in *Journal of the Acoustical Society of America* (107, 2256–2264; 2000), and in *Echolocation in bats and dolphins* (J.A. Thomas et al., eds.), pp. 531–535 (Chicago; 2004). In subsequent chapters of the latter volume F.T. Awbrey et al. (pp. 535–541) and W.E. Evans et al. (pp. 541–547) draw attention to vocalizations in Antarctic seals that may yet lead to the recognition of echolocation beyond the odontocetes.

226. See Z.-X. Luo and E.R. Eastman *Journal of Vertebrate Paleontology* 15, 431–442 (1995), and R.W. Fordyce *Smithsonian Contributions to Paleobiology* 93, 185–222 (2002); also J.M. Fahlke et al. *Proceedings of the National Academy of Sciences, USA* 108, 14545–14548 (2011).

227. See, for example, H. Brumm and S.A. Zollinger *Behaviour* 148, 1173–1198 (2011), who assess whether the Lombard effect has evolved independently in birds (where, for example, singing gets louder in noisy cities; see, for example, D. Luther and L. Baptista *Proceedings of the Royal Society of London, B* 277, 469–473; 2010) and mammals.

228. See M.M. Holt et al. *Journal of the Acoustical Society of America* 27, EL27–EL32 (2009), specifically with respect to killer whales. The Lombard effect has also been documented in nonecholocating whales; see, for example, S.E. Parks *Biology Letters* 7, 33–35 (2011).

229. Despite some reports (see M.J. Moore and G.A. Early *Science* 306, 2215 [2004], and further discussion on pp. 631–632, vol. 308 [2005]), it seems that the cetaceans possess mechanisms to avoid the "bends" (see A.S. Blix et al. *Journal of Experimental Biology* 216, 3385–3387 [2013]), notwithstanding impressively deep dives that approach 2 km (see P.L. Tyack et al. *Journal of Experimental Biology* 209, 4238–4253 [2006]). That, however, was not the case in some of the earlier whales, where bone damage indicates that independently odontocetes and mysticetes learned deep-diving, the former group having the head start (see B.L. Beatty and B.M. Rothschild *Naturwissenschaften* 95, 793–801 [2008]). Deep-diving is not, of course, restricted to the whales, and an interesting convergence is found in the myoglobins (specifically net surface charge) which also allows inferences about the likely diving depths of early whales (see S. Mirceta et al. *Science* 340, 1234192 [2013], and commentary on pp. 1293–1294 by E.L. Rezende); also M.F. Nery et al. *Journal of Molecular Evolution* 76, 386–387 (2013).

230. See his paper in *Bulletin of the American Museum of Natural History* 85, 1–350 (1945), p. 213.

231. See X.-M. Zhou et al. *Molecular Phylogenetics and Evolution* 61, 255–264 (2011).

232. See W.E. Barklow *Animal Behaviour* 68, 1125–1132 (2004).

233. See, for example, M. Orliac et al. *Proceedings of the National Academy of Sciences, USA* 107, 11871–11876 (2010); also Zhou et al. (2011); note 231.

234. See J.G.M. Thewissen et al. *Nature* 450, 1190–1194 (2007); also ensuing discussion by J.H. Geisler and J.M. Theodor in vol. 458 (2009) on pp. E1–E4 and reply by Thewissen et al. on p. E5. Phylogenetic discussion of the position of *Indohyus* has been developed by other authors, including Spaulding et al. *PLoS ONE* 4, e7062 (2009).

235. F.E. Fish (*Physiological and Biochemical Zoology* 73, 683–698; 2000) provides a valuable overview of the importance of extant semiaquatic mammals in indicating how the transformation of a furry mammal with limbs to a sleek whalelike creature was achieved.

184. See A.J. Corcoran et al. *Science* 325, 325–327 (2009).

185. See A.J. Corcoran et al. *Journal of Experimental Biology* 214, 2416–2425 (2011); also J.R. Barber and A.Y. Kawahara *Biology Letters* 9, 20130161 (2013).

186. As do, in an analogous way, the vocalization of birds. In the herons, for example, while some aspects of their vocalizations (which are innate and not learned) reflect their phylogenetic history (for example, syllabic structure), other aspects (including frequency range) are largely dependent on factors such as dense undergrowth and so much more subject to convergence (see K.G. McCracken and F.H. Sheldon *Proceedings of the National Academy of Sciences, USA* 94, 3833–3836 [1997]). In the regulid birds, where songs are learned (pp. 265–267), those acoustic traits involved with the learning process tend to be more convergent (see M. Päckert et al. *Evolution* 616–629 [2003]).

187. See M. Siemers et al. *Behavioral Ecology and Sociobiology* 50, 317–328 (2001).

188. See their paper in *Proceedings of the Royal Society of London, B* 274, 905–912 (2007).

189. Jones and Holderied (2007), p. 905; note 188.

190. See, for example, G. Neuweiler *Journal of Comparative Physiology, A* 189, 245–256 (2003).

191. See D. Vanderelst et al. *Journal of the Royal Society Interface* 9, 1100–1103 (2012).

192. See M. Kössl et al. *Journal of Comparative Physiology, A* 185, 217–228 (1999); also Jones and Teeling (2006); note 169.

193. For an overview, see M. Kössl and M. Vater in *Springer handbook of auditory research*, vol. 5 (Hearing in bats) (A. Popper and R. Fay, eds.), pp. 191–234 (Springer; 1995).

194. See G. Schuller and G. Pollak *Journal of Comparative Physiology* 132, 47–54 (1979).

195. See M. Kössl and M. Vater *Naturwissenschaften* 83, 89–91 (1996); also Kössl et al. (1999); note 192.

196. See G. Neuweiler *Physiological Review* 70, 615–641 (1990).

197. See Neuweiler (2003), p. 255; note 190.

198. Neuweiler (2003), p. 255; note 190.

199. Neuweiler (2003), p. 255; note 190.

200. Neuweiler (1990); note 196.

201. Schuller and Pollak (1979), p. 53; note 194.

202. See J.A. Simmons *Cognition* 33, 155–199 (1989).

203. See J.-E. Grunwald et al. *Proceedings of the National Academy of Sciences, USA* 101, 5670–5674 (2004).

204. See G.F. McCracken *Animal Behaviour* 45, 811–813 (1993).

205. See D. von Helversen and O. von Helversen *Journal of Comparative Physiology, A* 189, 327–336 (2003a).

206. Bat nectarivory is relatively uncommon, but in their review of this area Theodore Fleming and colleagues (*Annals of Botany* 104, 1017–1043; 2009) note that "bat pollination has evolved numerous times across angiosperm phylogeny" (p. 1033). Moreover, it is not without economic importance, not least the pol-lination of the agave cactus which provides tequila. Of the nectar-feeding bats best known are the glossophagines, but of these particularly remarkable is a species (*Anoura fistulata*) from Ecuador that has a tongue one and half times as long as its body. So immense is this tongue that the bat has to store it in a recess of the rib cage, known as glossal tube (see N. Muchhala *Nature* 444, 701–702 [2006]), and this invention most likely led to the co-evolution of ever longer tongues and ever deeper flowers (see N. Muchhala and J.D. Thomson *Proceedings of the Royal Society of London, B* 276, 2147–2152 [2009]). As the investigator Nathan Muchhala points out this has a parallel to Darwin's famous inference that a Madagascan orchid must be pollinated by a hawkmoth with a truly enormous proboscis (see C. Darwin *The various contrivances by which orchids are fertilised by insects*, 2nd ed. [Murray, 1899]; see pp. 162–166). For a historical overview, see J. Arditti et al. *Botanical Journal of the Linnean Society* 169, 403–432 (2012), a character that has evolved independently several times (see L.A. Nilsson *Trends in Ecology & Evolution* 13, 259–260 [1998]). Where bats define the pollination syndrome in general, the flowers and infloresences only open at night; what Fleming et al. refer to as a "chiropterophilous syndrome" is marked by a series of defining features such as dull-colored flowers "often located on branches and tree trunks . . . or suspended on long stalks, and tubular or radially symmetrical flowers, often of the 'shaving brush' type, that produce relatively large amounts of hexose-rich nectar" (p. 1020). In a way that recalls the remarkable carrion-flowers (pp. 162–163), bat-pollinated flowers often exude a rather unpleasant sour odor. This is due to sulfur compounds (see O. von Helversen et al. *Journal of Comparative Physiology, A* 186, 143–153 [2000]), and in the case of the New World it is clearly convergent (see J.T. Knudsen and L. Tollsten *Botanical Journal of the Linnean Society* 119, 45–57 [1995]; also S. Pettersson et al. *Biological Journal of the Linnean Society* 82, 161–168; 2004). Moths also employ crepuscular pollination, and there is evidence for convergence of floral volatiles. This is particularly notable among those visited by sphingophilous species with elongate probosces to probe deep flowers (see J.T. Knudsen and L. Tollsten *Botanical Journal of the Linnean Society* 113, 263–284 [1993]). So, too, among butterflies, despite the fact that a variety of compounds are produced, while many species are relatively indiscriminate as to which flowers they visit. Even so Susanna Andersson and colleagues (*Botanical Journal of the Linnean Society* 140, 129–153; 2002) remark, "A convergent pattern was revealed when the study focused on those compounds which were most frequently emitted and those both emitted in large relative amounts and of exclusively floral origin" (p. 149), although divergent patterns are also important.

207. See D. von Helversen et al. *Journal of Experimental Biology* 206, 1025–1034 (2003b). They also note, "At present, we do not understand how bats . . . manage roster of shrews is the venomous short-tailed shrew (see T.E. Tomasi *Journal of Mammalogy* 60, 751–759 [1979]; also B.M. Siemers et al. *Biology Letters* 5, 593–596; 2009). For a possible example in the early Tertiary mammal *Hyopsodus* see M.J. Orliac et al. *PLoS ONE* 7, e30000 (2012).

159. Ultrasound has, of course, evolved independently in a number of other groups, notably at least twice among the frogs; see A.S. Feng et al. *Nature* 440, 333–336 (2006), and V.S. Arch et al. *PLoS ONE* 4, e5413 (2009), as well as hummingbirds; see C.L. Pytte et al. *Journal of Comparative Physiology, A* 190, 665–673 (2004).

160. See N. Suga and P. H.-S. Jen *Journal of Experimental Biology* 62, 277–331 (1075); also M.B. Fenton et al. *Paleobiology* 21, 229–242 (1995).

161. See J.R. Speakman and P.A. Racey *Nature* 350, 421–423 (1991), and C.C. Voigt and D. Lewanzik *Journal of Comparative Physiology, B* 182, 831–840 (2012).

162. The convergences between the cochlea of the bats and the echolocating cetaceans are addressed on pp. 140–143. It is worth drawing attention to a remarkable frog-eating bat that necessarily has evolved the capacity to hear the low-frequency sounds of its prey, and correspondingly it shows one of the highest known concentrations of cochlear neurons; see V. Bruns et al. *Journal of Morphology* 199, 103–118 (1989).

163. See J. Zheng et al. *Nature* 405, 149–155 (2000).

164. See G. Li et al. *Proceedings of the National Academy of Sciences, USA* 105, 13959–13964 (2008); also commentary by E.C. Teeling *Trends in Ecology & Evolution* 24, 351–354 (2009).

165. For a comparable case of another auditory gene, *KCNQ4* in the bats, see Z. Liu et al. *PLoS ONE* 6, e26618 (2011).

166. For an overview of bat convergences, see M.B. Fenton *Current Zoology* 56, 454–468 (2010).

167. For an overview, see N. Ulanovsky and C.F. Moss *Proceedings of the National Academy of Sciences, USA* 105, 8491–8498 (2008).

168. See G.N. Eick et al. *Molecular Biology and Evolution* 22, 1869–1886 (2005); also E.C. Teeling et al. *Nature* 403, 183–192 (2000).

169. See G. Tsagkogeorga et al. *Current Biology* 23, 2262–2267 (2013); also commentary on pp. R999–R1001 by M.S. Springer.

170. Here frugivory has evolved independently of the other bats (collectively the microchiropterans), notably in the phyllostomids, although the methods of manipulation of the fruit differ substantially (see J.D. Vandoros and E.R. Dumont *Journal of Experimental Zoology, A* 301, 361–366 [2004]).

171. See, for example, H. Raghuram et al. *Folia Zoologica* 56, 33–38 (2007); the fruit bat *Eonycteris* evidently produces ultrasound by wing-clapping (see E. Gould *Journal of Mammalogy* 69, 378–379 [1988]).

172. See D.A. Waters and C. Vollrath *Acta Chiropterologica* 5, 209–219 (2003), and R.A. Holland *Journal*

of Experimental Biology 207, 4361–4369 (2004); also papers by Y. Yovel et al. in *Science* 327, 701–704 (2010), and *Journal of Comparative Physiology, A* 197, 515–531 (2011).

173. See Y. Yovel et al. *PLoS Biology* 9, e1001150 (2011).

174. See R.A. Holland and D.A. Water *Acta Chiropterologica* 7, 83–90 (2005).

175. This has led to yet another convergence (see Y.Y. Shen et al. *PLoS ONE* 5, e8838 [2010]) in the context for vision in dim-light, specifically between one of the opsins (RH1) in the fruit bats and a microchiropteran (the tomb bat). As noted, ultraviolet vision is evidently of particular importance (see Z.-B. Zhao et al. *Proceedings of the National Academy of Sciences, USA* 106, 8980–8985 [2009]), but of particular interest is that among those bats with highly specialized echolocation, the ultraviolet capacity is lost. Hua-bin Zhao and colleagues suggest that while "low-duty-cycle echolocators [continue to] augment their acoustic 'image' with short-wave vision" (p. 8984), the shift to Doppler-based echolocation means that these bats inhabit a world only seen through shouting. If they "see" any less clearly is quite another question.

176. See J. Ma et al. *Journal of Comparative Physiology, A* 192, 535–550 (2006).

177. See I. Ruczynski et al. *Journal of Experimental Biology* 210, 3607–3615 (2007), and commentary by G. Jones *Current Biology* 18, R34–R35 (2008).

178. See R.A. Suthers *Journal of Experimental Zoology* 158, 319–348 (1965).

179. See M.E. Bates et al. *Journal of Experimental Biology* 211, 106–113 (2008).

180. When it comes to more general or broadband noise, at least some bats show the so-called Lombard effect and increase their vocalizations (see J. Tressler and M.S. Smotherman *Journal of Comparative Physiology, A* 195, 923–934 [2009]), in much the same way that we and other animals do when the racket increases (see also notes 227 and 228).

181. Depending on relative proximity, these may be either symmetrical (i.e., both higher and lower frequencies are moved) or asymmetrical (only higher frequencies are shifted); see N. Ulanovsky et al. *Proceedings of the Royal Society of London, B* 271, 1467–1475 (2004). In addition to such examples of intraspecific JAR, not surprisingly JAR between species with similar broadcasts are also documented; see V. Necknig and A. Zahn *Journal of Comparative Physiology, A* 197, 469–473; 2011.

182. See E.H. Gillam et al. *Proceedings of the Royal Society of London, B* 274, 651–660 (2007).

183. JARs are not necessarily employed all the time. Sometimes it is evidently simpler to keep quiet and listen, especially when it is just the two of you (see C. Chiu et al. *Proceedings of the National Academy of Sciences, USA* 105, 13116–13121 [2008]). Not all bats employ JAR, because if they broadcast on a very narrow band, as for example the horseshoe bats, the risk of interference is much reduced.

NOTES 377

126. See M.C. Göpfert and D. Robert *Proceedings of the Royal Society of London, B* 267, 453–457 (2000).

127. See J.C. Jackson and D. Robert *Proceedings of the National Academy of Sciences, USA* 103, 16734–16739 (2006); also N. Mhatre and D. Robert (*Current Biology* 23, 1952–1957 [2013], with commentary on pp. R950–R952 by B. Geurten et al.), who document cochlealike behavior in the tympanic ear of a tree cricket.

128. So, too, in another dipteran, our old friend the fruit fly (see M.C. Göpfert and D. Robert *Proceedings of the National Academy of Sciences, USA* 100, 5514–5519 [2003]).

129. The title of his paper in *Comparative hearing: Insects* (R.R. Hoy et al., eds.), pp. 1–17 (Springer; 1998).

130. See P. Müller and D. Robert *Journal of Experimental Biology* 204, 1039–1052 (2001).

131. See their paper in *Current Opinion in Neurobiology* 12, 715–720 (2002).

132. Robert and Göpfert (2002), p. 717; note 131.

133. See their paper in *Science* 258, 1135–1137 (1992).

134. Robert et al. (1992), p. 1137; note 133.

135. See D. Robert et al. and D. Robert and U. Willi, respectively, in *Cell and Tissue Research* 284, 435–448 (1996), and 301, 447–457 (2000).

136. See R.R. Hoy and D. Robert *Annual Review of Entomology* 41, 433–450 (1996).

137. See D. Robert et al. *Journal of Experimental Biology* 202, 1865–1876 (1999), and R. Lakes-Harlan et al. *Proceedings of the Royal Society of London, B* 266, 1161–1167 (1999).

138. See R.S. Edgecomb et al. *Cell and Tissue Research* 282, 251–268 (1995).

139. Including from the fossil record, such as very well preserved examples from Eocene orthopterans; see R.E. Plotnick and D.M. Smith *Journal of Paleontology* 86, 19–24 (2012).

140. See, for example, A. Stumpner and D. von Helveren *Naturwissenschaften* 88, 159–170 (2001).

141. See A. Surlykke et al. *Journal of Experimental Biology* 206, 2653–2663 (2003).

142. See G. Boekhoff-Falk *Developmental Dynamics* 232, 550–558 (2005).

143. See B.P. Gupta and V. Rodrigues *Genes to Cells* 2, 225–233 (1997); also D. Jhaveri and V. Rodrigues *Development* 129, 1251–1260 (2002).

144. See A.P. Jarman et al. *Development* 121, 2019–2030 (1995). A significant parallel is found, for example, in the arrestin proteins involved in both vision and olfaction (see C.E. Merrill et al. *Proceedings of the National Academy of Sciences, USA* 99, 1633–1638 [2002]; also commentary on pp. 1113–1114 by A. Nighorn and J.G. Hildebrand).

145. See M.L. Pierce et al. *Evolution & Development* 10, 106–113 (2008).

146. See their overview in *Current Biology* 18, R869–R870 (2008).

147. Bechstedt and Howard (2008), p. R870; note 146.

148. Bechstedt and Howard (2008), p. R870; note 146.

149. See B. Nadrowski et al. *Current Biology* 18, 1365–1372 (2008), and commentary by Bechstedt and Howard (2008); note 146.

150. See F. Montealegre-Z. et al. *Science* 338, 968–971 (2012); also commentary on pp. 894–895 by R.R. Hoy.

151. In providing a succinct overview of the developmental biology of the Johnson's organ and notwithstanding its obvious derivation from the chordotonal organ, Daniel Eberl and Grace Boekhoff-Falk (see their paper in *International Journal of Developmental Biology* 51, 679–687 [2007]) are not embarrassed to draw attention to a curious fact. That is, the genes that otherwise inhibit the chordotonal organ are here required to implement the Johnson's organ, a fact that they refer to as "both astonishing and baffling" (p. 685). They, too, draw attention to the similarities that connect this remarkable piece of bio-engineering to the mammalian cochlea, and it will be interesting to see if deeper developmental principles emerge.

152. Curiously, within this group this complex arrangement has evolved twice, and even though there are a number of striking structural differences (revolving around the labial palp and labral pilifer) the aural capacity is very much the same; see M.C Göpfert and L.T. Wasserthal *Journal of Experimental Biology* 202, 909–918 (1999).

153. See J.R. Barber and W.E. Conner *Proceedings of the National Academy of Sciences, USA* 104, 9331–9334 (2007). In this paper they demonstrate not only examples of Batesian mimicry, but also the corresponding Müllerian mimicry where both "original" and "fake" species are toxic. See also J.R. Barber and A.Y. Kawahara (*Biological Letters* 9, 20130161; 2013), who document ultrasound production by a hawkmoth. A somewhat different sort of acoustic mimicry is found in the parasitic *Ormia* (p. 135), which runs the risk of being snapped up by a bat as it approaches its prey. Just as its ears are attuned to the cricket, when in flight it will show the same sort of startle response as its host if it happens to hear a bat; see M.J. Rosen *Journal of Experimental Biology* 212, 4056–4064 (2009).

154. See, for example, M.J. Novacek *Nature* 315, 140–141 (1985), and G.F. Funnell et al. *Palaeontologica Electronica* 5(2), 1–10 (2003).

155. See N.B. Simmons et al. *Nature* 451, 818–821 (2008).

156. See N. Veselka et al. *Nature* 463, 939–942 (2010).

157. See E. Gould *Proceedings of the American Philosophical Society* 109, 352–360 (1965).

158. See E. Gould et al. *Journal of Experimental Zoology* 156, 19–38 (1964); E.R. Buchler *Animal Behaviour* 24, 858–873 (1976); and K.A. Forsman and M.G. Malmquist *Journal of Zoology, London* 216, 655–662 (1988). B.M. Siemers et al. (*Biology Letters* 5, 593–596; 2009) suggest a principal function is to detect obstacles as they scurry from place to place. Among this and J.Z. Young in, respectively, vol. 306, pp. 158–189 (1984), and vol. 340, pp. 93–125 (1993).

89. Despite the body weight of mammals differing by seven orders of magnitude (shrews to blue whales), the size of these canals (in terms of their radii) hardly varies, and a compelling case exists that this system is not only robust and has little possibility to vary, but is optimally designed (see T.M. Squires *Physical Review Letters* 93, art. 198106 [2004]). Allometrically, however, the vestibular region of the cetaceans is conspicuously smaller than it "ought" to be, and with good reason.

90. See P.R. Stephens and J.Z. Young *Nature* 271, 444–445 (1978); also A.I. Arkhipkin and V.A. Bizikov *Journal of Zoology, London* 250, 31–55 (2000).

91. Much the same arrangement has evolved in the crabs (see C. Janse and D.C. Sandeman *Journal of Comparative Physiology* 130, 101–111 [1979]). Equally striking are the segmental series of fluid-filled canals located in the lumbrosacral region of the vertebral column of birds (see R. Necker *European Journal of Morphology* 37, 211–214 [1999]). These canals evidently serve as balancing organs, analogous to the semicircular canals and whose location makes good sense given the bird's bipedal hopping; see R. Necker *Journal of Comparative Physiology, A* 192, 439–448 (2006).

92. Stephens and Young (1978), p. 445; note 90.

93. See his paper in *American Naturalist* 125, 465–469 (1985).

94. Moynihan (1985), p. 465; note 93.

95. See M.A. Taylor *Nature* 323, 298–299 (1986).

96. See in particular A. Packard et al. *Journal of Comparative Physiology, A* 166, 501–505 (1990).

97. See K. Kaifu et al. *Fisheries Science* 74, 781–786 (2008).

98. See T.A. Mooney et al. *Journal of Experimental Biology* 213, 3748–3759 (2010).

99. Roger Hanlon gives a graphic description of panicking schools of squid disappearing just several seconds before the human observer could see (in clear water and excellent visibility) a fast-advancing pack of voracious carangid fish that, fortunately for the squid, make thumping noises when swimming quickly (see R.T. Hanlon and B.U. Budelmann, *American Naturalist* 129, 312–317 1987).

100. See J. Müller and L.A. Tsuji *P[LoS]ONE* 2, e889 (2007); this example is striking both because of its early appearance and unequivocal convergence based on the skull bones associated with the tympanic opening.

101. See G.A. Manley *Proceedings of the National Academy of Sciences, USA* 97, 11736–11743 (2000).

102. See G.A. Manley et al. *Journal of Comparative Physiology, A* 164, 289–296 (1989).

103. See C. Köppl *Hearing Research* 273, 65–71 (2011).

104. Manley et al. (1989), p. 295; note 102.

105. See, for example, M. Konishi *Scientific American* 268 (4, April), 34–41 (1993).

106. See C. Köppl et al. *Journal of Comparative Physiology, A* 171, 695–704 (1993).

107. See their paper in *Journal of Neurophysiology* 89, 2313–2329 (2003).

108. Köppl and Carr (2003), p. 2327; note 107.

109. See K.B. Payne et al. *Behavioral Ecology and Sociobiology* 18, 297–301 (1986); also M. Garstang *Journal of Comparative Physiology, A* 190, 791–805 (2004).

110. For an overview, see M. Vater and M. Kössl *Hearing Research* 273, 89–99 (2011).

111. Manley (2000), p. 11736; note 101. This was written, of course, before the wealth of new fossil evidence pointing to convergences in the evolution of the mammalian middle ear; see G.A. Manley *Hearing Research* 263, 3–8 (2010).

112. See L.P. Tatarinov *Paleontological Journal* 43, 1578–1599 (2009).

113. See T.H. Rich et al. *Science* 307, 910–914 (2005), with commentary on pp. 861–862 by T. Martin and Z.-X. Luo; also in vol. 309 criticism by G.S. Bever et al. (p. 1492a) and G.W. Rougier et al. (p. 1492b), and reply by Rich et al. (p. 1492c).

114. See his paper in *Nature* 450, 1011–1019 (2007).

115. Luo (2007), p. 1017; note 114.

116. This point is supported in more recent investigations (see Q. Ji et al. *Science* 326, 278–281 [2009]; and commentary on pp. 243–244 by T. Martin and I. Ruff) on the so-called Meckel's cartilage (see J. Meng et al. *Nature* 472, 181–185 [2011], and commentary on pp. 174–176 by A. Weil).

117. See, for example, S. Lange et al. *Naturwissenschaften* 94, 134–138 (2007).

118. This, of course, is only one of very many convergences that are found among the fossorial mammals (see, for example, E. Nevo *Mosaic evolution of subterranean mammals: regression, progression and global convergence* [Oxford; 1999]); also pp. 139–144 of my *Life's solution*.

119. See, for example, T. Lindenlaub et al. *Journal of Morphology* 224, 303–311 (1995).

120. See M. Müller et al. *Journal of Comparative Physiology, A* 171, 469–476 (1992).

121. See D.B. Webster and W. Plassmann in *The evolutionary biology of hearing* (D.B. Webster et al., eds.), pp. 633–636 (Springer; 1992); also M. Braun *Hearing Research* 78, 98–114 (1994).

122. See, for example, L.A. Miller *Journal of Morphology* 131, 359–382 (1970).

123. See J.E. Yack *Microscopy Research and Technique* 63, 315–337 (2004).

124. See, for example, M.J. van Staaden and H. Römer *Nature* 394, 773–776 (1998); also J.E. Yack and J.H. Fullard *Journal of Comparative Neurology* 300, 523–534 (1990).

125. See K.S. Boo and A.G. Richards *International Journal of Insect Morphology and Embryology* 4, 549–566 (1975).

47. See G.A. Doran and D.B. Allbrook *Journal of Mammalogy* 54, 887–899 (1973).

48. Even so, the distribution of these Pacinian corpuscles has its oddities: Why are the mesenteries of cats well endowed whereas ours (and dogs) have almost none?

49. See C.E. O'Connell-Rodwell et al. *Journal of the Acoustical Society of America* 108, 3066–3072 (2000).

50. See C.E. O'Connell-Rodwell et al. *Behavioral Ecology and Sociobiology* 59, 842–850 (2006).

51. See C.E. O'Connell-Rodwell et al. *Journal of the Acoustical Society of America* 122, 823–830 (2007).

52. See G.E. Weissengruber et al. *Journal of Anatomy* 209, 781–792 (2006).

53. In the subterranean habitat of the mole rat, seismic communication is also important. In at least one species the paws are an important conduit, and once again employ the Pacinian corpuscles to detect the seismic rumors (see T. Kimchi et al. *Journal of Experimental Biology* 208, 647–659 [2005]). Such seismic communication has, of course, evolved repeatedly among the vertebrates (see P.M. Narins in *Ecology of sensing* [F.G. Barth and A. Schmid, eds.], pp. 127–148 [Springer; 2001]), including even a frog (see E.R. Lewis and P.M. Narins *Science* 227, 187–189 [1985]; also Narins [2001]) and less surprisingly in the various burrow-dwellers, such as the kangaroo rats (see J.A. Randall and E.R. Lewis *Journal of Comparative Physiology, A* 181, 525–531 [1997]).

54. See D.M. Bouley et al. *Journal of Anatomy* 211, 428–435 (2007).

55. The highly tactile trunk has a number of functional similarities to the manatee and specifically its deployment of the facial hairs; see D. Bachteler and G. Dehnhardt *Zoology: Analysis of Complex Systems (ZACS)* 102, 61–69 (1999).

56. See G. Dehnhardt et al. *Zeitschrift für Säugertierkunde* 62 (Supplement, II; Proceedings of the 1ˢᵗ International Symposium on Physiology and Ethology of Wild and Zoo Animals), 37–39 (1997).

57. See L.E.L. Rasmussen and B.L. Munger *Anatomical Record* 246, 127–134 (1996).

58. See J.N. Hoffmann et al. *Anatomical Record, A* 281, 1138–1147 (2004).

59. See, for example, S.J. Lederman and R.L. Klatzky *Cognitive Psychology* 19, 342–368 (1987).

60. See B.H. Pubols and L.M. Pubols *Brain, Behavior and Evolution* 5, 342–366 (1972).

61. Hoffman et al. (2004), p. 1140; note 58.

62. Hoffman et al. (2004), p. 1145; note 58. Correspondingly a fovea for pain reception has also been identified in human fingertips; see F. Mancini et al. *Current Biology* 23, 496–500 (2013).

63. See M.W. Hamrick *Journal of Anatomy* 198, 683–688 (2001).

64. Hamrick (2001), p. 686; note 63.

65. See, for example, K.C. Catania *Journal of Comparative Physiology, A* 185, 367–372 (1999).

66. In the primates both sight and touch employ mechanisms of encoding neural information in an analogous fashion, pointing to some basic similarities that conceivably reflect convergent evolution (see J.M. Yau et al. *Proceedings of the National Academy of Sciences, USA* 106, 16457–16462 [2009]).

67. See P.D. Marasco and K.C. Catania *Journal of Experimental Biology* 210, 765–780 (2007).

68. In addition, the size of each receptive field is remarkably small (see R.N.S. Sachdev and K.C. Catania *Journal of Neurophysiology* 87, 2602–2611 [2002]).

69. See P.D. Marasco et al. *Proceedings of the National Academy of Sciences, USA* 103, 9339–9344 (2006).

70. The title of Catania (1999); note 65.

71. Catania (1999), p. 367; note 65.

72. See K.C. Catania and J.H. Kass *Journal of Comparative Neurology* 387, 215–233 (1997).

73. See K.C. Catania and F.E. Remple *Brain, Behavior and Evolution* 63, 1–12 (2004).

74. Catania and Remple (2004), p. 4; note 73.

75. Catania and Remple (2004), pp. 1–2; note 73.

76. See K.C. Catania and F.E. Remple *Nature* 433, 519–522 (2005).

77. While not identical, there are also some similarities in the epidermis of the hedgehog snout inasmuch as the receptors (of which there are three types) serve as an equivalent to the Eimer's organ; see L. Páč and L. Malinovský *Scripta medica* 54, 113–125 (1981).

78. See U. Proske and J.E. Gregory *Proceedings of the Linnean Society of New South Wales* 125, 319–326 (2004); also P.R. Manger and J.D. Pettigrew *Brain, Behavior and Evolution* 48, 27–54 (1996).

79. See, for example, M.J. Rowe et al. *Comparative Biochemistry and Physiology, A* 136, 883–893 (2003).

80. See D.A. Mahns et al. *Journal of Comparative Neurology* 459, 173–185 (2003).

81. See M.J. Rowe et al. *Proceedings of the Linnean Society of New South Wales* 125, 301–317 (2004); also Rowe et al. (2003); note 79.

82. Rowe et al. (2004), p. 301; note 81.

83. See M.J. Phillips et al. *Proceedings of the National Academy of Sciences, USA* 106, 17089–17094 (2009).

84. See D.B. Leitch and K.C. Catania (*Journal of Experimental Biology* 215, 4217–4230; 2012), who draw attention to the similarities of tactile systems of monotremes and moles.

85. They are the only cold-blooded (poikilothermic) mammals (see R. Buffenstein and S. Yahav *Journal of Thermal Biology* 16, 227–232 [1991]) and evidently depend on the stable thermal regime of their tunnels.

86. See S.D. Crish et al. *Brain, Behavior and Evolution* 62, 141–151 (2003), and T.J. Park et al. *Journal of Comparative Neurology* 465, 104–120 (2003).

87. See K.C. Catania and M.S. Remple *Proceedings of the National Academy of Sciences, USA* 99, 5692–5697 (2002); also E.C. Henry and K.C. Catania *Anatomical Record, A* 288, 626–645 (2006).

88. See B.U. Budelmann and T. Yu *Vie et Milieu* 47, 95–99 (1997); also papers in *Philosophical Transactions of the Royal Society of London, B* by B.U. Budelmann

Royal Society of London, B 226, 249–261 (1985). So, too, the setae on the antennules of the copepods have an extraordinary sensitivity (see D.M. Fields et al. *Marine Ecology Progress Series* 227, 173–186 [2002]) and probably play a key role in the remarkable success in capturing prey (see H. Jiang and G.A. Paffenhöfer in *Marine Ecology Progress Series* 373, 37–52 [2009], and T. Kiørboe et al. *Proceedings of the National Academy of Sciences, USA* 106, 12394–12399 [2009]).

14. See D.L. Feigenbaum *Canadian Journal of Zoology* 56, 536–546 (1978).

15. See L.M. Mejía-Ortiz and R.G. Hartnoll *Crustaceana* 79, 593–600 (2006).

16. See, for example, D.O. Elias et al. *Journal of Experimental Biology* 206, 4029–4039 (2003).

17. See D.O. Elias et al. *Journal of Experimental Biology* 209, 1074–1084 (2006).

18. See M. Virant-Doberlet and A. Cokl *Neotropical Entomology* 33, 121–134 (2004).

19. See I. Castellanos and P. Barbosa *Animal Behaviour* 72, 461–469 (2006).

20. See G.V. Broad and D.L.J. Quicke *Proceedings of the Royal Society of London B* 267, 2403–2409 (2000), and N. Laurenne et al. *Biological Journal of the Linnean Society* 96, 82–102 (2009).

21. See R.B. Rosengaus et al. *Naturwissenschaften* 86, 544–548 (1999).

22. This includes a regular reamplification as seen in both nervous transmission and human systems such as beacons (see A. Röhrig et al. *Insectes Sociaux* 46, 71–77 [1999]).

23. See T.A. Evans et al. *Proceedings of the National Academy of Sciences, USA* 102, 3732–3737 (2005).

24. See K.C. Catania *PLoS ONE* 3, e3472 (2008); also O. Mitra et al. *Biology Letters* 5, 16–19 (2008).

25. Catania (2008), p. 1 [Abstract]; note 24.

26. See J.H. Kaufman *Copeia* 1986, 1001–1004 (1986).

27. See N. Tinbergen *The herring gull's world: A study of the social behaviour of birds* (Collins [New Naturalist]; 1953); see pp. 32–35.

28. The latter example is particularly interesting because the gulls' equivalent of worm grunting is only observed in meadows (and is evidently engaged in by other birds such as lapwings). Darwin was well aware of the sensitivity of earthworms to vibrations, and in his book *The formation of vegetable mould through the action of worms, etc.* (Murray; 1883), he drew attention to a captive peewit that "used to stand on one leg and beat the turf with the other leg until the worms crawled out of their burrows, when they were instantly devoured" (pp. 28–29).

29. Tinbergen (1953), p. 34; note 27.

30. See D.S. Wethey and S.A. Woodin *Biological Bulletin* 209, 139–145 (2005).

31. The elongate bill of the enantiornithine *Longirostravis* suggests that such shore-feeding had evolved during the Cretaceous (see L.-H. Hou et al. *Naturwissenschaften* 91, 22–25 [2004]), and quite independently of the neorthines. Related forms have similar bills,

but it is likely that collectively they reflect a series of trophic specializations; see, for example, papers by J.-M.K. O'Connor et al. in *Journal of Vertebrate Paleontology* 29, 188–204 (2009), and *Acta Palaeontologica Polonica* 56, 463–475 (2011). For evidence of convergent activity by Jurassic pterosaurs see D.M. Unwin *Lethaia* 29, 373–386 (1997).

32. See T. Piersma et al. *Proceedings of the Royal Society of London, B* 265, 1377–1383 (1998); also S. Nebel et al. *Animal Biology* 55, 235–243 (2005).

33. Being tactile receptors, Herbst corpuscles find other functions in birds, not least in flight control (see W. Hörster *Journal of Comparative Physiology* 166, 663–673 [1990]). This finds a convergent counterpart in the mechanoreceptors found on the bat wing (see S. Sterbing-D'Angelo et al. *Proceedings of the National Academy of Sciences, USA* 108, 11291–11296 [2011]; also commentary by G. Jones *Current Biology* 21, R666–R667; 2011).

34. See papers by S.J. Cunningham et al. in *Auk* 127, 308–316 (2010), and *Journal of Avian Biology* 41, 350–353 (2010).

35. See C. Gutiérrez-Ibáñez et al. *Brain, Behavior and Evolution* 74, 280–294 (2009).

36. In either case the demands of a somatosensory enrichment reflect, respectively, the methods of filter-feeding and manipulation of seeds and fruit. This convergence, therefore, is more to do with tactility per se. What are almost certainly mechanoreceptors have also been identified in the Senegal parrots, albeit within the beak; see Z.P. Demery et al. *Proceedings of the Royal Society of London, B* 278, 3687–3693 (2011).

37. See A.N. Iwaniuk et al. *Zootaxa* 2296, 47–67 (2009).

38. Iwaniuk et al. (2009), p. 66; note 37.

39. See papers by S.J. Cunningham et al. in *Journal of Anatomy* 211, 493–502 (2007) and *PLoS ONE* 8, e80036 (2013).

40. See G.R. Martin et al. *PLoS ONE* 2, e198 (2007).

41. In this bird there is a peculiar emphasis on the heightened sense of smell linked to the nostrils; uniquely for birds they are located on the tip of the elongate bill. See F. Castro et al. *Journal of Avian Biology* 41, 213–218 (2010), who note how "Kiwi olfactory behaviours are reminiscent of mammalian sniffing" (p. 214); also S.S. Steiger et al. *Proceedings of the Royal Society of London, B* 275, 2309–2317 (2008), who demonstrate that the olfactory receptor genes in a variety of birds, including the kiwi (along with the kakapo and jungle fowl) approach in number those of mammals.

42. Martin et al. (2007), p. 1 [Abstract]; note 40.

43. See S.S. Seneviratne and I.L. Jones *Behavioral Ecology* 19, 784–790 (2008).

44. See, for example, A.K. McIntyre *Trends in Neurosciences* 3, 202–205 (1980).

45. See R. Saxod *Microscopy Research and Technique* 34, 313–333 (1996).

46. See J. Bell et al. *Progress in Neurobiology* 42, 79–128 (1994).

NOTES   373

remark how their study "reveals an unexpected history of expansion of these genes' subfamilies from only one or two genes in each insect order" (p. 2). One could observe that this is exactly what one should expect, and not to be surprised if these independent trajectories reveal further similarities.

108. See C. Kleineidam and J. Tautz *Naturwissenschaften* 83, 566–568 (1996); also C. Kleineidam et al. in vol. 88, pp. 301–305 (2001), for corresponding evidence for ventilation to ensure sufficient oxygen.

109. See C. Kleineidam et al. *Arthropod Structure & Development* 29, 43–55 (2000).

110. See C. Thom et al. *Journal of Chemical Ecology* 30, 1285–1288 (2004), and J. Goyret et al. *Proceedings of the National Academy of Sciences, USA* 105, 4565–4570 (2008).

111. See M.T. Gillies *Bulletin of Entomological Research* 70, 525–532 (1980); also T. Lu et al. *Current Biology* 17, 1533–1544 (2007).

112. See, for example, M. Enserink *Science* 298, 90–92 (2002).

113. Fruit flies, however, tell an interesting story. For reasons that are still not entirely clear, fruit flies avoid elevated levels of carbon dioxide (see G.S. Suh et al. *Nature* 431, 854–859 [2004]), which depends on two taste receptors but ones that are located in the antennae where the olfactory machinery is located (see J.Y. Kwon et al. *Proceedings of the National Academy of Sciences, USA* 104, 3574–3578 [2007], and W.D. Jones et al. *Nature* 445, 86–90 [2007], with commentary on pp. 30–31 by R.I. Wilson).

114. Quite possibly carbonic anhydrase plays a role, as it does independently in mice (see J. Chandrashekar et al. *Science* 326, 443–445 [2009]. Nor is the involvement of fruit flies limited to aversion to carbon dioxide, because it is also known that among their taste capacities is one for carbonated water (see W. Fischler et al. *Nature* 448, 1054–1057 [2007]). While the amount of carbon dioxide released when a can of soda beverage is opened can be so high as to make it detectable by smell, our interest is in the sugars. In the fruit fly it may seem paradoxical that they avoid carbon dioxide yet are tuned to carbonated water. Perhaps the resolution relies in the yeasts upon which they feed, which release carbon dioxide that in solution (as carbonated water) is a vital cue for the hungry insect.

CHAPTER 12

1. See, for example, papers in *Journal of Comparative Physiology, A* by B.U. Budelmann and H. Bleckman (vol. 164, pp. 1–5; 1988), and H. Bleckmann et al. (vol. 168, pp. 247–257; 1991); also S. Lenz et al. *Zoologischer Anzeiger* 234, 145–157 (1995).

2. H. Wilkens and U. Strecker (*Biological Journal of the Linnean Society* 80, 545–555; 2003) suggest that eye loss occurred multiple times. In addition, they present evidence to suggest that the genetic pathways

employed may have differed; also J.E. Niven *Current Biology* 18, R27–R29 (2008).

3. Caves, of course, are a rich source of insights into evolutionary convergence, with the recurrent evolution of many sorts of so-called troglomorphs; for an overview, see T.G. Langecker in *Ecosystems of the world 30: Subterranean ecosystems* (H. Wilkens et al., eds.), pp. 135–157 (Elsevier; 2000), and D.C. Culver et al. *Adaptation and natural selection in caves: the evolution of Gammarus minus* (Harvard; 1995). Best known are the cave fish but various arthropod groups (such as opilionid harvestmen; see S. Derkarabetian et al. *PLoS ONE* 5, e10388; 2010) are highly informative.

4. But also via their odontodes; as G. Haspal et al. (*Current Biology* 22, R629–R630; 2012) say, "By the teeth of their skin" [part of title].

5. Although this has been questioned by S.P. Windsor et al. *Journal of Experimental Biology* 211, 2950–2959 (2008).

6. If cave fish can navigate with perfect assurance in the pitch dark, it is scarcely surprising that the lateral-line organ has other versatilities. Consider, for example, the remarkable synchronicity of fish schools, a behavior that is almost certainly largely governed through the lateral line. So, too, detection of struggling prey or tracking it, and even sexual communication, are mediated by this versatile system.

7. See, for example, S. Coombs et al. *Philosophical Transactions of the Royal Society of London, B* 355, 1111–1114 (2000), and B. Ćurčić-Blake and S.M. van Netten *Journal of Experimental Biology* 209, 1548–1559 (2006).

8. See J. Goulet et al. *Journal of Comparative Physiology, A* 194, 1–17 (2008).

9. Nor do analogies with other sensory systems end here. Consider laterality in those blind cave fish. When it first approaches a novel object, initially there is a bias for using the lateral line on the right-hand side (see T. Burt de Perera and V.A. Braithwaite *Current Biology* 15, R241–R242 [2005]).

10. See C.A. York and I.K. Bartol *Journal of Experimental Biology* 217, 2437–2439 (2014)

11. For manatees see R.L. Reep et al. *Brain, Behavior and Evolution* 59, 141–154 (2002), and R.L. Reep and D.K. Sarko in *Evolution of nervous systems: A comprehensive reference*, vol. 3 (J.H. Kaas and L.A. Krubitzer, eds.), pp. 207–213 (Academic; 2007), where they note the body hairs "[constitute] a distributed three-dimensional somato-sensory array potentially capable of encoding the intensity and direction of water displacements and low frequency vibrations . . . this system may be used for touch at a distance, analogous to the lateral line system of fish" (p. 209). Also J.C. Gaspard et al. *Journal of Comparative Physiology, A* 199, 441–450 (2013). For seals see G. Dehnhart et al. *Nature* 394, 235–236 (1998).

12. See K.C. Catania *Brain, Behavior and Evolution* 56, 146–174 (2000), and *Anatomical Record, A* 287, 1038–1050 (2005).

13. See E.J. Denton and J. Gray *Proceedings of the*

trigger a cardiac response in a moth. Correspondingly, in *Volvox* (pp. 305–306, note 1) a glycoprotein is sufficient to induce sexual development at an amazingly low concentration of 6 x 10^{-17}M; see M. Sumper et al. *EMBO Journal* 12, 831–836 (1993).

70. There is a further twist to this story in that insects also employ ionotropic glutamate receptors for the detection of compounds such as alcohol; see for example U.B. Kaupp *Nature Reviews Neuroscience* 11, 188–200 (2010), and L. Abuin et al. *Neuron* 69, 44–60 (2011). Significantly these molecules are very ancient, but not only may they have arisen independently in plants and animals but among the latter group were evidently co-opted from a synaptic function; see V. Croset et al. *PLoS Genetics* 6, e100164 (2011).

71. See P. Sengupta et al. *Cell* 84, 899–909 (1996), and P.J. Clyne et al. *Neuron* 22, 327–338 (1999).

72. See R. Benton et al. *PLoS Biology* 4, e20 (2006), and M. Wistrand et al. *Protein Science* 15, 509–521 (2006); also P. Tsitoura et al. *PLoS ONE* 5, e15428 (2010).

73. See C. Lundin et al. *FEBS Letters* 581, 5601–5604 (2007).

74. See K. Sato et al. and D. Wicher et al. in vol. 452 (2008) of *Nature*, pp. 1002–1007 and pp. 1007–1011 (2008), respectively; also commentary by T.S. Ha and D.P. Smith *Cell* 133, 761–763 (2008).

75. This entails the use of various G-proteins in the signal cascade; see, for example, Y. Deng et al. *PLoS ONE* 6, e18605 (2011).

76. In this case transduction appears to be direct— see Sato et al. (2008); note 74; also M. Murakami and H. Kijima *Journal of General Physiology* 115, 455–466 (2000), in the related context of gustation of sugars. For a discussion of why there appear to be at least two distinct modes of signaling, see T. Nakagawa and L.B. Vosshall *Current Opinion in Neurobiology* 19, 284–292 (2009).

77. See their paper in *Insect Biochemistry and Molecular Biology* 38, 770–780 (2008).

78. See Y.T. Tang *Journal of Molecular Evolution* 61, 372–380 (2005).

79. Specific regions of the protein are central to the actual process of transduction; see A.S. Nichols and C.W. Luetje *Journal of Biological Chemistry* 285, 11854–11862 (2010).

80. Maybe a clue comes from the choanoflagellates, where apparently three times independently there has been a horizontal gene transfer from algae. A.M. Nedelcu et al. (*Journal of Evolutionary Biology* 22, 1882–1894; 2009) point out in each case the replacement genes usurp the function of the home gene, and given they are all involved in nitrogen metabolism, it is possible the imports are actually better at doing the job than the incumbents.

81. See the overview by P. Sengupta *Pflügers Archiv (European Journal of Physiology)* 454, 721–734 (2007).

82. See their paper in *Journal of Neurochemistry* 109, 1570–1583 (2009).

83. Spehr and Munger (2009), p. 1579; note 82.

84. See A.F. Silbering and R. Benton *EMBO Reports* 11, 173–179 (2010).

85. Silbering and Benton (2010; note 84) remark that this difference between insect and mammal could in principle "reflect a mere chance in evolution" (p. 177). Yet the employment of these ligand-gated ion channels can be traced to the gustatory apparatus of the water-flea (see D.C. Peñalva-Aruna et al. *BMC Evolutionary Biology* 9, e79 [2009]), which as a crustacean belongs to the group from which the insects arose.

86. See my overview in *EMBO Reports* 13, 281 (2012).

87. See X. Li et al. *PLoS Genetics* 1, e3 (2005).

88. See P.-H. Jiang et al. *Proceedings of the National Academy of Sciences, USA* 109, 4956–4961 (2012); these include a number of nonfelids and also in the marine realm dolphins and sea lions, which swallow their prey whole.

89. See R.F. Stocker *Cell and Tissue Research* 275, 3–26 (1994).

90. See their paper in *Current Biology* 14, 1065–1079 (2004).

91. Thorne et al. (2004), p. 1076; note 90.

92. See his paper in *Current Biology* 14, R560–R561 (2004).

93. The subtitle of Stocker (2004); note 92.

94. Stocker (2004), p. R560; note 92.

95. See, for example, K. Scott *Neuron* 48, 455–464 (2005).

96. See papers in *Proceedings of the National Academy of Sciences, USA* by H.M. Robertson et al. (vol. 100 [Supplement, 2], pp. 14537–14542; 2003), and K. Sato et al. (vol. 108, pp. 11680–11685; 2011); also H.-J. Zhang et al. *PLoS ONE* 6, e24111 (2011).

97. See A. Gardiner et al. *Biology Letters* 5, 244–247 (2009).

98. See their paper in *Cell* 139, 234–244 (2009).

99. Yarmolinsky et al. (2009), p. 240; note 98.

100. Yarmolinsky et al. (2009), p. 234 [Abstract]; note 98.

101. See, for example, H. Amrein and N. Thorne *Current Biology* 15, R673–R684 (2005).

102. See, for example, N. Meunier et al. *Journal of Neurobiology* 56, 139–152 (2003).

103. See, for example, Y. Jiao et al. *Proceedings of the National Academy of Sciences, USA* 104, 14110–14115 (2007), and J. Slone et al. *Current Biology* 17, 1809–1816 (2007).

104. See B. Gordesky-Gold et al. *Chemical Senses* 33, 301–309 (2008).

105. Gordesky-Gold et al. (2008), p. 306; note 104.

106. See A. Keller *Current Biology* 17, R77–R81 (2007).

107. Many other insects possess receptors for sugar. Most likely this capacity had a single origin, but the diversification (see L.B. Kent and H.M. Robertson *BMC Evolutionary Biology* 9, e41 [2009]) has proceeded independently from an ancestral basis of either one (in the flies and beetles) or two (in lepidopterans and hymenopterans) genes. Kent and Robertson also tory repertoire; see T. Kishida et al. *Biology Letters* 3, 428–430 (2007).

28. See M.R. McGowen et al. *Systematic Biology* 57, 574–590 (2008).

29. See their paper in *Genome Research* 20, 1–9 (2010).

30. Hayden et al. (2010), p. 2; note 29.

31. As noted (Thewissen et al. 2011; note 21), the mysticetes retain an olfactory capacity, but interestingly the odontocetes retain molecular machinery in the form of olfactory marker proteins that may serve for other neuronal functions; see T. Kishida and J.G.M. Thewissen *Gene* 492, 349–353 (2012). The loss of olfaction in the odontocetes was probably gradual, and squalodonts from the Miocene still possessed a high sensitivity that in due course was traded for echolocation (see S.J. Godfrey *Comptes Rendus. Palevol.* 12, 519–530 [2013]).

32. For animals that are more temporary inhabitants of water, an ingenious expedient has been arrived at whereby each nostril expels a bubble of air and then retracts it into the nose with its olfactory cargo. This has evolved independently in the star-nosed mole (see K.C. Catania *Nature* 444, 1024–1025 [2006]) as well as the tiny, but voracious and lightning-fast, water shrew (see K.C. Catania et al. *Proceedings of the National Academy of Sciences, USA* 105, 571–576 [2008]; also Catania [2006]).

33. See T. Kishida and T. Hikida *Journal of Evolutionary Biology* 23, 302–310 (2010).

34. See, for example, A. Keller and L.B. Vosshall *Current Biology* 14, R875–R878 (2004).

35. See R. Bentley *Chemical Reviews* 106, 4099–4112 (2006).

36. Keller and Vosshall (2004), p. R875, note 34.

37. See N. Sobel et al. *Nature*, 402, 35 (1999).

38. See R. Rajan et al. *Science* 311, 666–670 (2006); and also moles (see K.C. Catania *Nature Communications* 4, e1441 [2013]); probably sharks, including the hammerheads (see S.M. Kajiura et al. *Journal of Morphology* 264, 253–263 [2005]; also J.M. Gardiner and J. Atema *Current Biology* 20, 1187–1191; 2010); and maybe ants (see K. Steck et al. *Animal Behaviour* 79, 939–945 [2010]).

39. See M. Louis et al. *Nature Neuroscience* 11, 187–199 (2008).

40. See T.H.J. Burne and L.J. Rogers *Behavioural Brain Research* 133, 293–300 (2002); also G. Vallortigara and R.J. Andrew *Neuropsychologia* 32, 417–423 (1994).

41. See P. Letzkus et al. *Current Biology* 16, 1471–1476 (2006).

42. See her paper in *Brain, Behavior and Evolution* 59, 273–293 (2002).

43. The title of her paper; note 42.

44. See, for example, B.W. Ache and J.M. Young *Neuron* 48, 417–430 (2005).

45. Ache and Young (2005), p. 426; note 44.

46. See N.J. Strausfeld and J.G. Hildebrand *Current Opinion in Neurobiology* 9, 634–639 (1999); also R. Benton *Cellular and Molecular Life Sciences* 63, 1579–1585 (2006).

47. See their paper in *Seminars in Cell and Developmental Biology* 17, 433–442 (2006).

48. Kay and Stopfer (2006), p. 433; note 47.

49. See, for example, P. Pelosi et al. *Cellular and Molecular Life Sciences* 63, 1658–1676 (2006).

50. See D.R. Flower *Biochemical Journal* 318, 1–14 (1996); also M. Tegoni et al. *Biochimica et Biophysica Acta: Protein Structure and Molecular Enzymology* 1482, 229–240 (2000).

51. See, for example, L.A. Graham and P.L. Davies *Gene* 292, 43–55 (2002), and R.G. Vogt et al. *Journal of Experimental Biology* 205, 719–744 (2002); also Pelosi et al. (2006); note 49.

52. See, for example, I. Gaillard et al. *Cellular and Molecular Life Sciences* 61, 456–469 (2004).

53. See their paper in *Annual Review of Physiology* 70, 93–117 (2008).

54. Kaupp et al. (2008), subtitle of section, p. 101; note 53.

55. Kaupp et al. (2008); p. 101; note 53.

56. See W. Bialek *Annual Review of Biophysics and Biophysical Chemistry* 16, 455–478 (1987).

57. See D.A. Baylor et al. *Journal of Physiology* 288, 613–634 (1979); also J. Pahlberg and A.P. Sampath *BioEssays* 33, 438–447 (2011).

58. See his paper in *Interdisciplinary Science Reviews* 13, 27–39 (1988).

59. Autrum (1988), p. 31; note 58.

60. See his paper in *Hearing Research* 78, 98–114 (1994).

61. Braun (1994), p. 99; note 60.

62. See his review of "How the ear's works work," in *Nature* 341, 397–404 (1989).

63. Hudspeth (1989), p. 399; note 62; also A.C. Crawford and R. Fettiplace *Journal of Physiology* 364, 359–379 (1985), with comparable sensitivities in the more primitive turtle ear.

64. See W. Denk and W.W. Webb *Physical Review Letters* 63, 207–210 (1989); also F.-Y. Chen et al. *Nature Neuroscience* 14, 770–774 (2011).

65. Denk and Webb (1989), p. 209; note 64.

66. See H.M. Moir et al. *Biology Letters* 9, 20130241 (2013). While acknowledging that some bats may produce matching frequencies, these authors suggest it is a by-product of an ear geared to increased temporal acuity.

67. See M. Winklhofer et al. *Proceedings of the Royal Society of London, B* 280, 20130853 (2013).

68. See K.-E. Kaissling *Annual Review of Neuroscience* 9, 121–145 (1986); also V. Bhandawat et al. *Proceedings of the National Academy of Sciences, USA* 107, 18682–18687 (2010).

69. See A.M. Angioy et al. (*Chemical Senses* 28, 279–284; 2003), who report that less than six molecules can appropriate wavelengths of infrared (see H. Schmitz et al. *Nature* 386, 773–774 [1997], and H. Schmitz and H. Bleckmann *Journal of Comparative Physiology, A* 182, 647–657; 1998), and this is then released as thermal energy. The snag, of course, is that the rest of the insect cuticle has the same cuticular composition, and the movement of the sensor is very likely both subtle and complex (see D.X. Hammer et al. *Comparative Biochemistry and Physiology, A* 132, 381–392 [2002]).

190. See M. Müller et al. *Journal of Experimental Biology* 211, 2576–2583 (2008).

191. See his paper in *Annals of the Entomological Society of America* 98, 738–746 (2005).

192. Evans (2005), p. 740; note 191.

193. See H. Schmitz et al. *Journal of Comparative Physiology, A* 186, 543–589 (2000).

194. See W. Gronenberg and H. Schmitz *Cell and Tissue Research* 297, 311–318 (1999).

195. See A. Schmitz et al. *Naturwissenschaften* 95, 455–460 (2008).

196. See A. Schmitz et al. *Arthropod Structure & Development* 39, 17–25 (2010).

197. See H. Schmitz et al. *Naurwissenschaften* 89, 226–229 (2002).

198. See E.-J. Kreiss et al. *Arthropod Structure & Development* 34, 419–428 (2005).

199. See E. Kreiss et al. *Journal of Comparative Physiology, A* 193, 729–739 (2007).

200. In this case the beetles alight on the hot substrate, briefly run around, and then promptly disappear into the warm ash, probably to mate. The ensuing larvae in due course hatch, and most likely harvest the fungi that spring up from the charred remains.

201. It belongs, however, to another family and even if an infrared detector has evolved twice in the buprestid beetles and might be taken to indicate an evolutionary predilection, their completely different arrangements is noteworthy.

202. See T. Mainz et al. *Arthropod Structure & Development* 33, 419–430 (2004).

203. See H. Schmitz et al. *Naturwissenschaften* 87, 542–545 (2000); also E.S. Schneider and H. Schmitz *Arthropod Structure & Development* 42, 135–142 (2013).

204. See H. Schmitz and S. Trenner *Journal of Comparative Physiology, A* 189, 715–722 (2003).

205. Thus K. Ohtsu and S.-I. Uye (*Biological Bulletin* 221, 243–247; 2011) identify an infrared taxis in a jellyfish, suggesting heat-sensitive receptors are located on the rhopalia (which lack ocelli).

206. See S. Takács et al. *Proceedings of the Royal Society of London, B* 276, 649–655 (2009).

207. Takács et al. (2009), p. 654; note 206.

CHAPTER 11

1. See his paper in *Studies in History and Philosophy of Biological and Biomedical Sciences* 38, 807–819 (2007).

2. Shapiro (2007), part of title; note 1.

3. Shapiro (2007), p. 812; note 1.

4. See A. Briegel et al. *Proceedings of the National Academy of Sciences, USA* 106, 17181–17186 (2009).

5. Briegel et al. (2009), p. 17185; note 4.

6. See G. Morrot et al. *Brain and Language* 79, 309–320 (2001).

7. See R.A. Österbauer et al. *Journal of Neurophysiology* 93, 3434–3441 (2005).

8. See their paper in *Proceedings of the Royal Society of London, B* 271, S131–S133 (2004).

9. Curtis et al. (2004), p. S131; note 8.

10 See Curtis et al. *Medische Antropologie* 11, 143–158 (1999).

11. See, for example, Y. Zhang et al. *Nature* 438, 179–184 (2005); also C.H. Rankin *Current Biology* 16, R89–R91 (2006).

12. Zhang et al. (2005), p. 183; note 11.

13. See M. Rubio-Godoy et al. *Medical Hypotheses* 68, 61–66 (2007).

14. See A. Hay-Schmidt *Proceedings of the Royal Society of London, B* 267, 1071–1079 (2000); also S. Caverley et al. *Journal of Experimental Biology* 209, 4858–4868 (2006).

15. See, for example, E.C. Azmitia *Brain Research Bulletin* 56, 413–424 (2001).

16. See A.N.M. Bot *Ethology, Ecology and Evolution* 13, 225–237 (2001); also S. Ballari et al. *Journal of Insect Behavior* 20, 87–98 (2007).

17. Bot (2001), p. 235; note 16.

18. See J.A. Toronchuk and G.F.R. Ellis *Cognition and Emotion* 21, 1799–1818 (2007); see also ensuing discussion between J. Panksepp (pp. 1819–1828) and the authors (pp. 1829–1832).

19. See G. Nevitt et al. *Journal of Experimental Biology* 207, 3537–3544 (2004).

20. See G.A. Nevitt and I. Bonadonna *Biology Letters* 1, 303–305 (2005).

21. See J.G.M. Thewissen et al. *Marine Mammal Science* 27, 282–294 (2011).

22. See S. Kowalewsky et al. *Biology Letters* 2, 106–109 (2006).

23. See K.L.B. Wright et al. *Journal of Experimental Biology* 214, 2509–2511 (2011).

24. See P.W. Sorensen et al. *Current Opinion in Neurobiology* 8, 458–467 (1998).

25. See M.C. Stensmyr et al. *Current Biology* 15, 116–121 (2005); also S. Harzsch and B.S. Hannsson, *BMC Neuroscience* 9, e58 (2008), who address the question of olfactory capacities in the brain on a related land-dwelling hermit crab.

26. In terms of the specific molecules detected, however, the aquatic ancestry is more evident with a general reliance on water-soluble types; see A.-S. Kräng et al. *Proceedings of the Royal Society of London, B* 279, 3510–3519 (2012).

27. Not, however, for all marine animals; sea lions and sea turtles, for example, have a significant olfac-

158. It is, however, evident that in contrast to the general perception of warmth (note 148) and the corresponding capsaicin response, in at least the crotalids the temperature activation has a very different basis (see T.C. Pappas et al. *American Journal of Physiology. Cell Physiology* 287, C1219–C1228 [2004]), and so has been arrived at independently. The range of perception in terms of distance and more particularly the sensitivity to tiny differences in ambient temperate is remarkable. While less pronounced in the boids (see J. Ebert et al. *Journal of Zoology, London* 272, 340–347 [2007]), in the generally aggressive crotalids the sensitivity is very acute (see J. Ebert and G. Westhoff *Journal of Comparative Physiology, A* 192, 941–947 [2006]).

159. See E.O. Gracheva et al. *Nature* 464, 1006–1011 (2010).

160. Gracheva et al. (2010), p. 1010; note 159.

161. See S. Yokoyama et al. *Molecular Biology and Evolution* 28, 45–48 (2011).

162. See V. Moiseenkova et al. *American Journal of Physiology: Regulatory, Integrative and Comparative Physiology* 284, R598–R606 (2003).

163. See E.A. Newman and P.H. Hartline *Scientific American* 246(3), 98–107 (1982), and T. de Cock Buning *Journal of Theoretical Biology* 111, 509–529 (1984). In some species the forward overlap of the two pits suggests some sort of steropsis of the thermal image; see G.S. Bakka et al. *Journal of Experimental Biology* 215, 2621–2629 (2012).

164. See A.B. Sichert et al. *Physical Review Letters* 97, e068105 (2006).

165. See D.M. Berson and P.H. Hartline *Journal of Neuroscience* 8, 1074–1088 (1988).

166. K.C. Catania et al. (*Journal of Experimental Biology* 213, 359–367; 2010) demonstrate a comparable capacity in an aquatic piscivorous snake where information from extremely sensitive mechanoreceptors (located on rather extraordinary anterior tentacles) is evidently combined with the visual input.

167. See his paper in *Journal of Herpetology* 45, 2–14 (2011).

168. Goris (2010), p. 11; note 167.

169. Infrared capacities serve various functions, including thermoregulation, which may indeed have been its ancestral function; see A.R. Krochmal et al. *Journal of Experimental Biology* 207, 4231–4238 (2004).

170. See A.S. Rundus et al. *Proceedings of the National Academy of Sciences, USA* 104, 14372–14376 (2007); also commentary on pp. 14177–14178 by D.T. Blumstein.

171. See L. Kürten and U. Schmidt *Journal of Comparative Physiology, A* 146, 223–228 (1982), and L. Kürten et al. *Naturwissenschaften* 71, 327–328 (1984).

172. See E.O. Gracheva et al. *Nature* 476, 88–91 (2011); also commentary on pp. 40–41 by M.B. Fenton.

173. See R. Kishida et al. *Brain Research* 322, 351–355 (1984).

174. Blood sucking among arthropods has, of course, evolved multiple times. Not only are there convergences between the salivary nitrophorins of hemipterans (see J.G. Valenzuela and J.M.C. Ribeiro *Journal of Experimental Biology* 201, 2659–2664 [1998]), but more widely the common demands of persuading the host to dilate its blood vessels and fail to either coagulate the blood or induce inflammation have led to a striking reliance of salivary proteins that use the lipocalin fold. As William Montfort et al. (*Biochimica et Biophysica Acta: Protein Structure and Molecular Enzymology* 1482, 110–118; 2000) remark, despite the diversity of mechanisms to achieve the same aim, "The lipocalin fold appears to be a protein scaffold of choice for the development of antihemostatic functions in insects" (p. 116).

175. See C.R. Lazzari and J.A. Nunez *Journal of Insect Physiology* 35, 525–529 (1989).

176. See C.R. Lazzari and M. Wicklein *Memorias do Instituto Oswaldo Cruz* 89, 643–648; also in the same journal S. Catala, vol. 89, 275–277 (1994).

177. See G.B. Flores and C.R. Lazzari *Journal of Insect Physiology* 42, 433–440 (1996).

178. Curiously, even when starving, the bugs will not all rush toward the infrared source, but this arguably is a precautionary step, lest the tables are turned and the bug visitor turns into dinner, permanently (see H. Schmitz et al. *Journal of Insect Physiology* 46, 745–751 [2000]).

179. See J.V. Richerson et al. *Canadian Journal of Zoology* 50, 909–913 (1972).

180. See J.V. Richerson and J.H. Boden *Canadian Entomologist* 104, 1877–1881 (1972).

181. See T.U. Grafe et al. *Proceedings of the Royal Society of London, B* 269, 999–1003 (2002).

182. We might note the interesting analogies between natural wildfires and herbivory (see W.J. Bond and J.E. Keeley *Trends in Ecology & Evolution* 20, 387–394 [2005]), with the implication that flammability could determine the repeated evolution of, for example, serotiny in pines (where the cones must be heated to free their seeds; see D.W. Schwilk and D.D. Ackerly *Oikos* 94, 326–336; 2001).

183. See, for example, S. Archibald and W.J. Bond *Oikos* 102, 3–14 (2003); also Bond and Keeley (2005); note 182.

184. See S. Schütz et al. *Nature* 398, 298–299 (1999).

185. See his paper in *Journal of Economic Entomology* 36, 341–342 (1943).

186. Linsley (1943), p. 341; note 185.

187. See, for example, T. Vondran et al. *Tissue and Cell* 27, 645–658 (1995), and H. Schmitz and H. Bleckmann *International Journal of Insect Morphology and Embryology* 26, 205–215 (1997).

188. See A. Schmitz et al. *Arthropod Structure & Development* 36, 291–303 (2007).

189. It just so happens that various chemical bonds (notably C-H and O-H) within the molecules that make up the cuticle of the dome strongly absorb the

126. See, for example, P.W. Lucas et al. *Evolution* 57, 2636–2643 (2003).

127. See, for example, A.D. Melin et al. *Animal Behaviour* 73, 205–214 (2007).

128. See, for example, P. Riba-Hernández et al. *Journal of Experimental Biology* 207, 2465–2470 (2004), and A.K. Surridge et al. *Trends in Ecology & Evolution* 18, 198–205 (2003); also A.D. Melin et al. *Behavioral Ecology and Sociobiology* 62, 659–670 (2008) and E.R. Vogel et al. *Behavioral Ecology* 18, 292–297 (2007).

129. See, for example, E.R. Vogel et al. *Behavioral Ecology* 18, 292–297 (2007).

130. See, for example, A.K. Surridge et al. *Biology Letters* 1, 465–468 (2005).

131. See, for example, D. Osorio et al. *American Naturalist* 164, 696–708 (2004).

132. See A.A. Fernandez and M.R. Morris *American Naturalist* 170, 10–20 (2007).

133. See G.H. Jacobs et al. *Nature* 382, 156–158 (1996); also papers in vol. 38 (1998) of *Vision Research* by P.M. Kainz et al. (pp. 3315–3320) and B.C. Rega et al. (pp. 3321–3327).

134. See G.H. Jacobs and J.F. Deegan *Proceedings of the Royal Society of London, B* 268, 695–702 (2001).

135. See D.M. Hunt et al. *Vision Research* 38, 3299–3306 (1998).

136. See, for example, T. Kishida et al. *Biology Letters* 3, 428–430 (2007).

137. See Y. Gilad et al. *PLoS Biology* 2, e5 (2004).

138. See D.M. Webb et al. *Molecular Biology and Evolution* 21, 697–704 (2004).

139. See G.H. Jacobs et al. *Proceedings of the Royal Society of London, B* 236, 705–710 (1996).

140. See L. Peichl and K. Moutairou *European Journal of Neuroscience* 10, 2586–2594 (1998).

141. See, for example, G.H. Jacobs and J.F. Deegan *Journal of Comparative Physiology, A* 171, 351–358 (1992).

142. See G.H. Perry et al. *Molecular Biology and Evolution* 24, 1963–1970 (2007).

143. See S. Yokoyama et al. *Genetics* 170, 335–344 (2005).

144. See A. Reitner et al. *Nature* 352, 798–800 (1991).

145. See B.J. Bradley et al. *American Journal of Physical Anthropology* 139, 269–273 (2009).

146. See Y. Feldman *Physical Review Letters* 100, 128102 (2008).

147. See the commentary on Feldman et al. (2008); note 146 by P. Ball *Nature* 452, 676.

148. An ability to detect temperature differences is very widespread and is manifested in such behaviors as basking, navigation (as in the leaf-cutter ants [see C.J. Kleineidam et al. *Journal of Insect Physiology* 53, 478–487 (2007)]; also M. Ruchty et al. *Arthropod Structure & Development* 38, 195–205; 2009), and avoidance of danger. In flies and mammals the range of temperatures is detected using the same molecular equipment, specifically TRP (transient receptor potential) ion channels (see, for example, D.D. McKemy *Pflügers Archiv* (*European Journal of Physiology*) 454, 777–791 [2007]), and a similar principle may apply to temperature perception in plants (see S. Penfield *New Phytologist* 179, 615–628 [2008]). This example of the TRP ion channels reiterates the central paradox of deep. Employing the same tool kit obviously points to a strong evolutionary conservation, but did the common ancestor of flies and humans (which looked like neither and presumably was some sort of sluglike marine creature) also detect temperature differences in this way? Or, alternatively, was the tool kit largely ready and waiting for employment when the ancestors of flies (a crustaceanlike beast) and humans (a sprawling amphibian) dragged themselves onto land and so into a wildly fluctuating thermal environment? Another profoundly interesting point of temperature perception is that it reflects the qualia of the other senses, such as olfaction. This revolves around the psychophysics of sensation, with a fascinating correspondence between the activation of the different TRP ion channels and the sense of welcome coolness from menthol or (for some) exciting heat of chilies or cayenne (via the active agent capsaicin).

149. See R. Wehner and S. Wehner (*Physiological Entomology* 36, 271–281; 2011), who document "morphological, physiological and behavioural traits [that] form a characteristic 'thermophilia syndrome' that has evolved independently in the formicine genus *Cataglyphis* (Formicini) and the myrmicine genus *Ocymyrmex* (Pheidolini)" (p. 272), and among other items is evident from the convergent acquisition of extremely elongate legs (see S. Sommer and R. Wehner *Arthropod Structure & Development* 41, 71–77 [2011]), fast running speeds, impressive navigational skills, and a striking "getting a breather" or "respite" behavior.

150. See R. Loftus and G. Corbière-Tichané *Journal of Comparative Physiology* 143, 443–452 (1981).

151. See G. Corbière-Tichané and R. Loftus *Journal of Comparative Physiology* 153, 343–351 (1983).

152. See his paper in *Interdisciplinary Science Reviews* 13, 27–39 (1988).

153. Autrum (1988), p. 36; note 152.

154. Even so the eyes of some fish are evidently sensitive to near-infrared (see, for example, T. Matsumoto and G. Kawamura *Fisheries Science* 71, 350–355 [2005]), an ability that is evidently connected to their crepuscular activities; also D. Shcherbakov et al. *Zoology* 115, 233–238 (2012). Near-infrared sensitivity in cichlids is documented by D. Meuthen et al. (*Naturwissenschaften* 99, 1063–1066; 2012), but they do not exclude extraretinal reception.

155. See A.L. Campbell et al. *Micron* 33, 211–225 (2002); also H. Bleckmann et al. *Journal of Comparative Physiology, A* 190, 971–981 (2004).

156. See his paper in *Acta Biotheoretica* 34, 193–206 (1985).

157. de Cock Buning (1985), p. 204; note 156.

of sounds. What is so fascinating about this proposal is that it reinforces the likelihood that apparently disparate methods of sensory perception at a deeper level are much more similar than may be realized.

85. See, for example, S. Yokoyama and F.B. Radlwimmer *Genetics* 158, 1697–1710 (2001).

86. This includes the bats, which also employ ultraviolet vision; see D. Wang et al. *Molecular Biology and Evolution* 21, 295–302 (2004).

87. See, for example, M. Koyanagi et al. *Journal of Molecular Evolution* 66, 130–137 (2008).

88. See S. Yokoyama et al. *Genetics* 179, 2037–2043 (2008).

89. See S. Yokoyama *Gene* 300, 69–78 (2002).

90. See also O.S.E. Gustafsson et al. *Journal of Experimental Biology* 211, 1559–1564 (2008).

91. See B.S.W. Chang et al. *Molecular Biology and Evolution* 19, 1483–1489 (2002); also P.V. Nguyen *Trends in Neurosciences* 25, 549–550 (2002).

92. See S. Yokoyama et al. *Proceedings of the National Academy of Sciences, USA* 105, 13480–13485 (2008), and also commentary on pp. 13193–13194 by A.L. Hughes.

93. Importantly, as Yokoyama et al. (2008; note 92) note, the shifts in wavelength sensitivity "of most contemporary rhodopsins can be explained largely by a total of 15 amino acid replacements at 12 sites . . . [and] 4 of the 15 critical amino acid replacements occurred multiple times during rhodopsin evolution" (p. 13482).

94. Yokoyama et al. (2008), p. 13483; note 92.

95. Hughes (2008), p. 13194; note 92.

96. See L. Peichl et al. *European Journal of Neuroscience* 13, 1520–1528 (2001).

97. See D.H. Levenson and A. Dixon *Proceedings of the Royal Society of London, B* 270, 673–679 (2003).

98. See A.M. Mass and A.Ya. Supin *Anatomical Record* 290, 701–715 (2007); for other convergences in amphibious animals see J.G. Sivak in *Sensory Ecology* (M.A. Ali, ed.), pp. 503–519 (Plenum; 1978).

99. See M.F. Land in *Comparative physiology: Life in water and on land* (P. Dejours et al., eds.), pp. 289–302 (IX-Liviana Press; 1987).

100. See J. Ivanoff *Moken: Sea-gypsies of the Andaman Sea: Post-war chronicles* (White Lotus Press; 1997).

101. Mass and Supin (2007); note 98.

102. See A. Gislén et al. *Current Biology* 13, 833–836 (2003).

103. See A. Gislén et al. *Vision Research* 46, 3443–3450 (2006).

104. See, for example, A. Kelber and U. Hénique *Journal of Comparative Physiology, A* 184, 535–541 (1999).

105. See, for example, A. Balkenius et al. *Journal of Comparative Physiology, A* 192, 431–437 (2006).

106. See A. Kelber et al. *Nature* 419, 922–925 (2002).

107. See E.J. Warrant *Journal of Experimental Biology* 211, 1737–1746 (2008).

108. See H. Somanathan et al. *Journal of Comparative Physiology, A* 194, 97–107 (2008).

109. See H. Somanathan et al. *Current Biology* 18, R996–R997 (2008).

110. See H. Somanathan et al. *Journal of Comparative Physiology A* 195, 571–583 (2009).

111. So, too, in other groups; concerning the nocturnal capacity of some geckos see A. Kelber and L.S. Roth *Journal of Experimental Biology* 209, 781–788 (2006).

112. See, for example, L. Peichl *Anatomical Record, A* 287, 1001–1012 (2005).

113. See, for example, J.K. Bowmaker and D.M. Hunt *Current Biology* 16, R484–R489 (2006).

114. Tetrachromacy has evolved independently in some insects; see H. Koshitaka et al. *Proceedings of the Royal Society of London, B* 275, 947–954 (2008).

115. See W.L. Davies et al. *Current Biology* 17, R161–R163 (2007).

116. See C.P. Heesy and C.F. Ross *Journal of Human Evolution* 40, 111–149 (2001); also M.I. Hall et al. *Proceedings of the Royal Society of London, B* 279, 4962–4968 (2012).

117. See Y. Tan et al. *Proceedings of the National Academy of Sciences, USA* 102, 14712–14716 (2005).

118. For instance, the trichromacy of some primitive prosimians like Coquerel's sifaka and the red-ruffed lemur (see Y. Tan and W.-H. Li *Nature* 402, 36 [1999]; also S.D. Leonhardt et al. *Behavioral Ecology* 20, 1–12; 2009) may have emerged independently as each left the largely nocturnal, and dichromatic, world of their relatives. This conclusion, however, needs further testing, because another lemur that is active both in the day and night (that is, cathemeral) appears also to be trichromatic, possibly pointing to a deeper ancestry of this type of color vision (see C.C. Veilleux and D.A. Bolnick *American Journal of Primatology* 71, 86–90 [2009]).

119. See C.A. Arrese et al. *Current Biology* 12, 657–660 (2002), and *Proceedings of the Royal Society of London, B* 272, 791–796 (2005); also P. Sumner et al. *Journal of Experimental Biology* 208, 1803–1815 (2005).

120. See C.A. Arrese et al. *Gene* 381, 13–17 (2006); also J.A. Cowing et al. *Proceedings of the Royal Society of London, B* 275, 1491–1499 (2008).

121. See C.A. Arrese et al. *Current Biology* 16, R193–R194 (2006).

122. For an overview, see J.H. Jacobs *International Journal of Primatology* 28, 729–759 (2007).

123. See G.H. Jacobs et al. *Vision Research* 42, 11–18 (2002).

124. See S. Boissinot *Proceedings of the National Academy of Sciences, USA* 95, 13749–13754 (1998), and A.K. Surridge and N.I. Mundy *Molecular Ecology* 11, 2157–2169 (2002).

125. See, for example, A.C. Smith et al. *Journal of Experimental Biology* 206, 3159–3165 (2003); also S.-B. Lomáscolo and H.M. Schaeffer (*Journal of Evolutionary Biology* 23, 614–624; 2010), who explore the convergences of fruit color in terms of both primate and bird foraging.

K.F. Stranger-Hall et al. *Molecular Phylogenetics and Evolution* 45, 33–49 (2007), for evidence of convergences in signaling. Although examples of bioluminescence in terrestrial environments are most often thought of in the context of fireflies and glowworms (not to mention cockroaches; Vršanský et al. [2012]; note 31), the terrestrial pulmonate snail *Dyakia* is another example (see papers in *Malacologica* by J. Copeland and M.M. Daston [vol. 30, pp. 317–324 (1989)], and M.M. Daston and J. Copeland [vol. 35, pp. 9–19; 1993]). Some fungi also glow in the dark. *Armillaria*, the virulent honey-fungus (see M. Smith et al. *Nature* 356, 428–431 [1992]) is one of three groups of basidiomycete that have independently evolved bioluminescence (and apart from imparting an eerie glow to forests, also found service for travelers wishing to illuminate their jungle path); see D.E. Desjardin et al. *Photochemical & Photobiological Sciences* 7, 170–182 (2008).

57. See H.H. Seliger and W.D. McElroy *Proceedings of the National Academy of Sciences, USA* 52, 75–81 (1964), and papers by K.V. Wood et al. in *Science* 244, 700–702 (1989), and *Journal of Bioluminescence and Chemiluminescence* 4, 289–301 (1989); also D. Booth et al. *Journal of Experimental Biology* 207, 2373–2378 (2004).

58. See K.V. Wood *Photochemistry and Photobiology* 62, 662–673 (1995), and B.R. Branchini et al. *Biochemistry* 38, 13223–13230 (1999).

59. See their paper in *Photochemistry and Photobiology* 75, 22–27 (2002).

60. Viviani et al. (2002), p. 26, note 59.

61. See S.P. Leys et al. *Journal of Comparative Physiology, A* 188, 199–202 (2002).

62. See, for example, A.D. Briscoe and L. Chittka *Annual Review of Entomology* 46, 471–510 (2001).

63. The crustaceans, from which the insects arose, also possess this ability (see M.L. Porter et al. *Molecular Biology and Evolution* 24, 253–268 [2007]).

64. See A.G. Dyer et al. *Proceedings of the Royal Society of London, B* 279, 3606–3615 (2012).

65. See, for example, J. Spaethe and A.D. Briscoe *Journal of Experimental Biology* 208, 2347–2361 (2005).

66. See, for example, D.G. Stavenga and K. Arikawa *Arthropod Structure & Development* 35, 307–318 (2006).

67. See Y.S. Shi and S. Yokoyama *Proceedings of the National Academy of Sciences, USA* 100, 8308–8313 (2003).

68. See H.-B. Zhao et al. *Proceedings of the National Academy of Sciences, USA* 106, 8980–8985 (2009).

69. In one case, that of a glossophagine flower bat (p. 379, note 206), its vision is apparently otherwise monochromatic (see Y. Winter et al. *Nature* 425, 612–614 [2003]. The case of the bat being a monochromat is, however, questioned by Zhao et al. [2009]; note 68), but in many other bats color perception is retained. As Hua-bin Zhao and colleagues note, "UV color vision

is likely to play a considerably more important role in [their] sensory ecology than previously appreciated" (p. 8984).

70. See G.H. Jacobs *American Zoologist* 32, 544–554 (1992).

71. See, for example, D. van Norren and P. Schellenkens *Vision Research* 30, 1517–1520 (1990).

72. See J.W.L. Parry et al. *Biochemistry* 43, 8014–8020.

73. See A. Thorpe et al. *Vision Research* 33, 289–300 (1993).

74. See, for example, P.G. Hains *Experimental Eye Research* 82, 730–737 (2006).

75. See N. Shashar et al. *Biological Bulletin* 195, 187–188 (1998).

76. See E.P. Balskar and C.T. Walsh (*Science* 329, 1653–1656; 2010), who document the biosynthetic pathway in cyanobacteria and briefly note analogues in other groups; also note 7.

77. See J.M. Schick and W.C. Dunlap *Annual Review of Physiology* 64, 223–262 (2002).

78. See C.S. Cockell and J. Knowland *Biological Reviews* 74, 311–345 (1999).

79. As far as eyes are concerned, there are also some interesting exceptions. In the geckos, a return from nocturnality to a diurnal life has occurred independently at least three times (see B. Roll *Naturwissenschaften* 88, 293–296 [2001]). The evidence for this comes from the fact that these different geckos use separate crystallins in their lenses, and in one case (the i-crystallin) the protein has been co-opted to provide an ultraviolet filter (see P.J.L. Werten et al. *Proceedings of the National Academy of Sciences, USA* 97, 3282–3287 [2000]).

80. One idea is that with many different types of rhodopsin, mantis shrimps were extravagantly polychromatic. It seems, however, that these animals have a very distinct type of color vision that is linked to the scanning of eyes and the need for recognition and rapid decisions; see H.H. Thoen et al. *Science* 343, 411–413 (2014); also commentary on pp. 381–382 by M.F. Land and D. Osorio.

81. Cronin and Marshall in *Complex worlds from simpler nervous systems* (F.R. Prete, ed.), pp. 239–268 (MIT Press; 2004); see p. 258.

82. See T.-H. Chiou et al. *Current Biology* 18, 429–434 (2008); also commentary on pp. R348–R349 by M. Land.

83. Chiou et al. (2008), p. 433; note 82.

84. When it comes to the polychromatic vision, one possibility is that this actually collapses into a series of dichromatic channels. An alternative hypothesis is even more interesting because Cronin and Marshall (2004; note 81) ask, "Is the Mantis Shrimp's Eye an 'Ear'?" (p. 265). The basis of this idea is as follows. Just as the mantis shrimp eye has not only a wide spectral range but the specific sensitivities to form a series of well-defined peaks, so too the cochlea of the inner ear has hair cells capable of interpreting a wide frequency

15. Davenport (1998), part of the title; note 14.

16. See G.C. Hays et al. *Journal of Experimental Marine Biology and Ecology* 370, 134–143 (2009). Whether at least smaller sunfish are obligate consumers of gelatinous prey may, however, require revision; see J. Syväranta et al. *Journal of Fish Biology* 80, 225–231 (2012). This in turn was questioned in a subsequent issue (vol. 82; 2013) by J.M. Logan and K.L. Dodge (pp. 1–9), which earned a reply by C. Harrod et al. (pp. 10–16).

17. See the discussion by J.A. Freedman and D.L.G. Noakes (*Reviews in Fish Biology and Fisheries* 12, 403–416; 2002) as to why in terms of size the teleosts are outstripped by the sharks.

18. This famous example of aggressive sexual mimicry by firefly femme fatales was documented by J.E. Lloyd *Science* 187, 452–453 (1975), although F.V. Vencl et al. *Journal of Insect Behavior* 7, 843–858 (1994), have questioned the specificity of this mimicry.

19. See E.A. Widder *Science* 328, 704–708 (2010).

20. See, for example, T. Wilson and J.W. Hastings *Annual Review of Cell and Developmental Biology* 14, 197–230 (1998), and *Photochemistry and Photobiology* 62, 599–600 (1995). Nor should it be assumed that this system evolved originally to confer bioluminescence, and it is likely that primitively the luciferin-luciferase system was involved with oxygen detoxification; see G.S. Timmins et al. *Journal of Molecular Evolution* 52, 321–332 (2001).

21. See D.D. Deheyn and M.I. Latz *Invertebrate Biology* 128, 31–45 (2009).

22. Widder (2010), p. 707; note 19.

23. See P.J. Herring *Symposia of the Zoological Society of London* 38, 127–159 (1977).

24. See, for example, R.C. Guerrero–Ferreira and M.K. Nishiguchi *Cladistics* 23, 497–506 (2007).

25. See their paper in *Journal of Plant Growth Regulation* 19, 113–130 (2000).

26. Hirsch and McFall-Ngai (2000), p. 113; note 25.

27. These include the role of horizontal transfer, methods of securing the bacteria, and so-called auto-induction that ensure the growth of bacterial populations.

28. See P.V. Dunlap et al. *Cladistics* 23, 507–532 (2007); also P.V. Dunlap et al. *Marine Biology* 156, 2011–2020 (2009).

29. See I.A. Johnston and P.J. Herring *Proceedings of the Royal Society of London, B* 225, 213–218 (1985).

30. See P.J. Herring et al. *Journal of the Marine Biological Association, UK* 61, 901–916 (1981).

31. In a bioluminescent cockroach (*Lucihormetica*), the organ forms a lantern with a transparent cuticle modified as a type of reflector; see P. Vršanský et al. *Naturwissenschaften* 99, 739–749 (2012); see, however, D.J. Merritt in vol. 100, 697–698 (2013), who takes a more skeptical view.

32. See, for example, D. Pringgenies and J.M. Jørgensen *Acta Zoologica* 75, 305–309 (1994).

33. See M.K. Montgomery and M.J. McFall-Ngai *Journal of Biological Chemistry* 267, 20999–21003 (1992).

34. See W.J. Crookes et al. *Science* 303, 235–238 (2004).

35. See E.J. Denton and M.F. Land *Proceedings of the Royal Society of London, A* 178, 43–61 (1971).

36. See D. Tong et al. *Proceedings of the National Academy of Sciences, USA* 106, 9836–9841 (2009).

37. See W.D. Clarke *Nature* 198, 1244–1246 (1963).

38. See R.E. Young and C.F.E. Roper *Science* 191, 1046–1048 (1976).

39. Clarke (1963), p. 1244; note 37.

40. See, for example, J.A. Warner et al. *Science* 203, 1109–1110 (1979).

41. See R.D. Harper and J.F. Case *Marine Biology* 134, 529–540 (1999).

42. See papers in *Journal of Experimental Marine Biology and Ecology* by J.M. Claes et al. (vol. 388, pp. 28–32; 2010) and M. Renwart and J. Mallefet (vol. 448, pp. 214–219; 2013); intriguingly the latter suggests the system of bioluminescence may be completely novel.

43. See B.W. Jones and M.K. Nishiguchi *Marine Biology* 144, 1151–1155 (2004).

44. See N. Straube et al. *Molecular Phylogenetics and Evolution* 56, 905–917 (2010); they support the notion that bioluminescence evolved twice in the sharks—that is, etmopterids and dalatiids.

45. Claes et al. (2010), p. 28 [Abstract]; note 42.

46. See R.E. Young and F.M. Mencher *Science* 208, 1286–1288 (1980).

47. This is reminiscent of the release of green bioluminescent "bombs" by a deep-sea polychaete (see K.J. Osborn et al. *Science* 325, 964 [2009]).

48. See R.E. Young *Bulletin of Marine Science* 33, 829–845 (1983).

49. In the case of another deep-sea squid T. Kubodera et al. (*Proceedings of the Royal Society of London, B* 274, 1029–1034; 2007) suggest that the short flashes it emits from its arms might serve to blind its prey.

50. See B.H. Robison et al. *Biological Bulletin* 205, 102–109 (2003).

51. As does the polychaete *Chaetopterus*, by the rupturing of its photogenic gland; see N. Martin and M. Anctil *Biological Bulletin* 166, 583–593 (1984). For a similar example, but this time in the chaetognaths, see E.V. Thuesen et al. *Biological Bulletin* 219, 100–111 (2010).

52. See P.N. Dilly and P.J. Herring *Journal of Zoology, London* 186, 47–59 (1978).

53. See S.L. Bush and B.H. Robison *Marine Biology* 152, 485–494 (2007).

54. See his paper in *Hydrobiologia* 538, 179–192 (2005).

55. Oakley (2005), p. 187; note 54.

56. See, for example, V.R. Viviani *Cellular and Molecular Life Sciences* 59, 1833–1850 (2002), and J.M. Sivinski *Ecological Entomology* 7, 443–446 (1982); also mimicry of eyes seen in some cases, conspicuousness per se may be more important; see M. Stevens et al. *Behavioral Ecology* 19, 525–531 (2008). C. Blut et al. (*Entomologia Experimentalis et Applicata* 143, 231–244; 2012) suggest that the so-called sparkle effect in these eye-spots is far from accidental but adds further verisimilitude. For an overview, see U. Kodandaramaiah *Behavioral Ecology* 22, 1264–1271 (2011).

255. See his paper in *Animal Behaviour* 31, 1037–1042 (1983).

256. Davison (1983), p. 1037 [Abstract]; note 255.

257. Regarding the actual structural details of such feathers, I am not aware that direct information exists as to how the patterns are generated optically. I.M. Weiss and H.O.K. Kirchner (*Journal of Experimental Zoology: Ecological Genetics and Physiology, A* 313, 690–703; 2010) provide a detailed description of the peacock tail microstructure (especially the rachis) but do not address the ocelli.

258. On pp. 430–442 Darwin (1889; note 253) gives a detailed account of the patterning in the Argus and related pheasants, which anticipates the much later ruminations on biological patterning and self-organization such as B. Godwin's *How the leopard changed its spots* (Touchstone; 1996).

259. Strikingly similar arrangements have evolved independently in a number of groups of birds; see L. Auber *Proceedings of the Zoological Society of London* 129, 455–486 (1957).

260. See R. Bleiweiss *Proceedings of the National Academy of Sciences, USA* 101, 16561–16564 (2004).

261. See his paper in *Evolutionary Biology* 36, 171–189 (2009).

262. Bleiweiss (2009), p. 173; note 261.

263. See F.J. Ollivier et al. *Veterinary Ophthalmology* 7, 11–22 (2004).

264. Bleiweiss (2009), p. 180; note 261.

CHAPTER 10

1. A likely exception is the deepwater firefly squid (*Watasenia scintillans*), which has three visual pigments (see M. Seidou et al. *Journal of Comparative Physiology, A* 166, 769–773 [1990]). This animal has a banked retina reminiscent of some deepwater fish (see M. Michinomae et al. *Journal of Experimental Biology* 193, 1–12 [1994]), as well as the capacity to correct for chromatic aberration, which supports the notion of some sort of color discrimination (see R.H.H. Kröger and A. Gislén *Vision Research* 44, 2129–2134 [2004]).

2. This has evolved many times independently, and is probably best known in such insects as the bees. Celebrated for their ability to navigate to distant flowers as they zoom across the meadows, not only have they learned in the hive where to aim for (through the famous bee "dance"), but because the degree of light polarization across the sky varies, they use this information as a navigational vector.

3. See N. Shashar et al. *Marine and Freshwater Behaviour and Physiology* 35, 57–68 (2002).

4. See Shashar et al. (2002); note 3. This possibility of "private conversations" touches on a central question in biology. Life produces a continuous roar of communication, be it by sound, vision, vibration, electricity, or other modes. The dual trick, which has been solved in numerous different ways, is to work out how to communicate (say with a potential mate) without being eavesdropped on by somebody who would like to turn you (and perhaps your mate) into dinner. Correspondingly, if you are being hunted, how can you confuse or even jam the predator's communication system? Tricky.

5. See N. Shashar et al. *Vision Research* 40, 71–75 (2000).

6. See S. Johnsen, *Biological Bulletin* 201, 301–318 (2001).

7. Transparency, however, may be literally in the eye of the beholder. What to us is virtually invisible may become glaringly obvious once seen in the polarized light to which cuttlefish are sensitive (see N. Shashar et al. *Nature* 393, 222–223 [1998]; also Shashar et al. [2002]; note 3). Also, those animals that hunt in the ultraviolet part of the visual spectrum may find the transparency of their prey transformed, as far as the latter are concerned, to an unwelcome opacity, although in this case it is typically because the potential prey is employing organic compounds (notably the mycosporine-like amino acids and scytonemin, which are recruited from microbial sources; see, for example, R.P. Sinha and D.-P. Häder *Plant Science* 174, 278–289 [2008]. These help to protect the otherwise transparent animal from potentially dangerous ultraviolet radiation from which the body would in other circumstances provide an imperfect screen; see S. Johnsen and E.A. Widder *Marine Biology* 138, 717–730; 2001).

8. See G.W. Rouse and F. Pleijel *Polychaetes* (especially pp. 226–228; Oxford; 2001).

9. Even within the group to which the poeobids belong (the flabelligeridids), a pelagic mode of life has evolved independently twice (see K.J. Osborn and G.W. Rouse *Molecular Phylogenetics and Evolution* 49, 386–392 [2008]).

10. See B.H. Robison et al. *Journal of the Marine Biological Association, UK* 85, 595–602 (2005).

11. Nor are they alone in this regard; see, for example, M. Tatián et al. *Zoologica Scripta* 40, 603–612 (2011).

12. See M.N. Arai *Journal of the Marine Biological Association, UK* 85, 523–536 (2005).

13. See T.K. Doyle et al. *Journal of Experimental Marine Biology and Ecology* 343, 239–252 (2007).

14. See his paper in *British Herpetological Society Bulletin* 62, 4–8 (1998).

223. The squid (*Histioteuthus*) is rather odd, because one eye presumably serves for upward vision, and the other to peer below (see fig. 24 of M.F. Land in *Handbook of sensory physiology*, vol. 7/6B [H. Autrum, ed.], pp. 471–592 [Springer; 1981]), but otherwise they typify the convergent camera-eye arrangement of the cephalopods.

224. See J.D. Pettigrew in *Visual neuroscience* (J.D. Pettigrew et al., eds.), pp. 208–222 (Cambridge; 1986).

225. Pettigrew (1986; note 224) speculates that the panic-stricken rolling of eyes in frightened animals like horses might be a similar response.

226. See J.D. Pettigrew et al. *Current Biology* 9, 421–424 (1999), and commentary on pp. R286–R288 by M.F. Land.

227. See his commentary in *Nature* 399, 305, 307 (1999).

228. To paraphrase the subtitle of Land (1999); note 226.

229. Srinivasan (1999), p. 305; note 227.

230. Effectively, as the eye rotates, the position of the image shifts relative to the axis of rotation.

231. See J.D. Pettigrew et al. *Journal of Comparative Physiology, A* 186, 247–260 (2000).

232. Srinivasan (1999), p. 305; note 227.

233. Pettigrew et al. (1999); note 226.

234. See K.A. Fritscher and N.J. Marshall *Current Biology* 9, R272–R273 (1999).

235. As it happens the sandlance is not the only fish to have evolved independence of eye movement, and beyond the pipefish (relatives of the sea horse and distant from the sandlances), among the invertebrates the metaphorical ability of the mantis shrimp to watch two tennis matches at once (p. 12) further extends this convergence.

236. See, for example, V.U. Zhukov et al. *Acta Zoologica* 87, 13–24 (2006), reporting an instance in the freshwater viviparids.

237. See, for example, P.V. Hamilton et al. *Journal of Comparative Physiology, A* 152, 435–445 (1982), and J.-O. Seyer *Journal of Experimental Biology* 170, 57–69 (1992).

238. See M.V. Bobkova et al. *Invertebrate Biology* 123, 101–115 (2004).

239. See, for example, H.L. Gillary *Journal of Experimental Biology* 66, 159–171 (1977); H.L. Gillary and E.W. Gillary *Journal of Morphology* 159, 89–116 (1979); and J.-O. Seyer *Journal of Experimental Zoology* 268, 200–207 (1994).

240. See, for example, C.J. Berg *Behaviour* 51, 274–322 (1974), and L.H. Field *Pacific Science* 31, 1–11 (1977).

241. See, for example, R. Hesse *Zeitschrift für wissenschaftliche Zoologie* 68, 379–477 (1900), and M.J.F. Blumer *Zoomorphology* 119, 81–91 (1999).

242. See M.F. Land *Journal of Experimental Biology* 96, 427–430 (1982).

243. For a useful overview of these "eight-legged cats," see D.P. Harland and R.R. Jackson *Cimbebasia* 16, 231–240 (2000).

244. See R.S. Wilcox and R.R. Jackson, in *Animal cognition in nature: The convergence of psychology and biology in laboratory and field* (R.P. Balda et al., eds.), pp. 411–434 (Academic; 1998).

245. See, for example, M.F. Land *Journal of Experimental Biology* 51, 443–470 and 471–493 (1969); also A.D. Blest et al. *Journal of Comparative Physiology* 145, 227–239 (1981).

246. Not only is it complex, but as Mike Land (1969; note 245) notes, the organization of this remarkable visual system is "remarkably similar to mammals, and quite unlike most arthropods" (p. 488), inasmuch as the eye movements are effectively saccadal and serve to fix the most sensitive part of the eye on the object of interest. Another striking example of this scanning ability is found in the larvae of a diving beetle (dytiscids). Like the heteropod mollusks and salticid spiders, the eyes are tubular (and are known as stemmata; see G. Gilbert *Annual Review of Entomology* 39, 323–349; 1994). Although otherwise distinct from the usual compound eyes, rather remarkably the stemmata are bifocal, recalling the construction of some trilobite eyes (see A. Stowasser et al. *Current Biology* 20, 1482–1486 [2010]), but in the beetles the scanning is achieved by moving the entire head (see E.K. Buschbeck et al. *Journal of Comparative Physiology, A* 193, 973–982 [2007]). These larvae are effective predators, and while the main eyes are focused on the prey, peripheral eyes are ready to detect more victims lurking beyond the main field of vision. Just such an arrangement, of "spare" eyes busy investigating "blind spots," is found also in the salticids, while in the males of some margarodid aphids the compound eye has been reduced to stemmatalike structures (see E.K. Buschbeck and M. Hauser *Naturwissenschaften* 96, 365–374 [2009]); as Elke Buschbeck and Martin Hauser note in this case, there is a clear "trend toward evolving giant unicornal eyes" (p. 365).

247. This arrangement has evolved twice, and is employed in quite different ways of hunting (see K.F. Su et al. *Journal of Evolutionary Biology* 20, 1478–1489 [2007]).

248. See D.S. Williams and P. McIntyre *Nature* 288, 578–580 (1980).

249. See C.O. Hermans and R.M. Eakin *Zeitschrift für Morphologie und Tiere* 79, 245–267 (1974), and G. Wald and S. Rayport *Science* 196, 1434–1439 (1977).

250. See his paper in *Journal of the Marine Biological Association, UK* 85, 641–653 (2005).

251. Bone (2005), p. 641; note 250.

252. See S. Johnsen *Biological Bulletin* 201, 301–318 (2001).

253. See his *The descent of man, and selection in relation to sex* (Murray; 1889), pp. 398–400.

254. Eyespots have, of course, evolved multiple times and otherwise are probably most familiar in the butterflies, where they are widely held either to startle or deflect predators; see, for example, M. Stevens *Biological Reviews* 80, 573–588 (2005). Despite the striking analogy between the test of an entire urchin and a low-resolution compound eye" (p. 61).

193. See also M.P. Lesser et al. *Proceedings of the Royal Society of London, B* 278, 3371–3379 (2011), who identify the tube feet as photoreceptors and suggest that the terminal ossicles may serve as light collectors in a manner comparable to the ophiuroids (note 198). They also identify an opsinlike molecule, but do not rule out the possibility it is a novel protein comparable to the example in scallops (note 154).

194. See E.M. Ullrich-Lüter et al. *Proceedings of the National Academy of Sciences, USA* 108, 8367–8372 (2011).

195. Ullrich-Lüter et al. (2011), p. 8370; note 194. Study of the opsins (in fact, c-opsins) points in the same direction, especially in the echinoids where they occur in a variety of tissues (including the defensive pedicellaria), suggesting that the body as a whole is light sensitive; see E.M. Ullrich-Lüter et al. *Integrative and Comparative Biology* 53, 27–38 (2013).

196. See G. Hendler *Marine Ecology* 5, 379–401 (1984).

197. They also have an ability to detect polarized light; this may help to keep them clear of dangerously shallow water (see S. Johnsen *Journal of Experimental Biology* 195, 281–291 [1994]).

198. See G. Hendler and M. Byrne *Zoomorphology* 107, 261–272 (1987), and J. Aizenberg et al. *Nature* 412, 818–822 (2001); also J. Aizenberg and G. Hendler *Journal of Materials Chemistry* 14, 2066–2072 (2004).

199. See J.L.S. Cobb and G. Hendler *Comparative Biochemistry and Physiology, A* 97, 329–333 (1990); also J.L.S. Cobb and A. Moore *Marine Behaviour and Physiology* 14, 211–222 (1989).

200. The problem extends further because similar eyes have evolved also in the starfish (see P. Dubois and S. Hayt in *Echinoderm Research* [C. de Ridder et al., eds.], pp. 217–223 [Balkema; 1990], and C. Mah *Zoosystema* 27, 137–161 [2005]). Most likely they are somewhat less sophisticated, consisting of glass-clear tubercles studding the calcareous plate, but they evidently have arisen independently. A rather different arrangement occurs in *Linckia* where the compound eye (composed of approximately two hundred ommatidia) is located on the tip of a modified tube-foot; see A. Garm and D.-E. Nilsson *Proceedings of the Royal Society of London, B* 281, 20133011 (2014), and commentary by M.F. Land *Current Biology* 24, R200–R201 (2014).

201. Calcite has a very high refractive index, but interestingly an ostracod whose eyes employ a mirror system (pp. 101–102) also employs a cuticular lens, whose optical properties are close to those of calcite; see A. Andersson and D.-E. Nilsson *Protoplasma* 107, 361–374 (1981).

202. See D.I. Speiser et al. *Current Biology* 21, 665–670 (2011); also commentary on pp. R273–R274 by M.F. Land.

203. See my *Life's solution*, pp. 148–149. Long controversial, their closest relationship is with the beetles,

and accordingly this implies convergences in larval development; see O. Niehuis et al. *Current Biology* 22, 1309–1312 (2012).

204. See their paper in *Zoology* 111, 318–338 (2008).

205. Pohl and Beutel (2008), p. 334; note 204.

206. See, for example, G. Horváth et al. *Historical Biology* 12, 229–236 (1997).

207. See E.K. Buschbeck *Arthropod Structure & Development* 34, 315–326 (2005).

208. See H. Pohl et al. *Zoologica Scripta* 34, 57–69 (2005). Also Pohl and Beutel (2008); note 204.

209. See E.K. Buschbeck et al. *Journal of Comparative Physiology, A* 189, 617–630 (2003).

210. The exact evolutionary path from a usual compound eye is yet to be documented, but despite each lens being capable of providing a precise image, the structure of the retina suggests the overall resolution may be limited (see W. Pix et al. *Journal of Experimental Biology* 203, 3397–3409 [2000]; also Buschbeck et al. [2003]; note 209). It is a sort of echo of the box jelly camera eye (pp. 97–100) or the pinhole eyes of the mollusks, which serve to discern what needs to be discerned, and no more (see S. Maksimovic et al. *Journal of Experimental Biology* 210, 2819–2828 [2007]; also Pix et al. [2000]. For this reason, although the retinal structure of the schizochroal eye is conjectural, when it comes to making what is effectively a camera eye out of a compound eye (see E. Buschbeck et al. *Science* 286, 1178–1180 [1999]; also Buschbeck et al. [2003]; note 209), the long-extinct trilobites may have had the edge (see D.L. Bruton and W. Haas *Special Papers in Palaeontology* 70, 349–361 [2003]).

211. See, for example, A.D. Blest and M.F. Land *Proceedings of the Royal Society of London, B* 196, 197–222 (1977), and S. Laughlin et al. *Journal of Comparative Physiology* 141, 53–65 (1980).

212. See I.R. Schwab et al. *Transactions of the American Opthalmological Society* 99, 145–157 (2001).

213. See M.D.J. Sayer *Fish and Fisheries* 6, 186–211 (2005).

214. See S.K. Swamynathan et al. *FASEB Journal* 17, 1996–2005 (2003); also G.L. Owens et al. *PLoS ONE* 4(6), e5970 (2009).

215. See J.B. Graham and R.H. Rosenblatt *Science* 168, 586–588 (1970).

216. See O. Monk *Videnskabelige Meddelelser fra Dansk naturhistorisk Forening i Kjøbenhavn* 132, 7–24 (1969).

217. See J.B. Graham (*Marine Biology* 23, 83–91; 1973) and J. Nieder (*Journal of Fish Biology* 58, 755–767; 2001) for *Mnierpes* and *Dialommus*, respectively.

218. See E.R. Baylor *Nature* 214, 307–309 (1967).

219. See W.M. Saidel *Philosophical Transactions of the Royal Society, B* 355, 1177–1181 (2000).

220. See W.G. Pearcy et al. *Nature* 207, 1260–1262 (1965).

221. See Schwab et al. (2001), p. 154; note 212.

222. See M.F. Land *Philosophical Transactions of the Royal Society of London B* 355, 1147–1150 (2000).

161. Land (1984), p. 407; note 160.

162. See T.H. Oakley and C.W. Cunningham *Proceedings of the National Academy of Sciences, USA* 99, 1426–1430 (2002).

163. See T.H. Oakley *Integrative and Comparative Biology* 43, 522–530 (2004).

164. The embryology of the myodocopids is also strikingly distinct, and quite possibly the ostracods evolved more than once (see N. Wakayama *Journal of Zoology, London* 273, 406–413 [2007]). More fundamental, however, is the evidence that although the ancestral arthropod possessed some sort of eye, the evolution of a complex compound eye capable of precise focusing almost certainly occurred twice, specifically in the chelicerate-myriapod (broadly horseshoe crabs and centipedes) and crustacean-insect group (see D.E. Nilsson and A. Kelber *Arthropod Structure & Development* 36, 373–385 [2007]). An intriguing consequence of this dichotomy is that the eye structure of the latter group predisposes them to color vision and also the ability to detect polarized light, both functions put to effective use by many insects.

165. See F. Ducret *Zoomorphology* 91, 201–215 (1978); also T. Goto and M. Yoshida in *Photoreception and vision in invertebrates* (M.A. Ali, ed.), pp. 727–742 (Plenum; 1984).

166. See, for example, K.G. Helfenbein et al. *Proceedings of the National Academy of Sciences, USA* 101, 10639–10643 (2004), and D.Q. Matus et al. *Current Biology* 16, R575–R576 (2006).

167. See S.-X. Hu et al. *Palaeogeography, Palaeoclimatology, Palaeoecology* 254, 307–316 (2007); also my paper in *Acta Palaeontologica Polonica* 54, 175–179 (2009).

168. Mike Land and colleagues, for example, draw attention to the similarities between the crystalline cones in the euphausiid shrimps and the moths (M. Land et al. *Journal of Comparative Physiology* 130, 49–62; 1979). Thus they remark, "The resemblance between [these] cones . . . is uncanny . . . and it is hard to escape the conviction that this shape reflects a common optical design principle" (p. 52). Indeed the similarities extend to the entire eye (built on the refracting superposition arrangement), and subsequently Land (1984; note 159) wrote of the euphausiid eye, "The convergence between this design and that of a moth eye is quite extraordinary" (p. 427).

169. In some insects the so-called bright zone has greatly enhanced sensitivity and so serves as a sort of fovea; see, for example, J.H. Hateren et al. *Journal of Comparative Physiology, A* 164, 297–308 (1989). This structure has not only evolved independently, but evidently appeared very early in arthropod evolution; see M.S.Y. Lee et al. *Nature* 474, 631–634 (2011).

170. See S.J. Wilson and M.C. Hutley *Optica Acta* 29, 993–1009 (1982).

171. See C.G. Bernhard et al. *Zeitschrift für vergleichende Physiologie* 67, 1–25 (1970). Some cephalopods have also evolved corneal nipple arrays similar to arthropods, but because of differences in refractive indices underwater, they evidently have a different function; see C. Talbot et al. *Journal of Comparative Physiology, A* 198, 849–856 (2012).

172. See D.G. Stavenga et al. *Proceedings of the Royal Society of London, B* 273, 661–667 (2006); note that they discount the once popular idea that the corneal nipple arrays served to improve transmittance of light, as any such benefit would be minor.

173. See D.G. Stavenga and K. Arikawa *Arthropod Structure & Development* 35, 307–318 (2006).

174. See H. Peisker and S.N. Gorb *Journal of Experimental Biology* 213, 3457–3462 (2010).

175. See A. Yoshida et al. *Zoological Science* 14, 737–741 (1997).

176. Such wings are also found in the cicadas; see G.S. Watson and J.A. Watson *Applied Surface Science* 235, 139–144 (2004).

177. Stavenga et al. (2006); note 172.

178. Genomic information is providing new insights; see, for example, R.D. Burke et al. *Developmental Biology* 300, 434–460 (2006).

179. See his chapter (pp. 483–525) in *Nervous system of invertebrates* (M.A. Ali, ed.; Plenum; 1987).

180. Cobb (1987), p. 484; note 179.

181. For overviews, see J.L.S. Cobb in *The nervous systems of invertebrates: An evolutionary and comparative approach* (O. Briedbach and W. Kutsch, eds.), pp. 407–424 (Birkhäuser Verlag; 1995), and V.W. Pentreath and J.L.S. Cobb *Biological Reviews* 47, 363–392 (1972).

182. See his paper in *Zeitschrift für Biologie* 34, 298–318 (1896).

183. von Uexküll (1896), p. 317; note 182 (the German text reads, "auf Grand der Annahme eines Republik von Reflexen").

184. See J.E. García-Arrarás et al. *Journal of Experimental Biology* 204, 865–873 (2001).

185. See J.L.S. Cobb *Zeitschrift für Zellforschung und mikroskopische Anatomie* 108, 457–474 (1970).

186. Cobb (1987), p. 483; note 179.

187. See J.L.S. Cobb and A. Moore *Comparative Biochemistry and Physiology, A* 91, 821–825 (1988); see also D.S. Smith et al. *Proceedings of the National Academy of Sciences, USA* 82, 1555–1557 (1985).

188. See L.F. Jaffe *Biology of the Cell* 95, 343–355 (2003).

189. See papers in *Journal of Experimental Biology* by E. Blevins and S. Johnsen (vol. 207, pp. 4249–4253; 2004), and D. Yerramilli and S. Johnsen (vol. 213, pp. 249–255; 2010).

190. See, for example, S. Johnsen *Biological Bulletin* 193, 97–105 (1997).

191. See J.P. Woodley in *International Echinoderms Conference, Tampa Bay* (J.M. Lawrence, ed.), p. 61 [Abstract] (Balkema; 1982).

192. Woodley (1982; note 191) thus suggested "an

as Garm and Mori (2009; note 119) note, "came as a surprise and [is] admittedly not easily explained" (p. 3957).

121. Garm and Mori (2009), p. 3951; note 119.

122. As David Albert (*Neuroscience & Biobehavioral Reviews* 35, 474–482; 2010) provocatively asks with respect to the scyphozoan *Aurelia*, "What's on the mind of a jellyfish?" and demonstrates a range of behaviors that suggest these animals are far from being doddering robots.

123. See, for example, R. Quiring et al. *Science* 265, 785–789 (1994); also on pp. 742–743 commentary by C.S. Zuker.

124. See Z. Kozmik et al. *Developmental Cell* 5, 773–785 (2003).

125. See S. Plaza et al. *Journal of Experimental Zoology: Molecular and Developmental Evolution, B* 299, 26–35 (2003).

126. See, for example, V.J. Martin *Canadian Journal of Zoology* 80, 1703–1722 (2002).

127. See, for example, C. Weber *Journal of Morphology* 167, 313–331 (1981).

128. See H. Suga et al. *Proceedings of the National Academy of Sciences, USA* 107, 14263–14268 (2010).

129. Suga et al. (2010), p. 14267; note 128.

130. See Z. Kozmik *Brain Research Bulletin* 75, 335–339 (2008).

131. See, for example, D. Arendt and J. Wittbrodt *Philosophical Transactions of the Royal Society of London, B* 356, 1545–1563 (2001).

132. See Y.J. Passamaneck et al. *EvoDevo* 2, e6 (2011); also B.J. Eriksson et al. *BMC Evolutionary Biology* 13, e186 (2013), for an equivalent in arthropods.

133. See G.L. Fain et al. *Current Biology* 20, R114–R124 (2010).

134. Specifically rhabdomeric opsins are very close to the melanopsins, while the ciliary opsins form a distinct group that also include encephalopsins.

135. See H. Suga et al. *Current Biology* 18, 51–55 (2008); for an overview, see T.D. Lamb *Philosophical Transactions of the Royal Society of London, B* 364, 2911–2924 (2009).

136. See Z. Kozmik et al. *Proceedings of the National Academy of Sciences, USA* 105, 8989–8993 (2008a; with commentary on pp. 15376–15580 by M. Koyanagi et al.). The specifics include the employment of the so-called phosphodiesterase (PDE) system that employs the c-opsin in the PDE cascade, and melanin as the shielding pigment.

137. Piatigorsky et al. (1989, 1993); note 12.

138. See Z. Kozmik et al. *Evolution & Development* 10, 52–61 (2008b).

139. Kozmik et al. (2008a), p. 8991; note 136.

140. Kozmik et al. (2008a), p. 8992; note 136.

141. In terms of eye development, few examples are more spectacular than the redeployment of genes that play a central role in fly eye yet have been co-opted for avian muscle (see T.A. Heanue et al. *Genes & Development* 13, 3231–3243 [1999]). For a comparable example, consider a similar deployment to the vertebrate ear, see S.J. Silver and I. Rebay *Development* 132, 3–13 (2005). Even stranger, perhaps, are the remarkably simplified dicyemid metazoans (pp. 315–316; note 13) that, although effectively protistans, still retain *Pax6* (see J. Aruga et al. *BMC Evolutionary Biology* 7, e201; [2007]).

142. See, for example, J.A. Basil et al. *Journal of Experimental Biology* 203, 1409–1414 (2000).

143. See W.R.A. Muntz and U. Raj *Journal of Experimental Biology* 109, 253–263 (1984); also W.R.A. Muntz and S.L. Wentworth *Biological Bulletin* 173, 387–397 (1987), who draw attention to the complex structure of the so-called myeloid bodies in the retina.

144. See A.C. Hurley et al. *Journal of Experimental Zoology* 205, 37–43 (1978).

145. See P.V. Fankboner *Veliger* 23, 245–249 (1981).

146. See L.A. Wilkins *Biological Bulletin* 170, 393–408 (1986).

147. See M.F. Land *Proceedings of the Royal Society of London, B* 270, 185–188 (2002).

148. See Land (2002), p. 188; note 147.

149. See F.B. Krasne and P.A. Lawrence *Journal of Cell Science* 1, 239–248 (1966); also Kerneis (1975; note 49).

150. See D.-E. Nilsson *Philosophical Transactions of the Royal Society of London, B* 346, 195–212 (1994). Also Roth et al. (2011); note 76.

151. See E.K. Kupriyanova and G.W. Rouse *Molecular Phylogenetics and Evolution* 46, 1174–1178 (2009).

152. See R.S. Smith *Tissue and Cell* 16, 951–956 (1984).

153. Smith (1984), p. 955; note 152.

154. See, for example, M.F. Land *Journal of Physiology* 179, 138–153 (1965). A curiosity is that although a photoreceptor has a ciliary organization, it employs a novel opsin rather than the expected c-opsin; see D. Kojima et al. *Journal of Biological Chemistry* 272, 22979–22982 (1997).

155. See D.I. Speiser and S. Johnsen *Journal of Experimental Biology* 211, 2066–2070 (2008).

156. See J. Cable and R.C. Tinsley *International Journal for Parasitology* 21, 81–90 (1991), and C. Bedini and A. Lanfranchi *Acta Zoologica* 79, 83–90 (1998).

157. See H.-J. Wagner et al. *Current Biology* 19, 108–114 (2009); also commentary on pp. R78–R80 by M.F. Land. A somewhat similar arrangement is also found in the related *Rhynchohyalus*; see J.C. Partridge et al. *Proceedings of the Royal Society of London, B* 281, 20133223 (2014).

158. Wagner et al. (2009), p. 111; note 157.

159. See, for example, R. Elofsson *Arthropod Structure & Development* 35, 275–291 (2006); for a useful overview of crustacean eyes in general, see T.W. Cronin and M.L. Porter *Evolution: Education and Outreach* 1, 463–475 (2008).

160. See M.F. Land in *Photoreception and vision in invertebrates* (ed. M.A. Ali), pp. 401–438 (Plenum; 1984).

cal parallels because of the close relationship between propulsive and foraging modes" (pp. 284–285).

85. See J. Tibballs *Toxicon* 48, 830–859 (2006).

86. See, for example, R.A. Satterlie *Canadian Journal of Zoology* 80, 1654–1669 (2002); also M.O. Shorten et al. *Journal of Zoology, London* 267, 371–380 (2005).

87. See R.J. Larson in *Coelenterate ecology and behaviour* (G.O. Mackie, ed.), pp. 237–245 (Plenum; 1976).

88. See his article in *Natural History* 111(7), 72–75 (2002).

89. Seymour (2002), p. 72; note 88.

90. See B. Werner *Marine Biology* 18, 212–217 (1973); S.E. Stewart *Marine and Freshwater Behavior and Physiology* 27, 175–188 (1996); C. Lewis and T.A.F. Long *Marine Biology* 147, 477–483 (2005); and C. Lewis et al. *Publications of the Seto Marine Biological Laboratory* 40 (5/6), 1–8 (2008).

91. See, for example, J.S. Pearse and V.B. Pearse *Science* 199, 458 (1978); G. Laska and M. Hündgren *Zoologische Jahrbücher, Abteilung für Anatomie und Ontogenie der Tiere* 108, 107–123 (1982); and V.J. Martin *Canadian Journal of Zoology* 80, 1703–1722 (2002).

92. See D.E. Nilsson et al. *Nature* 435, 201–205 (2005).

93. See M. Coates et al. *American Zoologist* 41, 1640 (2001), and Nilsson et al. (2005), note 92.

94. See V.J. Martin *Hydrobiologia* 530/531, 135–144 (2004). This seems, however, to be less likely; see M. O'Connor et al. *Proceedings of the Royal Society of London, B* 277, 1843–1848 (2010); also M.M. Coates *Journal of Experimental Biology* 209, 3758–3765 (2006), and L.A. Gershwin and P. Dawes *Biological Bulletin* 215, 57–62 (2008).

95. One might, however, also draw attention to the counterintuitive ability of both salticid spiders (see T. Nagata et al. *Science* 335, 469–471 [2012]; also commentary on pp. 409–410 by M.E. Herberstein and D.J. Kemp) and squid (see W.-S. Chung and J. Marshall *Current Biology* 24, R64–R65) to engage in a defocusing that confers depth perception.

96. See E.J. Buskey *Marine Biology* 142, 225–232 (2003), and M.M. Coates *Integrative and Comparative Biology* 43, 542–548 (2003).

97. See A. Garm et al. *Journal of Comparative Physiology, A* 193, 547–557 (2007); see also O'Connor et al. (2010), note 94.

98. See A. Garm et al. *Current Biology* 21, 798–803 (2011).

99. See M. O'Connor et al. *Journal of Comparative Physiology, A* 195, 557–569 (2009).

100 O'Connor et al. (2009), p. 566; note 99.

101. As O'Connor et al. (2009; note 99) note, this example potentially "offers a unique insight into the early steps of lens evolution" (p. 567).

102. See W.M. Hamner et al. *Marine and Freshwater Research* 46, 985–990 (1995); also papers by A. Garm et al. in *Journal of Experimental Biology* (vol. 210, pp. 3616–3623 [2007]), and vol. 216, pp. 4520–4529; 2013).

103. See R.F. Hartwick *Hydrobiologia* 216/217, 171–179 (1991).

104. The box jellies are not the only animals with eyes but no brain. In the group known as the deuterostomes, and to which we belong, our close cousins the sea squirts (urochordates) show a number of surprising neurological and behavioral parallels with the sea anemones (see G.O. Mackie and R.C. Wyeth *Canadian Journal of Zoology* 78, 1626–1639 [2000]).

105. These have synaptic arrangements reminiscent of much more advanced animals (see G.C. Gray et al. *Biological Bulletin* 217, 35–49 [2009]).

106. See, for example, A. Garm et al. *Cell and Tissue Research* 325, 333–343 (2006).

107. See C. Skogh et al. *Journal of Morphology* 267, 1391–1405 (2006); also L. Parkefelt and P. Ekström *Journal of Comparative Neurology* 516, 157–165 (2009).

108. Their arrangement invites comparisons to ganglia; see R.A. Satterlie *Journal of Experimental Biology* 214, 1215–1223 (2011). Satterlie's overview of the cnidarian nervous systems is a salutary tonic against the idea that they are cripplingly primitive.

109. See L. Parkefelt et al. *Journal of Comparative Neurology* 492, 251–262 (2005).

110. Gray et al. (2009), p. 48; note 105.

111. In some species this can be surprisingly complex (see A. Garm and J. Bielecki *Journal of Comparative Physiology, A* 194, 641–651 [2008]).

112. See R. Petie et al. *Journal of Experimental Biology* 214, 2809–2815 (2011).

113. The larvae of these animals also possess eyes, albeit with a much simpler arrangement. What is more unexpected, however, is that their capacities as photoreceptors are conducted in the complete absence of a nervous system (see K. Nordström et al. *Proceedings of the Royal Society of London, B* 270, 2349–2354 [2003]).

114. See A. Garm et al. *Vision Research* 48, 1061–1073 (2008).

115. Unsurprisingly, genes for crystallins are present in these slit eyes, but more oddly the lens-less pit eye shows the same genetic expression (Kozmik et al. 2008a; note 136).

116. Consider, for example, the dragonflies with both magnificent compound eyes but also a median ocellus that now transpires to be much more sophisticated in its function than once thought (see R.P. Berry et al. *Vision Research* 47, 1394–1409 [2007]; also J. van Kleef et al. *Journal of Neuroscience* 28, 2845–2855 [2008]).

117. See their paper in *Journal of Comparative Physiology, A* 196, 213–220 (2010).

118. O'Connor et al. (2010), p. 213 [Abstract]; note 117.

119. See A. Garm and S. Mori *Journal of Experimental Biology* 212, 3951–3960 (2009); also Petie et al. (2011; note 112) and in *Journal of Comparative Physiology, A* 199, 315–324 (2013).

120. Rather oddly, the associated nervous tissue also seems to act as an extraocular photoreceptor, which,

49. This organelle has also been recruited in the evolution of the eyes and rather extraordinarily in some flatworms to form a lens (see, for example, N. Griesbach and U. Ehlers *Zoomorphology* 123, 199–202 [2004]; B. Sopott-Ehlers *Journal of Submicroscopic Cytology and Pathology* 31, 123–129 [1999]; and N.A. Watson *Australian Journal of Zoology* 46, 251–265 [1998]). This arrangement may well be convergent (see K. Rohde et al. *International Journal for Parasitology* 29, 511–519 [1999]), and the details of how the mitochondria are modified is quite variable. In some cases, however, the lens is highly refractile, and it is fascinating to speculate whether some mitochondrial enzyme has been recruited in an analogous manner to the other crystallins (see S. Tyler and M.D.B. Burt *Fortschritte der Zoologie* 36, 229–234 [1988]); B. Sopott-Ehlers (*Zoomorphology* 116, 95–101 [1996]) notes "a crystalline to lattice-like sub-structure" (p. 100) in eyes employing giant mitochondria. Nor are flatworms the only invertebrates in which the mitochondria play an optical role, because in the compound eyes of some annelids (p. 101) there is an extraordinary concentration of these organelles (see A. Kerneis *Journal of Ultrastructure Research* 53, 164–179 [1975]), which may serve to detect polarized light (see F.B. Krasne and P.A. Lawrence [1966]; note 149).

50. See D. Francis *Journal of Experimental Biology* 47, 495–501 (1967).

51. See, for example, E. Piccinni and P. Omodeo *Bollettino di Zoologica* 42, 57–79 (1975).

52. See his chapter (pp. 115–130) in *Photoreception and vision in invertebrates* (M.A. Ali, ed.; Plenum; 1984)

53. See Couillard (1984), p. 123; note 52.

54. See, for example, P. Hegemann et al. *Journal of Phycology* 37, 668–676 (2001); P. Gualtieri *Micron* 32, 411–426 (2001); and J.L. Spudich *Trends in Microbiology* 14, 480–487 (2006).

55. See, for example, C.L. Dieckmann *BioEssays* 25, 410–416 (2003).

56. G. Jékely *Philosophical Transactions of the Royal Society of London, B* 364, 2795–2808 (2009), see p. 2804.

. 57 For the case of fungi see, for example, G.M. Avelar *Current Biology* 24, 1234–1240 (2014).

58. See P. Albertano et al. *Micron* 31, 27–34 (2000); see also Gualtieri (2001), note 54.

59. Nor is the only example of versatility among the cyanobacteria. Some filamentous species have the curious capacity to weave themselves into quite substantial ropelike structures. This evidently enables them to colonize loose and shifting substrates, and has evolved independently several times (see E. Garcia-Pichel and M.F. Wojciechowski *PLoS ONE* 4 (11), e7801 [2009]).

60. Presumably this touches on not only our color vision but the vexed question of qualia of consciousness such as "redness."

61. See K.-H. Jung et al. *Molecular Microbiology* 47, 1513–1522 (2003).

62. See V.B. Bergo et al. *Journal of Biological Chemistry* 281, 15208–15214 (2006), and L. Vogeley et al. *Journal of Molecular Biology* 367, 741–751 (2007).

63. See L. Vogeley et al. *Science* 306, 1390–1393 (2004).

64. See O.A. Sineshchekov et al. *Journal of Biological Chemistry* 280, 14663–14668 (2005).

65. Spicules no doubt have several functions, including rendering the sponge decidedly unpalatable, as well as support.

66. See V.C. Sundar et al. *Nature* 424, 899–900 (2003), and J. Aizenberg et al. *Proceedings of the National Academy of Sciences, USA* 101, 3358–3363 (2004).

67. See F. Brümmer et al. *Journal of Experimental Marine Biology and Ecology* 367, 61–64 (2008).

68. See W.E.G.; Müller et al. *Biosensors and Bioelectronics* 21, 1149–1155 (2006).

69. See X.-H. Wang et al. *International Review of Cell and Molecular Biology* 273, 69–115 (2009).

70. Wang et al. (2009), p. 91; note 69.

71. See W.E.G. Müller et al. *FEBS Journal* 277, 1182–1201 (2010); also A.J. Rivera *Journal of Experimental Biology* 215, 1278–1286 (2012).

72. See A.R. Cashmore et al. *Science* 284, 760–765 (1999); also C.B. Green *Current Biology* 14, R847–R849 (2004).

73. See J.J. Wolken and R.G. Florida *Biological Bulletin* 166, 260–268 (1984).

74. Wolken and Florida (1984), pp. 266–267; note 73.

75. See M. Kawahara et al. *South African Journal of Marine Science* 20, 123–127 (1998).

76. J. Roth et al. (*Histochemistry and Cell Biology* 136, 11–23; 2011) suggest that the brush border on top of the compound eye of the ark shells also serves to guide light to the receptor.

77. See K. Franze et al. *Proceedings of the National Academy of Sciences, USA* 104, 8287–8292 (2007).

78. See their paper in *Zoology* 114, 1–10 (2011).

79. Adamksa et al. (2011), p. 4 [sub-heading]; note 78.

80. See I.G. Bebenek et al. *Development, Genes and Evolution* 214, 342–351 (2004).

81. See D. Hoshiyama et al. *FEBS Letters* 581, 1639–1643 (2007).

82. Specifically a spectral sensitivity in the larva of a haploclerid; see S.P. Leys et al. *Journal of Comparative Physiology, A* 188, 199–202 (2002).

83. See A. Hill et al. *Developmental Biology* 343, 106–123 (2010).

84. In addition to this type of jellyfish locomotion, there is an alternative mode more akin to a sort of rowing action. Both, however, are convergent (see J.H. Costello et al. *Invertebrate Biology* 127, 265–290 [2008]; also J.O. Dabiri *Journal of Experimental Biology* 213, 1217–1225; 2010). As John Costello and colleagues remark, "The developmental and structural means of arriving at either propulsive mode have varied between lineages, and parallel evolution has converged on the relatively limited array of functional solutions comprising the medusan morphospace. But these convergent solutions have also entailed ecologi-

1042–1054 (2012). See also C.E. Schnitzler et al. (*BMC Biology* 10, e107; 2012), who establish a link between opsins and photoproteins in the much more primitive ctenophores.

22. In any event, in terms of evolution this insect rhodopsin is closest to mammalian melanopsin. Melanopsin, of course, plays a key role in the maintenance of circadian rhythms and hence is of interest to the jet-lagged ("pass me the melatonin"), but it also occurs in the pigment patches (melanophores) of amphibians (see M.D. Rollag et al. *Methods in Enzymology* 316, 291–309 [2000]; also Provencio et al. [1998], note 18). Clearly melanopsins are related in one way or another to light (some indeed occur in the retina, but are not involved with image formation; see J. Bellingham et al. *PLoS Biology* 4, 1334–1343; 2006), although other opsins are found in the brain. These, aptly referred to as encephalopsins (see S. Blackshaw and S.H. Snyder *Journal of Neuroscience* 19, 3681–3690 [1999]; also S. Halford et al. *Genomics* 72, 203–208 [2001], where they refer to panopsin) and neuropsins (see E.E. Tarttelin et al. *FEBS Letters* 554, 410–416 [2003]), remain rather enigmatic. Neuropsin is also found, in an eyebrow-raising sort of way, in the testes (see Kumbalasiri and Provencio [2005]; note 17). Opsins also occur in the brains of invertebrates, such as the so-called pteropsins of the honeybee (see R.A. Velarde et al. *Insect Biochemistry and Molecular Biology* 35, 1367–1377 [2005]).

23. See D.C. Plachetzki et al. *PLoS ONE* 2, e1054 (2007), and R. Feuda et al. *Proceedings of the National Academy of Sciences, USA* 109, 18868–18872 (2012).

24. See A.K. Sharma et al. and J.L. Spudich in vol. 14 (2006) of *Trends in Microbiology*, pp. 463–469 and pp. 480–487, respectively.

25. See, for example, E.G. Govorunova et al. *Biophysical Journal* 86, 2342–2349 (2004); K.D. Ridge *Current Biology* 12, R588–R590 (2002); and O.A. Sineshchekov et al. *Biophysical Journal* 89, 4310–4319 (2005).

26. See, for example, L.S. Brown *Photochemical & Photobiological Sciences* 3, 555–565 (2004); M.M. Prado et al. *Current Genetics* 46, 47–50 (2004); and S.A. Waschuk et al. *Proceedings of the National Academy of Sciences, USA* 102, 6879–6883 (2005).

27. For an overview of these microbial proteins, see F. Zhang et al. *Cell* 147, 1446–1457 (2011).

28. See N.D. Larusso et al. *Journal of Molecular Evolution* 66, 417–423 (2008); also J. Soppa *FEBS Letters* 342, 7–11 (1994). E.L. Devine et al. (*Proceedings of the National Academy of Sciences, USA* 110, 13351–13355; 2013), however, dispute this convergence.

29. See E. Ebnet et al. *Plant Cell* 11, 1473–1484 (1999), and W. Deininger et al. *Trends in Genetics* 16, 158–159 (2000); also K.A. Mackin et al. *Molecular Biology and Evolution* 31, 85–95 (2014).

30. Larusso et al. (2008), p. 422 (emphasis added); note 28.

31. At least in the case of the fungi the sensory function has evidently evolved independently of the bacteria (see Y. Sudo and J.L. Spudich *Proceedings of the National Academy of Sciences, USA* 103, 16129–16134 [2006]).

32. See V. Busskamp et al. *Science* 329, 413–417 (2010).

33. See, for example, W. Gehring *Journal of Heredity* 96, 171–184 (2005).

34. See N. Shubin et al. *Nature* 457, 818–823 (2009).

35. These show examples of not just a secondary but a tertiary endosymbiosis of the plastids (see A. Bodyl and K. Moszczyński *European Journal of Phycology* 41, 435–448 [2006], and N.J. Patron et al. *Journal of Molecular Biology* 357, 1373–1382 [2006]).

36. See M.G. Levy et al. *Journal of Parasitology* 93, 1006–1015 (2007).

37. See, for example, C. Greuet in *The biology of dinoflagellates* (F.J.R. Taylor, ed.), pp. 119–143 (Blackwell; 1987); also Gehring (2005; note 33) who, in addition to illustrating these eyes (fig. 13), speculates that the warnowiid genes that code for the eye were transferred to the cnidarians, perhaps when symbiotically encapsulated in the tissue. It is an intriguing idea, open to testing, but I think it is as much an attempt to rescue his ideas of optical monophyly.

38. See F. Gómez *European Journal of Protistology* 44, 291–298 (2008).

39. See M. Hoppenrath et al. *BMC Evolutionary Biology* 9, e116 (2009).

40. See, for example, B.S. Leander *Trends in Ecology & Evolution* 23, 481–482 (2008).

41. For an account (pp. 482–484) of this extraordinary misunderstanding, see C.A. Kofoid and O. Swezy *The free-living unarmored Dinoflagellata; Memoirs of the University of California*, vol. 5 (1921).

42. See his paper in *Journal de l'anatomie et de la physiologie normales et pathologiques de l'homme et des animaux* 23, 87–112 (1887).

43. See his paper in *BioSystems* 13, 65–108 (1980).

44. Taylor (1980), p. 76; note 43.

45. See C. Greuet *Protistologica* 4, 419–422 (1968).

46. A rather similar eye, albeit somewhat simpler in its organization, has also been recorded from another group of dinoflagellates, known as the polykrikids (see J. Lecal *Bulletin de la Societie d'Histoire Naturelle de Toulouse* 108, 302–324 [1972]). Not everybody believes this report to be reliable (see M. Hoppenrath and B.S. Leander *Journal of Phycology* 43, 366–377 [2007]), but in any event not only may the polykrikids be related to the warnowiids, but these dinoflagellates are also instructive because of their convergent evolution of nematocysts.

47. See C. Greuet *Cytobiologie* 17, 114–136 (1978).

48. This is not only because of its ubiquity and widespread phototaxis in protistans, but the identification of rhodopsin in the dinoflagellate *Oxyrrhis*; see A.J. Hartz et al. *Journal of Eukaryotic Microbiology* 58, 171–178 (2011). Not only is this dinoflagellate a heterotroph, but is relatively primitive (see J.F. Saldarriaga et al. *International Journal of Systematic and Evolutionary Microbiology* 53, 355–365 [2002]).

230. Maeder (2002); note 227.

231. See F. Lucas et al. *Shirley Institute Memoirs* 28, 77–89 (1955).

232. See S.F. Jackson et al. in *Nature and structure of collagen* (J.T. Randall, ed.), pp. 106–116 (Butterworths Scientific Publication; 1953).

233. For example, a sabellariid polychaete anchors its tube by employing a cement whose amino acid composition (see R.A. Jensen and D.E. Morse *Journal of Comparative Physiology, B* 158, 317–324 [1988]) "is similar to arthropod silk, especially the glue-like component, sericin" (p. 323). To serve as a cement, stabilization of the molecular architecture is required. This is evidently achieved by employing a particular amino acid known as DOPA (short for 3, 4-dihydroxyphenylalanine), which also is employed in the adhesive plaques of the mussel byssus; see J.H. Waite *Journal of Comparative Physiology, B* 156, 491–496 (1986). Yet like a number of other molecules, DOPA is astonishingly versatile, with roles not only in molecular strengthening but also involvement with functions as diverse as the nervous and immune systems, and even ink (pp. 15–16); see J.H. Waite *Biological Bulletin* 183, 178–184; 1992.

234. The roster of "silk in unexpected places" does not end here, because it is also known that the outer layer of the eggs of carp are not only adhesive but contain abundant fibroinlike proteins; see Y.S. Chang and F.L. Huang *Molecular Reproduction and Development* 62, 397–406; 2002. It is also likely that in the same manner as spiders, the fish can produce a number of variants of the silk to meet different ecological settings where the eggs are laid.

235. See G.H. Dickinson et al. *Journal of Experimental Biology* 212, 3499–3510 (2009).

236. Dickinson et al. (2009), p. 3509; note 235.

237. Barnacles may be best remembered as the group upon which Darwin cut his evolutionary teeth, but they are also illustrative of convergences. These include the bonelike material of *Ibla* (p. 28), and numerous lineages showing a successive reduction in the number of parietal plates, evidently as a response to predation pressure imposed by drilling gastropods; see A.R. Palmer *Paleobiology* 8, 31–44 (1982).

CHAPTER 9

1. See pp. 186–189, 2nd ed. (Murray; 1860).

2. See my essay in *Current Biology* 19, R102–R103 (2009).

3. I refer, of course, to Paley's *Natural theology, on evidence of the existence and attributes of the deity* (Chambers; 1849).

4. Titled *Recollections of my life* (MIT Press; 1937).

5. Cajal (1937), p. 576; note 4.

6. See D.E. Nilsson and S. Pelger *Proceedings of the Royal Society of London, B* 256, 53–58 (1994).

7. See M. Eiraku et al. *Nature* 472, 51–56 (2011); also

commentary on pp. 42–43 by R.R. Ali and J.C. Sowden. For a similar story in terms of gut development, see T. Savin et al. *Nature* 476, 57–62 (2011).

8. In the fish *Anableps*, the transparency of the cornea evidently depends on gelsolin proteins. These are quite distinct from the other crystallins (see J. Kanungo et al. *Experimental Eye Research* 79, 949–956 [2004]. and S.-J. Jia et al. *FASEB Journal* 21, 3318–3328 [2007]).

9. See, for example, G. Wistow *Trends in Biochemical Sciences* 18, 301–306 (1993).

10. Their crystallins draw principally on three proteins: S-crystallins (based on so-called GST (glutathione S-transferases) enzymes, omega crystallins (based on aldehyde dehydrogenases), and O-crystallins (similar to some lipid-binding proteins). The latter two crystallins appear to be specific to the cephalopods. A useful overview is provided by S.I. Tomarev and J. Piatigorsky *European Journal of Biochemistry* 235, 449–465 (1996).

11. For an overview, see, for example, J. Piatigorsky *Journal of Biological Chemistry* 267, 4277–4280 (1992), and for a specific instance (gecko), see M.A.M. van Boekel et al. *Journal of Molecular Evolution*, 52, 239–248 (2001).

12. See J. Piatigorsky et al. *Journal of Comparative Physiology, A* 164, 577–587 (1989), and *Journal of Biological Chemistry* 268, 11894–11901 (1993).

13. There seems little rhyme or reason why particular crystallins are recruited, so it is all the more remarkable that an interesting case of convergent evolution has been identified in the promotor region of the genes responsible for the crystallins in scallop and vertebrates, even though the respective crystallins are completely unrelated; see E. Carosa et al. *Journal of Biological Chemistry* 277, 656–664 (2002).

14. See his article in *Current Opinion in Neurobiology* 10, 444–450 (2000).

15. Fernald (2000), p. 446; note 14.

16. See L. Barsanti et al. *Integrative Biology* 4, 22–36 (2012).

17. See, for example, T. Kumbalasiri and I. Provencio, *Experimental Eye Research* 81, 368–375 (2005), and A. Terakita *Genome Biology* 6(3), e213 (2005).

18. See I. Provencio et al. *Proceedings of the National Academy of Sciences, USA* 95, 340–345 (1998).

19. See W.L. Shen et al. *Science* 331, 1333–1336 (2011); also commentary on pp. 1272–1273 by B. Minke and M. Peters.

20. Given, however, that rhodopsin shows a very high thermostability (see K.W. Yau et al. *Nature* 279, 806–807 [1979]), which is just as well given its role in the retina, it is evident that the molecule must be integrated into a more complex system.

21. Shen et al. (2011; note 19) remind us that "it is interesting to consider the question as to whether the archetypal role for rhodopsin was in light sensation or in thermosensation" (p. 1336). Or maybe in some type of protosensory role to judge from its role in insect hearing; see P.R. Senthilan et al. *Cell* 150,

194. There is some evidence that some groups, such as the trichopterans and lepidopterans, have a common ancestor that spun silk, although the former group had to meet the further challenge of underwater spinning; see N. Yonemura et al. *Biomacromolecules* 7, 3370–3378; 2006.

195. See S.T. Case and J.R. Thornton *International Journal of Biological Macromolecules* 24, 89–101 (1999). Evidence for the convergence between the aquatic silks of these dipterans and the trichopterans is given by A. Papanicolaou et al. *Insect Biochemistry and Molecular Biology* 43, 1181–1188 (2013).

196. See D. Rubinoff and P. Schmitz *Proceedings of the National Academy of Sciences, USA* 107, 5903–5906 (2010).

197. See, for example, P. Schmitz and D. Rubinoff *Annals of the Entomological Society of America* 104, 1–15 (2011).

198. See D. Rubinoff and W.P. Haines *Science* 309, 575 (2005).

199. See B.J. Rathcke and R.W. Poole *Science* 187, 175–176 (1975).

200. This is not the only solution, because another caterpillar (of a noctuid moth) behaves more like a lawn mower, shaving the trichomes with its mandibles and disposing of the bundles of "clippings" with a sharp toss of the head. The trichomes are systematically removed in narrow strips, leaving the leaf defenseless (see P.E. Hulley *Ecological Entomology* 13, 239–241 [1988]).

201. Despite having a radically different body, this moth is a convincing mimic of jumping spiders (see J. Rota and D.L. Wagner *PLoS ONE* 1, e45 [2000]), as indeed are a number of unrelated insects as diverse as dipterans (see M.H. Mather and D.B. Roitberg and E. Greene in *Science* 236 [1987], respectively, pp. 308–310 and pp. 310–312) and homopterans (see, for example, A. Floren and S. Otto *Ecotropica* 7, 151–153 [2001]).

202. Conceivably these serve as either decoys or alternatively as landmarks (see A. Aiello and M.A. Solis *Journal of the Lepidopterists' Society* 57, 168–175 [2003]).

203. See J. Rota and D.L. Wagner *Animal Behaviour* 76, 1709–1713 (2008).

204. Rota and Wagner (2008), p. 1713; note 203.

205. Sometimes the same type of construction allows the tables to be turned, albeit by another group of insects, the ever-versatile ants. As is often the case, the Amazonian species in question has an intimate association with a plant, in this case *Hirtella*. Gathering bundles of trichomes, the ants then combine this with a network of fungi to serve as a trap for unwary insects; see A. Dejean et al. *Nature* 434, 973; 2005. Lurking beneath this cover, mandibles agape, they rapidly secure and then engulf the luckless visitor.

206. See B.B. Fulton *Annals of the Entomological Society of America* 34, 289–302 (1941).

207. See, for example, J.F. Jackson *American Midland Naturalist* 92, 240–245 (1974), and R.A. Broadley

and I.A.N. Stringer *Invertebrate Biology* 120, 170–177 (2001).

208. See T.D. Sutherland et al. *Insect Biochemistry and Molecular Biology* 37, 1036–1043 (2007).

209. See J.H. Young and D.J. Merritt *Arthropod Structure & Development* 32, 157–165 (2003).

210. See J.S. Edgerly et al. *Journal of Insect Behavior* 15, 219–242 (2002); also S. Mukerji *Records of the Indian Museum* 29, 253–282 (1927).

211. See D.-Y. Huang and A. Nel *Zoological Journal of the Linnean Society* 156, 889–895 (2009).

212. See their paper in *Insect Biochemistry and Molecular Biology* 39, 75–82 (2009); also M.A. Collin et al. *Zoology* 114, 239–246 (2011).

213. Collin et al. (2009), p. 81; note 212.

214. See T. Nagashima et al. *Cytologia* 56, 679–685 (1991).

215. The convergences of the webspinner extend yet further because in an Australian species not only does it produce an astonishingly fine silk (c. 65 nm), thinner than any other insect (see S. Okada et al. *International Journal of Biological Macromolecules* 43, 271–275 [2008]) and comparable to some spiders, but as Shoku Okada and colleagues note, the molecular structure provides "a striking example of convergent evolution" (pp. 271 [Abstract] and 274) with the silkworm.

216. See J.C. Chang et al. *Polymer* 46, 7909–7917 (2005).

217. See G.M. Gurr and M.J. Fletcher *Australian Journal of Entomology* 50, 231–233 (2011).

218. Perhaps the most remarkable example, however, is a parasitic wasp, the larva of which induces the host-spider to spin its protective silken cocoon; see W.G. Eberhard *Nature* 406, 255–256 (2000).

219. See T.D. Sutherland et al. *Genome Research* 16, 1414–1421 (2006).

220. This conclusion is based on their possessing a remarkable coiled structure (also found in the related hornets and solitary wasps); see T.D. Sutherland et al. *Molecular Biology and Evolution* 24, 2424–2432 (2007).

221. See K. Kronenberger et al. *Naturwissenschaften* 99, 3–10 (2012).

222. Kronenberger et al. (2012), p. 8; note 221.

223. See H. Zhao and J.H. Waite *Biochemistry* 45, 14223–14231 (2006).

224. For an overview, see J.H. Waite et al. *Philosophical Transactions of the Royal Society of London, B* 357, 143–153 (2002).

225. See K.J. Coyne et al. *Science* 277, 1830–1832 (1997).

226. See X.X. Qin et al. *Journal of Biological Chemistry* 272, 32623–32627 (1997); for an overview, see J.H. Waite et al. *Matrix Biology* 17, 93–106 (1998).

227. See F. Maeder and M. Halbeisen *Waffen und Kostumkunde* 43, 33–41 (2001), and F. Maeder *Archaeological Textiles Newsletter* 35, 8–11 (2002).

228. See his paper in *Ars Textrina* 29, 9–223 (1988).

229. The subtitle of his paper on "*Pinna* and her silken beard"; note 228.

157. See R. Gillespie *Science* 303, 356–359 (2004).

158. See A.C. Hawthorne and B.D. Opell *Journal of Experimental Biology* 206, 3905–3911 (2003).

159. See A.C. Hawthorne and B.D. Opell *Biological Journal of the Linnean Society* 77, 1–8 (2002); also F. Vollrath *Current Biology* 16, R925–R927 (2006).

160. The viscid silk is only one of several types of silk in a typical spider web, each produced by a separate gland and possessing the appropriate mechanical feature for its function (see, for example, T.A. Blackledge and C.Y. Hayashi *Journal of Experimental Biology* 209, 2452–2461 [2006]). The end product rightly excites our admiration, and the web has many advantages over its cribellate equivalent (see T.A. Blackledge and C.Y. Hayashi *Journal of Experimental Biology* 219, 3131–3140 [2006]). Not only is it stickier, but the construction of silk is more straightforward, the mechanical properties are improved, and even the reflectance from ultraviolet radiation (which many insects can perceive) is reduced.

161. See H.M. Peters in *Ecophysiology of spiders* (W. Nentwig, ed.), pp. 187–202 (Springer; 1987), and F. Vollrath and D.T. Edmonds *Nature* 340, 305–307 (1989).

162. See S. Zschokke *Nature* 424, 636–637 (2003), who speculates that this silk may be from a gumfoot web rather than an orb-web; also M. Brasier et al. *Journal of the Geological Society, London* 166, 989–997 (2009).

163. These adhesive threads, ejected by the velvet-worm, also employ viscous droplets (note 131).

164. See C. Kropf et al. (*Journal of Zoological Systematics and Evolutionary Research* 50, 14–18; 2012), who identify an organic coat on the spider body that prevents it from being ensnared.

165. See their paper in *Nature* 345, 526–528 (1990).

166. Vollrath et al. (1990), p. 527; note 165.

167. When levels of humidity are elevated, the silk undergoes marked contraction (see F.I. Bell et al. *Nature* 416, 37 [2002]). This has intriguing possibilities for a biomimetic application (see I. Agnarsson et al. *Journal of Experimental Biology* 212, 1990–1994 [2009]) because by inducing a cyclic behavior, not only can the silk apply very considerable forces, but it acts "as a high performance mimic of biological muscles" (p. 1990 [Abstract]).

168. Agnarsson et al. (2009), p. 1993; note 167.

169. See C. Dawson et al. *Nature* 390, 668 (1997).

170. See R. Elbaum et al. *Science* 316, 884–886 (2007).

171. See I. Agnarsson and T.A. Blackledge *Journal of Zoology, London* 278, 134–140 (2009).

172. Agnarrson and Blackledge (2009), p. 134 [Abstract]; note 171.

173. Such an arrangement has evolved multiple times, albeit for various functions; see M. Kuntner et al. *Biological Journal of the Linnean Society* 99, 849–866 (2010).

174. See W.G. Eberhard *Journal of Natural History* 9, 93–106 (1975), and M.K. Stowe *Journal of Arachnology* 6, 141–146 (1978).

175. A useful summary of this area is given by N.J. Vereecken and J.N. McNeil (*Canadian Journal of Zoology* 88, 725–752; 2010) as part of their excellent overview, "Cheaters and liars: chemical mimicry at its finest."

176. See C.K. Curtan and T. Miyashita *Biological Journal of the Linnean Society* 71, 219–235 (2000).

177. See K.V. Yeargan *Annual Review of Entomology* 39, 81–99 (1994).

178. See W.G. Eberhard *Psyche* 87, 143–169 (1980); see p. 144.

179. See M.K. Stowe et al. *Science* 236, 964–967 (1987), and K.V. Yeargan *Oecologia* 74, 524–530 (1988).

180. See J.-W. Zhu and K.F. Haynes *Journal of Chemical Ecology* 30, 2047–2056 (2004).

181. There is also evidence that the spider can adjust the proportions of false pheromones according to the times when different moths fly at night (see K.F. Haynes et al. *Chemoecology* 12, 99–105 [2002]). As it happens the juvenile females are too small to produce a bolas and they leap on their prey, but once again they evidently employ a chemical mimicry (see K.V. Yeargan and L.W. Quate *Oecologia* 106, 266–271 [1996]) to draw the small flies in to their doom. These spiders have other curiosities. One is a marked sexual dimorphism, and like the juvenile females the males never resort to using a bolas but rather retain a hunting mode, again employing a chemical mimicry to lure flies (see K.V. Yeargan and L.W. Quate *Oecologia* 112, 572–576 [1997]). Moreover, the behavior of the females has its oddities, including the release of a pungent odor if disturbed and having body markings suspiciously like a bird dropping.

182. See D.M. Brooks *Wilson Journal of Ornithology* 124, 345–353 (2012).

183. See K. Vahed *Biological Reviews* 73, 43–78 (1998).

184. See L.E. Costa-Schmidt et al. *Naturwissenschaften* 95, 731–739 (2008); see also M.J. Albo et al. *Journal of Zoology, London* 277, 284–290 (2009).

185. See T. Bilde et al. *Biology Letters* 2, 23–25 (2006).

186. The game continues, but this study confirms that death-feigning spiders are more successful in mating, and also enjoy longer copulations (see L.S. Hansen et al. *Behavioral Ecology* 19, 546–551 [2008]).

187. See T.D. Sutherland et al. *Annual Review of Entomology* 55, 171–188 (2010).

188. Sutherland et al. (2010), p. 172; note 187.

189. Sutherland et al. (2010), p. 182; note 187.

190. See, for example, B.L. Fisher and H.G. Roberston *Insectes Sociaux* 46, 78–83 (1999).

191. See R.N. Johnson et al. *Molecular Phylogenetics and Evolution* 29, 317–330 (2003), and S.K.A. Robson and R.J. Kohout *Australian Journal of Entomology* 44, 164–169 (2005).

192. See J.B. Wallace and R.W. Merritt *Annual Review of Entomology* 25, 103–132 (1980).

193. See M.C. Kim et al. *Journal of Entomological Science* 40, 178–185 (2005); also M.S. Engster *Cell and Tissue Research* 169, 77–92 (1976).

but in a later issue (104, 18595–18600; 2007), A.M. Peattie and R.J. Full argue in favor of the importance of specific evolutionary histories.

125. See J.S. Rovner *Journal of Arachnology* 8, 201–215 (1980).

126. In some ways the difference with other spiders is not so great because, as J.S. Rovner (see his paper in *Symposia of the Zoological Society of London* 42, 99–108 [1978]) remarks, in a sense the spiders simply carry their web substitute with them.

127. See J.O. Wolff et al. *PLoS ONE* 8, e62682 (2013), who note that these adhesive structures have evolved at least eight times.

128. See S.N. Gorb *Proceedings of the Royal Society of London, B* 265, 747–752 (1998).

129. See D. Li et al. *Journal of Zoology, London* 247, 293–310 (1999); W. Nentwig *Oecologia* 65, 284–288 (1985); and R.B. Suter and G.E. Stratton *Journal of Arachnology* 33, 7–15 (2005).

130. Li et al. (1999), p. 300; note 129.

131. In particular, the velvet worms (the onychophorans) have slime glands that eject sticky threads extremely rapidly to entrap prey (see M.V. St. J. Read and R.N. Hughes *Proceedings of the Royal Society of London B* 230, 483–506 [1987]). Although velvet worms are the most primitive of living arthropods, and as their name suggests soft-bodied, in point of fact they are a "formidable predator" (p. 502). The adhesive threads are similar to spider silk, both being composed of proteins with a high water content (see K. Benkendorff et al. *Comparative Biochemistry and Physiology, B* 124, 457–465 [1999]). Although in the velvet-worms the specific molecular structure is not at all silklike and consists of a distinctive proline-rich mixture of unordered proteins (see V.S. Haritos et al. *Proceedings of the Royal Society of London, B* 277, 3255–3263 [2010]), the mechanical properties (such as tensile strength) of this spit remain remarkable. Another convergence emerges when the velvet-worm threads are compared with the defensive adhesive glue produced by the frog *Notaden*, both in terms of material properties and biochemistry; see L.D. Graham et al. *Comparative Biochemistry and Physiology, B* 165, 250–259 (2013).

132. See F. Vollrath and D.P. Knight *Nature* 410, 541–548 (2001).

133. See J.M. Gosline et al. *Journal of Experimental Biology* 202, 3295–3303 (1999).

134. See J.C. Chang et al. *Australian Journal of Chemistry* 59, 579–585 (2006).

135. See A.M. Reynolds et al. *Biology Letters* 2, 371–373 (2006).

136. See J.T. Fingerut et al. *Oecologia* 150, 202–212 (2006).

137. See J. Gatesey et al. *Science* 291, 2603–2605 (2001); also N. Yonemura and F. Sehnal *Journal of Molecular Biology* 63, 42–53 (2006).

138. It seems likely that the use of silk in ancient civi-

lizations (China and the Indus valley) may have arisen independently; see I.L. Good et al. *Archaeometry* 51, 457–466 (2009).

139. See H.W. Kew *Proceedings of the Zoological Society of London* 1914, 93–111 (1914).

140. See, for example, P.D. Gabbutt *Journal of Zoology, London* 149, 337–343 (1966).

141. See S. Hunt *Comparative Biochemistry and Physiology* 34, 773–776 (1970).

142. See A. Hazan et al. *Comparative Biochemistry and Physiology, B* 51, 457–462 (1975).

143. See L.R. Jeppson et al. *Mites injurious to economic plants* (University of California Press; 1975).

144. Despite the nanometer dimensions of the serine-rich silk, it is remarkably strong; see M. Gobić et al. *Nature* 479, 487–492 (2011).

145. See, for example, E. Bini et al. *Journal of Molecular Evolution* 335, 27–40 (2004).

146. See P.A. Selden et al. *Proceedings of the National Academy of Sciences, USA* 105, 20781–20785 (2008).

147. See W.A. Shear et al. *Science* 246, 479–481 (1989).

148. See E. Penalver et al. *Science* 312, 1761 (2006).

149. See D. Dimitrov et al. *Proceedings of the Royal Society of London, B* 279, 1341–1350 (2012); also papers in vol. 24 (2014) of *Current Biology* by J.E. Bond et al. (pp.1765–1771) and R. Fernández et al. (pp. 1772–1777).

150. Other striking convergences also occur. For example, in the theridiid spiders, which spin a remarkable variety of cobwebs, there is evidence for rampant convergence; see W.G. Eberhard et al. *Systematics and Biodiversity* 6, 415–475 (2008).

151. See I. Agnarsson et al. *Journal of Zoological Systematics and Evolutionary Research* 51, 101–106 (2013).

152. See T.A. Blackledge et al. *Scientific Reports* 2, e833 (2012); these authors stress that this silk does not rival that of true orb weavers, where compliance and extensibility critically depend on a proline-rich protein (MaSp2). However, a relative of *Fecenia* does spin a silk with a similar protein (see A. Rising et al. *Insect Molecular Biology* 16, 551–561 [2007]; also D. Bittencourt et al. *Comparative Biochemistry and Physiology, B* 155, 419–426; 2010), suggesting that in due course the convergence may become yet closer.

153. For a comparable example from that other classic evolutionary laboratory, Galápagos, see C. de Busschere et al. (*Biological Journal of the Linnean Society* 106, 123–136; 2012), who document among a radiation of lycopsid spiders (*Hogna*) "phenotypic variation in non-genital traits [that] is strikingly identical in similar habitats, even between distantly related species" (p. 126), with those males in high-elevation habitats (one of three so distinguished) being "virtually indistinguishable" (p. 133).

154. See their paper in *Proceedings of the National Academy of Sciences, USA* 101, 16228–16233 (2004).

155. Blackledge and Gillespie (2004), p. 16232; note 154.

156. See R. Gillespie *Nature* 355, 212–213 (1992).

larger-winged insects where the cleaning capacity lies beyond the reach of the legs. This biological self-cleaning has found extraordinarily wide technological applications.

93. See M. Riedl et al. *Planta* 218, 87–97 (2003); also I. Scholz et al. *Journal of Experimental Biology* 213, 1115–25 (2010).

94. See L. Gaume et al. *New Phytologist* 156, 479–89 (2002).

95. See L. Gaume et al. *Arthropod Structure & Function* 33, 103–111 (2004); but see Scholz et al. (2010); note 93.

96. Furthermore, in at least some pitcher plants, beneath the upper layer of wax another surface can be exposed, and its foamlike constitution is designed to reduce the effective footprint of the sliding insect (see E. Gorb et al. *Journal of Experimental Biology* 208, 4651–62 [2005]).

97. See S. Poppinga et al. *Functional Plant Biology* 37, 952–961 (2010); also D. Bröderbauer et al. *Botanical Journal of the Linnean Society* 172, 385–97 (2013). The former authors point out that other entrapment mechanisms in carnivorous plants and so-called kettle-trap flowers include papillate epidermal cells, but these, too, are convergent.

98. Although the focus on entrapment has very much been on the slippery surfaces, it transpires that the "stomach" is not only equipped with digestive enzymes (note 80), but the fluid has rather remarkable visco-elastic properties (most likely due to polysaccharides) that serve to ensnare the prey. See L. Gaume and Y. Forterre *PLoS ONE* 2, e1185 (2007), and V. Bonhomme et al. *New Phytologist* 191, 545–554 (2011).

99. Although not a complete, sharp division, many species of *Nepenthes* evidently fall into a wax-glissade or aquaplaning mode, with the latter arrangement evidently evolving a number of times; see U. Bauer et al. *Journal of Evolutionary Biology* 25, 90–102 (2012).

100. See H.F. Bohn and W. Federle *Proceedings of the National Academy of Sciences, USA* 101, 14138–14143 (2004).

101. See U. Bauer et al. *Proceedings of the Royal Society of London, B* 280, 20122569 (2012), who document the same mechanism in the pitcher plant *Heliamphora*, arising independently of the earlier documentation in *Nepenthes* (note 99).

102. See U. Bauer et al. *Proceedings of the Royal Society of London, B* 275, 259–265 (2007).

103. Moreover, as Ulrike Bauer et al. (2007; note 102) remark, by retaining the initiative, the pitcher plant shows "interesting parallels to the strategies of animal predators" (p. 263).

104. In addition, the risk of drying out can be counteracted because the nectar secreted by the plant can facilitate the condensation of water. Not only does nectar serve as a lethal honeypot (in contrast to more benign access in other contexts), but it adopts a completely novel physical role.

105. See A. Jürgens et al. *Functional Ecology* 23, 875–887 (2009).

106. See K. Wells et al. *Journal of Tropical Ecology* 27, 347–353 (2011).

107. See B. Di Giusto et al. *Journal of Ecology* 98, 845–856 (2010); also G.D. Ruxton and H.M. Schaefer (*Journal of Ecology* 99, 899–904; 2011), who query this and related mimicry hypotheses.

108. See J. Ratsirarson and J.A. Silander *Biotropica* 28, 218–227 (1996).

109. Ratsirarson and Silander (1996); note 108.

110. Bohn and Federle (2004); note 100.

111. Moreover, in at least one case this symbiosis extends to a species of ant that not only can recover prey from the pool, but most likely feeds the pitcher plant with fragments of its food (and possibly feces); see V. Bonhomme et al. *Journal of Tropical Ecology* 27, 15–24 (2011); also M. Scharmann et al. *PLoS ONE* 8, e63556 (2013).

112. See W. Federle et al. *Oecologia* 112, 217–224 (1997), and W. Federle and T. Bruening in *Ecology and biomechanics: A mechanical approach to the ecology of animals and plants* (A. Herrel et al., eds.), pp. 163–183 (Taylor & Francis; 2006).

113. See W. Federle and F.E. Rheindt *Biological Journal of the Linnean Society* 84, 177–193 (2005).

114. See his chapter (pp. 430–433) in *Ant-plant interactions* (C.R. Huxley and D.F. Cutler, eds.; Oxford; 1991).

115. Harley (1991); title of chapter; note 114.

116. In addition, the ants benefit from an additional symbiotic association by virtue of a close association with scale-insects from which they obtain exudates (see H.-P. Heckroth et al. *Journal of Tropical Ecology* 14, 427–443 [1998]), but this intimate tripartite symbiosis evidently evolved later (see S. Ueda et al. *Proceedings of the Royal Society of London, B* 275, 2319–2326 [2008]) than the *Macaranga*-ant association (see S.-P. Quek *Evolution* 58, 554–570 [2004]).

117. One wonders, of course, how the ant keeps its grip, and it is clear that being able to run across the slippery wax does not confer an all-around advantage on smooth surfaces; could it be due to a subtle change in the adhesive fluid oozing from the pads (see W. Federle et al. *Journal of Experimental Biology* 203, 505–512 [2000])?

118. See S. Niederegger and S.N. Gorb *Journal of Comparative Physiology, A* 192, 1223–1232 (2006).

119. See A.B. Kesel et al. *Journal of Experimental Biology* 206, 2733–2738 (2003).

120. See D.E. Hill *Zoological Journal of the Linnean Society* 60, 319–338 (1977).

121. A.B. Kesel et al. *Smart Materials & Structures* 13, 512–518 (2004).

122. Kesel et al. (2004), p. 517; note 121.

123. Kesel et al. (2004), p. 516; note 121.

124. See E. Arzt et al. *Proceedings of the National Academy of Sciences, USA* 100, 10603–10606 (2003);

NOTES   351

60. See their paper in *Journal of Comparative Physiology, A* 192, 1169–77 (2006).

61. Irschick et al. (2006), p. 1175; note 60.

62. See W. Federle et al. *Proceedings of the National Academy of Sciences, USA* 98, 6215–20 (2001).

63. See R.G. Beutel and S.N. Gorb *Journal of Zoological Systematics and Evolutionary Research* 39, 177–207 (2001).

64. Here we find what appear to be examples of massive overengineering. In an American chrysomelid beetle, only a small proportion of the adhesive bristles are employed as it walks. When attacked by ants, however, it immediately clamps itself to the leaf and is effectively immovable (see T. Eisner and D.J. Aneshansley *Proceedings of the National Academy of Sciences, USA* 97, 6568–73 [2000]). Immune against ant attack, maybe, but holding on for grim death achieves little when the beetle is molested by a reduviid bug, which evidently injects a venom. As the authors explain, the chrysomelid goes limp within seconds and then can be easily lifted off.

65. See P. Drechsler and W. Federle *Journal of Comparative Physiology, A* 192, 1213–22 (2006).

66. See R.B. Singer *Annals of Botany* 89, 157–63 (2002).

67. See B.E. Juniper et al. *The carnivorous plants* (Academic; 1989); also E. Król et al. *Annals of Botany* 109, 47–64 (2012).

68. See V.A. Albert et al. *Science* 257, 1491–95 (1992); see also the special issue of *Plant Biology*, part 6, vol. 8 (2006), with an introduction on pp. 737–39 by S. Porembski and W. Barthlott.

69. See their paper in *Journal of Experimental Botany* 60, 14–42 (2009).

70. Ellison and Gotelli (2009), p. 19; note 69.

71. See Y. Yang et al. *Proceedings of the National Academy of Sciences, USA* 104, 8379–84 (2007).

72. See C.G. Pereira et al. *Proceedings of the National Academy of Sciences, USA* 109, 1154–58 (2012).

73. See C. Guisande et al. *Functional Plant Science and Biotechnology* 1, 58–68 (2007).

74. This is only one of many examples of explosive release; evidently the trade-offs in terms of the energy required to drain the bladders of water so that they can be employed for capturing prey is finely balanced and has led to intriguing and precise rearrangements in at least one of the respiratory enzymes (see L. Laakkonen et al. *Plant Biology* 8, 758–64 [2006]).

75. See their paper in *Proceedings of the Royal Society of London, B* 278, 2909–14 (2011).

76. Vincent et al. (2011), p. 2911; note 75.

77. See their paper in *Botanical Journal of the Linnean Society* 161, 329–56 (2009).

78. The full title is "Murderous plants: Victorian gothic, Darwin and modern insights into vegetable carnivory."

79. Chase et al. (2009), p. 332; note 77.

80. They are digested using enzymes secreted by special glands (see T.P. Owen and K.A. Lennon *American Journal of Botany* 86, 1382–90 [1999], and A.H. Thornhill et al. *International Journal of Plant Sciences* 169, 615–24 [2008]). These include enzymes we also employ (see, for example, Z.A. Tökés et al. and C.-I. An et al. in *Planta*, respectively, 119, 39–46 [1974], and 214, 661–67 [2002]), but not surprisingly enzymes capable of attacking the chitinous exoskeleton (chitinases) are also discharged (see H. Eilenberg et al. *Journal of Experimental Botany* 57, 2775–84 [2006]).

81. See L.-J. Chin et al. *New Phytologist* 186, 461–70 (2010).

82. See T.U. Grafe et al. *Biology Letters* 7, 436–39 (2011).

83. Nor are these the only carnivorous plants to obtain nutriments from feces, because in the South African *Roridula*, which traps prey on its sticky surface, a carnivorous hemipteran (*Pameridea*) with which it has an obligate association (see W.R. Dulling and J.M. Palmer *Systematic Entomology* 16, 319–28 [1991]) can walk across the leaves with impunity (see papers by D. Voigt and S. Gorb in *Journal of Experimental Biology* 211, 2647–57 [2008], and *Arthropod-Plant Interactions* 4, 69–79 [2010]) and so can eat the trapped prey; once converted into fecal pellets, these nourish the plant (see B. Anderson *Annals of Botany* 95, 757–61 [2005]; see also B.J. Ptachno et al. in vol. 104, pp. 649–54 [2009]). More generally, the total roster of carnivorous plants may be underestimated. As Chase et al. (2009; note 76) note, "We may be surrounded by many more murderous plants than we think" (p. 353), with obvious implications for both how these transitions are achieved and how many times.

84. See H.-Q. Li *Acta Botanica Gallica* 152, 227–34 (2005).

85. See M.A. Merbach *Nature* 415, 36–37 (2002).

86. The reason they perish in such inordinate numbers is that *Hospitalitermes* is remarkable among termites in resembling the army ants in forming foraging columns (see, for example, papers in *Insectes Sociaux* by D.T. Jones and F. Gathorne-Hardy [vol. 42, pp. 359–69 [1995] and T. Miura and T. Matsumoto [vol. 44, pp. 267–75 [1997]); also J.A. Moran et al. *Annals of Botany* 88, 307–11 (2001).

87. See, for example, K.F. Bennett and A.M. Ellison *Biology Letters* 5, 469–72 (2009).

88. Bennett and Ellison (2009), p. 470; note 86.

89. See their paper in *Biology Letters* 4, 153–55 (2008).

90. Part of the title of Schaefer and Ruxton (2008); note 89.

91. This interpretation, however, is contested by Bennett and Ellison (2009); note 87.

92. Such epicuticular wax also serves as an extremely effective method of self-cleaning, whereby water not only readily runs off but carries with it dust and pathogens. As W. Barthlott and C. Neinhuis (*Planta* 202, 1–8 [1997]) point out, not only is such self-cleaning found in many plants, such as the sacred lotus, but also in words of Santos et al. the arrangement is "strikingly similar to the insect smooth attachment pad in its organization" (p. 2561), it evidently differs in that the associated secretions from the soft disc have separate functions: one to adhere and the other to release.

31. See D.K. Riskin and P.A. Racey *Biological Journal of the Linnean Society* 99, 233–40 (2010).

32. See his chapter (pp. 225–56) in *Biological adhesives* (A.M. Smith and J.A. Callow, eds.; Springer [2006]).

33. Autumn (2006), p. 225; note 32.

34. See T. Gamble et al. *PLoS ONE* 7, e39429 (2012); they suggest eleven gains and nine losses.

35. See E.N. Arnold and G. Poinar *Zootaxa* 1847, 62–68 (2008); for a similar example from amber of Eocene age, see A.M. Bauer et al. *Journal of Zoology* 265, 327–32 (2005).

36. See their *Living reptiles of the world* (Hamish Hamilton; 1957).

37. Schmidt and Inger (1957), p. 70; note 36.

38. An idea touched upon by Autumn (2006; note 32), who also mentions an Indian legend of Shivaji and his Hindu warriors scaling the ramparts of the Deccan using attached lizards to attack the enemy forces.

39. The biomimetic possibilities of geckolike attachment are attracting wide attention; see, for example, A. Mahdavi et al. *Proceedings of the National Academy of Sciences, USA* 105, 2307–12 (2008).

40. Good descriptions can be found in Autumn (2006); note 32, and also K. Autumn and A.M. Peattie *Integrative and Comparative Biology* 42, 1081–90 (2002), and E.R. Pianka and S.S. Sweet *BioEssays* 27, 647–52 (2005).

41. See K. Autumn et al. *Proceedings of the National Academy of Sciences, USA* 99, 12252–56 (2002).

42. See C.J. Clemente et al. *Journal of Experimental Biology* 213, 635–42 (2010).

43. See W.R. Hansen and K. Autumn *Proceedings of the National Academy of Sciences, USA* 102, 385–89 (2005).

44. This is achieved by a distinctive rolling of the toes both upward and backward that changes the critical angle of the setae, breaks the molecular association, and so allows almost instant release (see Y. Tian et al. *Proceedings of the National Academy of Sciences, USA* 103, 19320–25 [2006]).

45. See Autumn (2006); note 32.

46. Pianka and Sweet (2005); note 40.

47. See A. M. Bauer *Journal of Morphology* 235, 41–58 (1998). Most probably this arrangement evolved more than once; in addition, the tails are prehensile. The geckos are instructive in other areas of convergence, not least in terms of more general aspects of foot structure (see A.P. Russell *Copeia* 1979, 1–21 [1979]). As is so often the case, however, these aspects are only recognized when molecular data overturn received phylogenetic wisdom and demonstrate that almost identical morphologies are unreliable indicators of relatedness. Thus, in even one such newly identified

clade (the trans-Atlantic phyllodactylids), three distinctive types of digit (including the classic padded arrangement) have evolved independently multiple times (see T. Gamble et al. *Zoologica Scripta* 37, 355–66 [2008]). While the loss of digits is commonplace in vertebrates, additions to the standard five (a feature known as hyperphalangy) are very uncommon. Nevertheless, in this clade, such has occurred twice, a feature that Tony Gamble and colleagues refer to as "particularly remarkable" (p. 362).

48. See T. Lamb and A.M. Bauer *Proceedings of the Royal Society of London, B* 273, 855–64 (2006).

49. Lamb and Bauer (2006), p. 862; note 48.

50. See T. Lamb et al. *Biological Journal of the Linnean Society* 78, 253–61 (2003); also E.N. Arnold *Journal of Zoology, London* 235, 351–88 (1995).

51. See W.C. Sherbrooke et al. *Zoomorphology* 126, 89–102 (2007).

52. See his paper in *Journal of Herpetology* 27, 265–70 (1993).

53. Withers (1993), p. 265; note 52.

54. See R. Ruibal and V. Ernst *Journal of Morphology* 117, 271–94 (1965).

55. See J. Losos et al. *Annals of the Missouri Botanical Garden* 93, 24–33 (2006).

56. See J.D. Lazell *Journal of Paleontology* 39, 379–382 (1965); also O. Rieppel *Nature* 286, 486–87 (1980).

57. See E.E. Williams and J.A. Peterson *Science* 215, 1509–11 (1982).

58. Their scientific name is *Prasinohaema virens*, which draws attention to a remarkable peculiarity: this lizard's lime-green blood. This coloration has no direct relation to the green pigment seen in the turaco birds (p. 206), and it is due to a product of the bile, the appropriately named biliverdin (see A.E. Greer and G. Raizers *Science* 166, 392–93 [1969]). The concentration of this generally toxic pigment is astonishing, being some 40 times higher than found in patients suffering from green jaundice (see C.C. Austin and K.W. Jessing *Comparative Biochemistry and Physiology, A* 109, 619–26 [1994]). Evidently it has evolved several times, and is found in three groups of fish, frogs, and insects, and also in another lizard, the shingleback (see M. Pennacchio et al. *Herpetological Review* 13, 161–65 [2003]). Why these groups have turned their blood bright green seems to be one of evolution's minor mysteries. Nor does it stand alone, because in animals the breakdown product of biliverdin is a related compound, bilirubin. In what Cary Pirone and colleagues (see their paper in the *Journal of the American Chemical Society* 131, 2830 [2009]) refer to as "an unexpected discovery" (p. 2830), the bright orange of some plants is also due to bilirubin. Whether the molecular pathways transpire to be convergent will be interesting to see.

59. As noted by Williams and Peterson (1982; note 57), in the skinks the setae appear to be elaborations of epidermal cell boundaries, whereas in geckos they are modified scales.

NOTES 349

415, 35–36 (2002); also T. Lingham-Soliar *Naturwissenschaften* 92, 231–36 (2005).

236. See D.A. Pabst *American Zoologist* 40, 146–55 (2000).

237. See G. Iosilevskii and D. Weihs *Journal of the Royal Society Interface* 5, 329–38 (2008).

CHAPTER 8

1. See W. Federle *Journal of Experimental Biology* 209, 2611–21 (2008).

2. See S.N. Gorb and R.G. Beutel *Naturwissenschaften* 88, 530–534 (2001); also S.N. Gorb *Advances in Insect Physiology* 34, 81–115 (2008).

3. Gorb and Beutel (2001), p. 533; note 2.

4. This convergence of smooth and hairy pad can even extend to a specific group, such as the dermapterans (the earwigs and their relatives); see F. Haas and S. Gorb *Arthropod Structure & Development* 33, 45–66 (2004).

5. See his article in *MRS Bulletin* 32, 479–485 (2007).

6. Barnes (2007), p. 479; note 5.

7. See A.D. Lees and J. Hardie and A.F.G. Dixon et al. *Journal of Experimental Biology* 136, 209–28 (1988), and 152, 243–253 (1990), respectively.

8. See L. Frantsevich et al. *Journal of Insect Physiology* 54, 818–827 (2008).

9. See J.M. Urban and J.R. Cryan *Molecular Phylogenetics and Evolution* 50, 471–484 (2009).

10. See P.J. Gullan and P.S. Cranston *The insects: An outline of entomology*, 3rd ed. (Blackwell; 2001), see p. 359.

11. See S.N. Gorb and E.V. Gorb *Journal of Experimental Biology* 207, 2917–2924 (2004).

12. See W. Federle et al. *Proceedings of the National Academy of Sciences, USA* 98, 6215–20 (2001).

13. Ingeniously these insects combine it with a frictional pad (see C.J. Clemente and W. Federle *Proceedings of the Royal Society of London, B* 275, 1329–36 [2008]). In addition, when climbing vertically, they employ dynamics that, as Daniel Goldman and colleagues (*Journal of Experimental Biology* 209, 2990–3000 [2006]) comment, is "remarkably similar to geckos, despite differences in attachment mechanisms, toe number and orientation, and leg number" (p. 2996).

14. See W. Federle et al. *Integrative and Comparative Biology* 42, 1100–1106 (2002). The disaster is not in terms of impact (the body mass of the ants is too low for them to risk damage), but finding their way back into the canopy colony from the forest floor without the benefit of an odor trail.

15. See J. Wojtusiak et al. *Tropical Zoology* 8, 309–18 (1995).

16. Wojtusiak et al. (1995), see fig. 4A (and also fig. 1); note 15.

17. See A.L. Ingram and A.R. Parker *Zoologischer Anzeiger* 244, 209–21 (2006); they document two types of pad—one of which they speculate acts in an analogous way to a car type on a wet surface or as a dewetting surface similar to that found in plants, while the second type is presumed to be simply frictional.

18. These are characterized by interdigital webbing, and in the larger species, surface adhesion is complemented by suction. D.M. Green and P. Alberch (*Journal of Morphology* 170, 273–82 [1981]) also note that the structure of the foot pads differs from those of arboreal frogs and reptiles, but while they respectively are leapers and rapid scurriers, the arboreal activities of these bolitoglossines is far more pedestrian. This feature has evolved several times (see P. Alberch *Evolution* 35, 84–100 [1981]), but interestingly in the case of the Neotropical species it is argued that only in one species does it specifically apply to arboreality (see M. Jaekel and D.B. Wake *Proceedings of the National Academy of Sciences, USA* 104, 20437–42 [2007]). When it comes to their European counterparts, however, things are more complicated; in essence, D.C. Adams and A. Nistri (*BMC Evolutionary Biology* 10, e216 [2010]) argue that in at least three species the ontogeny of the foot pads points to an adaptive explanation for climbing; as importantly the same evolutionary solution can be reached by different developmental pathways.

19. See D.M. Green and M.P. Simon *Australian Journal of Zoology* 34, 135–45 (1986).

20. See T.A. Ba-Omar et al. *Journal of Zoology, London* 250, 267–82 (2000).

21. See N. Crawford et al. *Journal of Experimental Biology* 215, 3965–72 (2012); this includes a self-cleaning mechanism that "bears many similarities to that in insects" (p. 3971).

22. See I. Scholz et al. *Journal of Experimental Biology* 212, 155–62 (2009).

23. Scholz et al. (2009), p. 155; note 22.

24. See W.J.P. Barnes et al. *Journal of Morphology* 274, 1384–96 (2013).

25. See S. Gorb et al. *Journal of Comparative Physiology* 186, 821–31 (2000).

26. See W.J.P. Barnes et al. *Journal of Comparative Physiology, A* 197, 969–78 (2011). These authors also draw attention to an arrangement of blood capillaries that are "strikingly similar to the blood sinuses that occur beneath the hairy adhesive pads of geckos" (p. 976), and suggest that here they are instrumental in serving as shock absorbers.

27. See W. Federle et al. *Journal of the Royal Society Interface* 3, 689–97 (2006).

28. See H.I. Rosenberg and R. Rose *Canadian Journal of Zoology* 77, 233–48 (1999).

29. See J.G.M. Thewissen and S.A. Etnier *Journal of Mammalogy* 76, 925–36 (1995).

30. Elsewhere in the oceans we find an arrangement for adhesion, this time in the sea urchins and their relatives that calls on the adhesive properties of their diagnostic tube feet (see R. Santos et al. *Journal of Experimental Biology* 208, 2555–67 [2005]; also P. Flammang *Echinoderm Studies* 5, 1–60 [1996]). Although in the oxygen utilization the gill structures dictate in favor of the tuna.

200. See D. Bernal et al. *Nature* 437, 1349–1352 (2005); they point out that while tuna does not show this temperature dependence, the muscle twitch is substantially higher.

201. See, for example, K.A. Dickson and J.B. Graham, *Physiological and Biochemical Zoology* 77, 998–1018 (2004).

202. It is, of course, no accident that endothermy has evolved in the metabolically active sphingid moths (pp. 200–201), and indeed independently in several other insect groups (pp. 169–70), the land vertebrates (notably birds and mammals) (226–28), and after a fashion, even plants (p. 163–64).

203. See D. Bernal et al. *Comparative Biochemistry and Physiology: Molecular & Integrative Physiology, A* 155, 454–63 (2010).

204. See J.C. Patterson et al. *Journal of Morphology* 272, 1353–1364 (2011); these workers also demonstrate that a comparable degree of endothermy is not present in the other two thresher sharks (bigeye, pelagic), which in turn poses a number of interesting phylogenetic questions, not least the possibility that the highly distinctive shape of the thresher could be convergent.

205. See C.A. Sepulveda et al. *Journal of Fish Biology* 73, 241–49 (2008).

206. See, for example, F.G. Carey et al. *Southern California Academy of Sciences, Memoirs* 9, 92–108 (1985). Unsurprisingly, this system of heat exchange is also found in the thresher sharks; see Q. Bone and A.D. Chubb *Journal of the Marine Biological Association, UK* 63, 239–241 (1983).

207. See F.G. Carey in *A companion to animal physiology* (C.R. Taylor et al., eds.), pp. 216–33 (Cambridge; 1982).

208. See C. Larsen et al. *Journal of Biological Chemistry* 278, 30741–47 (2003).

209. Carey (1982), p. 228; note 207.

210. See, for example, C.A. Sepulveda et al. *Journal of Fish Biology* 70, 1720–33 (2007).

211. See, for example, V.A. Tubbesing and B.A. Block *Acta Zoologica* 81, 49–56 (2000); also R.L. Alexander *Journal of Zoology, London* 245, 363–69 (1998).

212. See, for example, B.A. Block et al. *Science* 260, 210–14 (1993).

213. See R.L. Alexander *Zoological Journal of the Linnean Society* 118, 151–64 (1996).

214. See R.M. Runcie et al. *Journal of Experimental Biology* 212, 461–70 (2009).

215. See, for example, B.A. Block *Journal of Morphology* 190, 169–89 (1986).

216. See, for example, G. de Metrio et al. *Journal of Morphology* 234, 89–96 (1997).

217. See K.A. Fritsches et al. *Current Biology* 15, 55–58 (2005).

218. See A.G. Little et al. *Molecular Phylogenetics and Evolution* 56, 897–904 (2010).

219. Nor are they monophyletic inasmuch as *Pset-*

todes is excluded from the main clade, indicating a flatfish morph evolved twice; see M.A. Campbell et al. *Molecular Phylogenetics and Evolution* 69, 664–73 (2013); see also discussion by R. Betancur-R and G. Ortí (vol. 73, pp. 18–22 [2014]) and reply (vol. 75, pp. 149–53 [2014]). Even within the main pleuronectiform clade there is abundant polyphyly (see D.M. Roje *Molecular Phylogenetics and Evolution* 56, 586–600 [2010]), and as Dawn Roje notes, "asymmetrical jaws and specialized dentition have been gained, lost, and gained again and do not seem to have evolved sequentially from one to the other" (p. 597), thus knocking on the head any idea of a simple trend from generalized to specialized.

220. See M. Friedman *Nature* 454, 209–12 (2008), and commentary on pp. 167–70 by P. Janvier.

221. Little et al. (2010), p. 903; note 218.

222. See, for example, A.C. Dalziel et al. *Journal of Molecular Evolution* 62, 319–31 (2006).

223. See A. Tullis et al. *Journal of Experimental Biology* 161, 383–403 (1991).

224. See B.A. Block and C. Franzini-Armstrong *Journal of Cell Biology* 107, 1099–1112 (1988).

225. See C.D. Moyes et al. *Canadian Journal of Zoology* 70, 1246–1253 (1992).

226. As Christopher Moyes and colleagues (1992; note 225) note, this is not far off the "maximum degree of cristae packing that will still allow enough matrix space for two average-sized Krebs cycle enzymes, each in contact with an opposing membrane" (p. 1250).

227. So, too, in the billfish Ca^{2+}–ATPase evidently plays a key role (see, for example, D.C.F. Da Costa and A.M. Landeira-Fernandez *American Journal of Physiology: Regulatory Integrative and Comparative Physiology* 297, R1460–R1468 [2009]).

228. See P. Bartsch *Palaeontographica, Abt. A* 204, 117–226 (1988; in German, with English summary); also G. Arratia and P. Lambers in *Mesozoic fishes: Systematics and paleoecology* (G. Arratia and G. Viohl, eds.), pp. 191–218 (Verlag Dr Friedrich Pfeil; 1996).

229. See, for example, P.L. Forey *Bulletin of the British Museum (Natural History), Geology* 28, 123–204 (1977). Although almost entirely a set of detailed anatomical descriptions, he does note that "pachyrhizodontids were probably powerful swimmers and were certainly carnivorous. . . . The caudal fin of the American species of *Pachyrhizodus* is very deep and probably had a high aspect ratio suggesting that these species at least were capable of fast, sustained swimming" (p. 195).

230. See, for example, J. Alvarado-Ortega *Journal of Vertebrate Paleontology* 24, 802–13 (2004).

231. See M. Friedman *Proceedings of the National Academy of Sciences, USA* 106, 5218–23 (2009).

232. See J. Kriwet and M.J. Benton *Palaeogeography, Palaeoclimatology, Palaeoecology* 214, 181–94 (2004).

233. See M. Friedman et al. *Science* 327, 990–993 (2010); also commentary on pp. 968–69 by L. Cavin.

234. Friedman (2009), p. 5222; note 231.

235. See, for example, A.M. Boustany et al. *Nature*

166. This is on account of its rostrum extending far beyond the mandible (see C. McGowan *Nature* 322, 454–456 [1986]), while the appropriately named *Excalibosaurus* (see C. McGowan *Journal of Vertebrate Paleontology* 23, 950–956 [2003]) provides an intermediate between *Eurhinosaurus* and more orthodox forms.

167. Like many reptiles and birds these were supported by a sclerotic ring (see, for example, M.S. Fernández et al. *Journal of Vertebrate Paleontology* 25, 330–337 [2005]), but the equivalent ring in the teleost fish may have evolved independently (see T.A. Franz-Odendaal and B.K. Hall *Journal of Morphology* 267, 1326–1337 [2006]). Regarding the large eyes in the diving pinnipeds (see L.B. Debey and N.D. Pyenson *Marine Mammal Science* 29, 48–83 [2013]), but these mammals of course lack sclerotic rings.

168. See R. Motani et al. *Nature* 402, 747 (1999); also D.-E. Nilsson et al. *Current Biology* 22, 683–688 (2012), and commentary on pp. R268–R270 by J.C. Partridge; also D.-E. Nilsson et al. *BMC Evolutionary Biology* 13, e187 [2013]), who document the enormous eyes of giant squid and draw parallels with the ichthyosaurs.

169. See S. Humphries and G.D. Ruxton *Journal of Experimental Biology* 205, 439–441 (2002).

170. See R. Motani *Nature* 415, 309–312 (2002); also J.A. Massare *Paleobiology* 14, 187–205 (1988).

171. Motani (2005), p. 414; note 164.

172. See A. Bernard et al. *Science* 328, 1379–1382 (2010); also commentary on pp. 1361–1362 by R. Motani. What may be supportive evidence comes from evidence for melanosomes in the skin that may have assisted with thermal absorption and are convergent with fossil mosasaurs and leatherback turtles; see J. Lindgren et al. *Nature* 506, 484–488 (2014). These authors also note that ichthyosaurs may have lacked countershading, perhaps consistent with inferred deep diving.

173. Motani et al. (1996), p. 347; note 165.

174. See T. Lingham-Soliar and W.E. Reif *Neues Jahrbuch für Geologie und Paläontologie, Abhandlungen* 207, 171–183 (1998).

175. See T. Lingham-Soliar and G. Plodowski *Naturwissenschaften* 94, 65–70 (2007).

176. Both groups were, of course, highly effective predators, and one intriguing convergence concerns the teeth and specifically the cellular cementum; see E.E. Maxwell et al. *Journal of Morphology* 272, 129–135 (2011).

177. Such is evident from juveniles that were highly effective swimmers; see A. Houssaye and P. Tafforeau *Journal of Vertebrate Paleontology* 32, 1042–1048 (2012).

178. See M.R. Ross *Journal of Vertebrate Paleontology* 29, 409–416 (2009).

179. See B. Rothschild and L.D. Martin *Science* 236, 75–77 (1987).

180. See B.M. Rothschild and L.D. Martin *Netherlands Journal of Geology* 84, 341–344 (2005).

181. See B.M. Rothschild and V. Naples *Acta Palaeontologica Polonica* (in press) (2014); also B.M. Rothschild *Annals of the Carnegie Museum* 56, 253–258 (1987).

182. See, respectively, B.M. Rothschild and G.W. Storrs *Journal of Vertebrate Paleontology* 23, 324–328 (2003; here the damage is to the heads of the humerus and femur rather than the vertebrae; also A.A. Farke in vol. 27, pp. 724–726; 2007), and R. Motani et al. *Nature* 402, 747 (1999).

183. See B.M. Rothschild et al. *Naturwissenschaften* 99, 443–448 (2012). Rothschild (1987; note 181) suggests that in the case of a Cretaceous turtle the avascular necrosis may have originated from an escape maneuver from a mosasaur with rapid ascent.

184. See M.R. Ross *Journal of Vertebrate Paleontology* 29, 409–416 (2009).

185. See J.A. Massare *Journal of Vertebrate Paleontology* 7, 121–137 (1987).

186. This is the titanic species *Prognathodon currii* whose skull recalls that of the tyrannosaurids (see P. Christiansen and N. Bonde *Journal of Vertebrate Paleontology* 22, 629–644 [2002]).

187. See papers by J. Lindgren et al. in *Lethaia* 40, 153–160 (2007), and *Journal of Vertebrate Paleontology* 28, 1043–1054 (2008); also A.R.H. Le Blanc *Journal of Vertebrate Paleontology* 33, 349–362 (2013), who demonstrates the development of an akinetic skull consistent with high-speed pursuit of small prey.

188. Mosasaurs evidently survived right up to the K/T boundary; see W.B. Gallagher et al. *Journal of Vertebrate Paleontology* 25, 473–475 (2005).

189. Lindgren et al. (2007), p. 159; note 187.

190. See J. Lindgren et al. *PLoS ONE* 5(8), e11998 (2010); also T. Konishi et al. *Journal of Vertebtate Paleontology* 32, 1313–1327 (2012).

191. Lindgren et al. (2010), p. 1 [Abstract]; note 190; also J. Lindgren *Nature Communications* 4, e2423 (2013), for another mosasaur with a forked and highly asymmetric tail fin.

192. See also J. Lindgren et al. *Paleobiology* 37, 445–469 (2011), who focus on the trend toward a hypocercal tail and a shift to a fully oceanic habit.

193. See my *Life's solution*, pp. 223–224.

194. See S.L. Katz et al. *Nature* 410, 770–771 (2001).

195. For tuna, see J.B. Graham and K.A. Dickson, *Journal of Experimental Biology* 207, 4015–4024 (2004); for shark, see S. Gemballa et al. *Journal of Morphology* 267, 477–493 (2006).

196. See J.M. Donley et al. *Nature* 429, 61–65 (2004), and commentary on pp. 31–32 by A.P. Summers; also J.M. Donley et al. *Journal of Experimental Biology* 208, 2377–2387 (2005).

197. Donley et al. (2004), p. 61; note 196.

198. See D. Bernal et al. *Journal of Experimental Biology* 206, 2831–2843 (2003).

199. Thus, both depend on ram ventilation, but as N.C. Wegner et al. (*Journal of Experimental Biology* 215, 22–28 [2012]) point out, despite a series of convergent similarities in hydrodynamics, when it comes to

133. Ritzmann et al. (2004), p. 377; note 131.

134. See also K. Jung et al. *Bioinspiration & Biomimetics* 2, S42–S49 (2007).

135. See Y.-C. Yang et al. *Proceedings of the National Academy of Sciences, USA* 103, 18891–18895 (2006).

136. See M.A. MacIver et al. *IEEE Journal of Oceanic Engineering* 29, 651–659 (2004).

137. See R. Stafford et al. *BioSystems* 87, 164–171 (2007).

138. See A.E. Greer *Journal of Herpetology* 25, 166–173 (1991).

139. More general evolutionary principles emerge; rather remarkably it appears that limbs and digits can also reevolve, thereby confounding Dollo's "law" (p. 28).

140. See M.C. Brandley et al. *Evolution* 62, 2042–2064 (2008).

141. Brandley et al. (2008), p. 2056; note 140.

142. See also D. Liang et al. *BMC Evolutionary Biology* 11, e25 (2011).

143. Fossil evidence suggests they are related to the lacertid lizards, and the initial move to limblessness was connected to fossoriality and reinforcement of the head; see J. Müller et al. *Nature* 473, 364–367 (2011).

144. For example, the so-called shovel-headed morph (see M. Kearney and B.L. Stuart *Proceedings of the Royal Society of London, B* 271, 1677–1682 [2004]) has evolved three times independently. So, too, even among the Brazilian fauna (see T. Mott and D.R. Vieites *Molecular Phylogenetics and Evolution* 51, 190–200 [2009]) the structure of the tail and the ability to cast it off (autotomy) are, in the words of Tami Mott and David Vieites, exemplars of "extreme morphological homoplasy" (p. 190 [part of the title]).

145. See M. Wall and R. Shine *Biological Journal of the Linnean Society* 91, 719–727 (2007).

146. The main title of their paper in *Evolution* 55, 2303–2318 (2001).

147. So, too, they cast serious doubt on any simple change in the expression of key developmental genes (in this case the *Hox* genes), which too often are waved as a magic molecular wand to achieve the required transformation.

148. See J.J. Wiens et al. *Evolution* 60, 123–141 (2006); also R. Shine and M. Wall *Biological Journal of the Linnean Society* 95, 293–304 (2008), for the suggestion that snakes evolved from burrowing forms.

149. See M.W. Caldwell *Canadian Journal of Earth Sciences* 40, 573–588 (2003).

150. Caldwell (2003), p. 573 [Part of the title]; note 149.

151. See D.B. Wake *American Naturalist* 138, 543–567 (1991).

152. Wake (1991), p. 555; note 151.

153. See G. Parra-Olea and D.B. Wake *Proceedings of the National Academy of Sciences, USA* 98, 7888–7891 (2001).

154. Parra-Olea and Wake (2001), p. 7889; note 153.

155. Parra-Olea and Wake (2001), p. 7891; note 153.

156. Thus very much the same story of the "Lineatriton phenomenon" occurs in the Neotropical lizard *Sceloporus serrifer*; see N. Martínez-Méndez et al. *Zoologica Scripta* 41, 97–108 (2012).

157. In particular, the axial extension of vertebrae with ribs involves the necessary reorganization of the developmental *Hox* genes (see M.J. Cohn and C. Tickle *Nature* 399, 474–479 [1999]), which in itself is unsurprising given that these genes play a central role in axial differentiation. One might expect to see a simple expansion of the region of the *Hox* linear array that in typical vertebrates produces the thoracic region of the backbone (that is, with ribs; see J.M. Woltering et al. *Developmental Biology* 332, 82–89 [2009]). Yet this process turns out to be more complex than thought. Much more interesting, however, is that, as Joost Woltering and colleagues note, a comparable study "in a caecilian amphibian, which convergently evolved a deregionalized body plan, reveals a similar global collinear pattern of *Hox* expression" (p. 82 [Abstract]). The similarities between caecilian and snake are, therefore, not only underpinned by developmental convergence, but in a more subtle way than might be expected.

158. See R. Gaymer *Nature* 234, 150–151 (1971).

159. See J.C. O'Reilly et al. *Nature* 386, 269–272 (1997); also A.P. Summers and J.C. O'Reilly *Zoological Journal of the Linnean Society* 121, 65–76 (1997).

160. Like an earthworm one might expect the burrowing caecilians to be more or less colorless, yet about a third of species exhibit a variety of color (see K.C. Wollenberg and G.J. Measey *Journal of Evolution* 22, 1046–1056 [2009]). As Katharina Wollenberg and John Measey remark, this is surely "perplexing" (p. 1046 [Abstract]), not least because some taxa are "comparable with the most conspicuous frogs and salamanders" (p. 1047). Yet not only are there good adaptive reasons, revolving around aposematism (and possibly skin toxins) and crypsis, but importantly the "conspicuous yellow coloration has evolved in three convergent events" (p. 1052).

161. See J.C. O'Reilly *Nature* 382, 33 (1996).

162. Although the dolphins and many other odontocetes are usually considered strikingly hydrodynamic, it is worth recalling that the Pliocene *Odobenocetops* has some striking convergences with the walrus, including tusks; see C. de Muizon et al. *Smithsonian Contributions to Paleobiology* 93, 223–261 (2002), and C. de Muizon and D.P. Domning *Zoological Journal of the Linnean Society* 134, 423–452 (2002). Another delphinid convergence entails an older example in the form of *Australodelphis*, the skull of which converges on that of the beaked whales; see R.E. Fordyce et al. *Antarctic Science* 14, 37–54 (2002).

163. See R.L. Carroll *Special Papers in Palaeontology* 33, 145–155 (1985).

164. See R. Motani *Annual Review of Earth and Planetary Sciences* 33, 395–420 (2005).

165. See R. Motani et al. *Nature* 382, 347–348 (1996).

95. See T. Goto et al. *Ichthyological Research* 46, 281–287 (1999).

96. See P.A. Pridmore *Zoology, Analysis of Complex Systems (ZACS)* 98, 278–297 (1994/1995).

97. Pridmore (1994/1995), p. 285; note 96.

98. See M.I. Coates and R.W. Gess *Palaeontology* 50, 1421–1446 (2007).

99. See D.M. Koester and C.P. Spirito *Copeia* 2003, 553–561 (2003); also L.J. Macesic and S.M. Kajiura *Journal of Morphology* 271, 1219–1228 (2010). Some crabs can also punt; see M.M. Martinez et al. *Journal of Experimental Biology* 201, 2609–2623 (1998).

100. See L.O. Lucifora and A.I. Vassallo *Biological Journal of the Linnean Society* 77, 35–41 (2002).

101. See also R.J. Holst and Q. Bone *Philosophical Transactions of the Royal Society of London, B* 339, 105–108 (1993).

102. Lucifora and Vassallo (2002), p. 40; note 100.

103. Thus H.M. King et al. (*Proceedings of the National Academy of Sciences, USA* 108, 21146–21151 [2011]) document a sort of walking, principally employing the pelvic fins, in the lungfish, the sister group of the tetrapodomorphs. The mode of progression is, however, not similar, and given the highly specialized nature of this African species, one cannot help but wonder if it is convergent.

104. See S.M. Swift and P.A. Racey *Behavioral Ecology and Sociobiology* 52, 408–416 (2002).

105. Swift and Racey (2002), p. 411; note 104.

106. See W.A. Schutt and N.B. Simmons *Acta Chiropterologica* 3, 225–235 (2001); also D.L. Dietz *Journal of Mammalogy* 54, 790–792 (1973), who notices a link between walking and the bat's emotional state, even leading to "great excitement" (p. 792).

107. Evidently the limbs, especially the femora, are strong enough, but probably the musculature associated with the pectoral girdle is inappropriate; see D.K. Riskin et al. *Journal of Experimental Biology* 208, 1309–1319 (2005).

108. See E. Calvo *Journal of Biological Chemistry* 281, 1935–1942 (2006).

109. See, for example, A.Z. Fernandez et al. *Biochimica et Biophysica Acta* 1434, 135–142 (1999).

110. See A.M. Greenhall *Journal of Zoology, London* 168, 451–461 (1972).

111. See D.K. Riskin and J.W. Hermanson *Nature* 434, 292 (2005).

112. In addition, the running vampire has a distinctive aerial hop, related to a spectacular jumping behavior that initiates flight (see W.A. Schutt et al. *Journal of Experimental Biology* 200, 3003–3012 [1997]).

113. See D.K. Riskin et al. *Journal of Experimental Biology* 209, 1725–1736 (2006).

114. See, for example, M. Kennedy et al. *Molecular Phylogenetics and Evolution* 13, 405–416 (1999).

115. See B.D. Lloyd *Journal of the Royal Society of New Zealand* 31, 59–81 (2001).

116. Lloyd (2001), p. 68; note 115.

117. See his paper in *New Zealand Journal of Zoology* 6, 357–370 (1979).

118. Daniel (1979), p. 361; note 117.

119. Intriguingly, fossil evidence shows that some archaic mammals survived in New Zealand until at least the Miocene (c. 18 Myr) but then perished, perhaps due to climatic change (see T.H. Worthy *Proceedings of the National Academy of Sciences, USA* 103, 19419–19423 [2006]).

120. See S.J. Hand et al. *BMC Evolutionary Biology* 9, e169 (2009).

121. See P.D. Dwyer *Zoological Publications from Victoria University College of Wellington* 28, 1–28 (1962).

122. The convergences don't stop here, because like other bats this New Zealand mystacine houses a batfly, a sort of louselike guest (see B.A. Holloway *New Zealand Journal of Zoology* 3, 279–301 [1976]), but the fly in question is convergent with the typical batflies (see D.M. Gleeson et al. *Journal of the Royal Society of New Zealand* 30, 155–168 [2000[). The naked bulldog bats also have distinctive commensals, but this time in the form of earwigs (Schutt and Simmons 2001; note 106).

123. See G. Jones et al. *Journal of Experimental Biology* 206, 4209–4216 (2003); also J. McCartney et al. *New Zealand Journal of Zoology* 34, 227–238 (2007). The bat will also walk across flowers in search of nectaries (Daniel [1979]; note 117).

124. See P.D. Dwyer *Records of the Dominion Museum* 4, 77–78 (1962); in bats it is typically the canines that show greatest wear.

125. For a comparable example consider the so-called tree lobsters, members of the phasmatoid insects, and which show an example of "extreme convergence"; see T.R. Buckley et al. *Proceedings of the Royal Society of London, B* 276, 1055–1062 (2009). Loss of flight per se in insects has evolved multiple times; see T.F. Mitterboeck and S.J. Adomowicz *Proceedings of the Royal Society of London, B* 280, 20131128 (2013).

126. See my *Life's solution*, p. 218, for an overview of the peculiarities of the New Zealand fauna, also C.H. Daugherty et al. *Trends in Ecology & Evolution* 8, 437–442 (1993).

127. See M.J. Griffin et al. *New Zealand Journal of Ecology* 35, 302–307 (2011), for a trenchant review of this question.

128. See their paper in *Science* 311, 1575 (2006); also K.C. Burns *New Zealand Journal of Ecology* 30, 405–406 (2006); in addition, see subsequent discussion by M. Morgan-Richards et al. and K.C. Burns in vol. 32 (2008), respectively, pp. 108–112 and 113–114.

129. Duthie et al. (2006), p. 1575; note 128.

130. See my *Life's solution*, p. 284.

131. See R.E. Ritzmann et al. *Arthropod Structure & Development* 33, 361–379 (2004).

132. While typically thought of as scuttling, a South African species (*Saltoblattella*) shows a striking convergence with the jumping grasshoppers; see M. Picker et al. *Biology Letters* 8, 390–392 (2012).

L.J. Shapiro (*Journal of Human Evolution* 65, 253–265 [2013]) question habitual bipedality.

61. See M.W. Marzke and M.M. Shrewsbury *Journal of Human Evolution* 51, 213–215 (2006).

62. See E.E. Sarmiento and L.F. Marcus *American Museum Novitates* 3288, 1–38 (2000).

63. See L. Rook et al. *Journal of Human Evolution* 46, 349–356 (2004).

64. See their paper in *Proceedings of the National Academy of Sciences, USA* 94, 11747–11750 (1997).

65. Köhler and Moyà-Solà (1997), p. 11747 [Abstract]; note 64.

66. See E. Carnieri and F. Mallegni *Homo* 54, 29–35 (2003).

67. See S. Moyà-Solà and M. Köhler *Coloquios de Paleontologia, volumen Extraordinario (En honor al Dr Remmert Daams)* 1, 443–458 (2003); also Köhler and Moyà-Solà (1997); note 64.

68. See also I. Casanovas-Vilar et al. *Journal of Human Evolution* 61, 42–49 (2011), who discuss possible evidence from associated murids and their dentition.

69. See M.F. Williams *Bioscience Hypotheses* 1, 127–137 (2008).

70. Williams (2008), p. 133; note 69.

71. See her book *The aquatic ape: A theory of human evolution* (Souvenir; 1982).

72. I should say that intriguing as I find her arguments in the case of human bipedality (see also C. Niemitz *Naturwissenschaften* 97, 241–263 [2010]), as far as the fossil record is concerned, the aquatic connection looks tenuous, while concerning the much vaunted human nakedness I suspect this was entirely unrelated to bipedality and was the result of sexual selection along with the growing use of body color and ornaments.

73. See R.L. Bernor et al. *Bollettino della Societa Paleontologica Italiana* 40, 139–148 (2001); also S.D. Matson et al. *Journal of Human Evolution* 63, 127–139 (2012), who support this idea inasmuch as, despite a change in climate at the time of the disappearance of *Oreopithecus*, there is no evidence for a dramatic change in vegetation.

74. See B.G. Richmond and W.L. Jungers *Science* 319, 1662–1665 (2008), as well as commentary on pp. 1599 and 1601 by A. Gibbons.

75. See C.P.E. Zollikofer et al. *Nature* 434, 755–759 (2005).

76. See B.A. Patel et al. *Journal of Human Evolution* 57, 763–772 (2009).

77. See T.L. Kivell and D. Schmitt *Proceedings of the National Academy of Sciences, USA* 106, 14241–14246 (2009); also M. Dainton and G.A. Macho *Journal of Human Evolution* 36, 171–194 (1999). For arguments against this convergence see S.A. Williams *Journal of Human Evolution* 58, 432–440 (2010).

78. Evidence for the extinct *Sivapithecus* being a knuckle walker is given by D.R. Begun and T.L. Kivell *Journal of Human Evolution* 60, 158–170 (2011).

79. See C.M. Orr *American Journal of Physical Anthropology* 128, 639–658 (2005).

80. See R.H. Crompton et al. *Journal of Anatomy* 212, 501–543 (2008).

81. Crompton et al. (2008), p. 509; note 80.

82. See S.K.S. Thorpe et al. *Science* 316, 1328–1331 (2007), and commentary on pp. 1292–1294 by P. O'Higgins and S. Elton; also discussion in vol. 318 by D.R. Begun et al. (1066d) and reply by R.H. Compton and S.K.S. Thorpe (1066e).

83. O'Higgins and Elton (2007), p. 1293; note 82.

84. See his commentary in *Nature* 418, 133–135 (2002).

85. Wood (2002), p. 134; note 84.

86. See W.E.H. Harcourt-Smith and L.C. Aiello *Journal of Anatomy* 204, 403–416 (2004).

87. See, for example, R.M. Alexander *Journal of Anatomy* 204, 321–330 (2004).

88. Bipedality has evolved independently in the lemurs, specifically the sifakas (see R.E. Wunderlich and J.C. Schaum *Journal of Zoology, London* 272, 165–175 [2007]), although the method of walking is not that similar to ourselves.

89. Bipedality also appeared substantially earlier than the theropods, in the form of the bolosaurid *Eudibamus*, a herbivore living in the early Permian (see D.S. Berman et al. *Science* 290, 969–972 [2000], and commentary on p. 917 by E. Stokstad). As David Berman and his colleagues remark with respect to *Eudibamus*, "This form of locomotion is unique among Paleozoic tetrapods" (p. 969), but bipedal reptiles were going to evolve, come what may. This point is strikingly shown among the extant lizards (see R.C. Snyder *American Zoologist* 2, 191–203 [1962] and C.J. Clemente et al. *Journal of Experimental Biology* 211, 2058–2065; 2008), of which the most astonishing is the Basilisk lizard, which can run across water (see S.T. Hsieh *Journal of Experimental Biology* 206, 4363–4377 [2003]) by ingeniously striking the surface to form an air cavity (see J.W. Glasheen and T.A. McMahon *Nature* 380, 340–342 [1996]).

90. See C.L. Huffard et al. *Science* 307, 1927 (2005).

91. See G.S. Helfman et al. *The diversity of fishes* (Blackwell; 1997); also M. Jamon et al. *Journal of Experimental Zoology: Ecological Genetics and Physiology, A* 307, 542–547 (2007). In passing we can note that such fish fingers go beyond children's food because, as G.C. Jensen (*Biological Bulletin* 209, 165–167 [2005]) notes, the pectoral fins of the fourhorn poacher have free rays that serve as "fingers" to lift stones and shells in search of prey. Perhaps not surprisingly, this small fish seldom swims, but generally strolls around on the seafloor (using its pectoral and caudal fins).

92. Helfman et al. (1997), p. 105; note 91.

93. See A.P. Andriyashev *Journal of Ichthyology* 34, 149–155 (1994; originally published in *Zoologicheskiy Zhurnal* 72, 130–136 [1993]).

94. See J.L. Edwards *American Zoologist* 29, 235–254 (1989).

25. See P. Lemelin et al. *Journal of Zoology, London* 260, 423–429 (2003).

26. See P. Lemelin *Journal of Zoology, London* 247, 165–175 (1999).

27. See their paper in *American Journal of Physical Anthropology* 118, 231–238 (2002); also the chapter (pp. 329–380) by P. Lemelin and D. Schmitt in *Primate origins: Adaptations and evolution* (M.J. Ravosa and M. Dagosto, eds.; Springer; 2007).

28. Schmitt and Lemelin (2002), p. 235; note 27.

29. See their paper in *American Journal of Physical Anthropology* 141, 142–146 (2010).

30. Schmitt et al. (2010), p. 142 [Abstract]; note 29.

31. See their chapter (pp. 279–347) in *The biology of marsupials* (D. Hunsaker, ed.; Academic; 1977).

32. Hunsaker and Shupe (1977), p. 281; note 31.

33. Hunsaker and Shupe (1977), p. 281; note 31.

34. See D.T. Rasmussen *American Journal of Primatology* 22, 263–277 (1990); also P. Charles-Dominique in *Advances in the study of mammalian behavior* (J.F. Eisenberg and D.J. Kleiman, eds.), pp. 395–422. American Society of Mammalogists, Special Publication 7 (1983). See also S.M. Reilly et al. (*Journal of Morphology* 271, 438–450 [2010]), who point out that the Australian possum "*Trichosaurus* is clearly contrary to the fine branch primates in many ways and may be an informative model for parallels in the evolution of large branch/trunk and terrestrial primates" (p. 448).

35. Rasmussen (1990), p. 275; note 34.

36. See their chapter (pp. 775–803) in *Primate origins: Adaptations and evolution* (M.J. Ravosa and M. Dagosto, eds.; Springer; 2007).

37. Rasmussen and Sussman (2007), pp. 776; note 36.

38. Rasmussen and Sussman (2007), pp. 794–795; note 36.

39. See, in particular, J.D. Orkin and H. Pontzer *American Journal of Physical Anthropology* 144, 617–624 (2011), who draw attention to the locomotory similarities also seen in the squirrels but without other convergences with the primates. See also the interesting analysis of mice arboreality by C. Byron et al. (*Journal of Morphology* 272, 230–240 [2011]). Here mice when obliged to traverse fine branches not only managed a rudimentary grasping, but as the authors suggest, this combined with a small size and an ability to remold parts of the skeleton may parallel what turned out to be key ingredients of success in the first primates.

40. See D. Youlatos *Journal of Human Evolution* 55, 1096–1101 (2008).

41. Within the diprotodontian marsupials, for example, various species are either arboreal or terrestrial, and in the latter case may dig. Hands are central to all these ways of life, and unsurprisingly their anatomical arrangements reflect this (see V. Weisbecker and D.I. Warton *Journal of Morphology* 267, 1469–1485 [2006]). It seems likely that the forms now terrestrial descended from the trees, but independently in two major clades (the kangaroos and vombatiforms). The anatomy of the hand has evolved in a similar fashion, especially in the wrist, to meet this new challenge (see V. Weisbecker and M. Archer *Palaeontology* 51, 321–338 [2008]). The convergences between marsupial and placental mammals are, of course, one of the staples of evolution, but it is still important to note Vera Weisbecker and David Warton's words: "It appears that the relationship between diprotodontian hand anatomy and locomotion is essentially similar to that of placental clades" (p. 1483).

42. See A.N. Iwaniuk and I.Q. Whishaw *Trends in Neurosciences* 23, 372–376 (2000); also D. Sustaita et al. *Biological Reviews* 88, 380–405 (2013).

43. See D. Cassill et al. *Naturwissenschaften* 94, 326–332 (2007); also D.L. Cassill and D. Singh (*Annals of the Entomological Society of America* 102, 713–716; 2009) for a report on the ambidextrous nature of these ants.

44. See A. Manzano et al. *Journal of Anatomy* 213, 296–307 (2008); also L.A. Gray et al. *Journal of Experimental Zoology* 277, 417–424 (1997).

45. Manzano et al. (2008), p. 305; note 44.

46. Manzano et al. (2008), p. 306; note 44.

47. See, for example, S.T. Parker and K.R. Gibson *Journal of Human Evolution* 6, 623–641 (1977).

48. See, for example, A.N. Iwaniuk et al. *Australian Journal of Zoology* 48, 99–110 (2000).

49. See E. Pouydebat et al. *Journal of Evolutionary Biology* 21, 1732–1743 (2008); also various clarifications in vol. 22, pp. 2554–2557 (2009), and a joint letter by M.W. Marzke and E. Pouydebat in *Journal of Biomechanics* 42, 2628–2629 (2009).

50. Pouydebat et al. (2008), p. 1740; note 49.

51. Pouydebat et al. (2008), p. 1739 [Title of section]; note 49.

52. This topic is briefly addressed in my *Life's solution*; see pp. 269–270.

53. See L. Rook et al. *Journal of Human Evolution* 39, 577–582 (2000).

54. See, for example, E.E. Sarmiento *American Museum Novitates* 2881, 1–44 (1987).

55. See S. Moyà Solà and M. Köhler *Comptes Rendus de l'Academie des Sciences, Série IIa:* 324, 141–148 (1997).

56. See J.A. Finarelli and W.C. Clyde *Paleobiology* 30, 614–651 (2004).

57. Sarmiento (1987), p. 23; note 54.

58. See S. Moyà Solà et al. *Proceedings of the National Academy of Sciences, USA* 96, 313–317 (1999).

59. See L. Rook et al. *Proceedings of the National Academy of Sciences, USA* 96, 8795–8799 (1999).

60. See the two papers in *Journal of Human Evolution* by R.L. Susman questioning evidence for precision grip and bipedality in vol. 46, pp. 105–117 (2004) and vol. 49, pp. 405–411 (2005), respectively. So too, on the basis of the lumbosacral region, G.A. Russo and

Experimental Biology 144, 577–581 (1989); however, leg claws and teeth are reinforced with manganese.

268. See R.M.S. Schofield in *Scorpion biology and research* (P. Brownell and G. Polis, eds.), pp. 234–256 (Oxford; 2001).

269. See their paper in *Naturwissenschaften* 95, 257–261 (2008).

270. Berkov et al. (2008), p. 257; note 269.

271. These are remarkable structures; in the ichneumon *Megarhyssa* it extends for at least 5 cm in length yet is incredibly thin (c. 200 μm) (see J.F.V. Vincent and M.J. King *Biomimetics* 3, 187–201 [1995; Table 1]), who also describe how the drilling through the wood is achieved. In addition, their navigation depends on a remarkable steering mechanism that not only involves drilling through hard wood in search of victims to parasitize but a steering mechanism that is designed to allow bending and most probably is guided by sensillary hairs (also P.E. Brown and M. Anderson *Journal of Insect Physiology* 44, 1017–1025 [1998]) near the ovipositor tip. Most probably this mechanism has evolved at least twice; see D.L.J. Quicke and M.G. Fitton *Proceedings of the Royal Society of London, B* 261, 99–103 (1995).

272. See D.L.J. Quicke et al. *Zoological Journal of the Linnean Society* 124, 387–396 (1998); in the case of the mandibles the acquisition of zinc may have only been once, but it has, rather oddly, been lost in the vespid wasps. In addition to zinc, manganese is also employed in some groups.

273. In the squid, the remarkably tough, chitinous jaws have evolved similar mechanical properties without the use of either copper or zinc, but again the histidine (along with water content) evidently plays a key role (see A. Miserez et al. *Science* 319, 1816–1819 [2008], and on pp. 1767–1768 commentary by P.B. Messersmith; also the review by C.C. Broomell et al. *Journal of the Royal Society Interface* 4, 19–31 [2007]).

CHAPTER 7

1 Consider the camel and its distinctive pacing gait that has involved major changes to the anatomy of the foot (becoming secondarily digitigrade and developing a large pad), but as C.M. James et al. (*Journal of Vertebrate Paleontology* 22, 110–121 [2002]) demonstrate, this has evolved at least twice.

2. See, for example, K.J. Niklas *The evolutionary biology of plants* (Chicago; 1997), especially pp. 316–333.

3. See M. Maraun et al. *Proceedings of the Royal Society of London, B* 276, 3219–3227 (2009).

4. Maraun et al. (2009), p. 3226; note 3.

5. Maraun et al. (2009), part of title; note 3.

6. Those mites that have a more robust skeleton and, for reasons that are not really clear, tend toward sexual reproduction as against parthogenesis prove to

be the better candidates for arboreality. More recently evolved taxa are also more likely to shift to arboreality.

7. See J. Fröbisch and R.R. Reisz *Proceedings of the Royal Society of London, B* 276, 3611–3618 (2009).

8. Fröbisch and Reisz (2009), p. 3615; note 7. In due course we find this capacity in a late Triassic archosauromorph; see J.A. Spielman et al. *Rivista Italiana di Paleontologia e Stratigrafia* 111, 395–412 (2005).

9. See J.B. Losos et al. *Science* 279, 2115–2118 (1998); also commentary on p. 2043 by G. Vogel.

10. For convergence in grasping capacity between anolids and geckos, see G. Fontanarrosa and V. Abdala *Acta Zoologica* 95, 249–263 (2014).

11. See M.S. Fischer et al. *Zoology* 113, 67–74 (2010).

12. Prehensile tails have evolved numerous times, even in the fish (think of the sea horse; see M.E. Hale *Journal of Morphology* 227, 51–65 [1996]). Perhaps the most rampant example of this convergence is among the lizards where it has emerged multiple times. The case of the geckos is mentioned elsewhere, but prehensibility may be more familiar in the chameleons. Among the skinks, the prehensile tails depend on a strikingly different anatomical arrangement, but its system of muscles and tendons has some remarkable similarities to the sharks (see K.C. Zippel et al. *Journal of Morphology* 239, 143–155 [1999]).

13. Fischer et al. (2010), p. 67 [Abstract]; note 11.

14. See P. Legreneur et al. *Adaptive Behaviour* 20, 67–77 (2012).

15. Legreneur et al. (2012), p. 73; note 14.

16. See J.I. Bloch et al. *Proceedings of the National Academy of Sciences, USA* 104, 1159–1164 (2007).

17. See E.C. Kirk et al. *Journal of Human Evolution* 55, 278–299 (2008), and J.B. Hanna and D. Schmidt *American Journal of Physical Anthropology* 145, 43–54 (2011).

18. See M. Cartmill et al. *Zoological Journal of the Linnean Society* 136, 401–420 (2002).

19. See E. Larney and S.G. Larson *American Journal of Physical Anthropology* 125, 42–50 (2004); also Cartmill et al. (2002); note 18.

20. See D. Schmitt and P. Lemelin *Journal of Human Evolution* 47, 85–94 (2004); also I.J. Wallace and B. Dernes in vol. 54, pp. 783–794 (2008), for an analysis of a New World capuchin monkey (*Cebus*) where the diagonal sequence is employed when arboreal but a longitudinal sequence when on the ground.

21. See L.J. Shapiro and J.W. Young *Journal of Human Evolution* 58, 309–319 (2010).

22. See also D.A. Raichlen *Journal of Human Evolution* 49, 415–431 (2005).

23. See N.J. Stevens *Journal of Experimental Zoology: Comparative and Experimental Biology A* 305, 953–963 (2006).

24. See A.C. Delciellos and M.V. Vieira *Journal of Mammalogy* 90, 104–113 (2009).

and a type that has evolved at least twice (see L. Luiselli *Acta Oecologia* 30, 62–68 [2006]).

232. Such a dramatic shift to oophagy has interesting evolutionary consequences. In contrast to its venomous relatives, *Aipysurus* has effectively abandoned its toxin (see M. Li et al. *Molecular Biology and Evolution* 22, 934–941 [2005]), while *Emydocephalus* is now toothless.

233. In one particular case this has led to a further convergence. Thus, while snakes are close relatives of the lizards (as squamates), for all intents and purposes the former are not specifically territorial—except, that is, for a kukri snake from Taiwan, feeding on turtle eggs, where the females have reinvented this complex behavior (see W.-S. Huang et al. *Proceedings of the National Academy of Sciences, USA* 108, 7455–7459 [2011]).

234. See J.D. Scanlon and R. Shine *Journal of Zoology, London* 216, 512–528 (1988).

235. Scanlon and Shine (1988), p. 526; note 234.

236. See A. de Queiroz and J.A. Rodriguez-Robles *American Naturalist* 167, 684–694 (2006).

237. See C. Gans *Zoologica* (New York) 37, 209–244 (1952), for a detailed analysis of the method of swallowing and the manner in which ventral projections of the vertebrae (hypapophyses) serve to pierce and then rip open the egg. He also makes the telling point that, although almost edentulous the few tiny teeth are far from vestigial and in fact play an important role in securing the egg. Very much the same arrangement is found in one of the world's rarest snakes, the Indian *Elachistodon* (see C. Gans and M. Oshimi *American Museum Novitates* 1571, 1–16 [1952]), although the limited evidence suggests they are related (see also H.I. Rosenberg and C. Gans *Canadian Journal of Zoology* 54, 510–521 [1976]).

238. See G.E.A. Gartner and H.W. Greene *Journal of Zoology, London* 275, 368–374 (2008).

239. See, for example, Savitzky (1983); note 201.

240. See S.E. Vincent et al. *Journal of Evolutionary Biology* 22, 1203–1211 (2009).

241. Vincent et al. (2009), p. 1210; note 240.

242. See T.J. Hibbitts and L.A. Fitzgerald *Biological Journal of the Linnean Society* 85, 363–371 (2005); also J. Blicke et al. in vol. 88, pp. 73–83 (2006).

243. See K. Schwenk *Science* 263, 1573–1577 (1994); he also suggests the squamate forked tongue has evolved independently at least twice, possibly four times.

244. See S.B. McDowell *Evolutionary Biology* 6, 191–273 (1972).

245. McCarthy (1987); note 226.

246. See B.C. Groombridge *Journal of Natural History* 13, 661–680 (1979).

247. See, for example, B.A. Young *Journal of Natural History* 25, 519–531 (1991).

248. See A. Herrel et al. *Journal of Evolutionary Biology* 21, 1438–1448 (2008).

249. Herrel et al. (2008), p. 1446; note 248.

250. See, for example, R.J. Wealthall et al. *Journal of Morphology* 265, 165–175 (2005).

251. See, for example, E.D. Sone et al. *Crystal Growth & Design* 5, 2131–2138 (2005).

252. See J.A. Shaw et al. *Biological Bulletin* 218, 132–144 (2010).

253. See P. van der Wal et al. *Materials Science and Engineering, C* 7, 129–142 (2000).

254. See his paper in *Journal of Ultrastructure and Molecular Structure Research* 102, 147–161 (1989).

255. van der Wal (1989), p. 161; note 254.

256. van der Wal (1989), p. 161; note 254.

257. See G.W. Bryan and P.E. Gibbs *Journal of the Marine Biological Association, UK* 59, 969–973 (1979).

258. Not only is the use of zinc in teeth rampantly convergent, but in many cases there is clearly a specific association with histidine. Indeed, at least in the polychaete *Nereis*, we see almost the same coordination as carbonic anhydrase with the zinc associated with three histidines, although instead of water at the fourth site there is chlorine (see Lichtenegger et al. [2003]; note 29). Their specific comparison was with zinc-insulin, which also shows a zinc coordinated to three histidines. The occurrence of such halogens as chlorine is also a hallmark of the occurrence of zinc in teeth; see also H. Birkedal et al. *ChemBioChem* 7, 1392–1399 (2006).

259. See H.C. Lichtenegger et al. *Proceedings of the National Academy of Sciences, USA* 100, 9144–9149 (2003), and C.C. Broomell et al. *Journal of Experimental Biology* 209, 3219–3225 (2006).

260. See H.C. Lichtenegger et al. *Science* 298, 389–392 (2002); curiously, unlike the occurrences of zinc, at least some of the copper is in the mineral phase atacamite; also P.E. Gibbs and G.W. Bryan *Journal of the Marine Biological Association, UK* 60, 205–214 (1980).

261. See papers by C. Michel in *Cahiers de Biologie Marine*, respectively 7, 367–373 (1966), and 11, 209–228 (1970). There is, however, a difference to the classic snake delivery system inasmuch as the internal venom duct connects to the exterior by a series of small pores; this feature was first clearly seen in fossil material (see A. Charletta and P.S. Boyer *Micropaleontology* 20, 354–366 [1974]).

262. See F.A. Meunier et al. *EMBO Journal* 21, 6733–6743 (2002).

263. See Q. Bone et al. *Journal of the Marine Biological Association, UK* 63, 929–939 (1983).

264. See A.J. Edwards et al. *Cell Biology International* 17, 697–698 (1993) and R.M.S. Schofield et al. *Naturwissenschaften* 89, 579–583 (2002).

265. See two papers by B.W. Cribb et al. *Naturwissenschaften* 95, 17–23, and 433–441 (2008).

266. See J.E. Hillerton and J.F.V. Vincent *Journal of Experimental Biology* 101, 333–338 (1982); also D. Morgan et al. *Journal of Stored Products Research* 39, 65–75 (2003).

267. See R. Schofield and H. Lefevbre *Journal of*

190. Kuch et al. (2006), p. 86; note 189.

191. See A.R. Evans and G.D. Sanson *Biological Journal of the Linnean Society* 78, 173–191 (2003).

192. Evans and Sanson (2003), p. 184; note 191.

193. Evans and Sanson (2003), part of title; note 191.

194. Evans and Sanson (2003), p. 184; note 191.

195. See, for example, Z.-X. Luo et al. *Nature* 450, 93–97 (2007).

196. See, for example, J. Jernvall and M. Fortelius *Nature* 417, 538–540 (2002); also commentary on pp. 498–499 by J.M. Theodor.

197. See D. Cundall and H.W. Greene in *Feeding: Form, function, and evolution in tetrapod vertebrates* (K. Schwenk, ed.), pp. 293–333 (Academic; 2000).

198. See M. Hoso et al. *Biology Letters* 3, 169–172 (2007).

199. See G.J. Vermeij *Nature* 254, 419–420 (1975).

200. Hoso et al. (2007), p. 171; note 198.

201. See A.H. Savitzky *American Zoologist* 23, 397–409 (1983).

202. See Savitzky (1983), p. 403; note 201; also I. Sazima *Journal of Herpetology* 23, 464–468 (1989).

203. See T. Inoda et al. *American Naturalist* 162, 811–814 (2003).

204. Molluscan durophagy has evolved multiple times among the decapod crustaceans; see C.E. Schweitzer and R.M. Feldman *Palaios* 25, 167–182 (2010).

205. See J.B. Shoup *Science* 160, 887–888 (1968) and P.K.L. Ng and L.W.H. Tan *Crustaceana* 49, 98–100 (1985).

206. See G.P. Dietl and F.J. Vega *Biology Letters* 4, 290–293 (2008).

207. See G.P. Dietl and J.P. Hendricks *Biology Letters* 2, 439–442 (2006).

208. Dietl and Hendricks (2006), p. 440; note 207.

209. See R. Ueshima and T. Asami *Nature* 425, 679 (2003).

210. Ueshima and Asami (2003), p. 679 [figure explanation]; note 209.

211. See M. Hutchinson *The Beagle* 9, 61–70 (1992).

212. See D.A. Arena et al. *Proceedings of the Royal Society of London, B* 278, 3529–3533 (2011).

213. Arena et al. (2011), p. 3532; note 212.

214. See A.H. Savitzky *Science* 212, 346–349 (1981), and K. Jackson *Copeia* 1999, 815–818 (1999).

215. See their paper in *Journal of Zoology, London* 208, 269–275 (1986).

216. Patchell and Shine (1986), p. 270; note 215.

217. See J.A. Maisano and O. Rieppel *Journal of Morphology* 268, 371–384 (2007).

218. See T.H. Frazzetta *American Naturalist* 104, 55–72 (1970).

219. See D. Cundall and F.J. Irish *Journal of Zoology, London* 217, 569–598 (1989).

220. While the "broken jaw" of the bolyerine snakes has excited attention, it is not unique. Among the teleost fish, a break in the lower jaw (forming a so-called intramandibular joint) has evolved repeatedly, both among the coral reef fish (see N. Kernow et al. *Biological Journal of the Linnean Society* 93, 545–555 [2008]; also L.A. Ferry-Graham and N. Konow *Journal of Morphology* 271, 271–279 [2010]) and in freshwater habits. In their analysis of the latter case Alice Gibb and colleagues (see their paper in *Environmental Biology of Fishes* 83, 473–485 [2008]) remark how this "presents a striking example of functional convergence as the intramandibular joint has evolved independently . . . multiple times" (p. 474). While the exact arrangement varies in the many groups (at least six), typically the same bones (dentary against the angular and articular) are involved, and in general there is a strong correlation with feeding either on encrusted or attached food.

221. In its form as a large caudal paddle with vertebral supports, it has evolved independently twice; see K.L. Sanders et al. *Integrative and Comparative Biology* 52, 311–320 (2012).

222. In snakes alone, salt glands have evolved independently at least four times, and among the tetrapods as a whole a number more times, displaying multiple configurations and representing excellent examples of co-option; see L.S. Babonis and F. Brischoux *Integrative and Comparative Biology* 52, 245–256 (2012).

223. See A.R. Rasmussen *Symposia of the Zoological Society of London* 70, 15–30 (1997), and *Steenstrupia* 27, 47–63 (2003); also V. Lukoschek and J.S. Keogh *Biological Journal of the Linnean Society* 89, 523–539 (2006).

224. See S. Shetty and R. Shine *Australian Ecology* 27, 77–84 (2002); also J.C. Murphy *Integrative and Comparative Biology* 52, 217–226 (2012).

225. Moreover, the most lethal species of sea snake (*Enhydrina schistosa*) has evolved twice; see K.D.B. Ukuwela et al. *Molecular Phylogenetics and Evolution* 66, 262–269 (2013), which the authors regard as "an extreme case of convergent phenotypic evolution" (p. 266) and might reflect their peculiar propensity to tackle puffer fish (pp. 35–36).

226. This is a curious evolutionary turn (for their phylogenetic position, see, for example, K.L. Sanders et al. *Journal of Evolutionary Biology* 21, 682–695; 2008), but one that is also found in the closely related marbled sea snake (*Aipysurus*); see C.J. McCarthy *Journal of Natural History* 21, 1119–1128 (1987).

227. See H.K. Voris *Ecology* 47, 152–154 (1966).

228. See R. Shine et al. *Functional Ecology* 18, 16–24 (2004).

229. See R. Shine and J. Covacevich *Journal of Herpetology* 17, 60–69 (1983).

230. An Australian elapid, the death adder, extends this convergence to an African viper (*Bitis*). Even within a specific group of pit-vipers there is certainly a diversification, but it finds its counterpart in ecological convergences; see K.L. Sanders et al. *Journal of Evolutionary Biology* 17, 721–731 (2004).

231. These include some telling parallels in the whipsnake model, highly adept at running down lizards,

NOTES 339

evidently contain toxins. Not only are these used for self-anointing (p. 39) and evidently deter predators from approaching, but potentially they are also employed in biting; see L. Alterman in *Creatures of the dark: The nocturnal prosimians* (L. Alterman et al., eds.), pp. 413–424 (Plenum; 1995).

150. See W.C. Warren et al. *Nature* 453, 175–183 (2008).

151. See C.M. Whittington et al. *Genome Research* 18, 986–994 (2008).

152. Although the related echidna also has spurs, these are nonfunctional. Nevertheless, it is likely that the common ancestor of the echidna and platypus was venomous, and indeed there is reason to think yet more primitive mammals, again equipped with spurs, defended themselves from Mesozoic attackers (see J.H. Hurum et al. *Acta Palaeontologica Polonica* 51, 1–11 [2006]).

153. See G.H. Pournelle in *Venomous animals and their venoms*, vol. 1 (W. Bücherl et al., eds.), pp. 31–42 (Academic [1968]).

154. See M. Kita et al. *Proceedings of the National Academy of Sciences, USA* 101, 7542–7547 (2004).

155. See Y.T. Aminetzach et al. *Current Biology* 19, 1925–1931 (2009).

156. Aminetzach et al. (2009), p. 1925 [Abstract]; note 155.

157. Aminetzach et al. (2009), p. 1930; note 155.

158. See B.G. Fry et al. *Journal of Molecular Evolution* 68, 311–321 (2009a).

159. Aminetzach et al. (2009), p. 1930; note 155.

160. See B.G. Fry et al. *Annual Review of Genomics and Human Genetics* 10, 483–511 (2009b).

161. Fry et al. (2009a), p. 319; note 158.

162. See, for example, M.E. Ricci-Silva *Toxicon* 51, 1017–1128 (2008).

163. Fry et al. (2009b), p. 491; note 160.

164. Pournelle (1968), p. 31, note 153.

165. See his paper in *Pharmacology & Therapeutics* 53, 199–215 (1992).

166. Dufton (1992), p. 200; note 165.

167. An equivalent envenomation apparatus has been identified in Pleistocene shrews from Spain; see G. Cuenca-Bescós and J. Rofer *Naturwissenschaften* 94, 113–116 (2007); also M. Furió et al. *Journal of Vertebrate Paleontology* 30, 1294–1299 (2010). In her overview K.E. Folinsbee (*Comptes Rendus Paleovol* 12, 531–542 [2013]) casts some doubt on this suggestion, but more importantly points out that in mammals, envenomation is effectively restricted to the eulipotyphlans, where it has evolved three times.

168. See S.T. Turvey *Journal of Vertebrate Paleontology* 30, 1294–1299 (2010).

169. Turvey (2010), p. 1297; note 168.

170. See S. Peigné et al. *Geodiversitas* 31, 973–992 (2009).

171. Peigné et al. (2009), p. 973 [First part of title]; note 170.

172. Examples of venomous animals extend far beyond those given here. Perhaps the oldest example is among the conodonts, usually interpreted as primitive vertebrates, where in coniform elements like *Panderodus* an obvious groove may have been used for venom delivery; see H. Szaniawski *Acta Palaeontologica Polonica* 54, 669–676 (2009).

173. In particular, the identification of supposedly equivalent structures in some earlier mammals, notably a pantolestid (*Bisonalveus*) from the early Tertiary (see R.C. Fox and C.S. Scott *Nature* 435, 1091–1093 [2005]), has sparked controversy (see articles in vol. 27 [2007] of *Journal of Vertebrate Paleontology* by C.M. Orr et al. [pp. 541–546] and K.E. Folinsbee et al. [pp. 547–551]).

174. See H.-D. Sues *Nature* 351, 141–143 (1991); also A.B. Heckert et al. *Journal of Paleontology* 86, 368–390 (2012).

175. See J.S. Mitchell et al. *Naturwissenschaften* 97, 1117–1121 (2010).

176. See H.-D. Sues *Journal of Vertebrate Paleontology* 16, 571–572 (1996).

177. See V.-H. Reynoso *Journal of Vertebrate Paleontology* 25, 646–654 (2005).

178. See E.P. Gong et al. *Proceedings of the National Academy of Sciences, USA* 107, 766–768 (2010).

179. See B.G. Fry et al. *Proceedings of the National Academy of Sciences, USA* 106, 8969–8974 (2009).

180. See B.G. Fry et al. *Nature* 439, 584–588 (2006); also H.-Y. Yi and M.A. Norell *American Museum Novitates* 3767, 1–31 (2013).

181. See K. Jackson *Zoological Journal of the Linnean Society* 137, 337–354 (2003).

182. See K. Jackson *Toxicon* 49, 975–981 (2007).

183. Jackson (2007), p. 979; note 182.

184. A remarkably similar fang has evolved in a blenny (*Meiacanthus*). Here, however, its venomousness seems to be linked to predators swallowing this fish, then rapidly changing their mind; see L. Fishelson *Copeia* 1974, 386–392 (1974); also Szaniawski (2009); note 172, especially figs. 4A and 5D.

185. In this context see also D. Cundall (*Physiological and Biochemical Zoology* 82, 63–79; 2009) on the function of such "extreme teeth" (part of title).

186. Jackson (2007), p. 981; note 182.

187. Note F.J. Vonk et al. (*Nature* 454, 630–633 [2008]), while not completely ruling out convergence, argue the developmental similarities support effectively a single origin, but as ever when it comes to this sort of deep homology, the devil is in the details.

188. An interesting parallel is found in the pelican beak, which seems to have been effectively unchanged for 30 Ma. As A. Louchart et al. (*Journal of Ornithology* 152, 15–20 [2011]) note, "The morphology of the beak of pelicans may have been functionally quite optimal" (p. 19).

189. See U. Kuch et al. *Naturwissenschaften* 93, 84–87 (2006).

115. One such example is the convergence of the herbivorous lizard *Uromastyx* (see G.S. Throckmorton *Journal of Morphology* 148, 363–390 [1976]), especially the prismatic structure of the enamel (see J.S. Cooper and D.F.G. Poole *Journal of Zoology, London* 169, 85–100 [1973]; also J.E. Creech and G.M. Erickson *American Zoologist* 41, 1419 [Abstract]; 2001). One possibility is that this complex structure is a reflection of specialized diets (in the case of *Uromastyx* as a herbivore), and although enamel prism has been reported in various reptiles including theropods (see E. Buffetaut et al. *Naturwissenschaften* 73, 326–327 [1986]), it is evident that the occurrences are polyphyletic, and the only really convincing convergence with the mammals (and their therapsid ancestors; see A. Sahni *Scanning Microscopy* 1, 1903–1912 [1987]) is *Uromastyx* (see C.B. Wood and D.N. Stern in *Tooth enamel microstructure. Proceedings of the Enamel Microstructure Workshop 1994, University of Bonn* [W. von Koenigswald and P.M. Sander, eds.], pp. 63–83 [Balkema; 1997]).

116. Including rather startlingly an oviraptorosaurian dinosaur pretending to be a rodent (p. 415n142).

117. See R.L. Nydam *Journal of Paleontology* 81, 538–549 (2007).

118. See R.L. Nydam et al. *Journal of Vertebrate Paleontology* 20, 628–631 (2000).

119. Diet will help drive convergences, as is evident in another borioteiid (*Kuwajimalla*) from the Cretaceous of Japan, which is not only among the earliest of lizard herbivores but is interestingly convergent on living iguana including characteristically elongate multicuspate teeth (see S.E. Evans and M. Manabe *Palaeontology* 51, 487–498 [2008]). Not only has lizard herbivory arisen multiple times, but among the scleroglossans this convergence extends to being able to identify plant material on the basis of chemical cues; see W. Cooper *Journal of Zoology, London* 257, 53–66 (2002).

120. See J.M. Clark et al. *Science* 244, 1064–1066 (1989).

121. See P.M. O'Connor et al. *Nature* 466, 748–751 (2010).

122. See T.S. Marinho and I.S. Carvalho *Journal of South American Earth Sciences* 27, 36–41 (2009).

123. See G.A. Buckley et al. *Nature* 405, 941–944 (2000).

124. See, for example, E.M. Gomani *Journal of Vertebrate Paleontology* 17, 280–294 (1997), although she does not rule out occupation of previously constructed burrows.

125. See N.J. Kley et al. *Journal of Vertebrate Paleontology Memoir* 10 (Supplement, 6), 13–98 (2010).

126. Kley et al. (2010), p. 90; note 125.

127. See also R.V. Hill *Journal of Vertebrate Paleontology Memoir* 10 (Supplement, 6), 154–176 (2010). The osteoderms of *Simosuchus* find convergent counterparts in at least five other groups of vertebrates; see R.V. Hill *Systematic Biology* 54, 530–547 (2005).

128. While *Simosuchus* was busy reinventing itself as a pseudo-ornithischian, another Mesozoic crocodyliform known as *Dakosaurus* was certainly aquatic but had evolved a shortened skull with a dentition remarkably reminiscent of the carnivorous archosaurs. See Z. Gasparini et al. *Science* 311, 70–73 (2006); also D. Pol and Z. Gasparini *Journal of Systematic Palaeontology* 7, 163–197 (2009). The hypercarnivory of this crocodile and related forms such as *Geosaurus* is addressed by M.B. de Andrade et al. *Journal of Vertebrate Paleontology* 30, 1451–1465 (2010), noting in particular the multiple evolution of ziphodonty (teeth with additional denticles) and similarities to the theropod *Tyrannosaurus*.

129. Buckley et al. (2000), p. 944; note 123.

130. As X.-C. Wu and H.-D. Sues point out (in *Journal of Vertebrate Paleontology* 16, 688–702 [1996]) this is an important extension of what is otherwise a principally Gondwanan distribution.

131. See X.-C. Wu et al. *Nature* 376, 678–680 (1995); see also Wu and Sues (1996); note 130.

132. Wu and Sues (1996), p. 688; note 130.

133. Wu et al. (1995), p. 679; note 131.

134. See D. Pol *Journal of Vertebrate Paleontology* 23, 817–831 (2003).

135. Pol (2003), p. 826; note 134.

136. See F.E. Novas et al. *Journal of Vertebrate Paleontology* 29, 1316–1320 (2009).

137. Novas et al. (2009), p. 1319; note 136.

138. See P.H. Nobre et al. *Gondwana Research* 13, 139–145 (2008); also H. Zaher et al. *American Museum Novitates* 3512, 1–40 (2006).

139. See P.H. Nobre and I. de S. Carvalho *Gondwana Research* 10, 370–378 (2006).

140. See Wu and Sues (1996), p. 688 [Abstract]; note 130.

141. See A.G. Martinelli *Ameghiniana* 40, 559–572 (2003), part of title; also Pol (2003), p. 826; note 134.

142. See papers in *Neues Jahrbuch für Geologie und Paläontology, Abhandlungen* by A. Ősi et al. (vol. 243, pp. 169–177 [2007], and vol. 248, pp. 279–299 [2008]).

143. See A. Ősi and D.B. Weishampel *Journal of Morphology* 270, 903–920 (2009).

144. See A.R. Evans et al. *Nature* 445, 78–81 (2007).

145. Evans et al. (2007), p. 78; note 144.

146. While one usually thinks of envenomation in terms of hypodermic delivery, in fact most venomous animals employ a groove. This efficient system depends crucially on surface tension; see B.A. Young et al. *Physical Review Letters* 106, e198103 (2011).

147. See N. Hotton in *Origins of higher groups of tetrapods: Controversy and consensus* (H.-P. Schultze and L. Trueb, eds.), pp. 598–634 (Comstock; 1991).

148. For an overview of the many ways mammals defend themselves, see T. Stankowich *Adaptive Behavior* 20, 32–43 (2012).

149. Equally remarkable is the slow loris (or coucang). This primate has glands above its elbows that

1170–1181 (2011). In particular they note how in "*Varanus indicus* [it] has been described as approaching the truly labyrinthine condition seen in some Paleozoic amphibians" (p. 557). Despite the striking configuration of the dentine, adaptive explanations appear not to be so straightforward.

85. See M. Fabrezi and S.R. Emerson *Journal of Zoology, London* 260, 41–51 (2003).

86. Fabrezi and Emerson (2003), p. 41 (Abstract); note 85.

87. See R. Britz et al. *Proceedings of the Royal Society of London, B* 276, 2179–2186 (2009).

88. Their study is important in other ways, not least because sexual selection as well as its natural counterpart appears to have driven this adaptation. So, too, the ontogenetic pathways by which these fangs develop are also different, with each group choosing a particular trajectory to a highly effective solution.

89. See D.J.E. Murdock et al. *Nature* 502, 546–549 (2013); also commentary on pp. 457–458 by P. Janvier. This revolves around the conodonts, but the idea that they are vertebrates is questionable, and if so the evolution of teeth twice in the chordates is yet more interesting. For a possible third example, this time definitely among the vertebrates, see M.M. Smith and Z. Johanson *Science* 299, 1235–1236 (2003); also discussion in vol. 300, p. 1661 (www.sciencemag.org/sgi/content/fall/300/5626/1661b and reply /1661c). A contrary view taken by G.C. Young *Journal of Vertebrate Paleontology* 23, 987–990 (2003), is further addressed by Z. Johanson and M.M. Smith *Biological Reviews* 80, 303–345 (2005).

90. See, for example, J. Jernvall and M. Fortelius *Nature* 417, 538–540 (2002), and J.J. Flynn et al. *Palaeogeography, Palaeoclimatology, Palaeoecology* 195, 229–259 (2003).

91. See J. Damuth and C.M. Janis *Biological Reviews* 86, 733–758 (2011).

92. See G.M. Erickson et al. *Science* 338, 98–101 (2012). Among the tetrapods as a whole, herbivory has evolved multiple times, beginning in the late Carboniferous and most likely looks to gut bacteria to assist with digestion from an early stage; see H.-D. Sues and R.R. Reisz *Trends in Ecology & Evolution* 13, 141–145 (1998). For a convergence extending to a grass carp, see N.J.Gidmark et al. *Journal of Experimental Biology* 217, 1925–1932 (2014).

93. See B.F. Jacobs et al. *Annals of the Missouri Botanical Garden* 86, 590–643 (1999); also Y. Bouchenak-Khelladi et al. *Botanical Journal of the Linnean Society* 162, 543–557 (2010).

94. See G.D. Sanson et al. *Journal of Archaeological Science* 34, 526–531 (2007); also Damuth and Janis (2011); note 91.

95. This suggestion finds a correspondence in an African mole-rat (*Heliophobius*) that has found another route to hypsodonty—that is, by continuous replacement of the cheek teeth. As H.G. Rodrigues et al. (*Proceedings of the National Academy of Sciences, USA* 108, 17355–17359 [2011]) point out, its diet of tubers is hardly demanding, but its burrowing in very hard soils would challenge any tooth. Such continuous replacement of teeth is, of course, very unusual in mammals (unlike reptiles), but has evolved independently in the manatees and a marsupial (pygmy rock wallaby). All share three dental traits, but crucially individually any one of these traits (e.g., delayed eruption, as in humans) has a much wider distribution among the mammals.

96. See G. Billet et al. *Palaeogeography, Palaeoclimatology, Palaeoecology* 274, 114–124 (2009). While the hypsodontic notoungulates are quite well-known, as a group they were remarkably diverse and show various other parallels with mammals beyond South America; see N.P. Giannini and D.A. García-López *Journal of Mammalian Evolution* 21, 195–212 (2014).

97. See K.E.B. Townsend and D.A. Croft *Journal of Vertebrate Paleontology* 28, 217–230 (2008); see also Billet et al. (2009); note 96.

98. Billet et al. (2009), used in the title and text; note 96.

99. See F.J. Goin et al. *Naturwissenschaften* 99, 449–463 (2012).

100. See papers by D.W. Krause et al. (*Nature* 390, 504–507 (1997) and *Acta Palaeontologica Polonica* 48, 321–330 (2003).

101. See their paper in *Journal of Vertebrate Paleontology* 27, 521–531 (2007).

102. Wilson et al. (2007), p. 521; note 101.

103. See W. von Koenigswald et al. *Acta Palaeontologica Polonica* 44, 263–300 (1999).

104. See V. Prasad et al. *Science* 310, 1177–1180 (2007); also commentary on pp. 1126–1128 by D.R. Dolores and H.-D. Sues.

105. See J.F. Bonaparte *Journal of Vertebrate Paleontology* 6, 264–270 (1986).

106. See R. Pascual et al. *Journal of Vertebrate Paleontology* 19, 373–382 (1999); also Z. Kielan-Jaworowska et al. in pp. 517–519 (chap. 14) of their *Mammals from the age of dinosaurs: Origins, evolution, and structure* (Columbia; 2004).

107. See Y. Gurovich and R. Beck *Journal of Mammalian Evolution* 16, 25–49 (2009).

108. Wilson et al. (2007), p. 528; note 101.

109. Wilson et al. (2007), p. 528; note 101.

110. See G.G. Simpson *Journal of Mammalogy* 14, 97–107 (1933).

111. Simpson (1933), p. 106; note 110.

112. See Z.-X. Luo et al. *Nature* 450, 93–97 (2007).

113. Luo (2007), p. 1017; note 112.

114. Thus a primitive whale (*Janjucetus*) from the late Oligocene throws important light on the origin of the baleen whales but was itself a powerful predator showing convergences with Mesozoic pliosaurid reptiles (see E.M.G. Fitzgerald *Proceedings of the Royal Society of London, B* 273, 2955–2963 [2006]).

strength of its limb bones that a litoptern (a group of herbivores indigenous to South America, and some quite similar to horses) known as *Macrauchenia* was well adapted for abrupt swerving, notably when being chased by *Smilodon*.

67. See, for example, R. S. Feranec *Palaeogeography, Palaeoclimatology, Palaeoecology* 206, 303–310 (2004), also Therrien (2005; note 46) and Peigué and de Bonis (2003; note 47).

68. See W. D. Turnbull and W. Segall *Journal of Morphology* 181, 239–270 (1984).

69. See, for example, J. W. F. Reumer et al. *Journal of Vertebrate Paleontology* 23, 260–262 (2003), who point out on the basis of bones trawled from the North Sea that *Homotherium* was probably roaming in what was then dry land as recently as twenty-eight thousand years ago. See also M. Anton et al. *Quaternary Science Reviews* 24, 1287–1301 (2005), as well as the earlier paper by C.W. Marean and C.L. Ehrhardt *Journal of Human Evolution* 29, 515–547 (1995).

70. See M. Antón et al. *GeoBios* 42, 541–551 (2009).

71. See J. A. Holliday and S. J. Steppan *Paleobiology* 30, 108–128 (2004).

72. See also M. Sakamoto et al. *Journal of Evolutionary Paleobiology* 23, 463–478 (2010).

73. See P. Christiansen *PLoS ONE* 3, e2807 (2008).

74. See P. Christiansen *Journal of Morphology* 267, 1186–1198 (2006), and *Journal of Mammalian Evolution* 15, 155–179 (2008); also Therrien (2005; note 46). Christiansen stresses that the similarities of the clouded leopard are to the more primitive sabertooths notably *Paramachairodus*, and so might be instructive as to the killing techniques employed by its extinct analogues. This view of the clouded leopard being on the road to the reevolution of a sabertoothed cat is, however, contested by Slater and van Valkenburg (2008); note 52.

75. See P. Christiansen *Journal of Mammalogy* 89, 1435–1446 (2008).

76. See, for example, L.D. Martin *Science* 195, 981–982 (1977).

77. See his paper in *Science* 205, 1155–1158 (1979).

78. Adams (1979), p. 1155; note 77.

79. See R. Barnett et al. *Current Biology* 15, R589–R590 (2005). Barnett et al. also provide evidence that although the saber-toothed cats *Homotherium* and *Smilodon* are felids, they evidently diverged early and are not closely related to any living species.

80. Adams (1979), p. 1155; note 77.

81. Adams (1979), p. 1156; note 77. Note, though, that Sakamoto et al. (2010; note 72) identify a convergence in terms of biting performance with the lion.

82. Nor can we leave the felids without a corresponding mention of, first of all, the "honorary" felid *Cryptoprocta* (see N. Garbutt *Mammals of Madagascar* [Pica; 1999]; see pp. 128–130). Why so? As Egbert Leigh et al. (*Revue d'Écologie* [*La terre et la vie*] 62, 105–168 [2007]) remark in their overview of large islands, this animal is "Madagascar's largest predator . . . [and] was often classified as a cat. . . . After all, it has retractile claws, a cat-like skull and mode of killing, spotted kittens resembling small lion cubs, and it purrs during courtship" (p. 131), but in fact it is not closely related (see A.D. Yoder and J.J. Flynn in *The natural history of Madagascar* [S.M. Goodman and J.P. Benstead, eds.], pp. 1253–1256 [Chicago; 2003]). And then what about the viverrids? What's the news here? Although in the past this group and the felids have been closely allied, it now transpires that they are not particularly close (see P. Gaubert and P. Cordeiro-Estrela *Molecular Phylogenetics and Evolution* 41, 266–278 [2006]). This phylogenetic reorganization revealed, however, a completely unexpected convergence. It now transpires that while the African linsangs (*Poiana*) are certainly true viverrids, the remarkably similar Asian linsangs (*Prionodon*) are much closer to the cats (see P. Gaubert and G. Veron *Proceedings of the Royal Society of London, B* 270, 2523–2530 [2003]; also E. Barycka *Mammalian Biology* 72, 257–282 [2007]). Nor is this the only example, because in a subsequent paper P. Gaubert et al. (*Systematic Biology* 54, 865–894 [2005]) suggest that a viverridlike carnivore has evolved independently at least another two times, in the African palm civet (*Nandinia*) and the Malagsy euplerids (including *Fossa*). Even within the viverrids the convergences continue, including, for example, a striking similarity between the Asian bearcat (or binturong, *Arctictis*) and South American honeybear (or kinkajou, *Potos*); both are nocturnal, arboreal, and carnivorous, and both are equipped with a remarkably similar prehensile tail, comparable also to many primates (see D. Youlatos *Journal of Zoology* 259, 423–430 [2003]). Prehensile tails and arboreality go, so to speak, hand in hand. Consider, however, the paper by L.H. Emmons and A.H. Gentry (*American Naturalist* 121, 513–524 [1983]). Here they make the curious observation that gliders, which are otherwise rampantly convergent (p. 189–193), are absent from the tropical forests of South America, but here the prehensile tail rules in the arboreal habitat. They suggest the key factor is that in African forests, lianas are abundant and so provide an aerial roadway, whereas elsewhere the canopy lacks such bridges and so the prehensile tail and gliding come into their own. They also note that the African lianas are conspicuously tougher than their counterparts in other tropical forests and wonder if this represents adaptive pressure from elephants.

83. See, for example, C.N. Ciamaglio et al. *Sedimentary Research* (December 2005), 4–8 (2005).

84. Consider, for example, so-called plicidentine—where the dentine shows infolding around the pulp cavity. Among extant vertebrates it is only found in the varanid lizards, but its emergence in a variety of other groups points to convergence; see E.E. Maxwell et al. *Journal of Vertebrate Paleontology* 31, 553–561 (2011); also E.E. Maxwell et al. *Journal of Morphology* 272, these workers emphasize the craniodental adaptations, they also remind us that in terms of dentition (notably the presence of postcarnassial molars) the means of crushing bones were not identical. Their main emphasis, however, is to argue for social hunting in these borophagines, even though they note that some brain endocast data is less supportive of the proposed sociality.

38. Tseng and Wang (2011), p. 481; note 37.

39. Tseng and Wang (2011), p. 482; note 37.

40. Moreover, as Tseng and Wang (2011; note 37) note, the borophagines never manage to shorten their rostrum to the extent of the hyenids, but tantalizingly some specimens of the most derived borophagines "do show a very short rostrum comparable to the degree seen in hyaenids" (p. 482).

41. See B. van Valkenburgh *Paleobiology* 17, 340–362 (1991).

42. As with the hypercarnivorous cats, these canids show a recurrent vulnerability to extinction; see B. van Valkenburgh et al. *Science* 306, 101–104 (2004); also Tseng and Wang (2011); note 37.

43. See J.L. Gittleman *BioScience* 44, 456–464 (1994).

44. See B. Figueirido et al. *Journal of Evolutionary Biology* 23, 2579–2594 (2010).

45. Which show strikingly different growth rates; see R.S. Feranec *Palaios* 23, 566–569 (2008).

46. See, for example, F. Therrien *Zoological Journal of the Linnean Society* 145, 393–426 (2005).

47. See, for example, S. Peigué and L. de Bonis *Zoological Journal of the Linnean Society* 138, 477–493 (2003); also E. Barycka *Mammalian Biology* 72, 257–282 (2007).

48. See, for example, M. Morlo et al. *Zoological Journal of the Linnean Society* 140, 43–61 (2004).

49. See, for example, M. Anton et al. *Journal of Vertebrate Paleontology* 24, 957–969 (2004) and P. Christiansen *Cladistics* 29, 543–559 (2013).

50. See, for example, J. Morales et al. *Transactions of the Royal Society of Edinburgh: Earth Sciences* 92, 97–102 (2001).

51. See C.B. Schultz et al. *Bulletin of the University of Nebraska State Museum* 9, 1–31 (1970).

52. See G.J. Slater and B. van Valkenburgh *Paleobiology* 34, 403–419 (2008).

53. Slater and van Valkenburgh (2008), p. 415; note 52.

54. See L. D. Martin *Annales Zoologici Fennici* 28, 341–348 (1992).

55. See P. Christiansen *Zoological Journal of the Linnean Society* 151, 423–437 (2007).

56. See M. J. Salesa *Zoological Journal of the Linnean Society* 144, 363–377 (2005).

57. See M. Antón and R. García-Perea *Zoological Journal of the Linnean Society* 124, 369–386 (1998), for an interesting discussion as to how the face of *Smilodon* is reconstructed.

58. See L. D. Martin (1989); note 33.

59. Did they have slitlike vertical pupils in their eyes? Possibly, judging by their recurrent evolution in ambush predators as divergent as some felids and snakes; see F. Brischoux et al. *Journal of Evolutionary Biology* 23, 1878–1885 (2010).

60. See also J.A. Meachem-Samuels (*Paleobiology* 38, 1–14 [2012]), who explores the allometry of the forelimbs, again identifying convergences in form.

61. See B. van Valkenburg and F. Hertel *Science* 261, 456–459 (1993); also B. van Valkenburgh *Biological Journal of the Linnean Society* 96, 68–81 (2009).

62. In passing we should note how instructive claws are when it comes to evolution. Not only do bears (and tigers) have what are evidently optimally designed claws (see C. Mattheck and S. Reuss *Journal of Theoretical Biology* 150, 323–328 [1991]), but these structures, along with nails, show instructive convergences among the mammals (see M.W. Hamrick *Evolution & Development* 3, 355–363 [2001]). Most likely the mammalian claw is derived from a reptilian precursor, but when we descend to the amphibians, not only are claws far from universal but more likely convergent (see H.C. Maddin et al. *Journal of Experimental Zoology [Molecular and Developmental Evolution]* 308B, 259–268 [2007]). Some frogs, notably the African arthroleptids, are even more impressive. Not only do they possess sharp claws that can inflict impressive wounds—to the extent that although edible when hunting them the natives prefer to employ spears and machetes—but the claw is actually a piece of naked bone that can protrude from the foot (see D.C. Blackburn et al. *Biology Letters* 4, 355–357 [2008]). Another group of amphibians, in the form of a newt, go one better. Here the defensive array consists of spearlike ribs that, having rotated into position, then pierce the skin. Combined with a noxious skin secretion, this peculiar method of deterring predators is noteworthy—"Hurt yourself to hurt your enemy" being the opening words of the title in the paper by E. Heiss et al. *Journal of Zoology* 280, 156–162 (2010). In a different context L. Bonato et al. (*Zoomorphology* 130, 17–29 [2011]) note the independent evolution of pincerlike claws in the centipedes, commenting on the remarkable degree of derivation.

63. Martin (1989; note 33) points out that *Homotherium* appears to have an enlarged optic center in the brain; see also L. Radinsky *Brain, Behavior and Evolution* 11, 214–254 (1975) who discusses the evidence from brain endocasts of other saber-tooths.

64. P. Christiansen and J.S. Adolfssen *Zoological Journal of the Linnean Society* 151, 833–884 (2007), offer a cautionary note as to the extent these types are distinguishable.

65. See L.D. Martin et al. *Naturwissenschaften* 87, 41–44 (2000).

66. This is not to say the end result was always inevitable. Intriguingly R. A. Fariña et al. (*Ameghiniana* 42, 751–760 [2005]) speculate on an inferred transverse

138. See J.J. Meyers and A. Herrel *Journal of Experimental Biology* 208, 113–127 (2005).

139. See W.C. Sherbrooke and K. Schwenk *Journal of Experimental Zoology, A* 309, 447–459 (2008).

140. Meyers and Herrel (2005), p. 125; note 138.

CHAPTER 6

1. C. Darwin *On the origin of species, etc.* (Murray; 1860).

2. Darwin (1860), p. 79; note 1.

3. See F.R. Prete et al. *The praying mantids* (Johns Hopkins University Press; 1999).

4. See J.M. Carignan *Texas Journal of Science* 40, 111 (1988).

5. See, for example, C. Poivre *L'Entomologiste* 32, 2–19 (1976).

6. See H. Ulrich *Natur und Museum* 95, 499–508 (1965); also K. Kral et al. *Journal of Experimental Biology* 203, 2117–2123 (2000) and K. Kral *Physiological Entomology* 38, 1–12 (2013).

7. See T.C. Boyden *Journal of the New York Entomological Society* 91, 508–511 (1983).

8. See U. Aspöck and M.W. Mansell *Systematic Entomology* 19, 181–206 (1994); also U. Aspöck and H. Aspöck *Annalen des Naturhistorischen Museums in Wien, B* 99, 1–20 (1997).

9. See E. Haring and U. Aspöck *Systematic Entomology* 29, 415–430 (2004).

10. See G.C. Steyskal and L.V. Knutson in *Manual of nearctic diptera*, vol. 1, pp. 607–624 (J.F. McAlpine et al., eds.) (Agriculture Canada Monograph 27; 1981).

11. See O. Betz and R. Mumm *Arthropod Structure & Development* 30, 77–97 (2001).

12. See A. Cloarec *Journal of Experimental Biology* 120, 59–77 (1986).

13. See C. Weinrauch et al. *Cladistics* 27, 138–149 (2011).

14. See A.G. Sharov *Phylogeny of the Orthopteroidea* (Israel Program for Scientific Translations [Trudy Paleontologich-eskogo Instituta 118]; 1971).

15. See S.N. Gorb *Proceedings of the Royal Society of London, B* 265, 747–752 (1998).

16. As, more extraordinarily, is the caterpillar of a Hawaiian moth; see S.L. Montgomery *Entomologia Generalis* 8, 27–34 (1982).

17. See X.J. Nelson et al. *Biological Journal of the Linnean Society* 88, 23–32 (2006).

18. See C. de Muizon et al. *Nature* 389, 486–489 (1997) and C. de Muizon *Geobios* 32, 483–509 (1999). The short-tailed opossum (*Monodelphis*) makes a convincing pygmy version of a saber-tooth in terms of enormous canines, wide gape, and relatively weak bite (see R.E. Blanco et al. *Journal of Zoology, London* 291, 100–110 [2013]).

19. See Muizon (1999), note 18.

20. See D.H. Verzi and C.I. Montalvo *Palaeogeography, Palaeoclimatology, Palaeoecology* 267, 284–291 (2008).

21. See, for example, A.D. Rincón et al. *Journal of Vertebrate Paleontology* 31, 468–478 (2012).

22. See C. Argot *Alcheringa* 28, 229–266 (2004); see also F.J. Prevosti et al. *Ameghiniana* 47, 239–256 (2010; in Spanish, with English abstract), and S. Wroe et al. *PLoS ONE* 8, e66888 (2013) also A.M. Forasiepi and A.A. Carlini *Zootaxa* 2552, 55–68 (2010).

23. See A. Goswami et al. *Proceedings of the Royal Society of London, B* 278, 1831–1839 (2011).

24. See S. Wroe and N. Milne *Evolution* 61, 1251–1260 (2007). This does not, however apply to the marsupials as a whole; see V. Wiesbecker and A. Goswami *Proceedings of the National Academy of Sciences, USA* 107, 16216–16221 (2010).

25. See, for example, the chapters by M.E. Finch (pp. 553–561) and R.T. Wells et al. (pp. 573–585) in *Carnivorous marsupials*, vol. 2 (M. Archer, ed.; Royal Zoological Society of New South Wales [1982]).

26. See S. Wroe et al. *Proceedings of the Royal Society of London, B* 272, 619–625 (2005), and S. Wroe *Journal of Zoology, London* 274, 332–339 (2008).

27. See S. Wroe et al. *Australian Journal of Zoology* 47, 489–498 (1999).

28. Wroe et al. (1999), p. 495; note 27.

29. See their paper in *Zoology* 111, 196–203 (2008).

30. See S. Wroe et al. *Proceedings of the Royal Society of London, B* 274, 2819–2828 (2007).

31 Wroe et al. (2007), p. 2819; note 30.

32. See L. Werdelin *Australian Journal of Zoology* 34, 109–117 (1986).

33. See, for example, B. van Valkenburgh *Annual Review of Earth and Planetary Sciences* 27, 463–493 (1999), and L.D. Martin in *Carnivore behavior, ecology and evolution* (J. Gittleman, ed.), pp. 535–568 (Chapman & Hall; 1989); see also Goswami et al. (2011); note 23.

34. See B. Figueridio et al. *Paleobiology* 77, 490–518 (2011).

35. Further afield a convergent bone-crushing ability has been inferred in the enormous Eocene bird *Diatryma*; see L.M. Witmer and K.D. Rose *Paleobiology* 17, 95–120 (1991).

36. For overviews, see L. Werdelin in *Carnivore behaviour, ecology and evolution* (J.L. Gittleman, ed.), vol. 2, pp. 582–624 (Cornell University Press; 1996), and B. van Valkenburgh *Integrative and Comparative Biology* 47, 147–163 (2007). The latter, whose paper is appropriately titled, "Déja vu: the evolution of feeding morphologies in the Carnivora," also reviews the repeated evolution of the saber-tooths (as does Werdelin), as well as frugivores and a catlike ecomorph.

37. See their paper in *Paleobiology* 37, 470–489 (2011). B. van Valkenburgh et al. *Bulletin of the American Museum of Natural History* 279, 147–162 (2003), also draw attention to similarities between hypercarnivorous borophagine canids and the extant hyenas. While

111. See J.A. Goldbogen et al. *Marine Ecology Progress Series* 349, 289–301 (2007), who among other things point out the volume of water engulfed exceeds that of the entire body.

112. See D.J. Field et al. *Anatomical Record* 294, 1273–1282 (2011).

113. Field et al. (2011), p. 1279; note 112.

114. See G. Zweers et al. *Condor* 97, 297–324 (1995); also V. Mascitti and F.O. Kravetz *Condor* 104, 73–83 (2002).

115. See D.J. St. Aubin et al. *Canadian Journal of Zoology* 62, 193–198 (1985); they point out that in addition to the expected keratin (α and amorphous), there is also some hydroxyapatite, the standard phosphatic mineral found in bone.

116. See S.L. Olson and A. Feduccia *Smithsonian Contributions to Zoology* 316, 1–73 (1980), especially fig. 39.

117. Olson and Feduccia (1980), p. 64; note 116.

118. See S.L. Sanderson and R. Wassersug in *The skull*, vol. 3 (J. Hanken and B.K. Hall, eds.), pp. 37–112 (Chicago; 1993). The authors are dismissive of this flamingo/baleen whale convergence (see especially their fig. 2.12), drawing attention to various differences, but it remains the case that although baleen whales can feed in a variety of ways, one method is to draw water in by using the tongue, and using it again to expel water through the baleen curtain.

119. See P.A. Prince *Journal of Zoology, London* 190, 59–76 (1980).

120. See N.T.W. Klages and J. Cooper *Journal of Zoology, London* 227, 385–396 (1992).

121. Klages and Cooper (1992), p. 394; note 120.

122. In addition, the three basic methods of feeding found in the whales are correspondingly adopted by the prion birds, including a hydroplaning that serves to scoop up the surface plankton. In passing we should note that the loss of teeth in the birds and baleen whales is complemented by some five other transitions to toothlessness among the tetrapods, including the toads (albeit with the independent reinvention of fangs, p. 57), theropods (p. 415; note 42), and anteaters (see T. Davit-Béal et al. *Journal of Anatomy* 214, 477–501 [2009]) (again with striking convergences [see also my *Life's solution*, note 21 on pp. 353–354]). Tiphaine Davit-Béal et al. note that tooth loss "occurred independently at least four times during Avialae evolution" (p. 482; also A. Louchart and L. Viriot *Trends in Ecology & Evolution* 26, 663–673 [2011], who also draw attention to a convergence with pterosaurs), while among the reptiles the edentulous condition is most obvious in the turtles (see R.W. Meredith et al. *BMC Evolutionary Biology* 13, e20 [2013], and M. Tokita et al. *Evolution* 67, 260–273 [2013]) which like the birds have evolved a horny beak. To this roster it is worth drawing attention to the independent acquisition of a horny beak in the herbivorous synapsids known as the dicynodonts (see N. Hotton in *The ecology and biology of mammal-like reptiles* [N. Hotton et al., eds.], pp. 71–82 [Smithsonian; 1986]). Although characteristic of the Permo-Triassic, rather surprisingly they appeared to have survived until the Cretaceous (see T. Thulborn and S. Turner *Proceedings of the Royal Society of London, B* 270, 985–993 [2003]). Among the larger group of synapsids to which the dicynodonts belong, and known as the anomodonts (which include the arboreal *Suminia* [p. 67–68]), the complex system of mastication of plant material (referred to as propaliny) evolved twice (see K.D. Angielczyk *Paleobiology* 30, 268–296 [2004]), and so swells the roster of convergences among the synapsids. A wider parallelism also exists between the dicynodonts and ceratopsid dinosaurs, both equipped with powerful beaks, although the latter group have a battery of teeth to shear the vegetation (see S. Chrulew *Zentralblatt für Geologie und Paläontologie* 1976 [teil II], 291–292 [1976], and C.B. Cox *Palaeontology* 34, 767–784; [1991]).

123. See, for example, T.M. Sanchez *Ameghiniana* 10, 313–325 (1973); also P. Wellnhofer *The illustrated encyclopedia of pterosaurs* (Crescent; 1991, especially pp. 131–134), and A. Chinsamy et al. *Biology Letters* 4, 282–285 (2008).

124. See L.M. Chiappe and A. Chinsamy *Nature* 379, 211–212 (1996).

125. See R.T. Bakker *The dinosaur heresies* (Morrow; 1986), pp. 286–288; also Wellnhofer (1991), p. 159; note 123.

126. See A. Zemann et al. *Molecular Biology and Evolution* 30, 1041–1045 (2013).

127. See K.Z. Reiss in *Feeding: Form, function, and evolution in tetrapod vertebrates* (K. Schwenk, ed.), pp. 459–485 (Academic; 2000).

128. See her paper in *American Zoologist* 41, 507–525 (2001).

129. Reiss (2001), p. 517; note 128.

130. Reiss (2000), p. 469; note 127.

131. Nathan Muchhala (*Nature* 444, 701–702 [2006]) draws attention to the interesting convergence that arises when one needs to store an enormous tongue as in the pangolins and glossophagine nectar bats. Thus he notes how the anteating pangolins (p. 52) "also have a glossal tube; despite their different diets . . . [these] animals face similar evolutionary pressures for highly protrusible tongues, and . . . have independently converged on a similar solution" (p. 701).

132. See C. Charles et al. *Evolution* 67, 1792–1804 (2013), who also address vermivory.

133. See F. Delsuc et al. *Molecular Ecology* 23, 1301–1317 (2014).

134. See R.T. Hatt *Natural History* 34 (8), 725–732 (1934).

135. Hatt (1934), p. 728; note 134.

136. See E.R. Pianka and W.S. Parker *Copeia* 1975, 141–162 (1975).

137. See J.J. Meyers et al. *Biological Journal of the Linnean Society* 89, 13–24 (2006).

with the sheer diversity of cichlid species, also give a fascinating glimpse as to how the convergences continue to be unrolled through evolutionary time. Thus, factors such as feeding specializations or broad habitat serve as the primary guiding determinants, but with ever increasing specialization, so "trivial" factors (such as sexual selection) serve to determine not only color patterns but refine yet further the scope and scale of the convergences (see R.C. Albertson et al. *Proceedings of the National Academy of Sciences, USA* 96, 5107–5110 [1999] and P.D. Danley and T.D. Kocher *Molecular Ecology* 10, 1075–1086 [2001]).

83. See D. Kassam et al. *Molecular Phylogenetics and Evolution* 40, 383–388 (2006).

84. See N.B. Goodwin et al. *Proceedings of the Royal Society of London, B* 265, 2265–2272 (1998).

85. See K.O. Winemiller et al. *Environmental Biology of Fishes* 44, 235–261 (1995); also A.P. Martin and E. Bermingham *Molecular Phylogenetics and Evolution* 9, 192–203 (1998).

86. This particular study was by no means exhaustive, and a few species occupying unique ecological niches were identified. One such is the specialization for feeding on zooplankton, but in another comparison between a Brazilian cichlid and one from the Congo (ironically introduced into Brazil), the feeding styles again converge (see X. Lazzaro *Environmental Biology of Fishes* 31, 283–293 [1991]).

87. See C.G. Montâna and K.O. Winemiller (*Biological Journal of the Linnean Society* 109, 146–164 [2013]), who document convergences in ecomorphology, diets, and isotopes in cichlids from South America and centrarchids from Texas.

88. See K.R. McKaye *Environmental Biology of Fishes* 6, 361–365 (1981).

89. See M. Tobler *Journal of Fish Biology* 66, 877–881 (2005).

90. See F.Z. Gibran *Copeia* 2004, 403–405 (2004).

91. See p. 53 (spider) and p. 124 (lizard) in his *Journal of researches into the natural history and geology . . . Voyage of H.M.S. Beagle, etc.* (Nelson; 1890).

92. Unsurprisingly thanatosis has evolved repeatedly, both in vertebrates and invertebrates. Some instances of feigning death can have important adaptive consequences. In the highly aggressive fire ants the thin cuticle of the younger worker presumably makes them vulnerable so that when attacked they promptly collapse (see D.L. Cassill et al. *Naturwissenschaften* 95, 617–624 [2008]). In contrast, somewhat older workers run for their lives, and the mandible-to-mandible combat is left to the oldest workers.

93. See their chapter in *The biology of marsupials* (D. Hunsaker, ed.), pp. 279–347 (Academic; 1977).

94. Hunsaker and Shupe (1977), p. 314; note 93.

95. Although Peter Wainwright remarks that while there is "compelling circumstantial data" (p. 94; note 74) for this view, the specifics are more difficult to pin down.

96. We should also note that among the fishes an eel-like (anguilliform) ecomorph has evolved multiple times (see, for example, K.O. Winemiller *Ecological Monographs* 61, 343–365 [1991]), and even within relatively small groups such as the African catfish (the clariidids), an eel-like form has evolved at least four times (see G. Jansen et al. *Molecular Phylogenetics and Evolution* 38, 65–78 [2006], and S. Devaere et al. *Journal of Zoological Systematics and Evolutionary Research* 45, 214–229 [2007]; also J.M. Waters and R.A. McDowell for an example from the Australasian mudfishes [galaxids] in *Molecular Phylogenetics and Evolution* 37, 417–425 [2005]). Such convergences show that making an eel is unproblematic, but can be achieved in several different ways. Unsurprisingly, the principal method is to increase the number of vertebrae, typically either in the abdominal or tail region (see A.B. Ward and E.L. Brainerd *Biological Journal of the Linnean Society* 90, 97–116 [2007]). In different taxa among the true eels, either option for vertebral increase can be found (also R.S. Mehta et al. *Integrative and Comparative Biology* 50, 1091–1105 [2011]). In the moray eel the emphasis happens to be on increasing the number of abdominal vertebrae, and in all other respects they are typical eel ecomorphs. Thus the body is elongate, the pectoral and pelvic fins usually lost, and accordingly possesses a more or less continuous fin around the body.

97. See R.S. Mehta *Physiological and Biochemical Zoology* 82, 90–103 (2009); also J.S. Reece et al. *Molecular Phylogenetics and Evolution* 57, 829–835 (2010).

98. See R.S. Mehta and P.C. Wainwright *Journal of Morphology* 269, 604–619 (2008).

99. See R.S. Mehta and P.C. Wainwright *Nature* 449, 79–82 (2007).

100. See the commentary by M.W. Westneat (pp. 33–34) on Mehta and Wainwright (2007); note 99.

101. See T.J. Miller *Copeia* 1987, 1055–1057 (1987).

102. See R.S. Mehta and P.C. Wainwright *Journal of Experimental Biology* 210, 495–504 (2007).

103. See K. Nakaya et al. *Journal of Fish Biology* 73, 17–34 (2008).

104. To be sure, this analysis depends on the study of captured specimens, but it does suggest that despite its megamouth this shark is planktivorous rather than a suction feeder.

105. See M. Friedman *Proceedings of the Royal Society of London, B* 279, 944–951 (2012).

106. Friedman (2012), p. 949; note 105.

107. Friedman (2012), p. 949; note 105.

108. See T.A. Deméré et al. *Systematic Biology* 57, 15–37 (2008), and E.M.G. Fitzgerald *Biology Letters* 8, 94–96 (2012).

109. See T.A. Demére and A. Berta *Zoological Journal of the Linnean Society* 154, 308–352 (2008).

110. See P.F. Brodie in *Secondary adaptations of tetrapods to life in water* (J.-M. Mazin et al., eds.), pp. 345–352 (Verlag Dr Friedrich Pfeil; 2001).

54. See R. Aoki *Current Herpetology in East Asia* 1989, 17–21 (1989).

55. See G.H. Dalrymple *Journal of Herpetology* 13, 303–311 (1979).

56. See D.O. Mesquita et al. *Journal of Herpetology* 40, 221–229 (2006).

57. See C.D. Hulsey et al. *Evolution* 62, 1587–1599 (2008).

58. While there is by no means an exact correspondence, various durophagous convergences occur in representatives of the relatively closely related fish known as drums (sciaenids) and grunts (haemulids), as well as the more distant pompanos (carangids). As Justin Grubich (see his paper in *Biological Journal of the Linnean Society* 80, 147–165 [2003]) notes in this analysis, "the presence of multiple mechanisms that have evolved to direct crushing forces dorsally indicates a potential mechanical constraint for the task of molluscivory in marine teleosts" (p. 161). Further afield, durophagy evolved independently in the Devonian among some of the earliest of the actinopterygians; see L.C. Sallan and M.I. Coates *Zoological Journal of the Linnean Society* 169, 156–199 (2013).

59. See A.P. Summers et al. *Journal of Morphology* 260, 1–12 (2004), and D.R. Huber et al. *Journal of Experimental Biology* 208, 3553–3571 (2005).

60. See M.A. Edmonds et al. *Experimental Biology of Fishes* 62, 415–427 (2001).

61. See C.D. Wilga and P.J. Motta *Journal of Experimental Biology* 203, 2781–2796 (2000).

62. See D.R. Huber et al. *Journal of the Royal Society Interface* 5, 941–952 (2008).

63. See M.N. Dean et al. *Integrative and Comparative Biology* 47, 70–81 (2007).

64. See his online article in *Nature* (10 March, 2000; doi: 10.1038/news000316–2).

65. Huber et al. (2005); note 59.

66. See A.P. Summers *Journal of Morphology* 243, 113–126 (2000); also A.P. Summers et al. *Nature* 395, 450–451 (1998).

67. Although most familiar in mammalian bone, this arrangement has evolved convergently in the birds (see W.J. Bock and B. Kummer *Journal of Biomechanics* 1, 89–96 [1968]), while independently in the spines of the hedgehog and tenrec there is a striking trabecular arrangement that at least in the former group act as highly effective shock absorbers that probably explain a sometimes unappreciated head for heights (see J.F.V. Vincent and P. Owers *Journal of Zoology, London* 210, 55–75 [1986]). Trabecular architecture is also found among the invertebrates. These include a group of sea urchins known as the sand dollars that employ structurally analogous struts, but in this case composed of calcite (see A. Seilacher *Paleobiology* 5, 191–221 [1979]), while in the blue crab the arrangement of struts in the walking leg evidently serves to distribute stress (see E.M. Hecht *Arthropod Structure & Development* 39,

305–309 [2010]) and "is analogous in function to trabecular bone in the vertebrate skeleton" (p. 309).

68. See A.P. Summers et al. *Cell and Tissue Research* 312, 221–227 (2003).

69. See D.E. Sasko et al. *Zoology* 109, 171–181 (2006).

70. Summers (2000); note 66.

71. See M.N. Dean et al. *Journal of Morphology* 267, 1137–1146 (2006).

72. See, for example, C.P.J. Sanford and G.V. Lauder *Journal of Experimental Biology* 154, 137–162 (1990), and E.J. Hilton *Copeia* 2001, 372–381 (2001).

73. See N. Konow and C.P.J. Sanford *Journal of Experimental Biology* 211, 989–999 and 3378–3391 (2008).

74. See P.C. Wainwright in *Fish biomechanics* (vol. 23, Fish Physiology; R.E. Shadwick and G.V. Lauder, eds.), pp. 77–101 (Elsevier; 2006).

75. See, for example, K.F. Liem and S.L. Sanderson *Journal of Morphology* 187, 143–158 (1986).

76. See K. Mabuchi et al. *BMC Evolutionary Biology* 7, e10 (2007).

77. See M.W. Westneat et al. *Proceedings of the Royal Society of London, B* 272, 993–1000 (2005).

78. In Lake Tanganyika there has been recurrent evolution of coloration (facial stripes) (see N. Duftner et al. *Molecular Phylogenetics and Evolution* 45, 706–715 [2007]; also C.J. Allender et al. *Proceedings of the National Academy of Sciences, USA* 100, 14074–14079 [2003] for a similar example from Lake Malawi), and mouthbrooding by either parent and colonization of rocky habits (see S. Koblmüller et al. *Journal of Molecular Evolution* 58, 79–96 [2004]; also C. Clabaut et al. *Evolution* 61, 560–578 [2007]). In this lake also, repeated invasions of the same adaptive zones have resulted in the reinvention of both the same body shapes and feeding specializations (see L. Rüber and D.C. Adams *Journal of Evolutionary Biology* 14, 325–332 [2001]; also L. Rüber et al. *Proceedings of the National Academy of Sciences, USA* 96, 10230–10235 [1999]). An important corollary of this study, which has much wider implications, is that these ecomorphs track habitat by adaptive evolution and not by being trapped in some mysterious web of developmental constraints.

79. See T.D. Kocher et al. in *Molecular Phylogenetics and Evolution* 2, 158–165 (1993), and *Journal of Molecular Evolution* 16, 111–120 (1995); also M.L.J. Stiassny and A. Meyer *Scientific America* 280 (2, February), 44–49 (1999), and M. Muschick et al. *Current Biology* 22, 2362–2368 (2012), with commentary on pp. R71–R74 of vol. 23 (2013) by R.G. Gillespie.

80. See R.C. Albertson et al. *Proceedings of the National Academy of Sciences, USA* 100, 5252–5257 (2003); also their paper in vol. 102, 16278–16292 (2005).

81. See F. Duponchelle et al. *Proceedings of the National Academy of Sciences, USA* 105, 15475–15480 (2008).

82. Identification of these convergences depends, of course, on molecular phylogenies. These, together

20. See R.L. Mueller et al. *Proceedings of the National Academy of Sciences, USA* 101, 13820–13825 (2004).

21. See S.M. Deban et al. *Journal of Experimental Biology* 210, 655–667 (2007).

22. So, too, in the desmognathid salamanders we see how independently at least eight evolutionary lineages have made the transition from flowing to still waters; see D.A. Beamer and T. Lamb *Molecular Phylogenetics and Evolution* 47, 143–153 (2008).

23. See D. Bickford et al. *Current Biology* 18, R374–R377 (2008), and commentary on pp. R392–R393 by V.H. Hutchison.

24. See R.A. Nussbaum and M. Wilkinson *Proceedings of the Royal Society of London, B* 261, 331–335 (1995).

25. See M.H. Lake and M.A. Donnelly *Proceedings of the Royal Society of London, B* 277, 915–922 (2010).

26. See their paper in *Biological Journal of the Linnean Society* 62, 39–109 (1997); also *Zoological Journal of the Linnean Society* 126, 191–223 (1999).

27. Wilkinson and Nussbaum (1997), p. 91; note 26.

28. Among the plethodontids, the hemidactyllines remain permanently aquatic, not only retaining larval feature such as the gills but also the ability to engage in suction feeding.

29. See H.B. Shaffer and G.V. Lauder *Journal of Zoology, London* 216, 437–454 (1988).

30. See S.M. Deban and S.B. Marks *Zoological Journal of the Linnean Society* 134, 375–400 (2002).

31. It also seems likely that this transition is assisted if the process of metamorphosis from larva to adult is less abrupt, which also explains why the process of direct development has not only emerged but has occurred multiple times among the amphibians. Recall also that the plethodontids can reverse evolution and reinvent their larval stage, evidently because the hyobranchial apparatus is ready and waiting (p. 25); this may explain their success in colonizing still waters, which otherwise are dominated by salamanders that employ a direct development.

32. See also P.M. Sander et al. *PLoS ONE* 6, e19480 (2011), who described remarkable Chinese material of a shastasaurid with an edentulous and greatly reduced snout. This conclusion, however, is disputed by R. Motani et al. *PLoS ONE* 8, e66075 (2013). So, too, in terms of dramatically modified skulls and inferred suction feeding, a coeval amphibian in the form of the temnospondyl *Plagiosuchus* is also striking (see R. Damiani et al. *Zoological Journal of the Linnean Society* 155, 348–373 [2009]; also the related *Gerrothorax* (see F. Witzmann and R.R. Schoch *Journal of Morphology* 274, 525–542 [2013]).

33. See E.L. Nicholls and M. Manabe *Journal of Vertebrate Paleontology* 24, 838–849 (2004).

34. See N. Bardet et al. *PLoS ONE* 8, e63586 (2013).

35. See their paper in *Journal of Morphology* 233, 113–125 (1997).

36. Van Damme and Aerts (1997), p. 113, note 35.

37. See P. Lemell et al. *Journal of Experimental Biology* 205, 1495–1506 (2002).

38. Lemell et al. (2002), p. 1495 [Abstract]; note 37.

39. See C. Johnston and A. Berta *Marine Mammal Science* 27, 493–513 (2011); also V. Bouetel *Journal of Mammalogy* 86, 139–146 (2005).

40. See C.D. Marshall et al. *Journal of Experimental Biology* 211, 699–708 (2008).

41. Interestingly, as with the duck-billed platypus, the seal feeds with its eyes closed and appears to rely on the whiskers (or vibrissae). The latter structures must be very sensitive to touch. Keeping the eyes shut could simply be a protective measure, but possibly by turning off the vision, the tactile sensitivity is further enhanced.

42. Convergent, of course, and most familiar in the giraffes, but long necks have evolved widely, as in a variety of birds (e.g., ostrich, flamingo), plesiosaurs, and sauropod dinosaurs; see D.M. Wilkinson and G.D. Ruxton *Biological Reviews* 87, 616–630 (2012).

43. See C. Li et al. *Science* 305, 1931 (2004).

44. See letters in *Science* 308 (2005), respectively, by D. Peters, B. Demes and D.W. Krause (pp. 1112–1113), and reply by M. LaBarbera and O. Rieppel (p. 1113).

45. Motta et al. (2002); note 7.

46. See J.J. Day *Biological Journal of the Linnean Society* 76, 269–301 (2002); also T.M. Orrell and K.E. Carpenter *Molecular Phylogenetics and Evolution* 32, 425–434 (2004).

47. For example, in the wrasses, which along with the parrotfish belong to the important group known as the labrids, the range of feeding methods is very wide, but there is no neither a correlation to particular jaw morphologies nor a link between specialization of anatomy and food type; see D.R. Bellwood et al. *Proceedings of the Royal Society of London, B* 273, 101–107 (2006). Correlations do exist, however, between feeding and locomotion; see D.C. Collar et al. *Biology Letters* 4, 84–86 (2008).

48. Here the wrasses do show a correlation of feeding and anatomy; accordingly in terms of morphospace they occupy a much more discrete region (albeit not to the total exclusion of other types of fish).

49. See, for example, P.C. Wainwright *Ecology* 69, 635–645 (1988).

50. See, respectively, J.E. Martin *Geological Society of America, Special Paper* 427, 177–198 (2007); also D.A. Russell *Fieldiana, Geology* 33[13], 235–256 [1975]) and E.L. Nicholls et al. *Palaeontographica, Abt. A* 252, 1–22 (1999); also R. Motani *Journal of Vertebrate Paleontology* 25, 462–465 (2005).

51. See papers by J.M. Neenan et al. in *Nature Communications* 4, e1621 (2013); also e2284) and *Journal of Anatomy* 224, pp.603–613 (2014).

52. Martin (2007), p. 178; note 50.

53. See K. Carpenter and D. Lindsey *Journal of Paleontology* 54, 1213–1217 (1980); also C.A. Brochu *Journal of Vertebrate Paleontology* 19 (Memoir, 6), 9–10 (1999).

184. Oddly enough one of their most remarkable products, the extraordinary protein silk (p. 87–92), is relevant in this context. While most silks have an elastic component, this cannot be the overall design feature. Otherwise in the case of the spiders the web would simply act as a trampoline, catapulting the insect to safety. Some varieties, notably viscid (or flagelliform) silk, however, have a pronounced elasticity and in conjunction with a covering of sticky droplets (p. 88) this silk is perfectly designed to accommodate the struggling prey (see J.M. Gosline et al. *Endeavour* 10, 37–43 [1986]; also J.M. Gosline et al. *Nature* 309, 551–552 [1984]) until the web's owner appears, fangs poised, on the scene.

185. In these insects flight occurs at a relatively low frequency. Other insects beat their wings at a much higher frequency, and here a different arrangement is used.

186. In their own right, fleas provide some compelling examples of convergence; see the papers by R. Traub in *Bulletin of the British Museum* (*Natural History*), 23(10), 307–387 (1972), and in *Fleas* (R. Traub and H. Starcke, eds.), 33–67 (Balkema; 1980).

187. C. Neville and M. Rothschild aptly describe the flea as flying with its legs; see *Proceedings of the Royal Entomological Society of London, C* 32, 9–10 (1967).

188. See the two papers by M. Burrows in *Journal of Experimental Biology* 210, 3579–3589 and 3590–3600 (2007).

189. See D.P. Maitland (*Nature* 355, 159–161 [1992]), who describes how the larva of the notorious medfly draws itself into a tight loop, before sudden release allows it to catapult to safety (from predatory ants), the elastic energy being stored in the maggot cuticle. John Brackenbury (*Journal of Insect Physiology* 45, 525–533 [1999]) remarks how this example of an organism "breaking out of the design mold and evolving the ability to jump" (p. 528) is an important reminder not to assume it can't be done with a given body plan. Nor is the medfly the only drosophiloid larva to engage in leaping, as the related piophilids also engage in this activity (see R. Bonduriansky *Canadian Entomologist* 134, 647–656 [2002]). Russell Bonduriansky also notes that "larval leaping appears to have evolved independently in at least three lineages of Diptera" (p. 653).

190. See M. Burrows *Nature* 424, 509 (2003).

191. Unfortunately very little seems to be known about the composition of the elastic protein in flea-beetles, but its amino acids indicate that it is not a resilin (see Furth et al. [1983]; note 183) and presumably is yet another independent invention.

192. See P.R. Shewry et al. *Philosophical Transactions of the Royal Society of London, B* 357, 133–142 (2002).

CHAPTER 5

1. See P. Wainwright et al. *Integrative and Comparative Biology* 47, 96–106 (2007).

2. See R. Holzman *Journal of the Royal Society Interface* 5, 1445–1457 (2008).

3. Their fossil record can be traced to at least the Miocene, and interestingly this material shows little difference from extant taxa; see G. Carnevale et al. *Journal of Paleontology* 82, 996–1008 (2008).

4. See T.W. Pietsch and D.B. Grobecker *Science* 201, 369–370 (1978).

5. See D.B. Grobecker and T.W. Pietsch *Science* 205, 1161–1162 (1979).

6. See C.D. Wilga et al. *Integrative and Comparative Biology* 47, 55–69 (2007); also C.D. Wilga and C.P. Sanford *Journal of Experimental Biology* 211, 3128–3138 (2008).

7. See their paper in *Copeia* 2002, 24–38 (2002).

8. Motta et al. (2002), p. 36, note 7.

9. Cavitation also emerges in another shark—the thresher sharks, which use their tails to help capture prey and in the process of tail-slapping produce bubbles (see S.P. Oliver et al. *PLoS ONE* 8, e67380 [2013]). It is not clear if the bubble collapse induces shock waves that assist in disabling the prey.

10. Motta et al. (2002), fig. 2; note 7.

11. See M.J. Ajemian and C.P. Sandford *Journal of the Marine Biological Association, UK* 87, 1277–1286 (2007).

12. See B.C. Groombridge *Journal of Natural History* 13, 661–680 (1979); see also D. Cundall and H.W. Greene in *Feeding: Form, function, and evolution in tetrapod vertebrates* (K. Schwenk, ed.), pp. 293–333 (Academic; 2000).

13. See my *Life's solution*, pp. 112–115.

14. For some striking convergences between ephemopteran and plectopteran larva in Europe and South America, see J. Illies *Verhandlungen der Internationalen Vereinigung für theoretische und angewandte Limnologie* 14, 517–523 (1961).

15. In this context one normally thinks of the radula, and here, too, there are impressive convergences, such as those that have evolved among rock-scraping pulmonates; see A.S.H. Breure and E. Gittenberger *Netherlands Journal of Zoology* 32, 307–312 (1982).

16. See, for example, T. Geerinckx et al. *Journal of Morphology* 268, 805–814 (2007).

17. See W. Arens *Journal of Zoological Systematics and Evolutionary Research* 32, 319–343 (1994).

18. See S.M. Deban and W.M. Olson *Nature* 420, 41–42 (2002); also M.N. Dean *Copeia* 2003, 879–880 (2003); concerning pipid frogs, see also C.A. Carreño and K.C. Nishikawa *Journal of Experimental Biology* 213, 2001–2008 (2010).

19. This group is largely North American, but with representatives as far afield as Korea (see M.S. Min et al. *Nature* 435, 87–90 [2005]).

includes benzoquinones have various functions in the plants, such as antioxidant stress, they are an essential source for human vitamin E. So, too, in the case of the soldier beetle, which employs a highly effective chemical (8Z-dihydromatricaria acid), it would seem most likely that this was sequestered from either plants or fungi, but in fact its synthesis is de novo and thus convergent; see V.S. Haritos et al. *Nature Communications* 3, e1150 (2012).

149. See T. Nüchter et al. *Current Biology* 16, R316–R318 (2006), and S. Özbek et al. *Toxicon* 54, 1038–1045 (2009). Dinoflagellates (and also ciliates [see I.B. Raikov *BioSystems* 28, 195–201 (1992)] have also evolved a complex organelle with striking similarities to the stinging cells (nematocysts) of the cnidarians (see J.A. Westfall et al. *Journal of Cell Science* 63, 245–261 [1983]; also C. Greuet and R. Hovasse *Protistologia* 13, 145–149 [1977]). Technically the nematocysts are organelles, and as Gabriele Kass-Simon and Albert Scappaticci (*Canadian Journal of Zoology* 80, 1772–1794 [2002]) comment, "the complexity and ingenuity of their design, upon which their functioning depends, astonish" (p. 1778). An important part of the evolution of the cnidarian stinging cells was the importation of a synthase from a bacterial source (see E. Denker et al. *Current Biology* 18, R858–R859 [2008]). This is because the synthase probably plays a key role in the sudden shift in osmotic pressure that discharges the nematocyst (see J. Weber *Journal of Biological Chemistry* 265, 9664–9669 [1990]). It will be interesting to see if the dinoflagellates possess an analogous molecule. Might this analogy also extend to the employment of opsins? This is because in cnidarians opsins are employed (along with arrestins) in nematocyst discharge, representing an interesting co-option of the phototransduction cascade; see D.C. Plachetzki et al. *BMC Biology* 10, e17 (2012) and commentary by T.W. Holstein (e18). J.S. Hwang et al. (pp. 135–152 of *Evolutionary biology from concept to application* [P. Pontarotti, ed.; Springer; 2008]) also tentatively identify a possible minicollagen in the dinoflagellate nematocysts. These proteins are a characteristic component of the cnidarian nematocysts, but the authors favor some sort of common ancestry rather than independent recruitment.

150. The main title of the article by N.P. Money *Mycologia* 90, pp. 547–558 (1998).

151. The main title of the article by F. Trail *FEMS Microbiology Letters* 276, 12–18 (2007).

152. See A. Pringle et al. *Mycologia* 97, 866–871 (2005).

153. See, for example, N.P. Money (1998); note 150.

154. See X. Noblin et al. *Journal of Experimental Biology* 212, 2835–2843 (2009).

155. Noblin et al. (2009), p. 2836; note 154.

156. Noblin et al. (2009), p. 2835; note 154.

157. See C.S. Smith et al. *Applied and Environmental Microbiology* 54, 1430–1435 (1988).

158. See his Presidential Address in *Transactions of the British Mycological Society* 58, 179–195 (1972).

159. Ingold (1972), p. 145; note 158.

160. Trail (2007), p. 15; note 151.

161. See their paper in *PLoS ONE* 3, e3237 (2008).

162. Yafetto et al. (2008), p. 3; note 161.

163. Some specializations do exist. For example, in a pathogenic yeast, the end of the ascospore has a needlelike form that evidently assists in its penetration of the host tissue; see M.A. Lachance et al. *Canadian Journal of Microbiology* 22, 1756–1761 (1976).

164. See F. Trail (2007); note 151.

165. See M. Roper et al. *Proceedings of the National Academy of Sciences, USA* 105, 20583–20588 (2008).

166. See D. Evangelista et al. *Journal of Experimental Biology* 214, 521–529 (2011); in this case not only is there explosive discharge but the seed then drills itself into the ground.

167. And in a rather different way in the release of spores of the bryophyte *Sphagnum*. Here the collapsing capsule produces a vortex ring in which the spores are entrained, with a launch velocity that can almost reach 30 m/sec^{-1} (see D.L. Whitaker and J. Edwards *Science* 329, 406 [2010]). As Dwight Whitaker and Joan Edwards note, "vortex rings are commonly generated by animals such as jellyfish and squids. . . . Here we report vortex rings generated by a plant" (p. 406).

168. See J. Edwards et al. *Nature*, pp. 435, 164 (2005).

169. Much the same values are obtained for the remarkable hurling of pollen sacs on to the backs of bees by the Neotropical orchid *Catasetum*; see C.C. Nicholson et al. *Plant Signaling & Behavior* 3, pp. 19–23 (2008).

170. See their paper in *Functional Ecology* 21, 219–225 (2007).

171. Whitaker et al. (2007), p. 223; note 170.

172. See P.E. Taylor et al. *Sexual Plant Reproduction* 19, pp. 19–24 (2006).

173. Taylor et al. (2006), p. 23; note 172.

174. See their paper in *Science*, pp. 308, 1308–1310 (2005).

175. Part of the title of their paper; see note 174.

176. For an overview, see S.N. Patek et al. *Journal of Experimental Biology*, pp. 214, 1973–1980 (2011).

177. D.G. Furth reports subfossil material (*Israel Journal of Entomology*, pp. 12, 41–49; 1978) that is approximately eight hundred years old.

178. See D.G. Furth and K. Suzuki *Systematic Entomology*, pp. 17, 341–349 (1992).

179. See D.-Y. Ge et al. *Proceedings of the Royal Society of London, B* 278, 2133–2141 (2011).

180. Ge et al. (2011), p. 2138; note 179.

181. Ge et al. (2011), p. 2139; note 179.

182. See J. Brackenbury and R. Wang *Journal of Experimental Biology* 198, 1931–1942 (1995).

183. See D.G. Furth et al. *Journal of Experimental Zoology* 227, 43–47 (1983).

116. See L.S. Pereira et al. *Biochemical and Biophysical Research Communications* 352, 953–959 (2007).

117. See pp. 73–76 of Eisner et al. (2005), note 119.

118. If one wants an example of a convergence in terms of the accuracy of the defensive spray seen in the Bombardier beetle, a striking parallel is found in the uropygids, relatives of the spiders. The appropriately named vinegaroon blasts the opposition through the equivalent of a gun turret with an acidic spray, mostly composed of vinegar (acetic acid) (see Eisner et al. *Journal of Insect Physiology* 6, 272–298 [1961], where gripping accounts are given of various animals tangling with vinegaroons). A useful overview is given on pp. 4–6 of Eisner et al. (2005), note 119. Another parallel is found in the ground beetle *Galerita lecontei*, which ejects a spray principally composed of formic acid (also, of course, independently produced by some ants); see pp. 151–156 of Eisner et al. (2005), note 119 (and for formic acid production in ants, pp. 325–328).

119. See the wonderful survey of these and other arthropod defenses in *Secret weapons: Defenses of insects, scorpions, and other many-legged creatures* (Belknap; 2005) by T. Eisner et al. These authors have a succinct summary of the Bombardier beetle on pp. 157–162. They also note (p. 41) that the apparently remarkable frequency with which benzoquinones have been recruited by arthropods is actually less surprising because of their employment in the tanning process of the exoskeleton.

120. See X. Valderrama et al. *Journal of Chemical Ecology* 26, 2781–2790 (2000), and P.J. Weldon et al. *Naturwissenschaften* 90, 301–304 (2003).

121. Valderrama et al. (2000), p. 2783 (note 120); these authors suggest an antimicrobial role, but they also remark on the long-term risks to benzoquinone exposure, including potentially cancer.

122. For owl monkey, see M. Zito et al. *Folia Primatologica* 74, 158–160 (2003); for black lemurs, C.R. Birkinshaw *Folia Primatologica* 70, 170–171 (1999); and birds (the strong-billed woodcreeper), K.C. Parkes et al. *Ornitologia Neotropica* 14, 285–286 (2003). So, too, use of aromatic plants in birds' nests evidently confer protection against blood-sucking insects (see L. Lafuma et al. *Behavioural Processes* 56, 113–120 [2001]), as well as conferring other types of protection (see, for example, C. Peht et al. *Ecology Letters* 5, 585–589 [2002], and A. Mennerat et al. *Oecologia* 161, 849–856 [2009]).

123. See papers by B. Clucas et al. in *Animal Behaviour* 75, 299–307 (2008), and *Proceedings of the Royal Society of London, B* 275, 847–852 (2008).

124. B. Clucas et al. (*Journal of Evolutionary Biology* 23, 2197–2211 [2010]) present evidence that this antipredator behavior in ground squirrels evolved much earlier than their contact with rattlesnakes, suggesting that such anointing might be a primitive feature of rodents. This, however, does not address the different modes of anointing, nor why many rodents have abandoned such a useful expedient.

125. See T. Kobayashi and M. Watanabe *Ethology* 72, 40–52 (1986).

126. See Z.-J. Xu et al. *Journal of Mammalogy* 76, 1238–1241 (1995).

127. See their paper in *Proceedings of the Royal Society of London, B* 279, 675–680 (2012).

128. Kingdon et al. (2012), part of title; note 127.

129. Kingdon et al. (2012), p. 677; note 127.

130. See M. Arnaud *Comptes Rendus Hebdominaires des Séances de l'Académie des Sciences, Paris* 106, 1011–1014 (1888).

131. See his paper in *Animal Behaviour* 24, 68–71 (1976).

132. Brockie (1976), p. 68; note 131.

133. See E.D. Brodie *Nature* 268, 627–628 (1977).

134. See E. Gould and J.F. Eisenberg *Journal of Mammalogy* 47, 660–686 (1966).

135. See, for example, D.V. Andrade and A.S. Abe *Journal of Fish Biology* 50, 665–667 (1997).

136. See I.C. Castilla et al. *Proceedings of the National Academy of Sciences, USA* 104, 18120–18122 (2007).

137. See R.I. Fleming et al. *Proceedings of the Royal Society of London, B* 276, 1787–1795 (2009).

138. Fleming et al. (2009), p. 1787; note 137.

139. Moreover, these workers present evidence for the foams conferring antimicrobial protection, and so a role in the shielding against ultraviolet radiation is also possible.

140. See D.C. Hissa et al. *Journal of Experimental Biology* 211, 2707–2711 (2008).

141. See A. Cooper et al. *Biophysical Journal* 88, 2114–2125 (2005).

142. Fleming et al. (2009), p. 1794; note 137.

143. For an overview, see R.A. Rakitov *Zoologischer Anzeiger* 241, 117–130 (2002).

144. See M.L.S. Mello et al. *Insect Biochemistry* 17, pp. 493–502 (1987).

145. See U. Maschwitz et al. *Behavioral Ecology and Sociobiology* 9, 79–81 (1981).

146. See G.D.H. Carpenter and H. Eltringham *Proceedings of the Zoological Society, Series A* 1938, 243–252 (1938).

147. See T. Eisner et al. *Science* 172, 277–278 (1971).

148. Using repulsive ointments is only one of many examples of chemical defense. These include the widespread and convergent redeployment of chemicals, notably the alkaloids (pp. 34–35), which in the case of the pumiliotoxins found in poison frogs again act as insecticides (see P.J. Weldon et al. *Proceedings of the National Academy of Sciences, USA* 103, pp. 17818–17821 [2001]). Returning to the benzoquinones, note that they are also independently synthesized in plants (see S.E. Sattler et al. in *Plant Physiology* 132, 2184–2195 [2003], and also papers respectively in *Plant Cell* vol. 16, pp. 1419–1432 [2004] and vol. 18, pp. 3706–3720 [2006], where there is also evidence for convergence [see N.A. Eckardt in *Plant Cell* 15, 2233–2235 [2003]]). This, too, is of no small importance because while the end-products of the biosynthetic pathway that

84. Such an arrangement may owe its origins to the tetraodontiforms possessing a remarkably small number of vertebrae (Brainerd and Patek 1998; note 83), so in contrast to the usual flick of escape, these fish had to sit it out. It would be intriguing to think that restriction in vertebral number might, in some way, be linked to the famously tiny size of the genome (the smooth puffer fish has the smallest of any vertebrate), but this seems unlikely. In passing, however, it is worth noting that the decrease in the genome size of other tetraodontiform fish such as the trigger fish is evidently convergent, and even among the puffer fish this trend may have occurred independently in several lineages (see E.L. Brainerd et al. *Evolution* 55, 2363–2368 [2001]; see also D.E. Neufsey and S.R. Palumbi *Genome Research* 13, 821–830 [2003]).

85. And in at least one case also an actinomycete (see Z.-L. Wu et al. *Toxicon* 45, 851–859 [2005]).

86. See K. Miyuzawa and T. Noguchi *Journal of Toxicology—Toxin Reviews* 20, 11–33 (2001).

87. See T. Noguchi and Y. Hashimoto *Toxicon* 11, 305–307 (1973).

88. See J. Daly et al. *Toxicon* 32, 279–285 (1994).

89. See H.S. Mosher et al. *Science* 144, 1100–1110 (1964).

90. See V.C. Twitty and H.A. Elliott *Journal of Experimental Zoology* 68, 247–291 (1934).

91. See D.D. Sheumack et al. *Science* 199, 188–191 (1978).

92. See M. Yotsu-Yamashita et al. *Toxicon* 49, 410–412 (2007).

93. See E.V. Thuesen et al. *Journal of Experimental Marine Biology and Ecology* 116, 249–256 (1988).

94. See R. Ritson-Williams et al. *Proceedings of the National Academy of Sciences, USA* 103, 3176–3179 (2006).

95. See L.E. Llewellyn *Natural Products Reports* 23, 200–222 (2006) and S.A. Murray et al. *Molecular Biology and Evolution* 28, 1173–1182 (2011).

96. See J.D. Hackett et al. *Molecular Biology and Evolution* 30, 70–78 (2013); these authors point out that each group may have drawn on evolutionarily related proteins, but as ever their functions are not confined to saxitoxin synthesis.

97. Apart from a key role as one of the bases of DNA, guanine is widely employed in such matters as forming reflective surfaces in eyes (the tapetum) and gas-proofing fish swim-bladders (see my *Life's solution* pp. 112–113 and p. 355, respectively).

98. Llewellyn (2006); p. 212; note 95.

99. See K. Matsumara *Nature* pp. 378, 563–564 (1995).

100. See E.D. Brodie et al. *Journal of Chemical Ecology* 31, 343–356 (2005).

101. See S.L. Geffeney et al. *Nature* 434, 759–763 (2005); see also C.R. Feldman et al. *Proceedings of the Royal Society of London, B* 277, 3317–3325 (2010).

102. See C.R. Feldman et al. *Proceedings of the National Academy of Sciences, USA* 106, 13415–13420 (2009).

103. See their paper in *Proceedings of the National Academy of Sciences, USA* 109, 4556–4561 (2012).

104. Feldman et al. (2012), p. 4556 [Abstract]; note 103.

105. See V.M. Bricelj et al. *Nature* 434, 763–767 (2005).

106. See T.W. Soong and B. Venkatesh *Trends in Genetics* 22, 621–626 (2006).

107. See M.C. Jost et al. *Molecular Biology and Evolution* 25, 1016–1024 (2008); a similar story emerges among the tetrodotoxin-resistant snakes (see Feldman et al. 2012; note 103).

108. Jost et al. (2008), p. 1016 [Abstract]; note 107.

109. As Jost et al. (2008; note 107) point out, so also in clams, flatworms, and snakes.

110. See M.E. Arnegard et al. *Proceedings of the National Academy of Sciences, USA* 107, 22172–22177 (2010); also commentary on pp. 21953–21954 by E.D. Brodie.

111. See H.H. Zakon et al. in *Proceedings of the National Academy of Sciences USA* 103, 3675–3680 (2006); also commentary by P.S. Katz in *Current Biology* 16, R327–R330 (2006), and *Journal of Experimental Biology* 211, 1814–1818 (2008).

112. So, too, in the wider field of ion channels, Richard Copley (see his paper in *Trends in Genetics* 20, 171–176 [2004]) describes "remarkable examples of evolutionary convergence" (p. 171 [Abstract]), but here not in terms of mutational sites but in the process known as splicing (whereby the exons are inserted into the strand of DNA; see also V. Douris et al. [*Molecular Biology and Evolution* 27, 684–693 (2010)] for convergence of *trans*-splicing in disparate animals). What is important in this case is not only are the ion channels in question unrelated, but as he notes there appears to be "a general adaptive benefit to managing the regulation of ion channels in this way" (p. 175), with the important corollary that there might be "intrinsic benefits in modifying gene function by alternative splicing rather than by duplication and mutation" (p. 175). This example receives attention as part of a wider overview of "widespread recurrent evolution of genomic features," the title of the paper by E.I. Maeso et al. (*Genome Biology and Evolution* 4, 486–500 [2012]), but as they note, the example of recurrence of splicing "came as a profound surprise . . . The pattern indicates a rule that is at the same time extremely clear and poorly understood" (p. 496).

113. Among the most notorious are cyanogenic compounds, but it now transpires that the biosynthetic pathways in an insect (the moth *Zygaena*) and certain plants are not only identical but have evolved convergently; see N.B. Jensen et al. *Nature Communications* 2, e273 (2011).

114. See, for example, T. Eisner & D.J. Aneshansley, *Science*, 215, 83–85 (1982); also note 4 on p. 372 of my *Life's solution*.

115. See C.D. Derby *Biological Bulletin* 213, 274–289 (2007).

radiations of anurans with respect to their tadpole morphs show increasing levels of homoplasy.

46. See Y. Chiari et al. *Molecular Ecology* 13, 3763–3774 (2004).

47. See M. Vences et al. *Zoologischer Anzeiger* 236, 217–230 (1998).

48. See M. Vences et al. *Organisms, Diversity & Evolution* 3, 215–226 (2003).

49. See W. Takada et al. *Journal of Chemical Ecology* 31, 2403–2415 (2005), and R.A. Saporito et al. *Proceedings of the National Academy of Sciences, USA* 104, 8885–8890 (2007).

50. See V.C. Clark et al. *Proceedings of the National Academy of Sciences, USA* 102, 11617–11622 (2005).

51. See J.C. Santos et al. *Proceedings of the National Academy of Sciences, USA* 100, 12792–12797 (2003), and commentary on pp. 12533–12534 by K. Summers.

52. See C.R. Darst et al. *American Naturalist* pp. 165, 55–69 (2005).

53. Santos et al. (2003); note 51.

54. It has also allowed these frogs to move into the open, becoming diurnal and definitely beasts not to be tangled with. Indeed evidence exists that the toxins can also serve to deter insects, including malarial mosquitoes (see P.J. Weldon et al. *Proceedings of the National Academy of Sciences, USA* 103, 17818–17821 [2006]), and possibly even confer microbial defense (see C. Macfoy et al. *Zeitschrift für Naturforschung, C* 60, 932–937 [2005]).

55. See A. Rodríguez et al. *Biology Letters* 7, 414–418 (2011); see also discussion by G. Raspotnig et al. (pp. 555–556) and reply (p. 557) by M. Vences et al.

56. See J. W. Daly et al. *Journal of Natural Products* 68, 1556–1575 (2005).

57. See E. König and O.R.P. Bininda-Emonds *Peptides* 32, 20–25 (2011).

58. A particularly striking example of toxic convergence concerns two taxa of frog (*Litoria* and *Xenopus*), both of which employ the skin toxin caerulein, but have arrived at an identical peptide sequence independently (see K. Roelants et al. *Current Biology* 20, 125–130 [2010]). As Edmund Brodie (see his article in *Current Biology* 20, R152–R154 [2010]) remarks in his commentary, "these studies [and these include the equally striking convergence in shrew and lizard toxins] suggest a remarkable predictability of the biochemical details of evolution" (p. R153).

59. See I. van Bocxlaer et al. *Science* 327, 679–682 (2010).

60. van Bocxlaer et al., p. 681; note 59.

61. See her commentary on van Bocxlaer et al. (note 59) in *Science* 327, p. 633 (2010).

62. Pennisi (2010), 633; note 61.

63. Shall we employ snakes? Well, it is also worth noting an Asian snake that sequesters steroids from engulfed toads that are then stored in defensive glands, and are even passed on from the female to the young (see D.A. Hutchinson et al. *Proceedings of the*

National Academy of Sciences, USA 104, 2265–2270 [2007]).

64. See J.P. Dumbacher et al. *Science* 258, 799–801 (1992).

65. See J.P. Dumbacher et al. *Proceedings of the National Academy of Sciences, USA* 97, 12970–12975 (2000).

66. See E.X. Albuquerque et al. *Science* 172, 995–1002 (1971).

67. See S.-Y. Wang and G.K. Wang *Biophysical Journal* 76, 3141–3149 (1999).

68. See S. Bartram and W. Boland *ChemBioChem* 2, 809–811 (2001).

69. See J.P. Dumbacher et al. *Proceedings of the National Academy of Sciences, USA* 101, 15857–15860 (2004).

70. See J.P. Dumbacher et al. *Auk* 126, 520–530 (2009).

71. See J.P. Dumbacher and R.C. Fleischer *Proceedings of the Royal Society of London, B* 268, 1971–1976 (2001).

72. See J.P. Dumbacher et al. *Molecular Phylogenetics and Evolution* 49, 774–781 (2008).

73. See K.A. Jønsson et al. *Biology Letters*, 4, 71–74 (2008).

74. Dumbacher et al. (2008), p. 780; note 72.

75. These flocks are composed of several species, and this may represent a further convergence in the form of a Batesian mimicry, whereby the harmless species are mistaken for the poisonous *Pitohui*; see, however, E. Goodale et al. *Emu* 112, 9–16 (2012).

76. See J.C. Hagelin and I.L. Jones *Auk* 124, 741–761 (2007).

77. Dumbacher et al. (1992); note 64.

78. Here, as in the queen trigger-fish (see R.G. Turingan and P.C. Wainwright *Journal of Morphology* 215, 101–118 [1993]), the emphasis is on very powerful jaws and strong teeth, while in some puffer fish the jaws are also equipped with crushing plates, and although they possess pharyngeal jaws they are much reduced (see R.G. Turingan *Journal of Zoology, London* 233, 493–521 [1994]).

79. See, for example, E.L. Brainerd *Journal of Morphology* 220, 243–261 (1994).

80. See P.C. Wainwright and R.G. Turingan *Evolution* 51, 506–518 (1997).

81. Only these fish, which together comprise the smooth (tetraodontids) and spiny (diodontids) puffers, show full inflation, but significantly—in a related group known as the file fish—the genus *Brachaluteres* can also inflate its abdomen, while several other unrelated fish (a teleost [a goby] and the appropriately named swell shark) also show some capacity for inflation (Wainwright and Turingan 1997; note 80).

82. See pp. 27–28 in *Journal of researches into the natural history and geology . . . of H.M.S. Beagle, etc.* (Nelson; 1890).

83. See E.L. Brainerd and S.N. Patek *Copeia* 1998, 971–984 (1998).

793–801 (1991), and *Folia Primatologica* 62, 125–135 (1994); the latter paper is one of a thematic series on the aye-aye forming most of vol. 62.

18. See K.A. Phillips *Folia Primatologica* 74, 162–164 (2003).

19. I discuss this and related themes in my *Life's solution*, p. 372.

20. See D.R. Rawlins and K.A. Handasyde *Journal of Zoology, London* 257, 195–206 (2002). With respect to the marsupials, another sort of woodpecker analogue can be found in the yellow-bellied glider, which makes incisions in gum trees and so resembles the yellow-bellied sapsucker; see R.L. Goldingay *Oecologia* 73, 154–158 (1987), and *Wildlife Research* 27, 217–222 (2000).

21. See W. von Koenigswald et al. *Palaeontographica* A, 272, 1–169 (2005). The wider relationships of the apatemyids have been difficult to pin down, but exquisite fossil material seems to point to a connection with the euarchontoglires; see M.T. Silcox et al. *Zoological Journal of the Linnean Society* 160, 773–825 (2010).

22. See D.E. McCoy and C.A. Norris *Bulletin of the Peabody Museum of Natural History* 53, 355–374 (2012).

23. See R.M.D. Beck *Biological Journal of the Linnean Society* 97, 1–17 (2009).

24. Beck (2009), part of the title and elsewhere (e.g., pp. 2, 14); note 23.

25. Beck (2009), p. 14; note 23. Although not unequivocal, more recent finds have gone some way to support this supposition; see R.M.D. Beck et al. *Journal of Mammalian Evolution* 21, 127–172 (2014).

26. Erickson (1994); note 17.

27. See, for example, their paper in *Behavioral Brain Science* 2, 367–407 (1979; with commentaries on pp. 381–402).

28. Erickson (1994), p. 134; note 17.

29. For a critical overview, see P.J. Motta and K.M. Kotrschal *Netherlands Journal of Zoology* 42, 400–415 (1992).

30. See, for example, K.O. Winemiller *Ecological Monographs* 61, 343–365 (1991), and S.F. Norton et al. *Environmental Biology of Fishes* 44, 287–304 (1995); see also my *Life's solution*, p. 133. With respect to convergences in freshwater fish communities from tropical and temperate areas, see C. Ibanez et al. *Ecography* 32, 658–670 (2009).

31. See H. Fukami et al. *Nature* 427, 832–835 (2004).

32. See, for example, J.B. Losos et al. *Science* 279, 2115–2118 (1998), and *Annals of the Missouri Botanical Garden* 93, 24–33 (2006); also D.L. Mahler et al. *Science* 341, 292–295 (2013). See also my *Life's solution*, p. 125.

33. See N.B. Simmons et al. *Nature* pp. 451, 818–821 (2008).

34. See his paper in *Mammal Review* 31, 111–130 (2001).

35. The subtitle of the paper by Speakman (2001); note 34.

36. See N.B. Simmons *American Museum Novitates* 3077, 1–37 (1993).

37. See M. Weinbeer et al. *Biotropica* 38, 69–76 (2006).

38. Even within the genus *Myotis*, the distinctive ecomorph associated with prey capture over water turns out to be convergent, evolving independently in two distantly related species, respectively, from the Palaearctic (*M. daubentonii*) and the Americas (*M. lucifugus*); see M. Ruedi and F. Mayer *Molecular Phylogenetics and Evolution* 21, 436–448 (2001).

39. See N.U. Czech et al. *Acta Chiropterologica* 10, 303–311 (2008).

40. Ruedi and Mayer (2001), p. 436 [Abstract]; note 38.

41. Ruedi and Mayer (2001), p. 446; note 38.

42. As they are in areas of convergence as diverse as adhesion structures (pp. 182–185), gliding (pp. 189–193), and the reinvention of tadpoles (p. 25).

43. See F. Bossuyt and M.C. Milinkovich *Proceedings of the National Academy of Sciences, USA* 97, 6585–6590 (2000); also F. Bossuyt et al. *Systematic Biology* 55, 579–594 (2006).

44. As importantly, the investigators, Frank Bossuyt and Michel Milinkovich (2000; note 43) emphasize that while they cannot rule out functional or genetic constraints that forces adult and larva to evolve in tandem, they believe that this is much less likely than similar selective pressures serving to mold the parallel series of ecomorphs. For a more recent and comparable study see S. Moen et al. (*Proceedings of the Royal Society of London, B* 280, 20132156 [2013]). In this case evolutionary conservation can play a role, but the remarkable radiation in the Australian frog *Litoria* with independent evolution of ecomorphs remains striking. These emergences of repeated ecomorphs in even a local radiation underline an important point that even within given areas the inevitabilities play out. This applies to the radiation of the frogs in sub-Saharan Africa (see A. van der Meijden et al. *Molecular Phylogenetics and Evolution* 37, 674–685 [2005]; also F. Bossuyt et al. 2006; note 43), while in India striking adaptations that enable the larvae to hop across wet rocks have evolved twice (Bossuyt and Milinkovich 2000; note 43). Much the same story emerges if we move to New Guinea and consider the microhylids. Not only do they account for the majority of New Guinea frogs, but they have undergone a remarkable adaptive radiation with extraordinary strategies of parental care and even transport of the froglets. Yet this bizarre range of behaviors is evidently highly convergent, as indeed are the habitats that the microhylids have come to occupy (see F. Köhler and R. Günther *Molecular Phylogenetics and Evolution* 47, 353–365 [2008]).

45. See also K. Roelants et al. (*Proceedings of the National Academy of Sciences, USA* 108, 8731–8736 [2011]), where to the first approximation successive

84. See C.J. Meehan et al. *Current Biology* 19, R892–R893 (2009); also commentary on pp. R894–R895 by D.E. Jackson. See also R.R. Jackson et al. *Journal of Zoology, London* 255, 25–29 (2001).

85. See their paper in *Cimbebasia* 16, 231–240 (2000).

86. Harland and Jackson (2000), part of title; note 85.

87. See W. Nentwig *Oecologica* 69, 571–576 (1986).

88. See R.R. Jackson et al. *Proceedings of the National Academy of Sciences, USA* 102, 15155–15160 (2005); also F.R. Cross and R.R. Jackson *Proceedings of the Royal Society of London, B* 277, 3173–3178 (2010).

89. See F.R. Cross and R.R. Jackson *Biology Letters* 7, 510–512 (2011).

90. See R.R. Jackson *BioScience* 42, 590–598 (1992).

91. See M.A. Elgar and R.A. Allan *Naturwissenschaften* 91, 143–147 (2004).

92. See, for example, P.S. Oliveira *Biological Journal of the Linnean Society* 33, 1–15 (1988).

93. See K.P. Rajashekhar and K.P. Siju *Current Science* 85, 1124–1125 (2003).

94. This involves not only the salticids (and clubionids) but independently the neotropical aphantochilids (see P.S. Oliveira and I. Sazima *Biological Journal of the Linnean Society* 22, 145–155 [1984]). The spiders, however, show further evolutionary versatility: some such as a zodariid show a remarkable degree of specialization of ant prey (see S. Pekár et al. *Naturwissenschaften* 95, 233–239 [2008]).

95. See X.J. Nelson and R.R. Jackson *Proceedings of the Royal Society of London, B* 273, 367–372 (2006).

96. See A. Floren and S. Otto *Ecotropica* 7, 151–153 (2001).

97. Floren and Otto (2001), p. 151; note 96.

98. See R.R. Jackson and S.D. Pollard *Journal of Zoology, London* 273, 358–363 (2007).

99. See M.K. Wickstein *Crustaceana* 64, 314–325 (1993); also S.K. Berke and S.A. Woodin *Functional Ecology* 22, 1125–1133 (2008).

100 See J.L. Gressitt et al. *Nature* 217, 765–767 (1968).

101. See T. Eisner and M. Eisner *Proceedings of the National Academy of Sciences, USA* 97, 2632–2636 (2000); also M.R. Weiss in *Annual Review of Entomology* 51, 635–661 (2006), who reviews the many other uses to which insect feces can be employed.

102. See F. Nogueira-de-Sá and J.R. Trigo *Journal of Tropical Ecology* 21, 189–194 (2005).

103. See R.P. Duncan et al. *Proceedings of the Royal Society of London, B* 274, 3049–3056 (2007).

104. Duncan et al. (2007), p. 3055; note 103.

105. See his *Codebreaker in the far east* (Frank Cass; 1989).

106. Stripp (1989), p. 191; note 105.

CHAPTER 4

1. Such convergences are evident in terms of plumage (see, for example, W.S. Moore et al. *Biological Journal* of the Linnean Society 87, 611–624 [2006], and A.C. Weibel and W.S. Moore *Condor* 107, 797–809; 2005), including the New World orioles (see C.M. Hofmann et al. *Auk* 125, 778–789 [2008]) and seabirds (Why, for example, has dark plumage repeatedly evolved in tropical seabirds? See P.-A. Crochet et al. *Journal of Evolutionary Biology* 13, 47–57 [2000]). For evidence for convergence in the woodpecker *Picoides* see A.C. Weibel and W.S. Moore *Molecular Phylogenetics and Evolution* 22, 247–257 [2003].

2. Even as far as the remarkable New Caledonian crow (p. 275–278), on which G.R. Hunt (*Emu* 100, 109–114 [2000]) notes, "One individual used a distinct woodpeckerlike technique to remove wood around a hole containing a larva . . . lifted body and head (but not its feet) from the log then drove its slightly open bill into the wood with obvious force" (p. 112).

3. An excellent overview of island biotas, including Madagascar, and with an emphasis on convergence is given by E.G., Leigh et al. *Revue d'Écologie (La terre et la vie)* 62, 105–168 (2007).

4. Exactly the same applies to a radiation of songbirds in Madagascar that are now known to be monophyletic, but prior to molecular data their convergent similarities to three lineages (bulbuls, babblers, and warblers) had gone unrecognized; see A. Cibois et al. *Evolution* 55, 1198–1206 (2001).

5. See S. Yamagishi et al. *Journal of Molecular Evolution* 53, 39–46 (2001).

6. See their paper in *Proceedings of the National Academy of Sciences, USA* 109, 6620–6625 (2012).

7. Jønsson et al. (2012), p. 6624; note 6.

8. See, for example, U.S. Johansson et al. *Biology Letters* 4, 677–680 (2008).

9. See their paper in *Ibis* 138, 283–290 (1996).

10 For an overview, see Leigh et al. (2007); note 3.

11. See L. Pejchar and J. Jeffrey *Auk* 121, 548–556 (2004).

12. Fragments of amber in a Cretaceous bird suggest the independent acquisition of sap feeding in enantiornithines; see F.M. Dalla Vecchia and L.M. Chiappe *Journal of Vertebrate Paleontology* 22, 856–860 (2002).

13. And convergences, as for example in the supposed genus *Upucerthia*; see Chesser et al. *Molecular Phylogenetics and Evolution* 44, 1320–1332 (2007).

14. See R.G. Moyle et al. *Cladistics* 25, 386–405 (2009).

15. See J. Fjeldså et al. *Journal of Ornithology* 146, 1–13 (2005).

16. See G.W. Milliken et al. *Folia Primatologica* 56, 219–224 (1991), as to how this digit is independently controlled, and C. Soligo *Folia Primatologica* 76, 262–300 (2005); the latter author reviews the uses to which the finger can be put in addition to percussive hunting, including among young aye-aye tapping people the animal happens to meet.

17. See papers by C.J. Erickson *Animal Behaviour* 41,

36. See R.C. Bruce *Herpetological Review* 36, 107–112 (2005).

37. See their paper in *Herpetological Review* 36, 112–113 (2005); see also the following paper (pp. 113–117) by P.T. Chippendale and J.J. Wiens also supporting the reevolution argument.

38. See papers by H. Fukami et al. in *Nature* 427, 832–835 (2004) and *PLoS ONE* 3, e3222 (2008); also D.-W. Huang et al. *Molecular Phylogenetics and Evolution* 50, 102–116 (2009).

39. See their paper in *BMC Evolutionary Biology* 11, e37 (2011).

40. See, for example, A. Dinapoli and A. Klussman-Kollo *Molecular Phylogenetics and Evolution* 55, 60–76 (2010).

41. See M.L. Rosenzweig and R.D. McCord *Paleobiology* 17, 202–213 (1991).

42. See A.B. Ward and E. Azizi *Zoology* 107, 205–217 (2004).

43. See his paper in *Proceedings of the National Academy of Sciences, USA* 103, 1804–1809 (2006).

44. See G.J. Vermeij *Biological Journal of the Linnean Society* 72, 461–508 (2001).

45. See S.J. Adamowicz et al. *Proceedings of the National Academy of Sciences, USA* 105, 4786–4791 (2008).

46. Adamowicz et al. (2008), p. 4786; note 45.

47. Adamowicz et al. (2008), p. 4787; note 45.

48. Not to forget the multiple evolution of unisexuality among vertebrates; see W.B. Neaves and P. Baumann *Trends in Genetics* 27, 81–88 (2011).

49. See A.G. Himler et al. *Proceedings of the Royal Society of London, B* 276, 2611–2616 (2009) and C. Rabeling et al. *PLoS ONE* 4(8), e.6781 (2009).

50. See, for example, B.B. Normark et al. *Biological Journal of the Linnean Society* 79, 69–84 (2003).

51. See D. Fortaneto et al. *PLoS Biology* 5 (4), e87 (2007), and commentary in *Current Biology* 17, R543–R544 (2007), by D.M. Hillis.

52. See E.A. Gladyshev et al. *Science* 320, 1210–1213 (2008).

53. See R.J. Smith et al. *Proceedings of the Royal Society of London, B* 273, 1569–1578 (2006); but see the correspondence in *Nature* (vol. 453, p. 587 [2008]) from K. Nartens and I. Schon.

54. See K. Domes et al. *Proceedings of the National Academy of Sciences, USA* 104, 7139–7144 (2007).

55. See T.G. D'Souza et al. *Proceedings of the Royal Society of London, B* 271, 1001–1007 (2004).

56. See M.S. Zietara et al. *Hereditas* 143, 84–90 (2006).

57. See R.H. Cruickshank and A.M. Paterson *Trends in Parasitology* 22, 509–515 (2006).

58. See R. Poulin et al. *International Journal for Parasitology* 36, 185–191 (2006).

59. See W.S. Armbruster and B.G. Baldwin *Nature* 394, 632 (1998).

60. But not quite, because in a couple of Madagascan species a pedomorphic shift has resulted in the capac-

ity of buzz-pollination (p. 390; note 10) (see W.S. Armbruster et al. *Evolution* 67, 1196–1203 [2013]).

61. See, for example, P. Nosil and P. Nosil and A.Ø. Mooers in *Evolution*, respectively, vol. 56, pp. 1701–1706 (2002), and vol. 59, pp. 2256–2263 (2005); also A. Termonia et al. *Proceedings of the National Academy of Sciences, USA* 98, 3909–3914 (2001).

62. See N. Janz et al. *Evolution* 55, 783–796 (2001).

63. Janz et al. (2001), p. 793; note 62.

64. See M.D. Sorenson and R.B. Payne *Integrative and Comparative Biology* 42, 388–400 (2002).

65. See S.M. Lanyon *Science* 255, 77–79 (1992).

66. See S.I. Rothstein et al. *Behavioral Ecology* 13, 1–10 (2002).

67. See J.O. Stireman *Journal of Evolutionary Biology* 18, 325–336 (2005).

68. See N. Takebayashi and P.L. Morrell *American Journal of Botany* 88, 1143–1150 (2001); also H. Chapman et al. *International Journal of Plant Sciences* 164, 719–728 (2003).

69. See L. Werdelin *Journal of Zoology, London* 211, 259–266 (1987).

70. See R. Collin and R. Cipriani *Proceedings of the Royal Society of London, B*, 270, 2551–2555 (2003); also M. Pagel in *Trends in Ecology & Evolution* 19, 278–280 (2004).

71. See R. Collin and M.P. Miglietta *Trends in Ecology & Evolution* 23, 602–609 (2008).

72. Cruickshank and Paterson (2006), note 57.

73. Domes et al. (2007), note 54.

74. See R. Collin *Evolution* 58, 1488–1502 (2004), and R. Collin et al. *Biological Bulletin* 212, 83–92 (2007).

75. See S. Yamada *Biological Journal of the Linnean Society* 92, 41–62 (2007).

76. See R.J. Rainkov et al. *Condor* 81, 203–206 (1979).

77. See I. Agnarsson et al. *Evolution* 60, 2342–2351 (2006).

78. For an overview, see D.E. Jackson *Current Biology* 17, R650–R652 (2007).

79. See F. Vollrath *Behavioral Ecology and Sociobiology* 18, 283–287 (1986).

80. Although in another group of spiders, the stegodyphids, where sociality has probably arisen several times, the sociality appears to be more ancient (see J. Johannensen et al. *Proceedings of the Royal Society of London, B* 274, 231–237 [2007]).

81. See S.T. Kelley and B.D. Farrell *Evolution* 52, 1731–1743 (1998).

82. See B.M. Wiegmann et al. *American Naturalist* 142, 737–754 (1993).

83. In some ways mimicry, of which a number of examples are given in this book, exemplifies convergence with the often startling similarity between organisms. A huge area, but see A. Meyer *PLoS Biology* 4, e341 (2006), for a brief overview and H.D. Penney et al. *Nature* 483, 461–464 (2012, with commentary on pp. 410–411 by D.W. Pfennig and D.W. Kikuchi), on the controls on the preciseness of similarity.

the male, expend remarkable amounts of energy transporting rocks. The general notion is that they serve to provide a secure foundation to the nest (see F. Richardson *Ibis* 107, 1–20 [1965]), and this may indeed have been the original function. If so, it has been co-opted as a sign of sexual vigor, given only individuals in tip-top condition will lug kilograms of stone from one place to another (see J. Moreno et al. *Animal Behaviour* 47, 1297–1309 [1994]; see also Richardson [1965] who also suggested the rock collection could serve as a display function to the female).

116. See J.A. Mather *Journal of Zoology, London* 233, 359–363 (1994).

CHAPTER 3

1. See his *One long argument: Charles Darwin and the genesis of modern evolutionary thought* (Harvard; 1991).

2. See D. Griffiths *Journal of Animal Ecology* 49, 99–125 (1980).

3. See their paper in *Journal of Experimental Biology* 209, 3510–3515 (2006).

4. Fertin and Casas (2006), p. 3514; note 3.

5. See A.M. Hemmingsen *Entomologiske Meddelelser* 45, 167–188 (1977) [in Danish, with English summary].

6. See their paper in *Evolutionary Ecology* 23, 181–186 (2009).

7. Ruxton and Hansell (2009), title of paper; note 6.

8. See *Science* 143, 769–775 (1964).

9. His chapter (pp. 157–164), "Alone in a crowded universe" appears in *Extraterrestrials: Where are they?* (B. Zuckerman & M.H. Hart, eds.) (Cambridge University Press; 1995); for a similar outlook see chapter 12 in E. Mayr's *What makes biology unique?* Other aspects of the supposed improbability of aliens, addressed by other colleagues (e.g., Leonard Ornstein and George Beadle), are addressed on pp. 230–232 of my *Life's solution*.

10. See pp. 232–233 of *Life's solution*, drawing attention to the prescient views of Robert Bieri and Philip Morrison.

11. For an overview, see J.B. Losos *Evolution* 65, 1827–1840 (2011).

12. See J.W. Gruber's *A conscience in conflict: The life of St. George Jackson Mivart* (Columbia University Press [Temple Publications]; 1960).

13. Published by Macmillan; I consulted the second edition of 1871.

14. Mivart (1871), p. 85; note 13.

15. Mivart (1871), pp. 75–76; note 13.

16. Let me salute the prescient paper by C.F.A. Pantin *The Advancement of Science* 8, 138–150 (1951), which in a few pages brilliantly sketches out much of what I try to deal with at vastly greater length.

17. The most readily available edition is the 1969 reprint (MIT Press), with short introductions by

D'Arcy Thompson and (for this paperback) Theodosius Dobzhansky.

18. For a recent overview, see M. Debrenne and F. Debrenne *Comptes Rendus Palevol* 2, 383–395 (2003).

19. See the disturbing account by Peter Pringle in his *The murder of Nikolai Vavilov: The story of Stalin's persecution of one of the Twentieth century's greatest scientists* (JR Books; 2009), of Vavilov's energetic life and ultimate destruction.

20. See *Journal of Genetics* 12, 47–89 (1922).

21. Vavilov (1922), p. 87; note 20.

22. See his article in *Botanical Review* 39, 205–260 (1973).

23. See p. 203 (unnumbered), before Meyen (1973); note 22.

24 Meyen (1973), p. 253; note 22.

25. See his article in *Evolution* 5, 299–324 (1951). For a modern take on this classic paper see M. Chartier et al. *New Phytologist* (in press).

26. In a molecular context see also the "periodic table of coiled-coil protein structures"; the title of an interesting analysis by E. Moutevelis and D.N. Woolfson *Journal of Molecular Biology* 385, 726–732 (2009), who also draw attention to "expected" forms.

27. See his paper in *Evolution* 47, 341–360 (1993).

28. Skeletons are composed, of course, of biominerals, and here there is a rich field of convergences. Examples include evidence for the independent evolution of nacre (see M.J. Vendrasco et al. *Palaeontology* 54, 825–850 [2011]) and more extraordinarily the development of a lamellarlike bone in the shell-plates of the barnacle *Ibla* (see H.A. Lowenstam and S. Weiner *Proceedings of the National Academy of Sciences, USA* 89, 10573–10577 [1992]). More generally H. Ehrlich (*International Geology Review* 52, 661–699 [2010]) points out that very many examples of biomineralization revolve around either chitin or collagen as the fundamental template for crystal growth.

29. See his *On growth and form* (Cambridge University Press; 1942).

30. See especially his *Convergent evolution: Limited forms most beautiful* (MIT Press; 2011).

31. See his *Wonderful life: The burgess shale and the nature of history* (Norton; 1989).

32. See W.E. Duellman and L. Trueb *Biology of amphibians* (McGraw-Hill; 1986), pp. 23–28.

33. So too this is rampantly convergent; see J. Hanken in *The origin and evolution of larval forms* (B.K. Hall and M.H. Wake, eds.), pp. 61–108 (Academic; 1999).

34. And very probably independently in the hemiphractid or marsupial frogs (so called because the female carries the developing eggs on her back). See W.E. Duellman et al. *Copeia* 1988, 527–543 (1988); see also J.J. Wiens et al. *Evolution* 61, 1886–1899 (2007), who suggest that competitive release may be the trigger for this reacquisition.

35. See P.T. Chippindale et al. *Evolution* 58, 2809–2822 (2004).

et al. *BMC Evolutionary Biology* 11, e164 (2011). As these authors note, "Interestingly, gliding evolved independently at least four times through both convergent and parallel evolution, implying that there is strong positive selection for this life habit" (p. 6). This mode of life is also marked by a convergence in shell form and a corresponding recurrent occupation of a specific region of morphospace; see J.M. Serb et al. *Zoological Journal of the Linnean Society* 163, 571–586 (2011).

78. See A.G. Cole and B.K. Hall *Zoology* 107, 261–273 (2004).

79. Cole and Hall (2004), p. 267; note 78.

80. See A.G. Cole and B.K. Hall *Zoology* 112, 2–15 (2009).

81. See Cole and Hall (2004), note 78, who also draw attention to a wide variety of other invertebrates with cartilagelike connective tissue but do not fit into the strict category.

82. As the diameter of the axon doubles, the resistance halves, and the conduction speed goes up by the square root of the diameter.

83. See D.K. Hartline and D.R. Colman *Current Biology* 17, R29–R35 (2007).

84. See the overview in R.C. Eaton (ed.), *Neural mechanisms of startle behaviour* (Plenum [1984]).

85. For an overview, see B.U. Budelmann in *The nervous system of invertebrates: An evolutionary and comparative approach* (O. Breidbach and W. Kutsch, eds.), pp. 115–138 (Birkhäuser Verlag; 1995).

86. Budelmann (1995), p. 116; note 85.

87. See A. Giuditta et al. *Brain Research* 25, 55–62 (1971); also J.Z. Young *Proceedings of the Zoological Society of London* 140, 229–254 (1963).

88. Budelmann (1995), p. 129; note 85.

89. See, for example, the consecutive suite of papers in vol. 21 (1992) of *Journal of Neurocytology* (pp. 260–275 by M. Bundgaard and N.J. Abbott; pp. 276–294 by N.J. Abbott and M. Bundgaard; pp. 295–303 by N.J. Lane and N.J. Abbott; pp. 304–311 by N.J. Abbott et al.).

90. See B. Hochner et al. *Biological Bulletin* 210, 308–317 (2006).

91. Hochner et al. (2006), p. 309; note 90.

92. See J.Z. Young *Nature* 264, 572–574 (1976), and M.J. Hobbs and J.Z. Young *Brain Research* 55, 424–430 (1973).

93. Hochner (2008), p. R897; note 26.

94. See T. Shomrat et al. *Current Biology* 18, 337–342 (2008); see also R. Crook and J. Basil *Journal of Experimental Biology* 211, 1992–1998 (2008), for evidence for an approximately similar capacity in the more primitive *Nautilus*. Evidence for an episodic-like memory in cuttlefish would reinforce the likelihood of cognitive convergence with the vertebrates; see C. Jozet-Alves et al. *Current Biology* 23, R1033–R1035 (2013).

95. Shomrat et al. (2008), p. 341; note 94.

96. See J.A. Mather *Journal of Comparative Physiology, A* 168, 491–497 (1991).

97. See T. Moriyama and Y.-P. Gunji *Ethology* 103, 499–513 (1997).

98. See C. Alves et al. *Animal Cognition* 10, 29–36 (2007).

99. See her paper in *International Journal of Comparative Psychology* 19, 98–115 (2006).

100. Mather (2006), p. 101, note 99.

101. See L. Dickel et al. *Developmental Psychology* 36, 101–110 (2000); also Mather (2006), p. 101; note 99.

102. See J.A. Mather *Consciousness and Cognition* 17, 37–48 (2008a); also her paper in *American Malacological Bulletin* 24, 51–58 (2008b).

103. See, for example, J.A. Mather and R.C. Anderson *Journal of Comparative Psychology* 107, 336–340 (1993); also Mather (2008a, b); note 102. See, however, R. Pronk et al. (*Journal of Experimental Biology* 213, 1035–1041 [2010]) for evidence of episodic personality in the gloomy octopus (*Octopus tetricus*).

104. Indeed the evidence suggests that at least the same sort of personality resides in animals not otherwise noted for their mental prowess, such as hermit crabs (see M. Briffa et al. *Proceedings of the Royal Society of London, B* 275, 1305–1311 [2008]).

105. See D.L. Sinn et al. *Journal of Comparative Psychology* 115, 361–364 (2001); see also D.L. Sinn et al. *Animal Behaviour* 75, 433–442 (2008).

106. See the overview of octopus intelligence and playfulness, aptly titled, "Eight arms, with attitude," by J.A. Mather *Natural History* 116 (1, February 2007), 30–36 (2007).

107. For both instances see p. 105 in Mather (2006); note 99.

108. Mather (2008b), p. 55; note 102.

109. See J.A. Mather and R.C. Anderson *Journal of Comparative Psychology* 113, 333–338 (1999).

110. See Y. Ikeda *Japanese Psychological Research* 51, 146–153 (2009).

111. See R.A. Byrne et al. *Animal Behaviour* 64, 461–468 (2002); their study was based on the lateral asymmetry of eye use in *Octopus*.

112. See M. J. Kuba et al. *Journal of Comparative Psychology* 120, 184–190 (2006); see also Mather (2008a, b), note 102.

113. See D.J. Levey et al. *Nature* 431, 39 (2004), and for a more detailed assessment of rival hypotheses the paper by M.D. Smith and C.J. Conway *Animal Behaviour* 73, 65–73 (2007), who support prey attraction hypotheses but suggest that the use of dung may also serve to warn other owls that the burrow is already tenanted.

114. See J.K. Finn et al. *Current Biology* 19, R1069–R1070 (2009).

115. A somewhat similar case involves the yellowhead jawfish, which not only constructs a burrow but rearranges pebbles so that they ultimately serve as a well-defined roof to its home; see P.L. Colin *Copeia* 1973, 84–90 (1973). On land an equivalent situation is found in the black wheatear. These birds, especially

58. See C.D. Derby *Biological Bulletin* 213, 274–289 (2007).

59. See C. Kicklighter et al. *Proceedings of the National Academy of Sciences, USA* 108, 11494–11499 (2011). The authors confirm that as with the pigment these amino acids are sequestered from the red algae. They also make the important observation that these mycosporine-like amino acids have important parallels to such toxins as tetrodotoxin and saxitoxin (p. 34–38) inasmuch as, although being produced by a small number of typically microbial taxa, they are employed by an extraordinary variety of organisms and so serve as "keystone molecules."

60. See C.E. Kicklighter et al. *Animal Behaviour* 74, 1481–1492 (2007).

61. See J.F. Aggio and C.D. Derby *Journal of Experimental Marine Biology and Ecology* 363, 28–34 (2008).

62. See C.E. Kicklighter et al. *Current Biology* 15, 549–554 (2005); see also commentary on pp. R194–R196 by H.L. Eisthen and R. Issacs, who point out some intriguing parallels with other predatory-prey interactions.

63. For the aversive effect of this ink on blue crabs, see M. Kamio et al. *Animal Behaviour* 80, 89–100 (2010).

64. See E.T. Walters and M.T. Erickson *Journal of Comparative Physiology, A* 159, 339–351 (1986) for explanations as to the directionality of release.

65. See M. Nusnbaum and C.D. Derby *Animal Behaviour* 79, 1067–1076 (2010).

66. See C.D. Derby et al. *Journal of Chemical Ecology* 33, 1105–1113 (2007).

67. See E.A. Lapan *Comparative Biochemistry and Physiology, A* 52, 651–657 (1975).

68. See H. Furuya et al. *Journal of Morphology* 262, 629–643 (2004).

69. So, too, with respect to vertebrates and the equivalent structures in the insects, the malpighian tubules, despite their very disparate appearance (see E.D. Schejter and B.-Z. Shilo *Current Biology* 13, R511–R513 [2003]), there are "clear and striking parallels among developmental and functional attributes of insect and vertebrate renal organs" (p. R512), both in the way the tubular systems originate and the fact that the epithelial cells that go to form parts of the renal organ have different origins. Even within the protostomes, however, G. Mayer and S. Harzsch (*Journal of Comparative Neurology* 507, 1196–1208 [2008]) challenge whether the classic homology between the nephridia of arthropods and annelids is correct, not least because of the different manner of nervous innervation. On the other hand the genetics of development, notably in the all-important area of ultrafiltration of waste products from the blood, draw on the same tool kit (see H. Weavers et al. *Nature* 457, 322–326 [2009]). This means that the insect system may be informative about human renal disease, but on the other hand not only is the former remarkably complex (see K.W. Beyenbach et al. *Annual Review of Entomology* 55, 351–374

[2010]) but as so often the question of convergence of these otherwise disparate renal systems revolves around what was inherent in much more primitive systems.

70. See V.C. Barber and P. Graziadei *Zeitschrift für Zellforschung und mikroskopische Anatomie* 66, 765–781 (1965).

71. See M.-a. Yoshida et al. *Evolution & Development* 12, 25–33 (2010).

72. See J. Xavier-Neto et al. *Cellular and Molecular Life Sciences* 64, 719–734 (2007), with a thoughtful essay on the types of convergence in different types of heart; also A.W. Martin in *Hearts and heartlike organs* (G.H. Bourne, ed.), vol. 1, pp. 1–39 (Academic; 1980).

73. See, for example, R.E. Shadwick and J.M. Gosline in *Canadian Journal of Zoology* 61, 1866–1879 (1983) and *Journal of Experimental Biology* 114, 239–257 (1985).

74. See, for example, R.E. Shadwick and E.K. Nilsson *Journal of Experimental Biology* 152, 471–484 (1990).

75. A perfectly elastic material will deform under an applied force in a linear fashion, and when the force is removed, release all the energy without heat loss (thermodynamically impossible) and adopt its original shape. Some elastic proteins are astonishingly efficient, and the entire area of elastic behavior in organisms is deeply fascinating. An excellent introduction can be found in the books by S. Vogel, notably *Cat's paws and catapults: Mechanical worlds of nature and people* (Norton; 1998) and his earlier *Life's devices: The physical world of animals and plants* (Princeton [1988]).

76. See A.S. Tatham et al. *Biochimica et Biophysica Acta* 1548, 187–193 (2001), and papers in *Philosophical Transactions of the Royal Society of London, B* 357 (2002) by J. Gosline et al. (pp. 121–132) and A.S. Tatham and P.R. Shewry (pp. 229–234).

77. In the mollusks other than the cephalopods we find other striking convergences. Scallops resting on the seafloor will be startled into furious activity when they detect the approach of their principal natural predator, the starfish. We tactfully overlook our own insatiable desire for the meaty muscle (known as the adductor), but its rapid contractions flap the valves of the scallop, so it swims to safety. A key component is the antagonist to the adductor muscle (recalling muscles can only actively contract, and need an antagonist to restretch, hence our biceps and triceps serve as opposites), and in the bivalve mollusks this is provided by the ligament that joins the two valves and in the inner zone contains the elastic protein abductin (see P. Bochicchio et al. *Macromolecular Bioscience* 5, 502–511 [2005], and Q.P. Cao et al. *Current Biology* 7, R677–R678 [1997]). The group to which they belong, the Pectinidae, are also highly instructive when it comes to convergences in life habits, where at least seventeen recurrent transitions have been identified, including long-distance swimming/gliding; see A Alejandrino molecules go, simply enormous. (They are up to 450 kiloDaltons [kDa] in size, one Dalton being roughly equivalent to the mass of a hydrogen atom, so that is equivalent to about 450,000 atoms.) This means it is pretty cumbersome. Wouldn't it be far more sensible to have smaller units? In principle possibly so, but if the hemocyanin was simply dissolved in the blood, then the quantity required would be so substantial that the osmotic balance would be wrecked. So we can predict that not only will copper almost certainly be used for respiratory purposes on extraterrestrial planets, but the copper will be in intimate association with histidines and the proteins themselves will be among the largest known.

30. See I.K. Bartol *Biological Bulletin* 200, 59–66 (2001).

31. See J.T. Thompson et al. *Journal of Experimental Biology* 211, 1463–1474 (2008).

32. See T.P. Mommsen et al. *Proceedings of the National Academy of Sciences, USA* 78, 3274–3278 (1981); see also Q. Bone et al. *Journal of the Marine Biological Association, UK* 61, 327–342 (1981).

33. See A. Bairati et al. *Tissue and Cell* 35, 155–168 (2003).

34. These have their parallels in a number of other aquatic vertebrates (see W.M. Kier and A.M. Smith *Biological Bulletin* 178, 126–136 [1990]), and also oddly some bats.

35. See G. Sumbre et al. *Nature* 433, 595–596 (2005).

36. See R.A. Byrne et al. *Journal of Comparative Psychology* 120, 198–204 (2006).

37. See W.M. Kier and A.M. Smith *Biological Bulletin* 178, 126–136 (1990). Not only can the suckers readily manipulate fine objects, but they also have roles in locomotion, anchoring, chemotactility, display, and cleaning.

38. See her paper in *Advances in the Study of Behavior* 24, 317–353 (1995).

39. See Mather (1995), p. 321; note 38.

40. In this respect fish and cephalopods have another striking convergence in the form of both chromatophores and the reflective iridophores. See L.S. Demski *Brain, Behavior and Evolution* 40, 141–156 (1992), who emphasizes the various convergences, between the teleost fish and cephalopods. With respect to cephalopod chromatophores excellent reviews are given by A. Packard in *Cephalopod neurobiology* (N.J. Abbott et al., eds.), pp. 331–367 (Oxford [1995]) and J.B. Messenger *Biological Reviews* 76, 473–528 (2001). Cephalopods have never managed the transition to land, but among the terrestrial descendants of the fish chromatophores still operate in the amphibians and reptiles, but in the warm-blooded birds and mammals the skin largely vanishes beneath the pelage of feathers and fur. Where the skin remains exposed the often vivid coloration reveals another remarkable story of convergence (p. 206–207), but apart from the melanocytes, chromatophores become redundant. But not

completely, because they find a refugia in the iris of a number of birds (see L.W. Oliphant et al. *Pigment Cell Research* 5, 367–371 [1992]).

41. See M.E. Palmer et al. *Animal Cognition* 9, 151–155 (2006).

42. For a very helpful overview, see R. Hanlon *Current Biology* 17, R400–R404 (2007).

43. See L.M. Mäthger et al. *Journal of Comparative Physiology, A* 194, 577–585 (2008).

44. See their paper in *Journal of Experimental Biology* 211, 1757–1763 (2008).

45. Kelman et al. (2008), p. 1762; note 44.

46. See J.A. Mather and D.L. Mather *Journal of Zoology, London* 263, 89–94 (2004).

47. See K.V. Langridge et al. *Current Biology* 17, R1044–R1045 (2007).

48. See M.D. Norman et al. *Proceedings of the Royal Society of London, B* 268, 1755–1758 (2001).

49. R.T. Hanlon et al. (*Biological Journal of the Linnean Society* 93, 23–38; 2008) query some of these mimics, but they endorse the mimicry with the flatfish and also document presumably its independent occurrence in the Atlantic (see R.T. Hanlon *Biological Bulletin* 218, 15–24 [2010]), where they are "impressed with the contortions of the octopus body and the striking similarities in posture, speed, duration, and undulations to the flounder swimming" (p. 21).

50. The reverse situation is also worth noting, but this time involving a fish. Thus, the blue-striped fangblenny can temporarily change its color to mimic the juveniles of a cleaner-fish, and most likely relies on color vision (see K.L. Cheney et al. *Proceedings of the Royal Society of London, B* 276, 1565–1573 [2009]). In disguise it mingles among the innocuous throng before, to everybody's surprise, it lunges to attack mode (see the papers by I.M. Côté and K.L. Cheney *Nature* 433, 211–212 [2005], and, respectively, K.L. Cheney and I.M. Côte and Cheney et al. *Proceedings of the Royal Society of London, B* 272, 2635–2639 [2005] and 275, 117–122 [2008]). A somewhat more sceptical view of this interpretation is offered by M.L. Johnson and S.L. Hall in *Marine Biology*, 148, 889–897 (2006).

51. See R.T. Hanlon et al. *Nature* 433, 212 (2005).

52. See their paper in *Biological Bulletin* 192, 203–207 (1997).

53. Sauer et al. (1997), p. 203 [Abstract]; note 52.

54. See R.T. Hanlon et al. *American Naturalist* 169, 543–551 (2007), and commentary by E.J. Warrant *Current Biology* 17, R209–R211 (2007).

55. See, for example, M.T. Lucero et al. *Biological Bulletin* 187, 55–63 (1994).

56. Not entirely unrelated is the release by dwarf sperm whale of large quantities of apparently fecal material by a female and its calf, evidently for concealment; see M.D. Scott and J.G. Cordaro *Marine Mammal Science* 3, 353–354 (1987).

57. See, for example, J. Prince et al. *Journal of Experimental Biology* 201, 1595–1613 (1998).

456 [2011]) suggests that they may be fairly close to the scaphopods.

10. See R.K. O'Dor and D.M. Webber *Canadian Journal of Zoology* 64, 1591–1605 (1986).

11. See J.R. Voight et al. *Marine and Freshwater Behaviour and Physiology* 25, 193–203 (1994).

12. See V.A. Bizikov *Ruthenica* (Supplement, 3), 1–88 (2004).

13. See A. Packard and M. Wartz *Philosophical Transactions of the Royal Society of London, B* 344, 261–275 (1994).

14. The occurrence of dwarf males is a rampantly convergent occurrence among not only animals (see, for example, F. Vollrath *Trends in Ecology & Evolution* 13, 159–163 [1998]) but more remarkably also among the mosses (see L. Hedenäs and I. Bisang *Perspectives in Plant Ecology, Evolution and Systematics* 13, 121–135 [2011] and F. Rosengren and N. Cronberg *Biological Journal of the Linnean Society* 113, 74–84 [2014]) and even green algae (see H. Mei et al. *Plant Systematics and Evolution* 265, 179–191 [2007]).

15. Accordingly there is positive selection of numerous genes that are linked to the emergence of the cephalopod camera eye and associated nervous tissue; see M.-a. Yoshida and A. Ogura *BMC Evolutionary Biology* 11, e180 (2011).

16. See W.S. Jagger and P.J. Sands *Vision Research* 39, 2841–2852 (1999); also A.M. Sweeney et al. *Journal of the Royal Society Interface* 4, 685–698 (2007).

17. See H.-Y. Zhao et al. *Journal of Molecular Biology* 411, 680–699 (2011).

18. See A.R. Lindgren et al. *Integrative and Comparative Biology* 50 (Supplement, 1), e102 (2010). In this abstract they suggest the cornea has at least two originations in the cephalopods.

19. See D. Froesch *Zeitschrift für Zellforschungen und mikroskopische Anatomie* 145, 119–129 (1973); also R.H. Douglas et al. *Journal of Experimental Biology* 208, 261–265 (2005).

20. See A. McVean *Comparative Biochemistry and Physiology, A* 78, 711–718 (1984); see also papers by B.U. Budelmann and J.Z. Young in *Philosophical Transactions of the Royal Society of London, B*, respectively, in vol. 306, pp. 159–189 (1984) and vol. 340, pp. 93–125 (1993).

21. These include retinal sensitivity (see G. Groeger et al. *Vie et Milieu* 56, 167–173 [2006]) and migration of the screening pigment to adjust to ambient light levels that is controlled by dopamine (see I.G. Glendall et al. *Journal of Experimental Biology* 185, 1–16 [1993]), which, as Ian Glendall and colleagues remark, "provides yet another example of the remarkable physiological and anatomical analogues that exist between the cephalopods and vertebrates" (p. 2).

22. See A.M. Sweeney et al. *Integrative and Comparative Biology* 47, 808–814 (2007).

23. See J.Z. Young. *The anatomy of the nervous system of Octopus vulgaris* (Clarendon; 1971). This point is reemphasized by Mather (1995); see note 38.

24. See J.Z. Young *Philosophical Transactions of the Royal Society of London, B* 267, 263–302 (1974).

25. Young (1974), p. 299; note 24.

26. See his overview in *Current Biology* 18, R897–R898 (2008).

27. Hochner (2008), p. R898; note 26.

28. See O'Dor and D.M. Webber *Journal of Experimental Biology* 160, 93–112 (1991); note, however, that in their description of a deep-sea squid, T. Kubodera et al. (*Proceedings of the Royal Society of London, B* 274, 1029–1034 [2007]) point out that the unexpected agility of the animal is partly due to its fins, which they note serve a similar function to those of the rays.

29. Hemocyanin occurs not only in cephalopods but also chitons. More significantly, this protein is also found in many arthropods, including what may be the most primitive living representatives, the velvet worms (see K. Kusche et al. *Proceedings of the National Academy of Sciences, USA* 99, 10545–10548 [2002]). Both mollusks and arthropods are invertebrates, so is not the logical conclusion that this copper-based protein was present in a common ancestor that wafted its way through the sunlit seas more than 500 Ma ago? Not at all; hemocyanin provides a striking example of convergence. (For an overview of copper proteins and confirmation that the hemocyanins are convergent, deriving from α [mollusk] and β [arthropod] subclasses, see F. Aguilera et al. *BMC Evolutionary Biology* 13, e96 [2013]). The core of the convergence lies in the employment of two atoms of copper that serve to bind the oxygen and thereby allows it to be transported in the blood, where ultimately it will be released to the waiting tissue. Incidentally, when the oxygen is captured, the color of the blood turns to blue, which is of course the exact opposite to us. Here the deoxygenated blood turns from red to blue, and hence the bluish color of our veins and much more seriously the appropriately named cyanosis as a potentially lethal condition. In both arthropods (notably the crustaceans and spiders) and cephalopods (for an overview, see H. Decker et al. *Integrative and Comparative Biology* 47, 631–644 [2007]; see also C. Gatsogiannis et al. *Journal of Molecular Biology* 374, 465–486 [2007]), the active site of the protein involves not only the copper, but the amino acids associated (as ligands) with each atom are invariably histidines. Not surprisingly this convergence in molecular identity has been termed "remarkable," even "startling."

Given the "startling" similarity between the active sites of hemocyanin in arthropods and cephalopods, why should we suppose they have different evolutionary origins? Simply because the overall molecular structure is completely different. In arthropods it is based on sixfold units (hence known as hexamers), and these can combine in various multiples. In contrast, among the mollusks the fundamental arrangement is decameric and the overall shape like a cylinder. In both cases the resulting protein is, as these controversial. Immunological detection of serotonin (see R. Czaker *Cell and Tissue Research* 326, 843–850 [2006]) and components of the so-called extracellular matrix (including collagen; see R. Czaker *Anatomical Record, A* 259, 52–59 [2000]) invite the thought they must be exceedingly primitive, perhaps even "as intermediates between Protozoa and Metazoa" (p. 849). Yet the molecular evidence points to their being some sort of degenerate triploblastic animal. (Most likely they are lophotrochozoans; see, for example, M. Kobayashi et al. *Evolution & Development* 11, 582–589 [2009], and T.G. Suzuki et al. *Journal of Parasitology* 96, 614–625 [2010]). It is hardly surprising that they were regarded as parasitic, but the evidence points to a symbiotic role in the kidney in assisting both the flow of urine (see H. Furuya et al. *Journal of Morphology* 262, 629–643 [2004]) and the excretion of the ammonia (by lowering the pH; see E.A. Lapan *Comparative Biochemistry and Physiology, A* 52, 651–657; 1975). As importantly, these dicyemids find a series of striking convergences with a group of ciliates (including *Chromidina*; see papers by F.G. Hochberg in *Malacologia* 23, 121–134 [1982], and *Memoirs of the National Museum of Victoria* 44, 109–145 [1983]). Thus, both are vermiform, highly ciliated, and densely coat the renal surfaces. Intriguingly, in the pearly nautilus the "kidney" (the pericardial organ) employs two varieties of bacteria (β-proteobacteria and a spirochaete) that densely coat the surface and are evidently symbiotic with key roles in excretion; see M. Pernice et al. *Proceedings of the Royal Society of London, B* 274, 1143–1152 (2007); also M. Pernice and R. Boucher-Rodoni *Environmental Microbiology Reports* 4, 504–511 (2012). In the case of dicyemids and chromidinids each employs an attachment structure (known in the former as the calotte; see H. Furuya et al. *Journal of Zoology, London* 259, 361–373 [2003]) which, as Hidetaka Furuya and colleagues note, in the "Chromidinids exhibit configurations of the attachment end . . . that mirror shapes observed in the dicyemids" (p. 642). They also note how the chromidinid "life cycle [is] very similar to those of dicyemids" (p. 641), which Eric Hochberg points out it is "beautifully adapted to the requirements of their endoparasitic [*sic*] environment" (p. 132). Despite the fact that the metazoan has reinvented itself as an honorary ciliate there is one interesting distinction, inasmuch as the dicyemids typify the benthic cephalopods whereas their protistan counterparts lurk in equivalent cephalopods (see overview by F.G. Hochberg in *Diseases of marine animals*, vol. III [O. Kinne, ed.], pp. 47–202 [Biologische Anstalt Helgoland; 1990]) of the pelagic realm.

CHAPTER 1

1. Also referred to as the inebri-actometer; see J. Parr et al. *Journal of Neuroscience Methods* 107, 93–99 (2001).

2. See U. Heberlein et al. *Integrative and Comparative Biology* 44, 269–274 (2004).

3. Or for that matter *hangover*; see H. Scholz et al. *Nature* 436, 845–847 (2005).

4. See pp. 145–146 in his *The restaurant at the end of the universe* (Pan; 1980). Here Adams pokes gentle fun at the linguists, but also raises the possibility of convergence in language as each world stumbled on jynna-tonnyx (or gee-N'N-T'N-ix or jinond-o-nicks) independently. In fact, evidence for such phonetic convergences are few and far between, although grammar may be universal.

5. See his paper in *Science* 160, 1308–1312 (1968).

CHAPTER 2

1. Among H.G. Wells's books are *Evolution: Fact and theory* (Cassell; 1934) and *The science of life* (Cassell [1931]) (both written with Julian Huxley and G.P. Wells). With respect to the latter work Michael Coren gives some interesting details as to how the book came to be written (p. 186) in his *The invisible man: The life and liberties of H.G. Wells* (Bloomsbury [1993]).

2. This refers to the famous photograph of H.G. Wells embracing a gorilla skeleton, evidently in flattery (if that is quite the word) of his mentor and Darwin's pugnacious publicist.

3. See his *H.G. Wells and the culminating ape: Biological imperatives and imaginative obsessions* (Macmillan; 1996).

4. See their chapter (pp. 269–307) in *Complex worlds from simpler nervous systems* (F.R. Prete, ed.) (MIT Press; 2004).

5. Glendall and Shashar (2004), p. 307; note 4.

6. See A. Packard *Biological Reviews* 47, 241–307 (1972); also his earlier paper in *Archivio Zoologico Italiano* 51, 523–542 (1966). Not only do these touch on many of the issues discussed in this chapter, but in the literature of convergence they are classics of their kind, both in scope of analysis and judiciousness of approach.

7. Packard (1966), p. 524 (emphasis in original); note 6.

8. Technically this is the group known as the deuterostomes (literally "second mouth," a reference to its embryological origin in contrast to the protostomes ["first mouth"], where the blastopore goes to form the mouth), and also includes the hemichordates. Although this is a classic distinction and broadly consistent with metazoan molecular phylogeny, it is possible that in terms of the origins of these openings the mouth may have evolved first and the opposite opening evolved at least twice independently (see A. Hejnol and M.Q. Martindale *Nature* 456, 382–286 [2008]).

9. The origin of the cephalopods within the mollusks is still rather controversial, although one molecular phylogeny (see G.K.M. Kocot et al. *Nature* 477, 452–

photosynthesize, so is dependent on its host, but in return again provides nitrogen.

For an overview of ciliate symbiosis involving phototrophs, see M.D. Johnson (*Journal of Eukaryotic Microbiology* 58, 185–195 [2011]), and in particular a marine form known as *Mesodinium rubrum* (formerly *Myrionecta rubra*). This organism relies on photosynthetic plastids that are derived from a group of protists, known as cryptophytes, which the ciliate ingests. The snag is that once the plastids are ripped out of their original home they are isolated and have lost their original genetic instructions. The ciliate gets around this difficulty by obtaining nuclei from the cryptophytes that can maintain the plastids, although this has to be a more or less continuous process because the "stolen" nuclei only remain transcriptionally active for about a month (see M.D. Johnson et al. *Nature* 445, 426–428 [2007]). Evidently one nucleus can service up to eight of the plastids. Whether this is one step toward the acquisition of a photosynthetic organelle is not known, but the capability of juggling both the genomes of captured plastids and nuclei in order to maintain this complex symbiosis is impressive.

11. To build an animal one must call in various novelties, not least a nervous system. Arguably as important is a key structural protein known as collagen, with its diagnostic array of repeated amino acid triplets (notably glycine with proline and hydroxyproline) that are disposed in a triple helix. Genes that encode for collagen domains occur in the choanoflagellate *Monosiga* (see N. King et al. *Nature* 451, 783–788 [2008]). At first sight, to equip a unicellular choanoflagellate with a protein like collagen seems a trifle extravagant, until we remember the principle of co-option. There are more than hints of self-organizational properties in collagen (see, for example, M.M. Giraud-Guille et al. *Current Opinion in Colloid & Interface Science* 13, 303–313 [2008]), and given its relative molecular and structural simplicity, its evolution hardly seems fortuitous. So much is apparent in the emergence of collagenlike proteins in groups yet more distant from the choanoflagellates. Typically they occur as triplet repeats, with glycine (G) and two other sites (i.e., G-X-Y, and in many cases these sites are occupied by proline). In the fungus *Metarhizium* a collagenlike component plays an important role in allowing the invasive pathogen to disguise itself from the host's immune system. (See C.-S. Wang and R.J. St. Leger *Proceedings of the National Academy of Sciences, USA* 103, 6647–6652 [2006]; also M. Celerin et al. *EMBO Journal* 15, 4445–4453 [1996].) To explain the presence of this collagenlike protein in *Metarhizium*, Cheng-Shu Wang and Raymond St. Leger point out that its sporadic distribution in the fungi may not be due to repeated loss of a "collagen gene," but was the result of "convergent evolution, particularly as alignment of collagenous regions indicates high sequence divergence" (p. 6651). A protein highly reminiscent of collagen is also found

in a coccolithophorid (*Hymenomonas*); see H.D. Isenberg and L.S. Lavine in *Biological mineralization* (I. Zipkin, ed.), pp. 649–686 (Wiley [1973]). Even more remote are the collagenlike proteins found in some eubacteria. One, given the risks of bioterrorism, is particularly relevant because it concerns its presence in the outer wall of the lethal anthrax spores; see J.M. Daubenspeck et al. *Journal of Biological Chemistry* 279, 30945–30953 (2004), and L.N. Waller et al. *Journal of Bacteriology* 187, 4592–4597 (2005); also B.M. Thompson and G.C. Stewart *Molecular Microbiology* 70, 421–434 (2008). Nor is this the only example, because in streptococcal bacteria the protein has another hallmark of collagen (see Y. Xu et al. *Journal of Biological Chemistry* 277, 27312–27318 [2002]). No equivalents have been identified in the archaeal bacteria, and the very sporadic distribution of collagenlike proteins in the eubacteria and viruses make the idea of lateral gene transfer attractive (see M. Rasmussen et al. *Journal of Biological Chemistry* 278, 32313–32316 [2003]). Even more remarkable is the identification of equivalent proteins in viruses, notably in a mimivirus (see K.B. Luther et al. *Journal of Biological Chemistry* 286, 43701–43709 [2011]) and also in the tail fibers of some bacteriophages (see J. Engel and H.P. Bächinger *Proceedings of the Indian Academy of Sciences* [*Chemical Sciences*] 111, 81–86 [1999]; also correspondence in *Science* [vol. 279, p. 1834] by M.C.M. Smith et al. and response by J. Engel) and a serious scourge in farmed shrimps (white spot syndrome virus; see Q. Li et al. *Archives of Virology* 149, 215–223 [2004]).

Although the principal emphasis is on collagen as a structural biomolecule, this protein is also important in such roles as the adhesion of cells and their migration. Intriguingly this involves interactions with various receptors, including tyrosine kinases. (See J. Heino *BioEssays* 29, 1001–1010 [2007], and B. Leitinger and E. Hohenester *Matrix Biology* 26, 146–155 [2007].) So conceivably collagen was only subsequently co-opted for its structural role in such day-to-day tasks as building the Achilles tendon and giraffe necks, not to mention elephant trunks.

12. See my chapter (pp. 135–161) in *Complexity and the arrow of time* (C.H. Lineweaver et al., eds.) (Cambridge [2013]).

13. Of a possible niche, perhaps the most unexpected is in the kidney of the cephalopods. An evolutionary adventure ending up by being drenched in urine seems to be a bit of a short straw. It hardly seems the most propitious of environments, but here the experiment in metazoan simplification has been taken to an extreme in the form of the dicyemid mesozoans (see H. Furuya and K. Tsuneki *Zoological Science* 20, 519–532 [2003]). With a body composed of an axial cell and a series of about forty peripheral cells, entirely lacking anything like a gut and nervous system (or indeed any other organ system), it is scarcely surprising that the phylogenetic position of these animals has been to the chloroplasts. Importantly, study of its genome structure indicates parallels to other groups that have drastically slimmed the number of genes (see E.C.M. Nowack et al. *Current Biology* 18, 410–418 [2008]; and commentary on pp. R345–R347 by P.J. Keeling and J.M. Archibald), hinting that this cyanobacterium is also poised to explore the chloroplast road. Nor is it difficult to envisage how this association could have arisen, because other species of *Paulinella* lack the symbionts but ingest cyanobacteria (see P.W. Johnson et al. *Journal of Protozoology* 35, 618–626 [1988]). In *Paulinella chromatophora*, however, the cyanobacterium rather than being turned into dinner now plays a key role in the life of the amoeba, supplying it with sugars derived from its photosynthesis. Not only that, but it has taken key steps to becoming a fully fledged organelle, notably in the way that the division of host amoeba and guest cyanobacteria are synchronized. Such a coordination would be consistent with at least some genes having been transferred from the symbiont to the nucleus of *Paulinella*, and direct evidence exists for just such occurrences, including genes employed in photosynthesis (see T. Nakayama and K-i Ishida *Current Biology* 19, R284–R285 [2009]; E.C.M. Nowack et al. *Molecular Biology and Evolution* 28, 407–422 [2011]; and E.C.M. Nowack and A.R. Grossman *Proceedings of the National Academy of Sciences, USA* 109, 5340–5345 [2012], and on pp. 5142–5143 commentary by R.F. Waller; also A. Bodył et al. *Current Biology* 22, R304–R306 [2012]). As Takuro Nakayama and Ken-ichiro Ishida remark, "Another acquisition of a primary photosynthetic organelle is under way" (Nakayama and Ishida [2009], p. R284 [part of title]), but in terms of evolution this endosymbiosis is still in its infancy. The cyanobacterium has retained its genomic identity (see H.S. Yoon et al. *Current Biology* 16, R670–R672 [2006], and commentary on pp. R690–R692 by J.M. Archibald), but the protest by some workers (see U. Theissen and W. Martin *Current Biology* 16, R1016–R1017 [2006]) that this example is a symbiosis pure and simple and cannot qualify as a proto-chloroplast seems increasingly less likely. The closeness and permanency of this obligate association suggests that we are on the road to a new primary endosymbiosis. (See D. Bhattacharya and J.M. Archibald *Current Biology* 16, R1017–R1018 [2006]; also P. Mackiewicz and A. Bodył *Journal of Phycology* 46, 847–859 [2010], and T. Nakayama and J.M. Archibald *BMC Biology* 10, e35 [2012].)

Paulinella has understandably been the focus of attention, but the group to which it belongs (the cercozoans) has also yielded a rather remarkable form, powered by four flagellae and containing a cyanobacterial-like form that looks like yet another example of a primary endosymbiosis (see C. Chantangsi et al. *BMC Microbiology* 8, e123 [2008]). Nor are these the only such examples. S. Maruyama and E. Kim (*Current Biology* 23, 1081–1084 [also p. 1144 (2013)]; also

commentary on pp. R530–R531 by J.A. Raven) suggest that the mode of engulfment of bacteria by a prasinophyte might be instructive as to how the precursor chloroplasts were acquired. So, too, the long-term survival of cyanobacteria in vacuoles with cells of a colorless euglenophyte flagellate might also be instructive (see E. Schnepf et al. *Phycologia* 41, 153–157 [2002]). Another extraordinary instance of an intimate association involves another flagellate, the *Hatena arenicola* (which belongs to the group known as the katablepharids). This organism, unlike most of its relatives, which are swimmers, crawls across sand grains. It has struck up a striking symbiotic association with another protistan, this time belonging to the prasinophyceans. This latter organism comes equipped with a functional eye-spot, which presumably helps its host to navigate (see N. Okamoto and I. Inouye *Protist* 157, 401–419 [2006]). Otherwise the symbiont is mostly a photosynthetic plastid, and it too is conceivably heading toward organelle status. A key step in that direction is to synchronize cell division, but in *Hatena* the symbiont does not divide but stays in one daughter cell. The other cell, however, is not bereft, because without its photosynthetic partner it then begins to feed.

When we look at other symbiotic associations we see how flexibility and opportunism suggest that no evolutionary threshold is impregnable. A well-known example is the occurrence of green algae in the ciliate *Paramecium*, whose guests have evidently been acquired multiple times (see R. Hoshina and N. Imamura *Protist* 159, 53–63 [2008]). What may well be the beginnings of another primary endosymbiosis between a cyanobacterium and a dinoflagellate (see L. Escalera et al. *Protist* 162, 304–314 [2011]; also N. Gordon et al. *Marine Ecology Progress Series* 107, 83–88 [1994]) is complemented by other examples. One involves rhopalodiacean diatoms. Here a permanent endosymbiosis has been formed with other cyanobacteria (see J. Prechtl et al. *Molecular Biology and Evolution* 21, 1477–1481 [2004]), which, however, have lost all capacity for photosynthesis (see T. Nakayama et al. *Proceedings of the National Academy of Sciences, USA* 111, 11407–11412 [2014]) and there is corresponding evidence for genomic reduction (see C. Kneip et al. *BMC Evolutionary Biology* 8, e30 [2008]) This endosymbiotic association appears to have got under way quite recently (see T. Nakayama et al. *Journal of Plant Research* 124, 93–97 [2011]). Another such association is found in the prymnesiophytes (see A.W. Thompson et al. *Science* 337, 1546–1550 [2012]).Given they are photosynthetic, is it not rather extravagant to lasso a cyanobacterium that in principle is also photosynthetic? In actuality, in the case of the diatom it must wait for darkness to fall, because then the symbiont can engage in the reduction of nitrogen and so supply a key nutrient to the diatom. What of the prymnesiophyte? Here the cyanobacterium has lost its ability to

Review of Microbiology 61, 589–618 (2007)] stresses reveal "a striking evolutionary plasticity of cytoskeletal proteins" [p. 589]), but there is also overlap. This is because a bacterial actin is employed in the cell wall in much the same way as among the eukaryotes (see S.-Y. Wang et al. *Proceedings of the National Academy of Sciences, USA* 107, 9182–9185 [2010]), again suggesting that the emergence of a cytoskeleton is far from surprising. Nor are these the only proteins that once were thought to be effectively restricted to the eukaryotes. Central to their existence are the kinases, yet in various groups of bacteria (notably the myxobacteria) that show varieties of multicellularity, similar proteins play a key role in signaling (see J. Pérez et al. *Proceedings of the National Academy of Sciences, USA* 105, 15950–15955 [2008]).

These findings reinforce the point that while prokaryotes have many differences to the eukaryotes, it is a moot point whether they are "simpler." Not only does the former group possess some of the key building blocks to make a eukaryote, but other parallels exist. Among the most striking is a degree of social organization that finds clear analogues not only in microbial eukaryotes but in even more complex systems (see B.J. Crespi *Trends in Ecology & Evolution* 16, 178–183 [2001] and ensuing discussion on pp. 606–607; also S.A. West et al. *Annual Review of Ecology, Evolution, and Systematics* 38, 53–77 [2007]). Aspects of cell communication (and indeed the remarkable quorum sensing whereby bacterial aggregations arrive at joint decisions, a phenomenon that finds parallels among social insects such as in bee swarms; see T.D. Seeley and P.K. Visscher *Behavioral Ecology and Sociobiology* 56, 594–601 [2004]) and fish (see A.J.W. Ward et al. *Proceedings of the National Academy of Sciences, USA* 105, 6948–6953 [2008]) are, of course, a sine qua non for multicellularity. This, too, has evolved among the bacteria. In the cyanobacteria, for example, multicellularity evolved early in their history, and perhaps surprisingly unicellar forms have reevolved at least five times (see B.E. Schirrmeister et al. *BMC Evolutionary Biology* 11, e45 [2011]). As Bettina Schirrmeister and colleagues remark, the road to anatomical complexity can be reversed, but they also note how their study "clearly indicates that similar morphologies have been gained and lost several times during the evolutionary history of living cyanobacteria" (p. 9). Apart from the myxobacteria and cyanobacteria, another striking example involves magnetotactic bacteria. Forming small balls of cells, closely juxtaposed and with their locomotory flagella arranged across the outer surface, these bacteria are strikingly reminiscent of colonial green algae (see C.N.F. Keim *Journal of Structural Biology* 145, 254–262 [2004]) such as *Eudorina*.

10. What then of the chloroplasts? The general rules for genomic simplification in bacterial endosymbionts and pathogenic bacteria should again sound a warning with respect to the origin of the chloroplasts. If, as noted above, key genes associated with energy transduction must be retained while others are exported to the nucleus, then this could mean that every time chloroplasts evolve they are doomed to tread a similar evolutionary path. This is precisely the suggestion by John Stiller and colleagues (*Journal of Phycology* 39, 95–105 [2003]), who point out that as gene loss from the proto-chloroplast is far from random, then a convergent origin needs to be considered. This flies in the face of the almost universally accepted idea of a single origin, and indeed the evidence is far from straightforward (see J.D. Palmer *Journal of Phycology* 39, 4–11 [2003], and Stiller's reply on pp. 1283–1285). Monophyly remains the more popular option (see, for example, P.J. Keeling *American Journal of Botany* 91, 1481–1493 [2004], and N. Rodríguez-Ezpeleta et al. *Current Biology* 15, 1325–1330 [2005]). However, some indicators suggest that this may require rethinking, and new voices are urging caution. Tony Larkum and colleagues (*Trends in Plant Science* 12, 189–195 [2007]; also C.J. Howe et al. *Philosophical Transactions of the Royal Society of London, B* 363, 2675–2685 [2008]) point out, for example, that while chloroplasts are certainly derived from cyanobacteria, this probably happened more than 2 billion years ago, and no living cyanobacterium can be closely allied to the chloroplast genome (see also A. Criscuolo and S. Gribaldo *Molecular Biology and Evolution* 28, 3019–3032 [2011], who identify a very deep point of origination of the chloroplasts within the cyanobacteria). One idea, that chloroplasts are related to the so-called prochlorophyte cyanobacteria (not least because of possessing a particular type of chlorophyll), needs some sort of qualification given the prochlorophyte grade has evidently evolved independently at least three times (see E. Urbach et al. *Nature* 355, 267–270 [1992]; also the companion paper [pp. 265–267] by B. Palenik and R. Haselkorn). An interesting alternative is to seek the ancestral chloroplast in a cyanobacterium similar to *Nostoc* that bears cells (heterocysts) specialized for nitrogen fixation (see O. Deusch et al. *Molecular Biology and Evolution* 25, 748–761 [2008]). Somewhat similar symbioses in both the water fern *Azolla* and lichens employ the cyanobacteria, but the latter case is particularly interesting because the cyanobacterium is engulfed by a protrusion of the fungal hypha (see D. Mollenhaur et al. *Protoplasma* 193, 3–9 [1996]). Perhaps this recalls how the proto-chloroplast was assimilated into the life of the earliest eukaryotes, but we can go beyond speculation because the process is going on before our very eyes.

The story involves an amoeba known as *Paulinella chromatophora*, the specific name of which hints at the prominent green symbiont. The latter transpires to be cyanobacterium (see B. Marin et al. *Protist* 156, 425–432 [2005], and commentary by N. Rodríguez and H. Philippe *Current Biology* 16, R53–R56 [2006]), but one phylogenetically distant from the group that gave rise

Microbiology Reviews 32, 654–672 [2008]). Another striking example of a bacterial organelle are the so-called anammoxosomes, which are characteristic of some members of the planctomycetes and serve to oxidize ammonia (and so have an obvious relevance to treating polluted water). These organelles are complex with internal tubules and also a folded membrane that very much recalls the arrangement of mitochondria and presumably also serves to increase the surface area available for metabolism (see L. van Niftrik et al. *Journal of Structural Biology* 161, 401–410 [2008]). Processing ammonia is a ticklish business, not least because one of the intermediates in the reaction is hydrazine (N_2H_4) (see B. Kartal et al. *Nature* 479, 127–130; 2011) that otherwise finds employment as rocket fuel (see, for example, J. Schalk et al. *FEMS Microbiology Letters* 158, 61–67 [1998]). It is important to keep this highly toxic substance at a safe distance, and accordingly the membrane wall of the anammoxosome is exceptionally resistant to diffusion and employs specialized lipids (the "ladderanes") (see J.S. Sinninghe Damsté et al. *Nature* 419, 708–712 [2002], and on pp. 676–677 commentary by E.F. DeLong).

The planctomycetes have long been realized to be curiously eukaryote-like. For an overview, see J.O. McInerney et al. *BioEssays* 33, 810–817 (2011), albeit in what seems to be a somewhat tetchy review of those who would have the planctomycetes as intermediates between the eubacteria and eukaryotes. J.A. Fuerst and E. Sagulenko *Frontiers in Microbiology* 3, e167 (2012) set out to balance convergence as against homology with eukaryotes and tend toward the latter, although employing organisms such as *Giardia* for comparisons might raise an eyebrow. Not only do planctomycetes house internal membranes, but one taxon (*Gemmata*) encloses its DNA in a nuclear membrane, and this is almost certainly a convergence on the standard eukaryotic arrangement (see J.A. Fuerst and R.I. Webb *Proceedings of the National Academy of Sciences, USA* 88, 8184–8188 [1991]). Not only that but this bacterium exhibits an endocytosis-like behavior that would otherwise be taken as a hallmark of the eukaryotes (see T.G.A. Lonhienne et al. *Proceedings of the National Academy of Sciences, USA* 2010; see also commentary on pp. 12739–12740 by P. Forterre and S. Gribaldo; the authors do not rule out either a much deeper ancestry or acquisition by horizontal gene transfer, but an independent origin seems just as plausible). Another curiosity is that the cell wall is quite unbacterial and, instead of the customary peptidoglycans, is proteinaceous (see, for example, W. Liesack et al. *Archives of Microbiology* 145, 361–366 [1986]). Equally intriguingly, *Gemmata* is capable of making sterols (see A. Pearson et al. *Proceedings of the National Academy of Sciences, USA* 100, 15352–15357 [2003]), which in such forms as cholesterol and ergosterol play a central role in eukaryotes (see, for example, J.K. Volkman *Applied Microbiology and Biotechnology* 60, 495–506 [2003]) including membrane structure. Sterol synthesis is not restricted to this planctomycete, but occurs in a scattering of other bacteria (see H. Bode et al. *Molecular Microbiology* 47, 471–481 [2003]), notably *Methylcoccus* (see D.C. Lamb et al. *Molecular Biology and Evolution* 24, 1714–1721 [2007], and C. Nakano et al. *Bioscience, Biotechnology and Biochemistry* 71, 2543–2530 [2007]). This capacity to make sterols is almost certainly not the result of a lateral gene transfer from some eukaryote, but if it had a single origin in the bacteria it is not yet clear where the capacity originated. Nevertheless, Ann Pearson and colleagues (2003) also point out how the three bacterial lineages that have a sterol chemistry possess a number of peculiar features, including "Intra-cellular membranes, unusual cell walls and complex reproductive strategies" (p. 15357).

Planctomycetes are allied to a larger group of bacteria that includes the chlamydids (which include various venereal horrors) and another group, the Verrucomicrobia (see M. Pilhofer et al. *Journal of Bacteriology* 190, 3192–3202 [2008]). The latter employ the protein tubulin, which is of key importance in the cytoskeleton of the eukaryotes. In at least one verrucomicrobe (*Prosthecobacter*) this tubulin is evidently very ancient and although having a simpler configuration presumably has some sort of cytoskeletal function (see M. Pilhofer et al. *PLoS Biology* 9, e1001213 [2011]); while supporting a bacterial origin for this protein (as against earlier suggestions that it has obtained by horizontal gene transfer from a eukaryote) it may have been derived from an as yet unknown bacteria. It is clear that not only are these bacterial equivalents to tubulin (see also N. Yutin and E.V. Koonin [*Biology Direct* 7, e10 (2012)], who identify a tubulin in an archael bacteria that they propose is the closest to eukaryotic equivalents), but so too other proteins (actins and the so-called intermediate filament) that in eukaryotes are central in their cellular organization. In the bacteria, however, they play many roles, including the construction of the magnetosomes and the partitioning (see, for example, P.L. Graumann *Annual Review of Microbiology* 61, 589–618 [2007], and J. Pogliano *Current Opinion in Cell Biology* 20, 19–27 [2008]) of plasmids. It is important to emphasize two things. First, these proteins have very little sequence similarity to those in the eukaryotes, but the similarity of overall structure is generally thought to point to a common origin rather than convergence. (See, for example, E. Nogales et al. *Nature Structural Biology* 5, 451–458 [1998], and H.P. Erickson *Trends in Cell Biology* 8, 133–137 [1998]; see also, for example, S. Vaughan et al. *Journal of Molecular Evolution* 58, 19–29 [2004]). Eukaryotes, therefore, did not need to reinvent the essential (it should be noted that many of their functions in the prokaryotes find no counterpart in the eukaryotes and indeed as Peter Graumann [*Annual* that reduce the nitrogen while ensuring its availability for other energy-consuming components of this biochemical factory. All this depends quite crucially on the leghemoglobin (see T. Ott et al. *Current Biology* 15, 531–535 [2005]; and commentary on pp. R196–R198 by J.A. Downie). We see, moreover, a very precise set of molecular adaptations to the specific requirement of nitrogen reduction. In his review, Jonathan Wittenberg (*Gene* 398, 156–161 [2007]) not only draws attention to a series of striking convergences in the employment of the hemoglobins (and their equivalents) but emphasizes how these remarkably diverse systems are driven toward optimal solutions.

In terms of an ability to fix atmospheric nitrogen, this system has not only evolved in the legumes, but in other groups of flowering plants as well. When viewed phylogenetically they form a "nitrogen-fixing clade," and so evidently there is a predisposition to enter into symbiotic arrangements (see D.E. Soltis et al. *Proceedings of the National Academy of Sciences, USA* 92, 2647–2651 [1995]; also J.J. Doyle *Molecular Plant-Microbe Interactions* 24, 1289–1295 [2011]). It is very far from clear (see the overview by J.J. Doyle *Trends in Plant Science* 3, 473–478 [1998]), however, what underpins this propensity to form nodules that serve as homes to nitrogen-fixing bacteria. A clue of sorts, however, comes from a symbiosis that is extraordinarily widespread and involves plants and a group of fungi (glomeromycotans) that form a remarkable system of arbuscular mycorrhizae (p. 152). This intimate association, from which both sides evidently benefit, would also in principle potentially facilitate the introduction of the nitrogen-reducing bacteria, a notion that receives strong support from the fact that the methods of cell signaling (see C. Kistner and M. Parniske *Trends in Plant Science* 7, 511–518 [2002]) are the same in the root nodules and the arbuscular mycorrhizae (see M. Parniske *Nature Reviews Microbiology* 6, 763–775 [2008]). It shows how in principle a very ancient system connected to a signal pathway can be co-opted for symbiotic associations (see companion papers by S. Stracke et al. and G. Endre et al. on pp. 959–962 and pp. 962–966, respectively, in vol. 417 of *Nature* [2002], along with commentary on pp. 910–911 by H.P. Spaink). Nevertheless, as such this does not immediately explain the predisposition of this nitrogen-fixing clade, and even within this clade the actual occurrences of nitrogen fixation are generally very sporadic. It is evident that even if it is drawing repeatedly on a similar genomic tool kit (see, for example, R. Op den Camp et al. *Science* 331, 909–912 [2011]; also A. Tromas et al. *PLoS ONE* 7, e44742 [2012]), the system has evolved independently several times. This is most obvious because two quite distinct groups of bacteria are employed, the gram-negative rhizobia group and the gram-positive actinorhizal *Frankia*, and in at least the latter association there is evidence for multiple recruitment (see, for example,

S.C. Jeong et al. *Molecular Phylogenetics and Evolution* 13, 493–503 [1999]; also S.M. Swensen *American Journal of Botany* 83, 1503–1512 [1996]).

Nor is this the only way in which nitrogen fixation can be attained in a symbiotic context. There is, for example, an intriguing association in a lichen that has an endosymbiotic association (see M. Kluge et al. *Botanica Acta* 105, 343–344 [1992]) with the cyanobacterium *Nostoc* (lichens provide many other examples of a fungal-algal convergence but *Nostoc* also forms a curious association with a member of the Glomeromycota (see D. Mollenhauer et al. *Protoplasma* 193, 3–9 [1996]). So, too, among the primitive land plants known as the bryophytes (see D.A. Dalton and J.M. Chatfield *American Journal of Botany* 72, 781–784 [1985]; also D.G. Adams and P.S. Duggan *Journal of Experimental Botany* 59, 1047–1058; 2008), and also the much more advanced cycads (see J.K. Vessey et al. *Plant and Soil* 266, 205–230; [2004]) NB: This article is reprinted in vol. 274, 51–78 [2005], which often provide the vegetational backdrop to roaming dinosaurs, and as a group belong to gymnosperms), the associations are again achieved by a union with cyanobacteria. The convergent roster of symbiotic association between plant and cyanobacteria does not finish with the cycads (we should also note that in terms of a comparison with the rhizobial/actinorhizal associations and the cycads, Vessey et al. remark on an "interesting evolutionary convergence . . . of NiFe hydrogenase-uptake systems" [p. 209]). This is because such an association has also been arrived at independently among the flowering plants (in *Gunnera* [see B. Bergman in *Cyanobacteria in symbiosis* (A.N. Rai et al., eds.), pp. 207–232 (Kluwer [2002]); also B. Bergman et al. *New Phytologist* 122, 379–400 [1992]) and also a water fern (see G.M. Wagner *Botanical Review* 63, 1–26 [1997]) known as *Azolla*.

9. There is abundant evidence that some features key to eukaryote organization are present as harbingers among the prokaryotes. These include organellelike structures (for an overview of prokaryotic organelles see D. Murat et al. *Cold Spring Harbor Perspectives in Biology* 2, a000422 [2010]), notably the carboxysomes and other microcompartments (see T.O. Yeates et al. *Nature Reviews Microbiology* 6, 681–691 [2008]), gas vesicles (see A.E. Walsby *Microbiological Reviews* 58, 94–144 [1994]; also J.P. Ramsay et al. *Proceedings of the National Academy of Sciences, USA* 108, 14932–14937 [2011]), volutin granules (these are the bacterial version of the so-called acidocalcisomes that serve to store calcium and phosphates; they are almost universal and could well date back to the dawn of life (see M.J. Seufferheld et al. *Biology Direct* 6, e50[(2011]), but convergence cannot be ruled out (see R. Docampo and S.N.J. Moreno *Cell Calcium* 50, 113–119 [2011]), as well as the membranelike structures that serve to form the magnetosomes that enclose the crystals of bacterial magnetite (see D. Schüler *FEMS*

815–822 [2006]). Possibly they were exposed to new pathogens, but if so, first, relatively few target solely vertebrates, and second, one can invert the argument and propose that only certain types of pathogen have managed to insinuate themselves past the ramparts of vertebrate immunity. Another line of thought is to remember that the geological interval when adaptive immunity must have been first acquired more or less coincides with the Cambrian "explosion" so perhaps it was a corollary of the very rapid evolution of body plans?

This, however, would presumably apply to other groups, notably the arthropods, which still employ only innate immunity, albeit highly effectively. The issue is addressed by Jens Rolff (*Developmental and Comparative Immunology* 31, 476–482 [2007]). He raises the way in which an arms race between parasites, especially the group that includes such delights as the tapeworm and liver fluke (the platyhelminths) and their vertebrate hosts would drive the struggle to develop an effective immune system versus constant attempts to sabotage it. It is, however, another of his suggestions that is more intriguing. To some extent it is a chicken-and-egg problem, but the basic thesis is that both acceleration of metabolic rates and the concomitant development of the nervous system, including increasingly subtle brains, defined a new adaptive landscape, not least the increased risk of damage to DNA (by the free radicals produced during metabolism) and the spinning out of rogue cells (some cancerous)—hence the urgent need to develop a honed and focused immune system that could transcend the much more ancient innate immunity. In addition, there is some evidence that adaptive immunity is energy efficient in comparison to its innate equivalent (see I. Råberg et al. *Proceedings of the Royal Society of London, B* 269, 817–821 [2002]). The conclusion, which haunts this entire book, is that when evolution as a search engine breaks into new worlds, not only will this happen repeatedly, but previously unimaginable levels of biological complexity, such as adaptive immunity, will emerge again and again.

3. *Haplozoon* lives in the gut of a polychaete annelid, specifically a maldanid. Despite being a dinoflagellate (see S. Rueckert and B.S. Leander *European Journal of Protistology* 44, 299–307 [2008]) it shows a remarkable resemblance to the tapeworms (which are highly evolved flatworms). Tapeworms attach to the gut wall with suckers and hooks located on the scolex, while a string of segments (the proglottids) are successively budded off (in the process of strobilization) and ultimately detach to leave the host and continue the life cycle. And in *Haplozoon*? Once again attachment is achieved by a suctorial organ and adjacent stylets, while the remainder of the "body" forms a series of segments that include reproductive sporocytes that ultimately detach. A particular characteristic of the tapeworm is that the animal is covered in a dense

"fur" of microtrichs, and here, too, Brian Leander (*Journal of Eukaryotic Microbiology* 55, 59–68 [2008]) reports that "the thecal barbs of haplozoons are nearly identical in morphology and density" (p. 65).

Haplozoon, of course, is a gut-dwelling or enteric parasite, but when it comes to parasitism, more generally convergences are again rife. Robert Poulin (*Advances in Parasitology* 74, 1–40 [2011]) titles his overview, "The many roads to parasitism: A tale of convergence." As he notes, "parasites, despite their broad phylogenetic diversity, have converged on a limited number of adaptive peaks as reflected by analogies in morphological, ecological and epidemiological traits" (p. 3). In Chapter 14 I address the topic of parasitic plants, as does Poulin. His review provides an excellent summary, in which he concludes, "Parasitism has appeared well over 100 separate times, at least once in most phyla. . . . Yet this vast diversity boils down to only six general parasitic strategies, each characterised by an almost predictable suite of traits. From distinct phylogenetic origins and with vastly different biological properties at the start, separate lineages have inevitably gravitated across the adaptive landscape towards one of six stable peaks" (p. 32). This brief note can do little justice to the range of parasitic convergences, but among the innumerable examples, perhaps one of the most striking are those relatives of the cirripedes that have transformed themselves into endoparasites. Without the diagnostic cyprid larva their affinities would be difficult to guess because the adult forms of the aptly named rhizocephalans form an extraordinary set of rootlets that anastomose through the unfortunate decapodan host (see J. Bresciani and J.T. Høeg *Journal of Morphology* 249, 9–42 [2001]). Access to the host is gained via a puncturing stylet (see H. Glenner *Journal of Morphology* 249, 43–75 [2001]), and this releases a distinctive wormlike form that is motile and only later begins the process of arborescent invasion. Just the same arrangement is also found in the facetotectan cirripedes, but as Marcos Pérez-Losada et al. (*BMC Biology*, 7, e15 [2009]) demonstrate this is "a remarkable case of convergent evolution into extremely specialized endoparasitism" (p. 4).

4. See his article in *Journal of Theoretical Biology* 252, 1–14 (2008).

5. Stayton (2008), part of the title; note 4.

6. See my *Life's solution: Inevitable humans in a lonely universe* (Cambridge; 2003).

7. See his *Convergent evolution: Limited forms most beautiful* (MIT Press; 2011).

8. The nodules are packed with symbiotic bacteria that undertake the reduction of nitrogen, for effectively the same reasons as humans employ the Haber process to make agricultural fertilizers. In the leguminous nodules the bacterially mediated process is critically dependent on the precise control of oxygen supply, excluding it from the enzymes (nitrogenases) probably a recent adaptation in [such] holometabolous insects" (p. 176), given it is unknown in their more primitive hemimetabolous counterparts. (Well, not quite, because in the hemipterans a toll-like pathway exists; see, for example, A.E. Whitfield et al. *Insect Molecular Biology* 20, 224–242 [2011]). It will be interesting to see if this too evolved independently, but possibly not given the close phylogenetic proximity of the hemipterans to the holometabolous insects). They make, however, an even more important point that echoes Ausubel's insights when they raise the question as to "why evolution has retained a limited number of analogous regulatory modules in separate evolutionary lineages. It could be that the *intrinsic* properties of signaling modules are particularly well-suited to a specific function" (Leulier and Lemaitre [2008], p. 176 [emphasis added]).

Insect immune systems are not only complex but they may verge on the adaptive system found in vertebrates. As Joachim Kurtz and Sophie Armitage remark in their commentary on evidence for such a system in the mosquitoes (see their paper in *Trends in Immunology* 27, 493–496 [2006]; the commentary largely addresses the paper by Y.-M. Dong et al. [*PLoS Biology* 4, e229 (2006)]), it "should provide a new perception of invertebrate immunity that shows astounding analogies to the vertebrate adaptive immune system" (p. 495); but see C. Hauton and V.J. Smith *BioEssays* 29, 1138–1146 (2007) for a much more critical view. So it is that the insect immune system can show both a capacity for memory and a plasticity in the face of microbial threat (see T. Hibino et al. *Developmental Biology* 300, 349–365 [2006]). Even so it does not seem quite to match the extraordinary adaptive immunity possessed by the vertebrates. This superb defensive system effectively depends on the rapid identification of foreign molecules (antigens) that pose a threat. Based on recognition centers in the white blood cells (or lymphocytes, the well-known T-cells and B-cells derived, respectively, from the thymus and bone marrow), initial recognition is then followed by a clonal proliferation of antibodies to target the invasive antigens. Its versatility depends on the fact that the antibody genes can be rearranged in a colossal number of alternatives, almost one of which is certain to lock effectively onto the antigen. Nor only is it highly adaptive, but the system possesses a memory so that recurrent infections can be even more rapidly tackled.

Adaptive immunity only appears at about the time vertebrates acquired jaws (crudely sharks to humans). (It is very likely that as so often is the case the relevant genes have been co-opted, and indeed their discovery in the sea urchins (the echinoids), which certainly possess an immune system but apparently not one that is adaptive (see, for example, Y. Moret and M.T. Siva Jothy *Proceedings of the Royal Society of London, B* 270, 2475–2480 [2003]), points to a more ancient origin (see S.D. Fugmann et al. *Proceedings of*

the *National Academy of Sciences, USA* 103, 3728–3733 [2006]). The echinoids and the other echinoderms (such as starfish and sea cucumbers) are phylogenetically quite close to the vertebrates (collectively they all belong to the deuterostomes). Attempts to find the same immune system in the more primitive vertebrates, those without jaws, and probably best known in the form of the lamprey, were resoundingly unsuccessful. But they do not lack an adaptive immune system; rather they have invented their own independently, and it has been not only identified in the lamprey (see Z. Pancer et al. *Nature* 430, 174–180 [2004] and commentary on pp. 157–158 by M.F. Flajnik in the same issue; also M.N. Alder et al. *Science* 310, 1970–1973 [2005] and *Nature Immunology* 9, 319–327 [2008]) but also its revolting cousin, the hagfish (see Z. Pancer et al. *Proceedings of the National Academy of Sciences, USA* 102, 9224–9229 [2005]). Again the cells involved are similar to the lymphocytes, but this time the immunity depends on a system of proteins with leucine-rich repeats (LRR), which is of course the system we encountered in the plants. The similarities between jawed and jawless vertebrates extend to the latter possessing lymphocytes with analogous properties to the classic T (thymus) and B (bone marrow) cells (see Guo et al. *Nature* 459, 796–801 [2009]; also accompanying commentary by G.W. Litman and J.P. Cannon on pp. 784–786. See also J. Kasamatsu et al. *Proceedings of the National Academy of Sciences, USA* 107, 14304–14308 [2010]), even though Peng Guo and colleagues are driven to remark that the T analogues are "surprisingly similar" (p. 796). Guo et al. (2009) also note that "This unforeseen genotypic and functional division of lymphocyte differentiation offers an intriguing new piece to the puzzle of how the adaptive immune system may have evolved" (p. 799). And such "surprises" extend beyond this immune system to analogues of the thymus in the lamprey (see B. Bajoghli et al. *Nature* 470, 90–94 [2011], also commentary by M. Flajnik *Current Biology* 21, R218–R220 [2011]). These convergences suggest that there could well be very few solutions to adaptive immunity, and we should not be at all surprised to discover further parallels. The jawless vertebrate immune system is certainly adaptive and almost certainly also clonal. It may represent a strikingly different molecular pathway, but evolution steers it to very much the same destination. It should go without saying that having acquired by one route or another an adaptive immune system does not mean the vertebrates have discarded the more ancient innate immunity: far from it, and it remains the first line of defense. This, however, begs a question concerning the adaptive system: Why only in the vertebrates (but twice!), at about the same time, and under what sort of selective pressure? (As part of general overviews, these issues are discussed by Z. Pancer and M.D. Cooper *Annual Review of Immunology* 24, 497–518 [2006], and M.D. Cooper and M.N. Alder *Cell* 124, example is the protein known as flagellin, which as its name suggests occurs in the flagellar motor of the bacteria (p. 313). Once recognized this then prompts a response via a cascade of signals that mobilize the appropriate defense. Now, it will hardly come as a surprise that just as animals employ an innate immunity system, so do the plants. Given the profound and obvious differences between these two groups it is surely more remarkable that the two systems are strikingly similar but independently derived (see F.M. Ausubel *Nature Immunology* 6, 973–79 [2005]; see also J.-X. Yue et al. *New Phytologist* 193, 1049–63 [2012]). Or is it?

After all, immune systems face similar challenges: to detect hostile molecules and do something about it. So in both animals and plants we find molecules that span the cell wall, the so-called transmembrane receptors. In animals these are known as toll-like receptors (TLRs; an immunelike capacity using this system has also been identified in the social amoeba *Dictyostelium* [see G.-K. Chen et al. *Science* 317, 678–81 (2007)]), whereas in plants we find different proteins, representatives of an enormous family known as the kinases. So, too, the protein receptors located within the cell itself are not identical, but the overall similarities in organization are again striking (see also V. Bonardi et al. *Proceedings of the National Academy of Sciences, USA* 108, 16463–68 [2011]). This is because the structural arrangement follows a common tripartite scheme, including at the same end a domain rich in the amino acid leucine (technically the leucine-rich repeats [LRR]). One could, of course, argue that this similarity is simply a reflection of the common ancestor of plants and animals possessing an immune system. But this is most unlikely. A. Dievart and colleagues (*BMC Evolutionary Biology* 11, e367 [2012]) demonstrate how this leucine-rich repeat/receptor kinase system has evolved at least four times independently, including among the oomycetes (p. 151): not only are the actual proteins in either system markedly different (for instance, both transmembrane receptors respond to the bacterial flagellin but in different ways; see T. Nürnberger et al. *Immunological Reviews* 198, 249-66 [2004]). They also provide an absorbing review of innate immunity in animals and plants, with the subtitle, "Striking similarities and obvious differences"), but the details of the signaling pathways also differ. Frederik Ausubel (*Nature Immunology* 6, 973-979; 2005) puts the case in an interesting nutshell:

> Given the compelling case for convergent evolution of innate immune pathways [in plants and animals], an important issue is why evolution has chosen a limited number of apparently analogous regulatory modules in disparate evolutionary lineages. Does this reflect inherent biochemical constraints that result from a similar overall "logic" of how an effective immune system can be constructed? (p. 977)

Ausubel is careful to stress that the systems are not identical. Plants, for example, have a remarkable army of sensors tuned to various pathogenic signals, and it is not surprising that specific genes designed to resist disease (R genes) and are part of the innate immunity system also show convergence (see T. Ashfield et al. *Plant Cell* 16, 309–18 [2004] and J.M. McDowell *Trends in Plant Science* 9, 315–17 [2004]). Ausubel suggests that such versatility in the plants might be because not only are plants static but cannot rush cells to the collective defense; as he points out each and every cell is on the front line. This system also has another intriguing dimension because it transpires that the receptors of the signaling system employed in the extraordinary symbioses of arbuscular mycorrhizas and nitrogen-fixing root nodules also employ proteins with leucine-rich repeats (and thereby again similarities to the toll-like receptors). Here we face a fascinating implication that the roles of defense by innate immunity and symbiosis are far from divorced. After all, both involve "molecular conversations" so that what was foe now becomes first a valued and finally indispensable friend. Once again we are perhaps on the verge of discovering more general principles of evolution.

In addition to possessing an immune system, plants are also equipped with defensive compounds. These can illustrate the importance of evolutionary co-option, because in the oats the antimicrobial avenacin turns out to be recruited from an enzyme that otherwise has a completely different function (see X.-Q. Qi et al. *Proceedings of the National Academy of Sciences, USA* 103, 18848–53 [2006]). Plants, therefore, are well-protected, but what they do not appear to possess is that extra layer of protection, known as adaptive immunity. Nor as it happens do most animals; insects, for example, display an impressive system of innate immunity (see F.L. Watson et al. *Science* 309, 1874–78 [2005]; also commentary by L. Du Pasquires on pp. 1826–27) based on phylogenetically ancient immunoglobulins such as *dscam* (short for Down syndrome [DS] cell adhesion molecule; as its name suggests it was first identified in DS in connection with the classic trisomy of chromosome 21 and associated mental retardation [see K. Yamakawa et al. *Human Molecular Genetics* 7, 227–237 (1998)]). Equally vital to the operation of the insect immune system are the toll-like receptors. As proteins these are also very ancient, but among the insects clearly they have been independently recruited for the immune function because in comparison to the mammals there are significant differences in the pathways that are employed (see F. Leulier and B. Lemaitre *Nature Reviews Genetics* 9, 163–178 [2008]). In fact it is a familiar story. In each case there has been co-option followed by independent gene duplications. Using fly as a guide Françoise Leulier and Bruno Lemaitre suggest that employment of the toll-like receptors "is of fungi, N. Kienle et al. (*Biochemical Society Transactions* 37, 787–91 [2009]) confirm the low number of SNAREs in comparison with the animals, but unfortunately do not list the taxa they used.

That SNAREs may be a vital component in the emergence of multicellularity (or indeed other expressions of cellular complexity) reinforces the view that this biological property is an inevitability. As our knowledge of the genomic diversity of "primitive" organisms grows, we can be sure that further examples of how complexity is underpinned by preexisting systems will emerge. To take just one example, consider the molecular machinery associated with what is known as programmed cell death (PCD). Obviously this will be of vital (so to speak) importance in the operation of multicellular organisms. Yet it transpires that the complement of PCD molecules is remarkably diverse, among even the unicellular eukaryotes (see A.M. Nedelcu *Journal of Molecular Evolution* 68, 256–68 [2009]). There is further elaboration, but as Aurora Nedelcu notes, this diversity suggests "that the potential (i.e., the genetic basis) for the complex eukaryotic PCD machinery present in extant multicellular lineages has been established in their unicellular ancestors" (p. 266). It is particularly interesting that in both a unicellular green algae and choanoflagellate, respectively on the path toward green plants and animals, we find "such a complex PCD-related set of sequences . . . [as to suggest] that the evolution of the complex PCD machinery known in the multicellular lineages involved the co-option of sequences already present in their unicellular ancestors" (p. 258). Not, of course, that multicellularity is simply a result of the accumulation of SNAREs or PCDs. Much else will be involved, but no doubt more and more of the necessary building blocks will be seen to far predate their emergences.

2. Animals (and indeed all organisms) face an unceasing onslaught from potential pathogens. Not only have a variety of defense mechanisms evolved, but they have done so repeatedly. The same challenge is met by much the same solution. One line of protection is to employ antimicrobial compounds. These include natural antibiotics that play a central role in the agriculture of ants, but particularly important are the various antimicrobial peptides (see M. Zasloff *Nature* 415, 389–95 [2002]). These show a quite extraordinary diversity, including remarkably stable forms based on a circular arrangement that in the case of certain bacteriocins may, in the words of Marine Sáchez-Hidalgo and colleagues (see their paper in *Cellular and Molecular Life Sciences* 68, 2845–57 [2011]), be "close to perfection" (part of title). Moreover, despite the range of forms, general rules of engagement are still discernible, with the peptide structure arranged so as to ensure an effective molecular engagement with the enemy pathogen. The generality and the range of types also makes it very difficult for the bacteria to mount its own defense, which in these days of mounting resistance of bacteria to antibiotics is an observation that scarcely needs further emphasis.

The antimicrobial peptides are not, however, the only line of defense. Welcome to the peroxidases. Their central role, as the name suggests, is in managing oxidative stress, but it is important to stress that in at least plants they show an extraordinary versatility. As Filippo Passardi and colleagues (*Plant Cell Reports* 24, 255–65 [2005]) remark (p. 255 [Title]), they "have more functions than a Swiss army knife" (a useful catchphrase to indicate the versatility of molecular system; also S.G.J. Smith et al. [*FEMS Microbiology Letters* 273, 1–11 [2007]) for a similar example that addresses the bacterial Omp-A proteins), including roles in lignin formation and nodule-forming in nitrogenous legumes. Significantly Passardi et al. also note that despite a remarkable gamut of "functions [some of which] look paradoxical, the whole process is probably regulated by a fine-tuning that has yet to be elucidated" (p. 255 [Abstract]), epitomizing both the chutzpah of biological systems and their extraordinary fitness to purpose. Nor is it surprising that the peroxidases, and in particular those that bind to an iron-complex (heme), are ubiquitous. Indeed, just as with carbonic anhydrase it is difficult to see how biosphere could function in their absence. Yet they are patently convergent with a classic distinction (see A. Taurog *Biochemie* 81, 557–62 (1999); also K.G. Welinder *Current Opinion in Structural Biology* 2, 388–393 [1992]) between the two major groupings, one encompassing the animals while the convergent other is to be found in the prokaryotes along with the fungi. Evidence also points to at least two other originations, one in the form of yet another family (known as Tyr A, see C. Zubieta et al. *Proteins* 69, 234–43 [2007]) and the other in a thermophilic bacterium (see A. Ebihara et al. *Journal of Structural and Functional Genomics* 6, 21–32 [2005]). The capacity of peroxidases to create a lethal oxidative environment explains their effectiveness as antimicrobial agents, and independently nectar can launch a rocket-fueled knockout punch against microbes using hydrogen peroxide; see C. Carter and R.W. Thornburg *Trends in Plant Science* 9, 320–324 (2004), and C. Carter et al. *Plant Physiology* 143, 389–99 (2007). The crucial point about the peroxidases, however, is that they also play a role in what is known as the innate immune system (see M. Zamocky et al. *Proteins* 72, 589–605 [2008]) of the mammals. In this particular case, although, as one might expect, these peroxidases belong to the so-called animal group, the same type are also found in some bacteria. To be sure, their function in the bacteria is not clear, but it is difficult to avoid the conclusion that given the ancient origins of peroxidases, so waiting in the wings of the evolutionary theater is the immune response.

In detail this innate immunity is an extraordinarily complex system, but in essence it depends on the recognition of specific chemical threats. One such genomic information is inherent in the unicellular forms. As it happens a great deal (see S.E. Prochnik et al. *Science* 329, 223–26 [2010], also commentary on pp. 128–129 by E. Pennisi addressing the origin of multicellularity in the green alga *Volvox*). So Ichiro Nishii and Stephen Miller (*Current Opinion in Plant Biology* 13, 646–53 [2011]) note, it is evident "that the unicellular ancestor of *Volvox* carried a genetic toolkit that prepared it well for excursions into the higher levels of developmental complexity" (p. 651).

One key prerequisite of multicellularity is the need for adjacent cells to stick together, to adhere. Monika Abedin and Nicole King (*Trends in Cell Biology* 20, 734–42 [2006]) rightly stress, "the origin of cell adhesion holds the promise of further illuminating the roles of historical contingency, biological constraint, and chance during evolution" (p. 741). They point out there are several such paths, but so, too, "Cell adhesion mechanisms employed during [*Dictyostelium*] morphogenesis have striking similarities with animal cell adhesion" (p. 738), but in the case of this slime mold, "their cell adhesion molecules have distinct evolutionary origins" (p. 739).

In other respects, of course, animal multicellularity far outstrips that of these social amoeba. Such complex multicellularity as exemplified by not only the animals but also the plants and fungi (not to forget the laminarian brown algae and florideophyte red algae); see K.J. Niklas and S.A. Newman *Evolution & Development* 15, 41–52 (2013). For an overview of complex multicellularity albeit with an emphasis on plants and animals, see A.H. Knoll *Annual Review of Earth and Planetary Sciences* 39, 217–39 (2011). This is certainly less frequent, but once again convergences are the norm. Consider, for example, the systems that have evolved in animals and plants. As ever the programs drew on preexisting arrangements, and interestingly these may have had a special role in defense against viruses and only subsequently were recruited as a key agent in the formation of complex bodyplans (see S.S. Shabalina and E.V. Koonin *Trends in Ecology & Evolution* 23, 578–87 [2008]) to control both the proliferation of cells and their differentiation. Referred to respectively as geminin and GEM, Elena Caro and Cristanto Gutierrez (*Trends in Cell Biology* 17, 580–85 [2007]) exclaim, "it is remarkable that the functional convergence between geminin and GEM has been acquired using proteins without any apparent similarity in domain organization and in spite of the fundamental differences in organogenesis, body plan structure and developmental cues between plants and animals" (p. 584). These writers note that these respective proteins may be "amongst the most relevant for the acquisition and maintenance of complex forms of multicellular organization" (p. 584). So, too, what is often referred to as the "embryonic hourglass" in animals, whereby diverse embryologies are channeled through a similar phylotypic stage, finds a conver-

gent parallel in plants (see M. Quint et al. *Nature* 490, 98–101 [2012]). As Marcel Quint and colleagues note, "we speculate that such a mechanism may be required for enabling spatio-temporal organization and differentiation of complex multicellular life" (p. 100).

In the wider context of multicellularity its inherency is evident from the fact that molecular tool kits once thought to be central to multicellularity have now been widely identified among their unicellular ancestors (see, for example, A. Rokas *Current Opinion in Genetics & Development* 18, 472–78 [2008]). Molecular systems employed in such activities as cell adhesion and cell-cell communication may be central to the multicellularity, but they at least evolved much earlier (see, for example, A. Sebé-Pedrós et al. *Proceedings of the National Academy of Sciences, USA* 107, 10142–147 [2010]). With respect to cell signaling the receptor kinases are evidently of major importance. Intriguingly not only did they evolve independently in plants and animals (see J.M. Cock et al. *Current Opinion in Cell Biology* 14, 230–36 [2002]), but they may have played a crucial role in the convergent evolution of complex multicellularity, as in the brown algae (see J.M. Cock et al. *Nature* 465, 617–21 [2010]).

The receptor kinases were not alone, because it may transpire that the SNAREs (short for N-ethylmaleimide-sensitive factor attachment protein receptors) are also of particular importance. Like other membrane proteins in various eukaryotic lineages they have diversified and undergone gene duplications, but there is an interesting correlation between the expansion of this gene family and the emergence of multicellularity (see T.H. Kloepper et al. *Molecular Biology and Evolution* 25, 2055–68 [2008]; also J.B. Dacks and M.C. Field *Journal of Cell Science* 120, 2977–85 [2007]), including the plants (see, for example, A. Sanderfoot *Plant Physiology* 144, 6–17 [2007]). The correlation is not absolute because some single-celled protistans have numerous SNAREs, while in the fungi the number is relatively small. However, as Tobias Kloepper and colleagues (in *Molecular Biology of the Cell* 18, 3463–71 [2007]) point out, the abundance of SNAREs in *Paramecium* is perhaps less surprising given "its stunningly complex subcellular organization" (p. 3467). This point is echoed by Helmut Plattner (*Protist* 161, 497–516 [2010]) who notes that "the number of SNAREs in a *Paramecium* cell is rather high even when compared with some simple metazoans . . . [reflecting] their considerable structural complexity, particularly in the vesicle tracking system. . . . In other words there occurred a parallel evolution in the protozoan and metazoan" (p. 512). So, too, the data on fungi looks to the simpler forms (see N. Kienle et al. *BMC Evolutionary Biology* 9, e9 [2009]), and it would be interesting to know if the diversity of SNAREs has increased in histological complex forms, such as that extraordinary example of fungal artillery, the basidiomycete *Sphaerolobus*. In a survey of some seventy taxa

NOTES

A note on references: Space here precludes inclusion of titles and a full authorship list, but journal titles are given in full. Some journals are available only online, and, despite individual paginations, usually employ a prefix for the specific article. Prefixes vary, but for simplicity all are listed here as "e." So *PLoS ONE* 8, e69968 (2013) (on the absorbing topic of love darts, pp. 213–14).

INTRODUCTION

1. Richard Grosberg and Richard Strathmann (*Annual Review of Ecology, Evolution, and Systematics* 38, 621–54 [2007]) note this has happened at least twenty-five times, aßnd they remark how "the transition to multicellularity is relatively easy—a minor major transition" (p. 624). Easy they may have been, but apart from laboratory investigations (see W.C. Ratcliff et al., *Proceedings of the National Academy of Sciences, USA* 109, 1595–1600 [2012]) who induced multicellularity in yeast), for the most part they remain frustratingly inaccessible because these transitions to multicellularity occurred in the deep past and in a manner that would be very unlikely to be fossilized.

There are, however, exceptions. Consider the group of freshwater green algae known as the volvocaleans (see, for example, T. Nakada et al. *Molecular Phylogenetics and Evolution* 48, 281–91 [2008]). Here multicellularity has evolved at least three times (see, for example, T. Nakada et al. *Journal of Eukaryotic Microbiology* 57, 379–82 [2010]). Best known are the volvocines (exemplified by *Volvox*), and here multicellularity only evolved about 200 Ma ago (see M.D. Herron et al. *Proceedings of the National Academy of Sciences, USA* 106, 3254–358 [2009]). The beauty of this group is that among the extant forms practically every stage in the transition to the more complex multicellular forms is available. This is decidedly nontrivial: as Noriko Ueki and colleagues (*BMC Biology* 8, e103 [2010]) remark, "How 5000 independent rowers coordinate their strokes" (part of title) of the flagella represents a major shift in organization from the unicellular starting point. The many coordinated developments result in a new sort of entity, not least with respect to the emergence of oogamy (itself rampantly convergent; see D.L. Kirk *Current Biology* 16, R1028–R1030 [2006]) and the basis of the thrashing sperm and expectant egg, as well as the key step whereby the germ cells (that are technically immortal) become separated from their mortal counterparts, the somatic cells (see A. Hallmann *Sexual Plant Reproduction* 24, 97–112 [2011]). One should not, of course, fall into the trap of imaging this transition to ever-more complex multicellularity as being simply monotonic (see M.D. Herron and R.E. Michod *Evolution* 62, 436–451 [2008]). Not only do the intermediate steps themselves represent stable states (see, for example, A. Larson et al. *Molecular Biology and Evolution* 9, 85–105 [1992]), but as importantly even among the more advanced forms the trajectories toward particular systems, such as the development of reproductive cells, are convergent (see M.D. Herron et al. *Journal of Phycology* 46, 316–24 [2010]). Nowhere is this more apparent than in the formation of the embryos. Here the method of cellular differentiation at first results in the flagellate cells lining the interior, hardly a helpful orientation. The trick is to turn the embryo inside out, a process known as inversion. This has a striking similarity to the manner in which many animals engage in so-called gastrulation, whereby the ball of cells that defines the early embryo begins to infold and so define two layers of cells enclosing a cavity that will develop into the gut. Crucially, however, in different species of *Volvox* this process of inversion can be achieved in two quite different ways, but they lead to the same solution (see S. Höhn and A. Hallman *BMC Biology* 9, e89 [2011]; also commentary [e90] by R. Keller and D. Shook). The emergence of multicellularity in the volvocaleans brings into sharp focus how the potentially "selfish" interest of individual cells is melded into a cooperative whole. Influential in this regard is so-called multilevel selection theory (MLS); as Matthew Herron and colleagues note, while the specifics will vary from group to group, "there are basic principles of MLS that are relevant to all such transitions" (p. 3257). There is an equally important point revolving around emergence of multicellularity in *Volvox* and that is how much

He turned again to Mortimer: "Senza dub-bio, tuo amico pensa che tuo viaggio è finito. Qualchi anni qui, forse, e poi l'oblio. Ma si sbaglia totalmente. Io e tu sappiamo perfetta-mente che il viaggio è appena comminciato. Qualchi anni qui, forse, e poi l'oblio. Ma si sbaglia totalmente. Io e tu sappiamo perfetta-mente che il viaggio è appena comminciato" (No doubt your friend thinks that your jour-ney has now come to its end; a few years here perhaps, and then oblivion. How wrong can you be? You and I know perfectly well that the journey has only just begun).

protector, the deity Namagiri.[6] A simple rationalization? I rather doubt it; Ramanujan was stepping into much stranger territory. I don't remember every word Robert Kanigel wrote in his book, but this I do: 'Mystery, magic, and dark, hidden workings inaccessible to ordinary thought; it is these that Ramanujan's work invariably conjures up, a sense of reason butting hard up against its limits.'[7] Wigner says 'miracle,'[8] Kanigel says 'magician.'[9] Tut, tut, that won't please the good folk of North Oxford." Mortimer gazed across the lagoon and then continued, "Other things seem so commonplace that if they weren't actually very odd, then they'd hardly call for comment. Down to earth David Bellamy, you know that entertaining botanist, as a schoolboy sees his granny, large as life. Nothing wrong with that, except she had long been housebound, but there she is standing near the bus stop, in her Sunday black on a weekday. Our boy Bellamy knows straightaway what must have happened."[10]

I looked skeptical. "Well, she had to go at some point. Expect the young Bellamy was simply brooding."

My friend Mortimer arched his eyebrows. "Maybe so, but what about Wilfred Owen, turning up in his brother's cabin, off the African coast, thousands of miles from the Western Front. Harold could tell at once what had happened."[11]

Mortimer fell silent, and both of us simultaneously looked up to see the Milky Way arching over the lagoon. "Nonsense, surely. Just like Victor Goddard, the RAF chap, flying above that airfield near Edinburgh, that's right Drem, back in 1935. Well, our friend Victor had encountered some rough weather, his biplane flung all over the place. Then out in the clear, there he is above Drem airfield, fully functioning complete with mechanics in blue dungarees and not only some biplanes but also a couple of monoplanes, plain as you like. Only snag was that before the Second World War Drem was deserted, half-ruinous, part of a farm. Somehow he, and my experience is that pilots are more interested in staying alive than spinning yarns, had seen some four years into the future to what would be an important training base for the RAF. No Magister monoplanes in 1935, but Goddard saw them. Men in blue, yes in 1939, but when Goddard buzzed the airfield they should have been dressed in brown.[12] And I'll tell you my favorite. John O'Connor, ring any bells?"

"Of course." I replied. "He was the archetype of Father Brown and one of G.K. Chesterton's greatest friends, not that that gregarious saint lacked friends; even his enemies couldn't help liking him."

Mortimer raised his hand. "Spot on, but I think even Father O'Connor was more than a little surprised one Saturday morning to be summoned to the local lunatic asylum and ushered into the padded cell. To offer solace to the inmate? Not a bit of it. The warders pointed to a window, high in the wall, now smashed open. Our inmate had scarpered, not through the door like any mortal but somehow sailed out through that window."[13] Harold Owen on H.M.S. Astræa; Victor Goddard who rose to be an air marshal and to be knighted; Father O'Connor, a priest of transparent goodness, somehow to see them as accomplished liars seems just a little implausible."

"Ara, semo quasi arivai. Grassie, Gianni. Sì, proprio ea" (Well, almost here. Thank you, Gianni, yes, just over there). The boat glided against the jetty and in the shadows I saw the monastery of San Francesco del Deserto. A figure in Franciscan robes stepped forward and warmly grasped Mortimer's outstretched arm. "Ah, Professore Mortimer. Hai deciso e ti aspettavamo" (Ah, Professor Mortimer. You have decided, and we have expected you). He nodded to me and spoke in passable English. "Maybe one day you will trace the professor's footsteps?" He gestured, "Now the water-taxi awaits you. Soon you will be back in Venice."

CHAPTER 27

BACK TO THE LAGOON

IT WAS TIME to leave; the restaurant was almost empty. "Bepi, se no te dispiase, el conto par cortesia. Ottimo, meio che mai. Le vongoe gera squisite. Quatro cento neolire? Ze proprio poco. Sicuro?" (Bepi, would you mind, the bill, please? Excellent, better than ever. The vongole were superb. Four hundred Neoliras? It's really very little. Are you sure?) Mortimer gave another grin. "Well, we came by the ever-reliable Lagune-Norte, but I am afraid my plans are a little different." He obviously saw my surprise, if not alarm. "Don't worry, I have made all the arrangements and even with my little diversion the water-taxi—yes, yes, all paid—will not only take you direct to San Barnaba but much faster."

Leaving the restaurant we turned right beside the canal, crossed on the bridge, and in a couple of minutes were standing on the edge of the Lagoon, the campanile of San Martino looming above us. "Ah Gianni, come petavo, mai in ritardo. Se fa tragheto, o no?" (Ah, Gianni, as expected, and never late. Across the water, eh?), and here Mortimer gestured across the silky water, toward the south. Gianni was one of the experts, and with effective strokes Burano soon dropped away. For a few moments we were silent. "A long story, and even now I am very much feeling my way. To start with, I kept on thinking they must be just tall stories, to draw people round the inglenook, to stir their imaginations, a frisson of possibilities. And if not that, then simply misunderstandings, a trick

of the eye, that sort of thing. I'll freely grant you they make no overall sense,[1] but remember what I said over dinner? How in the final analysis the only interesting thing about evolution is that it is the mechanism by which our Universe becomes self-aware? Bubbling up all around us are other groups embarking on the same adventure—dolphins, crows, elephants." Mortimer fell silent again. "Of course we are one step ahead, on the threshold of previously unimaginable worlds. Yet what a curious notion it is that somehow we have now reached the end of the story. Mathematics alone is an obvious enough clue.[2] I have always been struck by what Eugene Wigner wrote. How did it go? Ah yes, he had just been discussing complex numbers,[3] and then he wrote, "It is difficult to avoid the impression that a miracle confronts us here, quite comparable in its striking nature to the miracle that the human mind can string a thousand arguments together without getting itself into contradictions."[4] Mortimer chuckled, "Can't say a photographic memory is always a blessing, but it can help. And you'll remember Ramanujan?"

"Wasn't he that totally extraordinary Indian mathematician," I replied, "the one who ended up in Cambridge, under Hardy's wing?"[5]

"Quite right," Mortimer replied, "but did his uncanny theorems just emerge from the end of his pencil? Not a bit of it; in his dreams the equations unfolded on scrolls held by his revolution."[144] Macklem is surely correct, and I would hope that the analysis of convergence and the narrow paths of evolutionary possibility will be integral to this renaissance.

Talk of "fields" and "emergence" begs many questions. As Macklem properly notes, "Consciousness is the most striking (and difficult) example of an emergent phenomenon."[145] A similar point was made by Polanyi when he remarked on "the obvious fact that consciousness is a principle that fundamentally transcends not only physics and chemistry but also the mechanistic principles of living beings."[146] Given that elsewhere he remarks "that the higher levels of life are [not] altogether absent in earlier stages of evolution [and] they may be present in traces long before they become prominent,"[147] it seems safer to conclude that Polanyi saw consciousness as somehow emergent. Nevertheless, it may be that the worldview articulated by Macklem and Polanyi (along with many others) has its limits. Darwin's genius was to explain the seamless nature of evolution, but extrapolations can only remain valid if the unfolding story remains embedded in the same world. When it came to the question of our mental capacities and how they were in any way reliable, Darwin found himself on much less secure ground. As he realized, if it was simply a process of extrapolation from material causes, how then can we trust any insight, let alone build vast architectures of thought? Consciousness remains the acid test. Is it not curious that the many and various explanations for consciousness, so convincing to some, have persistently slipped past the attention of the Nobel committee? Concede, however, that what we presently perceive is not all that there is—and indeed we may only be at the earliest stages of our exploration. Then we might make a little progress.

It is time to return to the lagoon.

ble of prevision, capable of retrospection, capable of wittolry.[139]

In the grand scheme of things, neither Muir nor Powys are particularly well-known. Muir is an appreciated poet, but possibly better remembered as the mentor of the gifted and engaging fellow Orcadian, George Mackay Brown. So, too, Llewelyn Powys is perhaps eclipsed by his brothers John Cowper and Theodore, of whom the latter's *Mr. Weston's Good Wine* remains a minor masterpiece.

Otherwise Muir and Powys would seem to have little in common. Neither professed a particularly orthodox religion, but each has articulated a paean to human uniqueness. To the committed materialist, of course, there is only one response: eyebrows arched to their absolute limits, before in accustomed style the head is lowered, held in the hands and slowly shaken from side to side, with accompanying murmurs to the effect that what one has just heard is unbelievable, absurd; indeed, words fail . . .

Beyond the Horizon

The various approaches taken in this, the penultimate chapter, explore the tension that somehow consciousness is emergent as against evolution being a process that not only allows it to navigate to recurrent cognitive solutions but in doing so discovers them to be preexistent realities. At first sight the first approach seems eminently sensible. After all, does not the "primitive intelligence" of the slime molds (p. 237) or hunting abilities of warnowiid dinoflagellates hint at a nascent awareness? Most investigators would hesitate to call such activities "conscious," but one might observe that if the brain is "only" a chemical machine, then it is difficult to deny some sort of continuum with ostensibly much more primitive organisms. The apparent elision in numerosity of sensory and cognitive processes might seem consistent with arguments for emergence, but in reality merely serves to blur a vital distinction upon which our rationality depends.

I would argue that our inability to provide any adequate explanation for the nature of consciousness is paradoxically quite encouraging. We are certainly dealing with unfinished business. Put simply, it is high time that just as Einstein transmogrified

the Newtonian world, so we need to move beyond evolution. Any more than gravity, neither the reality of evolution nor its importance is in any doubt, but we still need a new biology. In effect, the additive approach of the Darwinian paradigm has enjoyed immense success, but its reductionist program has now led us into the sands. Consider Michael Polanyi's[140] remarkable paper "Life's irreducible structure."[141] Here he not only suggests that "both machines and living mechanisms are irreducible to the laws of physics and chemistry,"[142] but as importantly he looks to a hierarchy of higher principles that enables life to transcend its origination. Quite how this is achieved has remained considerably more problematic. Polanyi invokes fieldlike powers, and forty years later Peter Macklem, writing an article titled "Emergent phenomena and the secrets of life,"[143] not only gave an experimental gloss to Polanyi's seminal tract but concluded, "Understanding life requires knowledge of how the design of living creatures and emergent phenomena, appearing spontaneously in self-ordered, reproducing, interacting, energy-consuming, non-linear, dynamic ensembles makes us what we are. I believe this will be the next biological then Shepard's comment, "more puzzling is the relatively sudden emergence of language competency in humans, and musical competencies in humans, birds, and whales,"[137] is all the more trenchant. As already noted with respect to music encountering Platonic worlds, mind may be universal and organized as a reflection of ultimately "abstract mathematical constraints," but the very fact it can understand these constraints points to mind being even stranger than most are prepared to admit.

Whether in terms of numbers, mirror self-recognition, qualia, or the incomprehensibility of death, each leads us toward worlds in which any self-respecting materialist should begin to feel deeply uncomfortable. To him they are tiresome intangibles that must ultimately be specters; they like everything else in the Universe do not actually *mean* anything. Do you really believe that? Let us insist that mind and consciousness are real and evolution is simply the search engine to enter new worlds. Who better to address this question than we humans?

A great deal hinges on who we are. Clearly we are animals, and so brothers to the great apes, and cousins, albeit many times removed, to the bacteria. Like all life we are mortal, but with a couple of possible exceptions, such as among the cetaceans and elephants (p. 293), seemingly only we know it. Yet is this so strange? Should not a dawning self-awareness be exactly what we would expect? The evolutionary landscape is littered with nascent states that we with hindsight see as the forerunners of complex endpoints. Yet this narrowest of demarcations is paradoxically also an immense gulf. Consider what Edwin Muir wrote:

Human beings are understandable only as immortal spirits; they become natural then, as natural as young horses;

they are absolutely unnatural if we try to think of them as a mere part of the natural world. . . . I do not have the qualifications to prove that man is immortal and that the soul exists; but I know that there must be such proof, and that compared with it every other demonstration is idle. It is true that human life without immortality would be inconceivable to me, though that is not the ground for my belief. It would be inconceivable because if man is an animal by direct descent I can see human life only as a nightmare populated by animals wearing top-hats and kid-gloves, painting their lips and touching up their cheeks and talking in heated rooms, rubbing their muzzles together in the moment of lust going through innumerable clever tricks, learning to make and to listen to music, to gaze sentimentally at sunsets, to count, to acquire a sense of humour, to give lives to some cause, or to pray.[138]

Nor is Muir alone. Here is what Llewelyn (Lulu) Powys wrote:

Let these devilish, badger-headed scientists reduce all matter to a series of revolving electrons, it still remains a sublime miracle of miracles that man, with brittle egg-shell skull, should have raised himself out of the dust. To open delicately contrived eyelids on this earth, on this fifth-smallest of the planets, which like a flock of frightened birds keep sailing about the sun, is surely a chance beyond all chances. It would be better to be a midget than a dead stone, it would be better to be a mud-eating lobworm under the ground than a dead stone, better to be a white-bellied beetle in Pitt Pond than a dead stone, and better, how much better, to be a cogitating mammal, firmly set upon his heels, capathree dimensions of representation of colors because there are, in fact, three corresponding degrees of freedom in the world."[129] We belong to a naturally illuminated world, and correspondingly in a nocturnal world drenched in sodium light we are not mistaken when we feel alienated.

To be sure, tricks of perception are familiar and can range from the masterpieces of trompe d'oeil to the vaguely disturbing, as is the case of some of Escher's ingenious impossibilities. Yet Shepard's seminal work provides a vital link between the natural world, our interpretative perception of it, and the accompanying mental processes. While pointing to inevitabilities of biological organization, avoiding the metaphysical implications is difficult. This is not to dispute that what we perceive reflects physical realities, such as the electromagnetic spectrum and euclidean space, and came about because of an adaptive evolutionary process. Nevertheless, we continue to hover on the edge of an apparently irreducible qualia of experience. In addition, as Shepard[130] points out, an empiricist position, exemplified by the philosopher Hume, of a world governed only by sensory experience runs into difficulties. Employing color-blind individuals, Shepard and Lynn Cooper[131] found the expected deficiency (mostly red-green). So far, so good, but in addition, they encountered a clear distinction between what they (and notably in the case of one individual) perceived as against their corresponding mental images of color. Believing, or rather knowing, is not necessarily the same as seeing.

The fact that the world about us is tractable in collapsing sense data into a minimum number of dimensions[132]—as for example the three that permit trichromacy—yet still allows in the words of Shepard a remarkable "structural richness and elegance of internalized knowledge,"[133] points to matters of profound significance. The links between their generality and the internalization of perceptions that allow us to know the real world point to principles of mind that are not only applicable to this planet but are universal. In a thrilling paragraph, Shepard[134] takes us almost to the limits of knowledge. To be sure, in his acknowledgments he thanks the editors for "this opportunity to indulge in unfettered speculation,"[135] and it would be rash of me to conclude that this quotation falls into such a category, but when one reads the following,

Holding even more certainly and under all possible conditions—including those inimical to life—are the abstract mathematical facts of logic, arithmetic, geometry, and probability. Far from being an arbitrary creation of the human mind, such mathematical facts have (in my view) universally held before the emergence of life, constraining what is possible in any world. Indeed, abstract mathematical constraints may have determined not only the form of the universe and its physical laws . . . but also the forms of evolutionary stable strategies, of sustainable social practices, and of the laws of individual thought, whenever and wherever life emerged.[136]

I at least am transported to a Darien Peak, a vast prospect where one might almost imagine two figures ascending from opposite directions and then greeting each other: Plato and Darwin embrace.

In moving toward the identification of universals in human cognition, but greatly doubting such universals can emerge from studying the brain, artificial "intelligence," or indeed cultural relativism (and I would suggest its demonic counterpart of deep political correctness), Shepard invites us to see the evolution of the mind in new ways. We should also note that even if complex cognitive states are inevitable and convergent, existing, but in doing so are independent of ourselves. If nervous (and, for all we know, other) systems are "tuned" to consciousness and qualia, then we can regard them as analogous to "antennae" in the same manner as Gray and colleagues propose for music.

The third line of inquiry that might encourage us to think that mental states are not simply emergent is to review the evidence that, in a very curious way, sensory systems transcend physics. Not that they themselves are anything but chemical machines, but as Roger Shepard[126] argues, the modes of perception "are guided by abstract general principles that reflect the invariant properties of the world in which we have evolved."[127] The basis for the identification of universals that might underlie perception (and so one might suppose extends to mental processes) depends on the recognition of general laws of psychological perception. In other words, while it is perfectly legitimate to look at sensory systems as simply being local adaptations that happen to rely on a given biological design, in reality they embedded in a predetermined architecture where the physical universe intersects with abstract realities.

Consider vision. Roger Shepard argues that this capacity is very deep-seated but is also an inevitable outcome of living in a world that is three-dimensional and euclidean. In other words, it comprises a planetary surface that is illuminated by a star and possesses an atmosphere. This explanation may sound straightforward, even trite, but Shepard's point is that this context governs our interpretation of the world about us and so bears on invariant mental realities. Consider colors. As he emphasizes, it is quite misleading to think that they somehow are "simple."[128] Why, for example, do we perceive the same color despite viewing them in very different types of illumination? So, too, despite red (long wavelength) and violet (short wavelength) being located at opposite ends of the spectrum (grading respectively into infrared and ultraviolet), why do we perceive them as much more similar than the intervening colors? In addition purest examples of a color are readily chosen, which appears to be a more or less universal ability of humans. But not all colors are natural. All agree that to see the world in the glare of sodium lights is pretty horrible, and in England at least we lament that at night our countryside, once a realm across which the mysterious was never very far away, is now more often than not daubed with pools of yellow vomit.

Put simply, Shepard proposes that we are not machinelike spectrometers, but rather engage in mental perception. He goes on to argue that this point can be explained by three chromatic components that can be determined either by cylindrical (thus hue, degree of saturation, and lightness) or rectangular coordinates. The latter may be more intuitively obvious and develop into distinctions between (a) light and dark; (b) longwavelength red versus the remainder of the spectrum, which averages as green; and (c) the opposite to (b)—that is, short-wavelength blue against the average of yellow. Effectively this is a consequence of solar illumination (that regularly fades to night), absorption of this starlight by the atmosphere (again the qualia of a red sunset), and scattering of the light, which is familiar as the blue sky. The evolutionary implications are rather remarkable. As Shepard emphasizes, trichromacy seems to engage with a fundamental order of the world and is not some arbitrary construct. Of course, even among humans there are varieties of color blindness, while in other groups, notably the insects and birds, the register of color vision extends to the ultraviolet. Nor is this to deny that color discriminations have selective value, but here we are dealing with deeper principles, where the physics of light meets perception. As Shepard notes, "the human visual system provides for their equations have a Platonic reality, Gray and colleagues ask, "Is there a universal music awaiting discovery, or is all music just a construct of whatever mind is making it—human, bird, whale? The similarities among human music, bird song, and whale song tempt one to speculate that the Platonic alternative may exist—that there is a universal music awaiting discovery."[121]

Let us then suppose that biological evolution is simply the search engine that allows the universe to become self-aware. Seen from one perspective it simply looks like emergence. By itself this qualifies as a narrative, but scarcely an explanation. We are left none the wiser as to how biological systems actually achieve any sort of mental state, let alone are in a position to make claims that it is comprehensible, let alone rational. If, however, we concede that what we perceive (at least at the moment) is not all there is—that is, there are deeper structures within the Universe—then we might make some progress. That animal cognition has begun to tap these Platonic worlds is consistent with such a view, but there seem to be limits to their intellectual capacities. It would be easy to attribute the differences to our enormous brain, but matters are not so simple.[122] Gerhard Roth and Ursula Dicke observe, "It remains open whether humans have truly unique cognitive properties,"[123] but it seems sterile to dwell on either the overwhelming similarities or seemingly minuscule differences. Either way we have transcended any animal mind, and however fortuitous or chosen they may be, some among us repeatedly provide penetrating insights into new worlds: Wolfgang Amadeus Mozart or Paul Dirac, Samuel Palmer or T.S. Eliot, take your choice.

If correct, this concept has a number of implications. First, however challenging the prospect, we will continue to uncover ever-deeper sets of order in the organization of the Universe. There will be no "End of Sci-ence,"[124] indeed very much the reverse.[125] Second, mental states are neither illusions nor fictions, and although necessarily interpreted through biological systems, they possess their own reality. Here we encounter a problem so fundamental that the usual response is to ignore it. At one level at least our mental world draws on information from the sensory realm. Studded with convergences, often with extraordinary sensitivities that take us to the limits of the physical universe, in whatever modality we choose to investigate—as familiar as color vision or remote as electrosensation—a fundamental contradiction seems to arise. In each case, what is utterly familiar elides into mysterious abstractions. Sensory systems tell us about the world, but whatever approach we adopt we are not a whit closer to providing an explanation for the corresponding qualia of experience.

One might respond that if the world is to be understood at all, it requires some sort of interpretation, but if that is all there is to the matter, then we would seem to have no warrant that what we perceive possesses any reality, let alone has any relevance beyond our species. One might also insist that sensory modalities of which we have no direct experience and can only "interpret" using scientific instruments, as is the case with electroperception and infrared detection, are effectively unknowable. Yet it will be evident that the many convergences in sensory systems, not least at the molecular level of transduction, point to a commonality of processes. I suspect that when a howler monkey sees the color red (p. 114) or a fruit fly tastes sugar (p. 127), they experience the same qualia. Equally I believe there are universes of qualia of which we can have no direct experience, but they too are real and not figments of an imagination. No neurobiological hypothesis exists that explains these qualia. If, however, the analogy of evolution as a search engine holds, then we can regard qualia as not only that this action most likely saved her life. He writes, however, of another instance where elephants broke into a shed in which remains of ears and feet were stored. They then partially buried them, prompting Spinage to remark, "What are we supposed to make of such curious behavior?"[108]

What indeed? As we have already seen in such contexts as the Von Economo neurons (p. 259), toolmaking (p. 279), and mirror self-recognition (p. 292), in their different ways all point to elephants having a far-above-average cognitive competence.[109] Anecdotes abound. Individual instances might raise an eyebrow, such as refusing to drive a pillar into a pit containing a sleeping dog,[110] engaging in painting,[111] plugging their bells with mud to avoid detection,[112] or even committing suicide,[113] but the collective evidence suggests that far from the primates, a mind is astir.[114] But what sort of mind? Benjamin Hart and colleagues[115] make what may be a more than throwaway remark. Thus, they comment as to how in elephants their undoubted capacity for memory does not obviously square with "conventional tests of cognitive performance . . . [and so] brings to mind studies on the rare savant syndrome in humans which is often associated with autism."[116]

The Universals of Perception

This has a tantalizing ring to it, even though it leads to areas that are not so much complex as simply opaque. So-called idiot savants may be exceedingly rare and far from stereotyped, but the astonishing cases of numeric or calendaric calculation, memorization, musicality, and artistic skill are profoundly fascinating.[117] How might we pursue this possible connection? Broadly there seem to be two strategies: one will be entirely familiar but ultimately may be incoherent; the other may sound mad but could be on the right track. The orthodox approach is to remind ourselves that both inherency (in this case for a nervous system) and convergence (specifically for a cognitive capacity) suggest that humans are "just another species." In other words, our undoubted cognitive prowess is effectively a seamless extrapolation from animal intelligence. It is difficult to argue with this point of view. Obviously we are a product of evolution, and the differences that separate us from the cognitive competence of corvids, New World monkeys, and perhaps even octopus are paper thin. "If only they could speak," we might murmur as we ponder our convergent counterparts. But this may be exactly the point. Our cognitive cousins might be able to vocalize, communicate, learn songs, and understand instructions, but language is a step too far. Have we simply traveled a few yards further along the same, if convergent, cognitive highway? Yes and no. We have, but I suggest in a rather unexpected way.

By way of explanation let me begin by saying how I have long been haunted by an essay on animal music written by Patricia Gray and colleagues.[118] They point out that in addition to the birds, animal music has evolved multiple times, perhaps most famously in the humpback whales.[119] In one sense such convergence is scarcely surprising, seeing that it typically involves vibrating columns of air. There is, however, a great deal more to animal music. Gray and colleagues remind us in the case of humpback whales that the "songs are constructed according to laws that are strikingly similar to those adopted by human composers,"[120] and similar remarks apply to the birds. Gray and colleagues then extend their observations in a quite startling direction. Just as many mathematicians believe note, it is unlikely that mirror self-recognition is a binary condition: either you can or you can't. There are what appear to be intermediary stages in both mammals[81] and birds,[82] suggesting "a possible cognitive continuum across animal taxa."[83] Such a capacity may have already been stirring much deeper in vertebrate history, at least if the responses of male cichlid fish are any guide.[84] Self-recognition itself seems to be unlikely, but as Julie Des-jardins and Russell Fernald note, "Clearly, the fish recognize something unusual about the mirror image."[85] Quite what is more difficult to establish, but as the authors also remark, "The mirror image presentation may induce fear . . . because it is a completely novel stimulus."[86] You don't need to be Alice to know mirrors *are* strange.[87] Perhaps with self-recognition comes the one fear that cannot be allayed. That's right, our own mortality.

THE FINAL FRONTIER?

In her autobiography[88] Rosemary Sutcliff hauntingly recalls how, following the death of one of the family dogs, for days afterward the survivor Mike searched high and low for his missing companion, "bewildered and inconsolable."[89] Anecdotes abound of animals showing distress at others' deaths, but few imagine they feel existentialist terror or perhaps deep curiosity as to what happens next. Here humans are almost alone. Certainly the Neanderthal burials[90] hint at the afterlife, but a solemn column of chimpanzees carrying a bier on which their alpha male reposes in a sea of flowers would be cause for comment. Yet if scattered reports, such as the interactions between a mother chimp and her dead infant,[91] are anything to go by, then perhaps there can be inklings of the ungraspable. Chimps are not alone, because among the dolphins in various species (including bottlenose,[92] Risso's,[93] and rough-toothed[94] dolphins), mothers interact with their dead calves, often pushing them to the surface. Significantly, others may be involved, exhibiting such behavior as intense vocalizations[95] or milling around.[96] Nor are these the only cetaceans to engage in this sort of activity, because among the belugas the ferrying of various objects, including bizarrely the best part of a caribou skeleton, may point to a surrogacy for lost calves.[97]

We sense an affinity for our sense of utter loss and helplessness at death, elsewhere articulated most obviously among the elephants. The immense importance of the matriarchs as the repositories of elephant know-how[93] is widely acknowledged. Their reactions to dying members of the clan, be they calves[99] or matriarchs,[100] are also well-known. When we read, therefore, of one instance where attempts were made to lift the dying elephant followed by intense interest in the corpse,[101] then it is difficult to disagree with Iain Douglas-Hamilton and colleagues when they suggest that these actions point to "a general awareness and curiosity about death."[102] All the more intriguing is, as Clive Spinage notes, "When a group of elephants become aware of a skeleton it often elicits great excitement."[103] As he reports, bits of the skeleton may be carted off, and although they might investigate any part of the skeleton, they have a particular fascination for the ivory.[104] Martin Meredith[105] comments as to how "Tusks excite particular interest and are sometimes passed from elephant to elephant."[106] Although elephant graveyards are a myth, Spinage also notes, "It has been known since ancient times that elephants will cover the dead bodies of other elephants and even men that they have killed."[107] And not always killed. He recounts an extraordinary story where elephants concealed a benighted woman under a mass of branches and was so well protected overnight from neural circuits [that] evolved to perform simpler behavior in animals."[70]

Yet, as indicated, numerosity and discriminatory judgments presumably represent the first steps to mathematics. The horse Clever Hans (p. 269) may be discredited, but some elementary mathematical tasks are within the capacity of apes.[71] Given the latter's evolutionary proximity to us, the case of the birds is arguably the more illuminating. Here, too, we see the glimmerings of a mathematical mind, at least in the justly celebrated male African grey parrot, Alex.[72] His capacity to count was first recognized during experiments[73] with another parrot, one Griffin, with whom Alex on previous occasions had had interactions, such as asking him to speak more clearly. Griffin was evidently in a truculent mood, or at least refused to answer when being tested for counting. Even though told to be quiet, Alex interrupted, and subsequent experiments revealed a capacity to do simple sums. Even more remarkable is that this parrot can formulate the concept of "none." Irene Pepperberg emphasized that while this cannot be simply equated with our concept of a mathematical zero (for example, Alex believes that 5 + 0 = 6), it is nevertheless very striking. For example, not only did Alex "invent" this concept without prompting, but it evidently entails a degree of abstraction and may be similar to the way a child understands the idea of "none."

The cognitive basis for this psittacine mathematics is not known, but despite having a brain the size of a walnut Alex seems to be hovering on the verge of a fully abstract world. Or does he? In a way analogous to the sensory qualia, the next step may in fact not be a step at all. In other words we need not doubt that a rudimentary mathematical capacity is a product of evolutionary processes,[74] yet there is a palpable unease as to whether this can even begin to explain an ability to formulate algebra. For example, individuals with brain damage that impairs numerosity may still retain higher mathematical functions. Once again we reach an impasse, with no obvious bridge between the natural processes and the emergence (or discovery) of abstractions. From the materialistic perspective it seems an awkward gap, perhaps to be papered over. To others, however, it simply points to unfinished business.

THROUGH THE LOOKING GLASS

Maybe another clue is when an individual looks in the mirror and says, "Good gracious, that's me!" Familiar enough to us: "Better shave, I suppose; another gray hair, drat it . . ." But what about elsewhere? When it comes to so-called mirror self-recognition, it is much the same story of various groups independently approaching this cognitive threshold and every now and again stepping across an invisible line to find themselves in a world of entirely new possibilities. If reports of mirror self-recognition in some squid (p. 19) are confirmed, this would be one more addition to their roster of convergences with the vertebrates. As far as the latter are concerned, apart from the great apes (and ourselves) the same capacity has evolved in the elephants,[75] and not only the dolphins[76] but other cetaceans.[77] Writing of the dolphins Diana Reiss and Lori Marino note how their "results represent a striking case of cognitive convergence."[78] The capacity for mirror self-recognition has also emerged in the corvids, specifically the magpie.[79] Echoing the point made elsewhere (p. 266), Helmut Prior and colleagues remind us how these reflective birds demonstrate that "a laminated cortex is not a prerequisite for self-recognition."[80]

Indeed not. Such convergences are inevitable, but as Joshua Plotnick aand colleagues erosity remain unknown. In the primates, however, the areas of the brain responsible for numerosity, notably the intraparietal sulcus,[65] have been well-mapped and are equivalent between human and monkey.[66] In the latter group, specifically the rhesus monkeys, the abstract and cognitive processes—ones that point to the pinnacles of human curiosity—appear to elide with a decidedly simpler set of neural principles—or so it would appear. The basis for these conclusions is found in a striking paper by Andreas Nieder and Earl Miller[67] and revolves around what are known as numerical judgments. Here the numerosity per se of the monkeys is not being tested (and in any event is not in dispute), but rather their powers to discriminate in terms of the so-called magnitude effects. In brief, these are separated into two components—numerical distance (that is, the separation between integers, say, 3 against 9 versus 8 against 9) as against numerical magnitude (that is, the size of the numbers, for example, 8 against 9 versus 35 against 40). Intuitively, one would imagine that such magnitude effects would call on both cognitive judgment and linguistic prowess: as already indicated, seeing numbers would not be expected to be equivalent to assessing relative values. When it came to the experiments, however, the ability to engage in both the discrimination of numerical distance and numerical magnitude was not constant; in other words, there was a systematic change in errors of judgment. In effect, as the numerical distance increased, so the ability to discriminate improved, whereas for numerical magnitude as the numbers increase so discriminatory powers worsened. So, 3 against 9 is easier to discriminate as a numerical distance than 8 against 9. Correspondingly, 35 against 40 is less easy to discriminate than 8 against 9.

At first glance, these results may seem to be little more than playing with numbers, but in fact they touch on a fundamental observation concerning sensory judgments, enshrined in the so-called Weber-Fechner law, which asks what is the smallest noticeable difference to enable a sensory process to detect the change. An example of holding a weight in one's hand may help clarify the point. It transpires that there is a systematic change such that 2 ounces are required to be added (or subtracted) from a weight of 80 ounces for me to sense any difference. If, however, the weight is reduced by a quarter—that is, it is now 20 ounces, an addition of ½ ounce will be sufficient, while the case of 40 ounces requires 1 ounce. In brief, the real world (in this case weight) meets sensory perception in such a way that heavier weights require proportionally greater additions to be detectable. In other words, the relationship is not linear, which would be one of simple proportionality, but exponential. To hold the world in your hand and then to be able to detect an additional weight would require proportionally vastly more than if you hold a sparrow.

This Weber-Fechner law is an illustration of how mind meets the material world—the field known as psychophysics. Crucially the monkey's capacities to discriminate numerical distance and numerical magnitude obey this psychophysical law. In the case of numerosity, the implication, therefore, is that far from being distinct there is a continuum of mental processes, from visual perception to cognitive judgments. The former are, like other sensory processes,[68] bound by the Weber-Fechner law, but so are the latter. In one way the fact that we are not dealing with independent systems is no more surprising than any other aspect of evolution, given there must be continuity. As Luiz Pessoa and Robert Desimone[69] write in their commentary on the seminal Nieder and Miller article (note 67), that this represents "one piece of the growing [weight of] evidence that even the most abstract, cognitive operations in humans may borrow basic building blocks abound, including not only those raiding monkeys (p. 281), but groups as disparate as social carnivores[52] and dolphins.[53] This tactic surely arose independently several times, but any doubt about its convergence is banished by the independent evolution of cooperative hunting among the birds. Examples include the loggerhead shrikes,[54] subantarctic skuas,[55] the crowlike pied currawongs,[56] the Harris's hawk,[57] as well as the corvids.[58] Among the last-named group, particularly striking examples are found among the brown-necked ravens. In one such instance, they drove off Egyptian vultures who had used stone tools to break open ostrich eggs (p. 274) to gain access to the contents.[59] More remarkable is where they hunt a large fossorial lizard, the mastigure. To guarantee success, however, one pair stands guard at the burrow entrance, while the remainder move in to attack.[60] Reuven Yosef and Nufan Yosef remark how their "observations suggest that the participating ravens have either communicated and formulated a strategy or have previously practised it to perfection."[61] Once again we dance within the circles of sentience. Plotting crows? These animals evidently know what they are doing, but do they actually understand?

The Convergence of Numbers

All that we see, or perceive by any other sensory modality, must pass through the doors of mentality. In this sense there are no raw data. Even though we interpret the world according to its natural structure, as Roger Shepard has indicated, ultimately these depend on deep abstractions. This view begs profound questions, not least as to the role of cognition and thought versus what we take to be "purely" sensory processes. In a way that seems intuitively obvious, we would want to draw a sharp distinction between the two. After all, even if both are embedded in the brain, surely thinking is not the same as seeing? At one level obviously not, yet it transpires that when we engage in the abstraction of numbers—the portal to mathematics—the processes of assessing such numerosity obey the same laws of psychophysics as do sensory processes.

The ability to differentiate numbers clearly has adaptive advantages: five lions may be worth more attention than one. For many years such capacities were overshadowed by the story of the supposed counting horse Clever Hans, although accusations on the part of the horse of "fraud," if not stupidity,

rather miss the point (p. 269). In fact the ability to assess numbers extends to deep phylogenetic levels. This in itself is tantalizing, not only because it bears on the origins of cognitive abstractions and intelligence, but such numerosity is evidently spontaneous (requiring no period of training) and also nonlinguistic. Just how deep this capacity extends is undetermined. In vertebrates it has been identified in both amphibians[62] and fish,[63] but not surprisingly their numerosity is relatively rudimentary. In salamanders, for example, the task was to choose between relative numbers of a favorite food—in their case fruit flies. Competently they can distinguish between two and three (a similar capacity occurs in year-old babies when choosing, for example, biscuits). In the mosquito fish, the females are highly social, and here too, as long as the numbers were not above four, the fish can discriminate different numbers in a shoal. It would be surprising if the cognitively adept cephalopods do not display at least an equivalent capacity, but among the invertebrates, numerosity is best characterized in the honeybees.[64]

In these cases the cognitive bases for num-

Humans, octopus, and perhaps even bees not only slumber but perhaps join in exploring dream landscapes. Given their tiny brains, the bees' descent into sleep might seem remarkable, but it chimes with their ability to recognize the faces of men.[37] Nor are they alone, because wasps can recognize each other.[38] This capacity not only has itself evolved independently a number of times[39] but has wider implications in terms of cognitive convergences, finding common ground between wasps and mammals.[40] Bees can also discriminate between presented faces with an impressive accuracy, even though these animals have "no evolutionary history

for this task."[41] In their analysis of memory in bees, Shaowu Zhang et al. touch on a central point when they write that irrespective of how much more sophisticated these processes are in the primates, "there seems to be a continuum . . . across the animal kingdom, rather than a sharp distinction between vertebrates and invertebrates."[42] This continuum also begs questions as to just how much neural equipment is required to achieve a high level of organization. Writing of the capacity for memory in the wasps, Michael Sheehan and Elizabeth Tibbetts[43] note, "Perhaps basic components of social cognition may not be as demanding"[44] as is often thought.

PLOTTING

That some sort of mind is astir in the insects, be it in terms of sleep or even tool use (p. 273), should not surprise us. This is neither to claim that their cognitive capacities are identical (although the degree of similarity may yet alarm us) nor that they have the same potentialities. To be sure, they have highly organized social systems, and the quorum sensing of honeybees where the scouts inform and then guide the swarm to their new home[45] is a sight to behold. Some insects can teach (p. 272), but perhaps to engage in plotting is for them even a step too far. But let us consider another cognitive yardstick: when animals go a-hunting.

While opportunistic cooperative hunting is quite widespread, when the enterprise involves two different groups, then this is surely astonishing. A remarkable example of such cooperation is found between two fish, specifically a coral trout (belonging to the genus *Plectropomus*) and a giant moray.[46] The coral trout actively solicits the eel, but either way it is bad news for the prey, because now open water (where the coral trout patrols) and coral reefs (where the moray lurks) are

equally lethal. This is not the only example of interspecific cooperation.[47] Better known is the bird—the greater honeyguide (with the appropriate binomial *Indicator indicator*)—that leads the honeybadger (also known as the ratel) to the bee nest.[48] The badger takes the honey, but the bird prefers the wax.[49] Not surprisingly, native hunters also employ the good services of this bird. But is the honeyguide just a robot? Herbert Friedman notes "that the entire performance is on a purely instinctive level,"[50] and in his view cannot be regarded as "purposeful." This does not square, however, with observations involving the Boran, nomadic people of Kenya, which suggest that information conveyed by the bird in various ways is laden with intentionality, perhaps even to the extent of misleading the Boran hunters if the prized nest is further than "expected."[51] If bird and human mind align, could this be also true for coral trout and moray eel? The roots of intentionality may draw on very deep phylogenetic soil.

If interspecific cooperations are known, it is scarcely surprising that co-specifics will collaborate. Among the mammals, examples our deep sleep. Stefan Sauer and colleagues[17] subsequently remark how "it would be most interesting to investigate whether nightly antennal movements in [bees] can also reflect daytime experience."[18] This need not equate with dreaming, but investigators have been repeatedly struck as to the parallels between the sleep of bees and the equivalents found in dozing mammals (including ourselves) and birds. Unsurprisingly, in bees this is manifest in similar responses to both sleep deprivation[19] (with interestingly a compensatory increase for "deep" sleep without antennal movements[20]) and arousal (at least when compared to cats[21]), and what may be individual idiosyncrasies reminiscent of those among us who get away with substantially less sleep than our snoring partners. Just how far these similarities extend remains to be seen, but they are one more pointer to the inevitability of the emergence of intelligence. Bees may have reached the cognitive pinnacle for the insects, but among the arthropods sleep is widespread. Sleep is known in the scorpions[22] and crayfish,[23] while further afield, pond snails are also known to nap.[24]

The genetic underpinnings of sleep in insects,[25] and the corresponding roles of certain proteins (including enzymes), have at least some similarities to those found in humans.[26] Indeed, the roots of serious disorders such as narcolepsy might be revealed by comparative studies between ourselves and the supposedly humble flies.[27] Nevertheless there is little doubt that as we, and most likely the bees, enter our respective dream worlds, this capacity has arisen independently. Of course, finding similar molecules and pathways—as is evidently the case in the sleep of insects, nematodes, and mammals—may well, as Birgitta Olofsson and Mario de Bono[28] point out, indicate "conserved mechanisms for sleep-like behaviors."[29] Until, however, we have a phylogeny of sleep (and such evidence as is available suggests that sleep is by

no means universal[30]) and know what, if any, molecules have been recruited from unrelated functions, then sleep may have the same architecture but still be convergent. While supporting the view that sleep is phylogenetically primitive, Ravi Allada and Jerome Siegel[31] make the important point that "While it is unlikely that plants or even unicellular organisms manifest the neurochemical and physiological machinery that underlies sleep seen in more complex organisms we cannot exclude the possibility that they may show biochemical precursors of sleep."[32] Such provides another example of that tension in evolutionary biology between inherency versus common descent. They seem almost opposite sides of the same coin, but inherency looks to a deeper structure in biology, where, come what may, the same building blocks will be employed and looking to very much the same instructional manual.

We should not be surprised if the same neuronal chemicals are recruited to ensure that we all get a good night's sleep. Again, to quote Allada and Siegel, "Sleep is largely regarded as a neurally driven phenomenon,"[33] and the similarities between our sleep and that of bees might be another strand of evidence to suggest that our respective cognitive worlds are far more similar than usually accepted. That message should hardly come as a surprise when we discover that the "honorary vertebrate," the octopus, periodically enters so-called quiet intervals that very probably are sleep.[34] The investigators note that in addition to the telltale signs in terms of electrical activity and responsiveness to being prodded, the pupils of the camera eyes reduce to narrow slits. Other investigators have also observed sleeplike states, and as Jennifer Mather[35] notes, "During the sleep time [of the octopus] they also found a specific color change not seen at other times, leading them to suggest a cephalopod equivalent of mammalian REM sleep."[36]

As the Eyes Close . . .

What does the word "cognition" conjure up, other than thoughts of eyes wide open, ears swiveling, noses lifted, all alertness and calculation. Yet one of its oddest manifestations is where much of the system shuts down and the animal nods off. The changes in brain state during sleep are profoundly interesting, not least the intervals of rapid eye movement (REM). One might think, however, that even if one has eyes, a lack of a brain might preclude anything like sleep. Not so; welcome back to the box jellies (or cubozoans), whose complexities we are yet to fully fathom, not least the astonishing evidence that they can sleep. As Jamie Seymour and colleagues[4] report, at night these otherwise lethal animals "lie motionless on the seafloor, with no bell pulsation occurring and with [their] tentacles completely relaxed."[5] If gently disturbed they briefly swim and then subside into inactivity. The fossil record of the box jellies goes back to the Carboniferous, about 300 Ma ago. In principle, to find such animals preserved at all would be surprising, except that these fossils (named *Anthracomedusa*[6]) occur in the extraordinary Mazon Creek biota, famous for its soft-part preservation, with not only box jellies and other jellyfish but a wonderful array of animals ranging from the familiar to the quite bizarre. *Anthracomedusa*, however, is strikingly similar to its living counterparts, and although its eyes do not appear to have fossilized, it seems very likely that at the time of the great Coal Forests, beneath a tropical sky, alert as ever, the box jellies scudded before night fell and then they slumbered as unknown constellations wheeled overhead.

As those constellations slowly reconfigured, so emerging sentiences became ever more acute in those separate descendants of the reptiles, the birds and mammals. Over the millions of years, each fleeting generation registered the passing of days and nights,[7] but so too they slept, but convergently. We and the birds both have independently developed the so-called slow wave sleep,[8] and also the intervals of REM that are typically associated with vivid dreams.[9] The similarities between sleep in birds and mammals are pronounced,[10] yet as Philip Low and colleagues note, the selective basis for "this remarkable similarity of characteristics"[11] remains obscure, although song (p. 266) may tantalizingly hold a clue. Moreover, the fact that slow-wave sleep and concomitant homeostasic regulation are convergent in birds and mammals[12] may be directly linked to members of either group possessing "large, heavily interconnected brains capable of performing complex cognitive processes."[13] Equally interesting, REM sleep evidently correlates with warm-bloodedness (p. 227). Given that this physiology has evolved independently in many insects and they also enjoy intervals of what researchers refer to as "profound rest," is it possible they also dream?

The evidence is tantalizing, especially in the honeybees, which display considerable cognitive competence.[14] Bees show all the hallmarks of sleep, both in terms of behavior and their physiology. Body temperature decreases, while not only do they enter a state of torpor but the bee is indifferent to bright lights. Muscle tone in the legs relaxes, so the bee sinks to the ground and eventually tilts over. One almost expects other bees to appear with pillows and blankets.[15] Most intriguingly, the near immobility of the bee is punctuated by intervals of antennal movement. As Walter Kaiser[16] notes, it is tempting to equate the periods when the antennae are moving spontaneously as equivalent to the intervals of light sleep in humans, and correspondingly times of antennal inactivity as the counterpart to

CHAPTER 26

THE FINAL STEPS

"WHAT," as so many children have asked, "is it *for*?" One such item deserving of scrutiny would be those dignified processions of academics in their glorious robes, with the spectator sport of spotting which political or intellectual monster is being lauded with an honorary degree. Or what about those wild-eyed enthusiasts who have taken the trouble to travel huge distances and present beautifully crafted talks, the central message of which is that all is without meaning. If you believe that, you'll believe anything. In the final analysis, the only reason why evolution is of the remotest interest is neither that it is true (which any idiot can see) nor that is fascinating (why else write this book?), but because it is the mechanism by which the Universe has become self-aware. In doing so it has allowed us to enter previously unimagined worlds. Poetry unstrings our hearts, music ravishes us, myths inspire us. What, though, of the final frontier, that "undiscovered country from whose bourn no traveller returns"? As Hamlet pondered death in his astonishing soliloquy so he murmurs, "To die, to sleep; To sleep, perchance to dream— ay, there's the rub."[1] As we know, sleep can be so deep it almost looks like death, but so too in slumber we visit other worlds that I am now sure are not only of our imagination.[2] Let us take the final steps, from sleep to self-awareness and then death, but this is not the end of the story as we will then see how the worlds of perception intersect with the fundamental architecture of the universe.

Grand stuff, but as mortals we need to know as much where we came from as against where we may be destined. Take intelligence, which, however imprecisely defined, is the obvious precondition of any metaphysical speculation. Simple questions, in part addressed earlier. How deep does intelligence extend, and might it manifest itself in unexpected ways? There are, after all, perfectly sane people who talk about plant neurobiology.[5] It is customary to speak of intelligence as "emerging," but if that is the case, where exactly are the thresholds—and if they exist, then are they gradual or abrupt? Or could this be completely back to front? Suppose mind is not only independent but also preexistent to matter? If that was the case, then evolution is simply the process to discover mind. How then might we distinguish the process of emergence as against discovery? The former seems far more sensible but appears to run into some intractable difficulties. If naturalistic explanations run into a dead end, then does evolution contain any clues as to how the Universe became self-aware? It is the topic of another book, but perhaps a start is to suggest that in the end the natural world lets us down: paradoxically what we don't know is more helpful than what we believe we do know. Let us begin this exploration by turning to a curious phenomenon, sleep.

play in "fish that leap, juggle, and tease."[225] As Burghardt documents, this literally includes "leaping (leapfrogging) of fish over objects floating in the water, such as plants, leaves, sticks, and other animals,"[226] and more unexpectedly, reports of how "great white sharks went into a play mode."[227] The thought of a great white rushing toward you because it is playtime might seem far-fetched, but is it? The cognitive world of sharks (p. 253), and indeed other fish, may be more sophisticated than is generally realized. Although aquatic locomotion is a spur to the evolution of play, so, too, can be intelligence. This may provide the link to remarkable reports of play in the elephantnose fish (or mormyriforms, p. 148). Drawing on the reports of a German doctor, one E. Meder, Burghardt recounts as to how

in the doctor's aquarium the fish not only pushed and pulled objects including snails but "one day, to Meder's astonishment, balanced one on its snout like a sealion. When the snail fell off, the fish would repeatedly rebalance it."[228] As he notes, "mormyrids may be the fish play champions,"[229] which is likely connected to the mormyrid's remarkable brain, not least an enormous cerebellum.[230] Burghardt is surely correct when he indicates that, however surprising, various studies provide evidence for "complex cognitive decision-making in fish suggest that the flexibility and plasticity of fish behavior may have been greatly underestimated."[231] Nor is this the only part of the cognitive realm that may have been underestimated.

destroying large objects,"[206] but when night falls the young hang around for "play and aggression. In one late-night interaction, for example, a sizable group of immature birds formed a screaming circle around a pair of keas that were involved in a particularly vicious brawl, much like gang members gathering around a knife fight."[207] These propensities are evidently underpinned by both the social context of keas and an almost compulsive drive to explore. Intriguingly Diamond and Bond also remark, "The present-day learning strategies of keas evolved during their isolation among the glaciers of the South Island,"[208] where "Their flexibility and cleverness, products of their evolution amid the unforgiving ice and snow, stood them in good stead in exploiting the erratic, patchy food supply of the cold grasslands."[209] Is there an echo here of how we became *Homo ludens*?

In the closely related (albeit flightless[210]) kakapo,[211] although they also exhibit social play it has nothing of the rough-and-tumble of the kea cavorting.[212] Diamond and Bond report how colleagues have "observed hand-reared kakapos to roll on their backs, waving their feet in the air as a play invitation to human handlers."[213] If kakapos resemble the family dog, so ravens ski or at least pass the time of day sliding down snowy roofs,[214] and as Bernd Heinrich and Rachel Smolker report, even doing so "on their backs with a stick held in the feet."[215] The playfulness of ravens and corvids is consistent with their obvious intelligence, but social play is also a hallmark of some other passerines, such as the Arabian babblers.[216] Not only are these birds highly inquisitive, but their play can involve a great deal of rough-and-tumble as well as mad chases. Orit Pozis-Francois and colleagues observed how in the case of a fledgling in company with his older siblings he often tried "to escape a play session that had turned violent, but his sisters rarely let him escape, and continued with what seemed

to us—and apparently to him—like bullying."[217] How sweet.

That birds and mammals both evolved complex play independently seems a reasonable supposition, but can we be sure that at least the kernel of such an ability did not evolve far deeper in the vertebrate tree? What of the reptiles? In their different ways, birds and mammals both evolved from the theropods (p. 193) and synapsids (p. 228), but among extant reptiles the sense of lethargy and cold-bloodedness does not readily conjure up images of them skipping around. But as we have seen, the monitor lizards have a number of intriguing convergences with the mammals. Burghardt not only suggests "that this group of lizards would be a prime candidate for play in reptiles,"[218] but provides a series of fascinating snapshots of the Komodo dragon romping around. Nor, as he points out, are the monitors alone. A report of a young dwarf caiman standing on its hindlegs in a shower[219] might be an eccentricity, but solid evidence exists for turtles at play, as in the pond (or emydid)[220] and Nile soft-shelled varieties.[221] In the latter case, the observations revolved around the behavior of a venerable individual affectionately known by his keepers as Pigface. And the significance of Pigface's disportings? As Burghardt and colleagues note, the evidence points toward his play as being "independently derived when a series of ecological, life history, and physiological factors coincide."[222] What might these coincidences be in the turtles? The key, as Matthew Kramer and Gordon Burghardt suggest, is that "Many turtles are aquatic, thus costs of locomotion are low . . . [and] Most are tolerant of conspecifics."[223]

A more general principle may also be lurking here; as Burghardt also notes, "Since swimming is less costly than terrestrial locomotion, play may be more common in aquatic animals."[224] This might help to explain the largely anecdotal but widespread evidence for winding past one's eyes for a second time: if we were to substitute an australopithecine for a capuchin, how closely would we have to look to see any real difference? Yet hominid tool use remained quite rudimentary for enormous periods of time, and it is a moot point as to when our technology finally pulled past an equivalent degree of complexity to that achieved by the New Caledonian crows. As importantly, tool use is only one milestone on the road to King Cortex. Tools may define our lives, but so does play.[196]

FOOLING AROUND

The exuberance of play—sometimes its edginess and irony, not to mention being helpless with laughter (a perhaps all-too-common occurrence in reading this book)—remind us that at heart the world is deeply comic. Play is also subversive. As Gordon Burghardt remarks, the rough-and-tumble play of children "is largely ignored in the child play literature . . . This may be because it is often too rambunctious for researchers and does not seem to be so connected with the cognitive skills that 'good' play is thought to instill."[197] Such is evident from the manic attempts of ambitious parents to ensure that every purchased game is "educational" before leading them to a sanitized play park where the first graze on the knee ensures highly remunerative litigation. Burghardt reminds us how all too often a "misunderstanding and mischaracterization of rough-and-tumble play by teachers and daycare workers . . . who are often sedentary females, leads to prohibitions against it in many contexts."[198] No matter—in a twinkling of an eye their young charges will be bungee jumping, surfing a widowmaker, climbing granite peaks without ropes, wing-suiting, and freediving. What other species would think these are sensible things to do? This catalogue of adrenalin-packed activities is one more hint that humans may be a whisker away from other mammals, but we now find themselves in new worlds. Even so the evolutionary range of play among mammals remains striking. Curiously among them, play (and here it is customary to consider it as either object-based, locomotory, or social) might be diagnostic, but some groups, such as the xenarthrans, are pretty quiescent. As Gordon Burghardt also mentions, "the wide variation in the ubiquity and complexity of play within these [mammalian] orders suggests play evolved and was later lost or greatly modified repeatedly during the evolution of placental mammals."[199] Just so, but what really matters is whether animal play has deep phylogenetic roots. Did it emerge from the mists of time when the first worm frolicked, or was it invented multiple times? As far as the mammals are concerned, "boxing" kangaroos,[200] and more tentative evidence for platypus play,[201] suggests that their clowning had already started in the Mesozoic. Much more likely is that complex and sophisticated play evolved multiple times in the mammals.[202] It certainly has in the birds.

As Judy Diamond and Alan Bond note in their overview,[203] "Social play in birds shares many characteristics with social play in mammals,"[204] but it is clearly convergent. For the most part it is associated with birds with a reputation for intelligence, notably the corvids, hornbills, and parrots. Among the last group, particularly striking is the New Zealand kea,[205] which not only epitomizes the convergence in terms of categories of play (such as mock wrestling, chasing, and tug-of-war), but as Diamond and Bond note, traits of juvenile delinquency. Not only do "young keas display a seemingly endless appetite for or to serve as knives,[173] not to mention aimed throwing,[174] the discovery that corresponding activities were going on in the wild without attendant humans came as something of a surprise. Nor was this outcome unreasonable, given that, like the other New World monkeys, the capuchins are thought of as primarily arboreal animals. But not only is their tool use spontaneous, but it is most characteristic of dry rain forests, such as the Caatinga[175] and Cerrado,[176] where terrestriality and seasonably available food are the likely spur for this innovation.[177] It is tempting to think that, with all its incalculable consequences, these capuchins are starting to explore the same convergent path as our hominid ancestors.[178]

In exploring these possible parallels Greg Westergaard reviewed not only the then preliminary observations on capuchin tool use, but in drawing attention to their decorative and related capacities pronounced, "I submit that the clay modification skills of capuchins provide additional evidence for evolutionary convergence between *Homo* and *Cebus*, even if mechanisms that underlie capuchin play with paint and clay differ from those that underlie abstract thought in humans."[179] While we tend to think of the "ascent of man the toolmaker" in the context of showers of arrows and thickets of spears, perhaps tool use can spur cognitive complexity in unexpected ways. In reviewing tool use by the Caatinga capuchins, Massimo Mannu and Eduardo Ottoni[180] drew attention to a rather surprising activity: "We also observed the capuchins using 'hammer' stones . . . to pulverize quartz pebbles embedded in conglomerate sedimentary rock. . . . Once the rock was broken the monkey licked . . . sniffed . . . and/or rubbing the face, chest and hands ....with the powder produced."[181] To be sure, about an equal number of pounding episodes failed to pulverize, and even when successful in only a small number of cases did the monkey actually dust itself. Nor is it clear what

this behavior means, but one wonders if, like the birds' "cosmetics" (p. 273), a symbolic mind is beginning to stir? Equally intriguing is that tool use in the Caatinga capuchins is quite diverse; not only do they use stones to dig plant tubers but sticks to retrieve bees and other insects. On one occasion, sticks were used for "threatening [an] opossum,"[182] although an earlier report describes drastic action meted out to a much more dangerous opponent: a venomous fer-de-lance.[183]

The hammering activity to crack open nuts has attracted particular attention. To see a capuchin standing bipedally[184] and about to release a large rock onto a designated anvil[185] is a powerful reminder that, at least in the hands of primates, sooner or later the world was going to change forever. As Elisabetta Visalberghi and colleagues[186] note, "nut-cracking in capuchins likely arose independently of similar behaviors in Hominoidea."[187] These workers also provided evidence not only for transport of stones and nuts[188] but consideration of useful escape routes up trees should the noise of banging attract unwelcome attention.[189] Given their body size, some of the stones the capuchins use are remarkably large.[190] The monkeys are also choosy in terms of deciding which nuts are weaker and which hammer is going to do the job most effectively,[191] not to mention how to employ the available anvils to best effect.[192] The effectiveness of individual capuchins in cracking open nuts varies widely,[193] and as Dorothy Fragaszy and colleagues note, "young capuchins persist at cracking many food items, including nuts, for years before succeeding routinely."[194] As they point out, such an ability can only arise in a social context, if not a tradition, and the importance of not only observation but tolerating scroungers[195] calls to mind the similarly relaxed attitudes among the New Caledonian crows (p. 276).

Seeing these monkeys at work, one has the uncanny sense of watching the tape of life size, feeding types (e.g., folivory), and social organization,[145] nonconvergence is the rule.[146] With the platyrrhines, among the "surprising" absences (or rarities) are socially solitary species (as exemplified by the orangutan) as well as folivores.[147] Peter Kappeler and Eckhard Heymann remind us, however, that for the most part the evolutionary trajectories taken are poorly known, and as importantly they "cannot exclude the possibility that some groups are still on their way towards convergence with other groups."[148] Three lines of evidence from the fossil record suggest that this may be the case. First, while not exact, a series of convergences unite the postcranial skeleton of a number of extant platyrrhines (notably *Cebus* and the atelines) to the Miocene Old World monkeys (catarrhines), with implications for both styles of arboreal locomotion and perhaps behavior.[149] Second, while the living platyrrhines are all relatively small, even in the very recent past, giant relatives in the form of *Caipora*[150] and *Protopithecus* [151] lived in Brazil.[152] In the latter case the animals were twice as large as any extant platyrrhine (c. 25 kg),[153] and conceivably were also folivores.[154] Third, it has been suggested that both *Caipora* and *Protopithecus* were not just arboreal, but at least in part terrestrial.[155] Nor were they necessarily alone. In a way perhaps reminiscent of *Oreopithecus* (pp. 70–71), the Caribbean islands housed a distinctive array of platyrrhines (the xenotrichinids), now all extinct.[156] Among them, *Paralouatta* from Cuba is of particular interest because, as Ross MacPhee and Jeff Meldrum[157] remark in comparison to the other xenotrichinids, this animal "is, if anything, even more distinctive—but in ways that converge to some extent on living Old World monkeys that spend significant amounts of time on the ground."[158]

All very interesting. But there is more. Our closeness to the chimpanzees raises many questions as to what might have been nascent in our common ancestor. One of the most disturbing comparisons concerns cases of incursions by the male chimps into neighboring territories to engage in lethal assaults (followed by embracing among the victors).[159] Although none of the observed encounters were lethal, spider monkeys employed very similar raiding parties.[160] In these cases brute force and teeth do the damage, but as the history of hominids shows, tools do come in handy. So do they in New World monkeys, and perhaps their capacity to crack open a nut is not so far removed from cracking open a skull?

That New World monkeys used tools at all took some time to be appreciated. Apart from a scattering of reports, such as spider monkeys using sticks to scratch themselves[161] and the rather curious account by Cécile Richard-Hansen and colleagues[162] of a howler monkey "using a stick to softly but repeatedly hit a two toed sloth,"[163] the focus of attention is now on the capuchins (*Cebus*).[164] Rightly so, because as Bill McGrew and Linda Marchant[165] note in their overview of tool use in the chimpanzees and capuchins, they "demonstrate some impressive convergences at the interface of elementary technology and laterality of function."[166] This report appeared during a relatively early stage of investigation, but it was already known that, for example, capuchins visited mangrove swamps to chip open oyster shells.[167] Other examples of tool use include bait fishing[168] and fishing for termites[169] with sticks,[170] but much of the research has focused on the capuchins' use of stone tools.

Anybody who has watched a piece of one's molar fly across the room having incautiously bitten on a nut will appreciate that tools do have their uses. So it is with the exceedingly tough palm fruits, some of which such as the piassava would otherwise be inaccessible to the capuchins.[171] Despite their capacities in captivity, including the production of stone flakes,[172] or using bones to break open walnuts living are concerned, a key problem is how to avoid the attention of blood-sucking insects. Commonly used techniques are shaking parts of the body or stamping, and of course in groups such as the bovids, tail twitching.[127] But Asian elephants use tools, specifically the branches, with the advantage that afterward one can always eat one's tool.[128] Not only do the young learn, but Benjamin Hart and colleagues[129] also note, "The modification of branches for fly switching potentially places Asian elephant tool use at the same level as that of . . . great apes."[130]

Most people associate nonhuman tool use with animals such as chimpanzees and orangutans. Yet so close are they to us in terms of evolution that they can hardly be informative as to the likelihoods of tool use evolving multiple times, especially in a context that might presage the emergence of our own lithic technologies. Here other groups of primates are highly instructive, especially the monkeys. At first sight this is rather surprising, as the words "monkey" and "tool use" are seldom associated. In fact, while sporadic, such occurrences represent a repeated reinvention.[131] One such example was an elderly female bonnet macaque who made tools and applied them to various parts of her body.[132] Eyebrows might be considerably more raised with "Monkeys opening oysters,"[133] but this is indeed the case for long-tailed macaques, which employ two types of stone hammer to access oysters.[134] As Michael Gumert and colleagues note, not only does one type involve the employment of a sort of axe hammer (or alternatively an elongate snail known as auger shells) to chip at the edges of the oyster shells, but it requires a degree of precision handling that leads these workers to draw attention to "the convergence . . . in the hand dexterity of macaques and capuchins, despite their difference in hand anatomy."[135] Such a convergence requires the necessary motor skills, which in the macaque involve regions in the brain

known as the parietal cortex (specifically labeled as areas 2 and 5). And what about the capuchins (*Cebus*)? Intriguingly the same two areas have evolved,[136] and as Jeffery Padberg and colleagues note with respect to area 2 in Old and New World monkeys, "aspects of its organization are remarkably similar."[137] This correspondence is all the more striking because although many primates can readily manipulate objects, this convergence of brain structure only finds one other example: ourselves.

Among the New World monkeys (platyrrhines), and notably *Cebus*, we find some of the most intriguing convergences in lithic tools. These monkeys make an abrupt appearance in the fossil record toward the end of the Oligocene (c. 26 Ma).[138] Most likely this was by rafting or island hopping across the then relatively narrow South Atlantic.[139] This diaspora, as Masanara Takai and colleagues note, "might have been a more inevitable event than imagined by many researchers."[140] Once safely on the shores of South America, there is not only evidence for two successive adaptive radiations[141] (with corresponding rounds of encephalization, p. 258), but a third and largely independent evolutionary adventure in the Caribbean.[142] In their new home these monkeys certainly seized the ecological opportunities, but so too particular solutions recurrently emerged. Although very largely arboreal the repeated evolution of locomotion involving either climbing or suspensory activities led to skeletal convergences,[143] of which one striking example is the independent evolution of a prehensile tail (p. 342n12) (in *Cebus* and the atelines).[144] Given their glorious isolation, of greater significance is to what extent did these New World monkeys tread the same evolutionary paths as their Old World cousins?

At first sight, not much. Taking not only the Neotropics but also Africa, Asia, and Madagascar, it seems that in terms of body evolutionary track to indicate how we actually got there? Of course, investigators are quick to point out that much of what we would like to believe is a lucid rationality is far from that, but there is a recurrent sense that the mental world of animals might largely depend on a nifty reconfiguration of long-established capacities. The example of the rooks is a reminder that a cognitive capacity cannot be automatically equated with toolmaking, but given that this capacity is rampantly convergent, it might help to constrain whether stirring in the brains of various groups is at least the inklings of a humanlike intelligence.[108]

Smashing Mammals

Not surprisingly the mammals are especially instructive. Tool use has evolved independently a number of times. Examples range from probable incidentals such as a lion using one thorn to extract another stuck in its paw[109] to the better-known case of some sea otters using an anvil and hammer to smash open seafood while swimming.[110] For the most part, an aquatic mode of life would seem to militate against tool use, but a remarkable exception is found in the dolphins. Best known from Shark Bay in Western Australia, the trick involves placing a sponge onto the rostrum so that it can serve as a glove while rummaging around in the seafloor.[111] Evidently this is not so much to avoid wear and tear, but because the gravels are bristling with a variety of venomous and toxic animals.[112] This capacity has most likely evolved independently on the other side of Australia,[113] but it qualifies as tool use for at least two reasons. First, it is evidently learned in a social context and thus is effectively an example of cultural transmission.[114] Second, it is largely an activity of the females (and their daughters) and seems to be especially prevalent in those dolphins ferreting around in the deepwater channels.[115] While rare,[116] such tool use is hardly surprising given dolphins' cognitive (and convergent[117]) capacities.

This finds a parallel in another group that one might not automatically associate with tools: the elephants. Such a capacity depends on their extraordinary trunk. Not only is it highly tactile (p. 131), but in its ability to pick up very small objects has parallels to the human hand.[118] The elephants are not the only mammals to have evolved a proboscis-like nose. While shorter and less developed in terms of functional versatility, the proboscis of the tapirs, relatives of the rhinos, is convergent.[119] In the tapir the development of a proboscis seems to be linked to olfaction, but in the case of the elephant the proboscis may have served as a snorkel[120] during its early aquatic history.[121] Like the tapir, the elephant trunk is a product of the nose and upper lip, but apart from special cases such as the woolly mammoth entombed in permafrost, it cannot fossilize. Nevertheless, its former presence can be inferred from the anatomy of the fossil skull, including evidence for muscle attachments and blood supply.[122] The living elephants represent only a fraction of their former diversity, but in reviewing their evolution, Emmanuel Gheerbrant and Pascal Tassy[123] note, "The shortening of the muzzle . . . and the flexibility of the typical elephant trunk took place at least five times in the Neogene, in an impressive parallelism."[124]

Whether these extinct proboscideans were also adept toolmakers is speculative, but although typically thought to be a human characteristic, elephants can fling objects with remarkable accuracy and effect.[125] Nor is that the only example of tool use. For elephants, tool use generally revolves around caring for their body, but sometimes it even involves burying the dead.[126] As far as the reminiscent of their natural capacities might help to decide the issue. Take Betty. Her triumph of constructing a hook (p. 275) was extended to test her capacities with not only novel materials but undertaking the reverse operation of unbending the wire. Learning quickly and engaging in complex tasks were characteristically impressive.[91] Alex Weir and Alex Kacelnik, however, remained well aware of the potential pitfalls in assessing cognitive competence and concluded that "it is not yet justified to assume that she possesses a full, human-like understanding of each task and that she uses it to plan and direct her behavior."[92] Other tests echo these conclusions, as experiments in the ability to select the appropriate tool are again impressive, but when new complications were introduced the performances took a nose-dive.[93] When the crows are taken to what must be their cognitive limits—when one tool must be employed to obtain another tool (i.e., a metatool) that will actually allow the bird to recover its reward—the evidence is tantalizing[94] rather than definitive.[95] There remains a divide between those who are unpersuaded that the undoubted successes truly represent examples of analogical reasoning[96] versus other groups who argue that in the crow's brain the problem is being handled in an abstract context of causal relations.[97]

The importance of this divergence of views can hardly be overestimated. All agree that there must be some sort of continuum in intelligences, and there are limits to the mental versatility—party tricks, if you will—of any species. The central question remains: Can these birds think themselves out of the box? Or, as Lucas Bluff and colleagues suggest, is the acquisition of tools effectively an ecological question, so that "the process of adaptation towards tool use may occur mostly through the evolutionary acquisition of motivational mechanisms, rather than by enhancing general intelligence."[98] In part, this

conclusion has been arrived at by considering the cognitive capacity of rooks. Laboratory tests reinforce the long-held supposition of their possessing considerable intelligence. On the other hand, while the string-pulling tests run by Bernd Heinrich and Thomas Bugnyar[99] are consistent with "some kind of understanding of means-end relationships, i.e. an apprehension of a cause-effect relation between string, food, and certain body parts,"[100] such a capacity was far from evident when other rooks were confronted by the trap-tube challenge.[101] More startling is that although rooks are not known to use tools in the wild, given the opportunity when captive they show impressive capacities. These include a confirmation of Aesop's fable, whereby the level of water in a container is raised by the rook dropping stones, thus bringing a worm within reach.[102] A parallel experiment involved dropping stones to release a trap, but in this set of investigations of using stones as metatools (one stone employed to release the necessary stone) and constructing hooks, the experiments seemed to point to impressive cognitive capacities.[103] Nevertheless, the accompanying commentaries raised some cautionary fingers. In his overview of the latter series of investigations, Alex Kacelnik[104] concluded, "The rooks' behavior is truly and unreservedly remarkable, but [the idea of] insight is, perhaps, best left alone."[105] Alex Taylor and Russell Gray echoed this notion,[106] pointing out that in the case of the rooks raising the water level, "At first glance, the way the birds solved these tasks seems remarkably insightful and human-like,"[107] but given their failure to "understand" that using bigger stones would achieve their objective more quickly, there evidently remains a mental lacuna.

So do we encounter other minds, or for some mysterious reason do we stand alone on a cognitive pinnacle that gives us an uninterrupted view of the foothills but no obvious defined along sexual lines (boys chucking spears, girls digging roots).

The convergent parallels between tool-making of New Caledonian crows and humans invite further questions. Do our avian analogues, for example, show any laterality in tool use? Can they employ so-called metatools, whereby one tool is employed in conjunction with another? The answers are intriguing, as they suggest that the corvid proto-technologists seem to be very close to us, but not quite. . . . So distinctive is human laterality in tool use that its evolutionary origins have provoked considerable debate. At first pass, while apes such as chimpanzees can show comparable lateralities, especially those involving cognitively challenging tasks,[75] a direct connection to humans is less obvious.[76] In the vertebrates, cerebral asymmetries can go very deep,[77] but it seems possible that embedded laterality goes hand in hand (so to speak) with sophisticated tool use. So what do the New Caledonian crows show? Laterality, certainly, but of a sort. When it comes to preparing a *Pandanus* tool, there is a very strong preference to use the left-hand edge of leaf,[78] but this applies to the stepped tool, whereas in the case of the simpler narrow and broad varieties, the preference is for the opposite side of the leaf.[79] When it comes to laterality of tool use—that is, according to which side of the head the proximal end of the probe lies— then in the case of both wild[80] and captive[81] crows there is a very pronounced preference in a given individual, albeit in favor of either the right or the left.

We seem to have an echo of a human capacity, but the crucial question remains: What exactly is the crow understanding? To a large extent insights depend on captive birds (naïve or experienced) and placing them in highly controlled experiments. Because of their self-evident skill in tool construction, it is appropriate to begin by considering how well New Caledonian crows fare in classic cognitive tests, such as the string-pulling task. Not surprisingly, in its simplest form the crows generally did well, but in more demanding tests, such as placing an obstruction in front of the bird, success rates fell considerably.[82] Alex Taylor and his co-experimenters remark that these performances are difficult to equate with a capacity of the crows constructing a mental image, writing that they "do not appear to have had any insight into the relation between the string and the reward."[83] What about the trap test? Here only half the crows solved the simple version, and even so it was a protracted process.[84] As the tests increased in difficulty, using so-called transfer experiments where the crows' success depends on their understanding the significance of colored cues, these birds reached their cognitive limit. Alex Taylor and colleagues concluded that while these crows "are capable of reasoning analogically,"[85] and so have some sense of cause and effect when compared to humans, "this reasoning was slower and may have operated at a lower abstract level . . . [suggesting that they] did not infer the existence of a distal causal mechanism (gravity)."[86]

Given the artificiality of such tests and the possibility that the crows may be importing neural routines designed for quite separate functions, establishing the humanlike extent of their cognitive capacity remains a challenging prospect.[87] Lucas Bluff and colleagues[88] remark, in their judicious overview of the problems of linking cognitive states to toolmaking capacities, "When observing fluent tool use by an adult New Caledonian crow, it is hard even for seasoned behavioral scientists to avoid interpreting the behavior in terms of planning and understanding,"[89] but confirming that within their brains there resides a capacity for genuine analogical reasoning is much less straightforward. It is, as Lucas and colleagues say, "an epistemological minefield."[90] Indeed it is. One might hope that experiments that draw on activities calls upon the skilled use of probes. In the case of the candlenuts, however, these trees are the result of human activity and thus a relatively recent introduction. As Christian Rutz and colleagues note,[55] their arrival may have "led to the rapid evolution of tool use in NC crows."[56] A probe, of course, would be little use against a nut, so here the crows employ a rather specific method of placing dried nuts in the fork of a tree before they fall onto the waiting anvil.[57]

Like other corvids, the New Caledonian crow is highly intelligent. They have strikingly large brains (p. 262), and their cerebrotype, with enlargement of specific areas (such as the mesopallium), is consistent with both fine motor control and enhanced cognitive capacities that are prerequisites for the making of complex tools.[58] Observations in the wild and laboratory-based experiments repeatedly provoke cries of admiration as to the skills of these birds in toolmaking and their subsequent employment.[59] This is not to deny that the rudiments of tool construction are innate. Such is evident from the behavior of naïve juveniles,[60] but despite this hereditary component, learning and innovation are clearly key to the tools actually working.[61] Interestingly, unlike some corvids the social structure[62] of the New Caledonian crow revolves around small family groups with both protracted pairbonds and the young receiving extended parental care,[63] not least in terms of feeding.[64] In such a way the tricky business of making effective tools is greatly facilitated. The parents are also notably tolerant with their offspring; while social learning does not seem to involve active demonstration, the juveniles are not only highly observant but spend a considerable amount of time experimenting and will tend to follow in their parents' footsteps.[65] Nor do all end up equally competent; not only is there evidence for a range of abilities, but any degree of success demands a high degree of skill.[66]

The net results can be highly impressive. Apart from the fine sculpting of hooks, the most intriguing activity revolves around the construction of the *Pandanus* probes. Not only does their manufacture involve a complex sequence of steps, but defective tools are rapidly discarded and the end product is remarkably standardized.[67] Even more strikingly the fibrous nature of the elongate monocot leaves makes them suitable for making three types of tool. Most likely the narrow and stepped variety arose as modified varieties of the more primitive wide variant.[68] This wide type shows a further innovation; some crows have learned to fold the tools, a procedure that might help to place the tip of the tool in the central field of vision.[69] Gavin Hunt and Russell Gray note that not only does this reinforce the evidence for "striking flexibility and innovation,"[70] but more crucially it can be identified as a "rudimentary cumulative technology"[71] that not only modifies an existing technique but most likely depends on social transmission. These workers also point out that the abundance of the stepped-tool type presumably reflects greater efficiency in terms of ease of manipulation and resistance to buckling. Hunt and Gray speculate that because the birds rip leaves "presumably attempting to capture prey concealed in the tightly packed leaves,"[72] this might have provided the "inspiration" for their being employed as probes. Nor is it an unlikely gambit if the crows were already using sticks for the same purpose. In one group of crows, not only were both tool types (sticks and leaves) in use (in the case of *Pandanus* employing the wide variety), but many individuals stuck to one type.[73] In other words, this preference, which most likely is socially transmitted, represents a "strong individual specialization."[74] As such it is equivalent to the division of labor connected to different modes of foraging, albeit unlike the typical human equivalent of being even be employed for drumming.[37] Perhaps because of the iconic role as Darwin's finches, tool use in the woodpecker finch (*Cactospiza pallida*) has become well-known. Using twigs and cactus spines, these birds are adept at prizing out nutritious insects, an ability that is evidently of crucial importance during the exigencies of the dry season.[38] The tools are modified before use, but when it comes to a cognitive assessment, while they are good at trial and error, there is no obvious flicker of real comprehension.[39]

THE NEW CALEDONIAN EXPERIMENT

So what is going on in their heads? Certainly there is a correlation between the size of the brain and tool use,[40] with Louis Lefebvre and colleagues noting a "strong concentration of true tool cases in Corvida and Passerida."[41] Regarding tool use among the corvids,[42] perhaps the most remarkable account is of a stand-off between a Steller's jay and a crow, with either bird using a pointed stick (first broken off by the jay) as a lance.[43] Other crows use probes,[44] and it is their use in which the New Caledonian crows not only excel but employ a tool kit, the construction of which seems to stand on the threshold of a convergent technology. Concerning the probes, most typical are stripped-out pieces of leaves from a monocot plant (*Pandanus*), but at higher altitudes twigs are employed.[45] Their principal role is to extract nutritious insects, notably beetle grubs, from the recesses of rotten wood. That these tools are the product of a flexible and innovative behavior is evident because, as Gavin Hunt and Russell Gray[46] observe, these crows "construct a similar functional product from a range of materials and thus suggests that their tool-making is goal-directed."[47] The range of materials drawn upon includes grass stems sufficient to obtain lizards.[48] In addition, many of the twig probes are hooked, and as Hunt and Gray observe, these crows "appear to be the only non-human species that manufacture and use hooked instruments."[49] Nor is this the only way New Caledonian crows employ hooked tools, and these workers describe the deft way in which the birds take a branch and by adept trimming prepare a hooked tool, with the final stage of preparation involving a careful sculpting.[50] These observations were made in the wild, but studies of the versatility of New Caledonian crows also involve captive birds. When it comes to making hooks, few cases are more celebrated than that of Betty, then a resident in Oxford, who after several trials apparently spontaneously bent a piece of wire, the necessary prerequisite to lift a concealed bucket containing appropriate morsels.[51]

So startling are these toolmaking abilities, and so pregnant are they as a possible guide to their convergent evolution in other groups, that they immediately beg two questions: Why did the New Caledonian crows develop this level of expertise, and do they actually know what they are doing? In terms of the former question, both the lack of competition and the fact that the food might be nutritious but not readily accessible may have been preceded by a more woodpeckerlike behavior.[52] Corvids are versatile in their range of diet, but when available will readily exploit carrion. New Caledonia, however, lacks any large native mammals that might be suitable as a source of carrion (although deer and pigs have been introduced), and it seems likely that this lack was one driving force in the evolution of toolmaking.[53] The principal sources of protein (and lipids) for these birds are the beetle grubs and candlenuts, although lizards are not insignificant.[54] All require extraction, which in the case of the grubs and lizards

100,000 years later it is more than likely that the ochre was used to paint bodies.[16] One supporting argument is that the brightest reds seem to be preferred.[17] In due course, pieces of ochre were incised in a pattern that look geometric,[18] while further evidence for symbolism comes from shells ornamented with this red pigment.[19] The oldest-known paintings by humans are younger still, but here too the birds seem to have arrived first, at least if one considers the artistic abilities of the satin bowerbird. Here the males arrange elaborate displays to attract the female, but the satin bowerbird shows a remarkable elaboration of its courtship by using a brush, in the form of a wad of bark, to apply a charcoal slurry to the walls of its bower.[20] The visiting girls seem to prefer fresh paint, but perhaps more for the smell than the sight.[21] Equally startling are those male bowerbirds that not only construct an avenue of stones to entice the females but so arrange them that their relative size and position shifts the entire perspective as viewed by the visitor.[22] Quite how the female is more impressed is yet to be determined, but as John Endler and colleagues note, not only does this hint at "the possibility of a previously unknown dimension of bird cognition,"[23] but it recalls one of the breakthroughs in Western art perhaps reaching its apotheosis in the genius of the Tiepolo family.

Give Us the Tools and . . .

If you get tired of painting, perhaps a spot of fishing? Again the birds are ahead of the game, taking fishing to new heights by the use of baits and lures—except, that is, when crocodiles return the compliment.[24] For example, when fishing, the green-backed heron uses insects and feathers.[25] Even more remarkably Hiroyoshi Higuchi[26] observed this heron deliberately placing a stick beneath its feet before breaking it into two pieces, one of which was then used for fishing. As he remarks, such behavior, which might do credit to a chimpanzee, is rather astonishing given that herons do not usually rate highly in terms of avian intelligence.[27] They are not alone, and of equal interest are examples where the bird literally spreads his bread on the water, with most satisfactory results.[28]

More familiar are those birds that have independently learned to drop stones out of the sky in order to break eggs.[29] Although more of a borderline case, various birds drop their prey.[30] While most familiar in the gulls and crows, Jeffery Boswell also draws attention to earlier reports "that tortoises were dropped in Greece by Golden eagles and were an important part of their diet."[31] Sometimes mistakes were made. As Stuart Kelly[32] recounts in ancient Sicily, one such dispatch had unforeseen consequences; having seized its tortoise, an eagle needed a suitable rock. Once released, then, as Kelly continues, "It was no rock that the tortoise hurtled towards, but the bald head of an elderly Greek named Aeschylus. He was killed outright. History does not record the fate of the tortoise."[33]

Indeed a tragedy, and echoed by those birds that have learned to throw stones, of which the most famous is the Egyptian vulture. While its congeners rootle around in noisome carcass (p. 208), this vulture chucks stones against otherwise inaccessible ostrich eggs.[34] Interestingly, as the only vulture to use tools, this bird also uses twigs, to engage in the time-honored task of wool gathering.[35] In one case the vulture was observed to sweep the wool into heaps using the twig before departing to the nest. The employment of twigs as tools by birds is quite widespread. While typically they are used as probes,[36] they can

CHAPTER 25

PLAYING WITH CONVERGENCE

ANTS ARE NOT humans, but not only can they farm and teach, they also use tools. Eyebrows raised? They shouldn't be, because tool use (and in many cases tool manufacture) is widespread among animals.[1] In ants, tool use revolves around using grains of sand or other debris to soak up liquid food (provided by the thoughtful investigator in the form of a small pool of honey).[2] Several species of ant display this capacity, and in at least the case of a forest ant, learning may be involved.[3] If rocks can be used to absorb valuable food, they can also be employed by a formicid ant (*Tetramorium*) to rain down on the heads of solitary halictid bees sheltering in their burrows, often leading to the death of the bee when they finally venture out.[4] Thus, George Schultz notes, "Upon discovering the bee, the ant normally paused several seconds at the rim of the nest, then wandered over the surrounding area, picked up a small piece of soil . . . [then] headed straight back to the nest entrance . . . hesitated for about 1 second, and then dropped the soil. The ant then waited there for several seconds before going quickly for another piece."[5] Funny what robots can do. Nor are the ants the only tool users among insects, and beyond the ants we find, for example, the employment of a fecal shield (p. 30), closing nests with a pebble,[6] and sand throwing.[7]

A DAB HAND

If ants, or for that matter fish[8] and octopus (p. 20), use tools, when it comes to their employment, birds and mammals win hands down. The range of techniques employed and levels of sophistication shown are a reminder not only of their repeated emergences in either group but the range of versatility. Fancy a bit of painting? A number of birds apply colors by either employing colored secretions from the skin (especially from a uropygidial gland) or deriving a fine powder from their feathers, but more interesting is where they daub themselves with iron oxides.[9] We can call them cosmetics, and as Kaspar Delhey and colleagues point out, "Given that the four main types . . . are not homologous, the use of cosmetics probably evolved at least four times in birds."[10]

Most remarkable are the bearded vultures that apply what is effectively ochre.[11] As J.J. Negro and colleagues say, this is "when art imitates life";[12] evidently this cosmetic helps to confer status, and intriguingly the "wild birds are secretive"[13] as to the location of their ochre baths. While some other animals apply cosmetics, in this department humans literally stand out.[14] It is all the more intriguing that iron minerals were being selected more than 250,000 years ago, and while their use in symbolism is somewhat speculative,[15] some

If the archerfish is the Spitfire of the animal world, how does it compute the various decisions that ensure such pinpoint accuracy? Acute vision is obviously essential, and the retina shows appropriate specializations,[163] but archerfish lack a visual cortex comparable to that of the mammals. Despite this, their visual processing and the all-important role of saliency—distinguishing the things that matter—show striking similarities with the way we obtain our visual representations.[164] Alik Mokeichev and colleagues point out that this capacity for saliency most likely depends on a "preexisting neural architecture,"[165] with the important implication "that no better alternatives have been found during the course of evolution."[166] Whether the archerfish have drawn on these neural functionalities independently remains to be established, but at least in the case of the C-type rapid start maneuver these fish employ a pre-existing program in which reticulospinal networks (the Mauthner cells) play a key role. Saskia Wöhl and Stefan Schuster point out, "It comes as no surprise, however, that the network does hold an enormous computational potential."[167] Indeed, the factors they identify, such as dendritic integration and distributing processing, are all the more plausible because the actual circuitry appears to be extraordinarily economical.[168]

Archerfish provide an object lesson in evolution. Highly specialized, they might be thought of as another of Nature's great one-offs. Not quite. The ability to shoot a jet of water has also been described in a gourami

(*Colisa lalia*),[169] a type of anabantoid fish that is effectively unrelated to the archerfish.[170] So, too, freshwater fish (*Brycon*) living in Costa Rican streams capture fruit dropping from overhanging trees in just the same way as the archerfish—detecting the fall, speeding to the point of impact by employing a C-type rapid start, and correcting for stream drift.[171]

The real importance of the archerfish is its flexible cognitive capacities that, although making it an astonishingly sophisticated ballistic predator, depend on a relatively simple neurology. This observation reinforces the view that mind begins to stir much deeper in the phylogenetic tree than is often suspected. Why are we so surprised? Consider the ant: their mastery of agriculture (p. 181) or gliding (pp. 192–93) are just two indicators of a considerable complexity. Thus to learn that ants can teach falls into a wider pattern. Being eusocial, ants must, of course, be effective communicators, but teaching is more specific because it involves an inexperienced individual (the "pupil") and transmission of useful information in either direction. Exactly this is achieved in so-called tandem running, whereby the pupil runs just behind the teacher, keeping in constant contact with its antenna. This approach enables the instructor not only to guide the inexperienced ant to a particular location but also to adjust mutual speeds.[172] Myrmecoid pedagogy goes further, because the teacher can also evaluate the competence of the pupil and the importance of getting to a particular goal.[173] Nor does the versatility of ants end here.

and many aspects of their behavior point to considerable cognitive capacities.[147] Such pointers would appear consistent with fish feeling pain and quite possibly being conscious.[148] Others, however, remain far more skeptical.[149] James Rose writes, "If fishes have consciousness, their consciousness must be so different from ours . . . we have no idea what it would be like."[150]

To say that this begs a question is an understatement, but the evidence indicates that fish have stepped into some form of the cognitive world. Blind cave fish are adept at forming a cognitive map of their surroundings (p. 128), which has a striking confirmation in bottom-dwelling fish that clearly memorize their surroundings.[151] Wolfgang Wickler notes how in the case of a freshwater blenny, if it is introduced into a new environment, it "shows a particular curiosity—exploring—behavior which is very similar to that of mice."[152] When it comes to recall, perhaps most striking is a reef-dwelling goby. This fish evidently memorizes the topography so that at low tide it can jump accurately[153] from one tidal pool to another, even though they are quite out of sight.[154]

Accuracy, again with cognitive overtones, plays an essential role in the extraordinary archerfish (*Toxotes*).[155] As their name implies, these tropical fish—inhabitants of mangrove lagoons as well as streams—project a jet of water that serves to topple their prey. Notwithstanding various handicaps, not least having to correct for the refractive differences between the water and air, their speed and accuracy are legendary.[156] This process is far from automatic; rather it is flexible and implies cognitive decisions including observation and learning, not to mention the capacity to generalize. As Stefan Schuster notes, the young fish "fire charmingly weak shots and improve their range and accuracy by continuous training."[157] To ensure the killing blow is effective, not only does

the archerfish produce a shot equipped with some ten times the energy actually needed to dislodge its victim, but the size of the shot is adjusted in direct proportion to the adhesive forces that attach their prey to the twig or other part of the tree.[158] This is ingenious, because given that the forces of adhesion follow a universal scaling law,[159] the archerfish can make the necessary allowance. In principle the fish could either increase the speed of the jet or the volume of water. It adopts the latter option, which proves a nifty trick: the amount of energy required to increase the mass of water in the jet depends upon a linear function, whereas if one employs a faster jet, then this is more expensive because the costs increase by the square.

Far from being trapped in an inflexible system, the archerfish makes the best possible use of the physical constraints. The flexibility of the archerfish is, however, most strikingly shown when it comes to being trained to not only hit moving prey, but at different heights. In other words, these fish achieve what many Royal Air Force (RAF) pilots found so difficult in the Battle of Britain—that is, to employ deflection shooting to ensure the enemy plane flew into the cone of bullets. This ability is all the more extraordinary because the fish has to extend its ballistic knowledge to not only three dimensions and a moving target, but make an allowance for the gravitational drag on the ascending jet.[160] Having nailed its prey, the archerfish then has to estimate exactly where it is going to fall and get there ahead of any other hungry fish. Normally, to obtain an express start that combines speed and accuracy, fish employ the so-called S-type rapid start. Counterintuitively the archerfish uses the alternative C-type rapid start, which as the name suggests entails the fish bending its entire body.[161] The net result is to provide an astonishing linear acceleration (up to 120 m s^{-2}) and allow the fish to head in exactly the right direction.[162]

and colleagues stress how this evidence for what appears to be a spontaneous solution combined with evidence for improved technique "make this behavior a likely candidate for insightful problem solving."[130] Yet Nathan Emery and Nicky Clayton point out that "it is not clear that this was the first instance of this behavior, thereby questioning whether this example is a true case of innovation."[131] In a commentary on experiments investigating the cognitive capacity of rooks, in which they rapidly solved a variant of the trap-tube problem but significantly only the female Guillem (out of seven rooks) cracked a more complex challenge,[132] Jackie Chappell[133] pointedly remarks, "But what exactly did this rook [Guillem] understand?"[134] Perhaps nothing? In the parallel case of the parrots, despite their undoubted cognitive capacities, when it comes to the trap-tube test, a variety of parrots (cockatoo, kea, and macaw) are all dismal failures.[135] Quite why is uncertain, but Jannis Liedtke and colleagues remind us that vocalization pathways in parrots differ from those of other birds (p. 266); despite having a large brain, this "enlargement might derive from different demands for neuronal processing capacity."[136]

That a particular neural architecture might preclude a specific cognitive solution is a sobering thought. Nevertheless, this emphatically is not an attempt to push these animals back into the realm of automata. Rooks and other corvids are far from stupid,[137] not least when it comes to tool use. Neither are elephants, but here, also, assumptions of causal reasoning may have to give way to what is known as associative learning.[138] Here the animal can link an action to a consequence (i.e., cause and effect) but is quite incapable of exporting this knowledge to a novel situation. Moti Nissani also points out that in the case of elephants, the capacity for associative learning "does not rule out—but still is not readily reconcilable with—the attribution to

elephants of . . . thinking . . . grief . . . suicide . . . and consciousness."[139] It is all so tantalizing. As we have seen, cognitive capacities show enormous individual variation. Not only are there geniuses like Guillem and duffers such as Fiona, but we need always to ask how much apparent creativity simply reflects a more innate behavior essential for the animal's survival. Amanda Seed and colleagues remind us that impressive as the rooks are, given that they never use tools in the wild, the trap-tube experiment may be less demanding because they "are the only corvids reported to cache food by digging a hole before placing the food inside it and then covering it."[140] Such a sequence of actions might predispose them to passing at least the first part of the examination with flying colors.

Have we then reached a stalemate? I suggest not: Once again the principles of inherency and convergence will serve us well. Both, of course, rely on a comprehensive knowledge of biology. The entertaining cartoon of a rat as the Pied Piper of Hamelin taking an excited throng of animal behaviorists to an uncertain destination[141] has long been superseded, as have the madder shores of behaviorism. All will admit, however, that many stones remain comfortably unturned. Consider the strange history of the raccoon. Investigation of this animal ran foul of various orthodoxies and suspicions,[142] yet the raccoon's playfulness and curiosity are evidently markers for an almost unprobed intelligence.[143] So, too, the phylogenetic roots of cognition raises some uncomfortable questions. Despite the behavior of monitors (p. 232), observations suggesting that in an anolid[144] "the cognitive abilities [of this lizard] are comparable with those of some endothermic species that are recognized to be highly flexible"[145] has proved controversial.[146] And "descending" the phylogenetic line of vertebrates, if lizards are suspect, then what about fish? Here we have already noted evidence for emotional states, the cognitive capacities of animals, investigators have been nervous about employing anecdotal evidence. Even so, criteria exist to judge the reliability of such "travelers' tales."[116] Also, and seldom emphasized, in experiment after experiment we find a range of capacities from the animal Einsteins to what Uta Seibt and Wolfgang Wickler[117] term "duffers." Their particular study involved the string-pulling task, whereby goldfinches and siskins had to recover a bucket of suspended food, a process requiring a highly coordinated action of beak and foot. These authors point out that at least part of this behavioral repertoire may draw on innate feeding methods, but apart from siskins outperforming the closely related goldfinches, Seibt and Wickler emphasized how the range of competences varied enormously. While some birds solved "the task immediately or within a few sessions . . . others seemed interested in the situation but incapable to lift the thimble."[118] Even among the parrots, which by general consensus are a peak in avian cognitive competence, in a similar string-pulling experiment employing spectacled parrotlets, there was still a spectrum of abilities.[119] Poor old Ginger took thirty-two actions (some admittedly repeated) until on the thirty-third he got his reward, while—sad to relate—Fiona and three others never got the hang of the business.[120]

We should, nevertheless, be careful not to dismiss these duffers. Consider the all-too-famous story of Clever Hans, the counting horse. His exposure as an equine mathematician plays the same role for initiates of animal behavior as Little Red Riding Hood does to overimpressionable children. Yet in assessing the vexed question of what it is (if anything) that separates us from other animals, Matt Cartmill[121] puts this hoary story into a refreshing context. To be sure, German psychologist Oskar Pfungst, who showed that Clever Hans relied on subtle cues provided by his trainer, is almost universally applauded for his unmasking of the equine fraud, however unwittingly perpetrated. Yet as Cartmill notes, "All this was widely regarded as a triumph of scientific observation. Yet oddly enough, it was regarded as a triumph for Pfungst and not for Hans."[122] Cartmill points out that Pfungst saw this animal as little more than an automaton geared to the consumption of bread and carrots, continuing, he comments: "it is gratifying to learn that Hans managed to bite Pfungst good and hard at least twice during the course of the study. . . . No doubt Hans mistook him for food."[123]

Horses and badgers, siskins and African greys—the questions of animal cognition remain a minefield of investigation. One difficulty is that among the birds, while corvids and parrots rule the cognitive roost, less endowed groups certainly can impress. Consider the house sparrows that learned how to activate automatic doors to gain access to a cafeteria, and then leave by the same route.[124] To be sure, the group to which the sparrows belong, the passeriforms, may outperform the dove and pigeon group (columbiforms).[125] What, however, are we to make of those pigeons that are dab hands at discriminating between a Monet and Picasso? And perhaps not to everybody's surprise, in the discrimination tasks they then failed when the Monet was hung upside-down, but not the Picasso?[126] In this particular case not everybody was convinced,[127] and in general when it comes to cognitive studies among the birds, the crows lead the field. Here ingenious tests, such as the string-pulling task or the trap-tube test,[128] attempt to establish whether animals actually *understand* what they are doing—that is, do they appreciate the causal chain?

Some evidence seems pretty clear-cut. Take the example of captive orangutans. In order to recover the proverbial peanut, the apes had to carry water in their mouths and then spit into the container so that eventually the morsel floated to accessibility.[129] Natacha Mendes songs are generated as against being heard[108] by different parts of the brain represents a "disassociation [that] has a striking parallel in humans."[109] Once again, obviously convergent, but as Michael Brainard and Allison Doupe remark, "there is little evidence that elements of song [in birds] can be used in a combinatorial manner to convey the rich and flexible semantic content that characterizes language."[110] So close, so very far. If vocalizations cannot provide the answer, perhaps we should turn to sheer cleverness.

CONVERGENT COGNITIVE LANDSCAPES

"Aren't they clever?" we exclaim, as the sheepdogs neatly corral their charges or dolphins engage in complex and coordinated pirouettes. Dogs and dolphins evidently enjoy our company; in some way we have a shared intelligence—or so we like to imagine. We are not too surprised, therefore, to see the glimmerings of mind in not only canids and cetaceans, but other mammals. Sometimes they definitely seem to be more than glimmerings. Consider the Giant otters (*Pteronura brasiliensis*), and specifically a matriarch of one group living on the margins of a lake in Peru (in Manu National Park). Once she was the generalissimo and a skilled hunter. However, as her vision failed, mobility problems began, and the last pups long born the inevitable decline did not lead to her exclusion and an early death, but to the contrary other members of her clan rallied round to provide regular assistance.[111] A clue as to this behavior comes from the fact that among the otters this species is unique in being cooperative breeders: cooperation and bigger brains are by no means unrelated.

Then what about the North American badgers, and specifically those that hunt ground squirrels? Some of their prey are caught in the open, but their burrows provide some protection. Even though badgers are powerful diggers, the burrow systems are well-suited for retreat, given that they form a complex of tunnels with a number of openings to the outside world. The badger's solution? Why, to plug these entrances, employing not only dirt (and even snow) but taking wooden blocks left by careless humans and jamming them into the openings.[112] Such tool use is, of course, one of the leitmotifs for animal intelligence; it might not only imply some sort of forethought and a sense of cause and effect, but it also opens the possibility of teaching and learning, even a primitive cultural transmission. In the case of these badgers, however, Gail Michener finds no evidence for either maternal demonstration or instruction being passed from generation to generation, commenting that "such techniques may remain idiosyncratic."[113]

This is what makes the mapping of cognitive landscapes in nonhumans such a tricky business. Some sort of correlation exists between tool use and intelligence, but other animals clearly possess comparable intelligences and get by perfectly well without any sort of tool.[114] Moreover, by necessity, many of the observations of the cognitive competences of various species depend on their being held in some sort of captivity. Despite the best intentions of the keepers and experimentalists to maintain neutrality, their subjects are also sentient. Some sort of bond is almost unavoidable. That much is apparent from an African grey parrot that had lost its feathers from a skin infection and was provided with a knitted bodice. Unsurprisingly the bird then employed various objects with which to scratch itself.[115] Does this sort of observation have a wider context? It is a fair question, because when trying to assess minor shifts in either the methods of sound production or possibly a genetic change that might mediate some neural change.

Regarding sound production, one important mechanism in the mammals involves the so-called air sacs. These are not to be confused with the ventilatory structures in birds, but they are balloonlike extensions arising from the larynx. These have evolved multiple times,[88] including in the primates,[89] where they can play an important role in vocalizations.[90] Although the australopithecines most likely possessed air sacs,[91] at some point in the hominin lineage they were lost. Could this be connected to our need to refine the process of sounds production, not least in terms of vowels?[92] So, too, presumably necessary was the evolution of our descended larynx? This is of more than passing interest, because this reconfiguration has not only been widely linked to the evolution of speech but considered to be unique—except it isn't. Next time you encounter a roaring lion, you must calmly explain that this impressive performance is only possible with a descended larynx.[93] If your companion is unimpressed with your sangfroid, why not continue the conversation by reminding her that unlike most cats, the lion and related felids cannot produce that soft, subdued, and pulsed vocalization we recognize as purring. As a further distraction from what is now near-panic, why not calmly remark that curiously something very like purring has evolved independently in a number of primates, notably a number of the New World monkeys,[94] where it may again be associated with a context of reassuring friendliness?[95]

It is no accident that just as the deep-throated roar of the lion requires a descended larynx, so its independent evolution in two species of deer[96] and gazelle[97] (and in terms of convergence, more strikingly the marsupial koala[98]) allows a series of booming vocalizations. Nor is it a coincidence that these vocalizations are restricted to the males, with the likelihood that the characteristic resonances (formants) generated in the vocal tract[99] are central to sexual selection, persuading your rivals that your voice matches your size and the attentive females that they are making the right choice. The same arguments have been extended to explain why the human larynx descended, and among the males so to produce what I am reliably informed is a delicious gruffness, unless, of course . . .

However speech emerged, it required not only a specific vocal apparatus and precise control of breathing[100] but a neural template. Could the breakthrough have depended on a genetic change, perhaps in itself trivial? Such has been widely held to be the case, with a fork-head box gene known as *FoxP2* and the dramatic effect that minor mutations have on the capacity for speech.[101] Given the striking molecular convergences of the *prestin* gene in the echolocating bats and whales (p. 143), one might expect a similar story between humans and those other animals that show complex learned vocalizations. Alas, the story is more complex.[102] In the bats, for example, there is a link between degree of mutation of *FoxP2* and their echolocation, as well as some intriguing convergences both within the bats and also with the echolocating whales, but overall no clear correlation exists.[103] Nor should we necessarily expect one. However dramatic the human mutation, when invoking the role of *FoxP2* there is recurrent risk of simplification.[104] The fact that the gene plays some role in birdsong is not in doubt,[105] but as Sebastian Haesler and colleagues remark, it is difficult to "distinguish between [induced] impairments in motor production and motor learning."[106] Whatever is happening with *Fox2P* genes in birdsong, there is little obvious link to any mutations.[107]

This is what makes the many parallels between birdsong and human speech so tantalizing. For example, the manner in which into the brains of birds.[58] Such vocal complexities emerge from a cortical architecture that is strikingly dissimilar in birds and mammals, but from which the same cognitive landscape emerges. So, too, with the vocalizations. In either case, of course, these depend not only on hearing and memory, but also precise motor control, be it of syrinx or larynx. In the birds, this intricate system of input, commands, and feedback is mediated within the brain via a series of so-called vocal nuclei, seven in total. These themselves do not have completely fixed functions, as is evident from the circuitry that allows the babbling.[59] Best documented in the songbirds,[60] much the same neural arrangement is also found in the parrots and hummingbirds. Each is convergent. This is evident from differences in the precise location of some of the nuclei in the parrots,[61] while in the hummingbirds not only are the nuclei somewhat differently arranged[62] but there are also histological distinctions.[63] All these bird brains derive, of course, from a common ancestor, and the recruitment of the vocal nuclei seems in part to have been governed by their proximity to regions of the brain central to the control of locomotion and other motor functions.[64] In order to sing, therefore, birds were constrained as to how they could deploy their brains, but songbirds, parrots, and hummingbirds all arrived at much the same solution independently. In one sense this explains why birdsong and human language converge; both employ similar neuronal pathways.[65] Yet as Michael Farries[66] remarks, "None of the pallial components of the [avian] song system . . . can be unequivocally regarded as homologues of mammalian isocortex."[67] A fascinating tension therefore exists between understanding the neural substrate and its emergent vocalizations, a tension defined by both striking differences and profound similarities.[68] Recall, however, that even if birds and mammals draw on the same neuronal substrate,

the actual pathways (and neuronal activity[69]) are clearly analogous. Erich Jarvis is correct to remind us that "vocal pathways may have evolved out of a pre-existing motor pathway that predates the ancient split from the common ancestor of birds and mammals."[70] The fact remains, however, that while many may be called, "few are capable of vocal learning."[71]

It is tantalizing to speculate why those few were chosen. Factors such as sexual selection and the capacity to mold sound production in different environments are certainly plausible.[72] Yet like language itself, it is still puzzling, as Erich Jarvis notes, that given the likelihood that "vocal learning evolved independently among birds and humans, then it did so under strong genetic constraints of a pre-existing basic neural network of the vertebrate brain,"[73] this capacity did not evolve many more times. Nevertheless, apart from its independent originations in the birds and humans, vocal learning has also evolved in some primates, cetaceans, seals, and perhaps more surprisingly the bats. Of these convergences, among the most interesting are the independent evolution of babbling (as in marmosets,[74] bats,[75] dolphins[76]), dialects (as in cetaceans[77] and seals[78]), mimicry and imitation (as in dolphins[79] and other cetaceans,[80] harbor seals,[81] and bats[82]), and perhaps most importantly complex songs (as in bats,[83] bowhead whales,[84] and most famously the male humpback whales[85]).

Recurrent themes to explain the emergence of these complex vocalizations revolve around cognitive prowess[86] or sociality. From this perspective, the emergence of human language would seem to be one more step. It may not be so simple. Despite the striking parallels with birdsong, Patricia Kuhl observes that "the mechanism that controls the interface between language and social cognition remains a mystery."[87] Would we do better to restrict our comparisons to the mammals? Maybe language arose from some relatively parroting. The fact that a Sri Lankan drongo is adept at imitating both predator and alarm calls[35] allows it, in the words of Eben Goodale and Sarath Kotagama, to "select sounds to mimic non-randomly and use these sounds with high context specificity,"[36] and by implication indicates considerable cognitive capacities. Mimicry is particularly familiar in birds such as captive African grey parrots,[37] where experiments with their musicality suggest a great deal more to their capacities than mere imitation.[38] This link to cognitive capacities is one indication of intriguing parallels between bird vocalizations and human language,[39] but ones that arise despite profound differences in both anatomy and neurobiology.

How much further can we pursue these parallels? Among the songbirds the tunes must not only be learned, but there is a critical interval when this is possible.[40] Just as with our language, while tutors are essential, the actual capacity to learn must be innate. Such is evident from the ease with which canaries isolated from other birds will memorize complex synthetic songs that no bird has heard before.[41] So, too, just as individual children go about mastering their language in different ways, young zebra finches use different avenues to gain their mastery of their song.[42] Wan-Chun Liu and colleagues remark that "we are struck by the fact that songbirds, with brains 1,000 times smaller than those of humans and with a very different evolutionary history, go about vocal learning, nonetheless, in ways that are not all that different from ours."[43] No more striking example of these parallels exists than the fact that just as our children go through an enchanting babbling stage before their language crystallizes, so too do the birds.[44] As Patricia Kuhl remarks in her commentary,[45] "At first glance, communication in babies and birds appears to have little in common . . . [but] a comparison of the ontogeny of communicative repertoires in human infants and avian song birds shows

striking parallels."[46] These acquisitions can only occur in a social context, but as any mother knows, the acquisition of language also involves nonvocal cues. In the cowbirds, for example, the females engage in activities such as wing striking that not only serve as an encouragement to song articulation[47] but have intriguing "Parallels to human speech development."[48] Given these similarities, it is no more surprising to find dialects in some birds.[49] More remarkable are when isolated populations of zebra finches at first produce only a primitive song, but generation by generation more complex variants emerge. As Olga Fehér and colleagues[50] remark, "Our findings resemble the well-known case of deaf children in Managua, Nicaragua, spontaneously developing sign language . . . as well as linguistic phenomena such as creolization."[51]

A popular suggestion is that before we humans spoke, we sang. Here the similarities between our speech and birdsong find a further important parallel. Unlike mammals that employ their larynx, birds rely on the syrinx.[52] This complex organ, in an analogous way to the larynx, both controls the flow of air and how it vibrates. Given its location deep in the bird, it has, however, proved far from easy to study.[53] Tobias Riede and Franz Goller note how the "vocal organs and respiratory systems of birds and humans are very different in their functional morphology. Despite these differences, however, strong convergent patterns emerge."[54] These include not only precise muscular control[55] and air flow, but at least in parakeets a degree of lingual articulation that allows specific resonances (formants).[56] As Gabriël Beckers and colleagues note, this finding "has most interesting consequences, for it indicates that the use of lingual articulation to create an extra dimension of vocal complexity may have evolved at least twice, once in humans and once in parrots."[57]

Just as language is somehow imprinted in our brains, so song is mysteriously woven of logic? Here the worlds that open up may be new, but emphatically they are not simply inventions of minds—human, dolphin, or otherwise. From such a perspective, evolution is not only a search engine by which the universe becomes self-aware but also one that perceives its deep order. As importantly, if such an order is invariant (p. 297), then it is hardly surprising that the routes of discovery turn out to be strikingly convergent. Curiously this is nowhere more evident than in song.

CONVERGENT SONG

For most of us, music is more than a pleasure; it is integral to being human. And other animals? In at least New World monkeys, experiments suggest they seem to prefer German techno music to a lullaby, but given the choice their preferred option is silence.[11] Birds do better, even if the process of discrimination is painfully slow.[12] When it comes to their own sound production, however, birds have few rivals. Almost automatically one thinks of their song,[13] but as ever, biology springs some surprises. Sonations—that is, sounds produced mechanically—have evolved several times. Fluttering of the tail feathers when Anna's hummingbird engages in a dive will produce a tonal chirping.[14] In the case, however, of another hummingbird, the streamertail (which, as its name suggests, has elongate tail feathers) the males' sonation is produced by its wing feathers.[15] These feathers also produce the remarkable sonations that characterize the Neotropical manakins.[16] Here sound production depends on the very rapid oscillation of specifically modified feathers,[17] the vibrations of which lead to a powerful resonance.[18] Such vibrations can only be produced by extremely rapid muscular contractions. Kimberly Bostwick and Richard Prum note that not only do these contractions rival those of hummingbirds, but significantly the "motor requirements for sound production in manakins have pushed them to physiological extremes."[19] This capacity for sonation has evolved multiple times[20] within this group of birds[21] (where it is evidently driven by sexual selection and courtship), but as Bostwick and Prum[22] also remark, "The stridulation mechanism used . . . to produce tonal sounds shows marked convergences[23] with sound production in many insects,[24] including the use of hardened integumentary appendages, extremely rapid limb vibration, and frequency multiplication through pick-and-file stridulation."[25]

What of those sounds that pour out of the syrinx as birdsong? Often harmonious and melodic, some can be extraordinarily complex. Consider, for example, the plain-tailed wren of the Neotropics. While bird duets are quite common, in this wren males and females both not only contribute to a highly synchronized antiphony but as each sex contributes to two of the four-part song, so individuals will drop out or join in almost perfect precision.[26] Nigel Mann and colleagues note, "This must be one of the most complex singing performances yet described in a non-human animal."[27] Related species of wren show a diversity of less complex songs,[28] but importantly in these wrens the songs are not innate (as is the case in some birds), but they must be learned. In the birds, however, this capacity has evolved independently three times.[29] Although best known in the oscines (or songbirds, that is, the corvids[30] and passerines[31]), song has emerged in the hummingbirds[32] and again in the parrots, including the remarkable New Zealand kea.[33] Where song is learned it is not surprising that mimicry is also found.[34] Clearly, however, that this is far from a clichéd

CHAPTER 24

CONVERGENT MINDS

ANIMALS SQUEAK, chirp, click, rasp, purr, grunt, and even—like some academics—boom. That this medley of vocalizations, not to mention other methods of sound production, have evolved independently and often convergently is hardly surprising. Humans also vocalize, but we do so by employing an extremely odd system: language. From whence comes this mysterious ability? There is no shortage of suggestions.[1] In part the solution must lie with social interactions and cognitive capacities. As Tecumseh Fitch[2] rightly stresses, given the importance of convergence in the context of the occupation of cognitive niches, this "leads us to expect convergent evolution of analogous cognitive mechanisms . . . in widely separated species that face similar cognitive problems."[3] Vocal learning and expression (not least as song) will thus emerge repeatedly. The similarities to human language are tantalizing, but also strangely elusive.

CONVERGENT LANGUAGE?

Strange indeed. In some sense there must have been a continuum. The fact, for example, that dolphins are syntactically and semantically competent when provided with a series of gestures (or sounds)[4] echoes the evidence that their characteristic signature whistles serve to convey information.[5] Vincent Janik and colleagues observe that the dolphins employ "these whistles as referential signals, either addressing individuals or referring to them, similar to the use of names in humans."[6] It is a tantalizing observation. Even so, as these workers remark, when presented with synthetic whistles to test their reactions, by no means did all of the dolphins pay attention. Robert Barton[7] points out in his commentary how "there is a danger of slippage . . . between accepting that dolphins can recognize and copy one another's whistles,

and the notion that they are using these calls to refer to individuals, either themselves or others."[8] There are hints that different whistle types convey meaning because they show a frequency of use in particular contexts that cannot but remind us of similar patterns in word use.[9]

Dolphins may be on the threshold of language, but as in the case of tool use (p. 279) an uncomfortable gap remains between undoubted skill and actual understanding. For the most part, biologists are content to leave the metaphysical implications to the philosophers, but the consequences could hardly be more momentous. Just as with their signature whistles, the ability of dolphins to classify objects might point to a capacity for abstraction.[10] But if so, do they then stand on the portals of rationality, not least in terms requires a new region, the subventricular zone (SVZ). That is exactly what some mammals have done,[154] but evidently independent of the birds.[155] Moreover, just as the SVZ is thicker in the primates, so it is in both parrots and songbirds, but in the latter birds once again this arose independently.[156]

One can also ask if specific sorts of complexity are convergent. The auditory cortex of birds has a strikingly mammal-like arrangement of laminations and radial columns. Homology seems the obvious option,[157] or is it? Possibly not, given that the cell circuitry of the respective cortices differs in various respects. Also, as Yuan Wang and colleagues point out, that bird and mammal share a specific type of neuron in one layer with a "unique somatodendritic morphology"[158] could be significant, but the undoubted convergence of Von Economo neurons (p. 269) is a reminder that specific neural demands can lead to precise neural solutions. One can make a strong argument that the road to cortical complexity in birds, and all that implies, was quite independent of the mammals. Onur Güntürkün adopted just such a stance.[159] He notes that the mammalian prefrontal cortex and its avian equivalent "show an astonishing degree of resemblance in terms of anatomical, neurochemical, electrophysiological and cognitive characteristics."[160] He sees this, however, through the lens of convergence,[161] suggesting that the "freedom to create different neural architectures that generate prefrontal architectures seems to be very limited,"[162] with the corresponding inference that it is "likely that there exist only very limited neural solutions for the realization of higher cognitive functions."[163]

These "higher cognitive functions" are remarkable (p. 276), but as with the mammals, the degrees of avian encephalization varies. One obvious constraint is the demands of flight on body weight, yet perhaps surprisingly among the flightless birds (which have evolved multiple times, p. 208), brains get no larger.[164] Indeed, in the remarkable New Zealand parrot, the kakapo, as well as the extinct Giant auk, the brains are conspicuously smaller.[165] Even the now-vanished Moa was decidedly average, with interesting implications for its intelligence and lack of sociality[166] that may have been less than helpful as the newly arrived Polynesians settled down to Moa Twizzlers and similar delicacies. Not so, however, in its flightless counterpart, the kiwi (p. 208). This palaeognathan bird shows a remarkable encephalization, independent of the neoavians, and perhaps linked to its nocturnal existence, sophisticated tactility (p. 130), and territorial nature[167] (thus recalling the monotreme counterparts in the form of the echidna).

In the case of the kiwi there is no obvious connection to cognitive functions, but among the convergent encephalizations of the neoavians, intelligence and big brains seem to march hand in hand.[168] The principal players need few introductions. Take the parrots. Andrew Iwaniuk and colleagues[169] note how the parrots (or psittaciforms) "tend to have significantly larger brain and telencephalic volumes than non-passerines,"[170] although they caution that not only do birds like the hornbills and owls have larger brains, but an increased size need not be linked automatically to cognitive prowess.[171] Nevertheless, in the parrots the correlation is convincing, and Georg Striedter and Christine Charvet[172] remark how they "closely resemble primates, which also tend to be highly intelligent, large-brained, and endowed with a proportionally large telencephalon."[173] Corvids match, and perhaps exceed, the parrots in intelligence (p. 292), and as a group they are encephalized, with the remarkable New Caledonian crow having a conspicuously large brain.[174] If a convergent mind is stirring anywhere among the birds, it is surely in the crows.

brain was established long before the amniotes, so the fundamental questions revolve around the paths of neural elaboration and how similar they were. In most respects the brains of birds and mammals differ markedly, but the traditional idea of the bird brain evolving by a massive elaboration of rather primitive areas has been thrown out of the window. In effect, most of the brain has been remapped, leading to a series of radical redefinitions.[140] Most significant in the birds is the cartography of the forebrain (the telencephalon), notably an area known as the dorsal ventricular ridge (encompassing a number of divisions such as the nidopallium). This has an arrangement quite different from the classical six-layered neocortex of the mammals. There is, however, no doubt that either region is functionally equivalent. As Anton Reiner and colleagues[141] note, "the avian hyperpallium, nidopallium and arcopallium, although not organized cytoarchitectonically into layers like mammalian neocortex, perform the same type of neural operations at the cellular level as mammalian neocortex. . . . Increase in the size and regional specialization of the neocortex in mammals and the pallium in birds appears to be the common underpinning of the emergence of complex vocal and cognitive abilities."[142]

At this point, opinion abruptly polarizes. Bird and mammal brains obviously differ. Does this, however, conceal a fundamental homology, or did the birds effectively evolve an independent neocortex?[143] The importance of this hardly needs emphasis, because on it hangs the question as to whether the cognitive landscapes of bird and mammal are convergent[144]—and, by implication, consciousness also.[145] At first glance, it seems a pretty open-and-shut case, but as ever with concepts of deep homology, the devil is in the details. When it comes to the expression pattern of key developmental genes, some correspondences are certainly striking.[146] Even when

equivalences are identified, such as between the Wulst area in the bird brain and corresponding mammalian neocortex, various similarities such as a lamination are almost certainly independently derived.[147] Overall, convergence appears to win the day. Expression patterns may be shared, but they are far from one to one.[148] Crucial in this regard is the role of a large extracellular protein. Known as *reelin* it is so named because in the mutant mouse, the animals stagger around or reel. In striking contrast to the birds, *reelin* in mammals plays a central role in determining the cortical architecture as well as the highly distinctive pyramidal neurons.[149] Tadashi Nomura and colleagues also note that other brain proteins (specifically *Er81* and *Brn2*) have "Expression patterns . . . in the quail pallium [that are] extremely different from those in mammals."[150] They remind us, however, that what amounts to profound differences in cortical architecture "might be provided by relatively small genetic differences effecting the timing or level of gene expression during early embryogenesis."[151]

We need also to remember that the various intermediates that are conjured up to connect birds and mammals, such as lizards and turtles, have been pursuing their own evolutionary paths for hundreds of millions of years. They are not necessarily keeping their brains in phylogenetic cold storage for our benefit. In the lizards, the employment of the protein *reelin* for cortical elaboration recalls the situation in mammals but finds no counterpart in other reptiles nor, as we have seen, the birds.[152] The protein *reelin* itself is far more ancient,[153] and while nobody doubts that the template for neuronal elaboration was established at the dawn of vertebrate history, what matters are the paths taken. Key to growing any sort of brain is the method of cell proliferation. Common to all vertebrate brains is the so-called ventricular zone, but if one wants an enlarged forebrain (or telencephalon), one bottlenose dolphins, "the cytoarchitectural patterns . . . are far more varied and complex than generally thought."[126] Given the importance of echolocation (p. 140), it is also plausible that brain enlargement, especially of the auditory area, was driven by these demands[127] and, as Stefan Huggenberger suggests, by the "need for greater precision and speed in processing sound due to the increased speed of sound in water compared to air."[128] Nor was this the only driver. Growing skills in com-munication and sociality probably provided further impetus, and Huggenberger suggests that dolphin encephalization can be viewed as "a neurobiological alternative to the situation in primates."[129] Here, in a milieu utterly different from ourselves, with a brain that has evolved in a radically different way, the path leads to the same cognitive solution. Mind as an attractor? If, however, you really wish to appreciate the power of such an attractor, take a look inside the skull of a bird.

BIRD BRAINS

From the time a dinosaur stood on its hind feet, shook its splendid scales, and called itself a theropod, birds were an evolutionary inevitability. It is also a commonplace that as far as cognitive capacities are concerned, some birds rival, and possibly outstrip, those of primates.[130] Indeed, is it time to turn the tables? Karin Isler and Carel van Schaik[131] ask, "Why are there so few smart mammals (but so many smart birds)?"[132] Big brains,[133] they argue, are so expensive that only if the mothers receive assistance with provisioning their insatiable young can the species escape the trap where the rates of growth (and so ultimately the ability to reproduce) would otherwise become so reduced that soaring death rates would make extinction a real threat. Be they the fathers or co-specifics, such cooperative breeding in the birds—that is, altricial—is not only the norm but evidently convergent from the more primitive precocial breeding.[134] Robert Ricklefs and Mathias Starck note that it has long been apparent that this cooperative context permits "a large postnatal growth increment of the brain and consequent large brain size in the adult."[135] Altricial development may also correlate with an enhanced degree of cerebellar foliation[136] (p. 255). At least in part the reason why there are "so few smart mammals" is that the altricial analogue of coop-erative breeding is uncommon. It provides, therefore, a clue as to why groups such as callitrichid monkeys or elephants (and even dogs) show enhanced cognition, but also has a bearing on us.[137]

Just as birds are an evolutionary inevitability, so too are some species with large brains and intelligence. The stirrings of such encephalization are evident among the preavian theropods (p. 198). Although scanty, evidence from brain endocasts suggests that by the Eocene,[138] birds such as *Prophaethon* had brains little different from those of living birds.[139] Such elaboration, however, arose from a dinosaurian template. The bird in your hand once sat beneath a bush, or more likely an enormous lycopod, because birds and mammals both trace a common ancestry to the very dawn of the amniotes. From about the late Carboniferous (c. 300 Ma), their evolutionary stories have been independent but rich in convergences. What applies to features such as warm-bloodedness (p. 227), tactility (p. 130), and even orgasms (p. 212) applies to brains. Or does it? How one views the evolution of the bird brain defines a major fault line in biology—shared ancestry or convergence, or if you prefer, inherency versus inevitability?

The basic architecture of the vertebrate

map onto the cognitive world of the elephant, but as Bob Jacobs and colleagues remind us, to a considerable extent, "the relationship between cortical architecture and cognition remains unclear."[98] One vital clue, however, lies in the recognition of highly distinctive neurons with a spindle shape, the so-called Von Economo neurons (VENs).[99] These large bipolar cells have attracted considerable attention; as Atiya Hakeem and colleagues note, they are evidently "part of the neural circuitry involved in social awareness and may participate in fast, intuitive decisions in complex and rapidly changing social situations."[100] As importantly, VENs are patently convergent and are found not only in us (as well as the great apes[101]), but a scattering of other mammals as well,[102] not least the cetaceans.[103]

The cognitive capacities of a number of cetaceans, most famously the dolphins, require little emphasis. How then does the evolutionary story play out with their brains?[104] Helmut Oelschläger[105] reminds us that when it comes to the dolphin brain, it presents "A challenge for synthetic neurobiology."[106] As he also notes, this brain "is characterized by a series of extreme adaptations, ranging from hypertrophy to massive reduction and loss of structures."[107] Either way one looks at it, the dolphin brain and that of the related toothed whales (odontocetes) is fascinating. From the vantage point of other giant brains, the similarities delineate the convergent pathways, not least at the molecular level.[108] Yet the differences are, in a way, more critical.[109] As we will see, irrespective of neurological architecture, the same cognitive landscape seems to emerge. At this higher level, and mysteriously through the lens of consciousness, cognitive convergences suggest that we all understand the world in the same way.

Among the toothed whales the dolphin brains have received the greatest attention. As in elephants the cerebellum is disproportionately large, which may reflect not only the demands for exquisite motor control[110] (think of dolphins remarkable coordinated swimming), but also the inputs from echo-location (p. 141) and cognitive processes.[111] In terms of convergence there is extensive development of the neocortex, but it is also extremely folded.[112] As Helmut Oelschläger and colleagues emphasize, it forms "deep and complicated gyri and sulci,"[113] as it does in other odontocetes such as the killer whales.[114] What of the differences? Particularly striking is that the area that serves to connect either hemisphere of the brain, the corpus callosum, is reduced.[115] This makes sense, because the dolphins exhibit so-called unihemispheric brain function, the most extraordinary capacity of which is for either side of the brain to take turns in sleeping.[116] Alternately one eye remains open and the other closed, perhaps to assist in the breathing cycle as well as keeping an eye out for friend or foe.[117]

Because unihemispheric sleep has evolved independently in some other aquatic mammals[118] (not to mention the birds[119]), it might be thought that the dramatic story of cetacean encephalization is largely linked to the return to the oceans in the early Tertiary. This is evidently not the case,[120] because the earliest whales (archaeocetes) have brains no different in size than the relatives they left standing on the shore.[121] Interestingly, encephalization was a two-stage process. The first major step was taken in the archaeocete-odontocete transition, which was followed by a further jump in the dolphins.[122] The latter increase was millions of years before that of the hominids,[123] but the idea has persisted that despite dramatic encephalization (and its cognitive implications), in a number of important respects the dolphin brain has remained rather primitive.[124] This latter view has obviously been difficult to square with their intelligence (p. 292). More recent investigations[125] by Patrick Hof and colleagues indicate, however, that at least in the

Simons and colleagues note, the cranium in question "provides yet another striking example of detailed morphological convergence between catarrhines and platyrrhines."[79] Conversely we find the counterpart of this convergence in the platyrrhines, whereby forms such as the Miocene *Chilecebus*[80] and *Homunculus*[81] also have small brains, yet in near contemporaries (as in *Kallikaike*) encephalization is well under way.[82] Indeed Richard Kay and colleagues[83] note how their studies "suggest that brain enlargement has occurred at least four times independently in Platyrrhini."[84] These parallels between catarrhine and platyrrhine encephalization have further ramifications, because in either case a common thread emerges. Although visual acuity appears to have evolved early on, the real driving force was how acute eyesight was employed for tasks such as either avoiding predators or growing manual dexterity. As Kay and colleagues remark, this convergence of encephalization among the monkeys and apes (collectively the anthropoids) can be seen as the "product of parallel selection for increased brain size in multiple evolutionary lineages."[85] It is difficult to avoid the conclusion that from the moment the primates emerged, big brains were an inevitability, with the curious consequence that one species would take an abiding interest not only in its own origins but in its many evolutionary cousins where intelligence repeatedly bubbles to surface.

What of the placentals other than primates? Here, too, there seem to be general rules of engagement.[86] Given multiple episodes of encephalization in the anthropoids, it would be surprising if in this respect they were alone. They are not. In the canids, for example, George Lyras[87] noted that "brain evolution followed three independent, yet convergent paths."[88] As each of the three groups underwent its successive radiation, so the most recent (the caninids) have the most complex brains. Among the hypercarnivores (pp. 54–56), there appears, however, to be a trade-off between brain size and accentuated jaw apparatus leading to a "stasis in brain evolution during the development of hypercarnivorous dietary adaptations."[89] Given the massive expense of running a nervous system (p. 241), if other parts of the body are also making huge demands, as has also been argued for brain size and testes in bats,[90] then this might be an important constraint on how large a brain can get.[91] Not, of course, that any single brain can do everything; even if some brains allow the universe to become self-aware they must still have evolved in specific functional contexts. A perhaps unexpected consequence is that the inevitable convergences lead to the discovery of the same cognitive worlds. Documenting these routes to extreme encephalization will draw us to the coda of this book. These occurrences may be highly sporadic, but repeatedly the metabolic burdens of running an enormous brain have been triumphantly resolved.

When it comes to giant brains beyond the primates, where better to start than with the elephants? As Jeheskel Shoshani and colleagues[92] point out, both the evidence for encephalization and their famous cognitive capacities support the notion "that similarities between human and elephant brains are due to convergent evolution."[93] Thus, convergence extends to molecular similarities revolving around the enormous energy demands of these brains, both in terms of oxygen requirements and reducing the danger of destructive radicals.[94] Apart from their large size, elephant brains are remarkable for their degree of folding (gyrification), as well as a conspicuous cerebellum.[95] Many gaps in our knowledge still remain. Clearly, however, although neuronal density is relatively low,[96] the neurons themselves show a striking diversity of form.[97] Of critical importance, of course, is how this brain structure might mates), yet it is not immediately obvious that echidna needs such cognitive equipment. One possibility is that its "enlarged frontal cortex may be involved in an enhanced spatial or olfactory memory of [its] . . . home range."[64] Hassiotis et al. freely admit this is speculative, but much more likely is their view that the "expansion of the brain must have occurred quite independently in monotremes and therians."[65]

Therians encompass the marsupials (or metatherians) and placentals (or eutherians), yet as documented above in the context of cortical structures and as John Johnson[66] observes, "the independent paths of marsupial and placental evolution have produced remarkably similar outcomes."[67] When it comes to marsupial encephalization,[68] these are also clearly convergent, be it in terms of animals like the Tasmanian wombat that "have relatively larger, gyrencephalic brains with a larger proportion of the brain devoted to the neocortex,"[69] or the strikingly large brain of the striped possum[70] which as Karlen and Krubitzer note has "an encephalization quotient that rivals that of some primates."[71]

That the primates exemplify encephalization needs no emphasis: If it hadn't happened, you certainly wouldn't be reading this book. But questions remain. Most important, and to be returned to below, is to ask, Are primates uniquely informative about encephalization? More parochially is to inquire, when did primate encephalization begin and what routes did it take? Paucity of fossil material and especially of brains in the form of endocasts lead us onto thin ice. The plesiadapiforms, however, provide important pointers, because this extinct group is the closest relative to the euprimates. Evidence from endocasts in a group of plesiadapiforms, the microsyopids, suggests that very early in primate history, encephalization was already under way.[72] Particularly tantalizing is what may have spurred this modest

increase, but a shift from olfaction to vision represents a plausible suggestion. Other proposals also command attention. The fossils remain largely mute, although recovery of endocasts from related groups living in the early Tertiary would solve many problems. More important is the growing realization that at least some of these changes occurred in parallel. In drawing comparisons with the euprimates Mary Silcox and colleagues note that not only are there indications of the "independent development of improvements to the visual system,"[73] but that microsyopids "have also independently evolved some caudal expansion of the cerebrum."[74] So, too, we need to reconsider not only the nature of yet more primitive brains (as in the tree shrews), but Silcox and colleagues crucially comment, "It makes no sense to talk about primate brain size increase as a single event. . . . It was a complex process, rife with parallelism."[75] They continue by reminding us that to try and find a single explanation for why primate brains mushroom in size is to follow a will o' the wisp and conclude by remarking that while "developments to the visual system were a critical part of the earliest phases of primate brain expansion, certainly other factors could be more relevant at other points in the history of the group."[76]

This must be correct, but the central question remains: How many routes are available? The nature of cerebrotypes indicates that not only do primates define their own brain space, but comparison between the Old World and New World monkeys (respectively, the catarrhines and platyrrhines) is instructive. Among the living platyrrhines, a number of forms, notably the capuchin monkey (*Cebus*), have strikingly high cranial capacities,[77] while the fossil record confirms that this must have arisen independently from the catarrhines. Among the latter, primitive representatives such as *Aegyptopithecus* had much smaller brains than once supposed,[78] but as Elwyn color vision (p. 114), the convergences in brain structure are genuine. As Sarah Karlen and Leah Krubitzer[54] observe, when one looks at the diversity of behaviors and ecology, then, "For each of these adaptations, marsupials have evolved an array of morphological, behavioral, and cortical specializations that are strikingly similar to those observed in placental mammals occupying similar habitats," indicative of how "constraints imposed on evolving nervous systems . . . result in recurrent solutions to similar environmental challenges."[55] Among the most striking instances these workers document are those concerning the cortical fields, notably the convergences in marsupials such as the possum between the whiskers (or vibrissae) and the so-called barrel field of rodents,[56] as well as other convergences in the visual cortex and apparently the auditory equivalent.

Not only are there obvious convergences between the cortical fields of marsupials and placentals, but parallels have evolved even within the placentals. Such is evident, for example, when we compare the visual cortex of monkey and cat.[57] A more important point is that given the constraints that ecology and behavior impose on the neural architecture of an animal, we can gain valuable insights into the likely starting point of one of the most startling of evolutionary adventures, the primates. Jon Kaas observes that although tree shrews (which are close relatives of the primates[58]) show obvious differences from the squirrels, the combination of arboreality and excellent vision define not only a convergent ecology but, as he remarks, "we are forced to conclude that the brain of a squirrel would serve well as a model of the early protoprimate brain, not because of a close phylogenetic relationship but simply because of convergences in brain organization driven by selection for somewhat similar niches."[59] Time to turn to the primates, but not just because they are exemplars of giant brains. A number of other groups embarked on much the same journey.

Giant Brains

Of all convergences, the most haunting is where encephalization takes off. So accustomed are we to possessing a brain approximately seven times larger than it "should" be, we forget the enormous metabolic, physiological, and computational challenges that big brains present and the important fact that the various strategies, be they of axon design or distributing neural functions more locally, fall short of the ideal.[60] Yet as with that walking disaster the enzyme RuBisCO, what to our eyes looks hopelessly compromised may still approach the very best. And convergences will be the norm. Such becomes apparent even if we look at what are regarded as the most primitive of living mammals, the monotremes—but primitive in only some senses.

In addition to being richly informative when it comes to convergences (p. 148), they can spring other surprises, not at least when it comes to the brain of the echidna. Although related to the iconic platypus, the evolution of the cerebral cortex has evidently followed a very different path.[61] While the echidna brain has its own neurological characteristics, it is striking both with respect to the degree of folding (gyrification) and overall encephalization.[62] As Maria Hassiotis and colleagues note, the echidna's encephalization "is comparable to the means for artiodactyls and carnivores and extends into the lower limits of the range for prosimians."[63] Extensive gyrification is the hallmark of intelligence (as in cetaceans, elephants and, of course, pricould be considered to possess a 'cognitive' cerebrotype. That is, the specialization of this cerebrotype is complex cognitive functions, such as problem solving, tool use and social behavior."[43] They stress how this is distinct from those cerebrotypes that have evolved to meet a given ecological challenge. Iwaniuk and Hurd continue, "In this regard, parrots and passerines are similar to primates,"[44] a point echoed by Fahad Sultan.[45] In addressing the evolution of the cerebellum in various birds, he remarks on how "The enlargement of specific visual and beak-related cerebellar parts in crows, parrots and woodpeckers fits well with their marked adeptness in using their beaks and/or tongues to manipulate and explore external objects. Their skills are even comparable to those of primates in using their hands. . . . some brains may have enlarged to solve similar problems by similar means during phylogeny."[46] Sultan's point is that as birds such as crows and parrots employ their beaks for "an increased active exploration and perception of the physical world, much as primates use of their hands to explore their environment,"[47] in doing so they may employ "a similar neuronal machine."[48] Sultan's claim serves to remind us of two central principles of brain evolution. The first, to which we return below, is the remarkable series of cognitive convergences that link primates, cetaceans, and birds. Second, however fascinating this convergence may be, it must be put in a larger context. How prevalent is convergence in the evolution of brains, and as importantly is massive encephalization an occasional fluke, or something else entirely?

Given the existence of cerebrotypes, then, in a sense the first question—that of convergence—is already answered. One such survey[49] looked specifically at bats, insectivores, and primates, and in each case the brain proportions accurately reflected ecology and behavior. In a sense each of these groups explored a particular dimension of "brain space," but in doing so reinvented the neurological wheel. Particularly striking in this regard are the bats for which different feeding modes (such as nectarivory [p. 419n65] and carnivory) have arisen independently and thus fall repeatedly into the same restricted region of brain space. Tellingly, this enabled Willem de Winter and Charles Oxnard to infer that one particular group of Neotropical bats, the murinines, must be nectarivorous, a hypothesis that fieldwork duly confirmed. As they remark, it has "an elegant external test of this prediction."[50]

One question to pose is as follows: How deep phylogenetically do these convergences extend? The birds provide a particularly interesting test case (p. 262), and as we have already seen, in their different ways both octopus (p. 18) and insects (p. 251) have some striking similarities with the mammals. But what of the mammals themselves? Relatively little is known of the monotremes (but see pp. 256–57), although comparisons between the brains of marsupials and placental mammals are once again instructive. Both groups possess the all-important hallmark of a cortex composed of six layers, but the developmental trajectories and style of cell proliferation show significant differences, not least in the marsupials the apparent absence of the so-called subventricular zone.[51] Nevertheless as Amanda Cheung and colleagues "stress the considerable differences in the elaboration of the cortical mitotic compartments . . . [and] the protracted development in opossum and wallaby does not result in an expanded cortical sheet."[52] Marsupial brains exhibit a number of important differences with their placental counterparts, not least that instead of mediating the interhemispheric connections via the classic corpus callosum, marsupials employ the so-called anterior commissures.[53] Given these contrasts, we can be confident that, just as with other aspects of marsupial evolution, such as

This animal has already caught our attention on account of its becoming a physiological reptile (p. 432n81). The peculiarities of its island existence also explain how *Myotragus* economized on its brain,[28] not least because the absence of predators removed the need for eternal vigilance.[29] As Meike Köhler and Salvador Moyà-Solà point out, "Changes in relative size and proportions of the brain of *Myotragus* strikingly parallel the reduction pattern under domestication."[30] Such reverse encephalization in domesticated animals such as horses[31] may be unsurprising, but the same holds for their domesticated

equivalents among the birds.[32] In his study of ducks, Peter Ebinger notes how these comparisons between birds and mammals suggest that there are "far-reaching similarities . . . in structural and associated functional changes, and thus the described alterations can be regarded as general trends."[33] Of equal importance is his emphasis on the speed of the process, given that domestication of animals probably only began about ten thousand years ago (and involved multiple independent events[34]), indicating that "these alterations are a remarkable example of the plasticity of the central nervous system."[35]

Mosaic Brains

This plasticity brings into sharp focus the extraordinary versatility of this evolutionary system, because not only does the brain have to integrate a multitude of functions, but at least in some species skittering through the neural labyrinth is the eerie property of consciousness. If changes in the size of the brain, especially in an upward direction, are to be managed coherently, then one might imagine the most straightforward expedient would be for the larger brain to be a scaled-up version of its ancestors'. Simple and sensible, but in reality brain evolution is mosaic, at least in mammals[36] and birds.[37] Mosaic evolution of the brain means that as it grows larger, rather than a uniform increase, different regions increase or decrease in relative size. This is not to say that the regions are completely independent, but the behavior and ecology of a given animal will make specific demands, not least in terms of sensory input. We can identify distinct cerebrotypes, and given that the behavioral and ecological milieu will impose very similar solutions, convergences will emerge.

Many such examples could be given, but among the most interesting are those that

cut across major taxonomic boundaries. In addressing mammalian cerebrotypes, Damon Clark and colleagues[38] draw attention to the strikingly large cerebellum of the echolocating bats (p. 137) and cetaceans[39] (p. 140). In extending this comparison to the mormyriform fish, they remind us, "The common factor of echolocation and electrolocation is the interpretation of subtle timing differences in echoes returned from transmitted pulses,"[40] hence the central role of a large cerebellum in processing these otherwise disparate sensory inputs. Avian cerebrotypes are not necessarily as clear cut as those of mammals, but the convergent linkages are again significant. Striking examples[41] include the cerebrotypes for prey capture, with its demands on acute vision, as well as the brain structures of owls and frogmouths (p. 205).

Most intriguing are the respective cerebrotypes for enhanced cognitive capacities. Thus, Clark and colleagues remark, "Cerebrotype shifts in primates are accompanied by significant expansion of the neocortical volume fraction,"[42] while in the birds Andrew Iwaniuk and Peter Hurd note that "parrots and passerines [which include the crows]

While the template for emotional complexity may well reach back to the earliest fish, as Broglio and colleagues remark, given that "during vertebrate phylogeny four major parallel radiations—agnathans, chondrichtyans, osteichtyans and tetrapods—have evolved independently . . . and increases in forebrain size and differentiation occur in members of all these radiations,"[10] then perhaps not only emotional intelligence but other cognitive capacities may have been an inevitability—but are they ones that emerged again and again? Such is evident from the identification of so-called cerebrotypes. This concept, while acknowledging the ground plan for all vertebrate brains, addresses the question as to why particular regions—say the cerebellum[11] or forebrain—are hypertrophied in particular groups. Among the sharks,[12] for example, such groups as the sphyrnids (the hammerheads) are notable for their disproportionate forebrains and foliated cerebellum.[13] Bigger brains, and here we can include the alopiids,[14] have emerged independently in the sharks (and indeed the rays[15]) and in either case evidently reflect living in complex worlds.[16] An apparent anomaly is that the relative energy consumption of the brain is markedly lower than other fish, a perennial puzzle, given that evolving large brains is metabolically enormously expensive.[17]

What might the cognitive capacities of these fish really be? We know extraordinarily little but suspect that the sharks may be considerably more intelligent[18] than their status as archaic hoodlums might indicate. In an earlier overview, Glenn Northcutt[19] not only was dismissive of sharks being no more than "primitive, robot like smelling and feeding machines,"[20] but pointed out that their neural capacities might well match those of some birds and mammals. Other clues that the sharks are advanced include the independent evolution of not only sociality (in the form of complex schooling)[21] and perhaps play (p. 285), but also giving live birth (viviparity) and even the development of a placenta (p. 219)—all that glorious biological complexity. Yet like the tuna (p. 81), many sharks are under severe pressure. As they slip toward extinction possibly an entire cognitive world will vanish. But who can complain? Shouldn't we as loyal Darwinians rejoice in the inevitabilities of evolution? There's the rub; evolution may be the way in which the Universe becomes self-aware, but although cognitive worlds are bubbling up all around us, only we actually *know*. And while it is understandable to seek out in the oceans and skies parallel examples of massive encephalization, we must always recall that brain evolution emphatically is not a one-way street.

BRAINS IN RETREAT

What of those animals that have turned their encephalization into reverse?[22] Even here, important principles emerge. As Kamran Safi and colleagues[23] remark, "Bigger is not always better: when brains get smaller."[24] While much of the context of their work involves bats and the delicate balance between having sufficient neural tissue to deal with the complexities of echolocation (p. 137) versus economies of weight reduction in a flying mammal, Safi and colleagues stress that "a reduction in brain size should be a general property of evolution."[25] They go on to say that to study either increase *or* decrease in brain size are "both pursuits [that] are equally rewarding when identifying fundamental processes shaping neural structures."[26]

Among the most fascinating examples of reduction in brain size[27] is a former inhabitant of Majorca, the extinct bovid *Myotragus*.

CHAPTER 23

THE ROAD TO "KING CORTEX"

JUST AS nervous systems are nascent in brainless sponges, so for the vertebrates their complex cognitive worlds are seeded in the fish, not least in the sarcopterygians,[1] whose descendants include ourselves. That hypercephalized species—the one that tackles algebra and enjoys Wagner—may be the constant point of reference, but recall that not only do a number of different groups stand on the threshold of cognitive complexity, but the evolution of brains holds a great deal more than the cliché of seemingly relentless encephalization.

FISHY BRAINS

We might do worse than start with the fish. It verges on parody to suppose that the story of vertebrate brain evolution is how a nubbin of a brain in a primitive fish, who only had trilobites to talk to, had successive layers of neural tissue plastered on top of it, ultimately to provide the crowning glory of the human neocortex. Not that anybody disputes that there has been striking elaboration, but this beguiling story of slow but relentless evolutionary improvement misses a deeper point. Consider the ray-finned fish (actinopterygians). A major stumbling block in comparing their forebrains (the telencephalon) with that of other vertebrates is a unique method of development (involving an eversion of the tissue instead of the customary evagination). At first sight the structure looks decidedly different,[2] but the difference is skin-deep. In identifying how various parts of the pallium of the fish forebrain have direct equivalents to mammal counterparts[3] (such as the lateral pallium to the hippocampus),[4] Cristina Bro-

glio and colleagues take this evidence to suggest "that the forebrain of vertebrates share a common, conserved pattern of basic organization."[5] This in turn might mean that various cognitive capacities have much deeper phylogenetic roots than once imagined. The proposed equivalence of the median pallium with the mammalian amygdala is consistent with fish[6] possessing an emotional intelligence.[7] So, too, with respect to the cerebellum, Fernando Rodriguez and colleagues[8] remark that the "functional involvement of the teleost cerebellum in learning and memory is strikingly similar to mammals and suggest that the cognitive and emotional functions of the cerebellum may have evolved early in vertebrate evolution."[9] Much of this work revolves around the teleosts, especially the goldfish whose emotional world is registered in terms of heartbeat and eyeblinks. You'll know the feeling.

We need to remember, however, that teleosts and not least goldfish are highly derived.

this invites a comparison to the mammals, whereby increases in computational ability are achieved by the corrugation of the brain into fissures (each a sulcus) and the corresponding gyri or intervening regions (each a gyrus). Given the scarabs have arrived at a gyrencephalic mushroom body, Farris and Roberts go on to remark how "Nonhomologous brain centers like the mushroom bodies and the cerebral cortex can thus adopt strikingly similar architectures when shared principles of cellular organization are acted on by similar selective pressures."[168] The strength of this analogy is reinforced by its independent evolution of gyrencephalic mushroom bodies in other groups of insects.[169] In addressing the elaboration of the mushroom body in the cockroach-termite transition[170] (p. 186), Sarah Farris and Nick Strausfeld draw our attention to the fact that "the increase in lobe size in the crown termitid groups is so extensive that the lobes have become folded into complex gyri and sulci that are rather reminiscent of mammalian cortex."[171]

Sarah Farris reiterates this theme[172] when she not only asserts "the high degree of convergence of insect mushroom bodies with the vertebrate cortex in terms of function, structure and developmental mechanisms,"[173] but emphasizes the increasing degree of compartmentalization and elaboration.[174] The words "insect" and "primate" do not often appear in the same paper on neurobiology, but Farris's comparisons command attention.[175] Nor is it the only way in which the cognitive capacities of insect and mammal convergently coincide, because as Sarah Farris and colleagues[176] point out, the manner in which the bee mushroom bodies increase in size when foraging has a wider significance. As they write, "The expansion of dendritic fields via lengthening and branching of the dendrites [of the intrinsic neurons] thus appears to be a phylogenetically widespread mechanism by which a brain may accommodate increasing levels of sensory information and learning, either as a result of increasing age or experience."[177]

Evidently stirring in the brains of insects is something akin to mind. Yet despite their immense complexity, arguably the cognitive worlds of the insects are close to the limits of what will ever be possible. Could this be also true for us? Welcome to the world of "King Cortex."

increase, so the reinvention of the wheel?

Time will tell, but time now to focus on the insects. Irrespective of possible homologies, their mushroom bodies reveal convergences at a more profound level. Why this must be so entails not only what any brain must do, but how it handles living in ever more complex worlds. As Nick Strausfeld reminds us, "In addition to their role in odor discrimination . . . mushroom bodies have another important role: that of integrating different sensory modalities. These include visual . . . tactile . . . and acoustic stimuli."[152] Embedded in this complex neuronal architecture is the capacity for learning and memory. Such is only possible if there is not only the capacity for increasing levels of computational sophistication,[153] but a neural plasticity, not least in terms of synaptic connections. In their investigation of the mushroom bodies of the honeybee, Benoît Hourcade and colleagues[154] note that "stable structural synaptic rearrangements, including the growth of new synapses, seem to be a common property of insect and mammalian brain networks involved in the storage of stable memory traces."[155] In the former case the key feature appears to be the density of microglomeruli. The fact, therefore, that the mushroom body is analogous to the mammalian hippocampus is, of course, consistent with the cognitive capacities of the honeybee (p. 288). Even though during its short life, the mushroom body of the honeybee shows some incremental increase, it is no coincidence that the foragers[156] have the largest volume of neuropil.[157] Nor is it surprising that other eusocial insects show similar trends. Thus, as the paper wasp moves from in-house duties to first working on the exterior and finally foraging far from the nest, so the neuropil in the calyx of the mushroom body expands.[158] Intriguingly that part of the mushroom body involved with visual input is neurologically stripped down before the foraging phase, perhaps to facilitate its remodeling and capacity

to handle the demands of flying to and from new territories.[159] Moreover, among the ants in *Camponotus*, the already hefty mushroom bodies increase in size yet further as they leave the nest to forage,[160] while the desert dweller *Cataglyphis*—famous for its feats of navigation—evidently relies on being able to process visual data through greatly enlarged mushroom bodies.[161]

Such increases in the size of the mushroom body are not restricted to the eusocial insects, but the fact that the sweat (or halictid) bees are facultatively social (p. 171) helps to explain the root causes for their spectacular changes in brain size. Thus, in a species of *Megalopta* (*M. genalis*), the driving force for the increase in the size of the mushroom body in the queen seems to be linked to her social context, not least in terms of dominance.[162] Such evidence chimes, of course, with the immensely influential arguments that brain size is intimately linked to social complexity.[163] This, however, may not be the only driver.[164] In a solitary bee (*Osmia*) the mushroom body again increases in size, but here sociality can obviously play no part; once again the need to forage successfully seems to provide the key.[165]

Learning to find one's way around the world may point to the deep commonalities that link the cognitive worlds of insects and mammals. Consider, for example, the scarab beetles. Like other insects they are equipped with mushroom bodies, but at first sight surprisingly the most complex are found in the phytophagous forms.[166] Evidently this reflects both the enormous range of potentially available plants and a choosiness as to which plant to attack, all of which demands considerable behavioral complexity.[167] Especially interesting is that this enhancement of the mushroom body revolves around the structure of the calyx—notably its cuplike shape, as well as internally the density of microglomeruli and the appearance of subdivisions. As Sarah Farris and Nathan Roberts point out, ple. As Ron Jenner[137] notes, "the position of cephalocarids remains very contentious . . . but no recent phylogenetic analysis places them as sister group to the remaining crustaceans";[138] the only reasonably secure link[139] is to another fascinating group, known as the remipedes.[140] In principle, "surprisingly" sophisticated brains may evolve in primitive representatives, but it is equally possible that supposedly primitive crustaceans such as the cephalocarids and remipedes have complex brains, because despite appearances they are actually advanced. Either way, in a manner reminiscent of the evolution of the insect mushroom body, in crustacean evolution there are evidently increasing levels of computational sophistication. The more advanced crustaceans have both elaborations in the deutocerebrum, and as Jeremy Sullivan and Barbara Beltz[141] note, it appears that "the hemiellipsoid body became an increasingly integral part of the olfactory pathway during the phylogeny of the Eumalacostraca."[142]

Yet despite the various similarities, at an earlier stage of this story Nick Strausfeld[143] noted, "the neural organization of olfactory pathways in insects and crustaceans differs markedly. Even [in the robber crab] . . . in which the hemiellipsoid body is situated within the protocerebral mass, the hemiellipsoid body bears little resemblance, if any, to a mushroom body."[144] But things have moved on. While the arrangements of the hemiellipsoidal bodies in the various crustaceans are far from uniform,[145] when we turn to the terrestrial hermit crab (*Coenobita*) the similarities to the mushroom body become particularly striking.[146] So case finally closed? Perhaps not. While some terrestrial crustaceans, notably the isopods (p. 165), have effectively lost their olfactory capacity,[147] this is certainly not the case for this hermit crab. Here the organization of the brain shows a major commitment to olfactory processing.[148] Not surprisingly the same is true for the closely related robber

crab (*Birgus*). The first thing to note is that the group to which these terrestrial crabs belong, the anomurans, are far from primitive but among the most derived of the decapod crustaceans.[149] Second, and even more significant, is the striking convergence between the olfactory organs of the robber crab (*Birgus*) and the insects[150] (see also p. 122). In retracing the steps onto land that marked the crustacean-to-insect transition hundreds of millions of years earlier, very much the same olfactory solution is arrived at. Might this explain why the hemiellipsoidal bodies of at least the terrestrial hermit crab are particularly similar to the mushroom body?

While keeping an open mind as to the likelihood of the convergences of mushroom bodies in different groups, Rudi Loesel and Carsten Heuer remark, "It could be argued that once a cerebral ganglion evolved to a certain level of neuronal complexity, similar architectural principles *have* to evolve accordingly due to specific computational needs and restrictions."[151] To be sure Loesel and Heuer prefer to contemplate whole-scale, if not irresponsible, loss of mushroom bodies, but I suspect that in the case of the annelids and arthropods we see a classic example of neurological convergence. To summarize: Despite crustaceans being principally aquatic and insects terrestrial, hemiellipsoidal bodies and mushroom bodies do similar things. Establishing, however, that they are homologous is much more challenging, and tellingly the most convincing comparisons involve crabs that just happen to have clambered onto land. But even if we grant that hemiellipsoidal and mushroom bodies share a fundamental identity, far more problematic is their occurrence in the phylogenetically remote annelids but not even a whiff in close relatives such as the mollusks. Which do you prefer? Reckless disposal in nearly every other group of ecdysozoan and lophotrochozoan, or as the demands of sophisticated computation innumerable clades, albeit for reasons best known to themselves, opted to discard this valuable computational unit. One might think that the identification of a mushroom body in the flatworms (or platyhelminthes) would go a long way to restore faith in this idea; the flatworms have long served as the totemistic first bilaterian. Not any more. With the important exception of the group known as the acoels (which are evidently genuinely primitive[117]—incidentally making platyhelminthes polyphyletic[118]—and whose brain[119] in any event lacks anything like a mushroom body), the remaining flatworms may have a "simple" anatomy but are highly derived and join the annelids in the lophotrochozoans. What then of their brains? While the planarians (or triclads) have enjoyed the scientific limelight,[120] their polyclad cousins have the most complex of flatworm brains.[121] Harold Koopowitz[122] notes that not only do they show an "unexpected" diversity of cell types, but "These cells are very similar to those found in the vertebrates."[123] This apparent sophistication is reflected in some taxa (such as *Notoplana*) possessing discrete clusters of globuli cells,[124] and reasonably these may be interpreted as being involved in the higher-order processing of sensory information.[125] Commenting on the histological and anatomical complexity of the polyclads, Larry Keenan and colleagues[126] remark on how these globuli masses "are considered the forerunners of mushroom bodies, *corpora pedunculata*, in higher invertebrate brains."[127] This identification seems decidedly forced. Not only do the anterior of the masses lie *outside* the main brain, but while their phylogenetic position is not entirely resolved, clearly the polyclads are not basal.[128] It is difficult to see what bearing these globuli masses have in determining the ancestry of the mushroom bodies.

But let's throw caution to the wind and stick doggedly to the notion that the mushroom bodies, scattered as they are across the metazoan Tree of Life, are so similar that they *must* all stem from a single progenitor. There is, however, a fly in the ointment, literally. Let us return to the insects. The first reason to be suspicious is that their most primitive representatives, the so-called archaeognathans, lack any trace of a mushroom body.[129] While there is little doubt that the remaining insects all draw on the one prototype,[130] significantly the primitive springtails only have what is interpreted as a simple mushroom body.[131] The second reason revolves around the fact that insects are not only closely related to the crustaceans but indeed derive from within them.[132] All, then, we need to do is to open up the brain of any crustacean and find the mushroom bodies—except they seem to be missing. Or are they? The question revolves of what in crustaceans are referred to as the *hemiellipsoidal bodies*. Are they mushroom bodies in disguise? The question is reasonable, because they do have some intriguing similarities to the insects. Thus olfactory information is channeled from the receptors (in the crustaceans known as esthetascs) to the deutocerebrum (specifically to the olfactory lobe and adjacent accessory lobe), neurons of which then project to the protocerebrum and in particular to these hemiellipsoidal bodies.[133] Not only that, but neurological similarities exist in the mode of operation of both hemiellipsoidal and mushroom bodies, notably in terms of the so-called parasol cells, which indicate that the former bodies also serve to integrate multimodal sensory data.[134]

While crustacean phylogeny remains something of a minefield,[135] animals such as *Hutchinsonella* might help to clinch the argument. Its brain possesses a multilobed complex in the protocerebrum that has some intriguing similarities to the mushroom body.[136] *Hutchinsonella* belongs to the cephalocarids and has a somewhat iconic status because it is widely regarded as an ur-crustacean. If only things were so sim-

and reflects their being employed for different functions.[95] Layering or stratification of nerve masses has evolved many times. It is tempting to compare[96] these to the equivalent columns in the mammalian cortex,[97] although in this case the analogy is not exact because of the varying roles in terms of sensory input.[98] Nevertheless, as Sarah Farris and Nick Strausfeld observe, in terms of the incremental addition of these laminae it "is reminiscent of a similar developmental progression of layered structures in the vertebrate brain, such as in the cerebral cortex and the cerebellum. . . . Common to all these structures are the obvious functional differences between laminae that are structurally heteromorphic in the adult, but that have arisen from an isomorphic pool of juvenile 'precursors.'"[99] As they go on to point out, this manner of "differentiation of cell types may have ancient origins and may be the most efficient way to build a layered brain region."[100]

This observation is especially pertinent to the mushroom bodies, because not only are they found in the insects, but also in a number of other arthropod groups (including the primitive onychophorans[101]) but intriguingly in the annelids and even the flatworms.[102] At first sight, all this seems unproblematic. In the case of the annelids, overall brain structure is certainly variable,[103] and the mushroom bodies display differences in size and structure. Despite the obvious differences in the modes of innervation of the mushroom bodies from the respective sensory organs,[104] overall the similarities between the annelids and their arthropodan counterparts are striking.[105] But things get a bit more complicated because, within the annelids as a group, mushroom bodies are by no means ubiquitous. Nor when they occur are they necessarily similar. Thus, a relative of the earthworm, a species of *Enchytraeus* (which is otherwise celebrated for its extraordinary regenerative powers[106]), does have mushroom body–like structures.[107]

They are, however, rather strange because they occur as "*four* bilateral pairs . . . in the dorso-anterior region of the brain,"[108] an arrangement that finds no counterpart elsewhere, even among other annelids. As far as the polychaetes are concerned, some taxa (such as *Nereis* and *Harmothoe*) do have well-defined mushroom bodies; in some the definition is far less precise, while in yet others there seems to be no trace, but perhaps this is on account of their sedentary existence.[109]

As long as annelids and arthropods were believed to be closely related, a common origin of the mushroom bodies was a sensible notion. Their phylogenetic union, in the so-called Articulata, has now, however, been exploded, making any idea of mushroom body homology deeply problematic.[110] Arthropods fall into a supergroup known as the Ecdysozoa, while annelids swell the ranks of the Lophotrochozoa, and so their relatives include the mollusks and platyhelminths. Cephalopods, representing the acme of molluskan neural architecture, do have vertical lobes in the brain that, as Binyamin Hochner[111] remarks, "somewhat resemble the mushroom bodies of the insect brain."[112] There is, however, no suggestion of a homology, and as Hochner also notes, "the vertical lobe . . . resembles the vertebrate hippocampus, both in its involvement in learning and memory and in its morphological organization."[113] Nor is there anything remotely like a mushroom body in the brain of other mollusks.[114] This is not to doubt that the acute olfactory systems of some mollusks, notably the terrestrials and slugs, show useful analogies to the mushroom body,[115] but as with the cephalopods, comparisons to the mammals seem even more fertile.[116]

Can we somehow still link the mushroom bodies of annelid and arthropod? Any such discussion is haunted by the possibility that, even if the mushroom body emerged at the dawn of bilaterian evolution, subsequently of differences that transpire to be skin-deep or as functional inevitabilities. In the next section we tackle the remarkable structures known as the mushroom bodies, as they provide a test case for neural convergence within the invertebrates. But they also have a wider significance. Insects and mammals both have brains, but even if they have similar functions, one might expect they are too different to pursue any detailed comparison. Not so; to quote Sarah Farris,[85] "This relationship among behavioral ecology, the sensory periphery and sensory representations in the mushroom bodies is strikingly convergent with observations of somatosensory specializations and their representations in the sensory cortices of mammals."[86] In fact, it has become a commonplace that the insects, most famously in the form of *Drosophila*, will reveal a functional architecture in their minute brains that will be directly applicable to ours.[87] This is not to underestimate the magnitude of the task. As Gilles Laurent asks, "Shall we even understand the fly's brain?"[88] but in asking if the problem will be cracked by 2106, he replies, "I am not so sure."[89] Nevertheless, the research program is promising. Reminding us of the inherency of nervous systems that extends deep into the Tree of Life, Laurent makes an important point entertainingly as he observes, "When it comes to computation, integrative principles, or 'cognitive' issues such as perception . . . most neuroscientists act as if King Cortex appeared one bright morning out of nowhere, leaving in the mud a zoo of robotic critters, prisoners of their flawed designs and obviously incapable of perception, feeling, pain, sleep, or emotions, to name but a few of their deficiencies. How nineteenth century!"[90] As we will see in the case of insect sleep (p. 287), one can't help but wonder that if Laurent's prognosis is not a little pessimistic. Can we envisage a day when a child, attempting to swat a fly, earns not applause but a severe reprimand? "Darling, you really mustn't. It's conscious." Time to turn to the arthropods.

ARTHROPOD BRAINS

Arthropods are particularly instructive. Just as the onychophoran brain is only bipartite, so its ventral nerve cord without ganglia suggests that the classic segmented "rope-ladder" arrangement seen in the more advanced arthropods is derived.[91] Correspondingly the very similar arrangement in the annelids must be convergent.[92] So, too, while the ventral nerve cord of insects shows a considerable diversity, the component ganglia are not only frequently fused but display configurations that are patently convergent.[93]

The mass of ganglia that in the arthropods go to form the anterior cerebral mass are all the more tantalizing because here we stand on the threshold of cognitive worlds, rooted in memory and learning. Central to this capacity in the insects are the remarkable neural structures, located in the protocerebrum, known as the mushroom bodies.[94] Their name is explained by their distinctive shape, which, although it shows considerable variation, at first glance consists of an expanded calyx linked to a "stalk" (the pedunculus) and a series of lobes. Essential in the operation of this complex neuropil are a large number of intrinsic neurons (the Kenyon cells, named after one of the early investigators of the bee brain) that surround the calyces in a densely packed array of tiny (globuli) cells. When it comes to the lobes, a striking feature in insects such as the cockroach are the alternating dark and light layers. This results from varying synaptic density simply reflect the inevitable co-option of the genetic toolbox.

This is not to claim that the evolution of nervous systems is a monotonic process. Brains can certainly decrease in size, and recall also groups such as the mesozoans (p. 315n13) that have entirely discarded their nervous systems. Rather, as Kiisa Nishikawa notes, "It is now clear that convergence is as pervasive in nervous systems as it is in the rest of the biological world."[70] This applies at all levels, be it within a given group or as yet wider binding principles. Recall, for example, the striking convergences in olfactory mechanisms (p. 123). In a similar vein Reinhard Wolf and colleagues[71] inquire, "Can a fly ride a bicycle?"[72] By this they mean that the coordination of motor control and visual input in the fly and human is such as to indicate that "the kind of visuo-motor coordination emerging from these studies may be universal in the animal kingdom."[73] So, do these deep-seated principles represent a quirk of history, even if the fly does possess the sensory-motor system needed to ride a bicycle, or are they the best solution?

BLOOD AND BRAINS

That complex brains have to work in much the same way should not be cause for surprise, but it also invites us to pursue the investigation of convergences. Something else that brains have in common is their sensitivity to environmental insult. A central line of defense is the so-called blood-brain barrier.[74] Key to its function of isolating the brain behind an impermeable membrane are so-called tight junctions as well as molecular transporters. This barrier is extremely effective at protecting the brain from chemical disturbance, but this comes with a cost; if the neural tissue becomes diseased, then delivery of the appropriate drugs can be highly problematical. In vertebrates such as ourselves, the barrier is derived from the blood vascular system and so employs endothelial cells. It appears, however, that this arrangement has evolved multiple times,[75] from a more primitive system that employs glial cells.[76] Such a transformation makes good sense; as Magnus Bundgaard and Joan Abbott point out, such a "division of labor between endothelium and glia . . . [allows the latter to become] specialized for regulating permeability and transport at the blood-brain interface, while glial cells became more involved with neu-

ronal support and induction/communication within the glio-vascular unit."[77]

The blood-brain barrier in the cephalopods is also convergent (p. 17), and as Swati Banerjee and Manzoor Bhat remark, its "absence in the lower invertebrates led investigators to suggest that a barrier is needed to perform complex integrative and analytical activities in the nervous system."[78] Such a selective advantage surely explains its evolution in the more neurologically sophisticated of the arthropods. The evidence for it evolving at least twice stems from its absence in the primitive onychophorans,[79] and that while among the chelicerates a blood-brain barrier is found in spiders[80] and scorpions,[81] no such equivalent is found in the more primitive limulids.[82] And what of the insects? Here the crustaceans provide the acid test. The absence of an effective blood-brain barrier in the crustaceans,[83] the group from which the insects evolved, suggests that as the latter group constructed the neural architecture to enable their cognitive world, so its protection became essential, even if this blood-brain barrier drew upon a preexisting tool kit.[84]

We hear the drumbeat of evolutionary convergence, irrespective of whether in terms will such brains be for behavioral flexibilities, necessarily underpinned by enhanced memory and the capacity to learn. Chittka and Niven note that various "types of higher cognitive functions . . . might indeed be too complex to be accommodated within the constraints of an insect head capsule."[51] As they say, "it is unlikely that in the insects we will find instances of flexible tool use, insight learning, theory of mind, etc."[52] Unlikely? Certainly. Impossible, perhaps not?

Either way, until some sort of enlarged brain emerges, then nobody is going to be able to think about evolution. If my overall thesis has any merit it will be more than helpful to document not only the many convergences in nervous systems, but as importantly the repeated evolution of big brains. Fortunately I am not alone. In championing the independent evolution of complex brains, Leonid Moroz[53] not only remarks how "multiple origins better explains the extant diversity of nervous systems and enormous plasticity in establishment of complex phenotypes"[54] and the associated developmental and neuronal machinery, but that such convergences are consistent with the recurrent recruitment of regulatory systems. Reminding us that we can envisage a "plausible scenario [whereby] neuron-like characteristics in different cell lineages evolved independently,"[55] so too the diversity of complex brains Moroz documents is impressive. To what extent, however, might this undeniable diversity conceal "skin-deep" differences? Especially tantalizing in this regard are the brains of cephalopods (p. 17); Hanno Jaaro and Mike Fainzilber[56] are prescient in asking whether there are "clear parallels in accelerated evolution, gene family expansion, and other evolutionary processes between cephalopods and vertebrates."[57]

These views seem to be seriously at odds with the prevailing paradigm of deep homology. Given that the developmental machinery in complex nervous systems, classically those of fly and mouse, operates in a strikingly similar fashion in terms of key genes and the cell biology that go to define the neural circuits,[58] then are not these apparently disparate systems of insect and vertebrate fundamentally the same? In reality there is no discrepancy, if perhaps a little confusion. First, any complex nervous system evidently evolved in one way or another from some sort of nerve net.[59] Such an arrangement, where the nerves form a plexus (as in cubozoans [p. 98] and echinoderms [p. 103]), does not look like any sort of brain but is still a sophisticated piece of neuronal machinery. Second, and more importantly, reliance on so-called model organisms usually involves looking at far-flung and highly derived animals with a fondness for the laboratory. If we expand the scope of investigation, a different story emerges. The central mantra of deep homology is that not only is the developmental machinery homologous,[60] but as a consequence the urbilaterian brain was necessarily tripartite.[61] It now transpires that despite its undoubted complexity the onychophoran brain,[62] which presumably serves as our best glimpse of a primitive arthropod, is effectively bipartite, and the posterior tritocerebrum must be a later addition.[63] When we look beyond the arthropods, to a much more basal group known as the chaetognaths, the organization of their brain is consistent with a central nervous system evolving at least twice.[64] Steffen Harsch and Andreas Wanninger[65] suggest that the "Urbilateria already was equipped with a genetic toolkit required to build a complex, concentrated central nervous system . . . although this was not expressed phenotypically."[66] Brains are inherent, they are inevitable, but they remain convergent.[67] From this perspective the undoubted similarities between the development of an annelid brain (as in a highly successful and active group, the nereids[68]) and that of a chordate,[69] separated as they are by an immense phylogenetic gulf, story of convergence is reinforced because the manner of formation of the copepod myelin is quite distinct;[33] more remarkable is that, in contrast to the other crustaceans, it does not have a glial origin.[34] In the case of the crab, however, James McAlear and colleagues noted, "the structure of myelinated nerves . . . is much more similar"[35] to the vertebrates than might be expected.[36] Although the arrangement of the myelin sheaths shows a variety of architectures, perhaps the most interesting is found in the axons of a shrimp. Here the nerve fiber is surrounded by a space filled with an electrolytic gel, in turn encased in the myelin sheath.[37] Effectively this combines the best of both worlds, recalling the giant axons but also employing myelin.[38]

Periodic interruptions in the sheath, another essential feature of myelination, are probably most familiar as the classic nodes of Ranvier seen in vertebrate nerves and where the all-important voltage-gated sodium channels (p. 37) are clustered.[39] Similar gaps, typically windowlike and so termed *fenestrae*, occur at regular intervals in the nerves of earthworms (the dorsal nodes[40]) and various crustaceans.[41] In their various ways these groups have independently arrived at much the same solution to ensure rapid as well as economical methods of nerve propagation.[42]

ENTER THE BRAIN

The evolutionary tropes of ancestral complexity, inherency, and convergence may be recurrent, if neglected, themes, yet if their importance crystallizes anywhere, then it is when we come to address not just the likelihood of nervous systems but of brains. A constant litany in this area is to eschew any idea that the evolution of brains and the emergence of cognition falls on any sort of *scala natura*. It doesn't, of course. But that neither precludes generalities nor inevitabilities. To begin with, as Lars Chittka and Jeremy Niven[43] remind us, "Almost all the molecular components of neurons . . . are present in both vertebrates and insects, and thus, presumably in their common ancestor."[44] It would be surprising if it was otherwise, yet Alain Ghysen[45] draws attention to "the unexpected sophistication of Urbilateria's neural equipment."[46] Such complexity began to emerge deeper still, in the nervous systems of cnidarians.[47] Here Hiroshi Watanabe and colleagues note how "much of [their] genetic complexity . . . is actually an ancestral feature,"[48] and as noted above the nerveless sponges are already proto-neuronal.[49]

Drawing attention to how the architecture of the brain is encoded at a deep phylogenetic level is all very well, but just as the nervous system is hemmed in by thermodynamic fences, then even more so do these strictures apply to a concentrated mass of nervous tissue. So intense do the energetic bounds appear to be that Lars Chittka and Jeremy Niven's apparently innocuous question, "Are bigger brains better?"[50] has a special force. In a thrilling overview, they remind us that, despite the extraordinary disparity in the brain size of an insect and vertebrate, the former, as exemplified by the honeybee, shows impressive cognitive capacities (p. 288) and a remarkable repertoire of behaviors. In some senses this is less surprising given that, in principle, a great deal can be achieved with surprisingly few neurons (p. 433n22). This is not to equate, of course, the brain of a bee and, say, a beaver. Larger bodies de facto have more neural demands, but more significantly when it comes to improved sensory perception, precision of movements, innovatory capacity, and toolmaking, then a larger brain is probably an essential prerequisite. So too

David Attwell and Simon Laughlin observe, a reduction in "the number of active synapses and ion channels to just greater than the level where synaptic and channel noise starts to destroy information, by fine tuning the properties of membranes, synapses and circuits."[12] The net result is that at one time much of the brain is remarkably quiet, with at most only a few percent of cortical neurons active.[13] Here we sense that, for all its intangibles, the brain is extremely finely honed, and like other biological systems, walking along a knife-edge of adaptive possibilities.

A Short Circuit?

Room for improvement would seem very limited. While one can identify a variety of strategies, such as improving blood flow or glucose/oxygen supply, their adoption ultimately leads to irreconcilable trade-offs.[14] So, too, there is a limit to which nervous tissue can be miniaturized,[15] and in axons smaller than about 0.1 μm, noise from the ion channels swamps the signal.[16] David Attwell and Alasdair Gibb note how any attempt either to increase the rates of action potentials or speed of propagation "carries a heavy metabolic price."[17] These fundamental constraints hedge in the operation of *any* nervous system, but there are still some loopholes. Think, for example, of the multiple evolution of the giant axon. There is an even more intriguing strategy when it comes to effective transmission: Why not insulate the nerve? In other words, employ some of the glial cells[18] to produce a fatty layer of myelin. Such myelinated fibers are well known in vertebrates, but they have evolved independently several times, notably in the earthworm and various other annelids,[19] a group of marine worms known as phoronids,[20] and crustaceans. Betty Roots[21] aptly notes in her overview of the phylogenetic distribution of myelin that these occurrences collectively provide "a remarkable example of convergent evolution,"[22] a fact underlined by the way in which the "myelin proteins . . . are [also] completely different."[23]

Of these examples the crustacean case proves to be particularly interesting because the capacity to evolve myelin has evolved several times. Examples include the claws of the snapping shrimps[24] (p. 143), a crab (*Cancer irroratus*),[25] and notably the copepods,[26] in which nerves with myelin only appear in the more evolved groups.[27] The copepods are particularly important because they are among the most abundant of animals on earth, occurring in vast numbers in the open ocean where they form the staple of the baleen whales and analogous feeders (p. 50). One reason for their success is not only an extraordinary tactile sensitivity[28] but extremely fast reaction times, where very high speeds of escape lead to a massive acceleration.[29] Given that the possession of a myelin-ensheathed axon is regarded as being of central importance in insulating the nerve and so allowing very fast transmission of the impulses, it may seem paradoxical that the more primitive copepods not only employ giant axons but have similar reaction times.[30] Nevertheless myelin appears to offer clear advantages when it comes to the general effectiveness in nervous transmission and the need to conserve energy, especially when food is scarce.

Sheathing a nerve in myelin can be achieved in several different ways. For example, in both the copepods and the vertebrates, the wrapping is tight and well-organized.[31] In contrast, in the former group the myelin layers have a concentric arrangement, whereas in the group to which we belong it has a spiral configuration, as it does in the annelids.[32] This

CHAPTER 22

CONVERGENT BRAINS

WITH A NERVOUS SYSTEM at hand, the universe emerges, blinking, to self-awareness but paradoxically also takes the observer to that chasm of incomprehension as we are invited to contemplate the apparently insoluble mystery of consciousness. Yet from one perspective it is surprising that a nervous system ever got off the ground.

RUINOUS EXPENSE?

The reason revolves around their stupendous expense; nervous systems simply gobble energy. To be sure our brains are something of an outlier (although by no means the extreme[1]), inasmuch as they account for 2 percent of our body mass but consume some 20 percent of our metabolic energy.[2] But even in a blowfly at rest, the retina alone (when in bright light) is consuming an extraordinary 8 percent of the entire metabolic budget.[3] Transmitting information involves costs that are far from negligible. As Simon Laughlin and colleagues observe, one needs "10^4 ATP molecules to transmit a bit at a chemical synapse,"[4] and even allowing for what is referred to as sparse neural coding, the energy required to power even a modest nervous system is substantial. Jeremy Niven and Simon Laughlin[5] observe how "there are high energetic costs incurred by neural tissue, including that of sensory systems, both while processing information and at rest. . . . There are also likely to be considerable energetic costs associated with the development and carriage of nervous systems."[6] The combination of maintaining ionic gradients, as well as making and transporting neurotransmitters, are only some of the factors that impose intense selective pressures to economize on how information can be coded and transmitted. In such a context, convergences[7] and streamlining (if not optimizations[8]) are both to be expected.

The convergences will unfold below, but what of the tricky area of optimization? Given both the legendary complexity of the brain and its black-box status, it might seem that here at last the attempt to establish general principles runs into the sand. Yet despite the diversity of nervous systems, not to mention the intangibles of perception and cognition, we can begin to tease out the universals. As noted, brain tissue is massively expensive, with the unavoidable penalty that it is peculiarly vulnerable to interruptions in blood supply.[9] As Simon Laughlin and Terrence Sejnowski[10] remark, "The more we learn about the structure and function of brains, the more we come to appreciate the great precision of their construction and the high efficiency of their operations."[11] This relentless drive for energetic efficiency promotes, as are usually combined as the animals with two germ layers—that is, the diploblasts. As Bernd Schierwater and colleagues remark, if their phylogenetic hypothesis proves correct,[82] then the common ancestor "already had the genetic capability and basic building blocks to build a nervous system, and that from here, the final build-up of the nervous system developed via independent, but parallel, pathways in diploblasts and Bilateria."[83] In a subsequent commentary,[84] they remark that not only may this dual origin "have been a simple step,"[85] but crucially the evolution of a nervous system "is hardly problematic and not much more than a morphological and physiological specialization of already existing proto-nerve cells."[86] Schierwater et al. conclude that "the invention of so-called nerve cells is anything but a major invention in metazoans."[87] What, then, are we waiting for?

tissue serves to instruct the animal to cease feeding, presumably when there is a risk of the aquiferous system becoming choked. This apparently strange arrangement has much wider implications, because electrical signaling without a nervous system is actually quite widespread, occurring in more advanced animals and intriguingly the plants. As Sally Leys and colleagues presciently remark, "We take the view that [nonnervous conducting tissue has] arisen repeatedly in evolution and for good reasons, serving efficiently as adjuncts or alternatives to nerves rather than being merely the vestiges of an earlier creation."[64]

Just as with the choanoflagellates, the evidence from the sponges points increasingly to a "nervous system in waiting." While lacking nerve cells and so necessarily synapses, nearly all the scaffold proteins employed in the postsynaptic region are expressed in the so-called flask cells of sponge larvae.[65] Signaling systems that are subsequently employed in neurogenesis have also been identified.[66] The extent to which synaptic molecules are available before they are employed in a nervous system[67] needs to be put into the wider context of the evolution of cellular machines.[68] With regards to what might be going on in animals that are within a whisker of evolving a nervous system, Kenneth Kosik[69] remarks, "The proto-synaptic structure may have been lifted from a metazoan ancestor, and with only minimal tinkering on the core structure added new gene products to create a novel cellular machine."[70]

That the apparent gulf between the nervous system and any evolutionary antecedent is as much in the mind as anywhere else also becomes apparent when we recall the parallels between chemosensory systems and neuronal synapses.[71] As Shia Shaham remarks, these systems show "interesting commonalities in the logic, organization and molecular machinery."[72] His review of what he intriguingly calls "functional logic"[73] touches on features such as the arrangement of the receptors, glial proteins, and the molecules employed. These are typically small, and among the most important are those involved with glutamate reception. As we saw earlier they play an important role in insect gustation and olfaction (p. 371n70). Their origins, however, go very deep,[74] and in the case of the animals, whether they were originally employed in a chemosensory or synaptic role is an open question.[75] At one level it hardly matters. Irrespective of whether the arrangement draws on an ancestral module or is convergent, Shaham's reference to the logic of organization is a powerful reminder that as and when a nervous system evolves, it is not going to have very much choice.

FINAL ORIGINS

The potentiality for a nervous system in the sponges[76] is therefore obvious, but the nerve itself is a further step. Here the first port of call is usually the cnidarians, yet even in this ostensibly primitive group[77] we need to remind ourselves that it is misleading to regard their nervous systems as simple nets capable of only the most rudimentary functions (p. 98). While the exact path by which the first nerve cells emerged is not entirely clear, it may well have entailed successive specialization of epithelial cells combined with a process of compartmentalization.[78] Moreover, although very controversial,[79] some analyses of early animal evolution suggest that the metazoan nervous system could have evolved twice.[80] The details of the debate revolve around the phylogenetic placement of such groups as the ctenophores and placozoans,[81] which together with the cnidarians

Even more extraordinary is that the plasmodium can anticipate a future event.[49] Thus, if subject to three pulses (at constant intervals) of an unfavorable kind—in this case, imposition of cooler and drier conditions—then even if the fourth pulse fails to arrive at the appropriate time, the plasmodium will react, specifically by markedly reducing its speed of locomotion.

As this team led by Toshiyuki Nakagaki emphasizes, even humans find it difficult to navigate mazes and choose the best pathways to connect scattered objects at minimum effort, let alone discern periodicities. The capacities displayed by the plasmodium allow Nakagaki to suggest that this organism is "smart or [has] something like a primitive intelligence."[50] Moreover, he stresses that the laboratory systems are far more rudimentary than what the plasmodium will experience in the wild. We may underestimate the intelligence of this curious organism, as is evident from the experiments that reveal its capacity to engage in some sort of anticipation and presumably memory. The implication is that the plasmodium possesses a "process of cellular computation,"[51] which in the absence of a central processing capability (as in a brain) is presumably analogous to parallel processing. The nature of the algorithms employed in this computation is, however, entirely enig-matic. As Nakagaki and colleagues remark, this "represents a challenging problem to be solved in the future."[52] Indeed it does, representing a fascinating intersection between the physics of computation and biological activity.

This diverse range of examples reminds us that evolution, and specifically what will emerge as a nervous system, are haunted by molecular potentialities. As we approach the animals, the same principles of inherency apply.[53] Let us, therefore, return to their closest relatives, the choanoflagellates. Given that a significant component of the genomic architecture of metazoans is present in the choanoflagellates, that key components in neurosecretion have been identified is hardly surprising.[54] These include various of the synaptic scaffold proteins,[55] and as Alexandre Alié and Michaël Manuel suggest, it is likely that "the ancestral function could have been to link calcium signalling and cytoskeleton regulation."[56] Echoing this theme, Tomás Ryan and Seth Grant[57] invoke "the protosynapse—that existed before the evolution of metazoans and neurons, and hence challenges existing views on the origins of the brain."[58] Choanoflagellates are certainly nerveless, and it might be thought a nervous system is just round the corner. There is, however, one more step. Consider the sponges.

Nerveless Sponges

Sponges are generally thought to be the most primitive of living animals, partly because they lack any sort of nervous system. They can, however, certainly "feel" inasmuch as some sponges show regular contractions.[59] These can be highly coordinated, to the extent that the sponges show a sort of "sneezing," with an explosive discharge of water.[60] This action makes good sense, because a perennial risk to sponges is a clogging of their aquiferous system[61] upon which their feeding depends.[62] Hence, "sneezing" to flush out the internal plumbing (or alternatively generating contractions that serve to close the body openings) will be of vital importance. More remarkable is that even though these sponges lack a nerve net,[63] there is evidence in the glass sponges (or hexactinellids, p. 96) for the generation of electrical impulses. Their propagation through the thin layer of conjuring process where things like nerves and brains appear out of thin air, then it is difficult to see how this Darwinian process could possibly otherwise work. What is much more important to establish is when in earth history the key components arose, to discover whether they did only once or many times, to ask why they are so specific in their newly recruited functions, and most tantalizingly to inquire how they interlock to act as superbly integrated systems. In a nutshell, when and how do evolutionarily complex things emerge?

Nowhere is this question more provokingly posed than with the evolution of intelligence. We tend, of course, to equate this with sentience, if not consciousness, and as such irrevocably connected to a highly organized nervous system. Nevertheless, the camera eye in the brainless box jellies (pp. 97–100), not to mention an equivalent "organ" in the warnowiid dinoflagellates (p. 95), should give us pause for thought. In the latter case the evidence for these single-celled and nerveless organisms engaged in hunting is greeted with a distinct nervousness. When reviewing the warnowiid eye, and echoing the remarks of Pierre Couillard (p. 95), Esther Piccinni and Pietro Omodeo[41] insist, "It is unthinkable that an apparatus of this sort, in spite of the sophistication of its design, can function like an image-forming eye."[42] Most likely they are correct, but when it comes to some sort of sentience in the protistans, I would be more inclined to keep an open mind. That this might be the case is based not only on these extraordinary dinoflagellates but from other lines of evidence, not least the observations of cooperative hunting in protistans such as amoeba.[43]

Related slime molds in the form of *Physarum polycephalum* are also instructive when seeking the evolutionary roots of intelligence. Part of their life cycle involves forming a ramifying network, the plasmodium. Although readily visible to the naked eye and packed with nuclei, it is effectively a single, giant cell. The central region is usually coalesced, but on the margins distinct tubes exhibit vigorous protoplasmic streaming. The plasmodium is adept at exploring its environment, especially in search of food, with the tubes either actively extending or being withdrawn from unpromising regions. In its own way it is a complex organism, and its ability to search the environment reveals, in the words of Toshiyuki Nakagaki and colleagues,[44] a remarkable capacity for "a primitive intelligence."[45] Here, perhaps, we stand on the border between the blindly automatic and intentionality.

To test an animal's intelligence, it is commonplace to design a maze. As with rats, so with the plasmodium. In the latter case, however, the maze is constructed from a plastic template (which on account of its dryness the plasmodium is usually reluctant to traverse) placed on a surface of agar jelly. Food, in the form of oatmeal flakes, is then strategically placed; the experiment begins—with surprising results. For the plasmodium the maze presents few difficulties, with the organism readily selecting the shortest available distance.[46] A variant on this type of experiment, whereby the plasmodium is "asked" to find several sources of food as effectively as possible, results in a highly efficient network that again finds the shortest routes.[47]

When it comes to maze experiments, the ingenuity of the investigator can be legendary, but the plasmodium rises to the occasion. Thus, its capacity to discover the shortest route presupposes that the journey is without risk. If, however, certain avenues are dangerous, then clearly further "thought" is required. For example, the plasmodium is strongly photophobic because light serves to damage its physiology. When faced with the challenge of navigating a maze with pools of illumination, the plasmodium will solve the problem and find the path of minimum risk.[48]

colleagues, though, remind us, "This demonstrates that complex systems like excitable tissues can evolve by coopting existing genes for new functions."[29] Once recruited, the evolutionary pathways of this channel in the vertebrates and invertebrates proceeded independently.[30] More importantly a sodium channel may have evolved independently at least three more times. Its employment (with a minor calcium component) in a single-celled protistan known as *Actinocoryne*[31] is evidently connected to this protistan's sensitivity to mechanical stimulation and capacity to contract with extreme rapidity.[32] A parallel situation is found in a diatom with a rapid Ca^{2+}/Na^+ action potential, although its present function is not clear.[33] As Alison Taylor notes, "The biophysical and pharmacological properties of the diatom Na^+ current are remarkably similar to those of invertebrates and vertebrate cardiac muscle."[34] Although she suggests that both animal and diatom derive this sodium channel from a common eukaryotic ancestor, it seems as likely this capacity was acquired independently. This is even more likely following the more surprising recognition of a voltage-gated sodium channel in a bacterium.[35] Although the mammalian equivalent is a more complex arrangement, the basic domain structure is strikingly similar, and here, too, this convergent sodium channel evidently evolved from a calcium precursor.[36] Given that bacteria have no nervous system, why then a sodium channel? Most likely it is a physiological response because this type of bacteria live in highly alkaline (i.e., high-pH) conditions, and the channel seems to be important in allowing the organism to follow chemical gradients (chemotaxis) and also to control its pH.[37]

Ion channels, in many and various ways, are central to the modus operandi of a cell—not only to run an immensely sophisticated chemical factory on a tiny scale but to be highly selective as to what is and is not allowed

to pass from the outside world. Migration of ions such as sodium and calcium is essential, but this requirement encounters an immediate paradox; the cell membrane is composed of a double layer of phospholipids and so forms an extremely effective barrier to these ions. Put simply, if ion channels did not exist it would be necessary to invent them: life has no choice. One might think that the gap between the complex domain structure of an ion channel—precisely arranged and specific in its operation—and any primitive precursor must remain unbridgeable, perhaps creaking under the weight of ad hoc hypotheses or, even worse, inviting the rictus grin of "intelligent design." Not so. Remarkably even simple peptides, consisting of a mere twenty-one amino acids and employing only leucine and serine, can serve as an ion channel (and intriguingly they also show similarities to the acetylcholine receptor).[38] To achieve this, each peptide must be configured as a helix, and up to six of these helices then brought into juxtaposition and positioned in a transmembrane location. This proto-channel is not very selective, but given that these helical arrangements can spontaneously arise, this strongly suggests that even the most primitive cell would be in a position to begin to shuttle ions. Some of the wider implications of this are explored by Andrew Pohorille and colleagues,[39] and as they remark, such "ion channels have a sufficiently universal architecture that they can readily adapt to the diverse functional demands arising during evolution."[40]

In terms of its capacity and function, the human brain is a truly astonishing product of evolution, but in large part the molecular scaffolding—be it the immune system, acetylcholine, or ion channels—arises from deep time, in some cases almost certainly more than a billion years ago. Nor should this be any occasion for surprise. Given that evolution emphatically is not some sort of neurotransmission. In nerve cells, such transmission depends on the synaptic system, and the molecule acetylcholine plays a key role as a neurotransmitter. But it is not restricted to the animals; rather its ubiquity across all the branches of life, even the bacteria, suggests that this molecule is very ancient.[9] Acetylcholine, therefore, does far more than serve in nervous systems. For example, it is found in both green algae[10] and in the higher plants, where it is important in such functions as cell proliferation. In bamboo, which is notorious for its stupendous capacity for growth, the growing tip of the plant can contain some eighty times more acetylcholine than the brain of a rat.[11] Acetylcholine occurs also in lichens (p. 152), where it may mediate cell interactions between the alga and fungus.[12] Even in the animals, acetylcholine has a range of nonneurogenic functions, including a role in the immune system.[13] Even if acetylcholine is highly conserved, other parts of this complex system are convergent. Thus the enzyme acetylcholinesterase, which serves to destroy the acetylcholine after it has performed its function, has evolved separately in plants and animals.[14]

So too the molecule GABA (that is γ-amino butyric acid[15]), which plays a central role in neurotransmission, not only occurs in substantial quantities in the ciliate *Paramecium*[16] but serves a number of functions. Paola Ramoino and colleagues note that when released in this protistan, it does so by "a synaptic vesicle-like exocytotic pathway."[17] In reviewing this area in the context of the origins of the nervous system, George Mackie[18] aptly referred to these as "Prophetic molecules in protists."[19] Earlier this point was echoed in more poetic form by J.B.S. Haldane[20] when he wrote how signaling between neighboring ciliates allowed "one to remark that the acetylcholine corresponded very well with the 'animal spirits' of Descartes," while longer-distance communication "employed more stable substances, the hormones, that correspond to the Cartesian 'vital spirits.'"[21] Descartes's immensely influential views on the nature of mind and whether animals were any more than automatons are perhaps superseded, but in their different ways Mackie and Haldane point to how deep the roots of consciousness reach. Other key components of the nervous system, notably hormones such as endorphin (as well as serotonin, p. 121), also have deep evolutionary origins.[22] These molecules often occur in low concentrations, and their exact functions remain somewhat conjectural. As with acetylcholine, however, we can be confident that they far predate any neural activity and presumably were only recruited for these functions at the onset of the Cambrian "explosion."

There is, however, one aspect of the evolution of the nervous system that might reasonably be thought to be a specific, and unique, innovation: the invention of voltage-gated sodium channels that play a central role in the rapid propagation of nervous signals. Of course, nothing is unusual about ion channels per se. They are ancient and ubiquitous. In fact, various methods of ion transport have evolved, but in the specific case of the ion channels the pore through which such cations as potassium, sodium, or calcium[23] must pass is defined by an array of protein domains that span the cell membrane. These not only display an extraordinary specificity for a given type of ion, but also have a near-optimal configuration[24] dependent on extraordinarily precise atomic distances.[25] The all-important sodium channel probably evolved from a calcium equivalent,[26] which in turn was derived from the still earlier and thus more primitive potassium channel.[27]

Although the sodium channel in the nervous system of animals appears to have a single origin, crucially the ancestry of this ion channel can be traced back to the nerveless choanoflagellates.[28] Benjamin Liebeskind and

CHAPTER 21

THE ROOTS OF SENTIENCE

"HEADMASTER WANTS to see you, Conway Morris." So you trudge past the misty playing fields; it is a vile day. The study is warm, but far from cosy. Punishment administered, and perhaps in its own way a salutary lesson for the future. Some things are best not done— or said. So if you want to enrage almost any evolutionary biologist, then why not sidle up and murmur, "Well, been thinking about it. Does seem to me that evolution has a remarkable directionality." If you are lucky all you'll need is a clean handkerchief to dab the spots of spittle, but sometimes the response is closer to foaming. One touches a raw nerve, but if mammalness is a genuine evolutionary property, then perhaps there are others? What about massive cognitive competence, another inevitability that is woven into the very fabric of evolution? It is not just that by the time a nervous system emerges we can have confidence as to the outcome. On the contrary, much of what you require to build a giant brain evolved billions of years before any such object appeared on the planet.

NERVELESS EUKARYOTES

Where shall we start? In the flatworm[1] nearly all of the relevant genes (or a near proxy) to the nervous system also occur in humans (110 out of 116, or 95 percent). This is scarcely surprising. More of a surprise, however, may be is that a significant proportion[2] of the genes employed in the development and functioning of this nervous system are also found in the brainless plants and the specialized fungi, specifically yeast.[3] In this study by Katsuhiko Mineta and colleagues, the genes were allocated to one of five functional categories—for instance, involvement in neurotransmission—and here, too, each category found its genetic equivalent among the plants and yeast.[4] Looking still deeper in the Tree of Life, some 14 percent of genes specific to the nervous system evidently had evolved before the common ancestor of yeast and human,[5] suggesting that components of the architecture of the nervous system were already emerging at least a billion years ago. None, of course, is specifically involved in nervous activity, and no doubt in some cases the genes have been redeployed.[6] Nevertheless, at first sight the fact that some third of the genes employed in the brains of animals occur in plants and yeast may seem peculiar, until we remind ourselves that the growth of the brain and its function are largely chemical processes.[7] Plants and fungi may be nerveless, but the transport and secretion of chemicals for multiple functions are essential.

In a study by Mineta and colleagues,[8] the largest number of shared genes was unsurprisingly in the functional category of the mantle of mammalness. Relatives of the crocodiles, the Cretaceous notosuchians display strikingly heterodont dentitions that converge on the early mammals. In terms of its approach to mammalian jaw movements, the Tanzanian *Pakasuchus* stands out.[201] As Patrick O'Connor and colleagues remark, "The complex morphology and high degree of occlusal precision of the cheek teeth in *Pakasuchus* shows a level of sophistication otherwise seen only in mammals."[202] Not only has it lost most of its reptilian body armor and is much more gracile than its crocodyliform relatives, but as O'Connor and colleagues

point out, the peculiarities of the Gondwana mammal faunas make it tempting to suggest that the innovations seen in *Pakasuchus* represent "an evolutionary-developmental experimentation by a clade in the absence of potentially competitive ecomorphs . . . that is, mammals."[203] The fact that neither monitor lizards nor notosuchian crocodyliforms[204] are or were going to evolve into mammals is to miss the point. What both exemplify is that biological form transcends phylogeny, and thus we can speak of the inherent nature of mammalness. That perspective leads us to some very interesting territory.

history . . . has been the remarkable, indeed all pervasive, convergence. . . . Repeatedly, members of the various subfamilies or families have developed morphologically identical or closely similar adaptations, perhaps in response to similar environmental challenges or opportunities."[179] There is even a curious link[180] between the kangaroos and the most celebrated of avatars, the marsupial mole (*Notoryctes*).[181]

More specifically the latter are convergent on placental diggers, especially the golden moles.[182] Not identical, of course, and a notable feature is the recruitment of the powerful tail to assist in the more or less continuous burrowing through the Australian sands, and in this respect this animal is more similar to the Namib Desert golden mole. These moles, however, did not begin their phylogenetic career in the deserts. Fossils from the Miocene[183] throw critical light on how they emerged[184] in a rain-forest environment and only subsequently, as Australia dried out, did they shift into the arid outback. Then there is the long-controversial *Necrolestes*, a Miocene denizen of Argentina, which had been regarded as a marsupial, with evidence for a molelike mode of life.[185] It now transpires, however, that it is a survivor of a nontherian lineage that flourished in the Mesozoic,[186] thus extending fossorial convergence further.

In their various ways, all these examples suggest that evolution is akin to an immense field of possibilities, but at widely scattered points there are deep wells to which biological forms are attracted. To trace the paths that evolution actually chooses to take is of great interest, but more fascinating still is to see how the recurrence of design points to deeper organizational principles. So we can talk of mammals and unpick the evolutionary threads that led, say, to a felid saber-tooth. But we can also visit the constructional well-labeled "saber-toothness" or even "mammalness." The latter is full of mammals, of course,

but also houses some intriguing analogues. Perhaps best-known is that "honorary mammal," the kiwi (p. 208). This bird is not alone, because at least two groups of reptiles have trodden similar paths to approximations of a mammal: the notosuchians and the varanid lizards. The point here is not to try and insist that one day either group (or the kiwi) will somehow become a mammal. They won't. What matters is that those features that define "mammalness" have evolved repeatedly.

Welcome first to the monitors (also known as the varanids). Not only do they show convergences with Permian synapsids such as *Varanops*,[187] but these lizards also exhibit a number of characteristics that are intriguingly mammalian. Well known is the Komodo dragon (*Varanus*), a truly venomous reptile (p. 62), but even larger was the fearsome *Megalania priscus* from the Pleistocene of Australia.[188] This extinct lizard seems to have emerged as a top carnivore, holding its own against its marsupial competitors.[189] So, too, in the greenhouse world of the Eocene of Myanmar, a giant acrodontan lizard (*Barbaturex*) weighed in as a convincing avatar of a mammalian herbivore.[190] As Jonathan Losos and Harry Greene[191] note, although monitors as a whole consume a variety of prey (including insects), "Some monitors do closely resemble certain endothermic predators. Giant forms such as *V. komodoensis* and *V. priscus* parallel large felids."[192] Even the smaller monitors have, for the lizards, rates of blood flow comparable to mammals[193] and elevated[194] metabolic rate,[195] all underpinned by a complex lung structure.[196] The mammal-like capacities of the monitors[197] extend to behavior, including large home ranges and acute vision, as well as sophisticated courtship and parental care.[198] Not only are monitors conspicuously active,[199] but they have a reputation for curiosity that is borne out in laboratory experiments showing these lizards to be fast learners.[200]

Nor are these the only reptiles to take on

can versus Northern Hemispheric origin.[157] Some convergences, such as the anteaters (pp. 51–52) in the form of afrotherian aardvarks and laurasiatherian pangolins, were already well-known. Even more striking is the similarity between the pangolins and the South American anteaters (p. 51),[158] the latter belonging to the third of the great mammal groups: the xenarthrans.[159] This latter group is highly instructive when it comes to documenting convergence, even though its principal representatives are as disparate as the sloth and armadillo.

Ponder the sloth.[160] With its hanging or suspensory locomotion and a pelt designed to retain water and encourage the growth of algae, these animals are the epitome of massive lethargy. They come in two varieties: two-toed and three-toed (respectively, *Choloepus* and *Bradypus*), and feed on leaves. In at least the case of the three-toed sloth this folivorous habit has converged on the more familiar rumination of the cows—that is, a fermentation of a mass of plant material[161] (p. 206). One might sense that apart from the number of toes, when you've seen one sloth you've seen them all, until one realizes that the two- and three-toed sloths are diphyletic. In the words of Timothy Gaudin,[162] they represent "a remarkable case of convergent evolution, perhaps the most remarkable in all of Mammalia."[163] Rummaging beneath the pendant sloths, who only descend to the forest floor about once a week to defecate and urinate,[164] are the armadillos, unexpectedly converging with the Jurassic mammal *Fruitafossor* (p. 229).

What of the afrotherians versus the laurasiatherians? Apart from those anteaters, are there any other convergences? The parallels are indeed striking. Just as there are true moles (including the remarkable star-nosed mole, p. 131), an afrotherian echo is found in the golden moles (chrysochlorids), which in turn show striking convergences with the

extinct palaeanodontans (e.g. *Xenocranium*), which were busy burrowing in the soils of Oligocene North America.[165] Chrysochlorids are closely related to the tenrecs of Madagascar, and among the latter we find both quasishrews and something very like a porcupine[166] or a hedgehog, while in West Africa the otter shrew[167] has some similarities to the otter.[168] In other cases the convergences between afrotherians and laurasiatherians are much less obvious—until, that is, one remembers to include the fossil record. The living afrotherian hyraxes, whose closest living relatives are the elephants, are very much the remnant of a former diversity. Among the fossil hyracoids, however, we find striking convergences with not only extant groups—such as the suids (pigs), horses, and tapirs—but other fossil[169] groups.[170]

When we consider the convergences among the marsupials, the pitfalls of the fossil record are arguably even more forceful.[171] What, one wonders, might still await one or another fortunate paleontologist? After all, whole swathes of the world once possessed rich marsupial faunas that have now vanished. Might we even find a marsupial bat?[172] Probably not, but as our knowledge grows, so will the convergences. Such confidence is not misplaced, given that we already possess a series of classic examples where both living and fossil representatives outline a recurrent story of repeated evolutionary solutions.[173] Even the very pouch, the marsupium, that serves to define the group has evolved[174] several times.[175] Consider the famous marsupial "wolf," the thylacines.[176] Are its similarities to the extinct borhyaenids of South America, which include the saber-toothed thylacosmilids (p. 54), a reflection of common ancestry or convergence? In this case, and others that include the various opossums, convergence is the answer.[177] Speaking of the iconic kangaroos Tim Flannery[178] commented, "the most striking aspect of [their] evolutionary its remains being discovered in the Morrison Formation, otherwise famous for its dinosaurs. Although a primitive form, it shows remarkable convergences with the armadillo (notably in its teeth and backbone). These are so striking that the authors are driven to insist that this fossil has no bearing on the evolutionary history of the xenarthrans. The similarities arise both because of diet and the capability for powerful digging. While *Fruitafossor* was busy rummaging in Colorado, across the world a related taxon, in the form of *Volaticotherium* (p. 190), was gliding from tree to tree. Other convergences include *Haldanodon*, which paralleled the desmans,[147] a group of semiaquatic moles. More remarkable is another semiaquatic docodont (*Castorocauda*), with a beaverlike tail and teeth convergent with those of fish-eating seals.[148] Not only, as Zhe-Xi Luo remarks, do these recent finds reveal "convergences to highly specialized extant mammals,"[149] but among this medley of forms there are convergences in the teeth that foreshadow a shift from a slicing action to tribosphenic crushing (p. 59).

Stirring in the Mesozoic are the potentialities for mammalian success. The orthodox view is that their chance only came when the great reptiles were kicked down the stairs by the end-Cretaceous (K/T) extinctions. Historically that is correct, but sooner or later the mammals were going to rise to dominance. All the K/T event did was accelerate the inevitable. If the asteroid had missed, readers of this book would have had to wait about another 50 Ma. But it didn't, so let us turn to the Tertiary.

Imagine we had a time machine. Having watched that mass extinction of the dinosaurs on our journey home, we decide to visit North America, stopping off roughly every 10 Ma. The mammalian faunas evolve, of course, but every time we step out of the machine there is a strong sense of déjà vu. There would be the saber-toothed cats, but

not only did this ecomorph evolve repeatedly, but effectively independently (p. 55). While we're busy avoiding one or another iteration of the saber-tooths, we must be careful that we don't get trampled by one or other of the recurrent hippo ecomorphs.[150] If we go to the other end of the scale and look at the shrews or the burrowing rodents or the larger herbivores, in each case the landscape will be populated by remarkably similar forms, but ones that had evolved iteratively.[151] In other words the entire mammalian community was evolving simultaneously, in much the same direction, summoning up the tried-and-tested ecomorphs. Extinction, perhaps mediated by climatic change, sweeps the ecological board largely clean, but when the dice are rolled again they are seen to be loaded. Echoing the example of a virtual saber-tooth lurking in the wings of the clouded leopard (pp. 56–57), so Larry Martin and Terence Meehan remark,[152] "Extinction may not be forever."[153] This cyclicity of ecomorphs is not, of course, an eternal return: the melody is indeed replayed, but new evolutionary harmonies are also successively introduced, and perhaps ultimately to allow the most sentient of species to hear the universal music. Yet despite time's arrow pointing to new, if not remarkable, destinations, the recurrence of forms is surely very striking. Just as the saber-tooths recur in North America, in the south the marsupial thylacosmilids navigated to a similar solution (p. 54). The drumbeat of convergence can be heard not only between the placentals and marsupial,[154] but also within either group.

The case for convergences within the placental mammals[155] only emerged recently as new molecular phylogenies began to radically revise some cherished assumptions.[156] In the new formulation four major groups are defined. Most notable are the Afrotheria and Laurasiathera, which, as their respective names indicate, broadly look to an Afri-

THE ROAD TO MAMMALS 229

solutions but because in both sarcopterygians and theropods we see multiple trends in the same direction. Give me a sarcopterygian or theropod, and I will conjure up a tetrapod or a bird. Why not the same for a synapsid? One possibility, to employ Kemp's imagery, is that in the case of the evolution of mammals the adaptive ridge is like a knife-edge, such that the degree of functional integration allows for little latitude. I suspect, however, that the story of mammal emergence will prove no different from any other major transformation, for this reason: In the scheme of correlated progression a series of organizational grades are identified. One of the first is marked by the emergence of the so-called sphenacodontid and therapsid grades. The mid-Permian beast *Raranimus*, however, not only helps us to close an important temporal gap,[134] but Jun Liu and colleagues note that it shows "a unique combination of therapsid and sphenacodontid features."[135] Sound familiar? So it should, and among the more advanced cynodonts, the discovery of *Ecteninion* led Ricardo Martinez[136] to remark,[137] "More than many eucynodonts [this taxon] demonstrates the 'mosaic' of primitive and derived morphology that characterizes much of cynodont evolution in the late Early and Late Triassic."[138]

One senses a tension between the desire to portray the emergence of the mammals as a neat and tidy sequence as against the likelihood that homoplasy is rife.[139] Not only is there homoplasy within particular groups of synapsid, but convergences across the group are also rampant.[140] With respect to the evolution of the foot and hand, and especially the so-called pharyngeal formula (which evolves from a primitive 2-3-4-5-3 to the mammalian 2-3-3-3-3), not only has this evolved multiple times,[141] but as Jim Hopson notes, "the rest of the appendicular skeleton shows the independent acquisition of derived more mammal-like morphologies in the different groups of therapsids."[142] Nor is this surprising. A more symmetrical foot and associated changes in the limbs and limb girdles are integral to a more effective mode of locomotion, from sprawling to a so-called parasagittal gait where the limbs are swung beneath the body. Moreover, the all-important bony secondary palate, which allows chewing and breathing to be effectively separated, arose at least three times in therapsids.[143] Not only can we reconstruct in ever-greater detail the evolutionary "Let's Make a Mammal" kit, but just as with the sarcopterygians and tetrapods or theropods and birds, we can be confident that at least by the time a synapsid wandered out of the wings of the Theatre of Evolution and edged toward the limelight then a mammal was an inevitability. And of course the story doesn't stop here.

Rerunning the Tape

We remain in the Mesozoic. Biologists might have a nodding acquaintance with dinosaurs, but when it comes to the mammals one is likely to face unexplored territory with a welter of unfamiliar names. They come from a fossil record that is still full of gaps, but never again can we think of these mammals as all-purpose, shrewlike animals cowering from the vast reptiles that dominated the Mesozoic landscape. Some grew to quite substantial sizes. The triconodont *Repenomamus* weighed up to 14 kg and returned the compliment by having juvenile dinosaurs for breakfast.[144] Equally striking is the previously unappreciated ecological diversity of these mammals, and with this come the convergences.[145] Consider *Fruitafossor*.[146] This animal lived in the late Jurassic of North America, mammals were already well on their way to colonizing the Mesozoic landscape. In a sense we are on the last lap, with a universe emerging into a slightly startled self-awareness, whereby the convergences of toolmaking and advanced intelligence will in due course ensure the telescope and geometry. How any of this would have been possible without endothermy is difficult to imagine. One might ponder how many of the convergences between bird and mammal are effectively determined by this shared, but independently derived, metabolic state. One line of inquiry looks to parental care and the need to pour food down the ever-open throats of the children, not to mention keeping them warm.[123] The ramifications by no means end here. Parental care and endothermy not only

presuppose parallel adaptive demands on a body capable of sustained exercise, but as Colleen Farmer also observes, "many other convergent features may also be explained . . . [including] exceptionally keen hearing and vocal communication."[124] In other respects the differences between birds and mammals remain obvious, and one might expect that in the ever-widening circles of similarity the convergences become ever more diffuse, if not slightly ridiculous. Nothing could be further from the truth. Seeded independently in both groups was the potentiality for emergence of complex cognitive states that provided the final steps to the universe becoming self-aware. As we have seen, the evolution of something like a bird is highly probable at least (p. 195), but what of the mammals?

Making a Mammal

To answer this question the fossil record is essential. If anything, this record of how the transition from a synapsid reptile to a fully furred mammal[125] was achieved is the best documented of all the major evolutionary transitions. A relative abundance of fossils from Permian and Triassic[126] sediments reveals a story of incremental change and functional integration, allowing Tom Kemp to formulate the powerful concept of correlated progression.[127] This paradigm helps to explain why the process was relatively protracted but, as Kemp[128] notes, was "of critical biological significance . . . involving as it did extensive reorganization of a low-energy, weak-jawed, sprawling-limbed, fully ectothermic, basal amniote into a radically new kind of organism."[129] Rapidly running, warm-blooded carnivores with ever stronger skulls,[130] housing larger brains and ears evolving to ever more acute hearing—here was a force to be reckoned with. Kemp explains how the concept of correlated progression

has a clear directionality and also "implies the existence of a long ridge in an adaptive landscape."[131] Such a trend does not preclude, of course, various excursions to provide oddities to delight the biologist, but the sense is that not only do functions remain integrated but they become increasingly sophisticated.[132] In contradistinction to the overriding theme of this book, Kemp argues, however, that the rise of the mammal-like reptiles "seems to have depended on a unique and fortuitous combination of mid-Permian biological and environmental circumstances,"[133] whereby both climate change and evolutionary changes (including physiology) set these animals on the route to success. But "unique and fortuitous"?

In the case of the sarcopterygian-tetrapod transition and the directly analogous case of the theropod-avian transformation (p. 193), one can argue that terrestrialization and flight, respectively, are inevitable—not only because they provide coherent functional otherwise is a most unusual mammalian trait; pigs may not fly, but they do build nests. The reason? In pigs, UCP1 has mutated and lost its function. In warm tropical environments this probably was manageable, but it presented a real challenge when in due course this group migrated to cooler regions.[93] Quite why pigs lost this capacity is less clear, but in general it is difficult to see how mammals would have achieved anything like the degree of success they enjoy if they had remained with a reptilian physiology. Vital as this novelty proved to be, nearly all the ingredients required for mammalian endothermy were waiting for their cue in the capacious wings of the evolutionary Spielhaus. In other words this family of proteins (which includes UCP2 and UCP3) is much more ancient,[94] and although UCP1 is central to so-called nonshivering thermogenesis,[95] even this specific protein goes deep into the evolutionary history of vertebrates.[96]

So what's the deal? Anything really new? When it comes to convergences, quite a lot.[97] First, for fish endothermy the physiological basis is most likely a "futile" cycling of calcium[98] (p. 80), and no role for UCP1 has been identified.[99] Second, thermogenesis in birds is evidently based in the flight muscle, which among other things explains the extreme energetics of the hummingbird (p. 200).[100] But UCP1 cannot be used for the simple reason that, along with UCP2, this protein was lost at some earlier point in bird evolution.[101] Heat is still generated across leaky[102] membranes,[103] and once again there is an uncoupling protein, known as ANT (adenine nucleotide translocator).[104] Birds also lack brown adipose tissue (BAT), but they employ analogous adipocytes.[105] As Mary Schweitzer and Cynthia Marshall[106] remark, "endothermy (more accurately heterothermy[107]) as expressed in modern birds is qualitatively and quantitatively different than the endothermy of mammals."[108] Just, however, as a much more efficient ventilation in birds (a pulmonary system with air sacs) and mammals conveys sufficient oxygen to feed the metabolic furnaces, so it is equally important that the hemoglobin in the red blood cells can effectively bind this oxygen. Crucial in this respect are a variety of phosphate compounds.[109] Mammals and birds achieve the same ends but through different means.[110] The former employ a phosphoglycerate (BPG), but birds on the other hand turn to phosphorylated inositols (IPs),[111] a representative of a remarkable and versatile molecule found across the Tree of Life.[112] In evolution, successful deployment is at least as important as novelty.

Albert Bennett and John Ruben[113] remind us that despite the enormous energetic investment, "mammalian and avian endothermy developed along essentially parallel lines among different groups of reptilian ancestors."[114] Correspondingly, in their overview of bird and mammal endothermy, Willem Hillenius and John Ruben[115] ask, "Who? When? Why?"[116] By no means have all these questions been solved. Fossils will never provide all the answers,[117] but despite the differences this convergence of endothermy ranks as a classic example. So striking are these similarities that, employing the irrefutable certainties of the cladistic methodology, Brian Gardiner suggested that far from birds being close allies to the crocodiles (and other archosaurs), they were in fact to be linked to the mammals as the so-called Haemothermia.[118] Tom Kemp[119] notes that although the "great majority of vertebrate biologists have dismissed Gardiner's hypothesis as simply outrageous . . . it must actually be taken seriously. . . . the list of characters supporting the taxon Haemothermia is indeed formidable."[120] Kemp then proceeds gently to dismantle the entire hypothesis, but the ghostly presence of "haemothermians" is a reminder of the many endothermic parallels[121] between birds and mammals.[122]

As birds spread their wings (p. 193), furry

into a tetrapod, so the apparently unpromising sprawling amphibian panting on the edge of the Devonian shore had the evolutionary inherencies that would lead to not only soaring swift and chattering ape, but also warm-bloodedness.

WARM BLOOD

If we had a time machine and dropped into points between the middle Permian and early Jurassic (and giving the mass extinctions a wide berth), we would find capturing the increasingly agile specimens of mammal ever more difficult. But once cradled in our hand—and avoiding those needle-sharp teeth—we would be struck by the warmth of their body. It is easy to see why endothermy has provided such a focus of attention in terms of mammalian evolution, even though Tom Kemp is correct in reminding us that it was part of an emerging functional complex that interbinds physiology, locomotion, thinking power, and stamina.[79] And sometimes it is forgotten that by no means all mammals have constantly high body temperatures. Many are heterothermic—that is, entering protracted periods of torpor, as in the hedgehoglike tenrecs[80] (p. 39). Even so, endothermy represents a whole new game. To feed the metabolic furnaces, food consumption soars, while precise thermoregulation calls on both hormonal and neuronal control.[81] It is an immense investment, but its rampant convergence suggests that it is a winning ticket that no planet can afford to do without.

Apart from its obvious relevance as to the emergency of avian endothermy (p. 227), the ongoing debate as to whether dinosaurs were endothermic[82] shows that fossil data are not always easy to read. Oddly enough, in these circumstances the best clue comes from the nose. Air drawn in will be typically cooler than the ambient body temperature, while correspondingly air leaving the lungs will be moisture rich, which in more arid environments might threaten the animal with excessive water loss and ultimately lethal dehydration. The trick is to channel the air over a series of highly folded bones (the maxilloturbinals[83]), where a countercurrent system ensures that incoming air is warmed while exiting air yields some of its moisture.[84] These bones are thin and delicate, and so seldom preserved, but their presence can still be inferred from associated ridges on the edges of the nasal cavity. This evidence points to endothermy emerging among relatively primitive therapsids, that is mammal-like reptiles, including the anomodont *Lystrosaurus*,[85] as well as the therocephalian *Glanosuchus*.[86] As John Ruben[87] remarks, "It is difficult to overstate the almost Rosetta Stone–like significance of incipient maxilloturbinals in therocephalians."[88]

Endothermy may well have been incipient in these early therapsids. It took perhaps 50 Ma[89] to reach its fully fledged state,[90] but the basic principle echoes the story of fish endothermy (p. 80). Once again the trick is to decouple the respiratory processes that operate across the inner mitochondrial membrane.[91] The net result is that instead of producing ATP, there is a constant flow of protons (H+) across this leaky membrane. This situation has to be rectified, and in doing so, heat is automatically generated. Central to this activity, notably in the lipid-rich brown adipose tissue (BAT), are the aptly named uncoupling proteins, notably UCP1.[92] The importance of this protein becomes self-evident when we look at pigs. Pigs? Why is that large porker shivering? The reason is that these animals are notoriously poor at thermoregulation. This incidentally explains what

Davis and colleagues suggest "that digit-like elements evolved at least twice within the Tetrapodomorpha."[64] This conclusion makes good sense in terms of *Sauripterus*'s life as a stealthy ambush predator, but it still raises some eyebrows. Even if it is "A fish with fingers?,"[65] the fin is encased in a stiff array of dermal bones (the lepidotrichia), meaning that although the internal skeleton would allow it to work like a limb, it is as if the fin was encased in plaster. Neil Shubin and Marcus Davis[66] remark, "The paradox of *Sauripterus* is that it has an exceedingly limblike endochondral pattern set in an entirely different structural, functional, and phylogenetic context from that of tetrapods."[67]

Even though this is another evolutionary experiment, is it really so different? In the canonical *Tiktaalik*, for example, the pectoral fin is well preserved; as Neil Shubin and colleagues[68] point out, "The lepidotrichia are solid rods that remain unjointed for most of their length and invest the endochondral bones dorsally and ventrally."[69] In this case, what is effectively a fin *and* a limb, and contributes so elegantly to its status as the *Archaeopteryx* of the sarcopterygians, may well have conferred a locomotory versatility, being "consistent with locomotion on the water bottom, along the water margins, and on subaerial surfaces."[70] So, too, and recalling their still fragmentary record, it will not be surprising if fossil material is recovered that shows how one or another rhizodontid also made the first steps onto land. Nor is this the end of the story. Just as fish like *Guiyu* and *Ptomacanthus* (p. 223) offer us tantalizing glimpses into what will ultimately be the deep foundations of the tetrapod body plan, we can also see that the potentiality for limbs goes deeper into the Tree of Life than just to rhizodontids. If one looks at the fins of the other sarcopterygians, as revealed by either the developmental patterning in a lungfish fin[71] or the deposition of fin bones in a fos-

sil coelacanth,[72] the fundamental similarities suggest that many of the components of what one needs to make a limb are already in place—and more than just a limb. Let us peer again into the very depths of evolution. In this developmental process, fundamental asymmetries suggest that one of the digits will, sooner or later, become a thumb.[73] Very handy.

The road to a limb need not only be revealed in fossil bones. In principle it could be encoded in the form of trace fossils, as trackways. Such indeed have been recorded from the Upper Devonian of Australia[74] and even more significantly (because they predate the record of body fossils) from the Middle Devonian of Poland.[75] Much more controversial, however, are other Australian trace fossils (located in the courtyard of a homestead) that have also been attributed to a tetrapod but are substantially older, perhaps even Upper Silurian.[76] Given both the indifferent preservation and questions of exact provenance, others have thought to invoke strolling arthropods.[77] Yet the matter is not closed, and is one of several hints that key aspects of tetrapod evolution may remain buried in Australia and other parts of the now dispersed Gondwana.[78]

With tetrapod feet now firmly on the ground, new vistas lay on the evolutionary horizons. Ironically one important road was straight back from whence they had come, to the oceans (p. 77). For other tetrapods, the trees beckoned to an increasingly agile arboreal existence. Once perched in the canopy, fresh horizons emerged as either gliders or ultimately 95 animals capable of powered flight. Sperm whale, capuchin monkey, flying lizard, and eagle, respectively, exemplify these various evolutionary destinations, but in each and every case the trajectories are haunted by striking convergences. Just as the sarcopterygian fish had lurking within them the latent potentialities to evolve inevitably a suitable starting point for limbs), but also a mobile head (and hence one day a neck), equipped with fangs that could engage in a crocodilelike lunge as against the more primitive mode of suction feeding[40] (p. 44).

Crucially this transformation happened several times, as emphasized by Emiliya Vorobyeva.[41] She notes that worrying as to whether the tetrapods have single or multiple origins is now irrelevant, given "the mosaic nature of evolution and the wide presence of parallelisms (homoplasies) in the phylogeny of various sarcopterygian groups";[42] she concludes that "different sarcopterygians acquire tetrapod features independently." Equally intriguingly she suggests that these parallelisms evidently reflect "common latent morphogenetic potentialities."[43] This becomes clear when we look at the phylogeny of these sarcopterygians. As Per Ahlberg and Zerina Johanson[44] remark, the tetrapods are "one of several similar evolutionary experiments,"[45] which is reflected in the evidence that "Tetrapod-like character complexes evolved three times in parallel within the Tetrapodomorpha."[46] In addition to the lineage that ultimately led to the tetrapods, two other groups are of interest: the tristichopterids[47] (which include the canonical fish *Eusthenopteron*[48]) and the rhizodontids.[49] In the context of evolutionary inevitabilities, the former group is perhaps the less compelling example simply because the tristichopterids are the sister group of the tetrapods and so are quite closely related. Nevertheless, not only do they show the characteristic mosaic evolution,[50] but Gael Clement[51] comments as to how the more derived tristichopterids also show "a remarkable parallelism [with the tetrapods and rhizodontids] in their dramatic increase in size, reduction or loss of their median fins, gain of diphycercal tails with a low aspect ratio, and development of a pair of fangs at the lower jaw symphysis."[52] This is echoed by Per Ahlberg and Zerina

Johanson,[53] who note how among the tristichopterids there is an evolutionary trend of moving from a life in open water to that of a stealthy ambush predator.[54] If the next meal depends on a sudden spurt followed by the lightning closure of the jaws, then a capacity to begin to move the head independently of the rest of the body would be advantageous. A neck is acquired convergently, as exemplified in the Australian tristichopterid *Mandageria*.[55] In their analysis of this fish Zerina Johanson and colleagues make the important point that this genus "presents an intriguing picture of 'piecemeal convergence' with the *Panderichthys*-tetrapod clade."[56] This point reinforces the already emphasized mosaicism, but the function of the neck joint is actually more similar to much more advanced tetrapods known as the temnospondyls.

Among the rhizodontids the convergences become more striking, simply because this group emerged from substantially lower in the sarcopterygian lineage. Although widespread in the mid-Paleozoic, with some representatives growing to an impressive size, overall the rhizodonts are not that well-known, with a limited number of genera and much of the material fragmentary.[57] Articulated material, of which the Devonian *Gooloogongia* is of particular note,[58] reinforces the evidence that the rhizodonts are following very much the same evolutionary trajectory as the tristichopterids and tetrapods in terms of body form and ecology.[59] Thus the form of the pectoral fin is not only consistent with life as an ambush predator,[60] but it too has a limblike arrangement.[61] This is particularly striking in a rhizodont from the Upper Devonian of Pennsylvania. In *Sauripterus*, at its far end the large fin is equipped with an array of narrow bones (the radials) that very much recall the diagnostic fingers of tetrapods.[62] These bones are an invention of the rhizodonts and so convergent with the equivalent region (known as the autopodium) of the tetrapods.[63] Marcus the Lower Devonian of China. This fish has been interpreted as sitting on a phylogenetic knife-edge, poised between the lungfish and the tetrapodomorphans and sharing features otherwise characteristic of either group.[31] Let us continue our descent of this part of the Tree of Life. Now we encounter a slightly older fish, again from China. Known as *Guiyu*, this animal shows a melange of characters between the sarcopterygians and the other major group of bony fish, the actinopterygians.[32] As Min Zhu and colleagues note, not only is there an "unexpected mix of postcranial features,"[33] but this mosaic evolution "suggests considerable parallelism between actinopterygians and sarcopterygians."[34] Deeper in this particular phylogeny? Consider the somewhat puzzling group of Palaeozoic fish known as the acanthodians. These are evidently key in the still-earlier diversification of jawed fish, and in the form of the Silurian *Ptomacanthus* the fish again displays a mosaicism that combines features of both the shark group (chondrichthyeans) and the bony fish.[35]

To the cladist this melange of characters is extremely tiresome, and so is expressed in terms such as "puzzling," "unexpected," and so on. At one level it is quite entertaining because every important new discovery tends to upset somebody's favorite apple cart.[36] Much more, however, is at stake. Far from being a local curiosity of the sarcopterygians and their friends, mosaic evolution is a universal. Even though this mosaicism is a megaphone directed at those wedded to the irrefutable certainties of the cladistic methodology, the jumble of advanced and primitive characters is very far from being the profound irritant at those who see evolution as little more than an exercise in Lego blocks (here's a blue one, where did I put the yellow one . . . ?). In reality, the universality of mosaic evolution highlights five things. First, organismal construction is far more

dynamic and flexible than often thought. Second, major groups do not appear by some mysterious saltatory method—no macroevolutionary hocus-pocus; on the contrary their emergence is gradual. Third, in the first stages of the diversification of a clade, species that with the benefit of hindsight will be identified as the ancestors of one major group (let us say a seagull, flying in as our representative tetrapod) will turn out to be almost indistinguishable from closely related species that ultimately will evolve into another major group (say, a herring, as a representative of the bony fish). In other words, the trajectories of sarcopterygian and actinopterygian certainly lead to what ultimately are radically different results (as the aerial seagull swallows the aquatic herring), but the common ancestor will inevitably show a mosaic of characters. Some will be deemed diagnostic of the sarcopterygians, and others, in hindsight, be taken as the hallmarks of actinopterygians. At each and every evolutionary bifurcation that leads finally to seagull and herring, there will be a mosaicism of characters. Fourth, evolution may be modular, but by no means all combinations are viable, and only a limited number of stable solutions will emerge, repeatedly. Finally, this sorting process is geologically rapid.[37]

Happily these various Paleozoic fish were unaware that they would be presenting evolutionary conundra to their tetrapod (if bipedal) descendants. Their evolutionary mosaicism makes sense when one views them as having the developmental capacity to evolve the morphology appropriate for a particular and often very specific ecology. The key to this story was the transition from a fish that was an open-water hunter to a lurking ambush predator[38] that skulked in a variety of aquatic habitats and seemed to have ranged from near-shore marine to lacustrine.[39] This entailed the evolution of fins suitable for precise maneuvering (and just happened to be surfaces."[9] Graham concludes, "air breathing has independently evolved among the fishes at least 38 times and perhaps as many as 67 times,"[10] a point that Karel Liem[11] echoes in his analysis of ABOs. He writes, "Synbranchids, clariids, and anabantoids realize closely similar functions in a periodically or chronically deoxygenated aquatic environment. Convergence in the evolution of the respiratory structure-function complex has resulted in the evolutionary development of strikingly similar structural novelties."[12]

THE FIRST FOOTPRINT

While some of these fish are effectively aquatic, others are denizens of intertidal zones,[13] and a few are capable of impressive terrestrial excursions—including not only those millet-eating clariids but the climbing gouramis (anabantids).[14] To see how, when, and even why it was that one group of fish, the sarcopterygians (the living descendants include the celebrated coelacanth as well as the lungfish[15]), hauled themselves onto land and so opened themselves to further evolutionary adventures such as invading the air (p. 189), reinvading the seas, and climbing trees (p. 67), we need to turn to a series of remarkable fossil discoveries. The greatest "surprise" is that much of what was "needed" to become terrestrial evolved in an aquatic milieu. Here we should salute the prescient remarks of Theodore Eaton,[16] who noted "that the walking movements of early tetrapods and the structure of their legs, feet and digits originated as a purely aquatic adaptation long before there was any occasion for, or possibility of, sustained locomotion on land."[17] Eaton sketched out what we can now confirm from the fossils, but Whiggish paleontology is evidently not quite defunct, given that the emphasis has very much been on the story of "from finned fish to fully limbed tetrapod." The latter are most famously represented by such Upper Devonian animals as *Acanthostega* and *Icthyostega*,[18] but this story also involves a series of truly fascinating intermediaries, such as the wonderful finds of *Tiktaalik*[19] and *Ventostega*.[20] The manner in which the various major transformations were achieved, such as the key area of hearing[21]—combined with the fact that the changes were not only incremental but sometimes had already emerged in more primitive sarcopterygians[22]—make this area a flagship for evolution. There are, however, some wider issues. Although touched on by some of the leading workers, we might offer a somewhat different perspective, one that chimes with both how complex forms emerge—be they cacti (p. 156–57) or birds (p. 193)—and what likelihoods accompany this process. In other words, is a tetrapod an evolutionary inevitability?

The first observation to make is that after a series of spectacular discoveries, new finds in the fossil record are unlikely to overturn the basic story, even if tantalizing but fragmentary material[23] (and controversial trackways, p. 225) suggests that further surprises will certainly occur.[24] Even with the material at hand, the style of evolution is strikingly mosaic. The fossils show an amalgam of advanced and primitive features, repeatedly provoking cries of "surprising" or "puzzling."[25] This applies not only to near-tetrapods such as *Panderichthys*[26] and *Livoniana*,[27] but more primitive sarcopterygians. In *Goganasus*,[28] for example, John Long and colleagues[29] noted that this fish "Unexpectedly ... shows a mosaic of plesiomorphic [primitive] and derived tetrapod-like features."[30] Unexpected? Let us look into the deeper ancestry of the sarcopterygians, specifically *Styloichthys* from

CHAPTER 20

THE ROAD TO MAMMALS

IN NATURE, if it can happen, it will. Less often appreciated is that what does happen is an infinitesimally small proportion of what in principle might evolve. Biologists are dazzled by the diversity, fecundity, and sheer chutzpah of biological systems—the apparent exuberance of the biosphere, of the rococo towering above the baroque. All good fun, but while encouraging us to enthuse on superficial splendors it has diverted inquiry from uncovering the architectural principles that underpin all life, the quest for biological properties.[1] If the roads of evolution are exceedingly narrow, then we need constantly to remind ourselves that practically all the alternatives are not worlds that might have been, but worlds that will never be. That applies as much to mammals—and us—as any other group. Neil Shubin writes of "Your inner fish,"[2] and it is equally fascinating to consider that series of successive palimpsests that successively reveal a primate or reptile, or indeed a fish. At each and every stage a set of extraordinary stories is to be told, but in some ways the most crucial question is this one: How did our remote ancestors first heave themselves onto land?

GASPING FOR AIR

Let us rewind the tape to what in the bad old days of Whiggish paleontology was known as the "Age of Fish" (although this is dwarfed by the deeply politically incorrect references to the "Age of Man"). We hardly need W.C. Fields's famous quip to remind us that fish live in water, but the fact that they were destined to invade the terrestrial habitat is not so surprising when we recall their examples of limblike locomotion (pp. 72–73). Of greater moment is the well-known equivalence of the swim bladder[3] and lung. So, too, the capacity of fish to breathe air has evolved multiple times.[4] Not only does this include the eponymous lungfish, but a remarkable variety of other fish. These include the labrisomids and butterfly fish, with their striking capacities for aerial vision (p. 105), as well as the clariids (or walking fish), of which the West African *Clarias* is reported to engage in nocturnal excursions to feed on millet planted by the locals.[5] Even a species of a mormyriform electric fish (p. 146) breathes air,[6] while its shocking Neotropical counterpart, the electric eel, is an obligatory air breather.[7] To be sure, the varieties of so-called air-breathing organs (ABO) are striking, but as Jeffery Graham notes,[8] "in spite of this diversity there has been a remarkable convergence in ABO structure owing to evolutionary constraints imposed on devices needed for air capture, for air storage space, and for adequate exchange independently in other groups of fish,[204] and the repeated emergence of these adaptive complexes is a further reminder, if it was needed, that evolution is neither "difficult" nor radically unconstrained.

One might still protest that the placentation of these fish has at best a tenuous connection to instances in other groups. Consider, for example, the so-called trophotaeniae. These are ribbons of tissue that extend from the embryo and serve to absorb the uterine nutrients.[205] Nevertheless, as John Wourms and Daniel Cohen remark the "Occurrence of trophotaeniae . . . is a remarkable example of convergent evolution,"[206] because these trophotaeniae have "a striking similarity to the mammalian strategy of embryonic nutrition, to that of oviparous onychophorans,[207] and to progressive provisioning of the brood in advanced social insects. *In all instances, the required nutrients are delivered upon demand rather than being sequestered* in toto, *prior to demand*."[208] And far-fetched as it might appear to link fish and mammal, consider the delightful sea horses. Among their peculiarities is that, like some frogs, the male is viviparous.[209] Not only do the fathers brood their young, but as Kai Stölting and Anthony Wilson explain, the extensive vascularization of the pouch demonstrates that the implanted embryo benefits from "far more than simple protection, and there is evidence for osmoregulatory, aerative, nutritive and possible immunoprotective roles."[210] Sea horses are not horses, and the viviparous modes have some obvious differences, yet Stölting and Wilson conclude that "mammalian pregnancy and seahorse reproduction exhibit compelling morphological and functional similarities."[211]

Given the diversity of placental fish, it might seem reasonable to suppose that when the sarcopterygian fish came onto land (p. 222), they were viviparous also. Such evidence as we have points in the opposite direction. To be sure, the living coelacanth is also placental,[212] and evidence for viviparity extends back to the Triassic,[213] but this is evidently another piscatorial convergence. Whether early amphibians evolved any sort of viviparity is conjectural, but certainly among the extant groups we find the usual riot of parental strategies, not least the employment of foam nests (pp. 39–40). Viviparity itself has evolved in all three major groups,[214] but is particularly characteristic of the caecilians, where it has evolved at least four times.[215] Although there is matrotrophy, in only one case might there be a placenta. This is a South American caecilian (*Typhlonectes compressicaudus*), which, in addition to a possible area of absorption on the ventral epidermis,[216] produces two enormous gills, richly endowed with blood vessels that make contact with the uterine wall.[217] The evidence that this is some sort of placenta is circumstantial,[218] but evidence certainly exists for oxygen exchange.[219] In this and many other cases the developing larvae derive nourishment from various sources, including scraping the uterus with special fetal teeth.[220] Among the more peculiar sorts of nourishment, the frogs leap to our attention, not least the male *Rhinaderma*, which both broods and nourishes the young in its vocal sacs.[221] Equally extraordinary are those frogs that house their young in chambers on their back, which has arisen independently in hylids and pipids. In the former, represented by *Gastrotheca*,[222] the poison glands in the skin are reduced, presumably to protect the young, while the proliferation of blood vessels points toward some sort of matrotrophy.[223] Much the same arrangement occurs in the viviparous pipids.[224] In this latter case, major nutrient transfer is thought to be less likely, but as Hartmut Greven and Susanne Richter note, "the nidation process, i.e., the embedment of eggs . . . strikingly resembles the embedding of the early embryo in the endometrium in mammals."[225] So to mammals we now turn.

placentation remains conjectural. The least surprising example involves the Cretaceous mosasaurs;[173] unsurprising because these impressive predators are close relatives of the varanid lizards[174] (and by implication had forked [and flicking] tongues[175]). So, too, were the wholly aquatic ichthyosaurs,[176] where viviparity is most evident from the spectacular material recovered from the Lower Jurassic Posidonia Shale.[177] As a group, however, they made their debut in the Lower Triassic; not only is there evidence for live birth but the exit of the young (head first) suggests that[178] viviparity arose in their terrestrial ancestors. Nor were they alone in the Mesozoic oceans. Their counterparts in the form of plesiosaurs are now known to have borne live young.[179] End of the convergent list? Not quite, because at least some pachypleurosaurids (*Keichousaurus*) were viviparous,[180] as most likely were the related nothosaurids.[181] All these examples are marine, but in freshwater deposits of Cretaceous age, a group of reptiles known as the choristoderans were viviparous.[182] Considerably more ancient is a possible example in the Permian mesosaurs,[183] while still further back in the Paleozoic we see the evolution of viviparity in the fish with its independent appearance[184] in the Devonian placoderms[185] and arthrodires.[186] This piscatorial roster of viviparity is swollen yet further by a number of more compelling examples, notably in teleosts such as the poeciliids,[187] along with the sharks and rays (chondrichthyean).

This latter occurrence is also very ancient[188] and has evidently evolved multiple times, albeit with occasional reversals to oviparity.[189] Echoing the remarks by Daniel Blackburn on squamate viviparity, Nicholas Dulvy and John Reynolds conclude their analysis of chondrichthyean viviparity by remarking, "Our data do not support a linear, irreversible progression toward a 'pinnacle' of maximum maternal input,"[190] and they reemphasize the lability of systems of paren-

tal care. All correct, of course, but both the strength of the trend and the sophistication of the endpoints may matter more. As with the lizards, the range of viviparous modes is considerable,[191] but significantly a number of shark species have evolved placentation.[152] Just how complex some of these placentae are seems to be a matter of debate. The spadenose shark (*Scoliodon*) has perhaps the most highly developed placenta,[193] with John Wourms suggesting that "placental viviparity has evolved . . . to a degree that rivals some eutherian mammals."[194] As is typical in these fish, matrotrophic nutrition is mostly via a uterine "milk" with its absorption mediated via specialized extensions of the umbilical cord (the appendiculae).[195] Wourms argues, however, that some nutrients are obtained via the blood supply (hemotrophically)[196] in the placenta.[197] Significantly it is evident that an immune system is also in operation,[198] further reinforcing the notion that the placenta is far from some sort of passive implant.

What of other fish? Successful as the sharks and rays remain, the waters of the Earth are dominated by the bony fish. These show a remarkable range of reproductive modes and parental care.[199] But all types of parental care have evolved repeatedly, and viviparity itself has emerged independently at least eight times. So too has placentation.[200] One such example is found in the four-eyed fish *Anableps*[201] (p. 105). This fish belongs to the large group of cyprinodontiforms, and it is in this diverse assemblage that placentation is particularly recurrent, notably among the poeciliids.[202] As David Reznick and colleagues stress, such examples of placentation are as important to evolutionary biology as other instances of the emergence of biological complexity (famously the eye), yet as these workers show not only has this organization evolved on multiple occasions but "placentas can evolve in 750,000 years or less."[203] Very similar types of placentation have evolved necessarily constrained to operate in a very similar manner, be it lizard or mammal.

Apart from the squamates the mammals may appear to be sine qua non of placentation, but even here convergences emerge. Thus, in the extinct multituberculate mammals, the structure of the pelvis (at least in the Cretaceous *Kryptobaatar*) points to a live birth, and although necessarily an inference would indicate that its viviparity had evolved independent of the therians.[157] If so, the neonate would have been tiny and would have been reminiscent of the equivalent in the marsupials. Even in the marsupials and eutherians the placentation may have arisen independently.[158] Such a view is shared by Anthony Carter,[159] who also points out that the fossil record is not quite silent with respect to placental evolution. This includes both the unborn[160] and, more intriguingly, the remains of the placenta itself.[161] Unfortunately in this latter case, a horse from the Miocene, histological details are not preserved. Most likely, like its descendants (and all other perissodactyls), it was of the epitheliochorial type (and so similar to the lizards), rather than either hemochorial or endotheliochorial. Which, however, came first? Even with advances in mammalian phylogeny it is not easy to decide. Perhaps more likely the hemochorial condition (where the maternal blood makes direct contact with the fetal chorion) is primitive, and if so, then the endotheliochorial placenta must have evolved independently[162] several times. In at least the hemochorial placenta, convergence can take some further unexpected turns. It transpires that certain genes (the syncytins) play a key role in placental development; not only are they derived from retroviruses, but among the mammals they have been recruited at least five times[163] independently.[164] The variety of mammalian placentae, which goes considerably beyond the three main types mentioned above, not only has proved evolutionarily problematic but has provoked the rhyme-and-reason sort of question. Is it all one vast muddle? Most likely not. Convergence of placental architectures and the molecular necessity of syncytins suggest that things are far from random. One explanation looks to the life histories, not least how fast the speed of the respective paces of life.[165]

There may also be wider principles in the evolution of viviparity and placentation, the most compelling of which revolves around the stark fact that the needs of the developing fetus and parent cannot coincide. By definition, the tissues of the former are not genetically the same as the adjacent maternal organs. Conflict is inevitable, not least in terms of demand for nutrition.[166] The embryo must also persuade its parent that it is not a foreign body, worthy of the attention of the immune system.[167] To avoid rejection and indeed encourage maternal investment, the embryo draws on an armory of hormones and other molecules. Cytokines such as the so-called interleukins play an important role, not only in mammals, but also placental lizards[168] and such fish[169] as sharks.[170] This conservation of molecular equipment begs the question of whether all these examples of viviparity and placentation can really be regarded as convergent as against reflecting a "deep homology." Convergence, however, provides the more valid perspective. First, the entire concept of so-called deep homology is deeply problematic. Related to this is the difficulty in identifying the ancestral function of a molecule as against its repeated co-option. So, for example, in the case of the interleukins,[171] their involvement in reproductive biology goes far beyond groups that happen to possess a placenta.[172]

In the final analysis, what matters is that something like a placenta is an evolutionary inevitability. Squamates show why this is so, but independently a number of extinct reptiles have evolved viviparity, although whether this was accompanied by any sort of reptiles, when the fertilized egg (the blastocyst) in lizards such as *Eulamprus* (*E. tympanum*) comes into contact with the uterus wall, its plasma membrane undergoes a series of major changes that permit its attachment.[133] In a subsequent paper,[134] Margot Hosie and colleagues note, "It may at first seem surprising that similar changes should take place in the plasma membrane of mammals and *Eulamprus tympanum*, when the uterine epithelium is penetrated by the blastocyst (hemochorial and endotheliochorial placentation) in many mammals, but not at all in *E. tympanum*."[135] They go on, however, to remind us that the type of placentation known as epitheliochorial[136] (p. 218), where by definition there is no breaching of the epithelium, does not in itself preclude the transformation of the plasma membrane.[137]

However, in the more complex types of squamate placenta, similarities to the eutherian mammals become the most pronounced. Here we see a complex interaction between the fetal and maternal tissues, notably via the so-called omphaloplacenta (where nutrients are supplied) and a respiratory chorioallantois. The best-known example among lizards with a type III placenta is *Chalcides* (*C. chalcides*),[138] where the secretory interactions between the maternal and fetal tissues led Carolyn Jones and colleagues[139] to remark that although this placenta has features that are unique, it "has evolved a true epitheliochorial placenta with many aspects in common with its therian counterparts."[140] In addition to the contact between mother and embryo being deeply interdigitated, Daniel Blackburn also notes, "Some of the mucosal folds of the placentome pull away from the uterine lining at the time of birth and are extruded with the chorioallantois and fetus. . . . This situation represents the only known case of deciduate placentation among squamates."[141] If *Chalcides* is the only lizard with an afterbirth, other African skinks such as *Eumecia*[142]

and *Trachylepis*[143] display equally complex and specialized placentae. With respect to the latter taxon, Daniel Blackburn and Alexander Flemming note how this lizard "shows an astonishing set of reproductive specializations that are convergent on those of certain eutherian mammals."[144] Here an endotheliochorial placenta[145] not only entails an intimate association of fetal and maternal tissues but also "represents the only known example of invasive implantation in any nonmammalian amniote."[146] An epitheliochorial placenta has evolved independently in the Australian skink *Pseudemoia*. This, however, shows a further refinement because the epithelium of the uterus is syncitial;[147] as Susan Adams and colleagues note,[148] "a similar placental morphotype has been described in a eutherian mammal, the American mole."[149]

Striking as these examples may be, pride of place goes to the Neotropical skink *Mabuya*. With its complex placenta[150] *Mabuya* has been repeatedly compared with eutherian equivalents.[151] This organization is consistent with a tiny egg[152] and protracted gestation time, with the embryo largely dependent on maternal input. While no single species of mammal displays all the specializations seen in the placenta of *Mabuya*, Martha Ramírez-Pinilla[153] notes, "All of these morphological specializations of the *Mabuya* allantoplacenta are very similar to those found in the placentae of different eutherian mammals, in the sense that they show an impressive convergence in tissue organization and cytological features of the endometrial and chorioallantoic placental components."[154] Not surprisingly, in the placentome the fetal/maternal boundary shows an intimate interdigitation, as well as specializations for the transfer of nutrients, notably in the form of pits[155] in the chorion known as areolae.[156] These convergences suggest that while multiple pathways to placentation and varying degrees of complexity exist, the most sophisticated arrangements will be insects (p. 175). To echo the thoughts of the poet David Jones, at first sight the topic of not only viviparity but parental care would seem to be little more than a "blooming, buzzing confusion." Nevertheless, as John Reynolds and colleagues[108] note, "many taxa show consistent trends in evolution of care,"[109] although the roles of either parent (or biparental care) show a considerable diversity. In squamates (i.e., lizards and snakes), for example, parental care has "undergone numerous transitions toward female inputs to the young from a state of no care and egg laying."[110] So, too, given that the range of viviparous groups embraces humans and tapeworms,[111] even a single trend emerging here would be surprising.

Nevertheless, when it comes to viviparity per se, it is not the mammals but the snakes and lizards (the squamates) that provide the gold standard.[112] Here viviparity is rampantly convergent, having evolved well over a hundred times.[113] In this respect of particular importance are the skinks[114] (as Daniel Blackburn remarks, "Saurian viviparity has reached a pinnacle in this large family"[115]) and the snakes (notably the colubrids and vipers),[116] but that it is an ancient invention is evident from a Cretaceous lizard containing the telltale embryos.[117] Of equal significance is that, once attained, viviparity is effectively irreversible,[118] with the oviparity[119] of a boid snake (*Eryx*) being one of the very rare exceptions where Dollo's "law" (p. 28) has been put into reverse. Not only do the adaptations to viviparity show a remarkable variation, but as Daniel Blackburn[120] has stressed, "a single [evolutionary] scenario cannot be assumed to account for the evolution of viviparity and matrotrophy. An important implication of this inference is that lineages may converge on viviparous matrotrophy by an array of different historical sequences."[121] This warning has particular force when Blackburn also writes, "Numerous reproductive similarities

of squamate reptiles and mammals make it feasible to compare viviparous representatives of the two groups,"[122] but he offers this as the opening lines to a section titled "Risks of Higher-level Generalizations."[123] As he points out, some features are derived from ancestral amniotes, if not still deeper, in the Tree of Life. Thus, the squamates themselves show a variety of preadaptations for viviparity, while lizards and mammals also display obvious differences. Blackburn, therefore, is correct to note that the diversity of viviparities and their tangled evolutionary paths militate against any simple model, but when he insists that "homocentric assumptions of an orthogenetic transformation toward the eutherian condition should be abandoned,"[124] we might gently protest that this is more of a straw man and ultimately misses the more important point of the emergence of particular biological properties.

Squamate viviparity[125] revolves around several features, but perhaps most notable is the independent evolution of a placenta.[126] By no means do all viviparous squamates show placentotrophy, and among those that do, classically four categories (I to IV)[127] are identified.[128] With the exception of the most complex variety—type IV in the skink *Mabuya*—each of these placental types has evolved repeatedly.[129] It would be a mistake to assume that these types can be arranged in a single sequential order of increasing complexity, but despite multiple historical pathways, a common theme in squamate placentotrophy is the capacity for fetal/maternal exchange. This culminates in the complex types III and IV,[130] but although they provide the most compelling similarities to the mammals, types I and II are not without interest.[131] For example, although there are "fundamental differences between female reproductive tracts of eutherians and squamates,"[132] which as Daniel Blackburn stresses precludes the resorption of the embryo in the these membranes evolved. The real innovations may have been connected more to the development of a membranous shell (often calcified) and the availability of substantial quantities of yolk that could nourish the embryo via an extensive set of blood vessels.[97] This mass of yolk situated in the eponymous yolk sac now means the amniotic egg is larger than the eggs of most amphibians, with the growing embryo effectively perched as a sort of plate on top of the yolk. One consequence is that the cell divisions no longer encompass a sphere but must be incomplete—that is, are meroblastic.[98] Only the mammals, with a tiny egg, have reevolved the so-called holoblastic cleavage, with complete cell divisions. Holoblasty, however, is also typical of the amphibians, and they offer a clue as to how the amniotic condition may have arisen.[99] In this context of particular interest is a Puerto Rican tree frog (*Eleutherodactylus coqui*) that, being terrestrial, has discarded the tadpole stage and so shows direct development[100] (p. 25). More important, however, is that unlike other frogs, where the yolk is engulfed during the process of gastrulation, in this species the body wall spreads across the yolk mass.[101] Of equal significance is that while some of the embryonic endoderm goes to form the gut, other cells are dedicated to absorbing the nutritious yolk.[102] No vascular system is involved,[103] but still it is

an echo of the yolk sac. As Daniel Buchholz and colleagues remark, "Although the movement of the body wall in *E. coqui* is superficially similar to that of the yolk sac, it is not homologous. . . . Nonetheless, the similarity suggests that the independent evolution of an enveloping layer can occur, and it could be used to surround an uncleaved yolk mass."[104] This frog, therefore, provides one pointer as to how an amniotic egg might have evolved.

Nevertheless, given the delicacy of the tissues, investigators have despaired that the original events leading to the birth of the amniotic egg will ever be revealed. Nevertheless, does this really matter? Let us don our evolutionary headphones and tune them to convergence, summon up the likelihood of mosaic evolution, and not be shy of seeking out analogies. In combination this strategy should allow us to not only predict the most likely path to the amniotic egg but in due course also show why it was an inevitability. Nor need the historical record, however imperfect, be dismissed. The recovery of fossil embryos in fish[105] (p. 220) and more notably in a primitive Permian reptile *Mesosaurus*[106] may mean that, one day (in combination with other examples of remarkable fossil preservation), direct evidence of the first amniotic egg will emerge.

Coming Out Alive

The case of *Mesosaurus* is all the more interesting because unlike nearly all reptiles, even in the Paleozoic, the young were conceivably born alive. In contrast, and as with the birds and even the monotreme mammals, the newborn's first glimpse of things to come is usually through a crack in the egg. Not always. In some cases the young slither alive from their mothers into the waiting world. In other words they are viviparous. At the heart

of viviparity are two key, if self-evident, features: retaining the egg in the reproductive tract (typically the oviduct)[107] and, by one means or another, nourishing (and in some groups this might involve one embryo consuming its siblings) the developing young—that is, the process of matrotrophy. Viviparity is, of course, just one aspect of the vast topic of parental care, examples of which we have already encountered among the lactating of the familiar, but also the very bizarre. As ever it is not unique: the story repeats itself. Once again the animals are hermaphrodites, but effectively unrelated to snails and slugs: the earthworms. The fact that these creatures fascinated Darwin[84] is a reminder that their familiarity might reveal some surprises. Blind maybe, but they engage in a courtship[85] that involves not only intense touching with the anterior of either animal but frequent prenuptial visits to each other's burrow.[86] Their hermaphroditic copulation is well-known, but less so is the fact that each partner will impale the other with its elongate chitinous chaetae,[87] to the extent that almost the entire bodywall of the worm is penetrated,[88] and in doing so, copious quantities of a mucuslike fluid are injected. The solution is convergent on the molluskan love darts. The penetra-

tive chaetae are strikingly similar to those found in some snails, being elongate, bearing grooves, and having a very sharp point.[89] As before, the mucus is secreted by adjacent glands,[90] and evidently injects a comparable allohormone.[91]

In my more wicked moments I wonder if above the beds of the ultra-Darwinists hangs a picture of that bearded genius. Darwin arranged for his earthworms to be serenaded by bassoon and piano,[92] but now does he look quizzically down on the evening's rompings? Nobody needs reminding that notwithstanding the wilder shores of animal reproduction, not to mention the zoo that represents human conjugal antics, all that really matters is a fertilized egg (or so it is claimed). But the egg itself turns out to be a most remarkable construction.

Cracking the Egg

Fish and amphibians lay eggs. Even though employing ingenious methods of brooding, their reproduction remains tied to water. All that changes with the arrival of the amniotic egg. Widely heralded as one of the great evolutionary innovations, this allowed full terrestrialization and so paved the way for the birds and mammals. Not all, however, are convinced. Here the argument is that the challenges of living on land have been exaggerated, and as importantly these views raise the possibility that supposedly key features of the amniote condition in reality have evolved elsewhere, including even in an aquatic milieu.[93] Thus, Joseph Skulan dismisses not only what he terms the "Haeckelian framework,"[94] where the evolutionary transition to the amniotic egg is regarded as little short of heroic, but goes on to argue that "the reproductive history of vertebrates is complex and has probably involved numerous reversals and convergences that likely have obliterated

any phylogenetic signal in the distribution of egg characters among vertebrates as a whole."[95] Given the prevalence of mosaic evolution, this would be expected and at the least suggests that reconstructing the details of the transition that resulted in the amniotic egg may be far from simple. Nevertheless, Skulan may be overlooking a wider point. The real question is not what was the exact route, but how likely is an amniotic egg per se? A fluke of history or, as with the analogous case of the seed, an inevitability?

As the name suggests, a key feature[96] is the appearance of the amnion, the membrane that encloses the embryo. It in turn forms one of the so-called extra-embryonic membranes. This also includes the chorion (the shell membrane) and allantois (for storing excretory material), which will play key roles in the subsequent evolution of the placenta (p. 217). It is, however, conjectural as to when in the emergence of the amniotic egg the hyenas), rodents, and bats, the baculum is also characteristic of the primates. We are an exception. Possibly even early species of *Homo* were so endowed, with the ensuing demise of the baculum linked to changes in reproductive behavior.[67] Nor are humans the only primates to have discarded the baculum. So have a number of New World monkeys, but frustratingly there are no obvious parallels in their sexual lives to provide further clues as to why the human baculum was lost. Let irony step forward, because among the uses humans have found for the baculum are as love tokens, and more mundanely tiepins (if small) and clubs (if somewhat larger).[68] From a human perspective the idea of a baculum has, at least for the male, perhaps a certain strangeness. In terms of evolution and convergence the arena of sexuality has yet wilder shores.

Consider the land snails. Even if they are hermaphrodites, some aspects of their behavior have a certain familiarity. For example, courtship involves not only a clockwise circling, but as Michael Landolfa[69] remarks, they "engage in bouts of mutual facial caressing, mouth-mouth and mouth-genital pore contact and biting."[70] So far, so predictable, but upon copulating, what are euphemistically termed "love darts" are deployed to forcibly stab the partner. As Joris Koene[71] dryly notes, it is "an extremely odd behavior."[72] The darts are not actually thrown, like a spear, but everted from the genital apparatus. By no means all achieve their function: some fail to penetrate and some simply miss, but a successful stab may lead to the tip of the dart protruding from the far side of the sexual partner.[73] In the appropriately named Samurai snail (the Japanese gastropod *Euhadra subnimbosa*), the activity reaches a frenzy: with about two stabs a second and in a foreplay (if that is quite the word) of about an hour, each partner inflicts more than three thousand stabs.[74] In this snail the dart is very small, and

indeed these copulatory daggers, which are typically calcareous, come in a remarkable variety of shapes.[75] Similar darts occur in the related slugs, and not only is it possible that they have evolved more than once, but they have certainly been lost several times.[76]

So what is the function of their decidedly curious behavior? Not surprisingly a popular idea has been an echo of some of the more arcane areas of human sexual behavior—that is, stimulation. Other ideas have revolved around some sort of nuptial gift, a dollop of donated calcium, or an indication of rude health.[77] The solution is very different. It revolves around the snail's hermaphroditism and their baroque reproductive complex, which includes not only the penis to deliver the sperm but a corresponding female unit. In the latter, however, there is a saclike structure (the so-called bursa copulatrix), which is a place no sperm would wish to end up; strong and one-way peristaltic movements drive the sperm into a literal dead end, and on arrival in the bursa copulatrix they are rapidly immobilized and digested.[78] This may seem self-defeating until one realizes that all things being equal, males will seek numerous inseminations, whereas the females need to be very choosy. By no means is all sperm up to scratch, which explains the darts, because as they leave their sac they pass special glands and are coated with a mucus rich in chemicals known as allohormones.[79] Their function is to sabotage the partner's sexual system, specifically serving to isolate the bursa copulatrix while, by enhanced peristalsis,[80] encouraging transfer of sperm in the right direction, toward the ovaries. This helps to explain not only the complexity of the dart, evidently to carry more mucus, but also the range of both dart numbers and intensity of stabbing.[81] Moreover, in the snail *Everettia* the dart functions as a syringe,[82] matching the independent evolution of hypodermic insemination.[83]

So this story of love darts has elements

retained or reevolved a penis? One hypothesis for their employment is the risk of water damage to the sperm, which could be consistent with possession of a penis in ducks and other waterfowl. In the Muscovy duck its eversion is truly explosive,[48] and as Patricia Brennan and colleagues note it takes "an average of 0.36 s, and [achieves] a maximum velocity of 1.6 m s^{-1}."[49] Not bad, not bad at all . . . but in the context of the truly heroic perhaps the prize should go to the male Argentine lake duck, the spiny penis of which can reach 20 cm, about half the body length.[50] As remarkable is the so-called phalloid organ, effectively a fake penis, in the buffalo weaver bird.[51] This striking and robust structure is located in front of the cloaca, but unlike a penis it remains rigid and contains no duct to transfer sperm. Its function has provoked some head scratching.[52] An earlier suggestion was the plausible thought that if inserted into the female oviduct it could serve to dislodge the sperm of a rival.[53] Now, however, a very different function is evident. Observations of its use in captivity, either using a taxidermist's model bird equipped with an artificial cloaca or stimulation of the phalloid organ, showed that it led to ejaculation. Evidently its primary employment is for male stimulation.[54] In the final stages of copulation, which can be quite protracted, Mark Winterbottom and colleagues report how "the male's wing beat slowed to a quiver and his entire body shook; with his leg muscles apparently in spasm [and] his feet clenched hold of the female."[55] Yes, orgasms are convergent.

Such birds as possess a penis show significant differences with those of mammals, because inflation is achieved using lymphatic fluid,[56] while the sperm is not conducted along a central duct but via longitudinal grooves. Nevertheless, the specific case of the mammalian penis does nevertheless find a striking convergence in the turtle equivalent. As Diane Kelly[57] has remarked, "the similarity of the anatomical designs used by [mammals and turtles] to produce penile stiffness is astonishing."[58] Neither is this surprising: to operate this hydrostatic structure[59] it must be inflated with blood, but success also depends on a combination of stiffness and nonbending, as well as a specific array (axial orthogonal) of collagen fibrils. This arrangement guarantees tumescence but guards against unfortunate aneurysms.

Turtle and mammal, and for that matter crocodile,[60] have independently navigated to a very satisfactory solution. Perhaps the mammals have the edge because in many groups the penis houses a free-floating bone. This is the baculum (sometimes the homologous clitoris has its diminutive equivalent, the baubellum), which has probably evolved more than once. In its own quiet way, echoing the question as to why the penis is so rare in the birds, it poses an evolutionary paradox: What is the baculum actually for? The old idea that it was more or less a by-product is pretty well abandoned, and with good reason. First, despite its penile constraints, it comes in a remarkable variety of shapes and sizes; indeed arguably it is the most diverse of all bone types. There are, however, patterns, and the recurrent evolution of particular forms strongly suggests functionality.[61] Add to this the cost of making this bone, and (ouch!) the real risk of fracture,[62] if the baculum is adaptive, then for what exactly? Support, rigidity, and load bearing[63] are one answer. Arguments also exist for protraction of copulation or induction of ovulation, but theory by and large does not accord with data. As Serge Larivière and Steven Ferguson[64] write, "the existence of the mammalian baculum remains one of the most puzzling enigmas of mammalian morphology."[65] Perhaps there is no simple answer, but broadly the solution must lie in sexual selection.[66]

There are, moreover, other puzzles. Along with such groups as the carnivores (except for the mammalian Y chromosome in terms of the genomic organization.[30] James Fraser and colleagues remark that, along with the plants, "These parallels reveal that similar mechanisms drive the evolution of sex determining regions in all three eukaryotic kingdoms."[31]

Where else in the world of sexualities do stimulating convergences emerge? What about so-called parental imprinting? What this means is that a particular gene expression is dependent on one or other parent, and yet as far as mammals and plants are concerned, this imprinting displays some important convergences.[32] As Robert Feil and Frédéric Berger remark, "The molecular regulation of imprinted expression . . . is remarkably similar in mammals and plants,"[33] not least in terms of the methylation (as the name suggests this refers to the addition of a methyl [CH₃] molecule) of DNA and histone. In both groups imprinting is closely involved in tissues related to nutrition of the next generation,[34] and as Claudia Köhler and colleagues[35] note with respect to this parental imprinting, "the independent recruitment of similar protein machineries for the imprinting of genes is a notable example of convergent evolution."[36] Nor is this the only example where molecular convergence meets the area of reproduction.[37]

PENISES AND LOVE DARTS

When it comes to the bedroom, even the most dedicated biologist might wish to put the question of parental imprinting or the XX-XY system firmly at the back of his (and I hope her) mind. And where better to start this topic than with the gorgeous birds? Courtships, as exemplified by the bowerbirds (p. 274), may be complex affairs, and what are we to make of Alec Chisholm's[38] remark of various species of Australian bird "flitting about, during courtship activities, with flowers in their bills"?[39] In the birds, however, the culmination usually involves a somewhat awkward copulatory maneuver of adpressing cloaca to cloaca, the male fluttering above the female.[40] This is something of an oddity, because the vast majority of birds have abandoned the time-honored employment of male intromittent organ.[41] Why should so many birds lack a penis? After all, practically all male terrestrial vertebrates possess some sort of penis, some of which are compellingly convergent? It is a neat biological puzzle because about 3 percent of bird species do have a penis, and it has evidently evolved or been lost several times. Patricia Brennan and colleagues[42] stress that "phallus evolution in birds is dynamic and complex and that so far we lack a hypothesis that explains the general pattern of phallus presence/absence."[43] One suggestion revolves around safe sex. The arrangement of the cloaca makes unavoidable the proximity of the sex duct and waste products, and so quite possibly pathogens. Given this configuration, then maybe using an intromittent organ is a shortcut to sexually transmitted disease? Here is another suggestion. With the famous exception of the swifts (after meeting midair in a way not so different to midair refueling, the pair usually descend in a shallow dive, the female gliding and the male flying[44]), could it be that given copulation is ground-based, then the risk of sudden interruption by a predator is much enhanced?[45] Nor should one assume that cloacal contact is always problematic.[46] Particularly extraordinary are the Vasa parrots, endemic to Madagascar (and nearby Comoros Islands). These birds possess large cloacal protrusions that interlock and allow copulation to proceed for well over an hour.[47]

What of those birds that have either

include haplo-diploidy and environmental cues (notably temperature), the latter being familiar among such reptiles as the crocodiles. These beasts are semiaquatic, but what happens when reptiles return to an oceanic existence, as spectacularly occurred in extinct groups like the ichthyosaurs (p. 77) and mosasaurs (p. 78)? In these cases, egg laying is hardly an option. Unsurprisingly such groups swell the roster of viviparity (p. 215), but this led presumably to the evolution of mechanisms of sex determination and was most likely a key to their remarkable adaptive radiations.[8] The inference concerning these Mesozoic behemoths reinforces the remarks by Melissa Wilson and Kateryna Makova[9] that "Sex chromosomes have arisen independently in a vast range of species."[10] It is generally agreed that sex chromosomes are derived from autosomes, but in the case of the birds and mammals[11] their respective systems arose independently.[12] Even with respect to groups that are phylogenetically much closer to the birds, again we find evidence for convergence of the sex chromosomes. Thus, in the snakes, which also employ the ZZ-ZW system, it evolved independently[13] of the birds.[14]

Another parallel among the birds, mammals, and snakes concerns the fate of the Y and W chromosomes. Eric Vallender and Bruce Lahn[15] note, "Even though the emergence of sex chromosomes has taken place independently in many different lineages, their fate seems to follow a similar evolutionary trajectory."[16] Notoriously in the mammals the Y chromosome is small, and the few genes it possesses are largely involved with male functions, such as spermatogenesis. During evolutionary time this Y chromosome has become increasingly degenerate, and indeed in the mole vole (*Ellobius lutescens*) it has disappeared.[17] Just the same degenerate trend occurs in the W sex chromosome of the birds. As significantly, evidence from the tinamou

suggests that in the flightless birds known as the ratites (p. 208), this has occurred independently of the more advanced birds that define the neognathans.[18] Judith Mank and Hans Ellegren[19] point out in their commentary, "What is most curious about this parallel divergence is the similarity of the tinamou and neognathous W [chromosome],"[20] and pregnantly they conclude that such data might allow us "to differentiate those aspects of avian sex chromosome evolution that are inevitable from those that are due strictly to chance."[21] My money is firmly on the first option.[22] The more adventurous might have also had a flutter on the possibility that not only would Y and W sex chromosomes show convergences, but so too would their counterparts (X and Z). And in birds and humans that is exactly what we see, with both independently arriving at an arrangement that not only has a lower gene density[23] but an abundance of testis-related genes.[24]

More surprisingly, investigations into the sex determination of the plants and fungi again reveal a series of convergences.[25] In the flowering plant *Silene*, for example, the XY system emerged geologically yesterday (c. 10 Ma ago), but already there is evidence for genomic degeneration in the Y sex[26] chromosome.[27] In the much more primitive bryophytes, specifically the liverwort *Marchantia*, an XY system has evolved completely independently.[28] Sachiko Okada and colleagues note how "the basic structure of the Y chromosome organization . . . is shared by bryophytes, higher plants, and animals . . . [and] could be a consequence of convergent evolution in response to the pressure to avoid frequent crossovers between the X and Y chromosomes."[29] Further afield among the fungi, their sexuality certainly can be very complex, but nevertheless with respect to the sex determination in a parasitic basidiomycete known as *Cryptococcus*, the so-called mating type shows some striking parallels to

CHAPTER 19

SEXUAL CONVERGENCE

As ROGER PAIN[1] memorably remarked, "If it is love that makes the world go round, then it is surely mucus and slime which facilitate its translational motion."[2] Do I see a flicker of interest cross my reader's face, something a little racy perhaps? Alas, for the prurient this chapter may be a little heavy going. Yes, orgasms and penises, all in a convergent framework of course, but hardly of interest to the lubricious, which may be just as well given that from one angle the sex act is deeply comic. Girl meets boy, sperm rushes to egg, and as fully signed-up mammals the male determines the sex of the progeny (the familiar XX-XY system). Hormones, and in many species pheromones, play their part, but in the end it all depends on the gonads. Among the vertebrates, however, the overall structure of the ovaries and testes is remarkably similar, but the ways at which this is arrived at are extraordinarily diverse in terms of cellular and molecular biology.[3] As Tony DeFalco and Blanche Capel remark, "It is striking that, despite the vast array of sex determination mechanisms, vertebrate adult testes and ovaries are astoundingly similar in structure. Yet the morphogenesis of these organs seems to be mechanistically different."[4] As part of the title of their paper proclaims, "Divergent means to a convergent end."[5] Hear! Hear!

PARENTS AND SEX

"Boy or girl?" asks the midwife brightly. Conversant with the sex determination in mammals being always male (so this is the heterogametic sex), she glances at the father in the certain knowledge that nine months before he played his part in the familiar XX-XY system of sex chromosomes. The nightingale singing outside the maternity ward, no doubt completely accidentally, is blissfully unaware that in all the birds sex determination lies with the female; heterogametically this is the ZZ-ZW arrangement. So nothing could be simpler? Not quite. With a review paper titled, "Weird Animal Genomes and the Evolution of Vertebrate Sex and Sex Chromosomes,"[6] one begins to get the idea. As Jennifer Graves notes, there is a "bewildering variety of sex-determining mechanisms just among higher vertebrates," but she immediately continues how this diversity actually "disguises a biochemical and histological commonality of the genetic pathway that induces gonad differentiation."[7] The central question is as follows: To what extent are the systems of sex determination ancestral as against being continuously reinvented?

While the paternal XX-XY and maternal ZZ-ZW systems revolve around the genetics of sex chromosomes, this is not the only avenue to sex determination. Alternatives not the only birds whose distribution has radically changed.[139] The oilbirds, remarkable for having independently evolved echolocation (p. 139), are now restricted to South America, but as fossils they occur in the English Eocene.[133] Most ironic are the New World vultures. In the Early Tertiary they were to be found from England to Mongolia,[134] but they are also splendidly convergent with the present-day Old World vultures![135] Their joint hallmark, of course, are prominent beaks and naked necks, but the convergences between the Old and New World vultures are striking in terms of scavenging.[136] In both cases, they divide into three categories: those birds that rip into the carcass, those that plunge into the viscera, and the third type feeding on the scraps. In each of the three cases skull and beak converge in function. In addition, the Egyptian vulture is well-known for its ability to crack open an ostrich egg with a stone, one of many examples of tool use by birds (p. 274).[137] Today the only place Old and New World vultures are likely to meet is in a zoo, but in the recent past there was certainly one place where they coexisted; their remains are mingled in the La Brea tar pits, famous also for its saber-toothed cats (p. 274). Here the gigantic *Teratornis*,[138] a condor with a wingspan of a small glider,[139] flapped past relatives of the Egyptian vulture.[140]

In speaking of redistributions, what could be more emblematic of the Argentinian pampas than the flightless rhea (or ostrich, as Darwin understandably described them during his South American travels[141])? As South American fossils they occur as far back as the Paleocene, but in the Eocene rheas could also be seen trotting around Europe. Nor are rheas alone, inasmuch as avian flightlessness has evolved multiple times.[142] Particularly spectacular are the multiple examples among the oceanic rails,[143] which evidently could abandon flight with extraordinary rapidity.[144] David Steadman suggests that, at the time of the Pacific diaspora, "from 500 to 1600 species of flightless rails inhabited Pacific islands,"[145] and even the upper estimate may be too low. Now nearly all are gone. These rails are a somber monument to the result of the arrival of humans, with their cargos of rats and dogs eagerly sniffing the new lands and their defenseless faunas.

Not only the rails faced termination on New Zealand the giant moa[146] was doomed, although the emblematic New Zealand bird, the kiwi, is endangered but survives. This remarkable bird definitely deserves the title of "honorary mammal,"[147] but it belongs to the primitive palaeognathans,[148] an important component of which are the flightless ratites. Ratites have been assumed to stem from a common ancestor that abandoned flying, but they are polyphyletic.[149] As John Harshman and colleagues note, "the many morphological and behavioral similarities among ratites [arise] by parallel or convergent evolution."[150] Birds, therefore, are endlessly fascinating but then so too is sex.

bright blue scrotum of the vervet monkey. It is hardly surprising to learn that these colors are produced by effectively identical collagen arrays in the skin.[123] The male mouse opossum is also equipped with a blue scrotum, and not only has this structural coloration evolved independently in the Old World primates and marsupials, but most probably at least twice in either group. In this analysis Richard Prum and Rudolfo Torres stress how these convergences arise "because collagen has several intrinsic features that *predispose* it to evolve"[124] structural coloration, including appropriate fiber arrays and a contrasting refractive index with the surround matrix of mucopolysaccharides. When we recall also collagen is a protein that effectively self-assembles (p. 315n11), then it is a fair guess that the glory of animal coloration on remote planets will have exactly the same basis.[125]

On such a visit we might also be struck by more unexpected parallels in skin coloration. Consider blushing, often thought to be a uniquely human characteristic. Not quite. As Juan Negro and colleagues[126] note, among the birds, facial flushing is documented in "12 different avian orders and at least 20 families,"[127] but as they also stress, the distribution is far from random and typifies large, dark, and tropical birds. The bright red color is striking and not surprisingly owes its origin to the skin being riddled with blood vessels.[128] Nor are these birds alone, because apart from massively embarrassed humans, a number of primates, notably the mandrill, owe the bright red color of their muzzle to extensive vascularization.[129] This caught the attention of Darwin, who devoted an entire chapter to the question of human blushing in his *Expression of the Emotions*.[130] Not that he supposed mandrills blushed, but just as other monkeys could go red with rage, so he noted, "When the Mandrill is in any way excited, the brilliantly colored, naked parts of the skin are said to become still more vividly coloured."[131]

Straying Convergence

Darwin's program of beginning the tricky work of unmasking our emotional antecedents was just one fertile line of inquiry that ultimately emerged from his *Origin*. Central to his argument of descent by natural selection was the importance of isolation with a concomitant explanation as to why faunas and floras in remote parts of the world are so obviously different. I cannot help but wonder if his musings on the curious fact that oceanic islands (like the St. Paul's Rocks he visited while on the *Beagle*) are always igneous in composition might not have given him the germ of the idea for what we call plate tectonics. The fact that mid-ocean volcanoes continue to erupt, combined with the fact that the outlines of South America and Africa are suspiciously congruent, might conceivably

have sparked a hypothesis of some sort of continental drift. After all, he certainly had the necessary genius, as is evident from his brilliant solution as to the origin of oceanic coral atolls. I suspect, however, that his fixist perception would be almost impossible to dislodge because it guaranteed the all-important principle of isolation. My guess is that Darwin would be surprised not only by plate tectonics but how, during geological time, biogeographical distributions prove to be extraordinarily dynamic.

As we saw earlier in this chapter, an Oligocene Englishman would not have been the least surprised to see an early hummingbird. Today they (the birds, not Englishmen) are entirely restricted to the Americas, especially the Neotropics. But the hummingbirds are wings were "rapidly and violently flapped."[104] Even more remarkable are the young, which climb up and down the trees using their wing claws.[105] Quelch gives a graphic description of the young hoatzins, which tenaciously hold onto the surrounding vegetation with wing claws, feet, and beak. Should they fall into water, they adroitly swim and can also dive to safety. Observers have been repeatedly struck by the almost reptilian appearance of these birds, and indeed the clambering on trees has invited comparison to the famous *Archaeopteryx* (p. 193), also equipped with prominent claws on its wings.[106] Neither turaco nor even the hoatzin are, however, primitive.

Their convergences extend, moreover, beyond arboreality. Most unusually for birds both are folivorous—the hoatzin feeding mostly on leaves while the turaco eats fruit, with a corresponding similarity in the bill and feeding apparatus.[107] The hoatzin, however, shows yet another twist in the story of convergence because it has independently cracked the problem of how to digest leaves—that is, it has become a ruminant.[108] That, in passing,

probably explains why the hoatzin's smell is regarded as repellent,[109] especially after death, thereby earning it the nickname of stinking pheasant.[110] Nor is the hoatzin the only bird to ruminate.[111] Fermentation of plant food in the foregut is well-known in such ruminants as the cow,[112] and has evolved independently in the colobine monkeys and the sloth,[113] while the convergence extends much deeper to the herbivorous fish.[114] The complex and diverse microbial floras in not only in the hoatzin[115] (which unsurprisingly includes typical rumen methanogens[116]) but other ruminants provide the key as to how they can process what is effectively indigestible cardboard. Here the evolutionary story takes a molecular turn; among the enzymes employed is one known as lysozyme. As its name suggests, its action of lysis—the breakdown of cell walls—means it plays a key antimicrobial role,[117] yet in the ruminants not only has it been repeatedly and independently recruited for a digestive role but evidence also exists for molecular convergence.[118]

CONVERGENT GLORY

Even if turacos are not ruminants, they are remarkable birds in their own right, not least because of their remarkably bright plumage. The turaco, however, is a sort of artist because its green (and red) coloration is based on a copper-based pigment,[119] whereas in most birds the colors are referred to as structural. The colors arise from so-called coherent light scattering (as against the familiar incoherent Rayleigh scattering that, among other things, explains why the sky, even in England, is blue), with the colors resulting from the light interacting with biological nanostructures in the feathers.[120] The topic of coloration in birds has, however, another interesting angle, which brings us back to the hoatzin. A striking feature of this bird is its blue[121] face, which in the same way as most plumage colors is the result of structural coloration arising from coherently scattered light. Once again it involves a biological nanostructure, this time employing special layers of collagen fibrils in the skin whose ordering results in the appropriate light scattering and so a particular color. Significantly, in the birds this arrangement is rampantly convergent, having evolved something like fifty times.[122]

Who also cannot fail to be impressed by the striking coloration in some mammals? Most familiar, perhaps, is the blue face of the male mandrill, while among the cognoscenti there is equivalent admiration for the

(p. 208) and hummingbirds. In any event, the parallel evolutionary tracks made by metavian and coronavian result in a whole series of striking avian correspondences.

In watery habitats the sunbittern stealthily wades to bring sudden death to its prey, but it stands in the convergent shadow of the booming bittern.[85] Continuing these comparisons, the more sedately metavian sandgrouse and its coronavian equivalent the seedsnipe both peck grain. So, too, out of the tropical sky in pursuit of their prey plunge coronavian boobies (close relatives of the gannets) and metavian tropicbirds.[86] The similarities are hardly surprising:[87] both forage across the open ocean, hunting out patchy resources, both plummeting after flying fish and squid. Hitting the surface of the ocean at high speed can be no fun, but boobies and tropicbirds both have air sacs around their face to act as shock absorbers. At night the owls hunt for nocturnal prey, and their metavian counterparts in the form of the frogmouth silently swoop toward their prey.[88] This convergence finds a neural counterpart because the cerebellum of the frogmouths shows significant differences from its relatives (which are placed with the goatsuckers [caprimulgiforms] and include the oilbirds, p. 139), but have corresponding similarities to the owls.[89] As a group the caprimulgiform birds[90] are notable for nocturnal types,[91] and an owl-like habit may have evolved independently in the owlet nightjars. The owl morph extends yet further because the *Elanus* kites not only ensure a silent descent on account of having the same feather structure, but they show a series of other owl-like convergences, including the asymmetric ears[92] that probably assist with prey location.[93]

Can we extend this convergence any further? "Can a primate be an owl?" asks Carsten Niemitz provocatively.[94] In the case of a tarsier and an owl from the forests of Borneo, an impressive array of convergences emerge, including not only the eyes and whiskerlike structures but even the methods of killing prey by perforation. Tarsiers don't fly, of course, but both are remarkably noiseless; despite obvious differences Niemitz concludes, "Although looking completely different the Bornean tarsier is almost an owl—almost."[95] One could, of course, simply explain these parallels as coincidences, and perhaps they are. But I am struck by the specific convergences;[96] once again one has a sense of biological forms almost "crystalizing." Of course, this analogy is imperfect; as repeatedly stressed the convergences are not identical, and not all coronavians have a metavian equivalent. But overwhelmingly my sense is that these nodes of biological stability are telling us something much deeper about the structure of life: evolution clearly navigates, but oddly enough to things not so different to archetypes.

Of these parallels, one of the strangest concerns the South American hoatzin. Its precise evolutionary position has proved remarkably hard to pin down, although it might have a remote connection to the doves.[97] Earlier molecular phylogenies had argued for a relationship to the African turaco (or touraco),[98] which seemed to support the many striking similarities between the two groups.[99] Once again we are dealing with a convergence: metavian hoatzin[100] and coronavian turaco.[101] This convergence is, however, worthy of particular attention. The hoatzin is highly arboreal, as are some species of turaco. In the case of the hoatzin, so arboreal is its way of life that very seldom do they descend to the ground, while on their underside an area of the sternum is devoid of feathers so that this dermal callosity provides a convenient cushion. Hoatzin and turaco are both poor fliers[102] but very agile among dense vegetation. The Victorian naturalist John Quelch[103] remarked how the hoatzin was hard-pushed to fly more than about forty meters, and in doing so the of intelligence (p. 269), technology (p. 276), and oddly orgasms (p. 212). In either group we find further fascinating examples. Among the mammals most celebrated are the classic comparisons between the placentals and marsupials. If these parallels remind us of the constraints of evolution, what do the birds show? Certainly in terms of avian body form, convergences are well-known.[69] Perhaps the most familiar example are those dark-colored speeding insectivores we call the swifts and swallows. The former are closely related to the hummingbirds, whereas the swallows belong to that vast group of birds known as the passerines. Like a number of other groups, swift and swallow are accomplished migrants, but so are some high-flying insects.[70] Jason Chapman and colleagues note how they have "flight behaviors [that] match the sophistication of those seen in migrant birds,"[71] but also that they "are analogous to Arctic-breeding shorebirds, which also select the fastest high-altitude winds."[72] This capacity to use favorable winds[73] is part of what Hugh Dingle[74] terms "a common migratory syndrome"[75] that convergently unites birds and such insects as hemipterans and lepidopterans (and also fish) in terms of metabolism (especially burning fat[76]) and hormonal regulation. The birds, however, remain the masters of migration, and while none can match the astonishing nonstop 11,000-km-long Pacific transits by the bar-tailed godwit,[77] among the more surprising migrants are the hummingbirds.[78]

The convergences that encompass the hummingbirds have been addressed, but just as their relatives, the swifts, closely resemble swallows, so in the birds we find a whole series of paired convergences. So highly adapted are the penguins to swimming that seeing them as birds can be difficult. Yet the powerful propulsive strokes of their wings find a convergent counterpart with the extinct plotopterids.[79] When it comes to foot-propelled

divers, we also know that the strikingly similar grebes and loons are in fact convergent, with the former close to the flamingos.[80] Yet when we look to the latter group, they are, of course, alarmingly similar to the spoonbills.[81] Not only are both colonial waders, both filter feeders, typically swing their bills from side to side and have similar displays, but the aptly named roseate spoonbill is gratifyingly a flamingo pink. Not only are the flamingos and spoonbills convergent, but they belong respectively to the two major clades of the so-called neoavians,[82] the Metaves and Coronaves. (At this juncture we need to note that few areas of phylogeny have created greater difficulties than the question of avian relationships, and indeed the concept of metavian and coronavian has been queried.[83] Even so, whatever scheme of avian phylogeny you choose, the convergence between flamingo and spoonbill seems to be secure as do many other avian examples, irrespective of whether one employs the metavian-coronavian dichotomy.)

As Matthew Fain and Peter Houde fascinatingly document,[84] a remarkable set of avian parallels echoes the more famous marsupial-placental convergence. Marsupials are in one sense ancient and highly divergent but show a reduced diversity. So, too, are the metavians. Conversely, like the coronavians the placentals are more diverse, both biologically and ecologically. Consider the proportions: there are about 19 families each of marsupial and metavian, but more than 100 placental (115) and coronavian (109) families. So, too, there are roughly ten species of coronavian for every metavian, and sixteen placental species for every marsupial. One explanation for this intriguing parallel is that marsupials are more or less of the southern hemisphere, if you like Gondwanan, whereas placentals are northern hemisphere. This hardly works for the birds, and an early primate in Europe would have enjoyed the company of New World vultures

A number of mammals are also avid nectar feeders. Particularly remarkable as a source of nectar is the Bertram palm of the Malaysian rain forest. This one smells like a brewery, and with good reason; yeast-based fermentation in the flower buds produces a copious nectar that rivals beer in alcoholic strength. This is enormously popular with many insects and arboreal mammals, but despite a substantial intake at least some of the mammals, notably a tree shrew, showed no sign of intoxication and presumably possess an appropriate physiology.[45] A few mammals, including the appropriately named honey possum, depend almost entirely on nectar.[46] Not surprisingly they have a remarkably modified skull,[47] effectively toothless and complete with brush-tipped tongue.[48] But in terms of hummingbirdoid convergences among the mammals, we need to turn to the nectar-feeding bats. Here the energetic costs of hovering are considerably lower than for either hawkmoth or hummingbird.[49] Yet in all cases the very ability to hover poses an extraordinary aerodynamic challenge, and this solution reveals convergences. Despite obvious differences between the wings of a bat[50] and a hawkmoth,[51] in both cases the process of flapping leads to the generation of a vortex along the leading edge of the wing that is a vital component in generating lift. In his commentary Michael Dickinson[52] illustrates the wind-tunnel experiments that revealed this vortex in either group and remarks how they "are uncannily similar."[53] What applies to bat and moth also finds a counterpart in the hummingbirds.[54] Here, too, there is employment of a leading-edge vortex, although in other respects the aerodynamics of the wing are less similar, at least in comparison to the moths.[55]

Dickinson also reminds us that even though bat and moth are more usually engaged in an unceasing warfare of detection by echolocation and desperate evasion (p. 138), as far as their flight is concerned they are "united by the laws of physics."[56] So, too, the nectar feeding of the hawkmoths finds a direct counterpart among the bats, where it has evolved several times.[57] Most obvious in this respect are examples among the New World phyllostomids,[58] as well as the so-called flying foxes of the Old World. Concerning this latter group (the so-called megachiropterans), Theodore Fleming and colleagues[59] remark on how "specialized nectarivory has evolved independently three times,"[60] and the specializations of skull and tongue have evolved independently several times as well.[61] This tendency is all the less surprising given that these megachiropterans are primitively fruit eaters, although as Elizabeth Dumont[62] notes, "New [i.e., phyllostomids] and Old World fruit bats exhibit many similarities in the organization of their frugivore assemblages."[63] Nor is it unexpected that in both this group and the phyllostomids have various convergences in the skull,[64] including reduction of the teeth, and an elongate tongue.[65] When it comes to the physiology of nectar metabolism,[66] as well as their ability to draw on the sugars as an emergency source almost immediately, the bats show significant convergences with the hummingbirds. Indeed this shared capacity means that these animals engage in the nearest thing to midair refueling,[67] reflecting physiological and biochemical convergences that link the assimilation of the sugars through the intestinal wall to its extraordinarily effective use in the flight muscles.[68]

SABER-TOOTH BIRDS?

These convergences between bat and hummingbird are only one of a series of examples

that link mammals and avians, among which we might think of the convergent evolution hand (so to speak)? Thus the Rat's tail (*Babiana ringens*), a type of iris endemic to South Africa, goes one better by specifically growing a perch for the malachite sunbirds to employ while engaged in nectar feeding.[31] The sunbird says a metaphorical thank-you and the plant receives its "reward" in terms of cross-pollination.[32] The sunbirds show a further convergence[33] with the hummingbirds, because as Steven Johnson and Susan Nicolson remark, the type and quality of the nectar obtained by these specialist African birds "almost exactly mirrors that in American plants pollinated by hummingbirds."[34]

Nectar-feeding birds are very widespread, and in terms of their geographical distribution, there is a parallel to the hummingbirds because, oddly, nectar-feeding birds are extremely uncommon in Europe.[35] Drawing on this observation, Hugh Ford suggested that nectar-feeding birds as a whole have been preempted by competition,[36] possibly with bees.[37] There are, moreover, important biogeographical distinctions, at least as far as they pertain to the nectarivorous bats and the birds.[38] Most obviously hummingbirds are restricted to the Neotropics, although interestingly when it comes to the Old World, tropical sunbirds Theodore Fleming and Nathan Muchhala note that "no striking differences between [these two groups] and their food plants emerge."[39] In general, however, quite marked contrasts separate the Neotropics, a region with a marked tendency toward evolutionary specializations, from the other tropical regions. So, too, distinctions can be made within the Old World (Paleotropics), specifically between Africa and Southeast Australasia. Evidently these differences stem from both phylogenetic histories and local ecologies (especially the predictability of floral resources), leading Fleming and Muchhala to conclude that together these factors "have limited the extent to which flower bird and bat niches have converged in the Neotropics and Palaeotropics."[40] All cor-

rect, but we need to ask ourselves: How may have things panned out if the hummingbirds had remained in their ancestral home?

In a similar vein, consider a stimulating article by Cris Cristoffer and Carlos Peres[41] titled "Elephants Versus Butterflies." Echoing Fleming and Muchhala, these workers emphasize the striking differences between the tropical forests of the Old and New Worlds. The former are notable for their large herbivores (e.g., elephants), whereas in the latter, insects (such as leaf-cutter ants, p. 181), birds (notably hummingbirds), and small mammals all flourish. Comparisons are difficult, especially in areas such as India and Malaysia, because we have to look through a filter of habitat destruction and extinction. Yet Cristoffer and Peres make a strong case that in the Old World the great herbivores have literally stamped the tropical forests into an ecology quite unlike that of the New World. In the geological past, however, the vertebrates of the tropical Old and New Worlds show convergences,[42] and the differences we see today are largely a result of how each area responded to the massive climate changes.[43] Similar evolutionary laboratories have now diverged, but as climates shift, areas become isolated or reconnect, so we can be confident that the same themes will be repeatedly revisited. It is, so to speak, learning how to see the woods for the trees. To most biologists a close-grained view provides a perennial fascination, with an emphasis on the particular and historical, even the fluke. Why, for example, has tool construction reached such sophistication among the crows in, of all places, New Caledonia (p. 275)? It is a perfectly fair point, but we need to decide whether such an emergence is ultimately a fluke of circumstance or an endpoint that sooner or later was bound to be reached. Thus, to find nectar-feeding evolving independently is certainly unsurprising, and to the first approximation among the birds it is one of similarity rather than convergence.[44]

mingbird are effectively identical. Moreover, Bartholomew notes, "The almost exact equivalence of the curves for mass-specific thermal conductance in sphingids, birds, and mammals is remarkable, particularly when one considers the structural differences among feathers, fur, and moth scales."[11] Just as feathers and fur confer essential insulation, so heat loss in hawkmoths[12] is controlled by "superb coats,"[13] and as Norman Church asserts it is "difficult to insulate a small body more efficiently than *Sphinx* is insulated, if the depth of insulation is taken into account."[14]

As Kenneth Welch and colleagues[15] emphasize, the biochemical and physiological innovations allowed the hummingbirds to occupy a niche once the exclusive reserve of insects, but to achieve this they had to follow an inevitable route. These convergences between hawkmoth and hummingbird are so striking[16] that one can be confident that in reaching independently the metabolic and functional limits of precise hovering flight[17] it will be almost inevitable that light years away hypnotically flitting from flower to flower will be an extraterrestrial hummingbirdoid. It hardly matters which hummingbird or hawkmoth evolved first. The earliest-known hawkmoth is Miocene,[18] whereas the hummingbirds appeared about 10 million years earlier (in the early Oligocene[19]) and were well adapted to hovering to obtain nectar.[20] What makes it more interesting is not only the inevitability of an aerodynamic endpoint we can call a "hummingbirdoid," but that to the first approximation it emerges at much the same time. The geological history of the hummingbirds has a further twist. The first fossils are found in Europe, but today the hummingbirds are confined to the Neotropics. Why they vanished from Europe and the rest of the Old World is obscure. Did competition with other birds, not to mention the bees, drive them to extinction? A haunting consequence is that a number of Old World flowers, such as some balsams and campanulas,

appear to be eminently suitable for hummingbird visitations. Gerald Mayr[21] intriguingly suggests that, in a manner of speaking, the flowers are still waiting for the now-vanished hummingbirds.[22] Next time you see a patch of balsam in a damp English meadow, look out of the corner of your eye for those ghostly hummingbirds.

Nectar-feeding has evolved numerous times in many other birds, notably the sunbirds and Neotropical honeycreepers.[23] Among the latter, nectarivory is blatantly convergent, having evolved something like seven times.[24] As so often, the Hawaiian archipelago proves especially instructive.[25] Among them are the only just extinct "honeyeaters"; one more victim of an ongoing mass extinction. In the wider scheme of things, however, this might seem to be less of a disaster given the close relationship of these birds to their Australasian counterparts, the meliphagids. It would be, except that generations of ornithologists have been entirely fooled.[26] Study of the DNA from museum specimens reveals that the Hawaiian birds represent "one of the most deceptive cases of convergent evolution in birds."[27] In fact these birds are related to the passerine birds such as the waxwings, but as Robert Fleischer and colleagues remark, the convergence to the meliphagids extends to "plumage . . . behavior, and song."[28] Interestingly, Fleischer et al. almost play down this extraordinary convergence by emphasizing how a divergence time of at least 10 million years ago gave these birds "ample opportunity to evolve the adaptations for nectarivory that make [them] so similar in gestalt to the distantly related Australasian honeyeaters."[29] This may be correct, but convergent solutions can emerge with surprising speed.

In terms of avian nectar feeders, the key difference is that only the hummingbirds, like the hawkmoths, have evolved hovering. The other nectarivores need to perch, at least to the first approximation.[30] But if your visitor has to perch, why not give a helping

CHAPTER 18

THE BIRDS CONVERGE

ONE CRITICISM of evolutionary convergence is to agree that while similarities exist—and notwithstanding the customary cries of astonishment—to ask if nevertheless they are anything other than superficial. A reasonable question, but let us join the Victorian explorer Henry Bates, onetime companion of Alfred Wallace, in the Amazon. Bates was a consum-mate naturalist and also a collector. Carefully he raises his rifle to shoot a hummingbird, but finds that rather than bagging another prize for one of his English connoisseurs he has just eliminated a hawkmoth (or sphinx moth and sphingids).[1] How idiotic! Not at all! For all his knowledge Bates joins that large and distin-guished club of those fooled by convergence.

HUMMINGBIRDOIDS

While in Brazil, Darwin[2] was equally struck by how in "their movements [bird and moth] are indeed in many respects very similar."[3] In the same fashion the locals were not in the least bit surprised. Natives (and "educated whites") knew perfectly well that moth could transmute into hummingbird; after all, do not caterpillars transform into butterflies?[4] While we smile (or as often sneer) at this apparent naïveté, the locals were not so wide of the mark. The similarities between bird and insect are not precise. Like other arthro-pods the hawkmoth is still constrained to use a compound eye, but it provides a quite extraordinary acuity.[5] Once again the appar-ently iron constraints of a biological design are triumphantly transcended, and in the case of the hawkmoth eye on the basis of an arrangement that at first sight looks distinctly unpromising.

Apart from the eyes, not only are the over-all shape and size, together with the ability to hover precisely while feeding on nectar, remarkably similar, but the resemblances go deeper. Thus, there are striking physio-logical, energetic, and metabolic similarities between moth and bird (and indeed bats, p. 203), not least in their ability to fuel their intensely energetic activities by direct use of the nectar sugars[6] rather than the customary metabolic burning of stored fats.[7] That the hummingbirds are endothermic is hardly surprising (even though they enter periods of torpor[8]), but so, too, are these moths. Like other insects the hawkmoth generates its heat using its flight muscles, as do the birds. As George Bartholomew[9] notes, when we come to compare the endothermic physiology of insect and bird they can be "strikingly sim-ilar and offer an almost unique example of convergent evolutionary responses of physio-logical functions to a common set of physical and physiological parameters operating in vastly different morphological structures,"[10] and in features such as oxygen consumption and energy costs the hawkmoth and humcial interest, however, are the troodontids. While taxa such as *Zanabazar* have a somewhat less impressive EQ,[204] *Troodon* itself is highly encephalized with a brain that would do credit to some birds and mammals.[205] This animal seems to be on the same cognitive trajectory that led to humans, as famously was captured by Dale Russell and Ron Séguin[206] in their speculation of how such an evolutionary journey might have ended in a troodontid equivalent to a humanoid. Usefully their ancestors already had binocular vision, and in some theropods its acuity might have rivaled that of a hawk.[207] In inviting their readers to take "cognizance of the ubiquity of the phenomenon of parallel evolution in the history of life on Earth,"[208] they remind us that if aerial avianlike and aquatic dolphinlike forms repeatedly emerged, then what Russell and Séguin term the dinosauroid could "represent a solution to the physical and physiological stresses imposed on the vertebrate organism by a greatly hypertrophied brain in a terrestrial environment."[209]

In one sense the experiment has been successfully carried out with the evolution of the close relatives of the troodontids, the birds; the many convergences between them and mammals are explored below. One obvious difference, feathers versus fur, may be much less profound than might appear. Both are composed of keratin, and their emergence from the integument relies on specific areas of dermal condensation. Ultimately these form papillae, and in both cases the same molecular machinery (specifically a Wnt/β-catenin pathway) is employed.[210] As Daniella Dhouailly points out, "hairs and feathers evolved independently, but in parallel."[211] The implications for this are intriguing, because among other possibilities, this suggests that the primitive arrangement in birds may have been to be covered with feathers, and indeed the fossil record of the theropods supports this view. If correct, then while evolutionary biologists like to point to the scales of a chicken as triumphant confirmation of their reptilian origins, oddly enough these arrangements might actually be secondary. This is not to dispute the origin of birds from theropod dinosaurs, but to remind ourselves that once a keratinous integument managed to evolve areas of dermal condensation, then the glory of a bower bird or the fur[212] of a mink were very much on the cards.

that a 'theropod-like body plan' developed in a group of crocodile-like archosaurs before it evolved in later theropod dinosaurs."[181] In both this paper and a companion publication Sterling Nesbitt and Mark Norell[182] do not hesitate to use the phrase "extreme convergence,"[183] while elsewhere Nesbitt[184] notes how "the skeleton of *Effigia* bears an uncanny resemblance to that of theropods and more specifically, ornithomimids."[185] Nor is this an isolated aberration, because a series of related taxa such as *Poposaurus*,[186] *Shuvosaurus*,[187] and *Tikisuchus*[188] are also theropodlike.[189] When first described Sankar Chatterjee concluded, "that *Shuvosaurus* is an ornithomimosaurid [dinosaur] appears inescapable from [its] cranial morphology,"[190] and in the same vein *Revueltosaurus* was confidently identified as an ornithischian dinosaur. It too, however, is a pseudosuchian,[191] and Nesbitt notes how the dentition of this reptile "shares an uncanny resemblance to the teeth of early ornithischians."[192] Nor are we likely to be at the end of this story, because as more fossils are found[193] so both the range of forms and their degree of specialization continues to grow.[194] In reviewing these various convergences between the pseudosuchian archosaurians and other reptiles, Ralph Molnar[195] makes a vital point when he writes, "Far from being a mere curiosity, the Triassic instances affect our whole understanding of the evolutionary origin of important components of the Mesozoic tetrapod faunae, and provide questions of why animals seemingly effectively occupying certain ecological niches were somehow replaced by phyletically distinct forms sufficiently similar to be mistaken for one another."[196]

Although many supposed Triassic dinosaurs are in reality pseudosuchian archosaurs, clearly toward the end of this period the two groups coexisted. It has long been assumed that the dinosaurs won out by a competitive process, assisted by major extinctions near the end of the Triassic, but analyses of their respective disparities during this critical interval do not support such a conclusion.[197] Stephen Brusatte and colleagues proceed to a properly orthodox formulation when they conclude that in the mass extinction the dice rolled against the pseudosuchians, and so they ultimately "died out by chance."[198] Quite so, but the wider perspective that seeks the deeper regularities in evolution will argue that it hardly matters who goes extinct and who survives: certain outcomes are inevitable, and in one sense how life got there hardly matters.

Yes, the dinosaurs enjoyed almost 150 million years of undisputed terrestrial supremacy, but as we have seen, in the theropodlike form was seeded the inherency for birds. Theropods like *Tyrannosaurus* and their quadrupedal relatives, the sauropodian behemoths, dominate the public imagination when they conjure up vistas of deep time, but arguably the real interest lies elsewhere. As a group the dinosaurs were cognitively challenged. Indeed when one considers brutes such as *Brachiosaurus*, this may well have been the case, although establishing precisely the ratio between the brain preserved as an endocast and the original body (the encephalization quotient or EQ) mass can be fraught with imponderables.[199] In this case, as well as such theropods as the oviraptorids,[200] the fact that the brain fits snugly in the cranial cavity means, however, that relative brain sizes can be calculated with some accuracy. More importantly, in well-preserved endocasts something of the brain structure can be inferred,[201] and among the theropods things get more interesting. To begin with, an increase in the size of the forebrain clearly began quite early in their history.[202] In addition, in an oviraptorid (*Conchoraptor*), the brain has a number of features that are more similar to extant birds than *Archaeopteryx*,[203] and in keeping with other features of this group are most likely convergent. Of spebird-sounding ornithischians form a distinct clade. These we have already briefly encountered in the form of the psittacosaurs (p. 196), which are instantly recognizable by their parrotlike beak. Although equipped with teeth, the cranial structure has a number of striking parallels to the parrots (psittaciforms), and their Cretaceous equivalents were probably equally adept at tackling nuts.[167] Looking at a distinctly different ornithischian, it is worth noting another convergence, concerning the stegosaurs. These are a highly successful group of Upper Jurassic dinosaurs and famous for their striking array of bony plates arrayed along the back. A close relative of *Stegosaurus*, a Portuguese beast known as *Miragaia*, shows a striking convergence with the long-necked sauropods.[168] This discovery is important not only because, as Octávio Mateus and colleagues remark, it reveals a "previously unsuspected level of morphological and ecological diversity among stegosaurs"[169] (and thereby a reminder that the reservoir of convergent examples in the fossil record must be far larger than currently appreciated), but because the manner in which the neck is elongated is very much in the same fashion as found in the mighty sauropods. In a broader context, despite the classic distinction between the pelvic arrangement in the ornithischians and saurischians, Matthew Carrano[170] demonstrates how "homoplasy is rampant in the evolution of the dinosaurian locomotor apparatus."[171] This revolves around such features as the orientation of the head of the femur, its associated musculature and the extension of the ilium. Nor are these convergences unexpected, because in each case they enable the reptile to move from a sprawling position to an erect posture, a necessary precondition for bipedality. Yet as Carrano also stresses, in the case of the hindlimb of the birds its evolution "represents a cumulative acquisition of characters, many of which were quite

far removed in time and function from the origin of flight."[172]

What of the story before the dinosaurs? Although it has long been appreciated that their rise to dominance began in the later stages of the Triassic, it transpires that a significant part of the record requires re-scrutiny. The reason? Convergence. It now transpires that while dinosaurs existed, they occurred neither in the diversity nor abundance once thought.[173] Much of the problem is that all too often the material is fragmentary, and as often as not, it occurs as teeth. But from whose jaw did they fall? Some from the dinosaurs,[174] but by no means all of them. We need to look at another group of archosaurians, the pseudosuchians.[175] It now transpires that to say, "It looks like a dinosaur tooth, so it must be a dinosaur tooth," won't wash. Nor is this surprising. In the case of carnivorous archosaurs as a whole, we see the repeated evolution toward larger and more powerful jaws, with fewer teeth but showing a greater degree of variation.[176] As Donald Henderson and David Weishampel note, their analysis suggests "that all these different predatory archosaurs were converging on a common predatory style for the attacking, dismembering, and ingesting [of] very large prey."[177] Just as the Tertiary was to host the saber-tooths, so one way or another the Mesozoic was going to be a time of blood and gore.

At the opposite end of the trophic spectrum is what Xing Xu and colleagues[178] refer to as "an extraordinary case of convergence among three higher archosaurian groups."[179] This entails two distinct groups of theropods (specifically a ceratosaur known as *Limusaurus* and the ornithomimosaurs) and the pseudosuchian *Effigia*, all of which have evolved striking similarities as a result of a herbivorous lifestyle. The theropodlike appearance of *Effigia* had already been identified by Sterling Nesbitt,[180] who noted, "Many convergences between *Effigia* and ornithomimids suggest so-called subarctometatarsus), and among the theropods this evolved independently five times.[147] So, too, while under some theropods the earth shook, others were much smaller. In fact miniaturization was frequent, pneumatic air sacs were already available,[148] and among the wealth of integumental filaments, vanes, and blades it is hardly surprising that feathers evolved.

All these reptiles, like their surviving descendants, laid eggs. It is intriguing that an oviraptorid from the Upper Cretaceous of Mongolia was discovered on top of its nest,[149] curled up over her eggs in a strikingly "bird-like posture."[150] Until we discover the nesting habits of more primitive oviraptorosaurians it will be difficult to decide if this protective brooding arose much earlier or was repeatedly reinvented. The former option might seem more likely given that not only extant crocodiles[151] but at least one ornithischian dinosaur (*Psittacosaurus*)[152] also exhibit parental care, but Qing-Jin Meng and colleagues are careful to remark that "homology of their parental care is debatable."[153] In addition, identification of similar behavior in a Permian varanopid "pelycosaur"[154] might, in the words of Jennifer Botha-Brink and Sean Modesto, be "completely unexpected in a Palaeozoic amniote," not least because of "the utter absence of evidence for any form of parental care in the Palaeozoic and Mesozoic synapsids."[155] It is as likely this trait emerged independently some 140 Ma in advance of the theropods.[156]

With the dromaeosaurids *Microraptor* and *Rahonavis* climbing into the sky in the company of birds, and with avianlike features emerging independently in the oviraptorids (and also among the alvarezsaurids[157]), we can say with confidence that by the time a maniraptorian emerged—that is, the clade that incorporates the therizinosaurians[158] (these include the bizarre *Mononykus*, a dinosaur convergent[159] on the anteaters[160]), oviraptorosaurians (some of which also appear to have converged on anteaters[161]), and the paravians—then birds were an inevitability. What if we look even deeper into dinosaur history?[162] Basal even to the tyrannosaurids are the allosauroids. This group of theropods was thought to have disappeared at least 20 Ma before the end-Cretaceous catastrophe. This now turns out to be incorrect[163] and significantly within the so-called megaraptorans the possession of "long, raptorial forelimbs, cursorial hindlimbs, appendicular pneumaticity [air sacs] and small size, [all represent] features acquired convergently in bird-like theropods."[164] As Roger Benson and colleagues go on to note, "Megaraptorans may have been 'coelurosaur parallels' in Cretaceous ecosystems."[165] Either way, an avianlike form seems inherent to the theropods as a group.

Do I see a hand raised, palm extended in a stop sign? "Enough of this talk of inherency and inevitability! What else would we expect? Consider! Theropods are bipedal, so forelimbs are free, yes, sometimes to dig ants or hold prey, but also to fly. Nor do I deny that at least some features that would prove key to successful flight appeared earlier in theropod evolution than generally thought, or indeed evolved several times. Give me a theropod and granted I will give you a bird, but can you give me a theropod?" The answer appears to be yes.

THE DINOSAURS CONVERGE

As ever the question of likelihoods revolves around convergence. As far as the dinosaurs are concerned, the theropods are closely allied to the herbivorous sauropods[166] (collectively they constitute the saurischians), while slightly confusingly the somewhat is a strong candidate[125]), and as we have seen not only has it evolved multiple times, but in at least some cases may be the necessary prerequisite to powered flight (p. 191). Moreover, there are some suggestions that *Archaeopteryx* itself may have been more a glider.[126] This is controversial,[127] and with *Microraptor* again we should not automatically assume that the current view is correct;[128] as John Hutchinson and Vivian Allen noted, the reconstruction of *Microraptor* as a biplane glider "is only one of several possibilities."[129]

The evidence shows that theropods took wing[130] at least three times, in the forms of *Microraptor, Rahonavis*, and *Archaeopteryx*.[131] This will come as no surprise to the student of convergence, but nor need it to the biological world as a whole. Numerous workers have emphasized (but not in these words) the inherency for a "bird" in the theropod tree, although more occasionally they have acknowledged that distinguishing between common ancestry and convergence is not always straightforward. The pattern of events is hauntingly similar to the story of tetrapod emergence from within the sarcopterygians. As Evgeny Kurochkin notes, "The mosaic distribution of avian characters in different theropod lineages suggest[s] that they developed in parallel. . . . By analogy with the process of mammalization, which developed in therapsids . . . such process can be named ornithization of theropod dinosaurs."[132] Not surprisingly much of the focus of attention has been on the evolution of feathers. Here, despite a reliance on exceptionally preserved fossils, as a group the theropods not only had a particular propensity for integumental elaboration, but this potentiality extended very deep into their history. Consider the tyrannosaurids: reconstructed as scaly behemoths they seem a far cry from their petite avianlike relatives, yet among the primitive forms *Dilong* is not only relatively small but equipped with a series of integumentary fil-

aments.[133] These are given the cognomen[134] of "protofeathers."[135] They come in a remarkable variety of forms,[136] and in the majority of cases the function cannot have been aerodynamic. Nevertheless, when the first "furry" theropod stretched forth its forelimbs, flight was on the agenda.[137]

From this perspective to find avianlike characters in the closely related dromaeosaurids and troodontids,[138] which together (as the deinonychosaurians) are the sister group to the avialians, is scarcely surprising. Deeper in the phylogeny of theropods we encounter the intriguing oviraptorosaurians. Their similarities to the birds have been taken to indicate a close relationship,[139] but in fact that the oviraptorosaurians were evolving birdlike characters independently.[140] This point became obvious with the discovery of a primitive form, *Incisivosaurus*,[141] from the Lower Cretaceous of China, which is distinctly less birdlike.[142] If, despite its remarkable specialized teeth, *Incisivosaurus* is more like a run-of-the-mill theropod, then the more derived species (such as the co-occurring *Caudipteryx*[143]) are distinctly avianlike. It should not come as a huge surprise if one day we find a fossil that demonstrates that independently an oviraptorid climbed to the top of a tree— and took off.[144] That this is on the cards is evident from an oviraptorid from the Upper Cretaceous of Inner Mongolia. To be sure it grew to an enormous size (standing 3.5 m high at the hip, and weighing in at a little under 1.5 metric tonnes), but in a number of respects this animal is strikingly birdlike.[145] If miniaturization[146] ever occurred, then flapping alongside the other theropods would be an oviraptorid.

Another prerequisite on the road to the skies is not only bipedality but a fleetness of foot, combined with an arrangement of ligaments that provides both an elastic rebound and a cushioning against impact. This entails a rearrangement of the metatarsal bones (as the of southern Germany[98] are exceedingly rare, and every new discovery is a sensation. New insights into this feathered theropod continue to emerge[99] and it is striking that even specimens that have received more than a century of study can still yield new data.[100] One such example is the "unexpected" discovery of feathers on the hindlimb of the second *Archaeopteryx* to be unearthed, the celebrated Berlin specimen.[101]

Iconic they may be, but their status is now being steadily eroded. First a whole series of remarkable discoveries of fossil birds from the Lower Cretaceous of Liaoning province, China, have thrown astonishing light on the next stage in the conquest of the air.[102] Of equal moment are the investigations into the evolutionary unfolding of the theropods as one of the most successful of dinosaur groups. Such research confirms beyond doubt that *Archaeopteryx* is a dinosaur with feathers, but in terms of such key features as its physiology[103] and brain structure,[104] this animal is effectively a reptile and not a bird. It is, however, in the wider context of theropod phylogeny[105] that the emerging framework points to exactly the same story as the terrestrialization of the sarcopterygian fish (p. 222): multiple trends in very much the same direction combined with a fascinating evolutionary mosaicism where not only are characters jumbled together but as they fall out of the evolutionary casino, then sooner or later a group comes up trumps.[106] Some of the most dramatic insights have come from the fossils of the Jehol Group in Liaoning,[107] but discoveries from other parts of the world[108] remind us that our knowledge of theropod diversity is far from complete.[109] New surprises, therefore, await us, but the main outline of theropod diversification is not only clear but reveals a series of remarkable stories in convergence.

First stop is Madagascar, where the Upper Cretaceous Maevarano Formation has yielded rich vertebrate spoils.[110] Above the dinosaurs flew several species of bird,[111] including *Vorona*[112] and *Rahonavis*,[113] except that the latter is not actually an avialian (that is, the birds and their immediate ancestors). Rather it belongs to a related group of theropods known as the dromaeosaurids,[114] and so are relatives of the troodontids. Unlike *Archaeopteryx*, *Rahonavis* is only known from its bones.[115] So how can we be so sure it flew? Evidence comes not only from its small body size, hollow bones, and elongate arms, but Catherine Forster and colleagues describe how part of the "ulna bears six, low slightly elongate papillae [that] . . . we interpret . . . to be quill knobs for the attachment of secondary flight feathers."[116] Just such a feature has been identified in the notorious *Velociraptor*. This animal was far too large to fly,[117] but is one line of evidence to indicate that feathers and similar integumentary structures evolved long before some theropods took to the air. Although much smaller, like any respectable dromaeosaur *Rahonavis* is equipped with a striking sicklelike claw on its foot, and scraps of fibrous material demonstrate that in life there was a keratinous sheath strikingly similar to that found in such living birds[118] as the emu.[119]

Rahonavis was not the only dromaeosaurid to become airborne. Independently, and earlier in the Cretaceous, we encounter *Microraptor gui* in the skies above China. Another star of the Jehol biota, not only does this dinosaur have the feathers preserved, but the fact that they arise from both arms and legs show this animal to be equivalent to a biplane.[120] The evidence for *Microraptor* being arboreal is strong,[121] but could it actually fly? Claims for the capacity for powered flight have been met with skepticism,[122] and biomechanical analyses suggests it was a very efficient glider.[123] That theropods evolved gliding is hardly surprising (and among the more primitive dromaeosaurids *Sinornithosaurus*[124] home in the canopy in about ten minutes.[85] It is evident that ants employ visual cues to stay on target.[86] Relatively few species of ant can glide, but for those that can it makes good sense, because not only are they arboreal, but they show features such as good eyesight and tendency to forage on the extremities of the trees. It may be an occupational hazard, but Stephen Yanoviak and colleagues propose that the gliding evolved originally because when the forests flood, an ant that happens to fall in to the water can expect a warm reception. The selective pressure to evolve gliding is obvious, and so too this capacity has emerged independently in African ants, notably *Cataulacus*,[87] which has many similarities to its distantly related cousins in South America.[88] Ants are by no means the only wingless insects gliding to safety. Among the much more primitive, bristle-tails, tree dwelling species in the tropics also control their descent, with a tail filament evidently acting as a rudder.[89]

If insects were the first to employ powered flight, in due course the skies were occupied by a series of vertebrates, including the pterosaurs, bats, and most familiar the birds. Nearly all look to the forelimb as the aerofoil, albeit in strikingly different configurations. Pterosaurs effectively employ an enormous finger, in the bats the equivalent membrane is stretched across a hand, whereas in the birds the end of the limb is more of a stump. Each,

therefore, wave part of an arm to achieve lift-off. Quite how the pterosaurs transformed themselves into the masters of the Mesozoic skies remains conjectural, although the first stage may have involved gliding. In the case of the birds, the path to the skies is clearer, but at first sight the convergences between at least the birds and bats might seem to be superficial. Both flap their wings, and that is the end of the story? Not quite, when it comes to digestive physiology they show some intriguing similarities.[90] The intense metabolic demands for flight are evidently also linked to a striking decrease in the size of the genome. This not only occurred independently in the birds[91] and bats,[92] but tellingly also in the extinct pterosaurs.[93] Like the bats the pterosaurs[94] employed a skin membrane as the wing, and in a way similar to that of the bats this is inferred to have had assisted with thermoregulation.[95] Correspondingly the pulmonary system of the pterosaurs is inferred to have been directly analogous to that of the birds.[96]

The bats and pterosaurs appear to have had a single origin, but I, for one, will not be the least surprised if some other extinct group of either mammal or reptile had a stab at becoming a bat or pterosaur. The fossil record will reveal if this actually happened, but just such a record shows that the birds did exactly that: not once, not twice, but at least three times they took to the skies.

MULTIPLE BIRDS

Perhaps that is just as well because it is difficult to imagine a world without birds. The received view, of course, is that birds are just one more evolutionary fluke. Feathered theropodan dinosaurs, emerging near the end of the Jurassic and exemplified by that "missing link" *Archaeopteryx*, the fossils provide fascinating insights into the transformation of

near-relatives of *Tyrannosaurs* to ultimately a swift. The story, however, is a little more complicated and a great deal more interesting. We now know much more than when the first specimen of *Archaeopteryx* was brought by a quarryman to the keen attention of the local doctor, Dr Carl Häberlein.[97] These celebrated fossils from the Solnhofen Limestone

It has been proposed that this rather strange animal may be close to pterosaurs,[58] but not all agree.[59]

This is something that the living lizard *Draco*, the "flying dragon," does to a consummate degree.[60] Once again it is the ribs that support the patagium. This structure can be readily folded and extended, and bands of elastin (p. 16) play an important role.[61] Although confined to the Indo-Malaysian forests *Draco* shows a remarkable diversity that includes a substantial range of body sizes. Because growth of the patagium is isometric, in principle the larger gliders have less gliding surface than they "require" and so potentially are at a disadvantage. So they are, but there are trade-offs in being large, and it is probably no coincidence that the big *Draco* are only found where a number of species[62] coexist.[63]

If *Draco* is the queen of reptile gliders,[64] at least three other groups show very considerable facility, not that this is always immediately apparent.[65] The West African lizard *Holapsis* is a competent glider, but it has no obvious adaptations for this aerial existence other than an ability to flatten its body.[66] This, combined with a very light body,[67] ensures not only a safe but graceful descent. In his analysis of the anatomy of *Holapsis* and the features that contribute to its effective gliding, Nick Arnold[68] is correct to remind us how it arose from "mosaics of cooption and adaptation."[69] To speak, however, of "multiple accidents of history"[70] begs the question. Just because *Holapsis* is the only gliding lacertid, and to say that it is "a lizard that glided by accident,"[71] may miss the wider point.[72] After all not only does it display skeletal lightening and a capacity to extend the skin, but its aerodynamic capacity is underpinned by various parts of the body bearing "modified scales . . . [forming] stiff lateral fringe-like extensions with little increase in weight . . . [and] are exactly paralleled in structure and function . . . in *Draco*, while *Ptychozoon* has

analogous cutaneous extensions along the length of the tail."[73] And what of the gecko *Ptychozoon*? Given their legendary capacity to grip (p. 84) it might seem surprising that this group of lizards not only learned to glide, but this ability has evolved more than once. *Ptychozoon* controls its descent using a patagium extending from the body margins, but this is a complex structure, and sensory structures[74] on the upper and lower surfaces may register airflow.[75]

Lizards and snakes are closely related, but the latter of course are legless, so a gliding snake would seem to be pressing the boundaries of biological credulity. Not in southern Asia, where the paradise tree snake (*Chrysopelea*) can launch itself in a variety of ways to become a glider.[76] It would seem an unpromising candidate, but nevertheless as it flattens out its ballistic drop, it extends its body and undulates, showing a considerable degree of aerial agility.[77] Its locomotion is, therefore, like that of its terrestrial counterparts, but it seems likely that the neuromuscular control during the descent is sophisticated.[78] If gliding snakes are improbable, then what about frogs? Hopping, yes, but gliding? Indeed so, and this capacity has evolved at least twice, specifically in the hylids and rhacophorids. The glide path is mediated by webbed feet (and additional flaps of skin attached to the limbs) that in combination with limb movements permit a controlled descent.[79] Although exhibiting a fair degree of maneuverability,[80] it is evident that the glide is not especially stable, and to lessen the risk from gusts and winds, much of their activity is nocturnal.[81] Descents can achieve a special urgency when the destination is a breeding aggregation.[82]

If a frog or snake gliding past you is an unexpected sight, can we draw the line at ants? No! Largely relying on hindlegs,[83] they too can be adept gliders.[84] In its controlled descent, the Neotropical *Cephalotes* travels abdomen first and having landed can be back assists with triangulation. The related sugar gliders are considerably more adept. In terms of lift and drag they are by no means aerodynamically identical to the flying squirrels, but the convergences[38] allow Kirstin Bishop to suggest how "we can begin to discover general rules for mammalian gliding."[39] The capacity for gliding has provoked speculation that this might provide an evolutionary route to fully powered flight, such as the bats. As Bishop trenchantly points out, given the radical differences between gliding patagium and bat wing, "How is it possible to move between two such highly specialized states without going through a phase in which the animal is not particularly well adapted for either function?"[40]

We face what might be called the "Darwinian trench," a nonadaptive chasm that seems impossible to bridge. Even though the fossil record is silent as to how bats first learned to fly (p. 137), Bishop suggests that one possibility is that fine manipulation of the patagium by moving the limbs could lead "to flapping behavior [that would provide] the means for overcoming this transition between aerodynamic regimes."[41] Such fine control is not only evident in the sugar glider but also their placental counterparts in the flying squirrels. What emerges is that even though the kinematics of launching can be traced with remarkably little change from their leaping cousins,[42] constant movements of the limbs and control of the wing surfaces during gliding impart a remarkable degree of maneuverability.[43] This is also reflected in what might seem unorthodox aerodynamics such as the aerofoil having a high angle of attack[44] and the final descent probably relying on deliberately stalling.[45] Intriguingly the cartilaginous strut that arises from the wrist serves as a winglet. This is probably not only central to controlling the vortices that arise from the leading edge but is directly comparable to the equivalent devised by NASA.[46]

Moreover, in the case of a Malaysian flying squirrel, W. Adams[47] reported that in Kuching he saw an individual "rise through the air with vigorous flapping movements of the skin between the fore and hind feet,"[48] landing about a meter higher than its departure point.

If mammals can do it, then so can reptiles. Among the extinct representatives, gliders took to the air as early as the Permian in the form of diapsid reptiles known as the coelurosauravians.[49] When first found, so improbable did the fossil appear to the German paleontologists that it was interpreted as a reptile that had come to lie on the fins of a coelacanth fish—not fins, but the elongate structures that once supported a wing-flap, although curiously in this case the bony rods supporting this aerofoil are of dermal origin rather than being the more standard elongated ribs.[50] The latter solution was employed in the succeeding Mesozoic, notably twice among the diapsid reptiles of the Triassic[51] and somewhat later in a Cretaceous lizard. Among the former, the best known are the English kuehneosaurids,[52] but earlier material from Poland[53] might give an unusual opportunity to see how the transition to gliding[54] was achieved. From Virginia *Mecistotrachelos* was certainly a glider but is remarkable for its long neck.[55] Given that in flight it could not recurve its neck (as in the manner of a heron) it is interesting to speculate how this reptile maintained its stability during its descent. Some 75 Ma later, as the flowering plants were spreading across the Cretaceous landscape, an acrodontan-like lizard from China was launching itself from the trees.[56] In this fossil (*Xianglong*) the skin of the patagium is preserved and was evidently strengthened by collagenous fibers. Such is also the case in an extraordinary glider known as *Sharovipteryx*, a contemporary of the kuehneosaurids, but from Kyrgystan. Its aerodynamics appear to have depended on a remarkable delta-wing.[57]

To think that the convergent evolution of gliding is a reflection of the sheer simplicity of the process is a misjudgment. Any expanded surface will in principle slow an inadvertent descent and so transform a potentially lethal mistake into an adaptive triumph. This must be correct. The diversity of gliders, from ants to geckos (and indeed seeds,[13] p. 389n10), and in some forms the capacity for impressive aerobatics that reflect sophisticated controls show, however, that evolutionary reality is more complex than meets the eye. Gliding makes good sense in terms of not only escape from predators but also the energetics of trudging across the forest floor.[14] Yet as Roman Dial observes in the case of mammals, "If gliding is such an efficient mode of travel, why are there only 60 gliding species worldwide?"[15] A real threat appears to be bird raptors, and this explains why these gliders are largely nocturnal. Nor is this the only constraint: gliding does not cross a level playing field. Among vertebrate gliders there is a remarkable diversity in the Indo-Malayan region, whereas equivalent areas in the tropics of Africa and South America are quite depauperate. A correlation seems to exist between the prevalence of lianas that can provide aerial highways and the employment of prehensile tails (p. 342n12) versus the capacity to glide.[16] This is unlikely to be the whole story, because while open and uncluttered forests will favor the glider,[17] the canopies of the Indo-Malaysian rain forests are also significantly taller than equivalent regions elsewhere in the tropics: with 80 meters or more of space beneath you, gliding looks increasingly attractive.[18]

Among the extant mammals the capacity to glide has evolved six times.[19] Three involve the marsupials (the feathertail gliders [Acrobates], petaurids such as the sugar glider and yellow-bellied glider, and the possumlike greater glider). The other three are placentals, evolving twice in the rodents (the fly-

ing squirrels[20] and African anomalurids[21]). Jonathan Kingdon also notes that the dwarf anomalure (Idiurus) "has an extraordinary convergence of form [with] . . . the South Australian Feather-tail glider, Acrobates,"[22] as well as the flying lemurs (or dermopterans) of Indo-Malaysia.[23] This by no means ends the roster of convergences, because extinct rodents known as the eomyids (relatives of New World pocket mice and pocket gophers[24]) had at least one Oligocene representative equipped with the characteristic aerofoil of skin (the patagium) suspended between the limbs.[25] In this eomyid the patagium is supported by a strut that arises from the elbow, a convergent arrangement seen in the anomalurids and the marsupial great glider,[26] whereas in the flying squirrels the strut arises from the wrist.[27] If the skies of the Tertiary were not already crowded enough, to this list of gliding mammals we can add a flying dormouse from the Miocene,[28] although a proposal that the extinct paromomyids were gliders (and related to the flying lemurs)[29] has been shot down.[30] Certainly arboreal,[31] this group of primates[32] have a dentition that is convergent with the marsupial sugar glider and suggests that it too gouged the bark to release the sap.[33] Mammals evolved long before the Tertiary, and at least one Mesozoic mammal was also a highly specialized glider, launching itself in the darkness above the heads of sleeping dinosaurs.[34]

The capacities of these extinct gliders must remain somewhat conjectural, but their living equivalents in the marsupial[35] and placental[36] lines reveal the degrees of aerial agility that can be achieved. Not that all species are equally adept. As Stephen Jackson observed in the case of the mahogany glider, the marsupial "often looked very hesitant at times prior to launch, appearing very fidgety when it came to gliding,"[37] and it also engaged in swaying of the head in the same manner as flying squirrels, a behavior that presumably

CHAPTER 17

THE ROAD TO THE SKY

COCOONED WITHIN a fuselage, gin and tonic at hand, for most of us the business of flying has long lost its romance. Yet now and again a pilot captures those extraordinary moments, perhaps vertiginously poised above a blue abyss or, as I vividly recall, flying down a fjord in North Greenland, walls of rock towering above our tiny plane. We, of course, are newcomers to the game. Once animals were on the land, then for all its insubstantiality it was inevitable that the air would hum with activity, filled with swooping and soaring forms.

TAKING OFF

Of all aerial convergences the most predictable is the capacity to glide. It has evolved at least thirty times independently,[1] and although usually thought of in terms of animals like the sugar glider or flying dragon, it is worth recalling that if some fish can walk (p. 72), others can glide. Visitors to the Triassic would have seen thoracopterid fish convergently gliding as they escaped the maws of marine reptiles,[2] but most familiar to us are the flying fish (the exocoetids). These come in both two-winged and four-winged types, the latter also employing their pelvic fins[3] and evidently representing a more advanced arrangement.[4] Not surprisingly the pectoral girdle is also enlarged, and the whole process of launching must be energetically expensive, which probably explains why these wonderful fish are confined to warmer waters.[5] As with their terrestrial counterparts it is likely that the capacity to travel[6] quite substantial distances is more than simply to escape their pursuers. In the Sargasso Sea, flying fish have been seen to glide from one wind-row of floating seaweed to another, and this accuracy is consistent with their capacity to see in both air and water (p. 105).[7] It is also likely that the evolution of gliding was considerably more complicated than simply leaping out of the water and finding the fins were working as wings. The origins evidently lay in high-speed plankton feeding, just beneath the surface of the ocean. With seabirds and dolphins in attendance, this is a very dangerous zone to inhabit; there was a logical step from taxiing to takeoff.[8] Nor are the flying fish alone, at least when we turn to freshwater.[9] Neither is it surprising that those masters of convergence, the squid (chapter 2), are also known to glide.[10] Liftoff is the result of jet propulsion, but in a case of a school of squid evading tuna off the coast of Australia, Silvia Maciá and colleagues noted, "some of the squid were observed jetting water while airborne."[11] These authors also observed other "squid rapidly undulated their lateral fins,"[12] while the posture of the arms suggests that the glide path was under active control.

example, the ability to chomp wood is found in some stag beetles (along with evidence for nitrogen fixation[229]), while among the vertebrates giant rodents such as the beaver and porcupine employ hind-gut digestion to the same effect.[230] Out to sea, the notorious shipworm, the bivalve *Teredo* and its relatives, causes havoc in submerged wooden structures as it bores through pilings and ships. Digesting wood all the time,[231] the metabolic challenge is met by enzymes such as glucanases.[232] It is probably not coincidental that although the guts of *Teredo* are not microbial-rich in the manner of termites, the gills of both these shipworms[233] and their deep-sea equivalents (in the form of the appropriately named *Xylophaga*[234]) possess a dense accumulation of gram-negative bacteria. These reside in bacteriocytes, recalling the xylophagous insects that faced an equally demanding diet. Wood crumbles and rots, is riddled with borings, and can readily burn. It also opens new evolutionary doors, not least providing jumping-off points for gliders.

shakers of the planet. Most obvious are the spectacular nests of some termites, ingeniously engineered to control temperature and humidity,[203] not to mention the construction of storage pits.[204] In addition to the consumption of vast quantities of vegetation (and other material), the subterranean excavations not only result in the movement of millions of tons of soil but significantly its enrichment in minerals.[205] This has a direct bearing on the apparently bizarre practice of geophagy. Just as pregnant women in Tanzania drink the water from "crab-ponds" (p. 167), so elsewhere in many parts of Southern Africa they consume so-called "pregnancy clays."[206] So, too, elephants will eat the soil of abandoned termite mounds.[207]

In the case of the termites it is, however, only the higher macrotermitines that cultivate fungi. Interestingly the fossil record of their fungus gardens suggest that this agriculture only evolved about 7 Ma ago[208]—that is, at much the same time as the leaf-cutter ants.[209] These fossils come from Chad, and like the human lineage this cultivation appears to have originated in the African rain forests.[210] The evidence indicates that the invasion of Asian habitats by these termites then occurred at least four times, but Duur Aanen and Paul Eggleton suggest that there were also multiple colonizations of the much drier savannahs. All this makes good sense: unwittingly the termites took their original rain-forest habitat with them—but as far as the fungi are concerned, as long as they can live in a nice warm, humid, and protected nest, then who cares!

Although the termites cultivate a fungus (appropriately *Termitomyces*), it is distantly related to the attine ants' crop and evidently has a single origin.[211] In contrast to the attine fungus it can produce fruiting bodies in the form of mushrooms that are highly nutritious. These are eagerly sought after[212] and can grow to an incredible size,[213] up to 1 m across.[214] By and large the specificity between

fungal cultivar and termite is low,[215] and presumably the mutualism is dynamic and flexible. Nevertheless, in a given nest the termites evidently propagate only one strain of *Termitomyces*, so the crop is a monoculture.[216] As Duur Aanen and colleagues note, not only is this convergent with the attine ants but "provides insights into the general principles that govern the stability of obligate ectosymbiotic mutualisms[, and] . . . it appears that the lifetime commitment between each farming society and a specific clonal crop is decisive for making both these fungus-farming mutualisms evolutionarily stable."[217] Unlike the attine ants, however, the macrotermitines have remained faithful to their *Termitomyces*, with one remarkable exception: the termite *Sphaerotermes*,[218] which extraordinarily cultivates a nitrogenous bacterium.[219]

With the exception of *Sphaerotermes*, about which little seems to be known, once established, the fungal gardens (the combs) are maintained within special chambers built of fecal pellets.[220] This serves to inoculate the organic-rich comb with the fungal spores.[221] Not only is the resultant fungal crop (forming nodules or mycotêtes) consumed, but also the basal parts of the comb. There is no doubt that this is an obligate mutualism, with the termites reliant on their agriculture. The exact nutritional role of the fungi[222] has, however, proved more elusive and may well vary among the macrotermitines.[223] Most likely the fungal breakdown of refractory plant material is important,[224] and in this context we see direct parallels with other systems of insect agriculture whereby both fungus and termite contribute appropriate and synergistic enzymes[225] that include both xylanases[226] and cellulases.[227] So, too, there is evidence for antimicrobial defenses.[228]

For those unacquainted with the termites, their destructiveness can be difficult to appreciate, but the capacity to digest wood (xylophagy) has evolved numerous times independently. Among the insects, for gardens and even show similar behavior in manipulation, so indicating, in the words of Rachelle Adams and colleagues, "that it has been convergently derived."[188] However, despite this, they fail to provide fresh substrate, and ultimately the ants are on the march again.

It is sometimes claimed that among the ants only the attines have achieved agriculture. Convergence would indicate otherwise. There are ants that not only harvest mushrooms[189] (in the Malaysian rain forest), but evidently process the collected fragments. This appears to result in a sort of fermenta-tion, but in any event it evidently prevents the food from going bad.[190] More intriguing is an association between an African ant and a plant (*Leonardoxa*). Such ant-plants have been addressed in other contexts of symbiosis (p. 86), but here the nesting cavities (known as domatia)[191] almost invariably contain a regularly arranged area of fungi.[192] The exact nature of this association is yet to be determined, but the evidence (including the fact that the ants supply the fungus with nutrients[193]) strongly points toward a form of agriculture.[194]

TERMITES AND BEYOND

One wonders how many more examples of insect agriculture remain to be discovered, or indeed may have evolved independently in the geological past. Considering the latter, what are we to make of some puzzling trace fossils from the Lower Cretaceous of Patagonia, consisting of ball-like structures permeated with roots?[195] Nailing their origin is not easy. Conceivably they were constructed by primitive attines, but if so they far predate their known geological history. Similar problems apply to otherwise interesting similarities to some termite nests, especially those constructed by the strange *Sphaerotermes*. As Jorge Genise and colleagues remark, these trace fossils could "represent a type of primitive fungus garden, associated with roots and/or the chamber wall, produced by a non-attine ant already present by the Lower Cretaceous."[196] If not an ant, then possibly a termite but either way arriving at this solution independently of its descendants. Certainly, of all the examples of insect agriculture, those that come closest to rivaling the attine ants are the fungus gardens of the termites.

Yet the emergence of the termites surely involves a lurking predictability.[197] Although often referred to as white-ants, in fact they arose from within the cockroaches.[198] As Daegen Inward and colleagues remark, given the profound differences between the two groups, this "may appear surprising to many people."[199] Among the cockroaches the closest relative to the termites is the wood-eating genus *Cryptocercus*.[200] This digestive capacity is a key prerequisite for the xylophagous termites, equipped as they are with a complex array of intestinal symbionts that assist with the breakup of the otherwise refractory cellulose. Elsewhere among the cockroaches another subsocial taxon (*Parasphaeria*) seems to be well on the road to evolving into a termitelike form, not least in terms of demonstrating brood care.[201] As these workers also stress, this convergence is not so surprising given that the capacity of these fungi to digest cellulose means that half the legwork has already been achieved as a preadaptation.

Once, therefore, the cockroaches arrived it was only a matter of time before the world welcomed the first termites, and in the event most probably this was in the early Cretaceous.[202] Although their overall diversity is low (approximately three thousand species), since their debut this extraordinary group has become literally one of the movers and day (except in heavy rain) and never returns to the nest to assist in foraging.[166] In one species the dumps are external and downslope of the nest (and well skirted by the foraging trails) so the waste workers are not aggressively handled by the occupants of the nest, whereas in another species of *Atta* the waste is put in dump chambers (in some taxa some large enough to house a human[167]) and however disposed of, dump workers are recruited from the elderly and suffer high rates of mortality.[168] And with good reason; the stuff they handle is toxic, laden with *Escovopsis* spores. Moreover, if in the fungal gardens themselves spore levels rise, then more workers are assigned to dump duties—and a short life.[169] It is hardly surprising, therefore, that when gardeners spread this waste material around the bases of vulnerable trees,[170] it provides an effective deterrent against the depredations of leaf-cutters.

Parallels with human agriculture and waste management have often been drawn, and certainly the nests are among the most intricate structures that are constructed by the social insects. In this context it is significant that the "leaf-cutter" queens enjoy multiple mating,[171] which means that these nests are far from being genetically uniform, staffed by robots.[172] Multiple mating, of course, increases genetic diversity and hence the propensity for these enormous societies to resist disease,[173] and in the case of the leaf-cutter ants this was evidently an abrupt transition that went hand in hand with a significant increase in the size of the metapleural glands.[174] New worlds now opened, but there are inevitably new threats.

As already noted the intricate symbioses that characterize the attine ants can be construed as the result of relentless competition as each "partner" attempts to exploit the others. As we have already seen in terms of the ant-actinomycete association, the evidence hardly points that way. So, too, the ants utilize enzymes derived from the fungi

that not only complement their own arsenal,[175] but help with both digestive capacities[176] and, by the application of fecal droplets, the maintenance of the fungal gardens.[177] The all-important business of nest-mate recognition may also be assisted and made more reliable by employing chemicals derived from the fungi.[178] Like other social insects the larvae, which are wholly dependent on the fungi, receive intensive care that includes frequent licking, feeding, and cleaning of the mouthparts.[179] As Julian Lopes and colleagues remark, not only does this reflect a complex behavior but it opens the possibility "that workers are to some extent able to assess the individual needs of larvae."[180] The flexibility and dynamism of these colonies is also reflected by the fact that in at least *Acromyrmex* the caste structure revolving around different types of workers is phenotypically plastic.[181] William Hughes and Jacobus Boomsma make the important point that the "convergent evolution of phenotypically plastic genetic influences on division of labour in leaf-cutting ants and honeybees suggests they may be an intrinsic property of complex, genetically diverse societies."[182] Combine this with evidence of the ants displaying behavioral plasticity and memory,[183] and our views on this complex set of symbioses might need to change.[184]

The attine farmers are an acme of evolutionary success, but one that is sometimes treated as a one-off among the ants. Not quite. Leaf-cutter ants will certainly defend their nests vigorously from attack by other ants, notably the army ant *Nomamyrmex* and where opposed sides both employ combat teams.[185] In the case of some so-called agropredators, however, the invading ants may induce panic before they plunder the garden,[186] but in other cases the invaders actually make a stab at agriculture.[187] In the latter instance (involving the ant *Megalomyrmex*) the adults appear to maintain the fungal structures, with associated glands that most likely are essential for the well-being of the guest actinomycetes.[146] The general assumption, therefore, has been that these bacteria are in the front line in the struggle against the principal threat to the fungal garden, the virulent *Escovopsis*. Whether matters are quite so simple is, however, less clear, and the idea of a close mutualistic association between ant and actinomycete may need reassessment. Ruchira Sen and colleagues[147] point out that the antibiotics secreted are unlikely to target specifically *Escovopsis*, and indeed can threaten the crops they are ostensibly protecting.[148] What is not in doubt is that these microbial associations are complex. For example, distinct clades of actinomycete can be recognized,[149] and particular strains of actinomycete may be associated with a given colony.[150] Clearly, considerable flexibility exists in these microbial associations with evidence of multiple recruitment.[151] This dynamism is not surprising, but it provides a sober check on the dream that if ants could manage to employ the same antibiotics for millions of years, then perhaps we too could learn how to escape the scourge of antibiotic resistance. As Ulrich Mueller and colleagues[152] dryly note, the hope that the actinomycete-ant association might render us "invulnerable to resistance evolution now appears to have been misquoted."[153] Rather than a relentless arms race, the interactions are more subtle, relying both on repeated recruitment from free-living populations of actinomycete and also employing different taxa.[154]

For the ants these natural antibiotics are by no means the only line of defense, and in the complex world of symbioses they can carry their own risk. Such is evident from the ant *Apterostigma*, where the action of the actinomycetes is potentially compromised by an association with a black yeast.[155] Indeed, among the attine ants there is evidence of a shift away from the use of antibiotics as an agent of biological control and to rely more on chemicals.[156] Here the metapleural glands play a central role. Although typical of ants as a whole, in the attines they secrete a variety of antimicrobial compounds.[157] This is associated with active grooming, and such selective application contrasts with the indiscriminate use of comparable antibiotics[158] by humans.[159] All this points to an effective system of immunity against ever-present pathogens, but it is more than likely that the complexity of this defensive array is even now not fully appreciated. In *Acromyrmex*, so-called allo-grooming increases as the pathogen threat rises,[160] providing what is effectively an adaptive social immunity. As Tom Walker and William Hughes remark, this arrangement "may be functionally analogous to the advanced physiological immune response of vertebrates"[161] (p. 307n2). Similar problems yield similar solutions, but the risks do not stop there.

One of the inevitable by-products of large and complex societies is waste, heaps of it. Seething with pathogens this is not material to have anywhere near the farm. The attine ants rise to the occasion, with intriguing differences between the more primitive groups and the leaf-cutters.[162] In the former, waste material is typically packed into a pocket adjacent to the mouth, and the resulting pellets may be either stacked near the nest or taken out to a more distant dump.[163] Where pellets are initially stacked, then, as Ainslie Little and colleagues note, "a separate caste of workers is devoted to the construction, management and eventual disposal of these piles."[164] Moreover, in *Trachymyrmex*, if infections rise, then pellet production soars. This infrabuccal pocket is far from being a wastepaper basket, and the gathered spores of *Escovopsis* are sterilized, most likely by antibiotics from resident actinomycetes.[165] In the leaf-cutters this can be taken to a remarkable degree of sophistication; in *Atta* a dedicated team works night and termed symbiont policing, whereby detection of alien strains provokes a strong reaction.[119] As Aniek Ivens and colleagues note,[120] their findings "add a novel layer of complexity to the sophistication of coadaptations in the attine ant-fungus symbiosis."[121]

A particular fascination with this agriculture is how pathogens are controlled, and as important how infected waste is safely disposed. Its relevance to some of our present-day predicaments will surely be obvious. Such protection begins from the day the garden is first planted by the queen. In addition to cleaning actions, methods to avoid contamination can include attaching the fungi to a rootlet (as in *Acromyrmex*[122]), while in some primitive attines the queen ingeniously employs her discarded wing as a platform for the fungus.[123] But the garden is always in danger of being overrun. Foremost in terms of threats is a virulent pathogenic fungus known as *Escovopsis*,[124] which specifically attacks the crop rather than the prepared mulch on which the fungi grow.[125] In general a particular strain of *Escovopsis* will track a specific crop, and indeed this pathogen is only found in association with the ant farms.[126] Like the crops, however, strains of the pathogen can switch,[127] as even happened when the ant *Apterostigma* began cultivating a completely new sort of fungus.[128] Unfortunately, *Escovopsis* is not the only threat. Not only are there a variety of other microfungal weeds,[129] but fresh plant material brought into the nest is potentially contaminated by fungi (and indeed prophylactic action[130] is taken[131]).

Another threat manifests itself in the form of so-called entomopathogenic fungi, and here we see disturbing parallels to human disease. One virulent fungus (known as *Ophiocordyceps*) may have only recently moved to the leaf-cutters,[132] although it is well known to attack other ants. Infection by this fungus produces one of the more lurid examples of behavioral control. David Hughes and colleagues[133] demonstrate how the ants effectively behave as zombies. Not withstanding the destruction by the fungus of much of the mandibular musculature, as the ant dies it clamps tightly to the plant[134] (the resultant scars can even be identified in the fossil record[135]). This gives time for the fungi to sprout a fruiting body on a long stalk from the ant's head (or other parts of the body).[136] Curiously the ants so infected may die close to each other, and these graveyards[137] are given a wide berth by the fit and healthy.[138] As far as the leaf-cutters are concerned if *Ophiocordyceps* is a looming threat, other pathogens are a very present danger. In particular the virulent fungus *Metarhizium* attacks not only ants but many arthropods (and so incidentally provides humans with a means of pest control). In the case of the ants, infection by this fungus presents a major challenge to its immune system. With its defenses weakened, another fungus (a species of *Aspergillus*), which is otherwise more or less harmless, suddenly becomes a major threat as it outcompetes the *Metarhizium*.[139] Sound familiar? As William Hughes and Jacobus Boomsma grimly note, this fungal interaction is "strikingly similar to that seen in immunocompromised vertebrates. . . . Indeed, infections of *Aspergillus* in humans are almost entirely limited to immunocompromised patients."[140]

These parallels to the threats all humans face, be it by infection or the risk of agriculture disaster, suggest that the attine ants may provide us with some timely lessons. Not least does this apply to antibiotics. Living in direct association with some of the gardens cultivated by *Atta*[141] we find a bacterium (*Buckholderia*) capable of secreting antibiotics. More remarkable is the recruitment of the actinomycetes.[142] These bacteria are housed on the ants[143] themselves and produce antibiotics[144] that contribute to keeping the pathogens at bay.[145] The associations can be complex, employing specialized cryptlike the folivorous howler monkeys are not identical, both show a strong preference for young leaves.[92] As Larry Rockwood and Kenneth Glander note, "while howling monkeys and leaf-cutting ants are not closely related in an evolutionary sense, they appear to have converged ecologically as generalist herbivores in neotropical forests."[93] So far as the ants are concerned, their success is not surprising given their organization (for example, in terms of "bucket-brigades"[94] and caching[95]), flexibility in foraging behavior,[96] trail construction,[97] and the extraordinarily high expenditure of energy by the leaf-cutting mandibles (albeit assisted by zinc impregnation, p. 66[98]) approaching that seen in that metabolic furnace, insect flight muscle.[99] So, too, while this agriculture had its origins in the tropics, these ants have made significant inroads into more temperate climes, largely by employing cold-resistant fungal symbionts.[100]

Higher attine ants rely on freshly cut vegetation, about which the individual workers appear to exert an "unexpected" degree of choice.[101] Among this group as a whole, however, the styles of agriculture reveal a story of increasing evolutionary sophistication toward more complex and apparently irreversible solutions. Attine ant agriculture appears to have had a single origin[102] some 50 Ma ago, and primitive taxa (such as *Mycetarotes*[103] and *Mycocepurus*[104]) give us glimpses as to how, in the words of Ulrich Mueller, this group "evolved the ability to grow their own food."[105] From this starting point, at least five important transitions can be recognized[106] with the debut of the most advanced leaf-cutting ants being "remarkably young"[107]—that is, only appearing some 10 Ma ago. This result depends on estimates using the molecular clock, but in age it is not far from the earliest known fossil nests built by attine ants in the late Miocene.[108] Although the crop usually depends on cultivating a group of basidiomycete fungi (known as the lepiotaceans), there

have been two significant shifts. One (in the ant *Cyphomyrmex*) is to the cultivation of yeasts[109]. In another ant (*Apterostigma*) there has been a major shift to growing gardens of a coral-mushroom (the pterulaceans),[110] a transition perhaps facilitated by the nests being associated with decomposing wood where these particular fungi flourish.[111]

Given this range of agricultural activities it is not easy to arrive at generalizations. Nevertheless, if recent research has achieved anything, it is to show that these associations are even more complex and dynamic than once envisaged. For example, although several shifts in crop type have been identified, the idea that the attine ants have maintained clonal monocultures for vast periods of time is now seen to be too simplistic. Rather we see evidence for both stability and the capacity for change. In the case of the cultivation of the coral-mushroom, which has long been regarded as an eccentric decision, phylogenetic analysis suggests that this domestication may have occurred more than once (although alternative scenarios remain equally possible).[112] Another insight comes from the ant *Mycocepurus*. This ant seems to be completely asexual[113] (which may seem rather astonishing until it is realized that such clonal reproduction has evolved multiple times among the ants[114]), and, as Anna Himler and colleagues remark, draws on an "unparallel diversity of cultivars."[115] This capacity to shift frequently "to novel, distantly related fungal crops"[116] may not be coincidental, and certainly this ant enjoys a very wide distribution.[117] With the more primitive attine ants it is likely that the free-living fungi provide a reservoir for potential cultivation. Among the higher attines, however, some evidence points to a less flexible system. Thus the incumbent fungal crop can itself chemically help to repel invading strains of potential competitors,[118] while in *Acromyrmex* when foraging the major workers exhibit what is and asphondylines), because here along with her eggs the female deliberately introduces fungal spores. In due course the larvae feed within the so-called ambrosial galls, on a mycelium of ascomycete fungi. Once again this is a highly integrated and interdependent mutualism,[75] and unsurprisingly the fungal spores are transported in pocketlike mycangia. These structures have evidently evolved independently at least twice, and in the asphondyline midge the mycangia are sensibly located in the ovipositor.[76] Many other insects, of course, employ such ovipositors for egg laying in wood (p. 66), and among the woodwasps (siricids and xiphydrids) we see a striking convergence to the beetles and dipterans. Here too the larvae live on fungi, this time a basidiomycete, and the spores are housed in the female wasp in mycangia-like structures.[77] This is evidently an intimate symbiosis, not least because as in other examples of arthropod agriculture the sterols[78] and at least two of the enzymes (which the larvae needs for digestion) are supplied by the fungus.[79]

THE MASTER CULTIVATORS

Wood wasps belong to the hymenopterans, and in this major group we encounter an insect agriculture that rivals human systems in complexity. Welcome to the attine ants. It is small wonder that this system has attracted so much attention.[80] The sophisticated underpinnings involve not only the careful cultivation of the fungi upon which the larvae are wholly dependent[81] but a series of related activities that are familiar to any gardener. These include weeding,[82] pruning,[83] application of fertilizers via fecal droplets,[84] and—paralleling the ambrosia beetles—the employment of antibiotics[85] to counter the ever-present danger of infection by pathogens. Key in this regard is specific preparation of the substrate to be employed in the fungal gardens. This may range from simply licking to what appears to involve inoculation with antibiotics.[86] The complex roles of antipathogenic bacteria is returned to below, but other bacteria (especially *Klebsiella*) play a key role in the fixation of nitrogen[87] (nor is this symbiotic association restricted to the attine ants[88]). Research into attine agriculture not only continues to underline the complexity of this agricultural system, but as importantly addresses questions central to evolution. How and when did this arrangement emerge? How has the balance between maintaining the stability of association and dealing with the dynamism of pathogenic threats actually been achieved? Even more important is to inquire whether the various interdependent, indeed obligate, mutualisms are in reality no more than a veneer for what in reality is a system ruled by ruthless competition and opportunism. Here in a microcosm is one of the central anxieties of the Darwinian formulation, revolving around the agency of blind, insensate forces engaged in cyclical viciousness as against the emergence of genuine mutualisms leading to genuine cooperation, even to transforming possibilities.

In the context of attine ant agriculture, the emphasis understandably has been on the classic leaf-cutter ants (including the well-known higher attines *Acromyrmex* and *Atta*). These, as Ted Schultz and Seán Brady[89] remind us, are a force to be reckoned with:[90] in the context of "a mature *Atta* colony [it is] the ecological equivalent of a large mammalian herbivore in terms of collective biomass, lifespan, and quantity of plant material consumed."[91] While the patterns of foraging that are shared by these ants and their neighbors the males emerging from unfertilized eggs and so necessarily are haploid.[56] Such an arrangement might be expected to slip into fully fledged eusociality, but to date is only known in one species from Australia.[57] In this beetle, however, both sexes are diploid (as are the termites, p. 171), indicating that although haplodiploidy may be key in such eusocial insects as the hymenopterans, what in some ways is a different route to eusociality[58] must have been employed.[59] A crucial factor may be that these Australian ambrosia beetles inhabit *living* gum trees, which for these insects is otherwise very unusual.[60] The advantage, however, is that whereas dead wood soon rots, here cavities can persist for decades. This leads Brian Farrell and colleagues to remark, "we predict that eusociality should be sought in other ambrosia beetles that attack living trees."[61]

As with other insect agricultures, sooner or later new colonies are founded, and they too need to be equipped with fungi. At its onset this may have been achieved by fortuitous adhesion of fungi. In the ambrosia beetles, however, nothing is left to chance because pure cultures are maintained in special pockets, the mycangia (also known as mycetangia). These are not only glandular but in being well supplied with respiratory trachea (and smaller tracheoles) the mycetangia are evidently sites of considerable metabolic activity.[62] These structures all serve the same function but are found in many separate locations in the different species of ambrosia beetle, provoking Alan Berryman to suggest that the mycangia "evolved simultaneously and independently within and between many of the modern genera."[63] On occasion, however, the mycangia have been lost. This is not due to sheer carelessness, but rather because at least three times independently ambrosia beetles have turned to stealing the fungus of other species.[64] Yet as Jiri Hulcr and Anthony Cognato note, "Ambrosia beetles are a true

showcase of the success of the fungus farming strategy: they dominate beetle communities in tropical forests . . . [and] are among the most frequent invasive species worldwide."[65]

They are, moreover, not alone. Another weevil (*Euops splendida*) not only possesses mycangia, but additional chambers serve respectively to incubate and nurture the fungal spores.[66] This symbiosis represents an independent agriculture, with the female constructing a larval cradle by cutting and then rolling a piece of leaf. Innoculation is achieved by yet another specialization in the form of a dedicated comb on the underside of the body. In some cases the fungi may render the leaf tissue more palatable,[67] but possibly they are more important in providing an antimicrobial defense.[68]

Weevils have become sophisticated fungal farmers, but probably the first group of beetles that turned to agriculture were the lymexylids.[69] Best known as the once-feared shipworm that were a major threat to wooden ships, the lymexylids typically eat fungi. In at least one case, the so-called leather-winged sailor (*Hylecoetus dermestoides*), their larvae flourish on ambrosial farms,[70] composed of an ascomycete fungus.[71] The larvae develop from eggs laid in the wood, with transmission of the fungus depending on pouchlike structures located adjacent to the ovipositor. The roster of ambrosial cultivation does not stop here. Some female stag beetles, for example, possess mycangia that house a form of yeast, which assists in the digestion of the wood by the larvae.[72]

Beyond the coleopterans,[73] a similar arrangement has evolved independently within the dipterans, specifically among the gall midges.[74] As their name suggests, they provoke gall formation in plants; for these midges the galls house the larvae, and in some species nutriment is obtained from the host plant. Not, however, in two groups of gall midges (specifically the lasiopterines snail *Littoraria irrorata* (which has evolved camera eyes [p. 106]), which deliberately wounds a marsh grass with its radula. These are then infected by an ascomycete fungi, which the snail nurtures by fertilizing with its fecal pellets. Although there is no direct evidence for inoculation by the snails, it is evidently an obligate mutualism whereby not only does the snail obtain food, but the fungus gains ready access to the plant tissues.[39] The extent of cultivation in the marine settings may be far more extensive than is generally realized.[40] Most seafloors are rid-dled with burrows and other constructions, and these survive in the rock record as trace fossils. Many types are recognized, among which the category of so-called graphoglyptid constructions are noteworthy for their complexity. Intriguingly these have been suggested to serve as farms,[41] whereby the builder (perhaps a crustacean) employs fungi or other microbes to attack organic material that would otherwise be difficult to digest.[42] Not only their organization, but the presence of chambers are further indicators that even in the abyss farming continues.[43]

Arthropod Farmers

Fungal farming has also evolved repeatedly on land among the insects, notably the beetles, termites, and those master agriculturalists, the leaf-cutter ants. At least as far as the beetles are concerned, the propensity of fungi, especially the ascomycetes, to adhere to the bodies of arthropods makes the emergence of a mutualistic agriculture all the more likely. But it is the weevils, and notably the so-called ambrosia beetles (comprising the platypodids and scolytids), which prove to be preeminent when it comes to fungal cultivation.[44] While related forms inhabit the bark and feed on the sugar-rich phloem (and thereby introduce such pernicious infections as Dutch elm disease), the ambrosia beetles live deep in the wood where they excavate a series of galleries. Most often they inhabit dead or decaying wood and seldom inflict much damage.[45] This agricultural system has evolved numerous times;[46] one study suggests at least ten times, first appearing some 50 Ma ago, and successive appearances being linked to times of global warming and the spread of tropical forests.[47] So, too, the fungi that are cultivated (often ophiostomatoids) are rampantly convergent.[48] The fungi are only one part of a microbial complex,[49] and the ambro-sial layers that grow in the galleries supply both the larvae and adults. The intricacy of this mutualism is remarkable. For example, many aspects of larval development specifically rely on a sterol (ergosterol, a fungal equivalent to our cholesterol) supplied by the fungus,[50] while oral secretions[51] and a bacterium (an actinomycete)[52] secrete antifungal compounds that deter invasive and pathogenic fungi (as independently occurs in the ants, p. 183).

The degree of social care and cooperation exhibited by the ambrosia beetles is also of a high order. Eggs, for example, may be placed in "cradles" of wood chips and supplied with the requisite fungi on which the larvae feed while the female removes waste products and reseals the fungal plug that serves to close the cradle.[53] Not only that, but in at least one species the larvae themselves assist with both gallery maintenance and the removal of waste.[54] The social cohesion is reflected in the colonies being haplodiploid, a system that itself is rampantly convergent, having evolved at least seventeen times.[55] In these beetles haplodiploidy evolved prior to the emergence of fungal agriculture and is marked by intense inbreeding, a predominance of females and

Fig. 48. The kea is a New Zealand parrot famous for its intelligence. (pp. 264, 283) [photograph: iStock]

Fig. 49. Despite its unassuming appearance, the archer fish is a master marksman. (p. 271) [photograph: iStock]

Fig. 50. The capuchin monkey is a tool-user. It arrived at this solution independently of other monkeys and apes. (p. 281) [photograph: iStock]

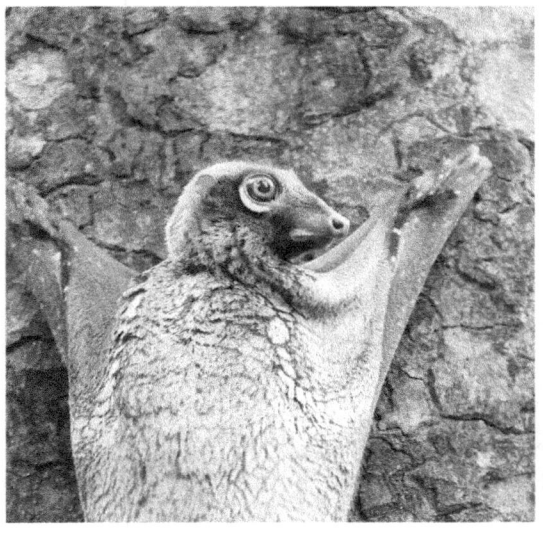

Fig. 45. The flying lemur is an inhabitant of Southeast Asia. It is one of the many mammals that have evolved gliding. (p. 190) [photograph: iStock]

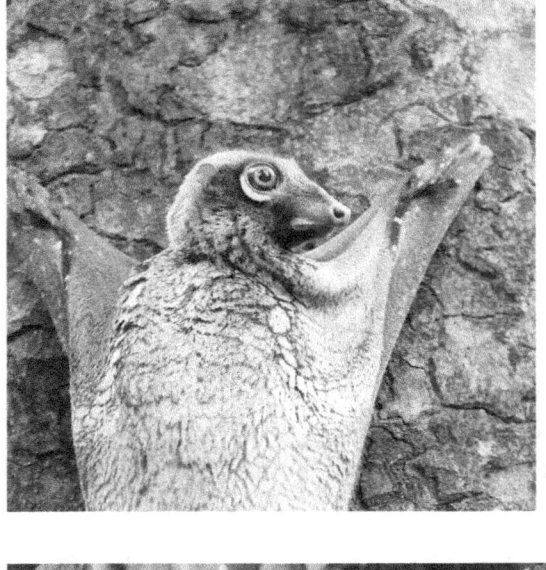

Fig. 46. The hoatzin is a South American bird that provides examples of convergence ranging from folivory to skin coloration. (pp. 205, 206) [photograph: iStock]

Fig. 47. The vulture is an iconic scavenger, but convergent. (p. 208) [photograph: iStock]

Fig. 42. The dung beetle: not only has this habit evolved several times, but these beetles are also endothermic. (p. 169) [photograph: iStock]

Fig. 43. Ants and aphids—from extraction of honeydew to taking on the role of shepherds. (p. 174) [photograph: iStock]

Fig. 44. The flying fish is one of many examples of the convergent evolution of gliding, but also with a direct counterpart in the Triassic. (p. 189) [photograph: iStock]

Fig. 39. The cactus is a classic xeromorphic plant, but exhibits a series of striking convergences. (p. 156) [photograph: iStock]

Fig. 40. Despite a flower of epic size, the plant *Rafflesia* is parasitic and also convergent. (p. 162) [photograph: iStock]

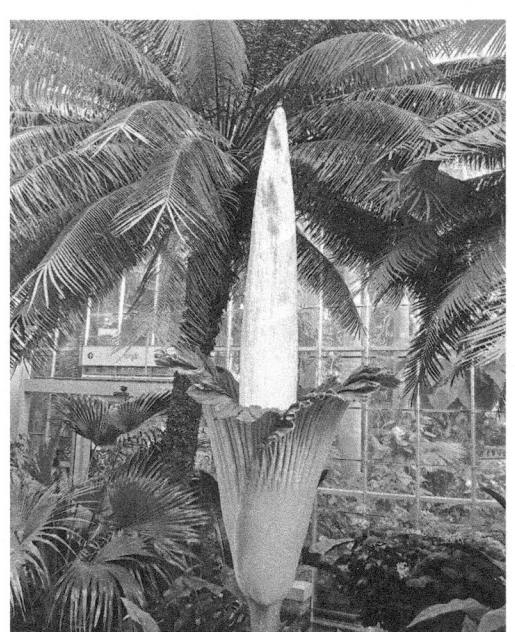

Fig. 41. The monstrous flower of this titan arum extends the roster of convergences to a repulsive smell and heat production. (p. 163) [photograph: iStock]

Fig. 36. The dolphin, an example of a toothed whale, exemplifies convergent evolution, from echolocation to cognition. (pp. 141, 259, 263, 279, 293)
[photograph: iStock]

Fig. 37. The lichen represents an extraordinary association between fungus and alga, and provides a rich source of convergences. (p. 152)
[photograph: iStock]

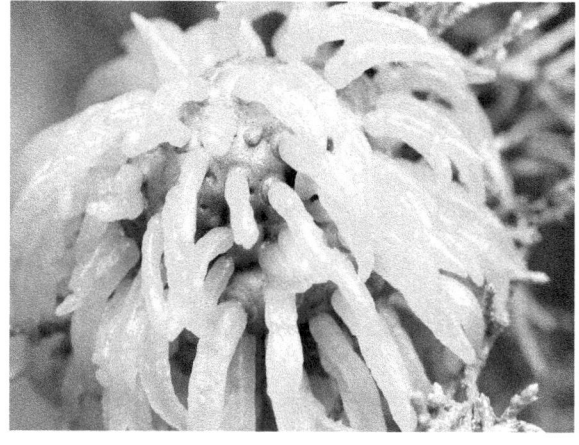

Fig. 38. The rust fungus. Similar fungi create startling mimics of flowers. (p. 153)
[photograph: iStock]

Fig. 33. The robber-crab is a crustacean that has made it onto land, and shows convergences of its sensory system. (pp. 122, 166, 249) [photograph: iStock]

Fig. 34. The termites, close relatives of the cockroaches, are independent inventors of agriculture. (pp. 129, 186) [photograph: iStock]

Fig. 35. The kiwi is the iconic flightless bird from New Zealand that provides rich insights into convergence. (pp. 130, 208, 262) [photograph: iStock]

Fig. 30. The firefly is not a fly but a beetle. Its bioluminescence finds many convergent counterparts. (p. 109) [photograph: iStock]

Fig. 31. The hawkmoth (or sphinxmoth) is an insect but is strikingly convergent with the hummingbirds. (pp. 113, 127, 169, 200) [photograph: iStock]

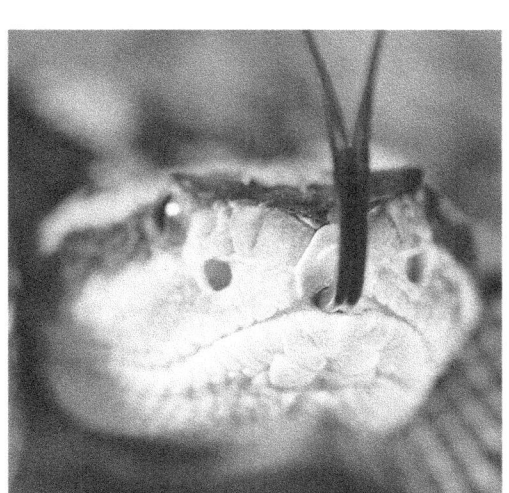

Fig. 32. The rattlesnake, justly feared, detects its warm-blooded prey with its infra-red pits. (pp. 116, 119) [photograph: iStock]

Fig. 27. Spider silk, familiar as the cob-web, is not only a highly effective trap but an object lesson in convergence. (p. 88) [photograph: iStock]

Fig. 28. The mosquito is the possessor of a compound eye, but has an olfactory system that shows a remarkable molecular convergence. (pp. 102, 123) [photograph: iStock]

Fig. 29. A brittle star possesses eyes of calcite, convergent with those of trilobites. (p. 103) [photograph: iStock]

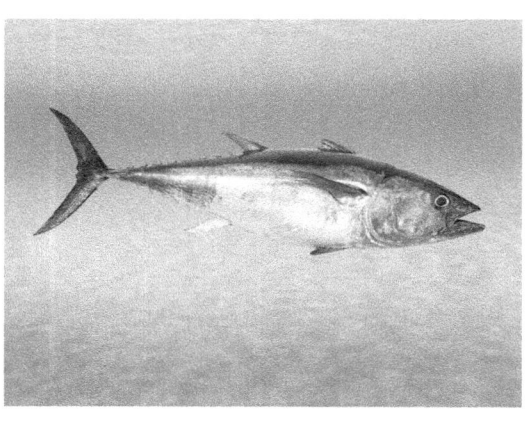

Fig.23. The cockroach's manner of walking is strikingly convergent with a dog. (pp. 74, 175, 186, 246) [photograph: iStock]

Fig. 24. The tuna is a hydrodynamic masterpiece, but also with a series of striking convergences. (p. 78) [photograph: iStock]

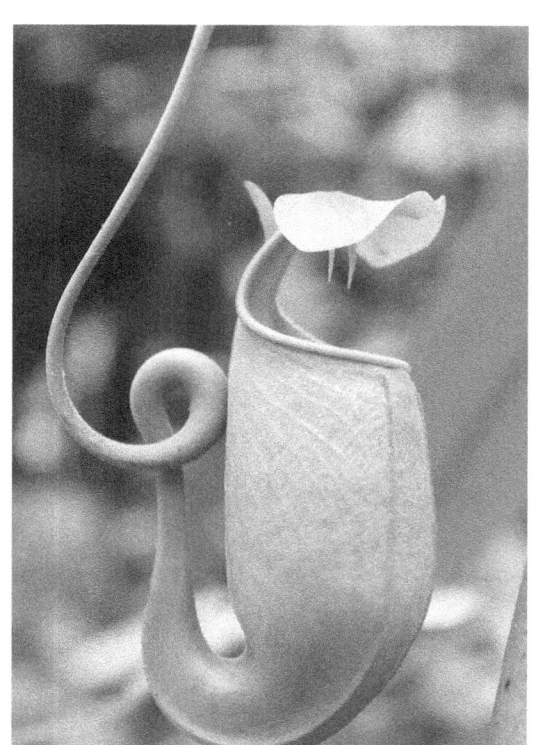

Fig. 25. The hummingbird is a marvel of evolution, but also strikingly convergent with the hawkmoth. (pp. 80, 200, 203, 207, 266) [photograph: iStock]

Fig. 26. A splendid example of evolutionary sophistication is the pitcher plant. There are also other convergences amongst carnivorous plants. (p. 85) [photograph: iStock]

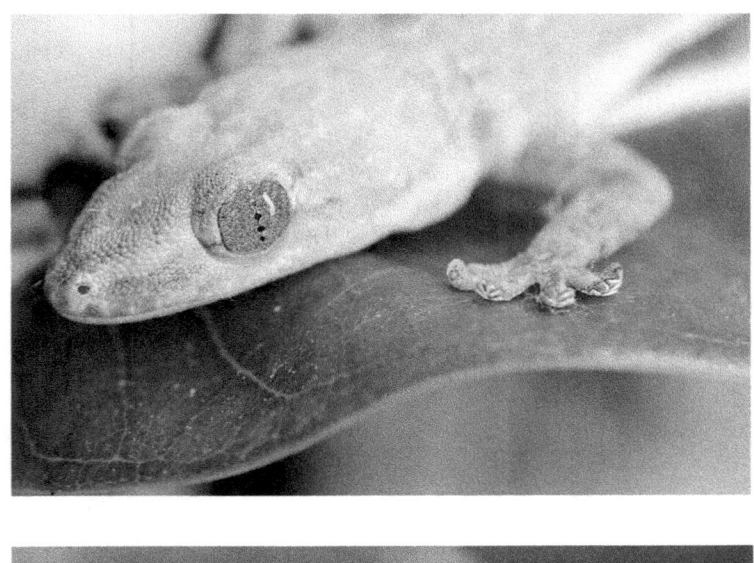

Fig. 20. The gecko is a lizard that exemplifies not only adhesive skill but other convergences as well. (pp. 68, 83, 175) [photograph: iStock]

Fig. 21. The chameleon. Its walking along branches closely parallels that of mammals. (p. 68) [photograph: iStock]

Fig. 22. The weta, a New Zealand grasshopper, qualifies as an honorary mouse. (p. 74) [photograph: iStock]

Fig. 17. The Komodo dragon is a type of monitor lizard that exemplifies convergences ranging from intelligence to toxins. (pp. 62, 232, 284)
[photograph: iStock]

Fig. 18. The sea-snake is an excellent example of the road back to the sea. It exemplifies a number of convergences. (pp. 64, 122)
[photograph: iStock]

Fig. 19. The leaf-cutter ant is the exemplar of the convergent evolution of agriculture, but also displays other convergences such as zinc-tipped teeth. (pp. 66, 121, 181)
[photograph: iStock]

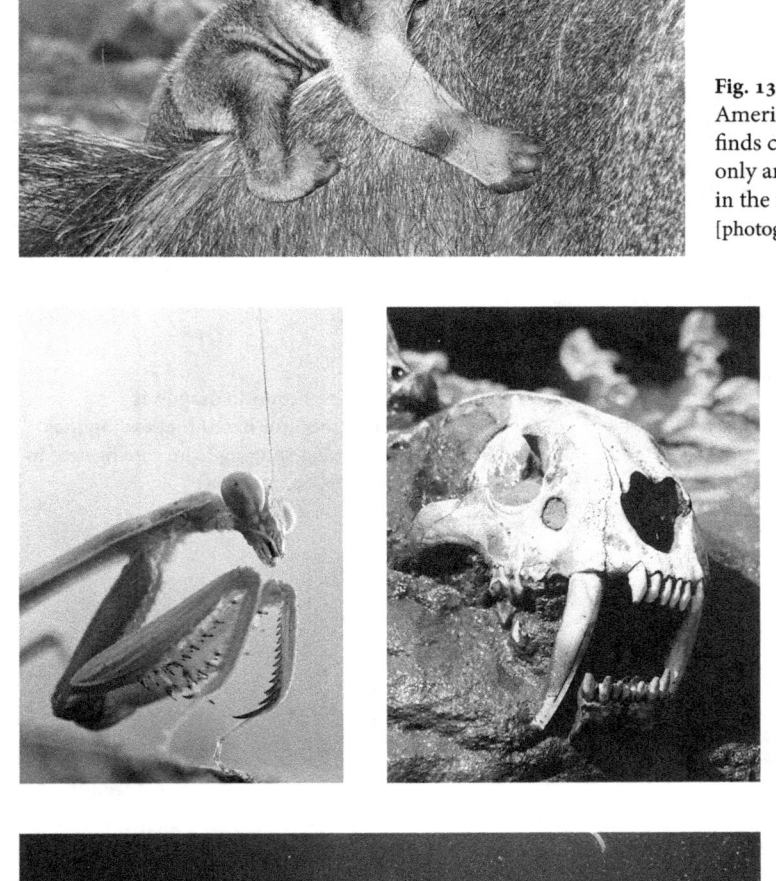

Fig. 13. The ant eater is a South American myrmecophage that finds convergent parallels, not only amongst mammals, but also in the reptiles. (pp. 51, 196, 231) [photograph: iStock]

Fig. 14. (Far left) The praying mantis is the epitome of raptorial effectiveness. (p. 53) [photograph: iStock]

Fig. 15. (Left) The saber-tooth cat is a superb predator and an object lesson in convergence. (pp. 55, 230) [photograph: iStock]

Fig. 16. The duck-billed platypus (monotreme) is a storehouse of convergences, from toxins to electrolocation. (pp. 61, 114, 32, 148) [photograph: iStock]

Fig. 10. The mosasaur is an extinct lizard and close relative of the Komodo dragon. These reptiles reveal convergences ranging from swimming and deep diving to ovoviviparity. (pp. 47, 78, 219)
[photograph: iStock]

Fig. 11. The cownose rays provide insights into the evolution of shell-crushing (durophagy). As a group the rays display other convergences, including walking and electroreception. (pp. 48, 73, 144)
[photograph: iStock]

Fig. 12. A fish to respect is the moray eel, especially with feeding habits similar to snakes. (p. 49)
[photograph: iStock]

Fig. 7. The bombardier beetle is a classic example of explosive evolution, but with unexpected links to the production of ink by the sea-hare. (p. 38)
[photograph courtesy of Maria L. Eisner]

Fig. 8. The mantis shrimp is not only an effective predator but is capable of smashing damage. (pp. 42, 112, 144)
[photograph: iStock]

Fig. 9. The nurse shark is equipped with teeth but adept at suction feeding. (pp. 44, 47)
[photograph: iStock]

Fig. 4. The bat is a test-case for the convergent evolution of echolocation. (pp. 33, 111, 137, 193, 203, 253, 267) [photograph: iStock]

Fig. 5. The dendrobatid frog is an object lesson in the convergent evolution of toxicity. (p. 34) [photograph: iStock]

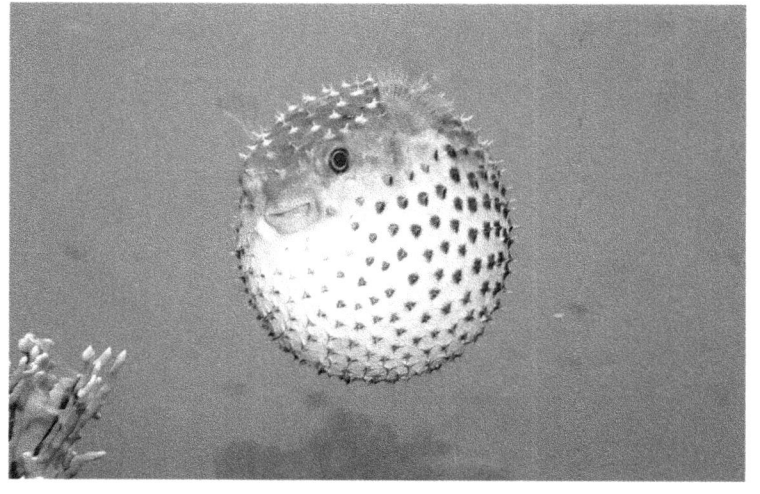

Fig. 6. The puffer fish is an anatomical marvel, but informative about toxic convergences. (pp. 36, 37) [photograph: iStock]

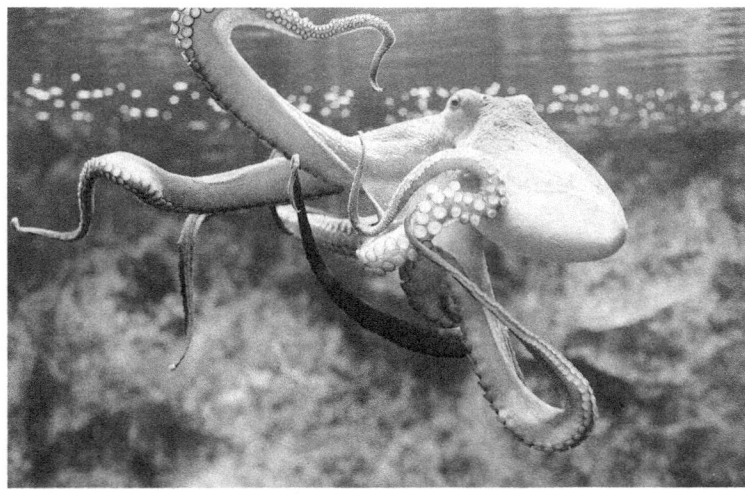

Fig. 1. The octopus is the cynosure of convergent evolution. (pp. 12, 72, 100, 288) [photograph: iStock]

Fig. 2. An octopus? Yes, but pretending to be a sea-snake. Such mimicry is rampantly convergent. (p. 15) [photograph: iStock]

Fig. 3. The sea-hare is apparently vulnerable; however, its protective ink clouds provide an object lesson in convergence. (p. 15) [photograph: iStock]

considerable impact on the types of sea grass. Of greater significance is the convergence between the dugongs and a group of Triassic marine reptiles known as the placodonts.[19] Long interpreted as durophagous shell crushers, Cajus Diedrich makes a compelling case when he writes, "A convergence of the cranial anatomy, the entire body shape and the body weight enhancements of placodontids is quite obvious when compared to the Lower Oligocene . . . *Halitherium*."[20] The convergences are not precise, but the similarities of ballasting (such as massive ribs),[21] a grinding jaw (oral pads versus flattened teeth), and possibly social behavior lead to an unexpected image as we envisage huge herds of these reptiles peacefully grazing over what is now Stuttgart. The sea grasses, of course, did not evolve until much later, and the efficient browsing of these placodontids must have been on banks of macroalgae.

We return to marine algal gardens because in terms of the invertebrates we find a series of intriguing examples in both polychaete annelids and various snails, notably the limpets.[22] How different, yet as George Branch and colleagues[23] note, "One of the striking features of [these] gardens is the similarity between the functional algal types present, even though the grazers responsible differ radically in terms of taxonomy, morphology, and mobility."[24] In nereid polychaetes the gardening involves attaching pieces of drifting algae to the anterior of the dwelling tube. Like all agriculture it confers a mutual benefit,[25] with the plant obtaining a secure anchorage while the worm has an immediate food source upon which to nibble.[26]

The mollusks are more versatile. As will be familiar from a garden snail or slug locomotion is achieved by a muscular foot lubricated by mucus, most obvious from the shiny trail as the brute heads toward the lettuces. Among some aquatic snails the mucus can be rapidly colonized by either bacteria[27] or microalgae.[28]

This is then grazed by the returning snail, and while consuming slime may not be to everyone's taste this presumably serves to offset the considerable metabolic expenses of secreting the mucus in the first place. The same argument applies to the diminutive foraminifera living on sea grasses. These too also produce a mucuslike trail that, once infested with bacteria, is grazed repeatedly.[29]

In the case of grazing the microalgae atop the mucus trail, the snails involved are the cap-shaped limpets. As Valerie Connor and James Quinn note, they are not only involved in "farming [the] algae for their exclusive use,"[30] but evidently go to considerable lengths not only to encourage algal growth but exclude other grazers.[31] Indeed, despite their dozy appearance, limpets have some surprising similarities to certain academics: tough carapaces, seemingly retiring unless provoked, when they become belligerent and aggressive. When provoked by a whelk,[32] some limpets will either bring their shell crashing down on the foot of the intruder, sometimes with sufficient force to cut off part of it, or subject it to "repeated battering [that] can become so violent that the whelk's shell is clipped."[33] So, too, those limpets that tend their algal gardens[34] leap to the defense. Most likely gardening has evolved several times, and it involves two main strategies. One is to construct a garden of red algae around the margins of the shell; the other is to maintain a more extensive algal farm over which they graze.[35] In the latter case the cultivation can cover about a square meter and is readily identifiable at a distance.[36] In other cases it appears that the limpet fertilizes the garden with nitrogenous excretion.[37] Researchers have emphasized the remarkable fertility of these gardens, with George Branch noting how they are "on a par with kelp forests: an astonishing production."[38]

All these marine examples employ algae, with one interesting exception. This is the

CHAPTER 16

Converging on the Farm

With their vast colony sizes and features, such as sophisticated pathogen control, as far as the complex eusocial worlds of the insects are concerned there can be no turning back. Nowhere is this more apparent than in the multiple inventions of agriculture among the insects. But before touching on these sophisticated systems, it is worth reminding ourselves that agriculture has sprung up across the biosphere even among the social amoeba (*Dictyostelium*), where some clones husband bacteria through their reproductive cycle.[1] Nobody doubts that this agriculture is primitive,[2] not least because as Debra Brock and colleagues report, these microbial "Farmers carry a variety of species of bacteria"[3] rather than cultivating a monoculture. Even so, they draw attention to "The striking convergent evolution between bacterial husbandry in social amoebas and fungus farming in social insects."[4] Among animal agriculturalists,[5] the crop is most often fungal but even here the fungi return the complement in the form of a morel (*Morchella*), which farms bacteria providing nourishment but also engaging in not only harvesting but transport of the "crop."[6] Yet if fungi are otherwise generally the target for cultivation, an alternative exists, in the form of algae.

Fishy Agriculture

And the farmers? Fish. Even within a single group there may be a striking range of feeding habits. Such is the case in the coral-dwelling damselfish (pomacentrids).[7] Not only do they show striking adaptive radiations,[8] but as Bruno Frédérich and colleagues note, there is "an overwhelming signature of convergence in [their] phenotypic traits,"[9] not least in terms of jaws (p. 340n220) and diet. Of particular relevance, however, is their capacity not only for herbivory,[10] but more remarkably the maintenance of algal farms. Not only are these highly productive submarine plots nurtured by weeding,[11] but the effectiveness of their aggressive defense[12] becomes apparent when the fish are excluded: the farm soon descends into rack and ruin.[13] While some species maintain farms with a variety of algae, others support what is effectively a monoculture,[14] as in a species of *Stegastes* (*S. nigricans*) and presumably independently in *Microspathodon* (*M. dorsalis*[15]). Moreover, in the monoculture farmed by *Stegastes* the red alga (a species of *Polysiphonia*) occurs in a form that is never found outside the farm, and evidently this crop represents an obligate mutualism.[16]

We should also note that herds of dugongs engage in what Anthony Preen[17] terms "cultivation grazing,"[18] and evidently they have a the particular groups of crustacean. Among their suggestions for potential domiciles were sponges. Riddled with cavities and chambers they offer shelter to many types of animal, and that is exactly where we find the eusocial synalpheid shrimps.[199] Synalpheid eusociality has evolved at least three times,[200] but at first sight it seems rather puzzling that although a variety of species make their homes in sponges, by no means all synalpheids have adopted this highly successful social organization. The immediate explanation seems to lie in the fact that the eusocial taxa do not have a planktonic stage that would allow them to drift away.[201] Emmett Duffy and Kenneth Macdonald note that "eusociality occurs only in shrimp species with non-dispersing larvae [and so] parallels the pattern in Hymenoptera."[202] Their habitat within the sponges is little different from the nest of many other eusocial animals, which means they are necessarily quite small (c. 1 cm). They show, of course, many of the hallmarks of eusocial behavior, not least aggressive defense against intruders.[203] Here the large males are active, and Emmett Duffy and colleagues remark how they "move around the sponge with an appearance of boldness."[204] If the intrusion is persistent, the colony displays a coordinated response much like that seen in other eusocial groups (albeit without mass attack), specifically by the snapping of their distinctive claws (the chelae). This action produces both

a jet of water and a loud noise[205] (p. 143). The snapping claws are probably one key to the emergence of eusociality, but in their versatility and multifunctionality were evidently the key preadaptation[206] in the remarkable success of the alpheid shrimps as a whole. Thus, the snapping claw has opened many other evolutionary doors, and unsurprisingly in these shrimps the same solution to particular problems has repeatedly emerged.

So these synalpheid shrimps nestle safely in sponges. Are they the only marine example of eusociality? Possibly not. Although the original query by Ehud Spanier and colleagues[207] is now answered, in their ruminations they pointed the finger at the isopods, noting how their wood-eating habits and employment of gut microbes had intriguing parallels to the termites. Once again we hear the drumbeat of evolutionary inevitabilities, because the origin of the termites lies within the cockroaches (p. 186). It is now clear that in this latter group the ability to digest wood has evolved at least twice,[208] and as with termites is reliant on a gut biota rich in flagellate protistans. Given that we have wood-eating isopods, should we be that surprised if perhaps one day in addition to the synalpheid shrimps yet another group of eusocial crustaceans will emerge, not this time from the recesses of a sponge but like the cockroaches from the recesses of a submerged and rotting log?

produce a living roof, while back in the ant nest the female mealybugs continue to give birth. While the milking of aphids and other hemipterans is very largely a prerogative of the ants, a striking convergence is found in the lizards, specifically a gecko living in Madagascar.[180] The reptile approaches the cicada with great care, but the rhythmic vibrations of the lizard, produced by bouncing its head against the bark, induce the insect to lift its abdomen. A drop of honeydew is placed on the snout, then to be licked off.

If exudates of one sort or another provide nourishment, surely we would draw the line at insects producing milk? Think again, not least with respect to the cockroaches. In some species the females secrete a "milk" in order to feed their young. This is most striking in one of the few insects that is a popular pet, the Madagascan hissing cockroach, where minutes after the ootheca is expelled from the female the hatchlings (neonates) are crowding around the extruded droplet, competing for access and sometimes stacked three deep.[181] This "milk" is almost certainly secreted by special integumentary glands located in the brood chamber. At least one other cockroach (*Diploptera punctata*) gives birth to live young (a process known as viviparity and one that is rampantly convergent, p. 215), and here the "milk" again derives from the brood sac and supplies the developing embryos with valuable nutrients.[182] Much the same arrangement has evolved independently in the viviparous tsetse fly,[183] although in this dipteran the milk gland has a somewhat different configuration.[184] Do the convergences end there? No, in at least some of the cockroaches the cycle of secretory activity is similar to that found in the mammary glands,[185] and there are also convergences at the molecular level. Thus a key protein in both the cockroach[186] and tsetse fly[187] are the lipocalins, but these are also found in mammalian milk. Geoffrey Attardo and colleagues point out, "The occurrence of lipocalins in lactation products of divergent organisms suggests that lipocalins have evolved to fill the role of milk proteins multiple times."[188] Why are they the "molecule of choice"? Most probably because they can serve to transport fats and similar hydrophobic molecules that are essential for nutrition. Long before the happy day when the baby insects are receiving their mother's milk, the lipocalins also help to initiate the process. This is because having used a pheromone (predictably called seducin[189]) to attract the opposite sex, at least some of the male cockroaches then secrete the same protein,[190] but this time as an aphrodisiac that the female licks up. Most likely it carries a pheromone, and it certainly seems to work.[191] Such a trick is too good to miss, and some mammals, including the hamster, have also recruited the lipocalins as aphrodisiacs, although this time the female[192] is in charge of the sexual chemistry.

THE FURTHER SHORES OF EUSOCIALITY

Eusociality is most familiar in the insects, and that perhaps underscores how it led to one of the few cases of a successful prediction in biology. This was the famous insight by R.D. Alexander[193] of the otherwise entirely unexpected recognition of a eusocial mammal, specifically the naked mole rats.[194] In fact, eusociality has arisen twice in the mole rats,[195] and quite possibly independently in some of the microtine voles.[196] So, too, in 1993 Ehud Spanier and colleagues[197] asked the question, "Why are there no reports of eusocial marine crustaceans?"[198] In a minor tour de force they identified both likely habitats and

This is dramatically confirmed in a third case, but this time deep inside a spittlebug where *Baumannia* has struck up an association with a β-proteobacterium known as *Zinderia*.[159] The similarities of the metabolic pathways are not precise, but as John McCutcheon and Nancy Moran note, "When examined at the level of essential amino acid production, the three systems . . . present a striking case of convergent evolution in the context of a common selection pressure to retain the capacity for a full complement of amino acids."[160] These authors note that *Zinderia* has a very small genome, but this is outstripped by the even tinier genome of another β-proteobacterium known as *Tremblaya*.[161] In this association, nurtured by a mealybug,[162] the intricacy of the symbiosis reaches new heights. Thus not only does *Tremblaya* actually house the partner γ-proteobacteria within its own cytoplasm,[163] but as John McCutcheon and Carol von Dohlen demonstrate, the degree of integration is such that "several essential amino acid pathways in the mealybug assemblage required a patchwork of interspersed gene products"[164] from the two bacteria and possibly the insect as well.[165]

The aphids and the related hemipterans emerge triumphant, albeit to our great cost to crops and other agriculture. Nevertheless, in the wider world the aphids are not the only beneficiaries. "Sugar in" means "exudate out." Dealing with this sticky extrusion has led to ingenious anal devices, including wax tubes, not to mention the delights of a forcible discharge.[166] The resultant honeydew can be produced in prodigious quantities, and is carefully harvested by a number of groups of ants.[167] This capacity has evolved many times;[168] just how many is impossible to judge because of the imprecise nature of aphid phylogeny, but either way it is clearly very labile. In at least some cases, however, this mutualism with the ants is evidently favored among those aphids with the longer mouthparts. This

potentially compromises their ability to withdraw it promptly if attacked, so having a fierce ant in attendance is of considerable benefit.[169] The mutual benefits to ant and aphid are well-rehearsed, although the wider ecological ramifications less so.[170] Striking examples of the former include the cleansing (to prevent fungal infection[171]) and storage of aphid eggs in the ant nest, which also offers shelter to the aphids themselves.[172] Even so protection is not always guaranteed:[173] specialized predators on the aphids employ hydrocarbons that allow them to slip by the tending ants[174] employing a chemical mimicry that David Lohman and colleagues identify as a "striking convergence in CHC [cuticular hydrocarbons]."[175] As importantly these workers conclude "that multiple paths to chemical mimicry may lead to a common result."[176]

Despite these striking mutualisms it would be unwise to draw too rosy a picture. In at least one association, involving a subterranean species, the ants not only feast on the honeydew but also the juvenile aphids themselves to obtain, so to speak, a balanced diet. As Aniek Ivens and colleagues[177] point out, this arrangement is not analogous to cattle husbandry inasmuch as the ants obtain both "milk" (as honeydew) and "meat" (from the luckless juveniles), but the "analogies between aphid husbandry in [the ant] and human cultural practices are quite striking."[178] Not only does this allow very large populations but the careful management of resources is another example of agriculture (chapter 16). That the ants remain masters of such associations is evident from a Malaysian ant. Utterly dependent on a pseudococcid insect as it is (and vice versa), nevertheless the ant has transformed itself into a migrating herdsman, a nomad with its flock of mealybugs.[179] The latter will only draw sap from young plants, so new pastures are also required. These involve organized mass processions, short-period depots, and when the rain pours the ants

THE ARTHROPODS SHOW THE WAY   173

what has been described as "molecular sabotage"[141] the saliva of the aphid prevents the equivalent of clotting.[142] Other challenges to the sapsuckers also find intriguing solutions. In tropical forests, finding young leaves that are both rich in nutrients such as nitrogen and have not had time to develop defenses is difficult and time-consuming (and indeed the evolution of protective measures adopted by those tropical plants show convergences, p. 394n6). Some treehoppers, however, have adopted communication based on vibrations from the scouts. These rapidly summon their siblings to the feast,[143] a mode of communication that is both convergent and versatile.

From the specifics of the aphids it is not so difficult to grasp the universals. It is surely likely that plants will evolve wherever starlight impinges on a suitable planetary surface. Photosynthesis commences, the sugars are synthesized, and the necessary transport of the sap will attract a stylet-bearing organism to tap the riches. Yet this food source poses a major dietary dilemma. Lacking proteins, as with the nectar diets of the hummingbirdoids (p. 418n34), this food is seriously short in vital nitrogen. As far as the aphids and other hemipterans are concerned, the balanced diet is only achieved by the repeated and independent recruitment of symbiotic bacteria.[144] Not only are these associations very ancient, but as ever they are deliciously intimate. For example, specific proteins may help to mediate the symbiosis between aphid and bacterium,[145] but as Shuji Shigenobu and David Stern point out, there is a "remarkable convergence of protein sequence composition and expression pattern for these genes between leguminous plants [p. 310n8] and aphids [that] suggests . . . there may be common principles underlying evolution of endosymbiosis in divergent taxa."[146] Moreover, such associations among the insects are widespread, and if the ladybirds are anything to go by, most likely we are only seeing the tip of this microbial iceberg.[147] Nevertheless, when we turn to the hemipterans, we find a rich evolutionary story that revolves around not only intricate symbioses and episodes of lateral gene transfer to enable carotenoid synthesis,[148] but also striking reductions in genome size. All this arises because of a diet that in any other context would be regarded as insane. Most extraordinary are those leafhoppers that tap not the sugary sap within the plant's phloem but the fluid that flows along the xylem.[149] It is remarkable that these insects can survive at all, because not only is the xylem extraordinarily dilute, meaning prodigious quantities have to be consumed, but it is a nutritional nightmare.[150] Without symbiotic bacteria (notably the γ-proteobacteria) they could not survive.[151] Are we surprised that such associations are strikingly convergent?[152] So it is that leafhoppers known as sharpshooters (cicadellinids) recruit a bacterium. Known as *Sulcia*,[153] it has an extraordinarily small genome, much of which is involved with the synthesis of amino acids that are essential for the insects. By itself, however, this bacterium could not manage, but a second symbiont in the form of another bacterium (the γ-proteobacteria *Baumannia*[154]) provides a remarkable metabolic complementarity in supplying other vital compounds such as vitamins.[155] Just the same complementary arrangement is found in the related cicadas, because the same demands of a xylophagous diet ensure convergence.[156] Instead of employing *Baumannia*, the cicadas look to an α-proteobacteria (known as *Hodgkinia*), which has an even tinier genome than *Baumannia*.[157] Once again *Sulcia* takes primary responsibility for supplying the "missing" amino acids, but what is extraordinary is that although the cicadas diverged from the sharpshooters close to the time the dinosaurs were gaining the upper hand, the respective genomes of *Sulcia* have retained exactly the same arrangement, suggesting rules of engagement that are not to be thwarted.[158]

because although they can become sexual,[122] typically they are clones.[123]

This has some interesting consequences, especially for those aphids that eschew the open skies and live in plant galls (a parallel found in many eusocial thrips that have also evolved soldiers in much the same circumstances as the aphids[124]). Gall-inducing behavior is rampantly convergent among the insects.[125] In their enclosed and snug environment all is well with the aphids—until, that is, either another clone or a predator turns up. What happens then? In the former case, as William Foster writes in his commentary on the consequences of clonal invasion,[126] the "soldier aphids go cuckoo."[127] It is all too similar to those guests who outstay their welcome ("Care to dry some dishes? . . . I think the liquor store is still open . . ."), the clonal "guests" don't lift a finger (or stylet) and get on with the serious business of growth and reproduction. That leaves the defense of the gall to the residents, but by no means is all lost. While the gall is in many ways an ideal home, it must maintain contact with the outside world, not least for waste disposal as well as migration, and so any opening is also a chink in the fortress. The soldiers[128] play a key role, and one line of defense is to activate the gall tissue to provide compensatory growth as a repair mechanism.[129] This capacity has evidently evolved independently in a Japanese aphid (appropriately *Nipponaphis*),[130] but here the repair reaches a spectacular apotheosis when the response to attack (for example, by a moth larvae) is a massive discharge of body fluids that by kneading serve to plaster the opening. In the best traditions of eusocial defense the aphids pay the full sacrifice, a token of which are their shriveled bodies embedded in the defensive ramparts.[131] That the aphids are instructive in terms of the evolution of social systems has never been in doubt, but earlier evidence of highly organized waste disposal[132] (including the production of wax that coats the drops of honeydew and so converts them into readily movable marbles[133]) and employment of alarm pheromones[134] (p. 170) when combined with the defense of the gall lead Mayako Kutsukake and colleagues to remark "that some social aphids can attain such a high level of social intricacy [to compare] with the nest building, repair and maintenance found in hymenopteran and isopteran social insects."[135] Those "small soft cushions of haemolymph" turn out, in their own particular way, to be marvels of evolution. It is only appropriate to round off this short narrative with one further insight. Wonders of biology they may be, but aphids are serious agricultural and garden pests.

Sucking the Sap

Aphids belong to the group of insects known as the hemipterans (which include the noisy cicadas who have their own convergences[136] and the treehoppers). Their ability to use the sap of plants, and thereby compound the nuisance value by introducing viruses and other pathogens into the plants, has evolved independently many times.[137] The stylets might be co-opted for fighting, but their primary function is to plunge into the plant to extract the sugar-rich[138] fluids. To say "plunge" is a massive simplification. In reality both the stylet and the mechanisms it employs are extremely complex.[139] Anybody who has seen an infestation of aphids and the consequent loss of sap will realize that the plant might have a view of this, akin as it is to the lapping up of the lifeblood by a vampire bat (p. 73).[140] In normal circumstances the probing by the aphid would provoke a reaction by the plant, but in

"Other eusocial insect lineages such as termites, wasps and bees also developed in the Early Cretaceous. . . . The fact that this phenomenon [eusociality] apparently occurred in a relatively short period during the Early Cretaceous in insect clades *of different ecologies* remains largely unexplained".[101] Just a coincidence?

It certainly looks as if long ago the advanced eusocial groups passed the point of no return and signed up to a permanent eusociality. Such groups dominate the stage, but waiting in the wings of the evolutionary theater are other groups that throw invaluable light on how once-solitary insects embarked on the road to eusociality. Most notable are the sweat bees (or halictids). These bees show a full spectrum in social arrangements, from solitary to fully eusocial. It appears that the latter arrangement has evolved independently at least three times, although within each clade there have also been reversals to more solitary states.[102] Of particular interest is that the three transitions to eusociality were more or less simultaneous, about 20 Ma ago and during an episode of global warming.[103] Cause and effect in evolution can be difficult to disentangle, but this is a reminder that extrinsic factors may well act as a spur (or brake) to a crowd of evolutionary actors who, when the prompt comes, surge onto the stage, but all repeating pretty much the same lines.

All things being equal, even though the halictid bees still have an escape clause from eusociality it may well be they are on the same convergent path as the corbiculate bees, the ants, the termites, or for that matter the wasps, where in the vespids we see again not only eusociality but a system that evolved twice.[104] Along with the independent eusociality so we see the repeated evolution of such features as cognitive capacities[105] (p. 289), social sanctions[106] and worker policing,[107] coordinated defense,[108] nest thermoregulation,[109] disease prevention and hygiene[110] (pp. 184–185), and agriculture (p. 181). The focus of attention remains on the hymenopterans, but they are not alone.[111] Best known in this roster of eusociality are the highly successful termites. A strong argument can be made for the near-inevitability of the evolution, but as Daegan Inward and colleagues[112] note, the termite route to eusociality is different from what we see in the hymenopterans. Still other groups of insects that have ventured into the world of eusociality include the gall thrips,[113] beetles,[114] and the aphids.

Soft Cushions of Hemolymph

The aphids are particularly interesting. As William Foster[115] graphically writes of this seemingly defenseless group, are not they little more than "small soft cushions of haemolymph [huddled] on leaves and stems, ready to be ripped apart by ladybirds, sucked dry by lacewings, hoovered up by hoverflies and punctured by wasps?"[116] No, because the seemingly innocuous aphids can fight back. One way is thanks to the evolution in two groups of specialized soldiers.[117] Here accentuated claws or stylets, and in one group (the cerataphidinids) even horns,[118] serve to engage with mischief-makers. Aphid soldiers[119] have evolved independently a number of times,[120] but the real peculiarities of the aphids are twofold. First, while many eusocial groups inhabit subterranean nests or other secluded nooks and crannies, these aphids spend part of their life out in the open.[121] More extraordinary is that while eusocial animals typically share considerable genetic similarity—after all, they share one mother—the aphids take this to an extreme hornets. If not swiftly repelled, an en masse attack by up to thirty hornets will begin. As they say, this is no laughing matter. A single hornet can dispatch forty bees a minute, and at that rate of attrition the beleaguered colony succumbs to the onslaught, leaving the hornets to pillage the protein bank of the bee larvae for their own young. However, the marking by a scout hornet with a pheromone that serves to guide its colleagues to the proposed scene of the pillage also galvanizes the residents. A visiting hornet is engulfed in a living ball, composed of more than five hundred honeybees. The interior temperature swiftly rises (with the bees on the outside evidently providing the thermal insulation[90]), and although the lethal limit of the bees (which is

about 48°C) is only a couple of degrees higher than that of the hornet, the latter is roasted.[91] It is not only heat that dispatches the visiting hornet, because within the bee ball levels of carbon dioxide[92] also rise to lethal levels.[93] Stingless bees do it differently, but to similar effect. In what Mark Greco and colleagues[94] call the "Pharaoh approach,"[95] unwelcome visitors in the form of invasive beetles are effectively mummified in a goo composed of "batumen" (a concoction of mud, resin, and wax). These workers conclude, "The convergent evolution of live mummification of nest parasites in stingless bees and social encapsulation in honeybees is another striking example of evolution between insect societies and their parasites."[96]

BEE SOCIETY

The eusociality of honeybees provides a benchmark in sophistication, but such an arrangement has evolved many times. How exactly did it come about, and what might be the key adaptations? Here the bees are particularly instructive. Evolutionary acceleration in a number of genes provides some revealing contrasts between the complex eusociality of the honeybees and their relatives with more primitive socialities.[97] This roster includes genes linked to the development of glands, important for secreting pheromones and so forth, but as Hollis Woodard and colleagues also note, "Genes associated with carbohydrate metabolism appear to have been a particularly strong target of selection during eusocial bee evolution."[98] Not surprisingly, given such demands as thermoregulation and intense foraging, the honeybee shows further genetic changes related to carbohydrate utilization, but Woodard and colleagues make an important point when they write "that the multiple independent evolutionary paths to eusociality may have each been shaped by dif-

ferent combinations of extrinsic and intrinsic factors, and perhaps also via different forces of selection."[99]

Their work points to a necessary genomic substrate that enables the emergence of eusociality, and that lineages need not have identical histories. One also needs to ask not only how many times did it happen, but whether this trajectory arrives at an irreversible state of evolution? This question is difficult to address because in the classic systems we see in such groups as the corbiculate bees (these include the honeybees and stingless bees) and ants, their eusociality evolved tens of millions of years ago and the intermediate stages have been irretrievably lost. There is, of course, a fossil record, but it is rather sporadic and for the most part not helpful in documenting the precise staging posts toward full eusociality. It does, however, reveal some uniquely interesting information. In their documentation of key evidence for eusociality in ants preserved in Cretaceous ambers, Vincent Perrichot and colleagues[100] conclude by noting,

THE ARTHROPODS SHOW THE WAY   169

multiple times, and typically is reliant on the flight muscles.[65] The most striking example are the sphinx moths, whose energetics rival that of the hummingbirds (pp. 200–201). So, too, thermogenesis in an enormous Neotropical fly (*Pantophthalmus*) is linked to not only flight readiness but effective flight itself.[66] More surprisingly, given their smaller size (and so larger area-to-volume ratio), syrphid flies are also endothermic.[67] Here instant readiness to fly may be linked to lekking (where males congregate to impress the females), and this example is also notable because of the striking mimicry these thermogenic flies have with a variety of bees and wasps.[68] More examples, please? Yes, endothermy has also evolved independently in some beetles. In the tropical dynastine scarabs, more familiar as the rhinoceros beetles, thermogenesis in terrestrial activities (as against flying) can result in a metabolic rate equivalent to that of a mammal like a shrew.[6] In other dynastines not only is the beetle endothermic, but it has a pollination strategy (canthophily, p. 163) involving a thermogenic plant[70] (p. 163). Nor are they the only beetles, because endothermy has evolved in the African dung beetles.[71] To us, rolling balls of dung[72] might seem to verge on the comic (or worse), but in fact a vast heap of elephant dung is a very valuable resource[73] to be quickly located and removed. Nor is the setting of the sun any obstacle, because at night dung beetles can use the Milky Way as a guide.[74] As George Bartholomew and Bernd Heinrich also note, there is a "premium on rapidity of ball building and speed of rolling it from areas of high beetle density in and around the dung."[75] Lest this be thought an exaggeration, give a thought to those luckless individuals in areas such as India where the beetles don't bother to wait outside and—to put it delicately—engage in rectal excursions.[76] Given the intensity of competition it is neither surprising that dung feeding has evolved several times,[77] nor that these lead to

specializations.[78] Consider dense tropical forests, teeming with arboreal mammals, such as monkeys. Little chance here of the descending dung hitting the ground, and so we see the independent evolution of arboreal specialists that roll the feces to the edge of the plant and then fall with it to the forest floor.[79]

The link between locomotion and endothermy, be it the exquisite hovering of a sphinx moth or the frantic rolling of a dung beetle, makes good sense, but other areas of frenetic activity are also powered in this way. Consider the evolution of endothermy in some hemipteran cicadas. Not only may this capacity play a role in evaporative cooling (facilitated by feeding on the contents of the xylem, p. 173),[80] but in some species it also assists in their famous sound production:[81] "Hot-blooded singers,"[82] in the words of Allen Sanborn. As is the case in many other thermogenic organisms the convergence extends to the cellular level.[83] If one is endothermic, an effective insulation is also a good idea. It is no coincidence that such methods of controlling convective heat loss have evolved among the insects.[84] Sphinx moths have a thick, furry coat,[85] as do bumblebees. In these and other social insects, endothermy is not only linked to active lifestyles but in the thermoregulation of the nests,[86] including, when necessary, cooling such as by fanning. This is essential for the well-being of the brood, but it can also assist in unexpected ways. Just as the raging temperature we "enjoy" when suffering from an infection is actually beneficial with respect to the triggering of the immune system (p. 307n2),[87] in what Philip Starks described[88] as "a striking example of convergent evolution between [the hive] and other fever-producing animals,"[89] fever in the honeybee colony is instrumental in destroying a lethal fungal pathogen.

In a different way, thermal regulation again comes to the rescue, this time for the Japanese honeybee when threatened with attack by that the selective regime is far from obvious, and they wonder if there might be "an innate tendency."[51] Its basis might be developmental, but its recurrence could reflect an all-purpose adaptability. Some support for this comes from the fact that the crab roster does not finish here. In a study of the extinct and somewhat enigmatic crustaceans[52] known as the cycloids, Fred Schram and colleagues[53] summarized their morphology and likely habits. They wrote that the cycloids "bear striking, convergent similarities to the body plan of crabs!"[54] and suggested that these cycloids might have preempted opportunities for the crabs.[55]

When it comes to terrestrial competence the crabs certainly deserve a silver medal, but among the arthropods the insects are the clear winners. Many convergences link them to their rivals the vertebrates, but self-evidently they are generally substantially smaller. Often this has been linked to the constraints imposed by a respiratory system based on trachea.[56] Things, however, are a bit more complicated. Not only do some insects have a discontinuous exchange of gases (which has evolved at least five times[57]), but by using synchrotron X-ray imaging, Mark Westneat and colleagues[58] demonstrated how pressure changes within the insect body lead to "Tracheal compression . . . [that] functions as a mechanism of air convection much like that of vertebrate lungs."[59] Not only that, but some insects, specifically caterpillars, have evolved lungs where tufts of trachea project into the interior of haemocoel[60] with a branching pattern that invites direct comparison to our pulmonary system.[61] In due course we will encounter some alarming similarities in terms of cognitive competence, but whatever else insects have achieved, among the most striking are their social organizations.

HIVES AND FORTRESSES

The integration of the insect societies has excited endless admiration, sometimes tinged with fear. Consider, for example, the nomadic army ants. Not only do they carry the queen with them and construct bridges to cross water, but if a pothole obstructs the advance of a foraging column then it is rapidly filled with living ants while the rest of the army marches over and on.[62] Those who fail to flee the advancing horde are rapidly dismembered, and there is more than an echo of the calamity of the Mongol invasions. Nobody is suggesting exact parallels, but the repeated evolution of insect eusociality is the exemplar of the emergence of complex systems. So too arguably they are locked in to an endpoint that can never be unraveled. Like us, for better or worse, there is no turning back. Along with the termites and wasps, perhaps the most familiar example of insect eusociality are the honeybees. As is the norm in such systems each colony possesses its reproductive queen, vast numbers of sterile workers (as in all hymenopterans [ants, bees, and wasps] the females), and a smattering of male drones. Their moment of glory comes when they encounter the new queen on her nuptial flight. The aerial engagement with the drones involves multiple matings,[63] before the queen descends to her hive, packed with a lifetime's supply of sperm to generate an immense progeny.

The astonishing society of the honeybee is no less fascinating when seen through scientific spectacles. Perhaps most striking are the cognitive powers of the bee (p. 289), but complexity is evident in many other areas of their life. Consider, for example, the elevated body temperature of the bees.[64] In the insects alone, such "warm-bloodedness" has evolved

The similarities of this crab lung to that of the vertebrates are not only striking,[38] but as A.J. Innes and colleagues note, the structure of the branchial lungs with their "anastomosing primary and secondary airways . . . bear striking structural and functional resemblances to the parabronchi and air capillaries respectively of the avian lung, while the crab 'bellows' system is analogous to the air sac system of birds."[39] Here we have a lung that is "A quantum leap in . . . evolution."[40] Not only is there pumping of air but also gas exchange that rather than being dependent on simple diffusion has moved to active perfusion, as in the pulmonary arrangement in the vertebrates. This crab may represent an especially striking instance of how to live far from the oceans, but an understanding of terrestrialization in a wider context is important,[41] not only in terms of the many functional adaptations, but also because of the emergence of complex, and convergent, behaviors.

Consider a grapsid crab (*Metopaulias depressus*) that lives in dry areas of Jamaica and exhibits an extraordinary degree of maternal care.[42] Analogous to many amphibians, the larval stages of this crab still require water, but the reservoir of choice is restricted to small ponds. These are not scattered across a beneficent landscape but located in the axils of an epiphytic plant belonging to the bromeliads. It is hardly a propitious place, the water being quite acidic, depressingly low in oxygen, and lacking calcium, something of a disadvantage given that it is essential for the crabs' skeletons. The mother, however, is definitely in charge. Not only does she clear the pond of debris, but she maintains oxygen levels and never at a loss supplies calcium by the expedient of dumping snail shells that make the nursery more alkaline.[43] Moreover, while she will collect the shells when she comes across them, if the conditions deteriorate, then extra shipments are organized.[44] Given this degree of maintenance it is unsurprising that maternal care extends to feeding the young and protecting them. It all makes good sense, yet in experiments where the young are transferred to another pond the mother finds them or the young may trek back to their original home. As Rudolph Diesel notes these comprise a set of "surprising results suggest[ing] that the mother cares for the young, and the young seek maternal care."[45] As this worker points out, parallels exist with a Neotropical poison frog (p. 34) that also employs bromeliad ponds and exhibits care of its young, including feeding.[46] Elsewhere in Africa other freshwater crabs (potamoid brachyurans) have arrived at very much the same solution, notably in the forests of East Usambara in Tanzania.[47] It is the last place littoralists would look for a crab, but in water-filled boles these crabs conduct their lives in parallel to their cousins in Jamaica. While less well-known they keep the pond free of debris, add snail shells, and most likely engage in parental care. Nor is the water of value to the crabs alone because it is also used in local medicine, a sovereign protection against miscarriages. In these Tanzanian crabs, and presumably the convergent examples from Liberia, an unsurprising adaptation are elongate legs that are helpful for clambering around trees.[48] Nor are these the only crabs to have taken up an arboreal life.[49]

The grapsid crabs seem to be especially suited to various terrestrial adventures, and this may point to a more general principle. Thus, among the decapod crustaceans (the lobster being a familiar example) the crabmorph, effectively defined as tucking the abdomen beneath a well-calcified carapace, has evolved four more times (in total: once in the brachyurans and four times in the anomurans).[50] In some ways this presents a puzzle because of the extraordinary range of habitats in which these crabs flourish: from abyssal trenches to forests. C. L. Morrison and colleagues draw our attention to the fact are the isopods,[9] familiar as the wood-louse. In achieving terrestriality they provide a treasure trove of convergences.[10] This transition has involved significant changes in the sensory apparatus,[11] and some taxa (*Hemilepistus*) have not only managed to colonize desert environments but have a complex social system of monogamy, biparental care, and tight kin recognition.[12] So, too, isopods have evolved lungs multiple times.[13] This entails not only closure mechanisms in the form of spiracles,[14] but even tracheal-like extensions into the body cavity.[15] Tracheae are too good a trick to miss, and we find convergent counterparts in the arachnids (such as spiders[16] and harvestmen[17]), a group that also reveals an interesting marine-terrestrial transition.[18]

Among the various experiments in terrestrialization (and indeed the invasion of freshwaters), the crabs (especially among the brachyurans[19]) provide some unexpected insights. A popular suggestion is that the drive to land was in response to the intense levels of predation in the adjacent seas. Jumping from the frying pan has its own dangers, especially if humans are involved. In a delicious irony the bamboo traps that natives employ to trap land crabs in the Caribbean (Trinidad and Tobago) and the Philippines (Cebu and adjacent islands) have converged in the same highly effective design.[20] Indeed, most likely such hunting was responsible for the disappearance of the land crabs on Hawaii.[21] As with the flightless birds, the indigenous land crabs of tropical islands seem to be highly vulnerable to introduced competitors. And not just rats, pigs, and humans, but even ants, as is apparent in what Dennis O'Dowd and colleagues[22] term "Invasional 'meltdown'"[23] of the Christmas Island ecosystem after supercolonies of the crazy ant got a grip. Gustav Paulay and John Starmer remark how this vulnerability "may restrict crabs to the island they evolved on, rendering them evolutionary dead ends from a global perspective."[24]

This may well be correct, but in their own context many of these land-crabs are far from unsuccessful. Reports speak of crabs being so abundant as to result in a "literally reddening of Clipperton Island."[25]

Among the most celebrated of the terrestrial crabs are the robber (or coconut) crabs. Not only is this because of a dramatic convergence in the olfactory system (p. 122), but these robber crabs (and related terrestrial forms that still employ shells as hermit crabs, collectively the coenobitids) have invented lungs.[26] Although they still retain their ancestral gills,[27] practically all the uptake of oxygen is now pulmonary, while exchange of carbon dioxide[28] is mostly across the gills.[29] Impressive as this group is in terms of terrestrial adaptations,[30] the robber crabs are by no means the only crabs to have trekked on to land. In shifting to terrestrial respiration the gills are transformed from soft, feathery, and collapsible structures to well-supported lamellae[31] whose capacity to enact gaseous exchange is again augmented by lungs[32] that display an extensive degree of vascularization.[33] E.W. Taylor and A.J. Innes[34] point out, "The evolution of air-breathing in crustaceans closely parallels that of vertebrates with a transition from aquatic, diffusion-limited gill breathing, via bimodal, diffusion-limited 'skin'/gill breathing to [ultimately] perfusion-limited lung breathing."[35] This last refers to one of the most extraordinary cases of respiratory convergence and concerns a crab (*Pseudothelphusa garmani*) that lives in the mountains of Trinidad. Like other crustaceans it employs the respiratory protein hemocyanin (p. 13). Despite the oft-quoted inferiority of this protein, in this crab it employs a partial pressure of oxygen "which is close to that of *Homo sapiens*."[36] Moreover, it manages an acid-base balance in its blood such that like vertebrates it is "in a state of compensated respiratory acidosis with respect to their aquatic counterparts."[37]

CHAPTER 15

THE ARTHROPODS SHOW THE WAY

SUCH IS THE diversity and success of the arthropods, from farming ants (p. 181) to bolus-wielding spiders (p. 89), that common themes might seem elusive. Yet in areas as disparate as walking (p. 74) and silk (p. 89), convergences reveal that the number of evolutionary solutions is limited. These two particular examples are relevant in another context, that of terrestriality. While the oceans are awash with arthropods—think of krill—they, along with the vertebrates, have been by far the most successful colonizers of land. Geerat Vermeij and Robert Dudley[1] remind us that the total number of groups managing this transition was surprisingly small, and indeed vice versa.[2] One of evolution's favorite chestnuts is why, despite their overwhelming terrestrial success, so few insects managed to return to the oceans. Various ingenious ideas have been presented.[3] It may be, however, that although numerous insects have an aquatic phase, the marine environment simply lacks the facilities necessary for completion of the life cycle,[4] specifically any way of "launching" the adult flying stage inasmuch as even fresh-water insects employ vegetation or some sort of substrate as they metamorphose. By far the most successful colonizers of the marine realm are relatives of the pond-skaters, notably *Halobates*,[5] and these gerromorph insects have invaded the marine realm at least fourteen times.[6]

Such exclusions remind us that not all things are possible in all groups, although when we meet dinoflagellates that have transformed themselves into tapeworms (p. 310n3) and animals that have become honorary protistans (p. 315n13), the versatility and flexibility of evolution impresses rather than the supposed constraints. Fascinating as the marine arena may be, the center of attention is necessarily the terrestrial realm. Why? Simply because despite only occupying a third of the surface area, the overall diversity of terrestrial life far outstrips the marine world.[7] This rise to dominance, however, is geologically relatively recent (c. 100 Ma) and may well have been triggered by the emergence of the flowering plants.[8]

FIRST THE CRABS

With the terrestrial realm as the cockpit of evolution it might seem sensible without further ado to turn to the insects and vertebrates. So we shall, but our agenda is to seek the general principles, both in terms of recurrences along with determining the limits of what is actually possible. In the context of terrestrialization, another group of arthropods is remarkably instructive, the crustaceans. Most successful in this regard stupendously large and far from fragrant *Rafflesia*[196] (and the related *Rhizanthes*[197]) as well as the holoparasitic *Hydnora*, whose repellent smell evidently serves to attract dermestid beetles.[198]

To suggest that the warmth of a human and stinking arum have much in common other than an elevated temperature might seem to be risible, but when we look to molecular mechanisms, convergent similarities arise. In essence any type of thermogenesis requires a diversion of respiratory pathways and a process of "futile" cycling, which in turn leads to the inevitable employment of enzymes and mitochondria.[199] This is not to say that only one biochemical route to thermogenesis exists; there are several, but as ever the convergent result is what counts. Mammalian endothermy looks to so-called uncoupling proteins (UCPs). As the name suggests these proteins derail the oxidative processes in the mitochondria, specifically by the futile cycling of protons across a leaky inner mitochondrial membrane. Intriguingly UCPs are found in some thermogenic plants, such as the dead-horse arum, but the protein is not restricted to the thermogenic tissue.[200] Nor is this surprising: UCPs are effectively ubiquitous in plants[201] and among other functions play a key role in photosynthesis.[202] At least some thermogenic plants, however, employ UCP, and in a philodendron the metabolic substrate is a lipid (as in mammals).[203] In the same paper the investigators Kikukatsu Ito and Roger Seymour demonstrate that another thermogenic plant (the voodoo lily) employs a different enzyme (AOX, or alternative oxidase), which most likely draws on a carbohydrate substrate. Just as with the UCPs, so AOX shows a diversity of functions linked to regulation of oxygen,[204] but in many thermogenic plants AOX appears to be a key factor in generating heat.[205] This cannot be the whole story,[206] and in some cases, such as the skunk cabbage, AOX and UCP may work in tandem,[207] while as in other examples of endothermy the density of the mitochondria and the style of activity[208] again serve to link the thermogenesis of plant and animal. And to the animals we now turn, beginning with a group intimately associated with the plants. I speak, of course, of the arthropods.

hamburger or an accident left by my dog is no more coincidental, because the odors wafting from this fungus are evidently convergent on the putrid equivalents from carrion flowers; certainly they attract the flies.[173] In many of these plants, including the mosses, the cues are not only olfactory but also visual. Charles Davis and colleagues catch this sense of the revolting with their vivid writing, noting how these carrion flowers not only often are colored a dark, liverish red emblazoned with light splotches, but in addition may display "a tangled mat of hairs, festering pustules and darkened orifices [that] further add to the visual and tactile sensation of the model."[174] The carrion flies may regard these monstrous blossoms as the nearest thing to a rose, but the repulsion exuded by the often massive flowers[175] may also help to deter wandering herbivores.[176]

Flowers smelling of cat pee might be bad enough, but in many cases these revoltingly smelling plants have the capacity to provoke a further shudder of disgust: they are warm to the touch. Not only has endothermy evolved multiple times among the animals (pp. 226–27), but so too there are thermogenic plants.[177] Roger Seymour[178] notes, "some plants produce as much heat for their weight as birds and insects in flight,"[179] while among the arums, well known for their thermogenic capacity, *Philodendron* can reach an astonishing 46°C. Appropriately referred to by Ingolf Lamprecht and colleagues[180] as "Flower ovens,"[181] a routine method of scientific investigation involves infrared thermography. This capacity to generate heat is remarkable because, of course, there can be no central regulatory system. Even so, a number of these plants exhibit thermoregulation, such as in the arum *Symplocarpus foetidus*.[182] As both the specific designation and its popular name of skunk cabbage indicate, this arum smells pretty bad. A recurrent suggestion for such plants being thermogenic is to increase the volatility of the horrible odors, and so enhance yet further the attraction to pollinators.

The thermogenic arums have attracted considerable attention, including the truly massive infloresence of the appropriately named titan arum.[183] Growing up to 3 m high, its distribution of the thermogenic areas suggests that it does indeed serve to waft cadaverous odors across the jungle; while speaking of the dead-horse arum, Anna-Maria Angioy and colleagues[184] remark, "Blowflies find the horrific smell, and possibly the fleshy colored infloresence, irresistible."[185] These carrion flies and equally beetles treat these thermal beacons as magnets, but at least in the case of beetle pollination (cantharophily) thermogenesis is also important as the flower provides its visitors with an effectively endothermic environment.[186] Nor are the benefits restricted to the pollinating insects, because in a skunk cabbage, which flowers in the early spring,[187] the heat generated is crucial in allowing the pollen to germinate and then assist the pollen tube on its epic journey[188] toward the expectant ovule (p. 154).

Cantharophily is rampantly convergent, and it is no coincidence that both this type of pollination and plant thermogenesis have evolved repeatedly in tandem.[189] Thus the cones of some cycads are thermogenic,[190] while among the angiosperms heat production is particularly characteristic in some of the more basal representatives,[191] including the famous Amazon waterlily[192] and the sacred lotus.[193] With respect to the latter plant Roger Seymour and Paul Schultze-Motel note that not only is it an exceptional thermoregulator, but the infloresences can increase "the rate of heat production in proportion to the decrease in ambient temperature, a pattern identical to that in homeothermic birds and mammals."[194] The roster of thermogenic flowers goes considerably further and includes not only the palms[195] and the above-mentioned arums, but also parasitic plants. These include the plants the enzyme RuBisCO, which is central to photosynthesis, is retained.[146] Or is it so strange? RuBisCO may be a walking disaster in terms of biochemical efficiency, yet like so many biomolecules it shows an unexpected versatility and in the oil-seed rape plays a key role in the synthesis of vegetable oils.[147] Whether this applies to the parasitic plants is yet to be determined, but whatever its role (or more likely roles) one can be sure that its retention reflects a vital function.

The habits of many parasitic angiosperms leave something to be desired, although the so-called Maltese mushroom (*Cynomorium*) is widely employed for herbal medicine.[148] The dodder group (*Cuscuta*)[149] is particularly pernicious as an agricultural pest,[150] notably of legumes. Given its capacities this is hardly surprising, because this obligate parasite employs a complex haustorium.[151] This attaches to the host by an ingenious combination of a cement and highly pliable epidermal cells,[152] while the extractive cells[153] are hyphae-like.[154] Dodder selectively forages for hosts,[155] evidently depending on the plant being able to detect volatiles released by its victim.[156] Justin Runyon and colleagues remark how "this system [is] similar to that previously described for foraging insect herbivores . . . and thus [reveals] an unexpected convergence in the host-location strategies used by disparate natural enemies of plants."[157]

On the scale of remarkability among parasitic plants, surely nothing rivals the astonishing *Rafflesia*.[158] With a flower that can be a meter across, yet bursting out the host vine where the rest of the parasite consists of only threadlike tissue, is not *Rafflesia* emblematic of the sheer strangeness of life, another one-off, never to be repeated? Not quite. It is now clear that the larger group of plants referred to as the Rafflesiales are blatantly polyphyletic.[159] *Rafflesia* and its closest relatives fall in the enormous group known as the Malpighiales and appear to be close to the spurges

(euphorbiaceans).[160] The cytinaceans, which again have a fungal-like body with only the flowers erupting from the host,[161] turn out to be close to the Malvales, while the supposed rafflesiacean known as *Mitrastema*[162] is a relative of plants that include tea and the heathers (Ericales).

It may be that the colossal flower of *Rafflesia* has reached the ultimate size limit of any flower, yet it is clear that this remarkable evolutionary trajectory not only represents an astonishing acceleration but one underpinned by trivial molecular differences with its more modestly attired relatives.[163] Todd Barkman and colleagues remark that given that these plants might be taken as an endpoint in angiosperm experimentation, so one might expect "that large flowers would have evolved over long periods of time and probably only once during the history of the genus." But they continue: "Contrary to expectations these large flowers appear to have evolved rapidly, recently, and repeatedly."[164] In the wider context of plant evolution, such floral gigantism (i.e., larger than 30 cm) has evolved independently at least nine times,[165] and they also exhibit recurrent pollination syndromes. *Rafflesia* exemplifies one type because notoriously it smells of rotting meat and relies on pollination by carrion flies.[166] This is disgustingly convergent because various groups of flowering plants have independently evolved odors[167] that recall not only corpses, but feces,[168] urine, rotting fish,[169] and more benignly even yeast.[170] In the case of the appropriately named dead-horse arum (or lily), its nauseous odor is a molecular mimic of the smell emanating from meat well past its sell-by date.[171] Nor is this stench confined to flowering plants, because the splachnacean mosses have arrived independently at the same solution, attracting insects to disperse their sticky spores by producing volatiles that mimic herbivore dung.[172] The fact that a stinkhorn smells something like an overripe gent.[121] Jonathan Leake notes that "despite the very diverse range of families and genera [the seeds of] these plants show some quite remarkable parallel evolutionary forms,"[122] notably a tiny size.[123] He adds, "Even more remarkable is the convergent evolution of virtually identical adaptations of seed morphology and seed coat structures among the taxonomically disparate genera."[124] Not surprisingly the leaves are often vestigial or completely wanting. So, too, in many taxa, "stems are exceptionally slender and thread-like,"[125] and the vascular strands typically are greatly reduced. Just because these so-called mycoheterotrophic plants lack chlorophyll does not mean they escape the attention of herbivores, and in one way or another they seek to deflect unwanted attention. In an ericacean (*Monotropsis*), concealment is achieved by bracts that arise from the stem and allow the plant to blend effortlessly with the surrounding leaf litter.[126] Jonathan Leake notes how mycoheterotrophy has evolved in "more than 400 species of vascular plants [and] in 87 genera."[127] It is, however, by no means the only route plants can take to a life of parasitism.

It is easy to see why parasitic plants[128] provide a frisson of the alien: pale and etiolated, seeking attachment to their hosts with a blind intentionality, draining the sap via intricate haustoria,[129] and flowers—sometimes of a monstrous size and smelling of corpses—erupting from the host. In their various ways parasitic plants exemplify the strange and bizarre. Molecular phylogenies have led to some radical reconfigurations of this part of the tree of life, with supposedly trustworthy morphologies transpiring to be rampantly convergent. In the flowering plants, parasitism has evolved at least eleven times, and of these, eight are holoparasitic and so have lost all capacity for photosynthesis[130] (in contrast to the hemiparasites, such as the familiar mistletoe,[131] which remain green). Intriguingly among the plants, several groups seem

to have a particular tendency toward parasitism. Thus, as Todd Barkman and colleagues note, "within lamiids alone (ca. 12 percent of angiosperms [and an enormous group of plants that include the potato family and gentians]), parasitism has independently evolved 3 times."[132] On the other hand, "monocots, campanulids, and caryophyllids [accounting for about 40 percent of angiosperm diversity] have never evolved parasitism,"[133] or if they ever did, it was among extinct representatives. So why this propensity?[134]

Not only has parasitism evolved in the flowering plants, but also in a New Caledonian gymnosperm[135] and even more remarkably in the much more primitive liverworts,[136] specifically the appropriately named ghostwort. Subterranean, and of course incapable of photosynthesis, this bryophyte effectively inserts itself into an existing mycorrhizal association.[137] This incapacity to photosynthesize does not mean, however, that the plastids have been lost. Even in green plants, plastids—located, for example, in the roots—are involved in a series of important metabolic functions unconnected to any interaction with sunlight.[138] In a way that echoes pathogenic and endosymbiotic (pp. 173–74) bacteria, the plastids show a dramatic decrease in the size of their genome (the so-called plastome).[139] In the beechdrop (appropriately named *Epifagus*, and once again a member of the lamiids), all the genes connected to photosynthesis,[140] as well as others involved with RNA activity, have been lost.[141] Not only that, but the story of gene loss in nonphotosynthetic plants[142] finds striking similarities[143] in the relict chloroplast of the group of protists known as the apicomplexans.[144] Yet in the case of the plants, despite what Kenneth Wolfe and colleagues describe as "massive pruning,"[145] remarkably the order of genes around the circular chromosome has hardly altered and the plastid is still functional. What seems particularly surprising is that in many parasitic duction of potent ecdysteroids that could be used to fight off insect predators."[103] Their potency lies in the fact that the hormone can disrupt larval development and so act as a natural insecticide.[104] Such examples are a focus of attention in developing chemicals that might assist us in protecting vital crops from the insect onslaught. Different types of insecticide target one or other aspect of insect biochemistry or physiology. Correspondingly the intense selective pressure in fields drenched in chemicals has led not only to adaptive countermeasures on the part of the arthropods,[105] but so too a serious problem of resistance to a range of insecticides. Darwinian enthusiasts like to trot out such examples as "evolution in action." Tellingly less often is it emphasized that not only does insecticide (and antibiotic) resistance demonstrate the reality of evolution, but the process is more predictable than often thought. One striking instance concerns the organochlorine cyclodiene insecticides such as dieldrin. These aim to disable a standard neurotransmitter (GABA, or gamma-aminobutyric acid).[106] Not only has the resistance evolved independently in numerous groups of insects,[107] but also within particular groups such as that serious pest the flour beetle,[108] and malarial vectors such as mosquitoes.[109] Almost invari-

ably it involves a substitution at a single site in the protein, usually occupied by the amino acid alanine. Replace this, typically with serine, and the insect acquires resistance.

Similar stories emerge with other insecticides: DDT and pyrethroids should knock out a sodium channel, but typically only two mutations will confer resistance. These have also evolved repeatedly in important pests such as aphids[110] and other hemipterans like the whitefly.[111] Other insecticides, such as the organophosphates (OPs) designed to clobber that part of the insect nervous system that employs an esterase enzyme (acetylcholinesterase), again depend on a specific substitution[112] and so follow the path of convergence.[113] When Carol Hartley and colleagues[114] write of this system in flies as providing "evidence for biochemically precise convergent evolution [that suggests] the options for evolving esterase-based metabolic resistance to OPs are tightly constrained,"[115] this would apply with equal force to other insecticide resistances. Convergent resistance has also emerged in fungicides, such as those that attack vineyards,[116] and herbicides designed to tackle serious agricultural threats such as wild oat.[117] In both cases it is again the same very specific sites in the protein that repeatedly change.[118]

FLOWERS SMELLING OF CORPSES

The range of plant-insect interactions is enormous, but of these perhaps the strangest are the carnivorous plants (p. 85). This, however, is not the only case where the plant retains its photosynthetic capacity but manages to obtain some of its carbon from another source. If not trapping insects, then why not become friends with the so-called mycorrhizal fungi in what is termed a mixotrophic association? This has evolved at least twice in some orchids, as well as a group of ericace-

ans known as the pyroloids,[119] but many more convergences are likely to be identified not only in the angiosperms but other groups of plants.[120] Confidence that this must be correct is based on the observation that mixotrophy is only a staging post to fully fledged parasitism. Here the plant loses all photosynthetic capacity and relies entirely on some sort of mycorrhizal association. Representing a dramatic shift from autotrophy to heterotrophy, these associations are rampantly converof the plant. In flowering plants they draw upon distinct developmental pathways,[81] and more generally have proved to be rampantly convergent.[82] A striking instance is found in a Carboniferous lianalike seed fern, *Blanzyopteris*. Such a mode of growth looks to the same adaptive strategies for effective attachment[83] that are found in the flowering plants where again this habit has evolved multiple times.[84] Not only did *Blanzyopteris* festoon the coal forests of France, but interestingly it possessed two types of trichome. Of these the glandular variety employs a strikingly similar mechanism to that seen in the touch-sensitive trichomes of various flowering plants, notably the cucurbitids such as the musk cucumber.[85] As their name suggests, glandular trichomes function by breaking open and releasing a sticky exudate that stops insects like aphids in their tracks. This is not the only way in which these pests can get bogged down. A famous example entails the various resins that solidify into the golden tombs that as amber provide so many remarkable insights into ancient life, not least in terms of spider silk (p. 88) or predatory fungi (p. 150). Amber had appeared by the Carboniferous. Curiously it has a chemistry typical of flowering plants,[86] even though they did not evolve until much later. This leads David Grimaldi[87] to exclaim that not only is this an "astonishing evolutionary convergence at the molecular level"[88] but an important reminder that early plants were well equipped to deter unwanted invaders as well as seal wounds.

Plants synthesize an extraordinary array of metabolites, not least those involved in sexual mimicry[89] as well as cyanogenic compounds. In their overview, Eran Pichersky and Efraim Lewinsohn[90] explain how in some cases much the same function is achieved using quite different chemicals, whereas in other cases unrelated plants have succeeded in synthesizing the same compounds, but as they also note "overall convergent evolution in plant specialized metabolism is surprisingly common."[91] Many of these compounds, of course, are employed for defensive purposes. Particularly potent is the repeated evolution of specialized tubes known as lactifers[92] that house various sorts of latex. As Jillian Hagel and colleagues[93] comment, such "specialized cells . . . occur in over 20 families in several unrelated angiosperm orders."[94] Their contents are of more than passing interest because latex encompasses products such as rubber and opium, plus the fact that latex has obviously evolved numerous times.[95] Nor for that matter is latex restricted to plants,[96] because it has evolved independently in the fungi. Best known are the aptly named milkcaps, notably *Lactarius*,[97] but so too (and presumably independently) latex can be found oozing from the fruiting bodies of some species of *Mycena*.[98] This is only one aspect of the chemical armory that fungi employ to defend themselves, and although much remains to be learned there are at least some parallels to the plants.[99] In these fungi the latex evidently contains a potent brew of chemicals, and although we associate the latex with a sticky entrapment for incautious insects, it also houses various poisons.[100] Some insects, however, flourish on lactiferous plants by the simple expedient of cutting open the veins to drain the latex.[101]

Latex is, of course, only one part of an immense chemical armory that plants employ to discourage chewing, shredding, and chomping by the legions of herbivorous insects. The trick, of course, is to find some sort of chink in the armor. Consider, for example, the striking convergences in the metabolic pathways for the synthesis of steroids in plants and their insect equivalent, the so-called ecdysteroids.[102] Both can serve as hormones, but as Carl Thummel and Joanne Chory note, "In a further twist, plants [have] . . . diverted part of their steroid biosynthetic pathway toward the pro-only may the internal cavities (aerenchyma) have assisted in carbon dioxide concentration, but the "low [atmospheric] CO_2 excursion coincides . . . with the greatest diversity of arborescent lycopsids."[64]

THE PLANTS FIGHT BACK

The landscape of evolution might shimmer with vibrant life-form, but so too it resonates with the echoes of vanished denizens. Particularly haunting are those lands, notably Madagascar and New Zealand, both of which represent evolutionary laboratories of the first order. Here almost within living memory they housed giant flightless birds such as the elephant bird (which weighed twice as much as a grizzly bear) and the moa. Almost certainly hunted to extinction to provide a toothsome fricassee, the plants have longer memories. It has been suggested, for example, that striking changes in the leaf shape and color of the growing *Pseudopanax* (endemic to New Zealand) might have served to camouflage[65] the plant from the unwanted attentions of the moa.[66] It may not be coincidental that the leaves also bear spines, perhaps as further defense against herbivorous attack. It is in the wider context of so-called anachronistic adaptations that we repeatedly see features that will never be needed again. Consider the architecture of the so-called wire plants.[67] These nicely show how effectively unrelated plants can arrive at the same solution, because in each case by drawing on such features as narrow, wiry stems, and widely spaced leaves, they evidently frustrated the giant herbivorous birds. In Madagascar and New Zealand, of course, large browsing ungulates are absent,[68] which was just as well, because against these mammals the wire plants would be so much fodder.

But when the ungulates are roaming, the plants are far from helpless. That is why we see the repeated evolution of thorns and spines.[69] These structures are capable of administering much more than a sharp stab, because they harbor a wide variety of pathogenic bacteria that can lead to necrosis, tetanus, and gangrene.[70] The story of plant deterrence goes further. In animals, obvious spines with a warning coloration (this is known as aposematism) has evolved innumerable times.[71] Think of the porcupine[72] and the porcupine fish,[73] not to mention those hedgehog spines dripping with toad venom (p. 39). Plants have evolved the same trick,[74] and even more remarkably take the art of mimicry to greater heights by the expedient of either a sort of color printing to give the impression of thorns or structures that look just like thorns but are not.[75] Warning coloration in plants has explored other ingenious avenues, with some species adopting either an anomalous color[76] or a striking pattern[77] (in this context, recall also poison frogs [p. 34] and snakes) that metaphorically shout danger. Equally intriguing are plants that pepper themselves with structures that look suspiciously like ants or aphids or adopt a color pattern similar to that of toxic caterpillars, thus discouraging the inquisitive visitor.[78] This by no means exhausts the armory of plant defenses. Consider the trembling aspen, yet the leaf structure in this and a variety of other plants that allows movement in even the lightest breezes may be a protection against unwanted herbivores and pathogens.[79]

So, too, among the plants there is an entire battery of other physical (e.g., mineral deposits or hairy surfaces) and chemical armaments.[80] Particularly well known are the trichomes. These are hairlike structures that play an important role in the protection (or indeed *any* complex form) the question revolves not only around the availability of a slate of available traits but the combinatorial possibilities that ultimately lead to their functional integration. For cacti this would include the evolution of succulent tissue with mucilage cells for water storage as well as the need to postpone the development of a constraining bark (or periderm).[38] It is significant that among the cacti and their relatives, these key characters are seen to be both labile, and in some cases can show, as Matthew Ogburn and Erika Edwards note, "very high levels of homoplasy."[39] In some ways evolution appears to be more like a melting pot whereby the characters can be assembled in a variety of combinations that make good functional sense. So too they are likely to evolve repeatedly, much to the chagrin and frustration of those biologists wedded to the atomistic certainties of the cladistic methodology. Importantly Ogburn and Edwards note that if such characters can evade adaptive scrutiny "and are also evolving independently from one another, then it follows that higher evolutionary lability in these traits will increase the likelihood that one lineage will evolve the 'right' character states in all of them."[40] As they point out, this will promote both evolutionary innovation and ultimately its integration into a stable solution. To invoke homoplasy as a creative force in evolution may raise eyebrows in some quarters, but surely it resonates with the theme of this book.

Among the curiosities of the cacti is an abundance of crystals (druses) of calcium oxalate.[41] Biologically these are widespread,[42] and to us are horribly familiar in the form of kidney stones. Their function in the cacti is debated, but in some plants one function appears to control stomatal closure,[43] which in general is dependent on calcium.[44] Such a control takes on a special significance in the cacti, because water loss must be kept at a minimum. If, however, the stomata remain shut, then water will not escape, but neither will the air with its carbon dioxide be able to enter the plant to be available for photosynthesis. The solution is to be like an old-fashioned English pub—that is, restrict the opening times. In the case of the stomata they only open during the cool hours of the night. This in turn is linked to an ingenious system of then managing to store the carbon dioxide until the sun rises and photosynthesis can start up. We see, therefore, another method of concentration of carbon dioxide[45] (p. 311n11), but this time employing large cellular vacuoles that house an organic acid, typically malic acid.[46] This method of carbon dioxide storage underpins the so-called Crassulacean Acid Metabolism (or CAM) photosynthesis.[47] Just like C_4 photosynthesis,[48] with which it shares a number of biochemical similarities,[49] CAM is rampantly polyphyletic.[50] The demands of this type of physiology, not least the capacity to store the carbon dioxide in the malic acid until it can be remobilized as daylight returns, has led to anatomical convergences.[51]

CAM is known in ferns[52] and gymnosperms,[53] as well as a diverse range of flowering plants that includes the bromeliaceans[54] and orchids.[55] Moreover, while CAM photosynthesis typifies the cacti[56] and other stem succulents,[57] its versatility extends far beyond the xerophytes.[58] Most surprising in this regard are a variety of aquatic plants (including the remarkable Andean fern *Stylites*, whose stomata are confined to its roots[59]) that independently employ CAM. Here, too, the fundamental driving force is the need to utilize nocturnal abundances of carbon dioxide.[60] One possibility is that the CAM system originally emerged from ponds, and this could be consistent with the inference of a CAM-like metabolism in the giant lycopsids (as characterizes their modern-day descendants, the aquatic quillworts (*Isoetes*])[61] that were such a feature of the Coal Measure swamps.[62] As Walton Green points out,[63] not draws two important conclusions. First, not only has this capacity been frequently achieved, but remarkably rapidly. Second, and more intriguingly for Darwinians, these steps to convergence do not necessarily imply intermediate forms. After all, if the developmental machinery can be turned on or off, then the convergent character will flicker in and out of existence. When the developmental biology is investigated,[20] as Hearn notes, his research "reveals remarkable parallels in structure and development between separate evolutionary origins of storage parenchyma in stems and roots of species of *Adenia*."[21] Be it in either stem or root, the crucial factor is the coordination of the vascular system and parenchymatous tissue. Hearn also draws our attention to similarities in the storage roots of more distantly related plants, not least the carrot, concluding how "these studies present multiple instances of parallel evolution and development of storage tissue in stems and storage roots across distantly related lineages of angiosperms."[22]

When it comes to water storage and convergences in the flowering plants,[23] among the other xerophytes some of the most striking parallels are those found between the cacti and euphorbiaceans.[24] Both groups have taken xeromorphy to remarkable heights. One key to their success is the possession of succulent tissue. Such tissue has, of course, evolved repeatedly, but in terms of the history of diversification in the cacti and euphorbaceans[25] (along with the agaves)[26] the crucial factor appears to have been the major onset of aridity (along with falling levels of atmospheric CO^2; and hence the emergence of C_4 plants, p. 392n48) during the Miocene and Pliocene.[27] The story of succulent evolution is yet to be fully told, but the cacti offer some useful insights. Indeed even within the portulacineans, the group of plants that encompass the cacti, not only has succulence evolved multiple times but as significantly the three

major ways in which any plant can become a succulent life-form are all found in this one group.[28] So far as the cacti are concerned, a key step in their evolution was, however, the introduction of vascular bundles into the cortex,[29] thus allowing the development of a massive stem.[30] What is important to appreciate is that when the sequence of events that culminated in the cacti are analyzed, it appears that when evolution shuffles this particular pack, sooner or later the aces will play out.

This conclusion depends on knowing what features are the necessary prerequisites for any sort of cactus to emerge and in turn demands a knowledge of the wider evolutionary relationships of the cacti.[31] In this context it is a primitive representative of the cacti, in the form of *Pereskia*,[32] that takes on a particular relevance. This is because *Pereskia* demonstrates, not surprisingly, that the key to becoming a fully spined cactus[33] lies in its capacity for water regulation. As Erika Edwards and Michael Donoghue[34] succinctly note, this plant "may not look like a cactus, but it behaves like one."[35] What this means is that the capacity to live in drought-stricken environments emerged "long before [the cacti] evolved the anatomical specializations"[36] that we otherwise tend to associate automatically with such a mode of life. Once again there is a sense of evolutionary inherency, of a "cactus-in-waiting," with Edwards and Donoghue remarking how in this wider scheme of things "the evolution of leaflessness and stem succulence does not seem quite so extraordinary."[37] Might we shift the discussion from the prevailing Darwinian view of endpoints being deeply improbable to at least their being likely, perhaps even inevitable?

Cacti, therefore, seem to be waiting in the wings of the evolutionary theater. Accordingly we need to remind ourselves that to understand the plot we should pay attention not only to the actor but also to the prompt. Central to the eventual emergence of the cacti

CHAPTER 14

THE ROAD TO PLANTS

UNTIL THE CONTINENTS turned green, the first animal visitors to dry land would have been only day trippers, even though these brief excursions onto land would, one day, lead to the first steps on the Moon. It is, therefore, to the cockpit of plant evolution that we must first turn. Indirect evidence suggests that some sort of plant colonization was under way in the late Precambrian,[1] with the important corollary that from then on levels of atmospheric oxygen began to rise:[2] one vital prerequisite for ultimately terrestrial animals.[3] The nature of these first plants is conjectural, although lichens (p. 152) may have been an important component.[4] So, too, plausibly would have been bryophytes, familiar as the mosses and representing the most primitive of the living land plants.[5] This group is highly instructive as to the inevitable routes a plant must take to gain a roothold on land, but bryophytes themselves also provide informative convergences.[6] Nevertheless, full-scale terrestrialization—forest and flowers and that sort of thing—might be thought to involve a whole series of deeply improbable evolutionary coincidences. No trees, then no arboreal marsupials (p. 69); no fruits, then no trichromatic monkeys choosing the ripest examples (p. 114). From almost any perspective it is difficult to imagine a world remotely similar to ours without its comparable flora. Convergence, however, suggests that we can be pretty sure what to expect. If space permitted, one might turn to a series of compelling examples that range from wood[7] (as well as xylem[8]) to leaves[9] and pollination,[10] as well as more specific instances such as mangroves,[11] myrmechory,[12] fruits[13] (and their colors[14]), and even autumnal colors.[15] Moreover, in crucial innovations such as the seed, one can argue that they are evolutionary inevitabilities.[16] But let us turn to a truly succulent example.

SUCCULENT CONVERGENCE

For most plants if there are two central factors, these are sunlight and water. Concerning the former we can ask how inevitable is the process of photosynthesis,[17] or whether it might be optimized.[18] Water may be crucial to the survival of any plant, not least in times of drought, but once again important evolutionary lessons can be learned. A telling example, given by David Hearn,[19] concerns *Adenia*, an African relative of the passionflower. The hundred or so species show a remarkable variety of growth forms, ranging from trees to herbs and lianas. For *Adenia* a key evolutionary gambit is storage of water, which it achieves by either developing succulent stems or underground tubers. Both have evolved multiple times and call on the same developmental machinery. Hearn the insects fooled by these fungal pseudo-flowers, so also were her botany students.[57] These "flowers" do not specifically resemble those of the host, but with their bright yellow color and abundant sugary "nectar"[58] they lure insects to a tissue largely made of fungi and from which their spores can be dispersed as a pseudo-pollination. For some insects the attraction is evidently visual, but for others it is olfactory,[59] and intriguingly the emerging scent mimics chemicals that moths find irresistible.[60] This arrangement is convergent, because in other plants the rust *Puccinia* again induces "flowers" (actually altered leaves) that in their yellow color, secretion of nectar, and scent draw in the flies.[61] The nature of some of these fungal mimics verges on the extraordinary.[62] For example, a fungus (*Monilinia vaccinii-corymbosi*) that causes a highly destructive blight in the blueberry (the so-called mummy berry) follows the tradition by secreting sugars and smelling pleasant but also alters the host's leaves so that they become reflective to ultraviolet radiation, which also serves to attract insects.[63] All par for the course, but the fungus goes one better. When its spores germinate on the stigmata of the flowers (in the so-called secondary infection) they show a series of striking parallels to the way in which the pollen germinates and then grows as the pollen tube (similarities that extend to the modes of adhesion).[64]

This busy transfer of "pollen" in pursuit of sexual reproduction is analogous to a sexually transmitted disease, and indeed, in the case of smut in pinks,[65] Jacqui Shykoff and Erika Bucheli bluntly refer to the transmission of this fungus as "a venereal disease of plants."[66] Parasitism can, however, yield to a more benign mutualism. In an intimate threeway association between a grass, a fungus (known as *Epichloë*, an ascomycete and thus distinct from the basidiomycete rusts and smuts) that infects the grass and a fly (aptly named *Botanophila*) the latter two mutually benefit by assisting with spore dispersal and providing food for the larvae. So curiously does the grass; the volatile chemicals (an alcohol) released by the fungus not only attract the fly but also serve to protect the plant from unwanted attention by other microbes.[67] As Florian Schiestl and colleagues note, the production of these volatiles is suggestive of "similar yet convergent evolutionary pathways [leading] to interspecific communication signals in both fungi and plants,"[68] and this is by no means the only[69] example.[70] Fungi and plants have more in common, therefore, than meet the eye, but now it is time to turn to the latter and let them take center stage.

other principal cyanobacterium has long assumed to have been the taxon *Scytonema*, but while this may account for some associations, many now clearly involve the genus *Rhizonema*.[43] This may seem of arcane relevance, of interest only to that sturdy band, the lichenologists, but in fact this association has a much wider relevance. First, it has clearly been arrived at multiple times. As importantly the various groups of fungi concerned are highly selective as to this choice so that it is likely that once acquired this cyanobacterium has been "shared" among other lichens. Robert Lücking and colleagues note how it is that if "unrelated lichen mycobionts inadvertently 'cooperate' in the selection, 'domestication,' and distribution of photobionts in the lichen community by proliferating particular strains that increase the success of the lichen association [then] This phenomenon is not unlike human crop domestication."[44] In the latter case, high-yield varieties are shared widely, but Lücking and colleagues also note, "The notion of fungi cultivating photobionts invites comparison with leaf-cutter ants,"[45] famous for their cultivation of fungi (p. 181).

The lichenous association is at least as old as that involving arbuscular mycorrhizae, with examples from the famous Rhynie Chert of Lower Devonian age[46] and possibly even extending to the dawn of animal life.[47] In any event, lichens have evolved many times independently[48] in both ascomycetes and basidiomycetes, but there is also an association between a glomeromycote (otherwise central to the arbuscular mycorrhizae) and a cyanobacteria.[49] Given the range of both fungi and photosynthetic companion, the style of association varies considerably. Despite significant differences in the ascomycete and basidiomycete symbioses, we see convergences[50] in either group, while in the peltigeralean lichens (that have a typical association with cyanobacteria), multiple and independent acquisitions of green algae have led to tripartite symbiosis.[51] Although only accounting for a small proportion of the lichens, those involving the basidiomycetes[52] are still blatantly polyphyletic, with at least five independent instances of lichenization.[53] To give just one example, revolving around the description of a new basidiolichen from tropical Africa, Damien Ertz and colleagues acknowledged that "morphological convergence is relatively common in the Agaricomycetes [(that is mushroom-forming), but] finding that the basidiolichens forming coral-like fruiting structures are not monophyletic was entirely unexpected . . . this is the only case of two unrelated basidiolichen clades with morphologically indistinguishable fruiting structures."[54] One wonders how long this will remain an "only case," but we may never know, as many of these lichens are found in areas under increasing environmental threat.

FOOLING FLOWERS

If plants can engage in all sorts of clever tricks, sometimes they can also be taken for a ride. There is no doubt that some fungal associations, especially the mutualistic (and convergent) myccorhizae, are essential for the health of the plant. In other cases, however, the relationship is very much more one-sided. Most remarkable are the cases of parasitic fungi, notably the rusts that engage in ingenious sorts of plant mimicry.[55] A rust fungus (*Puccinia*) can radically transform the morphology of the host plant (members of the mustard family) and not only prevent it from flowering but induce false equivalents.[56] In identifying this extraordinary floral mimicry the investigator remarked that not only were

At the heart of evolutionary convergence is the repeated evolution of an adaptive solution, in this case how to be a particular type of plant pathogen. In the final analysis the route employed is of less importance. Evolution may describe the details of the journey and the type of vehicle employed, but it is silent as to the map of life. Space does not allow any detailed consideration of the main groups of fungi, be they safe or dangerous, but they provide numerous insights into convergence not least in terms of the fruiting bodies,[27] notably the mushrooms.[28] Rather we need to turn first to the lichens.

INTIMATE FUNGI AND THE ROAD TO LICHENS

Many fungal associations are benign. Most notable are the basidiomycetes, which form so-called ectomycorrhizal associations, a symbiosis whereby sugar from the plants and minerals from the fungi confer mutual benefit. Both the multiple origins of this association and the variety of plants with such fungal symbionts point to this being a highly dynamic association. This view is reinforced by evidence for perhaps as many as nine reversals from ectomycorrhizy to the primitive state for fungal existence, that is saprotrophy.[29] Important as these symbioses are,[30] they are overshadowed by the arbuscular mycorrhizal associations (which also appear to have played a central role in the evolutionary origin of the nitrogenous root nodules). Arbuscular mycorrhizae are especially characteristic of a group of fungi known as the glomeromycotans.[31] As the name suggests the arbuscules are characteristic branching structures that proliferate in the host cells and across which exchange of materials occurs.[32] While this symbiosis is extraordinarily widespread among the land plants, there have evidently been multiple shifts to ectomycorrhizal partnerships.[33] The mycorrhizal association is also very ancient. It can be traced back to the dawn of land plants in the early Devonian,[34] while glomeromycotan fungi are known as far back as the Ordovician.[35] Given that this association dates back to the earliest land plants, these fungi may have played a key role in the invasion of land.[36]

If so much of plant evolution revolves in one way or another around the fungi, be it via mycorrhizal associations or mycoheterotrophy (p. 161), then it is hardly surprising that they display other skills of intimacy. Consider the lichens. With the fungi and algae (both the eukaryotic and cyanobacteria) in close association, these represent one of the best exemplars of symbiosis, and it has been reinvented multiple times. Conrad Schoch and his sixty-three colleagues note that although a precise figure is difficult to arrive at, "a conservative interpretation of our data [is] that lichenization evolved multiple times in Ascomycota."[37] When the partner is a eukaryote it is usually a green algae, but remarkably a lichenous association is found between a marine brown alga (*Petroderma*) and an ascomycete (*Verrucaria*).[38] Similar symbiotic associations are known and form so-called mycophycobioses,[39] but these fall short of forming a true lichen. In contrast the *Petroderma-Verrucaria* association has "cytological characteristics [that] are quite similar to those described in other ascomycete lichens,"[40] specifically in terms of an intimate symbiosis with marked changes in the growth properties of both partners. If a lichen looks to the cyanobacteria, then the association is often with *Nostoc*.[41] Nor is this surprising, because this filamentous cyanobacteria forms a wide variety of symbioses with different groups of plants, including a specific association with the water-fern *Azolla*.[42] The source of nitrogen, and it is more than likely that in either nutrient-poor soils or among those fungi that subsist on wood, such prey provides a vital supplement. Naturally one's mind turns to an intriguing analogy with the carnivorous plants.[14] Ying Yang and colleagues remind us that, as with the pitcher plant (p. 85) and comparable forms, so the "predatory fungi have the ability to capture and to absorb nutrients from their prey, [presenting] a fascinating example of convergent evolution."[15] Perhaps these parallels go further. As in the carnivorous plants, the nematophagous fungi have digestive enzymes, including those capable of destroying collagen and chitin,[16] as well as serine proteases.[17]

FUNGAL IMPOSTORS

Mention of the ascomycetes, basidiomycetes, and zygomycetes in the context of seizing, if not strangling or impaling, hapless nematodes also serves to introduce the three most important divisions of fungi. Each shows a series of fascinating convergences, but first it is necessary to escort an impostor to the door. Welcome (or farewell) to a group known as the water molds (or oomycetes). Yet another obscure group, of interest to otherworldly researchers? Not quite. Among the notorious water molds are species of *Phytophora* responsible for the blight that led to the catastrophe of the Irish potato famine as well as the ongoing disaster of Sudden Oak Death. Other oomycetes are an equal scourge for fish and crayfish. Water molds may be strikingly similar to the fungi, especially those that are invasive pathogens of the plants, but they are clearly convergent and are only distantly related.[18] In addition to general similarities, including the molecular strategies the invaders employ to overwhelm the innate immunity (p. 307n2) of the plant,[19] there are also key convergences. These revolve around the way in which the defenses are breached by both enzymatic attack and more especially concentrated application of turgor pressure.[20] This convergence, however, takes a very interesting twist because a number of the genes, notably those involved with osmotrophy and hence turgor control, have clearly arrived in the water molds from ascomycete fungi by lateral gene transfer.[21] In addition, enzymes known as cutinases—which, as their name suggests, work by attacking the resistant cuticle of the host cells (and employ the classic and convergent catalytic serine triad,[22])— were also transferred[23] laterally.[24]

These cases of lateral gene transfer between fungi and water molds (and bacteria) that just happen to show striking convergences could imperil the central thesis of this book. After all, if genes are being swapped all over the place, then maybe what we perceive as convergences are simply the genes for the same thing doing the same thing. This, however, is too simplistic. First, even among the eukaryotes, lateral gene transfer is commonplace,[25] but it has no obvious connection to convergences. Next, consider the osmotrophy of the fungi and oomycetes. Given that they are both composed of the threadlike cells (hyphae) and filled with fluid, turgor pressure is central to their existence. In an investigation of how turgor pressure is regulated in an ascomycete (which as noted are evidently the source of the osmotrophic genes in the oomycetes) and an oomycete, it emerged, however, that despite significant similarities in the hyphal tip and rates of growth, the actual mechanisms employed in either group are probably different.[26] In each case one needs to distinguish specific needs (thus in water molds the role of osmotrophy and cutinases) as against the functioning of a highly integrated organism.

CHAPTER 13

THE ROAD TO MUSHROOMS

THE FUNGI represent one of the great eukaryotic kingdoms, and it will be a strange planet that doesn't have mushrooms from the fields of Farmer Maggot, sinister toadstools to dispense death to the unwary,[1] not to mention yeasts that any intelligent species will seize upon to ferment alcohol. Is it only their toxic reputation or crepuscular emergence that sends shivers down the spines of nonmycologists? They deserve better, because not only do they have a fascinating biology that makes them central to the economy of the planet but they provide a rich source of insights into convergent evolution. Let us begin our exploration of this fascinating group by examining what might be the least expected feature of any fungus: the development of carnivory, or if you prefer, a taste for meat.

HUNTING FUNGI

Fungi are, like us, heterotrophs, taking complex compounds as food and then breaking them down to yield both energy and components to rebuild bodies. Fungi, however, do not have guts. Typically they grow across nutritious substances (and thus are saprotrophic), which as often as not are found lurking in remote corners of a neglected fridge. Extraordinarily, however, some fungi have convergently learned how to trap animal prey. This capacity to eliminate nematodes,[2] which can be serious agricultural pests, has excited the attention of biotechnologists. Generally referred to as nematophagous fungi, the abundance of nematodes in soils provides a ready source of nutrition, but many other groups[3] fall victim, including rotifers and tardigrades.[4] Best documented in the ascomycetes (notably the hyphomycetes[5]), examples are also known in the basidiomycetes[6] (the most familiar of the fungi on account of their reproductive bodies we call mushrooms and toadstools) and even the third and more primitive zygomycetes (the appropriately named *Zoophagus*).[7] Methods of entrapment are also convergent. Most typically they entail such structures as adhesive knobs[8] or nets, but in some ascomycetes an extraordinary arrangement is found where a ring of cells serves to ensnare the prey, sometimes by a sudden constriction.[9] This lasso mechanism, which is very rapid[10] and depends on a sudden increase in turgor pressure, has also been identified in fossil fungi embedded in amber of Cretaceous age.[11] This latter instance, however, most likely arose independently,[12] because associated with the hyphae bearing the lasso (which here is composed of only a single cell) are yeastlike cells.[13]

From the perspective of evolution the nematophagous fungi invite wider similarities. Animal tissue is, of course, a convenient nae,[412] but John Pettigrew notes how in the former case, "The highly directional response of the platypus head saccades suggests that it must be performing very sophisticated signal processing of the electric image over large numbers of electroreceptors."[413] Pettigrew and Wilkins also remark that, notwithstanding "The remarkable anatomical and behavioral similarities between these two independent electroreceptive systems"[414] that unite paddlefish and platypus, there are important differences. First, the former has no counterpart to the mechanoreceptors of the platypus. Despite possessing electroreceptors the platypus appears to lack specifically dedicated neurons,[415] so in some way the platypus evidently combines the tactile and electric signals.[416] This is reflected in the brain, where, as John Pettigrew and colleagues[417] note, we find an "elaborate cortical structure, where inputs from these two sensory arrays are integrated in a manner that is astonishingly similar to the stripe-like ocular dominance array in primate visual cortex, that integrates input from the two eyes."[418]

The platypus is a highly effective hunter, and one might then assume that its electrosensory capacity was intimately connected to its aquatic life. Not quite. Just as the related echidna has equivalent mechanosensors (p. 132), the long-beaked echidna possesses electroreception that assists in its exploratory foraging in the moist soils of upland New Guinea.[419] More surprisingly, however, there are electroreceptors on the short-beaked equivalent.[420] Uwe Proske and colleagues, however, remark that despite this echidna foraging in generally drier soils, not only do they "have an uncanny ability to detect prey when it is buried in moist soil,[421] but "Perhaps the reason why the echidna always has a runny nose[422] is to maintain a low resistance pathway between sources of electric current in the soil and the electroreceptors."[423] Even so, given that the echidnas had an aquatic ancestry,[424] their possession of electroreceptors is less puzzling. The ability to electrolocate probably goes deep into monotreme history,[425] but among the vertebrates they were not alone. Still earlier, in the Permian, both primitive reptilomorphs[426] and more advanced theriodonts[427] possess characteristic bony pits suggesting that pursuit of their aquatic prey employed more than eyes.

So the senses close, and now we must seek new roads.

an auditory world or the JAR of the weakly electric fish. In this context fish and barn owl are not so different. Masakazu Konishi notes, "The fish algorithm [for JAR] is thus remarkably similar to that of the barn owl, even though the problems that are solved, the sensory systems involved, the sites of processing in the brain and the species are different. The similarities suggest that brains follow certain general rules for information processing that are common to different sensory systems and species."[393] So it is scarcely surprising that the avoidance of jamming has not only evolved independently in the mormyriforms and gymnotiforms,[394] but most likely twice in the latter group.[395] The systems are not identical, not least because the computational center for comparisons of the timing of the electrical signals is located in the hindbrain of mormyriforms, but in the midbrain of their Neotropical counterparts.[396] Yet as Masashi Kawasaki notes, "Despite this difference, the internal organization of the phase comparison circuitry within the structures is strikingly similar."[397] The point of JAR, of course, is to prevent interference between two adjacent fish, so it is equally striking that the algorithm each group employs to compute the necessary shift in frequency is identical.[398] Not only that but the sensitivity of the systems is extraordinary, with the hyperacuity evident in both terms of timings (in the order of nanoseconds) and minuscule differences in frequencies.[399]

That the JARs are so effective is predictable given that these fish live in worlds where interference in electrical signals, be it for effective navigation[400] or social communication,[401] is hardly likely to be helpful. The tables can, however, be turned, as in a gymnotiform that employs jamming in aggressive interactions.[402] These signals can also be turned to a cooperative use, as in the rather remarkable case of group hunting by small packs of mormyriforms in Lake Malawi. This study did not find specific evidence for cooperation in flushing out their prey (p. 289), but as Matthew Arnegard and Bruce Carlson[403] point out, this activity points toward an "unexpected behavioral richness."[404] This chimes with the link between the impressive diversification of the mormyriforms, a process whereby an ability to discriminate different signals will be paramount, and the relative size of that part of the brain that engages in communication.[405] While among the fish the encephalization of the sharks is the most striking (p. 253), the gymnotiforms are certainly above average.[406] When we turn to mormyriforms we are almost in for a shock when dissection reveals an astonishingly large and complex cerebellum.[407] Is it so surprising, then, that some mormyriforms even engage in play (p. 285)?

It would be easy to dismiss these extraordinary fish as just one more arcane example of evolution, and one that from our perspective at least is entirely alien. Yet not only do we find striking parallels to other sensory systems, not least in terms of their extreme sensitivities and the still mysterious methods of neuronal computation, but the world of electroreception not only extends beyond the fish but even to the mammals. Given its extraordinary biology one could almost guess the mammal in question is the platypus. On its iconic bill there is a tactile array (p. 132) of electroreceptors,[408] a system that finds a series of striking convergences with the paddlefish.[409] These similarities embrace not only the electroreceptors but, as John Pettigrew and Lon Wilkens note, "Given the similar orientation of the rostrum and bill, it is perhaps not surprising that both paddlefish and platypus have evolved a short-latency head saccade."[410] If a platypus can do it, then why not a dolphin? Sure enough, on the rostrum of the Guiana dolphin we discover ampullary electroreceptors.[411] Not only do platypus (and dolphin) possess analogous electric antenthese weakly electric teleosts and the rays. Not only do the latter also employ undulatory fins, but as the pioneer of electroreception, Hans Lissman[368] notes in each case that the body is more or less rigid, an advantage if the electroreceptors are to remain in alignment with the electric field.

Using electroreception to locate prey is by no means the prerogative of the weakly electric teleosts, but has evolved independently in the paddlefish whose striking rostrum serves as an electric antenna to detect zooplankton.[369] The sensitivity of these systems is extraordinary, and it is customary to infer that at least the weakly electric teleosts form some sort of "electric image."[370] Even though the differences with vision, not least the inability to focus the electrical "image," remain self-evident, in at least a mormyriform the fish can distinguish an impressive variety of shapes.[371] Gerhard von der Emde and Steffen Fetz remark how "even in complete darkness, they can perceive parameters such as the volume, size, 3-D shape, contour, material and possibly the orientation of an object."[372] Even when the object is presented as a wire outline, recognition is possible. It is conjectural, of course, as to what the fish is actually "seeing," and as von der Emde and Fetz also comment these mormyriform "probably have to carry out some complex neural computations in order to acquire information about object shape."[373]

Quite so. Perhaps what Christian Graff and colleagues[374] term "electroperception"[375] has closer parallels to vision (and other sensory modalities) than might be expected. As enthusiasts for convergence we can hardly be surprised with the identification in both gymnotiforms[376] and mormyriforms[377] of specific regions richly endowed with the tuberous electroreceptors—that is to say, an electric fovea. These invite immediate comparisons with not only visual fovea,[378] but convergent equivalents such as the tactile fovea of

the star-nosed mole (p. 132).[379] An apparent curiosity is that in these fish there appear to be two fovea. Here, however, they are not alone. This is because various birds (including pigeons[380] and independently diurnal raptors[381]) and anolid lizards[382] also possess a pair of fovea (one more or less central, the other [or temporal] to one side). As Katherine Fite and Bradford Lister note, "Despite major differences in evolutionary history, taxonomy, and ecology, *Anolis* lizards and diurnal birds of prey appear to provide an unusual example of convergent evolution with respect to bifoveal vision."[383] In at least the case of the birds their function appears to be to dedicate the temporal fovea for close-up work pecking food, while the central fovea keeps an eye on more distant threats.[384] A direct parallel exists among such mormyriforms as *Gnathonemus*, because one of the electric fovea is located on the nasal region while the other is found on the prominent anterior "chin," appropriately named the Schnauzenorgan.[385] Like the birds it appears that each fovea has a distinct function: the nasal fovea to confer a long-distance and wide-angle acuity, while that on the Schnauzenorgan is employed in foraging.[386] The latter structure can be moved at remarkably high speeds,[387] and its shape serves to funnel the electrical field so that, in the words of Roland Pusch and colleagues, it acts "as an electric searchlight."[388] Such rapid movements of a foveal surface recall, of course, the situation in the star-nosed mole (p. 132), and Gerhard von der Emde[389] notes how this movement occurs "in a saccade-like manner while the fish moves forward."[390]

In terms of convergence one could also draw attention to other sensory fovea, not least the auditory equivalent of bats (p. 138). They and the weakly electric fish are, however, bound by an even more striking convergence in the form of the jamming avoidance response (JAR). As Catherine Carr[391] aptly observes, "Timing is everything,"[392] be it in physiologically similar, the tuberous electro-receptors found in Mormyriformes and Gymnotiformes are anatomically quite distinct . . . and represent another extraordinary example of convergence."[345] Nor does the story end here. Gymnotiforms are closely related to the catfish (or siluriforms), and conceivably their common ancestor was already electrorecep-tive.[346] However, as Alves-Gomes points out, it is not unreasonable to assume that among the catfish "there were four independent evolutionary events of electric organ appear-ance."[347] Catfish possess ampullary electrore-ceptors,[348] albeit of considerable diversity,[349] yet a blind South American catfish has inde-pendently evolved tuberous electroreceptors. Most likely these are to assist detection of its prey, ironically the gymnotiforms.[350] The same grisly and convergent fate awaits those mormyriforms that broadcast their presence to the African sharptooth catfish.[351]

Just as gymnotiforms have electric cous-ins, so do the mormyriforms in the guise of the African knife-fish (xenomystids),[352] again complete with ampullary receptors.[353] Curiously their Asian counterparts lack this capacity,[354] although why it was discarded[355] seems no clearer than why the first teleosts dispensed with their electroreception. The repeated invention of electroreception among the teleosts has, however, allowed the reoccu-pation of an extraordinary sensory domain. Yet Theodore Bullock and colleagues muse that, with respect to the crossing of this evolu-tionary threshold, "Remarkably, there are no clearly borderline or transitional fishes, with respect to these [electrogenic] characters, physiologically or anatomically,"[356] be it in terms of either the generation or detection of the electric signals or indeed in the necessary modifications of the brain (notably with the emergence of the all-important region known as the electrosensory lateral-line lobe).

The many striking convergences between the mormyriforms and gymnotiforms, which

extend beyond their shared electrosensory prowess,[357] do not mean that the systems are identical.[358] For example, in both groups a region of the midbrain (known as the torus circularis) houses electrosensory neurons, but as Thomas Finger and colleagues note, this area "is laminated in S [Siluriformes] and G [Gymnotiformes] but has a nuclear organization in M [Mormyriformes]."[359] This is just one of some seventeen differences[360] (in seven categories) that they identify, but these workers not only compile an even more impressive list of similarities[361] but remark, "One might expect that the independent evo-lution of electroreception in different teleost lineages would result in many significant dif-ferences. . . . Instead, however, one is struck by the extraordinary similarity of these [elec-trosensory] systems in diverse teleosts."[362]

This similarity has allowed either group of fish to enter a sensory world of remark-able complexity. Generation (by the electric organ) and interpretation (in the brain) of the electric field not only allows navigation in total darkness,[363] but in different species of gymnotiform the signals are tuned to local environmental factors such as flow regime and modes of foraging.[364] Feeding styles result not only in convergences between gymnotiform and mormyriforms in terms of a distinctive grasp-suction mechanism,[365] but these weakly electric fish are also renowned for their maneuverability.[366] This capacity largely depends on elongate fins (dorsal in mormyriforms and ventral in gymnotiforms) that allows them to swim forward or back-ward with equal facility. In some gymnoti-forms this is put to good effect because by propelling itself backward the potential prey can first be detected by the posteriorly located electric organ before the mouth comes within striking distance.[367] Propulsion using elon-gate fins is, of course, widespread, but in the context of electrolocation and body form we find an unexpected convergence between behaviors. These include not only hunting and courtship, but even the capacity for the embryos in their characteristic egg cases to "freeze" when avoiding detection.[324] Not only is this system multifunctional, but as Ryan Kempster and colleagues[325] note, through time it "has become more complex and highly specialized,"[326] perhaps conferring enhanced resolution. Certainly its sensitivity can be extraordinary, with measurements suggesting detection capacities in the range of nanovolts.[327] As Hansjochem Autrum[328] notes, this is equivalent to "the field that would be obtained by switching on a torch battery with its two poles 10,000 km apart."[329] Equally intriguing is that although the distinctive wings of the skates are automatically associated with their swimming,[330] in at least some forms their shape appears to be linked to the distribution of the electroreceptors and thus an enhanced tuning of the electric field.[331] That these systems can detect a thousand-millionth of a volt or help determine a shape as iconic as that of a skate[332] are—in their different ways—consistent with my thesis that evolution is precisely regulated.

Fish are well-equipped with various sensory modalities, not least their lateral-line system (p. 128), but in a number of groups electrosensitivity is clearly important, and in forms such as the extraordinary paddlefish[333] it is more or less vital. So as we clamber up the phylogeny of fish, from the relatively primitive elasmobranchs and chondrosteans (which include the paddlefish and sturgeon[334]), so I, the narrator, will enthrall you with a series of spellbinding accounts of how the more advanced bony fish (or teleosts) drew on this deep electric ancestry. Except they didn't. Rather at some point the teleosts threw away this capacity. Not only did they then have to reenter the electrical world, but this occurred at least twice and possibly four separate times.[335]

Strangest are the fish appropriately named stargazers, ambush predators that spend most of their time buried in sand. Some taxa possess an electric organ, which as usual is of muscular derivation. Its location, however, is somewhat odd. In a way comparable to the heater organ in scombrid fish (p. 80), the electric organ is based on extrinsic eye muscles,[336] and it remains far from clear what might be the function of the electrical signals. A role in prey capture seems unlikely[337] but conceivably they advertise to passing rays that they are crossing occupied territory.[338] Equally curious is the fact that while the stargazers can generate electrical pulses, they have no electroreceptors, so no means of detecting them.

In other teleosts, however, systems of both electrogeneration and electroreception have evolved. These are remarkable not only for their sophistication, but on account of a series of striking convergences that intriguingly emerged at much the same time in the Cretaceous.[339] Indeed the parallels between the mormyriform (or elephant-nose) fish of Africa and their electrical counterparts in the gymnotiforms of the Neotropics provide an example of convergent evolution[340] that easily rivals that of the camera eye of the octopus. As Carl Hopkins[341] notes, along with other electric fish, this "provides an excellent example of convergent evolution: not simply that electric fish evolved electric organs several times over, but that the mechanisms of generating complex discharges converge on different solutions to the same problem. . . . Not only are time codes important in the independently evolved groups, but the sensory processing of temporal cues has converged on similar computational algorithms."[342]

Unsurprisingly the mormyriforms and gymnotiforms have independently reinvented ampullary electroreceptors, but crucially an additional set of so-called tuberous receptors. In contrast to the former array these distinctive receptors are responsible for detecting high-frequency electrical signals.[343] As Jose Alves-Gomes[344] notes, "Despite being not only the release of a flash of light, but very locally a temperature rise of above 5,000°K.[305] Exactly this has been observed in the mantis shrimps, which independently have evolved cavitation,[306] not as a signal but as a component of their legendary ability to smash open the shells of their prey. This is achieved by using a lethal chelate hammer, which imposes an immense physical force, followed by the cavitation shock as the bubble generated by the strike rebound (combined with a complex flow pattern) is followed by its implosion.[307] One might well wonder how the mantis shrimp avoids damage to itself, and indeed the exoskeleton shows pitting that probably is the result of cavitation.[308] Cavitation is familiar to marine engineers because the repeated implosions readily erode a propeller and can seriously damage it. The exoskeleton of the mantis shrimp, however, has a distinctive microstructure that might alleviate damage, and the carapace is also discarded every few months and a new one secreted ready to take the punishment. The shock tactics of the mantis shrimps are one measure of their evolutionary success, and is clearly dependent on a sophisticated attacking appendage[309] that can function not only as a hammer but also as a spear. The force and speed of this mantis leg far exceeds what is possible by using muscles alone, and not surprisingly it employs a classic and convergent method of energy release, albeit by an unusual and superbly designed elastic spring.[310]

STUNNING CONVERGENCE

Echolocation may have at least some resonance with our sensory world, but some modalities seem almost impossible to imagine. Consider the many animals that live in an electrical world. We may live in a world permeated with electricity, but just beyond living memory were Western societies to which even a lightbulb or telephone were revelations. So, too, the stunning blow delivered by an electric eel[311] can come as quite a shock.[312] The fact remains, however, that electricity is intrinsic to all life. Bernd Kramer[313] reminds us, "Any living tissue generates electric fields associated with the regulation of its ionic balance. . . . In animals, electric fields also arise from normal nerve or muscle-cell activity."[314] Weak as these electric fields may be, once detectable they allow animals to enter a world of electrical perceptions, even if to us they represent a mysteriously orthogonal projection of more familiar sensory realms. So analogues of eyes and ears are required, but remarkably in some groups not only is there a capacity for electroreception[315] but also electrogeneration.

To inhabit an electric world one needs a conducting medium, so usually this means water. Until recently this was thought to be the prerogative of the vertebrates, but electroreception has evolved in several groups of invertebrates.[316] Most notable are freshwater crayfish[317] where it may serve as an alarm system[318] although how the electrical field is actually detected is more elusive.[319] It would not be surprising if other invertebrates independently have evolved electroreception, but to date the most striking examples are found among the fish. This capacity is very ancient,[320] so it is, for example, unsurprising to find evidence for electroreception not only in the lungfish[321] but also amphibians.[322]

Electroreception typically depends on specialized cells or ampullae. In the elasmobranchs—that is, the sharks and rays—they were recognized in the seventeenth century, hence the ampullae of Lorenzini, although their electroreceptive capacity is a far more recent discovery.[323] The scope of the electric world that the elasmobranchs inhabit is evident from the range of documented phins employing their biosonar[290] can easily discriminate a wide variety of objects. Also, the information garnered by one dolphin can be gleaned by another's eavesdropping.[291] This is all consistent with these animals living in an echolocatory world. While bats certainly have eyes, in the case of the cetaceans they generally appear to have excellent vision, and it is all the more intriguing that object recognition garnered from their echolocation can be combined with visual data.[292]

By no means identical, but still a series of striking similarities link bats and whales. Does that exhaust our catalogue of convergences? Not quite. As already noted in the bats, a particular protein (*prestin*) plays a key and convergent role in the amplification of sounds. When we turn to the echolocating toothed whales, and specifically the dolphins, to learn that they also use the *prestin* gene is unexpected. More remarkable, however, is that not only is there a striking molecular convergence[293] at a number of specific sites in the protein,[294] but convergences are also identified in two other auditory genes.[295] As Gareth Jones[296] remarks, these results provide "a stunning insight into how gene and protein sequences can be subject to convergent adaptive evolution in similar ways to morphological characters."[297] In the case of the bats and echolocating whales, just such a morphological convergence is found in the structure of the cochlea.[298] Although no cetacean appears to possess a specific acoustic fovea, in the case of the dolphin cochlea[299] the enormous number of ganglion cells points to superb pitch discrimination.[300] Moreover, just as in the bats, the structure of the cochlea (notably the basilar membrane) is central in preventing the system suffering from auditory overload.[301]

The Click of Evolution

Biological systems can provide guidance into technological challenges as diverse as adhesion (p. 82) or self-cleaning surfaces (p. 83). But it was not always so. Famously the inventors of sonar in World War II, a crucial element in the defeat of the Nazi U-boats, could only conclude there had been a serious breach in security when they heard biologists happily and publicly discussing the discovery of the evolutionary equivalent in the form of bat sonar. And it was only after this war that the convergent equivalent was identified among the cetaceans. Marine sonar now finds many uses, but its employment is often frustrated by a background roar from the seabed. Much of this noise[302] is due to myriads of snapping shrimps, flicking together their claws and in at least one species of alpheid shrimp the sound is produced in a remarkable way. The very abrupt juxtaposition of an extension (known as the plunger) on the articulating "finger" (the dactyl) against the rest of the claw produces a jet of water traveling at a high velocity. So fast does this jet move (at about 25 meters a second!) that the local pressure in the water drops dramatically. This enables a small bubble to form, in a physical process known as cavitation.[303] The bubble is, of course, short-lived, but its implosion is prompt and violent, releasing a very substantial amount of energy and so a loud noise. Hence the sonar-defeating racket generated by the snapping shrimp. This principle was applied in antisonar warfare whereby the Nazis developed their *Pillenwerfer*, a device used by the U-boat to operate clouds of bubbles that in a way analogous to "window" in radar aimed to baffle the pursuer.[304]

In cavitation the implosion of the collapsing bubble has other dramatic effects, including the dolphin itself is moving at high speed, are considerable. Just as the head is equipped with impressive sonar capabilities, so are the receivers. These are in the form of a variety of channels (including the mandible with its acoustic fat) that deliver the impulses via two pathways to the waiting ear.[262] There may also be further surprises. Even if dolphins cannot smile, all are familiar with the sort of grin imparted by strikingly regular teeth. Catching prey is a self-evident function,[263] but in principle this regularity might also impose a filter as the sound travels along the jaw, enhancing some frequencies and greatly diminishing others.[264] It is an intriguing suggestion, and on this basis one cannot help but wonder if other aquatic vertebrates with so-called homodont dentition, notably the ichthyosaurs, might have possessed some sort of echolocation?

If they ever did, it might have been rather rudimentary, but in the case of the odontocetes their echolocation is both extraordinarily sensitive and flexible in its use.[265] Its effectiveness in hunting prey is not in doubt, and at least a river dolphin[266] employs a stealth approach that gives it a margin of safety between the bursts of sonar.[267] So, too, they can eavesdrop. Given their complex social organization, which typically revolves around the rampantly polyphyletic fission-fusion system,[268] such eavesdropping is hardly surprising. This appears to play an important role in the remarkable synchronized swimming that dolphins can display,[269] and in the case of a tightly coordinated pod of rough-toothed dolphin, only one individual was actually echolocating.[270]

The echolocatory abilities of dolphins and bats arose completely independently, and given both the phylogenetic and ecological contexts, important differences separate the two groups.[271] Even so, although in one case it is a 20g animal nailing a moth in a forest while the other entails a 20 tonne cetacean tracking

a squid in a submarine canyon off Norway,[272] their respective systems of echolocation converge in various ways. Dolphins exert control on the width of their sonar beam[273] and compensate for the strength of the returning echo as they approach their prey,[274] as can bats.[275] Dolphins, like bats,[276] can engage in eavesdropping. Bats most likely have a wider range of acoustics, but just as they show convergence in their signals (p. 138), cetaceans as different as the dwarf sperm whale,[277] a riverine dolphin,[278] and various delphinids[279] (notably *Cephalorhynchus*[280]) also have independently evolved highly distinctive, narrowly tuned and high-frequency calls that are also relatively weak. The most likely factor in common is the generation of a sort of sonar that will not draw the very unwelcome attention of the killer whales.[281] An additional factor may be that the frequencies these whales employ correspond to a low noise "window" in the oceans.[282]

While such acoustic crypsis may defeat a potential predator, cetacean biosonar is of course central to find tiny targets in immense volumes of seawater. Here, too, we see convergences with the bats. This entails not only being able to adjust their hearing as they approach their prey,[283] but just as in the bats the final moments before contact with the insect prey are marked by a diagnostic buzz. The same acoustics have evolved independently in porpoises,[284] sperm whales,[285] and a beaked whale.[286] In the last case it is possible that this buzzing assists in persuading the scattered prey to clump as a school,[287] but Peter Madsen and colleagues[288] are surely correct when they remark, "It is striking to note how two very different groups of mammals in functional convergence have evolved the same basic acoustic behavior and movements . . . during echolocation and capture of prey in aquatic and aerial habitats."[289] Given these similarities, it is no more surprising that just as bats "see" through their ears, so dolof sound. One striking change involves the shrinking of the vestibular organ with its semicircular canals. Thus, in the blue whale they are little different in size from those of a human, and the fact that this reduction had already occurred among the primitive pakecetids indicates that these morphologically transitional forms were fully aquatic.[237] A possible explanation for this remarkable reduction in size is that it reflects the extraordinary agility and acrobatics of many whales; as Darlene Ketten[238] remarks, this arrangement might "be highly adaptive for cetaceans, permitting high-flying spins without 'space-sickness' side-effects."[239] In any event, as Fred Spoor and his colleagues note, by the time of the pakicetids, "the modification of the semicircular canal represented a crucial 'point of no return' event in early cetacean evolution."[240] This must apply to the rest of the hearing system, where the fossil record reveals many of the profound modifications that occurred.[241] The end result is an astonishing piece of biological engineering, especially among the odontocetes, whose biosonar still outstrips our technology.[242]

Toothed whales produce a variety of sounds, but the famous whistles (p. 263) are probably a fairly recent invention.[243] They are also of relatively low frequency,[244] whereas those used for sonar not only have much higher frequencies,[245] but can be very loud. The click of a sperm whale can reach a shattering 236 decibels and requires an enormous power input (c. 4 kW).[246] All this revolves around a radical reorganization of the vocalization system,[247] and one that is best known from the dolphins. How radical is evident from the fact that laryngeal production has been abandoned. This is hardly surprisingly given that the animal will be submerged for protracted intervals but still needs to engage in intense vocalizations. So no point in repeatedly opening the mouth and attempting to drown. The solution is to relocate sound production deep inside the body,

using the so-called phonic lips.[248] These are linked to a complex set of air-filled cavities[249] with vibrations being generated by a passing flow of air,[250] in a way not dissimilar to the reed of a musical instrument.[251]

If the lips are nasally located, how then do the sonar pulses leave the animal? Not, of course, via the mouth. Instead various parts of the head serve both to transmit the sonar beam and once returned as an echo not only to channel the sounds back to the ears,[252] but to achieve this in a way that allows an extraordinary sensitivity in hearing.[253] Pinna, invaluable for bats[254] (not to mention cats[255]), are out, and not only for hydrodynamic reasons. At first glance the streamlined head of most cetaceans hardly looks promising for sound reception. Look a little closer. In fact for both transmission of the sonar and the hearing of the returning echoes, so-called acoustic fats play a key role, not least in a structure located above the jaws that evidently serves as an acoustic lens.[256] Known as the melon, it is intriguing not only in terms of its function[257] but also composition. The lipids concerned (such as isovaleric acid) have very unusual compositions[258] that in other circumstances can be highly toxic, as is evident from the fact that even starving dolphins do not draw on this potential reservoir. It is widely supposed that the melon serves to refract and collimate[259] the sonar beam in a manner reminiscent of a condensing lens.[260] The ability to serve as an acoustic lens depends not only on its varying composition, but a series of tendons allow changes of shape that may assist propagation, not least to allow part of the posterior melon to act as a sort of megaphone.[261] Incidentally, that explains why, for all their intelligence and playfulness, dolphins cannot employ facial expressions such as smiling, because the relevant muscles have been radically redeployed.

Having projected the sound, what about the returning signal? The challenges of receiving a scatter of weaker pulses, while structures in the mid-brain[220] seem to be specifically modified.

The echolocatory capacities of these birds do not rival the bats. They are in any event largely connected to navigation, although a role in social vocalizations in at least the swiftlets is probably important.[221] Oilbirds and swiftlets can fly confidently in pitch-dark caves, and while the latter group recall the fruit bats in being mostly diurnal, like some of their relatives (such as the frogmouths, p. 105), the oilbirds pursue a crepuscular existence in search of nocturnal fruit. Their echolocation probably plays a role in these activities, but the visual capacity of these birds is also extraordinary.[222] Unsurprisingly the retina possesses relatively few cone cells, but the rod cells are not only remarkably small

(indeed they approach the theoretically minimal size), but also occur at an astonishing density (about a million per square millimeter—more than twice that in the retina of the falcons. Graham Martin and colleagues stress how the oilbirds are "pushing at the limits of sensitivity,"[223] but with the prize not going to a razor-sharp vision but a remarkable capacity to collect any available light.[224]

In the air and on the ground, and principally for nocturnal activities, the pulse of echolocation has evolved independently something like six or seven times. What of the oceans? It is intriguing to speculate whether any of the great marine reptiles in the Mesozoic evolved echolocation, but as far as the sea is concerned the convergence of this sensory modality is firmly anchored in the whales.

OCEAN ECHOES

Not, however, all whales. Although the mysticetes actively vocalize (and sing, p. 266), echolocation among the marine mammals[225] is the prerogative of the toothed whales (odontocetes). This capacity may be very ancient. By the late Oligocene (c. 25 Ma)[226] the oceans were probably astir with the clicks, buzzes, and creaks that are now an inescapable part of the din beneath the waves, but one to which maritime traffic may alarmingly be making an unwelcome contribution. Just as we end up shouting at a packed cocktail party (manifesting the so-called Lombard effect[227]), so today whales find it necessary to bolster their sonar output.[228] Like their terrestrial counterparts, cetacean echolocation depends, of course, on transmissions (vocalizations) and receivers (ultimately the inner ear), but a moment's thought will reveal the immense evolutionary challenges that had to be met in the transition from land back to the sea.

So well adapted are the whales to aquatic life[229] that even one of the greatest biolo-

gists of the twentieth century, G.G. Simpson, despaired of finding the correct part of the mammalian tree to attach them.[230] He need not have worried, but it still came as a considerable surprise that among their close relatives are cows and, more precisely, a fellow artiodactyl in the form of the hippopotamus.[231] It may be no coincidence that, although hippos do not echolocate, they are able to make sounds under water.[232] Not only is the fossil record of the so-called cetartiodactyls good enough to be integrated with the all-important molecular data,[233] but one can obtain insights of how in the Eocene these animals began to wade into the water[234] before embarking on a series of transformations[235] into fully oceanic animals that not only completely remolded the body but were achieved in a remarkably short time.[236]

Along with the loss of the hindlimbs and evolution of flippers, there are also profound changes in the hearing system that reflect the very different auditory environment that water presents, not least the far faster speed if his appeal to "an accidental non-functional mishap"[199] reveals a reluctance to concede the true nature of the evolutionary landscape. Far from being an accident, is not evolution once again in the grip of a sort of "attractor"? This convergence between horseshoe and mustached bats extends beyond an acoustic fovea because in the brain there is also a dichotomous arrangement of the auditory cortex for, respectively, processing the FM and CF signals as well as a major dedication of cortical tissue to the interpretation of these frequencies.[200] Not only is the frequency range in these bats extraordinarily narrow, but a key aspect of the response is a so-called Doppler shift compensation whereby the bat can adjust for changes in the frequency of the echolocatory pulses as it closes in at high speed on the luckless insect. This, too, is convergent in horseshoe and mustache bats, and as significantly Gerd Schuller and George Pollak remark that "the acoustic system of [horseshoe bats] bears a marked resemblance to the visual system: eye movements (i.e. echoes) focused on the [acoustic] fovea,"[201] thus achieving a constant and precise tracking.

This observation begs the question of just how close the analogy is between echolocation and vision. The bats not only display extraordinary sensitivities, but by combining temporal and spectral responses it seems reasonable to infer that the acoustic image is for all intents and purposes interpreting a spatial world.[202] Such is evident from evidence that bats can evaluate different types of texture,[203] not to mention the extraordinary capacity of a Mexican bat to relocate its pup in vast roosts, which implies a capacity to memorize the topography of the pitch-dark cave.[204] Striking as well is the case of nectar-feeding bats (p. 203). Not only can they readily distinguish flowers at different stages of opening (with implications for nectar rewards),[205] but because the associated pollination syndrome[206] leads to the evolution of distinc-

tive flower types in at least the case of those with a characteristic bell shape, the bats are highly attuned to their recognition.[207] They are evidently assisted by the plant providing an acoustic mirror or beacon, either by the specific arrangement of the petals[208] or a dish-shaped leaf that is remarkably effective at returning the signal across a wide range.[209] This feature is echoed (so to speak) in the capacity of bats to discriminate even between two hollow bowls of identical depth and diameter, but respectively with a hemispherical and paraboloid shape.[210] In total darkness evidently the bats see.

So, too, do some birds.[211] Echolocation is best known in the so-called oilbirds of South America,[212] but it has evolved independently in the swiftlets. These birds have a very wide distribution across Asia, with many species inhabiting the deep recesses of caves with their saliva-encrusted nests forming the basis for the ever-popular bird's nest soup. In a way that is reminiscent of the bats, the molecular phylogenies for the swiftlets[213] are finely balanced between identifying a single origin of echolocation that was followed subsequently by several losses, as against this sensory capacity evolving at least twice, but the evidence seems to be pointing in favor of the latter possibility.[214] Meanwhile their Brazilian counterparts may also employ echolocation,[215] and if confirmed would presumably be independent of the Asian swiftlets. In both the oilbirds and swiftlets the auditory clicks,[216] which are audible to humans, are not made by the tongue (as in the fruit bats, pp. 137–38), but are generated in the syrinx, the source of song (p. 265). Nevertheless, despite the considerable similarities in sound production, the actual arrangements of the syrinx in the oilbird[217] and swiftlet[218] show significant differences. Although the powers of echolocation must have a neural counterpart, the details remain somewhat elusive, and neither the middle ear[219] nor relevant auditory the case in other echolocators such as the tenrecs and swiftlets). Although the echo-location of the rousettids is often regarded as somewhat rudimentary, these fruit bats show capacities comparable to other bats in terms of avoiding obstacles[172] and target-approach.[173] Their highly mobile ears may also play an important role.[174] In this group, vision remains important, but more sur-prisingly, notwithstanding their crepuscular existence, the same applies to many of their microchiropteran counterparts.[175]

"Blind as a bat" misses the point, but their echolocation is justly admired. Most often the image that comes to mind is their agile swooping in even highly cluttered environ-ments, but echolocation is also used in com-munication. The vocalizations of horseshoe bats show a remarkably rich repertoire of sounds,[176] while in noctule bats the capacity to eavesdrop on their neighbors is impor-tant when trying to find the safety of a tree hole in which to roost.[177] When fish-catching bats find themselves on a collision course, one or both emits a honk that is followed by immediate evasive action.[178] In dense swarms of bats, such as those that pour forth from cave systems, air traffic control is at a premium but will be compromised if the echolocatory signals jam each other.[179] The solution is to employ a jamming avoidance response (JAR)[180], whereby frequencies are not only shifted,[181] but with an extremely rapid[182] response time.[183] Midair collisions are best avoided, but the primary function of bat echolocation is to catch nocturnal prey, not least moths. The military mind might suggest broadcasting warning signals, but wouldn't it be better to simply jam the bat sonar? At least one tiger moth has learned to do just that.[184] By emitting protracted bursts of ultrasonic clicks, the moth disorientates the bat's sonar in those closing moments that would otherwise end in a score of Bat 1, Moth 0.[185]

Echolocation may be central to their hunt-ing (and sometimes its frustration), but it would be mistaken to imagine among the bats a one-size-fits-all approach. Once again, con-vergences crowd forward. Bats that occupy similar niches but live in very different parts of the world, such as Neotropical myotisids and European pipistrellids, show convergen-ces[185] in their echolocatory repertoires.[187] In a helpful overview, Gareth Jones and March Holderied[188] note "how signal design in bats provides remarkable examples of convergent evolution, with distantly related taxa evolving similar signal designs independently in the face of similar environmental challenges."[189] The opportunities for convergence are also widened because several types of echolo-cation have evolved,[190] revolving around varying employment of constant frequency (CF) versus frequency modulation (FM). Most striking are those bats that employ CF echolocation. Arguably this finds an acme in terms of convergences among the Old World horseshoe bats (rhinolophids) with an aston-ishingly sophisticated echolocation. Their strange-looking noseleaves serve to focus the emissions,[191] which are broadcast on a very narrow frequency and have a striking evolu-tionary convergence with a New World mor-moopid, the greater mustached bat.[192] Because the frequencies are so sharply tuned, a signif-icant section of the cochlea[193] is dedicated to their specific reception. In other words, we have an acoustic fovea, which has evolved independently in the horseshoe bats[194] and mustached bat,[195] and so is analogous to the visual fovea.[196]

Gerhard Neuweiler identifies this as "one of the most striking examples of convergent evolution,"[197] but in his discussion he almost seems to lose his nerve. So he doubts that "evolutionary driving forces"[198] are capable of producing such a superb biological system. As a leading authority on bats he may well be correct, yet one cannot help but wonder purpose, but they are modified mouthparts.[152] And to whom might the ability to hear ultrasound literally be a matter of life or death? Why, those that need to escape animals that have evolved (convergently, of course) the capacity for echolocation. Perhaps the best trick is to jam the incoming signal, but given that some moths are laden with toxic chemicals, another solution is for potentially palatable relatives to emit much the same set of ultrasonic signals in a form of acoustic mimicry.[153] And who might be so misled? I speak of those masters of nocturnality, the bats.

SHOUTING TO SEE

Fossil bats from as far back as the Eocene (c. 50 Ma) show compelling evidence, notably in the form of an enlarged cochlea, for fully fledged echolocation.[154] Bats may have learned to fly before they exchanged vision (perhaps also assisted by hearing and smell) for echolocation,[155] but this capacity cannot be ruled out in even the earliest-known bats.[156] Any idea, however, that either flight itself was the necessary prerequisite for echolocation, or that this sensory modality involved some significant evolutionary hurdle, can be safely dismissed. First, although not widely appreciated, a number of other terrestrial mammals have independently evolved echolocation. These include the foaming tenrecs (p. 39) from Madagascar,[157] as well as a variety of shrews.[158] Given that the ecology of these animals is probably not far removed from that of the proto-bats, in principle there seems no reason for echolocation not to have preceded flight. Second, despite the intensity of high-frequency sounds[159] generated (incidentally necessitating means of avoiding self-deafening[160]), echolocation is evidently metabolically cheap and presents no serious barrier to its emergence.[161] Echolocation, of course, isn't going to work without ears that have the capacity to deal with ultrasound (in whales it is a different story, p. 141). This is evident not only in terms of prominent pinnae but also with respect to the cochlea,[162] where the sounds impinging on the hairs (stereocilia) of the basilar membrane are subject to transduction (so the electrical signals can be transmitted to the auditory centers of the brain). Part of the transduction process involved with the amplification of the sound draws on a specific protein (an anion channel, prestin[163]), and importantly genomic changes have arisen independently[164] among the echolocating bats.[165]

Just as the molecular phylogeny of the myotisid bats revealed what were described as deterministic ecomorphs (p. 39), comparable investigations that address the bats as a whole uncover convergences,[166] not least in terms of echolocation.[167] Despite advances, very seldom can a given phylogeny be regarded as watertight. With the bats, currently two alternatives need to be entertained. As is often the case the choice is finely balanced,[168] but either way convergence still wins. One proposal is that echolocation evolved twice, specifically in the group that includes the horseshoe bats (rhinolophids) and a large assemblage of other microbats.[169] This phylogeny, however, does not automatically rule out a single origin for echolocation, and in this case it would have been subsequently lost in the group that ultimately evolved into the megachiropteran fruit bats,[170] only to be reinvented in the genus Rousettus. One reason to think this approach could have been the pathway is that in contrast to the microchiropterans, which employ the larynx, in the rousettid fruit bats the ultrasound is produced by clicking the tongue[171] (as is also hearing.[137] Each is an object lesson in the evolution of an important novelty,[138] but each also recapitulates very much the same story across the insects. Ears evolved repeatedly, but in case after case we see the recruitment of the chordotonal organ, a thinning of the cuticle to provide an "eardrum" (tympanum), and the development of an air-filled acoustic cavity, formed by a judicious enlargement of the tracheal system that is otherwise employed for respiration. Not only do we find tympanic ears in many different insects[139] and many different parts of the body, but in locations such as the legs and wings that might in principle be compromised by vibrations,[140] but sometimes in the most unexpected of places—such as the internal tympanic ears of the hooktip moths.[141]

Just as with the eye (pp. 99–100), the fact that the tympanic ears repeatedly draw on the same developmental tool kit may appear to beg the question of what we mean by convergence. This potential objection appears to achieve even greater force when we consider the developmental biology of ears in general and especially the genes employed. In many ways this discussion exemplifies an important tension within evolutionary biology: Are my ears *really* the same as a mosquito's Johnston's organ? In part the answer is one of perspective. Certainly it is evident there are important developmental parallels between the ears of insects and vertebrates.[142] The story involves many genes, but among the most important is one known as *atonal*. But *atonal* has considerable versatility; not only is it employed in the developing ear but in the fruit fly also for olfaction[143] and vision.[144] So really the question is pushed to a much deeper level of evolution, because if an ear must draw on neurosensitive cilia (as indeed do olfactory systems and many eyes), then it is scarcely surprising to find molecular conservation.[145] Convergence never said that each and every building block has to be rein-

vented, simply that as and when you want to reinvent an evolutionary wheel, you might as well employ the same master craftsmen. In point of fact, this redeployment is the norm: genes are repeatedly recruited, and it is not always obvious which role actually came first.

At a more macroscopic level it is scarcely surprising if ostensibly entirely different organs draw on an inherited cellular architecture. Thus, in the scolopidial cells of the insect chordotonal organ, a key element in the process of transduction involves a ciliary unit. In addition, in the vertebrate cochlea there is an array of ciliated neurosensory cells. These cilia, however, are just one component of an exceedingly complex system that draws on a specific anatomy and molecular machinery to enable highly effective hearing. As Susanne Bechstedt and Jonathan Howard[146] point out, given the obvious contrasts between insect and vertebrate ears, one might expect them to "operate by very different molecular and biophysical mechanisms,"[147] but as they note, "the *Drosophila* ear contains a transduction apparatus that functions almost identically to that in hair cells"[148] of vertebrates. At a quite remarkable level of detail, either group has ended up with very much the same solution.[149] In a comparable way, in a katydid insect the ear might be minute, but the manner in which the airborne sounds are transmitted to the fluid-filled "inner" ear via a lever system provides a striking analogy to the vertebrates.[150] The series of exquisite (if not excruciating) descriptions of such biologically complex structures may be the bread and butter for the specialist, but perhaps they run the risk of making us overlook just how extraordinary these products of evolution actually are.[151]

The sensitivities of this and other types of insect ear have already been stressed, as should their range of aural perceptions. Particularly striking are those insects that can detect ultrasound. In the hawkmoths (pp. 200–201), for example, the ears are employed for just such a dent evolution of a low-frequency acoustic fovea?[120] Nor are low-frequency sounds unimportant above ground, as is apparent from the series of convergences, including a specialized cochlea, as well as enormous tympanic membranes and cavities, that link the nocturnal desert rodents of the Old and New Worlds.[121]

The auditory worlds of many of these mammals would surely seem rather alien to us, but not nearly as much as those of the insects where ears have evolved at least twenty times. To find an ear in an insect, however, one can look almost anywhere: legs, abdomen, antennae, mouthparts, and even wings[122] are all sites for audition. Despite this extraordinary diversity, all draw upon the common agency of a particular type of mechanoreceptor, the chordotonal organ.[123] This, it may be recalled, is also the basis for an infrared sensor in insects (p. 118), but in the case of the ear the evolutionary antecedents are even clearer because more generally the chordotonal organs play a vital role of assessing the degree of deformation and stretching of the cuticular exoskeleton. It was, therefore, an easy evolutionary step to begin to respond to external vibrations.[124]

A small step maybe, but now an insect can enter a world full of sounds, of which the most important is the detection of friend, foe, or lover. Consider, for example, the extraordinary acoustic acuity of the male mosquito. In this insect and many other dipterans, the ear is typically found in the antenna, in the form of the so-called Johnston's organ. As ever the chordotonal organ is in attendance, and the sensory units (known as the scolopedial cells), which act as the transductors of sound into nervous impulse, can total up to thirty thousand.[125] Such a number implies an astonishing sensitivity; in terms of equivalences the minute deflection of the antenna necessary to trigger a response is equivalent to deflecting the top of the Eiffel Tower by less

than a millimeter.[126] Such sensitivity is essential because the male must be attuned to the sound of the female wingbeats, which given their tiny size requires the best ear possible in order to detect a most welcome approach. The number of sensory cells rivals those found in our cochlea, and a number of striking parallels exist. In essence, acute hearing not only demands tuning to frequencies but also so-called nonlinear responses whereby signals can be amplified so that the very faint but important sounds can be more readily identified. As in our cochlea, so in the Johnston's organ[127] of the male mosquito.[128]

"Acute as a bug's ear" as Ronald Hoy[129] aptly puts it in his accessible introduction to hearing in insects, and in this context another dipteran again stands out: the tachinid fly known as *Ormia*. This insect displays astonishingly acute hearing in being able to judge both the direction and distance of a sound source,[130] which is, in the words of Daniel Robert and Martin Göpfert,[131] "strikingly reminiscent of the barn owl's perception of acoustic targets."[132] In contrast to the mosquito, in this fly the sensitivity is the prerogative of the female; she unerringly navigates toward the sound of a male cricket, who unwittingly is broadcasting his sexual music, in order to lay her eggs that will subsequently parasitize the cricket. In this context an ear in the guise of a Johnston's organ will not suffice, and as Daniel Robert and colleagues[133] remark, "for a fly to act like a cricket, it must hear like one."[134] To achieve this, the ormiid fly has independently evolved a tympanic ear.[135]

In doing so it is in good company because this type of ear has evolved independently many other times and across at least seven orders of insects,[136] including of course those doomed crickets. Nor is this fly alone, because independently in another group, the sarcophagids (and notably *Emblemasoma*), the same functional solution has been arrived at and again confers precise directional adaptive pressure to improve the sensitivity of hearing has led to recurrent solutions within the cochlea.[101] One of the most striking is the evolution of specialized hair cells: one series has the primary role in actual hearing along the available frequencies, while the other set is central in the process of amplification. Such an arrangement has evolved quite independently in the birds and mammals. It is an important test case for convergence because at first glance the configurations have little in common. Look a little closer.[102] Within the mammalian cochlea we distinguish a single row of inner hair cells and several rows of outer hair cells, whereas in birds the hair cells divide into tall and short. Even though the latter do not occur as regular rows, patently they have the same functions (respectively, of hearing and amplification) and accordingly important differences in their styles of innervation.[103] As Geoffrey Manley and colleagues write, "Considering the independent evolutionary origin of these two vertebrate classes from different groups of reptile ancestors, it is remarkable to find indications of a clear functional parallel between avian and mammalian hair-cell populations. The parallel evolution of such hair-cell populations suggests that there are fundamental properties of hair cells which lend themselves to a particular division of labour."[104]

That birds have excellent hearing is scarcely surprising given their songs (pp. 264–66), but in terms of sheer acuity the owls have few rivals.[105] Just as in some bats, they possess an auditory fovea that confers intense sensitivity at the all-important higher frequencies.[106] Not only that, but a comparison of the physiological responses in the cochlear nucleus in the brain of an owl shows many parallels to those of mammals.[107] Christine Köppl and Catherine Carr conclude, "Considering the fundamental morphological differences between the cochlear nuclei of birds and mammals, the great similarity of response types was

somewhat surprising,"[108] and indeed these contrasts in brain structure pose equally compelling conundra with respect to other convergences, not least those of vocalizations (p. 265).

As a boy I could hear the high-frequency cries of bats, but no longer. Nor at any time would I have been able to catch the infrasonic communications between elephants.[109] Collectively, therefore, mammals register a vast range of acoustic frequencies,[110] but central to their success is the transmission mechanism across the middle ear in the form of three bones. While the stapes harks back to the dawn of tetrapod evolution, the recruitment of the incus and malleus from the lower jaw of our reptilian ancestors is one of the iconic examples of evolutionary transformation. It has been described as "a quirk of phylogeny,"[111] and while it is a complex story that revolves around the transition from the therapsid reptiles, it may well have involved convergences.[112] Not only is there evidence that the iconic arrangement of the three ear bones rose independently in the marsupials and placentals (that is, the therians) and monotremes,[113] but as far as the Mesozoic mammals are concerned, Zhe-Xi Luo[114] notes[115] that is "beyond doubt that the last step in the transformation of the definitive mammalian middle ear occurred homoplastically in some Mesozoic lineages."[116] Convergences also emerge among the mammals. The independent evolution of high-frequency biosonar in bats and cetaceans is addressed below, but what about the other end of the spectrum? Within the burrow systems constructed by mammals (especially rodents), hearing is necessarily shifted toward the detection[117] of low-frequency sounds.[118] Among the most striking of these convergences are the recurrent adaptations for hearing the low-frequency sounds that characterize the subterranean world of the mole-rats.[119] Are we then surprised that this convergence extends to the indepenpair of incisors can move independently of one another. Even more sensational is that almost a third of the somatosensory cortex is dedicated to the input from these incisors, making it very likely that these animals enjoy exquisite tactility via their teeth.[87]

A Balancing Act

Animals may need to see, hear, and touch or draw on a rich series of other sensory modalities. As any disorientated pilot flying in thick clouds will confirm, they also need instruments to say which way is up. Gravity rules, and there is an essential need for precise balance. Again and again this has been achieved by using tiny grains (or statoliths) whose relative movements convey the vital information. In the cephalopods we find that the statoliths are housed in a small sac known as the statocyst. Linked to the eyes, their movement (in what is termed the oculomotor reflex) is strikingly similar in its arrangement to the vertebrates,[88] including ourselves. Shake your head from side to side, and unless there is something wrong with your inner ear—notably a fault in the fluid-filled semicircular canals—you will still easily read this page.[89] In other words, the angular acceleration of your head is being compensated. Many cephalopods are fast moving, so a capacity to deal with angular acceleration is obviously important and results in a remarkable convergence. This depends on the complex arrangement of the statocyst that serves to direct the fluid flow[90] in much the same way as the semicircular canals.[91] As P.R. Stephens and J.Z. Young note, "It is clear that these devices for canalising flow in the statocyst have the same function as the semicircular canals of vertebrates. . . . We must, therefore, add one more to the long list of parallels between the functional organisation of cephalopods and vertebrates."[92]

Oddly cephalopods have never evolved ears, which is all the more curious because the latter are rampantly convergent. As Martin Moynihan[93] remarks, "Their deafness is so remarkable that it needs to be explained in functional and evolutionary terms."[94] Ingeniously he develops an idea that deafness protects the cephalopods from being stunned by staggeringly noisy pulses emitted by whales (p. 141). The snag here is that cephalopods without ears evolved long before the oceans filled with echolocators.[95] Nor does this necessarily mean cephalopods are actually deaf.[96] Various lines of evidence point to sound perception at lower frequencies. Mediated by the statocysts,[97] the net result is that, despite the obvious differences, cephalopods hear in much the same way as fish.[98] This point could be of vital importance because however acute cephalopod eyes might be in murky or nocturnal conditions, not to mention the inky blackness of the abyss, failure to detect the approaching enemy could result in disaster.[99]

Hearing Evolution

Like any other sensory modality, the ability to hear has evolved many times independently. General sensitivity to vibrations is one thing, but the prerequisite for hearing prey and chatting (that is, impedance-matching hearing) involves a tympanic ear. This has evolved at least three times among extant vertebrates (in lizards, birds, and mammals), and perhaps even more strikingly in Permian representatives of the so-called parareptiles.[100] So, too, impressions that calculations suggest the star-nosed mole is capable of scanning an extraordinarily large area in a short period of time.

As the story of "A nose that looks like a hand and acts like an eye"[70] first broke, Kenneth Catania correctly emphasized "The many surprising parallels"[71] with the visual systems. These have now transpired to be even more exact than perhaps was first anticipated. Among the most striking of these discoveries is the recognition of a tactile fovea[72] on the so-called tentacle 11 (there are twenty-two in total), and here the size of the receptive fields is smaller than those of the other tentacles. As significantly, just as our eyes can only encompass a small part of the visual field and so must jerk rapidly to provide all-around vision, the equivalent jerks (or saccades) in the star-nosed mole[73] are, as Kenneth Catania and Fiona Remple note, "surprisingly similar to primate eye movements."[74] This tactile fovea also shows a corresponding increase in dedicated brain tissue, and these authors tellingly remark on "a remarkable degree of convergence in the design of sensory systems that handle large volumes of complex information. Visual, auditory, and somatosensory specialists have all hit upon the same solution of dividing the sensory system into a small high resolution area for detailed analysis and a larger, low resolution area for scanning a sensory scene."[75] A final point deserves emphasis: there is a good argument that the star-nosed mole has reached the limits of design, at least as far as the processing of the tactile information is concerned. Just as the eye and ear have reached their respective physical limits, so the tactile world of the star-nosed mole simply cannot work any faster.[76]

The exquisite sensitivity of the star-nosed mole depends on receptors that are housed in the so-called Eimer's organs. These have a characteristic domelike exterior and are found in nearly all the moles. The overall structure of an Eimer organ is quite complex, yet it is strongly convergent[77] with the mechanoreceptors possessed by the monotreme mammals, especially the duck-billed platypus.[78] Both this animal and the related echidna are intensely tactile.[79] Interestingly, just as in the star-nosed mole, so for the echidna enhanced tactility also characterizes the forepaws,[80] which play a central role in digging for ants. In the duck-billed platypus we see other parallels. These include an analogous tactile fovea, once again equipped with remarkably small receptive fields, as well as a striking allocation of cortical tissue in the brain for this tactile processing.[81] In the words of Mark Rowe and colleagues, the mechanoreceptors in the bill of the platypus and the star-nose of the eponymous mole show a "remarkable resemblance."[82] These workers, however, also make a point of the greatest importance in reminding us that labeling the monotremes as "primitive" may have a certain applicability (e.g., it is a mammal but it still lays eggs), but that this has by no means constrained the possibilities of evolutionary exploration and sophistication, including not only the Eimer analogues but in the case of the echidna (which like the elephant [p. 131] evidently had an aquatic ancestry[83]) possession of a rather remarkable brain (pp. 258–59).

The platypus and star-nosed mole (and indeed crocodiles[84]) are highly adept tactile hunters, but the mechanosensory worlds have further surprises to spring. Consider the naked mole rats. Most famous because of their independent evolution of eusociality, their subterranean life is one where eyes are more or less useless and the rows of sensory hairs along the otherwise naked body[85] evidently serve a very similar role to the more familiar whiskers (vibrissae).[86] Even more extraordinary, however, are their enormous incisors. These are employed for a wide variety of tasks: digging, carrying the young, feeding, and so on, and rather remarkably the lower

Pacinian corpuscles, which are highly attuned to vibratory stimulation.[44] They are almost certainly convergent on the Herbst corpuscles.[45] These corpuscles are constructed like a minute onion, with successive membranes that enclose fluids that serve to transmit the pressure or vibration to the nerve cell.[46] These corpuscles are widespread: they (or something very similar) are found, for example, in the tips of the tongues of pangolin[47] anteaters[48] (p. 52). They also take a central role in the elephants, where seismic communication is very important, with their low-frequency "rumbles" being transmitted through the air and at a slower speed via the ground.[49] These sounds play an important role in the behavior of elephants,[50] not least in allowing them to discriminate very subtle differences in alarm calls.[51] It has long been suspected that the giant feet, which are effectively huge cushions with masses of fat and fibers of elastin (p. 16), provide highly sensitive structures,[52] and both the distribution of the Pacinian corpuscles and the manner in which the elephants position their feet strongly suggest that these remarkable animals are able to receive seismic communications[53] via these colossal pads.[54] The emblematic trunk,[55] which is highly sensitive to touch[56] and probably can also pick up vibrations, is equipped with somewhat unusual Pacinian corpuscles,[57] but not their counterparts in the form of the so-called Meissner corpuscles.[58]

Meissner corpuscles, however, play an important role in the tactility of the skin, and in the primates are a vital component for an arboreal life as branches are grasped and food felt. The focus of attraction is, of course, the hands,[59] although in such monkeys as the New World spider monkey the highly prehensile tail (p. 342n12) has a conspicuous smooth (glabrous) pad that is highly sensitive to touch.[60] Even so, in the spider monkey the hand is superior in tactility. Of particular interest, however, is that just as the eye has an area known as the fovea—which is ultrasensitive both on account of the number of photoreceptors and a corresponding dedication of cortical tissue in the brain—so as Joscelyn Hoffmann and colleagues note, the "motor control and magnified cortical representation of digits I-III . . . are consistent with the notion that monkey digits may function as a tactile eye, having a small behavioral focus, or fovea, at the center."[61] They suggest that such tactile sensitivity may be important in assessing the texture of fruit, and thus be an important complement to their trichromatic vision (pp. 114-15). As these workers emphasize, the potential for a "tactile foveation in anthropoids is intriguing."[62] Tactility also plays an important role in some marsupials, notably the didelphid *Chironectes*. In contrast to many of its relatives, which are skillful climbers (p. 69), this animal is not only an adept swimmer but seeks its prey with outstretched fingers that seem to be exceptionally tactile.[63] As Mark Hamrick also notes, "The radial arrangement of epidermal projections on the digital skin of *Chironectes* resembles somewhat the finger-like projections surrounding the snout of the star-nosed mole."[64] To these remarkable animals we now turn.

These animals are equipped with a series of prehensile and highly tactile tentacles that surround the nose. As is now well known,[65] the neural arrangement of these tentacles has a number of striking similarities to the eye,[66] but their tactility is extraordinarily sensitive[67] and can assess surface textures to a remarkable degree of resolution.[68] "Reading" the surfaces is not, however, achieved in the way a Braille script is brushed, but by a series of frequent (often more than ten per second) and exceedingly brief contacts. Given the constant contact with often rough surfaces, the sensitivity not only encompasses touch but also an array of pain sensors,[69] which no doubt reduce the dangers of general wear and tear. So fast and frequent are the tactile digging moles."[25] So successful is this technique, and one that is handed down generation to generation, that the world of the worm grunter is now hemmed in by permits. In the world of convergence the stomping of turtle feet[26] and paddling of herring gulls[27] achieve much the same result.[28]

TACTILE EVOLUTION

In considering apparently similar behavior by birds feeding on the shorelines, Niko Tinbergen drily observes, "There are no moles in the intertidal zone,"[29] yet seabeds are a hive of seismic activity with news of hunters and mates, neighbors and competitors, all traveling through the sediment.[30] Who, for example, could possibly mistake the diagnostic vibrational signal of a defecating lugworm, and to whom might it be of more than passing interest? The answer is such shorebirds (or waders) as the sand pipers (which belong to a group known as the scolopacids) whose probing for buried prey as they patrol intertidal mud-flats is a familiar sight. Given this type of feeding[31] it is unsurprising to learn that the tip of the elongate bill houses characteristic pits that house an abundance of mechanosensory cells.[32] These are the Herbst corpuscles,[33] and they serve to detect the pressure gradients within the sediment. Just the same arrangement is found in the ibises and most likely evolved independently.[34] Linked to this tactile capacity there is a corresponding enlargement in the part of the brain (PrV, short for principal sensory trigeminal nucleus) innervated by the trigeminal nerves. An increase, although less dramatic, has arisen independently[35] in some waterfowl and parrots.[36] The enormous trigeminal nerve in a now-extinct bizarre flightless duck (*Talpanas*) from Hawaii suggests it, too, may have largely lived in a tactile world.[37] As Andrew Iwaniuk and colleagues suggest, "*Talpanas* may be viewed as a duck that evolved into a kiwilike niche."[38]

A similar combination of inferred flightlessness and a reliance on nonvisual modalities also exemplifies the kiwi, with its employment of an elongate beak for tactile probing.[39] This New Zealand bird is a palaeognathan (see p. 208) and so ostensibly more primitive than the neognathans already mentioned (including the scolopacids), but among its other remarkable features is the size of its brain (p. 262). In the immediate context of mechanosensory convergences, however, it transpires that while olfaction was thought to be paramount in this nocturnal animal, in fact the kiwi has striking parallels to the elongate bills of the shorebirds, with again terminal pits housing the Herbst corpuscles. The similarity extends to the corresponding sensory regions within the brain, and as with the shorebirds, the PrV is enlarged.[40] Graham Martin and colleagues also note how the shift to an olfactory[41] and tactile capacity at the expense of vision "is markedly similar to the situation in nocturnal mammals that exploit the forest floor. That kiwi and mammals evolved to exploit these habitats quite independently provides evidence for convergent evolution in their sensory capacities."[42] In mammals, of course, facial tactility is usually associated with the vibrissae (whiskers), but the bristlelike feathers of a crevice-dwelling seabird, the appropriately named whiskered auklet, are evidently very useful in its underground life, and most likely a similar mechanosensory function has evolved in several other groups of birds equipped with facial bristles.[43]

In the mammals a vital component in sensing pressure changes depends on the so-called detection of underwater sounds? Somewhat surprisingly, cephalopods lack ears. Nevertheless not only have they jury-rigged a convergent substitute (p. 133), but independently they have evolved a strikingly convergent lateral-line organ.[10] The anatomy is well-documented, but unfortunately less is known of its function and versatility. If the many other convergences are anything to go by, then one can be pretty certain that this system will approach, if not rival, that seen in the vertebrates. As aquatic denizens, the cephalopods and fish do not stand alone. Intriguing analogues to the lateral-line system have been identified in several aquatic mammals, notably the manatees and seals,[11] as well as the semiaquatic moles known as the desmans.[12] Is that the end of this particular roster? No, because marine crustaceans known as the penaeids cleverly employ their antennae for just this purpose,[13] while the small but ferocious arrow-worms (the chaetognaths) may well use body hairs in a similar way.[14]

THE VIBRATIONS OF EVOLUTION

In one way or another all these hydrodynamic detection systems depend on the deflection of tiny hairs that trigger a nervous signal. When, like the fish, decapod crustaceans invade caves, eyes become useless, but a variety of hairlike structures (setae) spout from between the redundant lenses to confer a tactile (or chemosensory) function.[15] Such tactility reaches extraordinary heights in other arthropods, especially the insects and spiders. Here vibrational communication is both highly attuned and diverse in function. In the spiders, apart from their silk webs, vibrational exchanges play a crucial role in the courtship of the salticids[16] and wolf spiders.[17] The diversity of vibrational communication among the insects is also remarkable,[18] as the following three (of potentially very many) examples indicate. One type of caterpillar larvae can discriminate the vibrations made by inoffensive herbivores as against potentially lethal predators. If the latter is detected, then the caterpillar will lower itself to safety by a silk drag-line, extending itself still farther if the threat is identified as a flying wasp.[19] Second example: Parasitic wasps, before they start drilling through the wood with their zinc-hardened ovipositors (p. 66), need to locate their hapless prey. This is achieved by knocking, employing a modified antennal hammer, which has evolved multiple times.[20] Even more remarkable is the case of a termite. If it finds itself heavily contaminated with the spores of a fungal pathogen, which might spell doom to their agricultural system (pp. 186–87), it will vibrate a message of alarm from which the unaffected will flee.[21] The vibrational world of termites shows other versatilities, including chains of drumming soldiers to communicate alarm signals[22] and vibrational cues to assess the size of pieces of wood that will be shredded for food.[23]

The modality of sensory vibrations is likely far more widespread and convergently evolved than generally realized. What, for example, are we to make of the *International Festival of Worm Charming*, or indeed the whole topic of worm grunting? More than might meet the eye. In some parts of America, earthworms (needed for bait) are brought to the surface by the expedient of vibrating a wooden stake, which once driven into the ground is activated by sliding a piece of metal (say, a car suspension spring) across the top.[24] The earthworms have reason to be worried, because, as Ken Catania points out, the worm grunters (also known as fiddling and snoring, as well as charming) "unknowingly mimic

CHAPTER 12

(IN)TANGIBLE EVOLUTION

OUR WORLD IS flooded with photons and buzzes with tasty and smelly molecules. It also hums with vibrations, and to some animals electrical signals are an entirely familiar sensory modality. We humans might possess first-rate eyes and exquisitely tuned ears, but we only know about such sensory modalities as echolocation or electroreception via our scientific instrumentation. These systems tend not only toward extraordinary sensitivities but, of course, convergences. So first let us have a peek beneath the waves.

THE PRESSURE OF EVOLUTION

Underwater sounds and stimulation by vibration are effectively the same, so any distinction is somewhat academic. Fish have ears, but a key component of their sensory arsenal (as well as the descendant amphibians) is a pressure-sensitive device known as the lateral-line organ system.[1] These consist of elongate channels, especially around the head, housing tufts of cilia that respond to incoming signals. Known as neuromasts these tufts are typically embedded in a jellylike substance (forming the so-called cupola). Pressure changes serve to bend the cilia, electrical signals are fired, and so the fish swims in a world full of information. Like at least some other sensory systems the lateral line is probably closer to an "eye" than might be appreciated. So much is apparent from fish, such as the Mexican cave fish (*Astyanax*). These are habitually blind (indeed eye loss[2] is convergent[3]) and are largely dependent on the information received through the lateral-line system.[4] At first sight this cave fish faces a truly daunting challenge because the zone of sen-sitivity of the lateral-line system is orders of magnitude smaller than the ecological space it inhabits. The solution lies in the extreme sensitivity of the lateral line by which means obstacles can be readily navigated. Counterintuitively when a cave fish is placed in a novel environment, typically it swims faster, but for good reasons. At higher velocities the amplitude of the sensory response generated by the lateral line is increased; in addition the so-called boundary layer (the "sticky" skin of water that adheres to the body of the fish) is diminished, again improving sensitivity.[5] To describe the lateral-line organ as an "eye" might raise eyebrows,[6] yet one can speculate how closely the "hydrodynamic image"[7] provided by the lateral-line system might approximate to seeing. What is not in doubt is the remarkable degree of sophistication[8] of this sensory modality.[9]

Fish could not work unless they were equipped with a lateral-line organ, but how do their honorary counterparts in the form of cephalopods manage when it comes to the

THE SMELL AND TASTE OF EVOLUTION   127

how convergence points to central principles of organization. Of course, the gustatory complexes are not identical. In mammals,[95] for example, the taste cells are not themselves neurons, but serve to activate neurotransmitters that then start a cascade of information toward the brain. So, too, the family of taste receptors are organized rather differently, but both employ the standard seven-helical transmembrane proteins. However, the gustatory proteins in the insects are related to the olfactory battery.[96] They too must be convergent on the standard array, but despite their common origin as cation-channel proteins, the gustatory component shows considerable divergence that reflects important functional differences.[97]

If cats don't like sugar, fruit flies certainly do. One of the additional surprises in terms of gustatory convergence is not only the similarities—the logic—of the overall design, but when it comes to capacities for taste, some striking parallels exist between humans and fruit flies. As David Yarmolinksy and colleagues[98] note, "Given that the fruit fly taste system has evolved quite independently from that of mammals, it detects a remarkably similar range of tastants."[99] They identify "clear parallels in the organization and coding logic between the two systems."[100] Both insect and mammal enjoy a series of discrete modalities,[101] with a notable overlap in both the ability to taste bitter substances[102] (such as quinine, but many of which are potentially toxic) and more particularly receptors for scrumptious sugars.[103] Remarkably the fruit fly is also receptive to many artificial sweeteners. Unsurprisingly they also provoke responses in our close cousins the great apes, but in many cases elicit no response in the monkeys.[104] As Beth Gordesky-Gold and colleagues remark, "In this regard, *Drosophila* responses are more human-like than are those of many mammals, including New World monkeys."[105] A delicious irony, but the reason

is not so strange: both of us originate from Africa, not only cut our teeth (so to speak) on sweet African fruits and retain an avidity for huge sticky cakes, but are also remarkably catholic in our omnivorous tastes. Given these similarities, it is less surprising that the fruit flies, especially the archetypal *Drosophila melanogaster*, are one of our closest commensals,[106] as indeed Mortimer pointed out (p. 9).[107]

We have already seen in the case of the cats that what we taste depends on the receptors available. So there is a spectrum of tastes and by implication qualia. In humans, for example, the umami sensibility explains the extraordinary effectiveness of monosodium glutamate as an "enhancer" of flavor. Among the more remarkable of tastes in insects is the ability to detect carbon dioxide, which only becomes evident to us at very high concentrations (nose to soda can!). For the insects this gas provides an important sensory modality, but it is employed in strikingly varied ways. In the fungus-farming ants (pp. 181–85), for example, detection of carbon dioxide levels in the nest[108] is of key importance.[109] In the outside world newly opened flowers can produce conspicuously higher concentrations of carbon dioxide, providing an important cue for the hawkmoth (pp. 200–201) in search of nectar.[110] Notoriously the plume of exhaled breath, rich in carbon dioxide, is one magnet for mosquitoes[111] (although socks help). If one could interfere with this navigational aid, then an important step against the transmission of malaria[112] might be achieved.[113] That a seven-helical transmembrane protein can act as a transductor for not only photons, olfactory molecules, and taste compounds but even an atmospheric gas is surely remarkable, and it will be fascinating to discover precisely how this process is achieved.[114] But this is not the end of the story, as we now move into the realms of the intangible.

activities. In one sense this discovery very much echoes the convergent evolution of bacteriorhodopsin,[79] but to my mind it remains decidedly puzzling.[80] The canonical worm, *C. elegans*, is quite closely related to the insects (all belong to the ecdysozoans), but its chemosensory capacity evidently draws on the classic receptors used by other animals.[81] As importantly, given that the insects have visual opsins, and in other groups the olfactory proteins are related to the opsins (all having arisen from a common ancestor), then why on earth not simply employ the same method in olfaction? As Marc Spehr and Steven Munger[82] remark, when comparing insect and mammalian systems,

"A remote possibility is that the diversity of receptor families [in the two groups] is simply a byproduct of evolutionary chance. This seems unlikely, as there is strong evidence for adaptive evolution."[83] An intriguing possibility[84] revolves around the configuration of the noses, antennal in insects versus being buried deep in the mammalian skull, along with the very different dynamics whereby the cow leisurely inhales air while the rapidly buzzing fly intersects narrow plumes of odor separated by clear air.[85] Far more importantly it demonstrates that even if one discards the obvious solution there is very little choice in the matter.[86] Very much the same solution still emerges: it is as if the world was made of twins.

THE TASTE OF EVOLUTION

Other than the mysterious worlds of synesthesia, where music may be seen as bright green or smelled as sandalwood, the nearest conjunction to smell are the gustatory sensations. Once again, striking instances of convergence are found. So, too, the gustatory world touches on the seemingly evanescent sensations of qualia. At the same time we should remind ourselves that however expansive access to a given sensory world may be, there will inevitably be lacunae. Just as we cannot perceive ultraviolet, the taste receptors available will determine the range of gustatory worlds. Cats, for example, are obligatory carnivores; indeed, the domestic variety flourishes on a rich and varied diet of songbirds. They are also notoriously indifferent to sweet things, and this is due to a key receptor gene being immobilized.[87] Nor are they alone, because independently a number of other carnivores have been stripped of this capacity.[88]

When one compares the gustatory organs of ourselves (as mammals) with the best-documented alternative, the insects (especially the fruit fly), at first it is difficult to see

anything in common. We employ taste buds, but the insects use their hairlike sensilla (as do they in infrared detection [pp. 117–19] and olfaction). Moreover, while some are located around the mouth, others occur on the legs, wings, and even the female genitalia.[89] The last example is most likely involved with the decision as to the most propitious place to lay her eggs, but having such a dispersed gustatory system means that the insect can taste but avoid poisoning itself. Despite these obvious differences, convergence emerges triumphant. As Natasha Thorne and colleagues[90] stress, the perception of taste has a "Similar Logic" so that we see "a remarkable convergence of anatomical as well as molecular features of gustatory systems."[91] This point is echoed by Reinhard Stocker,[92] who in describing the fruit fly as "A model of good taste"[93] also writes of the "remarkable parallels with the mammalian gustatory system . . . [that] may reflect crucial constraints in the design of taste detection systems."[94] Talking of "logic" in the context of evolution is surely unfamiliar, but it is a reminder of

THE SMELL AND TASTE OF EVOLUTION   125

in a mad steeplechase to the egg, in point of fact, very few are capable of fertilization. Their guidance relies on a finely honed chemotaxis, hence the employment of olfactory molecules. While the overall principle is obvious, Benjamin Kaupp and colleagues[53] stress the system is not only complex but by no means completely understood. However, these workers point out how sperm, olfactory sensory neurons, and indeed photoreceptors display "An Illuminating Kinship"[54] by exhibiting a fascinating series of functional parallels. These involve intriguingly similar schemes of molecular signaling that "confers absolute sensitivity."[55]

At this point we touch on what is the most unremarked feature of sensory systems: that is, their limits of capacity seem to run into the buffers not as a result of some intrinsic failure of biological design but because of the physical boundaries of the universe itself. The sensitivities of the various sensory systems are astonishing.[56] Thus, the rods in a retina can respond to a single photon,[57] and as Hansjochem Autrum[58] notes, "The photoreceptors of vertebrates have also developed to the limit of what is physically possible."[59] When it comes to hearing, then, as Martin Braun[60] notes, "The mammalian ear has reached an absolute detection threshold which is hard to imagine. It has been calculated that a stapes [one of the middle-ear bones] displacement of a tenth of the hydrogen atom's diameter leads to a hearing sensation."[61] So, too, within the inner ear. As James Hudspeth[62] explains, "The threshold of hearing occurs at a sinusoidal hair-bundle deflexion near ±0.3 nm. . . . This motion of ±0.003° corresponds to a displacement of the pinnacle of the Eiffel Tower by only a thumb's breadth."[63] Not surprisingly the limits of transduction within the inner ear are at a level of sensitivity that is almost that of thermal noise.[64] As Winifried Denk and Watt Webb note, this latter system is "as sensitive as the physical nature of the stimulus will

permit."[65] Equally remarkable is the capacity for the tympanic ear of the greater wax moth (*Galleria*) to hear at very high frequencies (c. 300 kHz), some twenty times past the human threshold and also significantly beyond any echolocatory pulse of a bat.[66] As we will see, fish have electrosensitivities at the level of nanovolts (p. 145), while birds can adapt to magnetic intensities about a tenth weaker than they normally employ for navigation.[67]

Evolution, therefore, has navigated to the very edges of perception. This applies also to olfaction, because even a single molecule of an odorant can trigger a nervous impulse.[68] Such work explains, for example, the remarkable sensitivity of some insects to sexual pheromones.[69] Given what we have already learned, it would then be a secure prediction that given that insects have eyes, and thus necessarily opsins, so the equivalent molecules (the classic seven helical transmembrane proteins) would be found serving for the equivalent olfactory transduction.[70] Indeed they are,[71] but stunningly it now transpires that they have a completely different origin.[72] The hallmarks of convergence are obvious because not only is there no sequence similarity, but more importantly the entire molecule is effectively back-to-front so that the amino (N)–terminus of the protein is inside the cell membrane and the carboxy (C)–terminus is outside the membrane, even though the ends are separated by the canonical seven helices.[73] These molecules not only have a very different origin (evidently they are a new class of nonselective cation channel proteins[74]), but the modes of transduction are complex and employ systems similar to other animals,[75] as well as others that are evidently novel.[76]

Close runners-up to this exciting discovery, Renee Smart and colleagues[77] support the idea that here we have a novel receptor, but go on to suggest it may be related to the so-called PAQR proteins.[78] These are evidently very ancient and are implicated in receptor the brain crystallizes as "yes, yes, violets, and, yes, now lime blossom . . ." As Barry Ache and Janet Young remark, "There are striking similarities between species in the organization of the olfactory pathway from the nature of the odorant receptor proteins, to perireceptor processes to the organization of the olfactory CNS [central nervous system], through odor-guided behavior and memory . . . [spanning] a phylogenetically broad array of animals. . . . Such conservation also implies that there is an optimal solution to the problem of detecting and discriminating odors."[45] Like Eisthen these authors are careful to point out that the evidence of convergence as against a common ancestry needs to be weighed carefully.

The preferences of Ache and Young lie toward the similarities being convergent, but a definitive answer is difficult because to date only a handful of species have been studied in any detail. In principle, therefore, features that appear convergent may, as more and more species are investigated, transpire to stem from some common and ancestral arrangement. The evidence, nevertheless, points ever more strongly in exactly the opposite direction. Not only that, but the solution arrived at reflects a remarkably limited freedom in design but paradoxically also the best of designs. Most striking in this regard is the degree of convergence between the olfactory systems of insects, notably *Drosophila*, and the mammals.[46] Strongly echoing the aforementioned remarks of Ache and Young, so Leslie Kay and Mark Stopfer[47] stress how convergence operates at the structural, functional, and physiological levels. Indeed, for them it represents "a beautiful case of convergent evolution," and so suggests "there may be one best way to process information about odorants."[48]

Of the various aspects of this convergence, a number of features stand out. First, the proteins that initially bind the odorants in the fluid above the olfactory receptors have significant similarities. These include being of a small size, occurring in very high concentrations close to the olfactory neurons, and of course having a strong affinity for various chemicals. Yet their origins are completely different.[49] In the mammals they are lipocalin proteins,[50] whereas those found in insects have a quite separate molecular architecture.[51] This is curious because lipocalins are ubiquitous, and in the insects they have a variety of functions, including their role in the convergently evolved milk (p. 175). Processing of the olfactory data, including gain control of the signal, shows many parallels as well. In addition, the way in which the neurons are distributed to the glomeruli is strikingly similar, and it comes as no surprise to learn that these distinctive spherical structures have themselves almost certainly evolved independently. For olfaction to work there must also be appropriate receptors that act as transductors—that is, changing the molecular signal into an electrical impulse.[52] Central to this operation are proteins with a distinctive structure, effectively consisting of seven helices that span the cell membrane (7TM). Such transmembrane proteins are very widely employed in sensory transduction and have already been encountered in the visual "molecule of choice," the opsins. Nor is it surprising that opsins and their olfactory cousins derive from a common ancestral form.

Just as opsins crop up in locations very far removed from the eyes (p. 94), so we find the equivalent olfactory proteins lurking in out-of-the-way places, such as the testes. Bizarre as it might seem to find olfactory proteins in this fertile location, such a co-option (and here we hope, for reasons I can't quite put my finger on, that the evolutionary journey was nose to testicle rather than vice versa) makes a great deal more sense when we recall that any self-respecting sperm needs to know where it is heading. Contrary to the popular idea that once ejaculated the sperm engage into words. Most haunting is the cues they can provide to memory, triggering intense recollections and uncapping emotional wells that otherwise might be better left untapped. This gulf between our perception versus the mechanics of molecular intersection and transmission is, as with other sensory perceptions, all too often neatly swept under the materialist carpet (dull beige, much tattered, and no discernible pattern), yet Andreas Keller and Leslie Vosshall are right to insist that "How the activation of populations of olfactory sensory neurons is translated in the brain into a discretely perceived odor quality is still completely mysterious."[36]

While the contrasts between olfaction and vision are vivid, there are also some parallels. One time I was with a colleague who had brought along a bottle of ancient wine, specifically a crippled Vouvray. Slightly to my surprise on opening it, the cosmologist held it to one nostril and inhaled deeply. All was well; although almost twenty years old the wine had retained its complex structure. Perhaps this preprandial maneuver made good biological sense because not only does the airflow into our nostrils periodically switch, but more particularly each side seems to be sensitized to different odors and so delivers an olfactory "image" that is slightly different.[37] More extraordinary is that rats not only have very acute smell, but can smell in stereo.[38] Even though each nostril is only separated by a few millimeters, each has a separate airflow, and in a couple of sniffs the source of interest is precisely located, presumably by integration of the olfactory "images" in the brain. Such a capacity may be convergent because even in the olfactory world of the fruit-fly larvae, equipped with only a handful of olfactory neurons along either side of the body, either side can function, but the accuracy of perception is improved when both sides are brought to bear.[39] This study also revealed a slight bias in favor of the right-hand side of the larvae. How curious, or is it? Such lateralities seem inherent to the world, and in the birds (or at least the ubiquitous chick) there is evidence that olfaction is lateralized. The right-hand nostril has the advantage, with the corresponding part of the brain having the greater role in olfactory memory and preferences.[40] The convergences roll on, because in the bees, as in other insects, olfaction is mediated via the antenna. Which one has by far the greatest sensitivity? Why, the right-hand antenna.[41]

The similarities between olfaction in the insects and the vertebrates might be thought to end there: What could be more different than antenna and nose? Convergence trumps this assumption. At almost every level what at first sight looks utterly dissimilar conceals not only a striking identity, but one that points to the inevitability of design specifications. This is not to deny that there are alternatives, but strikingly we find some of the most compelling similarities among the most sophisticated systems. When invoking this or any other convergence, one needs to be careful to refute the seemingly more parsimonious, if not plausible, option that the identity of form is simply because each olfactory solution is basically the same because it has descended from a common origin (and so deep homology). As Heather Eisthen[42] asks, "Why are olfactory systems of different animals so similar?"[43] In this thoughtful review she also points out that even if a system is convergent, then could it be that with the constraints in terms of organization there is effectively little choice in the matter? Impressively, for example, the surfaces of the olfactory membranes, be they sniffing in air or under water, have the same fundamental arrangement.[44] Thus, the neurons (which are bipolar) sprout hairlike projections (usually cilia) into a fluid-filled compartment, with the impinging odorant molecules triggering the appropriate receptors to unleash a cascade of chemistry that in as the Antarctic prion (p. 51), as they do in their ecological counterparts, the mysticete whales[21] (along with predators like harbor seals[22] and penguins[23] attracted to narrow zones of high productivity). Such sensitivities are evidently acute, and even if humans do not inhabit an odor landscape we are not "blind." While the human olfactory repertoire is diminished, it is by no means vestigial. People suddenly bereft of smell (a variety of anosmia) can find it disorientating, and in its own way a permanent loss can be as serious as blindness. Think of leaking cooking gas or the horrible risks of a putrid sausage. More constructively, connoisseurs of tea and perfumes need to make a living, while in those sumps of consumerism, the supermarkets, olfactory wafts of freshly baked bread from nonexistent bakeries are artfully directed as the shopper dithers over fifty types of bread that all taste the same (if indeed of anything).

In terms of olfactory aptitude, probably the most familiar territory is that of pheromone detection. This includes not only remarkable examples of chemical mimicry (p. 393n89), but striking parallels between the pheromonal cues employed by fish and insects.[24] This example emphasizes that the aquatic world is potentially as olfactorily rich as that encountered by air-breathing animals, yet the transition to a terrestrial way of life poses a major challenge. The recurrent evolutionary question as to whether evolution embarks on the route of innovation or perforce is compelled to adopt the same solution as everybody else is compellingly answered in favor of convergence in at least the case of the robber or coconut crabs.[25] Apart from the larval stage these monsters are entirely terrestrial and inquisitive scavengers, and despite being closely related to the other aquatic crustaceans the olfactory organs show striking convergences with the insects in terms of anatomy, behavior, and physiology.[26]

As with the robber crab, many animals possess an acute sense of smell. But not all. Humans most likely traded olfactory prowess for trichromatic vision (p. 115), but a return to the oceans also leads to a similar loss.[27] Thus, in terms of olfactory repertoire the whales have suffered massive reduction in capacity, to the extent that the toothed whales evidently have no sense of smell.[28] But they are not alone. As Sara Hayden and colleagues[29] demonstrate, there are "spectacular examples of OR [olfactory receptor] gene losses in three independent lineages of aquatic and semi-aquatic mammals, yet convergent, selection of similar functional OR families."[30] Thus, not only do the toothed whales show a loss of olfactory capacity,[31] but so to a considerable extent do the manatees (and to some extent aquatic carnivores such as the sea otter).[32] Given this, it is less surprising to discover an impoverishment of the OR genes in the viviparous sea snakes, but not among their egg-laying counterparts that necessarily still have to make return trips to land.[33]

The smells of parched land after a sudden shower, a good claret, or an avenue of lime trees in flower . . . all very familiar, yet consider how strange olfaction actually is. Unlike vision there is no ordered electromagnetic spectrum; in contrast the nasal epithelia must receive and distinguish a riot of molecules, of widely varying specificity and affinity. Not only that, but the odors are delivered in plumes of turbulent air so that local concentrations can vary enormously.[34] Olfaction, therefore, is in many respects quite unlike vision, yet they invoke not only the qualia of smell, but ones that remain equally elusive: How is it that we translate a molecular interaction into a conscious experience? Molecules that are identical except for being mirror images of each other (these are known as enantiomers) may smell completely different (in some cases one enantiomer will even be odorless),[35] while others may reveal levels of subtlety yet are notoriously difficult to put combination of color and aroma can enhance the occasion further still. Yet there is a darker side. Consider the gut-wrenching list offered by Val Curtis and colleagues:[8] "faeces, vomit, sweat, spit, blood, pus, sexual fluids . . . toenail clippings, rotting meat, slime, maggots . . .";[9] all provoke disgust, especially when combined with other departments of perception, such as the tactile. These workers make a compelling case that this disgust is a powerful evolutionary agency to avoid infection and disease.[10] This all-too-familiar repulsion is universal among human cultures, yet it bears also on aesthetic judgments. The basis for disgust goes very deep. Nematode worms, to take just one example, employ their olfactory capabilities to avoid pathogenic bacteria.[11] Yun Zhang and colleagues euphemistically refer to "intestinal distress," as they note the parallels in the "conditioned taste aversion . . . in mammals, snails, cuttlefish and fish in which animals avoid foul flavours."[12] A key molecule in this response is serotonin. Although well known as a neurotransmitter, this molecule is extraordinarily versatile. Significant quantities are generated in the gut tissues, and its role in disgust is important in the induction of vomiting, which however unpleasant is a vital conduit to dispose of toxins and pathogens. Intriguingly there are also possible links to the immune system,[13] which is implicated in aspects of brain development.

Serotonin occurs in practically all animals[14] and most likely is very ancient, given that it occurs also in plants.[15] It is a rather simple molecule (technically 5-hydroxytryptamine), and given its derivation from the ubiquitous amino acid tryptophan, its synthesis is conceivably convergent. As far as animals are concerned, the avoidance of potentially lethal food is at one level primitive, yet the more advanced aversions have most likely evolved repeatedly and are closely associated with learning. An interesting convergence emerges in a comparison of the waste management in two complex social groups, humans and leaf-cutter ants.[16] The latter are famous for their independent evolution of agriculture (p. 181), and as with us a vital component is the exclusion of waste products, including corpses, with their load of pathogens. The ants have designated workers, and they not only engage in a series of specific behaviors (including acting as undertakers) but are vigorously excluded from the nest. In this unattractive environment, mortality rates are high, but as Adrianne Bot writes, "There are a number of interesting parallels between waste management in agricultural societies of ant and human. . . . natural selection has probably led to a series of general attitude of disgust towards different types of waste product. . . . Apparently, natural selection and evolution have produced convergent methods."[17] At what point, however, does disgust achieve an emotional dimension?[18] We stray again into the vexed question of qualia: disgust and repulsion are not, I would suggest, simply some tendentious extension of evolutionary psychology. Who knows if a leaf-cutting ant posted to guard a foraging entrance feels disgust as a reeking, corpse-handling waste-management worker stumbles toward her, but I would not be too surprised. At the least I can insist that if I see glory in a painting by Titian, a near-emetic response is my reaction to Gilbert and George (not to mention the buzz of approval from the chattering classes). Not all that smells, however, revolts.

THE SMELL OF EVOLUTION

For most animals the sense of smell is vital. For example, chemicals such as dimethyl sulphides and pyrazins play a key role in foraging[19] (and navigation[20]) of oceanic birds such

CHAPTER 11

THE SMELL AND TASTE OF EVOLUTION

BEING SUCH A visual species, the other sensory modalities we possess sometimes seem to take a backseat, although those who, for example, lose their sense of taste or smell often need to make profound adjustments. Not surprisingly, convergence is not only ubiquitous in the eyes but all other sensory systems. Of particular interest are those to which we are blind, such as echolocation, infrared sensitivity, and electroreception. As significant, not least because it touches perhaps on the roots of sentience, are the sensory capacities of plants and microbes, which play vital roles in assessing the worlds they inhabit. As James Shapiro[1] remarks, "Bacteria are small but not stupid."[2] He is unafraid to write (and I unembarrassed to read), "Once properly oriented in our thinking, we can find cognitive and computational phenomena

in many of the classical bacterial systems,"[3] and he goes on to remind us of the parallels between neurally based sensory systems and bacterial chemotaxis. Not only are bacteria highly adept at chemotaxis, but this entails their receptors being disposed in a hexagonal array.[4] As Ariane Briegel and colleagues note, "The universal hexagonal architecture and secondary structure of chemoreceptor arrays we observed in diverse bacterial species"[5] points to strong evolutionary conservation. No doubt this is correct, but the many examples of bacterial convergence not only give us at least a moment's pause for thought, but specifically in the context of sensory systems as a whole—again and again—we find that there really is a best way to do things. So with reluctance, let us leave the bacteria and gingerly swivel our noses toward the first topic.

DISGUSTING EVOLUTION

Ironically, our immersion in chromatic worlds too often etches our appreciation of exactly why we see and more importantly *know* we see, not least in the completely mysterious qualia of consciousness. Look again at a painting—say, Titian's utterly astonishing *Assumption of the Virgin* on the main altar of the Frari in Venice. Smears of oil paint on a canvas? We might fool ourselves that the electromagnetic spectrum of visible light, photons, even quantum mechanics somehow weave the rainbow and allow us

to address the quiddity of visual qualia. Of course they don't! This is brought even more sharply into focus (so to speak) when we consider olfaction. On the mammalian scale of snozzles humans are sadly blunted and vision can notoriously usurp smell. People—well, French undergraduates—will insist that a white wine is red when the former is adulterated with an innocuous dye (grape anthocyanins),[6] and similar tricks can be played with other drinks like fruit juice[7] and food. Just as Mortimer found in his restaurant, the

the beetles but in heteropterans, specifically a Pyrophilus flat-bug (*Aradus*). Here, too, we see a dome-shaped detector, lamellate wall, and associated dendrite.[195] Most likely this infrared organ has evolved several times in the aradids. Although the sensitivity may be less than in the melanophilids,[196] this similarity is telling because it provides powerful evidence that infrared detectors may be designed in several different ways, but each time the given solution is very much on the cards.

Despite the convergence of function, the arrangement of the infrared detectors in these insects and the snakes is markedly different, although just as the arthropodan detectors derive from mechanoreceptors, those of snakes probably originate from nerves once associated with touch and pressure. Other convergences, however, are closer. In another beetle, known as *Acanthocnemus*, the infrared receptors are arranged as microbolometers and have a design strikingly similar to those of the crotalid snakes. As in *Melanophila* the detectors are located on the thorax, but this time in front of the first pair of legs and more importantly with a completely different design.[197] Here the sensory disc is attached to the rest of the body by a narrow stalk, and like the crotalids it is stiffly suspended above an air-filled cavity. This sensory disc is not only isolated and insulated, but richly endowed with sensory dendrites and packed with mitochondria.[198] Studded with peglike sensilla, each has an electron-dense rod that may serve to channel the incoming infrared radiation. Notwithstanding the differences with other insects, these acanthocnemids are also fire beetles,[199] but operate over shorter ranges and with a different behavior.[200]

Just as this appropriately named little ash beetle converges on the crotalid snakes, so the final example among the beetles has some interesting similarities to the boid snakes. This beetle, *Merimna atracta*, is also a buprestid.[201] They are attracted to fires above which they mate so that the larvae can develop in the newly burned wood. Its infrared detectors are different again. This time they are located on the abdomen, although the variability in the number of detectors suggests that they may be a relatively recent innovation and one that is yet to stabilize (and might indeed give us further clues as to how these thermal eyes originate).[202] In detail the detectors show a honeycomblike structure associated with a slight depression.[203] Like the little ash beetle and the snakes, they consist of a mass of dendrites, associated with very large mitochondria. These similarities extend to their physiology, and while the response pattern is much less sensitive than that of the snakes, the tolerance to the intensity of input is remarkable: even using a 20 mW infrared laser, the beetle's detector[204] failed to saturate.

This story of infrared detection is by no means complete and rather extraordinarily even extends into the aquatic realm.[205] Although the examples have largely involved heat being generated by either furry animals or forest fires, these are not the only sources of infrared radiation. Plants can absorb solar radiation differentially so that some parts become much brighter in this part of the spectrum. In the case of the conifers this is evidently an important cue for hemipterans, specifically the Western conifer seed bug as it seeks out warmer cones. Quite independently this insect has evolved infrared organs that are similar to the merimnids in location and microstructure.[206] Stephen Takács and colleagues note that a similar infrared (IR) capacitance may extend to some moths and midges, and they pregnantly conclude how "IR perception [may be] a potentially widespread foraging cue in . . . frugivorous insects and the invertebrate and vertebrate pollinators."[207] And these are by no means the only doors to sensory perception. What's that in the wind?

sensitivity of this animal is extraordinary because it can detect forest fires from many miles away. An important component is evidently a smoke detector located in the antennae that seems to be extremely sensitive to volatiles, such as phenols, produced by burning.[184] Thus the beetles will home in on all sorts of fire, such as smelter plants, and the irritation of spectators in big football games in Berkeley was thought by Gorton Linsley[185] to be due to these insects being "attracted by the smoke from some twenty thousand (more or less) cigarettes which on still days sometimes hangs like a haze over the stadium."[186]

Pity these beetles as they fly into such a deeply politically incorrect atmosphere, but in the natural world happily forest fires are very infrequent and localized (that is, without the assistance of pyromaniacs). In its natural habitat there must be very strong selective pressure to enable *Melanophila* to rush toward the conflagration. As far as the capacity for infrared detection is concerned it resides in an extraordinary pair of organs located on either side of the thorax and a little above the insertion point of the middle leg. Each consists of a battery of sensory domes, typically numbering fifty to a hundred.[187] They originate from the so-called mechanoreceptors, which in themselves are of great interest because they provide the ears of insects (p. 135). In the fire beetles, however, they have been transformed into a complex structure, and even now not every step in their evolution from a mechanoreceptor is understood.[188] The overall structure of each sensory unit, a sensillum, is a globelike unit attached to a nerve cell. It is generally agreed that the impingement of the infrared radiations heats the dome, and its expansion then triggers the nervous response.[189] What may be of key importance is the precise mechanical properties of the cuticle, so that thermal expansion, in conjunction with adjacent pockets of fluid, exerts the necessary pressure to activate the nerve cell.[190] The tip of the dome contains an electron-dense channel, and intriguingly William Evans[191] has proposed that it might be "a waveguide acting like the light-guiding cones and rhabdoms of compound eyes,"[192] as well as controlling internal reflection in a manner very much like fiber optics (pp. 96–97). The degree of engineering of this detector is remarkable, reflected in its extraordinary sensitivity to detect heat across great distances.

While it had been thought that the detector might employ opsins for the transduction of the infrared radiation, it is now clear that the action is as follows: infrared → heat → mechanical → electrical. Given the location of these organs on the sides of the insect, the animal evidently has to tack toward the fire, and indeed the exact angle of approach would lead to different sensilla being activated. An ability to discriminate different frequencies of the infrared, to provide an equivalent to color vision, seems less likely as the sensilla are much the same size.[193] Nevertheless the likelihood is that just as in the crotalid snakes in the central nervous system of the beetle, some sort of image is created. It is also likely that the input from these sensory batteries is combined with visual and olfactory cues,[194] although most likely the infrared capacity is of central importance as the beetles fly among the flames and then settle on the hot bark.

Several other beetles have independently evolved infrared detection. The respective organs show a considerable diversity, even if sometimes they display similarities to comparable detectors in the snakes. It might seem, therefore, that the domal arrangement seen in *Melanophila* is another one-off, albeit showing remarkable sensitivities. Given that infrared organs are rather characteristic of the buprestid beetles, even if such an arrangement did evolve more than once in this group. In itself this might be less of a surprise. Remarkably, however, just such an arrangement has evolved, but in a completely different group of insects. This evolved not in

THE COLOR OF EVOLUTION   117

evidently confer an infrared capacity, and just as with the "five-site rule" in vision (p. 112), we not only find three critical substitutions but these are identical in the boids and crotalids.[161]

At first sight the infrared sense of the crotalids surely seems decidedly alien: we readily sense heat but cannot envisage it. The differences, however, turn out to be rather superficial. To begin with, although the pit organs are designed to intercept the infrared, they possess a much wider sensitivity that evidently extends to the ultraviolet.[162] More importantly the crotalid infrared pit is convergent with the pinhole camera eye (p. 101),[163] although it suffers the same drawback of such eyes because the image must be blurred. There is, however, reason to think that neuronal processing can lead to a significant sharpening.[164] Even more intriguing is the evidence that this image is combined with the input from the eyes[165] and most likely is assessed at a high neural level in the forebrain.[166] Indeed Richard Goris[167] insists[168] that "it is fallacy to consider the pit organs as an independent sixth sense evolved solely for the purpose of detecting and acquiring prey. Not so. What the pits do is improve vision for their owners."[169] For these snakes a key role is the detection of warm-blooded mammals, but the lunge of the snake need not spell doom. Some squirrels anoint themselves with rattlesnake scent (p. 39), while another well-known maneuver is in the form of tail waving. But there are snakes and then there are snakes, and only when the squirrel sees the rattlesnake does it pump blood into its tail so that its potential hunter sees a bright infrared flag as a further dissuasion.[170]

The attractions of warm blood can take other routes, such as the siphoning off of somebody else's blood. Once again infrared detection appears to play a role, with an obvious candidate: the vampire bats. Adjacent to the so-called nose-leaf, which is remarkably mobile, three pits are kept substantially cooler and act as thermosensors.[171] They employ a protein (TRPV1) already known for its heat sensitivity, but modifying it specifically for the needs of the bat.[172] The proteins used by snake and bat, therefore, differ, but in the brain of the vampire bat there is a structure that has significant parallels to the neural center for infrared detection in snakes.[173] If one escapes the attention of a vampire bat—a good idea, given that they can carry rabies—there is always the chance of an insidious visit by bedbugs (hemipterans, belonging to the reduviids) in search of a blood meal.[174] These also offer the risk of infection, but this time by some protistan horror such as the trypanosome responsible for Chagas disease. To enjoy their blood supper, infrared perception again plays a role.[175] Evidently this involves a specialized structure located in the antennae,[176] and their waving movements[177] seem to play a key role in orientation.[178]

Apart from a parasitic wasp, equipped with a bizarre antennal sensory structure that may serve for infrared detection[179] of potential hosts—that, although living in wood, leak sufficient metabolic heat to be detected[180]—among the arthropods the real masters are the beetles. Here this capacity has evolved independently at least three times, but in each case to detect fire. This is, of course, something from which nearly all animals will flee, and there is even a frog that is attuned to the crackling sound of burning[181] and will head for safety.[182] Even if some plants can withstand the passing flames,[183] at first sight it would seem to be lunacy for the beetles to be heading for the conflagration. For them, however, the aim is reproduction and paradoxically to provide a haven for their larvae. The heat provides the lure, and in each case the infrared organs are located in the cuticle. Despite having the same functions they show a remarkable diversity of form and in some cases an astonishing complexity.

The latter applies in particular to a buprestid beetle known as *Melanophila*. The sensory been touched upon. At yet shorter wavelengths one moves into X-rays: certainly not a zone in which to linger. Detection in the opposite direction, however, presents some decidedly more interesting possibilities. Rather extraordinarily, human skin can act as a two-dimensional antenna sensitive in the extremely high-frequency region where radio waves merge into microwaves.[146] This apparent oddity is due to the conductive layers of the skin combined with the helical structure of the sweat glands. Not surprisingly, the response is dramatically enhanced after a bout of intense jogging, and while these observations might find use in detecting emotional states[147] there is no implication that the skin is actually serving as a sort of "radio eye."

When we move back toward the visible spectra, and specifically the infrared, then indeed new (and of course convergent) sensory worlds open up. For the most part the infrared spectrum is associated with the detection of warmth.[148] This ability is very widespread and can lead to some remarkable convergences, such as those thermophilic ants traversing deserts that to anybody else would prove lethal.[149] Detection methods can also involve some extraordinary sensitivities. Particularly remarkable are the larvae of a cave beetle (*Speophyes*), whose sensory hairs (sensilla)[150] are capable of responding to minute shifts of temperature.[151] Lacking eyes, this capacity presumably prevents the animal from straying, but as Hansjochem Autrum[152] notes, "Of course, no adaptation to the customary environment, however well developed, will serve if overridden by lust for sensual gratification. Overripe Roquefort cheese is one means of luring the beetles away from their normal temperature range."[153] Apart from cheese, these beetles react to both heat and cold, but as far as the former is concerned, a specific ability to detect not warmth in general but specifically infrared radiation

is a more specialized art. The snag, of course, is that the levels of available energy are low and for the most part no transduction system based on the opsins can operate at these frequencies.[154] Nevertheless, in at least one of the eight or so times in which the infrared sense[155] has evolved, the organ concerned does very much act as an eye.

When it comes to infrared detection, best known are the snakes, with sensory pits located on the head. This arrangement has evolved twice, in the boids (the pythons) and the crotalids (familiar in the form of the rattlesnakes). As Tjeerd de Cock Buning[156] remarks, "it will be obvious that the peripheral differences between the heat receptors and pit structures in pythons and crotalids represent two different solutions for the same functional demands."[157] Their independent origins are evident both from the different arrangement of the pits (in the crotalids they are restricted to a pair located between eyes and nostrils) and more particularly their anatomy. Whereas pythons employ a straightforward pit, the crotalids are more ingenious. Here a very thin sensitive membrane forms a sort of false floor and is suspended above a lower air-filled cavity that serves to insulate the detector. Such an arrangement is known as a bolometer, and it also forms the basis for many artificial heat sensors. The actual detection of the infrared radiation is mediated by specialized nervous array comprising a mass of dendrites and mitochondria, the latter indicating that the perception is a highly energetic process.[158] Note that the detection of infrared radiation does not depend on a sort of phototransduction. To the contrary, it employs a heat-sensitive ion channel protein (TRPA1),[159] with Elena Gracheva and colleagues noting how the boids and crotalids "have independently adopted [this ion channel] as an infrared sensor through convergent evolution."[160] This is dramatically confirmed in terms of the amino acid substitutions that acid sites in the opsin molecule.[135] Howlers, therefore, presumably see the world in very much the same way as we do, yet with at least one striking difference. In contrast to many mammals, our olfactory capabilities are crippled, with our genome peppered with a litter of pseudogenes. A plausible explanation is that trichromacy (and hearing) has trumped smell.[136] Now while in comparison with other New World monkeys the howlers have lost a substantial olfactory capability,[137] unlike humans they have retained the capacity (via the vomeronasal organ) to detect pheromones.[138] And maybe habitat explains this: as we ran free on the savannah rejoicing in the color of the setting sun and welcoming firelight, the howlers remained in the jungle, sniffing each other.

While the howlers saw their world brighten, another group in the New World, the owl monkeys (*Aotus*), sank back into the crepuscular world and became monochromatic,[139] as indeed independently have several other mammalian groups, ranging from rodents[140] to carnivores such as the kinkajou.[141] All are nocturnal, and echoing the puzzle of polymorphic color vision in the New World monkeys, it is far from clear why other nocturnal mammals retain color vision—and in at least the case of the aye-aye (p. 32), the eyes can evidently function in moonlit conditions.[142] Among humans, red/green color blindness is not uncommon among males, and if it is any consolation is strikingly convergent with the dichromatic world of elephants.[143] True monochromatism is very rare, but even so, a sort of color vision is still found.[144] Let us conclude with another sort of ocular convergence. Blue eyes can make a startling impact, yet among mammals as a whole they are very unusual. A striking exception is found in a subspecies of black lemur (*Eulemur macaco flavifrons*), and its eye color also depends on a genetic mechanism. Not, however, the same as ours, so here the Sinatra syndrome is convergent.[145]

The Heat of Evolution

The chromatic range of vision is striking, but ultimately constrained by the realities of the electromagnetic spectrum. Humans, of course, cannot see past violet, and other than by using detectors or admiring the dramatic fluorescence in some minerals when irradiated by an ultraviolet source, we are blind to this region of the electromagnetic spectrum. Blind we may be, but even so our color vision, along with sounds—say, the Good Friday music of *Parsifal*—and olfaction—"a glass of the old Petrus, anybody?"—provide pretty impressive sensory palates. It is a worthwhile journey because such sensory worlds swiftly lead to metaphysical doors sliding noiselessly open as we touch on the enigma of qualia, epitomized by the vividness of the kingfisher or the solemnity of the setting sun. Scientists instinctively shy away from these questions, and certainly while the qualia of redness is difficult enough to grapple with, how can we possibly imagine "seeing" ultraviolet?

Physics, however, provides some constraints. Consider what happens as we travel beyond the visible range of the electromagnetic spectrum, which of course is only the narrowest of bands covering a tiny fraction of the range of frequencies from gamma rays to radio waves. A favorite speculation by science fiction writers has been whether evolution in alien environments could lead to biological antennae that could eavesdrop in the wavelengths in the high energies beyond the ultraviolet or alternatively toward radio frequencies. As far as the former is concerned, the dangers of ultraviolet vision have already dichromatic,[112] having lost two of the pigments that are found in most other vertebrates,[113] which are, therefore, tetrachromatic.[114] Interestingly the most primitive of mammals, the monotremes (or at least the duck-billed platypus), still hold on to one of these pigments,[115] but as eggs were abandoned so the chromatic world of the mammals narrowed. In the primates, however, trichromacy was reinvented. A popular supposition is that the world took on new colors as the early primates abandoned the nocturnal life,[116] but alternative views suggest that the primates were trichromatic from the start and returned repeatedly to the dark.[117] Deciding between these alternatives is not straightforward and revolves around questions as to the phylogeny of early primates, nocturnal adaptations such as a reflective tapetum, and in extinct species the size of the eye orbits. Nor is it necessarily out of the question that both nocturnality and trichromacy are convergent within the primates.[118] Nevertheless, while the invention of mammalian trichromacy has largely focused on the primates, some marsupials (such as the nectar-feeding honey possum) have clearly arrived at this solution.[119] This capacity is possibly retained from yet more primitive mammals, but it is just as likely to be convergent. In any event it extends to some specific molecular convergences in terms of key amino acid sites,[120] but unlike primate trichromacy can include an ultraviolet component.[121]

It does seem certain, however, that the trichromacy of the monkeys and apes stems from a common ancestor. But even here are unexpected twists and turns. In the New World monkeys, which have enjoyed evolutionary isolation for about 30 Ma and are a storehouse of instructive convergences, the trichromacy is polymorphic. The reason is the sex linkage in some of the opsin genes, so that the males are dichromatic (as are the homozygous females), whereas the het-erozygous females are trichromatic.[122] The system relies on three alleles and most probably evolved only once, although interestingly very much the same arrangement has emerged independently in a lemur, Coquerel's sifaka.[123] In the New World monkeys, polymorphic color vision is very ancient,[124] and such stability over millions of years makes almost certain that it is of adaptive value. Why should the boys be dichromatic and most of the girls trichromatic? Popular hypotheses have revolved around the discrimination of colored fruits[125] and leaves,[126] as well as the ability to detect insects.[127] Yet there is still a distinct air of puzzlement over this issue.[128] The fact that the polymorphism is almost certainly adaptive does not mean that one should be too procrustean and end up regarding these monkeys as little more than arboreal color monitors. After all, they live different lives, in light that fluctuates between tropical glare and jungle crepuscularity and may conceal various predators.[129] The eyes of these monkeys do not show identical spectral sensitivities; and even within a single species may look for different types of food[130] (not least in times of shortage[131]). One potential spin-off, however, is that among the apes and monkeys that show gregarious mating (as against those species that are monogamous/solitary), there is a striking association with either a red pelage or skin, suggesting a link between trichromacy and sexual selection.[132] Curiously, among early hominids we find the widespread use of ochre. Is there perhaps a connection?

New World monkey polymorphism is almost ubiquitous, with one vital exception: the howler monkeys (*Alouatta*). These have independently evolved a full and very acute trichromacy,[133] even though their close relatives, the spider (*Ateles*) and woolly (*Lagothrix*) monkeys remain in a polymorphic world.[134] The return to trichromacy is marked by a molecular convergence of key amino tionarily represent a dynamic system, with gains and losses. One apparent oddity is that as one descends into deeper water, the longer wavelengths of light (the reds) are quickly filtered out, so the shorter blues predominate (this incidentally explains why a wounded diver sees green blood oozing from the cut). Yet independently both the whales and seals have lost color perception in this part of the visual spectrum.[96] They have the necessary genes, but they have been rendered impotent (as so-called pseudogenes). The fact that the loss of blue vision is convergent makes it less likely that this is simply an accident. Because light is more effectively scattered in the shorter wavelengths (hence the sky is blue), by way of explanation it is supposed that in the blue depths whales and seals traded sensitivity of perception for a broader spectrum of visibility.[97] These are not the only adaptations to aquatic vision, and in particular the whales show some striking changes in design, which find parallels in the fish and to some extent other aquatic mammals, notably the seals.[98]

When we dive without a face mask, our vision is, of course, severely blurred, which is hardly surprising seeing that some two-thirds of the eye's refractive power is lost.[99] Yet the appropriately named sea-gypsies, the Moken people,[100] inhabiting the littoral of Burma and Thailand, have taken an important step toward aquatic vision. In particular, the children hunt for food across the seafloor and enjoy a remarkable underwater acuity, more than twice that of European children. How is this achieved? A key adaptation is a dramatic constriction of the pupil (echoing what is seen in whales and seals,[101] although here related to the transition from brightly sunlit shallows to hadeal gloom), combined with accommodation of the eye by adjusting the shape of the lens.[102] Although strongly adaptational, it evidently is not genetic, because Europeans can be trained to match this acuity in a matter of weeks.[103]

The blur of underwater swimming is not the only area where human vision fails. While in daylight we enjoy acute trichromatic vision, as night falls we enter a monochromatic world. But this is not true of all animals. Apart from the echolocating oilbirds (pp. 139–40), perhaps the most astonishing are the hawkmoths (pp. 200–201). Daylight species enjoy excellent trichromatic vision, which like many insects extends into the ultraviolet.[104] Some species, however, are completely nocturnal, yet as "hummingbirdoids" (p. 200) they still need to seek out flowers. Olfaction may well play a part,[105] but incredibly at least one nocturnal hawkmoth can discriminate colors in the equivalent to dim starlight.[106] Other insects, notably some bees and wasps, also possess nocturnal vision, a capacity that has evolved independently in both groups.[107] Such an ability is all the more intriguing because the type of compound eye (known as appositional) is inherently less sensible than the alternative (superpositional), which is in fact employed by such creatures as the nocturnal moths. Even so, similar optical adaptations are employed, and while many of these hymenopterans are crepuscular, an Indian carpenter bee (*Xylocopa tranquebarica*) not only flies on moonless nights[108] but employs color vision.[109] Like other nocturnal insects, the rhabdoms of this compound eye are very wide, but in general there is little in the structure of the eye to explain its extraordinary sensitivity.[110] Perhaps the photoreceptors are more sensitive, but there seems little doubt that additional neuronal processing must dramatically improve the signal-to-noise ratio and ensure that the insects we usually associate with sunlit meadows can fly with confidence under the stars.

In hawkmoths,[111] nocturnal stages have played an important part in their evolution, and it is generally thought that this also has been central in the development of mammalian vision. In general, mammals are fish[73] to mammals,[74] and as might be expected in the cephalopods.[75] Among the compounds employed are the mycosporine-like amino acids[76] (p. 15) and 3-hydroxykynurenine. These have been widely recruited in numerous groups to provide other sorts of protection against ultraviolet radiation,[77] although alternative strategies[78] have also been arrived at convergently.[79]

Among the animals, color vision has evolved repeatedly, but perhaps most strikingly in the mantis shrimps.[80] Employing an equally extraordinary pair of compound eyes, they evidently can operate independently of each other; as Thomas Cronin and Justin Marshall remark, it "may be more like two people watching the same tennis match."[81] In addition, this animal can not only detect polarized light but more extraordinarily the eyes are sensitive to circular polarized light. This capacity may be employed for communication (as a sort of "private channel" invisible to other animals), but it could also assist the animal in peering through sediment-laden water.[82] Human technology employs some comparable systems, but Tsyr-Haei Chiou and colleagues emphasize that the mantis shrimp system had a head start of hundreds of millions of years; as they say, "nature got there first."[83] With this optical capacity, the visual world of the mantis shrimp seems almost incomprehensible to us.[84]

Nevertheless, the story of color vision as it applies to us and the vertebrates is in its way equally fascinating. Color vision itself is not only very ancient, but some general rules apply. Most notable, perhaps, is the so-called five-site rule whereby red/green vision is very largely dependent on substitutions of amino acids at a mere five critical sites[85] in the rhodopsin molecule.[86] This is neither to say that this "rule" applies to all animals[87] nor to deny that certain synergies may be operating between some of the key amino acids.[88] Nevertheless, it reinforces the view that color per-

ceptions from ultraviolet to dim light (which in deeper water is blue-shifted) often depend on only a limited number of mutations at key sites.[89] These sensitivities have two implications. First, one can make a reasonable inference as to what the amino acid sequence must have been in the first vertebrates, jawless creatures gliding across a Cambrian seabed. By extrapolation one can design an ancestral gene and make a fair guess as to how some extinct animals could see.[90] For example, knowing the structure of opsins among the living vertebrates and specifically their phylogenetic connections to the ancestral archosaurs, predecessors to the great dinosaurs, resurrection of their opsin genes allows one to infer that these ancient carnivores could see well in dim light,[91] possibly bad news for the newly emerging mammals. A similar exercise, when applied to the capacity for dim-light vision by fish that range in habitats between the sunlit shallows and the eternal darkness of the deep sea, demonstrates that molecular adaptations to different levels of available light evolved multiple[92] times.[93] Note, however, that not only is the evolutionary process reversible so that fish now living in shallow water had deep-sea ancestors, but "similar functional changes can be achieved by different amino acid replacements."[94] There is also the lurking possibility that at least sometimes these arrangements are arrived at by "Non-Darwinian Mechanisms,"[95] inasmuch as the requisite mutations may require several uncoordinated changes that, at first, were accidentally arrived at by genetic drift. Only subsequently, it has been suggested, was there a selective "decision" that enabled the necessary shift in wavelength sensitivity. It is a contrarian thought, but in the world of convergence we need to be constantly aware that coordinations that permit complex solutions may turn out to be far more frequent than realized and very far from random.

Visual perception and color vision evoludetection by hunting eyes, sensitive to polarized light (p. 364n7).

Bioluminescence is so convergent that one can be confident that on any earthlike planet this strangely beautiful light will be flickering from ocean abysses to fungi in rain forests; certainly in both habitats on this planet the arthropods shine forth. Speaking specifically of the myodocopid ostracods and the evidence that bioluminescence evolved at least twice, Todd Oakley[54] drew upon this example as one that might "reveal [the] generalities of evolution,"[55] as well as his emphasizing the likelihood that the luciferases have been co-opted from quite different functions. Among the other arthropods light production is well-known in the insects.[56] Not only that,

but in the fireflies we find intriguing examples of a sort of reverse color vision, where the light organs emit a variety of spectra producing specific colors that range from green to orange.[57] Moreover, just as in color vision (p. 112), the colors produced are controlled by very specific substitutions of amino-acid sites,[58] and in certain cases the shift depends on a single key site. By no means are all insects bioluminescent, and among the dipterans only the fungus gnats are bioluminescent. One might expect, therefore, that at least here it had a single origin. But no. As Vadim Viviani and colleagues[59] demonstrate, not only are the light-emitting organs quite distinct, but the biochemical systems are "completely different."[60]

THE COLOR OF VISION

Bioluminescence presupposes that somebody is watching, but determining what frequencies of light can actually be perceived is second only in importance to vision itself. Even within the eyeless sponges (pp. 96–97), we find one such example, because some swimming larvae show a marked sensitivity to blue light.[61] Different animals are adept at exploring the entire range of visible light (and even in a different way infrared radiation, p. 116), but in the higher energy range of the spectrum, from violet to the ultraviolet, the potentialities and challenges of evolution are well displayed. Many insects[62] possess color vision,[63] and among those plants that rely on insect pollination, the color of the flowers can be crucial. In hymenopterans such as the bees, the match between their discriminatory ability and the color spectra of the flowers is not only strikingly matched but in the endemic world of Australia has arisen independently of that in the Old World.[64] Bees are trichromatic, but in contrast to humans have acquired ultraviolet vision,[65] as have inde-

pendently the butterflies.[66] So have the vertebrates.[67] Among the most curious instances are the bats. To be sure, among those bats that have acquired highly specialized forms of echolocation[68] there has been a loss of ultraviolet vision, but despite their associates also employing echolocation and typically being nocturnal, ultraviolet vision can play an unexpected role.[69]

The advantages of employing the ultraviolet part of the spectrum,[70] which might include an expansion of the spectral window and enhanced capacity for detection of prey or flowers, need to be set against the potential dangers of such high-energy radiation, including damage to the retina.[71] This probably explains why many vertebrates appear to have shifted their sensitivity to the less energetic violet part of the spectrum, and even within the mammals this seems to have occurred independently several times.[72] Pigments in the eye lens capable of blocking dangerous ultraviolet radiation have also evolved independently in vertebrates ranging from

"What do a squid and a legume have in common?"[26] Considerably more than meets the untutored eye. The roots of legumes play host to rhizobial bacteria that assist with nitrogen fixation (p. 310n8), and despite the obvious differences between soil and ocean, the respective bacteria in legume and squid show a number of intriguing similarities, especially in the early stages of symbiotic association.[27]

As far as the bioluminescent bacteria are concerned, very much the same symbiotic arrangement has also evolved multiple times in the teleosts,[28] including the extraordinary angler fish. Moreover, at least one scopelarchid fish[29] (*Benthalbella*) and some squid[30] convergently employ an ingenious "reverse eye," arranged so that light shines through transparent portals.[31] In both cases this reverse eye is based on modified muscle, and at least in the case of the squid not only is there a lens[32] but it employs crystallins.[33] Reverse eye it may be, but not only is it equipped with reflective tissue employing platelets of a novel protein (appropriately named reflectins) that are convergent[34] on the more usual guanine-based tapeta,[35] but it also possesses a retina and opsins.[36] Using its symbiotic bacteria the eye shines, but it evidently also monitors its own performance.

Such illumination most likely serves several functions, and in many cases where the light source points downward it serves as an equivalent to countershading[37] (familiar in fish with a silvery underside and darker dorsum) by helping to diminish the otherwise telltale silhouette.[38] Given that a variety of mesopelagic groups are bioluminescent, it is neither surprising that many employ ventral photophores, nor, as William Clarke duly notes, that they may have "evolved in . . . phylogenetically diverse groups . . . [but] are remarkable in their structural similarity."[39] This convergence of counterillumination encompasses crustaceans,[40] teleost fish,[41] and

sharks,[42] as well as the aforementioned squid, of which another example is the bobtail squid.[43] As Julie Claes and colleagues remark in the context of the shark *Etmopterus*,[44] this recurrent employment of counterillumination "illustrates how evolution can take different routes to converge on identical complex behavior."[45] Obviously an effective system, but what happens as the strength of the illumination (be it from the sun or moon) changes, especially as the squid migrates toward the surface? Ingeniously the animal shifts the frequency of its bioluminescence, but evidently judges this by the temperature of the water as it moves from the cold depths to the warmer surface.[46]

If one cannot disappear by counterillumination then an alternative seen in some jellyfish is either to release a mass of luminescent particles that rapidly turn on or off,[47] or alternatively as a brightly illuminated secretion that sticks to whatever it touches.[48] In the vampire squid there is not only a series of light organs,[49] but the tips of the arms can secrete a viscous fluid that is swept into a glowing cloud around this deep-sea animal.[50] This is not the only squid to produce a luminous secretion,[51] but in the case of *Heteroteuthis dispar* it flows out of a sac-like gland that is located immediately beside one of the hallmarks of the cephalopods, the remarkable ink sac. Although the production of this ink is typically associated with shallow sunlit waters,[52] in the inky depths of the deep sea the cephalopods not only produce both smoke screens and pseudomorphs to assist with either flight or act as a decoy, but they can employ the ink in surprisingly versatile ways.[53] For example, in one case the ink forms a long ropelike structure, perhaps to imitate a venomous deep-sea siphonophore (a relative of the jellyfish). In another case the mantle cavity of the squid expands and fills with ink. This curious behavior may assist in rendering this otherwise transparent species immune to matches that of the animal). The net result of this gradient in the refractive index is that the surface reflectivity is reduced dramatically.[7]

Becoming transparent can lead to quite bizarre reorganizations of anatomy, such as the strange-looking poeobid[8] polychaetes.[9] Somewhat more familiar are the jellyfish, among which are the remarkable box jellies (pp. 97–98). With their contractile bell these animals are familiar enough, but in this regard they are not alone. Particularly striking is a pelagic tunicate, belonging to the doliolids, that in migrating into deeper water has turned itself into a medusoid, specifically resembling a craspedote hydromedusa.[10] In doing so it has abandoned the customary mode of filter feeding and become a carnivore.[11] This does not prevent these and other gelatinous animals from becoming prey,[12]

even though their calorific value as food is depressingly low.[13] It is, therefore, all the more extraordinary that, in the words of John Davenport,[14] the endothermic leatherback turtle (p. 433n121) flourishes "on a diet of cold jelly."[15] To achieve this it has to swallow prodigious quantities of gelatinous food, while its esophagus not only shreds the food but its enormous length allows it to eat yet more and probably warm the jelly. When it comes to gobbling jellyfish they are not alone, because so does the ocean sunfish (*Mola*). This dietary convergence for the gelatinous[16] has a further paradox because not only is the food distressingly unnutritious, but coincidentally or not, ocean sunfish[17] and leatherback are also the largest living representatives of the bony fish and turtles, respectively.

EERIE CONVERGENCE

Some animals do their best to become invisible, and apart from transparency there are innumerable other avenues to camouflage. In other cases, however, they go out of their ways to advertise themselves. Nowhere is this more apparent than in the evolution of that cold light we call bioluminescence. Best known among the insects, such as the fireflies and glowworms (both actually beetles), typically it is used for sexual communication. It can also be employed, however, in the more sinister context of sexual confusion. One striking instance involves a firefly (*Photuris*) that emits flashes of light—as they should, except that the signals are dangerously similar to the females of various other species, and rather than hinting at a delicious engagement serve to lure the males to their doom.[18]

Such a capacity for bioluminescence is rampantly convergent.[19] Although typically it involves the action (by the oxidation of a substrate labeled luciferin) of the appropri-

ately named luciferase, this enzyme is convergent[20] and has evolved at least thirty times independently. Moreover, although the luciferase system is generally regarded as the universal, bioluminescence in at least one syllid polychaete (relatives of which release glowing "bombs," p. 365n47) may involve a different type of photoprotein.[21] Edith Widder notes, "bioluminescence is estimated to have evolved independently at least 40 times," but, she continues, "Remarkably, not only is there evidence of independent origins within taxa . . . but even within individual species."[22] Examples range from the fungi to the insects, as well as many deep-sea animals, including cephalopods.[23] Some of these squid possess specific light organs that employ light-producing vibrionacean bacteria (members of the γ-proteobacteria),[24] which have provided a series of classic insights into symbiosis. Very much in the spirit of this book, Ann Hirsch and Margaret McFall-Ngai[25] inquire,

CHAPTER 10

THE COLOR OF EVOLUTION

EYES ARE FOR seeing, but exactly what will be seen and with what acuity depends critically on the available optical apparatus. Both may be mollusks, but nobody imagines that the world seen in the mirror eyes of a scallop looks the same as through the camera eye of a squid. We might, however, have more confidence that as a sperm whale grapples with a giant squid, what they convergently see is much more similar. Despite living in the oceans, paradoxically the whales have lost the ability to see the blue part of the visible spectrum (p. 113), but otherwise very few mammals lack some sort of color vision. In comparison, and despite their spectacular capacities in camouflage, cephalopods seem to lack color vision.[1] On the other hand, they enjoy a type of vision that humans can only manage when they enter the cool zone and put on their sun glasses—that is, an ability to detect polarized light.[2] The cephalopods, of

course, live in a very different environment, and here we find further subtleties in their employment of polarized light.[3] Cephalopods are not only able to change body patterns with quite remarkable facility and speed, but in addition their ability to see polarized light (and also produce it via reflective structures known as iridophores) means that a cuttlefish can communicate with a potential rival or transmit sexual messages. Not only that, but because nearly all of its potential predators, especially fish, are blind to polarized light, the cuttlefish can conduct a "private conversation."[4] The compliment, however, is returned with a vengeance. Typically fish have a reflective undersurface, which in its own way is an ingenious way of camouflaging the body against the bright "sky." Unfortunately for the fish, the light reflecting off this silvery surface is partially polarized, which the hunting cuttlefish is able to detect.[5]

INVISIBLE CONVERGENCE

Oh dear, to escape this dilemma perhaps it would be more sensible simply to become invisible—that is, to render the body transparent. In principle an excellent idea, because numerous pelagic animals have independently learned how to become transparent. In fact, the evolutionary trick of turning oneself into the equivalent of the Invisible Man has been solved in several different ways,[6] but per-

haps the most ingenious method is to ensure that the surface of the animal is not smooth but studded with minute pimples. Why so? Because in the same way as some insects (p. 102), according to where on this surface the light falls, the refractive index changes seamlessly from the tips of the pimples (where the index is the same as the seawater) to their base (where the refractive index ming around independently without any corporeal link."[251] The fact that eyes are pigmented is potentially highly compromising for animals that are otherwise transparent, although some of the pelagic mollusks ingeniously sidestep this problem by employing natural mirrors as a camouflage that—in reflecting only the surrounding ocean—thereby render the eye invisible.[252]

We started this chapter with Darwin, so in a way let us conclude with him. If his all-encompassing biological curiosity is evident anywhere, it is in the realm of sexual selection. He was particularly struck with the male Argus pheasant[253] and its striking eye-like structures (ocelli) arrayed across the feathers.[254] Writing of the ocelli in the related Malaysian peacock, G.W.H. Davison[255] remarked how in displays to the females not only was the plumage arranged around the actual eye, but in terms of comparison the similarities included "color, convexity, iridescence, roundness and concentricity."[256] This worker also noted that the ocelli stood proud of the rest of the feather, and it would be interesting to know the structural configuration that confers this arrangement.[257] In the case of the Argus pheasant, Darwin was greatly struck by the ball and socket appearance of the ocelli,[258] but this is an optical illusion. Not, however, in all feathers. In the so-called ocular feathers[259] of the blue-winged mountain tanager, which contribute to the striking blue plumage of this Neotropical bird,[260] there is a series of chambers. As Robert Bleiweiss[261] demonstrates, these are responsible for a "color mechanism whose architecture, microstructure, and reflectance parallels that of chambered (pin-hole, mirror, camera) image-forming eyes—to a remarkable degree."[262] There is, of course, nothing like a retina, but in terms of features such as domed portals to receive the light and especially a striking analogue to the reflective tapetum lucidum (such eye-shine is, of course, strikingly convergent in vertebrate eyes), then these ocular feathers are aptly named.[263] As Bleiweiss stresses, "The structural and optical parallels between eyes and feathers can be traced ultimately to certain basic physical principles that apply to any light-handling structure,"[264] but in this imaginative and wide-ranging paper he is careful to emphasize how we might identify the basic organizational principles that are associated with the absorbance and reflection of light. As he indicates, seeing and signaling may have more in common than meets the eye.

of design principles."[229] In particular apart from the capacity to track prey using a single eye,[230] the focus depends not on the lens but on very rapid adjustments of the cornea, entailing employment of a strikingly similar musculature.[231] Srinivasan writes of the "extraordinary parallels"[232] between the sandlance and chameleon, with John Pettigrew and colleagues[233] listing some thirteen, including a remarkably deep fovea in each animal. The sandlance shows a further peculiarity: unlike the universal tendency of the eye to jerk continuously from position to position (as saccades), in these fish typically the eye slowly drifts, perhaps to maintain a panorama or rendering it less conspicuous[234] and so helping the fish to avoid the fate it so readily delivers to others.[235]

Apart from the vertebrates, when it comes to camera eyes the exemplars of convergence are found among the mollusks. While the normal point of reference are the cephalopods (pp. 12–13), among the gastropods—probably most familiar from the garden snail and slug—we find a particular recurrence of this design. Thus, a cameralike eye has arisen in the so-called prosobranchs (probably more than once[236]), most strikingly among the winkles or littorinids,[237] as well as the pulmonate snails.[238] The visual capacities of these snails vary, but at least in the case of the littorinids there is little doubt that the eye is a finely honed piece of optics and able to provide a sharp image. However, among the strombid gastropods (the conchs) and the remarkable pelagic heteropods, the convergences are the most striking. The case of the strombids has long excited attention because these snails are slow-moving herbivores, and possession of a remarkably sophisticated camera eye[239] thus stands in apparent contradiction to the rule of thumb that animals with such eyes tend to be fast-moving, predatory, and intelligent. Strombids might be dozy vegetarians, but they can readily detect predators (including

other snails, such as the lethal cone shells) and in response engage in a series of rapid kicks that propel them to safety.[240]

The large camera eye of the actively swimming carnivores known as the heteropods[241] almost certainly has a vertebratelike acuity, but the retina shows an interesting arrangement; instead of being arranged as the usual cup, it forms a narrow ribbon. In principle the animal should be peering through a narrow slit, but this potential limitation is overcome by a scanning movement whereby the eye quickly flicks down and then slowly returns to its original position.[242] It is a striking arrangement, but not unique because independently several other groups have arrived at the same solution. Best known are the jumping spiders (salticids). These have an acute vision[243] that not only enables them to launch themselves accurately toward their prey but engage in complex signaling that suggests an otherwise unsuspected cognitive sophistication.[244] Here, too, the retina consists of a narrow strip of receptors, but because the lens is anchored to the carapace of the spider, the retina is scanned by an elaborate[245] set of muscles.[246] The range of visual capacities in the jumping spiders varies, but at its most acute employs a complex retinal ultrastructure that is astonishingly effective at collecting light.[247] Not only that, but part of the retina is organized to magnify the image, effectively providing a telephoto capacity.[248]

So does that exhaust the roster of camera eyes among the invertebrates? Not quite. Consider the alciopids, a group of marine polychaetes that are also equipped with prominent camera eyes.[249] These are beautiful pelagic creatures, and like many denizens of this realm are almost completely transparent. The effectiveness of this transparency is tellingly conveyed by Quentine Bone[250] when he writes, "All that can be seen of some Alciopid polychaetes . . . are a pair of crimson eyes and a single row of black spots apparently swim-so is typical of amphibious fish.[213] Just as our vision is blurred underwater (p. 113), so fish are myopic in air. Unlike us the cornea does not normally assist in the focusing of light (effectively because it has the same refractive index as water), so the convergent adaptations revolve around the corneal architecture. In this gallery of four-eyed fish, perhaps the best known is *Anableps*. Here the dorsal and ventral eyes[214] each have their own pupil and are separated by a ridge, with the aerial upper eye having a thicker and flatter cornea. Division of the cornea, but this time more or less vertically, is instrumental for aerial vision in two closely related labrisomid fish from the eastern Pacific (*Mnierpes*[215]) and the Galápagos (*Dialommus*[216]), which engage in intertidal excursions in pursuit of prey. The eyes of these fish, therefore, are equipped with two corneal windows,[217] but another aerial tourist, the flying fish (p. 189), goes one step further; its cornea has a pyramidal structure that defines three fields of view and evidently allows vision in both air and water.[218] Also, in the African fish *Pantodon*, which is a ballistic leaper rather than a flying fish (p. 410n9), the visual field has a tripartite structure with the lower section dedicated to aerial vision.[219] Curiously the fish that has taken the four-eyed solution to its logical endpoint is a deepwater fish (*Bathylychnops*) where either eye has two lenses, each with its own retina.[220] As Ivan Schwab and colleagues note, these represent "evolutionary adaptations so unusual as to seem unbelievable."[221] When one looks at the wider picture, then, as they note, commonalities emerge. In this fish the larger eye points upward, the other downward. Just such an arrangement is found, albeit in different ways, in deepwater crustaceans and cephalopods and represents the need[222] to detect silhouetted forms overhead and bioluminescence below.[223]

Across the biosphere, therefore, the eyes of fish and other vertebrates scan, gaze, exam-ine, and patrol their surrounding environments. Of particular value, of course, is if they are positioned to provide stereoscopic vision, so recurrently we find forward-facing eyes that can confer binocular vision. In birds this arrangement is almost certainly convergent with the mammals, and in the case of the frogmouths (p. 205), this capacity rivals that found in the owls.[224] The frogmouths, however, show a particular curiosity, because the eyes can readily unlock from a binocular condition, thus rendering the bird monocular with each eye scanning a separate panorama. In humans the equivalent squint (or strabismus) where each eye follows a separate gaze is usually involuntary, but in the frogmouth it is readily deployed when these well-camouflaged birds feel themselves under threat.[225]

More generally the ability to rotate each eye independently of its opposite number has evolved several times. Among other vertebrates, surely the most remarkable is the convergence in both optical systems and behavior that joins a diminutive fish known as the sandlance to the more familiar chameleon.[226] In each case the eyes are housed in mobile turrets and readily move completely independently. Typically as one eye scans the world in search of prey, the other remains motionless. Every now and again the eyes swap over, and when a victim is detected, the strike of either fish or reptile is extraordinarily fast and accurate. In the sandlance the lunge involves the entire body, which, as Mandyam Srinivasan[227] points out, is effectively equivalent in action to the protrusible tongue of the chameleon, with eyes added. Despite the radical differences in both size and habitat of these tropical animals, with the sandlance lurking in coral sands whereas the chameleon dwells among vegetation, these vertebrate eyes not only "break all the rules"[228] but, as Srinivasan remarks, they have evidently "converged on a common set as light enters the crystal lattice, it then splits into two rays. Unless viewed along the so-called optic axis, any image will be duplicated. Hardly conducive to unhindered vision, and so in both arthropod and echinoderm, the optic axis is perpendicular to the surface of each lens in the compound eyes (although in the mollusks known as the chitons, the eyes employ the other polymorph of calcium carbonate, aragonite, but here the birefringence of the lens may allow two images, one for use underwater and the other when the animal is exposed[202]). In the case of the trilobites a reasonable assumption is that they had a brain with optic centers. Nevertheless, the schizochroal arrangement is rather strange because in principle each of the large lenses can act as a separate eye, capable of forming an individual image, whereas each lens of a compound eye typically acts more like a point source for the light.

There is, however, an intriguing convergence with the eyes of a curious group of insects known as the twisted-wing insects or strepsipterans.[203] Only the males need the eyes, because in their very short active phase—during which, in the words of Hans Pohl and Rolf Beutel,[204] they represent little more than "flying spermatophores"[205]—they fly in search of the females. The latter typically spend their entire life as the parasitic "guest" of other insects such as wasps. The compound eyes of the strepsipterans are, however, very strange. Quite unlike any other insect, the lenses are enormous and separated by expanses of tough cuticle that typically bear a thicket of hairs around each lens. Overall their appearance is strikingly similar to the schizochroal eyes of trilobites,[206] but as far as the insects are concerned, this odd arrangement bears on several important questions. How and why did it evolve from the typical compound eye of most insects? This question is by no means trivial, because it is far from clear (so to speak) what any plausible intermediate would look like. One escape clause is to argue that in fact the giant lenses seen in the strepsipterans are derived from the simple eyes of the larval form (thus similar to the stemmata in the ferocious diving beetles, p. 246). While this cannot be ruled out, the evidence is in favor of their derivation from the normal compound eye.[207] Unfortunately, in this case the fossil record is of little help, because although a fair number of examples are known, including a strikingly primitive type, they display the classic strepsipteran arrangement.[208] The clue may come when we remember the giant headlamplike eyes of some of the crepuscular hunting spiders, a link that is consistent with the evidence that the first strepsipterans[209] were also nocturnal.[210]

A NEW ORBIT

Overall, the compound eye is nowhere as effective as the camera eye, and not surprisingly several groups of arthropods have made a fair try at evolving something like the camera eye. Other than the jumping spiders, most striking in this regard are the appropriately named ogre-face spider, whose giant eye (specifically the posterior median) is evidently of key importance for its crepuscular hunting by conferring excellent nighttime vision.[211] When it comes to the genuine article, however, the vertebrates provide a central reference point.

They also show various curiosities, not least among several groups of fish that independently have doubled up their eyes, from two to four.[212] Most often this is associated with the need to see in both air and water, and because in some insects a very similar array is found on the wings, including a hawkmoth, which like a number of related forms has glasslike[175] wings.[176] Here, too, most likely they help the animal to disappear in a remarkably sophisticated form of camouflage. Such anti-glare arrays have even found technological applications, not least for window panes and cell-phone displays.[177]

Think of a compound eye and most likely an arthropod swims into view, but as indicated, a variety of animals should come to mind. Few, however, would expect this convergent roster to extend to the echinoderms, perhaps most familiar in the form of the spiny sea urchin and five-armed starfish. Being deutero-stomes, echinoderms are closely related to us, but presumably as a result of having thrown away their heads, most of the nervous system is rather diffuse—and in comparison with many other groups is rather poorly known.[178] Indeed, writing in 1987 James Cobb[179] dryly noted how earlier his external examiner had advised him to become "a world author-ity on something obscure."[180] So Cobb duly did, but the nervous system of echinoderms remains remarkably unexplored.[181] While often described as forming a nerve net, to the extent that the celebrated zoologist Jakob von Uexküll[182] suggested these animals were little more than a "republic of reflexes,"[183] the ner-vous system houses hidden complexities.[184] Some sections form distinct nerve cords,[185] but nevertheless Cobb did not hesitate to describe them as "brainless."[186] Another curi-osity is that in contrast to most metazoan ner-vous systems, the action potentials are based on calcium,[187] with correspondingly slow rates of propagation (p. 436n23).[188]

Nevertheless it has long been appreciated that echinoderms such as the echinoids (sea urchins) are sensitive to light,[189] which is con-sistent with their possessing the "molecule of choice"[190] for vision, rhodopsin (pp. 93–94). Given that a typical sea urchin is constructed

as a sort of globe with spines, is it possible that the entire animal could function as a sort of cyclopean eye?[191] This prescient sug-gestion by Jeremy Woodley[192] has been dra-matically (and elegantly) confirmed by the identification of two sets of photoreceptors, one located near the tips of the numerous tube feet[193] and the other set tucked in near the base.[194] Crucially the skeleton serves to shade incoming light, allowing a direction-ality in vision; as Esther Ullrich-Lüter and colleagues write, "we propose the sea urchin visional photoreceptor system to function rather like a huge compound eye."[195] Nor among the echinoderms are the sea urchins alone. Consider the delicate brittle stars (the ophuiroids). The fact that some are capable of striking color changes[196] certainly hints at some sort of visual capacity.[197] One brittle star, known as *Ophiocoma wendtii*, however, took many by surprise when it was found to pos-sess compound eyes.[198] Like all echinoderms the skeleton is composed of calcareous plates, but in this brittle star some of those on the upper surface of the arms form a striking array of calcite lenses. Not only can each lens correct for spherical aberration, but adjacent cells contain pigments that if necessary can extend over the lenses and thus can control the amount of light received. Compound eyes in an animal without a brain; what exactly are the brittle stars seeing? Certainly a bright light provokes a nervous response,[199] but like the box jellies, the nature of an image pro-cessed outside a brain is conjectural.[200]

Compound eyes of echinoderms are strik-ing on their own account, but especially in the case of the brittle stars because they are con-vergent with the calcitic eyes[201] of the extinct trilobites, notably the so-called schizochroal design found in the group known as the pha-copids. An important feature found in both the eyes of the brittle stars and phacopids is related to the optical properties of calcite, and in particular its high birefringence—that is, mirror optics have been regarded as the pre-rogative of the invertebrates, and however superbly engineered, the camera eye of verte-brates is widely assumed to be highly conser-vative, trapped in a given design. Not so! In the deep-sea fish *Dolichopteryx*, the tubular eyes house a secondary retina in a diverticu-lum, which receives focused light via a mirror composed of ingeniously arranged reflective plates.[157] This arrangement no doubt helps to augment the little available light, but in addi-tion to the sophisticated optical arrangement an important evolutionary point emerges. As Hans-Joachim Wagner and colleagues conclude, although this is "an entirely novel image-forming vertebrate eye . . . [it] demon-strates that image formation in vertebrates is not constrained to refraction, despite the evo-lutionary pathway taken, or retained, by most vertebrates."[158]

Arguably the cynosure of mirror optics is found in the crustaceans.[159] They show vari-ous arrangements, and sometimes the mir-rors operate in conjunction with lenses. Of these the most extraordinary is found in the ostracod *Gigantocypris*. As the name implies, its body size—for an ostracod—is very large, as are its eyes, which consist of enormous par-abolic mirrors.[160] Such an arrangement uses no lenses, and indeed Michael Land exclaims, "Anything less like a conventional eye is hard to imagine."[161] Evidently they serve their pur-pose, because this animal is a denizen of the deep sea, and in ways comparable to noc-turnal spiders (p. 104) the eyes are geared to high sensitivity rather than resolving power. Most ostracods are, of course, more like the great majority of arthropods because they employ compound eyes, a defining charac-teristic of this group. Again the ostracods spring a surprise because it appears that the compound eyes have evolved independently in the group known as the myodocopids.[162] Understandably this development has raised the collective eyebrow, given that within the

crustaceans compound eyes are more or less ubiquitous, but the alternative argument[163]—that there must have been repeated loss of such eyes (and for whatever reason) within the ostracods—seems weaker.[164]

Compound eyes are clearly convergent, and in addition to the examples already given, they have also evolved in some of the marine arrow-worms (or chaetognaths), notably in the taxon *Eukrohnia*, where they are strik-ingly arthropodlike.[165] This convergence is all the more notable because of this group's rela-tive phylogenetic isolation.[166] Not only are the chaetognaths very ancient,[167] but they remain highly effective predators equipped with not only a fearsome array of feeding spines but a decidedly toxic venom (p. 36). Compound eyes are therefore an evolutionary success story and have not only evolved multiple times, but even within particular sorts of compound eye, more specific examples of convergence[168] are found.[169]

Like other types of eye, those of a com-pound design are used both to help find food and to ensure that the animal is not itself turned into dinner. Among the insects, how-ever, the eyes can give the game away, as and when the serried ranks of lenses reflect the incoming light in a concentrated glare. An ingenious solution is at hand, convergently drawing on exactly the same principle that renders an animal transparent (pp. 108–9). Just as in the case of those animals that fade into transparency, so—by bearing an array of nipplelike protuberances—the surface of the cornea serves to equalize, in a progressively gradual fashion, the transition between the refractive index of the air and the eye.[170] Such an arrangement has evolved several times[171] and is best known in the moths[172] (as well as some butterflies[173]). Not only do these nipple arrays serve to mask glare, but in a way analo-gous to plants, they are also self-cleaning and so prevent the lenses from becoming filthy.[174] This arrangement is not restricted to the eyes, and the squid—have very much the same arrangement. It is not, however, universal among the cephalopods, because in the beautiful pearly nautilus the eye is based on the simpler pinhole eye design. This has attracted attention inasmuch as there is a sense that the nautilus "ought" to possess the same sort of eye as its close relatives. Yet the animal lives in deep water and may largely rely on olfaction.[142] Perhaps it uses its eyes to detect bioluminescence, and in any event for this, it hardly needs a lens. We should, however, not underestimate the sophistication of this system, given possession of a high-quality retina comparable to that of octopus,[143] as well as the ability for the "pinhole" to act as an adjustable pupil, albeit slowly.[144]

The pinhole camera has evolved more than once, and among its more remarkable manifestations is an infrared eye in some snakes (p. 117). In terms of the visible spectrum, however, the pinhole has evolved not only in the nautilus but in other mollusks,

this time among the giant clams.[145] They, of course, are entirely sedentary, nestling within coral reefs. At first sight, having more than a thousand eyes arranged along the edge of the shell and embedded in mantle margin looks rather extravagant. However, the giant clam has to perform something of a balancing act. Its shells must gape to allow feeding, but more importantly the mantle margin is also teeming with symbiotic algae that need to be exposed to the sunlight for their photosynthesis. On the other hand, exposed, potentially succulent flesh is an obvious draw for predators. The eyes, therefore, form an alarm system.[146] As Mike Land emphasizes, a pinhole arrangement is just what is required,[147] serving as a minimal optical system. If beefed up and complete with a lens, this would certainly improve the eye's sensitivity and resolution, but then the clam would spend most of its life in near-panic responding to each and every visual stimulus and hence "would spend most of its life shut."[148]

COMPOUNDING VISION

A similar alarm system has evolved a number of other times in sedentary animals. Here, too, the optical system meets similar requirements to the giant clam, but instead of a pinhole camera we see the recurrent evolution of compound eyes, notably in a group of tube-dwelling polychaete annelids (the sabellids)[149] and the bivalve mollusks known as the ark shells.[150] Moreover, and no doubt with a similar function, compound eyes have evolved in relatives of sabellids,[151] in another sedentary group of annelids known as the serpulids.[152] These eyes appear to be the most complex yet identified and, as Richard Smith notes, "are probably the closest rivals to the compound eyes of arthropods."[153] In these worms, with their beautiful fans of feeding tentacles, the eyes are highly effective, judging from both

their sensitivity and the speed of the animal's retraction into their sheltering tubes.

These are not the only animals to keep a wary lookout, as is evident from the extraordinary eyes of a familiar group of bivalve mollusks, the scallops. They, however, explore yet another realm of eye evolution, employing a system of mirror optics.[154] Arrayed around the shell margins, they have classically been thought to act as an alarm system. Their visual acuity, however, suggests a possible capacity to detect predators—and even more remarkably the size and speed of passing suspended particles upon which, of course, the scallop must subsist.[155] Many telescopes employ mirror optics, and unsurprisingly this trick has also been employed by other animals, including even the flatworms.[156] Nevertheless, evolved. Such might be implicit in the insistence by Hiroshi Suga and colleagues when they write, "It is thus very likely that for eye development and/or maintenance, three distinct animal lineages [i.e., cubozoans, hydrozoans, bilaterians] use three distantly related *Pax* genes [respectively *PaxB, PaxA, Pax6*]."[129] They go on to argue that notwithstanding such a distant relationship, the case for the monophyly holds, given that all these genes belong to a single family. In one way this is correct, but once again we may be able to identify deeper principles at work. In the end, what is an eye? To be sure, lenses are widely employed, but they per se are not essential. In essence, any eye is a bipartite structure. Thus, it comprises a photosensitive unit with opsins together with a shielding pigment to ensure directionality of the incoming light. Accordingly this may explain why a paired-box gene is necessarily co-opted, because it has two corresponding binding domains.[130] In other words, lurking in the eyeless sponges is the inherent capacity for an eye.

Eyes, therefore, are convergent. Nowhere is this more apparent than when we stare across the phylogenetic gulf and back into the eye of the cubozoan. The similarities are far more than skin-deep. In vertebrates, much has been made of the fact that the membranes of the photoreceptors are based on a ciliary system, whereas invertebrates tend to have the alternative rhabdomeric photoreceptors. Neat as this dichotomy might be, the reality is considerably more complicated.[131] To give just one example: in terms of so-called cerebral eyes (and, as we shall see, eyes crop up all over the place), it had been thought that the ciliary system was restricted to vertebrates, but more recently they have been identified in the larvae of brachiopods.[132] In the end, either system has particular advantages. In the rhabdomeric arrangement it revolves around sensitivity to illumination, whereas energy efficiency is the hallmark of their ciliary counterpart, but all in all the systems are probably broadly comparable.[133] An important consequence is that not only are there a wide range of opsins in animals, but the opsins of either photoreceptor type (so respectively termed c- and r-opsins) are only distantly related,[134] and each is associated with a distinct phototransduction cascade.[135]

What then of the cubozoans? As they lack a brain, let alone a head, their camera eyes can hardly be said to be cerebral. As in the vertebrates, the eyes are, however, based on the ciliary design. Indeed it transpires that the entire phototransduction cascade is assembled using the same molecular machinery as the vertebrates.[136] While this might reflect a common ancestry, it is much more likely to have arisen independently. As noted,[137] the cubozoan crystallins are convergent on those of vertebrates, but here too the regulatory systems are closely similar, except of course they employ *PaxB* rather than the canonical *Pax6*.[138] As Zlynek Kozmik and colleagues note, their "work reveals surprising similarities in the genetic components used for visual system development in vertebrates and cubozoan jellyfish."[139] Surprising? Hardly, not least because Kozmik and colleagues refer to an "apparent evolutionary 'promiscuity' of developmental cascades"[140] whereby these pathways can be co-opted for entirely new functions.[141]

The World through a Pinhole

I stare into the eye of an octopus, and she glances back; both myself and my aquatic companion are employing a camera eye. The relatives of the octopus—the cuttlefish one that shows a bilateral organization.[109] As Clark Gray and colleagues note, "the integration of visual information in cubomedusae may be more sophisticated than previously thought."[110] This information, which includes a pacemaker signal,[111] is then transmitted to the so-called nerve ring that encircles the bell; in ways that are not well understood, this is translated into the adroit maneuvers[112] of this "honorary fish."

One senses that the cubozoans may yield other secrets. To date there is no evidence that the pairs of camera eyes on each rhopalia are integrated, but these clublike structures bear in total another sixteen eyes. Adjacent to the lower eye there is a pair of slit eyes and on either side of the upper eye there is yet another pair, the pit eyes. While these eyes are considerably simpler,[113] especially in the case of the pit eyes, those with a slitlike arrangement not only possess photoreceptors (rich in mitochondria) but cells with a vitreous content that evidently act as a sort of lens,[114] albeit lacking the power of refraction.[115] While it is not unusual for animals to possess two (or even more) types of eyes,[116] an animal equipped with twenty-four eyes, eight of which are highly sophisticated camera eyes, does seem a little excessive. One possibility is that each type has its specific function. Megan O'Connor and colleagues[117] draw attention to the slow temporal resolution of the camera eyes being consistent with a mode of life that has less concern with small and rapidly moving objects, but as importantly they propose that "the visual system of box jellyfish is a collection of special purpose eyes."[118] Even so, there is evidence that the camera eyes and pit eyes work together to control, via the pacemaker,[119] the rate and direction of swimming.[120] The slit eyes perhaps serve some other role and may be sensitive to vertical movements. As Anders Garm and S. Mori also note, "What exact information the eyes register and how it is being processed by the rhopalial nervous system is largely unknown."[121] Should we ask whether we are in danger of underestimating the sophistication of these "primitive" animals?[122] Is it even conceivable that the combined eyes provide a sort of superintegrated system, a type of adaptive optics?

So what has all this to do with the evolution of eyes in the wider context, not least the iconic role of the *Pax6* gene?[123] The complex optical system of cubozoans obviously points to a certain degree of developmental sophistication, so given what we know of the evolution of camera eyes, it might be expected that as in nearly all other cases cubozoans will call on *Pax6*. Except they don't. Rather they employ the *PaxB* gene,[124] the one that had already evolved in the more primitive and eyeless sponges (p. 97). First loophole for the deep homology argument: Perhaps *PaxB* is simply the precursor to *Pax6*? No, in this particular case the evolutionary connection with *Pax6* seems to lie with another of the cnidarian *Pax* genes, the one referred to as *PaxC*. So, do cubozoans possess *PaxC*? Yes, but this gene has no involvement in cubozoan eye development.[125] Nor is this the end of a somewhat complex story. Among the cnidarians, eyes[126] have not only evolved in the cubozoans, but have emerged fully fledged in the jellyfish stage of hydrozoans.[127] And which of the *Pax* genes do they employ? *PaxB* as in the cubozoans? Failing that, what about the possible precursor of *Pax6*, the *PaxC* gene? The answer is: Neither! They use the third of the *Pax* genes available to a cnidarian: *PaxA*.[128]

To the outsider, this blizzard of *Pax* genes may seem arcane, but on it hinges an issue central to convergence. After all, if the same gene is employed in *all* eyes, then this monophyletic origin hardly squares with multiple evolutionary inventions. Much is at stake here, and not only with respect to eyes. If the proponents of deep homology are correct, then one could argue that however stunning the evolutionary end products, without the necessary genes they might never have cnidarians), but these extraordinary animals have moved into a new realm. They almost qualify as "honorary fish." Many observers have remarked on their agility and speed of movement, with an easy maneuverability that includes sudden turns.[86] Prey, such as fish, once caught are conveyed toward the highly prehensile lips and rapidly swallowed.[87] In at least some species, the principal digestive area is not the stomach but canals within each tentacle. These, as Jamie Seymour[88] notes, have linings that "bear an uncanny resemblance to the villi that line vertebrate digestive systems."[89] With a metabolic rate ten times that of other jellyfish, it is no more surprising that they are unique among the cnidarians in displaying a sort of courtship and subsequent copulation.[90]

Such a life would hardly be possible without eyes, the animals being equipped with a visual battery carried on four clublike extensions (the rhopalia) that hang from the bell. The centerpiece are the camera eyes,[91] two on each rhopalium, the upper pointing skyward and the other downward. The lower eye is more remarkable because it has a pupil. Moreover, in both eyes the lens is not only spherical but just like those found in cephalopods and vertebrates they show a graded refractive index that corrects for spherical aberration.[92] This correction is best developed in the upper eye, but the ability to produce a precisely focused image runs into an extraordinary paradox because the actual point of focus lies well *behind* the retina.[93] "Honorary fish" they may be, but if the box jellies live their life in a blur, why do they possess such sophisticated eyes? It had been thought the camera eyes were used in hunting, and even the possibility of color vision has been raised.[94] In fact this strange defocusing appears to be an adaptation to allow the animal to navigate safely without constant distractions (in a way analogous to the pinhole eyes of the giant clam, p. 101).[95] This is most obvious in those species that live among mangrove

roots, hunting dense clouds of swimming copepods.[96] These arthropods aggregate in the sunlit shafts that speckle the mangroves, and into these areas the box jellies precisely navigate and at the same time avoid potentially damaging encounters as they move among the thicket of submerged roots.[97] It is also important that the jellyfish do not stray beyond the confines of the mangroves. Here that upwardly pointing eye plays a vital role, because it can detect the trees, employing them as terrestrial cues.[98] It is also likely that, as with other groups, these box jellies show a range of optical capacities. In at least one species, the capacity to focus is almost disastrous,[99] and in the specific case of the lower eye, Megan O'Connor and colleagues note, "the alarming conclusion is that the lens only has a marginal effect on spatial resolution,"[100] so that replacing the lens with seawater would make little difference. In some respects this eye is closer to the pinhole design (pp. 101–2), but it touches on the apparent Darwinian dilemma that if a lens is practically useless, why bother to evolve it at all? One possibility is that originally the role of the lens was to protect the retina and only subsequently was it co-opted for an optical function.[101]

Box jellies have, therefore, a remarkable visual capacity, but seemingly of a rather special sort. They are visually adept, and in laboratory experiments can not only readily avoid obstacles,[102] but have been seen to navigate precisely to a submerged twine before docking using curious adhesive pads.[103] So the question is repeatedly raised: How can an animal like this see without a brain?[104] Like other cnidarians, the nervous system is more like a net and has no concentration of neurons to form anything like a brain. It might, however, be a mistake to underestimate the neural capacities of these animals. Thus, it is likely that visual processing begins in the retina[105] and then continues in the rhopalia.[106] The latter structures not only have quite a complex[107] nervous system,[108] but ing, because these proteins are very widely employed in light sensitivity (for example, in plants and for circadian rhythms), but this capacity is convergent.[72]

Remarkable as these sponge spicules are, this is not the only time a biological optical fiber has evolved. In a syllid annelid, which has a complex eye (and is well on the way to forming a camera eye), the lens is traversed with a striking array of rods.[73] These, Jerome Wolken and Robert Florida suggest, may be "a fiber-optics system serving as an efficient light collector for the photoreceptors,"[74] spe-

cifically to detect bioluminescence radiated by their glowing mates, hence the appropriately named fireworm. The connection between bioluminescence and a fiber-optic system has also evolved in the cephalopods, specifically in the firefly squid, where photophores located beneath the eye have a striking array of light guides that not only channel the light but most likely also polarize it.[75] Fiber optics find still other roles,[76] because it turns out that a specific component of the vertebrate retina, the so-called Müller cells, have just such a capacity.[77]

Eyes, but No Brains

While the optic-fiber systems in these various eyes are elegant adaptations of an already complex system, given the absence of a nervous system in sponges, the idea of their spicules playing a similar role might seem a step too far. Surely their entire anatomy is too simple? The genes, however, tell a different story, because in contrast to the anatomy it is evident that as Maja Adamska and colleagues[78] note, "The developmental genetic toolkit of sponges is surprisingly complex."[79] Embedded in the sponge genome are genes involved in eye development. Not any old gene, but ones like $Six 1/2$[80] and Pax[81] that play a canonical role in forming the eye. Each, however, is only represented by one gene, and although a role in the photosensitivity of sponges[82] cannot be ruled out, this seems less likely.[83] In terms of straws in the wind, these genes are, however, surely a harbinger of things to come. Indeed the sponge Pax gene, known as $PaxB$, is involved in the eye formation of a jellyfish. All neat and tidy—a classic case of deep homology? If only it was so simple.

To explain why, we need to turn back to that Darwinian epitome, the camera eye—not in its many iterations among the vertebrates and mollusks, but in its strangest manifes-

tation in a group of jellyfish, specifically the box jellies (the cubozoans). This distinctive group of jellyfish have, as their name suggests, a more or less quadrate body. Despite this distinctive body, like other jellyfish the animals employ a sort of jet propulsion. This mode of propulsion, which can be contrasted to a sort of rowing action found in other jellyfish,[84] is typical of ambush predators. Such swimming involves propulsive spurts that allow the jellyfish to move away from potential threats and toward new foraging areas where they then passively drift . . . waiting. As ambush predators, the box jellies are not to be tangled with lightly; one species (Chironex fleckeri) is regarded as one of the most venomous animals on the planet. Contact with the stinging cells has agonizing consequences that unless quickly treated can lead to death (although for immediate first aid, use vinegar!). While the related carybdeid box jellies may be less lethal, they induce the extremely painful Irukandji Syndrome.[85]

These potentially fatal encounters are accidental inasmuch as, although the box jellies are hunters, they pursue small prey. They may be close relatives of such animals as the sea anemone and coral (collectively the but here we find an extraordinary anticipation of color vision.[60] This is because the rhodopsin is not only sensory[61] (and in showing distinct differences from other bacteriorhodopsins converges on the rhodopsin of vertebrates[62]), but it adopts stable states that are sensitive, respectively, to orange and blue illumination.[63] Why be color sensitive? Most likely the cyanobacteria possess a feature known as chromatic adaptation, whereby the two wavelengths of light stimulate the production of different types of pigment that are most suitable for photosynthesis in that particular part of the spectrum.[64]

EYES, BUT NO NERVES

The ability to detect light has, therefore, extremely ancient origins, but most people would insist on some sort of distinction between sensitivity to light and actually being able to see an image. The latter, they would go on to argue, surely depends on a nervous system that interprets the electrical signals. It might be reasonably concluded that an eye without nerves is useless, but as we have already seen with the warnowiid dinoflagellates, this may not be the case. As we move into this problematic evolutionary territory, we may need to rethink what is perceived and how it is interpreted. We should be careful not to automatically dismiss organisms that lack a nervous system as somehow being irredeemably "primitive," even "stupid." Here the glass sponges may be unexpectedly informative.

A hallmark of the sponges are their spicules.[65] In the glass sponges, as the name suggests, the spicules are composed of silica, but they also effectively define the animal. This is because the thin tissue that constitutes the body is little more than a smear and literally festooned from its glassy cage. Some of the spicules are elongate and serve to anchor the sponge to the seafloor. Remarkably, however, these spicules also are closely similar to the optical fibers used in the telecommunication industry, with a core of solid silica surrounded by a series of concentric shells that show changes in the refractive index, allowing them to act as highly efficient light guides.[66] For what purposes, and if it is not an accidental similarity, in which direction is the light being channeled, and for whose benefit? In sponges living in shallow sunlit waters, the spicules may serve to channel light deep into the body, most probably for the benefit of symbiotic algae.[67] In the glass sponges the mystery deepens because they typically are found far below the sunlit (or photic) zone of the ocean. Speculation now takes over, but one possibility is that the spicules act more like a "reverse eye" (p. 110), serving to broadcast the bioluminescence generated by the sponge and so helping to attract shrimps that end up living symbiotically in the sponge cavity. Some support for this idea comes from the fact that the frequencies of light that can be transmitted are rather precisely controlled,[68] so maybe the sponges act as beacons?

This need not, however, be the only possibility. The spicule length is typically in the order of 10 to 15 cm, but in *Monorhaphis chuni* they can reach an astonishing 3 m. These, too, act as highly effective optical fibers,[69] but here Xiao Hong and colleagues suggest that these "sponges are provided with an unusual, (perhaps) unrecognized photoreception system,"[70] where the complete absence of a nervous system does not necessarily preclude a sort of vision. It would be tempting to invoke the opsins, but evidence from another group of sponges with siliceous spicules suggest that the photosensitivity may depend on so-called cryptochromes.[71] Nor is this surprisnot least with the near universal employment of the *Pax6* gene. Does not my insistence on convergence now begin to crumble? After all, given such a commonality, is not the natural conclusion that all eyes must have a common origin?[33] Not so fast. Despite the seductive appeal of *Pax6* on further analysis, this (or any other) example of so-called deep homology[34] soon begins to unravel. In evolution it is better to take the long view. Let us begin by reminding ourselves that eyes are by no means restricted to animals. I speak of the dinoflagellates.

Although these protistans are usually photosynthetic and so employ chloroplasts,[35] some types root themselves to fish as parasites (a phenomenon that is almost certainly convergent[36]) while others have transformed themselves in a radically new direction. Welcome to the warnowiids; despite being a single cell the organism carries an extraordinary bulbous eye[37] (or occasionally two[38]). Its similarity to the eyes of the animals, even to the extent of having an irislike structure,[39] has been repeatedly stressed.[40] Indeed an earlier investigator, one Carl Vogt, insisted that it was the result of a protistan ciliate having accidentally swallowed the eye (ocellus) of a jellyfish.[41] Others were not fooled, but as long ago as 1887 Georges Pouchet[42] insisted that if we didn't know it was attached to a dinoflagellate then we would have no hesitation in attributing it to an animal. This point was echoed by F.J.R. Taylor,[43] who emphasized how they had "an uncanny parallelism to the structure of metazoan eyes."[44] Moreover, although most workers have emphasized the similarity to a camera eye, in one species of warnowiid it has taken a step toward a compound eye, because in this case up to fifteen discrete crystalline structures occupy the upper surface,[45] much like the facets of an arthropod eye.[46]

Even though the optical system of the warnowiids functions like any other eye, its origins are quite unlike that of any animal, with the lower part being derived from a now-redundant chloroplast.[47] It will, however, be surprising if rhodopsin is not employed.[48] Interestingly, the lens of the warnowiid eye is surrounded by mitochondria,[49] although the precise origin of the lens itself is not clear. Even so it is clearly capable of precise focusing,[50] which makes sense because the warnowiids have become hunters. For some investigators the possibility that these single-celled organisms can actually see is a step too far.[51] As Pierre Couillard[52] remarks, "If an image is really formed on the retinoid, we fail to see, in the present state of knowledge, how an integrative computer could exist to analyze it within the cell."[53] Who, I wonder, is failing to see?

Although these dinoflagellate eyes represent a remarkable acme of protistan complexity, eyespots occur in many single-celled protistans.[54] Of these, the eyespot of the green alga *Chlamydomonas* has received the most attention,[55] but as Gáspár Jékely notes, among the eukaryotes a capacity for phototaxis has "evolved at least eight times independently,"[56] including even some fungi[57]. Yet the roots of a visual capacity go much deeper than the single-celled protistans, at least as far as the cyanobacteria. Like many prokaryotes they employ bacteriorhodopsin, but when it comes to light detection, then once again this cyanobacterial opsin turns out to be the "molecule of choice." Rather remarkably one type of filamentous cyanobacterium (*Leptolyngbya*) possesses a distinct eyespot orange in color and housing a rhodopsinlike protein.[58] Given that they are photosynthetic, it might seem strange that any cyanobacteria requires an eyespot, until one learns that this variety was found living on the damp frescoes of the famous Golden House (Domus Aurea) in Rome. Here in the dim light the cyanobacteria employ their eyespots to move, very slowly, toward the optimum illumination.[59]

In terms of photosensitivity, other cyanobacteria show a more generalized capacity, aptly referred to their employment in the context of vision as "irresistible";[15] they are indeed the "molecules of choice." Their special features are noteworthy: not only do they show intense absorption and an ability to operate across the entire visible spectrum (and also ultraviolet)—and the reaction is rapid, precise, and reliable (along with a quick recovery)—but the all-important retinal component is derived from a common vitamin A (as β-carotene).[16] While being central to the transduction of photons, as is so often the case we find that such molecules are also remarkably versatile[17] and although often located in the eye also have nonvisual functions. Melanopsin is probably the best-known example.[18] One major surprise was the discovery[19] (in the larvae of fruit fly) that rhodopsin is also involved in thermosensation[20] (pp. 116–17). Why and how it was recruited[21] remain unsolved questions.[22]

Molecules of choice opsins may be, but convergence remains the guiding principle. The absence of functional opsins in the most primitive animals suggests that the variety (so-called type II) employed in eyes is an animal invention.[23] Paradoxically, however, rhodopsin has an effectively universal distribution. It occurs not only in other microbes,[24] notably in algae[25] and fungi,[26] but also bacteria where it is generally referred to as bacteriorhodopsin (or type I opsins).[27] The overall structure of this bacterial protein as seven helices inserted between two membranes (thus in the trade known as seven-helix transmembrane protein or 7-TM) is strikingly similar to the rhodopsin in our eyes, but it is almost certainly convergent.[28] In reviewing the three main lines of evidence (lack of sequence similarity, important structural differences, and absence of opsins in primitive animals), Nicholas Larusso and colleagues note that while there are some dissenting voices,[29] the evidence very much points to this being "an amazing convergent evolution of the two opsin types . . . [implying] that the 7-TM structure common to both families of opsins is *vital* to the sensory functionality of the protein."[30]

In the archael bacteria the bacteriorhodopsin has long held a classic status as a proton pump that is driven by light energy. This is only the beginning of the story, because this pump can be readily converted into a sensory receptor. Given that the microbial rhodopsins are evidently derived from bacteriorhodopsin, it is hardly surprising to find them employed in algal eyespots or fungal[31] phototactile responses. Nor should the aficionados of convergent evolution raise much of an eyebrow when we discover that (in the form of archael halorhodopsin) this protein can rescue retinal cones subject to the destructive condition of retinitis pigmentosa.[32] Perhaps one day even the blind will see. From an evolutionary perspective, rest assured that wherever a sun shines, so rhodopsin will be there to convert photons into information. So just one more evolutionary fact? Perhaps not, because in such molecules we have extraordinary potentialities, not least the capacity for color vision with the implication that the qualia of a red sunset are as vivid and beautiful for us as any sentient extraterrestrial. Suddenly we stand on the edge of mysterious intangibles, as I am torn between the mundane sense of "Well, what on earth did you expect?" and a cosmic thrill that "This is truly extraordinary!"

EYES, BUT NOT EVEN IN ANIMALS

The molecular machinery of crystallins and opsins underpins the operation of any eye, but so too it transpires that their developmental systems show startling similarities,

CHAPTER 9

WHEN EVOLUTION BEGINS TO SEE

READING DARWIN, one doesn't usually get the impression he was overendowed with a sense of humor. On the other hand it is difficult to avoid the thought that his famous rumination in chapter 6 of the *Origin*[1] on the eye as an organ of extreme perfection involved more than a little leg-pulling. Even if it was, one should never forget Darwin's steely purpose. Whatever the *Origin* set out to do, one of its principal purposes was the assassination of William Paley's watchmaker.[2] Darwin undertook this with consummate intelligence, but perhaps Darwin never could quite free himself of Paley's seductive arguments?[3] Certainly others found the spell difficult to break. Consider the musings by the Spanish genius and founding father of neurobiology Ramón y Cajal in his autobiography.[4] Writing of the retina, he remarks how

life never succeeded in constructing a machine so subtilely devised and so perfectly adapted to an end as the visual apparatus. . . . I must not conceal the fact that in the study of this membrane I for the first time felt my faith in Darwinism (hypothesis of natural selection) weakened, being amazed and confounded by the supreme constructive ingenuity revealed not only in the retina and in the dioptric apparatus of the vertebrates but even in the meanest insect eye. There, in fine, I felt more profoundly than in any other subject of study the shuddering sensation of the unfathomable mystery of life.[5]

Should we join y Cajal in his vertiginous experience, or might we transform this precipice into sunny grassy slopes where the Darwinian children run uncaring?

DARWIN'S SHUDDER

Much is made of the potential ease of transformation of a simple eyespot into a camera eye.[6] Equally striking is the capacity of the optic cup to form by self-organization,[7] reinforcing the sense that making an eye is far from difficult. Perhaps less often is it remembered that the two vital prerequisites for any sort of eye—that is, the maintenance of a transparent tissue and a mechanism capable of transducting photons into an electrical signal—draw on molecular systems that evolved long

before any eye. To confer transparency, eyes almost always look to crystallins, although not invariably.[8] The crystallins are rampantly polyphyletic, being recruited from a variety of proteins (often enzymes) whose original function was to deal with a variety of physiological stresses.[9] Even among the camera eyes of cephalopods[10] and vertebrates,[11] not to mention the cubozoan jellyfish,[12] the crystallins have entirely separate origins.[13]

What of the opsins? Russell Fernald[14] has

different from typical mussel byssus,[232] and it is a fair bet that sea silk is richly endowed with silklike units. Just as various bivalve mollusks employ a molecular construction that employs silklike motifs, so other invertebrates that need to anchor themselves to the seafloor[233] look to similar solutions.[234] While attachment is mediated in many cases by silklike molecules, in the case of the barnacles the cements are proteinaceous, but interestingly the process of polymerization is strikingly similar to that of blood clotting, in particular with the employment of serine proteases.[235] Gary Dickinson and colleagues go so far as to suggest that this cementation is "a specialized form of wound healing,"[236] given that a cut finger and the limpetlike adherence of a barnacle[237] look to the same enzymatic mechanisms. But let us not remain rooted to convergence; let us rather begin to look further.

this group extends to the only hemipteran (a leafhopper, *Kahaono montana*) known to spin silk[216] (to construct shelters to provide defense against predators[217]). The hymenopterans represent yet another excursion into spinning silk,[218] and in addition to the ants, so too among the bees their larvae[219] produce what is a completely novel silk.[220]

So closely is silk associated with the spiders and such insects as the silkworm—both, of course, members of the arthropods—that it is understandable if one concludes that the evolution of this quite remarkable protein must have been contingent on the emergence of this particular phylum. This would not explain, however, its multiple and independent origins, nor is it obvious that the ancestral arthropod was biochemically favored. The fact that there is a variety of adhesive fibrous proteins can be seen as simply one glimpse into the diversity of nature. Yet such a catalog will overlook both recurrence of form and the search for general biological properties—outcomes of evolution to be sure, but inevitable manifestations of organic order. One clue that this may be correct comes from the recognition of a marine silk, specifically to bind together the agglutinated tube of an amphipod crustacean.[221] An arthropod to be sure, and one that also appears to draw the silk from the pores located on the tips of the legs, but crucially the threads are composed of a mixture of proteins and mucopolysaccharides. The former evidently confer to flexibility, while the adhesive properties are linked to the mucopolysaccharides. In terms of protein structure, the crustacean silk has obvious differences to its terrestrial counterparts, yet as Katrin Kronenberger and colleagues note, "The comparison of independent amphipod and spider silk gland processing systems suggests certain generic underlying processing principles for a silk."[222] Such principles also apply to adhesion.

Consider, for example, the remarkable adhesive threads, known as the byssus, that serve to anchor such bivalve mollusks as the mussels to the substrate. It seems a far cry from the apparent fragility of a spider's web to animals capable of withstanding the surge of an Atlantic storm. It will come as no surprise that the byssal threads are superb examples of bioengineering. The adhesion to the rock is via a natural glue, and interestingly the junction between fiber and plaque is very tough. Its basis? Proteins rich in the amino acid histidine, and with a remarkable number of binding sites to copper.[223] Another oddity? No, recall that just such an arrangement is found in the astonishingly robust jaw of a polychaete worm (p. 66). The byssal threads themselves are a brilliant compromise, between a section that has "give" and more distally a crystalline length that is very durable.[224] Interestingly, the former region includes collagens, which unlike its more typical equivalents elsewhere, display a pronounced extensibility and indeed converge on the elastins (p. 16).[225] The collagen that helps to confer strength is equally interesting because it contains two units that are strikingly similar to the fibroin sequences in spider silk, specifically the dragline whose hallmark is again remarkable strength.[226]

These similarities reach a curious acme in the textile known as sea silk, woven to provide such garments as hats and cravats and once very highly prized for its fineness, suppleness, and color.[227] Now largely forgotten, one might assume it to be the silk from some marine arthropod, perhaps one of the pseudoscorpions? No, sea silk is obtained from the byssus of the fan mussel. How long sea silk has been used is debated. Daniel McKinley[228] gives a peppery and entertaining overview in his "foray into historical misappropriations,"[229] dismissing such favorites as the Argonauts' Golden Fleece, yet an example is known from fourth century Roma Aquincum (Budapest).[230] While some of its mechanical properties are not like silk,[231] an earlier analysis revealed a molecular structure very silk function."[189] Thus it is no accident that various features of silk, not least the amino acids employed, are so recurrent. Many hymenopterans use silk. Among these, some ants are particularly proficient,[190] where silk finds its most famous employment in weaving and in this capacity most likely has originated several times.[191] Such weaving is, however, typically employed in nest construction, and not to build silken webs. Nevertheless, at least two groups of aquatic insects employ nets to trap food.[192] These are the relatively well-known caddisflies[193] (belonging to the trichopterans),[194] and also the larvae of midges (chironomids).[195] Silk is also used as tie-lines in fast-flowing streams, but this example is more remarkable because it involves caterpillars of the Hawaiian moth *Hypsomocona*, in which at least three lineages have become aquatic.[196] This genus has undergone a striking set of adaptive radiations,[197] and the caterpillars of another species of this moth have moved from the familiar territory of chomping plants to hunting snails, which the caterpillar first enmeshes in a silk net before devouring.[198]

Nor is the limit in terms of versatility of silk use among lepidopteran larvae. The larvae of an ithomiid butterfly show a remarkable adaptation because collectively they spin a silk scaffolding over the leaf surface that is bristling with defensive trichomes[199] (pp. 158–59), allowing them to crawl over the leaf with impunity.[200] More striking are the caterpillars of the choreutid moth, *Brenthia*.[201] Here the caterpillar constructs a silk roofing on the leaf, which may be supported by "stalactites" composed of fecal pellets.[202] The silk, however, provides more than just a barrier, because it also serves as an extraordinary extension of the caterpillars' sensory field. Contact is mediated by conspicuously long setae, allowing not only extraordinarily rapid reactions but employment of a previously chewed escape hatch leading to the safety of the underside of the leaf.[203] The very frightened caterpillar keeps in touch with the news from above via two especially elongate setae. Using silk as a sensory field recalls, of course, a spider's web, but this strategy has evidently evolved several times independently in the lepidopterans. As Jadranka Rota and David Wagner[204] note, "The multiple origins and similarities in associated behaviors provide a compelling testament to the selective advantages that accrue to the practitioners of this disappearing art."[205]

Among the dipterans, in addition to the aquatic chironomid midges, the larvae of fungus gnats (mycetophilids) also spin adhesive webs. Some cover the ground,[206] but more spectacular are those that suspend silken ropes from the web to engage in a sort of aerial fishing.[207] As far as the dipterans and silk are concerned, the examples of convergence do not end here. Consider the dance flies (or hilarinids, belonging to the empidids, which in another group have evolved a mantidlike raptorial forelimb [p. 53]) which independently have evolved a silk.[208] As their name suggests, these flies form mating swarms, and the silk plays a key role. It is only produced by the males, and like the pisaurid spiders, it serves as a nuptial gift. The silk is produced by glands housed in a leg segment (the basitarsus) that connect to hollow spines.[209] The arrangement has many similarities with the aptly named webspinners (embiopterans),[210] which have probably been spinning their silk since the Jurassic.[211] Here, too, we see that the silk, which as Matthew Collin and colleagues[212] note, "has remarkably convergent similarities [at a molecular level] and surprising differences [in terms of tensile strength]"[213] to that spun by other arthropods, is weaved into domiciles derived from glands packed into a swollen segment of the front leg.[214] Webspinners and dance flies are distantly related, and this distinctive way of producing silk is surely of independent origin.[215] Webspinners are regarded as quite primitive insects, and the roster of convergences among stickier it is. This leads Ingi Agnarsson and Todd Blackledge to conclude that "orb webs function optimally when threads are able to detach and adhere repeatedly to struggling prey."[172] While the trapping abilities of the spider's web are certainly remarkable, some insects still manage to escape the net. Particularly notable are the moths whose bodies and wings are covered with tiny scales. When they fly into a web the scales readily detach, but the story is not entirely one-sided. One strategy adopted by the spider *Scolodecus* is to extend the web into a remarkable vertical ladder.[173] When the moth blunders into the trap it slides down the ladder, leaving a trail of scales until eventually enough are detached for the viscid silk to capture the prey.[174]

Even more remarkable are the so-called bolas spiders.[175] Evidently they evolved from spiders similar to the cyrtarachnids that successively reduced their web.[176] The female bolas spiders still spin a sort of web that effectively provides a trapezelike platform, but as their name suggests they employ a single line of silk that ends in a sticky droplet. Either swung as a pendulum or in rapid whorls, it readily adheres to a flying moth[177] (it is no accident that one species of the best-known bolas spider *Mastophora* is named *dizzydeani* by William Eberhard after "one of the greatest baseball pitchers of all time, Jerome 'Dizzy' Dean"[178]). The silk droplet is very liquid and so can flow between the scales of the moth, while the center of the bolas contains a folded thread that might help to secure the prey. Once trapped the moth never escapes, although notwithstanding "Dizzy" Dean the success rate is not that high (a couple of moths per night might be average). The chances of intercepting a moth would seem very low, until one realizes that the bolas spider is capable of a remarkable molecular convergence; it produces chemical counterparts to the sex pheromones[179] that are used to attract the male moth, luring him to a sticky end rather

than an expectant female. This mimicry is all the more remarkable because the bolas spider can attract several species of moth,[180] each of which produces a specific pheromone.[181]

To see a spider's web en-jewelled on a misty autumnal morning may distract attention from its more sinister function. Recalling the monstrous Shelob it is difficult not to feel a frisson of horror as the captured fly (or even bird[182]) is efficiently en-parcelled in silken chords. But in some spiders the food is being gift-wrapped to provide a so-called nuptial gift. While such gifts (of many varieties) are a commonplace among the insects,[183] including the dance flies (p. 90), it is very rare among the spiders. Best known in the so-called nursery web spiders (the pisaurids), it has evolved convergently in a Neotropical group (*Paratrechalea*)[184] and in both cases has a very important function that may not be entirely unfamiliar to some of my readers: placating the female. As with people, courtship in spiders is not without its risks, especially for the male. In the spiders that present nuptial gifts, the pattern of sexual behaviors has significant similarities. These are not identical, however; perhaps the most important difference is that, in some pisaurid spiders, if the female becomes aggressive and the going gets rough, the male promptly drops "dead."[185] Not only is the male completely motionless but still holding on to the gift; the female may drag the "corpse" around. Evidently she loses interest, begins to enjoy her wedding breakfast, and presumably fails to notice that the "corpse" has very cautiously revived.[186]

Among the insects alone the capacity to make silk has evolved multiple times.[187] In their crisp review, Tara Sutherland and colleagues identify some twenty-three "silk lineages,"[188] and note that "Given the diversity of insect silks, it is remarkable that certain features are common among silk production systems. Convergent features in independent lineages likely represent key properties for blundering into spiderwebs, and from Lower Cretaceous (c. 135 Ma) amber there is exquisite evidence not only of silk but entrapped insects.[148]

This silk may well have formed an orb web, which is probably the most familiar example of silk in the natural world. Given its distinctiveness one might expect the orb web to have a single origin, and so much is indicated by the genomics of silk.[149] But even if the silken orb evolved only once,[150] in a distantly related spider (*Fecenia*), an orblike web has evolved independently[151] and its dragline silk is markedly more robust than that of its relatives.[152] Consider also the tetragnathids. These are a widespread group of spiders, but those that arrived in Hawaii underwent a series of striking adaptive radiations, and not just in morphology.[153] Thus, in what Todd Blackledge and Rosemary Gillespie[154] term "ethotypes," it transpires that three highly distinctive orb web constructions evolved independently across the islands, with each type arising twice. As the authors remark, "Thus, evolution can act in surprisingly predictable ways even upon complex behavioral traits."[155] And not all these spiders spin webs. Others have moved to active hunting, including one remarkable tetragnathid that impales passing insects on an enormous claw.[156] This lineage has strikingly long legs, and not only are there four distinct ecomorphs, but in one or another combination these have also evolved several times independently, thus imposing a recurrent community structure.[157]

The evolution of the silk in webs has its own fascinations. The most primitive type forms simple cylinders, and its adhesion to captured prey depends on the same van der Waals forces[158] that the gecko feet employ. This type of silk is known as cribellar because as it is drawn out of the spider it is passed over a spinning plate (the cribellum) located on the abdomen. More usually this silk has pufflike structures, and these evidently employ a form of capillary or hygroscopic force that further enhances the adhesion.[159] This use of moisture possibly explains a remarkable breakthrough in silk design, because in the majority of spiders the capture thread is now wet and viscid.[160] The trick was to add a liquid coating, and because this is unstable it spontaneously organizes itself into a series of sticky droplets[161] (and these are beautifully preserved in Cretaceous amber[162], and also find a parallel in the velvet-worms[163]) that ingeniously the spider itself does not fall foul of as it races across the web.[164] Having an aqueous solution hanging from the equivalent of a washing line is, of course, highly problematic, given the likelihood of evaporation. In the spiders, at least, the solution is brilliantly arrived at by the redeployment of molecular compounds that otherwise are associated with the control of osmotic stress—that is, the movement of water from less concentrated to more concentrated solutions. Now they serve to stabilize the droplets, and as Fritz Vollrath and colleagues[165] note, "The composition of the solution is as good as any that can be devised to prevent evaporation and yet not interact with the fibres."[166] Water also plays an important role in the so-called dragline silk.[167] Ingi Agnarsson and colleagues suggest that the mechanisms by which the water is added or extracted from the silk "is strikingly similar to the mechanism proposed to explain how plant tissues can act as motors,"[168] such as the way in which pine cones open[169] or more remarkably how similar hygoscopic forces act on the awns to drive wheat seeds into the ground.[170]

Evidence for the best design in silk also emerges in the potentially conflicting demands between the strength of the thread and its degree of stickiness.[171] After all, the more adhesive the glue, the more likely the prey will be held—but as it violently struggles, it may snap the thread. There is a direct correlation: the stronger the thread, the but what they label "Spider-Post-Its"[122] possess "attachment systems . . . [that] show astounding similarities"[123] to those of the geckos. In part, at least, this similarity arises because spiders are generally substantially larger than many insects. In addition, the scaling of hairy adhesion applies over a rather remarkable six orders of magnitude,[124] and so we find an overlap between the spiders and lizards. These geckolike spiders are not only adept at climbing, but are hunters like the bird and wolf spiders; rather than employing a silk web, they use the eight adhesive legs to grapple with their prey[125] and at the same time hold on to the substrate.[126]

The use of adhesive hairs by hunting spiders,[127] and indeed beetles,[128] is fascinatingly complemented by a very different method adopted by the scytodid spiders. They spit, forcibly and directionally. The prey, often salticids, find themselves enmeshed with a sticky mixture composed of saliva, silk, and venom.[129] If the first spurt, which comes out of enormous venom glands via a nozzle at the base of the fang, fails to entrap the prey, then there will be repeated discharges, each with "a convulsive shudder."[130] Once enmeshed, then the fangs descend. Meeting your death in saliva is an unattractive thought, but needless to say it has evolved more than once.[131]

A Silken Convergence

When it comes to prey capture and adhesion, the example of silk springs to mind. This remarkable material has attracted attention because although it is stored in a liquid form, when extruded via the nozzle (classically the spinnerets of a spider), it solidifies into a fine thread with a complex molecular structure that consists of both crystalline and noncrystalline domains.[132] The mechanical properties[133] can be quite extraordinary, with some spider silk approaching the tensile strength of steel but being many times lighter. Typically silk is characterized by particular motifs of amino acids, often drawing on alanine and glycine, and they in part determine the silk's physical properties. Functions show an extraordinary range, from not only the familiar capture webs of spiders and cocoons of the silk moth, but applications that encompass the construction of waterproof tents,[134] assistance with ballooning,[135] and aquatic draglines.[136]

Such versatility calls upon a complex molecular repertoire, often with distinctive amino acid compositions. Nevertheless, there is good evidence that in at least some cases

the amino acid motifs have remained effectively invariant for tens of millions of years, providing powerful evidence not only for stabilizing selection but the fact that the motifs are far from accidental.[137] Silk has evolved several times independently in the arthropods.[138] Even among the arachnids, spinning is almost certainly independent of spiders in such groups as the pseudoscorpions,[139] which build nests[140] with a very distinctive silk more reminiscent of keratin,[141] and the aptly named spider-mites[142] that are serious agricultural pests[143] and spin an extraordinarily fine silk.[144] Moreover, although various differences are apparent between the threads spun by spiders and insects, important similarities in overall design also exist.[145] The geological evidence for silk production is ancient. Primitive spiders (the aptly named *Attercopus*) from the Middle Devonian (c. 380 Ma) are equipped with rows of spigots located on a ventral plate.[146] Given that flying insects evidently did not make their debut for another 60 Ma the silk was probably not used in a web, but may have served to line a burrow.[147] It was not long before the early flying insects must have been their doom because the trichomes on the margin of the pitcher apparently resemble their food.[86] Insects (and the shrews) are drawn by the sugar-rich nectaries, in what Katherine Bennett and Aaron Ellison[87] aptly refer to (in the context of the pitcher plant *Sarracenia*) as "the sweet shop of horrors."[88] In at least one case, as Martin Schaefer and Graeme Ruxton[89] note, the plants also "roll out the red carpet"[90] because the red color of the pitchers provides a strong lure.[91] However they arrive, the insects then find themselves on a lethal glissade because the rim of the pitcher is coated with tiny crystals of aldehyde wax[92] that provide no purchase for the adhesive pads.[93] Nor does it make any difference whether the insect pad is smooth or hairy; in either case the insect finds it almost impossible to obtain any traction.[94] Evidently what happens to at least the hairy adhesive pads is that the setae either become clogged with broken wax crystals[95] or some amorphous substance.[96] Not all such trips, however, are one way. Many plants must trap pollinating insects that subsequently escape with their cargo of pollen, but as they, too, slide toward the waiting reproductive organs, their journey is assisted by convergent epicuticular waxes.[97]

Nor is this the only way insects may slither to their doom.[98] In some pitcher plants the rim lacks a waxy surface but is extraordinarily wettable.[99] Rain droplets quickly spread out to form a glossy surface across which the insect aquaplanes.[100] This method is highly effective and convergent,[101] but it presupposes that the surface remains wet. It stops raining, the clouds clear, and as the rim of the pitcher plant dries out, it ceases to function? Is this not a serious design fault? Should the natural engineer be sacked? Not necessarily. The very unpredictability of when the rim is wet and slippery[102] may allow the plant to retain the element[103] of surprise.[104] In other cases the plant releases chemicals, either as volatiles that

in one way or another may lure the unwary insect[105] (or indeed defecating mammals[106])— or in the case of a pitcher plant from Borneo, the scents produced mimic those of flowers.[107] More remarkable is evidence for the release of other compounds, possibly alkaloids, which appear to anesthetize the insect. Investigators noted that after several hours working among the pitcher plants they began to suffer from headaches and dizziness.[108]

The pitcher plant is not always the winner. Ants can cut holes in the side of the plant to draw the fluid and catch the contents.[109] More particularly, despite the generally lethal rim, species of ant that enjoy a symbiotic association with the pitcher plant can cross this surface with impunity,[110] as well as swimming and diving in the digestive fluid.[111] This situation finds a fascinating counterpart in some plants, such as certain species of the tropical euphorbiacean *Macaranga* in Southeast Asia. Here, too, the stems are slippery, with wax crystals across which only one species of ant can walk.[112] In *Macaranga* this feature has evidently emerged more than once,[113] but what Ray Harley[114] aptly terms "the greasy pole syndrome"[115] has independently evolved in various other plants. Harley points out that this exclusion makes sense, because ants more often than not steal nectar. In the case of *Macaranga*, however, both sides benefit, with the plant providing sugars,[116] in return for which the ants not only deter herbivorous insects but also prune the surrounding area to discourage competition.[117]

What of other arthropods with adhesive pads? Among the most notable are the spiders, where the pads are known as scopula and act as a dry adhesive.[118] Most likely they rely on the van der Waals forces[119] and are of central importance in the highly agile salticids, the jumping spiders.[120] As Antonia Kesel and colleagues[121] remark, not only does the ultrastructural arrangement of branched setules provide an extremely effective grip

Anolids and geckos are functionally almost identical, yet the latter group has the edge in proficiency. The adaptive advantage evidently lies in the geckos being able to fold and unfold their foot pads, which allows a lightning shift from a firm grip to disengagement. This finds an echo in the alternative adhesive system, where such hymenopterans as the ants and bees can rapidly adhere and with equal facility detach their limbs.[62]

If lizards can walk up walls, so can insects. Unsurprisingly they, too, have evolved hairy adhesion pads several times independently,[63] albeit drawing on chitinous setae rather than using the keratin of lizards.[64] A further difference is that whereas the adhesion in lizards is completely dry, in insects small amounts of fluid assist in adhesion, as indeed they do in the equivalent smooth pads.[65] This can lead to a situation in which insects become seriously unstuck. Elsewhere I mention the ingenious (and convergent) sexual mimicry of the orchids (p. 393n89), but in at least one case not only is there pseudocopulation but the flower is so arranged that its waxy surface means the bee loses its grip and once trapped makes contact with the pollen mass.[66] Here the bee eventually escapes, but in many other cases when an insect visits certain plants, it is, I am afraid to say, a one-way journey.

Consider those plants that have transformed themselves into stomachs, the carnivorous plants.[67] This syndrome is rampantly convergent, having evolved at least six times.[68] As Aaron Ellison and Nicholas Gotelli[69] remark, "the degree of morphological and physiological convergence across carnivorous taxa is remarkable."[70] Nor is this surprising, because capture of animals may often be a vital supplement (notably of nitrogen) in nutrient-starved soils; as a result, we see some intriguing parallels among the fungi.[71] Familiar examples may include the Venus fly-trap and the sticky sundews, but one genus (*Philcoxia*) of plantains even operates underground and

traps nematodes[72] (albeit in a different way to those ingenious fungi, pp. 150–51). Others are aquatic; consider that triumph of biological engineering in the form of the bladderwort (*Utricularia*).[73] This carnivorous plant is equipped with a remarkable spring trap,[74] which as Olivier Vincent and colleagues[75] demonstrate involves a buckling process that ranks "among the fastest movements generally known in the plant kingdom."[76]

Lurking in ponds, the bladderworts can be easily overlooked, but coming across a stand of pitcher plants can induce an alien shiver of appreciation. As Mark Chase and colleagues[77] note in their informative review titled "Murderous plants,"[78] accustomed as we are to think of plants as vegetables, "there is something deeply unnerving about the thought of carnivorous plants."[79] As their name suggests, pitcher plants have a vaselike form holding a water trap into which insects fall.[80] Well, usually. In Borneo they can provide jungle lavatories where some enormous pitcher plants serve as latrines when tree shrews choose to visit.[81] The posture they adopt ensures that the nitrogen-rich feces (and perhaps urine) falls into the well. An even more intimate association has been identified in aerial pitchers[82] where the bats find the pitcher a convenient roosting site while the plant evidently benefits from the droppings.[83] In this world it seems anything will do. Nor should we be surprised to discover pitcher plants on some distant planet because on dear old Earth they have evolved three times independently, with the group known as the sarraceniaceans dating back to the time of the dinosaurs.[84] In each case they form the pitcher by an inrolling of the outer markings of the leaf (a process known as epiascidiation), but the entire plant forms a remarkable functional complex. Frequent victims are ants, but also termites. In the latter context, particularly remarkable is the entrapment of huge numbers of *Hospitalitermes*[85] that are foragers and so led to

Given the speed at which they skitter around, it is not surprising that the toe pads can very quickly detach.[44] Apart from elegance, one of the most surprising findings is that the adhesive power of the gecko appears to be massively overengineered; only a few percent of the pad area is required to provide the necessary grip, and in theory the lizard could support the weight of two humans.[45] So why the excess? Sometimes not all of the adhesive pad may be correctly orientated. This, combined with the fact that natural surfaces may often be rough and crumbling, may go some way to providing an explanation. But possibly the most important reason comes from when the animal falls, inadvertently or otherwise. Eric Pianka and Samuel Sweet[46] provide a gripping instance of an Amazonian gecko that parachuted down from the forest canopy and then, while several meters from the ground, used one foot to break its descent as it held a leaf. After a tropical cyclone in Mauritius the geckos had been beaten to death by the lashing of leaves, but the corpses were still hanging on. Finally, while one might think that four adhesive pads per gecko were enough, the carphodactyline geckos of New Caledonia and New Zealand have tails with adhesive pads built to the same design.[47]

While primarily arboreal, a variety of gecko groups have become fully terrestrial. Moving into sandy deserts, however, presents a particular challenge given the adhesive role of the ancestral feet.[48] Among these desert dwellers those that burrow show striking convergences in the structure of the feet, with successive reduction of the pads and development of a distinctively spiny surface. This leads Trip Lamb and Aaron Bauer to identify how "repeated evolution (pedal traits, particularly) in similar dune environments corroborates selection's role in shaping observed morphological similarities."[49] Desert environments are, of course, a productive source of insights into many areas of convergence,

but in the immediate context of desert-dwelling lizards, the dune ecomorphs provide some compelling examples.[50] Deserts are also adaptively challenging in terms of water, which also leads to a rich area of convergence, such as the repeated evolution of succulent plants (p. 155-57). Among the lizards we find a notable convergence between an Australian agamid (the thorny devil) and a North American iguanid (horned lizard). Independently they have developed a remarkable capacity to wick water rapidly across the body and into the corner of the mouth, employing channel-like structures situated at the base of the scales.[51] Writing of the thorny devil, Philip Withers[52] remarked how this lizard "will readily absorb water by a 'blotting-paper' action"[53] and evidently is even able to extract water from damp sand.

Lizards provide many insights into evolutionary convergence—from limblessness (p. 75-76) to viviparity (pp. 216–17)—but one might expect that the extraordinary capacity of geckos to maintain a grip is a one-off. Not so; admirable as is its design, it is one that has been arrived at repeatedly. Even within the reptiles this design has evolved twice more. The first is among the anolid lizards[54] (which in the Caribbean islands also offer insights into convergent evolution in terms of recurrent ecomorphs[55]), and specimens preserved in amber suggest arboreal agility.[56] The second convergence involves skinks,[57] living in the forests of Papua New Guinea and the Solomon Islands.[58] The adhesive arrangement in these skinks is closely similar to that of geckos, but the setae appear to be differently derived.[59] Duncan Irschick and colleagues[60] aptly point out how it demonstrates that one can arrive at "the same functional endpoint along very different trajectories."[61] In emphasizing this threefold convergence within the lizards they stress that the skinks are less adept climbers, which can be attributed to the simpler structure of the adhesive pads.

shark, and so the pandarid crustaceans have independently evolved adhesive pads.[17]

What of the vertebrates? Among the arboreal tree frogs (and also salamanders[18]), we see the repeated evolution of adhesive toe pads,[19] even if their similarity has occasioned surprise.[20] Not only are they highly effective,[21] with nanometer-sized pillars evidently of key importance,[22] but as Ingo Scholz and colleagues note, not only are frog pads "an exceptionally good example of convergent evolution [but] . . . there really does appear to be a best 'design' for a toe pad."[23] In addition to a highly structured surface on the nanoscale, these toe pads have a strikingly recurrent hexagonal arrangement.[24] Just such a pattern is also found in some orthopteran insect pads.[25] The convergence extends further, because in both cases the structure is such that while the pad is extremely soft, it is capped by a more resistant exterior that confers resistance against abrasion.[26] As with nearly all the other animals with smooth pads, an extremely thin coating of fluid evidently plays a key adhesive role.[27] In frogs the fluid is a mucus, secreted by glands located between the hexagons, but in the feathertail glider, the adhesive pads call upon sweat.[28] Nor is this marsupial the only mammal that has navigated to an adhesive solution, because in some vespertilionid bats the thumbs are equipped with adhesive fluid.[29] Given that mammals usually rely on their claws, this mode of attachment is unusual enough, but *Thyroptera* uses a device closely similar to an octopus sucker.[30] Independently the sucker-footed bats of Madagascar have evolved adhesive capacity. Despite their name, they also employ an adhesive, and the convergences extend to roosting; in both *Thyroptera* and *Myzopoda*, the bats are most unusual because they roost head-up.[31]

Remarkable as these adhesive contrivances are in their phylogenetic range and similarity, arguably even they are left hanging when it comes to their hairy counterparts, most famously exemplified by the geckos. As Kellar Autumn[32] notes, these lizards can run "with seeming reckless abandon on vertical and inverted surfaces."[33] Not that all geckos have adhesive feet, and it appears such an arrangement has been lost (as in the burrowing forms [p. 84]) almost as many times as it has evolved.[34] But in the world of grip, the geckos have an iconic role. To judge from a specimen preserved in amber, such a capacity dates back to at least the Cretaceous, and the animal in question (*Cretaceogekko*) was probably arboreal.[35] Their companions, the dinosaurs, departed, but the geckos maintained their grip; today these diverse and successful reptiles are a familiar feature in tropical rooms. Their clinging ability can moreover be put to mischievous use. Karl Schmidt and Robert Inger[36] relate how in Malaysia the boys fasten a string onto a gecko and lower it from a high window or the roof onto the hat of a passing pedestrian, which the gecko immediately grabs, and at that point lizard and hat are drawn upward.[37] How convenient might it be to have adhesives that never left a trace but might be used by the new pioneers of mountain climbing:[38] as you grapple with the ropes, belays, cams, and all the other paraphernalia, the rival team passes with a nonchalant nod and continues up an overhang with their gecko-inspired pads providing a perfect grip.[39]

What is the origin of this extraordinary grip? The answer lies with the pads on the feet, each one of which bears a huge number of fine setae. These in turn have a myriad of tiny branches, so the whole structure is decidedly tuftlike,[40] thus providing an extraordinarily effective adhesive surface. Its basis relies on the van der Waals force,[41] a type of intermolecular attraction that includes an electrostatic component. The pad is also self-cleaning (as convergently is the case in the insects[42]), so dust and other debris do not clog the pad.[43]

CHAPTER 8

STICKING TO CONVERGENCE

THE METAPHOR of "skin-deep differences" may find its classic application in the tuna/lamnid shark convergence, but the principle resonates across evolution. While the thunniform convergence is specific to the oceanic environment, what matters is the repeated emergence of sophisticated, adaptational complexes. The previous chapters are replete with examples, whether in terms of feeding or locomotion. In the latter context, for one reason or another sometimes the animal in question may need to grip hard. The solutions are neat—and of course convergent.

THE GRIP OF EVOLUTION

Sometimes the need to keep a foothold is imperative, such as when it comes to strolling across a ceiling. In reality there are two separate ways in which animals have learned how to keep just such an impressive grip. One can employ either smooth pads or hairy[1] pads, and both are rampantly convergent. In some groups, notably the insects, both types of pad are found,[2] but in both cases the convergences are plausibly explained in the words of Stanislav Gorb and Rolf Beutel as the "restricted number of design principles"[3] that can be drawn upon.[4] Despite the range of animals that have learned to cling with smooth pads, a common identity exists inasmuch as typically the surface is very soft and draws on adhesive fluid. So striking is this convergence that Jon Barnes[5] is persuaded to write that it represents "clear evidence of an optimum design."[6] Among the insects, we find this arrangement in the aphids[7] and the celebrated fulgorids[8] (which are instructive with respect to various convergences,[9] with one taxon [*Fulgora laternaria*] evolving an extraordinary mimicry because it "bears an uncanny resemblance to that of a caiman"[10]), heteropterans,[11] hymenopterans (notably the ants and bees[12]), and cockroaches.[13] The adhesive grip of some of the weaver ants is particularly noteworthy, given that in their arboreal habit losing their footing could have catastrophic consequences[14] (although other ants have convergently learned to glide [p. 192–93]). Having a firm grip also comes in useful among those weaver ants that combine forces to haul home remarkably large objects, including dead birds and rodents, which in the former case can involve climbing up a tree.[15] There is little doubt that the collective adhesion is due to the well-developed foot pads (or arolia), although even a single worker has been observed clinging on to a dead bird suspended in space.[16] Although all these examples involve terrestrial arthropods, the principle also helps if you wish to keep your footing as an ectoparasite on a first on the block. Swarming through the Mesozoic were a variety of fast-swimming predatory fish, along with a group known as the pachycormids that have some striking similarities to the tuna.[228] However, when the bad times arrived at the end of the Cretaceous (the K/T event), these and similar groups of fish (such as the pachyrhizodontids[229] and ichthyodectiforms[230]) were just as vulnerable as the dinosaurs.[231] The Tertiary oceans, however, were soon repopulated by thunniform groups (and similar stories occur among the sharks[232] and the giant planktivorous vertebrates[233] [p. 50]), but as Matt Friedman remarks with respect to the thunniforms, "Ironically, the very same groups that seem to have diversified into emptied ecospace at the dawn of the Cenozoic now face the greatest risks of extinction from overexploitation."[234] Tuna, and for that matter shark, may be immensely successful, but as convenient cylinders of high-quality protein both have long been of interest to another cognitive species. Nobody needs to be reminded that the future of both, at least outside the problematic fish farms, is looking bleak.

Notwithstanding this rather pessimistic view, a wider, if not universal, point deserves to be made. On any planet, fast, aggressive, thunniform predators will emerge, and with impressive oceanic ranges.[235] Although the emphasis has been on ichthyosaurs, mosasaurs, lamnid sharks, and scombrid teleosts, important convergences also exist between the latter group and such cetaceans as the dolphin in terms of musculature and tendons.[236] Impressive as the speeds that such animals can reach, interestingly universal limits in the larger thunniforms are governed by cavitation (p. 143) at the propeller.[237] Convergence gives us a useful tip. Next time you visit another planet, just make sure your speedboat can go faster than 15 meters a second, and then you'll be quite safe.

deep dives), there is obviously a premium on this countercurrent being efficient. But how efficient? In fact, with respect to the effectiveness of heat exchange, one calculation suggested an astonishing 98.3 percent efficiency.[207] Yet, ingenious as this system might be, it poses a problem. Warm blood stores less oxygen, so in principle the incoming cooler blood will "steal" oxygen away from the metabolic furnace where it is required. Hardly a satisfactory state of affairs, but in tuna the solubility of oxygen is directly linked to blood temperature (the so-called reverse temperature effect). The lamnid sharks have independently arrived at the same solution,[208] and as Frank Carey notes, tuna and sharks (specifically bluefin and porbeagle) both "achieved a remarkable temperature stability in the oxygenation of their haemoglobin, but have evolved quite different molecular mechanisms for doing so."[209]

This capacity to both maintain and regulate elevated body temperatures is not only a feature of the propulsive core, but is also found in the brain and eye of tuna[210] and sharks.[211] Such cranial endothermy provides striking convergences in other groups of fish.[212] Among the chondrichthyeans, these include the mobulid rays,[213] while the roster of teleost examples lists a predatory lampridiform (the opah[214]) as well as the so-called billfish, which include the marlins[215] and swordfish.[216] These, too, are fast-moving ocean dwellers, and in the latter case there is evidence that the eye heater greatly improves the swordfish's vision.[217] At first sight, identifying convergent endothermy in the billfish would seem to be relatively trivial, given that they are generally regarded as scombroids, and so close to the tuna. Not at all. It now transpires that they are close relatives of the flatfish (pleuronectiforms),[218] a group justly celebrated by evolutionary biologists[219] for their strikingly asymmetric skulls and the migration of one eye across the head.[220] As A.G. Little and colleagues emphasize, the misassignment of the billfishes to scombroids demonstrates "the power of similar selective pressures in driving the evolution of nearly identical complex adaptations in distantly related groups . . . [and] also show[s] how varying selective pressures can spur rapid and extreme divergence within a single monophyletic group."[221]

The underlying molecular physiology of fish endothermy reveals convergences and inevitably the capacity to draw on existing systems.[222] Unsurprisingly the oxidative capacity of the musculature is greatly enhanced, and in the case of the billfish even exceeds that of hummingbird flight muscle.[223] In this case, the ultrastructure of the thermogenic eye muscle is dramatically modified, showing both an abundance of mitochondria and smooth membranes.[224] Nor is this surprising. The mitochondria are, of course, the powerhouses of the cell, and an important way to bolster their contribution is to increase the surface area of the internal membranes (cristae) where the oxidative reactions take place. In the thermogenic skeletal muscle of tuna, the surface area of the cristae[225] can reach an extraordinary 70 square meters per cubic cm.[226] As in other cases of endothermy (pp. 226–27), these membranes are crucial to thermogenesis; typically they employ a so-called futile cycling whereby transport of protons or ions (such as Ca^{2+}) results in the processes that would otherwise be involved with oxidative metabolism being cleverly sidetracked. Instead of producing "useful" energy-rich molecules (typically ATP), "waste" heat is generated.[227]

In the birds and mammals, warm-bloodedness opens up a whole range of evolutionary possibilities. What of fish? Both lamnid sharks and tuna are superb ocean-roving creatures that exemplify not only high-performance biology but they also seem to be very close to the functional limits of what is evolutionarily possible. But they weren't the puffer fish (p. 36), to do this they call upon two types of muscle. For normal cruising, only red (or aerobic) muscle is employed, but for exceptional bursts of activity the white (or anaerobic) muscle is brought into play. The latter, however, is used at a cost, because as its alternative name indicates, it rapidly builds up an oxygen debt that then needs to be paid off. Much the same happens when we run fast (and just possibly are not at the very summit of fitness) and the leg pains are a consequence of a buildup of lactic acid, a metabolic product that forms when the oxygen cannot be supplied quickly enough. Tuna, being a fish, has both red and white muscle, but the key development is not only to shift the red muscle to deep inside the body (recall that in most fish it is on the margins, so that their contractions can throw the body into a sinous wave), but also effectively to decouple it so that it can contract independently.[194]

The next trick is that when the red muscles contract, the resultant force is transmitted to the large tail by a system of tendons. It is a most ingenious system, but an almost identical arrangement has evolved independently in the lamnid shark. As ever, however, the evolutionary trajectory leaves a series of footprints. For example, although both tuna and shark employ tendons to transmit the force, their actual origin is different.[195] Yet the overall degree of convergence is remarkable; as Jeanine Donley and colleagues[196] remark, the "mechanical design between the distantly related lamnids and tunas is much more than skin deep."[197] As far as these active, wide-ranging oceanic predators are concerned, from an evolutionary perspective they are accidents waiting to happen. In these reiterations of convergence it is never intended, however, striking the similarities to insist on an exact identity. Thus, in some respects tuna are arguably "superior" to sharks—for example, in terms of muscle structure and associated details such as blood capillaries,[198]

as well as the ability to use oxygen.[199] On the other hand, in the salmon shark (a lamnid), the muscles are actually more mammal-like; if the body temperature drops, the muscles become ineffective.[200] Tuna, in contrast, can continue to function at depressed body temperatures.

What is all this about body temperatures? Surely these fish are cold-blooded? Not only are their biomechanics convergent, but another key feature is the independent emergence[201] in both lamnid shark and tuna of warm-bloodedness. In shark and tuna, endothermy is clearly a key component to their success, but it is a convergence that not only extends to other groups of fish but far beyond.[202] Thus, in addition to the lamnids, there is evidence that despite a markedly different body shape, the common thresher shark (a member of the alopiids) has convergent similarities in the deep location of the red muscle and its dynamics[203] and has also arrived independently at warm-bloodedness.[204] Along with the tuna, this physiology evidently goes deep into their history,[205] and it may well have been initiated as they invaded cold-water niches, not only at high latitudes but at depth. Just as the tendon systems of tuna and lamnid shark are very similar but have different origins, so the source of heat generation is associated with different regions of the alimentary canal, but in both cases a crucial necessity is a so-called countercurrent arrangement. This is a rampantly convergent biomechanical device, perhaps most familiar in a long-legged bird standing on ice whereby a closely associated network of blood vessels (the rete) ensure that heat descending the leg is captured and returned to the body. In the same way, in tuna and lamnid shark,[206] the heat generated in the core region is constantly returned to maintain a stable central body temperature rather than being dissipated on the margins. Living in water that may be cold (especially during

Ichthyosaurs could evidently also cruise at appreciable speeds. In this respect they probably rivaled the tuna with their swimming capacity underpinned by an elevated basal metabolic rate.[170] This raises the possibility that forms such as "*Stenopterygius* had basal metabolic rates between typical mammalian and reptilian levels at 20°C,"[171] but this may be an underestimate because analyses employing oxygen isotopes (a routine method for establishing paleo-temperatures) recovered from the bones of ichthyosaurs (and also plesiosaurs) suggest body temperatures as high as 35°C.[172] While ichthyosaurs are, therefore, clearly tunalike, how precise is this convergence? Comparing the range of forms Ryosuke Motani remarks how "sharks provide the best analogue for ichthyosaurs in overall body shape and locomotion, although differing in details,"[173] a point reinforced by exceptional fossil preservation of the peduncular region indicative of a tendon sheath (although the tail is more tunalike).[174] So, too, the structure of the fins is consistent with life in the fast lane.[175]

While ichthyosaurs exemplify hydrodynamic agility and so stand (or swim) in contrast to the medley of other Mesozoic aquatic reptiles, consider the mosasaurs. A highly successful group, close relatives of the snakes and probably warm-blooded (albeit perhaps more akin to gigantothermy; see p. 433n121), otherwise for the most part they seem far removed from any sort of ichthyosaur.[176] Not so fast. First, like ichthyosaurs, at least some mosasaurs were viviparous (p. 219), pointing to a wholly aquatic mode of life.[177] This can also be inferred from the streamlining of the scales,[178] as well as bone damage (avascular necrosis) arising from too rapid ascent from a deep dive,[179] provoking the notorious "bends" (or Caisson disease) of human divers (but not whales, p. 380n229). This capacity for deep diving evidently evolved several times among the mosasaurs, and correspondingly some

forms (such as the durophagous *Globidens* [p. 47]) remained in the shallow waters.[180] Nor were they the only Mesozoic reptiles at risk; avascular necrosis has also been identified in the turtles,[181] plesiosaurs, and ichthyosaurs,[182] although in the last case the first representatives were evidently splashing around in the shallows and deep diving only commenced as the risks of predation escalated.[183] Having invaded the oceans, the mosasaurs underwent a series of spectacular adaptive radiations.[184] This is especially apparent in terms of feeding and dentition,[185] with one mosasaur even resembling a tyrannosaurid.[186] Consider, in particular, the remarkable *Plotosaurus*. Here the body form has evolved from the more typical snakelike form to fusiform, with reliance on the tail for propulsion, and is thus beginning to converge on the ichthyosaurs.[187] Johan Lindgren and colleagues note that was it not for the end-Cretaceous extinctions,[188] "with time, the *Plotosaurus* lineage might even have yielded fully thunniform animals."[189] Evidence that this conclusion is by no means fanciful comes from *Platecarpus*. Although well-known from skeletal material, a superb specimen with soft-tissue preservation[190] shows that the transition to "a streamlined body plan and crescent-shaped caudal fin"[191] was already well under way prior to the appearance of *Plotosaurus*.[192] So why not a Tertiary ichthyosauroid? Just how close the ichthyosaurs themselves came to the living thunniforms in details of muscle structure and physiology will probably remain circumstantial, although if a time machine were set for 140 Ma ago, I'd be happy to take bets with the crew before we began our dissection of one of these Jurassic monsters.

Such confidence becomes apparent when we look at the tuna and lamnid shark. Both, of course, are superb high-performance swimmers,[193] but they are only distantly related. All fish swim, of course, by muscular action, and excluding rather remarkable forms like the

These salamanders are not the only amphibians to have disposed of their limbs. In another group, the caecilians, we see an effectively universal loss of legs and by the usual multiplication of vertebrae the adoption of a snakelike form.[157] The convergences do not end there. In some caecilian taxa (notably *Dermophis*), there has been a remarkable transformation into what is effectively a worm.[158] In burrowing this amphibian decouples the spine from the skin and associated musculature so that, remarkably for a vertebrate body plan, it employs a hydrostatic skeleton that is distinctly reminiscent[159] of many invertebrates, such as the earthworm.[160] Among the other oddities of the caecilians is that although they are fossorial, with the result that many either have reduced eyes or are even blind, in one case the eyes are actually protrusible.[161] A unique feature among the vertebrates, but of course reminiscent of the snails.

From opossums teetering along branches to running vampire bats, the apparent diversity of locomotory styles soon reveals itself to be the recurrent appearance of a limited number of evolutionary solutions. Apparent differences are only skin-deep. Even underwater examples of walking convergences are much less surprising when the functional context is understood. When we think, however, about aquatic vertebrates, most usually it is synonymous with swimming. Nothing strange about that—until, that is, we turn to the torpedoes of the oceanic realm. Masters of hydrodynamics, the convergent similarities go much deeper than simply streamlining.

More Than Skin Deep

It has long been appreciated that large, fast-moving marine animals are usually fusiform with prominent arcuate tails and stabilizing fins. Indeed, in zoological popularity of convergences, close behind the comparison of the cephalopod and vertebrate camera eye (pp. 12–13) is the long-appreciated similarity between the extinct ichthyosaurs (p. 47) and the extant thunniform swimmers in the form of the scombrid fish (tuna), lamnid sharks, and the dolphins.[162] Long-distance, fast swimming demands not only a fusiform body but, for the propulsive unit, a large tail capable of flicking backward and forward and so is only attached to the rest of the body by a narrow "neck" or peduncle. This thunniform locomotion is quite different from the sinuous locomotion of most fish, most fully exemplified in the anguilliform eels. In the thunniforms the trick is to transmit the muscular forces down to the tail, which calls on a special arrangement.

The convergences are indeed striking, but at the same time they are in principle entirely predictable. Although a wide variety of Mesozoic reptiles became aquatic (and these can be divided into a series of categories[163]), the ichthyosaurs were evidently the adaptive pinnacle, and with their streamlined shape, flippers, a powerful caudal tail, and prominent dorsal fin,[164] they exemplify the thunniform solution. The earliest forms had a rather undulatory style of swimming, but as they headed for deeper water their characteristic body plan soon became apparent.[165] Their adaptive radiations produced a medley of forms: an early form was evidently durophagous (p. 47), while the Jurassic *Eurhinosaurus* is strikingly reminiscent of the living swordfish.[166] Among the more remarkable features of ichthyosaurs is their viviparity (p. 129), as well as truly enormous eyes.[167] In forms like *Opthalmosaurus* these eyes are consistent with both deep diving and the capacity to hunt in very dim light,[168] but they may have possessed exceptional acuity of vision.[169]

would seem an entirely plausible hypothetical sequence of first elongation of the body, then reduction in the limbs, and finally loss of the digits is not what actually happens; on the contrary it is a more or less entire and simultaneous transformation.[147] In this work they identify two distinctive snakelike ecomorphs, one for surface dwelling and one for burrowing, and subsequently they show that this has a much wider generality among the squamates.[148] Paradoxically they go on to argue that in reality we should be relatively unimpressed by the twenty-five or so transformations into a snakelike body, because if competitive exclusion were relaxed, it would have happened hundreds of times.

Having clambered onto land, disposing of one's limbs might seem a regressive maneuver. The squamate reptiles prove the opposite, and neither are they alone. Consider the amphibians,[149] which as Michael Caldwell aptly remarks, can be "Without a leg to stand on."[150] Among the most successful of the salamanders, the plethodontids (p. 25) and specifically a subgroup known as the bolitoglossines, extreme elongation of the body and reduction of the limbs is correlated with a fossorial mode of life and may be a key factor in these Neotropical animals' successful invasion of the lowlands from cooler upland habitats.[151] In one such taxon, known as *Oedipina*, elongation is achieved in a "sensible" way— that is, by increasing the total number of vertebrae in the trunk. This recalls, of course, the manner of making an eel (p. 49). In contrast, another form known as *Lineatriton* has adopted a "giraffe-neck" solution to the problem of how to become elongated[152]—that is, it has retained the smaller number of vertebrate but extended them. Two solutions to the same problem, the former perhaps more "logical" and one found in other elongate bolitoglossines, the other an eccentric "one-off." Except it isn't, because molecular phylogeny reveals that not only is *Lineatriton* closely related to a rather generalized taxon (known as *Pseudo-*

eurycea), but this elongate morph has evolved at least twice independently to the extent that separate populations both assigned to the species *Lineatriton lineolus* are more closely related to different species of *Pseudoeurycea* than they are to each other![153] The investigators draw attention to what is in effect a double convergence, both in terms of general bolitoglossine fossoriality and at the level of a single taxon. They cannot, however, resist exclaiming, "This is extraordinary because the morphology . . . is extreme in its degree of specialization . . . and has been considered [in *Lineatriton*] to be unique in the combination of characters,"[154] but patently has arisen independently from generalized ancestors.

Perhaps in the way of many convergences there is something "innate" about those pseudoeurycean ancestors that preordained the *Lineatriton* condition. But the surprise of Gabriela Parra-Olea and David Wake is not only genuine, but perhaps explains a revealing comment at the end of their paper. Here they write that the so-called "'Lineatriton' phenomenon appears to be local and limited . . . [and] can only be considered [as] terminal twigs in the bolitoglossine radiation."[155] "Twigs of a tree" is a standard evolutionary trope, and in its own way is perfectly legitimate. These may, however, fail to do full justice to both the ubiquity of convergence and its nested arrangement. In brief, suppose that at each and every bifurcation[156] in the Tree of Life, the options are much more limited than might be thought. This is, in fact, evident from the abundance of "parallelisms" (or homoplasies) that *all* phylogenies display at any scale. To focus on the twigs is, therefore, to miss the point. The "Lineatriton phenomenon" is far from trivial; it is a vital clue to the likelihood that far from being a sprawling arborescence, the geometry of the Tree of Life is far more determinate. Darwinian evolution is the mechanism, of course, but at each and every level, convergence hints at the template for the Map of Life.

the kinematics of movement rather than anatomy;[131] in each case the animal needs to balance the propulsive force against stability. Thus, in insects (specifically cockroaches[132]) as well as mammals, the push forward comes from the force exerted by the relatively simple operation of the posterior pair of limbs. In the case of insects there is a difference, inasmuch as they possess an additional pair of midlimbs that serve both in propulsion and braking so as to assist with overall balance. The movements of the anterior limbs, however, are considerably more complex because they have to combine a forward movement that must come down at the most appropriate spot, followed by a braking maneuver. Ritzmann and coworkers write of how "Remarkable similarities are found in insects and vertebrates, in spite of the fact that legs evolved independently in these two groups."

They continue, "Engineers are well advised to pay attention to these common solutions,"[133] and with good reason, because a major thrust of their article is the applicability of this work to robotics.[134] The study of biological solutions, and not least the fact that evolution has repeatedly arrived at the same destination, is providing biomimetic inspiration. These include such areas as underwater detection that might draw on analogues of the lateral line system[135] (p. 128), or those fish that sense the surrounding environment by generating an electric field and are also capable of precise maneuvering (pp. 146–47),[136] not to mention a potential collision detector based on insect neurobiology.[137] The prospects of a new technology inspired by evolutionary solutions to given problems, and ones that are almost invariably convergent, are indeed tantalizing.

SLITHERING TO CONVERGENCE

Is there any planet with life where limbs have failed to evolve? I doubt it. Yet in many groups of vertebrates, once evolved they have subsequently been completely discarded, most obviously in the snakes. Certainly within the squamate reptiles, building a snake from a lizard ancestor seems remarkably straightforward, given that a reduction in the limbs has occurred at least twenty-six times independently.[138] Not only has a snakelike form been achieved in much the same way by coordinated evolution of body parts,[139] but also in markedly different ecological settings.[140] Matthew Brandley and colleagues pose the key question: "What might explain the generally similar patterns of body-form evolution across squamates?"[141] and conclude that on balance it is adaptation rather than developmental constraints that determine this elegant rerunning of the tape.[142]

The specific examples among the liz-

ards provide many fascinating insights. The appropriately named worm-lizards (the amphisbaenians) are notable because, as their name suggests, nearly all are limbless.[143] Yet their phylogeny indicates that the limbs were lost at least three times, while the distinctive cranial specializations essential for burrowing fall into four main categories, but here, too, we see convergences.[144] Other groups of lizards are also instructive. We have already encountered the skink-swallowing Burton's legless lizard with its collapsible teeth (p. 64), and while it does not employ venom (or indeed engage in constriction) its method of lunging and dealing with a potentially very irritated prey shows a series of striking (so to speak) parallels to the snakes.[145] In explaining "How lizards turn into snakes," John Wiens and Jamie Slingluff[146] draw on the anguid lizards. They show that not only has this happened multiple times, but significantly what skillful fliers, employing echolocation and roosting in trees, they have robust legs. As B.D. Lloyd remarks, these bats "are remarkably fast-moving and agile on the ground and when climbing."[116] These words are echoed by M.J. Daniel[117] when he notes, "The manner of folding the wings and protecting them from injury and the robust hind limbs and feet give this bat astonishing rodent-like agility on the ground,"[118] and he also remarks on their unbatlike ability to chew their way through timber to provide cavities and tunnels. The usual explanation for these bats turning to a terrestrial existence is that they evolved in an environment where normal ground predators such as snakes and other mammals are absent.[119] The discovery, however, of Miocene relatives in Australia that evidently also walked suggests that it was not lack of predators that spurred this convergent development (even if the surviving New Zealand mystacines have taken full advantage of their situation).[120] In many ways the short-tailed bat is like a mouse. It has velvetlike fur,[121] spends a substantial proportion of time on the ground where it displays considerable agility,[122] and when walking folds its wings tightly. Seeking out arthropod prey,[123] the bat employs hearing and smell, and there is a corresponding shift to the incisors showing the maximum wear.[124]

The absence of mammals from New Zealand has other interesting consequences for convergence, not least in terms of the evolution of plants (p. 158). The celebrated weta, a giant grasshopper,[125] has long been regarded as another sort of honorary mouse.[126] How precise these comparisons really are has been a matter of debate,[127] but one key test would be if the weta stepped beyond any known insect by following the tracks of many mammals that feed on fruits, swallow the seeds, and thereby assist in the dispersal of the plants. Such, in fact, is the case. Catherine Duthie and colleagues[128] demonstrate how the "weta form mutualistic partnerships with fleshy-fruited plants, characteristic of relationships between plants and small mammals, providing a striking example of ecological convergence between unrelated organisms."[129] They go on to remark that ironically the very influx of introduced species of mammals to New Zealand threaten many of the weta.

One might think that, in other respects, comparisons between insect and mammal would be entirely superficial. Consider, for example, their walking. This is such a generalized system that to describe it as convergent would seem to be excessively vague. Yet when it comes to such insects as the cockroach and a typical mammal—ponder the greyhound—then we find a much more striking degree of convergence. Mammals stand in contrast to the more sprawling gait of most land vertebrates because the limbs are swung under the body, to their considerable locomotory advantage. Moreover, in the case of the forelimb (composed of the humerus and ulna/radius), the shoulder blade (or scapula) also becomes an integral part of the structure, while in the hindlimb the accentuation of the ankle joint (with the femur and tibia/fibia) again produces a limb built on three components. The crucial point is that the kinematics of locomotion in both insect and mammal converge, especially with respect to the construction of the forelimbs where the distal segment of the insect leg (known as the coxa) acts very much in the same way as the scapula. This convergence is all the more notable because the system of muscles and skeletons is effectively reversed, and in addition the insects employ six legs (hence the widely used alternative name of hexapods; an arrangement that is also convergent[130]) in contrast to the usual four of nearly all mammals.

Here is another case of skin-deep differences and has nothing to do with a hypothetical six-limbed mammal or similar. As Roy Ritzmann and colleagues demonstrate, the fundamental identity revolves around for example, employs both sets of fins[95] to stroll in a surprisingly tetrapodlike way over its reef habitat.[96] To study these convergent kinematics, the investigator Peter Pridmore was interested to see whether terrestrial locomotion was also a possibility. As he noted, when the individual sharks were placed on a "sandpaper-covered board [they] exhibited considerable displeasure at being removed from the water; nonetheless each [shark] progressed with facility when placed venter down on the test surface."[97] Nor are they alone, because a Carboniferous shark (*Onychoselache*) has a convergent pectoral fin structure suggesting that it also strolled across an ancient Scottish lagoon.[98] Equally impressive are examples from among the skates. Here we see not only limblike extensions that enable the beast to punt,[99] but also a remarkable convergence with the independent evolution of the anterior part of the pelvic fin into what is effectively a leg.[100] Skates have independently arrived at bipedal locomotion.[101] As always, biological structures evolve for good reasons, and Luis Lucifora and Aldo Vassallo suggest that not only might this be more energetically favorable than flapping the large pectoral fins, but it could also enable a more stealthlike approach to prey. These workers, like Pridmore, also remind us of the striking parallels with the tetrapod limbs, which enabled the sarcopterygian fish to clamber on to land, remarking that the "legs" of the skate "argue for a limited number of structural designs responding to functional requirements."[102]

The ability of sarcopterygian fish to leave the water is scarcely surprising given that their legs originally evolved in their aquatic home (pp. 222–25), and this capacity may go deeper into their history than realized.[103] Limbs, of course, depend upon a system of articulating rods to enable them to walk and run. Even when vertebrates learn to fly, the legs are there as undercarriage, available for any terrestrial strolling. There are, however, some interesting exceptions, especially among the bats. As well as being adept fliers and largely relying on the convergent acquisition of echolocation (pp. 137–39), typically they roost. There are a few exceptions. A myotisid bat typically gleans ground-dwelling arthropods.[104] As S. M. Swift and P. A. Racey note, in captivity it was "adept at quadrupedal locomotion and frequently caught prey by running after it,"[105] and most likely the same occurs in the wild. A molossian, the sturdily built naked bulldog bat (*Cheiromeles*), enjoys arboreal locomotion and is equipped with something suspiciously like an opposable "thumb" on its hindleg.[106] Otherwise, for the most part, bats are very poor walkers.[107]

Consider now the predicament of the vampire bats, the only mammals to seek blood as effectively its sole food (although such hematophagy has repeatedly evolved in such groups as the leech and many insects, and in the latter led to convergent strategies to overcome host defenses[108]) and that employs an anticoagulant glycoprotein appropriately known as draculin.[109] Before making an incision, not only do they have to move around sniffing their prey to find a suitable spot (or reopen an old wound) but should the animal roll over (as the mules in Trinidad are wont to do when a vampire lands on their back),[110] then an ability to run is obviously desirable. That is exactly what has happened: the vampires have reinvented running.[111] There is one difference: instead of the hind limbs, now the power unit depends on the front pair, which as in all bats are the more powerful because they drive the wings.[112]

This reinvention of locomotion among the bats is not unique to the vampires. In New Zealand the short-tailed bat has arrived at very much the same solution.[113] The relationships of these mystacine bats have been controversial, but they are probably relatives of the Neotropical noctilionids,[114] and they have attracted considerable attention.[115] Although that our ancestors were knuckle walking. Evolutionary proximity sometimes needs to be handled carefully, and in the great apes knuckle walking may have evolved two[77] or three[78] times (as well as independently in the anteaters[79] [p. 52]). More importantly the biomechanical transition to hominids remains deeply problematic.[80] As Robin Crompton and colleagues explain, "Remarkably . . . we find bipedal kinematics and kinetics closest to those of modern human bipedalism in the locomotor behavior of the most arboreal of the great apes."[81] So we come full circle. Unless the early primates had not become fine-branch specialists, and subsequently our nearer ancestors locomoted across the trees in the manner reminiscent of the highly arboreal orangutans—with a tree-based bipedalism and the hands grasping branches[82]—then we would not be standing here.

If this arboreal scenario is correct, then as Paul O'Higgins and Sarah Elton point out,

"bipedal walking might have evolved independently in various early hominins."[83] This point is echoed by Bernard Wood,[84] who reminds us that in an adaptive radiation, "because of the independent acquisition of similar shared characters (homoplasy), key hominid adaptations such as bipedalism, manual dexterity and a large brain are likely to have evolved more than once."[85] Nor, when we consider the range of foot structures in fossil hominids, should we be surprised to find a range of bipedal capacities.[86] Some will still allow arboreality, but in other cases to be a tree climber verges on the eccentric. One can still make much of the fact that human bipedality per se is unique.[87] If, however, one takes the wider view, then evolution suggests a different narrative: threading its way through primate history is the inherent inevitability that while some would remain quadrupedal, others sooner or later would rise on their two feet.[88]

The Convergent Walkers

Our bipedality looks less like a fluke than is often thought. What of looking further afield? Bipedality is familiar in such animals as the theropod dinosaurs (and of course their descendants, the birds), but it has evolved a number of other times among the reptiles.[89] What about beyond the vertebrates? Consider again the octopus. Here their similarity with the vertebrates takes another turn, with the recognition that some octopus move across the seabed on what are effectively two legs—that is, they show bipedal locomotion. To be sure, progression is more in the form of a rolling gait, and each of the two arms effectively unfurls across the seafloor. And the reason? Christine Huffard and her colleagues suggest that with the six other arms in the air (so to speak) the octopus camouflages itself, with one species possibly resembling

a rolling coconut while to the unobservant predator the other species is seen as "a clump of algae tiptoeing away."[90] Nor are the octopus the only marine animals that have learned to walk.

Many bony fish (the teleosts) have adopted this locomotory style,[91] not least the sea robins that "give the appearance of someone tiptoeing on many moving fingers."[92] Also remarkable are those fish that show a tetrapodlike locomotion, including the pigfish Zanclorhynchus[93] and the antennariids (also known as anglerfish or frogfish). Usually these fish lurk immobile, but when they move they can even manage a strange gallop, albeit at a snail's pace.[94] Among the other major group of fish, the chondrichthyeans (encompassing the sharks and rays), walking has also evolved. The epaulette shark (Hemiscyllium), pose that *Oreopithecus* is considerably more primitive and is effectively an ape close to such forms as *Dryopithecus*[55] (in which case the convergences become very significant). Its geological age (Miocene) can be read in support of either possibility, but in various ways this ape appears to be "out of time," and the lineage may have appeared substantially earlier.[56] Wrestling with an incomplete fossil record and a complex mosaic of characters make all this very tantalizing. But convergence may be the elephant in the room.

If *Oreopithecus* ultimately transpires to be a close relative of the hominids, then Esteban Sarmiento's analysis of the arm structure and his conclusion—"Because *Oreopithecus* and hominoids have arrived at the same morphological solutions to the mechanical problems imposed by climbing behaviors, convergence is a very unlikely supposition"[57]—will be fully vindicated. This author concedes, however, that alternative hypotheses remain open, and the repeated references to the peculiar mosaic of characters in *Oreopithecus* suggest that appearances may be deceiving. This matters because if *Oreopithecus* is more distant from hominids, then evidence for a precision grip[58] and bipedality[59] suggest that the trajectories of evolution are constrained. Given the controversial position of *Oreopithecus*, these proposals have generated debate,[60] not least because inferences as to how ligaments and muscles were originally arranged are far from easy.[61] Even if *Oreopithecus* was more than capable of arboreal excursions,[62] the evidence for both a precision grip and bipedality seem convincing.

Nor are we asked to imagine this ape as striding across the savannah. While the ear structures, notably the semicircular canals, certainly indicate agility,[63] Meike Köhler and Salvador Moyà-Solà[64] argue that although "bipedal activity made up a significant part of the positional behavior of [*Oreopithecus* . . . it was] functionally intermediate between apes

and early hominids."[65] Progress was probably more like a shuffle as against a brisk walk, let alone sprinting. In addition, a precision grip need not be equated with a grinning face above the rock that comes crashing down on the defenseless tortoise. In fact, the dentition points to plant eating,[66] and the convergent employment of bipedality and a precision grip can be put in the context of an effective method of harvesting plant material.[67] It has been proposed that not only was *Oreopithecus* an effective vegetarian but it waded through swamps[68] chewing aquatic plants and perhaps the odd invertebrate.[69] Does this strike a chord? Marcel Williams reminds us that "Alister Hardy proposed that the origin of hominin bipedalism and power precision grips were aquatic locomotive and manual adaptations,"[70] an idea made more famous by Elaine Morgan.[71] In support of the "aquatic ape" she weaves an intriguing skein of evidence, and it is fair to say that the academic reception has been decidedly frosty.[72] So too, in a manner of speaking, was it for *Oreopithecus*. At the time this Miocene ape was wandering around, these islands were a zone of safety, with few predators. When connections were established with larger land masses by the ceaseless tectonic activity within the Mediterranean, then the carnivores moved in and *Oreopithecus* was doomed.[73]

Speculation continues as to how the early bipedal hominids managed to avoid a similar fate. A tool-less, diminutive ape on two legs in Pleistocene Africa hardly seems an encouraging prospect. Yet the evidence suggests that bipedality evolved very early, as is argued for such key forms as *Orrorin*[74] and *Sahelanthropus*.[75] Nor were they only terrestrial excursionists, because a start in this direction was achieved earlier, in forms such as the Miocene ape *Equatorius*.[76] In the case of humans, focus remains on our nearest relatives, the chimps and gorilla, with the dilemma as to how we managed to become bipedal given observation revolves around the fact that with propulsion largely concentrated in the hindlimbs, the forelimbs are freed up. This is not only to test the stability of the next step, but to reach out and, assisted by accurate vision, grasp a fruit or an insect. What was once a paw is becoming a hand, and convergences are inevitable.[41] Instead of thundering feet and raking claws, the forelimbs become increasingly manipulative. This capacity, however, may go deep into tetrapod history, although distinguishing between homoplasy or homology depends crucially on certain phylogenetic assumptions.[42] Beyond the vertebrates, examples of manipulative ability range from the octopus (pp. 274–83) to various insects, including fire ants.[43] When these capacities extend to a precise grip, then various arboreal frogs (p. 14) offer an instructive example of convergence. Particularly striking is a specialized arborealist known as *Phyllomedusa*.[44] This frog can not only can walk along the equivalent of very narrow branches but, as Adriana Manzano and colleagues note, it employs "a diagonal gait typical of primates and other arboreal mammals when walking on narrow substrates."[45] These workers also remark how one unexpected result was that "*Phyllomedusa* is mechanically capable of executing a precision grip, known only from higher primates and so characteristic of human manipulative skills."[46]

The capacity for a precise grip and skilled manipulation has invited tempting correlations to both the evolution of tool use (p. 34) and enhanced intelligence.[47] Although both are also strikingly convergent, simple correlations are seldom to be trusted.[48] Even when one looks at the capacity for grasping among the anthropoids (that is, humans, apes, and monkeys), the well-known distinction between a power grip and a precision grip involves a complex evolutionary story. It is one that can neither be reduced to a story of ever increasing precision nor a single mechanism, let alone a straightforward link to cognitive capacity.[49] Rather, given the variety of hand anatomies, styles of grip vary, but importantly convergences emerge. As Emmanuelle Pouydebat and colleagues write, "precision grip is far more widespread than previously thought. . . . These movements may not be homologous and represent convergent evolution of motor patterns that superficially resemble reaching."[50] These workers also comment on the "Unexpected similarities between capuchin and human prehensile behaviors."[51] This has a bearing on the convergent evolution of tool use by capuchins (pp. 281–83), but as importantly on whether a precision grip might have evolved independently in extinct groups. Such indeed appears to be the case.

FROM FOUR TO TWO

A most intriguing case emerges in *Oreopithecus*,[52] a Miocene ape (c. 7 Ma) that inhabited Mediterranean islands that subsequently became part of Tuscany and Sardinia.[53] Insularity is often equated with endemicity. *Oreopithecus* very much fits into this mold, because it shows a decidedly puzzling mixture of morphological features. In its own way this is another test case for convergence,

although one of particular importance because the specifics revolve around features that are widely agreed to be central in the emergence of hominids. In effect, the phylogenetic position is polarized between those workers who argue that the striking similarities to the hominids are simply a reflection of evolutionary proximity[54] (so convergence is hardly an issue), as against those who pro-a series of striking convergences are found in the marsupials, and specifically the opossums. Like their primate counterparts the ability to climb and the amount of time spent on the ground vary from species to species.[24] In the case of the woolly opossum (*Caluromys*), however, their gait pattern, including a diagonal sequence of footfalls and emphasis on the hindlimbs, is strikingly convergent with the primates,[25] especially such groups as the cheirogaleid lemurs.[26] Daniel Schmitt and Pierre Lemelin[27] stress that the "gait characteristics of woolly opossums are essentially identical to those of the majority of primates."[28] When the compliance of the forelimbs is compared, then as Daniel Schmitt and colleagues[29] also note, "These data add another convergence trait between arboreal primates, *Caluromys*, and other arboreal marsupials and support the argument that all primates evolved from a common ancestor that was a fine-branch arborealist."[30]

Caluromys is instructive in a wider context. Writing of the didelphid opossums as a group and especially the Virginian opossum, Don Hunsaker and Donald Shupe[31] remind us that "it is very easy for one to picture the entire family as a slow moving, primitive group of beasts with remarkable dentition and some objectional personal habits that include biting the investigator. If unable to make tooth contact, they are just as liable to defecate and spray him with evil smelling fluids. In contrast to this species, however, the active, curious, and intelligent acting Woolly opossums (*Caluromys*) are highly reminiscent of lemurs bounding from limb to limb investigating and manipulating every unusual thing they encounter."[32] These authors were optimistic

that the didelphids "have a great potential for studies in convergent evolution and comparative behavior."[33] In terms of such features as brain size, rates of development, small litters, and basal metabolic rate, the woolly opossums are not only instructive in transcending the supposed marsupial norm but their convergences might provide an object lesson in the origins of primates.[34] Tab Rasmussen remarks, "*Caluromys* . . . makes a good model of what primitive prosimians might have been like."[35] Nor, it must be stressed, is *Caluromys* uniquely valuable in this respect, because more generally the opossums show a number of striking parallels with the primates. In their overview Tab Rasmussen and Robert Sussman[36] remind us that if we wish to know more about primate origins, we should "find and examine as many parallel independent evolutionary runs as possible."[37] Focusing on the phalangeroid marsupials (which include the sugar glider and acrobatids [p. 190], the honey possum [p. 434n154], and woodpecker mimic *Dactylopsila* [p. 32]), they produce an impressive roster of parallels. Among the various taxa these include orbital convergence, prehensile tail, enhanced encephalization, and most strikingly grasping hands and feet. Certainly phalangeroids show a striking diversity of form, but as Rasmussen and Sussman note, "The greatest convergence of primate-like traits occurs in the small-bodied omnivores . . . [and] clearly offer broad support for the idea that primates, too, may have diversified as terminal-branch-foraging omnivores."[38] Important though these comparisons[39] are, it might be thought that as a general exercise in evolutionary predictability we are grasping at straws. That is precisely the point.

GETTING A GRIP

When comparing the gaits in primates and marsupials like the woolly opossum, and this

includes convergence as to the manner in which the foot grasps the branch,[40] a central digit may have acted in an opposable, thumblike fashion.[7] Jorg Fröbisch and Robert Reisz duly note how "*Suminia* independently evolved grasping and clinging abilities long before (approx. 30 Ma) the evolution of these characters in any other tetrapod."[8] Among the extant reptiles, the many examples of arboreality include the Caribbean anolids (with a series of corresponding ecomorphs in other habitats[9]), the geckos[10] with their adhesive pads (pp. 83–84), and the chameleons. The latter group are particularly adept arborealists, confidently navigating slender branches. They owe their success to a series of skeletal modifications,[11] including suppressing lateral undulations, rendering the shoulder girdle much more mobile, and employing prehensile feet (as well as a tail[12]). Martin Fischer and colleagues note, "All these evolutionary novelties parallel very similar modifications in the evolution of the locomotor apparatus of therian mammals. We propose that the convergent 'invention' of dynamic stability and a compliant gait seem to be responsible for the locomotor similarities between chameleons and mammals."[13] Integral to arboreality is not only a secure grip but, when the occasion demands, the ability to launch oneself confidently into the void, perhaps to escape the unwanted attention of a predator. Particularly adept are some anolids and lemurs. Despite the enormous phylogenetic gulf that separates lizard and primate, the mechanics of leaping are strikingly convergent,[14] leading Pierre Legreneur and colleagues to "suggest that general laws of locomotion apply even in distantly related clades, despite their dissimilar morphological designs"[15] and physiologies. And we now turn to the primates because good arguments exist to indicate that if our ancestry had not taken us to the trees, then ultimately we would have neither held a pen to write about it or walked upright to the library shelf to check some niggling fact.

IN THE TREES

Analysis of the hands of very primitive primates, including the plesiadapiforms, suggests not only an arboreal existence[16] but one that may well have involved moving into the swaying zone, to walk along the thinnest possible of branches.[17] In this teetering environment, paradoxically we can trace the road to human bipedality. Of equal importance, however, is that when it came to achieving a precarious foothold, these primates were not alone: other mammals were converging on the same solution. In parallel to the arboreal primates walk the opossums.

To stroll along very slender branches, with the attendant advantages of moving across the forest canopy and leaving behind baffled predators, the arboreal primates typically employ a walking gait whose footfalls of left (L) and right (R)/fore (F) and hind (H) define a diagonal pattern (so LF→LH→RF→RH, etc).[18] Integral to this pattern of walking is the arrangement of knees and elbows to provide a compliant gait, combined with the hindlimbs providing much of the support and propulsive force and freeing the forelimbs to test the reliability of the next step forward.[19] It is important to stress that this arrangement is not invariable; in the slender loris the gait is again diagonal, but the forelimbs take more of the force,[20] while in its arboreal locomotion the sugar glider (*Petaurus*) walks in a somewhat different way (in part this may be because of a significantly smaller size).[21] There is also a risk to seeing these gaits in a rather static way. In reality the gait is a highly dynamic process revolving around the musculature[22] and kinematics of locomotion.[23] Nevertheless, when it comes to agile behavior high in the trees,

CHAPTER 7

WALKING (AND SWIMMING) TO CONVERGENCE

AS HEROIC WALKERS, occasional tree climbers, and even swimmers, humans can otherwise only admire the grace and fluidity of animal locomotion.[1] At first sight it is the diversity of ways of getting around that impresses. After all, is there anything more peculiar than human bipedality? Certainly it is distinctive, but if we can understand how we learned to stand on our own two feet, then what looks like an oddity turns out to be threaded with inevitabilities.

ARBOREAL EXCURSIONS

Living in trees?

Stand we may, but we wouldn't be here unless our ancestors knew all about living in trees. Not just our ancestors, because many other groups engage in arboreal excursions for which the sine qua non is the existence of trees (they, too, are rampantly convergent[2]). Animal arboreality was an inevitability, with convergent examples encompassing many groups of vertebrates (including frogs, p. 192) and invertebrates. Among the latter are tree-dwelling crabs (p. 167), as well as a group of mites (they, together with the revolting ticks, constitute the acarines) known as the oribatids.[3] At first sight, not least because of their tiny size, oribatid arboreality might seem to be an arcane example. This microcosm of acarine evolution is, however, a neat test case for the primacy of convergence. As their popular name, the "moss mites," suggests, typically they are soil-dwellers, but their ascent into the trees has taken place at least fifteen times. Interestingly, a sensory appendage (or sensilla) has been transformed so as to register wind speed (anemometer), but importantly not all oribatids are destined to become arboreal. The dice are loaded. Mark Maraun and colleagues conclude that the "ecological forces swamp chance events, such as drift and historical contingencies during evolution, supporting the 'adaptationist program.'"[4] As these authors stress, their work points to "the primacy of ecology,"[5] but equally this oribatid story is underpinned by function and evolutionary likelihood.[6] What then of the vertebrates?

For most people, climbing in trees either conjures up visions of reckless children or graceful gibbons. Long before any mammal evolved, other vertebrates were exploring the arboreal world, but the earliest known example is a synapsid reptile, the group from which the mammals arose. The creature in question, *Suminia*, is a late Permian anomodont. Its skeletal configuration unequivocally points to this animal not only being an agile climber but possessing a hand whose first be instructive in our technologies,[253] and as Paul van der Wal[254] remarks with respect to the chitons and limpets, "Although the minerals are different, some of the most crucial properties are strikingly similar."[255] These include the overall hardness and the manner in which the leading edge is kept sharp; he continues, "The fact that this combination of traits has evolved twice in nature may show that the tooth is optimally designed for its function."[256]

Not only iron finds such employment. Repeatedly and independently, zinc has also been recruited as a way of hardening teeth in a variety of invertebrates. When first detected in the jaws of the polychaete *Nereis* one possibility considered was marine pollution.[257] It certainly isn't, and the zinc[258] clearly plays a major structural role.[259] In *Glycera*, copper[260] is employed; this may be linked to the need to pierce prey, because this polychaete has independently evolved a method of venom injection[261] using a highly effective neurotoxin.[262] Not only is zinc used to harden the teeth of worms, but also other marine creatures, such as the predatory arrow-worms (and perhaps the extinct conodonts, distant relatives of the vertebrates),[263] while on land we encounter zinc in the mandibles of leaf-cutter ants (p. 182)[264] and some termites (p. 187),[265] as well as other insects.[266] In another great division of the arthropods known as the chelicerates, this metal reinforces the fangs of spiders[267] and similar structures in the scorpions, while in the sting the concentration far exceeds mineral ores with up to a quarter of its dry weight being zinc.[268] As noted, fangs or harpoons or stings that can deliver toxins to immobilize prey or discourage attention are widespread, yet as ever the degree of convergence (and its location) can be surprising. In a cerambycid beetle from Peru, Amy Berkov and colleagues[269] describe the antenna, which bears at its tip a sophisticated sting "almost identical to that found in the stinger of a deadly buthid scorpion."[270] Not only is the antenna highly maneuverable, but most likely it can engage in multiple stings, although at least for the humans the toxin is far less potent. Quite likely the sting is hardened with zinc. It is not only teeth and stings that are so hardened: zinc is also found in the ovipositors of parasitic wasps that drill through wood,[271] which has evidently evolved at least four times.[272] Of course, although strikingly convergent, zinc is only one way of hardening jaws and other structures. While one might intuitively think of the employment of zinc as the evolutionary equivalent of cutlery, in point of fact the greatly enhanced strength is more likely linked to molecular cross-links.[273]

wider significance? In the eyes of some, snake oophagy underlines the contingencies of evolution, given that the immediate ancestors of these snakes tend to feed on the animals that lay the eggs.[236] In one such analysis, particular stress is laid on a remarkable African colubrid (*Dasypeltis*) that only eats bird eggs,[237] yet in at least this case, the snakes' oophagy is scarcely accidental, given the adaptive possibilities that presented themselves when these ground-dwelling birds underwent a major radiation.[238]

If you don't fancy swallowing entire eggs, what about an aquatic mode of life hunting fish? In snakes this also has evolved multiple times,[239] and here the natricine snakes are especially informative.[240] Shawn Vincent and colleagues remark this is a potentially rich field for "even more powerful . . . opportunities [that] exist in . . . other snake clades [such as the homolapsines] to test for phenotypic convergence."[241] Yet in the stream-dwelling natricines themselves, the convergent functional complex is highly instructive in terms of common hydrodynamics and methods of foraging.[242] As ever an apparently mundane example of convergence opens a much wider field of inquiry. Thus, in striking contrast to most vertebrates, the aquatic snakes have, with one exception, never evolved suction

feeding. Presumably this is because of the elaboration of their iconic forked tongue for chemoreception[243] as against its assisting with swallowing and thereby necessitating the modification of the hyomandibular complex,[244] an essential component in the sudden enlargement of the oral cavity in suction feeding (p. 44). And the exception? Egg-eating sea snakes that have managed to equip themselves with a novel jaw muscle. This evidently serves to depress the floor of the oral cavity,[245] and also finds a direct counterpart in a group of primitive snakes known as scolecophidians that spend their lives hoovering up ants and termites.[246] No other aquatic snakes have ever learned this trick. They continue to strike and employ either a sideways or forward movement.[247] Each has hydrodynamic consequences. In the case of the natricines, the mode of frontal striking leads to convergences,[248] with Anthony Herrel and colleagues concluding that "underwater striking could be a model system for studying convergence and the constraints therein. Underwater striking has arisen independently several times in snakes. . . . Our data . . . suggest that the outcome of evolutionary divergence may be predictable in the face of severe physical constraints on performance."[249]

ARMORED TEETH

This chapter opened with Darwin's musings on swiftness of death, as maws gape, teeth descend, and death struggles turn to limpness. Wherever one looks, teeth like poniards, stilettos, and dirks hover in eager anticipation, and as often as not dripping with venom. It is a formidable prospect, but some animals go one step further. Why not armor one's teeth? Most celebrated are a group of mollusks known as chitons. Like many members of this phylum they possess a long ribbon of teeth,

the radula. In their case, however, scraping of the algal films on rocky surfaces is much enhanced by the addition of iron oxide, specifically magnetite, to produce a highly effective file.[250] But they are not alone, because independently another group of mollusks, the gastropods in the form of limpets, also employ an iron oxide—but this time employing the mineral goethite.[251] While wear pattern in each group differs,[252] they clearly help confer excellent cutting properties that might crack open the shell.[211] It is not alone, because an extinct Miocene marsupial from the same area has arrived at almost exactly the same solution,[212] which as Derrick Arena and colleagues note represents "an extraordinary example of convergence between a lizard and a mammal."[213]

What about even more robust prey, such as a scaly lizard? Delicious they may be, but how is a snake to tackle what is effectively an armored ball? Enter the fully collapsible tooth, ingeniously hinged near its base rather than the usual arrangement of a firm fusion to the jaw. The teeth are no longer fanglike, but shorter and with surfaces well designed to grip a tough and unyielding prey. A splendid solution, so we will not be surprised to learn that it has evolved independently at least three or four times.[214] Below I briefly address the many times the lizards have evolved a snakelike form, and so too when we look at one such lizard, Burton's legless lizard (*Lialis*), we find that it also preys on other hard-bodied lizards. In the words of Frederick Patchell and Richard Shine[215] the solution stands as "a remarkable example of convergent evolution in the dentition of scincivorous [skinks] snakes and lizards,"[216] because it too has evolved hinged teeth.

This is not the only way that snakes choose to grapple with the biological equivalent of an oil barrel. Consider, for example, the remarkable bolyerine snakes (notably *Casarea*) from the tiny Round Island, close to Mauritius. They are a minor evolutionary icon because, alone of all vertebrates, the upper jaw (the maxilla) possesses a joint about which anterior and posterior sections can articulate.[217] Given the difficulty of imagining an intermediate form (either the maxilla is "broken" or it isn't), this snake may be one of the closest, and very few, approaches to an idea championed by Richard Goldschmidt, that is a "hopeful monster."[218] Maybe so, but its function is linked to a particular style of feeding that is suitable for engulfing hard-bodied lizards. A collapsible jaw, rather than collapsible teeth, but this snake shows a suite of behaviors,[219] notably in searching for and then handling its prey, that are strikingly similar to those found in the lizard *Lialis*.[220]

Nobody needs reminding how dangerous many snakes can be, and we find particular groups of snakes repeatedly slithering toward the same solution, or in the case of the sea snakes, swimming to a functional complex that predictably includes a flattened paddlelike tail[221] and salt glands.[222] All are elapids, but evidently the transition from a terrestrial form has evolved three times independently,[223] and convergences can also be found with other water snakes.[224] Most sea snakes are feared for their aggressive venomousness,[225] but curiously the turtle-headed sea snake (*Emydocephalus*[226]) hunts out eggs[227] and has adopted a mode of foraging similar to herbivorous browsing mammals (as well as some lizards).[228] Nor is this the only convergence within the elapids, because the Australian taipan (and small-scale snake) shows a whole series of striking parallels[229] to the equally deadly African Black mamba.[230]

The list of recurrently evolving ecomorphs continues.[231] While some sea snakes consume small eggs,[232] their terrestrial counterparts have developed a taste for reptile eggs multiple times.[233] The case of an Australian coral snake (*Simoselaps*) is particularly interesting.[234] As John Scanlon and Richard Shine note, not only is "The dentitional similarity between the oophagous *Simoselaps* and various oophagous genera of colubrid snakes . . . impressive,"[235] but in contrast to typical colubrids, their Australian counterpart is venomous. This convergence—revolving around a reduction in maxillary teeth and a corresponding development of a bladelike tooth to puncture eggs—is all the more noteworthy because only some species of *Simoselaps* have navigated to this specific solution. Is there a

THE TOOTH OF PERFECTION

Perfection? Not a word biologists often use. But the snakes are not alone. Consider the teeth of mammals[191]—specifically the designs based on the cutting teeth of the carnivores (carnassials) and the multiconed tribosphenic (protoconoids) arrangement found in many insectivores. These invite a comparison against ideal models, leading to the conclusion "that these teeth have the best shape for their function."[192] Each is indeed "The tooth of perfection."[193] As importantly Alistair Evans and Gordon Sanson note that "Dental morphology has managed to escape whatever developmental constraints apply to it to achieve essentially optimal functional tooth designs."[194] This no doubt helps to explain the many examples of convergence among mammalian carnassials and protoconoids,[195] but even in cases where the teeth are not obviously ideal (as in molars), convergence[196] again points to a limited number of solutions that may reflect the constraints of diet. In their different ways, therefore, at least some mammalian teeth and the fangs of snakes suggest that this is the best there will ever be. Moreover, while we typically associate snakes with dripping fangs, some exceptions are intriguing.

Fangs may be just the trick when the snake is lunging at the warm, soft mammal, but what happens when the snake might be better equipped with a pair of nutcrackers? In principle, as any self-respecting Frenchman will confirm, snails are a splendid addition to the diet, and indeed some pareatine snakes from Southeast Asia specialize on these gastropods.[197] The problem is that snails typically coil dextrally, and given that the jaws of this snake are too weak to break the shell, extraction of the flesh might be problematic. However, like those weird pieces of cutlery that one finds in dinner parties in the Sixth Arondissment, the pareatine snakes have

markedly asymmetrical mandibles that facilitate removal of the soft parts.[198] Masaki Hoso and colleagues point out the otherwise enigmatic abundance of such sinistral snails in this part of the world[199] "may be attributable to 'right-handed' predation" by the snakes.[200] As significant is the likelihood that the same syndrome has evolved independently in the Neotropical dipsadine snakes;[201] as Alan Savitzky notes, "The dipsadines and pareines resemble each other so closely that many previous authors considered them a single subfamily."[202]

These convergences on asymmetric feeding occur elsewhere. The insectivorous larvae of water beetles, for example, have prominent mandibles but are asymmetrical to assist with tackling dextral snails.[203] Crabs are notorious predators[204] on snails, and in members of the oxystomatids such as *Calappa*, one claw (normally the right one) is transformed into a massive structure with a scissorlike action that is highly effective in peeling open the snail.[205] Not only has this evolved independently in a Cretaceous crab (*Megaxantho*),[206] but among more recent marine fossils there is again evidence for sinistral snails enjoying a selective advantage.[207] Gregory Dietl and Jonathan Hendricks do not conceal their puzzlement when they conclude, "If selectively advantageous, why is left-handedness then so rare, especially in tropical seas where crab predation is ecologically important?"[208] A clue comes from the intricacies of snail mating (p. 213), such as in a Japanese land snail (*Euhadra*) where a dextral morph has evolved independently several times,[209] but when dextral meets sinistral this leads to what Rei Ueshima and Takahiro Asami gently refer to as "Genital mismatch."[210] Nor is this the only avenue to grappling with a snail. An Australian lizard (*Cyclodomorphus*) has an extraordinary tooth that acts as a hammer to fanciful, given that the shrew in question (the water shrew, a species of *Neomys*) is venomous. Conceivably on the vast tracts of once-undrained land, farm stock may have been bitten, even if there was more danger from a subsequent infection.

Shrews that are venomous[167] presumably rely on the puncture wounds of the needle-like teeth to impart the venom, but the endangered Haitian *Solenodon* has grooved incisors that evidently act as venom ducts. Nor is it the only insectivore to have evolved a venom delivery system, because notwithstanding unresolved phylogenetic questions, it is very likely that envenomation of another West Indian resident, the recently extirpated nesophontids, is convergent.[168] Samuel Turvey notes that this arrangement "required the evolution of a suite of soft tissue characters (glands, ducts, neurovasculature) unknown in any living mammals,"[169] but one that also evolved independently in various reptiles. The potential for the delivery of toxic saliva has also been identified in an Oligocene mammal (*Siamosorex*), but this insectivore is even more remarkable because it has reinvented itself as a carnivore.[170] Stéphane Peigné and colleagues titled their paper, "An astonishing example of convergent evolution towards carnivory." [171] Once again the capacity for venom delivery is inferred from grooved teeth, but while grooved canines have evolved several times independently, by no means all of these mammals[172] also dribble venom.[173]

The reference point for venomous vertebrates are, of course, the snakes. But long before they (and the lizards—collectively the squamates) had, other reptiles had invented venom delivery systems. In the Triassic of the United States, archosaurlike animals possessed bladelike teeth equipped with either deep grooves[174] or ducts that were entirely enclosed,[175] strikingly similar to the arrangement in the elapid snakes.[176] Later in the Mesozoic a sphenodont (today known from

the famous tuatara "lizard" of New Zealand) once again evolved grooved teeth, arguably immobilizing its prey with venom.[177] Perhaps surprisingly, the same sort of evidence is used to infer the capacity for venom delivery in a theropod dinosaur, specifically the dromeosaurid *Sinornithosaurus*.[178]

One automatically equates living venomous reptiles with the snakes, but a number of lizards also employ venom, including (perhaps surprisingly given its large size) the famous Komodo dragon.[179] While it has been thought that snake and lizard envenomation arose independently, current evidence points toward a single origin.[180] Nevertheless, convergences abound. The arrangement most to be feared is tubular fangs connected to large and complex venom glands and associated compressor muscles. This ensures massive and rapid delivery and has evolved independently three times (in the atractaspidids, elapids, and viperids).[181] If the complex venom glands arose independently, then did also the tubular fangs?[182] It is an obvious inference, but as Kate Jackson notes, "There is nothing in the morphology or development of the tubular fangs of the three groups of venomous snakes to give any indication of homoplasy."[183] Jackson offers us four scenarios, but she plumps for the supposedly primitive condition of a grooved fang[184] being actually the derived condition. This, of course, begs the question as to why such a superb hypodermic delivery system[185] was abandoned. In remarking, however, that the proposed uniqueness of tubular fangs means that this "system does not represent the pinnacle of adaptive perfection,"[186] one cannot help but wonder if this will yet transpire to be another "surprising" convergence.[187] Significantly, the design of these fangs has remained effectively unchanged[188] for at least 20 Ma,[189] leading Ulrich Kuch and colleagues to argue that here is a structure "brought close to perfection."[190]

of recurrent fascination to biologists—mammals but laying eggs, secreting milk but lacking nipples, even possessing electroreception (pp. 148–49). Equally remarkably, however, the males are equipped with a spur on their hind feet that can inject poison quite capable of killing a dog.

With the unlocking of the platypus genome,[150] the venom naturally was one focus of interest. As is often the case in toxins, the platypus venom depends on very small proteins, but in this case the defensinlike peptides are convergent[151] with those employed by snakes and lizards.[152] The latter, of course, employ fangs, but so unexpectedly do some mammals. Concerning this mode of delivery there is no doubt that several types of shrew secrete a venom into their saliva; a useful trick that most likely significantly extends their range of prey.[153] Of these the venom of the short-tailed shrew (*Blarina brevicauda*), which has also evolved echolocation independently (p. 137), is particularly potent. It has been known for some time that this shrew venom employs a protein (specifically a serine protease with a catalytic triad) that is similar to some reptile venoms.[154] It now transpires, however, that with respect to the venom produced by the Mexican beaded lizard, the convergence is much more precise,[155] arising as Yael Aminetzach and colleagues point out, through "almost identical functional changes."[156] These entail a series of relatively minor changes around the active site of the serine protease that transform it from what is otherwise an innocuous protein to one with pronounced toxicity. Aminetzach and colleagues ask the central question: "How repetitive, and thus predictable, is evolution?"[157] and answer that, as far as not only this shrew and lizard example is concerned but in other toxins,[158] then the evidence "suggests that adaptation may be highly predictable."[159] Notwithstanding the fact that these

various toxins draw on a wide variety of proteins,[160] Bryan Fry and colleagues are surely correct to emphasize that "The remarkably similar biochemical composition of the cephalopod glandular secretions and the complex venoms across the Animal Kingdom suggest that there are structural and/or functional constraints as to what makes a protein suitable for recruitment."[161] Among the proteins recruited for toxicity are the lipocalins[162] (p. 175), but doffing their hats to evolutionary versatility, "Several groups of hematophagous [blood-sucking] arthropods have independently recruited this protein family"[163] to prevent in one way or another the clotting of their victim's blood.

Given their diminutive size, it might seem faintly comic that notwithstanding their venomous capacity, shrews could ever be regarded as a threat. Yet as George Pournelle mentions in his overview of these venomous insectivores, as a child living on the Gulf Coast of Florida he was repeatedly warned to "watch out for cottonmouths [a very dangerous snake] and don't pick up any of those long-nosed mice [i.e. shrews]!"[164] In a more recent assessment of venomous mammals,[165] Mark Dufton reminds us how our predecessors took a decidedly different view. Even in the time of Shakespeare, not least his *The Taming of the Shrew*, to be a "'shrew' was associated with matters of depravity, wickedness, evil, ill omen and malignancy."[166] Dufton also draws attention to the "shrew ash," where the bole of the ash tree had a hole augured, prior to the entombment of a luckless shrew. Branches taken from such a shrew ash were, it was insisted, most efficacious in banishing malign spirits and curing cattle, while children with whooping cough were also steered in this direction. Let us all smirk most patronizingly, but an association between shrews and a reputation for malevolence (or its cure) may not have been entirely mechanism,[134] but as Diego Pol remarks, its "feeding mechanisms show similarities with either non-mammalian cynodonts . . . or therian mammals . . . denoting singular cases of convergent evolution."[135] These examples of heterodonty serve to underline the diversity of notosuchian dentitions. In the case of *Yacarerani* the teeth include both insiciviforms and molariforms, but, as importantly, evidence from wear patterns and inferred jaw musculature indicate complex food processing.[136] This included tooth-to-tooth occlusion and anterior-posterior movements of the jaws, leading Fernando Novas and colleagues to conclude that the observed "Diversity in dental patterns may reflect different dietary habits, and [they] depict notosuchians as a very versatile group that filled a vast range of ecological niches during the Cretaceous."[137] Indeed it does; the dentition of *Mariliasaurus* suggests this crocodyliform may have rootled in the group in the manner of pigs,[138] while the arrangement in *Adamantinasuchus* points to scavenging if not carnivory.[139] Considering the close relationship of these notosuchians to the crocodiles and alligators, which although capable of durophagy (p. 47) are mostly thought of in terms of a smiling dentition

of rather uniform conical teeth, it is hardly surprising that the mammal-like heterodonty of these extinct crocodylians has provoked words like "singular"[140] and "bizarre."[141] This, however, is not the only time such heterodonty has evolved, because a somewhat less primitive group, the eusuchians, a basal form from Hungary (*Iharkutosuchus*), has complex molariform teeth[142] and a complex chewing mechanism.[143]

Critics of evolutionary convergence may complain of yet another catalog, yet it is equally important to draw attention to more general principles of organization. This has been strikingly demonstrated when comparisons are drawn between overall tooth complexity and diet (ranging from hypercarnivore to herbivore) in the carnivorans and rodents.[144] Notwithstanding striking differences in the body sizes, methods of mastication, and digestion, in the words of Alistair Evans and colleagues, "both mammalian groups have independently and repeatedly evolved different dietary specializations covering most of the range from animal to plant foods,"[145] allowing evolution to transcend both scaling and phylogeny.

VENOMOUS EVOLUTION

It is bad enough being stabbed, sliced, and chewed, but when the teeth serve to deliver venom as well, this might be regarded as the last straw. Often it is, but it is also convergent. So, shall we turn to the snakes? Not so fast. What about the mammals?

Now, apart from literary reviews by the deeply envious, to think of any mammal as venomous usually requires a considerable stretch of imagination. In fact, it is possible that a venom delivery system had appeared at the very dawn of mammal evolution, because

a Permo-Triassic synapsid (*Euchambersia*) not only had canines with grooves[146] but a pit in front of the eye that may well have housed a gland full of poison.[147] If indeed this was the primitive arrangement, then evidently it was lost in favor of teeth and claws as the primary mode of defense,[148] but as ever this does not prevent the reinvention of a venomous wheel. We have already encountered the hedgehog draped with toxins[149] (p. 39). And what about the monotremes? These animals, and especially the platypus, remain a source mortar action. In the so-called tribosphenic dentition, the protocone is opposed to talonid basin to form a highly effective crushing mechanism. Yet among the Mesozoic mammals,[112] this arrangement is evidently convergent. As Zhu-Xi Luo notes, not only is this dental evolution much more labile than had previously been realized, but it demonstrates how there is "more than one pathway to combine slicing and crushing functions, as exemplified by tribosphenic and pseudotribosphenic molars."[113]

Other examples of teeth evolution can also be very instructive,[114] not least among the reptiles. While the generality in this group is for a rather uniform dentition consisting of rows of conical teeth, reptiles can in fact show considerable variability. Such is evident, for example, among the durophagous mosasaurs (p. 47) and ichthyosaurs (p. 47). It is, however, the reptilian convergences with the mammals that are particularly informative,[115] not least because of the independent evolution of so-called heterodonty[116] exemplified by anterior canines or incisors along with a set of posterior molars. In a late Cretaceous borioteiid[117] lizard (*Peneteius aquilonius*), the posterior teeth are not only molariform but evidently operated in much the same way as an insectivorous mammal and show a tribosphenic-like[118] arrangement.[119]

Considerably more remarkable are extinct relatives of the crocodiles, known as the notosuchians. Abundant in the Cretaceous across Gondwana, not only are these crocodyliforms striking because of a mammal-like dentition,[120] but one form (*Pakasuchus*[121]) seems to be well on the road to becoming an honorary mammal (p. 233). The diversity of forms points to considerable ecological success, with the appropriately named *Armadillosuchus* from Brazil with its bands of bony plates making a fair approximation to its xenarthran successor.[122] Even more noteworthy is the late Cretaceous *Simosuchus* from Madagascar.

With its blunt head and powerful neck muscles it was probably an effective burrower,[123] as indeed were other notosuchians.[124] Its teeth, moreover, are most uncrocodile-like and possess five specific characters that unequivocally point toward a diet of plants.[125] As Nathan Kley and colleagues note, "These five dental characters, in varying combinations and degrees of expression, have been documented in an extraordinarily wide range of herbivorous vertebrates . . . including some teleost fish . . . pareiasaurs . . . several clades of both ornithischian and saurischian dinosaurs . . . iguanine lizards . . . and anomodont synapsids."[126] There are wider implications, because as Gregory Buckley and his colleagues also remark, "Other features of *Simosuchus* [also] appear to be convergent with those of ankylosaurs [a group of ornithischian dinosaurs], such as a broad compact body, extensive dorsal and ventral shielding,[127] bony protection of the skull above the supratemporal fenestrae and orbits, and a deep cranium with a broad, short snout." Fascinatingly, they continue, "A crocodyliform convergent upon an ornithischian dinosaur[128] is intriguing in the light of the apparent absence of the latter from the Late Cretaceous of Madagascar."[129]

The ankylosaur-mimic was not the only notosuchian to turn to herbivory, because the Chinese[130] *Chimaerasuchus* has the appropriate teeth, but this time the convergence is toward the mammals.[131] In fact as Xiao-Chun Wu and Hans-Dieter Sues note, "The molariform teeth . . . are so unusual that its crocodyliform affinities were not recognized for many years, and the fossil was initially considered a possible multituberculate mammal."[132] Crocodyliform it is, but the convergence remains, and as Xiao-Chun and colleagues noted, the "maxillary teeth of *Chimaerasuchus* strikingly resemble the upper postcanines of the Tritylodontidae, highly derived non-mammalian synapsids."[133] Much the same applies to *Sphagesaurus*, which not only has a complex feeding dinosaurs not only evolved a similar grinding dentition but employed even more types of dental tissue.[92] The received wisdom links hypsodonty to more effective grazing and the emergence of grasslands[93] (in which in due course hominids had a walk-on part), and in particular the abrasion suffered by the teeth as they encountered the siliceous phytoliths in the grass. This intuition falls foul of the observation that the teeth are tougher than the phytoliths, and the real culprit for grinding down the teeth is the grit and dust[94] swept up while cropping the lawn.[95] Nor is the correlation with grazing and hypsodonty necessarily straightforward. In the extinct notoungulates, endemic to South America, hypsodonty was not only a highly successful strategy but it evolved at least three times.[96] Nevertheless, while grazing, including that of grasses, might in part explain the rise of hypsodonty, the issue is still very much open. Indeed, hypsodonty might as plausibly reflect these notoungulates being browsers and also having to cope with volcanic dust.[97] Guillaume Billet and colleagues refer to this example of hypsodonty as "precocious,"[98] indicating that its evolution in South America substantially predated the better-known examples in other mammals.

But the notoungulates weren't the first on the block. Welcome to the gondwanatherians. As their name suggests, these Cretaceous mammals (which survived into the Eocene[99]) are a characteristic component of ancient Gondwana.[100] As Greg Wilson and colleagues[101] note, they "have garnered attention as the only Mesozoic mammals to evolve high-crowned teeth (i.e., crown hypsodonty)."[102] But this convergence has further implications. First, given the widespread assumption that grasses are an innovation of the Tertiary it seemed reasonable to think of gondwanatherians living somewhat like a beaver, tackling roots and perhaps semi-aquatic.[103] With the discovery of grass phy-

toliths in dinosaur coprolites from India, it seems reasonable to suppose that not only the giant sauropods but co-occurring gondwanatherians were grazing the first grasses.[104] The other fascinating question revolves around the wider relationships of this group. So strange are the gondwanatherians that when first discovered, even a relationship to the xenarthrans was contemplated.[105] A more realistic idea is a connection to the multituberculates, but here it is very difficult to rule out convergence,[106] not least because much of the fossil material exists as teeth and jaws.[107] It could give the word "circularity" a bad name. As Wilson and colleagues also note, "Perhaps the most striking example of Gondwana endemism is the gondwanatherian radiation,"[108] which reinforces the view they most "likely filled ecological roles similar to those of 'Laurasian' multituberculates."[109]

Nor is this latter group without interest when it comes to the transformation of one or more of the cheek teeth in the lower jaw into the equivalent of the bread knife, albeit with the jaw-stretching name of plagiaulacoid dentition. This is indeed one of the hallmarks of the multituberculates, but the evolution of a large, laterally compressed tooth with a sharp serrated edge is rampantly convergent, having evolved at least three more times in the marsupials and primates.[110] George Gaylord Simpson notes how this instance is "not confined to the denticulate shearing teeth, but extends to greater or lesser extent throughout the dentition. It is a complex and thoroughgoing example of convergence."[111] And this is evident from a suite of convergent dental characters, involving not only procumbent incisors but typically a diastema, all to assist in a polyphyletic attack on tough and fibrous vegetation. Of all these examples, possibly the most important is one that occurred early in the history of the mammals and concerns the molars. Here a key mammalian arrangement is to configure what is effectively a pestle-and-seems well on the way, and we might even go so far as to pinpoint the place of origin as Sunda Island, where Diard's clouded leopard is even closer to the saber-toothed morph than its mainland cousins.[75] But will it happen? The answer is probably "no," not because evolution is not predictable—convergence shows it is—but along with the future of all the great cats, the clouded leopard may not be with us much longer. When the hominids appeared all bets were off.

Spectacular as the saber-toothed convergences are, even within the cats they are by no means the only example. Consider, for instance, the American cheetah (*Miracinonyx*). Although now extinct it has long been realized to be remarkably similar to the Old World cheetah.[76] Indeed Daniel Adams[77] writes, "The points of similarity are so extensive and of such a complex nature that a hypothesis attributing their origin to other than common genetic descent would require pushing the concept of parallel evolution to an unprecedented extreme."[78] So what does the more recent molecular evidence show? One could almost guess; the American

cheetah evolved from a pumalike ancestor and so is convergent with the African cheetah.[79] This particular episode is minor enough, but it epitomizes the whole topic of convergence. Again and again molecular phylogenies, even if not foolproof, demonstrate how evolution almost goes in circles—reinventing the wheel, so to speak. Disbelief at the extent of convergence goes hand in hand with adjectives of surprise: it is no accident that in naming the American cheetah *Miracinonyx* (and of course believing it was but a subgenus of *Acionyx*, the cheetah), the prefix is the Latin for "surprising, amazing."[80] It is easy to say that both cheetahs are cats, and thus the similarity is better described as a parallelism rather than a convergence. Yet this statement would miss the point, because the similarity is almost exact—be it the characteristic domed head or extraordinary dentition where the upper and lower teeth occlude so precisely that it is effectively "an enormous pair of carnassials."[81] Is it so daft to think that cheetah morph "out there" toward which the real cats inevitably navigate?

TEETH

The cats[82] and their marsupial avatars, both with their dripping canines and ferocious bite, understandably form a focus of attention. Among vertebrates as a whole, however, convergence of teeth is the norm.[83] Much of the emphasis has been on the mammals, but the roster is much wider.[84] For example, not only are some frogs equipped with fangs, but this arrangement has evolved independently at least four times.[85] Significantly investigators Marissa Fabrezi and Sharon Emerson suggest that for a frog to inflict a bite, these "Fangs seem to represent the single best design solution."[86] It is interesting to see much the same story in the appropriately named *Danionella dracula*. Despite its diminutive size,

this cyprinid fish reevolved impressive fangs even though the group as a whole lost their teeth[87] millions of years ago.[88] These fangs, or odontoids, are not "true" teeth, but in the vertebrates, teeth per se seem to have evolved at least twice.[89] And while the first teeth seem to have been largely employed in grasping and stabbing, they have shown a series of evolutionary trajectories that, however extraordinary (think of the narwhal), are relentlessly convergent, not least among the mammals.

In this context the evolution of so-called hypsodonty (high cheek teeth) is perhaps the most familiar.[90] In the ungulates alone it has evolved at least seventeen times,[91] while beyond the mammals the hadrosaur able to adjust their grip by retracting their claws. Nevertheless, the fact that the mauling inflicted by a bear is also achieved without claw retraction suggests the thylacosmilids more than managed. Nature red in tooth and claw, as Tennyson reminds us.[62]

And what of the scimitar cats? Evidently rather than skulking in the undergrowth, they were the faster runners,[63] chasing down their prey and inflicting repeated slashes that led to successive hemorrhaging and the final fatal stumble. While dirk and saber are, to judge by their recurrence, sensible solutions,[64] a third way is found in *Xenosmilus*.[65] This cat, living in the early Pleistocene of Florida, was a sort of saber-toothed amalgam, evolving a dirk-tooth arrangement and lurking in dense vegetation, but using a killing bite much more like a scimitar cat.

The mayhem the thylacosmilids and their feline equivalents inflicted on countless generations of prey[66] can be augmented by other details, and also some possibly wild speculation. In both cases the way of life was probably largely solitary as against any sort of pack hunting, but on the other hand the technicalities of hunting with precision lunges almost certainly meant the young needed nurture and training, possibly for several years.[67] And bringing down the prey in a welter of gore and roars (if not subsequent purring, p. 267) was all very well, but how did they detect prey? Clearly vision played a role, but at least in the thylacosmilids binocular vision was quite restricted. There is, however, a very intriguing convergence in the ear region,[68] and although the functional significance is debated, one possibility is that the bones help to enhance the reception of low-frequency sounds. Is it even possible that in these otherwise ferocious beasts, the giant canines were gently inserted into the soil and so could act as sensory "antennae." How unlikely . . . until we remember that other mammals, such as the mole rats, use their enlarged incisors for just such a function (p. 133).

Although the saber-toothed morph evolved at least four times, all are extinct, although doubtless they were all too familiar to our ancestors[69] (yet no convincing example has been found of representation in Paleolithic art[70]). Do their immense canines simply provide another telling example of an evolutionary dead end, a solution that might indeed have been very successful—after all, why else did it repeatedly reevolve?—but in the wider panorama of evolutionary "huntin' and fishin'" was doomed to ultimate failure. Certainly if one compares the disparity of the saber-toothed morphs—that is, the amount of morphospace they occupied—with that of their near relatives, then it appears that with the exception of nimravids it was more restricted.[71] Nor is this particularly surprising: the saber-tooths are evidently a highly adaptive complex, effectively locked into a particular solution.[72] Indeed in the specific case of the machairodont saber-tooths the evolution of the skull and lower jaw has followed a distinct evolutionary trajectory in response to very specific functional demands.[73] In effect, like the other felids, biting ideally employs as wide a gape as possible in combination with very powerful muscles, but these are effectively impossible to reconcile: the cat can either have an immense gape with concomitant giant canines—the saber-toothed solution—or massive jaw muscles but correspondingly a more restricted gape. Daggers versus strength, but in their different ways equally effective and both employing a precision bite.

What the saber-tooths did, they did very well: a passing hominid would have been astonished to see them chewing watercress. But their story is finished, isn't it, never to reevolve? Possibly not, because the clouded leopards (*Neofelis*), among the most enigmatic of the great cats, show many of the key features that presage not only the emergence of a new saber-tooth,[74] but what would be a first-time hit for the feline felids. Indeed it there are certainly differences, "The overarching pattern of convergence is quite clear";[38] among the borophagines the "derived ecomorphs do appear to share functional advantage in bite force and skull strength."[39] What is particularly fascinating is, in general, how similar the evolutionary trends are in the two groups.[40] So, too, ironically some dogs (the canids)[41] have evolved into catlike forms.[42] And what about those carnivores that have opted to turn themselves into herbivores, as exemplified by the convergence of the distantly related giant and red pandas, both of which consume vast quantities of bamboo?[43] The shift is dramatic, and by no means are all things possible from this starting point, but convergences still emerge.[44]

In terms of hypercarnivores, of all the examples, the saber-tooths surely spring to mind. As far as the placental saber-toothed cats are concerned there were broadly two types,[45] respectively, those with somewhat shorted scimitarlike canines as against the "classic" elongate sabers of the so-called dirk-tooths:[46] the two well-known genera *Homotherium* and *Smilodon* exemplify this contrast. Also, even within the placentals the saber-toothed type has evolved at least three times: in the relatively primitive nimravids,[47] the barbourofelids,[48] and the machairodont felids.[49] Nor has this recognition of convergence been straightforward, because the very similarities serve to confound phylogenetic analysis, repeatedly persuading investigators of a close relationship that on closer inspection turns out to be more distant than supposed.[50] The development of the hypertrophied canines, which in the nimravid *Barbourofelis* are estimated to have reached at least 22 cm in length,[51] involve substantial reconfiguration of the skull.[52] Not only is this achieved by a mosaic evolution, but as Graham Slater and Blaire van Valkenburgh note, "the similarity in the evolutionary pathways seen in [the skull of] nimravids and felid sabertooths

is striking and suggests that the transition to this extreme morphology makes functional demands that can be met in a limited number of ways."[53] Yet despite these extremes of predatory adaptation, a range of specializations exists. Among the nimravids, not only did a miniature saber-tooth evolve, little bigger than a domestic cat,[54] but more significantly the saber-morph evidently evolved very rapidly. Does this not indicate that this evolutionary route was "waiting to be discovered," so that we can hardly be surprised that this destination was then repeatedly revisited? It is not proposed, of course, that any of these evolutionary trajectories was instantaneous, and in the case of the machairodont felids, the discovery of more primitive representatives allows us to see what key characters, linked to an effective slicing action (the shear bite[55]) and stabilizing the descending bite, heralded the route to the superb cats *Homotherium* and *Smilodon*,[56] complete with whiskers.[57]

The two variants, the scimitar- and dirk-toothed, also evolved several times independently, each time representing a particular ecological solution.[58] The dirk-tooths were evidently ambush predators,[59] lurking in dense vegetation, and using not only the immensely powerful jaws to subdue their greatly surprised prey but also employing their massive forelegs.[60] Death was quick as the canines and well-developed incisors inflicted gaping wounds. This ecological solution was also adopted by the marsupial thylacosmilids, although with one crucial advantage in comparison to its placental counterparts; the canines showed a continuous growth, so that should they be snapped off in the fray, then ultimately they would regrow. The same situation for the placentals was far more serious, and indeed such fractures seem to have been commonplace.[61] There were also some interesting differences. In grappling with their prey, the thylacosmilids had the relative disadvantage of not being diverse as grinding pavements of teeth or long sticky tongues spell their own varieties of doom, but being mammals ourselves, perhaps most graphic to our way of thinking is a mouthful of teeth designed to slice and impale. And what could be more emblematic than a saber-toothed cat, but in its convergent form as a marsupial. Welcome to the thylacosmilids—now extinct, but for millions of years they stalked the landscapes of South America. These cat mimics belong to the group known as the borhyaenids, whose origins within the marsupials are somewhat obscure but appear to be fairly close to the didelphids, that is, the opossums.[18] The borhyaenids were broadly doglike, but among the roster of forms the thylacosmilids stand out not only as an extreme morphological excursion but one that just happened to navigate to something very much like a saber-toothed cat. Not only were they evidently successful, but they are the last of the borhyaenids to disappear, surviving until the late Pliocene (about 2 Ma ago).[19] Why they disappeared is uncertain; possibly it was a result of a changing climate, but in any event it seems unlikely to have been the result of the influx of mammals from North America, an invasion that started as a trickle at least 6 Ma ago[20] and as the Panamanian isthmus solidified became a flood. Thus, as the placental saber-tooths spread across South America,[21] they never snarled face-to-face with the thylacosmilids but only padded past their bones.

Convergence is never precise, so the identity between the thylacosmilids and saber-toothed cats is not exact.[22] As I like to point out, however, it seems pretty academic whether you were to fall victim to either the saber-toothed marsupial or placental, even if you could console yourself that the first was far more closely related to a koala "bear" and the latter to a hedgehog than they were to each other. And although, like the saber-toothed cats, the thylacosmilids are extinct,

in both cases we can make some deductions as to the last minutes of their prey. This convergence between thylacosmilids and saber-tooths is, of course, only one of a wider roster of similarities between marsupial and placental, and even within the carnivores there are other examples of at least an approximate matching.[23] This is not to deny, however, important phylogenetic constraints as well as a relatively smaller brain in the marsupials that may be reflected in correspondingly more massive jaw muscles (and correspondingly a trade-off in the placental carnivores between bigger brains and a correspondingly weaker bite).[24] Notably in the other crucible of marsupial evolution, Australia, the extinct marsupial "lion"[25] was immensely powerful[26] and used its incisors, rather than the canines, to stab its prey. Despite these and other differences, when compared to the saber-toothed *Smilodon*, there is an intriguing list of similarities[27] that, as Stephen Wroe remarks, are "strongly suggestive of niche correspondence."[28] In their paper titled, "How to Build a Mammalian Super Predator" Wroe and his colleagues[29] continue to explore these convergent correspondences and point out that these robustly built hypercarnivores also have some intriguing similarities with the bears. Similarly the correspondence between the marsupial "wolf," the also extinct thylacine, and the dogs is not exact, but an analysis of the stresses experienced by either skull when biting are striking.[30] Again as they note there was probably "considerable ecological overlap."[31] So, too, the bone-cracking hyenas have similarities to the marsupial Tasmanian devil.[32] Among the other mammalian carnivores, convergence is again widespread,[33] although history certainly plays a role.[34] For instance, a hyenalike ecomorph that specializes in bone cracking[35] has evolved independently perhaps six times.[36] As Zhijie Jack Tseng and Xiaoming Wang[37] note in their comparison of hyenid and borophagine canid skulls, although

CHAPTER 6

BITING CONVERGENCE

CONCLUDING HIS third chapter of the *Origin*,[1] the one devoted to the struggle for existence, Darwin ended on an almost optimistic note. Thus he wrote how notwithstanding his overview of this murderous world, where the weak always went to the wall, nevertheless "we may console ourselves with the full belief, that the war of nature is not incessant, that no fear is felt, that death is generally prompt, and that the vigorous, the healthy, and the happy survive and multiply."[2] Perhaps it was a sunny morning when Darwin penned this consolatory passage, and whatever eyebrows it might raise, one can concede that sometimes death at least comes as a surprise when with lightning speed the predator strikes.

RAPTORIAL DOOM

Consider the praying mantis:[3] famous as predators and using a stunningly designed raptorial appendage that can even ensnare a hummingbird.[4] In the insects, this arrangement is strikingly convergent and has evolved independently at least six times.[5] These include the mantispids[6] (who themselves mimic wasps[7]) and the rhachiberothids[8] (although they and the mantispids are neuropterans, the raptorial leg has evidently evolved independently[9]), the empidid flies (dipterans),[10] staphylinid beetles (coleopterans),[11] ranatrid water-stick insects (heteropterans),[12] phymatine bugs (hemipterans),[13] and the extinct titanopterans (orthopteroids).[14] This is a striking roster. The example from the beetles is interesting, not only because of the rapidity of the strike, but contact with the prey is enhanced by the adhesive hairs of the raptorial foreleg, which are equipped with a glandular system very similar to that seen in the adhesive foot pads (p. 82) of the brachycerid flies.[15] The beetle is an ambush predator[16] but can hunt in complete darkness, evidently relying on its antennae as a sensory probe. This is a reminder that sensory modalities extend far beyond the eye, although other raptorial insects tend to be visual. The mantids are highly visual, and their acuity can be confirmed in an interesting manner. Recall the mimicry between the salticid spiders and the ants they stalk (p. 30). Investigations reveal that not only do the mantids avoid the aggressive ants but they also decline to tangle with the fakes.[17]

THE DESCENDING CANINE

Be it the strike of the mantis, the fall of the merlin or the gobble of the piranha, the fact that death arrives in a myriad of ways needs little emphasis. As we have seen, systems as is evidently "a gradation of specializations among ant-eating taxa,"[129] but intriguingly animals such as the anteater and pangolin not only represent extremes of specialization but are strikingly convergent. In particular these animals have a highly protrusible and elongate tongue. As Reiss notes, "The bulk of the tongue of echidnas, pangolins, and anteaters is an unusual muscle known as the sternoglossus . . . also present in some nectarivorous bats,"[130] and like the bats inserted deep in the thorax.[131] Other convergences among myrmecophages entail modification and even loss of teeth,[132] what appears to be an enhanced olfactory capacity but reduced taste buds, sticky saliva, powerful digging arms, and perhaps no more surprisingly the gut microbiome.[133] Opening up a nest with thousands of aggressive occupants is a challenging proposition, although some reports[134] speak of pangolins trapping and crushing the ants beneath their scales and then, as Robert Hatt writes, "To complete the act, the crafty pangolin walks into the water, raises its scales,

and thus permits the numerous dead bodies to float to the surface, where they are picked up and eaten."[135]

These mammals are not the only vertebrates to specialize in ants for food. Among the reptiles there is a specialization in this direction by the horned lizards (p. 84).[136] This myrmecophagous trend is registered in various ways, including reduction in the teeth and weaker jaws.[137] The tongue is also protrusible,[138] and significantly these lizards produce copious quantities of mucus to immobilize a very angry lunch.[139] Although fairly distantly related, the Australian thorny devil lizard is also a mymecophage. Its method of capturing ants differs in various ways, but as Jay Meyers and Anthony Herrel note, these lizards show "several convergent aspects,"[140] including an understandably rapid rate of consumption. Further afield in both phylogeny and time, there is the extraordinary theropod dinosaur *Mononykus* (p. 196), and if its gut contents are ever found, most likely they, too, will be termites.

sive gape connected not only with a flexible pouch but an elongate and toothless mandible.[110] And it transpires that while the volume of water ingested by the whale is vastly greater,[111] the biomechanics of these methods of engulfment both in terms of the pouch and the strength of the mandible are strikingly similar.[112] As Daniel Field and colleagues note, "our results suggest that convergence between rorquals and pelicans is deeper than previously recognized: the selective pressures imposed by similar feeding mechanics have resulted in the convergent evolution of an effective jaw design for resisting bending, and of similarly extensible buccal tissue enabling engulfment of large volumes of water."[113]

While the methods of feeding are not identical, the baleen whales also show some intriguing convergences with the flamingo. To obtain their food, both draw in water and then expel it using a muscular tongue. In the flamingos there is a complex series of horny lamellae on the bill that serve to strain the water[114] in an analogous way to the baleen (which being composed of keratin is akin to hair[115]). Indeed, Storrs Olson and Alan Feduccia[116] remarked that "the general appearance of the head [of the baleen whale] is so like that of flamingos . . . as to be one of the more outstanding examples of convergence in the animal kingdom."[117] They emphasize that the convergence also extends to skeletal features and suggest that the famous bent bill of the flamingo is similar to the baleen whales because in both cases it allows the entire jaw to open with the minimum of movement, and that the peculiar inverted feeding of the bird is a later feature.

These similarities are not precise,[118] but it is worth recalling a group of oceanic seabirds, the broad-billed prions. They are also known as the whale birds, and for good reason. These birds feed voraciously on planktonic crustaceans, including krill.[119] Norbert Klages and John Cooper[120] point out that despite being a thousand times smaller than a fin whale, "Their feeding methods are remarkably similar."[121] Just how similar is evident not only in terms of the lamellae situated along the bill, but also a large muscular tongue that serves to draw in and expel the seawater, as well as a distensible pouch that can hold proportionally larger volumes of water.[122] A further extension of this type of convergence is to another group of flying reptiles, the extinct pterosaurs. This involves the quite extraordinary pterosaur *Pterodaustro*[123] from the Cretaceous of Argentina. This reptile has not only a recurved skull but a striking array of baleenlike extensions (although in this case they are not, of course, hair but actually very elongate teeth[124]). They evidently served as a filter and, it is argued, are convergent on the flamingo.[125]

A MOUTHFUL OF ANTS

If baleen whales and flamingos are adept at scooping up zooplankton, so other animals have turned to hoovering up ants and termites even though rootling through their nests guarantees a warm reception. This capacity, technically known as myrmecophagy, has evolved not only in the pangolins (also referred to as the manids), but independently in a number of other mammals.

These include the spiny anteater (or echidna), marsupials in the form of numbats,[126] primitive mammals belonging to the egg-laying monotremes, and in the Neotropics anteaters like the eponymous *Myrmecophaga* (and along with the sloths belonging to the xenarthrans).[127]

In reviewing the mammalian myrmecophages, Karen Reiss[128] explains how there of prey by snakes.[99] While the process is not identical,[100] the moray eel also shows a racheting mechanism whereby the pharyngeal and oral jaws alternately grip the prey and assist its backward passage. The moray eels show other intriguing parallels to the snakes, such as knotting when engaged in feeding.[101] This highly effective method of feeding most

likely evolved in response to the characteristic dwelling of morays in crevices and other recesses.[102] If their prey can keep clear of rocks and reef, then they should be relatively safe, except of course in the more open water other predators are lurking and this in turn has led to an extraordinary instance of cooperative hunting (p. 289).

FILTERING EVOLUTION

To talk of feeding usually conjures up descending canines and the grinding maw within the mouth, but of course some prey are too small to be chomped, shredded, sliced, or otherwise dismembered. Such typifies the capture of zooplankton. The fish have proved adept here also, not least the electrifying paddlefish (p. 147). Even among the sharks, where we tend to think in vivid terms of slicing teeth, the feeding frenzy, and spreading gore, among this group there are the docile plankton sievers in the form of the basking and whale shark, while the aptly named megamouth shark is also planktivorous. Its mode of feeding, which involves extraordinarily extensible skin around the mouth, seems to have a much closer parallel to some of the mysticetes, especially the great balaenopterans[103] such as the humpback whale.[104]

Nor is it any coincidence that when it comes to processing enormous quantities of seawater for food, there is a striking and recurrent tendency to gigantism. Yet one is entitled to ask: Does this represent much the same adaptive peak arrived at by a similar pathway? Here the extinct pachycormiforms are instructive.[105] Ecologically this group of Mesozoic fish are diverse, with some representatives converging on the tuna (pp. 78–79). Others, however, evolved into oceanic suspension feeders, and crucially an intermediate in the form of *Ohmdenia* shows how these pachycormiforms and "mysticete

whales converge on a common region of morphospace from different ancestral conditions."[106] Nor did they tread this pathway alone, because as Matt Friedman also points out, not only does this example "suggest a tendency for large-bodied marine suspension feeders to emerge from within clades of moderately large- to large-bodied pelagic predators . . . [but] additional evidence for this assertion comes from basking and megamouth sharks . . . [while] putative examples of extinct suspension feeders (e.g. the Devonian 'placoderm' *Titanichthys* . . .) imply a comparable evolutionary trajectory."[107]

The pachycormiforms, however, are particularly important because the fossil record outlines a pathway of not only loss of teeth but a dramatic change in the jaws, especially the development of an elongate mandible. Among the baleen whales, the basis of their dramatic success also revolves around a radical reorganization of the skull, allowing the animal to process vast volumes of seawater. This profound evolutionary transition can be traced in a step-by-step fashion from their toothed ancestors,[108] with an archaic group (known as the aetiocetids) being strikingly transitional in having both teeth *and* the beginnings of baleen.[109] The end result is, of course, one of the glories of the planet, and it might also be thought a unique specialization. Not so. Consider the rorqual whale and brown pelican. Both are capable of an impresregarded as an evolutionary cause celebré on account of their remarkable adaptive radiations, both in the great lakes of Africa and also across the Atlantic in Central America. As ever there is strong overprint of convergence. We have already seen this with respect to the durophagous heroines (p. 47), but many more examples are known, even from a single lake.[78] Striking as these convergences certainly are, they are overshadowed by a series of remarkable parallels between the cichlids of the different African lakes, with perhaps the classic instance between those inhabiting Lake Malawi and Lake Tanganyika.[79] Here *within* each lake the methods of feeding have dramatically diverged, but correspondingly they have converged *between* the lakes.[80] Among the most remarkable are those that revolve around life histories, including the independent evolution[81] of a specific style of mouthbrooding.[82] No wonder the cichlids of the great African lakes show so many convergent forms.[83] When we look farther afield and compare the African cichlids with those of the Neotropics, especially Central America, we tread familiar ground. On this wider compass it is evident that mouthbrooding has evolved more than ten times, while a shift from biparental care of the young to a responsibility of the female alone perhaps as many as thirty times.[84] These and other flexibilities in behavior must also be an important contributor to the remarkable success of the cichlids, with implications for both the evolution of their brains and cognitive capacities. While the research emphasis has very much been on the lake-dwelling cichlids, among the river-dwelling forms there have been largely unremarked evolutionary radiations. Although the faunas on either side of the Atlantic are by no means identical in terms of morphological diversity or range of diets, nevertheless each divergence has led to widespread ecomorphological[85] convergences.[86] Nor do the convergences end here, because they extend to the centrarchid fish.[87]

Perhaps the strangest of the cichlid convergences involves the feigning of death. In these cichlids the curious should be very careful because independently in Lake Malawi[88] and Mexico[89] the fish lies motionless on the bottom, sometimes having very slowly sank, and awaits a scavenger whose hunger exceeds caution. The Lake Malawi example is even more remarkable because to make it even more appealing the cichlid appears to have begun to decay. A similar behavior, although more akin to appearing to be on the point of death or just looking very ill, has also been identified in a marine setting, specifically in a grouper (comb grouper).[90] Such thanatosis, "playing possum," was noticed of course by Darwin,[91] and finds its archetypal expression in the marsupial opossums.[92] As Don Hunsaker and Donald Shupe[93] remark in this regard, "It appears that the animal has definite opinions as to what a dead opossum is supposed to look like, and will assume these positions if changed by an investigator. If the opossum's eyes are closed, he opens them."[94]

Cichlids present as powerful an example of evolution as one could ask for. Among these fish this potential seems to have been largely realized by the versatility of their pharyngeal jaws.[95] Cichlids by no means exhaust the evolutionary potential of the pharyngeal jaws. Consider the moray eels.[96] Their mouth is quite small and has poor powers of suction, but these eels are notorious for their aggressive behavior and capacity to grapple with sizeable prey. Morays are most often thought of as broadly piscivorous (and can hunt cephalopods), but some are durophagous, and this habit may be convergent.[97] In either case, how is this achieved? The solution involves the pharyngeal jaws. Instead of remaining well back to process the incoming food, they show a remarkable series of protrusive movements that engulf the prey and then drag it back to the gullet.[98] This is remarkable enough, but it is strikingly convergent in terms of function and behavior to the capture material that provides the underpinning for our noses. In the words of Henry Gee, to think of rays "chewing clams with cartilaginous jaws must be like felling a tree with a custard-filled sock."[64] In the cownose ray, however, the crushing teeth are reinforced with both layers of calcified cartilage (as are horn sharks)[65] and also a series of fine internal struts, the trabeculae, that are strikingly convergent with trabecular bone.[66] In this biomechanical arrangement, strength is combined with economy of material.[67] Not only have the cownose rays reinvented bone, but the tendons that are associated with the powerful muscles that bring the crushing plates together have striking convergences in terms of both morphology and ultrastructure to mammalian tendon.[68]

Like other sharks the cownose ray employs suction to loosen the sand around its prey, while extensions of its pectoral fins that form the prominent cephalic lobes adjacent to the mouth are not only highly sensory but serve to manipulate the food in a way that recalls the activity of the mammalian dugongs.[69] Despite a relatively uniform morphology the rays extend styles of feeding far beyond durophagy. The magnificent manta rays, for example, are closely related to the cownoses but are plankton sievers. Yet their jaws are also reinforced with trabeculae, which might reflect the enormous size of these rays and the need to prevent the jaws from buckling.[70] So too the lesser electric ray has independently evolved jaws reinforced with mineralized trabeculae, but in this case the animal is not durophagous. Rather it hunts its prey by a ballistic protrusion of the jaws, and evidently the trabeculae reinforce the areas of the jaw under the greatest loadings.[71]

In terms of biomechanics, durophagy is highly sophisticated, but it still looks to brute force, such as the crushing teeth that often form a pavement to immensely powerful jaws and associated muscles. This arrangement can be achieved in a number of ways, and

in the teleosts involves converting the pharyngeal jaws into powerful mills. Clearly a mouth with a double set of jaws is an interesting proposition, but other groups of fish have gone even further by interpolating a third set between the pharyngeal and oral set. Known as the tongue-bite apparatus, it comprises a complex array of toothplates and associated ligaments that allow the fish to engage in highly effective raking movements.[72] All very impressive and intricate, but it is convergent having evolved independently in the salmon group and also an assemblage known as the osteoglossomorphs.[73]

The pharyngeal jaws of bony fish have attracted the most attention, and most likely their sheer versatility underpins much of teleost success. This arguably reaches its acme in the freshwater cichlids and marine labrids. A combination of pharyngeal tooth form, fusion of the lower jaw bones, a specific articulation of the upper jaw against the skull, and complex musculature has endowed these teleosts with a striking array of possibilities for feeding, ecological diversification, and speciation.[74] The similarities in the pharyngeal jaws of the cichlids and labrids has long been taken to indicate a close phylogenetic relationship,[75] but one could almost guess: not only is this arrangement convergent in the two groups of fish, but the evolutionary novelty is relative because the assembly of this complex feeding structure draws on an existing anatomical repertoire.[76] In the labrids the function of the jaws and the rest of the skull shows a pattern that is central to the argument of this book: While at the lower levels of labrid taxonomy we see a recurrent divergence of form, at the higher taxonomic levels convergence emerges as the global phenomenon.[77] In other words, locally there will be a plethora of evolutionary experiments, but on the broader scale they are encompassed by broader constraints that confer an inevitability on the evolutionary process.

What of the cichlid fish? Rightly they are

and physiology of long-extinct animals. Irrespective as to whether this protorosaur was a hoover or a grabber, it does have prominent teeth. But this does not rule out suction feeding because many such feeders possess teeth, and in the nurse shark[45] they can be employed

to hold their prey. In the sharks, predatory feeding is almost certainly the most primitive arrangement, whereas in the bony fish, suction feeding may take evolutionary precedence over the use of teeth.

Crunching Evolution

Among those fish that employ teeth, the barracuda and piranha being two fearsome reminders, we gain fresh insights into evolution. Even within quite small groups of bony fish, such as the porgies (the sparids), among approximately thirty genera there is not only a notable variety of feeding methods but also convergences in the evolution of the various specializations.[46] And in underpinning this versatility of feeding types and a corresponding evolutionary success we find in a number of groups of the bony fish a rather remarkable arrangement whereby in addition to the standard oral jaws there is a complementary pharyngeal set located at the back of the oral cavity. We might expect, therefore, that given the general correlation between tooth shape and the corresponding diet, this principle could be readily extended to the bony fish.

Things, however, are not quite so simple,[47] but some general rules do emerge. Consider the ability to crush hard-shelled prey (such as mollusks). This is the process known as durophagy.[48] Evolving into the equivalent of a gravel crusher is a tall order but the rewards are considerable,[49] and in evolving multiple times durophagy provides compelling insights into convergence. While the fish provide the lion's share of examples, among other aquatic vertebrates the mosasaurs (in the form of the Cretaceous globidentids), some Triassic ichthyosaurs,[50] and placodonts[51] possess obvious crushing dentitions. In the case of the mosasaurs, James Martin

confirms it, noting how "The rib-cage area [is] filled with densely packed bivalve fragments, principally large, flat inoceramids that had a large visceral content."[52] Just as we otherwise associate mosasaurs and ichthyosaurs with rows of sharp, gleaming teeth, so also in extinct turtle-crunching crocodiles[53] and some living examples (including the Chinese alligator),[54] molariform dentition and a concomitant crushing ability have evolved a number of times. Terrestrial durophagy is also found. Various squamates tackle hard-bodied prey (p. 64), with the crushing dentition of the teiid lizard *Dracaena*[55] echoing the common theme, although interestingly its prey is primarily aquatic snails.[56] But in terms of molluskivory the real masters of this convergence are the fish. Among the heroine cichlids of Central America alone it has evolved six times,[57] but this convergence is also found among larger groupings of fish.[58]

Beyond the teleosts we again find examples of durophagy, especially among the sharks and their relatives, the rays (collectively the chondrichthyeans). This capacity has evolved multiple times, including the horn shark[59] (which also employs suction[60]), a hammerhead shark known as the bonnet head,[61] the ratfish (or chimaeras),[62] and several times among the rays.[63] The solutions employed are certainly diverse, but of these examples perhaps the most interesting are the cownose rays. As with all living chondrichthyeans the skeleton is cartilaginous, the same of the projectile tongue and its co-option of the hyobranchial would have been a later development. Yet because the larval forms are aquatic[28] suction feeders, the transition to land as an adult ought to represent a major evolutionary challenge, not least because of a complete shift in mode of feeding. Should we appeal to some sort of "hopeful monster"? Not at all. The transition from water to land involves a largely seamless transformation and for the most part little change in the kinematic architecture.[29] The potential importance of this for envisaging what appear to be profound transitions scarcely needs emphasis, not least because it also revolves around a central evolutionary question—that is, the relative ease of terrestrialization. Yet given the process is one of evolution, it cannot be flawless. Thus, at the time of larval metamorphosis, the highly adapted system of suction feeding has to be closed down[30] and so no feeding occurs.[31]

When we look beyond the amphibians to groups that were once fully terrestrial but have now returned to the ocean, we see that suction feeding has also been reinvented. In the extinct ichthyosaurs, although the majority were highly effective predators, one Triassic form was not only enormous, but effectively edentulous and may have been a suction feeder[32] relying on a massive expansion of the gular region.[33] At a more parochial level, when other groups of reptiles return to the water, suction feeding again reemerges. In the chelonian turtle *Ocepechelon*, from the upper Cretaceous of Morocco, the elongate snout points to suction feeding as well as convergences with the pipe fish and beaked whales.[34] Among the living turtles, the Australian snake-necked pleurodiran *Chelodina* has also taken this route, and with interesting consequences. Not only is the feeding apparatus necessarily transformed from operating in a terrestrial context, but more significantly, as Johan Van Damme and Peter Aerts[35] remark,

despite "largely different structural solutions, optimal feeding conditions as deduced for suction in feeding fishes are also employed by *Chelodina*."[36] Another pleurodiran, the South American fringed turtle (or matamata), also shows some striking adaptations for suction feeding. These include a streamlined skull (that helps to eliminate a bow-wave), large hyoid, and an enormously distensible esophagus.[37] Patrick Lemell and colleagues note how this adaptive complex allows one to see this "specialized suction feeder . . . as one endpoint in the feeding evolution of aquatic reptiles."[38] Once again we seem to see that the evolutionary solution is probably about as good as it possibly can be, and that the striking parallels in the kinematics of feeding reflect the immutable realities of hydrodynamics. This explains also the convergences we see in the suction feeding of such aquatic mammals as the whales[39] and the bearded seal.[40] In the latter case, precise control of the shape of the mouth using the rostral lips is directly analogous to the employment of comparable labial structures in the suction-feeding amphibians and fish. The mechanism of rapidly expanding the oral cavity depends of course on a mammalian anatomy (in this case the hyolingual complex), but the net result is very much the same.[41]

Hydrodynamics is evidently of key importance in the evolution of suction feeding and in one sense makes its convergence less remarkable (and conversely more predictable). Suction feeding has likely also evolved in groups that have vanished from the face of the earth. Such may be the case in an aquatic reptile belonging to the protorosaurs. These animals have remarkably long necks,[42] and while they are most often depicted as darting after fish, Triassic representatives from China may have employed suction feeding.[43] The ensuing debate[44] that arose from this suggestion certainly revealed the difficulties in inferring both the functional morphology anatomical modification.[11] If, however, it was found as a fossil, no doubt the interpretative diorama would have a lurid display of its terrorizing the ocean.

Unsurprisingly, suctorial feeding is only effective in aquatic animals. The air is not nearly viscous enough, and perhaps the only terrestrial exception is the scolecophidian snakes (p. 65).[12] Also in a parallel way, despite the desperate attempt to conjure up Fortean bladders, suspension feeding on land is similarly a biological impossibility.[13] As with many other aspects of the move from the oceans to dry land—the process of terrestrialization—when it comes to suction feeding the group that epitomizes the straddling of this major transition, the eponymous amphibians, can provide some useful insights. Their aquatic manifestation is probably most familiar in the form of the tadpoles (p. 25), which typically graze algal films and the like. This in itself is a complex procedure, and such grazing has evolved multiple times, notably in aquatic insect larva,[14] mollusks,[15] and even catfish.[16] Algal films offer various advantages as a food source, but require specialized equipment not only to remove from the substrate but to collect and often squeeze out excess water. Nature has responded by evolving five distinct types of tool, often brushlike or gouging, but these have resulted in a series of spectacular convergences.[17] While the great majority of tadpoles also employ a rasping action, despite its tiny size *Hymenochirus* has not only become a predator but adopted a mode of suctorial feeding that finds no counterpart in any other frog but is strikingly similar to the arrangement in the teleost fish.[18]

In the related salamanders the larval forms also employ suctorial feeding, and this has also been arrived at independently. Among the salamanders, the most important group is the plethodontids.[19] As adults they are terrestrial, and their striking success is in large part due to an extraordinary projectile tongue. This can function in several different ways and show some striking convergences,[20] not least in the ballistic arrangement that has evolved independently three times in the plethodontids and is also similar to the famous tongue of the chameleons.[21] The operation of this tongue, however, comes with an apparent penalty; a central component known as the hyobranchial, which is involved in the very rapid protraction in the salamanders, is otherwise employed for the ventilation of the lung. But it is evidently a worthwhile trade-off, because although the plethodontids are lungless, their ecological diversity is notable, including remarkable burrowing forms.[22]

Within the amphibians, lunglessness is not restricted to these plethodontids. It has evolved independently in the frogs[23] and twice among the Neotropical caecilians, one being a very small terrestrial taxon (*Caecilata*)[24] and the other an aquatic caecilian referred to as *Atretochoana*.[25] In their description of the latter animal, which is a member of the typhlonectids, Mark Wilkinson and Ronald Nussbaum[26] described its many exceptional features both as a caecilian and tetrapod. They remarked how this animal "demonstrates the unpredictability of evolution."[27] Yet as they and others note, amphibians might have a propensity to lunglessness because of their capacity for cutaneous respiration. Nor is it any accident that this lungless caecilian and the frog, from Brazil and Borneo, respectively, inhabit fast-flowing and cold streams where the amount of dissolved oxygen is elevated. In addition, the potential buoyancy function of the lung is removed. In passing, so to say, in both these discoveries the scientists note that not only are these amphibians exceedingly rare, but continuing environmental destruction of their habitats does not bode well.

The origins of the lunglessness in the plethodontids are widely believed to have a similar explanation, and if so the evolution

CHAPTER 5

SWALLOWING CONVERGENCE

IN DUE COURSE we will see that while the world swarms with venomous creatures, and let us not forget embittered literary reviewers, for most people the point of reference remains the snakes. If it is not the fangs dripping venom, then it is the gaping jaws, sudden strike, and then serpentine engulfment ("Anybody seen the gerbil?") that engender an atavistic shudder. This style of feeding is, of course, highly coordinated and a specialized process, but it is also one that finds a series of remarkable parallels in the moray eels. But to explain the nature of this striking similarity (pp. 49–50) we need first to remind ourselves how, in general, fish feed. In doing so inevitably we discover many other convergences.

THE SUCK OF EVOLUTION

Among the bony fish—that is, the teleosts—the prevalent mode of feeding is by suction.[1] Typically this means shaping the mouth (and often accompanied by its protrusion)[2] and very rapidly expanding the oral cavity: as the water rushes in, so the prey disappears into the maw. Of these perhaps the most remarkable are the antennariid anglerfish.[3] These live as cryptic ambush predators and possess a remarkable array of lures; in one species it is almost identical to a small fish.[4] Antennariids can expand the size of their oral cavity by up to twelve times, engulfing the victim in four milliseconds.[5] Suctorial feeding has evolved independently many times, even in the sharks. These include the horn shark, and more particularly in a group known as the orectolobiformids, where in the bamboo and nurse sharks suctorial feeding is the rule.[6] When compared with the teleosts the arrangement of the jaw and surrounding anatomy shows important differences, yet as Philip Motta and colleagues[7] note with respect to their study of the nurse shark, it "comes close to functionally replicating the feeding mechanism of an inertial suction feeding bony fish . . . and it does it with a similar suite of modifications to the prey capture apparatus."[8] On occasion the prey is repeatedly blown out and sucked in again, evidently either to reorientate the item or make it smaller. The investigators also noted that not only was the suction force very powerful, but speculated that the loud popping noises they heard in particularly vigorous suction might represent the process of cavitation[9] (pp. 143–44). Photographs of the nurse shark with its tubelike mouth[10] are striking, but the relatively small cat shark provides a cautionary tale, especially for paleontologists. The group to which it belongs, the important scyliorhinids, are relatively primitive, and most are active predators. The cat shark, however, evidently largely relies on suction feeding but shows little in the way of curious but important by-product: the elasticity of the gluten is an essential component of dough and the kneading before it is baked to provide the food of the gods—bread.

To conclude, convergence opens new ways of looking at evolution, fresh possibilities that may yet transcend the oxymoronic triumphal aridity of the ultra-Darwinists who continue to rejoice in both the complete meaningless of the universe while taking a special, if not gleeful, pleasure in perversely magnifying our utter cosmic insignificance. An emphasis on convergence as the key to a deeper structure in biology is certainly not meant to imply that *everything* is convergent, although as we have seen and will see, apparently disparate organisms can still show surprising similarities. To identify examples of biological uniqueness, therefore, by no means undermines the general thesis. Thus, not only might convergence delineate the map of life, but by implication also provide a predictive framework. Can you swallow that?

the departing pollen is more than half that of the speed of sound.[172] Moreover, Philip Taylor and colleagues stress that not only does the discharge of this pollen exceed "all previously recorded values in biology for time motion, initial velocity and mass specific power output," but continue by noting that this system "approached the theoretical physical limits for hydraulic movements in plants, and exceeded the performance of all other plants and fungi within the category dominated by 'explosive fracture.'"[173] This category not only is clearly demarcated from generally slower movements that depend on fluid transport across very small distances, but in the numerous cases of explosive release (and similar so-called snap buckling) that are found in plants and fungi, almost invariably the mechanism depends on an abrupt shift in turgor pressure. In defining these fields, Jan Skotheim and L. (Maha) Mahadevan[174] rightly refer to "design principles,"[175] and it is striking how much of the potential space is occupied.

One can hardly leave the field of explosive release without a nod to the elastic systems found in animals,[176] not least that of the remarkable mantis shrimps. While these animals are adept at delivering truly stunning blows (pp. 143–44), their terrestrial counterparts in the form of the insects employ elasticity in various contexts, not least in jumping. Consider the aptly named flea beetles (alticinids). They employ an ingenious spring located in an enlarged segment (metafemora) of their hind legs. This robust structure[177] has a hooklike form, and among the coleopterans this arrangement has evolved independently a number of times.[178] Its effectiveness in the flea beetles is self-evident, but even within this group not only has it evolved on four occasions,[179] but De-yan Ge and colleagues note how the group shows "an inherent propensity to acquire this specialized jumping apparatus."[180] Equally crucially they refute the assumption that such "structural and func-

tional complexity . . . [can be] considered as evidence against their convergent evolutionary history."[181] And what enables this impressive leaping capacity, which in one species provides an acceleration of more than 2,500 meters per second and a force equivalent to 270 g?[182] Given its location, not surprisingly the spring is composed of chitin and evidently its fibrous arrangement in conjunction with a protein confers the necessary elasticity.[183]

And when it comes to elastic proteins, the prizes must go to the arthropods.[184] It is the appropriately named resilin that is the key not only to the success of flight in such insects as the locust and dragonfly,[185] but also in explaining how the flea[186] catapults itself away from the poised fingertips of doom.[187] But both the fleas and flea beetles, as well as such insects as the leafhoppers[188] (not to mention jumping maggots[189]), are in turn completely outclassed by the froghoppers. By using a novel catapult they can accelerate at up to 4,000 meters per second^{-2}, and lift the equivalent of more than 400 times their body weight.[190]

Elastic proteins have, therefore, repeatedly evolved, and in serving diverse functions show a wide range of structural properties, although overall abductin, elastin, and resilin are remarkably similar. Typically an elastic protein will have domains that confer elasticity interspersed with other domains that serve to reinforce the protein by cross-links and prevent it behaving like jelly. The key point, however, is that the characteristic amino acid sequence of each elastic protein has virtually no similarity to any other.[191] As ever with convergent evolution it is not how you got there, but how it works that matters. Perhaps the strangest example of an elastic protein is that found in wheat, specifically the gluten of the seed.[192] The elasticity certainly has nothing to do with bouncing around, and it is suggested it is effectively a by-product of the way the proteins need to be stored in the endoderm of the seed. There is, however, another very

THE INEVITABILITY OF FORM   41

sheds surface energy, which is exchanged for kinetic energy and momentum. The energetic efficiency of this process is remarkable,[154] and as Xavier Noblin and colleagues demonstrate in their detailed analysis of the overwhelming importance of surface tension, the entire process "reveals an exquisite fine-tuning of the different stages."[155] Intriguingly they also note that although this ballistic mechanism may be unique to the fungi, it does have "a surprising similarity with the mechanics of jumping in animals,"[156] inasmuch as both revolve around a redistribution of mass (I crouch), a sudden release (I spring), and a push against a firm substrate (with luck I lift off). Of course, in other respects the ballistic world of fungal spores remains very different because of the enormous differences in scaling and the negligible role of gravity. Thus, launch time is less than 10 microseconds and the acceleration probably exceeds 12,000 g. One may wonder how the spore survives such a colossal acceleration, but one needs to recall that the individual spores are tiny, in the order of picograms (10^{-12} g)[157] and only travel a short distance before gravity takes over and they are entrained in any passing wind.

Fungal ballistics are by no means restricted to the ballistospores. In what Terence Ingold[158] refers to as "a clear record for fungal gunnery,"[159] the striking case of a gastromycete known as *Sphaerobolus* is addressed. Here by virtue of a complex anatomy and very rapid changes in the osmotic pressure, the cup that houses a large mass of spores suddenly everts and can hurl it more than 5 meters. The fungal cannons of a completely different group of fungi, known as the ascomycetes, and probably most familiar as the yeasts, also employ turgor pressure for explosive dispatch. These fungi take their name from the flasklike structures (the asci) that contain the sexual spores. As Frances Trail aptly notes, "Ascospore discharge is no small feat of evolution,"[160] involving as it does precise control of

turgor that enables the tip of the ascus to blast open. No small feat it may be, but in a subsequent investigation of blastoff by fungi (both ascomycetes and zygomycetes) growing in the romantic medium of animal dung, not only were some of the accelerations of spores extraordinary (in one ascomycete species, equivalent to 180,000 g), but as the investigators Levi Yafetto and colleagues[161] noted, while these are "remarkable feats of natural engineering," in reality the "explosive mechanisms of spore discharge do not require any extraordinary mechanisms of osmolyte accumulation, nor the elaboration of any specialized cell wall structures to maintain this pressure before discharge."[162] As ever, timing is all, to ensure near instant rupture and launch.[163] The change in turgor pressure seems to be mediated by an influx of ions (H^+, Cl^-), and again we encounter incredible accelerations, in the order of 30,000 g.[164] Unsurprisingly the shape of the spores is typically consistent with the minimization of drag, with many of the species studied having a shape that in terms of drag was no more than 1 percent from the optimal shape.[165]

Hurling reproductive bodies at vast speeds obviously has its advantages, and some interesting parallels can be found in the catapulting of seeds[166] and more particularly pollen.[167] In the latter case the mechanisms revolve around changes in turgor pressure and the sudden release of elastic energy stored in the stamens, specifically the slender filaments that support the pollen-filled anthers. In a *Cornus* (bunchberry dogwood), the action has been compared to the medieval siege engine known as the trebuchet, and here too we see not only massive accelerations (equivalent to 2,400 g)[168] but a capacity to launch the pollen[169] that Dwight Whitaker and colleagues[170] interpret as an "optimality of launching speed."[171] Much the same mechanism has evolved independently in the white mulberry, where the initial velocity of biological froth also shows a series of remarkable physical properties. Not only are they very stable, but even though they look similar to the white of a beaten egg the concentration of proteins is about one hundred times less.[140] They also possess striking surfactant properties. The ranaspumin proteins are evidently novel,[141] but again to quote Fleming and colleagues the frog foam "is not only a remarkable biomaterial . . . but also . . . shows aspects of convergent evolution for the defense of reproductive and dispersal stages of a wide range of organisms"; they continue how these are "of intrinsic interest, not least for the general principles [such foams] may illuminate in terms of the evolution and adaptability of these unusual biological materials."[142]

It is always helpful to think in terms of general principles, and in the context of foaming perhaps the most familiar revolve around the insects. Probably most familiar are the highly destructive spittlebugs (hemipterans), whose nymphs typically encase themselves in a foam that is evidently a product of the malpighian tubules mixed with air.[143] This is presumably a protective measure, and although the fibers that serve to stabilize the bubbles have been compared to silks (pp. 89–90), this seems a rather distant comparison; much remains to be understood as to the nature of very complex secretion.[144] Other insects, however, produce a defensive foam that either entangles or repels potential attackers, typically employing a chemical secretion that is aerated by air being blown through the respiratory spiracles. Most striking, perhaps, are some Malaysian ants that spray long ropes of foam, evidently employing their venom glands.[145] So, too, moths and grasshoppers have independently learned to produce foul-smelling froth,[146] and in one instance the defensive chemical[147] may have been sequestered from a herbicide![148]

BALLISTIC CONVERGENCE

If such animals as the Bombardier beetle employ a chemistry familiar to rocket science (as indeed do some bacteria), so elsewhere in evolution do we encounter other extraordinary methods of explosive delivery. Some of these involve quite astonishing accelerations. In the cnidarians a combination of a molecular spring and very high pressure allow the nematocyst to discharge in about 700 nanoseconds and accelerate to an equivalent of more than 5 kg, sufficient to puncture the cuticle of its arthropod prey.[149] Other extraordinary examples are to be found among the fungi. With articles titled "More g's than the Space Shuttle"[150] or "Fungal cannons,"[151] one gets the sense that the polymathic biologist will expect his colleagues to be conversant not only with mushrooms but also the elements of ballistics. And the ballistospores found in the basidiomycetes (these are the most familiar group of fungi and include the mushrooms and toadstools) are aptly named. So rapid is the discharge of their spores that only recently has ultrafast photography managed to capture the process and, in doing so, triumphantly verify earlier observations.[152] Remarkably unlike natural springs or turgor pressure, both of which can also produce some impressive ballistic performances, the ballistic departure of these spores depends on a catapult that employs surface tension.[153] Just adjacent to the spore, a minute sugary drop is secreted, which being hygroscopic allows water to condense. At this point the droplet spreads extremely rapidly around the spore and in doing so shifts its center of mass. Newtonian mechanics kicks in, because as the droplet spreads around the spore it insects such as beetles and termites but also millipedes—have independently evolved benzoquinones for defensive purposes.[119] In the case of the millipedes their role in defense is taken one stage further because capuchin monkeys have learned to rub them on their bodies, evidently to confer protection against mosquitoes.[120] Somewhat oddly, given the intensely irritating nature of the chemical's release, the capuchins have also been observed to place the millipede in the mouth, whereupon, as Ximena Valderrama and colleagues remark, "they drool copiously and their eyes appear to glaze over."[121] Nor are capuchins the only animals to self-anoint themselves with millipedes to deter insects.[122] And deterrence can take other remarkable routes. For many small mammals, snakes are a perennial hazard. The risk is compounded given infrared detection for warm little bodies has evolved within the snakes two times (p. 116), although as we will see, ground squirrels employ an ingenious thermal beacon in response (p. 117). However, squirrels have other lines of defense, which include chewing pieces of discarded rattlesnake skin and then anointing themselves.[123] This olfactory camouflage is evidently highly effective and has evolved independently in other rodents.[124] These include the Siberian chipmunk, which applies chunks of snake flesh,[125] and the Rice-field rat, whose preference is for the secretions from the anal gland of weasels.[126] Camouflage is one route, deterrence another. Consider the African crested rat, which as Jonathan Kingdon and colleagues[127] note, delivers "A poisonous surprise."[128] This is achieved by chewing the bark of the poisonous *Acokanthera* and then applying the glycoside (ouabain) to the hairs. As they note, this exemplifies a "unique, exceptional optimization and economy of structure for sequestration and delivery of toxins."[129] Given that hunters on the east coast of Africa apply this poison to their arrows,[130] it is hardly sur-

prising that when dogs tangle with this rat, they come off second best.

Nor are these rats quite alone. The hedgehog provides another benchmark, given, as Robert Brockie[131] remarks, its "curious habit of plastering itself with its own saliva."[132] These animals will chew almost anything and literally foaming at the mouth duly apply the frequently pungent solution. The reek of hedgehog may be important in the breeding season but is so ubiquitous it could on occasion be no more than a displacement activity. But in certain instances it is evidently defensive because the hedgehog chews toad skin (the rest of the animal is eaten) and then anoints its prickles with a toxic application,[133] with an effect not so different from other poisonous spines. What is particularly impressive is that among the various Madagascan tenrecs, some are not only morphologically convergent with the hedgehogs (not least with their spines), remarkably this convergence extends to their behavior. Thus tenrecs like *Echinops* also show massive salivation followed by anointing, albeit with their forepaws.[134] Not only that, but their mode of attack is also convergent.

These are not the only animals to engage in foaming (and here I exclude potential reviewers), and a number of interesting examples can be found elsewhere. Various groups of animals, for example, have independently evolved nests of foam. These include freshwater fish,[135] while in the marine environment a group known as the tunicates (or sea-squirts) utilize masses of algal foam to facilitate the fertilization and retention of the larvae.[136] Better known are the foam nests constructed by frogs, employing the appropriately named ranaspumins.[137] As Rachel Fleming and colleagues note, "This style of nesting appears to have evolved independently in several lineages,"[138] and they list the Neotropical leptodactylids, Australasian myobatrachids, and Afro-Asian rhacophorids.[139] But this compromise the action of the sodium channel. Very impressive, and one might reasonably point out that these toxins are, to put it mildly, powerful agents for selection. Dealing with tetrodotoxin simply leaves life with very few choices, yet this example surely points to deeper principles. Once again we see that convergence is not simply a question of navigation along well-traveled roads but at least in this case more like an elephant walking confidently along a gossamer-thin tightrope.

Some sense of these sensitivities also comes from the weakly electric fish, known as the gymnotiforms and mormyriforms (pp. 145-147). When it comes to electrogeneration and electroreception they themselves provide a striking example of convergent evolution (p. 146), but so far as their sodium channels are concerned they provide another interesting test case. Typically the electric organs are derived from muscle tissue, and perhaps unsurprisingly it transpires that one of the duplicated genes (specifically $Na^v1.4a$) is not only lost from the muscle of the fish but is now expressed in the electric organ of both groups, perhaps because of a shift from contractability to the generation of an electric current. More importantly, however, an associated series of amino acid replacements at key sites within the sodium channel of both mormyriforms and gymnotiforms[110] occurs to ensure that this channel is silenced in the muscle, whereas in its new role in the electric organ the rate of its evolution has been accelerated as it adapts to its new function.[111] Given also that in a number of cases the amino acid substitutions are at sites adjacent to ones that in human mutations provoke disease, again there is little doubt that these changes are very far from random.[112]

Exploding Evolution

If tetrodotoxin and saxitoxin demonstrate nature at its most lethal, there are of course many other avenues of chemical defense.[113] Consider the Bombardier beetle.[114] To produce a scalding (100°C) and toxic spray—not to mention aiming with remarkable accuracy and employing an antidrip mechanism (Coanda effect)—by the combination of benzoquinones and hydrogen peroxide (and catalysis by oxidant enzymes) is wholly remarkable. After all, rockets and explosives depend on a similar chemistry. Or is it so remarkable? Recall first the convergent ink of sea hares, a product that proves highly effective at deterring predators (pp. 15–16). An important part of the sea hare chemistry involves the combination of an enzyme (escapin, a variety of oxidase) with a particular substrate (in this case L-lysin). And as in the Bombardier beetle in the sea hare each ingredient is stored separately (respectively, in the ink and opaline glands) and can only mix and react when both glands discharge—a sort of Bombardier beetle in slow motion.[115]

What about the hydrogen peroxide component in the Bombardier beetle? So far as defense in the arthropods is concerned, employment of this highly reactive molecule appears to be unique to these beetles, but this compound is also generated by some tarantula spiders (where interestingly it has an antimicrobial function).[116] While the Bombardier beetles justifiably attract considerable attention, when it comes to blasting defense they are by no means alone. For example, a viviparous cockroach (p. 175) has glands that secrete benzoquinones, and these are expelled by bursts of air from its breathing tubes (the trachea).[117] In this case, however, the expulsion is more as a fine mist and is not specifically aimed.[118]

Numerous other arthropods—not only

origin may have been bacterial, followed by lateral gene transfer, it is perhaps more likely that the biosynthetic pathways for saxitoxins in cyanobacteria and dinoflagellates arose independently.[96] Saxitoxin is like tetrodotoxin in being potentially lethal, and although their respective molecular structures are not that similar, in both cases their effectiveness depends on a derivative of guanine.[97] Their grim effectiveness lies in their extraordinarily precise ability to block the voltage-gated sodium channels (pp. 37–38) essential for nervous and muscular activity. No wonder paralysis is the result.

The gruesome nature of these toxins tends to overshadow remarkable insights into some evolutionary principles, central to both the sensitivity of biological systems and convergence. Even for the toxic molecule to locate a sodium channel in the first place is remarkable. As Lyndon Llewellyn graphically writes, the saxitoxin "must force its way through water which is energetically demanding, find the Na^+ channel among a forest of other extracellular receptors, orient itself and deform its structure, if necessary, to access the binding site . . . [and] do all this while being enveloped in an enormous torrent of Na^+ ions."[98] Yet this binding is extraordinarily effective and medically almost impossible to treat. And, of course, it begs the question of how those animals like the fugu fish do not end up dead themselves; indeed in the puffer fish, tetrodotoxin can serve as a male-attracting pheromone.[99] One clue comes from the newt *Taricha*. While for nearly all purposes it is invulnerable, the newt can readily fall prey to some snakes, notably a garter snake known as *Thamnophis*. Not only has this capacity evolved several times,[100] but in this snake substitutions by amino acids at critical points in the protein that constructs the sodium channel render the tetrodotoxin harmless.[101] Here, too, the critical mutations necessary to confer defense have not only evolved independently

in different species of this snake[102] but, as Chris Feldman and colleagues[103] go on to note, when one looks further afield among three lineages of snake, there is "remarkably constrained convergence during the evolution of resistance to . . . tetrodotoxin," which reflects the fact that "only a small fraction of the experimentally validated mutations"[104] actually serve to confer resistance. So, independently, some clams, relying on a single amino acid substitution, have managed to enormously increase their resistance to saxitoxin.[105]

And this explains how in the fugu fish and most likely the other animals with a similar cargo of tetrodotoxin the substitutions of amino acids at key sites in the sodium channel protein prevent the toxin from binding and thus dramatically increase the resistance, often by many thousandfold.[106] Not only is this a striking case of molecular adaptation, but when a series of puffer-fish species are compared, we see strikingly similar substitutions of amino acids at key points in each of the four (I to IV) domains that constitute the sodium channel.[107] These are not identical, but Manda Jost and colleagues note that while these fish have a variety of types of sodium channel, they "all confer varying degrees of toxin resistance, yet show remarkable convergence among genes and phylogenetically diverse species."[108]

What is particularly striking is that in one domain (IV), not only does the same type of replacement occur independently, but in usual circumstances this region of the protein is crucially important in maintaining the selectivity for sodium ions. The importance of this is difficult to exaggerate. The sodium channel is extraordinarily precisely engineered; by all means tamper with key sites, but it is at your peril. Yet not only can tetrodotoxin and saxitoxin bind with great efficiency, but the countermeasures at the very same sites that are employed by the fugu fish[109] are not only astonishingly effective but do not habit (pp. 47–48).[78] Along with the capture and manipulation of the prey by the oral teeth, frequent use is made of blowing water out the mouth (and corresponding suction). This is a particular characteristic of the tetraodontiforms, and may help to explain the origins of one of the most bizarre behaviors on the planet, the transformation of a fish into a globular ball. This requires a remarkable anatomy, including a stomach that can distend with seawater, the absence of various bones (including ribs), and an extraordinarily extensible skin.[79] Thus it is suggested that the general capacity to first suck and then blow out jets of water—for example, to wash sediment away from concealed prey—was the key requirement[80] to allow the ultimate evolution of the puffer fish.[81]

A fully inflated spiny puffer fish is an extraordinary sight and is nearly impregnable. Characteristically they caught the attention of Charles Darwin, and in his account of his voyage on the *Beagle* he not only makes his own acute observations but relates accounts of puffer fish being found alive and well in the stomachs of sharks, and if finding their confinement somewhat irksome calmly chewing their way through the body wall to freedom.[82] Indeed, in the tetraodontiforms there is a remarkable emphasis on the defensive. For example, constructing a carapace by suturing large scales (reminiscent of a tortoise) has evolved independently at least four times,[83] and the generally boxlike shape with propulsion depending on the beating of the fins is in marked contrast to the sinuosity of most fish, culminating in the eels (pp. 49–50).[84]

Inflated and bristling with spines, or enclosed in a tanklike carapace, the tetraodontiforms might be thought to be pretty secure, but these fish have a further trick nestling in their livers and ovaries. Why is the diner lying inert on the floor; has the fugu fish not been properly prepared? The culprit is a fearsome molecule, tetrodotoxin, that by binding to the sodium channels (pp. 37–38) of the muscles and nerves induces paralysis and often death. A remarkable series of evolutionary stories unfold. Although first identified in the tetraodontid puffer fish, hence tetrodotoxin, the molecule is actually synthesized by a wide variety of bacteria.[85] Either by symbiosis in the tissue or sequestration through the diet tetrodoxin is employed by an extraordinary range of animals.[86] These include other fish (such as a goby),[87] and among the amphibians a dendrobatid poison frog[88] (that oddly does not appear to employ the customary alkaloids) and a newt known as *Taricha*.[89] In the newt its discovery was curiously serendipitous because in an experiment where the eye vesicles of *Taricha* were transplanted into the embryo of a tiger salamander, to the investigators' surprise (and possibly also the salamander's) development continued normally but the animal remained in a state of complete paralysis.[90] As they say, "never forget to add the eye of newt." While the tetrodotoxin evidently provides a defensive role in these and many other animals, it has also been employed as a venom as in the potentially lethal blue-ringed octopus[91] (although the tetrodotoxin occurs in many parts of the body, notably the arms),[92] the predatory chaetognaths,[93] and even the flatworms.[94]

Tetrodotoxin is, therefore, best given a very wide berth, and not just in the notorious fugu fish. And, oh dear, across the restaurant another diner has just collapsed. She decided not to risk the fugu and went for the clam chowder. A serious mistake as the paralytic shellfish poison takes its fearful grip. Here the culprit is saxitoxin, and although best known from the clams (the toxin takes its name from the genus *Saxidomus*) that feed on the "red tides" of dinoflagellates that synthesize the molecule, in fact saxitoxin has a wide distribution among animals and can be produced not only by dinoflagellates but other groups, including some cyanobacteria.[95] While its evidently evolved at least three[57] times.[58] And then there are the toads. In terms of toxicity, particularly notorious are the cane toads. Yet its generally unwelcome success as an introduced species in Australia is much less surprising given that its toxicity is just one of a suite of characters (including size, the capacity to move away from water, and large numbers of eggs) that are evidently instrumental in the capacity to engage in major expansions of range and so provide the key to global colonization.[59] Yet as Innes van Bocxlaer and colleagues note, having conquered the Old World "several lineages on different continents independently attained more specialized ecomorphs,"[60] a point echoed by Elizabeth Pennisi,[61] who comments that "this quintessential 'toad' form has emerged multiple times on multiple evolutionary branches."[62] The success of something like the cane toad now makes much better evolutionary sense, even if active measures are under way to tackle the menace.[63]

Apart from the frogs, among the other vertebrates perhaps the most striking example for the convergent sequestration of defensive chemicals are the toxic birds of New Guinea, specifically *Pitohui*[64] and *Ifrita*.[65] And the convergences these birds show with the dendrobatid frogs are intriguing because they also employ alkaloids known as batrachotoxins. These are highly effective in disrupting neuromuscular transmission by blocking sodium channels,[66] and recall of course the similar effectiveness of tetradotoxin (p. 36). Resistance to batrachotoxin can also be conferred by point mutations in the sodium channel.[67] Just as the frogs (and puffer fish,

p. 36) are insensitive to their own toxin, a molecular parallel may well exist in these birds.[68] So, too, the alkaloids are evidently sequestered from the diet, and in these birds (and at least some frogs) melyrid beetles may be an important source of the batrachotoxins.[69] The feathers and skin are particularly enriched in the toxin, which is consistent with the birds being repulsive to potential predators and perhaps also ectoparasites.[70] In addition, the birds are brightly colored, possibly as a defensive aposematism, also showing elements of convergence in *Pitohui*.[71] Not only that but this genus turns out to be polyphyletic,[72] and while there may be a general propensity toward toxicity in this group,[73] closely related forms are benign. Not only has this trait evolved multiple times, but as John Dumbacher and colleagues point out, "the six *Pitohui* species appear to have evolved convergently in a number of important characteristics,"[74] including not only neurotoxicity and overall morphology, but interestingly as leaders of flocks of birds.[75] In any event there is little doubt that, as with the dendrobatids, their toxicity is defensive,[76] and the locals fondly refer to the Pitohuis bird as the "rubbish bird"[77] because of its inedibility and an acrid, sour smell. Birds that taste disgusting and highly toxic frogs, of which the most famous are those dendrobatids employed by some Indian tribes to give lethal effect to the darts of their blowpipes, are only a small part of nature's armory of lethal compounds. Most familiar, of course, are the various convergent venoms (p. 61), but first let us turn our attention to a group of fish known as the tetraodontiforms. Never heard of them?

DEADLY CONVERGENCE

In fact they are probably most familiar in the form of the bizarre puffer fish, but this group displays an impressive range of both body

forms (including the world's largest teleost, the sunfish) and feeding strategies, including the independent evolution of a durophagous step it will be accompanied by convergence. Every time the evolutionary process probes (or more likely discovers) a new area of biological hyperspace, the clade in question will find the novelty to be relative—because if evolution is a deterministic process, then each solution has a high probability of evolving several times.

Telling as the myotisid ecomorphs are, this example from the bats is only the tip of an ecomorphological iceberg. In this regard the frogs are particularly instructive.[42] Yet at first sight given what appears to be a largely invariant body form, any attempt to find ecomorphological convergences among the frogs would seem decidedly unpromising. Such, however, is not the case. Consider, for example, the frogs of Madagascar and Asia (which in the latter case is largely based on frogs living in India). In each region there has been a striking adaptive radiation, not least among the burrowing and arboreal frogs.[43] Understandably, when morphology was our only guide, the respective forms were thought to be closely related, but molecular phylogeny unequivocally shows that they form a series of convergent ecomorphs. What is especially striking is that in these ecomorphs typically the convergence embraces both adult and larval[44] forms.[45]

But is this ecomorphological series complete? After all, however striking the parallels may be between these pairs of ecomorphs any frog expert will quickly point to a glaring lacuna. Thus, in Madagascar the frog repertoire includes the remarkable poison mantellids. In terms of the endangered Malagasy biodiversity, these frogs have a near-iconographic status on account of their sometimes vivid coloration, which serves as a warning aposematism, and in these mantellids this coloration is clearly convergent.[46] Their poisonous nature, however, depends on toxic skins that store disgusting alkaloids. These are not manufactured by mantellids

themselves, but are sequestered from the consumption of ants and other small arthropods.[47] It is a striking arrangement, but as no such ecomorph is found in Asia, surely now the principle of predictability fails? No, because when we look to the Neotropics, and specifically at the dendrobatid poison frogs, we find a series of striking convergences with the mantellids. Here too there are examples of vivid aposematism, which again within the dendrobatids themselves are convergent.[48] So, too, the skin toxicity depends on devouring small insects (as well as mites)[49] containing alkaloids. We see some rather remarkable convergences here also, but only a small number of species, notably among the ants, are suitable sources of alkaloids. It is evident, however, that the ants of the Neotropics are not closely related to those living in Madagascar, but have evolved comparable biosynthetic pathways for this alkaloid synthesis.[50] In the dendrobatids, not only has this evolved independently a number of times, but at least twice this has led to dietary specialization,[51] presumably allowing yet a further concentration of the vile-tasting alkaloids.

While the association between aposematic warning and toxicity is not absolutely invariant,[52] the similarities between dendrobatids and mantellids remain striking. In the former group it is also clear that convergences are very dynamic,[53] with some emerging in ancient times and others more recently.[54] Rendering oneself vile in terms of taste is too good a trick to be used by the dendrobatids and mantellids alone; independently the tiny eleutherodactylid frogs have learned how to sequester alkaloids from mites.[55] But the defensive repertoire of the amphibians by no means ends here. Not only are there a remarkable number of alkaloids employed, and these point again to chemical convergence,[56] but several other amphibians are similarly poisonous. As far as frogs are concerned, although the antimicrobial defenses are peptide-based, these have

Ecomorphs

Woodpeckeroids may provide a useful test case for the likelihood that evolution is strewn with inevitabilities, but the aim here is to demonstrate that at whatever level of biology one considers there will be loci of persistent biological stability that will act as irresistible attractors. One such line of evidence revolves around ecomorphs—recurrent anatomical configurations that answer the call of particular ecological needs.[29] Here, too, we seem to view evolution through a much more deterministic lens than current fashion generally allows. Many such examples could be cited, including fish[30] (not least the eels, pp. 49–50), corals,[31] and perhaps best known the anolid lizards of the Caribbean.[32] Any more? What about those remarkable creatures, the bats?

Crepuscular dwellers, but aerodynamic masters, silent to our ears, but living in a world of screaming echoes, these animals are surely closer to the stuff of legend than nearly any other animal. From an evolutionary perspective they are richly informative. Such as is evident from the sanguinary vampire bats that have independently evolved not only running (p. 73) but also infrared detection (p. 117). So, too, other bats have learned to hover and lap up nectar (p. 202), while in New Zealand the mystacinids are almost "honorary mice" (pp. 73–74). Most likely the first bats were agile arborealists,[33] and when they flew into the nocturnal realm, then, as John Speakman[34] remarks, the evolution of their echolocation (pp. 137-139) was "another leap in the dark."[35] In becoming aerial the terrestrial locomotion of bats has been largely compromised, but a combination of roosting and flight has led to other curiosities, not least the evolution of pubic nipples that like the more orthodox thoracic pair also lactate.[36]

Once launched, the story of bat evolution since then has been one of extraordinary success, with divergences into a series of specialized niches. Some bats, for example, are peculiarly adept at capturing prey over open water and show a corresponding series of specific adaptations in morphology and behavior. Indeed this and other divergences would seem to undermine the central thesis of this book; after all, is not the imagination of any biologist continuously overwhelmed not only by the immensity of the Tree of Life (albatrosses to aphids, giardia to giraffes) but also its apparently endless diversity? Yet this one example of water-gleaning bats actually reveals a very different pattern. In fact this mode of feeding has evolved independently a number of times among the bats, and even the fact that it has done so only once in the ecologically diverse leaf-nosed bats (phyllostomids, the group that includes the vampire bats),[37] is the exception that proves the rule.[38] Others glean from the ground, and interestingly several species of myotisid bat have independently evolved a series of whiskers along the margin of the tail membrane (or uropatagium) that evidently are tactile and potentially an important adjunct to their echolocation.[39] Nor are these isolated examples, because in the myotisid bats, at least two more instances of parallel ecomorphs can be identified. As Manuel Ruedi and Frieder Mayer remark in the context of bat ecomorphs, "Independent adaptive radiations [have] produced strikingly similar evolutionary solutions in different parts of the world,"[40] and even more strikingly they refer to "deterministic evolution."[41] The phylogeny that underpinned their conclusions employed molecular data and in doing so revealed that even an almost identical morphology is no guide to phylogenetic closeness; rather in a given biogeographic realm there is a strong likelihood that a given ecomorph will evolve. The implications are surely obvious. Yes, evolution is divergent, but at each and every

Hawaii is also famous for its bird diversity, albeit with the usual litany of extinctions and other species hanging in by their toes and claws. So too the rare 'akiapola'au[11] fills a woodpeckerlike niche, drilling holes in search of sap much like the woodpecker known as the yellow-bellied sapsucker.[12] Yet the most striking of all the parallels to the true woodpeckers is not found on an island but in the Neotropics among the furnariids. Like the vangids this group of birds shows an astonishing range of adaptations,[13] among which some woodcreepers, notably *Glyphorynchus* and the ovenbird *Xenops*,[14] are remarkably similar to the woodpeckers.[15] A key feature in hammering tree trunks is to have a skull that can take the punishment, and woodpeckers and these woodcreepers have both arrived at the same cranial solution of sturdy ossification and associated skeletal structures (such as with the nasal bars) to accommodate rapid percussive impacts.

Thus, in an alien wood maybe we would be less surprised to hear the sound of tapping. Yet reminding ourselves that we are ultimately interested in the emergence of biological properties it is not so surprising that woodpeckerlike habits have emerged from beyond the birds. In Madagascar the remarkable aye-aye (*Daubentonia*), a lemur famous for its extraordinarily elongated middle finger[16] that it uses for percussive foraging[17] (a behavior that finds a convergence in the capuchin monkeys[18] of the New World), has often been regarded as an analogue of the woodpecker.[19] This arrangement is evidently too good an invention to use only once, and striking parallels also exist with the marsupial possum *Dactylopsila*.[20] Peering out of the fossil record are three more examples of honorary woodpeckers. Best known are the extinct Eocene apatemyids,[21] but more recently strong cases have been made for examples from the Miocene, especially a notoungulate (*Hegetotherium*) from South America[22] and

a marsupial (*Yalkparidon*) from Australia.[23] Robin Beck aptly refers to this example of "a mammalian woodpecker."[24] Entering into the spirit of this game, despite incomplete skeletal material he is willing to predict that not only will evidence emerge for an arboreal habitat (as with its convergent counterparts) but "at least one of its manual digits may be elongate."[25]

Irrespective of how woodpeckerlike these different groups of mammals transpire to be, the behavior of especially the aye-aye opens different vistas concerning the emergence (and convergence) of cognitive systems. Naturalists are repeatedly impressed by the speed and efficiency of this nocturnal animal's percussive foraging. Carl Erickson suggests that via its fingertip the aye-aye not only constructs a precise cognitive map of the hidden galleries housing the insect prey in a way analogous to ultrasound or seismic profiling,[26] but drawing on earlier work by Sue Parker and Kathleen Gibson[27] noted that this capacity for precise extractive foraging could be "an important step in the evolution of increased brain size and cognitive capabilities."[28] Rather little appears to be known about the intelligence of aye-aye, but certainly accounts of it meeting travelers and tapping them inquisitively with its famous digit, as well as the superstitious fear in which some indigenes hold it, points to some capacity. Erickson's point, although specific to the aye-aye, is nevertheless prescient in terms of several themes of this book. Convergences of sensory systems, including tactile and seismic, are striking, as is their evident contribution to the construction of cognitive maps that in some way underpin (but probably do not explain) consciousness. So, too, precision of action, be it by the middle finger of the aye-aye or the elephant trunk (p. 131), may give surprising insights into the emergence of cognitive complexity.

CHAPTER 4

THE INEVITABILITY OF FORM

THESE PRELIMINARY observations suggest that the compass of evolution points true. Yet one might complain that it is all very well to trot out a series of examples, but do these turtles and spiders transcend the anecdote? Indeed they do, because as the remainder of this book sets out to show, when one looks at either the functionality of biological solutions or the roads taken, then the choices are restricted, if not inevitable. This becomes apparent also when we consider the nature of biological form. Is nothing unique? What about the woodpeckers?

WOODPECKEROIDS

If there is a cynosure for evolutionary uniqueness, it must be the woodpecker, not least because Jared Diamond has used this group of birds as the proxy for a similar human uniqueness. To be sure, within the woodpeckers themselves we find some interesting examples of convergence,[1] but as a group they evidently all originate from a single ancestral species—that is, they are monophyletic. Yet birds much like a woodpecker have evolved, repeatedly.[2] One such example concerns the vangids, an astonishingly diverse assemblage of birds found only in Madagascar.[3] So different in morphology and habit are the vangids that it came as a considerable surprise, when molecular data became available, that they all derive from a single ancestor,[4] although we surely know enough about evolution to realize how the arrival of a group in a "new world" swiftly leads to the dramatic occupation of the "ecological barrel" and almost invariably the reinvention of the evolution-

ary wheel as the same challenges are met by very much the same solutions. For whatever reason the woodpeckers never flew to Madagascar, but one of the vangids, known as *Falculea*, stepped in to take much the same role.[5] This bird, along with similar woodcreepers, is among the most derived of the vangids, but as Knud Jønsson and coworkers[6] note, "This 'woodcreeper' key innovation may have been so advantageous that the clade was able to radiate significantly, even after the vangids as a whole had reached an ecological limit."[7] In the meantime, other vangids radiated wildly in other directions,[8] as is clear from the extraordinary variety of bill types that Satoshi Yamagishi and Kazuhiro Eguchi entertainingly and directly compare to the many different types of plier.[9]

Any oceanic island, especially in the diverse tropics, can serve as an evolutionary laboratory, and the Hawaiian islands provide many fascinating insights into convergence.[10]

Salticids also transform their bodies into extraordinary imitations of their potential dinner,[92] as indeed others have come to resemble wasps.[93] The ingenuity of this mimicry is almost breathtaking.[94] What of those salticids that are strongly sexual dimorphic, for example, when the male has prominent jaws? Can it still be a mimic? Of course! Can't you see it is an ant carrying a parcel in its jaws?[95] It is a common query, of course, as to whether the mimicry and camouflage are only in the eye of the beholder. Ultimately this touches on profoundly interesting questions of universalities of perception, yet at least in the case of the salticid spider mimics, we can be confident that it is a visual mimicry.

Ironically the disguise can also work in the opposite direction. Particularly extraordinary is a homopteran belonging to the fulgoroids. Here the wings are transparent—apart, that is, from each bearing the image of a salticid spider![96] As Andreas Floren and Stefan Otto note, "This imitation is so realistic that the insect was erroneously sorted out as a spider in the first instance."[97] In a different way some assassin bugs decorate themselves with the corpses of ants that evidently deter the jumping spider,[98] either because in many cases these spiders avoid ants or perhaps the response is simply, "Doesn't look like any bug I've ever seen. . . ."

Camouflage achieved by taking foreign objects and attaching them has, of course, evolved in many other contexts. Think of the crabs that decorate themselves with all manner of invertebrates (and algae).[99] Perhaps even more remarkable camouflage is achieved by some weevils, which carry a complete garden of moss and algae on their backs,[100] although more repellent are those chrysomelid beetles whose larvae are equipped with a sort of pitchfork that provides the anchor for a defensive thatch of feces.[101] To some the sight might be repulsive, but more importantly deterrence also involves a chemical component.[102] And in a happily different way the bodies of some desert spiders bear a specific type of setae whose adhesive properties (which evidently operate in much the same way as gecko pads, pp. 83–84) enable the animal to coat itself with sand grains and so meld into the background.[103] And of course it is convergent, with Rebecca Duncan and colleagues noting that they may "represent a general design principle for particle capture and retention."[104]

The camouflage of the natural world has, of course, provided inspiration for humans, not least in time of war. Let me wrap up, however, with a short anecdote that concludes Alan Stripp's fine book on code breaking.[105] Noting that the story might be apocryphal, he writes how "the Germans painstakingly built a decoy airfield, with dummy aircraft, fuel bowsers, hangars and control all beautifully made of wood, to distract attention from the real thing. On the day it was completed the RAF, who had been keeping a quiet eye on what was happening, showed their appreciation by dropping one wooden bomb on it."[106] Time to move on, I think.

emerge are rampantly convergent. Consider the evolution of sociality in the spiders.[77] This is remarkably rare but also blatantly convergent. Of the twenty-six[78] known species of social spider (out of a total of about thirty-nine thousand species!) rather remarkably about eleven of these instances have arisen among a group known as the theridiids, with one case evidently verging on eusociality (pp. 170–71).[79] Even within this clade of spiders, however, their phylogenetic distribution is very scattered. It is most likely, therefore, that these specialized excursions in sociality are short-lived experiments,[80] but

the vital point remains that they arise independently: as one group goes to the wall, another emerges. Talking of "dead ends" in this context is, therefore, relative, and it is very likely that this example from the spiders and others such as is seen in bark beetles,[81] and the phytophagous (note 61, this chapter) and parasitic[82] insects are the tip of a convergent iceberg: many specializations may indeed lead nowhere particularly interesting (at least from our perspective), but as the theridiid spiders nicely demonstrate, even here the tape of life is replayed at a high level of fidelity.

How Clear Is the Signal?

The spiders allow us to complete this introductory survey of convergence. Not in terms of silk (pp. 87–89) or eyes (p. 104). They come later. Rather, the spiders permit us to consider all too briefly what many would regard as the exemplification of convergence. Now you see it, now you don't. I speak, of course, of mimicry.[83] Like nearly all spiders the remarkable jumping spiders (or salticids) are carnivores, even if curiously one has transformed itself into a herbivore.[84] Not only are they carnivores but highly sophisticated hunters, which Duane Harland and Robert Jackson[85] aptly refer to as "eight-legged cats."[86] As is often the way in hunting, different groups of spiders tend to show trends toward specialization in their diet,[87] and salticids are no exception (in captivity they are known to enjoy sausages). This specialization reaches an acme with a salticid that consumes the blood of vertebrates: not by plunging its fangs into a startled ungulate but by the simpler expedient of picking on blood-engorged mosquitoes.[88] And just as the mosquito employs olfactory cues to find the human, so too experiments (employing socks) demonstrate that the spider has converged on the same saliency.[89]

For many spiders the last thing they see is a salticid hurling itself at them. Should their prey live on a web, then this presents no obstacle because like those ants that skitter across the otherwise lethal rim of a pitcher plant (pp. 85–86), so the salticids can walk with impunity across the otherwise sticky silk webs built by their prey.[90] Rather remarkably, however, apart from tracking down other spiders, there is a pronounced tendency among the salticids to target eusocial insects like the ants, even though one is in danger of stirring up the proverbial hornet's nest in terms of vigorous, venomous, and coordinated defense. In order to slip thorough the defenses we encounter some classic cases of mimicry. To hunt ants, and to avoid their aggressive response, salticids have, therefore, evolved various disguises. Some, for example, obtain chemicals (hydrocarbons) that the ants, specifically weaver ants, themselves employ for recognition of nest mates against intruders. The source of the hydrocarbons is the larvae of the colony, and while it is not entirely clear how these chemicals are actually transferred, the spider doubtlessly acquires an effective chemical disguise to infiltrate the colony.[91]

parasitism,[58] is whether evolution marches to a drummer whose beat is insistently to drive the generalist, that jack-of-all-trades, to what is only too familiar, the specialist, the species that is truly superb at doing only one thing very well. If so, is the drummer choosing a road that may ultimately end in disaster? That view is also widely held, but in fact the evidence is much more open ended. In certain cases, becoming a generalist is not so surprising. Consider the vines known as *Dalechampia*.[59] On the African mainland their pollination is highly specialized, depending on an association with resin-collecting bees. Yet on that laboratory of evolution, in the form of the off-shore island of Madagascar, pollination is generalized. Evidently *Dalechampia* had no choice, because when it dispersed to Madagascar its specialized pollinators missed the boat.[60] On arrival the plant had two choices: become extinct or generalize. It took the obvious route. Yet in many other cases, although the route from generalist to specialist is the more common, it is clearly a finely balanced matter and numerous journeys occur in the opposite direction. Such, for example, appears to be the case in the phytophagous insects.[61] For example, writing on the plant-feeding associations of the larvae of nymphalid butterflies, Niklas Janz and colleagues[62] emphasized that specialization "is not a path of no return."[63] So, too, in other instances it may well be that primitively the arrangement was one of specialism and subsequently it became increasingly generalized. Consider brood parasitism in the birds. This is probably most familiar with the cuckoo (which has an intriguing mimicry with the hawks [p. 442n137]), where the host unwittingly rears the nestlings of a different species. Unsurprisingly this arrangement is rampantly convergent, having evolved at least seven times in the birds and including even an example from the ducks.[64] Apart from the cuckoo, another well-known

example is the cowbird, and here it is argued that in its earliest form it was a highly specific association but subsequently became much more generalized.

In any one of these discussions, the conclusion as to where evolution is really heading is only as reliable as the phylogeny one cares to employ. In the case of the parasitic cowbirds, the argument[65] that brood parasitism moved from specific to general is finely balanced, and in the case of the cuckoos (where interestingly brood parasitism evolved independently several times), the direction does seem to be toward greater specialism.[66] Biologists, like any scientists, like neat conclusions, but it is clear that in examples as disparate as parasitoid flies[67] and self-fertilization in plants,[68] clearly no firm conclusions can yet be reached as to whether the generalist or specialist is the endpoint.

Nevertheless, the existing evidence clearly shows that in many cases evolution is reversible and so confounds Dollo's "law," an articulation that the tape of life is irreversible; that is to say, once an evolutionary structure is lost it cannot be regained. One famous example involves the supernumerary teeth of the lynx,[69] while another looks to the gastropods. Classically recognized by their coiled shell, independently many times they have developed lineages with uncoiled shells. These have been seen as classic evolutionary dead ends. Yet now we know that coiling, in the shell of the calyptraeid *Trochita*, has indeed reevolved.[70] Indeed case studies in the breaking of Dollo's "law" have turned into a minor industry,[71] encompassing both some of the examples given above (e.g., parasitism,[72] sexuality)[73] and others that range from marine larvae (again involving the calyptraeid snails)[74] and ostracodes,[75] even to bowerbirds.[76]

But are we not beginning to flounder again; where please are our general principles? There are two points to make. The first is that in many cases the specializations that suggests that it will always be outflanked. As Michael Rosenzweig and Robert McCord emphasize, the notion of evolutionary progress deserves a much fairer hearing than it has generally received. How many ears are open remains to be seen. Why at least some attention should be given is evident from an intriguing analysis of evolving complexity in the crustaceans.[45] These present several advantages, not only in terms of a good fossil record, but to the outside world a bewildering array of appendages that often show striking specializations along the length of the body (the phenomenon known as *tagmosis*). And as such, they provide an excellent test case for some of the most famous of evolutionary problems. Nobody can fail to be impressed by the diversity of extant forms; recall that Darwin cut his teeth on the barnacles. But how did this arise? As Sarah Adamowicz and colleagues remark, this "could have resulted from nothing more than random diffusion from a minimum boundary . . . or from the differing fortunes of a small number of clades that happen coincidentally to differ in complexity."[46] And this is exactly the mantra chanted by many evolutionary biologists: diversification, of course, but little more than a random muddle punctuated by the odd fluke. The data speak otherwise. Not only is there clear evidence for increasing morphological complexity in the crustaceans but crucially this occurs "within many independent lineages."[47] Intriguingly these authors also point out that, as far as tagmosis is concerned, the theoretical limit has not yet been reached, although this may be more theoretical than practical.

Such trends of increasing complexity imply for others either extinction or at best retreat into refugia. Yet even as the masters of the evolutionary universe are toppled from their pedestals, does this not also imply evergreater specializations and the sense that however superbly designed, each and every lineage is ultimately doomed to explore what

sooner or later will transpire as a dead end? And if evolution is littered with wrecks of beautifully designed but fragile organisms, then surely a sense of predictability vanishes: yes, the history of life is a coruscating marvel of forms, but as they flicker in and out of existence, how can there possibly be a deeper pattern?

There is certainly a very widespread sense that to become evolutionarily specialized is to walk along an ever-narrowing path of successive adaptation, ending in the sheer drop of extinction. And in certain cases this appears to be the case. A particular focus in the study of life's blind alleys is the evolution of asexuality. One of the most surprising[48] is in a fungus-farming ant[49] (pp. 181–85), but more famous are those examples in some ostracods and rotifers, which appear to be very ancient.[50] In some instances, such as the bdelloid rotifers, it may really be that the lineage has irrevocably abandoned sexuality. However, this does not mean that these animals form a massive clone. A species structure can evidently still emerge,[51] and in addition the genome is remarkably rich in genes imported from other kingdoms.[52] Conversely, in the darwinulid ostracods, the very odd male can be found lurking.[53] Moreover, in many other cases an escape clause is evidently built in, so in groups as diverse as oribatid mites,[54] flatworms,[55] and monogenean platyhelminthes,[56] sexuality is recovered. The last group happens also to be parasitic, and here too there is a pervasive view that the extraordinarily complex associations such animals have with the hosts and the equally intricate adaptations that are associated with parasitism study would make evolutionary reversals all but impossible. Evolution turns out (as ever) to be more labile and inventive, and indeed it is clear that parasitism is by no means irreversible.[57]

Closely linked to this question, and involving an inquiry that extends far beyond morphology, which had served not only to define a phylogeny but also identify single originations, on the contrary turn out to be repeatedly reinvented.

The Compass of Evolution

It seems, therefore, more plausible that evolution is unlikely to be thwarted in its exploration of all potentialities, including as it happens even the emergence of intelligence. This is not to deny that entrenched positions exist and incumbents may indeed take a disproportionately long time to be dislodged. Consider the turtles. Inhabiting what is effectively a mobile bone box represents a highly effective defensive strategy, with a key corollary being the ability to withdraw the head in time of danger. Unsurprisingly this has evolved at least twice, independently in the so-called cryptodire and pleurodire turtles with the respective distinction that the flexure of the withdrawn neck is either in a vertical or lateral plane. Being unable to retract the head might be labeled the "executioner's-block syndrome": it is surely a distinct adaptive disadvantage.[41] Yet the primitive turtles (which are referred to as the amphichelydians) could not withdraw their heads, perhaps because when they first evolved the levels of risk were relatively low. Yet in a changing world, to survive, if not prosper, the turtles would do well to evolve head retraction. And so they do, yet extraordinarily those that never learned to do so only disappeared from the planet on the edge of historical memory, some 150 Ma after the cryptodire turtles first began to retract their heads. Mass extinctions probably helped to accelerate the process, and the unfolding story of the decline of the amphichelydians follows separate timetables in different parts of the world, with Australasia slouching to the slowest tempo of expiration. For the protruding necks of the amphichelydians, the writing was on the wall for a long time, but independently on continent after continent the inevitable tape of life was played out and each time they shuffled off the evolutionary stage into the black side wings of extinction.

Getting out of trouble is likely, therefore, to become increasingly selective, and so, too, other ways of escape will not only evolve but converge. Thus, just as turtles learned to withdraw their heads, so among the fish and amphibians we see an escape maneuver that employs a very specific type of head retraction. This has evolved independently at least six times and seems to be particularly associated with living either in burrows or a very cluttered environment.[42] In these and many other cases the evidence suggests both ever increasing levels of predation and ever more complex defensive measures, be they venomous shrews (pp. 61–62) or toxin-laden fish (pp. 36–37). Multiple threats lead to multiple responses, and unsurprisingly this has led to rampant convergence. In a fascinating essay, Geerat Vermeij, for example, addresses evolutionary contingency and the supposed uniqueness of biological innovation.[43] As he points out, counterexamples abound; for example, in the rise of predatory snails the so-called labral tooth (a spinose projection arising from the shell margin that plays a key role in carnivory, including forcible entry or apparently conferring stability during attack) represents a key adaptation that has evolved independently at least fifty-eight times.[44] In his analysis Vermeij also emphasizes the likely importance of incumbency to explain why, despite its rampant convergence, snails already possessing the predator high ground may be difficult to displace.

Maybe so. Evolutionary incumbency is certainly real, but ultimately the evidence

Turning Back?

Such a notion, arguing how much is inevitable in terms of evolutionary outcomes, sits uneasily with much of neo-Darwinian thinking. Here, of course, the emphasis is on randomness, be it at the level of mutation or mass extinction. It is, however, already possible to show that at least some mutational pathways are far more constrained than might be thought, while arguably the only thing mass extinctions achieve is an acceleration of a preordained process. Yet in the case of convergence it is possible to take the view that however many numerous and convincing the examples, the direction of evolution is still dependent on prior conditions and critical transitions. After all, evolution does not work in a vacuum, and if the necessary ingredient hasn't materialized, then a given route is forever barred.

The question of constraint in evolution is understandably central in determining what might never be possible as against the very likely, if not inevitable. In effect, what happens when a given constructional kit has failed to materialize? One view of life would be to suggest that starting points critically determine any outcome, as was popularized in Stephen J. Gould's metaphor of rerunning the tape of life.[31] The prevalence of convergence, of course, points in exactly the opposite direction. But even if initial conditions are not determining factors, is it not likely that once an organism has dug itself a metaphorical adaptive hole it becomes extremely difficult to clamber out and start some new evolutionary adventure? Thus we need to consider not only constraints on what is possible but also the topic of evolutionary irreversibility.

Consider the curious case of the tadpole. The contrast with the adult frog is, of course, striking, and the transition is defined as a metamorphosis from a wholly aquatic animal with gills and a tail to the frog with legs and lungs. Frogs, of course, belong to the amphibians, which as their name implies occupy a halfway house between the water and dry land. Like many other groups, therefore, they are on the road to full terrestrialization (pp. 221–26), and unsurprisingly, not only have all the major groups of amphibians become terrestrial, but also many times independently.[32] Obviously, once on land a tadpole is redundant, so rather than following the indirect route of their aquatic cousins the egg develops in a direct fashion, bypassing both tadpole stage and subsequent metamorphosis.[33]

But what happens when history goes into reverse and the amphibian follows the whale road and returns to the water? Having thrown it away, surely it is extraordinarily difficult to reevolve the tadpole stage? Nevertheless, within the plethodontid salamanders[34] this is exactly what has happened.[35] But why in the plethodontids, but apparently not in other salamanders? Paul Chippindale and colleagues (note 35, this chapter) suggest that the retention of certain larval structures, notably the hyobranchial apparatus, in the developing egg leaves the door open for the reinvention of the tadpole. Not surprisingly, such a dramatic reversal has been met with skepticism,[36] yet in response, as Ronald Bonett and colleagues also note, our attitude should not be one of surprise.[37] Astonishment and claims of incredulity arise in many other instances of convergence that had passed unrecognized. The catalyst has been the reinvigoration of phylogenetic analysis on the basis of molecular data. Again and again, be it examples from groups as disparate as corals[38] (not for nothing do Danwei Huang and colleagues refer to "Bigmessidae")[39] or pulmonate snails,[40] the molecular data convincingly show that other words, morphological laws cannot be reduced to history, function and adaptation, just as biology itself cannot be reduced to physical and chemical factors."[24] Meyen was also embedded in dialectic materialism and had little time for Darwin.

Yet this approach is reflected to some extent in George Ledyard Stebbins's prescient analysis[25] of angiosperms, and especially the structures of the diagnostic flowers and fruits, embedded in a matrix of all possibilities. In contrast to Vavilov, however, Stebbins's analysis was emphatically Darwinian, emphasizing selection and adaptation. Significantly, and in a way that echoes a much more recent and similar analysis of enzyme active sites,[26] Stebbins found that in his matrix of the 256 possible combinations, only 86 were realized, and of those only 49, or less than 20 percent, with any frequency. So, too, Roger Thomas's definition of "skeleton space"[27] again draws on a matrix, this time of constructional possibilities, and broadly concludes that just about everything has been invented and usually multiple times.[28] The sense that evolution might be predictable and thereby conceivably reveals a deeper structure, perhaps amenable to mathematical description, extends far beyond the prescient work by Vavilov, Stebbins, and Thomas, all of which look to the matrix of possibilities and at least by implication forms that are either inevitable or, equally relevant, impossible. Parallel strands of inquiry look either to geometry of form, articulated most brilliantly by D'Arcy Thompson,[29] or interest in how stability emerges out of chaotic systems via attractors, another compelling metaphor for both the stability of biological systems and their repeated "rediscovery."

Unsurprisingly these approaches have sometimes been explicitly anti-Darwinian, which is doubly unfortunate as their portrayal may be seen as antithetical and thereby giving unnecessary (and erroneous) support to exponents of intelligent design and other such theological mischief. The question is not whether evolution is Darwinian, which it certainly is, but whether it alone provides a complete explanation. Ultra-Darwinists protest that it does, but claims and reality may be far apart. More fruitfully, and as is already happening, we can ask if a more complete theory of evolution, post-Darwinian for want of a simpler label, is in the offing. My sense is that convergence suggests exactly this, whereby evolution perforce must follow a metaphorical map of highly restricted possibilities. The corpus of Darwinian theory—that is, adaptation arising via natural selection—is unchanged (and uncontroversial), but to use an analogue is as unlikely to explain the outcomes of evolution as ionic bonding is the periodic table. Such a view has been intriguingly extended by George McGhee.[30] To my way of thinking the insights of Vavilov (and McGhee) not only point to a deeper structure of evolution but allow the map of life to extend much further than reductionist materialists can allow.

The plausibility of this claim remains to be established and might seem to founder immediately inasmuch as it would be a rash claim to argue that everything is convergent and nothing is unique. Indeed, notwithstanding Vavilov's optimism that his method heralded a systematic approach to biology that had long been familiar to chemists, biologists broadly divide. There are those who relish the cavalcade of examples, the sheer particularity of an instance. Yet so, too, as often they might throw up their hands at life's fecund exceptionality, perhaps even muttering, "What a dog's breakfast." And on the other side of this divide? Here the eye attempts to seek a deeper order—that many differences are really only skin-deep, and constraint is the order of the day.

So too when he earns any attention it is as much mingled with contempt, but in his *On the Genesis of Species*,[13] his imagination was clearly caught by what are now seen as classic examples of convergence. In speaking of those which involved the cephalopods versus vertebrates he wrote, "It would be difficult to calculate the odds against the independent occurrence and conservation of two such complex series of merely accidental and minute haphazard variations."[14] Mivart's arguments, revolving around what we would now identify as the immense hypervolume of biological space and the question as to how such a minute number of workable variations are found, is of course part of the intelligent design mantra and need not detain us. For Mivart, and ourselves, the reality of evolution was—and is—not in question. However, when Mivart noted that there might "by the concurrence of some other and internal natural law or laws co-operating with external influences and with 'Natural Selection' in the evolution of organic forms"[15] that together served to govern among other things convergence, then we would be unwise to dismiss this possibility any more than do physicists or chemists.[16] Evolutionary convergence may be just that sort of clue, and although articulated in an imprecise fashion, Mivart's suggestion was only the first of several attempts to arrive at a nomothetic synthesis.

So, too, drawing on the topic of evolutionary convergence there is Leo Berg's *Nomogenesis or Evolution as Determined by Law*,[17] published in the early days of Soviet entrenchment and rapidly published in English, but now largely forgotten. Not only is it a fascinating inventory of convergence, but it too is stridently anti-Darwinian. His countryman, Nikolai Vavilov,[18] an earlier supporter of the notorious Lysenko but ultimately imprisoned in Saratov where he died of malnutrition in 1943,[19] is better remembered for his law

of homologous series, which attempted to establish a predictive framework of "expected forms."[20] While a botanist, and largely focused on plants such as wheat and barley, Vavilov was fully aware of the implications of his work to both animals and fungi, and to the topic of convergence as a whole. Thus he regarded his studies as directly analogous to those of a chemist, writing, "For a systematist is not a man who knows all the curiosities of nature but one who grasps the order and sense of it all."[21] Unsurprisingly for the time, he added a Marxist coda of his approach being "historically necessary and inevitable." In due course Vavilov paid the inevitable price of "bourgeois heresy," one "insignificant" victim of the thanatocratic twentieth century where the great materialist heresies found fertile roots as the very few decreed the fate of the very many. Today Marxism may be at a low ebb, skulking behind the piles of corpses and a miserific vision as the ever-glib apologists grin center stage, but it would be entirely mistaken to think that evolution is somehow immune to hijacking by forces that promise to be yet more malign.

The attempts by Vavilov to find order and predictability in the plant kingdom were recapitulated by Sergei Meyen,[22] who also emphasized the contributions by Berg and other Soviet biologists. Interestingly his article was preceded by a gently worded health warning ("free flow of ideas . . . unfamiliar to most English-speaking botanists . . . foster progress"),[23] and indeed much of the language remains alien to the present day. Yet Meyen (and his predecessors) still deserve a hearing, and as he writes, the "morphological laws in plants cannot be learned by way merely of historical, functional and adaptational (ecological analysis) of plant forms and structures. We have to study the laws as such (just as in the art sciences Roman pottery may well be studied outside its utility). In unyielding realities of physics and chemistry, then much hangs on this conclusion. As has been repeatedly stressed, if, according to received neo-Darwinian wisdom, human intelligence is a mere fluke of circumstance, then the search for an extraterrestrial equivalent is most likely a complete waste of time. Such a view has been articulated forcibly by George Gaylord Simpson in his essay bluntly titled, "The Nonprevalence of Humanoids,"[8] and with equal conviction by Jared Diamond.[9] An important argument used by Diamond is that given unique evolutionary solutions such as the woodpecker, we should not expect anything similar elsewhere, including of course human intelligence. The argument for the inevitability of intelligence is returned to later, as is the evidence that woodpeckers are far from unique (pp. 31–32). Yet if human intelligence is very much likely to happen,[10] then this reopens the question posed by Enrico Fermi as to why we appear to have no glimmer of extraterrestrial intelligence.

Equally important to the argument that life's trajectories are effectively indeterminate and open-ended is the emphasis on randomness. Most typically this is expressed in terms of the molecular dice, that is mutation, or the historical turn-of-event, notably mass extinctions. Yet, as we shall see, the molecular architecture upon which mutations are imprinted—be it enzyme structure or gene networks—is rampantly convergent and greatly constrains what is possible. Similarly megadisasters such as massive flood volcanism, oceanic crises, extreme global warming, and bolide impacts undoubtedly perturb (to put it mildly) the biosphere, but it is far less obvious that they radically redirect the course of evolution as against simply accelerating the inevitable.

When we begin to consider the implications of evolution there is indeed a lot to play for. Indeed it is no accident that just as earlier versions of Darwinism were seized upon by some unhealthy customers, notoriously Ernst Haeckel with his celebration of Prussian hegemonies, so too scientific world-pictures speaking to evolution will be accommodated, often in an almost comically procrustean way, to meet the cultural zeitgeist. It is surely no coincidence that aspects of neo-Darwinism continue to be wrenched out of any legitimate scientific context to prop up one or other dubious metaphysics. So if we choose to live in a world wedded to relativism, then there is little surprise if counterfactual evolutionary stories find a resonance. Correspondingly, if we can demonstrate a deep structure to evolution, unsurprisingly other metaphysics will be more fruitful.

A Brief History of Convergence

Evolutionary convergence[11] is certainly more interested in the generalities of biological properties rather than the coruscating and as often bewildering cascade of organisms as different as bacteria and bees. The history of biology has, however, yet to provide an overall account of the part that convergence has played in our understanding of evolution. In fact, ironically the boot has often been on the other foot, because in this area there has been a recurrent emphasis on anti-Darwinian thinking.

This emerged very early in the ultimately tragic figure of St. George Mivart. He is a neglected figure who made the cardinal mistake of falling foul first of the Darwin/Huxley camp and then of the Catholic Church to which he converted at the age of sixteen.[12]

CHAPTER 3

CONVERGENCE: HOW CLEAR IS THE SIGNAL?

THE CONVERGENCES in the octopus (and the related cephalopods) are striking in their range and also sometimes the subtleties of the similarities. It is hardly surprising that the convergence in the camera eyes (see also p. 106) holds a classic status in biology. At the other end of the octopus spectrum, the striking similarity between their tentacular arms and those of a human is a reminder that convergences may be much more than skin-deep. Convergence is ubiquitous; the number of possibilities in evolution in principle is more than astronomic, but the number that actually work is an infinitesimally smaller fraction. A bold claim and short of stumbling across an extraterrestrial biosphere, how will we ever know if this is correct? So far as we are concerned, n = 1. What we see may be only local, and combined with the lingering suspicion that some steps, such as inventing the eukaryotic cell, are very difficult, then the case for evolution being predictable seems to fall at the first hurdle. I argue otherwise, and to paraphrase Ernst Mayr, the Runes of Evolution is "One long argument."[1] Here is a small test case. A number of animals have invented methods of entrapment: cunning humans (p. 166) and web-weaving spiders immediately come to mind. But there are alternatives. Consider the conical pit built by the larval ant lion, an insect belonging to the lacewings (neuropterans). Lurking at the base, envenomed mandibles poised, it is

ready to seize luckless visitors that tumble down the sandy sides.[2] The pits are simplicity itself in construction, but as Arnold Fertin and Jérôme Casas[3] remark, it "remains a mystery why such simple traps have so rarely been adopted by the animal kingdom."[4] Even though a dipteran larva (the worm lion) has arrived at the same solution,[5] Graeme Ruxton and Mike Hansell[6] are correct to ask, "Why are pitfall traps so rare in the natural world?"[7] The answer, as they point out, might be straightforward. Simple they may be, but these pits suffer from a string of restrictions, ranging from specific soil types to vulnerability to both a downpour and probing predators.

Wherever one looks it seems as if evolution has very little choice. If, then, we can discover the details of the metaphorical "map" across which life must navigate to the very few available solutions, then are we not on the threshold of a predictive biology? This view of life cuts cleanly across one of the central areas of neo-Darwinian thinking, an area that insists on the randomness of evolution and the unpredictability of the outcomes. That might explain why the adjectives associated with descriptions of evolutionary convergence are so often exclamatory; "remarkable," "surprising," "astonishing," "uncanny," and "stunning" are all routinely employed. And perhaps such surprise is justified, because if evolution is fundamentally nonpredictive, an iridescent form skittering across the dung by burrowing owls. While this material is employed in the construction of the underground nest it is also scattered liberally around the entrance. Its function? Suggestions that from a human perspective sound a little odd, such as mate attraction or what is euphemistically referred to as "olfactory camouflage" are certainly possible, but the evidence strongly points to the dung serving to attract insects that unwittingly provide dinner for the owls.[113] Deliberate collection and placing of lumps of dung qualify as tool use.

So tools come in surprising contexts, but what about the octopus? Most striking are veined octopus (*Amphioctopus*) that employ coconut halves in lagoons off Indonesia, extracting them, carrying them, and then employing them as a most convenient lair.[114] An echo of this sort of activity is also found in the octopus den. Often this is a crevice or other recess, and as such may well not be fit for the purpose. If so, the octopus will not only construct a wall (employing sometimes a remarkable range of items including pieces of glass and shell as well as rocks),[115] but evidently show choice, selection, and planning consistent with cognitive complexity that allows the connecting and anticipation of unrelated pieces of information.[116] Just one more clue that stirring in the brain of the octopus is yet another approach to a universal mind.

the temperaments of the octopus named, respectively, Emily Dickinson and Lucretia McEvil.[106] Among the attributes of the latter was an unwelcome tendency to trash the support system in her tank. The propensity of the octopus to escape from its tank is also well known. While this may lead to a sad demise on the floor, in the Brighton Aquarium one octopus was well-known to pay visits to the next-door tank to feast on lump-fish, and not because it was hungry.[107] The notion that all this reflects a consciousness can generate a distinct nervousness in some biologists. But, not only does the evidence seem strong, but as Jennifer Mather also reminds us, the arms of the octopus are very richly endowed with nervous tissue, including ganglia. As she remarks, "Our mammalian heritage leads us to think of central planning as complex and peripheral as simple, but these divisions may need re-thinking for octopuses."[108] Indeed so, and in due course when we encounter visually alert jellyfish (pp. 97–100) and maze-solving slime molds (pp. 237–38), we will see that intelligence may be even more curious than we realize.

A central theme of this book is, of course, how evolution rediscovers the inevitable, thus hinting at lawlike features that confer predictability, a central goal for all scientists. Thus, when we assess the cognitive sophistication of the octopus it should not come as a surprise (but probably will) that they exhibit not only traits of personality but one of the hallmarks of advanced intelligence: playfulness. In one species such activity was relatively undramatic, inasmuch as play was more like some sort of bathtime activity with the octopus deliberately directing jets of water at a floating bottle. However, of the eight octopuses on parade, only two manifested play, which the investigators suggested was personality-based.[109] All the more remarkable, therefore, is *Octopus vulgaris*, where nine out of the fourteen animals played with

Lego bricks. Octopus not only play but are evidently curious. Rather remarkably, some cephalopods, such as the oval squid, will react to seeing themselves in a mirror,[110] and if this is actually self-recognition then they will be in rare company indeed (pp. 292–93). And this finds parallels in studies[111] that note the amount of time the animals spent looking out of the tank, noting a marked increase when there is a stimulus. Interestingly, however, the degree of curiosity differed widely among the individuals, again consistent with personality. A male, nicknamed Calimero, became far more interested in the outside world when there was a stimuli, but sad to relate Celeste rather let the side down because she spent about equal amounts of time looking beyond the tank irrespective of stimuli. You'll know the type. But even if not all octopus engage in play or have much of an attention span, there is no doubt that Michael Kuba and colleagues are right to emphasize the connection between cognitive complexity and behavior.[112] Even more importantly is their suggestion "that there might be a common principle for how animals and humans interact playfully with objects" (p. 188). Indeed so, and when we examine the one-to-one matching between types of social play in some birds and mammals, which again is convincingly convergent (pp. 283–84), then we can be increasingly confident that not only is intelligence evolutionarily inevitable, and in like manner tool-making and ultimately technology, but the sentient extraterrestrials will also be playing.

Given this roster of cognitive sophistications it is less surprising that a type of tool use has also been identified in the octopus. This needs some explanation because among behavioral biologists the identification of tool use goes far beyond the usual notion of femur head crashing onto the antelope skull, let alone well-hafted spear thudding into the side of a short-tempered ungulate. Consider, for example, the use of pieces of the regions referred to respectively as the peduncle and vertical lobes). Indeed, in the case of the vertical lobes, not only are there intriguing similarities to the mammalian hippocampus, but as Binyamin Hochner also notes they "somewhat resemble the mushroom bodies of the insect brain,"[93] not least in terms of a remarkable neuronal density. It is not surprising, therefore, that the capacity for learning (and correspondingly memory) are so highly developed in the octopus. We also see here intriguing similarities with the vertebrates, such as a clear distinction between brain areas given over to long- versus short-term memory.[94] As these investigators, led by Tal Shomrat, point out, parallels also exist with the insects, and crucially they propose that "short- and long-term memory of nonreflexive behaviors . . . appear to have a universal organizational principle."[95] Octopus can thus readily learn by imitation—as well as showing other aspects of learning that encompass such features such as discriminating shapes, employing spatial memory to recognize landmarks and so navigate effectively (with the implication that the octopus constructs some sort of cognitive map)[96]—and engaging in so-called autonomous learning whereby the animal does not behave as a robot but decides when navigating a maze the best course of action in terms of tradeoffs of speed as against actual distance traveled.[97] The octopus, therefore, evidently lives in a rich cognitive environment, and there is little doubt that among the cephalopods they are not alone in this respect. Cuttlefish, for example, not only show spatial learning but the investigators emphasize the parallels with the vertebrates.[98] And when reviewing how the behavior of cephalopods changes through their life, Jennifer Mather[99] remarked, "what is surprising is the clear parallel between cephalopods and mammals,"[100] showing how a preprogrammed stage gives over to learning and a corresponding flexibility. It is not

surprising to discover that cuttlefish brought up in "interesting" tanks (with various objects and a sandy floor) memorized more readily and grew faster than those stuck in featureless ("boring") tanks.[101]

Just how rich, and perhaps also to an underappreciated extent, is the cognitive world of the octopus and its relatives is evident also in at least rudimentary expressions of personality, play, tool use, and perhaps even sleep (pp. 287–89). All these are widely taken to be hallmarks of advanced intelligence and what is almost certainly some form of consciousness.[102] Evolution, and especially the study of animal behavior, is burdened by terms that are not only anthromorphic but carry a moral freight: *selfish genes, suicidal bacteria, cheating birds,* and so forth. For the most part these terms are recognized for the metaphors they are. When, however, we start to invoke terms such as *personality*[103] in octopus,[104] one can treat their expression as at least a rudimentary reflection of human complexity. Entirely reasonable, but in fact this regression of inquiry presents some formidable problems, not least when we come to consider the wider problems of consciousness. So, too, if as appears to be the case, evolutionary convergence indicates that the available solutions are very limited, then we should not be surprised that just as toolmaking and prototechnology have evolved independently, so have other aspects of the cognitive landscape, including personality. In essence, the question of animal personality— entirely uncontroversial to those familiar with such animals as dogs and horses (cats I leave to the relevant expert)—so it is possible to assess the behavior of not only octopus but other cephalopods such as the dumpling squid in terms of temperamental traits that are both idiosyncratic and individualistic.[105] As I have mentioned elsewhere, it may not be completely daft to call an individual octopus by name, and no prizes are offered to guess

Octopus Nerves

When we move to the nervous system, matters become increasingly interesting as we read the runes of evolution. In the realms of physiology the nervous system of the squid has long held central place because the appropriately named giant axons are particularly suitable for experimental work, not least the classic studies of ion channels. As their name suggests, these axons are relatively enormous, which means that resistance to the electrical current is greatly reduced and the speed of conduction proportionally increases.[82] As Daniel Hartline and David Colman note, axonal gigantism is "the easy solution."[83] Giant axons are ideal for rapid response, and so we see their independent evolution in groups as diverse as the annelidan tube worms, crayfish, and flies.[84] In each case the giant axons facilitate the rapid startle response, be it the fan worm zipping back into its protective tube or the crayfish flicking its tail and vanishing backward.

Giant axons are obviously a sensible solution for getting out of trouble quickly. By and large, vertebrates do it differently. Their solution is to wrap the nerve in a fatty layer known as the myelin sheath. The myelinated solution has evolved multiple times (pp. 242–43), but for whatever reasons, the cephalopods have not chosen this evolutionary route. So, too, at first sight, their brains being built on a molluskan ground plan[85] are strikingly unlike any vertebrate. As Bernd Budelmann remarks, not only do "cephalopods have the largest of all the invertebrate nervous systems,"[86] with some 100 million neurons,[87] but their nervous system "manifests the highest degree of centralization ('cerebralization') for any mollusk."[88] Yet, not only are there convergences (including the all-important blood-brain barrier[89] [p. 245]), but more intriguingly their cognitive world appears to be much less alien than might at first be supposed. Not that

evolutionary convergence is meant to imply an absolute identity, and it would be a duller world if such was the case. For example, the octopus employs both visual and chemotactile senses, but as we will see, evidence for their integration suggests not only a correspondingly richer sensory world but more importantly some fundamental similarities of how, ultimately, consciousness may be constructed (or accessed . . .). In any event it is no accident that the octopus falls within more stringent ethical guidelines when it comes to experimentation than practically any other invertebrate. In their excellent overview,[90] Binyamin Hochner and colleagues emphasize not only their "sophisticated and extraordinary ability to adapt their behavioral repertoire to the current environmental circumstances,"[91] but how an octopus newly introduced into captivity is within a few days transformed from a fearful individual to a friendly and attentive petlike creature. Given the continuing emphasis on the convergences between octopus (and other cephalopods) with the vertebrates, it should be less surprising that they extend to the brain and its cognitive abilities.

Or is it? As a number of commentators have stressed, the high degree of intelligence in at least the octopus a priori is rather odd because these animals are solitary and short-lived. Hardly worth the expense? In fact, as we will see, the evolution of intelligence is considerably more subtle than simply organizing the emergence of a troop of crafty apes. Thus, these convergences deserve particular emphasis, not only because they are one strand pointing toward universal mentalities that can sprout from radically different starting points, but more specifically because there are striking similarities between specific regions of the vertebrate brain, notably the cerebellum,[92] and hippocampus, and equivalent areas in the octopus brain (specifically might seem at first sight that this convergence of inks between cephalopod and gastropod is relatively superficial, not least because in contrast to the former the sea hares are slow-moving and thus the more reliant on chemical deception. Nevertheless, the evidence now suggests that the antipredatory chemistry of the sea hare finds its convergent counterpart in the cephalopods.[66]

ELASTIC EVOLUTION

Wherever one looks in the cephalopods, the convergences pop out. Not only eyes and tentacles, but across their anatomy again we find many striking similarities. Perhaps some are more subtle than realized? Take the so-called kidneys. This renal complex has little similarity to its vertebrate counterpart, and although it is by far the most complex arrangement among the mollusks its invertebrate origins are obvious. But these "kidneys" also house a remarkable convergence in terms of two quite separate groups of vermiform symbionts (p. 317n13). Given the evidence that they might help the excretion of ammonia[67] (standard among aquatic animals on account of its solubility) as well as contribute to the smooth flow of the urine on account of their dense covering of beating cilia,[68] then is it possible that symbiont and "kidney" are more complex, even more vertebratelike than realized?[69]

Other anatomical convergences are more obvious. Unlike most other mollusks, the blood system is closed, entirely confined to vessels that even have the analogue of vertebrate capillaries.[70] This is to be expected, because as a group the cephalopods are typically active with enhanced metabolic activity. An integral part of these vessels is a lining of so-called endothelial cells, and unsurprisingly the developmental machinery, especially a growth factor receptor, is convergently employed in both cephalopods and vertebrates.[71] Not only is this receptor employed in the vascular system but also in the development of the heart. Not surprisingly the hearts are chambered and convergent with those of

vertebrates.[72] What is even more interesting is that the structure of the aorta is remarkably similar to that of the mammals.[73] Both must manage the rapidly fluctuating pressure as the blood surges out of the heart, so the primary function of the aorta is not only to be resilient and not rip open in an aneurysm but to smooth out the pressure fluctuations.[74] The shared secret lies in the use of proteins that act elastically[75]—that is, they store energy and then release it in a controlled fashion. In itself this is an intriguing example of convergence at the molecular level because elastic proteins have evolved independently many times.[76] In the aortas the protein of choice is elastin, which in the vertebrates occurs very widely, allowing the elasticity of our skin and lungs, not to mention that towering monument to convergence, the penis (p. 212).[77]

As far as internal anatomy is concerned, the convergences include not only the aorta but also a cartilaginous skeleton.[78] This is strikingly similar to that of the vertebrates, most notably in the cranial cartilage of cephalopods, which as Alison Cole and Brian Hall remark, demonstrates an "extreme histological convergence."[79] Not only is the cephalopod cartilage strikingly reminiscent of the hyaline equivalent in vertebrates, but the areas that develop first in the embryo show further striking similarities in terms of differentiation by cellular condensation[80] (cartilage has evolved independently twice more, in the fan-worms [sabellid polychaetes] and limulids [aquatic chelicerate arthropods, relatives of the spiders]).[81]

may serve to confuse the enemy,[47] a signal analogous to the squirrels that employ their tails as flaming beacons (p. 117). Even more remarkable, however, is an octopus that puts its camouflage on an offensive footing by providing astonishing imitations of a poisonous sea snake, a toxic flatfish, as well as a variety of other venomous animals that should all be given a wide berth. This mimicry is highly dynamic, including not only patterning but other aspects of impersonation. Mark Norman and colleagues[48] suggest the octopus[49] makes a positive decision as to which Walter Mitty mode to adopt.[50]

But among the cephalopods even more outrageous is the example of sexual mimicry documented by Roger Hanlon and others.[51] It involves a giant cuttlefish, the females of which aggregate for sex but are extraordinarily fussy. The premium on being a successful male is, therefore, very high, and accordingly the females are closely guarded. The smaller males, however, can literally instantaneously mimic the females, approach, and, with considerable frequency, achieve success. And the attendant "official" male? He can indeed be distracted by another presumptuous intruder, but in some cases there is no distraction, but still he notices nothing. In passing we should also note that the reproductive strategies among the cephalopods show a considerable variety, including what Warwick Sauer and colleagues[52] report in some South African squid of a nuptial dance. This, they note, is similar to a lek and is "unexpectedly complex, rivaling those of mammals and birds."[53] The female squid evidently choose the larger males, but sad to say once again the younger chaps will sneak in uninvited, the escort remaining oblivious. In these South African squid, the fun often continues after sunset, whereas in the Australian giant cuttlefish the signaling and deception cease when dusk falls. Then the ability to camouflage themselves takes on a different urgency, because remarkably at night these cuttlefish alter their body patterns to match the surrounding rocks and so avoid the attentions of nocturnal predators such as fish and dolphins.[54] This implies, of course, that the nighttime vision of these cuttlefish is also acute.

And cephalopods are masters of disguise in another way, by the celebrated employment of their ink. While we associate this ink in visual terms of concealment and bafflement, there are some indications that its defensive role extends to the chemical realm.[55] And this is certainly the case in the convergent evolution of inking[56] by another mollusk, a swimming gastropod known as the sea hare (*Aplysia*). In this case two glands work in tandem, one to produce the colored ink (with a pigment derived from eating red algae)[57] and the other (from the so-called opaline gland), a clear but viscous substance. Intriguingly the chemistry of mixing recalls that seen in the bombardier beetle (p. 38), albeit in slow motion.[58] The chemical activity that is invoked is certainly complex. Thus, the opaline secretion serves to warn nearby individuals of attack, and this alarm cue employs particular amino acids (mycosporine-like)[59] that otherwise play an important role in ultraviolet screening (p. 112). The ink also broadcasts alarm cues,[60] but even more remarkably in one of the main enemies of the sea hare, the spiny lobster, the chemicals either induce the crustacean to flee (and here one of the products of this chemical reaction, hydrogen peroxide, is important)[61] or fool it into feasting on the ink instead of the retreating sea hare.[62] In this so-called phagomimicry it may be that the viscous secretions from the opaline gland assist in a tactile deception. Crustaceans[63] are not the only enemies, but in the case of fish the strategies employed by the sea hare and its squirting ink[64] are somewhat different.[65] It of rods articulate via a series of joints. With its flexible tentacles with effectively unlimited freedom of movement it is difficult to imagine anything much more dissimilar—until, that is, we look at the octopus more closely, using a high-speed camera. Then, rather extraordinarily, the tentacles turn out to be far more like our arms than might have been expected. The tentacle is transformed into an arm! To be sure there are no fingers as the food is grasped by the highly tactile suckers.[34] What happens, however, is that waves of muscular contraction roll along the arm, and when they collide form a pseudo-joint. The net result is a quasi-arm, defined by three segments, with most of the rotation taking place at the "elbow." The position of the articulatory "joints" are not precisely fixed, but interestingly the proportion between the lengths of the two main segments usually remains remarkably constant. Germán Sumbre and his colleagues take pains to note that this striking functional convergence may at first sight be surprising but it actually suggests that when evolution needs an arm, then there really is

an "optimal design."[35] That's the way the world works. Moreover, although each of the eight arms is effectively identical (apart from those octopus with an add-on penis [p. 425n59]), the octopus evidently knows the difference because particular arms appear to be chosen to undertake specific tasks.[36] But if the arms of the octopus are less alien than imagined, surely the suckers of the arms—traditionally clamped to the face of the doomed diver— have no useful counterpart elsewhere? In one sense this is correct, but in many ways they are a sort of analogue of the elephant's trunk and its versatility (p. 131). In fact, the octopus suckers have multiple functions, ranging from display to locomotion. But of these, perhaps the most extraordinary is the way an individual sucker can fold into two halves and, acting like a mittened hand, is quite capable of grasping a strand of fishing line.[37] Not only that, but as Jennifer Mather[38] also notes, the manipulative ability of arms extends to being able to "untie knots in the finest surgical silk."[39]

EVOLUTION IN DISGUISE

One of the most startling features of the cephalopods is the extraordinary patterns that can flicker across their body as they deploy the so-called chromatophores.[40] They serve many functions, not least as a remarkably dynamic method of communication. Thus, even though human observers are quite incapable of distinguishing sexes of cuttlefish, the females appear to have no such difficulty.[41] It comes as no surprise, therefore, that many cephalopods are also masters of camouflage[42]—for example, not only matching themselves to the surrounding seafloor, but despite being color-blind (p. 115) even managing a chromatic correspondence with

surrounding substrate.[43] All this is consistent with their acute powers of vision. Indeed Emma Kelman and colleagues[44] remark how they "have been struck by the similarity of the cuttlefishes' camouflage behavior to human object recognition,"[45] not least in terms of pictorial depth. Yet their abilities to imitate more than gravel, let alone rolling coconuts and tiptoeing algae, still provide constant surprises. One trick adopted by an octopus is to generate a "passing cloud" across its body, evidently in an attempt to startle prey while all the time the octopus actually remains motionless, waiting to pounce.[46] So, too, among the cuttlefish, eyespotlike structures details of the retina.[21] Striking as this check-list of convergences might be, evolutionary convergence neither means the structure is identical nor that all the solutions discovered are optimally equivalent. Thus, the many correspondences between the octopus and vertebrate eye should not allow us to overlook that in various respects the retina of the latter is more sophisticated and is involved with signal processing in a way not found in any cephalopod. But is the cephalopod really so disadvantaged? Recall first that it has the more "sensible" arrangement of the retina being uppermost, unlike the vertebrates, where in principle (p. 97) the light has to travel through a layer of cells before it impinges on the retina. Nobody doubts their acuity of vision, and in some squid it can rival that of humans.[22] Less surprising, therefore, is to see that the outer cortex (or deep retina) of the cephalopod optic lobe, which is juxtaposed to the eye and connected of course by the optic nerves, is convergent with the vertebrate retina.[23]

In extending his study to the optic lobe of a squid, J. Z. Young[24] remarked how his analysis brought "out further remarkable similarities between the pattern of organization of the visual system in cephalopods and vertebrates. The general resemblance between the retina profunda and the vertebrate retina is obvious enough, but similarity extends even into details."[25] As Binyamin Hochner[26] notes, this "similarity [is] all the more striking as the octopus's typically invertebrate mechanisms of transduction and physiological responses to light are quite different from those of vertebrates."[27] Recalling also that the vertebrate retina is an outgrowth of the brain—and so in terms of evolutionary ancestry is crucially different from the mode of origination of the cephalopod retina, which is formed by the infolding of the skin—in the final analysis the proclaimed differences in the modes of visual processing in fact are achieving the same result by different routes.

Getting a Grip

If squid are "honorary fish," then we would expect them to be highly effective swimmers. And so they are, but unlike the fish, which wave a tube of muscle from side to side, cephalopods employ a form of jet propulsion. Although this system is inherently less efficient[28] (not least in the employment of a cumbersome respiratory protein known as haemocyanin)[29] and makes extraordinary respiratory demands, there is a striking analogy in fish and squid between those muscles used for sudden bursts of speed as against normal cruising.[30] Not only is it likely that in the squid this arrangement has evolved independently several times,[31] but the similarity extends to the cellular level. The reason is that the corresponding two types of muscle fiber found in fish that exhibit markedly different

oxidative metabolisms find a direct counterpart in the squid.[32] Effectively, therefore, we find an equivalent to the white and red muscles in fish, which in turn reveals an extraordinary story of convergence when we come to look at the tuna (pp. 78–80). Of course, muscle needs to be enclosed in an integument, a skin. Mollusks they may be, but once again we find that the cephalopods have departed from such close relatives as the bivalves and snails, because now they have evolved an integument much more similar to that of the vertebrates.[33]

So, "honorary fish" they remain, but if their apparent alienness appears to have somewhat receded, what about those famous tentaclelike arms? I put down my fountain pen, my friend lifts her book. In either case at its simplest a set females of *Ocythoe*.[13] In the latter case, why not their companions? Presumably because they occur as dwarf males.[14] Not surprisingly, therefore, investigators have been repeatedly struck by not only the versatility of the cephalopod body plan, but the way they have pushed it to some quite extraordinary limits.

This is true, for example, of their circulatory system and even more so in terms of nervous development. So far as the latter is concerned, cephalopods now stand on the threshold of new worlds: stirring in their nervous system is a mind.

THE GAZE OF THE MOLLUSK

Look at the octopus, and it looks back. In both cases the eye is built somewhat like a camera. Indeed this independent evolution of the camera eye is rightly celebrated as one striking example of convergence.[15] In fact, as we will see (p. 106), the camera eye has evolved independently at least six more times. Evidently it is an excellent arrangement—one might say *design* (but with no implication of so-called intelligent design). By definition the lens must be transparent, which this is achieved by using proteins that, in the context, are appropriately known as *crystallins*. They are a compelling example of molecular convergence, so it is quite unsurprising to learn that both octopus and vertebrate have equally transparent lenses and have crystallins that function in exactly the same way, but in reality the crystallins in either group have completely different evolutionary origins (p. 93).

Typically the lens in cephalopods and vertebrates is spherical, but this presents a special problem when it comes to how the light travels through it. As is well known, when light moves from one medium to another, usually the path it must take is slightly bent. Lower a perfectly straight stick into water, and it will appear bent (the degree of bending being determined by what is called the *refractive index*). In the case of a lens this is potentially very serious because if the bending of the light is not properly corrected, the image will not be precisely focused on the ret-

ina. The result: A blurred image and potential disaster for the octopus. The solution to this problem is for the octopus to progressively change the refractive index from the center to the edge of the lens so as to correct precisely the spherical aberration. This degree of correction is very close to the ideal a physicist would employ. So, too, it is almost identical to what we see in the equivalent vertebrate, in this case the example chosen being a fish (trout) as it is also aquatic. Not only are the camera eyes convergent, but so is the correction for spherical aberration.[16] This refractive correction is, of course, molecular, but not only is it based on remarkably high concentrations of the crystallin proteins but specific amino acid substitutions that find intriguing parallels between vertebrate and cephalopod.[17] Now in one way all this is entirely predictable: after all, if the lens cannot focus properly it is pretty useless. Nevertheless, it is a reminder that, as often as not, convergences are far from superficial; not only are they subtle, but they suggest the constraints are far from being accidental. Rather they point to universal principles.

So far as the camera eyes of octopus and vertebrate are concerned, the convergences extend far beyond the overall anatomy, employment of crystallins, and correction for spherical aberration. Rather, the roster of similarities extends to the cornea,[18] pupil,[19] the so-called extraocular muscles (responsible for rotating the eyeball),[20] as well as

CHAPTER 2

CONSIDER THE OCTOPUS

THE OCTOPUS HAS a remarkable hold on our imaginations. It is certainly no accident that in *The War of the Worlds* H.G. Wells peopled (if that is the word) the sinister tripods that stalked across a terrified and panic-struck England with octopoidlike occupants. Although best-known as a consummate writer, Wells was more than biologically literate,[1] having not only studied under the great Thomas Henry Huxley but literally aped him.[2] Moreover, as Peter Kemp reminds us,[3] Wells had long been fascinated in the octopus, with their apparently inhuman and groping tentacles (but see below), not to mention a weakness for human flesh. Nor is this fascination simply the purlieu of gifted novelists, however competent they might be in biology. Ian Gleadall and Nadav Shashar[4] echoed this in their nice review on octopus vision when they wrote, "Cephalopods are among the most exotic and alien life forms imaginable."[5]

Yet for all its alien attributes, in reality the octopus shows many remarkable similarities with the vertebrates, even though in terms of the animal kingdom the evolutionary gap could scarcely be wider.[6] But as Andrew Packard remarks, "ours is a functional and democratic age: no longer is genealogy of primary importance. And cephalopods functionally *are* fish."[7] Vertebrates, which include fish and such descendants as the mammals (and thereby us), are closely related to the starfish (and the other echinoderms).[8] The octopus (and other cephalopods such as the squid

and cuttlefish), however, are close cousins of the oyster (and other mollusks).[9] This evolutionary gulf means that the many similarities we see between the octopus and vertebrate have clearly been arrived at independently. There are two reasons for thinking this must be correct. First, the common ancestor lived about 550 million years ago and may have approximated a sluglike beast. In any event, what this remote form could not possibly have possessed are the many complex features that the two groups now share. Second, while the convergences are indeed very striking, in each case there is a clear "footprint" that clearly points to a completely separate evolutionary trail.

In many ways cephalopods like the squid can be justly labeled "honorary fish." The similarities are certainly not exact (convergence never claims they are), and in such features as locomotion, respiration, and digestion the squid clearly takes second place.[10] Nevertheless, they remain immensely successful and in the deep oceans arguably outclass the fish. Ingeniously in this environment they store the waste product ammonia, which, having a low density, imparts buoyancy. A nice trick, and one that in this group has evolved independently several times.[11] Nor is this the only way cephalopods can achieve neutral buoyancy. In fish the standard method is to employ the swim-bladder, but just such a structure has evolved independently in the octopus, specifically the deep-sea *Alloposus*[12] and the like Douglas Adams was right. Remember that episode in *The Hitchhiker's Guide to the Galaxy* where jynna-tonnyx turns out to be the galactic-wide refreshment?[4] And what's true for alcoholism, and for all we know gin and tonic, applies across biology: from enzymes to penis, from insecticide resistance to toolmaking and technology, from love darts to music, everywhere you look evolution is hedged in by convergence. By no means everything is possible; in fact nearly everything is impossible.

"At last I do believe we might be on the threshold of a general theory for biology. The implications are"—and here Mortimer paused and looked at the first stars that had begun to shine above the lagoon—"well, I think they might be rather considerable. After all, what is it about life? I'm no vitalist, but you can't help being amazed by the intricacies of that tiny chemical factory we call the cell. Why, it far outstrips even our most advanced nanotechnology. And then think of the way the embryo develops—not only the wonderful way in which it unfolds, but how robust it is against insult and injury. Polanyi was quite right: Life is much more than just a sort of souped-up chemistry and physics.[5] And there is so much to play for: I have a

hunch that convergence could even give us some new clues on that mystery of mysteries: what consciousness is, but if anybody thinks mind is entirely material they are barking up the wrong tree. And aren't these conclusions positively cosmic? When I look at the night sky then like Pascal those infinities terrify me. Where on earth are they? You know—the extraterrestrials! Fermi was on to something, and now we know about convergence we really can predict what these extraterrestrials will look like, and as importantly how they'll think. Physicists know that the elements they see in the farthest reaches of the Universe are exactly the same as on Earth. So if life is universal, which incidentally might not be quite as secure an assumption as is usually thought, and the roads of evolution are indeed narrow and inevitable, then the differences between us and an alien will be trivial. But as I said: Where on earth are they? Is life just a fluke? Could there be something very peculiar about the Earth? Whichever way you look at it, something doesn't add up. . . ."

In the silence the next dish was served. "Splendid," remarked Mortimer. "What better example of convergence? Consider the octopus."

CHAPTER 1

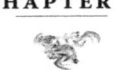

DINNER ON THE LAGOON

MORTIMER HAD TAKEN ME to the best restaurant he knew, and we were not disappointed. The food was a sensation, a visual triumph, appetizing aromas and exquisitely balanced flavors: eyes, noses, and taste buds all jostled for attention. Beyond the quay the lagoon shimmered, with vast thunderheads illuminated by the setting sun. We had much to discuss and would only just catch the last vaporetto. "Do you recall that lecture by Gould? How insistent he was that evolution had no directionality, how the history of life veered erratically and unpredictably, clobbered by catastrophe after catastrophe, supereruption followed by cometary impact. To think evolution was like either physics or chemistry, that it is with a deep structure, real predictabilities, was absurd. Rerun the tape of life, as ironically he repeated so often, and the outcome will be entirely different. Humans? Yet another accident of evolution, as interesting in their own way as a tapeworm, or, if you prefer, a tulip. Of course, it was perfectly clear that, like any scientist, Gould had based his stance on deep, if hardly articulated, metaphysical roots. His worldview made no sense unless humans were utter flukes of circumstance. Only then could we construct the world as we wished it to be, rather than as it actually is." Mortimer emptied his wineglass, and then grinned. "Now we know that he was gloriously wrong, and indeed the evidence stares us in the face."

Mortimer gestured at the rim of his wineglass. Just visible were a couple of fruit flies, a stock-in-trade for molecular biologists (where they are just "fly" or *Drosophila*). Finding them hovering around the wine was scarcely surprising. In the wild they seek out fruit, which of course soon rots and, by fermenting, so dinner for the fruit fly is laced with alcohol. As he explained, Mortimer filled both glasses, emptied the bottle, and signaled to the waiter. "Tricky stuff, alcohol," and taking an appreciative sip continued, "and highly poisonous. Pity the fruit fly; yes it, too, becomes inebriated, and not surprisingly the device biologists use is the inebractometer.[1] No, no, I am not joking. Well, you see the fascinating thing is, if you study how the flies get drunk—sorry? Oh yes, you see how fast their little legs are moving, well, would you believe it but their behavior is remarkably similar to the way we get drunk?[2] Wonderful example of evolutionary convergence, and you can hardly be surprised if the genes for alcoholism in a fruit fly have names like *barfly*, not to mention *cheapdate* and *amnesiac*.[3]

"Sounds trivial, but it isn't. Think about it. How different biology appears to be from either physics or chemistry. Where are the grand principles, the major laws, the equivalent of the periodic table or the theory of general relativity? Of course, there is evolution, but that doesn't *predict* anything! Just as Gould claimed was the case: everything a fluke. But that story of the fruit flies getting intoxicated is just one example of convergence. It looks of pure speculation. As the Universe has become self-aware, we as one of its products can appreciate its deep beauty. But suppose— and this is by no means a novel idea—mind is not so much self-realized as brains increase in size and complexity but rather the brain serves as conduit. In this way it encounters the abstract realms of mathematics, music, and language, all of infinite potentiality. Such realms are familiar, but they also hint at the existence of orthogonal worlds. Perhaps our journey has only just begun?

mals. Choanoflagellates are too small for collagen to have a structural function, and in animals evidently it has been co-opted for its new role. Nestling in these tiny protistans is one component essential to the elephant. More striking still is the case of the nervous system. Unique to animals (although a dual origin is possible) and, more importantly, a significant part of both its genetic basis and molecular machinery (especially with respect to synaptic functions) evolved long before any nervous system materialized. Deep in the history of eukaryotes lies the potentiality for a brain.

Evolutionary inherency is no more surprising than convergence, but it does pose unsolved problems as to the versatility of molecular systems. How are they recruited, and, as importantly, how do the various components build on each other to give systems of increasing complexity? In the history of life, things not only change but they get decidedly more interesting. One might observe, "Once there were bacteria, now there is New York."[12] We need to be careful, however, not to equate "simple" with "primitive." In fact, however far you go down the Tree of Life, really simple things are pretty elusive. Molecular phylogenies allow us to infer with some accuracy what the common ancestor possessed, and in the case, for example, of the earliest eukaryotes they turn out to be "surprisingly" complex. And this is the general rule. So, too, organisms of apparent simplicity turn out not only to have been derived from much more complex predecessors, but they remain highly sophisticated. Among the most astonishing are minute animals known as dicyemids. They consist of less than a hundred cells, have no real organs, and spend their lives drenched in urine, living as they do on the surface of cephalopod kidneys. Despite their simplicity, dicyemids clearly derive from advanced animals. But there is more. First, they are strikingly convergent on a group of pro-

tistans known as the ciliates, specifically the chromodinids, that also infest cephalopod kidneys. Second, long thought to be parasites (another treasure trove for convergence), it is more likely that they contribute to renal function[13] and so are a vital ingredient in allowing the cephalopods to transcend their molluskan origins and become "honorary fish."

This honorific status is matched by the cephalopods having remarkably large brains, indicative that independently in these mollusks a mind is stirring. For Darwin the mystery of mysteries was the origin of species, but for us it is the nature of mind. Convergence helps us to stake out the territory. Giant brains and cognitive sophistication have evolved multiple times. Obvious manifestations are learned vocalizations, toolmaking, and social play. Less tangibly, sleep, mirror self-recognition, and even an awareness of death are as much tantalizing as informative. All are patently a product of evolution, and the differences that separate us from apes, crows, dolphins, and maybe even octopus are paper thin. This is what Darwin taught, but now we stand alone. For nearly all biologists this is simply a matter of seamless extrapolation. By Darwinian descent this is patently correct, but is it true of mind? Yes, we share the same nervous system, the same types of brain (even if convergent) and a cognitive architecture that confers memory, learning, manipulative skill, and so on. But do our animal counterparts actually *understand* what they are doing? Their ingenuity, not least in the New Caledonian crow's toolmaking, is indeed remarkable. Cleverly devised experiments, however, are beginning to suggest severe limits to animal intelligence, and when it comes to rationality perhaps they remain in the dark. If this is the case, then maybe rationality is not so much a question of emergence but one of discovery?

The book is a long journey, but at the end we return to a lagoon and a short chapter they discarded a perfectly good system and by recruiting a convergent protein reinvented the wheel is less clear. It is, however, powerful evidence that biology travels through history but ends up at much the same destination.

The aim of this book is far wider than simply to show the ubiquity of evolutionary convergence. Such does indeed suggest a metaphorical landscape where the adaptive peaks are very few relative to the immensity of territory that remains permanently inaccessible, except of course to our imaginations. And here is the nub of the problem. How is it that we have imaginations that can conjure up counterfactual possibilities? More generally, how is it that the Universe became self-aware? Convergence, especially of cognitive systems, shows us that it certainly happens, but in what context?

In this discussion, one stumbling block needs to be dynamited straightaway: the notion that while convergence per se is unremarkable, there still remain key transitions. If, so the argument goes, any of these were frustrated, then this would forever preclude one or another evolutionary adventure. This is not to contest the fact that our understanding of many of these major transitions is woefully incomplete. Nevertheless, straws in the wind suggest likelihoods of outcome. Consider the eukaryotes. Without them, then no fungi, nor animals, nor plants. Yet we need to take the wider view. First it is increasingly clear that not only are there interesting convergences between eukaryotes and bacteria, but significant components of the eukaryotic condition have evolved in bacteria and are available for co-option.[9] So, too, key steps—such as the acquisition of chloroplasts from once free-living bacteria—are replicated today, suggesting that, as revolutionary as this step was, it was, by no means inherently improbable.[10]

The neglected concept of inherency—that is, how waiting in the wings of the evolu-

tionary theater are the players for the next act—reinforces the sense of evolutionary inevitability. Evolutionary co-option and horizontal gene transfer are well enough known. On the other hand, the extent to which the molecular equipment required for a key innovation actually evolved long before it was recruited to its new role is far more pervasive than sometimes is appreciated. The proteins that confer transparency to the lens and cornea of animal eyes, the so-called crystallins, represent a classic example of co-option. Be it via the familiar modalities of vision, smell, or tactility to the seemingly alien worlds of infrared perception, echolocation, and electrosensation, each and every sensory system is fascinating in many other regards. This is true both in terms of individual convergences and also unexpected commonalities, such as not only optical foveae but striking tactile, auditory, and electrical analogues. What also emerges is the astonishing sensitivity of these (and many other) evolutionary systems. Repeatedly we find a breathtaking precision of operation, be it the operation of the Johnston's organ (a sort of ear) of the mosquito or the infrared detector of the buprestid fire-beetle. One can make a general argument that in their different ways these sensory systems have effectively reached the limits of the physical universe, at least as far as biology is concerned. Science, of course, opens inaccessible windows of observation to which the universe would otherwise be blind.

Animals not only need to see and smell but among other things also support themselves and transmit instructions. With respect to the former, the protein collagen plays a key structural role. Yet not only is this molecule convergent, evolving independently in both fungi and bacteria, but more importantly collagen itself evolved in the group of protistans known as the choanoflagellates.[11] Their appearance far predated that of aniso it is equally legitimate to conjure up "hummingbirdoids" with evolutionary analogues like the sphinx moth flickering into convergent existence.

Ecomorphs remind us that many convergences are most easily explored in terms of functional morphology. Nevertheless, a complaint recurs that sometimes the concept of convergence is pushed too far with similarities that are really quite superficial. Would we bother, for example, to discuss topics such as nesting or gliding? Yet, even here, there can be subtleties, such as foam nests, and surprises: Who would have predicted a gliding snake? More important, however, are where the apparently superficial similarities conceal a more interesting story. Such skin-deep differences are exemplified in the tuna and lamnid sharks, both fish but, respectively, as a teleost and chondrichthyean not at all closely related. It is hardly rocket science to notice that both are superbly hydrodynamic. But look a little closer: quite independently each has evolved a complex musculotendinous system to provide powerful thrusts to the large tail, but each has also developed a warm-bloodedness that represents a quantum leap in predatory effectiveness and range of operation.

We would hope that any ichthyologist would have sufficient command of the area to appreciate not only these skin-deep differences but inquire also whether extinct thunniform counterparts (think of the Mesozoic pachycormid fish and ichthyosaurs) might provide further insights. Yet, as a whole, taxonomists necessarily specialize: she on cycads, he on ascomycetes, myself on priapulids. Inevitably with such focus on a particular group (and as often a lifetime's dedication of study), the recognition of deeper patterns and commonalities in biology may be frustrated. Yet the prospect of a more general theory of biology will depend in teasing out what unites form rather than divides it. Who, for example, would have expected that the

manner in which a cockroach (with incidentally its convergent milk production) strides across the floor is strikingly similar in terms of the biomechanics to a mammal? So, too, the neural circuits that fire the walking programs are intriguingly convergent. But this example is overshadowed by the extraordinary way in which the octopus can transform its arm into a simulacrum of a vertebrate arm. These general rules of operation extend far beyond biology because they are now having a major influence on the design of robots and other biomimetic systems.

Convergence brings into focus a profound tension between the basic rules of organization, if you like the geometry of life, and the innumerable historical pathways that paleontologists in particular revel in discovering. How in other words does history mesh with the immutable realities of the Universe? One hotly debated area is to what extent evolution is any more than a vast exploration of dead ends, biological cul-de-sacs where organisms "run out of things to do." This question has profound implications for the possible limits of biological exploration, but we should never underestimate the versatilities of evolution. Consider the bats. Superb fliers (and independently they have evolved echolocation), but notoriously poor walkers and once landed they usually roost. Vampire bats face a problem. They fly to their luckless host, but once perched then how do they get around and choose the best place to slice open the skin and start lapping the blood? Simply by reinventing the capacity to walk and run. Perhaps, however, the most startling example of reinvention is linked to the capacity of insects to taste and smell. In the same way as vision, these sensory systems employ a canonical protein (opsins and the like) with a diagnostic series of helices (seven in total) that span the cell membrane. Just the same arrangement is found in the insects, except that here it is completely independent. Why that something very like a tapeworm has evolved in a group of protistans known as the dinoflagellates?[3] Should we rise to the challenge of Tristam Stayton's[4] question, "Is convergence surprising?"[5] Yet his inquiry is rhetorical; simulations employing four separate metrics lead Stayton to conclude that evolution is far from random. The real world tells the same story, as outlined earlier by myself[6] and George McGhee.[7]

In this book, however, the net been thrown much wider than before, and as a consequence a whole series of neglected evolutionary questions arise. Some are remarkably general. Why, for example, are convergences such as parasitism, carnivory, and nitrogen fixation in plants[8] concentrated in particular taxonomic hot spots? Why do certain groups have a particular propensity to evolve toward particular states? Another chestnut involves what is referred to as "deep homology." If the underlying genetic mechanisms (such as the famous example of the paired-box gene *Pax6* in eyes) are the same in quite different groups, then how can it be legitimate to speak of convergence? The reality is rather different and points to some much more interesting principles. Other questions are more specific but lead to unexpected evolutionary insights. If bees sleep (as they do), do they dream? Why is that insect copulating with an orchid? Why have sponges evolved a system of fiber optics? What do mantis-shrimps and submarines have in common? If mosasaurs had not gone extinct, what would have happened next? Will a saber-toothed cat ever reevolve?

With this blizzard of questions we must trust to Ariadne's thread as we pace the labyrinths of this book. Yet not only is the journey punctuated by regular shafts of light but as importantly the local, even the anecdotal, melds into the general. Like a tapestry one has to stand back to see how the figures and buildings are set in a wider landscape where the recurrent themes of the narrative reveal the deeper patterns of evolution. If we are going to start this journey, then where better than in a restaurant and out in the Venetian lagoon? Episodes in the lagoon open and close this book, but at the beginning it is my shadowy colleague Mortimer who, turning from his wine, exclaims "Consider the octopus!" What better starting point? Of course, with its camera eyes the octopus is a totem of convergent evolution, but most intriguing are the cognitive convergences that, toward the end of this book, open dramatic new territory.

Whatever other significance convergence might have, it is central in the study of evolution because it confirms the power of adaptation. Most familiar are the forms of mimicry. Apart from the classic instances of Batesian and Mullerian mimicry, a more mysterious world shadows these famous examples in the form of chemical mimicry, not least those putrid "carrion" flowers with their loathsome scents that some insects find irresistible. And what about the remarkable bolas spider and their whirling balls of sticky silk that intercept moths lured by fake sexual signals? Equally striking in the context of convergence are so-called ecomorphs—that is, recurrent body forms that are spawned by similar environmental pressures. Justly celebrated are the anolid lizards of the Caribbean islands, but perhaps even more startling are ecomorphological convergences between the poison dart frogs of the Neotropics and the mantellid frogs of Madagascar. Is there not, however, a hitch in this argument? Are not some ecomorphs unique? With its percussive skull, sticky tongue, and distinctive stance, the woodpecker is confirmed by both ornithologists and molecular biologists to have a single—that is, monophyletic—origin. This book insists, however, that we need to take the wider view. Woodpeckers are unique, but "woodpeckeroids," both among other birds and further afield, most certainly are not. And just as one can invoke woodpeckeroids,

INTRODUCTION

LONG BOOKS DEMAND short introductions. Yes, another book on evolution, but one with a difference. Not to dispute the realities of evolution, or for that matter the primacy of the Darwinian explanation. Neither would it dream of contesting the self-evident observation that we humans are just one endpoint of, to paraphrase the science fiction writer Brian Aldiss, a 3-billion-year evolutionary spree. But here is a book that is prepared to be heterodox, and not before time. So what is it all about?

Even among the mammals, let alone the entire Tree of Life, humans represent one minute twig of a vast (and largely fossilized) arborescence. So, it would be very poor form were we to demand center stage. Nor is that my intention. Every living species is a linear descendant of an immense string of now-vanished ancestors, but evolution itself is the very reverse of linear. Rather it is endlessly exploratory, probing the vast spaces of biological hyperspace. Indeed this book is a celebration of how our world is (and was) populated by a riot of forms, a coruscating tapestry of life.

But only humans understand this, and so while there is no narrative of human origins, no "Monad to Man," threaded through this book are some of the staging posts that led from a eukaryotic cell to animals, including fish that clambered onto land, furry reptiles that transformed into mammals, and in one branch evolved into primates, great apes, and ultimately ourselves. If from unicell to human every generation is taken as a stepping-stone then they would total billions. Even a tally of really key steps as to how the first eukaryote evolved into the biological form writing these lines would run into the hundreds, if not the thousands. We need to be selective, and while far from imagining this list to be exhaustive, among the most significant might be: multicellularity,[1] tissues (including a nervous system), sensory systems (not least vision), limbs (including those that grasp), toolmaking, and intelligence. Others, such as an immune system (notably an adaptive one[2]), placentas (and even a penis), agriculture, and sleep might be less obvious but in one way or another each was biologically seismic. All are part of a much longer list that collectively entails becoming human. They have, however, a wider significance. Each of the features just listed is an essential component in a much broader argument that forms the core of this book. Each and every one of them is either demonstrably convergent or inherent in terms of prior history, that is, drawing on building blocks that have already evolved (and as often as not for a quite separate purpose).

Of these two concepts, evolutionary convergence is the more familiar. It is the otherwise uncontroversial observation that from very different starting points in the Tree of Life very much the same solution has evolved multiple times. Did you know, for example,

The Runes of Evolution

Acknowledgments

Like many other things in my life this book needs to start with an apology. Its genesis lay in the generous invitation to deliver the 2007 Gifford Lectures at the University of Edinburgh. As this work grew, however, it became clear that it had diverged from my original theme by too great an extent. It is still my firm intention to publish the Gifford lectures and it is incumbent on me to thank the many people in Edinburgh—most especially Wilson Poon, Isabel Roberts and Susan Manning—for this memorable time marked by their many kindnesses. I am in debt to many other people. First is Vivien Brown who has handled manuscript preparation with panache, verve—and patience. Next are the Map of Life team, Chloë Cyrus-Kent and Verena Dietrich-Bischoff as well as the web-designer Francis Rowland who collectively not only delivered what I believe is a very effective website (www.mapoflife.com) but drew innumerable facts to my attention. The project was supported generously by the John Templeton Foundation and I warmly thank them for their support. More recently my wits have been sharpened by discussions with Sylvain Gerber, Jen Hoyal-Cuthill, Victoria Ling and Mags Pullen. Then there are those who gave their time to read various sections, especially Rob Asher, Nick Davies, William Foster, Peter Grubb, Chris Howe, David Norman and Nick Strausfield. All offered acute criticisms and by no means agreed with all I said; the faults remain firmly with me. Many others also alerted me to recently published papers, and I especially thank Pablo de Felipe and Ken McNamara. Practically every word written draws on the painstaking work of the "invisible college" and I hope that the research and conclusions drawn by them are dealt with honestly. In places I use short quotations and these fall under the rubric of fair usage. Librarians across Cambridge assisted the chasing of references and I especially thank Sarah Humbert. Mr R. Garry has also sent me a remarkable range of books and other material. My agent Barbara Levy zealously guarded my interests and offered constant encouragement and support. In addition my brother Roderick kindly provided the translations in the last chapter. Finally I warmly thank my Department and St. John's College for providing a secure basis where long-term projects can reach some sort of fruition. Finally, as ever, I thank Zoë for her love and forbearance even if the conversation seemed to revolve around the number of bottles consumed, ink that is.

ILLUSTRATIONS

Following page 178

1. Octopus
2. Octopus
3. Sea-hare
4. Bat
5. Dendrobatid frog
6. Puffer fish
7. Bombardier beetle
8. Mantis shrimp
9. Nurse shark
10. Mosasaur
11. Cownose rays
12. Moray eel
13. Myrmecophage
14. Praying mantis
15. Saber-tooth cat
16. Duck-billed platypus
17. Komodo dragon
18. Sea-snake
19. Leaf-cutter
20. Gecko
21. Chameleon
22. Weta
23. Cockroach
24. Tuna
25. Hummingbird
26. Pitcher plant
27. Spider silk
28. Mosquito
29. Brittle star
30. Firefly
31. Hawkmoth
32. Rattlesnake
33. Robber crab
34. Termites
35. Kiwi
36. Dolphin
37. Lichen
38. Rust fungus
39. Cactus
40. *Rafflesia*
41. Titan arum
42. Dung beetle
43. Ants and aphids
44. Flying fish
45. Flying lemur
46. Hoatzin
47. Vulture
48. Kea
49. Archer
50. Capuchin monkey

21. The Roots of Sentience 234

22. Convergent Brains 241

23. The Road to "King Cortex" 252

24. Convergent Minds 263

25. Playing with Convergence 273

26. The Final Steps 286

27. Back to the Lagoon 301

Notes 305

General Index 463

Index of Genera 489

CONTENTS

Acknowledgments xiii

Introduction 3

1. Dinner on the Lagoon 9

2. Consider the Octopus 11

3. Convergence: How Clear Is the Signal? 21

4. The Inevitability of Form 31

5. Swallowing Convergence 44

6. Biting Convergence 53

7. Walking (and Swimming) to Convergence 67

8. Sticking to Convergence 82

9. When Evolution Begins to See 93

10. The Color of Evolution 108

11. The Smell and Taste of Evolution 120

12. (In)tangible Evolution 128

13. The Road to Mushrooms 150

14. The Road to Plants 155

15. The Arthropods Show the Way 165

16. Converging on the Farm 177

17. The Road to the Sky 189

18. The Birds Converge 200

19. Sexual Convergence 209

20. The Road to Mammals 221

Bernard of Chartres used to compare us to [puny] dwarfs perched on the shoulders of giants. He pointed out that we can see more and farther than our predecessors, not because we have keener vision or greater height, but because we are lifted up and borne aloft on their gigantic stature.

The Metalogicon of John of Salisbury,
Book III [Translated by D.D. McGarry
(University of California Press; 1962) p. 167]

To the memory of my parents

Templeton Press
300 Conshohocken State Road, Suite 500
West Conshohocken, PA 19428
www.templetonpress.org

© 2015 by Simon Conway Morris

All rights reserved. No part of this book may be used or reproduced,
stored in a retrieval system, or transmitted in any form or by any means,
electronic, mechanical, photocopying, recording, or otherwise,
without the written permission of Templeton Press.

Designed and typeset by Gopa & Ted2, Inc.

Library of Congress Cataloging-in-Publication Data on file

Printed in the United States of America

15 16 17 18 19   10 9 8 7 6 5 4 3 2 1

THE RUNES OF EVOLUTION

How the Universe Became Self-Aware

Simon Conway Morris

Consciousness: a process by which the universe experiences & understands itself. This is where unrelated species develope similar traits

TEMPLETON PRESS

THE RUNES OF EVOLUTION

* "runes" was the metaphor of ancient writing to suggest that these patterns in evolution are a meaningful message about the inevitability of certain life forms arising.
* convergence: act of things moving together. unrelated species develop similar traits.